Volume Editor

Tal Rabin
IBM T.J.Watson Research Center
Hawthorne, NY, USA
E-mail: talr@us.ibm.com

Library of Congress Control Number: 2010931385

CR Subject Classification (1998): E.3, G.2.1, F.2.1-2, D.4.6, K.6.5, C.2, J.1

LNCS Sublibrary: SL 4 – Security and Cryptology

| | |
|---|---|
| ISSN | 0302-9743 |
| ISBN-10 | 3-642-14622-8 Springer Berlin Heidelberg New York |
| ISBN-13 | 978-3-642-14622-0 Springer Berlin Heidelberg New York |

Typesetting: Camera-ready by author, data conversion by Scientific Publishing Services, Chennai, India
Printed on acid-free paper        06/3180

# Lecture Notes in Computer Science        6223

*Commenced Publication in 1973*
Founding and Former Series Editors:
Gerhard Goos, Juris Hartmanis, and Jan van Leeuwen

Tal Rabin (Ed.)

# Advances in Cryptology
# CRYPTO 2010

30th Annual Cryptology Conference
Santa Barbara, CA, USA, August 15-19, 2010
Proceedings

 Springer

# Preface

CRYPTO 2010, the 30th Annual International Cryptology Conference, was sponsored by the International Association for Cryptologic Research (IACR) in cooperation with the IEEE Computer Society Technical Committee on Security and Privacy and the Computer Science Department of the University of California at Santa Barbara. The conference was held in Santa Barbara, California, during August 15-19, 2010, in conjunction with CHES 2010 (Workshop on Cryptographic Hardware and Embedded Systems). Zulfikar Ramzan served as the General Chair.

The conference received 203 submissions. The quality of the submissions was very high, and the selection process was a challenging one. The Program Committee, aided by a 159 external reviewers, reviewed the submissions and after an intensive review period the committee accepted 41 of these submissions. Three submissions were merged into a single paper and two papers were merged into a single talk, yielding a total of 39 papers in the proceedings and 38 presentations at the conference. The revised versions of the 39 papers appearing in the proceedings were not subject to editorial review and the authors bear full responsibility for their contents. The best-paper award was awarded to the paper "Toward Basing Fully Homomorphic Encryption on Worst-Case Hardness" by Craig Gentry.

The conference featured two invited presentations. This year we celebrated 25 years from the publication of the ground-breaking work of Shafi Goldwasser, Silvio Micali and Charles Rackoff "The Knowledge Complexity of Interactive Proof-Systems." We had the privilege of having "GMR" give the first invited talk of the conference. The second invited talk was in a joint session with CHES. The topic was "Is Theoretical Cryptography Any Good in Practice?" and the talk was jointly given by Ivan Damgård and Markus Kuhn. The program also included a Rump Session, chaired by Daniel J. Bernstein and Tanja Lange, featuring short informal talks on new and in-progress results.

I am in debt to the many people who contributed to the success of the conference, and I apologize to those I have forgotten. First and foremost I thank the authors who submitted their papers; a conference is only as good as the submissions that it receives. The Program Committee members made a great effort contributing their time, knowledge, expertise and taste and for that I am grateful. I also thank the large number of external reviewers who assisted in the process. (The Program Committee and sub-reviewers are listed in the following pages.) The submission and review process used the software that Shai Halevi designed and I received a lot of help from him in running it.

And always, I want to thank my friends at IBM Research, Rosario Gennaro, Craig Gentry, Shai Halevi, Charanjit Jutla, Hugo Krawczyk and Vinod Vaikuntanathan – being part of this group makes everything so much more worthwhile.

June 2010 — Tal Rabin

# CRYPTO 2010

The 30th International Cryptology Conference

August 15–19, 2010, Santa Barbara, California, USA

Sponsored by the
International Association for Cryptologic Research (IACR)

in cooperation with
IEEE Computer Society Technical Committee on Security and Privacy,
Computer Science Department, University of California, Santa Barbara

## General Chair

Zulfikar Ramzan       Symantec

## Program Chair

Tal Rabin       IBM Research

## Program Committee

| | |
|---|---|
| Michel Abdalla | ENS, France |
| Adi Akavia | Weizmann Institute, Israel |
| Amos Beimel | Ben-Gurion University, Israel |
| Xavier Boyen | Université de Liège, Belgium |
| Christian Cachin | IBM Research, Zurich, Switzerland |
| Serge Fehr | CWI, The Netherlands |
| Johan Håstad | Royal Institute of Technology, Sweden |
| Carmit Hazay | Weizmann Institute and IDC Herzelia, Israel |
| Susan Hohenberger | Johns Hopkins, USA |
| Thomas Holenstein | ETH, Switzerland |
| Yael Tauman Kalai | Microsoft Research - New England, USA |
| John Kelsey | NIST, USA |
| Eike Kiltz | CWI, The Netherlands |
| Eyal Kushilevitz | Technion, Israel |
| Tanja Lange | Technische Universiteit Eindhoven, The Netherlands |
| Yehuda Lindell | Bar-Ilan University, Israel |
| Ilya Mironov | Microsoft Research, USA |
| Tal Moran | Harvard, USA |

| | |
|---|---|
| Jesper Buus Nielsen | University of Aarhus, Denmark |
| Eiji Okamoto | University of Tsukuba, Japan |
| Pascal Paillier | Gemalto, France |
| Rafael Pass | Cornell University, USA |
| Giuseppe Persiano | University of Salerno, Italy |
| Thomas Peyrin | Ingenico, France |
| Leonid Reyzin | Boston University, USA |
| Matt Robshaw | Orange Labs, France |
| Palash Sarkar | Indian Statistical Institute, India |
| abhi shelat | University of Virginia, USA |
| Vinod Vaikuntanathan | IBM Research, USA |
| Brent Waters | University of Texas, Austin, USA |
| Hoeteck Wee | Queens College, CUNY, USA |
| Andrew Yao | Tsinghua University, China |

## Advisory Members

Shai Halevi (CRYPTO 2009 Program Chair) - IBM Research
Phil Rogaway (CRYPTO 2011 Program Chair) - University of California,
  Davis

## External Reviewers

| | | |
|---|---|---|
| Divesh Aggarwal | Anne Canteaut | Maria Dubovitskaya |
| Shweta Agrawal | Claude Carlet | Leo Ducas |
| Jae Hyun Ahn | David Cash | Dejan Dukaric |
| Joel Alwen | Nishanth Chandran | Orr Dunkeman |
| Benny Applebaum | Donghoon Chang | Sebastian Faust |
| Gilad Asharov | Melissa Chase | Matthias Fitzi |
| Aslan Askarov | Sanjit Chatterjee | Manuel Forster |
| Jean-Philippe Aumasson | Lily Chen | Pierre-Alain Fouque |
| Roberto M. Avanzi | Victor Chen | David Freeman |
| Steve Babbage | Nathan Chenette | Georg Fuchsbauer |
| Daniel J. Bernstein | Cline Chevalier | Thomas Fuhr |
| Luk Bettale | Christophe Clavier | Benjamin Fuller |
| Rishiraj Bhattacharyya | Jean-Sébastien Coron | Steven Galbraith |
| Sanjay Bhattacherjee | Scott Coull | Clemente Galdi |
| Niek Bouman | Giovanni Di Crescenzo | Sharon Goldberg |
| Elette Boyle | Dana Dachman-Soled | Prasant Gopal |
| Zvika Brakerski | M. Prem Laxman Das | Dov Gordon |
| Eric Brier | Blandine Debraize | Louis Goubin |
| Dan Brown | Cécile Delerable | Aline Gouget |
| Jan Camenisch | Yevgeniy Dodis | Vipul Goyal |
| Sbastien Canard | Chandan Dubey | Matthew Green |
| Ran Canetti | Renaud Dubois | Iftach Haitner |

# Table of Contents

## Key Exchange, OAEP/RSA, CCA

## Attacks

# Composition

# Computation Delegation and Obfuscation

# Multiparty Computation

# Pseudorandomness

# Quantum

# Circular and Leakage Resilient
# Public-Key Encryption
# under Subgroup Indistinguishability
## (or: Quadratic Residuosity Strikes Back)

Zvika Brakerski[1] and Shafi Goldwasser[2]

[1] Weizmann Institute of Science
zvika.brakerski@weizmann.ac.il
[2] Weizmann Institute of Science and Massachusetts Institute of technology
shafi@theory.csail.mit.edu

**Abstract.** The main results of this work are new public-key encryption schemes that, under the quadratic residuosity (QR) assumption (or Paillier's decisional composite residuosity (DCR) assumption), achieve key-dependent message security as well as high resilience to secret key leakage and high resilience to the presence of auxiliary input information.

In particular, under what we call the *subgroup indistinguishability assumption*, of which the QR and DCR are special cases, we can construct a scheme that has:

- **Key-dependent message (circular) security.** Achieves security even when encrypting affine functions of its own secret key (in fact, w.r.t. affine "key-cycles" of predefined length). Our scheme also meets the requirements for extending key-dependent message security to broader classes of functions beyond affine functions using previous techniques of Brakerski et al. or Barak et al.
- **Leakage resiliency.** Remains secure even if any adversarial low-entropy (efficiently computable) function of the secret key is given to the adversary. A proper selection of parameters allows for a "leakage rate" of $(1 - o(1))$ of the length of the secret key.
- **Auxiliary-input security.** Remains secure even if any sufficiently *hard to invert* (efficiently computable) function of the secret key is given to the adversary.

Our scheme is the first to achieve key-dependent security and auxiliary-input security based on the DCR and QR assumptions. Previous schemes that achieved these properties relied either on the DDH or LWE assumptions. The proposed scheme is also the first to achieve leakage resiliency for leakage rate $(1 - o(1))$ of the secret key length, under the QR assumption. We note that leakage resilient schemes under the DCR and the QR assumptions, for the restricted case of composite modulus product of safe primes, were implied by the work of Naor and Segev, using hash proof systems. However, under the QR assumption, known constructions of hash proof systems only yield a leakage rate of $o(1)$ of the secret key length.

T. Rabin (Ed.): CRYPTO 2010, LNCS 6223, pp. 1–20, 2010.

# 1   Introduction

The "classical" definition of *semantic secure* public-key encryption by Goldwasser and Micali [16], requires that an efficient attacker with access to the public encryption-key must not be able to find two messages such that it can distinguish a random encryption of one from a random encryption of the other. Numerous candidate public-key encryption schemes that meet this definition have been presented over the years, both under specific hardness assumptions (like the hardness of factoring) and under general assumptions (such as the existence of injective one-way trapdoor functions).

This notion of security, however (as well as other commonly accepted ones), does not capture certain situations that may occur in the "real world":

- Functions of the secret decryption-key can be encrypted and sent (note that semantic security only guarantees security with respect to messages which an efficient attacker can find).
- Information about the secret key may leak.
- The same secret key may be used in more than one application, or more generally the attacker can somehow obtain the value of a hard-to-invert function of the secret key.

In recent years, extensive research effort has been invested in providing encryption schemes which are provably secure even in the above settings. Such schemes are said to achieve *key-dependent message (KDM) security, leakage-resilience*, and *auxiliary-input security* in correspondence to the above real world settings. To date, we know of: (1) Candidate schemes which are KDM secure under the decisional Diffie-Hellman (DDH) and under the learning with errors (LWE) assumptions; (2) Candidate schemes that are resilient to key leakage of rate $(1 - o(1))$ (relative to the length of the secret key), under the LWE assumption and under the DDH assumption. In addition, candidate scheme achieving some leakage resilience under a general assumption: the existence of universal hash-proof systems, with a leakage rate depending on the hash proof system being used; (3) Candidate schemes that are auxiliary input secure under the DDH assumption and under the LWE assumption.

In this work, we present an encryption scheme that achieves all of the above security notions simultaneously and is based on a class of assumptions that we call *subgroup indistinguishability assumptions*. Specifically, this class includes the quadratic residuosity (QR) and the decisional composite residuosity (DCR) assumptions, both of which are related to the problem of factoring large numbers. In addition, our schemes have the following interesting property: the secret key consists of a randomly chosen binary vector independent of the group at hand. The instantiation of our scheme under QR enjoys the same useful properties for protocol design as the original [16] scheme, including re-randomization of ciphertexts and support of the XOR homomorphic operation over the $\{0, 1\}$ message space, with the added benefit of leakage resilience.

To best describe our results, we first, in Section 1.1, describe in detail the background for the new work, including the relevant security notions and previous results. Second, in Section 1.2, we describe in detail the new results and encryption schemes. Then, in Section 1.3, we describe the new techniques. Section 1.4 discusses some additional related works and Section 1.5 contains the paper organization.

## 1.1 Background

**Key-dependent messages.** The shortcoming of the standard security definition in the case where the plaintext to be encrypted depends on the secret key was already noticed in [16]. It was later observed that this situation is not so unlikely and may sometimes even be desirable [9,1,21]. Black, Rogoway and Shrimpton [5] formally defined KDM-security: the attacker can obtain encryptions of (efficient) functions of its choosing, taken from some specified class of functions $\mathcal{F}$, applied to the secret key. The requirement is that the attacker cannot tell if all of its queries are answered by encryptions of some constant symbol 0, instead of the requested values. This definition is extended to the case of many (say $n$) users that can encrypt each others' secret keys: the attacker's queries now contain a function to be applied to *all* secret keys, and an identity of the user whose public key should be used to encrypt. This latter case is referred to as KDM$^{(n)}$-security while the single-user case is called KDM$^{(1)}$-security.

Boneh, Halevi, Hamburg and Ostrovsky [6] constructed a public key encryption scheme that is KDM$^{(n)}$ secure w.r.t. all affine functions,[1] under the decisional Diffie-Hellman (DDH) assumption, for any polynomial $n$. This first result was followed by the work of Applebaum, Cash, Peikert and Sahai [3] who proved that a variation of Regev's scheme [25] is also KDM secure w.r.t. all affine functions, under the learning with errors (LWE) assumption.

More recent works by Brakerski, Goldwasser and Kalai [8] and by Barak, Haitner, Hofheinz and Ishai [4] presented each general and different techniques to extend KDM-security to richer classes of functions. In [8], the notion of *entropy-$\kappa$ KDM-security* is introduced. A scheme is entropy-$\kappa$ KDM-secure if it remains KDM-secure even if the secret key is sampled from a high-entropy distribution, rather than a uniform one. They show that an entropy-$\kappa$ KDM-secure scheme implies a scheme that is KDM-secure w.r.t. roughly any pre-defined set of functions of polynomial cardinality. In [4], the notion of *targeted public-key encryption* is introduced. A targeted encryption scheme can be thought of as a combination of oblivious transfer and encryption: it is possible to encrypt in such a way that the ciphertext is decryptable only if a certain bit of the secret key takes a predefined value. They show that a targeted encryption scheme implies a KDM-secure scheme w.r.t. all functions computable by circuits of some predefined (polynomial) size. These two results achieve incomparable performance. While in the former, the public key and ciphertext lengths depend on the size of the function class (but not on its complexity) and are independent of the number of users $n$, in the latter the public key size does not depend on the function class, but

---

[1] More precisely "affine in the exponent": the secret key is a vector of group elements $g_1, \ldots, g_\ell$ and the scheme is secure w.r.t. functions of the form $h \cdot \prod g_i^{a_i}$.

the ciphertext length is linear in the product of $n$ times the complexity of the functions.

**Leakage resiliency.** The work on cold boot attacks by Halderman et al. [17], gave rise to the notion of public-key encryption resilient to (bounded) memory leakage attacks, presented by Akavia, Goldwasser and Vaikuntanathan [2] and further explored by Naor and Segev [22]. In their definition, security holds even if the attacker gets some information of its choosing (depending on the value of the public key) on the scheme's secret key, so long as the total amount of information leaked does not exceed an a-priori information theoretic bound. More formally, the attacker can request and receive $f(sk)$ for a length-restricted function $f$.[2] [2,22] presented public-key encryption schemes that are resilient to leakage of even a $1 - o(1)$ fraction of the secret key (we call this the "leakage rate"). In particular, [2] showed how this can be achieved under the LWE assumption, while [22] showed that this can be achieved under the DDH (or $d$-linear) assumption. It is further shown in [22] that some leakage resilience can be achieved using any universal hash proof system (defined in [10]), where the leakage rate depends on the parameters of the hash proof system. This implies secure schemes under the the QR and DCR assumptions as well. However, using the known hash proof systems, the leakage rate achievable under the QR assumption was only $o(1)$ — much less than the desired $1 - o(1)$. Based on the DCR assumption, a leakage rate of $(1 - o(1))$ was achievable [22,10,11].

**Auxiliary input.** Dodis, Kalai and Lovett [13] and Dodis, Goldwasser, Kalai, Peikert and Vaikuntanathan [12] considered the case where the leakage is not restricted information theoretically, but rather *computationally*. In the public key setting, the attacker is allowed to access any information on the secret key, with the following computational restriction: as long as recovering the secret key $sk$ from said information $f(pk, sk)$, for $f$ of the attackers choosing, is computationally hard to a sufficient extent (see discussion of several formalizations in [12]). This notion of security was termed *security in the presence of auxiliary input* (or *auxiliary-input security*, for short). Public-key auxiliary-input secure encryption schemes under the DDH and LWE assumptions were recently presented in [12].

### 1.2    New Results

Let us define a generalized class of assumptions called *subgroup indistinguishability* (SG) assumptions. A subgroup indistinguishability problem is defined by a group $\mathbb{G}_U$ ("the universe group") which is a direct product of two groups $\mathbb{G}_U = \mathbb{G}_M \times \mathbb{G}_L$ (interpreted as "the group of messages" and "the language group") whose orders, denoted by $M, L$ respectively, are relatively prime and where $\mathbb{G}_M$ is a cyclic group. Essentially, the *subgroup indistinguishability assumption* is that a random element of the universe $\mathbb{G}_U$ is computationally indistinguishable from a random element in $\mathbb{G}_L$. In other words, the language $\mathbb{G}_L$

---

[2] To be more precise, the requirement is that the min-entropy of the secret $sk$ drops by at most a bounded amount, given $f(sk)$.

is hard on average in the universe $\mathbb{G}_U$. The precise definition is a little more involved, see Section 3 for details.

Two special cases of the subgroup indistinguishability assumptions are the quadratic residuosity (QR) assumption on Blum integers and Paillier's decisional composite residuosity (DCR) assumption. This is easily seen for QR as follows. Let integer $N = p \cdot q$, where $p, q$ are random primes of equal bit-length, $\mathbb{Z}_N^* = \{x \in \mathbb{Z}_N : \gcd(x, N) = 1\}$, $\mathbb{J}_N$ denote the group of Jacobi symbol $(+1)$ elements of $\mathbb{Z}_N^*$, and $\mathbb{QR}_N = \{x^2 : x \in \mathbb{Z}_N^*\}$ denote its subgroup of quadratic residues. The *quadratic residuosity* (QR) assumption is then, that the uniform distributions over $\mathbb{J}_N$ and $\mathbb{QR}_N$ are computationally indistinguishable. Taking $N$ to be a *Blum integer* where $p, q = 3 \pmod 4$ (otherwise the orders of $\mathbb{G}_L, \mathbb{G}_M$ we define next will not be relatively prime) and setting $\mathbb{G}_U = \mathbb{J}_N$, $\mathbb{G}_L = \mathbb{QR}_N$ (which is of odd order), and $\mathbb{G}_M = \{\pm 1\}$ (which is cyclic and has order 2), the QR assumption falls immediately into the criteria of subgroup indistinguishability assumptions.

We are now ready to describe the new encryption scheme for a given subgroup problem $(\mathbb{G}_U, \mathbb{G}_M, \mathbb{G}_L)$ where $h$ is a generator for $\mathbb{G}_M$. In general, we view the *plaintext message space* as the elements $h^m \in \mathbb{G}_M$ (sometimes the exponent $m$ itself can be viewed as the message). For the case of QR, the plaintext message space is $\mathbb{G}_M = \{\pm 1\}$.

A word on the choice of parameters is in order. All parameters are measured as a function of the security parameter $k$. As customary, in the QR and DCR cases, think of the security parameter as the size of the modulus $N$ (i.e. $k = \lceil \log N \rceil$). We let $\ell$ denote a parameter whose value is polynomially related to $k$,[3] selected in accordance to the desired properties of the scheme (KDM security, amount of leakage resilience etc.).

## The Encryption Scheme for Subgroup Problem $(\mathbb{G}_U, \mathbb{G}_M, \mathbb{G}_L)$ with Parameter $\ell$:

- *Key generation.* Set the secret key to a random binary vector $\mathbf{s} = (s_1, \ldots, s_\ell)$ of length $\ell$. Set the public key to be the tuple $(g_1, \ldots, g_\ell, g_0)$ where $g_1, \ldots, g_\ell$ are uniformly chosen elements of $\mathbb{G}_L$ and $g_0 = \prod g_i^{-s_i}$. (For the QR assumption, the public key thus consists of $\ell$ random squares, followed by a product of a random subset of them, selected by the secret key $\mathbf{s}$).
- *Encryption.* On input message $h^m$,[4] sample a uniform integer $r$ from a large enough domain and output the ciphertext $(g_1^r, \ldots, g_\ell^r, h^m \cdot g_0^r)$. (For the QR assumption case, encryption is of single bits $\{\pm 1\}$, and the ciphertext is the tuple of squares in the public key, raised to a random power, where the last one is multiplied by the plaintext message.)
- *Decryption.* On ciphertext $(c_1, \ldots, c_\ell, c_0)$, compute $h^m = c_0 \cdot \prod c_i^{s_i}$. (For the case of QR, $m = c_0 \cdot \prod c_i^{s_i}$.) In general, recoverability of the exponent $m$ depends on whether taking discrete logs in base $h$ of $h^m$ is easy.

We remark that the basic structure of our construction is strikingly similar to [6], where the public key also contains $\ell$ independent "random" elements and

---

[3] More precisely, $\ell$ is a polynomial function $\ell(k)$.

[4] Recall that $h$ is a generator of $\mathbb{G}_M$, which is a part of the description of $\mathbb{G}_U$.

an additional element that is statistically close to uniform, but in fact is a combination of the previous ones. The difference and challenge is in how to prove security. This challenge is due to the fact that the subgroup indistinguishability assumptions seem inherently different from the DDH assumption. In the latter, for cyclic group $\mathbb{G}$ where DDH is assumed, the assumption implies that the distribution $(g_1, g_2, g_1^r, g_2^r)$ is computationally indistinguishable from $(g_1, g_2, g_1', g_2')$ giving complete re-randomization (a similar property follows for LWE). Such re-randomization does not follow nor is it necessarily true from subgroup indistinguishability. Rather, we will have to use the weaker guarantee that $(g_1, g_2, g_1^r, g_2^r)$ is indistinguishable from $(g_1, g_2, h^{r'} \cdot g_1^r, h^{r''} \cdot g_2^r)$, giving only "masking" of the message bits.

Similarly to the scheme of [6], our scheme is lacking in efficiency. This is most noticeable in our QR-based scheme, where the encryption of one bit requires a ciphertext containing $\ell + 1$ group elements, each of size roughly the security parameter $k$. The situation is somewhat better when relying on DCR: there each such ciphertext encrypts $\Omega(k)$ bits. Improved efficiency can be achieved by using the same values $g_1, \ldots, g_\ell$ with many vectors $\mathbf{s}$, however this makes KDM security hold only with respect to a less natural function class (this is similar to the more efficient LWE based scheme of [3]) and significantly reduces leakage resiliency. Coming up with more efficient KDM secure or leakage resilient schemes remains an interesting open problem.

We prove the following properties for the new encryption scheme.

**Property 1: KDM-Security.** First, we prove that the scheme is KDM$^{(1)}$-secure w.r.t. affine functions of the secret key. To show this for QR case, we show that for any affine function specified by $a_0, \ldots, a_\ell$, the encryption of $(-1)^{a_0 + \sum_i a_i s_i}$ is indistinguishable from the encryption of $(-1)^0$. For the general case, it is more natural to view KDM$^{(1)}$ with respect to the affine functions "in the exponent": for any $h_0, h_1, \ldots, h_\ell \in \mathbb{G}_M$ where $h_i = h^{a_i}$, for the generator $h$, we show that an encryption of $h_0 \cdot \prod h_i^{s_i} = h^{a_0 + \sum_i a_i s_i}$ is indistinguishable from an encryption of $h^0$.

Second, we prove that for any polynomial value of $n$, the above encryption scheme satisfies KDM$^{(n)}$ security, if $\ell$ is larger than, roughly, $n \log L$. We note thus that the public key size and ciphertext size grow with $n$ to achieve provable KDM$^{(n)}$ security. Interestingly, in the works of [6,3], $\ell$ did not need to grow with $n$. This seems difficult to achieve without the complete "re-randomization" property discussed above which does follow from the DDH and LWE assumptions, but not from ours.

Finally, we can also show that our scheme can be used to obtain KDM security for larger classes of functions than affine function: The scheme is *entropy-$\kappa$ KDM-secure* (for proper values of $\ell$), as required in [8] and therefore implies a scheme that is secure w.r.t. functions of the form $a_0 + \sum_i a_i f_i(sk)$ for (roughly) any set of polynomially-many efficiently computable functions $\{f_1, \ldots, f_\ell\}$. Our scheme also implies a *targeted encryption scheme*, as required in [4], and therefore implies that for any polynomial bound $p$, there is a scheme that is secure w.r.t. all functions computable by size-$p$ circuits.

**Property 2: Improved Key-Leakage Resiliency.** We prove that the new scheme is resilient to any leakage of a $(1 - o(1))$ fraction of the bits of the secret key. Stated differently, if one specifies in advance the amount of leakage $\lambda$ (a polynomial in the security parameter) to be tolerated, we can choose $\ell$ to obtain a scheme that is secure against a leakage of $\lambda$ bits. The growth of $\ell$ is additive in $\lambda$ (i.e. $\ell = \ell_0 + \lambda$) and therefore we can select the value of $\ell$ to obtain schemes that are resilient to leakage of a $(1 - (\ell_0/\ell)) = (1 - o(1))$ fraction of the secret key.

We emphasize that while schemes with the above guarantees were known under LWE [2] or DDH [22], and even (implicitly) under DCR [22,10], this was not the case under QR. Previous results with regards to QR-based leakage resiliency [22,10] could only approach a leakage rate of $1/k = o(1)$ (recall that $k$ is the security parameter, or the bit-length of the modulus $N$), compared to $(1 - o(1))$ in our scheme.

In addition, previous constructions of QR and DCR based hash proof systems required that the modulus used $N = p \cdot q$ is such that $p, q$ are *safe primes*. We do not impose this restriction. In the QR case we only require that $p, q = 3 \pmod 4$ (i.e. $N$ is a *Blum integer*) and in the DCR case we only require that $p, q$ have the same bit-length.

**Property 3: Auxiliary Input Security.** We prove that our schemes remain secure when the attacker has access to additional information on the secret key $sk$, in the form of $f_{pk}(sk)$, where $f_{pk}$ is a polynomial time function (which may depend on the public key) that is evaluated on the secret key $sk$. First, we consider the case where $f$ is such that the transition $(f_{pk}(sk), pk) \to sk$ is computationally hard. Namely, that retrieving the secret key $sk$ given the public key $pk$ and the auxiliary information $f_{pk}(sk)$, is sufficiently hard. This notion was termed *weak auxiliary-input* security in [12]. In turn, [12] show how to leverage weak auxiliary-input security to achieve security when the requirement on $f$ is weaker: now, only the transition $f_{pk}(sk) \to sk$ needs to be hard. The latter is called *auxiliary-input* security.

We conclude that for all $\delta > 0$, we can select the value of $\ell$ such that the scheme is auxiliary-input secure relative to any function that is hard to invert (in polynomial time) with probability $2^{-\ell^\delta}$. We note that the input to the function is the secret key – a length $\ell$ binary string, and therefore we measure hardness as a function of $\ell$ (and not of the security parameter $k$).

## 1.3   Our Techniques

The circular security, leakage resiliency and auxiliary-input security properties of our scheme are proved using a new technical tool introduced in this work: the *interactive vector game*. This proof technique can also provide an alternative proof for the KDM$^{(1)}$-security, leakage resiliency and auxiliary-input security of (known) public-key encryption schemes based on DDH and LWE, thus providing an alternative, more generic proof for some of the results of [6,3,22,12].[5]

---

[5] In this work, the interactive vector game is defined only for our subgroup indistinguishability assumptions, but it easily extends to other assumptions.

This suggests an alternative explanation to the folklore belief that the three notions are related: that it is the proof technique that is related in fact. Namely, the proof techniques for each property can be generalized to interactive vector games which, in turn, imply the other properties.

We proceed to overview the proofs of security for the various properties of our scheme. Again, let us consider the groups $\mathbb{G}_U = \mathbb{G}_M \times \mathbb{G}_L$ with $h$ being a generator for $\mathbb{G}_M$, such that the subgroup indistinguishability assumption holds.

To best explain the ideas of the proof, let us consider, as a first step, a simple semantically secure encryption scheme (which is a generalization of the Goldwasser-Micali scheme [15]). An encryption of 0 is a random element $g \in \mathbb{G}_L$ and an encryption of 1 is $h \cdot g$ (in the QR case, the encryption of $(+1)$ is a random quadratic residue and the encryption of $(-1)$ is a random quadratic non-residue). The two distributions are clearly indistinguishable (consider the indistinguishable experiment where $g$ is uniform in $\mathbb{G}_U$). In order to decrypt, one needs some "trapdoor information" that would enable to distinguish between elements in $\mathbb{G}_L$ and $\mathbb{G}_U$ (such as the factorization of the modulus $N$ in the QR (and DCR) case).

The first modification of this simple idea was to fix $g$ and put it in the public key, and set the ciphertext for $h^m$ to $h^m \cdot g^r$ for $r$ large enough. Note that the sender does not know the order of $\mathbb{G}_U$: Indeed, in the QR case, knowing the order of the group $\mathbb{J}_N$, which is $\frac{\varphi(N)}{2}$, enables to factor $N$. For the QR case, this modification still amounts to encrypting $(+1)$ by a random square, and $(-1)$ by a random non-square.

The second modification does away with the need of the secret key owner to distinguish between elements in $\mathbb{G}_L$ and $\mathbb{G}_U$ (e.g. with the need to know the factorization of $N$ in the QR case), by replacing the "trapdoor information" with a secret key that is a uniform binary vector $\mathbf{s} = (s_1, \ldots, s_\ell)$. Holding the secret key will not enable us to solve subgroup indistinguishability, but will enable us to decrypt as in [6]. We take a set of random elements $g_1, \ldots, g_\ell \in \mathbb{G}_L$ and define $g_0 = \prod g_i^{-s_i}$. If $\ell$ is large enough, then the leftover hash lemma implies that $g_0$ is almost uniform. As the ciphertext is $(g_1^r, \ldots, g_\ell^r, h^m \cdot g_0^r)$, one can recover $h^m$ using $\mathbf{s}$. Recovering $m$ itself is also possible if the discrete logarithm problem in $\mathbb{G}_M$ is easy, as is the case in the QR scenario.

The crux of the idea in proving security is as following. First, we note that the distribution of $g_0$ is close to uniform in $\mathbb{G}_L$, even given $g_1, \ldots, g_\ell$ (by the leftover hash lemma). Recall that in a DDH-based proof, we could claim that $((g_1, \ldots, g_\ell, g_0), (g_1^r, \ldots, g_\ell^r, g_0^r))$ is computationally indistinguishable from $((g_1, \ldots, g_\ell, g_0), (g_1', \ldots, g_\ell', g_0'))$ (where $g_i'$ are uniform). However, based on subgroup indistinguishability, a different method is required: Consider replacing $g_0$ with $g_0' = h \cdot g_0$, the distribution $((g_1, \ldots, g_\ell, g_0), (g_1^r, \ldots, g_\ell^r, g_0^r))$ is computationally indistinguishable from $((g_1, \ldots, g_\ell, h \cdot g_0), (g_1^r, \ldots, g_\ell^r, h^r \cdot g_0^r))$ under the subgroup indistinguishability assumption. The crucial observation now is that since the orders of $\mathbb{G}_M$ and $\mathbb{G}_L$ are relatively prime, then in fact $g_0'^r = h^{r'} \cdot g_0^r$, where $r'$ is independent of $r$. Combined with the fact that $\mathbb{G}_M$ is cyclic, we get that $((g_1, \ldots, g_\ell, g_0), (g_1^r, \ldots, g_\ell^r, g_0^r))$ is indistinguishable from $((g_1, \ldots, g_\ell, h \cdot g_0), (g_1^r \ldots g_\ell^r, h' \cdot g_0^r))$, for a random $h' \in \mathbb{G}_M$. Semantic security now follows.

To address the issues of circular security, leakage resiliency and auxiliary-input, we generalize the idea presented above, and prove that the distributions $((g_1, \ldots, g_\ell), (h^{a_1} \cdot g_1^r, \ldots, h^{a_\ell} \cdot g_\ell^r))$ and $((g_1, \ldots, g_\ell), (g_1^r, \ldots, g_\ell^r))$ are indistinguishable. We provide an interactive variant of this claim, which we call an *interactive $\ell$-vector game*, where the values of $a_1, \ldots, a_\ell \in \mathbb{Z}$ are selected by the distinguisher and can depend on $(g_1, \ldots, g_\ell)$, and show that the above is hard even in such case. The interactive vector game will be employed in the proofs of all properties of the scheme.

For key-dependent message security, we consider the ciphertext $(g_0^r, g_1^r, \ldots, h \cdot g_i^r, \ldots, g_\ell^r)$. This ciphertext will be decrypted to $h^{s_i}$ and in fact can be shown (using an interactive vector game) to be computationally indistinguishable from a legal encryption of $h^{s_i}$. Key-dependent message security follows from this fact.

Proving $\text{KDM}^{(n)}$-security for our scheme is more complex. To illustrate this, we contrast it with the ideas in the proof of [6]. They used homomorphism and re-randomization to achieve $\text{KDM}^{(n)}$-security: Their scheme is shown to have *homomorphic* properties that enable to "shift" public keys and ciphertexts that are relative to a certain secret key, into ones that are relative to another secret key. In order to apply these "shifts", one only needs to know the relation between the original and final keys (and not the keys themselves). In addition, their scheme is shown to have *re-randomization* properties that enable to take a public key (or ciphertext) and produce an independent public key (or ciphertext) that corresponds to the same secret key (and message, in the ciphertext case). These two properties enable simulating the $\text{KDM}^{(n)}$-security game using only one "real" secret key, fabricating the $n$ required keys and ciphertexts using homomorphism and re-randomization. In [3], similar ideas are employed, but the re-randomization can be viewed as implicit in the assumption (the ability to generate independently looking vectors that are in fact linearly related).

Our scheme can be shown to have such homomorphic properties, but it doesn't enjoy as strong re-randomizability as required to use the above techniques. As an example, consider a public key $pk = (g_0, g_1, \ldots, g_\ell)$ corresponding to a secret key $sk = (s_1, \ldots, s_\ell)$, i.e. $g_0 = \prod g_i^{-s_i}$. Let $j \in [\ell]$ and consider $\hat{pk} = (\hat{g}_0, \hat{g}_1, \ldots, \hat{g}_\ell)$ defined as follows: for all $i \notin \{j, 0\}$, set $\hat{g}_i = g_i$; for $j$, set $\hat{g}_j = g_j^{-1}$; and finally set $\hat{g}_0 = g_j \cdot g_0 = \hat{g}_j^{-(1-s_j)} \cdot \prod_{i \neq j} \hat{g}_i^{-s_i}$. We get that $\hat{pk}$ is a properly distributed public key corresponding to the secret key $\hat{sk} = sk \oplus e_j$ ($sk$ XORed with the $j^{\text{th}}$ unit binary string). Namely, we were able to "shift" a public key to correspond to another (related) secret key, without knowing the original key. However, the joint distribution of $pk, \hat{pk}$ is easily distinguishable from that of two independent public keys. What we lack is the ability to re-randomize $\hat{pk}$ so that it is distributed as a public key for $\hat{sk}$ which is independent of $pk$.

Intuitively, this shortcoming requires us to use more "real randomness". Our proof simulates the $\text{KDM}^{(n)}$-security game using only one "real" secret key, as in the idea presented above. This secret key is used to fabricate $n$ secret and public keys. However, when we want to apply the leftover hash lemma to claim that the $g_0$ components of all $n$ fabricated public keys are close to uniform, we need the one real secret key to have sufficient entropy. This requires a secret key whose size is linear in $n$. These ideas, combined with the ones used to prove $KDM^{(1)}$ security, give our final proof.

The property of entropy-$\kappa$ KDM-security requires that the scheme remains secure even when the secret key is sampled from a high-entropy (but not necessarily uniform) distribution. This is shown to hold using the leftover hash lemma, since $\prod g_i^{s_i}$ is a 2-universal hash function. A targeted encryption scheme is obtained similarly to the other constructions in [4], by using the fact that we can "fabricate" ciphertexts that correspond to affine functions of the secret key without knowing the secret key itself.

Leakage resiliency and auxiliary-input security are proven by an almost identical argument: consider a case where we replace the ciphertext $(h^m \cdot g_0^r, g_1^r, \ldots, g_\ell^r)$ with a computationally indistinguishable one: $(h^{-\sum \sigma_i s_i} \cdot h^m \cdot g_0^r, h^{\sigma_1} \cdot g_1^r, \ldots, h^{\sigma_\ell} \cdot g_\ell^r)$, where $\sigma_i \in \mathbb{Z}_M$ are uniform. Computational indistinguishability (even for a known secret key) follows from the interactive vector game mentioned above. For leakage-resilience, the leftover hash lemma implies that so long as there is sufficient entropy in $\mathbf{s}$ after the leakage, $\sum \sigma_i s_i$ will be close to uniform and will "mask" the value of $m$. For auxiliary input we use the *generalized Goldreich-Levin* theorem of [12] to show that $\sum \sigma_i s_i$ is close to uniform in the presence of a function of $\mathbf{s}$ that is hard to invert, even given the public key. Thus obtaining *weak* auxiliary-input security. In the QR case, the inner product is over $\mathbb{Z}_2$ and therefore we can use the "standard" Goldreich-Levin theorem [14], which implies better parameters. We use leveraging (as used in [12]) to obtain the full result.

### 1.4   Other Related Work

Cramer and Shoup [10] presented the notion of *hash proof systems*, which are similar to subgroup indistinguishability assumptions. Their implementations from QR and DCR also do not require the factorization of $N$ in order to decrypt. However they use the discrete logarithm of (their analog to) the $g_i$'s as a secret key for the system. Our scheme can be seen as taking another step towards "stripping" the secret key of all structure: in our scheme, it is just a uniform sequence of bits (resulting in a weaker form of a hash proof system that is "universal on average").

Hemenway and Ostrovsky [19] show how to construct lossy trapdoor functions (see [24] for definition) from the QR and DCR assumptions (among other assumptions). Similar ideas can be used in a straightforward manner to construct lossy trapdoor functions from subgroup indistinguishability assumptions with special properties.

### 1.5   Paper Organization

Due to space constraints, this extended abstract only discusses the construction based on the QR assumption. In addition, some of the proofs are omitted. We refer the reader to the full version of this paper [7] for the complete presentation, including all details.

Preliminaries and definitions are presented in Section 2. The definition of subgroup indistinguishability assumptions and instantiations from QR and DCR appear in Section 3.

Our QR-based encryption scheme is presented in Section 4, followed, in Section 5, by introduction of the interactive vector game: a central technical tool to be used for the analysis throughout the paper. KDM-security is discussed in Section 6, leakage-resilience in Section 7 and auxiliary-input security in Section 8.

## 2 Preliminaries

We denote scalars in plain lowercase ($x \in \{0,1\}$) and vectors in bold lowercase ($\mathbf{x} \in \{0,1\}^n$). The $i^{\text{th}}$ coordinate of $\mathbf{x}$ is denoted $x_i$.

For vectors $\mathbf{g}, \mathbf{h} \in \mathbb{G}^n$, where $\mathbb{G}$ is a multiplicative commutative group, we denote by $\mathbf{g}^r$ the vector whose $i^{\text{th}}$ coordinate is $g_i^r$. We denote by $\mathbf{h} \cdot \mathbf{g}$ the vector whose $i^{\text{th}}$ coordinate is $h_i \cdot g_i$. Note that this does *not* denote an inner product. For a group element $g \in \mathbb{G}$ and a vector $\mathbf{x} \in \mathbb{Z}$, we let $g^{\mathbf{x}}$ denote the vector whose $i^{\text{th}}$ coordinate is $g^{x_i}$.

Let $X$ be a probability distribution over a domain $S$, we write $x \xleftarrow{\$} X$ to indicate that $x$ is sampled from the distribution $X$. The uniform distribution over a set $S$ is denoted $U(S)$. We use $x \xleftarrow{\$} S$ as abbreviation for $x \xleftarrow{\$} U(S)$. An ensemble $X = \{X_k\}_k$ is $\epsilon = \epsilon(k)$-*uniform* if for all $k$, $X_k$ is within statistical distance $\epsilon(k)$ from the uniform distribution. Statistical and computational indistinguishability are defined in the standard way. We write negl($k$) to denote an arbitrary *negligible* function, i.e. one that vanishes faster than the inverse of any polynomial.

We use the following simple lemma.

**Lemma 2.1.** *Let $T, N \in \mathbb{N}$ and let $x \xleftarrow{\$} [T]$, then $x \pmod{N}$ is $(N/T)$-uniform in $\mathbb{Z}_N$.*

We use the following lemma which is an immediate corollary of the leftover hash lemma and explicitly appears in [6, Lemma 2].

**Lemma 2.2.** *Let $H$ be a 2-universal hash family from a set $X$ to a set $Y$. Then the distribution $(h, h(x))$ where $h \xleftarrow{\$} H$, $x \xleftarrow{\$} X$ is $\sqrt{\frac{|Y|}{4|X|}}$-uniform in $H \times Y$.*

The following lemma states the properties of a class of hash functions that we use.

**Lemma 2.3.** *Let $\mathbb{G}$ be any finite commutative group and let $\ell \in \mathbb{N}$. Then the set of functions $H = \{h_{g_1,\dots,g_\ell} : \{0,1\}^\ell \to \mathbb{G}\}_{g_1,\dots,g_\ell \in \mathbb{G}}$ where $h_{g_1,\dots,g_\ell}(\mathbf{x}) = \prod_{i \in [\ell]} g_i^{x_i}$, is 2-universal.*

We use the standard definitions of KDM security, leakage resilience and auxiliary input security as appear, e.g., in [6,22,12], respectively.

# 3 Subgroup Indistinguishability Assumptions

We present the class of subgroup indistinguishability assumptions in Section 3.1 and then discuss instantiations under the QR and DCR assumptions in Section 3.2.

## 3.1 Definition of a Subgroup Indistinguishability (SG) Problem

Let $\mathbb{G}_U$ be a finite commutative multiplicative group, such that $\mathbb{G}_U$ is a direct product of two groups: $\mathbb{G}_U = \mathbb{G}_M \times \mathbb{G}_L$ (interpreted as the "message group" and the "language group"), where $\mathbb{G}_M$ is cyclic of order $M$, $\mathbb{G}_L$ is of order $L$ (and is not necessarily cyclic) and $\mathbb{G}_U$ is of order $M \cdot L$ (we abuse notation and use $M, L$ to index the groups and to denote their orders). We require that $\gcd(M, L) = 1$. Let $h$ be a generator for $\mathbb{G}_M$ such that $h$ is efficiently computable from the description of $\mathbb{G}_U$. We require that there exists an efficient algorithm $OP_{\mathbb{G}_U}$ to perform group operations in $\mathbb{G}_U$, and also that there exist efficient sampling algorithms $S_{\mathbb{G}_M}, S_{\mathbb{G}_L}$ that sample a random element from $\mathbb{G}_M$, $\mathbb{G}_L$ respectively. We further require that an upper bound $T \geq M \cdot L$ is known.

We stress that as always, all groups described above are in fact families of groups, indexed by the security parameter $k$. To be more precise, there exists a polynomial time randomized algorithm that given the security parameter $1^k$, outputs $I_{\mathbb{G}_U} = (OP_{\mathbb{G}_U}, S_{\mathbb{G}_M}, S_{\mathbb{G}_L}, h, T)$. We refer to $I_{\mathbb{G}_U}$ as an instance of $\mathbb{G}_U$.

For any adversary $\mathcal{A}$ we denote the subgroup distinguishing advantage of $\mathcal{A}$ by

$$\mathrm{SGAdv}[\mathcal{A}] = \left| \Pr_{x \xleftarrow{\$} \mathbb{G}_U} [\mathcal{A}(1^k, x)] - \Pr_{x \xleftarrow{\$} \mathbb{G}_L} [\mathcal{A}(1^k, x)] \right| .$$

That is, the advantage $\mathcal{A}$ has in distinguishing between $\mathbb{G}_U$ and $\mathbb{G}_L$. The subgroup indistinguishability (SG) assumption is that for any polynomial $\mathcal{A}$ it holds that for a properly sampled instance $I_{\mathbb{G}_U}$, we have $\mathrm{SGAdv}[\mathcal{A}] = \mathrm{negl}(k)$ (note that in such case it must be that $1/L = \mathrm{negl}(k)$). In other words, thinking of $\mathbb{G}_L \subseteq \mathbb{G}_U$ as a language, the assumption is that this language is hard on average. We define an additional flavor of the assumption by

$$\mathrm{SG}'\mathrm{Adv}[\mathcal{A}] = \left| \Pr_{x \xleftarrow{\$} \mathbb{G}_L} [\mathcal{A}(1^k, h \cdot x)] - \Pr_{x \xleftarrow{\$} \mathbb{G}_L} [\mathcal{A}(1^k, x)] \right| .$$

It follows immediately that for any adversary $\mathcal{A}$ there exists an adversary $\mathcal{B}$ such that $\mathrm{SG}'\mathrm{Adv}[\mathcal{A}] \leq 2 \cdot \mathrm{SGAdv}[\mathcal{B}]$.

## 3.2 Instantiations

We instantiate the SG assumption based on the QR and DCR assumptions.

For both instantiations we consider a modulus $N$ defined as follows. For security parameter $k$, we sample a random *RSA number* $N \in \mathbb{N}$: this is a number of the form $N = pq$ where $p, q$ are random $k$-bit odd primes.

We note that our instantiations work even when the modulus $N$ is such that $\mathbb{QR}_N$ is not cyclic.

**Instantiation Under the QR Assumption with Any Blum Integer.** Consider a modulus $N$ as described above. We use $\mathbb{J}_N$ to denote the set of elements in $\mathbb{Z}_N^*$ with Jacobi symbol 1, we use $\mathbb{QR}_N$ to denote the set of *quadratic residues* (squares) modulo $N$. Slightly abusing notation $\mathbb{J}_N, \mathbb{QR}_N$ also denote the respective groups with the multiplication operation modulo $N$. The groups $\mathbb{J}_N, \mathbb{QR}_N$ have orders $\frac{\varphi(N)}{2}, \frac{\varphi(N)}{4}$ respectively and we denote $N' = \frac{\varphi(N)}{4}$. We require that $N$ is a *Blum integer*, namely that $p, q = 3 \pmod 4$. In such case it holds that $\gcd(2, N') = 1$ and $(-1) \in \mathbb{J}_N \setminus \mathbb{QR}_N$.

The *quadratic residuosity* (QR) assumption is that for a properly generated $N$, the distributions $U(\mathbb{J}_N)$ and $U(\mathbb{QR}_N)$ are computationally indistinguishable.[6] This leads to the immediate instantiation of the SG assumption by setting $\mathbb{G}_U = \mathbb{J}_N$, $\mathbb{G}_M = \{\pm 1\}$, $\mathbb{G}_L = \mathbb{QR}_N$, $h = (-1)$, $T = N \geq 2N'$.

**Instantiation Under the DCR Assumption.** The *decisional composite residuosity* (DCR) assumption, introduced by Paillier [23], states that for a properly generated RSA number $N$, it is hard to distinguish between a random element in $\mathbb{Z}_{N^2}^*$ and a random element in the subgroup of $N^{\text{th}}$-residues $\{x^N : x \in \mathbb{Z}_{N^2}^*\}$. The group $\mathbb{Z}_{N^2}^*$ can be written as a product of the group generated by $1 + N$ (which has order $N$) and the group of $N^{\text{th}}$ residues (which has order $\varphi(N)$). This implies that setting $\mathbb{G}_U = \mathbb{Z}_{N^2}^*$, $\mathbb{G}_L = \{x^N : x \in \mathbb{Z}_{N^2}^*\}$ and $\mathbb{G}_M = \{(1+N)^i : i \in [N]\}$ provides an instantiation of the SG assumption, setting $h = (1+N)$ and $T = N^2$. It is left to check that indeed $\gcd(N, \varphi(N)) = 1$. This follows since $p, q$ are odd primes of equal length: assume w.l.o.g that $p/2 < q < p$, then the largest prime divisor of $\varphi(N) = (p-1)(q-1)$ has size at most $(p-1)/2 < p, q$ and the claim follows.[7]

## 4   Description of the Encryption Scheme

We now present our QR-based scheme $\mathcal{E}[\ell]$.

**Parameters.** The scheme is parameterized by $\ell \in \mathbb{N}$ which is polynomial in the security parameter. The exact value of $\ell$ is determined based on the specific properties we require from the scheme.

The message space of $\mathcal{E}[\ell]$ is $\mathcal{M} = \{0, 1\}$, i.e. this is a bit-by-bit encryption scheme.

**Key generation.** The key generator first samples a Blum integer $N$. We note that the same value of $N$ can be used by all users. Furthermore we stress that no entity needs to know the factorization of $N$. Therefore we often refer to $N$ as a public parameter of the scheme and assume that it is implicitly known to all users.

---

[6] The QR assumption usually refers to random RSA numbers, which are not necessarily Blum integers. However, since Blum integers have constant density among RSA numbers, the flavor we use is implied.

[7] If greater efficiency is desired, we can use a generalized form of the assumption, presented in [11].

The key generator also samples $\mathbf{s} \xleftarrow{\$} \{0,1\}^{\ell}$ and sets $sk = \mathbf{s}$. It then samples $\mathbf{g} \xleftarrow{\$} \mathbb{QR}_N^{\ell}$ and sets $g_0 = (\prod_{i \in [\ell]} g_i^{s_i})^{-1}$. The public key is set to be $pk = (g_0, \mathbf{g})$ (with $N$ as an additional implicit public parameter).

**Encryption.** On inputs a public key $pk = (g_0, \mathbf{g})$ and a message $m \in \{0,1\}$, the encryption algorithm runs as follows: it samples $r \xleftarrow{\$} [N^2]$,[8] and computes $\mathbf{c} = \mathbf{g}^r$ and $c_0 = (-1)^m \cdot g_0^r$. It outputs a ciphertext $(c_0, \mathbf{c})$.

**Decryption.** On inputs a secret key $sk = \mathbf{s}$ and a ciphertext $(c_0, \mathbf{c})$, the decryption algorithm computes $(-1)^m = c_0 \cdot \prod_{i \in [\ell]} c_i^{s_i}$ and outputs $m$.

The completeness of the scheme follows immediately by definition.

## 5   The Interactive Vector Game

We define the *interactive $\ell$-vector* game played between a challenger and an adversary. We only present the QR-based game and refer the reader to [7] for full details.

**Initialize.** The challenger samples $b \xleftarrow{\$} \{0,1\}$ and also generates a Blum integer $N$ and a vector $\mathbf{g} \xleftarrow{\$} \mathbb{QR}_N^{\ell}$. It sends $N$ and $\mathbf{g}$ to the adversary.

**Query.** The adversary adaptively makes queries, where each query is a vector $\mathbf{a} \in \{0,1\}^{\ell}$. For each query $\mathbf{a}$, the challenger samples $r \xleftarrow{\$} [N^2]$ and returns $(-1)^{\mathbf{a}} \cdot \mathbf{g}^r$ if $b = 0$ and $\mathbf{g}^r$ if $b = 1$.

**Finish.** The adversary outputs a guess $b' \in \{0,1\}$.

The advantage of an adversary $\mathcal{A}$ in the game is defined to be

$$\mathrm{IV}_{\ell}\mathrm{Adv}[\mathcal{A}] = |\Pr[b' = 1|b = 0] - \Pr[b' = 1|b = 1]| .$$

Under the QR assumption, no poly($k$)-time adversary (where $k$ is the security parameter) can obtain a non-negligible advantage in the game, as formally stated below.

**Lemma 5.1.** *Let $\mathcal{A}$ be an adversary for the interactive $\ell$-vector game that makes at most $t$ queries, then there exists an adversary $\mathcal{B}$ for QR such that*

$$\mathrm{IV}_{\ell}\mathrm{Adv}[\mathcal{A}] \leq 4t\ell \cdot \mathrm{QRAdv}[\mathcal{B}] + 2t\ell/N .$$

*Proof.* A standard hybrid argument implies the existence of $\mathcal{A}_1$ which is an adversary for a 1-round game ($t = 1$ in our notation) such that $\mathrm{IV}_{\ell}\mathrm{Adv}[\mathcal{A}] \leq t \cdot \mathrm{IV}_{\ell}\mathrm{Adv}[\mathcal{A}_1]$.

We consider a series of hybrids (experiments). For each hybrid $H_i$, we let $\Pr[H_i]$ denote the probability that the experiment "succeeds" (an event we define below).

---

[8] A more natural choice is to sample $r \xleftarrow{\$} [|\mathbb{J}_N|]$, but since $|\mathbb{J}_N| = 2N' = \frac{\varphi(N)}{2}$ is hard to compute, we cannot sample from this distribution directly. However, since $r$ is used as an exponent of a group element, it is sufficient that $(r \bmod 2N')$ is uniform in $\mathbb{Z}_{2N'}$, and this is achieved by sampling $r$ from a much larger domain. We further remark that for the QR case, it is in fact sufficient to use $r \xleftarrow{\$} [(N-3)/4]$.

**Hybrid $H_0$.** In this experiment, we flip a coin $b \xleftarrow{\$} \{0,1\}$ and also sample $i \xleftarrow{\$} [\ell]$. We simulate the 1-round game with $\mathcal{A}_1$ where the challenger answers a query $\mathbf{a}$ with $(g_1^r, \ldots, g_{i-1}^r, (-1)^{b \cdot a_i} \cdot g_i^r, (-1)^{a_{i+1}} \cdot g_{i+1}^r, \ldots, (-1)^{a_\ell} \cdot g_\ell^r)$. The experiment succeeds if $b' = b$.

A standard argument shows that

$$\frac{\mathrm{IV}_\ell \mathrm{Adv}[\mathcal{A}_1]}{2\ell} = \left| \Pr[H_0] - \frac{1}{2} \right| .$$

**Hybrid $H_1$.** In this hybrid we replace $g_i$ (which is a uniform square) with $(-g_i)$. We get that there exists $\mathcal{B}$ such that $|\Pr[H_1] - \Pr[H_0]| \le 2 \cdot \mathrm{QRAdv}[\mathcal{B}]$.

We note that in this hybrid the adversary's query is answered with

$$(g_1^r, \ldots, g_{i-1}^r, (-1)^{b \cdot a_i} \cdot (-g_i)^r, (-1)^{a_{i+1}} \cdot g_{i+1}^r, \ldots, (-1)^{a_\ell} \cdot g_\ell^r) .$$

**Hybrid $H_2$.** In this hybrid the only change is that now $r \xleftarrow{\$} \mathbb{Z}_{2N'}$ (recall that $N' = \frac{\varphi(N)}{4}$) rather than $U([N^2])$. By Lemma 2.1 it follows that $|\Pr[H_2] - \Pr[H_1]| \le 1/N$. We note that while $N'$ is not explicitly known to any entity, this argument is statistical and there is no requirement that this hybrid is efficiently simulated.

We denote $r_1 = (r \bmod 2)$ and $r_2 = (r \bmod N')$. Since $N'$ is odd, the Chinese Remainder Theorem implies that $r_1, r_2$ are uniform in $\mathbb{Z}_2, \mathbb{Z}_{N'}$ respectively and are independent. The answer to the query in this scenario is therefore

$$(g_1^r, \ldots, g_{i-1}^r, (-1)^{b \cdot a_i} \cdot (-g_i)^r, (-1)^{a_{i+1}} \cdot g_{i+1}^r, \ldots, (-1)^{a_\ell} \cdot g_\ell^r) =$$
$$(g_1^{r_2}, \ldots, g_{i-1}^{r_2}, (-1)^{b \cdot a_i + r_1} \cdot g_i^{r_2}, (-1)^{a_{i+1}} \cdot g_{i+1}^{r_2}, \ldots, (-1)^{a_\ell} \cdot g_\ell^{r_2}) .$$

However since $r_1$ is a uniform bit, the answer is independent of $b$. It follows that $\Pr[H_2] = \frac{1}{2}$. Thus $\mathrm{IV}_\ell \mathrm{Adv}[\mathcal{A}_1] \le 4\ell \cdot \mathrm{QRAdv}[\mathcal{B}] + 2\ell/N$, and the result follows. $\qquad\square$

## 6 KDM Security

In this section, we discuss the KDM-security related properties of our QR-based scheme (for the general discussion and full details, see full version [7]). We prove the KDM$^{(1)}$-security of $\mathcal{E}[\ell]$, for $\ell \ge \log N + \omega(\log k)$, in Section 6.1. Then, in Section 6.2, we state that for $\ell \ge n \cdot \log N + \omega(\log k)$, $\mathcal{E}[\ell]$ is also KDM$^{(n)}$-secure. Finally, extensions beyond affine functions are stated in Section 6.3.

We define $\mathcal{F}_{\mathrm{aff}}$ to be the class of affine functions over $\mathbb{Z}_2$. Namely, all functions of the form $f_{a_0, \mathbf{a}}(\mathbf{x}) = a_0 + \sum a_i x_i$, where $a_i, x_i \in \mathbb{Z}_2$.

We use KDM$_{\mathcal{F}}\mathrm{Adv}[\mathcal{A}]$ to denote the advantage of an adversary $\mathcal{A}$ in distinguishing between a case where it gets legal encryptions of functions in $\mathcal{F}$ and the case where it gets encryptions of the constant message 0.

## 6.1  KDM$^{(1)}$-Security

The intuition behind the KDM$^{(1)}$-security of $\mathcal{E}[\ell]$ is as follows. Consider a public key $(g_0 = \prod g_i^{-s_i}, \mathbf{g})$ that corresponds to a secret key $\mathbf{s}$, and a function $f_{a_0,\mathbf{a}} \in \mathcal{F}_{\text{aff}}$. The encryption of $f_{a_0,\mathbf{a}}(\mathbf{s}) = (-1)^{a_0 + \sum a_i s_i}$ is

$$(c_0, \mathbf{c}) = ((-1)^{a_0 + \sum a_i s_i} \cdot g_0^r, \mathbf{g}^r) = ((-1)^{a_0} \cdot \prod((-1)^{a_i} \cdot g_i^r)^{-s_i}, \mathbf{g}^r) \ .$$

We notice that if $\mathbf{s}, a_0, \mathbf{a}$ are known, then $c_0$ is completely determined by $\mathbf{c} = \mathbf{g}^r$. Therefore, if we replace $\mathbf{g}^r$ with $(-1)^{\mathbf{a}} \cdot \mathbf{g}^r$ (an indistinguishable vector, even given the public key, by an interactive vector game), we see that $(c_0, \mathbf{c})$ is indistinguishable from $(c_0', \mathbf{c}') = ((-1)^{a_0} \cdot g_0^r, (-1)^{\mathbf{a}} \cdot \mathbf{g}^r)$, even when the secret key and the message are known. Applying the same argument again, taking into account that $g_0$ is close to uniform, implies that $(c_0', \mathbf{c}')$ is computationally indistinguishable from $(g_0^r, \mathbf{g}^r)$, which is an encryption of 0. A formal statement and analysis follow.

**Theorem 6.1.** *Let $\mathcal{A}$ be a KDM$^{(1)}_{\mathcal{F}_{\text{aff}}}$-adversary for $\mathcal{E}[\ell]$ that makes at most $t$ queries, then there exists an adversary $\mathcal{B}$ such that*

$$\text{KDM}^{(1)}_{\mathcal{F}_{\text{aff}}}\text{Adv}[\mathcal{A}] \leq 4t(2\ell + 1) \cdot \text{QRAdv}[\mathcal{B}] + \sqrt{N \cdot 2^{-\ell}} + O(t\ell/N) \ .$$

The theorem implies that taking $\ell = \log N + \omega(\log k)$ is sufficient to obtain KDM$^{(1)}$-security.

*Proof.* The proof proceeds by a series of hybrids. Let $b'$ denote $\mathcal{A}$'s output.

**Hybrid $H_0$.** In this hybrid, the adversary gets the public key, queries functions $f_{a_0,\mathbf{a}} \in \mathcal{F}_{\text{aff}}$ and gets legal encryptions of the functions of the secret key.

**Hybrid $H_1$.** In this hybrid, we change the way the challenger answers the adversary's queries. Recall that in hybrid $H_0$, the query $f_{a_0,\mathbf{a}} \in \mathcal{F}_{\text{aff}}$ was answered by $(c_0, \mathbf{c}) = ((-1)^{a_0 + \sum a_i s_i} \cdot g_0^r, \mathbf{g}^r)$. In hybrid $H_1$, it will be answered by $(c_0, \mathbf{c}) = ((-1)^{a_0} \cdot g_0^r, (-1)^{\mathbf{a}} \cdot \mathbf{g}^r)$.

We prove that

$$\left| \Pr_{H_1}[b' = 1] - \Pr_{H_0}[b' = 1] \right| \leq \text{IV}_\ell \text{Adv}[\mathcal{A}'] \leq 4t\ell \cdot \text{QRAdv}[\mathcal{B}_1] + O(t\ell/N) \ ,$$

for some $\mathcal{A}', \mathcal{B}_1$, even when $\mathbf{s}$ is fixed and known.

To see this, we notice that in both hybrids $c_0 = (-1)^{a_0} \cdot \prod_{i \in [\ell]}((-1)^{a_i} \cdot c_i^{-1})^{s_i}$ and $g_0 = \prod_{i \in [\ell]} g_i^{-s_i}$. Therefore an adversary $\mathcal{A}'$ for the interactive $\ell$-vector game can simulate $\mathcal{A}$, sampling $\mathbf{s}$ on its own and using $\mathbf{g}$ to generate $g_0$ and "translate" the challenger answers. Applying Lemma 5.1, the result follows.

**Hybrid $H_2$.** In this hybrid, we change the distribution of $g_0$, which will now be sampled from $U(\mathbb{QR}_N)$. By Lemma 2.3 combined with Lemma 2.2, $(g_0, \mathbf{g})$ is $\sqrt{\frac{N'}{2^{\ell+2}}} \leq \sqrt{\frac{N}{2^{\ell+2}}}$-uniform. Thus

$$\left| \Pr_{H_2}[b' = 1] - \Pr_{H_1}[b' = 1] \right| \leq \sqrt{\frac{N}{2^{\ell+2}}} \ .$$

**Hybrid $H_3$.** In this hybrid, we again change the way the challenger answers queries. Now instead of answering $(c_0, \mathbf{c}) = ((-1)^{a_0} \cdot g_0^r, (-1)^{\mathbf{a}} \cdot \mathbf{g}^r))$, the challenger answers $(c_0, \mathbf{c}) = (g_0^r, \mathbf{g}^r)$. The difference between $H_2$ and $H_3$ is now a $t$-query interactive $(\ell+1)$-vector game and thus by Lemma 5.1,

$$\left| \Pr_{H_3}[b' = 1] - \Pr_{H_2}[b' = 1] \right| \le 4t(\ell+1) \cdot \mathrm{QRAdv}[\mathcal{B}_2] + O(t\ell/N) \ ,$$

for some $\mathcal{B}_2$.

**Hybrid $H_4$.** We now revert the distribution of $g_0$ back to the original $\prod_{i \in [\ell]} g_i^{-s_i}$. Similarly to $H_2$, we have

$$\left| \Pr_{H_4}[b' = 1] - \Pr_{H_3}[b' = 1] \right| \le \sqrt{\frac{N}{2^{\ell+2}}} \ .$$

However, hybrid $H_4$ is identical to answering all the queries of the adversary by encryptions of 0. Summing the terms above, the result follows. □

## 6.2 KDM$^{(n)}$-Security

A formal statement for the QR case follows.

**Theorem 6.2.** *Let $\mathcal{A}$ be a $\mathrm{KDM}^{(n)}_{\mathcal{F}_{\mathrm{aff}}}$-adversary for $\mathcal{E}[\ell]$ that makes at most $t$ queries, then there exists an adversary $\mathcal{B}$ such that*

$$\mathrm{KDM}^{(n)}_{\mathcal{F}_{\mathrm{aff}}} \mathrm{Adv}[\mathcal{A}] \le 4nt(2\ell+1) \cdot \mathrm{QRAdv}[\mathcal{B}] + (N \cdot 2^{-\ell/n})^{n/2} + O(nt\ell/N) \ .$$

Thus, taking $\ell = n \cdot \log N + \omega(\log k)$ is sufficient for $\mathrm{KDM}^{(n)}$-security.

## 6.3 Beyond Affine Functions

Two building blocks have been suggested in [8,4] to obtain KDM-security w.r.t. a larger class of functions. Our scheme has the properties required to apply both constructions, yielding the following corollaries (that can be generalized to any SG assumption, see full version [7]).

The first corollary is derived using [8, Theorem 1.1]. A set of functions $\mathcal{H} = \{h_1, \ldots, h_\ell : h_i : \{0,1\}^\kappa \to \{0,1\}\}$ is *entropy preserving* if the function $f(x) = (h_1(x)\| \cdots \|h_\ell(x))$ is injective (the operator $\|$ represents string concatenation).

**Corollary 6.1.** *Consider $\mathcal{E}[\ell]$ and let $\kappa$ be polynomial in the security parameter such that $\kappa \ge \log N + \omega(\log k)$. Then for any entropy preserving set $\mathcal{H} = \{h_1, \ldots, h_\ell : h_i \in \{0,1\}^\kappa \to \{0,1\}\}$ of efficiently computable functions, with polynomial cardinality (in the security parameter), there exists a $\mathrm{KDM}^{(1)}$-secure scheme under the QR-assumption w.r.t. the class of functions*

$$\mathcal{F} = \left\{ f(\mathbf{x}) = a_0 + \sum a_i h_i(\mathbf{x}) : (a_0, \mathbf{a}) \in \mathbb{Z}_2 \times \mathbb{Z}_2^\ell \right\} \ .$$

The second corollary is derived using [4, Theorem 4.1].

**Corollary 6.2.** *Based on the* QR *assumption, for any polynomial p there exists a* KDM[(1)]*-secure encryption scheme w.r.t. all functions computable by circuits of size p(k) (where k is the security parameter).*

# 7   Leakage Resiliency

We prove that the scheme $\mathcal{E}[\ell]$ (our QR based scheme) is resilient to a leakage of up of $\lambda = \ell - \log N - \omega(\log k)$ bits. This implies that taking $\ell = \omega(\log N)$, achieves $(1 - o(1))$ leakage rate.

Intuitively, to prove leakage resiliency, we consider the case where instead of outputting the challenge ciphertext $((-1)^m \cdot g_0^r, \mathbf{g}^r)$, we output $((-1)^m \cdot (-1)^{\sum \sigma_i s_i} \cdot g_0^r, (-1)^\sigma \cdot \mathbf{g}^r)$, for a random vector $\sigma \xleftarrow{\$} \mathbb{Z}_2^\ell$. The views of the adversary in the two cases are indistinguishable (by an interactive vector game).[9] Using the leftover hash lemma, so long as $\mathbf{s}$ has sufficient min-entropy, even given $g_0$ and the leakage, then $\sum \sigma_i s_i$ is close to uniform. In other words, the ciphertexts generated by our scheme are computationally indistinguishable from ones that contain a strong extractor (whose seed is the aforementioned $\sigma$), applied to the secret key. This guarantees leakage resiliency.[10] The result in the QR case is formally stated below, where $\text{Leak}_\lambda \text{Adv}[\mathcal{A}]$ denotes the advantage of an adversary $\mathcal{A}$ in breaking the security of the scheme using $\lambda$ bits of leakage.

**Theorem 7.1.** *Let $\mathcal{A}$ be a $\lambda$-leakage adversary for $\mathcal{E}[\ell]$. Then there exists an adversary $\mathcal{B}$ such that*

$$\text{Leak}_\lambda \text{Adv}[\mathcal{A}] \le 8\ell \cdot \text{QRAdv}[\mathcal{B}] + \sqrt{N \cdot 2^{\lambda - \ell}} + O(\ell/N) \ .$$

# 8   Auxiliary-Input Resiliency

As in previous work, we start by stating weak auxiliary-input security in Lemma 8.1 below and then derive general auxiliary-input security for sub-exponentially hard functions in Corollary 8.1.

A function $f$ is $\epsilon$-weakly uninvertible if for any efficient $\mathcal{A}$, $\Pr[\mathcal{A}(1^k, pk, f_k(sk, pk)) = sk] \le \epsilon(|sk|)$.

**Lemma 8.1.** *Let $\epsilon(\ell)$ and $f$ be such that $\epsilon$ is negligible and $f$ is $\epsilon$-weakly uninvertible function (more precisely, family of functions). Then under the* QR *assumption, the scheme $\mathcal{E}[\ell]$ is secure even with auxiliary input $f(sk)$.*

We note that the above may seem confusing since it appears to imply auxiliary-input security, and thus also semantic security, regardless of the value of $\ell$.

---

[9] Of course the latter ciphertext can only be generated using the secret key, but the indistinguishability holds even when the secret key is known.

[10] In the spirit of [22], we can say that our scheme defines a new hash proof system that is universal with high probability over illegal ciphertexts, a property which is sufficient for leakage resiliency.

However, we recall that if $\ell$ is too small, then we may be able to retrieve $\mathbf{s}$ from $pk$ without the presence of any auxiliary input. Therefore the value of $\ell$ must be large enough in order for $f$ to be weakly uninvertible.

We can then derive the following corollary.

**Corollary 8.1.** *Assuming that a subgroup indistinguishability assumption holds, then for any constant $\delta > 0$ there is an encryption scheme that is resilient to auxiliary input $f(sk)$ any function $f$ is hard to invert with probability $2^{-\ell^\delta}$.*

**Acknowledgments.** The authors wish to thank Gil Segev for illuminating discussions. The first author wishes to thank Microsoft Research New-England, for hosting him at the time of this research.

# References

1. Adão, P., Bana, G., Herzog, J.C., Scedrov, A.: Soundness of formal encryption in the presence of key-cycles. In: di Vimercati, S.d.C., Syverson, P.F., Gollmann, D. (eds.) ESORICS 2005. LNCS, vol. 3679, pp. 374–396. Springer, Heidelberg (2005)
2. Akavia, A., Goldwasser, S., Vaikuntanathan, V.: Simultaneous hardcore bits and cryptography against memory attacks. In: Reingold, O. (ed.) TCC 2009. LNCS, vol. 5444, pp. 474–495. Springer, Heidelberg (2009)
3. Applebaum, B., Cash, D., Peikert, C., Sahai, A.: Fast cryptographic primitives and circular-secure encryption based on hard learning problems. In: Halevi (ed.) [18], pp. 595–618
4. Barak, B., Haitner, I., Hofheinz, D., Ishai, Y.: Bounded key-dependent message security. In: Gilbert, H. (ed.) EUROCRYPT 2010. LNCS, vol. 6110, pp. 423–444. Springer, Heidelberg (2010)
5. Black, J., Rogaway, P., Shrimpton, T.: Encryption-scheme security in the presence of key-dependent messages. In: Nyberg, K., Heys, H.M. (eds.) SAC 2002. LNCS, vol. 2595, pp. 62–75. Springer, Heidelberg (2003)
6. Boneh, D., Halevi, S., Hamburg, M., Ostrovsky, R.: Circular-secure encryption from decision diffie-hellman. In: Wagner, D. (ed.) CRYPTO 2008. LNCS, vol. 5157, pp. 108–125. Springer, Heidelberg (2008)
7. Brakerski, Z., Goldwasser, S.: Circular and leakage resilient public-key encryption under subgroup indistinguishability (or: Quadratic residuosity strikes back). In: Cryptology ePrint Archive, Report 2010/226 (2010), http://eprint.iacr.org/
8. Brakerski, Z., Goldwasser, S., Kalai, Y.: Circular-secure encryption beyond affine functions. In: Cryptology ePrint Archive, Report 2009/485 (2009), http://eprint.iacr.org/
9. Camenisch, J., Lysyanskaya, A.: An efficient system for non-transferable anonymous credentials with optional anonymity revocation. In: Pfitzmann, B. (ed.) EUROCRYPT 2001. LNCS, vol. 2045, pp. 93–118. Springer, Heidelberg (2001)
10. Cramer, R., Shoup, V.: Universal hash proofs and a paradigm for adaptive chosen ciphertext secure public-key encryption. In: Knudsen, L.R. (ed.) EUROCRYPT 2002. LNCS, vol. 2332, pp. 45–64. Springer, Heidelberg (2002)
11. Damgård, I., Jurik, M.: A generalisation, a simplification and some applications of paillier's probabilistic public-key system. In: Kim, K.-c. (ed.) PKC 2001. LNCS, vol. 1992, pp. 119–136. Springer, Heidelberg (2001)

12. Dodis, Y., Goldwasser, S., Kalai, Y., Peikert, C., Vaikuntanathan, V.: Public-key encryption schemes with auxiliary inputs. In: Micciancio, D. (ed.) TCC 2010. LNCS, vol. 5978, pp. 361–381. Springer, Heidelberg (2010)
13. Dodis, Y., Kalai, Y.T., Lovett, S.: On cryptography with auxiliary input. In: Mitzenmacher, M. (ed.) STOC, pp. 621–630. ACM, New York (2009)
14. Goldreich, O., Levin, L.A.: A hard-core predicate for all one-way functions. In: STOC, pp. 25–32. ACM, New York (1989)
15. Goldwasser, S., Micali, S.: Probabilistic encryption and how to play mental poker keeping secret all partial information. In: STOC, pp. 365–377. ACM, New York (1982)
16. Goldwasser, S., Micali, S.: Probabilistic encryption. J. Comput. Syst. Sci. 28(2), 270–299 (1984)
17. Halderman, J.A., Schoen, S.D., Heninger, N., Clarkson, W., Paul, W., Calandrino, J.A., Feldman, A.J., Appelbaum, J., Felten, E.W.: Lest we remember: Cold boot attacks on encryption keys. In: van Oorschot, P.C. (ed.) USENIX Security Symposium, pp. 45–60. USENIX Association (2008)
18. Halevi, S. (ed.): CRYPTO 2009. LNCS, vol. 5677. Springer, Heidelberg (2009)
19. Hemenway, B., Ostrovsky, R.: Lossy trapdoor functions from smooth homomorphic hash proof systems. In: Electronic Colloquium on Computational Complexity (ECCC), vol. 16(127) (2009), http://eccc.uni-trier.de/report/2009/127/
20. Kiltz, E., Pietrzak, K., Stam, M., Yung, M.: A new randomness extraction paradigm for hybrid encryption. In: Joux, A. (ed.) EUROCRYPT 2009. LNCS, vol. 5479, pp. 590–609. Springer, Heidelberg (2010)
21. Laud, P., Corin, R.: Sound computational interpretation of formal encryption with composed keys. In: Lim, J.-I., Lee, D.-H. (eds.) ICISC 2003. LNCS, vol. 2971, pp. 55–66. Springer, Heidelberg (2004)
22. Naor, M., Segev, G.: Public-key cryptosystems resilient to key leakage. In: Halevi (ed.) [18], pp. 18–35
23. Paillier, P.: Public-key cryptosystems based on composite degree residuosity classes. In: Stern, J. (ed.) EUROCRYPT 1999. LNCS, vol. 1592, pp. 223–238. Springer, Heidelberg (1999)
24. Peikert, C., Waters, B.: Lossy trapdoor functions and their applications. In: Ladner, R.E., Dwork, C. (eds.) STOC, pp. 187–196. ACM, New York (2008)
25. Regev, O.: On lattices, learning with errors, random linear codes, and cryptography. In: Gabow, H.N., Fagin, R. (eds.) STOC, pp. 84–93. ACM, New York (2005)

# Leakage-Resilient Pseudorandom Functions
# and
# Side-Channel Attacks on Feistel Networks

Yevgeniy Dodis and Krzysztof Pietrzak

New York University and CWI Amsterdam

**Abstract.** A cryptographic primitive is leakage-resilient, if it remains secure even if an adversary can learn a bounded amount of arbitrary information about the computation with every invocation. As a consequence, the physical implementation of a leakage-resilient primitive is secure against every side-channel as long as the amount of information leaked per invocation is bounded.

In this paper we prove positive and negative results about the feasibility of constructing leakage-resilient pseudorandom functions and permutations (i.e. block-ciphers). Our results are three fold:

**1.** We construct (from any standard PRF) a PRF which satisfies a relaxed notion of leakage-resilience where (1) the leakage function is fixed (and not adaptively chosen with each query.) and (2) the computation is split into several steps which leak individually (a "step" will be the invocation of the underlying PRF.)

**2.** We prove that a Feistel network with a super-logarithmic number of rounds, each instantiated with a leakage-resilient PRF, is a leakage resilient PRP. This reduction also holds for the non-adaptive notion just discussed, we thus get a block-cipher which is leakage-resilient (against non-adaptive leakage).

**3.** We propose generic side-channel attacks against Feistel networks. The attacks are generic in the sense that they work for any round functions (e.g. uniformly random functions) and only require some simple leakage from the inputs to the round functions. For example we show how to invert an $r$ round Feistel network over $2n$ bits making $4 \cdot (n+1)^{r-2}$ forward queries, if with each query we are also given as leakage the Hamming weight of the inputs to the $r$ round functions. This complements the result from the previous item showing that a super-constant number of rounds is necessary.

# 1 Introduction

Traditional cryptographic security definitions only give the adversary black-box access to the primitive at hand. For example, a function $\mathsf{F} : \Sigma^k \times \Sigma^m \to \Sigma^n$ ($\Sigma \stackrel{\text{def}}{=} \{0,1\}$) is pseudorandom if no efficient adversary given oracle access to a function $\mathcal{O} : \Sigma^m \to \Sigma^n$ can tell whether the oracle is a uniformly random function or instantiated with $\mathsf{F}(K,.)$ for a random key $K \in \Sigma^k$.

T. Rabin (Ed.): CRYPTO 2010, LNCS 6223, pp. 21–40, 2010.

Unfortunately, this model does not capture many attacks in the real-world where adversaries can attack concrete *implementations* of cryptosystems which potentially leak information about their internal secret state during computation. Attacks exploiting such leakage are called side-channel attacks. Popular side-channels that have been exploited for cryptanalytic attacks include running-time [28], electromagnetic radiation [39,20] or power consumption [30].

**Countermeasures.** Side-channel attacks are a very real threat for systems used in practice. Not surprisingly, much research has concentrated on developing countermeasures against such attacks. This research is mostly done by practitioners (i.e., the cryptographic hardware community) who are also active in finding and exploiting new side-channels, [37] gives an overview of this research. The countermeasures proposed are usually ad-hoc, in the sense that they aim to protect against some particular, known attack, and are backed up by heuristic security arguments. This is fundamentally different from the provable security approach taken by modern cryptography, where one requires that a scheme is *proven* secure against a class of resource bounded (e.g. polynomial time) adversaries and not only particular attacks. This situation is very unsatisfying; after all, what is a provably secure cryptosystem good for, if ultimately its security hinges on an ad-hoc side-channel countermeasure? Nonetheless, until recently there was almost no input from the theory community on side-channel countermeasures as it was believed that this is a practical problem, and theory can only be of limited use in this context. Fortunately, recent results indicate that this view was much too pessimistic. In an early influential paper, Micali and Reyzin [35] propose the "physically observable cryptography" framework which adapts the concept of cryptographic *reductions* to the context of side-channel attacks. Only very recently direct constructions of cryptographic schemes were proposed which are provably secure against general classes of side-channel attacks. We'll discuss several such modes below.

**Leakage-Resilient PRFs.** A cryptographic primitive is *leakage-resilient* if it remains secure even if the adversary can – with each invocation – learn a bounded amount of arbitrary information about the computation. This notion was introduced in [17], and is formally modelled by allowing the adversary to choose (besides the regular input, if there is any) a leakage function $g$ with bounded range $\Sigma^\lambda$ for some leakage parameter $\lambda$.[1] After the invocation the adversary gets – besides the regual output – the leakage $g(\tau)$ where $\tau$ is all data accessed by the primitive during this invocation (that is, the part of the secret state that was accessed and – if the primitive is probabilistic – any random coins used). We will take a more "fine-grained" view and split an invocation into $t > 1$ sequential steps, where the adversary is allowed to learn a bounded amount of information

---

[1] The basic idea to consider adversaries who can learn any (sufficiently compressing) function $g(S)$ about the secret state $S$ goes back to Maurer's bounded storage model [32,15,42]. The bounded retrieval model [14,8] adapts this to the computational setting.

$g_1(\tau_1), \ldots, g_t(\tau_t)$ about every step. Here $\tau_i$ denotes absolutely all information that is accessed in the $i$-th step.

As a consequence, the physical implementation of a leakage-resilient cryptosystem will remain secure in the presence of any side-channel attack, as long as the information exploited by this attack can be modelled by adaptively chosen leakage functions as just described. A sufficient (but not necessary) condition on the side-channel is to require that (1) the amount of information leaked per invocation (or, in the fine-grained approach, per step) is at most $\lambda$ bits and (2) "only computation leaks information", which means that parts of the memory which are not accessed during an invocation (or step) will not leak.

*Remark 1 (On "Only computation leaks information").* "Only computation leaks information" is an assumption about the physical properties of cryptodevices, and was originally put forward as one the "axioms" in the physically observable cryptography framework of Micali and Reyzin [35]. As just mentioned, devices adhering to this axiom are captured by the model of leakage resilience, but this is only a sufficient condition and by no means necessary. For example, [38] explains why the mathematical model of leakage-resilience also captures certain physical attacks which explicitly violate this axiom, like "cold-boot attacks" [22] or when considering memory that is subject to static leakage.

**Limitations of Current Techniques.** The only leakage-resilient primitives that were constructed so far *in the standard model* are stream-ciphers [17,38] and signature schemes [19]. A leakage-resilient public-key encryption scheme has been constructed, but only in the idealised generic group model [27]. A central open problem is this line of research is the construction of pseudorandom functions (PRFs) and permutations (PRPs, or equivalently, block-ciphers). Block-ciphers are the work horses of crypto. Not surprisingly, they are also a favourite target of side-channel cryptanalysts.

In this work we consider the problem of constructing leakage-resilient PRFs and PRPs. The techniques used in the construction of leakage-resilient streamciphers and signature schemes crucially rely on *key evolution*. For example, in a stream-cipher the key evolves naturally, while for signatures one can sample a fresh public/secret key pair with each signature query and sign the new key with an old key. Unfortunately it is not clear how to evolve the key of a PRF/PRP. The same difficulty arises with public-key encryption, so the leakage-resilient PKE scheme from [27] does not rely on evolution, but rather on sharing the secret key. The sharing is rerandomized after each invocation. In order to decrypt using the shares of the secret key without actually reconstructing it, one exploits the homomorphic property of the group. Thus, even aside from the reliance on idealised generic groups [27], this technique is not an option to construct leakage-resilient PRFs/PRPs if we do not want to use inefficient techniques and assumptions (like DDH) that are used in public-key cryptography.

**Our PRF Results.** As leakage-resilient PRFs seem out of reach with our current techniques, we will consider a relaxed notion of leakage-resilience, where the leakage function is not adaptively chosen by the adversary before each invocation, but is fixed. This notion still captures all side-channel attacks where the adversary will always measure (almost) the same leakage if she performs exactly the same computation. This for example captures timing and to some extent power-analysis attacks[2], but not probing attacks (where different wires can be probed on different invocations on the same input.) We construct a PRF which is secure under this relaxed notion from any standard PRF. The construction, as illustrated on the left in Figure 1, can be seen as a hybrid of the GGM construction [21] (which constructs a PRF from any PRG) and the leakage-resilient stream cipher from [38].

*Related Work.* The idea to only consider non-adaptive leakage functions and that this could be useful in the context of the GGM construction goes back at least to Micali and Reyzin [35].[3] A similar point for a particular leakage function (power analysis) was made by Kocher [29]. The idea to consider leakage-resilience but to fix the leakage function is due to Standaert et al. [41]. They suggest that the GGM construction is secure in this setting if the PRG is modelled as a uniformly random function and the leakage function is fixed.[4]

**Side-Channel Attacks on Feistel.** A pseudorandom permutation (PRP) $\mathsf{F} : \Sigma^k \times \Sigma^n \to \Sigma^n$ is defined like a PRF, except that one requires that for every key $K \in \Sigma^k$, $\mathsf{F}(K, .)$ is a permutation. A super PRP (sPRP) satisfies a stronger notion where the adversary can also make inverse queries. The additional structural properties of permutations are often useful as they allow for better efficiency and/or security. Block-ciphers, which are strong PRPs, are the "work horses" of cryptography and a favourite target of side-channel cryptanalysts.

PRPs seem to be much more complicated objects than PRFs, but in a classical paper, Luby and Rackoff [31] prove that a simple 3 round Feistel network (cf. Definition 6) instantiated with PRFs, is a PRP. With one round more one even gets a sPRP. More recently, [7] prove that a six round Feistel network instantiated with

---

[2] If the power-analysis just leaks the number of wires set to 1, then this is captured, but if the power-analysis leaks the number of wires that "switch" from 0 to 1, then this is no longer possible.

[3] From [35]: *Our definitions allow for repeated computation to leak new information each time. However, the case can be made (e.g., due to proper hardware design) that some devices computing a given function f may leak the same information whenever f is evaluated at the same input x. This is actually implied by making the leakage function deterministic and independent of the adversary measurement. Fixed-leakage physically observable cryptography promises to be a very useful restriction of our general model (e.g., because, for memory efficiency, crucial cryptographic quantities are often reconstructed from small seeds, such as in the classical pseudorandom function of [21]).*

[4] The model considered is basically the random oracle model, but it is conceptually used in a different way. In the RO model, a uniformly random function is accessible to all parties, and security proofs only exploit the fact that a random oracle allows to efficiently access an exponential amount of true randomness. In contrast, in [41] the security proof exploits the fact that the adversarial leakage functions cannot *query* the random oracle.

random functions is *indifferentiable* [34] from a uniformly random permutation. These results suggest that a Feistel network with some small constant number of rounds instantiated with *leakage-resilient* PRFs, would yield a *leakage-resilient* PRP.

Unfortunately, this is not true. We show very simple side-channel attacks against Feistel networks where the round functions can be *arbitrary*, and the only leakage is some (simple) function $g(.)$ of the inputs to the round functions. We identify a simple property of leakage-functions function $g(.)$ – which we call "reconstructible" (cf. Definition 7) – that is sufficient for our attack to work. This property is shared by many simple and natural leakage functions (like the Hamming weight or the identity function with very high noise). Thus our attacks are quite practical. We explain these attacks in detail in Section 3 (which is self contained and can be read independently of the rest of this paper), here only giving the brief summary. We show that getting leakage from any reconstructible leakage function $g(.)$ is sufficient to allow the side-channel attacker to invert the Feistel network on any input using a number of *forward* queries which is exponential in the number of rounds (and, thus, in polynomial time for any fixed constant number of rounds). This breaks the security of any fixed-round Feistel network as a PRP.

For readers familiar with the notion of Indifferentiability [34,6], it might seem that our attacks contradict the beautiful result of Coron et al. [7] showing that a six round Feistel network with random functions is indifferentiable from a random permutation. The reason this is not a contradiction is that the indifferentiability simulator $\mathcal{S}$ is allowed to make arbitrary *additional* forward/backward queries to the random permutation when trying to "fake" the six random round functions, as opposed to the queries made by the distinguisher (which the simulator does not even see). For example, for our attack making only forward queries, the simulator will be "smart enough" to figure out the backward query we are "computing" using our forward queries, and will make such a query in advance to avoid any inconsistencies. Translated to the setting of leakage, the indifferentiability framework will imply the following much weaker notion of security than the one we are aiming for: after making $q$ block-cipher queries and observing the leakage, all but specially chosen $poly(q)$ input/outputs of the block cipher will "look random". In contrast, we will ensure that *every* un-queried input/output pair will "look random".

We also mention that [12] defined a notion of "honest but curious indifferentiability". As observed by [12,7] this notion is *incomparable* to standard indifferentiability. On one hand, it is stronger because the simulator $\mathcal{S}$ is not allowed to make any queries to $\mathbf{P}$ or $\mathbf{P}^{-1}$ (but only sees the queries made by the distinguisher). But it is also weaker, as the distinguisher is not allowed to query intermediate round functions, but only the entire Feistel network (or its simulation) together with all the inputs/outputs of the internal round functions. This notion is much closer to the setting of side-channel attacks, except with side-channels we allow a much richer class of leakage functions (e.g., those that depend on the key). In fact, the side-channel attacks we propose generalize (and strengthen) a lower bound from [12] which basically corresponds to our attack for the special case where the leakage contains the entire inputs to the round functions.

**Leakage-Resilient PRPs.** In light of the results discussed in the previous section, the best we can hope for is that an $r$-round Feistel network $\Psi_r$, instantiated with leakage-resilient PRFs, is secure against adversaries who make at most an exponential (in $r$) number of queries. In Section 4 we show (again using techniques from [12]) that this is indeed the case: the $r$-round Feistel network is a secure leakage-resilient super PRP as long as the number of queries is bounded by $q \leq 1.38^{r/2-1}$.

We notice that the leakage-resilient sPRP, as just described, is secure in an attack scenario where the adversary with every query to $\Psi_r$ gets to see all the inputs[5] to the $r$ round functions and *also* leakage from every round function (as computed by any leakage function for which the underlying leakage-resilient PRF is secure). Also, the reductions works for other notions of leakage-resilience, in particular for the original notion of leakage-resilience where the leakage-function is chosen adaptively. Thus, although our current PRF constructions only give us "non-adaptive-leakage" sPRPs, future advances in leakage-resilient PRFs would immediately translate to stronger leakage-resilient sPRPs.

In contrast, when proposing attacks, we want to consider a setting where the adversary is as limited as possible. As explained in the previous section, the side-channel attacks we propose against Feistel require a very limited setting where the only leakage the adversary gets is some simple function (e.g. Hamming weight) of the inputs to the round functions. The attack works no matter what the round functions are, they can be leakage-proof PRFs or even uniformly random functions.

**More Related Work.** We shortly discuss some work on *provable* side-channel security not already covered in the introduction. The more practical work on this topic is too extensive to cover here, [37] gives an overview of this research.

*Private Circuits.* Ishai et al. [25,24] consider a model where the adversary can choose some wires in the circuit on which the cryptographic algorithm is run, and then learns the values carried by those wires during the computation (This can be seen as a generalisation of exposure resilient cryptography [13], where the adversary was restricted to learn some bits of the *input*.) They were the first to *prove* how to implement *any* algorithm secure against an interesting side-channel, i.e. probing attacks. This work uses techniques from general multiparty computation (MPC).[6] Recently Faust et al. [18] extended this result to significantly more general classes of leakage, in particular, they give a construction

---

[5] The outputs of the round functions can be computed from the input: the output of the $i$th round functions is the XOR of the inputs of rounds $i-1$ and $i+1$.

[6] Formally, Ishai et al. prove the following: let $t \geq 0$ be some constant and let $[X]$ denote a $(t+1)$ out of $(t+1)$ secret sharing of the value $X$. They construct a general compiler, which turns every circuit $G(.)$ into a circuit $G_t(.)$ (of size $O(t|G|)$) such that $[G(X)] = G_t([X])$ for all inputs $X$, and moreover one does not learn any information on $G(X)$ even when given the value carried by any $t$ wires in the circuit $G_t(.)$ while evaluating the input $[X]$. This transformation uses multiparty-computation, which is quite different from all other approaches we discuss here.

(also based on general MPC) which remains secure given leakage computed by any function from a low complexity class like $AC_0$. The main drawback of those constructions is that the amount of leakage that can be tolerated is very small: to tolerate $t$ bits leakage, the circuits must be blown up by a factor of at least $t$. Moreover the construction from [18] requires (albeit very simple) completely leakage proof components.

*(Continuous) Memory Attacks.* A cryptographic scheme is secure against memory attacks, if it remains secure even if a bounded amount of information about the secret key is given to the adversary. In this model [1,36,4] construct public-key encryption schemes and [26,2] construct signature schemes, identification schemes and key exchange protocols.[7] Unlike leakage-resilience, here the leakage function gets the *entire* secret state as input, and not only what was accessed. On the downside – unlike leakage-resilience or private circuits – memory attacks are a "one-shot" game where the total amount of leakage cannot be larger than the length of the secret key. Very recently [10,5] extended the model of memory attacks to the continuous setting. In their model the secret key gets periodically updated (using local randomness and without changing the public key), and a bounded amount about of information about the secret key can leak in-between every two updates. The update phases can also leak, but only a logarithmic amount. In this model, [10] construct identification, signature and authenticated key agreement schemes, [5] construct signatures and PKE.

*Auxiliary Input.* [11] introduce the notion of security against auxiliary input, where one requires the scheme to be secure even if the adversary is given some leakage $g(K)$ about the secret key as long as $g(.)$ is uninvertible. That is, $K$ cannot be inverted given $g(K)$ but with very small probability. In this model private-key [11] and public-key [9] encryption schemes have been constructed.

## Notation & Basic Definitions

- $\Sigma^t$ denotes $\{0,1\}^t$, i.e. all bitstring of length $t$. $\Sigma^{\leq t} \overset{\text{def}}{=} \bigcup_{i=0}^t \Sigma^t$ denotes all bitstrings of length at most $t$, including the empty string $\varepsilon$.
- $[a,b]$ denotes the interval $\{a, a+1, \ldots, b\}$, $[b]$ is short for $[1,b]$.
- Sequential composition of functions is denoted with $g \circ f(x) \overset{\text{def}}{=} g(f(x))$.
- Concatenation of two strings $x, y$ is denoted $x\|y$, or, if no confusion is possible, simply $xy$.
- $w_H(x)$ denotes the number of 1's (i.e. Hamming weight) in $x$.
- $\mathbf{R}_{m,n}$ denotes a uniformly random function $\Sigma^m \to \Sigma^v$, $\mathbf{P}_n$ a uniformly random permutation over $\Sigma^n$.
- For $X \in \Sigma^n$ we denote with $X_{|i}$ the $i$ bit prefix of $X$.

---

[7] Let us mention that PRFs and PRPs (i.e. the primitives considered in this paper) that are secure against memory attacks do not even exist. E.g. we can trivially distinguish $F(K, X)$ (here $K$ is the key and $X$ is any fixed input to the PRF $F(.,.)$) from uniform with advantage $1 - 2^{-\lambda}$ given as leakage the first $\lambda$ bits of $F(K, X)$.

- $\mathsf{pre}(X) = \bigcup_{i=0}^{n} X_{|i}$ denotes the set of all prefixes of $X$, including the empty string $\varepsilon = X_{|0}$ and the entire $X = X_{|n}$.
- We sometimes write $X^q$ to denote a sequence $X_1, \ldots, X_q$ of values.
- For a set $\mathcal{X}$, $X \xleftarrow{*} \mathcal{X}$ denotes that $X$ is assigned a value sampled uniformly at random from $\mathcal{X}$.
- We denote with $\delta^{\mathsf{D}}(X;Y)$ the advantage of a circuit $\mathsf{D}$ in distinguishing the random variables $X, Y$, i.e.: $\delta^{\mathsf{D}}(X;Y) \stackrel{\mathrm{def}}{=} |\Pr[\mathsf{D}(X) = 1] - \Pr[\mathsf{D}(Y) = 1]|$. With $\delta_s(X;Y)$ we denote $max_{\mathsf{D}} \delta^{\mathsf{D}}(X;Y)$ where the maximum is over all circuits $\mathsf{D}$ of size $s$.

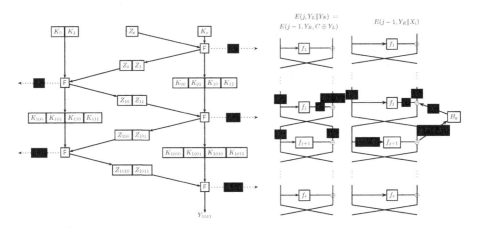

**Fig. 1. Left:** Illustration of the NALR-secure PRF $\Gamma^{\mathsf{F},m} : \Sigma^{3k+n} \times \Sigma^m \to \Sigma^{4k+2n}$ (here shown for $m = 4$ and input $1011 \in \Sigma^m$) from any standard (weak) PRF $\mathsf{F} : \Sigma^k \times \Sigma^n \to \Sigma^{4k+2n}$. We consider adversaries who with each such query $X$ can get leakage $\Lambda_I$ for every $I \in \mathsf{pre}(X)$ which is defined as $\Lambda_I \stackrel{\mathrm{def}}{=} g(K_I, Z_I, I)$, where $g$ is any function of bounded size $s$ and range $\lambda$. And moreover all the $Z_I, I \in \mathsf{pre}(X)$.
**Right**: Illustration of the second Claim from the proof of Theorem 2.

## 2   Leakage-Resilient PRFs

Figure 1 (left) illustrates our construction of a PRF $\mathsf{F} : \Sigma^k \times \Sigma^m \to \Sigma^n$ for which we will show that it satisfies a relaxed notion of leakage-resilience where the leakage function is a priori fixed (and not adaptively by the adversary with every query). Recall the standard definitions of (weak) PRFs.

**Definition 1 (PRF/weak PRF).** $\mathsf{F} : \Sigma^\kappa \times \Sigma^m \to \Sigma^n$ *is an* $(\epsilon_{\mathsf{prf}}, s_{\mathsf{prf}}, q_{\mathsf{prf}})$-*secure pseudorandom function (PRF) if no adversary of size* $s_{\mathsf{prf}}$ *can distinguish* $\mathsf{F}$ *(instantiated with a random key) from a uniformly random function, i.e. for any* $\mathcal{A}$ *of size* $s_{\mathsf{prf}}$ *making* $q_{\mathsf{prf}}$ *oracle queries we have*

$$\Pr_K[\mathcal{A}^{\mathsf{F}(K,\cdot)} \to 1] - \Pr_{\mathbf{R}_{m,n}}[\mathcal{A}^{\mathbf{R}_{m,n}(\cdot)} \to 1] \leq \epsilon_{\mathsf{prf}}$$

F *as above is a* $(\epsilon_{\mathsf{prf}}, s_{\mathsf{prf}}, q_{\mathsf{prf}})$-*secure* <u>*weak*</u> *PRF if the above only holds for randomly (and not adversarially) chosen inputs, i.e. for* $K \xleftarrow{*} \Sigma^\kappa$ *and*

$$\text{for } i = 1, \ldots, q_{\mathsf{prf}}: \quad X_i \xleftarrow{*} \Sigma^m \quad Y_i \leftarrow \mathsf{F}(K, X_i) \quad R_i \leftarrow \mathbf{R}_{m,n}(X_i)$$

*we have* $\quad \Pr[\mathcal{A}(X^{q_{\mathsf{prf}}}, Y^{q_{\mathsf{prf}}}) = 1] - \Pr[\mathcal{A}(X^{q_{\mathsf{prf}}}, R^{q_{\mathsf{prf}}}) = 1] \leq \epsilon_{\mathsf{prf}}$

Definition 2 below specifies what we mean by a PRF F being leakage-resilient w.r.t. to a class of leakage functions $\mathcal{L}$. Informally, we consider an adversary $\mathcal{A}$ with access to two oracles. Initially, we sample a key $K \xleftarrow{*} \Sigma^k$. The first oracle then takes as input some $X \in \Sigma^m$ and outputs the output of the PRF $Y \leftarrow \mathsf{F}(K, X)$ on this input and the leakage $\Lambda \leftarrow g(K, X)$ (where $g$ is any function from the class $\mathcal{L}$). The second oracle is either a uniformly random function $\mathbf{R}_{m,n}$, or the PRF $\mathsf{F}(K, .)$ (using the same key as $K$ the first oracle.). We require that no efficient $\mathcal{A}$ can distinguish these two cases. Of course we have to require that $\mathcal{A}$ never queries the two oracles on the same input $X$, as otherwise distinguishing becomes trivial.

The practical implication of this definition is as follows. Consider an adversary who can launch a side-channel attack against $\mathsf{F}(K, .)$, where for every query $\mathsf{F}(K, X)$ made she can measure some leakage $\Lambda(K, X)$. If F is $\mathcal{L}$ resilient, and the leakage $\Lambda(K, X)$ can be modelled as $\Lambda(K, X) = g(K, X)$ for some $g \in \mathcal{L}$, then for all inputs $X'$ on which $\mathsf{F}(K, .)$ has not yet been queried, the output $\mathsf{F}(K, X')$ will be indistinguishable from random.

**Definition 2 ($\mathcal{L}$-resilient PRF/PRP/sPRP).** F $: \Sigma^\kappa \times \Sigma^m \rightarrow \Sigma^n$ *is a* $(\epsilon_{\mathsf{prf}}, s_{\mathsf{prf}}, q_{\mathsf{prf}})$-*secure* $\mathcal{L}$-*resilient pseudorandom function if for every adversary* $\mathcal{A}$ *of size* $s_{\mathsf{prf}}$ *and every* $g \in \mathcal{L}$

$$\Pr_K[\mathcal{A}^{\mathsf{F}^g(K,.),\mathsf{F}(K,.)} \rightarrow 1] - \Pr_{K,\mathbf{R}_{m,n}}[\mathcal{A}^{\mathsf{F}^g(K,.),\mathbf{R}_{m,n}(.)} \rightarrow 1] \leq \epsilon_{\mathsf{prf}} \qquad (1)$$

*Here* $\mathcal{A}$ *can make a total of* $q_{\mathsf{prf}}$ *queries (arbitrarily scheduled) to his two oracles, but the queries to the first and second oracle must be disjoint. The first oracle* $\mathsf{F}^g(K, .)$ *takes as input* $X \in \Sigma^m$ *and outputs* $\mathsf{F}(K, X), g(K, X)$.

$\mathcal{L}$-*resilient pseudorandom permutations (PRP) are defined similarly, except that now for every* $K$, $\mathsf{F}(K, .)$ *has to be a permutation and the random function* $\mathbf{R}_{m,n}$ *in eq.(1) is replaced with a random permutation* $\mathbf{P}_m$. *A* $\mathcal{L}$-*resilient* **super** *PRP (sPRP) is defined the same way, except that now we additionally allow the adversary to make inverse queries. Here* $\mathcal{A}$ *is also not allowed to make an inverse (forward) query* $Y$ *to one oracle, if* $Y$ *has been received as output to a forward (inverse) query from the other oracle.*

**Definition 3 (NARL security).** *We say that a PRF* F *(same for PRP,sPRP) is* non-adaptive leakage-resilient *if the computation of* $\mathsf{F}(K, X)$ *can be split into* $t \geq 1$ *steps, and* F *is* $\mathcal{L}$-*resilient w.r.t. to a class* $\mathcal{L}$ *which can leak, for every of the* $t$ *steps, arbitrary* $\lambda$ *bits of information about all the data that is accessed in this step.*

Below we define our construction $\Gamma^{F,m}$ of a function as illustrated is Figure 1 for which we will prove that it is NARL secure if instantiated with any standard weak PRF F. This construction can be seen as a hybrid of the GGM construction [21] and the leakage-resilient stream-cipher from [38].

**Definition 4 (Construction $\Gamma^F$).** *For a functions* $F : \Sigma^k \times \Sigma^n \to \Sigma^{4k+2n}$, *we denote with* $\Gamma^F$ *a function* $\Sigma^{3k+n} \times \Sigma^m \to \Sigma^{4k+2n}$ *defined as follows (cf. Figure 1). The secret key $K$ consists of the four values* $Z_\varepsilon \in \Sigma^n, K_\varepsilon, K_0, K_1 \in \Sigma^k$. *The output on input* $X \in \Sigma^m$ *is* $Y_X \leftarrow F(K_X, Z_X)$ *where* $Z_I, K_I$ *for* $I \in$ pre$(X)$ *are recursively defined as*

$$(Z_{I0}, Z_{I1}, K_{I00}, K_{I01}, K_{I10}, K_{I11}) \leftarrow F(K_I, Z_I)$$

*Figure 1 illustrates this construction for $m = 4$ on input $X = 1011$.*

Theorem 1 below states that $\Gamma^F$ is NARL secure. Or more precisely, $\mathcal{L}$-resilient, where $\mathcal{L}$ contains all functions that leak $\lambda$ bits of arbitrary information about every invocation of F. How large $\lambda$ can be depends on the security of F. Roughly, if F cannot be broken with advantage $2^{-w}$, then we can leak $\lambda = w/6$ bits with each of the $n$ invocations of F. (and thus $nw/6$ bits in total.)

NARL security requires that the leakage in each of the $m + 1$ steps (i.e. the invocations of the underlying F) can depend on absolutely all data that is accessed during this step. For step $i$ $(0 \le i \le m)$ this means $Z_I, K_I$, where $I = X_{|i}$ is the $i$ bit prefix of the input $X$, but also the last two bits of $I$ itself, as this bits specify which part of the state[8] must be accessed in this step. We will even give the entire $I$ as input to the leakage function.

**Theorem 1.** *If F is a weak PRF, $\Gamma^{F,m}$ is a NARL super-PRP, where each invocation of the underlying F is considered a step as in Def. 3. If the PRF cannot be distinguished from random with advantage more than $\epsilon_{prf}$, then we can tolerate leakage of $\lambda = \log(\epsilon_{prf}^{-1})/6$ bits per step. The precise quantitative statement is given below.*

*Assume* $F : \Sigma^k \times \Sigma^n \to \Sigma^{4k+2n}$ *is a* $(\epsilon_{prf}, s_{prf}, n/\epsilon_{prf}^2)$ *secure weak PRF (where* $\epsilon_{prf} \ge n \cdot 2^{-n/3}$ *and* $n \ge 20$) *and let* $\lambda = \log(\epsilon_{prf}^{-1})/6$. *Then* $\Gamma^{F,m} : \Sigma^{3k+n} \times \Sigma^m \to \Sigma^{4k+2n}$ *is a* $(\epsilon'_{prf}, s'_{prf}, q'_{prf})$ *secure* $\mathcal{L}_{s,\lambda}$-*resilient PRF for any* $q'_{prf}$ *and*

$$s'_{prf} = s_{prf}\epsilon_{prf}^2/2^{\lambda+2}(n+k)^3 - s \cdot m \cdot q'_{prf} \qquad \epsilon'_{prf} = 8 \cdot q'^2_{prf} \cdot m \cdot \epsilon_{prf}^{1/12}$$

*where the class $\mathcal{L}_{s,\lambda}$ contains all functions $\mathcal{L}_g$ indexed by a function $g : \Sigma^{k+n+m} \to \Sigma^\lambda$ of size at most $s$ defined as (with $K_I, Z_I$ as in Definition 4)*

$$\mathcal{L}_g(K, X) = \{\Lambda_I, Z_I \; : \; I \in \text{pre}(X)\} \qquad \Lambda_I \stackrel{\text{def}}{=} g(K_I, Z_I, I)$$

Recall that a random variable $X$ has min-entropy $k$, denoted $H_\infty(X) = k$, if $\Pr[X = x] \le 2^{-k}$ for any $x$ in the support. In the proof, we will extensively use a computational version of this notion called HILL-pseudoentropy [23,3].

---

[8] Let $I_d$ denote $I$ where the last $d$ bits deleted. Then before step $I$ the state is $Z_{I_10}Z_{I_11}, K_{I_200}, K_{I_201}, K_{I_210}, K_{I_211}$.

**Definition 5 (HILL-pseudoentropy[23,3]).** *We say $X$ has* HILL *pseudoentropy $k$, denoted by $\mathsf{H}^{\mathsf{HILL}}_{\epsilon,s}(X) \geq k$, if there exists a distribution $Y$ with minentropy $\mathsf{H}_\infty(Y) = k$ where $\delta_s(X;Y) \leq \epsilon$.*

*Proof (of Theorem 1).* Our construction $\Gamma^{\mathsf{F},m}$ is inspired by the construction of the leakage-resilient stream-cipher from [38], and also the proof is very similar. We will use several technical results from [38,17] which for space reasons are moved to Appendix A.

It will be convenient to consider an adversary which is stronger than what is actually required in the proof. We consider an adversary $\mathcal{A}$ who can adaptively "explore" the tree structure underlying the $\Gamma^{\mathsf{F},m}$ construction. This is modeled by giving her access to two oracles $\mathcal{O}_K(.)$ and $\mathcal{O}_K^b(.)$. These are initialised with a random key $K$ (as used in $\Gamma^{\mathsf{F},m}$), a random bit $b$ and a uniformly random function $\mathbf{R}$. The $\mathcal{O}_K^b$ oracle takes inputs from $\Sigma^m$ and outputs either random outputs (if $b = 1$) or the output of $\Gamma^{\mathsf{F},m}$ (if $b = 0$). The $\mathcal{O}_K$ oracle allows to "explore" the tree structure of $\Gamma^{\mathsf{F},m}$.

$$\mathcal{O}_K(I) \rightarrow \begin{cases} Z_{I0}, Z_{I1}, \Lambda_I & \text{if } I \in \Sigma^{\leq m-1} \\ Y_I, \Lambda_I & \text{if } I \in \Sigma^m \end{cases} \qquad \mathcal{O}_K^b(I) \rightarrow \begin{cases} Y_I & \text{if } b = 0 \\ \mathbf{R}(I) & \text{if } b = 1 \end{cases}$$

We put the additional restriction on the order in which queries can be made: $\mathcal{A}$ can only make a query $I$ to $\mathcal{O}_K$ or $\mathcal{O}_K^b$, if the $|I| - 1$ bit prefix of $I$ has already been queried (the first query can only be $\varepsilon$). Wlog. we assume that $\mathcal{A}$ never makes the same query twice. $\mathcal{A}$ can never make the same query $I \in \Sigma^m$ to both oracles (which would trivially allow to distinguish the cases $b = 0$ and $b = 1$.)

A $q'_{\mathsf{prf}}$-query adversary $\mathcal{A}'$ who breaks the $\mathcal{L}_{s,\lambda}$ security of $\Gamma^{\mathsf{F},m}$ with advantage $\epsilon$ can be turned into an adversary $\mathcal{A}$ of almost the same size who has advantage $\epsilon$ in distinguishing the cases $b = 0$ and $b = 1$ in the experiment just described: A query $X$ to $\Gamma^{\mathsf{F},m}(X)$ can be simulated by making the queries $\mathsf{pre}(X)$ to $\mathcal{O}_K$. A query $X$ to the second oracle can be simulated the same way, except that the query $X$ is forwarded to $\mathcal{O}_K^b(.)$. This $\mathcal{A}$ makes at most $(m-1)q'_{\mathsf{prf}}$ and $q'_{\mathsf{prf}}$ queries to the first and second oracle respectively. Thus it remains to upper bound

$$\Pr_K[\mathcal{A}^{\mathcal{O}_K(.),\mathcal{O}_K^0(.)} \rightarrow 1] - \Pr_{K,\mathbf{R}}[\mathcal{A}^{\mathcal{O}_K(.),\mathcal{O}_K^1(.)} \rightarrow 1]$$

This means we must show that the outputs of the oracle $\mathcal{O}_K^0 : I \rightarrow \mathsf{F}(K_I, Z_I)$ are pseudorandom even given access to $\mathcal{O}_K$, and thus cannot be distinguished from the uniformly random outputs of $\mathcal{O}_K^1 : I \rightarrow \mathbf{R}(I)$. Let $\mathtt{view}_i$ denote the view of $\mathcal{A}$ after the $i$th query, the initial view is $\mathtt{view}_0 = \{Z_\varepsilon\}$. We say that $I \in \Sigma^{\leq m}$ is a "potential query" if $\mathcal{A}$ did not yet make the query $I$ but all its prefixes $\mathsf{pre}(I) \setminus I$. The following facts hold (with high probability) after the $i$th query and for any potential query $I$. (We ignore the precise bounds on HILL pseudoentropy, writing only $\mathsf{H}^{\mathsf{HILL}}$ to denote $\mathsf{H}^{\mathsf{HILL}}_{\epsilon,s}$ for "small" $\epsilon$ and "large" $s$.)

1. $K_I$ and $Z_I$ are independent given the view $\mathtt{view}_i$ of $\mathcal{A}$.
2. $\mathsf{H}^{\mathsf{HILL}}(K_I | \mathtt{view}_i) = k - 2\lambda$ and $\mathsf{H}^{\mathsf{HILL}}(Z_I | \mathtt{view}_i \setminus Z_I) = k - 2\lambda$.

3. If $K_I, Z_I$ satisfy fact 1 & 2 then
   (a) $\mathsf{F}(K_I, Z_I)$ is pseudorandom given $\mathtt{view}_i$.
   (b) $\mathsf{H}^{\mathsf{HILL}}(\mathsf{F}(K_I, Z_I)|\Lambda_I, \mathtt{view}_i) = |\mathsf{F}(K_I, Z_I)| - 2\lambda$.

Note that fact 3.(a) implies that a query $I$ to $\mathcal{O}_K^0$ will result in a pseudorandom value $\mathsf{F}(K_I, Z_I)$. As just described, this establishes the theorem. The lemmata below are given in Appendix A.

Fact 1 follows from Lemma 3 (originally from [16], also given as Lemma 5 in [38]). The only reason we add $Z_{I0}Z_{I1}$ to the output of $\mathcal{O}_K(I)$ (and not only the leakage $\Lambda_I$) is so we can apply this lemma.

Fact 3.(a) follows from Fact 2 using Lemmata 4 and 5, which state that the output $\mathsf{F}(K, Z)$ of a weak PRF is pseudorandom as long as $K$ and $Z$ are independent and have sufficiently high pseudoentropy.

Fact 3.(b) follows from Fact 3.(a) and Theorem 2 from [17] (or, independently [40]), which states that a pseudorandom value like $\mathsf{F}(K, Z)$ has high pseudoentropy, even if a bounded amount of information about the seed (in our case $K, Z$) is leaked. The precise quantitative statement of Fact 3.(b) is given as Lemma 6 (which is Lemma 6 from [38]).

Finally, Fact 2 holds by induction over the queries that $\mathcal{A}$ makes using Fact 3.(b). To see this, note that Fact 2 holds initially for $i = 0$ as $K_0, K_1, K_\varepsilon, Z_\varepsilon$ are independently and uniformly sampled. Now assume it holds after the $i$th query, and $\mathcal{A}$ makes the query $I$ (where $|I| < m$), then by Fact 3.(b) the newly computed values $Z_{I0}, Z_{I1}, K_{I00}, \dots, K_{I11} \leftarrow \mathsf{F}(K_I, Z_I)$ will also satisfy Fact 2.

So far we have only established the qualitative statement that $\Gamma^{\mathsf{F},m}$ is a NARL secure PRP but said nothing about the exact security as claimed in the proof. The HILL-pseudoentropy in the facts above must be quantified, e.g. in fact 2. above $\mathsf{H}^{\mathsf{HILL}}(K_I|\mathtt{view}_i) = k - 2\lambda$ can be expressed as $\mathsf{H}^{\mathsf{HILL}}_{\epsilon,s}(K_I|\mathtt{view}_i) = k - 2\lambda$ for some $\epsilon, s$. One then has to do some bookkeeping bounding how this parameters get worse (i.e. how $s$ decreases and $\epsilon$ increases) during the run of the experiment. As this is not very instructive we omit this calculations. The bounds we get here are exactly the same bounds that are proven for the leakage-resilient stream-cipher in [38] (when using the same $\mathsf{F}$ and the number of invocations to the underlying $\mathsf{F}$ is the same). In fact, minor adaptions of the proof from [38] give us the claimed bounds. The only difference is that the advantage $\epsilon'_{\mathsf{prf}}$ in this paper is a factor $q'_{\mathsf{prf}}$ larger, the reason is that our $\mathcal{A}$ can make $q'_{\mathsf{prf}}$ "challenge queries" to the $\mathcal{O}_K^b$ oracle, whereas in [38] only one challenge query is considered.    □

# 3   Side-Channel Attacks on Feistel

In this section we put forward generic side-channel attacks on Feistel networks. As Feistel networks (and minor variations thereof) are the only generic constructions of PRPs from PRFs known, this indicates that constructing leakage-resilient PRPs from leakage-resilient PRFs might be significantly harder than constructing PRPs from PRFs in the normal (non-leakage) setting. Below we first define the Feistel network.

**Definition 6 (Feistel, $\mu$).** *For a function $f : \Sigma^n \to \Sigma^n$, we denote with $\Psi[f]$ the permutation over $\Sigma^{2n}$ defined as $\Psi[f](x_L, x_R) \stackrel{\text{def}}{=} f(x_L) \oplus x_R \| x_L$. $\Psi[f_1, \dots, f_r]$ denotes $\Psi[f_r] \circ \dots \circ \Psi[f_1]$.*

*We define $\mu$ as $(R_0, \dots, R_{r+1}) \stackrel{\text{def}}{=} \mu(\Psi[f_1, \dots, f_r], R_1 \| R_0)$ where for $i \geq 1$ : $R_i \stackrel{\text{def}}{=} R_{i-1} \oplus f_{i-1}(R_{i-1})$, so $R_i$ is the input to the ith round function on input $X = R_1 \| R_0$.*

In a classical paper, Luby and Rackoff prove that the advantage of any $q$-query distinguisher in distinguishing $\Psi_3 \stackrel{\text{def}}{=} \Psi[f_1, \dots, f_3]$ from a uniformly random permutation over $\Sigma^{2n}$ is upper bounded by[9] $q^2/2^n$ if the $f_i : \Sigma^n \to \Sigma^n$ are uniformly random functions.[10] This in particular implies that no adversary who can query $\Psi_3$ in forward direction can invert $\Psi_3$ on a random $Y \in \Sigma^{2n}$, unless she makes $q = \Theta(2^{n/2})$ queries.

We consider a setting where the adversary not only can make queries to some Feistel network $\Psi_r \stackrel{\text{def}}{=} \Psi[f_1, \dots, f_r]$, but with each query $X$, besides the output $Y \leftarrow \Psi_r(X)$, also gets some "leakage" about the intermediate values.

We will consider different leakage functions $g : \Sigma^n \to \Sigma^*$, our attack will work for any functions which allow "reconstruction" as defined below

**Definition 7 (reconstructible).** *A function $g : \Sigma^n \to \Sigma^*$ is $(k, \delta)$ reconstructible, if there exists an efficient algorithm $B_g$ such that $\Pr[C' = C] \geq \delta$ in the experiment below:*

1. *Sample a random challenge $C \stackrel{*}{\leftarrow} \Sigma^n$.*
2. *$B_g$ can adaptively make $k$ queries $X_1, \dots, X_k$ to an oracle which on input $X_i$ outputs $g(C \oplus X_i)$.*
3. *$B_g$ outputs $C'$.*

*If $g$ is probabilistic, then it is $(k, \delta)$ reconstructible if there exits a single $B_g$ such that the expectation (over the randomness of $g$) of the probability $\mathsf{E}[\Pr[C' = C]]$ is at least $\delta$. Two examples of reconstructible functions are given below.*

**Hamming-weight:** The Hamming-weight function $g : \Sigma^n \to \Sigma^{\lceil \log n \rceil}$, $g(X) \stackrel{\text{def}}{=} w_H(X)$ is $(n, 1)$ reconstructible: For $i \in [n]$ let $B$ ask for $\Lambda_i = g(X \oplus e_i)$, where $e_i = 0^{i-1}10^{n-i-1}$ for $i = 1, \dots, n$. Note that $\Lambda_i$ can only take two values, $w_H(X) - 1$ or $w_H(X) + 1$, which is the case if the ith bit of $X$ is 1 and 0 respectively.[11]

---

[9] With one round more, the same result holds even if the distinguisher is allowed to make inversion queries.

[10] This then implies that $\Psi[f_1, \dots, f_3]$ is a pseudorandom permutation if the $f_i$'s are pseudorandom functions. In fact, Luby-Rackoff proved this latter result directly, but as advocated e.g. in [33], the detour via uniformly random objects is cleaner and easier.

[11] If all $\Lambda_i$ are the same then $X = 1^n$ or $0^n$, which is the case can be deduced from $\Lambda_1$ (which is $n - 1$ or 1 in those cases).

**Noise:** For some $\gamma > 0$ consider the probabilistic function $g_\gamma : \Sigma^n \to \Sigma^n$ which flips every bit of its input with probability $1/2 - \gamma$ (and each bit of every input is flipped independently.) For any $k$, $g_\gamma$ is $(k, 1 - n \cdot e^{-2 \cdot k \cdot \gamma^2})$ reconstructible: $B_{g_\gamma}$ uses any sequence $X_1, \ldots, X_k$ of distinct inputs, and guesses that the $i$th bit of $C$ is 0 iff the majority of the $i$th bits in $g_\gamma(C \oplus X_1), \ldots, g_\gamma(C \oplus X_k)$ is 0. By the Chernoff bound, the probability that the $i$th bit is guessed wrong is at most $e^{-2 \cdot k \cdot \gamma^2}$, taking the union bound over all $n$ bits we get the bound as claimed.

**Theorem 2.** *For some $r \geq 3$ and any $f_1, \ldots, f_r : \Sigma^n \to \Sigma^n$, consider the $r$ round Feistel network $\Psi_r = \Psi[f_1, \ldots, f_r]$ and some leakage function $g : \Sigma^n \to \Sigma^*$ which is $(k, \delta)$ reconstructible. Then there exists an attacker $\mathcal{A}$ which can invert $\Psi_r$ on any value $Y$ with probability $\delta^{(k+1)^{r-2}}$, where $\mathcal{A}$ makes $4(k+1)^{r-2}$ forward queries to $\Psi_r$, and with each query $X$ learns the output $\Psi_r(X)$ and leakage $g(R_1), \ldots, g(R_{r-1})$ about the inputs to the round functions $(R_0, \ldots, R_{r+1}) \leftarrow \mu(\Psi_r, X)$. The running time of $\mathcal{A}$ is $O((k+1)^{r-3}|B_g|)$ where $|B_g|$ is the running time of $B_g$ as in Definition 7.*

In the theorem we only consider the case $r \geq 3$, for $r = 0, 1$ or 2 one can trivially invert with probability 1 making $0, 1$ or 4 forward queries respectively. This theorem generalizes Theorem 3.1 from [12], who consider the case where the adversary gets all the $R_i$'s. (or equivalently, where $g$ is $(1, 1)$ reconstructible.)

*Remark 2.* Note that we don't have to leak $g(R_i)$ for $i \in \{0, 1, r, r+1\}$ as for those $i$ the entire $R_i$ is already contained in the input or output. The above theorem can also be proven (with worse bounds: $(k+1)^r$ queries and probability $\delta^{(k+1)^r}$) in a weaker setting where the adversary does not even get to see the output $\Psi_r(X) = R_r \| R_{r+1}$, but instead gets the leakage $g(R_r), g(R_{r+1})$.

*Remark 3.* The success probability $\delta^{(k+1)^{r-2}}$ drops very fast in $k$ and $r$. This is not an issue for leakage functions where $\delta = 1$ like Hamming weight. But this also is good enough for noisy leakage, where we get a success probability of $(1 - n \cdot e^{-2 \cdot k \cdot \gamma^2})^{(k+1)^{r-2}} \geq (1 - n \cdot e^{-2 \cdot k \cdot \gamma^2} \cdot (k+1)^{r-2})$ which approaches 1 exponentially fast in $k$.

*Proof (of Theorem 2).* The proof by induction on the number of rounds $r$. For $j \in [r]$ let $\Psi_j \stackrel{\text{def}}{=} \Psi[f_1, \ldots, f_j]$ denote the first $j$ rounds of $\Psi_r$. For any $j, 1 \leq j \leq r$, we let $E(j, Y_j) \stackrel{\text{def}}{=} \Psi_j^{-1}(Y_j)$, that is, the input $Z$ such that the intermediate value after $j$ rounds in the computation $\Psi_r(Z)$ is $Y_j$. It will be convenient to define $E'(j, Y_j) = \{Z, \Psi_r(Z), g(R_1), \ldots, g(R_r)\}$ where $(R_0, \ldots, R_{r+1}) \leftarrow \mu(\Psi_r, Z)$. We show that

*Claim.* $E'(1, Y_L \| Y_R)$ can be computed (with probability $\delta$) making $k+1$ forward queries to $\Psi_r$.

*Proof (of Claim).* As $Z \stackrel{\text{def}}{=} E(1, Y_L \| Y_R)$ is $Y_R \| f_1(Y_R) \oplus Y_L$, to get $Z$ it is sufficient to learn $C \stackrel{\text{def}}{=} f_1(Y_R)$. To get $E'(1, Y_L \| Y_R)$ we then make one more $\Psi_r$ query $Z$. Let $B_g$ be as in Definition 7, we will use it to reconstruct $C$ as follows: For every query $X_i$ asked by $B_g$, we make the query $Y_R \| X_i$ to $\Psi_r$. The answer will contain the leakage $\Lambda_2 = g(C \oplus X_i)$, which is exactly what $B_g$ expects as answer to his query $X_i$. Thus after $k$ queries we learn $C$ with probability $\delta$.    $\square$

*Claim.* For $j \in [2, r-2]$, $E'(j, Y_L \| Y_R)$ can be computed (with probability $\delta$) making $k+1$ queries to $E'(j-1, .)$.

*Proof (of Claim).* The proof of this claim is illustrated in Figure 1. The idea is similar as in the previous claim; We will use $B_g$ to reconstruct $C \stackrel{\text{def}}{=} f_j(Y_R)$ (as explained below) and then we get $E'(j, Y_L \| Y_R) = E'(j-1, Y_R \| C \oplus Y_L)$ with one more $E'(j-1, .)$ query.

To reconstruct $C = f_j(Y_R)$, for every query $X_i$ made by $B_g$, we ask for $E'(j-1, Y_R \| X_i)$ which includes the leakage $\Lambda_{j+1} = g(C \oplus X_i))$ as expected by $B_g$. Thus after $k$ queries $X_1, \ldots, X_k$, $B_g$ outputs $C = f_j(Y_L)$ with probability $\delta$.

*Claim.* For $j \in \{r-1, r\}$, $E'(j, Y_L \| Y_R)$ can be computed making 2 queries to $E'(j-1, .)$.

*Proof (of Claim).* We ask for $E'(j-1, 0^n \| Y_L) = \{Z, \Psi_r(Z), \ldots\}$, here $\Psi_r(Z)$ contains $f_j(Y_L)$ in the clear (it's the left part of $\Psi_r(Z)$ for $j = r-1$ and right part for $j = r$). Make one more $E'(j-1, .)$ query to get $E'(j, Y_L \| Y_R) = E'(j-1, Y_R \| f_j(Y_L) \oplus Y_L)$.    $\square$

Let us for now assume that $\delta = 1$ (i.e. $B_g$ always reconstructs correctly) and let $T_{j,r}$ denote the number of forward queries to $\Psi_r$ one has to make in order to compute $E'(j, .)$. By the above claims

1. $T_{1,r} = k+1$
2. $T_{i,r} = (k+1)T_{i-1,r}$ for $i \in [2, r-2]$.
3. $T_{i,r} = 2 \cdot T_{i-1,r}$ for $i = r-1$ or $i = r$.

For $i \leq r-2$, the relations 1. and 2. are satisfied by

$$T_{i,r} \leq (k+1)^i$$

So $T_{r-2,l} = (k+1)^{r-2}$, with 3. this gives

$$T_{r,r} = 4(k+1)^{r-2}$$

As claimed in the theorem. We just have to verify the success probability, the error $\delta^{(k+1)^{r-2}}$ comes up as follows: by the first claim, we can compute $E(1, .)$ with probability $\delta$. For $E(j, .)$ $(1 < j \leq r-1)$ we need $k+1$ invocations of $E(j-1, .)$, thus the error exponentiates with $k+1$. For $j = r-1$ and $j = r$ no extra error is introduced.    $\square$

## 4   Leakage-Resilient PRPs

Theorem 3 below states that an $r$ round Feistel network, instantiated with $\mathcal{L}$-resilient PRFs, is a $\mathcal{L}'$-resilient super PRP. Here $\mathcal{L}'$ contains all leakage functions which for every round round $i \in [r]$ leak $g_i(K_i, R_i)$ where $g_i \in \mathcal{L}$ is an admissible leakage function for the leakage-resilient PRF used in the round functions. Moreover the round function inputs $R_i$ are leaked entirely. Thus, if the PRF is NALR secure, so is the super PRP. The number of queries a distinguisher can make is exponential in $r$, thus for super-logarithmic $r$ we get security against any polynomial distinguisher.

**Theorem 3.** *An $r$ round Feistel network instantiated with NARL secure PRFs is a NARL secure super PRP for $q$-query distinguishers satisfying $q \leq 1.38^{r/2-1}$.*

*More precisely, let $\mathsf{F} : \Sigma^k \times \Sigma^n \to \Sigma^n$ be a $(\epsilon_{\mathsf{prf}}, s_{\mathsf{prf}}, q)$-secure $\mathcal{L}$-resilient PRF and $\Psi_r = \Psi[f_1, \ldots, f_r]$ denote an $r$ round Feistel network instantiated with $f_i = \mathsf{F}(K_i, .)$. Then $\Psi_r$ (whose key is $K \stackrel{\text{def}}{=} \{K_1, \ldots, K_r\}$) is a $(\epsilon, s, q)$ $\mathcal{L}'$-resilient super-PRP for*

$$q \leq 1.38^{r/2-1} \qquad s = s_{\mathsf{prf}} - |F| \cdot q \cdot r \qquad \epsilon = (2 + q \cdot r) \cdot \epsilon_{\mathsf{prf}} + \frac{q^6 r^6}{5! \cdot 2^n} + \frac{q^2}{2^n}$$

*Where $\mathcal{L}'$ contains, for every $g_1, \ldots, g_r \in \mathcal{L}$, the function $g'$ defined as*

$$g'(K, X) = \{g_1(K_1, R_1), \ldots, g_r(K_r, R_r), R_0, \ldots, R_{r+1}\}$$

*with $(R_0, \ldots, R_{r+1}) \leftarrow \mu(\Psi_r, X)$.*

We will prove this theorem using a combinatorial lemma from [12]. Consider an adversary $\mathcal{A}$ making $q$ queries (forward or inverse) to $\Psi_r = \Psi[f_1, \ldots, f_r]$. Let $R[i, j]$ denote the input to the $j$th round function on the $i$th query. We say $R[i, j + 1]$ (resp. $R[i, j - 1]$) is "freshly generated" if the $i$th query is a forward (resp. inverse) query where $R[i, j]$ is fresh in the sense that $R[i, j] \neq R[k, j]$ for all $k < j$ (and thus $f_j$ has not been invoked on $R[i, j]$ before). We say that for this sequence of queries the 5-XOR condition holds, if some freshly generated value can be expressed as the bitwise XOR of 5 previously computed round function inputs. In [12] the following Lemma is proven

**Lemma 1 (Lemma 4.1 from [12]).** *Let $\Psi_r$ be any $r$ round Feistel network. For any $s \leq r/2$, if after making $q \leq 1.38^{s/2}$ forward/inverse queries to $\Psi_r$ the 5-XOR condition does* not *hold, then there is no collision on the input to the $j$th round function for any $j \in [s, r - s]$.*

Next we show that it is hard to provoke the 5-XOR condition in $\Psi_r$.

**Lemma 2.** *Assume an adversary $\mathcal{A}$ of size $s$ can satisfy the 5-XOR condition with probability $\epsilon$ making $q$ queries to $\Psi_r(K, .)$ as in Theorem 3 (with each query $X$ also getting the leakage $g'(K, X)$ for some $g' \in \mathcal{L}'$.) Then $\mathsf{F}$ is* not *a $(s_{\mathsf{prf}}, \epsilon_{\mathsf{prf}}, q)$-secure $\mathcal{L}$-resilient PRF where $s_{\mathsf{prf}} = s + |F| \cdot q \cdot r$ and $\epsilon_{\mathsf{prf}} = \frac{\epsilon}{q \cdot r} - \frac{q^5 r^5}{5! \cdot 2^n}$.*

*Proof.* We define an adversary $\mathcal{A}'$ (which will use $\mathcal{A}$ as a black-box) against the $\mathcal{L}$-resilience of F. As in Definition 2, $\mathcal{A}'$ has access to $F^g(K, .)$ (Where $g \in \mathcal{L}$ and $F^g(K, X) \stackrel{\text{def}}{=} [F(K, X), g(K, X)].)^{12}$ and $\mathcal{O}(.)$, and has to guess whether $\mathcal{O}(.)$ is a random function or $F(K, .)$.

$\mathcal{A}'$ first guesses a random query $i$ and round $j$ $(1 \leq i \leq q, 1 \leq j \leq r)$. Then it simulates an attack of $\mathcal{A}$ on $\Psi_r$, where for the first $i$ queries it uses its first oracle $F^g(K, .)$ as the function for the $j$th round, and samples the round keys for the other $r - 1$ rounds at random.

On the $i$th query, if the input to the $j$th round function is not fresh or the 5-XOR conditions already holds, $\mathcal{A}'$ outputs 0 and stops. Otherwise it uses its second oracle $\mathcal{O}(.)$ to compute the output, which gives a "freshly generated" value $R$. If this value can be expressed as the XOR of 5 previous round values, $\mathcal{A}'$ outputs 1 and 0 otherwise.

Assume $\mathcal{O}(.)$ is a uniformly random function, then the probability that $\mathcal{A}'$ outputs 1 is at most $q^5 r^5 / (5! \cdot 2^n)$ as the output of $\mathcal{O}(.)$ is uniformly random, and there are at most $q^5 r^5 / 5!$ possible values (i.e. each subset of 5 queries specifies one possibility) which will trigger the 5-XOR condition.

Now assume the other case, where $\mathcal{O}(.)$ is $F(K, .)$. If $\mathcal{A}$ will provoke the 5-XOR condition (which holds with prob. $\epsilon$), and $\mathcal{A}'$ guessed which fresh query will satisfy this condition for the first time (with happens with prob $1/(q \cdot r)$), then $\mathcal{A}'$ will output 1. Thus in this case $\mathcal{A}'$ outputs 1 with prob. $\epsilon/(q \cdot r)$.

By definition, the gap $\epsilon/q \cdot r - q^5 r^5 / (5! \cdot 2^n)$ between those two probabilities is $\mathcal{A}'$ advantage in breaking the $\mathcal{L}$-resilience of F. $\qquad \square$

*Proof (of Theorem 3).* Consider an adversary $\mathcal{A}$ of size $s$ against the $\mathcal{L}'$-resilience of $\Psi_r$ as specified in Definition 2. This $\mathcal{A}$ has access to two oracles, the first being $\Psi_r^{g'}(K, .) : X \to [\Psi_r(K, X), g'(K, X)]$ and the second being either $\Psi_r(K, .)$ or a uniformly random permutation $\mathbf{P}_n(.)$ (we call this the real and random experiment). By Lemma 2, in the real experiment the inputs to the functions in round $w \stackrel{\text{def}}{=} \lfloor r/2 \rfloor$ and $w + 1$ will be distinct with probability at least $1 - \epsilon'$ where $\epsilon' = q \cdot r \cdot \epsilon_{\text{prf}} + q^6 r^6 / (5! \cdot 2^n)$. Conditioned on this, the output of the right oracle in the real experiment is pseudorandom and thus cannot be distinguished from the output of the right oracle $\mathbf{P}_n(.)$ in the random experiment but with probability $2 \cdot \epsilon_{\text{prf}} + q^2 / 2^n$, here the $2\epsilon_{\text{prf}}$ accounts for the output only being *pseudo*random, and the $q^2 / 2^n$ accounts for the fact that even if those values were uniform, the distribution would still be slightly off from what the oracle $\mathbf{P}_n$ in the random experiment outputs (we omit the details here.) Thus, $\mathcal{A}$ cannot distinguish the two experiments better than with probability $\epsilon' + 2 \cdot \epsilon_{\text{prf}} + q^2 / 2^n$. $\qquad \square$

---

[12] The following reduction also works for the original notion of leakage-resilience where the leakage-function can be adaptively chosen. For this one must consider the oracle $F^{\mathcal{L}}$ (instead $F^g$) defined as $F^{\mathcal{L}}(K, X, g) \stackrel{\text{def}}{=} [F(K, X), g(K, X)]$ (where $g \in \mathcal{L}$). Thus, although our current PRF constructions only give us "non-adaptive-leakage" sPRPs, future advances in leakage-resilient PRFs would immediately translate to stronger leakage-resilient sPRPs.

# References

1. Akavia, A., Goldwasser, S., Vaikuntanathan, V.: Simultaneous hardcore bits and cryptography against memory attacks. In: Reingold, O. (ed.) TCC 2009. LNCS, vol. 5444, pp. 474–495. Springer, Heidelberg (2009)
2. Alwen, J., Dodis, Y., Wichs, D.: Leakage-resilient public-key cryptography in the bounded-retrieval model. In: Halevi, S. (ed.) CRYPTO 2009. LNCS, vol. 5677, pp. 36–54. Springer, Heidelberg (2009)
3. Barak, B., Shaltiel, R., Wigderson, A.: Computational analogues of entropy. In: RANDOM-APPROX, pp. 200–215 (2003)
4. Brakerski, Z., Goldwasser, S.: Circular and leakage resilient public-key encryption under subgroup indistinguishability (or: Quadratic residuosity strikes back). In: Rabin, T. (ed.) CRYPTO 2010. LNCS, vol. 6223, pp. 1–20. Springer, Heidelberg (2010)
5. Brakerski, Z., Kalai, Y.T., Katz, J., Vaikuntanathan, V.: Cryptography resilient to continual memory leakage. Cryptology ePrint Archive, Report 2010/278 (2010), http://eprint.iacr.org/
6. Coron, J.-S., Dodis, Y., Malinaud, C., Puniya, P.: Merkle-Damgård revisited: How to construct a hash function. In: Shoup, V. (ed.) CRYPTO 2005. LNCS, vol. 3621, pp. 430–448. Springer, Heidelberg (2005)
7. Coron, J.-S., Patarin, J., Seurin, Y.: The random oracle model and the ideal cipher model are equivalent. In: Wagner, D. (ed.) CRYPTO 2008. LNCS, vol. 5157, pp. 1–20. Springer, Heidelberg (2008)
8. Di Crescenzo, G., Lipton, R.J., Walfish, S.: Perfectly secure password protocols in the bounded retrieval model. In: Halevi, S., Rabin, T. (eds.) TCC 2006. LNCS, vol. 3876, pp. 225–244. Springer, Heidelberg (2006)
9. Dodis, Y., Goldwasser, S., Kalai, Y.T., Peikert, C., Vaikuntanathan, V.: Public-key encryption schemes with auxiliary inputs. In: Micciancio, D. (ed.) TCC 2010. LNCS, vol. 5978, pp. 361–381. Springer, Heidelberg (2010)
10. Dodis, Y., Haralambiev, K., Lopez-Alt, A., Wichs, D.: Cryptography against continuous memory attacks. Cryptology ePrint Archive, Report 2010/196 (2010), http://eprint.iacr.org/
11. Dodis, Y., Kalai, Y.T., Lovett, S.: On cryptography with auxiliary input. In: STOC, pp. 621–630 (2009)
12. Dodis, Y., Puniya, P.: Feistel networks made public, and applications. In: Naor, M. (ed.) EUROCRYPT 2007. LNCS, vol. 4515, pp. 534–554. Springer, Heidelberg (2007)
13. Dodis, Y., Sahai, A., Smith, A.: On perfect and adaptive security in exposure-resilient cryptography. In: Pfitzmann, B. (ed.) EUROCRYPT 2001. LNCS, vol. 2045, pp. 301–324. Springer, Heidelberg (2001)
14. Dziembowski, S.: Intrusion-resilience via the bounded-storage model. In: Halevi, S., Rabin, T. (eds.) TCC 2006. LNCS, vol. 3876, pp. 207–224. Springer, Heidelberg (2006)
15. Dziembowski, S., Maurer, U.M.: Tight security proofs for the bounded-storage model. In: 34th ACM STOC, pp. 341–350. ACM Press, New York (2002)
16. Dziembowski, S., Pietrzak, K.: Intrusion-resilient secret sharing. In: FOCS, pp. 227–237 (2007)
17. Dziembowski, S., Pietrzak, K.: Leakage-resilient cryptography. In: 49th FOCS, pp. 293–302. IEEE Computer Society Press, Los Alamitos (2008)

18. Faust, S., Rabin, T., Reyzin, L., Tromer, E., Vaikuntanathan, V.: Protecting circuits from leakage: The computationally-bounded and noisy cases. In: Gilbert, H. (ed.) EUROCRYPT 2010. LNCS, vol. 6110, pp. 135–156. Springer, Heidelberg (2010)

19. Faust, S., Kiltz, E., Pietrzak, K., Rothblum, G.N.: Leakage-resilient signatures. In: Micciancio, D. (ed.) TCC 2010. LNCS, vol. 5978, pp. 343–360. Springer, Heidelberg (2010)

20. Gandolfi, K., Mourtel, C., Olivier, F.: Electromagnetic analysis: Concrete results. In: Koç, Ç.K., Naccache, D., Paar, C. (eds.) CHES 2001. LNCS, vol. 2162, pp. 251–261. Springer, Heidelberg (2001)

21. Goldreich, O., Goldwasser, S., Micali, S.: How to construct random functions. Journal of the ACM 33, 792–807 (1986)

22. Halderman, J.A., Schoen, S.D., Heninger, N., Clarkson, W., Paul, W., Calandrino, J.A., Feldman, A.J., Appelbaum, J., Felten, E.W.: Lest we remember: Cold boot attacks on encryption keys. In: USENIX Security Symposium, pp. 45–60 (2008)

23. Håstad, J., Impagliazzo, R., Levin, L.A., Luby, M.: A pseudorandom generator from any one-way function. SIAM Journal on Computing 28(4), 1364–1396 (1999)

24. Ishai, Y., Prabhakaran, M., Sahai, A., Wagner, D.: Private circuits II: Keeping secrets in tamperable circuits. In: Vaudenay, S. (ed.) EUROCRYPT 2006. LNCS, vol. 4004, pp. 308–327. Springer, Heidelberg (2006)

25. Ishai, Y., Sahai, A., Wagner, D.: Private circuits: Securing hardware against probing attacks. In: Boneh, D. (ed.) CRYPTO 2003. LNCS, vol. 2729, pp. 463–481. Springer, Heidelberg (2003)

26. Katz, J., Vaikuntanathan, V.: Signature schemes with bounded leakage resilience. In: Matsui, M. (ed.) ASIACRYPT 2009. LNCS, vol. 5912, pp. 703–720. Springer, Heidelberg (2009)

27. Kiltz, E., Pietrzak, K.: How to secure elgamal against side-channel attacks (2009) (manuscript)

28. Kocher, P.C.: Timing attacks on implementations of Diffie-Hellman, RSA, DSS, and other systems. In: Koblitz, N. (ed.) CRYPTO 1996. LNCS, vol. 1109, pp. 104–113. Springer, Heidelberg (1996)

29. Kocher, P.C.: Design and validation strategies for obtaining assurance in countermeasures to power analysis and related attacks. In: Proceedings of the NIST Physical Security Workshop (2005)

30. Kocher, P.C., Jaffe, J., Jun, B.: Differential power analysis. In: Wiener, M. (ed.) CRYPTO 1999. LNCS, vol. 1666, pp. 388–397. Springer, Heidelberg (1999)

31. Luby, M., Rackoff, C.: How to construct pseudorandom permutations from pseudorandom functions. SIAM Journal on Computing 17(2) (1988)

32. Maurer, U.M.: A provably-secure strongly-randomized cipher. In: Damgård, I.B. (ed.) EUROCRYPT 1990. LNCS, vol. 473, pp. 361–373. Springer, Heidelberg (1991)

33. Maurer, U.M.: Indistinguishability of random systems. In: Knudsen, L.R. (ed.) EUROCRYPT 2002. LNCS, vol. 2332, pp. 110–132. Springer, Heidelberg (2002)

34. Maurer, U.M., Renner, R., Holenstein, C.: Indifferentiability, impossibility results on reductions, and applications to the random oracle methodology. In: Naor, M. (ed.) TCC 2004. LNCS, vol. 2951, pp. 21–39. Springer, Heidelberg (2004)

35. Micali, S., Reyzin, L.: Physically observable cryptography (extended abstract). In: Naor, M. (ed.) TCC 2004. LNCS, vol. 2951, pp. 278–296. Springer, Heidelberg (2004)

36. Naor, M., Segev, G.: Public-key cryptosystems resilient to key leakage. In: Halevi, S. (ed.) CRYPTO 2009. LNCS, vol. 5677, pp. 18–35. Springer, Heidelberg (2009)

37. European Network of Excellence (ECRYPT). The side channel cryptanalysis lounge, http://www.crypto.ruhr-uni-bochum.de/en_sclounge.html
38. Pietrzak, K.: A leakage-resilient mode of operation. In: Joux, A. (ed.) EUROCRYPT 2009. LNCS, vol. 5479, pp. 462–482. Springer, Heidelberg (2010)
39. Quisquater, J.-J., Samyde, D.: Electromagnetic analysis (ema): Measures and counter-measures for smart cards. In: E-smart, pp. 200–210 (2001)
40. Reingold, O., Trevisan, L., Tulsiani, M., Vadhan, S.P.: Dense subsets of pseudorandom sets. In: FOCS, pp. 76–85 (2008)
41. Standaert, F.-X., Pereira, O., Yu, Y., Quisquater, J.-J., Yung, M., Oswald, E.: Leakage resilient cryptography in practice. Cryptology ePrint Archive, Report 2009/341 (2009), http://eprint.iacr.org/
42. Vadhan, S.P.: Constructing locally computable extractors and cryptosystems in the bounded-storage model. Journal of Cryptology 17(1), 43–77 (2004)

# A   Technical Lemmata

**Lemma 3 ([16]).** *Let $A_0, B_0$ be independent and $\phi_1, \phi_2, \ldots$ be any sequence of functions. Let $A_1, A_2, \ldots, B_1, B_2, \ldots$ and $V_1, V_2, \ldots$ be defined as*

$$((A_{i+1}, V_{i+1}), B_{i+1}) := (\phi_{i+1}(A_i, V_1, \ldots, V_i), B_i) \quad \textit{if } i \textit{ is even}$$
$$(A_{i+1}, (V_{i+1}, B_{i+1})) := (A_i, \phi_{i+1}(B_i, V_1, \ldots, V_i)) \quad \textit{otherwise}$$

*Then $B_i \to \{V_1, \ldots, V_i\} \to A_i$ (and $A_i \to \{V_1, \ldots, V_i\} \to B_i$) is a Markov chain (or equivalently, $A_i$ and $B_i$ are independent given the $V_1, \ldots, V_i$)*

**Lemma 4 ([38]).** *For any $\alpha > 0$ and $t \in \mathbb{N}$: If $\mathsf{F} : \{0,1\}^\kappa \times \{0,1\}^n \to \{0,1\}^m$ is a $(\epsilon_{\mathsf{prf}}, s_{\mathsf{prf}}, q_{\mathsf{prf}})$-secure wPRF (for uniform keys), then it is a $(\epsilon'_{\mathsf{prf}}, s'_{\mathsf{prf}}, q'_{\mathsf{prf}})$-secure wPRF even if the keys are only sampled from a distribution with min-entropy $\kappa - \alpha$ with*

$$q_{\mathsf{prf}} \geq q'_{\mathsf{prf}} \cdot t \qquad s_{\mathsf{prf}} \geq s'_{\mathsf{prf}} \cdot t \qquad \epsilon_{\mathsf{prf}} \leq \epsilon'_{\mathsf{prf}}/2^{\alpha+1} - \frac{q^2_{\mathsf{prf}}}{2^{n+1}} - 2 \cdot \exp\left(-\frac{t \cdot \epsilon'^2_{\mathsf{prf}}}{8}\right)$$

**Lemma 5 ([38]).** *Let $\beta > 0$, then if $\mathsf{F} : \{0,1\}^\kappa \times \{0,1\}^n \to \{0,1\}^m$ is a $(\epsilon_{\mathsf{prf}}, s_{\mathsf{prf}}, 1)$-secure wPRF (for uniform inputs), it's also a $(\epsilon'_{\mathsf{prf}}, s'_{\mathsf{prf}}, 1)$-secure wPRF if the input is chosen from a distribution with min-entropy $m - \beta$, where for any $t \in \mathbb{N}$*

$$s_{\mathsf{prf}} \geq s'_{\mathsf{prf}} \cdot 2t \qquad\qquad \epsilon_{\mathsf{prf}} \leq \epsilon'_{\mathsf{prf}}/2^{\beta+1} - 2 \cdot \exp\left(-\frac{2 \cdot t \cdot \epsilon'^2_{\mathsf{prf}}}{64}\right)$$

**Lemma 6 ([38]).** *Let $\mathsf{F} : \{0,1\}^\kappa \times \{0,1\}^n \to \{0,1\}^m$ be a $(\epsilon_{\mathsf{prf}}, s_{\mathsf{prf}}, n/\epsilon^2_{\mathsf{prf}})$-secure wPRF. Let $K \in \{0,1\}^\kappa$ and $X \in \{0,1\}^n$ be independent where $H_\infty(K) = \kappa - 2\lambda$ and $H_\infty(X) = n - 2\lambda$ and let $f : \{0,1\}^{\kappa+n} \to \{0,1\}^\lambda$ be any leakage function, then for $\lambda \leq \log(\epsilon^{-1}_{\mathsf{prf}})/6$*

$$\Pr_{X,Y}[\mathsf{H}^{\mathsf{HILL}}_{\epsilon', s'}(\mathsf{F}(K,X)|X, f(K,X)) \geq m - 2\lambda] \geq 1 - 2^{-\lambda/2+1}$$

*with $\epsilon' = 2^{-\lambda/2+2}$ and $s' = s_{\mathsf{prf}}/2^{\lambda+3}(n+\kappa)^3$.*

# Protecting Cryptographic Keys against Continual Leakage

Ali Juma and Yevgeniy Vahlis[*]

Department of Computer Science, University of Toronto
{ajuma,evahlis}@cs.toronto.edu

**Abstract.** Side-channel attacks have often proven to have a devastating effect on the security of cryptographic schemes. In this paper, we address the problem of storing cryptographic keys and computing on them in a manner that preserves security even when the adversary is able to obtain information leakage during the computation on the key.

Using any fully homomorphic encryption with re-randomizable ciphertexts, we show how to encapsulate a key and repeatedly evaluate arbitrary functions on it so that no adversary can gain any useful information from a large class of side-channel attacks. We work in the model of Micali and Reyzin, assuming that only the active part of memory during computation leaks information. Our construction makes use of a single "leak-free" hardware token that samples from a distribution that does not depend on the protected key or the function that is evaluated on it.

Our construction is the first general compiler to achieve resilience against polytime leakage functions without performing any leak-free computation on the protected key. Furthermore, the amount of computation our construction must perform does not grow with the amount of leakage the adversary is able to obtain; instead, it suffices to make a stronger assumption about the security of the fully homomorphic encryption.

## 1 Introduction

Leakage-resilient cryptographic constructions – constructions that remain secure even when internal state information leaks to the adversary – have received much recent interest. Traditionally, security models have treated such internal state information as perfectly hidden from the adversary. However, the development of various side-channel attacks has made it clear that this traditional view is inconsistent with physical reality. In a side-channel attack, an adversary obtains information about the internal state of a device by measuring such things as power consumption, computation time, and emitted radiation.

Cryptographic primitives with long term keys, such as encryption and signature schemes, are often targeted by such attacks. An adversary observing information leakage from computation on the key can potentially accumulate enough data over time to compromise the security of the scheme. Consequently, storing

---

[*] Supported by the Natural Sciences and Engineering Research Council of Canada (NSERC).

T. Rabin (Ed.): CRYPTO 2010, LNCS 6223, pp. 41–58, 2010.

keys and computing on them in adversarial environments has been an important goal both in theory and practice. Indeed, many operating systems provide cryptographic facilities that allow programs to access keys only through designated functions, such as signing and encrypting. Smart cards provide a similar interface in hardware. In both cases, the goal is to limit any adversary to interacting with the scheme through designated channels. Nevertheless, information leakage through physical side-channels is often sufficient to overcome such barriers and break the scheme.

In this paper, we propose an approach for protecting cryptographic keys and computing on them repeatedly in a manner that preserves the secrecy of the key even when information about the state of the device continuously leaks to the adversary. Towards this goal, we define a new primitive called a *key proxy*, which encapsulates a key $K$ and provides a structured way of evaluating arbitrary functions on $K$. This allows, for example, the conversion of any pseudorandom function, signature scheme, or public-key encryption scheme into a leakage-resilient variant of itself. Our construction withstands a bounded amount of leakage per invocation (where an invocation occurs each time a function is evaluated on $K$), but the total amount of leakage is unbounded. Previously, only stream ciphers, signature schemes, and identification scheme have been made resilient to an unbounded total amount of leakage.

For our construction, we make use of the recently achieved fully homomorphic encryption [12,4], and an additional "leak-free" component. We describe two ways of instantiating this component, and in both cases the component samples from a globally fixed distribution that does not depend on $K$.

*Leakage-resilient cryptography.* The problem of executing code in an adversarial environment has always been on the minds of cryptographers. Still, most cryptographic schemes are designed assuming that the hardware on which they will be implemented is a black box device, and information is accessible to the adversary only through external communication channels. Goldreich and Ostrovsky [13] consider the problem of protecting software from malicious users, and define the concept of an oblivious RAM – a CPU that is capable of evaluating encrypted programs using a constant amount of leak-free memory and an unbounded amount of memory that is fully visible to the adversary. The oblivious RAM is initialized with a secret key, which is used to decrypt encrypted instructions, execute them, and re-encrypt the output. The encrypted state of the program is stored in the clear. Oblivious RAMs provide the strong security guarantee that even if an adversary can keep track of the memory locations accessed by the computation, she is still unable to gain any additional information about the program over what would normally be revealed through black box access.

Since the work of Goldreich and Ostrovsky, the focus in leakage-resilient cryptography has been steadily shifting towards allowing the adversary ever-growing freedom in observing the *computation* of cryptographic primitives. Ishai, Sahai, and Wagner [17] introduce "private circuits" – a generic compiler that transforms any circuit into one that is resilient to probing attacks. In a probing attack, the adversary selects a subset (of some fixed size) of the wires of the circuit and obtains the values of these wires. Goldwasser, Kalai, and Rothblum [15] define one-time programs – programs that come with small secure hardware tokens, and can be executed a bounded number of times without revealing anything but the output, even if the adversary observes the entire computation. The secure tokens are the hardware equivalent of oblivious transfer – each token stores two keys and reveals one of them upon request, while the second key is erased.

Micali and Reyzin [20] outline a framework for defining and analyzing cryptographic security against adversaries that perform side channel attacks. They introduce an axiom: only computation leaks information. That is, at any point during the execution of an algorithm, only the part of memory that is actively computed on may leak information. This allows for convenient modeling of leakage: an algorithm is described as a sequence of procedures and the set of variables that is accessed by the procedure. The adversary may then obtain leakage separately from the contents of each set of variables as they are accessed during the execution of the algorithm. The only-computation-leaks model (OCL) has since been used to obtain stream ciphers [9,21] and signature schemes [10] that remain secure even if the adversary obtains leakage from the active state each time the primitive is used, and the total amount of leakage is unbounded. We refer to such leakage as "continuous leakage" for the rest of the paper.

Faust *et al* [11] propose an alternative restriction on side-channel adversaries: restricting the computational power of the leakage function but allowing leakage on the entire state. Faust *et al* describe a circuit transformation that immunizes any circuit against leakage functions that can be described as $AC^0$ circuits[1]. The transformed circuit can leak information from the entire set of wires at each invocation, and makes use of a polynomial number of leak-free components that generate samples from a fixed distribution that does not depend on the computation of the circuit. We make use of a similar leak-free component, although the distribution generated by our component is significantly more complex than the one in [11] due to the fact that we must defend against leakage functions that are not restricted to circuits of small depth.

Very recently, specific leakage-resilient cryptographic primitives have been constructed under even more general continuous leakage models. Dodis, Haralambiev, Lopez-Alt, and Wichs [7] have constructed several primitives, including signature schemes and authenticated key agreement protocols, that remain secure even if the entire state (and not just the active part) leaks information continuously. The public key of the scheme remains fixed throughout the lifetime of the system. Brakerski, Kalai, Katz, and Vaikuntanathan [3] construct a public-key encryption scheme that allows continuous leakage on the entire state, and does not require a leak-free key update procedure. [3] also construct signature schemes and identity based encryption under slightly different leakage models. As in our work, both above works provide protection against leakage that can be described by arbitrary polynomial-time computable functions with sufficiently short output.

---

[1] $AC^0$ circuits have constant depth and unbounded fan-in.

In addition to the recent work on cryptographic constructions that are resilient to continuous leakage, there has been significant progress [1,2,22,19] on obtaining resilience to "memory attacks" – side channel attacks where the adversary obtains a bounded amount of information about the memory contents of the device throughout its lifetime. Perhaps due to the bounded nature of this type of leakage, constructions secure against memory attacks tend to be quite efficient and do not require the algorithm to maintain a state.

*Concurrent work of Goldwasser and Rothblum.* In a concurrent paper [16], Goldwasser and Rothblum construct a general compiler that achieves resilience to polynomial time leakage. Their construction relies on a linear number of leak-free components, while ours relies on a single component. On the other hand, they rely on the standard Decisional Diffie Hellman assumption, whereas we rely on fully homomorphic encryption.

*On testable leak-free components.* When constructing leakage-resilient cryptographic primitives, one has to take care in the nature and amount of components that are assumed not to leak any information. It is preferable, but may not always be possible, to avoid such components altogether. For example, one can protect any functionality against leakage given an arbitrary number of leak-free gates that can decrypt a ciphertext, perform a logical operation on the plaintext, and re-encrypt the result. Such a component can be used to evaluate the circuit $F$ on $K$ gate by gate, keeping all intermediate values encrypted, and thereby rendering leakage useless. However, building such leak-free components may be as difficult as constructing a leak-free computer and forgetting all about side-channels. Consequently, the focus of research in this area has always been to reduce the power and amount of computation that is assumed to be a-priori insulated from side-channel attacks.

Our construction uses a leak-free component that produces random encryptions of some fixed message (in our case – $\bar{0}$) under a given public key in the fully homomorphic encryption scheme. More specifically, the leak-free component we use is a randomized component that, given $pub$, produces two random encryptions of $\bar{0}$. Consequently, the computation performed by this component does not depend on any user or adversarially supplied inputs, and in particular does not depend on the key $K$ or the function $F$ that is evaluated on $K$. We call such a component *testable* because it can be accurately simulated in a controlled environment – all one has to do is feed the component random bits and randomly generated public keys and observe its behavior. More generally, we say that a component is testable if its inputs come from a globally fixed distribution that is independent from other inputs to the system.

We propose testability as a rule of thumb for secure hardware components in leakage resilient cryptography. All hardware components leak at least *some* information such as timing (every computation takes time) and power consumption. Therefore, the best we can hope for is that the information leaked by the components that we assume to be leak-free is useless to the adversary. Testability gives us the ability to observe the leakage from the secure component – as it will happen during actual usage – and estimate whether the component is safe to use. We note that the components used by [11] and [16] are testable.

In contrast to previous general compilers that achieve leakage resilience, we use only one leak-free component, regardless of the size of the circuit that is evaluated on $K$, or the amount of information leakage per invocation. Thus, our construction does not require the number of leak-free components to grow with the amount of leakage.

*Our contributions.* We study the problem of computing on a cryptographic key in an environment that leaks information each time a computation is performed. We show that in the OCL model with a single leak-free randomized token, a cryptographic key can be protected in a manner that allows repeated computation on it while making sure that the adversary gains no information from side-channel information leakage.

More precisely, we propose a tool which we call a *key proxy* – a stateful cryptographic primitive that is initialized once with a key $K$, and then given any circuit $F$ computes $F(K)$. Any leakage obtained by an adversary from the computation of the key proxy can be computed given just $F$ and $F(K)$. Using any *fully homomorphic encryption* (FHE) we construct a key proxy with the following properties:

*Resilience to adaptive polynomial time leakage.* During each invocation of the key proxy, we allow the adversary to adaptively select leakage functions that are modeled as arbitrary circuits with a sufficiently short output. The exact amount of round leakage that our construction can withstand depends on the level of security of the underlying FHE. Assuming the most basic security for the FHE (i.e. against polynomial time adversaries) permits security against $O(\log n)$ bits of leakage each time a function is evaluated on $K$. More generally, given a $2^{l(n)}$-secure FHE, our construction can withstand roughly $l(n)$ bits of leakage per invocation.

*Independent complexity.* The starting point of leakage-resilient cryptography is that *computation leaks information*. It does not require a large leap of faith to suspect that *more* computation leaks more information. In fact, to the best of our knowledge, this is indeed the case for many side-channel attacks in practice. The amount of computation performed by our key proxy construction does not depend on the amount of leakage that the adversary obtains per invocation. Instead, to get resilience to larger amounts of leakage, a stronger assumption about the security of the underlying fully homomorphic encryption is used. This allows us to avoid a circular dependency where, in order to obtain resilience to larger amounts of leakage one must build a more complex device, which in turn leaks more information.

*One-time programs with efficient refresh.* The one-time programs of [15] can be implemented without leak-free one-time memory tokens by storing the contents of the tokens in memory, and then accessing only the needed values during computation. The one-time programs can then be refreshed occasionally in a secure environment to allow continuous use. Currently, the refresh procedure performs as much computation as the evaluation of the program that it protects. If one is willing to trade resilience against complete exposure of the active memory (achieved by [15]) for resilience length bounded leakage then by pre-computing the outputs of the leak-free tokens in our construction and storing them in memory we obtain one-time programs with an update procedure of fixed complexity that does not depend on the protected program.

*Our approach.* The underlying building block for our construction is fully homomorphic encryption. An FHE is a public-key encryption scheme that allows computation on encrypted data. That is, given a ciphertext with corresponding plaintext $M$, the public key, and a circuit $F$, there is an efficient algorithm that computes an encryption of $F(M)$.

For our construction, we partition the state of the key proxy into two parts, $A$ and $B$ (or equivalently two devices). Given a key $K$, the key proxy is initialized as follows. An FHE key pair $(pri, pub)$ is generated and is stored in memory $A$. Then, a random encryption $C$ of $K$ under $pub$ is computed and is stored in memory $B$. To evaluate a function $F$ (described as a circuit) on $K$, the following actions are performed. First, a new pair of keys $(pri', pub')$ is generated and stored in memory $A$, and an encryption $C_{pri} = \mathsf{Enc}_{pub'}(pri)$ of the old private key is written to a public channel. Then, computing on memory $B$ and the public channel, the following two ciphertexts are generated homomorphically from $C$ and $C_{pri}$: an encryption $C_{\mathsf{res}}$ of $F(K)$ and a fresh encryption $C_{\mathsf{key}}$ of $K$. Note that both $C_{\mathsf{res}}$ and $C_{\mathsf{key}}$ are encryptions under the new public key $pub'$. The ciphertext $C_{\mathsf{res}}$ is then sent back to memory $A$ where it is decrypted, and $F(K)$ is returned as the output of the program. This basic approach is described in Figure 1.

It is clear that without leakage, the above construction is secure. Of course, the main difficulty is showing that leakage does not provide the adversary with any useful information. Below we provide an informal description of two main technical issues that arise.

*Leakage on private keys and ciphertexts.* It is easy to see that without refreshing the encryption $C$ of $K$, a leakage adversary will eventually learn all of $K$ by gradually leaking all of $C$ and $pri$ and then simply decrypting. Therefore, it is clear that an update procedure is necessary. The algorithm described in Figure 1 performs such an update: After each invocation, memory $A$ contains a freshly generated private key and memory $B$ contains an encryption of $K$ under the corresponding public key. However, we cannot directly claim that this refreshing procedure provides the necessary level of security. The main difficulty stems from the fact that the adversary obtains leakage on the private key in memory $A$ both before and after she obtains leakage on the encryption $C$ of $K$ under the corresponding public key. In particular, if the adversary could obtain the entire ciphertext $C$, she would be able to hardcode it into the second leakage function that is applied to the private key. The leakage function would then decrypt $C$ and leak bits of information about $K$.

This requires us to make use of the fact that the adversary obtains only a bounded amount of leakage on the ciphertext $C$, and never sees it completely.

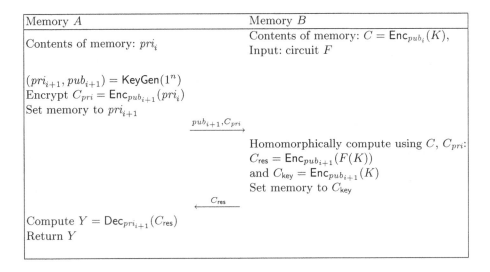

**Fig. 1.** Informal description of the construction

We argue that any leakage function that provides enough information about the ciphertext in order to later learn something about the plaintext given the private key, essentially acts as a distinguisher and can be used to break the semantic security of the FHE.

*Randomizable FHE.* Ciphertexts produced by fully homomorphic encryption schemes may carry information about the homomorphic computation that was performed to obtain them. For instance, it is possible that the ciphertext $C_{res}$ is actually first decrypted to a string of the form $(F(K), K)$ and then the decryption algorithm ignores the second element in the pair. In this case, the adversarial leakage function is clearly not forced to follow the honest decryption algorithm and can make use of the intermediate values of the decryption process to leak information about $K$. Similarly, the ciphertext $C_{key}$ may contain information about the function $F$ that was evaluated on $K$. For some applications, such as encryption where $F$ encodes in plain text the message to be encrypted, this is undesirable since the adversary may use future leakage functions to gain information about the message.

Fortunately, the homomorphic encryption schemes of Gentry [12] and of van Dijk *et al* [4] have the following additional property: given any encryption $C$ of a message $M$ and a random encryption $C'$ of $M'$, the ciphertext $C + C'$, where the addition is performed over the appropriate group of ciphertexts, is a random encryption of $M + M'$. Consequently, to address the issue described above, we randomize both $C_{res}$ and $C_{key}$ by adding random encryptions of zero to both ciphertexts. In order to make use of the property described above, the encryptions of zero need to be generated without leakage; otherwise, the leaked information maintains a correlation between the randomized ciphertext and the history of the computation that was used to produce the original ciphertext.

We note that in the FHE schemes of [12] and [4], $C'$ has to be generated in a special way in order to have enough noise to annihilate any dependence between $C + C'$ and the computation history of $C$. For simplicity of exposition we ignore this distinction, and instead remark that the randomization procedures of both FHE schemes satisfy the properties needed for our construction.

*Function privacy in key proxies.* In the above description of key proxies, we require that the leakage obtained by the adversary can be simulated given just $F$ and $F(K)$. However, in some applications, such as private-key encryption, the function $F$ itself also needs to be hidden. In the case of encryption, $F$ contains the message $M$, so an adversary can break semantic security simply by leaking information about $F$, ignoring $K$ completely. This raises a subtle modeling issue: the message $M$ must exist somewhere as plaintext, and if the adversary obtains leakage on that computation, she will trivially break semantic security. Therefore, irrespective of the definition of leakage-resilient key proxies, semantic security cannot be achieved when every invocation of every algorithm leaks information.

There are several ways in which this issue can be addressed. One solution is to weaken the definition of semantic security by requiring that the plaintexts have high pseudo-entropy[2] given the leakage obtained by the adversary. We avoid this approach both because it leads to complex definitions, and because it does not seem to have a clear advantage over the following much cleaner solution. Instead, we allow the adversary to obtain leakage both before and after the challenge ciphertext is generated, but not on the computation of the challenge ciphertext itself. This essentially means that while leakage can compromise individual encryptions, the long-term key remains safe. Under this restriction, our definition of key proxies provides the needed level of security. This approach is consistent with previous definitions of leakage-resilient semantic security (see e.g. [9,22,8,6]), and allows us to avoid additional complexity in our definition. This is desirable especially given the fact that for some applications of key proxies, such as signature schemes, function privacy is not necessary.

We mention briefly that another option is to define a leakage model for private-key encryption which allows the encryption algorithm to perform some leak-free pre-processing that is independent of the key. Then, the encryptor can generate an encrypted version of the circuit $F$, which can be safely given to the adversary without compromising security.

*Organization.* In Section 3, we describe the computational and leakage models that we use, and define a leakage-resilient key proxy. In Section 4, we provide our main construction, and analyze its security. In Section 5, we describe several variants of our model and construction, and provide several applications of leakage-resilient key proxies.

---

[2] A distribution has pseudo-entropy $\geq k$ if it is computationally indistinguishable from some distribution with min-entropy $\geq k$.

## 2   Preliminaries

*Notation.* We write PPT to denote Probabilistic Polynomial Time. When we wish to fix the random bits of a PPT algorithm $M$ to a particular value, we write $M(x; r)$ to denote running $M$ on input $x$ and randomness $r$. We write $time_n(M)$ to denote the running time of algorithm $M$ on security parameter $n$. We use $x \in_R S$ to denote the fact that $x$ is sampled according to a distribution $S$. Similarly, when describing an algorithm we may write $x \leftarrow_R S$ to denote the action of sampling an element from $S$ and storing it in a variable $x$.

It is common in cryptography to describe probabilistic experiments that test the ability of an adversary to break a primitive. Given such an experiment Exp, and an adversary $A$, we write $A \leftrightarrows \mathsf{Exp}$ to denote the random variable representing outcome of Exp when run with the adversary $A$.

### 2.1   Fully Homomorphic Encryption

The main tool in our construction is a fully homomorphic public-key encryption (FHE) system. Intuitively, such a system has the usual semantic security properties of a public-key encryption (PKE) scheme, but in addition, can perform arbitrary computation on encrypted data. The outcome of this computation is of course also encrypted. The first construction of FHE was given by Gentry in [12], and is based on ideal lattices. Recently another construction was proposed by van Dijk *et al* [4].

We do not go into the details of the FHE constructions, but rather present the result with respect to an arbitrary FHE with an additional randomization property, which is satisfied by both constructions.

**Definition 1.** *Let* FHE $=$ (KeyGen, Enc, Dec, EncEval, Add, Subtract) *be a tuple of PPT algorithms, and let* $l : \mathbb{N} \to \mathbb{N}$. *We say that* HPKE *is an* $l(n)$-*secure fully homomorphic public key encryption scheme if the following conditions hold:*

1. *The triple* (KeyGen, Enc, Dec) *is a public-key encryption scheme. We assume without loss of generality that the private key is always the random bits of* KeyGen.
2. *The algorithm* EncEval($pub, \mathbf{C}, F$), *where pub is a public key,* $\mathbf{C} = (C_1, \dots, C_n)$ *is a vector of ciphertexts with plaintexts* $(m_1, \dots, m_n)$, *and $F$ is a circuit on $n$ inputs, outputs a string $C'$ which is a valid encryption of* $F(m_1, \dots, m_n)$.
3. *The algorithms* Add *and* Subtract *have the following properties:*
    (a) *For all pri, for pub $=$ KeyGen(pri), for all messages $M_1$ and $M_2$, for a random encryption $C_1$ of $M_1$ under pub and for every encryption $C_2$ of $M_2$ under pub,* Add($pub, C_1, C_2$) *is distributed identically to* $\mathsf{Enc}_{pub}(M_1 + M_2)$, *and* Subtract($pub, C_1, C_2$) *is distributed identically to* $\mathsf{Enc}_{pub}(M_1 - M_2)$.
    (b) *For all ciphertexts $C_1$ and $C_2$,* Add($pub$, Subtract($pub, C_2, C_1$), $C_1$) $= C_2$. *That is, subtracting a ciphertext is the inverse of adding it.*
4. *For every probabilistic adversary $A$ running in time at most $l(n)$, the advantage of $A$ in breaking the semantic security of* FHE *is at most* $1/l(n)$.

*Remark 1.* The algorithms Add and Subtract may be implemented as addition and subtraction over the space of ciphertexts, though we do not require this. In some fully homomorphic encryption schemes, Add and Subtract may not achieve the exact requirement of step 3 above. Specifically, Add and Subtract may produce an encryption that cannot be computed on homomorphically using EncEval. We note that this is not a problem for our construction since we only use EncEval on encryptions of *pri*, which are ephemeral and never the output of Add or Subtract. We avoid formalizing this issue to improve exposition.

# 3   Models and Definitions

In this section, we present the definition of a leakage-resilient key proxy (LRKP). We start with a syntactic description of the primitive, and then describe the security experiment and the leakage model.

*Stateful Algorithms.* Due to the continuous nature of side-channel attacks, it is necessary for an LRKP to maintain a state in order to achieve security. We model stateful algorithms by considering algorithms with a special input and output structure. A stateful randomized algorithm takes as input a triple $(x; R, S)$ where $x$ is the query to the algorithm, $R$ is a random string, and $S$ is a state (when $R$ is clear from context we omit it, and denote the input by $(x; S)$). It then outputs $(y, S_{\mathsf{new}})$ where $y$ is the reply to the query, and $S_{\mathsf{new}}$ is the new state.

**Definition 2.** *A* key proxy *is a pair* $KP = (\mathsf{KPInit}, \mathsf{KPEval})$, *where* KPInit *is an algorithm, and* KPEval *is a stateful algorithm. For fixed* $c \in \mathbb{N}$ *and for all* $n \in \mathbb{N}$, $K \in \{0,1\}^{n^c}$, $\mathsf{KPInit}(1^n, K)$ *outputs an initial state* $S$. *For every circuit* $F : \{0,1\}^{|K|} \to \{0,1\}^n$, *and random coins* $R$, *the stateful algorithm* $\mathsf{KPEval}(1^n, F; R, S)$ *outputs* $F(K)$.

We now describe the security experiment of LRKPs. This experiment is parameterized by the leakage structure on a single invocation of the KPEval algorithm. However, for clarity we start with the description of the general experiment, and then provide details on the leakage that occurs at each invocation. We model the the leakage resilience of a key proxy by requiring the leaked information to be simulatable. That is, we require the existence of a simulator Sim that, given $F$ and $F(K)$, can simulate the leakage and messages obtained by the adversary during the computation of $\mathsf{KPEval}(1^n, F; R, S)$. No efficient adversary should be able to tell whether she is getting actual leakage and messages, or interacting with a simulator. We now describe the real and ideal security experiments:

Let $KP = (\mathsf{KPInit}, \mathsf{KPEval})$ be a key proxy. Let $A$ and Sim be PPT algorithms, $n \in \mathbb{N}$, and consider the following two experiments:

ExpReal **(Real Interaction).** The interaction of the adversary with the key proxy proceeds as follows:
1. A key $K$ is chosen by the adversary, and $\mathsf{KPInit}(1^n, K)$ is used to generate an initial state $S$.

2. The adversary repeats the following steps an arbitrary number of times:
   (a) The adversary submits a circuit $F$, which is evaluated on $K$ by KPEval. During the computation, the adversary acts as a single invocation leakage adversary (described below in Definition 5) for KPEval.
   (b) At the end of the computation of KPEval, the adversary is given $F(K)$.
3. After the adversary is done making queries, it outputs a bit $b$.

ExpIdeal **(Ideal Interaction).** The interaction of the adversary with simulated leakage proceeds as follows:

1. The adversary submits a key $K$, which is not revealed to the simulator.
2. The adversary then repeats the following steps an arbitrary number of times:
   (a) The adversary submits a circuit $F$, and Sim is given $F$ and $F(K)$. The adversary then acts as a single invocation leakage adversary according to Definition 5, except that the leakage functions are submitted to the simulator, which returns simulated leakage values and messages.
   (b) Eventually the adversary stops submitting leakage functions, and is given $F(K)$.
3. After the adversary is done making queries, it outputs a bit $b$.

**Definition 3.** *We say that* KP *is a* Leakage-Resilient Key Proxy *if for every PPT $A$ there exists a PPT $S$ and a negligible function $neg(\cdot)$ such that*

$$|\Pr[(A \leftrightarrows \mathsf{ExpReal}) = 1] - \Pr[(A \leftrightarrows \mathsf{ExpIdeal}) = 1]| \leq neg(n)$$

The above definition describes the security of an LRKP relative to some unspecified procedure which allows the adversary to obtain leakage during each invocation of KPEval. The exact procedure for a single-invocation leakage depends on the leakage model and on the structure of the implementation of KPEval. Below we formalize the structure of our solution, and describe the leakage obtained by the adversary during a single invocation of KPEval.

Our construction of KPEval is described as a protocol between two parties EvalA and EvalB that leak information separately, and where the messages between EvalA and EvalB are public. In this format, our construction requires two flows between the parties: one from EvalA to EvalB and one from EvalB to EvalA. The following definition formalizes this structure.

**Definition 4.** *A 2-round* split state key proxy $KP = (\mathsf{KPInit}, \mathsf{KPEval})$ *is a key proxy such that the state $S$ is represented as a pair $S = (\mathsf{MemA}, \mathsf{MemB}) \in (\{0,1\}^{n^d})^2$ for some fixed $d \in \mathbb{N}$, and the algorithm KPEval is described as four algorithms $(\mathsf{LeakFree}, \mathsf{EvalA}_1, \mathsf{EvalB}, \mathsf{EvalA}_2)$, each running in time polynomial in $n$, where*

1. *$\mathsf{EvalA}_1$ takes as input $\mathsf{MemA}$, $\mathsf{OutLF}_A$, and randomness $\mathsf{RandA}$, and outputs an updated state $\mathsf{MemA}' \in \{0,1\}^{n^d}$ and a message $M_{AB}$ to $\mathsf{EvalB}$.*

2. LeakFree *takes as input message* $M_{AB}$ *and randomness* RandLF, *and outputs string* OutLF.

3. EvalB *takes as input* MemB, *randomness* RandB, OutLF, *the message* $M_{AB}$, *and a circuit* $F : \{0,1\}^{|K|} \to \{0,1\}^n$ *of arbitrary size. It then outputs an updated state* MemB$' \in \{0,1\}^{n^d}$ *and a message* $M_{BA}$ *to* EvalA.

4. EvalA$_2$ *takes as input* MemA$'$, *the message* $M_{BA}$ *and outputs an updated state* MemA$''$ *and the result* $F(K)$.

*The output of* KPEval *is* $F(K)$, *and the updated state is* (MemA$''$, MemB$'$).

Recall that our construction requires a leak-free component. This leak-free component is modeled by algorithm LeakFree above. A crucial point here is that LeakFree receives only randomness and a public message as input, and, in particular, receives neither $F$ nor the saved state (MemA, MemB) as inputs; therefore, regardless of the actual construction, the above definition prevents LeakFree from carrying out the evaluation of $F$ on $K$, which would make the construction trivial.

We are now ready to describe the leakage structure on a single invocation of a 2-round split state key proxy. The leakage model we use, commonly known as "only computation leaks information" (OCL), lets the adversary obtain leakage only on the active part of memory during each computation.

**Definition 5.** *Let* $l : \mathbb{N} \to \mathbb{N}$ *and let KP be a 2-round split state key proxy. A single invocation leakage adversary in the only-computation-leaks model chooses a circuit* $f_1$, *then sees* $f_1(\text{MemA}, \text{RandA})$ *and* $M_{AB}$, *chooses circuit* $f_2$, *then sees* $f_2(\text{MemB}, \text{OutLF}, \text{RandB})$ *and* $M_{BA}$, *chooses a circuit* $f_3$, *and finally sees* $f_3(\text{MemA}')$. *The adversary is* $l$-*bounded if for all* $n$ *the range of* $f_1, f_2, f_3$ *is* $\{0,1\}^{l(n)}$.

Note that in the above definition, the leakage functions can compute any internal values that appear during the computations of EvalA$_1$, EvalB, and EvalA$_2$. This means, for example, that it is unnecessary to explicitly provide $M_{AB}$ to $f_1$ or $M_{BA}$ to $f_2$.

*History freeness.* In Definition 3 we allow information about the functions $F_i$ that are evaluated on $K$ to leak to the adversary. In particular, it is possible that during some invocation $j$ the adversary can obtain, through leakage, information about some previously queried function $F_i$. In the introduction we mentioned that leakage-resilient variants of some applications, such as private-key encryption, are defined to allow leakage both before and after the generation of the challenge ciphertext, but not on the challenge itself. However, if the state of LRKP keeps a history of some of the functions that were applied to $K$, then by leaking on it after the challenge was computed, the adversary may be able to break the semantic security of the encryption. We note that the above definition is sufficient to obtain security in the presence of what we call "lunch-time leakage" attacks – where the adversary obtains leakage only before the challenge ciphertext is generated, but not after.

To address the above issue, and allow full leakage in applications such as encryption, we introduce an additional information theoretic property that requires that the state of the LRKP is distributed identically after all sequences of functions that are evaluated on $K$. This property is satisfied by our construction, and prevents the above mentioned "history attack".

**Definition 6.** *An LRKP* (KPInit, KPEval) *is called* history free *if for all $n \in \mathbb{N}$ and all $K \in \{0,1\}^{poly(n)}$, there exists a distribution $D$ over the states of the LRKP such that for all $j \in \mathbb{N}$, all sequences of functions $F_1, \ldots, F_j$ : $\{0,1\}^{|K|} \rightarrow \{0,1\}^n$, and all sequences of random tapes $R_0, \ldots, R_{j-1}$, the random variable $\{S_{j+1} | S_1, \ldots, S_j\}$ over $R_j$ is distributed according to $D$, where $S_1 =$ KPInit$(1^n, K; R_0)$ and $S_i$ is the updated state after* KPEval$(1^n, F_{i-1}; R_i, S_{i-1})$.

## 4   Leakage-Resilient Key Proxies from Homomorphic Encryption

Given a fully homomorphic public-key encryption scheme FHE = (KeyGen, Enc, Dec, EncEval, Add, Subtract) we construct a leakage-resilient 2-round split state key proxy LRKP = (KPInit, KPEval).

KPInit$(1^n, K)$: The algorithm KPInit$(1^n, K)$ first runs KeyGen$(1^n)$ to obtain a public-private key pair $(pub_1, pri_1)$ for the FHE. It then generates a ciphertext $C_{key} = \text{Enc}_{pub_1}(K)$ and assigns MemA $\leftarrow pri_1$ and MemB $\leftarrow C_{key}$. The output is an initial state that consists of two parts (MemA, MemB).

KPEval$(1^n, F; (\text{MemA}, \text{MemB}))$: The algorithm KPEval consists of four subroutines: $\langle$LeakFree, EvalA$_1$, EvalB, EvalA$_2\rangle$ that are used as follows: on input circuit $F$ first generate $(\text{OutLF}_A, \text{OutLF}_B) \leftarrow_R \text{LeakFree}(1^n)$. Then, follow the protocol described in Figure 2 by computing

$$(M_{AB}, \text{MemA}') \leftarrow_R \text{EvalA}_1(\text{MemA}, \text{OutLF}_A);$$
$$(M_{BA}, \text{MemB}') \leftarrow_R \text{EvalB}(\text{MemB}, \text{OutLF}_B, M_{AB});$$
$$Y \leftarrow \text{EvalA}_2(\text{MemA}', M_{BA})$$

The final state after one evaluation of KPEval is $(\text{MemA}', \text{MemB}')$, and the output is $Y$.

We now describe the subroutines $\langle$LeakFree, EvalA$_1$, EvalB, EvalA$_2\rangle$ of KPEval:

LeakFree$(pub)$: Parse randomness as $(r_{LF1}, r_{LF2})$, and compute

$$C_{R0} = \text{Enc}_{pub}(\bar{0}; r_{LF1})$$
$$C_{R1} = \text{Enc}_{pub}(\bar{0}; r_{LF2})$$
$$\text{OutLF} = (C_{R0}, C_{R1})$$

and output OutLF.

The subroutines $\mathsf{EvalA_1}$, $\mathsf{EvalB}$, and $\mathsf{EvalA_2}$ are described in Figure 2 as a two round two party protocol where $\mathsf{EvalA_1}$ and $\mathsf{EvalA_2}$ specify the actions of party $A$ and $\mathsf{EvalB}$ specifies the actions of party $B$. In the definition of $\mathsf{EvalB}$ we use subroutines $\mathsf{Evaluate}$ and $\mathsf{Refresh}$ that are defined as follows:

$$\mathsf{Evaluate}(F, C, pri): \text{Compute and output } F(\mathsf{Dec}_{pri}(C))$$

$$\mathsf{Refresh}(C, pri): \text{Compute and output } \mathsf{Dec}_{pri}(C)$$

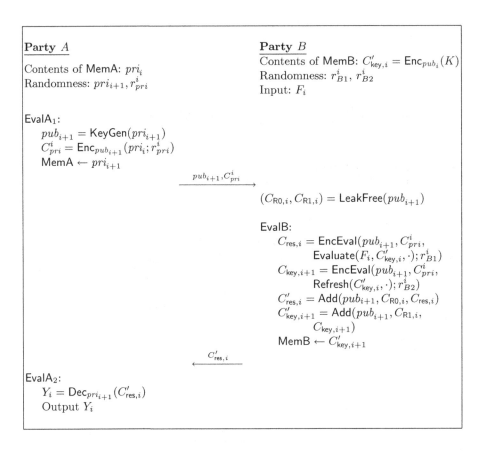

**Fig. 2.** The algorithm KPEval in its $i$th invocation

The correctness of this construction follows in a straightforward manner from the correctness of the underlying FHE. We also note that our construction is *history free* according to Definition 6. This is due to the fact that the values assigned to MemA and MemB at the end of KPEval are independent from the function $F$. In particular, MemA is simply a random private key, and MemB contains an encryption of $K$ which was obtained by a homomorphic evaluation of Refresh on the previous contents of MemB and an encryption of the previous private key, neither of which depends on $F$.

The bulk of the analysis is in showing that our construction is in fact leakage-resilient according to Definition 3, where during each invocation the leakage structure on the computation of KPEval is given in Definition 5. We now state our main theorem.

**Theorem 1.** *Let* LRKP *be the 2-round split state key proxy described in the above construction, and let* $l : \mathbb{N} \to \mathbb{N}$. *If* FHE *is a* $2^{O(l(n))}$*-secure fully homomorphic encryption then* LRKP *is leakage-resilient against all* $O(l(n))$*-bounded adversaries in the OCL model.*

The theorem follows as a corollary from the following lemma:

**Lemma 1.** *Consider the experiment* $ExpReal$ *instantiated using scheme* $LRKP$. *Then, for every function* $\varepsilon(n) > 0$, *every* $d > 0$, *every* $l : \mathbb{N} \to \mathbb{N}$, *and every* $l$*-bounded PPT adversary* $Adv$ *that makes* $n^d$ *queries and gets leakage according to the only-computation-leaks model, there exists a PPT simulator* $S$ *such that if*

$$|\Pr[(Adv \leftrightarrows ExpReal) = 1] - \Pr[(Adv \leftrightarrows S) = 1]| \geq \varepsilon(n)$$

*for infinitely many* $n$, *then for every function* $\varepsilon'(n) > 0$ *there exists an adversary* $Adv'$ *that runs in time*

$$\frac{2^{3l(n)+7}}{\varepsilon'(n)^2}\left(3l(n) + 4 + \log\frac{1}{\varepsilon'(n)}\right) \cdot time_n\,(LRKP \leftrightarrow Adv)$$

*and breaks the semantic security of* (KeyGen, Enc, Dec) *with advantage*

$$\frac{\varepsilon(n)}{3 \cdot 2^{2l(n)}(n^d + 1)} - 2\varepsilon'(n)$$

*for infinitely many* $n$. *Specifically,* $S$ *runs in time* $time_n(LRKP \leftrightarrow Adv)$.

### 4.1 Proof Approach for Lemma 1

Let $Adv$ be a PPT adversary according to Definition 3 that makes $n^d$ function evaluation queries and gets leakage according to the only-computation-leaks model described in Definition 5. We define a sequence of experiments where the initial experiment is the real security experiment ExpReal, and the final experiment is such that the leakage obtained by the adversary for each KPEval query $F$ can be simulated given only $(F, F(K))$. Specifically, the final experiment involves instantiating our construction with key $\bar{0}$ instead of $K$. We show that if $Adv$ can distinguish the initial experiment and the final experiment, we can construct an adversary $Adv'$ that, roughly speaking, distinguishes variants of these experiments that consist of only two rounds. We then show how pairs of the leakage queries of $Adv'$ can be combined into a single query (of larger output length) using a guess-and-check approach: when the adversary would normally make the first of the pair of leakage queries, it instead guesses an output and verifies this guess when it makes the second leakage query; when the guess is wrong, the adversary outputs a randomly chosen bit. Repeatedly combining queries in this manner yields an adversary that just makes a single leakage query and (essentially) distinguishes encryptions of $K$ and $\bar{0}$. To finish the proof, we use an observation of Akavia $et$ $al$ [1] that every $2^{O(\ell(n))}$-semantically-secure public-key encryption scheme remains secure when the adversary gets $O(\ell(n))$ bits of leakage on KeyGen. We defer the details to the full version of this paper [18].

## 5 Extensions and Applications

Below we describe some variants and applications of our scheme.

**Resilience against simultaneous leakage.** In Definition 5, the adversary is only allowed to see leakage from the part of memory where computation is occurring. Our construction is also secure under an alternative leakage model where the adversary is allowed to see *independent* leakage from *both* parts of memory each time it makes a leakage query. The basic idea is to first show that our construction is secure under a variant of Definition 5 where the adversary sees an additional leakage $f_4$ on memory $B$. Under this variant of Definition 5, the adversary's leakage queries strictly alternate between memory $A$ and memory $B$. We then use an observation of Pietrzak [21] that simultaneous but independent leakage on two pieces of memory can be perfectly simulated by strictly alternating leakage (of twice the output length) on these two pieces of memory.

**Resilience against complete compromise.** Our scheme can be viewed as a protocol between two devices that communicate over a public channel. The key remains hidden even if the memory contents of one of the devices are leaked completely (for example, in a cold boot attack), provided that the compromise is detected and no further computation is performed using the counterpart device.

**One-time programs.** Our construction can be modified to work without any leak-free components by pre-computing a large number of tuples of the form $(pri, pub, C, C')$ where $C$ and $C'$ are encryptions of 0 under $pub$, and storing the tuples in memory. Then, at each invocation, one such tuple is used (first $pri$ and $pub$ are used by EvalA$_1$, and then $C, C'$ are used by EvalB). Assuming that only computation leaks information, the remaining tuples remain hidden until they are accessed. Therefore, security is obtained following essentially the same argument as the proof of Theorem 1. The number of invocations in this case is bounded by the number of pre-computed tuples. This approach provides a weaker security guarantee than the one time programs of [15] (i.e. only security against leakage), but has the advantage that the pre-computing phase is independent from the functionality that is being protected.

**Concurrent composition.** We have shown that an adversary interacting with a single instance of our construction gains no information about the underlying key. However, for some applications, such as private-key encryption where several parties compute on the same agreed upon key, this may not suffice. It is quite possible that the adversary is performing side-channel attacks on several parties simultaneously, and is coordinating his leakage functions adaptively. In the full version of this paper, we show that an adversary interacting concurrently with several instances of our construction still gains no information through leakage.

**Leakage-resilient private-key encryption.** Extending the traditional notions of semantically secure encryption to the leakage setting is non-trivial. In particular, suppose that every invocation of the encryption algorithm leaks information. Then, since the plaintext of the adversary's challenge message is an input to the encryption algorithm, the adversary can trivially break semantic security by leaking even a single bit about this message. To deal with this problem, several works [9,22,8,5] adopt the approach that the computation of the encryption of the challenge is *not* allowed to leak. We follow this approach, and show how to obtain semantically-secure private-key encryption in the leakage setting using LRKPs. The details are deferred to the full version of this paper.

**Leakage-resilient public-key encryption.** Constructions of public-key encryption schemes that are resilient to an a-priori bounded amount of leakage were recently given by [22,2,5]. However, no constructions are known of PKEs that remain secure under Chosen Ciphertext Attack (CCA), if the adversary can obtain leakage during each decryption query. LRKPs provide a convenient way to achieve such a construction. Specifically, given a CCA-PKE (KeyGen, Enc, Dec), we construct a new PKE (KeyGen', Enc, Dec') where the encryption algorithm stays the same; the key generation KeyGen' runs KeyGen to obtain $(pub, pri)$ and then initializes an LRKP with $pri$. The public key is $pub$, and the private key is the initial state $state_1$ of the LRKP. The decryption algorithm is stateful, and to decrypt a ciphertext $C$, Dec' generates a circuit $H(x)$ that computes that function $Dec_x(C)$, and then uses KPEval to evaluate it on the private key $pri$.

*Acknowledgements.* We thank Charles Rackoff for many hours of discussion.

# References

1. Akavia, A., Goldwasser, S., Vaikuntanathan, V.: Simultaneous hardcore bits and cryptography against memory attacks. In: Reingold, O. (ed.) TCC 2009. LNCS, vol. 5444, pp. 474–495. Springer, Heidelberg (2009)
2. Alwen, J., Dodis, Y., Wichs, D.: Leakage resilient public-key cryptography in the bounded retrieval model. In: Halevi, S. (ed.) CRYPTO 2009. LNCS, vol. 5677, pp. 36–54. Springer, Heidelberg (2009)
3. Brakerski, Z., Kalai, Y., Katz, J., Vaikuntanathan, V.: Cryptography resilient to continual memory leakage (2010) (manuscript)
4. van Dijk, M., Gentry, C., Halevi, S., Vaikuntanathan, V.: Fully homomorphic encryption over the integers. In: Gilbert, H. (ed.) EUROCRYPT 2010. LNCS, vol. 6110, pp. 24–43. Springer, Heidelberg (2010)
5. Dodis, Y., Goldwasser, S., Kalai, Y., Peikert, C., Vaikuntanathan, V.: Public-key encryption schemes with auxiliary inputs (2009)
6. Dodis, Y., Goldwasser, S., Kalai, Y.T., Peikert, C., Vaikuntanathan, V.: Public-key encryption schemes with auxiliary inputs. In: Micciancio, D. (ed.) TCC 2010. LNCS, vol. 5978, pp. 361–381. Springer, Heidelberg (2010)

7. Dodis, Y., Haralambiev, K., Lopez-Alt, A., Wichs, D.: Cryptography against continuous memory attacks. Cryptology ePrint Archive, Report 2010/196 (2010), http://eprint.iacr.org/

8. Dodis, Y., Kalai, Y.T., Lovett, S.: On cryptography with auxiliary input. In: STOC 2009: Proceedings of the 41st Annual ACM Symposium on Theory of Computing, pp. 621–630. ACM, New York (2009)

9. Dziembowski, S., Pietrzak, K.: Leakage-resilient cryptography. In: FOCS 2008: Proceedings of the Annual IEEE Symposium on Foundations of Computer Science, pp. 293–302. IEEE Computer Society, Washington (2008)

10. Faust, S., Kiltz, E., Pietrzak, K., Rothblum, G.N.: Leakage-resilient signatures. In: Micciancio, D. (ed.) TCC 2010. LNCS, vol. 5978, pp. 343–360. Springer, Heidelberg (2010)

11. Faust, S., Rabin, T., Reyzin, L., Tromer, E., Vaikuntanathan, V.: Protecting against computationally bounded and noisy leakage. In: Gilbert, H. (ed.) EUROCRYPT 2010. LNCS, vol. 6110, pp. 135–156. Springer, Heidelberg (2010)

12. Gentry, C.: Fully homomorphic encryption using ideal lattices. In: STOC 2009: Proceedings of the 41st Annual ACM Symposium on Theory of Computing, pp. 169–178. ACM, New York (2009)

13. Goldreich, O., Ostrovsky, R.: Software protection and simulation on oblivious rams. J. ACM 43(3), 431–473 (1996)

14. Goldwasser, S., Kalai, Y., Peikert, C., Vaikuntanathan, V.: Robustness of the learning with errors assumption. In: Proceedings of the 1st Innovations in Computer Science Conference, ICS 2010 (2010)

15. Goldwasser, S., Kalai, Y.T., Rothblum, G.N.: One-time programs. In: Wagner, D. (ed.) CRYPTO 2008. LNCS, vol. 5157, pp. 39–56. Springer, Heidelberg (2008)

16. Goldwasser, S., Rothblum, G.: Securing computation against continuous leakage. In: Rabin, T. (ed.) CRYPTO 2010. LNCS, vol. 6223, pp. 59–79. Springer, Heidelberg (2010)

17. Ishai, Y., Sahai, A., Wagner, D.: Private circuits: Securing hardware against probing attacks. In: Boneh, D. (ed.) CRYPTO 2003. LNCS, vol. 2729, pp. 463–481. Springer, Heidelberg (2003)

18. Juma, A., Vahlis, Y.: Protecting cryptographic keys against continual leakage. Cryptology ePrint Archive, Report 2010/205 (2010), http://eprint.iacr.org/

19. Katz, J., Vaikuntanathan, V.: Signature schemes with bounded leakage resilience. In: Matsui, M. (ed.) ASIACRYPT 2009. LNCS, vol. 5912, pp. 703–720. Springer, Heidelberg (2009), http://dx.doi.org/10.1007/978-3-642-10366-7

20. Micali, S., Reyzin, L.: Physically observable cryptography. In: Naor, M. (ed.) TCC 2004. LNCS, vol. 2951, pp. 278–296. Springer, Heidelberg (2004)

21. Pietrzak, K.: A leakage-resilient mode of operation. In: Joux, A. (ed.) EUROCRYPT 2009. LNCS, vol. 5479, pp. 462–482. Springer, Heidelberg (2009)

22. Segev, G., Naor, M.: Public-key cryptosystems resilient to key leakage. In: Halevi, S. (ed.) CRYPTO 2009. LNCS, vol. 5677, pp. 18–35. Springer, Heidelberg (2009)

23. Standaert, F.X., Malkin, T., Yung, M.: A unified framework for the analysis of side-channel key recovery attacks. In: Joux, A. (ed.) EUROCRYPT 2009. LNCS, vol. 5479, pp. 443–461. Springer, Heidelberg (2009)

24. Standaert, F.X., Pereira, O., Yu, Y., Quisquater, J.J., Yung, M., Oswald, E.: Leakage resilient cryptography in practice. Cryptology ePrint Archive, Report 2009/341 (2009), http://eprint.iacr.org/

# Securing Computation against Continuous Leakage

Shafi Goldwasser[1,*] and Guy N. Rothblum[2,**]

[1] Weizmann Institute of Science and MIT
[2] Princeton University

**Abstract.** We present a general method to compile any cryptographic algorithm into one which resists side channel attacks of the *only computation leaks information* variety for an unbounded number of executions. Our method uses as a building block a semantically secure subsidiary bit encryption scheme with the following additional operations: key refreshing, oblivious generation of cipher texts, leakage resilience re-generation, and blinded homomorphic evaluation of one single complete gate (e.g. NAND). Furthermore, the security properties of the subsidiary encryption scheme should withstand bounded leakage incurred while performing each of the above operations.

We show how to implement such a subsidiary encryption scheme under the DDH intractability assumption and the existence of a simple secure hardware component. The hardware component is independent of the encryption scheme secret key. The subsidiary encryption scheme resists leakage attacks where the leakage is computable in polynomial time and of length bounded by a constant fraction of the security parameter.

## 1 Introduction

Modern cryptographic algorithms are designed under the assumption that keys are perfectly secret, and computations done within one's computer are opaque to the outside. Still, in practice, keys do get compromised at times, and computations are not fully opaque for a variety or reasons. A particularly disturbing loss of secrecy is as a result of *side channel attacks.*

These attacks exploit the fact that every cryptographic algorithm is ultimately implemented on a physical device and such implementations enable "observations" that can be made and measured on computations which use secret data and secret keys, or on the secret keys and data directly. Such observations can and have lead to complete breaks of systems which were proven secure, without violating any of the underlying mathematical principles. (see [KJJ99, RCL] for just two examples). Recently, a growing body of research on *side-channel-resilient cryptography* aims to build general mathematical models of realistic side

---

\* This research is supported in part by ISF710267, BSF710613, NSF6914349 and an internal Weizmann KAMAR grant.
\*\* Research supported by NSF Grants CCF-0635297, CCF-0832797 and by a Computing Innovation Fellowship.

T. Rabin (Ed.): CRYPTO 2010, LNCS 6223, pp. 59–79, 2010.

channel attacks, and to develop methods grounded in modern cryptography to provably resist these attacks.

Modeling side channel attacks on a cryptographic algorithm so as to simultaneously capture real world attacks and achieve the right level of theoretical abstraction, is an intriguing and generally controversial question. Indeed, the number of answers seems to be nearly as high as the number of papers published on the topic. Perhaps the only universally agreed on part of the modeling is that each physical measurement should be modeled as the result of computing an adversarially chosen but computationally bounded function $\ell$ (the so called "leakage" function) on the "internal state of the computation". We find the most important modeling questions to be:

– *How should we characterize which leakage functions $\ell$ can be measured?*
– *How many measurements occur and how often?*
– *Are all portions of the computation's internal state subject to measurement at the same time? Namely, what is the input to the leakage function $\ell$?*
– *Can we use secure hardware components, and if so which ones are reasonable to assume as building blocks to achieve side channel security?*

*"Only computation leaks information" and the question of granularity.* Micali and Reyzin, in their pioneering work [MR04], set forth a model of physical security, which takes a particular approach at these modeling questions. One of the axioms in their model was that any computation *but only computation* leaks information (OC attack model). In other words, every time a computation step of a cryptographic algorithm "touches" data which may contain portions of (but not necessarily the entirety of): cryptographic secret keys, internally generated randomness, and results of previous computations done on cryptographic keys, a measurement on this data can be made by an adversary. However, data which is not "touched" by a computation step of an algorithm, can not be measured at this time (and thus does not leak). Stated in terms of leakage functions, this means that a leakage function can be computed in each computation step, but each such function is restricted to operate only on the data utilized in that computation step. Within this model, various constructions of cryptographic primitives [GKR08, DP08, Pie09, FKPR09] such as stream ciphers and digital signatures, have been proposed and proved secure for certain leakage function classes and under various computational intractability assumptions.

This is the model of attacks which we focus on in this paper. Our main result addresses *how to run any cryptographic algorithm* (i.e an algorithm which takes as input secret keys and uses secret randomness) securely in this model for an unbounded number of executions.

Implicit in using this model, is the view of program execution (or computation) as preceding in discrete 'sub-computation steps' $S_1$, $S_2$... Each sub-computation $S_i$ computes on some data $d_i$ (which is a combination of secret and public data and randomness). At each $S_i$, the side-channel attack adversary can request to receive the evaluation of a new leakage function $\ell_i$ on $d_i$. The choice of $\ell_i$ to be evaluated at step $S_i$ may depend on the results of values attained by previous $\ell_1, ..., \ell_{i-1}$, but $\ell_i$ can only be evaluated on the $d_i$ used in step $S_i$.

An important question in evaluating results in the OC attack model emerges: what constitutes a sub-computation step $S_i$, or more importantly what is the input data $d_i$ to $S_i$ available to $\ell_i$ in this sub-computation? Let us look for example at the beautiful work of Dziembowski and Pietrzak [DP08] which construct secure *stream ciphers* in the OC model. Initialized with a secret key, their stream cipher can produce an unbounded number of output blocks. In [DP08], the $i$-th sub-computation is naturally identified with the computation of the $i$-th block of the stream cipher. The input $d_i$ to this sub-computation includes a pre-defined function of the original input secret key. The class of tolerated leakage functions $\ell_i$ (each computed on $d_i$) are (roughly) restricted to a length shrinking function whose output size is logarithmic in the size of the security parameter of the stream cipher[1]. Another example is in the work of Faust *et al.* [FKPR09] which construct secure randomized digital signature scheme which can generate an unbounded number of signatures in the OC attack model. The $i$-th sub-computation is identified with the computation of the $i$th signature, and $d_i$ is (essentially) fresh randomness generated for the $i$-th sub-computation. Coupled with one-time signatures of [KV09], the class of leakage functions $\ell_i$ tolerated are length shrinking functions whose output size is a constant fraction of the size of the security parameter of the signature scheme, under the intractability of DDH and various lattice problems.

An interesting practical as well as theoretical question is what *granularity (i.e. size of sub-computations)* is reasonable to consider for general cryptographic computation. Certainly, the larger the granularity (and the sub-computations), the better the security guarantee. For security, ideally we'd prefer to allow the leakage to work on the entire memory space of the computation. However, the assumption that physical leakage is "local" in time and space, and applies to small sub-computations as they happen, still encapsulates a rich family of attacks. Carried to the extreme, one might even model leakage as occurring on every single gate of a physical computation with some small probability, and even this model may be interesting.

In this work, we advocate the approach of allowing the programmer of a cryptographic computation, the freedom to divide the computation into arbitrary sub-computations, and then analyzing security by assuming that leakage is applied to each sub-computation's input independently (i.e. only computation leaks information). In particular, this will mean that the total amount of leakage from a computation can grow with the complexity of the computation (as well as the number of executions), as it well should, since indeed in practice the possibility of leakage increases with the complexity (length of time) of the computation. General approach aside, our positive results are much stronger: we work with granularity that is a polynomial in a security parameter.

---

[1] Alternatively stated, their construction of is based on an exponential hardness assumption (where the assumption degrades as a function of the amount of leakage tolerated).

## 1.1   The Contributions of This Work

In this work we focus on general cryptographic computations in the OC attack mode, and address the challenge of how to run *any cryptographic algorithm* securely under this attack model, for any polynomial number of executions.

Our contributions are twofold. First, we show a reduction. Starting with a subsidiary semantically secure bit encryption scheme $E$, which obeys certain additional homomorphic and leakage-resilience properties (see below and Section 3), we build a compiler that takes any cryptographic algorithm in the form of a Boolean circuit, and probabilistically transforms it into a functionally equivalent probabilistic stateful algorithm. The produced algorithm can be run by a user securely for an unbounded number of executions in the presence of continuous OC side-channel attacks. Second, we show how to implement such a subsidiary encryption scheme $E$ under the DDH intractability assumption and using a secure hardware component. The hardware component (see Section 1.1) samples from fixed polynomial time computable distribution (it does not compute on any secrets of the computation). The security assumed about the component is that there is no leakage on the randomness it uses or on its inner workings.

*The execution and adversary model:* We start with a cryptographic algorithm $C$ and its secret key $y$ ($C$ is a member of a family of poly($n$)-size Boolean circuits $\{C_n\}$ and $y \in \{0,1\}^n$). In an initial *off-line stage when no side-channel attacks are possible*, $C(y, \cdot)$ is converted via a probabilistic transformation to an algorithm $Eval_C$ with state – which is updated each time $Eval_C$ is executed, and is functionally equivalent to $C$, i.e $Eval_C(\cdot) = C(y, \cdot)$. After this initial off-line stage, $Eval_C$ is computed on an unbounded number of public inputs $x_1, x_2,...$ which can be chosen by the adversary in the following manner. The computation of $Eval_C(x_i)$ is divided into sub-computations $C_{i,1}, ..., C_{i,n}$ each of which are evaluated on data $d_{i,1}, ..., d_{i,n}$ respectively. At this stage, for each sub-computation $C_{i,j}$, the OC side-channel adversary is allowed to request the result of evaluating leakage function $\ell_{i,j}$ on $d_{i,j}$. The leakage functions we tolerate can be chosen adaptively based on the result of previously evaluated leakage functions, and belong to the class of polynomial time computable length shrinking functions. We emphasize that after the initial off-line stages all computations of $Eval_C$ (including its state update) are subject to OC side-channel attacks.

*The security guarantee:* is that even under the OC side-channels and adversarially chosen inputs, the adversary learns no more than the outputs of $C(y, \cdot)$ on the chosen inputs (formally, there is a simulation guarantee). In particular, it is important to distinguish between leakage incurred on the cryptographic algorithm $C(y, \cdot)$ being protected, and the leakage on the subsidiary cryptographic scheme. There is constantly leakage on the subsidiary scheme's secret keys, and the specific scheme we use can handle this. On the other hand, for the algorithm $C(y, \cdot)$ *there is no leakage at all on $y$*. Only its black-box behavior it exposed.

For example, if we think of $C$ as the decryption algorithm for *any* public-key scheme, and $y$ as its secret decryption key (which is completely unrelated to the secret keys of the subsidiary cryptosystem we use as a tool!), then an adversary who wants to decrypt a challenge ciphertext $x_1$, and has OC leakage access to the evaluation of $Eval_C(y, x_2)$ for decrypting another ciphertext $x_2$, still cannot break the security of $x_1$ and in particular cannot decrypt it. This is a qualitatively different security guarantee from the setting of memory-bound leakage [AGV09] or even in the more recent work of Brakerski *et al.* [BKKV10] on public key encryption under continual leakage. In these works, no security is guaranteed for a challenge ciphertext that is known to the adversary when it chooses the leakage function.

*The granularity of our sub-computations:* We let a subsidiary security parameter $\lambda$ govern the granularity of the computation steps as follows. The computation of $Eval_C$ is divided into sub-computations each of which consist of performing a basic cryptographic operations (e.g. encrypt, decrypt, key generate, etc.) of a subsidiary encryption scheme $E$ with security parameter $\lambda$. Essentially, $E$ is used as a tool to emulate a secure executions of $C$, in such a way that a constant number of cryptographic operations of $E$ emulate the evaluation of each gate of $C$. Thus the complexity of $Eval_C$ is $O(poly(\lambda) \cdot |C|)$. In accordance with the OC attack model, leakage functions are assumed to apply to each input of the cryptographic operations of $E$ separately. The main idea behind obtaining the leakage resilience for any algorithm $C$, is that whereas how $C$ works is out of our control (as it is a given), we can choose an $E$ for which we are able to continually refresh its keys. As each key will be utilized as input for only a constant number of cryptographic operations, only a bounded number of leakage functions (measurements) can be made on each key. Indeed, for an appropriately chosen $E$, we can tolerate any polynomial time computable leakage functions, whose output length is up to a constant fraction of the security parameter $\lambda$. Note that the security parameter $\lambda$ is chosen for the purposes of side-channel security, and may be chosen to be different than the security parameter $n$ of the cryptographic algorithm $C$, by the implementer.

*Leakage grows with the complexity of $Eval_C$:* The total amount of leakage that our method can tolerate per execution of $Eval_C$ is $O(\lambda \cdot |C|)$ whereas and the complexity of $Eval_C$ is $O(poly(\lambda) \cdot |C|))$. Thus, our method tolerates more measurements and leakage as the computation time increases. This is in contrast with previous general compilers (see Section 1.2), where the size of the transformed circuit grows as a function of the total amount of leakage tolerated.

**Main tool: a subsidiary cryptosystem.** The subsidiary cryptosystem utilized by our compiler is a semantically secure bit encryption scheme with the following special properties (even in the presence of OC side channel attacks). See Section 3 for full definitions of these properties.

• **Semantic Security under Multi-source Leakage.** We require semantic security to hold even against an adversary who can (measure) receive leakage both from the secret key *and the cipher-texts which we attempt to protect, and are encrypted under this secret key.* Note that we depart here from the [AGV09] model in considering leakage also on the challenge ciphertexts, and not only on the keys. A priori, this might seem impossible. The reason it is facilitated is that due to the OC nature of our attacks an adversary can never apply a leakage function to the ciphertext and the secret-key at the same time (otherwise it could decrypt); furthermore the leakage length bound ensures that the adversary will not learn enough of the ciphertext to be useful for him at a later time when it can apply an adaptively chosen leakage function to the secret key (otherwise, again, it could decrypt the ciphertext).

• **Key Refreshing.** It should be possible to "refresh" secret keys in the scheme, changing them into new keys, via a randomly generated *correlation value*. In addition, we require that using the correlation value alone and *without knowledge of the secret key*, one can also refresh old ciphertexts under the old secret key to new ciphertext under the new secret key. Intuitively, this property is useful for taking secret keys on which there has already been a large amount of leakage, and transforming them into new keys on which there is less leakage (i.e. with more entropy). The requirement that refreshing on ciphertexts must not use the secret key, is due to the fact that otherwise a leakage function could be evaluated on the ciphertext and key (which are computed on at the same time) simultaneously and used to decrypt the ciphertext! The fact that the correlation value alone can be used to refresh ciphertexts avoids attacks of this type.

• **Oblivious Ciphertext Generation.** It should be possible to generate fresh encryptions of random bits. Even an OC adversary should not be able to tell anything about the plaintexts in these new obliviously generated ciphertexts. For example, the Goldwasser-Micali [GM84] cryptosystem naturally has this property (by generating a random Jacobi symbol 1 element).

• **Leakage Resilience Regeneration.** It should be possible to "re-generate" leakage resilience on ciphertexts and keys: i.e., to take a ciphertext and secret key and repeatedly generate a new "random-looking" ciphertext and key pair, encrypting the same value. The security requirement is that even after many such regenerations (with accumulated ciphertext and key OC leakages), as long as the amount of leakage between two successive regenerations is bounded, an adversary cannot tell whether the original ciphertext was an encryption of 0 or of 1. Intuitively, this property is useful for taking old ciphertexts and keys, on which there has been previous leakage, and re-generating them into new ones that are more secure (i.e. injecting new entropy).

• **Blind Homomorphic NAND.** It should be possible to take three ciphertexts $c_i, c_j, c_k$, encryptions of $b_i, b_j, b_k$ (respectively), and output a data string $hc$ (a "homomorphic ciphertext") which can later be decrypted (using the secret key) to yield $b = (b_i \text{ NAND } b_j) \oplus b_k$.[2] Moreover, we require a "blinding" property: that the encrypted outcome $hc$ contains no more information beyond the plaintext outcome $b$, even w.r.t an adversary who can launch OC attacks on the homomorphic evaluation, and who is later given OC access to the decryption of $hc$ (which also computes on the secret key). In particular, such an adversary should not be able to learn anything about the values of $b_i, b_j, b_k$ beyond $b$. Note that we do not require that $hc$ itself be a ciphertext or support any further homomorphic operations: i.e. we require only "one-shot" homomorphism.

---

[2] Actually, we require homomorphic evaluation of a slightly more complex functionality that also takes some plain-text inputs, see Section 3.

**Instantiating the subsidiary cryptosystem.** A slight modification of the encryption scheme of Naor and Segev [NS09] and Boneh et al[BHHO08], amplified with a simple secure hardware device, satisfies all of these properties. Here we highlight some of the novel challenges and ideas. See Section 3 for details.

We do not specify the scheme [NS09] in its full detail here, but only recall that operates over a group $G$ of order $q$ where the Decisional Diffie Hellman Problem (DDH) is hard. The secret key is a vector $s \in \mathrm{GF}[q]^m$ for some small $m > 0$ (for our parameters $m = 10$ suffices) and the public key is $g^s, g$ for a generator $g$. To encrypt $b \in \mathrm{GF}[q]$, the scheme masks $g^b$ by multiplying it by a group element whose distribution is indistinguishable (under DDH) from $g^{\langle s,r \rangle}$, where $r \in \mathrm{GF}[q]^m$ is uniformly random. We note further that the scheme supports homomorphic addition (over $\mathrm{GF}[q]$) and scalar multiplication.

*Semantic security under multi-source leakage.* We need to prove that semantic security holds when an adversary can launch a "multi-source" leakage attack *separately* on the secret key and the cipher-texts which we attempt to protect (encrypted under this secret key), a (constant fraction) of leakage is computed on each. The proof of security uses ideas from theory of two source extractors. In particular a theorem of Chor and Goldreich [CG88], showing how to extract statistically close to uniform bits from two independent min-entropy sources. We argue (assuming DDH) that the adversary's view is indistinguishable from an attack in which the plaintext $b$ is masked by $g^{\langle s,r \rangle}$, where $r$ is a uniformly random vector in $\mathrm{GF}[q]^m$. Given the adversary's separate leakage functions from the key $s$ and the ciphertext $r$, $s$ and $r$ will have sufficient *entropy* (because the amount of leakage is bounded) and are *independent* random sources (because the leakage operates separately on key and ciphertext). Using [CG88] we conclude that $\langle s, r \rangle$, and also $g^{\langle s,r \rangle}$, are statistically close to uniform. This is all in an attack where $r$ is uniformly random, but this attack is (under DDH) indistinguishable from the real one, and so semantic security holds. No secure hardware is used here.

*Key refresh.* Key refresh is enabled by the homomorphic properties of the Naor-Segev cryptosystem. In particular, choosing a correlation value $\pi \in GFq^m$, we can add this value to the secret key and update the public key and any ciphertext accordingly in a homomorphic manner, without accessing the secret key. No secure hardware is used here.

*Secure hardware.* The secure hardware device **CipherGen** (see Section 3.1) that is used in this work is simple. The device receives as input a public key and mode of operation $mode \in \{0, rand\}$. In mode $mode = 0$ it computes and outputs a fresh encryption of 0, and in mode $mode = rand$ it chooses a uniformly random bit $b \in \{0, 1\}$ and computes and outputs a fresh encryption of $b$. I.e. it only runs public key operations. We assume that when this device is invoked, there is leakage on its input and output, but not on its internal workings or randomness. It is interesting to compare this device to the device used by Faust *et al.* [FRR$^+$09]. Their device samples a random string whose XOR is 0. This can be viewed as a string "encrypting" the bit 0. The adversary, who is bounded to $AC^0$

bounded length leakage functions, cannot determine the XOR, or "decryption", of the string that was generated. We also note that in several works addressing continual leakage for particular functionalities, it is assumed that during parts of the computation either there is *no leakage* from the computation's internal randomness [DHLAW10], or that leakage from the internal randomness is very limited [BKKV10].

*Oblivious Generation and Leakage-Resilience Regeneration.* These two properties are satisfied almost immediately by the **CipherGen** secure hardware device. Activating the device in mode *rand* generates opaquely a ciphertext encrypting a random plaintext bit — giving immediately an oblivious generation procedure. For ciphertext and key regeneration we first use key refreshing to regenerate the secret key (injecting new entropy). We then use mode 0 of **CipherGen** to generate a fresh encryption of 0, and add it to the ciphertext. This effectively regenerates the randomness of the ciphertext, injecting new entropy.

*Homomorphic blinded masked NAND.* Perhaps the most challenging obstacle in constructing the subsidiary cryptoscheme is coming up with a procedure for computing blinded homomorphic masked NAND, i.e. given ciphertext $c_1, c_2, c_3$ encrypting plaintexts $b_1, b_2, b_3 \in \{0, 1\}$, computing a homomorphic blinded ciphertext containing $(b_1 \text{ NAND } b_2) \oplus b_3$.

Suppose for a moment that $b_3 = 0$ (i.e. we are computing the NAND of $b_1$ and $b_2$). We could homomorphically add the three ciphertexts, obtaining an encryption $d$ of a plaintext $\gamma$, where $\gamma$ is either 0,1 or 2 (it is important the homomorphic addition here is over $\mathbb{GF}[q]$ only). Here $\gamma = 2$ means that $b_1 = b_2 = 1$ and the NAND is 0, and $\gamma \in \{0, 1\}$ outcomes imply that the NAND is 1. We note however that the exact value of $\gamma \in \{0, 1\}$ leaks information about the input $b_1$ and $b_2$, which we will need to "blind".

There are two main ideas in blinding. The first is to use mode *rand* of **CipherGen** to generate an encryption $u$ of a random bit in $\mu \in \{0, 1\}$. We can then homomorphically compute an encryption of $\gamma - \mu - 2$, which will always be non-zero if the NAND is 1, and will be zero w.p. $1/2$ (over the ciphertext generated by **CipherGen**) if the NAND is 0. Similarly, for the case where $b_3 = 1$ we can compute an encryption of $\gamma - \mu$ which will have the same distribution depending on the value of the masked NAND. In conclusion, if we compute homomorphically an encryption of $b_1 + b_2 - \mu - 2 \cdot (1 - b_3)$ we obtain an encryption of a non-zero value when the NAND is 1, or a zero value w.p. $1/2$ when the NAND is 0. Repeating this several times, for different $u$, all the homomorphic decryptor needs to do is check whether any of these homomorphic computations resulted in a zero plaintext (in which case the output is 0) or not (there is a negligible error probability of incorrect decryption). We emphasize, that for each ciphertext generated in the above procedure, being an encryption of a zero or non-zero plaintext exposes *no information about the inputs* (beyond the output). This is because even an OC leakage adversary cannot tell whether **CipherGen** generated an encryption of 0 or 1. In a different context and cryptosystem, similar ideas for blinding were used by Sander, Young and Yung [SYY99].

Still, another idea is necessary, as the specific non-zero plaintext value (e.g. 1 rather than 2) might leak information about the inputs. An initial observation is that homomorphic multiplication by a random scalar $e$ leaves zero ciphertexts as encryptions of zero, but completely randomizes the plaintext values of non-zero ciphertexts. This can blind the ciphertexts while maintaining (for correctness) their plaintext being zero or non-zero (respectively). Unfortunately, in the presence of OC leakage there will be leakage on the value $e$, and this blinding will not be secure. We handle the OC leakage using a more complicated blinding procedure, which essentially homomorphically multiplies the plaintext by an inner product of two vectors $e$ and $f$ of random scalars. We use ciphertext regeneration (mode 0 of **CipherGen**) in-between homomorphic sub-steps to ensure that the leakage from each scalar is independent (or rather indistinguishable under DDH from an experiment with independent leakage). In the end, even given the leakage, the scalar $\langle e, f \rangle$ by which we multiply the ciphertext is indistinguishable from uniform, even to an OC leakage adversary, and blinding is guaranteed.

**Main result: the compiler.** The main contribution of this paper is a compiler which takes any cryptographic algorithm in the form of a Boolean circuit, and transforms it into a functionally equivalent probabilistic stateful algorithm. In this overview we assume an intuitive understanding of the subsidiary encryption scheme $E$ and its properties and letting $(pk_j, sk_j)$ denote public and secret key pairs of $E$. See Section 4 for details. We emphasize that in the description that ensues there is a distinction between a user who is executing the evaluation algorithm and an adversary whose view of this execution (which proceeds by a sequence of sub-computations) is only through the results of leakage functions applied on secret data as sub-computations are actually performed on this data.

The *input* to the compiler is a *secret* input $y \in \{0,1\}^n$, and a *public circuit C* of size $poly(n)$ that is known to all (compiler and adversary alike). The circuit takes as inputs $y$ and also public input $x \in \{0,1\}^n$ (which may have been chosen by the adversary), and produces a single bit output.[3] Without loss of generality, the circuit $C$ is composed of NAND gates with fan-in and fan-out 2, which are organized in layers. The inputs of layer $i$ arrive from the outputs of layer $i-1$. The *output* of the compiler on $C$ and $y$ is a probabilistic evaluation algorithm $Eval_C$ with state (which will be updated during the run of $Eval_C$) such that for all $x$, $C(y, x) = Eval_C(x)$. The compiler is run once at the beginning of time and is not subject to side-channels. See Section 2.2 for a formal security definition.

The idea of the evaluation algorithm is that in its state it keeps the value $v_j$ of each wire $j$ of the original input circuit $C(y, x)$ in the following secret-shared form: $v_j = a_j \oplus b_j$. The invariant for every wire is that the $a_j$ shares are public and known to all whereas $b_j$ are secret and kept encrypted by the subsidiary encryption algorithm $E$ under a secret key $sk_j$ (i.e. there is a key-pair for every wire). We emphasize that the OC side-channel adversary does not actually ever see even the cipher-text of plain text $b_j$ – let alone $b_j$ itself – in their entirely, but rather only the result of a leakage function on these cipher-texts at the time when they are involved in a sub-computation.

---

[3] We focus on single bit outputs, the case of multi-bit outputs also follows naturally.

At the outset of computing $Eval_C$, for all input wires corresponding to the $y$-input, $a_j = 0$; for all input wires corresponding to the $x$ input, $b_j = 0$; for all the other wires $b_j$ are chosen uniformly at random independently of the input; This generation of random ciphertexts containing the $b_j$ value is done using the oblivious generation procedure of $E$. Finally, for the circuit's output wire, $b_{output} = 0$. As the user selects an input $x$ to run $Eval_C$ on, he sets $a_j$ on the input wires of the $x$-input by the value of the bits of $x$, and is now ready to start updating the shares $a_j$ on the internal wires and compute $a_{output} = C(y, x)$.

The crux of the idea is to show how the user can compute the public shares corresponding to the internal wires of $C(y, x)$. Here is where we use the fact that encryption scheme $E$ can support a blinded homomorphic evaluation of a single NAND gate. Say, the user already computed the values of $a_j$ of all wires $j$ on layer $s$ (starting from the input wires this will hold inductively). Then, for each pair of wires $i, j$ into a gate on layer $s + 1$ with output wire $k$, the user will compute the public share of the output wire $a_k$ via a sequence of sub-computations as follows: first, transform the cipher texts of $b_i, b_j$ (using the key-refresh property) to encryptions of the same plaintexts under the secret key $pk_k$; second, homomorphically using $a_i, a_j$ and the cipher texts of $b_i, b_j, b_k$ all under $pk_k$ compute a (blinded) ciphertext $hc_k$ of $a_k$ under $pk_k$ (note that $a_k = ((a_i \oplus b_i) \text{ NAND } (a_j \oplus b_j)) \oplus b_k)$[4] and finally, decrypt the blinded $hc_k$ using secret key $sk_k$ to obtain $a_k$. Note that this is one place the "only computation leaks information" assumption is of essence. For example, if the leakage function would have taken the inputs to the first sub-computation as well as to the third sub-computation, it could have used $sk_k$ to decrypt $b_i$ and discover in full the value of $v_i$, which of course would destroy the security of the entire construction (since it is non black-box information about the computation being performed). It is also important to note here that we will set the leakage parameter $\lambda$ to be such that the adversary cannot even see enough of the ciphertexts corresponding to secret shares $b_j$ under any key (and in particular under $sk_k$). Otherwise, the adversary could "remember" these ciphertexts and then adaptively choose a future leakage function applied on $sk_k$ to decrypt it. Proceeding inductively, finally the user will compute $a_{output}$ and since $b_{output}$ was set initially to 0, the user has obtained $v_{output} = a_{output}$.

Finally, to prepare for another execution of $Eval(x')$ for a new $x'$, all ciphertexts and keys containing secret shares of the bits of the secret input $y$ are regenerated. This effectively "resets" the amount of leakage that has happened on these ciphertexts and keys. In the next execution we again (from scratch) choose a new oblivious encryption of a random $b_j$ for each internal wire $j$.

*Summary:* One of the main advantages of the above construction was that it let us go from a procedure for blinded OC-secure homomorphic evaluation of a single (NAND) gate, and obtain an evaluation mechanism for an arbitrary functionality (using several other properties of the subsidiary cryptosystem). We note that if the subsidiary cryptosystem supports more complex homomorphic computations,

---

[4] Note that, in terms of leakage, this sub-computation may itself be separated into smaller sub-computations.

we may hope to use the same framework to obtain a more efficient construction, operating at the level of larger computations rather than gate-by-gate (perhaps with improved granularity). We also note that the above construction should be viewed mainly as a proof-of-concept, we did not attempt here to make it practical enough for implementation.

## 1.2   Related Work

Our work is inspired by many beautiful classical techniques in the field of cryptography. For one, the central idea of our compiler may be thought of as a cross between the garbled circuit method originated by Yao [Yao82] and the pioneering idea of Goldreich, Micali, and Wigderson [GMW87] of computing on data by keeping it in a secret shared form and computing on the shares. Using limited homomorphic properties of encryption schemes in order to perform reduced round oblivious circuit evaluation was proposed in the work of Sander, Young, and Yung [SYY99]. Secure hardware was proposed in many prior works in the context of achieving provable security, starting with work of Goldreich and Ostrovsky [GO96] on software protection which assumes a universal secure leak-free processor. Most importantly, our work should be compared to results in other side channel attack models. We note that in the random oracle model other works have appeared (we do not cover all these results here).

The pioneering work of Ishai, Sahai, and Wagner [ISW03] first considered the questions of converting general cryptographic algorithms (or circuits) to equivalent leakage resistant circuits. They treat leakage attacks which leak the values of an a-priori fixed number of wires of the circuit, and produce leakage resistent circuits which grow in size as a function of the total bound on number of wires which are allowed to leak. The work applies to an unbounded number of executions of the circuit, assuming leakage attacks only apply per execution. Stated differently, the assumption is that the history of all past executions is erased. This is closely inspired by the model of proactive security. In quantitative terms, they place a global bound $L$ on the number of wires whose values leak, compile any circuit $C$ into a new circuit of size roughly $C \cdot L^2$ which is resilient to leakage of up to $L$ wire values (in our work the leakage bound grows with the complexity of the transformed circuit).

Faust, Tromer, Rabin, Reyzin, and Vaikuntanathan [FRR+09] also address converting general cryptographic algorithms (or circuits) to equivalent leakage resistant circuits extending [ISW03] significantly. They allow a (measurement) a side channel attack on an execution to receive the result of a leakage function which takes as input the entire existing (non-erased) state of the computation (rather than values of single wires), but in return restrict the leakage functions $\ell_i$ that can be handled to $AC^0$. Quantitatively, as in [ISW03], they place a fixed bound $L$ on the amount of leakage, and blow up the computation size by a factor of roughly $L^2$. [FRR+09] require a secure hardware component as well.

The bounded *memory leakage model* [AGV09] has received much attention. Here one allows $\ell$ to be defined on the *entire contents of memory* including all stored cryptographic secret keys, all previous computation done on the secret key results, and internally generated randomness. Obviously, in this strong setting, if no erasures are incorporated in the system, one must bound *the total amount* of information that measurements can yield on the original cryptographic keys, or else they will eventually be fully leaked by the appropriate adversarial choice of $\ell$. This is the model used in the works of [AGV09, NS09]. In contrast, in our work, we are interested in the continuous leakage question. Namely, the cryptographic algorithm initialized with secret cryptographic keys is invoked again and again for a (not specified in advance) polynomial (in the size of the initial cryptographic keys) number of times; each time the side-channel adversary continues to get some information on the secrets of the computation. Thus, the total amount of information that the adversary gets over the life time of the system will unbounded.

Coming back to the OC attack model, the ideas of Goldwasser, Kalai, and Rothblum [GKR08] in the work on one-time programs provide another avenue for transforming general cryptographic circuits to equivalent leakage resistant algorithms. The resulting leakage resistant algorithm will be secure in the OC attack model if it is *executed once*. To obtain an unbounded number of executions of the original circuit, one can resort to an off-line/on-line per-execution model where every execution is preceded by an off line stage in which the circuit conversion into a leakage resistent algorithm is performed a-new (obviously using new randomness). This is done prior to (and independently from) the choice of input for the coming execution. Surprisingly, the produced circuits are secure even if *all* data which is touched by the computation leaks. Namely, in presence of any polynomial time leakage functions including the identity function itself!

A recent independent work published in this proceedings is by Juma and Vahlis [JV10]. They also work in the OC attack model and address the question of how to run general computations in this model. They use as a tool a fully homomorphic encryption scheme and a leakage free hardware component independent from the functionality being computed. In terms of granularity, they divide each activation into two parts: one of which is large (a homomorphic computation of the entire circuit), and the second of which is small (a decryption). Quantitatively, To tolerate a leakage bound of $L$ bits in total, they transform the computation into one of size $C \cdot \exp(L)$. Under stronger assumptions (e.g. sub-exponential security of the fully homomorphic encryption) the transformed computation can be of size $C \cdot \operatorname{poly}(L)$.

## 2    Security Definitions

### 2.1    Leakage Model

*Leakage Attack.* A leakage attack is launched on an algorithm or on a data string. In the case of a data string $x$, an adversary can request to see any efficiently computable function $\ell(x)$ whose output length is bounded by $\lambda$ bits. In the case of an algorithm, we divide the algorithm into disjoint sub-computations. We assume that *only computation leaks information*, and so the adversary can request to see a bounded-length function of each sub-computation's input (separately).

**Definition 1 (Leakage Attack $\mathcal{A}[\lambda : s](x)$).** *Let $s$ be a source: either a data string or a computation. We model a $\lambda$-bit leakage attack of adversary $\mathcal{A}$ with input $x$ on the source $s$ as follows.*

*If $s$ is a computation (viewed as a boolean circuit with a fixed input), it is divided into $m$ disjoint and ordered sub-computations $sub_1, \ldots, sub_m$, where the input to sub-computation $sub_i$ should depend only on the output of earlier sub-computations. A $\lambda$-bit Leakage Attack on $s$ is one in which $\mathcal{A}$ can adaptively choose PPTM functions $\ell_1, \ldots \ell_m$, where $\ell_i$ takes as input the input to sub-computation $i$, and has output length at most $\lambda$ bits. For each $\ell_i$ (in order), the adversary receives the output of $\ell_i$ on sub-computation $sub_i$'s input, and then chooses $\ell_{i+1}$. The view of the adversary in the attack consists of the outputs to all the leakage functions.*

*In the case that $s$ is a data string, we treat it as a single subcomputation. A $\lambda$-bit leakage attack of $\mathcal{A}$ on $s$ is one in which $\mathcal{A}$ adaptively chooses $\lambda$ single-bit functions of the string in its entirety.*

*Multi-Source Leakage Attacks.* A multi-source leakage attack is one in which the adversary gets to launch concurrent leakage attacks on several sources. Each source is an algorithm or a data string. The leakages from each of the sources can be interleaved arbitrarily, but each leakage is computed as a function of a single source only.

**Definition 2 (Multi-Source Leakage Attack $\mathcal{A}[\lambda : s_1, \ldots, s_k](x)$).** *Let $s_1, \ldots, s_k$ be $k$ leakage sources (algorithms or data strings, as in Definition 1). We model a $\lambda$-bit multi-source leakage attack on $[s_1, \ldots, s_k]$ as follows. The adversary $\mathcal{A}$ with input $x$ runs concurrently $k$ separate $\lambda$-bit leakage attacks, one attack on each source. The attacks can be interleaved arbitrarily and adaptively. The attacks on each of the sources separately form a $\lambda$-bit leakage attack as in Definition 1. It is important that each leakage function is computed as a function of a single sub-computation in a single source (i.e. the leakages are never a function of the internal state of multiple sources). It is also important that the attacks launched by the adversary are concurrent and adaptive, and their interleaving is controlled by the adversary.*

***Simulated** Multi-Source Leakage Attacks.* For security definitions, we will occasionally want to replace the adversary's access to one or more source in a multi-source leakage attack with a view generated by a *simulator*. To facilitate composition, we view some sources as fixed: these are outside of the simulator's control. Both the adversary and the simulator get leakage access to these fixed sources (these are analogous to the environment in the UC framework [Can01]). Access to all of the other sources is simulated by the simulator.

**Definition 3 (Simulated Multi-Source Leakage Attack).** *Let $s_1, \ldots, s_k$ each be either a special symbol $\perp$ or a leakage source (algorithm or data string, as in Definition 1). Denote by $s'_1, \ldots, s'_\ell$ the subset of $s_1, \ldots, s_k$ that are not $\perp$. A simulated $\lambda$-bit multi-source leakage attack $(\mathcal{A}[\lambda : s_1, \ldots, s_k](x), \mathcal{S}[\lambda' : s'_1, \ldots, s'_\ell](x'))$ on $[s_1, \ldots, s_k]$ is defined as follows.*

*A with input x runs concurrently k separate λ-bit leakage attacks, one attack on each of its k sources, as in Definition 2. The difference here is that the sources which are ⊥ are all under the control of the simulator S. The simulator S, which itself has an input x′ and can launch a λ′-bit multi-source leakage attack on $[s'_1, \ldots, s'_\ell]$, produces the answers to all of the adversary's queries to all of the sources (including the ⊥ sources).*

*As in Definition 2, the adversary's (and the simulator's) access to its sources can be interleaved arbitrarily. The only difference is that the adversary's leakage queries to some of the sources are answered by the simulator. The simulator's answers may also be adaptive and depend on its prior view, which includes all of the adversary's past queries to simulated sources.*

As discussed above, the motivation for including sources that are outside the simulator's control is to facilitate composition between different components that are each (on their own) resilient to multi-source leakage attacks. Throughout this work, it will be the case that $\lambda' \geq \lambda$, and so it is "easy" for the simulator to answer $\mathcal{A}$'s queries to the "non-⊥ sources" (by making the same query itself). The challenge is answering $\mathcal{A}$'s queries to the "⊥-sources".

## 2.2   Continuous Side-Channel Secure Compiler

We divide a side-channel-secure compiler into two parts: the first part, the *initialization* occurs only once at the beginning of time. This procedure depends only on the circuit $C$ being compiled and the private input $y$. We assume that during this phase there are no side-channels. The second part is the *evaluation*. This occurs whenever the user wants to evaluate the circuit $C(\cdot, y)$ on an input $x$. In this part the user specifies an input $x$, the corresponding output $C(x, y)$ is computed, and side-channels are in effect.

**Definition 4 (λ(·)-Continuous Side-Channel Secure Compiler).** *for a circuit family $\{C_n(x, y)\}_{n \in \mathbb{N}}$, where $C_n$ operates on two n-bit inputs, we will say that a compiler $(Init_C, Eval_C)$ offers λ(·)-security under continuous side-channels, if for every integer $n > 0$, every $y \in \{0, 1\}^n$, and every security parameter κ, the following holds:*

- *Initialization: $Init_C(1^\kappa, C_n, y)$ runs in time $poly(\kappa, n)$ and outputs an initial state $state_0$*
- *Evaluation: for every integer $t \leq poly(\kappa)$, the evaluation procedure is run on the previous state $state_{t-1}$ and an input $x_t \in \{0, 1\}^n$. We require that for every $x_t \in \{0, 1\}^n$, when we run: $(output_t, state_t) \leftarrow Eval_C(state_{t-1}, x_t)$, with all but negligible probability over the coins of $Init_C$ and the t invocations of $Eval_C$, $output_t = C_n(x_t, y)$.*
- *λ(κ)-Continuous Leakage Security: for every PPTM (in κ) leakage-adversary $\mathcal{A}$, there exists a PPTM simulator S s.t. the view of $\mathcal{A}$ when adaptively choosing inputs $(x_1, x_2, \ldots x_T)$ while running a continuous leakage attack on the evaluation procedure, is indistinguishable from the view generates by S which only gets the inputs-output pairs $((x_1, C(x_1, y)), \ldots, (x_T, C(x_T, y)))$.*

*Formally, the adversary repeatedly and adaptively, in iterations $t \leftarrow 1, \ldots, T$, chooses an input $x_t$ and launches a $\lambda(\kappa)$-bit leakage attack on $Eval_C(state_{t-1}, x_t)$ (see Definition 1). The view $view_{\mathcal{A},t}$ of the adversary in iteration $t$ includes the input $x_t$, the output $output_t$, and the leakages. The complete view of the adversary is $view_{\mathcal{A}} = (view_{\mathcal{A},1}, \ldots, view_{\mathcal{A},T})$, a random variable over the coins of the adversary, of the $Init_C$ and of the $Eval_C$ procedure (in all of its iterations).*

*We note that modeling the leakage attacks requires dividing the $Eval_C$ procedure into sub-computations. In our constructions the size of these sub-computations will always be at most polynomial in the security parameter.*

*The simulator's view is generated by running the adversary with simulated leakage attacks. In each iteration $t$ the simulator gets the input $x_t$ chosen by the adversary and the circuit output $C(x_t, y)$. It generates simulated side-channel information as in Definition 3. It is important that the simulator sees nothing of the internal workings of the evaluation procedure. We compute:*

$$state_{\mathcal{S},0} \leftarrow \mathcal{S}(1^\kappa, C_n), x_t \leftarrow \mathcal{A}(view_{\mathcal{S},1}, \ldots, view_{\mathcal{S},t-1}),$$

$$(state_{\mathcal{S},0}, view_{t,\mathcal{S}}) \leftarrow \mathcal{S}(1^\kappa, x_t, C(x_t, y), view_{\mathcal{S},t-1})$$

*where $view_{\mathcal{S},t}$ is a random variable over the coins of the adversary when choosing the next input and of the simulator. The complete view of the simulator is $view_{\mathcal{S}} = (view_{\mathcal{S},1}, \ldots, view_{\mathcal{S},T})$.*

*We require that $view_{\mathcal{S}}$ and $view_{\mathcal{A}}$ are computationally indistinguishable.*

# 3  Subsidiary Cryptosystem and Hardware

We now present the subsidiary cryptosystem and hardware device we will use to instantiate our main construction. We also define the properties we need from the subsidiary cryptosystem. We omit the full formal details of the instantiations of these properties by the subsidiary cryptosystem for lack of space, but direct the reader back to Section 1.1 for an overview of these properties and how they are instantiated.

## 3.1  The Naor-Segev/BHHO Scheme and Secure Hardware

Security is based on the Decisional Diffie-Hellman (DDH) Assumption: Let $Gen$ be a probabilistic group generator, s.t. $G \leftarrow Gen(1^\kappa)$ is a group of order $q = q(\kappa)$. We will take $G$ to be $\mathbb{GF}[q]$, i.e. the field of prime order $q$ (which also supports addition operations). The DDH assumption for $Gen$ is that the ensembles below are computationally indistinguishable:

$$(G, g_1, g_2, g_1^r, g_2^r) : G \leftarrow Gen(1^\kappa), g_1, g_2 \in_R G, r \in_R \mathbb{GF}[q]$$

$$(G, g_1, g_2, g_1^{r_1}, g_2^{r_2}) : G \leftarrow Gen(1^\kappa), g_1, g_2 \in_R G, r_1, r_2 \in_R \mathbb{GF}[q]$$

The cryptosystem has the following algorithms (we take $m = 10$, this choice is arbitrary and effects the constant in the fraction of leakage we can tolerate):

- *KeyGen*($1^\kappa$): choose $\boldsymbol{g} = (g_1, \ldots, g_m) \in_R G^m$ and $\boldsymbol{s} = (s_1, \ldots, s_m) \in_R \mathbb{GF}[\mathsf{q}]^m$. Define: $y = \prod_{i=1}^m g_i^{s_i}$. Output $pk = (\boldsymbol{g}, y)$ and $sk = \boldsymbol{s}$.
- *Encrypt*($pk, b \in \{0,1\}$): parse $pk = (\boldsymbol{g}, y)$ and choose $r \in_R \mathbb{GF}[\mathsf{q}]$. Output: $c \leftarrow (g_1^r, \ldots, g_m^r, y^r \cdot g_1^b)$
- *Decrypt*($sk, c$): parse $sk = \boldsymbol{s}$ and $c = (f_1, \ldots, f_m, h)$. Compute $h' = \prod_{i=1}^m f_i^{s_i}$. Output 1 if $h = g_1 \cdot h'$ and output $\bot$ otherwise.

**CipherGen** *Secure Hardware.* This device will be used to realize additional useful properties for the subsidiary cryptosystem. We assume that when this device is invoked, there is leakage on its input and output, but not on its internal workings or randomness. The device receives as input a public key and mode of operation $m \in \{0, rand\}$. In mode $m = 0$ it computes and outputs a fresh encryption of 0, and in mode $m = rand$ it chooses a uniformly random bit $b \in \{0, 1\}$ and outputs a fresh encryption of $b$.

## 3.2   Homomorphic and Leakage-Resilient Properties

**Definition 5 (Semantic Security Under $\lambda(\cdot)$-Multi-Source Leakage).** *An encryption scheme (KeyGen, Encrypt, Decrypt) is semantically secure under multi-source leakage attacks if for every PPTM adversary $\mathcal{A}$, when we run the game below, the adversary's advantage in winning (over 1/2) is negligible:*

1. *The game chooses a key pair $(pk, sk) \leftarrow KeyGen(1^\kappa)$, chooses uniformly at random a bit $b \in_R \{0,1\}$, and generates a ciphertext $c \leftarrow Encrypt(pk, b)$.*
2. *The adversary launches a multi-source leakage attack on $sk$ and $c$, and outputs a guess $b'$ for the value of $b$:*

$$b' \leftarrow \mathcal{A}[\lambda(\kappa) : sk, c](pk)$$

*The adversary wins if $b' = b$.*

**Lemma 1.** *The Naor-Segev cryptosystem, as defined in Section 3.1, is semantically secure under $(\lambda = mq/3)$-multi-source leakage.*

**Definition 6 (Key Refreshing).** *An encryption scheme supports key-refreshing if it has additional algorithms with the following properties:*

1. *The key refresh procedure $Refresh(1^\kappa)$ outputs a "correlation value" $\pi$ every time it is run.*
2. *The key correlation procedures output new secret and public keys $pk' \leftarrow PKCor(pk, \pi)$ and $sk' \leftarrow SKCor(sk, \pi)$. Here $pk'$ is a public key corresponding to $sk'$. We require that even for fixed $sk$, the new $sk'$ (as a function of a randomly chosen $\pi$) is uniformly random.*
3. *The ciphertext correlation procedure transforms an encryption from one key to the other. I.e. if $c' \leftarrow CipherCor(pk, c, \pi)$, then $Decrypt(sk, c) = Decrypt(sk', c')$.*
4. *The key linking procedure outputs a correlation value linking its two input secret keys. I.e. if $\pi \leftarrow KeyLink(sk, sk')$, then $sk' = SKCor(sk, \pi)$.*

5. *A correlation-inverter CorInvert such that $\pi^{-1} \leftarrow CorInvert(\pi)$ satisfies that if $sk' = SKCor(sk, \pi)$, then $sk = SKCor(sk', \pi^{-1})$. Also for the corresponding public keys $pk = PKCor(pk', \pi^{-1})$.*

**Definition 7 ($\lambda(\cdot)$-Leakage Oblivious Generation).** *An encryption scheme (KeyGen, Encrypt, Decrypt) supports oblivious generation if there exists a randomized procedure OblivGen such that:*

1. *OblivGen outputs the encryption of a random bit:*

$$\forall b \in \{0, 1\} : \Pr_{c \leftarrow OblivGen(pk)}[Decrypt(sk, c) = b] = 1/2$$

2. *The security requirement is that there exists a Simulator $\mathcal{S}$ such that for every bit $b_1 \in \{0, 1\}$ and every PPTM adversary $\mathcal{A}$, when we run the game below, the real and simulated views are indistinguishable:*
   (a) *The game chooses a key pair $(pk, sk) \leftarrow KeyGen(1^\kappa)$.*
   (b) *In the real view, $\mathcal{A}$ launches a $\lambda(\kappa)$-bit multi-source leakage attack:*

$$\mathcal{A}[\lambda(\kappa) : sk, c_0 \leftarrow OblivGen(pk), c_0](pk)$$

   *In the simulated view, the game encrypts bit $b_1$: $c_1 \leftarrow Encrypt(pk, b_1)$, and we run $\mathcal{A}$ with a simulated $\lambda(\kappa)$-multi-source leakage attack:*

$$(\mathcal{A}[\lambda(\kappa) : sk, \perp, c_1](pk), \mathcal{S}[\lambda'(\kappa) : sk, c_1](pk))$$

   *I.e., here the leakage attacks on the oblivious generation procedure are simulated by $\mathcal{S}$. We require that $\lambda'(\kappa) = O(\lambda(\kappa))$ (the simulator may get access to a little more leakage than the adversary).*

**Definition 8 ($\lambda(\cdot)$-Leakage Ciphertext Regeneration).** *An encryption scheme (KeyGen, Encrypt, Decrypt) supports oblivious generation if it has a procedure Regen such that:*

1. *When we run $(pk', sk', c') \leftarrow Regen(pk, sk, c)$, it is the case that $Decrypt(sk', c') = Decrypt(sk, c)$.*
2. *The security requirement is that for every PPTM adversary $\mathcal{A}$ that runs for $T$ repeated regenerations, every bit $b \in \{0, 1\}$ (determining whether the input ciphertext is an encryption of 0 or 1), the view generated by the adversary in the game below is indistinguishable.*
   (a) *The game chooses a key pair $(pk_0, sk_0) \leftarrow KeyGen(1^\kappa)$ and generates a ciphertext $c_0 \leftarrow Encrypt(pk, b)$.*
   (b) *The adversary $\mathcal{A}$ launches $\lambda(\kappa)$-bit multi-source leakage attack on $T$ repeated regenerations:*

$$\mathcal{A}[\lambda(\kappa) : sk_0, c_0, (pk_1, sk_1, c_1) \leftarrow Regen(pk_0, sk_0, c_0),$$
$$sk_1, c_1, (pk_0, c_0 pk_2, sk_2, c_2) \leftarrow Regen(pk_1, sk_1, c_1),$$
$$\ldots,$$
$$sk_{T-1}, c_{T-1}, (pk_T, sk_T, c_T) \leftarrow Regen(pk_{T-1}, sk_{T-1}, c_{T-1})](pk_0, \ldots, pk_T)$$

*We further require that the input to each sub-computation in the Regen procedure depends either on the input secret key or the input ciphertext, but never on both.*

*Homomorphic Masked NAND.* A homomorphic masked NAND computation is given three ciphertexts $c_1, c_2, c_3$ encrypted under the same key and with corresponding plaintexts $b_1, b_2, b_3 \in \{0, 1\}$, and two plain-text values $a_1, a_2 \in \{0, 1\}$. It should compute homomorphically (without using the secret key) compute a "blinded" (see below) ciphertext $hc$ that can later be decrypted to retrieve the value $((a_1 \oplus b_1) \text{ NAND } (a_2 \oplus b_2)) \oplus b_3$.

**Definition 9 ($\lambda(\cdot)$-Leakage Blinded Homomorphic NAND).** *An encryption scheme (KeyGen, Encrypt, Decrypt) supports blinded homomorphic masked NANDs if there exist procedures HomEval and HomDecrypt such that:*

1. *When take $hc \leftarrow HomEval(pk, a_1, a_2, c_1, c_2, c_3)$, for the secret key $sk$ corresponding to $pk$ w.h.p. it holds that $HomDecrypt(sk, hc) = ((a_1 \oplus b_1) \text{ NAND } (a_2 \oplus b_2)) \oplus b_3$.*

2. *The result should be "blinded". There exists a Simulator $\mathcal{S}$ such for every PPTM adversary $\mathcal{A}$, PPTM ciphertext generators $G_1, G_2, G_3$,[5] and plaintext values $a_1, a_2 \in \{0, 1\}$, the real and simulated views in the game below are indistinguishable:*

   (a) *The game chooses a key pair $(pk, sk) \leftarrow KeyGen(1^\kappa)$ and generates ciphertexts $c_1 \leftarrow G_1(pk), c_2 \leftarrow G_2(pk), c_3 \leftarrow G_3(pk)$ using random strings $r_1, r_2, r_3$ for $G_1, G_2, G_3$ respectively.*

   (b) *In the real view, the adversary $\mathcal{A}$ launches a multi-source leakage attack on the homomorphic evaluation and decryption:*

   $$\mathcal{A}[\lambda(\kappa) : sk, c_3 \leftarrow G_3(r_3),$$
   $$hc \leftarrow HomEval(pk, a_1, a_2, c_1, c_2, c_3),$$
   $$a_3 \leftarrow HomDecrypt(sk, hc)](pk, a_1, a_2, r_1, r_2)$$

   *In the simulated view, the simulator does not get any access to homomorphic evaluation or decryption, but rather gets only the output $a_3$ of the homomorphic decryption:*

   $$(\ \mathcal{A}[\lambda(\kappa) : sk, c_3 \leftarrow G_3(r_3), \perp, \perp](pk, a_1, a_2, r_1, r_2),$$
   $$\mathcal{S}[\lambda'(\kappa) : sk, c_3 \leftarrow G_3(r_3)](pk, a_1, a_2, r_1, r_2, a_3))$$

   *We require that $\lambda'(\kappa) = O(\lambda(\kappa))$.*

---

[5] In the security proof for our construction these generation procedures will be the *OblivGen* or *Regen* procedure.

# 4   A Continuous-Leakage Secure and Compiler

The compiler can be based on any subsidiary cryptosystem with the properties of Section 3. We refer the reader to Section 1.1 for the construction overview and preliminaries, and to Section 2.2 for the security definition. The initialization and evaluation procedures are presented below in Figure 1. The evaluation procedure is separated into sub-computations (which may themselves be separated into sub-computations of the cryptographic algorithms). For each such sub-computation we explicitly note which data elements are computed on ("touched") in the sub-computation. We defer the proof of security to the full version.

---

**Initialization** $Init_C(1^\kappa, C, y)$

For every input wire $i$, corresponding to bit $j$ of the input $y$, generate new keys: $(pk_i, sk_i) \leftarrow KeyGen(1^\kappa)$ and compute an encryption $c_i = Encrypt(pk_i, y_j)$. $state_0 \leftarrow \{(pk_i, sk_i, c_i)\}_i : i$ is a $y$-input wire

**Evaluation** $Eval_C(state_{t-1}, x_t)$

1.  *Generate keys and ciphertexts for all wires of $C$ except the $y$-input wires.*
    For the $x$ input wires, generate fresh keys and encryptions of 0.
    Proceed layer-by-layer (from input to output). For each gate $g$ with input wires $i$ and $j$ and output wire $k$: (repeat independently for gate $g$'s second output wire $\ell$)

    (a)  Generate a random correlation value $\pi_{i,k} \leftarrow Refresh(1^\kappa)$. Apply this value to wire $i$'s keys to get a new key pair for wire $k$: $pk_k \leftarrow PKCor(pk_i, \pi_{i,k}), sk_k \leftarrow SKCor(sk_i, \pi_{i,k})$. Derive a correlation value specifying the correlation between the keys of wires $j$ and $k$: $\pi_{j,k} \leftarrow KeyLink(sk_k, sk_j)$. Store the keys and correlation values. "Computed on" keys, correlation values

    (b)  Generate a ciphertext encrypting the share $b_k$ for wire $k$: for internal wires, use the oblivious generation procedure to generate an encryption of a random bit $c_k \leftarrow OblivGen(pk_k)$.
    For the output wire $o$, generate an encryption $c_o \leftarrow Encrypt(pk_o, 0)$.
    Store the ciphertexts. "Computed on" ciphertexts

2.  *Compute the value of $C(y, x_t)$.*
    Proceed layer by layer (from input to output). For each gate $g$ with output wire $k$ and input wires $i, j$, the previous gate evaluations yield the shares $a_i, a_j \in \{0, 1\}$ of the gate's input wires. Compute an encryption of $a_k$: (do the same independently for gate $g$'s second output wire $\ell$):

    (a)  First transform the ciphertexts $c_i$ and $c_j$ to be encryptions under $pk_k$: $c'_i \leftarrow CipherCor(pk_i, c_i, \pi_{i,k})$ and $c'_j \leftarrow CipherCor(pk_j, c_j, \pi_{j,k})$. "Computed on" ciphertexts and correlation values.

    (b)  Run the blinded homomorphic evaluation procedure: $hc_k \leftarrow HomEval(pk_k, a_i, a_j, c'_i, c'_j, c_k)$. "Computed on" ciphertexts.

    (c)  Compute $a_k \leftarrow HomDecrypt(sk_k, hc_k)$. "Computed on" $hc_k$ and the secret key.
    Taking $o$ to be the output wire, the output is $output_t \leftarrow a_o$.

3.  *Generate the new state.*
    For each $y$-input wire $i$ regenerate wire $i$'s keys and ciphertext: $(pk_i, sk_i, c_i) \leftarrow Regen(pk_i, sk_i, c_i)$.
    The new state is $state_t \leftarrow \{(i, pk_i, sk_i, c_i)\}_i : i$ is a $y$-input wire.

---

**Fig. 1.** $Init_C$, performed off-line without side channels, and $Eval_C$, performed on input $x_t$ in the presence of side-channel attacks

**Theorem 1.** *Let $(KeyGen, Encrypt, Decrypt)$ be a subsidiary encryption scheme with security parameter $\kappa$ and with the properties specified in Definitions 6 (key refreshing), 5 (multi-source leakage resilience), 7 (oblivious generation), 8 (leakage resilience regeneration), and 9 (homomorphic masked NAND), all with*

$\lambda = \Omega(\kappa)$-*leakage resilience.*[6] *Then the* ($Init_C$, $Eval_C$) *compiler specified in Figure 1 offers* $\Omega(\kappa)$-*leakage security under continuous side-channels as in Definition 4.*

# References

[AGV09]    Akavia, A., Goldwasser, S., Vaikuntanathan, V.: Simultaneous hardcore bits and cryptography against memory attacks. In: Reingold, O. (ed.) TCC 2009. LNCS, vol. 5444, pp. 474–495. Springer, Heidelberg (2009)

[BHHO08]   Boneh, D., Halevi, S., Hamburg, M., Ostrovsky, R.: Circular-secure encryption from decision diffie-hellman. In: Wagner, D. (ed.) CRYPTO 2008. LNCS, vol. 5157, pp. 108–125. Springer, Heidelberg (2008)

[BKKV10]   Brakerski, Z., Kalai, Y.T., Katz, J., Vaikuntanathan, V.: Cryptography resilient to continual memory leakage. Cryptology ePrint Archive, Report 2010/278 (2010)

[Can01]    Canetti, R.: Universally composable security: A new paradigm for cryptographic protocols. In: FOCS, pp. 136–145 (2001)

[CG88]     Chor, B., Goldreich, O.: Unbiased bits from sources of weak randomness and probabilistic communication complexity. SIAM J. Comput. 17(2), 230–261 (1988)

[DHLAW10]  Dodis, Y., Haralambiev, K., Lopez-Alt, A., Wichs, D.: Efficient public-key cryptography in the presence of key leakage. Cryptology ePrint Archive, Report 2010/154 (2010)

[DP08]     Dziembowski, S., Pietrzak, K.: Leakage-resilient cryptography. In: Annual IEEE Symposium on Foundations of Computer Science, pp. 293–302 (2008)

[FKPR09]   Faust, S., Kiltz, E., Pietrzak, K., Rothblum, G.: Leakage-resilient signatures. Cryptology ePrint Archive, Report 2009/282 (2009), http://eprint.iacr.org/2009/282

[FRR+09]   Faust, S., Rabin, T., Reyzin, L., Tromer, E., Vaikuntanathan, V.: Protecting against computationally bounded and noisy leakage (2009) (manuscript)

[GKR08]    Goldwasser, S., Kalai, Y.T., Rothblum, G.N.: One-time programs. In: Wagner, D. (ed.) CRYPTO 2008. LNCS, vol. 5157, pp. 39–56. Springer, Heidelberg (2008)

[GM84]     Goldwasser, S., Micali, S.: Probabilistic encryption. J. Comput. Syst. Sci. 28(2), 270–299 (1984)

[GMW87]    Goldreich, O., Micali, S., Wigderson, A.: How to play any mental game or a completeness theorem for protocols with honest majority. In: STOC, pp. 218–229 (1987)

[GO96]     Goldreich, O., Ostrovsky, R.: Software protection and simulation on oblivious rams. J. ACM 43(3), 431–473 (1996)

[ISW03]    Ishai, Y., Sahai, A., Wagner, D.: Private circuits: Securing hardware against probing attacks. In: Boneh, D. (ed.) CRYPTO 2003. LNCS, vol. 2729, pp. 463–481. Springer, Heidelberg (2003)

[JV10]     Juma, A., Vahlis, Y.: On protecting cryptographic keys against continual leakage. Cryptology ePrint Archive, Report 2010/205 (2010)

---

[6] We mean that there exists an explicit constant $0 < c < 1$ s.t. we allow leakage of $c \cdot \lambda$ bits.

[KJJ99]    Kocher, P., Jaffe, J., Jun, B.: Differential power analysis. In: Wiener, M. (ed.) CRYPTO 1999. LNCS, vol. 1666, pp. 388–397. Springer, Heidelberg (1999)

[KV09]    Katz, J., Vaikuntanathan, V.: Signature schemes with bounded leakage resilience. In: Matsui, M. (ed.) ASIACRYPT 2009. LNCS, vol. 5912, pp. 703–720. Springer, Heidelberg (2009)

[MR04]    Micali, S., Reyzin, L.: Physically observable cryptography (extended abstract). In: Naor, M. (ed.) TCC 2004. LNCS, vol. 2951, pp. 278–296. Springer, Heidelberg (2004)

[NS09]    Naor, M., Segev, G.: Public-key cryptosystems resilient to key leakage. In: Halevi, S. (ed.) CRYPTO 2009. LNCS, vol. 5677, pp. 18–35. Springer, Heidelberg (2009)

[Pie09]    Pietrzak, K.: A leakage-resilient mode of operation. In: Joux, A. (ed.) EUROCRYPT 2009. LNCS, vol. 5479, pp. 462–482. Springer, Heidelberg (2009)

[RCL]    Boston University Reliable Computing Laboratory. Side channel attacks database, http://www.sidechannelattacks.com

[SYY99]    Sander, T., Young, A., Yung, M.: Non-interactive cryptocomputing for $nc^1$. In: FOCS (1999)

[Yao82]    Yao, A.C.: Theory and application of trapdoor functions. In: Symposium on Foundations of Computer Science, pp. 80–91 (1982)

# An Efficient and Parallel
# Gaussian Sampler for Lattices

Chris Peikert*

Georgia Institute of Technology

**Abstract.** At the heart of many recent lattice-based cryptographic schemes is a polynomial-time algorithm that, given a 'high-quality' basis, generates a lattice point according to a Gaussian-like distribution. Unlike most other operations in lattice-based cryptography, however, the known algorithm for this task (due to Gentry, Peikert, and Vaikuntanathan; STOC 2008) is rather inefficient, and is inherently sequential.

We present a new Gaussian sampling algorithm for lattices that is *efficient* and *highly parallelizable*. We also show that in most cryptographic applications, the algorithm's efficiency comes at almost no cost in asymptotic security. At a high level, our algorithm resembles the "perturbation" heuristic proposed as part of NTRUSign (Hoffstein *et al.*, CT-RSA 2003), though the details are quite different. To our knowledge, this is the first algorithm and rigorous analysis demonstrating the security of a perturbation-like technique.

## 1 Introduction

In recent years, there has been rapid development in the use of *lattices* for constructing rich cryptographic schemes.[1] These include digital signatures (both 'tree-based' [13] and 'hash-and-sign' [8, 6]), identity-based encryption [8] and hierarchical IBE [6, 1], noninteractive zero knowledge [19], and even a fully homomorphic cryptosystem [7].

The cornerstone of many of these schemes (particularly, but not exclusive to, those that 'answer queries') is the polynomial-time algorithm of [8] that samples from a so-called *discrete Gaussian* probability distribution over a lattice $\Lambda$. More precisely, for a vector $\mathbf{c} \in \mathbb{R}^n$ and a "width" parameter $s > 0$, the distribution $D_{\Lambda+\mathbf{c},s}$ assigns a probability proportional to $\exp(-\pi\|\mathbf{v}\|^2/s^2)$ to each $\mathbf{v} \in \Lambda + \mathbf{c}$ (and probability zero elsewhere). Given $\mathbf{c}$, a basis $\mathbf{B}$ of $\Lambda$, and a sufficiently large $s$ (related to the 'quality' of $\mathbf{B}$), the GPV algorithm outputs a sample from a

---

* This material is based upon work supported by the National Science Foundation under Grant CNS-0716786. Any opinions, findings, and conclusions or recommendations expressed in this material are those of the author(s) and do not necessarily reflect the views of the National Science Foundation.

[1] A lattice $\Lambda \subset \mathbb{R}^n$ is a periodic 'grid' of points, or more formally, a discrete subgroup of $\mathbb{R}^n$ under addition. It is generated by a (not necessarily unique) *basis* $\mathbf{B} \subset \mathbb{R}^{n \times k}$ of $k$ linearly independent vectors, as $\Lambda = \{\mathbf{Bz} : \mathbf{z} \in \mathbb{Z}^k\}$. In this paper we are concerned only with *full-rank* lattices, i.e., where $k = n$.

distribution statistically close to $D_{\Lambda+\mathbf{c},s}$. (Equivalently, by subtracting $\mathbf{c}$ from the output, it samples a lattice point from a Gaussian distribution centered at $-\mathbf{c}$.) Informally speaking, the sampling algorithm is 'zero-knowledge' in the sense that it leaks no information about its input basis $\mathbf{B}$ (aside from a bound on its quality), because $D_{\Lambda+\mathbf{c},s}$ is defined without reference to any particular basis. This zero-knowledge property accounts for its broad utility in lattice-based cryptography.

While the sampling algorithm of [8] has numerous applications in cryptography and beyond, for both practical and theoretical purposes it also has some drawbacks:

- First, it is rather *inefficient*: on an $n$-dimensional lattice, a straightforward implementation requires exact arithmetic on an $n \times n$ matrix having $\Omega(n)$-bit entries (even ignoring some additional $\log n$ factors). While approximate arithmetic and other optimizations may be possible in certain cases, great care would be needed to maintain the proper output distribution, and the algorithm's essential structure appears difficult to make truly practical.
- Second, it is *inherently sequential*: to generate a sample, the algorithm performs $n$ adaptive iterations, where the choices made in each iteration affect the values used in the next. This stands in stark contrast to other 'embarrassingly parallelizable' operations that are typical of lattice-based cryptography.

## 1.1   Contributions

We present a new algorithm that samples from a discrete Gaussian distribution $D_{\Lambda+\mathbf{c},s}$ over a lattice, given a 'high-quality' basis for $\Lambda$. The algorithm is especially well-suited to '$q$-ary' integer lattices, i.e., sublattices of $\mathbb{Z}^n$ that themselves contain $q\mathbb{Z}^n$ as a sublattice, for some known and typically small $q \geq 2$. These include NTRU lattices [10] and the family of random lattices that enjoy 'worst-case hardness,' as first demonstrated by Ajtai [3]. Most modern lattice-based cryptographic schemes (including those that rely on Gaussian sampling) are designed around $q$-ary lattices, so they are a natural target for optimization.

The key features of our algorithm, as specialized to $n$-dimensional $q$-ary lattices, are as follows. It is:

- *Offline / online*: when the lattice basis is known in advance of the point $\mathbf{c}$ (which is the norm in cryptographic applications), most of the work can be performed as offline precomputation. In fact, the offline phase may be viewed simply as an extension of the application's key-generation algorithm.
- *Simple and efficient*: the online phase involves only $O(n^2)$ integer additions and multiplications modulo $q$ or $q^2$, where the $O$-notation hides a small constant $\approx 4$.
- *Fully parallelizable*: for any $P$ up to $n^2$, the online phase can allocate $O(n^2/P)$ of its operations to each of $P$ processors.
- *High-quality*: for random bases that are commonly used in cryptographic schemes, our algorithm can sample from a Gaussian of essentially the same

'quality' as the prior GPV algorithm; this is important for the concrete security of applications. See Section 1.2 below for a full discussion.

We emphasize that for a practical implementation, parallelized operations on small integers represent a significant performance advantage. Most modern computer processors have built-in support for "vector" instructions (also known as "single instruction, multiple data"), which perform simple operations on entire vectors of small data elements simultaneously. Our algorithm can exploit these operations very naturally. For a detailed efficiency comparison between our algorithm and that of [8], see Section 1.2 below.

At a very high level, our algorithm resembles the "perturbation" heuristic proposed for the NTRUSign signature scheme [9], but the details differ significantly; see Section 1.3 for a comparison. To our knowledge, this is the first algorithm and analysis to demonstrate the theoretical soundness of a perturbation-like technique. Finally, the analysis of our algorithm relies on some new general facts about 'convolutions' of discrete Gaussians, which we expect will be applicable elsewhere. For example, these facts allow for the use of a clean *discrete* Gaussian error distribution (rather than a 'rounded' Gaussian) in the "learning with errors" problem [20], which may be useful in certain applications.

## 1.2   Comparison with the GPV Algorithm

Here we give a detailed comparison of our new sampling algorithm to the previous one of [8]. The two main points of comparison are the width ('quality') of the sampled Gaussian, and the algorithmic efficiency.

**Gaussian Width.** One of the important properties of a discrete Gaussian sampling algorithm is the *width* $s$ of the distribution it generates, as a function of the input basis. In cryptographic applications, the width is the main quantity governing the concrete security and, if applicable, the approximation factor of the underlying worst-case lattice problems. This is because in order for the scheme to be secure, it must hard for an adversary to find a lattice point within the likely radius $s\sqrt{n}$ of the Gaussian (i.e., after truncating its negligibly likely tail). The wider the distribution, the more leeway the adversary has in an attack, and the larger the scheme's parameters must be to compensate. On the other hand, a more efficient sampling algorithm can potentially allow for the use of larger parameters without sacrificing performance.

The prior sampling algorithm of [8], given a lattice basis $\mathbf{B} = \{\mathbf{b}_1, \ldots, \mathbf{b}_n\}$, can sample from a discrete Gaussian having width as small as $\|\widetilde{\mathbf{B}}\| = \max_i \|\widetilde{\mathbf{b}}_i\|$, where $\widetilde{\mathbf{B}}$ denotes the Gram-Schmidt orthogonalization of $\mathbf{B}$.[2] (Actually, the width also includes a small $\omega(\sqrt{\log n})$ factor, which is also present in our new algorithm, so for simplicity we ignore it in this summary.) As a point of comparison, $\|\widetilde{\mathbf{B}}\|$ is always at most $\max_i \|\mathbf{b}_i\|$, and in some cases it can be substantially smaller.

---

[2] In the Gram-Schmidt orthogonalization $\widetilde{\mathbf{B}}$ of $\mathbf{B}$, the vector $\widetilde{\mathbf{b}}_i$ is the projection of $\mathbf{b}_i$ orthogonally to $\mathrm{span}(\mathbf{b}_1, \ldots, \mathbf{b}_{i-1})$.

In contrast, our new algorithm works for a width $s$ as small as the *largest singular value* $s_1(\mathbf{B})$ of the basis $\mathbf{B}$, or equivalently, the square root of the largest eigenvalue of the Gram matrix $\mathbf{B}\mathbf{B}^t$. It is easy to show that $s_1(\mathbf{B})$ is always at least $\max_i \|\mathbf{b}_i\|$, so our new algorithm cannot sample from a narrower Gaussian than the GPV algorithm can. At the same time, any basis $\mathbf{B}$ can always be efficiently processed (without increasing $\|\widetilde{\mathbf{B}}\|$) to guarantee that $s_1(\mathbf{B}) \leq n \cdot \|\widetilde{\mathbf{B}}\|$, so our algorithm is at worst an $n$ factor looser than that of [8].

While a factor of $n$ gap between the two algorithms may seem rather large, in cryptographic applications this *worst-case* ratio is actually immaterial; what matters is the relative performance on the *random* bases that are used as secret keys. Here the situation is much more favorable. First, we consider the basis-generation algorithms of [4] (following [2]) for 'worst-case-hard' $q$-ary lattices, which are used in most theoretically sound cryptographic applications. We show that with a minor modification, one of the algorithms from [4] outputs (with overwhelming probability) a basis $\mathbf{B}$ for which $s_1(\mathbf{B})$ is only an $O(\sqrt{\log q})$ factor larger than $\|\widetilde{\mathbf{B}}\|$ (which itself is asymptotically optimal, as shown in [4]). Because $q$ is typically a small polynomial in $n$, this amounts to a cost of only an $O(\sqrt{\log n})$ factor in the width of the Gaussian. Similarly, when the vectors of $\mathbf{B}$ are themselves drawn from a discrete Gaussian, as in the basis-delegation technique of [6], we can show that $s_1(\mathbf{B})$ is only a $\omega(\sqrt{\log n})$ factor larger than $\|\widetilde{\mathbf{B}}\|$ (with overwhelming probability). Therefore, in cryptographic applications the performance improvements of our algorithm can come at almost no asymptotic cost in security. Of course, a concrete evaluation of the performance/security trade-off for real-world parameters would require careful analysis and experiments, which we leave for later work.

**Efficiency.** We now compare the efficiency of the two known sampling algorithms. We focus on the most common case of $q$-ary $n$-dimensional integer lattices, where a 'good' lattice basis (whose vectors having length much less than $q$) is initially given in an offline phase, followed by an online phase in which a desired center $\mathbf{c} \in \mathbb{Z}^n$ is given. This scenario allows for certain optimizations in both algorithms, which we include for a fair comparison.

The sampling algorithm from [8] can use the offline phase to compute the Gram-Schmidt orthogonalization of its given basis; this requires $\Omega(n^4 \log^2 q)$ bit operations and $\Omega(n^3)$ bits of intermediate storage. The online phase performs $n$ sequential iterations, each of which computes an inner product between a Gram-Schmidt vector having $\Omega(n)$-bit entries, and an integer vector whose entries have magnitude at most $q$. In total, these operations require $\Omega(n^3 \log q)$ bit operations. In addition, each iteration performs a certain randomized-rounding operation, which, while asymptotically poly$(\log n)$-time, is not especially practical (nor precomputable) because it uses rejection sampling on a value that is not known until the online phase. Lastly, while the work within each iteration may be parallelized, the iterations themselves must be performed sequentially.

Our algorithm is more efficient and practical in the running time of both phases, and in the amount of intermediate storage between phases. The offline phase first computes a matrix inverse modulo $q^2$, and a 'square root' of a matrix whose entries have magnitude at most $q$; these can be computed in $O(n^3 \log^2 q)$ bit operations. Next, it generates and stores one or more short integer 'perturbation' vectors (one per future call to the online phase), and optionally discards the matrix square root. The intermediate storage is therefore as small as $O(n^2 \log q)$ bits for the matrix inverse, plus $O(n \log q)$ bits per perturbation vector. Optionally, the offline phase can also precompute the randomized-rounding operations, due to the small number of possibilities that can occur online. The online phase simply computes about $4n^2$ integer additions and multiplications ($2n^2$ of each) modulo $q$ or $q^2$, which can be fully parallelized among up to $n^2$ processors.

Lastly, we mention that our sampling algorithm translates very naturally to the setting of *compact* $q$-ary lattices and bases over certain rings $R$ that are larger than $\mathbb{Z}$, where security is based on the worst-case hardness of *ideal lattices* in $R$ (see, e.g., [16, 21, 14]). In contrast to GPV, our algorithm can directly take advantage of the ring structure for further efficiency, yielding a savings of an $\tilde{\Omega}(n)$ factor in the computation times and intermediate storage.

### 1.3   Overview of the Algorithm

The GPV sampling algorithm [8] is based closely on Babai's "nearest-plane" decoding algorithm for lattices [5]. Babai's algorithm takes a point $\mathbf{c} \in \mathbb{R}^n$ and a lattice basis $\mathbf{B} = \{\mathbf{b}_1, \ldots, \mathbf{b}_n\}$, and for $i = n, \ldots, 1$ computes a coefficient $z_i \in \mathbb{Z}$ for $\mathbf{b}_i$ by iteratively projecting ('rounding') $\mathbf{c}$ orthogonally to the nearest hyperplane of the form $z_i \mathbf{b}_i + \mathrm{span}(\mathbf{b}_1, \ldots, \mathbf{b}_{i-1})$. The output is the lattice vector $\sum_i z_i \mathbf{b}_i$, whose distance from the original $\mathbf{c}$ can be bounded by the quality of $\mathbf{B}$. The GPV algorithm, whose goal is instead to sample from a discrete Gaussian centered at $\mathbf{c}$, uses *randomized rounding* in each iteration to select a 'nearby' plane, under a carefully defined probability distribution. (This technique is also related to another randomized-rounding algorithm of Klein [11] for a different decoding problem.)

In addition to his nearest-plane algorithm, Babai also proposed a simpler (but somewhat looser) lattice decoding algorithm, which we call "simple rounding." In this algorithm, a given point $\mathbf{c} \in \mathbb{R}^n$ is rounded to the lattice point $\mathbf{B}\lfloor \mathbf{B}^{-1}\mathbf{c} \rceil$, where each coordinate of $\mathbf{B}^{-1}\mathbf{c} \in \mathbb{R}^n$ is independently rounded to its nearest integer. With precomputation of $\mathbf{B}^{-1}$, this algorithm can be quite practical — especially on $q$-ary lattices, where several more optimizations are possible. Moreover, it is trivially parallelized among up to $n^2$ processors. Unfortunately, its deterministic form it turns out to be completely insecure for 'answering queries' (e.g., digital signatures), as demonstrated by Nguyen and Regev [18].

A natural question, given the approach of [8], is whether a *randomized* variant of Babai's simple-rounding algorithm is secure. Specifically, the natural way of randomizing the algorithm is to round each coordinate of $\mathbf{B}^{-1}\mathbf{c}$ to a nearby integer (under a discrete Gaussian distribution over $\mathbb{Z}$, which can be sampled efficiently), then left-multiply by $\mathbf{B}$ as before. Unlike with the randomized nearest-plane algorithm, though, the resulting probability distribution here is unfortunately not spherical, nor does it leak zero knowledge. Instead, it is a 'skewed' (elliptical) Gaussian, where the skew mirrors the 'geometry' of the basis. More precisely, the *covariance* matrix $E_{\mathbf{x}}[(\mathbf{x}-\mathbf{c})(\mathbf{x}-\mathbf{c})^t]$ of the distribution (about its center $\mathbf{c}$) is approximately $\mathbf{BB}^t$, which captures the entire geometry of the basis $\mathbf{B}$, up to rigid rotation. Because covariance can be measured efficiently from only a small number of samples, the randomized simple-rounding algorithm leaks this geometry.[3]

Our solution prevents such leakage, in a manner inspired by the following facts. Recall that if $X$ and $Y$ are two independent random variables, the probability distribution of their sum $X + Y$ is the *convolution* of their individual distributions. In addition, for continuous (not necessarily spherical) Gaussians, covariance matrices are additive under convolution. In particular, if $\Sigma_1$ and $\Sigma_2$ are covariance matrices such that $\Sigma_1 + \Sigma_2 = s^2\mathbf{I}$, then the convolution of two Gaussians with covariance matrices $\Sigma_1$, $\Sigma_2$ (respectively) is a spherical Gaussian with standard deviation $s$.

The above facts give the basic idea for our algorithm, which is to convolve the output of the randomized simple-rounding algorithm with a suitable non-spherical (continuous) Gaussian, yielding a spherically distributed output. However, note that we want the algorithm to generate a *discrete* distribution — i.e., it must output a lattice point — so we should not alter the output of the randomized-rounding step. Instead, we *first* perturb the desired center $\mathbf{c}$ by a suitable non-spherical Gaussian, then apply randomized rounding to the resulting perturbed point. Strictly speaking this is not a true convolution, because the rounding step depends on the output of the perturbation step, but we can reduce the analysis to a true convolution using bounds related to the "smoothing parameter" of the lattice [17].

The main remaining question is: for a given covariance matrix $\Sigma_1 = \mathbf{BB}^t$ (corresponding to the rounding step), for what values of $s$ is there an efficiently sampleable Gaussian having covariance matrix $\Sigma_2 = s^2\mathbf{I} - \Sigma_1$? The covariance matrix of any (non-degenerate) Gaussian is symmetric positive definite, i.e., all its eigenvalues are positive. Conversely, every positive definite matrix is the covariance of some Gaussian, which can sampled efficiently by computing a 'square root' of the covariance matrix. Since any eigenvector of $\Sigma_1$ (with eigenvalue $\sigma^2 > 0$) is also an eigenvector of $s^2\mathbf{I}$ (with eigenvalue $s^2$), it must be an eigenvector of $\Sigma_2$ (with eigenvalue $s^2 - \sigma^2$) as well. Therefore, a necessary and sufficient condition is that all the eigenvalues of $\Sigma_1$ be less than $s^2$. Equivalently, the algorithm works for any $s$ that exceeds the largest singular value of the given basis $\mathbf{B}$. More generally, it can sample any (possibly non-spherical) discrete Gaussian with covariance matrix $\Sigma > \Sigma_1$ (i.e., $\Sigma - \Sigma_1$ is positive definite).

---

[3] Given the above, one might still wonder whether the covariance $\mathbf{BB}^t$ could be simulated efficiently (without any privileged knowledge about the lattice) when $\mathbf{B}$ is itself drawn from a 'nice' distribution, such as a discrete Gaussian. Indeed, if the vectors of $\mathbf{B}$ were drawn independently from a *continuous* Gaussian, the matrix $\mathbf{BB}^t$ would have the so-called *Wishart distribution*, which can be generated 'obliviously' (without knowledge of $\mathbf{B}$ itself) using the *Bartlett decomposition*. (See, e.g., [12] and references therein). Unfortunately, these facts do not quite seem to carry over to discrete Gaussians, though they may be useful in another cryptographic context.

In retrospect, the high-level structure of our algorithm resembles the "perturbation" heuristic proposed for NTRUSign [9], though the details are quite different. First, the perturbation and rounding steps in NTRUSign are both *deterministic* with respect to two or more bases, and there is evidence that this is insecure [15], at least for a large polynomial number of signatures. Interestingly, randomization also allows for improved efficiency, since our perturbations can be chosen with offline precomputation (as opposed to the deterministic method of [9], which is inherently online). Second, the signing and perturbation bases used in NTRUSign are chosen independently, whereas our perturbations are carefully chosen to conceal the statistics that would otherwise be leaked by randomized rounding.

## 2  Preliminaries

### 2.1  Notation

For a countable set $X$ and a real-valued function $f$, we write $f(X)$ to denote $\sum_{x \in X} f(x)$. A nonnegative function $f \colon \mathbb{N} \to \mathbb{R}$ is called *negligible*, written $f(n) = \mathrm{negl}(n)$, if it vanishes faster than any inverse polynomial, i.e., $f(n) = o(n^{-c})$ for every constant $c \geq 0$. A function $g \colon \mathbb{N} \to [0,1]$ is called *overwhelming* if it is $1 - \mathrm{negl}(n)$. The *statistical distance* between two distributions $X$ and $Y$ (or two random variables have those distributions, respectively) is defined as $\Delta(X, Y) = \sup_{A \subseteq D} |X(A) - Y(A)|$. When $D$ is a countable set, we have $\Delta(X, Y) = \frac{1}{2} \sum_{d \in D} |X(d) - Y(d)|$.

We use bold lower-case letters (e.g., $\mathbf{x}$) to denote vectors in $\mathbb{R}^n$, for an undetermined positive integer dimension $n$ that remains the same throughout the paper. We use bold upper-case letters (e.g., $\mathbf{B}$) for ordered sets of vectors, and identify the set with the matrix having the vectors as its columns. We frequently use upper-case Greek letters such as $\Sigma$ to denote (symmetric) *positive (semi)definite* matrices, defined below. In contexts where a matrix is expected, we sometimes use a scalar $s \in \mathbb{R}$ to denote $s \cdot \mathbf{I}$, where $\mathbf{I}$ is the identity matrix of appropriate dimension. We let $\|\mathbf{B}\| = \max_i \|\mathbf{b}_i\|$, where $\|\cdot\|$ denotes the Euclidean norm.

### 2.2  Linear Algebra

A symmetric matrix $\Sigma \in \mathbb{R}^{n \times n}$ is *positive definite*, written $\Sigma > \mathbf{0}$, if $\mathbf{x}^t \Sigma \mathbf{x} > 0$ for all nonzero $\mathbf{x} \in \mathbb{R}^n$. Equivalently, its spectral decomposition is

$$\Sigma = \mathbf{Q}\mathbf{D}^2\mathbf{Q}^{-1} = \mathbf{Q}\mathbf{D}^2\mathbf{Q}^t,$$

where $\mathbf{Q} \in \mathbb{R}^{n \times n}$ is an orthogonal matrix (i.e., one for which $\mathbf{Q}^t\mathbf{Q} = \mathbf{Q}\mathbf{Q}^t = \mathbf{I}$) whose columns are eigenvectors of $\Sigma$, and $\mathbf{D}$ is the real diagonal matrix of the square roots of the corresponding eigenvalues, all of which are positive. We have $\Sigma > \mathbf{0}$ if and only if $\Sigma^{-1} > \mathbf{0}$. We say that $\Sigma$ is positive *semi*definite, written $\Sigma \geq \mathbf{0}$, if $\mathbf{x}^t \Sigma \mathbf{x} \geq 0$ for all $\mathbf{x} \in \mathbb{R}^n$; such a matrix may not be invertible. Positive (semi)definiteness defines a partial ordering on symmetric matrices: we say that

$\Sigma_1 > \Sigma_2$ if $(\Sigma_1 - \Sigma_2) > \mathbf{0}$, and likewise for $\Sigma_1 \geq \Sigma_2$. It is the case that $\Sigma_1 \geq \Sigma_2 > \mathbf{0}$ if and only if $\Sigma_2^{-1} \geq \Sigma_1^{-1} > \mathbf{0}$, and likewise for the analogous strict inequalities.

For any nonsingular matrix $\mathbf{B} \in \mathbb{R}^{n \times n}$, the symmetric matrix $\Sigma = \mathbf{BB}^t$ is positive definite, because

$$\mathbf{x}^t \Sigma \mathbf{x} = \langle \mathbf{B}^t \mathbf{x}, \mathbf{B}^t \mathbf{x} \rangle = \|\mathbf{B}^t \mathbf{x}\|^2 > 0$$

for nonzero $\mathbf{x} \in \mathbb{R}^n$. We say that $\mathbf{B}$ is a *square root* of $\Sigma > \mathbf{0}$, written $\mathbf{B} = \sqrt{\Sigma}$, if $\mathbf{BB}^t = \Sigma$. Every $\Sigma > \mathbf{0}$ has a square root $\mathbf{B} = \mathbf{QD}$, where $\Sigma = \mathbf{QD}^2\mathbf{Q}^t$ is the spectral decomposition of $\Sigma$ as above. Moreover, the square root is unique up to right-multiplication by an orthogonal matrix, i.e., $\mathbf{B}' = \sqrt{\Sigma}$ if and only if $\mathbf{B}' = \mathbf{BP}$ for some orthogonal matrix $\mathbf{P}$. A square root of particular interest is given by the *Cholesky decomposition* $\Sigma = \mathbf{LL}^t$, where $\mathbf{L}$ is a (unique) lower-triangular matrix. Given $\Sigma$, its Cholesky decomposition can be computed efficiently in fewer than $n^3$ multiplication and addition operations (on real numbers of sufficient precision).

For a nonsingular matrix $\mathbf{B}$, a *singular value decomposition* is $\mathbf{B} = \mathbf{QDP}^t$, where $\mathbf{Q}, \mathbf{P} \in \mathbb{R}^{n \times n}$ are orthogonal matrices, and $\mathbf{D}$ is a diagonal matrix with positive entries $s_i > 0$ (called the singular values) on the diagonal, in non-increasing order. Under this convention, $\mathbf{D}$ is uniquely determined, and $s_1(\mathbf{B}) = \max_{\mathbf{u}} \|\mathbf{Bu}\| = \max_{\mathbf{u}} \|\mathbf{B}^t \mathbf{u}\|$, where the maximum is taken over all unit vectors $\mathbf{u} \in \mathbb{R}^n$. Note that

$$\Sigma = \mathbf{BB}^t = \mathbf{QDP}^t \mathbf{PD}^t \mathbf{Q}^t = \mathbf{QD}^2 \mathbf{Q}^t,$$

so the eigenvalues of $\Sigma$ are the squares of the singular values of $\mathbf{B}$.

### 2.3  Gaussians

The $n$-dimensional Gaussian function $\rho : \mathbb{R}^n \to (0, 1]$ is defined as

$$\rho(\mathbf{x}) \overset{\Delta}{=} \exp(-\pi \cdot \|\mathbf{x}\|^2) = \exp(-\pi \cdot \langle \mathbf{x}, \mathbf{x} \rangle).$$

Applying a linear transformation given by a nonsingular matrix $\mathbf{B}$ yields the Gaussian function

$$\rho_{\mathbf{B}}(\mathbf{x}) \overset{\Delta}{=} \rho(\mathbf{B}^{-1}\mathbf{x}) = \exp\left(-\pi \cdot \langle \mathbf{B}^{-1}\mathbf{x}, \mathbf{B}^{-1}\mathbf{x} \rangle\right) = \exp\left(-\pi \cdot \mathbf{x}^t \Sigma^{-1} \mathbf{x}\right),$$

where $\Sigma = \mathbf{BB}^t > \mathbf{0}$. Because $\rho_{\mathbf{B}}$ is distinguished only up to $\Sigma$, we usually refer to it as $\rho_{\sqrt{\Sigma}}$.

Normalizing $\rho_{\sqrt{\Sigma}}$ by its total measure $\int_{\mathbb{R}^n} \rho_{\sqrt{\Sigma}}(\mathbf{x})\, d\mathbf{x} = \sqrt{\det \Sigma}$ over $\mathbb{R}^n$, we obtain the probability distribution function of the (continuous) *Gaussian distribution* $D_{\sqrt{\Sigma}}$. It is easy to check that a random variable $\mathbf{x}$ having distribution $D_{\sqrt{\Sigma}}$ can be written as $\sqrt{\Sigma} \cdot \mathbf{z}$, where $\mathbf{z}$ has spherical Gaussian distribution $D_1$. Therefore, the random variable $\mathbf{x}$ has *covariance*

$$\mathop{\mathrm{E}}_{\mathbf{x} \sim D_{\sqrt{\Sigma}}} \left[\mathbf{x} \cdot \mathbf{x}^t\right] = \sqrt{\Sigma} \cdot \mathop{\mathrm{E}}_{\mathbf{z} \sim D_1} \left[\mathbf{z} \cdot \mathbf{z}^t\right] \cdot \sqrt{\Sigma}^t = \sqrt{\Sigma} \cdot \frac{\mathbf{I}}{2\pi} \cdot \sqrt{\Sigma}^t = \frac{\Sigma}{2\pi},$$

by linearity of expectation. (The $\frac{1}{2\pi}$ covariance of $\mathbf{z} \sim D_1$ arises from the independence of its entries, which are each distributed as $D_1$ in one dimension, and have variance $\frac{1}{2\pi}$.) *For convenience, in this paper we implicitly scale all covariance matrices by a $2\pi$ factor, and refer to $\Sigma$ as the covariance matrix of $D_{\sqrt{\Sigma}}$.*

The following standard fact, which will be central to the analysis of our sampling algorithms, characterizes the product of two Gaussian functions.

**Fact 1.** *Let $\Sigma_1, \Sigma_2 > 0$ be positive definite matrices, let $\Sigma_0 = \Sigma_1 + \Sigma_2 > 0$ and $\Sigma_3^{-1} = \Sigma_1^{-1} + \Sigma_2^{-1} > 0$, let $\mathbf{x}, \mathbf{c}_1, \mathbf{c}_2 \in \mathbb{R}^n$ be arbitrary, and let $\mathbf{c}_3 \in \mathbb{R}^n$ be such that $\Sigma_3^{-1}\mathbf{c}_3 = \Sigma_1^{-1}\mathbf{c}_1 + \Sigma_2^{-1}\mathbf{c}_2$. Then*

$$\rho_{\sqrt{\Sigma_1}}(\mathbf{x} - \mathbf{c}_1) \cdot \rho_{\sqrt{\Sigma_2}}(\mathbf{x} - \mathbf{c}_2) = \rho_{\sqrt{\Sigma_0}}(\mathbf{c}_1 - \mathbf{c}_2) \cdot \rho_{\sqrt{\Sigma_3}}(\mathbf{x} - \mathbf{c}_3).$$

### 2.4   Gaussians on Lattices

A *lattice* $\Lambda$ is a discrete additive subgroup of $\mathbb{R}^n$. In this work we are only concerned with full-rank lattices, which are generated by some nonsingular *basis* $\mathbf{B} \in \mathbb{R}^{n \times n}$, as the set $\Lambda = \mathbf{B} \cdot \mathbb{Z}^n = \{\mathbf{B}\mathbf{z} : \mathbf{z} \in \mathbb{Z}^n\}$. When $n \geq 2$, every lattice has infinitely many bases, which are related by unimodular transformations: $\mathbf{B}'$ and $\mathbf{B}$ generate the same lattice if and only if $\mathbf{B}' = \mathbf{B}\mathbf{U}$ for some unimodular $\mathbf{U} \in \mathbb{Z}^{n \times n}$. The *dual lattice* of $\Lambda$ is defined as $\Lambda^* = \{\mathbf{w} \in \mathbb{R}^n : \langle \mathbf{x}, \mathbf{w} \rangle \in \mathbb{Z} \; \forall \, \mathbf{x} \in \Lambda\}$. (We only need this notion for defining the smoothing parameter of a lattice; see below.)

Let $\Lambda \subset \mathbb{R}^n$ be a lattice, let $\mathbf{c} \in \mathbb{R}^n$, and let $\Sigma > 0$ be a positive definite matrix. The *discrete Gaussian distribution* $D_{\Lambda+\mathbf{c}, \sqrt{\Sigma}}$ is simply the Gaussian distribution restricted so that its support is the coset $\Lambda + \mathbf{c}$. That is, for all $\mathbf{x} \in \Lambda + \mathbf{c}$,

$$D_{\Lambda+\mathbf{c}, \sqrt{\Sigma}}(\mathbf{x}) = \frac{\rho_{\sqrt{\Sigma}}(\mathbf{x})}{\rho_{\sqrt{\Sigma}}(\Lambda + \mathbf{c})} \propto \rho_{\sqrt{\Sigma}}(\mathbf{x}).$$

We recall the definition of the *smoothing parameter* from [17].

**Definition 1.** *For a lattice $\Lambda$ and positive real $\epsilon > 0$, the smoothing parameter $\eta_\epsilon(\Lambda)$ is the smallest real $s > 0$ such that $\rho_{1/s}(\Lambda^* \backslash \{\mathbf{0}\}) \leq \epsilon$.*

Observe that if $\Lambda_1$ is a sublattice of a lattice $\Lambda_0$, then $\eta_\epsilon(\Lambda_1) \geq \eta_\epsilon(\Lambda_0)$ for any $\epsilon > 0$, because $\Lambda_0^* \subseteq \Lambda_1^*$ and hence $\rho_{1/s}(\Lambda_0^* \backslash \{\mathbf{0}\}) \leq \rho_{1/s}(\Lambda_1^* \backslash \{\mathbf{0}\})$ by positivity of $\rho_{1/s}$.

Note that the smoothing parameter as defined above is a scalar; in this work we need to extend the notion to positive definite matrices.

**Definition 2.** *Let $\Sigma > 0$ be any positive definite matrix. We say that $\sqrt{\Sigma} \geq \eta_\epsilon(\Lambda)$ if $\rho_{\sqrt{\Sigma^{-1}}}(\Lambda^* \backslash \{\mathbf{0}\}) = \rho(\sqrt{\Sigma} \cdot \Lambda^* \backslash \{\mathbf{0}\}) \leq \epsilon$, i.e., if $\eta_\epsilon(\sqrt{\Sigma^{-1}} \cdot \Lambda) \leq 1$.*

**Lemma 1 (Corollary of [17, Lemma 4.4]).** *Let $\Lambda$ be any $n$-dimensional lattice. For any $\epsilon \in (0,1)$, $\Sigma > 0$ such that $\sqrt{\Sigma} \geq \eta_\epsilon(\Lambda)$, and any $\mathbf{c} \in \mathbb{R}^n$,*

$$\rho_{\sqrt{\Sigma}}(\Lambda + \mathbf{c}) \in [\tfrac{1-\epsilon}{1+\epsilon}, 1] \cdot \rho_{\sqrt{\Sigma}}(\Lambda).$$

*Proof.* Follows directly by applying $\sqrt{\Sigma^{-1}}$ as a linear transform to $\Lambda$, and by $\eta_\epsilon(\sqrt{\Sigma^{-1}} \cdot \Lambda) \leq 1$.

**Lemma 2 (Special case of [17, Lemma 3.3]).** *For any $\epsilon > 0$,*

$$\eta_\epsilon(\mathbb{Z}^n) \leq \sqrt{\ln(2n(1+1/\epsilon))/\pi}.$$

*In particular, for any $\omega(\sqrt{\log n})$ function, there is a negligible $\epsilon = \epsilon(n)$ such that $\eta_\epsilon(\mathbb{Z}^n) \leq \omega(\sqrt{\log n})$.*

## 3 Analysis of 'Convolved' Discrete Gaussians

In this section we prove some general facts about 'convolutions' of (possibly non-spherical) discrete Gaussian distributions, which are important for the conception and analysis of our sampling algorithm; we expect these facts to have other applications as well. (Strictly speaking, the probabilistic experiments that we analyze are not true convolutions, because we are adding random variables that are not formally independent. However, the spirit of the experiment and its outcome are entirely 'convolution-like.')

Because the proof of the theorem is rather technical, the reader who is interested in applications may wish to skip ahead to the next section after understanding the theorem statement.

**Theorem 1.** *Let $\Sigma_1, \Sigma_2 > 0$ be positive definite matrices, with $\Sigma = \Sigma_1 + \Sigma_2 > 0$ and $\Sigma_3^{-1} = \Sigma_1^{-1} + \Sigma_2^{-1} > 0$. Let $\Lambda_1, \Lambda_2$ be lattices such that $\sqrt{\Sigma_1} \geq \eta_\epsilon(\Lambda_1)$ and $\sqrt{\Sigma_3} \geq \eta_\epsilon(\Lambda_2)$ for some positive $\epsilon \leq 1/2$, and let $\mathbf{c}_1, \mathbf{c}_2 \in \mathbb{R}^n$ be arbitrary. Consider the following probabilistic experiment:*

*Choose $\mathbf{x}_2 \leftarrow D_{\Lambda_2 + \mathbf{c}_2, \sqrt{\Sigma_2}}$, then choose $\mathbf{x}_1 \leftarrow \mathbf{x}_2 + D_{\Lambda_1 + \mathbf{c}_1 - \mathbf{x}_2, \sqrt{\Sigma_1}}$.*

*The marginal distribution of $\mathbf{x}_1$ is within statistical distance $8\epsilon$ of $D_{\Lambda_1 + \mathbf{c}_1, \sqrt{\Sigma}}$. In addition, for any $\bar{\mathbf{x}}_1 \in \Lambda_1 + \mathbf{c}_1$, the conditional distribution of $\mathbf{x}_2 \in \Lambda_2 + \mathbf{c}_2$ given $\mathbf{x}_1 = \bar{\mathbf{x}}_1$ is within statistical distance $2\epsilon$ of $\mathbf{c}_3 + D_{\Lambda_2 + \mathbf{c}_2 - \mathbf{c}_3, \sqrt{\Sigma_3}}$, where $\Sigma_3^{-1} \mathbf{c}_3 = \Sigma_1^{-1} \bar{\mathbf{x}}_1$.*

*If $\mathbf{x}_2$ is instead chosen from the continuous Gaussian distribution $D_{\sqrt{\Sigma_2}}$ over $\mathbb{R}^n$, the marginal distribution of $\mathbf{x}_1$ is as above, and the conditional distribution of $\mathbf{x}_2 \in \mathbb{R}^n$ given $\mathbf{x}_1 = \bar{\mathbf{x}}_1 \in \Lambda_1 + \mathbf{c}_1$ is within statistical distance $2\epsilon$ of $\mathbf{c}_3 + D_{\sqrt{\Sigma_3}}$. (In this setting, the lattice $\Lambda_2$ and the hypothesis $\sqrt{\Sigma_3} \geq \eta_\epsilon(\Lambda_2)$ are unneeded.)*

*Proof.* We start by analyzing the joint distribution of $\mathbf{x}_1 \in \Lambda_1 + \mathbf{c}_1$ and $\mathbf{x}_2 \in \Lambda_2 + \mathbf{c}_2$. Let $\bar{\mathbf{x}}_1 \in \Lambda_1 + \mathbf{c}_1$ and $\bar{\mathbf{x}}_2 \in \Lambda_2 + \mathbf{c}_2$ be arbitrary, and let $\Sigma_3^{-1} \mathbf{c}_3 = \Sigma_1^{-1} \bar{\mathbf{x}}_1$. Then we have

$$\Pr[\mathbf{x}_1 = \bar{\mathbf{x}}_1 \wedge \mathbf{x}_2 = \bar{\mathbf{x}}_2]$$

$$= D_{\Lambda_1 + \mathbf{c}_1 - \bar{\mathbf{x}}_2, \sqrt{\Sigma_1}}(\bar{\mathbf{x}}_1 - \bar{\mathbf{x}}_2) \cdot D_{\Lambda_2 + \mathbf{c}_2, \sqrt{\Sigma_2}}(\bar{\mathbf{x}}_2) \tag{1}$$

$$= \frac{\rho_{\sqrt{\Sigma_1}}(\bar{\mathbf{x}}_2 - \bar{\mathbf{x}}_1) \cdot \rho_{\sqrt{\Sigma_2}}(\bar{\mathbf{x}}_2)}{\rho_{\sqrt{\Sigma_1}}(\Lambda_1 + \mathbf{c}_1 - \bar{\mathbf{x}}_2) \cdot \rho_{\sqrt{\Sigma_2}}(\Lambda_2 + \mathbf{c}_2)} \tag{2}$$

$$\propto \frac{\rho_{\sqrt{\Sigma}}(\bar{\mathbf{x}}_1) \cdot \rho_{\sqrt{\Sigma_3}}(\bar{\mathbf{x}}_2 - \mathbf{c}_3)}{\rho_{\sqrt{\Sigma_1}}(\Lambda_1 + \mathbf{c}_1 - \bar{\mathbf{x}}_2)}. \tag{3}$$

Equation (1) is by construction; Equation (2) is by definition of $D_{\Lambda+\mathbf{c}}$ and by symmetry of $\rho_{\sqrt{\Sigma_1}}$; Equation (3) is by Fact 1. (The $\rho_{\sqrt{\Sigma_2}}(\Lambda_2 + \mathbf{c}_2)$ term in the denominator of (2) is the same for all $\bar{\mathbf{x}}_1, \bar{\mathbf{x}}_2$, so we can treat it as a constant of proportionality.)

We now analyze the marginal distribution of $\mathbf{x}_1$. For any $\bar{\mathbf{x}}_1 \in \Lambda_1 + \mathbf{c}_1$, let $\Sigma_3^{-1}\mathbf{c}_3 = \Sigma_1^{-1}\bar{\mathbf{x}}_1$ as above; then we have

$$\Pr[\mathbf{x}_1 = \bar{\mathbf{x}}_1]$$

$$= \sum_{\bar{\mathbf{x}}_2 \in \Lambda_2 + \mathbf{c}_2} \Pr[\mathbf{x}_1 = \bar{\mathbf{x}}_1 \wedge \mathbf{x}_2 = \bar{\mathbf{x}}_2] \qquad \text{(by construction)}$$

$$\propto \rho_{\sqrt{\Sigma}}(\bar{\mathbf{x}}_1) \cdot \sum_{\bar{\mathbf{x}}_2 \in \Lambda_2 + \mathbf{c}_2} \frac{\rho_{\sqrt{\Sigma_3}}(\bar{\mathbf{x}}_2 - \mathbf{c}_3)}{\rho_{\sqrt{\Sigma_1}}(\Lambda_1 + \mathbf{c}_1 - \bar{\mathbf{x}}_2)} \qquad \text{(Equation (3))}$$

$$\in \rho_{\sqrt{\Sigma}}(\bar{\mathbf{x}}_1) \cdot [1, \tfrac{1+\epsilon}{1-\epsilon}] \cdot \frac{\rho_{\sqrt{\Sigma_3}}(\Lambda_2 + \mathbf{c}_2 - \mathbf{c}_3)}{\rho_{\sqrt{\Sigma_1}}(\Lambda_1)} \qquad (\sqrt{\Sigma_1} \geq \eta_\epsilon(\Lambda_1), \text{ Lemma 1})$$

$$\subseteq \rho_{\sqrt{\Sigma}}(\bar{\mathbf{x}}_1) \cdot [\tfrac{1-\epsilon}{1+\epsilon}, \tfrac{1+\epsilon}{1-\epsilon}] \cdot \frac{\rho_{\sqrt{\Sigma_3}}(\Lambda_2)}{\rho_{\sqrt{\Sigma_1}}(\Lambda_1)} \qquad (\sqrt{\Sigma_3} \geq \eta_\epsilon(\Lambda_2), \text{ Lemma 1})$$

$$\propto \rho_{\sqrt{\Sigma}}(\bar{\mathbf{x}}_1) \cdot [\tfrac{1-\epsilon}{1+\epsilon}, \tfrac{1+\epsilon}{1-\epsilon}].$$

It follows that

$$\Pr[\mathbf{x}_1 = \bar{\mathbf{x}}_1] \in [(\tfrac{1-\epsilon}{1+\epsilon})^2, (\tfrac{1+\epsilon}{1-\epsilon})^2] \cdot \frac{\rho_{\sqrt{\Sigma}}(\bar{\mathbf{x}}_1)}{\rho_{\sqrt{\Sigma}}(\Lambda_1 + \mathbf{c}_1)} \subseteq [1 - 16\epsilon, 1 + 16\epsilon] \cdot D_{\Lambda_1 + \mathbf{c}_1, \sqrt{\Sigma}}(\bar{\mathbf{x}}_1),$$

because $\epsilon \leq 1/2$. The claim on the marginal distribution of $\mathbf{x}_1$ follows by definition of statistical distance.

When $\mathbf{x}_2$ is chosen from the continuous Gaussian $D_{\sqrt{\Sigma_2}}$, the analysis is almost identical: we simply replace the summation over $\bar{\mathbf{x}}_2 \in \Lambda_2 + \mathbf{c}_2$ with integration over $\bar{\mathbf{x}}_2 \in \mathbb{R}^n$. Because $\int_{\bar{\mathbf{x}}_2} \rho_{\sqrt{\Sigma_3}}(\bar{\mathbf{x}}_2 - \mathbf{c}_3) \, d\bar{\mathbf{x}}_2 = \sqrt{\det \Sigma_3}$ is independent of $\mathbf{c}_3$, there is no need to invoke Lemma 1 a second time.

The analysis of the conditional distribution of $\mathbf{x}_2 \in \Lambda_2 + \mathbf{c}_2$ proceeds similarly; due to space restrictions, we omit here it an refer the reader to the full version.

# 4   Discrete Gaussian Sampling Algorithms

In this section we define and analyze some new discrete Gaussian sampling algorithms. We start in Section 4.1 by defining and analyzing some important randomized-rounding subroutines. In Section 4.2 we describe a general-purpose (but unoptimized) sampling algorithm, then in Section 4.3 we describe a highly optimized sampling algorithm for $q$-ary lattices.

## 4.1   Randomized Rounding

We first need to define and analyze some simple 'randomized-rounding' operations from the reals to lattices, which are an important component of our sampling algorithms.

We start with a basic rounding operation from $\mathbb{R}$ to the integers $\mathbb{Z}$, denoted $\lfloor v \rceil_r$ for $v \in \mathbb{R}$ and some positive real 'rounding parameter' $r > 0$. The output of this operation is a random variable over $\mathbb{Z}$ having distribution $v + D_{\mathbb{Z}-v,r}$. Observe that for any integer $z \in \mathbb{Z}$, the random variables $\lfloor z + v \rceil_r$ and $z + \lfloor v \rceil_r$ are identically distributed; therefore, we sometimes assume that $v \in [0, 1)$ without loss of generality. We extend the rounding operation coordinate-wise to vectors $\mathbf{v} \in \mathbb{R}^n$, where each entry is rounded independently. It follows that for any $\mathbf{v} \in \mathbb{R}^n$ and $\bar{\mathbf{z}} \in \mathbb{Z}^n$,

$$\Pr\left[\lfloor \mathbf{v} \rceil_r = \bar{\mathbf{z}}\right] \propto \prod_{i \in [n]} \rho_r(\bar{z}_i - v_i) = \exp(-\pi\|\bar{\mathbf{z}} - \mathbf{v}\|^2/r^2) = \rho_r(\bar{\mathbf{z}} - \mathbf{v}).$$

That is, $\lfloor \mathbf{v} \rceil_r$ has distribution $\mathbf{v} + D_{\mathbb{Z}^n - \mathbf{v}, r}$, because the standard basis for $\mathbb{Z}^n$ is orthonormal.

The next lemma characterizes the distribution obtained by randomized rounding to an arbitrary lattice, using an arbitrary (possibly non-orthonormal) basis.

**Lemma 3.** *Let $\mathbf{B}$ be a basis of a lattice $\Lambda = \mathcal{L}(\mathbf{B})$, let $\Sigma = r^2 \cdot \mathbf{BB}^t$ for some real $r > 0$, and let $\mathbf{t} \in \mathbb{R}^n$ be arbitrary. The random variable $\mathbf{x} = \mathbf{t} - \mathbf{B}\lfloor \mathbf{B}^{-1}\mathbf{t} \rceil_r$ has distribution $D_{\Lambda + \mathbf{t}, \sqrt{\Sigma}}$.*

*Proof.* Let $\mathbf{v} = \mathbf{B}^{-1}\mathbf{t}$. The support of $\lfloor \mathbf{v} \rceil_r$ is $\mathbb{Z}^n$, so the support of $\mathbf{x}$ is $\mathbf{t} - \mathbf{B} \cdot \mathbb{Z}^n = \Lambda + \mathbf{t}$. Now for any $\bar{\mathbf{x}} = \mathbf{t} - \mathbf{B}\bar{\mathbf{z}}$ where $\bar{\mathbf{z}} \in \mathbb{Z}^n$, we have $\mathbf{x} = \bar{\mathbf{x}}$ if and only if $\lfloor \mathbf{v} \rceil_r = \bar{\mathbf{z}}$. As desired, this event occurs with probability proportional to

$$\rho_r(\bar{\mathbf{z}} - \mathbf{v}) = \rho_r(\mathbf{B}^{-1}(\mathbf{t} - \bar{\mathbf{x}}) - \mathbf{B}^{-1}\mathbf{t}) = \rho_r(-\mathbf{B}^{-1}\bar{\mathbf{x}}) = \rho_{r\mathbf{B}}(\bar{\mathbf{x}}) = \rho_{\sqrt{\Sigma}}(\bar{\mathbf{x}}).$$

*Efficient rounding.* In [8] it is shown how to sample from $D_{\mathbb{Z}-v,r}$ for any $v \in \mathbb{R}$ and $r > 0$, by rejection sampling. While the algorithm requires only $\text{poly}(\log n)$ iterations before terminating, its concrete running time and randomness complexity are not entirely suitable for practical implementations.

In this work, we can sample from $v + D_{\mathbb{Z}-v,r}$ more efficiently because $r$ is always fixed, known in advance, and relatively small (about $\sqrt{\log n}$). Specifically, given $r$ and $v \in \mathbb{R}$ we can (pre)compute a compact table of the approximate cumulative distribution function of $\lfloor v \rceil_r$, i.e., the probabilities

$$p_{\bar{z}} := \Pr[v + D_{\mathbb{Z}-v,r} \leq \bar{z}]$$

for each $\bar{z} \in \mathbb{Z}$ in an interval $[v - r \cdot \omega(\sqrt{\log n}), v + r \cdot \omega(\sqrt{\log n})]$. (Outside of that interval are the tails of the distribution, which carry negligible probability mass.) Then we can sample directly from $v + D_{\mathbb{Z}-v,r}$ by choosing a uniformly random $x \in [0, 1)$ and performing a binary search through the table for the $\bar{z} \in \mathbb{Z}$ such that $x \in [p_{\bar{z}-1}, p_{\bar{z}})$. Moreover, the bits of $x$ may be chosen 'lazily,' from most- to least-significant, until $\bar{z}$ is determined uniquely. To sample within a $\text{negl}(n)$ statistical distance of the desired distribution, these operations can all be implemented in time $\text{poly}(\log n)$.

## 4.2  Generic Sampling Algorithm

Here we apply Theorem 1 to sample from a discrete Gaussian of any sufficiently large covariance, given a good enough basis of the lattice. This procedure, described in Algorithm 1, serves mainly as a 'proof of concept' and a warm-up for our main algorithm on $q$-ary lattices. As such, it is not optimized for runtime efficiency (because it uses arbitrary-precision real operations), though it is still fully parallelizable and offline/online.

---

**Algorithm 1.** Generic algorithm $\mathsf{SampleD}(\mathbf{B}_1, r, \Sigma, \mathbf{c})$ for sampling from a discrete Gaussian distribution.

---
**Input:**
  *Offline phase:* Basis $\mathbf{B}_1$ of a lattice $\Lambda = \mathcal{L}(\mathbf{B}_1)$, rounding parameter $r = \omega(\sqrt{\log n})$, and positive definite covariance matrix $\Sigma > \Sigma_1 = r^2 \cdot \mathbf{B}_1 \mathbf{B}_1^t$.
  *Online phase:* a vector $\mathbf{c} \in \mathbb{R}^n$.
**Output:** A vector $\mathbf{x} \in \Lambda + \mathbf{c}$ drawn from a distribution within $\mathrm{negl}(n)$ statistical distance of $D_{\Lambda+\mathbf{c}, \sqrt{\Sigma}}$.
  *Offline phase:*
1: Let $\Sigma_2 = \Sigma - \Sigma_1 > \mathbf{0}$, and compute some $\mathbf{B}_2 = \sqrt{\Sigma_2}$.
2: Before each call to the online phase, choose a fresh $\mathbf{x}_2 \leftarrow D_{\sqrt{\Sigma_2}}$, as $\mathbf{x}_2 \leftarrow \mathbf{B}_2 \cdot D_1$.
  *Online phase:*
3: **return**  $\mathbf{x} \leftarrow \mathbf{c} - \mathbf{B}_1 \lfloor \mathbf{B}_1^{-1}(\mathbf{c} - \mathbf{x}_2) \rceil_r$.

---

**Theorem 2.** *Algorithm 1 is correct, and for any $P \in [1, n^2]$, its online phase can be executed in parallel by $P$ processors that each perform $O(n^2/P)$ operations on real numbers (of sufficiently high precision).*

*Proof.* We first show correctness. Let $\Sigma, \Sigma_1, \Sigma_2$ be as in Algorithm 1. The output $\mathbf{x}$ is distributed as

$$\mathbf{x} = \mathbf{x}_2 + (\mathbf{c} - \mathbf{x}_2) - \mathbf{B}_1 \left\lfloor \mathbf{B}_1^{-1}(\mathbf{c} - \mathbf{x}_2) \right\rceil_r,$$

where $\mathbf{x}_2$ has distribution $D_{\sqrt{\Sigma_2}}$. By Lemma 3 with $\mathbf{t} = (\mathbf{c} - \mathbf{x}_2)$, we see that $\mathbf{x}$ has distribution $\mathbf{x}_2 + D_{\Lambda+\mathbf{c}-\mathbf{x}_2, \sqrt{\Sigma_1}}$. Now because $\Lambda = \mathcal{L}(\mathbf{B}_1) = \mathbf{B}_1 \cdot \mathbb{Z}^n$, we have $\sqrt{\Sigma_1} = r\mathbf{B}_1 \geq \eta_\epsilon(\Lambda)$ for some negligible $\epsilon = \epsilon(n)$, by Definition 2 and Lemma 2. Therefore, by the second part of Theorem 1, $\mathbf{x}$ has distribution within $\mathrm{negl}(n)$ statistical distance of $D_{\Lambda+\mathbf{c}, \sqrt{\Sigma}}$.

To parallelize the algorithm, simply observe that $\mathbf{B}_1^{-1}$ can be precomputed in the offline phase, and that the matrix-vector products and randomized rounding can all be executed in parallel on $P$ processors in the natural way.

## 4.3  Efficient Sampling Algorithm for $q$-ary Lattices

Algorithm 2 is an optimized sampling algorithm for $q$-ary (integral) lattices $\Lambda$, i.e., lattices for which $q\mathbb{Z}^n \subseteq \Lambda \subseteq \mathbb{Z}^n$ for some positive integer $q$. These include

**Algorithm 2.** Efficient algorithm $\mathsf{SampleD}(\mathbf{B}_1, r, \Sigma, \mathbf{c})$ for sampling a discrete Gaussian over a $q$-ary lattice.

**Input:**

    *Offline phase:* Basis $\mathbf{B}_1$ of a $q$-ary integer lattice $\Lambda = \mathcal{L}(\mathbf{B}_1)$, rounding parameter $r = \omega(\sqrt{\log n})$, and positive definite covariance matrix $\Sigma \geq r^2 \cdot (4\mathbf{B}_1\mathbf{B}_1^t + \mathbf{I})$.

    *Online phase:* a vector $\mathbf{c} \in \mathbb{Z}^n$.

**Output:** A vector $\mathbf{x} \in \Lambda + \mathbf{c}$ drawn from a distribution within $\mathrm{negl}(n)$ statistical distance of $D_{\Lambda+\mathbf{c},\sqrt{\Sigma}}$.

    *Offline phase:*

1: Compute $\mathbf{Z} = q \cdot \mathbf{B}_1^{-1} \in \mathbb{Z}^{n \times n}$.
2: Let $\Sigma_1 = 2r^2 \cdot \mathbf{B}_1\mathbf{B}_1^t$, let $\Sigma_2 = \Sigma - \Sigma_1 \geq r^2 \cdot (2\mathbf{B}_1\mathbf{B}_1^t + \mathbf{I})$, and compute some $\mathbf{B}_2 = \sqrt{\Sigma_2 - r^2}$.
3: Before each call to the online phase, choose a fresh $\mathbf{x}_2$ from $D_{\mathbb{Z}^n, \sqrt{\Sigma_2}}$ by letting $\mathbf{x}_2 \leftarrow \lfloor \mathbf{B}_2 \cdot D_1 \rceil_r$.

    *Online phase:*

4: **return** $\mathbf{x} \leftarrow \mathbf{c} - \mathbf{B}_1 \left\lfloor \frac{\mathbf{Z}(\mathbf{c}-\mathbf{x}_2)}{q} \right\rceil_r$.

---

NTRU lattices [10], as well as the family of lattices for which Ajtai [3] first demonstrated worst-case hardness.

Note that Algorithm 2 samples from the coset $\Lambda + \mathbf{c}$ for a given *integral* vector $\mathbf{c} \in \mathbb{Z}^n$; as we shall see, this allows for certain optimizations. Fortunately, all known cryptographic applications of Gaussian sampling over $q$-ary lattices use an integral $\mathbf{c}$. Also note that the algorithm will typically be used to sample from a *spherical* discrete Gaussian, i.e., one for which the covariance matrix $\Sigma = s^2\mathbf{I}$ for some real $s > 0$. As long as $s$ slightly exceeds the largest singular value of $\mathbf{B}_1$, i.e., $s \geq r \cdot (2s_1(\mathbf{B}_1) + 1)$ for some $r = \omega(\sqrt{\log n})$, then we have $\Sigma \geq r^2 \cdot (4\mathbf{B}_1\mathbf{B}_1^t + \mathbf{I})$ as required by the algorithm.

**Theorem 3.** *Algorithm 2 is correct, and for any $P \in [1, n^2]$, its online phase can be implemented in parallel by $P$ processors that each perform at most $\lceil n/P \rceil$ randomized-rounding operations on rational numbers from the set $\{\frac{0}{q}, \frac{1}{q}, \ldots, \frac{q-1}{q}\}$, and $O(n^2/P)$ integer operations.*

When the width of the desired Gaussian distribution is much less than $q$, which is the case in all known cryptographic applications of the sampling algorithm, all the integer operations in the online phase may be performed modulo either $q$ or $q^2$; see the discussion following the proof for details.

*Proof.* First observe that because $\Lambda = \mathcal{L}(\mathbf{B}_1)$ is $q$-ary, i.e., $q\mathbb{Z}^m \subseteq \Lambda$, there exists an integral matrix $\mathbf{Z} \in \mathbb{Z}^{n \times n}$ such that $\mathbf{B}_1\mathbf{Z} = q \cdot \mathbf{I}$. Therefore, $\mathbf{Z} = q \cdot \mathbf{B}_1^{-1} \in \mathbb{Z}^{n \times n}$ as stated in Step 1 of the algorithm. We also need to verify that $\mathbf{x}_2 \leftarrow \lfloor \mathbf{B}_2 \cdot D_1 \rceil_r$ has distribution $D_{\mathbb{Z}^n, \sqrt{\Sigma_2}}$ in Step 3. Let $\mathbf{w} \in \mathbb{R}^n$ have distribution $\mathbf{B}_2 \cdot D_1 = D_{\sqrt{\Sigma_2 - r^2}}$. Then $\mathbf{x}_2$ has distribution

$$\lfloor \mathbf{w} \rceil_r = \mathbf{w} + (-\mathbf{w} + \lfloor \mathbf{w} \rceil_r) = \mathbf{w} + D_{\mathbb{Z}^n - \mathbf{w}, r},$$

by Lemma 3 (using the standard basis for $\mathbb{Z}^n$). Then because $r \geq \eta_\epsilon(\mathbb{Z}^n)$ for some negligible $\epsilon = \epsilon(n)$ and $\mathbf{B}_2 \mathbf{B}_2^t + r^2 = \Sigma_2$, by Theorem 1 we conclude that $\mathbf{x}_2$ has distribution $D_{\mathbb{Z}^n, \sqrt{\Sigma_2}}$ as desired.

We now analyze the online phase, and show correctness. Because $\mathbf{B}_1^{-1} = \mathbf{Z}/q$, the algorithm's output $\mathbf{x}$ is distributed as

$$\mathbf{x}_2 + (\mathbf{c} - \mathbf{x}_2) - \mathbf{B}_1 \lfloor \mathbf{B}_1^{-1}(\mathbf{c} - \mathbf{x}_2) \rceil_r.$$

We would like to apply Lemma 3 (with $\mathbf{t} = \mathbf{c} - \mathbf{x}_2$) and Theorem 1 (with $\Lambda_1 = \Lambda$, $\Lambda_2 = \mathbb{Z}^n$, $\mathbf{c}_1 = \mathbf{c}$, and $\mathbf{c}_2 = \mathbf{0}$) to conclude that $\mathbf{x}$ has distribution within $\mathrm{negl}(n)$ statistical distance of $D_{\Lambda + \mathbf{c}, \sqrt{\Sigma}}$. To do so, we merely need to check that the hypotheses of Theorem 1 are satisfied, namely, that $\sqrt{\Sigma_1} \geq \eta_\epsilon(\Lambda)$ and $\sqrt{\Sigma_3} \geq \eta_\epsilon(\mathbb{Z}^n)$ for some negligible $\epsilon = \epsilon(n)$, where $\Sigma_3^{-1} = \Sigma_1^{-1} + \Sigma_2^{-1}$.

For the first hypothesis, we have $\sqrt{\Sigma_1} = 2r \cdot \mathbf{B}_1 \geq \eta_\epsilon(\Lambda)$ because $\Lambda = \mathbf{B}_1 \cdot \mathbb{Z}^n$, and by Definition 2 and Lemma 2. For the second hypothesis, we have

$$\Sigma_3^{-1} = \Sigma_1^{-1} + \Sigma_2^{-1} \leq 2 \cdot \left(2r^2 \mathbf{B}_1 \mathbf{B}_1^t\right)^{-1} = \left(r^2 \mathbf{B}_1 \mathbf{B}_1^t\right)^{-1}.$$

Therefore, $\sqrt{\Sigma_3} \geq \eta_\epsilon(\Lambda) \geq \eta_\epsilon(\mathbb{Z}^n)$, as desired. This completes the proof of correctness.

For the parallelism claim, observe that computing $\mathbf{Z}(\mathbf{c} - \mathbf{x}_2)$ and the final multiplication by $\mathbf{B}_1$ can be done in parallel using $P$ processors that each perform $O(n^2/P)$ integer operations. (See below for a discussion of the sizes of the integers involved in these operations.) Moreover, the fractional parts of the $n$-dimensional vector $\frac{\mathbf{Z}(\mathbf{c} - \mathbf{x}_2)}{q}$ are all in the set $\{\frac{0}{q}, \ldots, \frac{q-1}{q}\}$, and rounding these $n$ entries may be done independently in parallel.

*Implementation notes.* For a practical implementation, Algorithm 2 admits several additional optimizations, which we discuss briefly here.

In all cryptographic applications of Gaussian sampling on $q$-ary lattices, the length of the sampled vector is significantly shorter than $q$, i.e., its entries lie within a narrow interval around 0. Therefore, it suffices for the sampling algorithm to compute its output modulo $q$, using the integers $\{-\lfloor \frac{q}{2} \rfloor, \ldots, \lfloor \frac{q-1}{2} \rfloor\}$ as the set of canonical representatives. For this purpose, the final multiplication by the input basis $\mathbf{B}_1$ need only be performed modulo $q$. Similarly, $\mathbf{Z}$ and $\mathbf{Z}(\mathbf{c} - \mathbf{x}_2)$ need only be computed modulo $q^2$, because we are only concerned with the value of $\frac{\mathbf{Z}(\mathbf{c} - \mathbf{x}_2)}{q}$ modulo $q$.

Because all the randomized-rounding steps are performed on rationals whose fractional parts are in $\{\frac{0}{q}, \ldots, \frac{q-1}{q}\}$, if $q$ is reasonably small it may be worthwhile (for faster rounding) to precompute the tables of the cumulative distribution functions for all $q$ possibilities. Alternatively (or in addition), during the offline phase the algorithm could precompute and cache a few rounded samples for each of the $q$ possibilities, consuming them as needed in the online phase.

# 5    Singular Value Bounds

In this section we give bounds on the largest singular value of a basis $\mathbf{B}$ and relate them to other geometric quantities that are relevant to the prior sampling algorithm of [8].

## 5.1    General Bounds

The *Gram-Schmidt orthogonalization* of a nonsingular matrix $\mathbf{B}$ is $\mathbf{B} = \mathbf{QG}$, where $\mathbf{Q}$ is an orthogonal matrix and $\mathbf{G}$ is right-triangular, with positive diagonal entries $g_{i,i} > 0$ (without loss of generality). The Gram-Schmidt vectors for $\mathbf{B}$ are $\widetilde{\mathbf{b}}_i = g_{i,i} \cdot \mathbf{q}_i$. That is, $\widetilde{\mathbf{b}}_1 = \mathbf{b}_1$, and $\widetilde{\mathbf{b}}_i$ is the component of $\mathbf{b}_i$ orthogonal to the linear span of $\mathbf{b}_1, \ldots, \mathbf{b}_{i-1}$. The Gram-Schmidt orthogonalization can be computed efficiently in a corresponding iterative manner.

Let $\mathbf{B} = \mathbf{QG}$ be the Gram-Schmidt orthogonalization of $\mathbf{B}$. Without loss of generality we can assume that $\mathbf{B}$ is *size-reduced*, i.e., that $|g_{i,j}| \leq g_{i,i}/2$ for every $i < j$. This condition can be achieved efficiently by the following process: for each $j = 1, \ldots, n$, and for each $i = j - 1, \ldots, 1$, replace $\mathbf{b}_j$ by $\mathbf{b}_j - \lfloor g_{i,j}/g_{i,i} \rceil \cdot \mathbf{b}_i$. Note that the size reduction process leaves the lattice $\mathcal{L}(\mathbf{B})$ and Gram-Schmidt vectors $\widetilde{\mathbf{b}}_i = g_{i,i} \cdot \mathbf{q}_i$ unchanged. Note also that $\|\mathbf{g}_i\| \leq \sqrt{n} \cdot \max_i g_{i,i}$, by the Pythagorean theorem.

**Lemma 4.** *Let $\mathbf{B} \in \mathbb{R}^{n \times n}$ be a size-reduced nonsingular matrix. We have*

$$s_1(\mathbf{B}) \leq \sqrt{n} \cdot \sqrt{\sum_{i \in [n]} \|\widetilde{\mathbf{b}}_i\|^2} \leq n \cdot \|\widetilde{\mathbf{B}}\|.$$

The lemma is tight up to a constant factor, which may be seen by considering the right-triangular matrix with 1s on the diagonal and $1/2$ in every entry above the diagonal.

*Proof.* Let $\mathbf{B}$ have Gram-Schmidt orthogonalization $\mathbf{B} = \mathbf{QG}$. We have

$$s_1(\mathbf{B}) = \max_{\mathbf{x}} \|\mathbf{B}^t \mathbf{x}\| = \max_{\mathbf{x}} \|\mathbf{G}^t \mathbf{x}\| \leq \sqrt{\sum_{i \in [n]} (\sqrt{n} \cdot g_{i,i})^2} = \sqrt{n} \cdot \sqrt{\sum_{i \in [n]} g_{i,i}^2},$$

where the maxima are taken over all unit vectors $\mathbf{x} \in \mathbb{R}^n$, the second equality uses the fact that $\mathbf{Q}$ is orthogonal, and the first inequality is by Cauchy-Schwarz.

## 5.2    Bases for Cryptographic Lattices

Ajtai [2] gave a procedure for generating a uniformly random $q$-ary lattice from a certain family of 'worst-case-hard' cryptographic lattices, together with a relatively short basis $\mathbf{B}$. Alwen and Peikert [4] recently improved and extended the construction to yield asymptotically optimal bounds on $\|\mathbf{B}\| = \max_i \|\mathbf{b}_i\|$ and $\|\widetilde{\mathbf{B}}\| = \max_i \|\widetilde{\mathbf{b}}_i\|$. Here we show that with a small modification, one of the constructions of [4] yields (with overwhelming probability) a basis whose largest singular value is within an $O(\sqrt{\log q})$ factor of $\|\widetilde{\mathbf{B}}\|$. It follows that our efficient Gaussian sampling algorithm is essentially as tight as the GPV algorithm on such bases. Due to space restrictions, we state the main result here; the proof may be found in the full version.

**Lemma 5.** *The (slightly modified) construction of [4, Section 3.2] outputs a basis* $\mathbf{B}$ *such that* $s_1(\mathbf{B}) = O(\sqrt{\log q}) \cdot \|\widetilde{\mathbf{B}}\|$ *with overwhelming probability.*

### 5.3   Gaussian-Distributed Bases

Here we show that for a lattice basis generated by choosing its vectors from a discrete Gaussian distribution over the lattice (following by some post-processing), the largest singular value $s_1(\mathbf{B})$ of the resulting basis is essentially the same as the maximal Gram-Schmidt length $\|\widetilde{\mathbf{B}}\|$ (with high probability). Such a bound is important because applications that use 'basis delegation,' such as the hierarchical ID-based encryption schemes of [6, 1], generate random bases in exactly the manner just described.

Due to space restrictions, we state the main theorem here; the proof may be found in the full version.

**Theorem 4.** *With overwhelming probability, the* RandBasis *algorithm of [6] outputs a basis* $\mathbf{T}$ *such that* $\|\widetilde{\mathbf{T}}\| \geq s \cdot \Omega(\sqrt{n})$, *and for any* $\omega(\sqrt{\log n})$ *function,* $s_1(\mathbf{T}) \leq s \cdot O(\sqrt{n}) \cdot \omega(\sqrt{\log n})$. *In particular,* $s_1(\mathbf{T})/\|\widetilde{\mathbf{T}}\| = \omega(\sqrt{\log n})$.

*Acknowledgments.* The author thanks Phong Nguyen and the anonymous CRYPTO'10 reviewers for helpful comments.

# References

[1] Agrawal, S., Boneh, D., Boyen, X.: Efficient lattice (H)IBE in the standard model. In: Gilbert, H. (ed.) EUROCRYPT 2010. LNCS, vol. 6110, pp. 553–572. Springer, Heidelberg (2010)

[2] Ajtai, M.: Generating hard instances of the short basis problem. In: Wiedermann, J., Van Emde Boas, P., Nielsen, M. (eds.) ICALP 1999. LNCS, vol. 1644, pp. 1–9. Springer, Heidelberg (1999)

[3] Ajtai, M.: Generating hard instances of lattice problems. Quaderni di Matematica 13, 1–32 (2004); Preliminary version in STOC 1996 (1996)

[4] Alwen, J., Peikert, C.: Generating shorter bases for hard random lattices. In: STACS, pp. 75–86 (2009)

[5] Babai, L.: On Lovász' lattice reduction and the nearest lattice point problem. Combinatorica 6(1), 1–13 (1986)

[6] Cash, D., Hofheinz, D., Kiltz, E., Peikert, C.: Bonsai trees, or how to delegate a lattice basis. In: Gilbert, H. (ed.) EUROCRYPT 2010. LNCS, vol. 6110, pp. 523–552. Springer, Heidelberg (2010)

[7] Gentry, C.: Fully homomorphic encryption using ideal lattices. In: STOC, pp. 169–178 (2009)

[8] Gentry, C., Peikert, C., Vaikuntanathan, V.: Trapdoors for hard lattices and new cryptographic constructions. In: STOC, pp. 197–206 (2008)

[9] Hoffstein, J., Howgrave-Graham, N., Pipher, J., Silverman, J.H., Whyte, W.: NTRUSIGN: Digital signatures using the NTRU lattice. In: Joye, M. (ed.) CT-RSA 2003. LNCS, vol. 2612, pp. 122–140. Springer, Heidelberg (2003)

[10] Hoffstein, J., Pipher, J., Silverman, J.H.: NTRU: A ring-based public key cryptosystem. In: Buhler, J.P. (ed.) ANTS 1998. LNCS, vol. 1423, pp. 267–288. Springer, Heidelberg (1998)

[11] Klein, P.N.: Finding the closest lattice vector when it's unusually close. In: SODA, pp. 937–941 (2000)

[12] Kshirsagar, A.M.: Bartlett decomposition and Wishart distribution. The Annals of Mathematical Statistics 30(1), 239–241 (1959), http://www.jstor.org/stable/2237140

[13] Lyubashevsky, V., Micciancio, D.: Asymptotically efficient lattice-based digital signatures. In: Canetti, R. (ed.) TCC 2008. LNCS, vol. 4948, pp. 37–54. Springer, Heidelberg (2008)

[14] Lyubashevsky, V., Peikert, C., Regev, O.: On ideal lattices and learning with errors over rings. In: Gilbert, H. (ed.) EUROCRYPT 2010. LNCS, vol. 6110, pp. 1–23. Springer, Heidelberg (2010)

[15] Malkin, T., Peikert, C., Servedio, R.A., Wan, A.: Learning an overcomplete basis: Analysis of lattice-based signatures with perturbations (2009) (manuscript)

[16] Micciancio, D.: Generalized compact knapsacks, cyclic lattices, and efficient one-way functions. Computational Complexity 16(4), 365–411 (2007); Preliminary version in FOCS 2002 (2002)

[17] Micciancio, D., Regev, O.: Worst-case to average-case reductions based on Gaussian measures. SIAM J. Comput. 37(1), 267–302 (2007); Preliminary version in FOCS 2004 (2004)

[18] Nguyen, P.Q., Regev, O.: Learning a parallelepiped: Cryptanalysis of GGH and NTRU signatures. J. Cryptology 22(2), 139–160 (2009); Preliminary version in Eurocrypt 2006 (2006)

[19] Peikert, C., Vaikuntanathan, V.: Noninteractive statistical zero-knowledge proofs for lattice problems. In: Wagner, D. (ed.) CRYPTO 2008. LNCS, vol. 5157, pp. 536–553. Springer, Heidelberg (2008)

[20] Regev, O.: On lattices, learning with errors, random linear codes, and cryptography. J. ACM 56(6) (2009); Preliminary version in STOC 2005 (2005)

[21] Stehlé, D., Steinfeld, R., Tanaka, K., Xagawa, K.: Efficient public key encryption based on ideal lattices. In: Matsui, M. (ed.) ASIACRYPT 2009. LNCS, vol. 5912, pp. 617–635. Springer, Heidelberg (2009)

# Lattice Basis Delegation in Fixed Dimension and Shorter-Ciphertext Hierarchical IBE

Shweta Agrawal[1], Dan Boneh[2,*], and Xavier Boyen[3]

[1] University of Texas, Austin
[2] Stanford University
[3] Université de Liège, Belgium

**Abstract.** We present a technique for delegating a short lattice basis that has the advantage of keeping the lattice dimension unchanged upon delegation. Building on this result, we construct two new hierarchical identity-based encryption (HIBE) schemes, with and without random oracles. The resulting systems are very different from earlier lattice-based HIBEs and in some cases result in shorter ciphertexts and private keys. We prove security from classic lattice hardness assumptions.

## 1 Introduction

Hierarchical identity based encryption (HIBE) is a public key encryption scheme where entities are arranged in a directed tree [HL02, GS02]. Each entity in the tree is provided with a secret key from its parent and can delegate this secret key to its children so that a child entity can decrypt messages intended for it, or for its children, but cannot decrypt messages intended for any other nodes in the tree. This delegation process is one-way: a child node cannot use its secret key to recover the key of its parent or its siblings. We define HIBE more precisely in the next section.

The first HIBE constructions, with and without random oracles, were based on bilinear maps [GS02, BB04, BW06, BBG05, GH09, Wat09]. More recent constructions are based on hard problems on lattices [CHKP10, ABB10] where the secret key is a "short" basis $B$ of a certain integer lattice $L$. To delegate the key to a child the parent creates a new lattice $L'$ derived from $L$ and uses $B$ to generate a random short basis for this lattice $L'$. In all previous constructions the dimension of the child lattice $L'$ is larger than the dimension of the parent lattice $L$. As a result, private keys and ciphertexts become longer and longer as one descends into the hierarchy.

**Our results.** We first propose a new delegation mechanism that operates "in place", i.e., without increasing the dimension of the lattices involved. We then use this delegation mechanism to construct two HIBE systems where the lattices used have the same dimension for all nodes in the hierarchy. Consequently, private keys and ciphertexts are the same length for all nodes in the hierarchy. Our

---

* Supported by NSF and the Packard Foundation.

T. Rabin (Ed.): CRYPTO 2010, LNCS 6223, pp. 98–115, 2010.

first construction, in Section 4, provides an efficient HIBE system in the random oracle model. Our second construction, in Section 5, provides selective security in the standard model, namely without random oracles. We prove security of both constructions using the classic learning with errors (LWE) problem [Reg09].

To briefly explain our delegation technique, let $L$ be a lattice in $\mathbb{Z}^m$ and let $B = \{b_1, \ldots, b_m\}$ be a short basis of $L$. Let $R$ be a public non-singular matrix in $\mathbb{Z}^{m \times m}$. Observe that the set $B' := \{R b_1, \ldots, R b_m\}$ is a basis of the lattice $L' := RL$. If all entries of the matrix $R$ are "small" scalars then the norm of the vectors in $B'$ is not much larger than the norm of vectors in $B$. Moreover, using standard tools we can "randomize" the basis without increasing the norm of the vectors by much. The end result is a random short basis of $L'$. This idea suggests that by associating a public "low norm" matrix $R$ to each child, the parent node can delegate its short basis $B$ to a child by multiplying the vectors in $B$ by the matrix $R$ and randomizing the resulting basis. Note that since the dimension of $L'$ is the same as the dimension of $L$ this delegation does not increase dimensions.

The question is whether delegation is one way: can a child $L'$ recover a short basis of the parent $L$? More precisely, given a "low norm" matrix $R$ and a random short basis of $L'$, can one construct short vectors in $R^{-1}L'$? The key ingredient in proving security is an algorithm called SampleRwithBasis that given a lattice $L$ (for which no short basis is given) outputs a "low norm" matrix $R$ along with a short basis for the lattice $L' = RL$. In other words, if we are allowed to choose a low norm $R$ then we can build a delegated lattice $L' = RL$ for which a short basis is known even though no short basis is given for $L$. Algorithm SampleRwithBasis shows that if it were possible to use a random short basis of $L'$ to generate short vectors in $L$ then it would be possible to solve SVP in any lattice $L$ — generate an $L' = RL$ with a known short basis and use that basis to generate short vectors in $L$. More importantly, the algorithm enables us to publish matrices $R$ so that during the HIBE simulation certain private keys are known to the simulator while others are not. The key technical challenge is showing that these simulated matrices $R$ are distributed as in the real system.

**Comparison to other lattice-based HIBE.** Table 1 shows how the HIBE systems derived from our basis delegation mechanism compare to existing lattice-based HIBE systems. In the random oracle model the construction compares favorably to other lattice-based HIBE in terms of ciphertext and private key size. In terms of computation time, the encryptor in our system computes an $\ell$-wise matrix product when encrypting to an identity at depth $\ell$, which is not necessary in [CHKP10]. However, this product is not message dependent and need only be computed once per identity.

Our construction in the standard model treats identities at each level as $k$-bit binary strings. Table 1 shows that the construction is only competitive with existing HIBEs [CHKP10, ABB10] in applications where $k < \ell$ (such as [CHK07] where $k = 1$). When $k > \ell$ the construction is not competitive due to the $k^2$ term in the ciphertext length (compared to $k\ell$ in [CHKP10] and $\ell$ in [ABB10]). Nevertheless, this HIBE is very different from the existing HIBEs and the techniques of [ABB10] can potentially be applied to improve its performance.

**Table 1.** A comparison of lattice-based HIBE schemes

| selective secure HIBE | ciphertext length | secret key length | pub. params. length | lattice dimension | security $n/\alpha$ |
|---|---|---|---|---|---|
| [CHKP10] with RO | $\tilde{O}(\ell n d^2)$ | $\tilde{O}(\ell^3 n^2 d^2)$ | $\tilde{O}(n^2 d^2)$ | $\tilde{O}(\ell d n)$ | $\tilde{O}(d^d n^{d/2})$ |
| Sec. 4 with RO | $\tilde{O}(n d^2)$ | $\tilde{O}(\ell n^2 d^2)$ | $\tilde{O}(n^2 d^2)$ | $\tilde{O}(dn)$ | $\tilde{O}((dn)^{\frac{3}{2}d})$ |
| [CHKP10] no RO | $\tilde{O}(k\ell n d^2)$ | $\tilde{O}(k^2\ell^3 n^2 d^2)$ | $\tilde{O}(kn^2 d^3)$ | $\tilde{O}(k\ell d n)$ | $\tilde{O}(d^d (kn)^{d/2})$ |
| [ABB10] no RO | $\tilde{O}(\ell n d^2)$ | $\tilde{O}(\ell^3 n^2 d^2)$ | $\tilde{O}(n^2 d^3)$ | $\tilde{O}(\ell d n)$ | $\tilde{O}(d^d n^{d/2})$ |
| Sec. 5 no RO | $\tilde{O}(k^2 n d^2)$ | $\tilde{O}(k^3\ell n^2 d^2)$ | $\tilde{O}(k^3 n^2 d^3)$ | $\tilde{O}(kdn)$ | $\tilde{O}((kdn)^{kd+\frac{d}{2}})$ |

The table compares the lengths of the ciphertext, private key, and lattice dimension. We let $n$ be the security parameter, $d$ be the maximum hierarchy depth (determined at setup time), and $\ell$ be the depth of the identity in question. When appropriate we let $k$ be the number of bits in each component of the identity. The last column shows the SVP approximation factor that needs to be hard in the worst-case for the systems to be secure (the smaller the better). We focus on selectively secure HIBE since for all known adaptive lattice HIBE security degrades exponentially in the hierarchy depth.

**Relation to bilinear map constructions.** The recent lattice-based IBE and HIBE systems are closely related to their bilinear map counterparts and there may exist an abstraction that unifies these constructions. While the mechanics are quite different the high level structure is similar. The construction and proof of security in [CHKP10] resembles the tree construction of Canetti et al. [CHK07]. The construction and proof of security in [ABB10] resembles the constructions of Boneh and Boyen [BB04] and Waters [Wat05]. The constructions in this paper have some relation to the HIBE of Boneh, Boyen, and Goh [BBG05], although the relation is not as direct. Waters [Wat09] recently proposed dual-encryption as a method to build fully secure HIBE systems from bilinear maps. It is a beautiful open problem to construct a lattice analog of that result using either the basis delegation in this paper or the method from [CHKP10]. It is quite possible that two lattice-based dual-encryption HIBE systems will emerge.

# 2   Preliminaries

**Notation.** Throughout the paper we say that a function $\epsilon : \mathbb{R}_{\geq 0} \to \mathbb{R}_{\geq 0}$ is negligible if $\epsilon(n)$ is smaller than all polynomial fractions for sufficiently large $n$. We say that an event happens with overwhelming probability if it happens with probability at least $1 - \epsilon(n)$ for some negligible function $\epsilon$. We say that integer vectors $v_1, \ldots, v_n \in \mathbb{Z}^m$ are $\mathbb{Z}_q$-linearly independent for prime $q$ if they are linearly independent when reduced modulo $q$.

## 2.1   Hierarchical IBE

Recall that an Identity-Based Encryption system (IBE) consists of four algorithms [Sha85, BF01]: Setup, Extract, Encrypt, Decrypt. The Setup algorithm generates system parameters, denoted by PP, and a master key MK. The Extract algorithm uses the master key to extract a private key corresponding to a given identity. The encryption algorithm encrypts messages for a given identity and the decryption algorithm decrypts ciphertexts using the private key.

In a Hierarchical IBE [HL02, GS02], identities are vectors, and there is a fifth algorithm called Derive. Algorithm Derive takes an identity $\text{id} = (\text{id}_1, \ldots, \text{id}_k)$ at depth $k$ and a private key $\text{SK}_{\text{id}|\ell}$ of a parent identity $\text{id}_{|\ell} = (\text{id}_1, \ldots, \text{id}_\ell)$ for some $\ell < k$. It outputs the private key $\text{SK}_{\text{id}}$ for the identity id which is distributed the same as the output of Extract for id.

**Selective and Adaptive ID Security.** The standard IBE security model of [BF01] allows an attacker to adaptively choose the identity it wishes to attack. A weaker notion of IBE called selective security [CHK07] forces the adversary to announce ahead of time the public key it will target. We use both notions, but restrict the adversary to chosen-plaintext attacks.

**Security Game.** We define HIBE security using a game that captures a strong privacy property called *indistinguishable from random* which means that the challenge ciphertext is indistinguishable from a random element in the ciphertext space. This property implies both semantic security and recipient anonymity, and also implies that the ciphertext hides the public parameters (PP) used to create it. For a security parameter $\lambda$, we let $\mathcal{M}_\lambda$ denote the message space and let $\mathcal{C}_\lambda$ denote the ciphertext space. The selective security game, for a hierarchy of maximum depth $d$, proceeds as follows.

**Init:** The adversary is given the maximum depth of the hierarchy $d$ and outputs a target identity $\text{id}^* = (\mathsf{I}_1^*, \ldots, \mathsf{I}_k^*), k \leq d$.

**Setup:** The challenger runs $\mathsf{Setup}(1^\lambda, 1^d)$ (where $d = 1$ for IBE) and gives the adversary the resulting system parameters PP.

**Phase 1:** The adversary adaptively issues queries on identities $\text{id}_1, \text{id}_2, \ldots$ where no query is for a prefix of $\text{id}^*$. For each query the challenger runs algorithm Extract to obtain a private key $d_i$ for the public key $\text{id}_i$ and sends $d_i$ to the adversary.

**Challenge:** Once the adversary decides that Phase 1 is over it outputs a plaintext $M \in \mathcal{M}_\lambda$ on which it wishes to be challenged. The challenger chooses a random bit $r \in \{0, 1\}$ and a random ciphertext $C \in \mathcal{C}_\lambda$. If $r = 0$ it sets the challenge ciphertext to $C^* := \mathsf{Encrypt}(\mathsf{PP}, \text{id}^*, M)$. If $r = 1$ it sets the challenge ciphertext to $C^* := C$. It sends $C^*$ as the challenge to the adversary.

**Phase 2:** The adversary issues additional adaptive private key queries as in phase 1 and the challenger responds as before.

**Guess:** Finally, the adversary outputs a guess $r' \in \{0, 1\}$ and wins if $r = r'$.

We refer to such an adversary $\mathcal{A}$ as an INDr–sID-CPA adversary and define its advantage in attacking $\mathcal{E}$ as    $Adv_{\mathcal{E},\mathcal{A},d}(\lambda) = \left| \Pr[r = r'] - 1/2 \right|.$

**Definition 1.** *A depth d HIBE system $\mathcal{E}$ is selective-identity, indistinguishable from random if for all INDr–sID-CPA PPT adversaries $\mathcal{A}$ the function $Adv_{\mathcal{E},\mathcal{A},d}(\lambda)$ is negligible. We say that $\mathcal{E}$ is INDr–sID-CPA secure for depth d.*

We define the adaptive-identity counterparts to the above notions by removing the Init phase from the attack game, and allowing the adversary to wait until the Challenge phase to announce the identity $id^*$ it wishes to attack. The adversary is allowed to make arbitrary private-key queries in Phase 1 and then choose an arbitrary target $id^*$ as long as he did not issue a private-key query for a prefix of $id^*$ in phase 1. The resulting security notion is defined using the modified game as in Definition 1, and is denoted INDr–ID-CPA.

## 2.2   Statistical Distance

Let $X$ and $Y$ be two random variables taking values in some finite set $\Omega$. Define the *statistical distance*, denoted $\Delta(X;Y)$, as

$$\Delta(X;Y) := \frac{1}{2} \sum_{s \in \Omega} \left| \Pr[X = s] - \Pr[Y = s] \right|$$

We say that $X$ is $\delta$-uniform over $\Omega$ if $\Delta(X;U_\Omega) \leq \delta$ where $U_\Omega$ is a uniform random variable over $\Omega$. Two ensembles of random variables $X(\lambda)$ and $Y(\lambda)$ are statistically close if $d(\lambda) := \Delta(X(\lambda);Y(\lambda))$ is a negligible function of $\lambda$.

## 2.3   Integer Lattices

**Definition 2.** *Let $B = \begin{bmatrix} b_1 & | & \cdots & | & b_m \end{bmatrix} \in \mathbb{R}^{m \times m}$ be an $m \times m$ matrix whose columns are linearly independent vectors $b_1, \ldots, b_m \in \mathbb{R}^m$. The m-dimensional full-rank lattice $\Lambda$ generated by B is the set,*

$$\Lambda = \mathcal{L}(B) = \left\{ y \in \mathbb{R}^m \quad s.t. \quad \exists s \in \mathbb{Z}^m , \quad y = B s = \sum_{i=1}^{m} s_i b_i \right\}$$

Here, we are interested in integer lattices, i.e, when $L$ is contained in $\mathbb{Z}^m$. We let $\det(\Lambda)$ denote the determinant of $\Lambda$.

**Definition 3.** *For $q$ prime, $A \in \mathbb{Z}_q^{n \times m}$ and $u \in \mathbb{Z}_q^n$, define:*

$$\Lambda_q(A) := \left\{ e \in \mathbb{Z}^m \quad s.t. \quad \exists s \in \mathbb{Z}_q^n \ where \ A^\top s = e \pmod q \right\}$$
$$\Lambda_q^\perp(A) := \left\{ e \in \mathbb{Z}^m \quad s.t. \quad A e = 0 \pmod q \right\}$$
$$\Lambda_q^u(A) := \left\{ e \in \mathbb{Z}^m \quad s.t. \quad A e = u \pmod q \right\}$$

Observe that if $t \in \Lambda_q^u(A)$ then $\Lambda_q^u(A) = \Lambda_q^\perp(A) + t$ and hence $\Lambda_q^u(A)$ is a shift of $\Lambda_q^\perp(A)$ .

## 2.4   The Gram-Schmidt Norm of a Basis

Let $S$ be a set of vectors $S = \{s_1, \ldots, s_k\}$ in $\mathbb{R}^m$. We use the following standard notation:

- $\|S\|$ denotes the $L_2$ length of the longest vector in $S$, i.e. $\max_{1 \leq i \leq k} \|s_i\|$.
- $\tilde{S} := \{\tilde{s}_1, \ldots, \tilde{s}_k\} \subset \mathbb{R}^m$ denotes the Gram-Schmidt orthogonalization of the vectors $s_1, \ldots, s_k$ taken in that order.

We refer to $\|\tilde{S}\|$ as the Gram-Schmidt norm of $S$.

Micciancio and Goldwassser [MG02] showed that a full-rank set $S$ in a lattice $\Lambda$ can be converted into a basis $T$ for $\Lambda$ with an equally low Gram-Schmidt norm.

**Lemma 1 ([MG02, Lemma 7.1]).** *Let $\Lambda$ be an $m$-dimensional lattice. There is a deterministic polynomial-time algorithm that, given an arbitrary basis of $\Lambda$ and a full-rank set $S = \{s_1, \ldots, s_m\}$ in $\Lambda$, returns a basis $T$ of $\Lambda$ satisfying*

$$\|\tilde{T}\| \leq \|\tilde{S}\| \quad and \quad \|T\| \leq \|S\| \sqrt{m}/2$$

Ajtai [Ajt99] and later Alwen and Peikert [AP09] show how to sample an essentially uniform matrix $A \in \mathbb{Z}_q^{n \times m}$ with an associated basis $S_A$ of $\Lambda_q^{\perp}(A)$ with low Gram-Schmidt norm. The following follows from Theorem 3.2 of [AP09] taking $\delta := 1/3$. The theorem produces a matrix $A$ statistically close to uniform in $\mathbb{Z}_q^{n \times m}$ along with a short basis. Since $m$ is so much larger than $n$, the matrix $A$ is rank $n$ with overwhelming probability and we can state the theorem as saying that $A$ is statistically close to a uniform rank $n$ matrix in $\mathbb{Z}_q^{n \times m}$.

**Theorem 1.** *Let $q \geq 3$ be odd and $m := \lceil 6n \log q \rceil$. There is a probabilistic polynomial-time algorithm $\mathsf{TrapGen}(q, n)$ that outputs a pair $(A \in \mathbb{Z}_q^{n \times m}, S \in \mathbb{Z}^{m \times m})$ such that $A$ is statistically close to a uniform rank $n$ matrix in $\mathbb{Z}_q^{n \times m}$ and $S$ is a basis for $\Lambda_q^{\perp}(A)$ satisfying*

$$\|\tilde{S}\| \leq O(\sqrt{n \log q}) \quad and \quad \|S\| \leq O(n \log q)$$

*with all but negligible probability in $n$.*

**Notation:** We let $\tilde{L}_{\mathrm{TG}} := O(\sqrt{n \log q})$ denote the maximum (w.h.p) Gram-Schmidt norm of a basis produced by $\mathsf{TrapGen}(q, n)$.

## 2.5   Discrete Gaussians

**Definition 4.** *Let $L$ be a subset of $\mathbb{Z}^m$. For any vector $c \in \mathbb{R}^m$ and any positive parameter $\sigma \in \mathbb{R}_{>0}$, define:*

$$\rho_{\sigma,c}(x) = \exp\left(-\pi \frac{\|x - c\|^2}{\sigma^2}\right) \quad and \quad \rho_{\sigma,c}(L) = \sum_{x \in L} \rho_{\sigma,c}(x)$$

*The discrete Gaussian distribution over L with center c and parameter $\sigma$ is*

$$\forall y \in L \quad , \quad \mathcal{D}_{L,\sigma,c}(y) = \frac{\rho_{\sigma,c}(y)}{\rho_{\sigma,c}(L)}$$

For notational convenience, $\rho_{\sigma,0}$ and $\mathcal{D}_{L,\sigma,0}$ are abbreviated as $\rho_\sigma$ and $\mathcal{D}_{L,\sigma}$. When $\sigma = 1$ we write $\rho$ to denote $\rho_1$.                                               □

The distribution $\mathcal{D}_{L,\sigma,c}$ will most often be defined over the lattice $L = \Lambda_q^\perp(A)$ for a matrix $A \in \mathbb{Z}_q^{n \times m}$ or over a coset $L = t + \Lambda_q^\perp(A)$ where $t \in \mathbb{Z}^m$.

**Properties.** The following lemma from [Pei] captures standard properties of these distributions. The first property follows from Lemma 4.4 of [MR07]. The last two properties are algorithms from [GPV08].

**Lemma 2.** *Let $q \geq 2$ and let $A$ be a matrix in $\mathbb{Z}_q^{n \times m}$ with $m > n$. Let $T_A$ be a basis for $\Lambda_q^\perp(A)$ and $\sigma \geq \|\widetilde{T_A}\| \, \omega(\sqrt{\log m})$. Then for $c \in \mathbb{R}^m$ and $u \in \mathbb{Z}_q^n$:*

1. $\Pr\left[\, x \sim \mathcal{D}_{\Lambda_q^u(A),\sigma} \; : \; \|x\| > \sqrt{m}\,\sigma \,\right] \;\leq\; \mathrm{negl}(n)$.
2. *There is a PPT algorithm* SampleGaussian$(A, T_A, \sigma, c)$ *that returns $x \in \Lambda_q^\perp(A)$ drawn from a distribution statistically close to $\mathcal{D}_{\Lambda,\sigma,c}$.*
3. *There is a PPT algorithm* SamplePre$(A, T_A, u, \sigma)$ *that returns $x \in \Lambda_q^u(A)$ sampled from a distribution statistically close to $\mathcal{D}_{\Lambda_q^u(A),\sigma}$, whenever $\Lambda_q^u(A)$ is not empty.*

**Randomizing a basis:** Cash et al. [CHKP10] show how to randomize a lattice basis (see also [GN08, Sec. 2.1]).

RandBasis$(S, \sigma)$:
On input a basis $S$ of an $m$-dimensional lattice $\Lambda_q^\perp(A)$ and a gaussian parameter $\sigma \geq \|\widetilde{S}\| \cdot \omega(\sqrt{\log n})$, outputs a new basis $S'$ of $\Lambda_q^\perp(A)$ such that

- with overwhelming probability $\|\widetilde{S'}\| \leq \sigma\sqrt{m}$, and
- up to a statistical distance, the distribution of $S'$ does not depend on $S$. That is, the random variable RandBasis$(S, \sigma)$ is statistically close to RandBasis$(T, \sigma)$ for any other basis $T$ of $\Lambda_q^\perp(A)$ satisfying $\|\widetilde{T}\| \leq \sigma/\omega(\sqrt{\log n})$.

We briefly recall how RandBasis works:

1. For $i = 1, \ldots, m$, let $v \leftarrow$ SampleGaussian$(A, S, \sigma, 0)$ and
       if $v$ is independent of $\{v_1, \ldots, v_{i-1}\}$, set $v_i \leftarrow v$, if not, repeat.
2. Convert the set of vectors $v_1, \ldots, v_m$ to a basis $S'$ using Lemma 1 (and using some canonical basis of $\Lambda_q^\perp(A)$).
3. Output $S'$.

The analysis of RandBasis in [CHKP10] uses [Reg09, Corollary 3.16] which shows that a linearly independent set is produced in Step (1) w.h.p. after $m^2$ samples from SampleGaussian$(A, S, \sigma, 0)$. It is not difficult to show that only $2m$ samples are needed in expectation.

## 2.6   Hardness Assumption

Security of all our constructions reduces to the LWE (learning with errors) problem, a classic hard problem on lattices defined by Regev [Reg09].

**Definition 5.** *Consider a prime $q$, a positive integer $n$, and a distribution $\chi$ over $\mathbb{Z}_q$, all public. An $(\mathbb{Z}_q, n, \chi)$-LWE problem instance consists of access to an unspecified challenge oracle $\mathcal{O}$, being, either, a noisy pseudo-random sampler $\mathcal{O}_s$ carrying some constant random secret key $s \in \mathbb{Z}_q^n$, or, a truly random sampler $\mathcal{O}_\$$, whose behaviors are respectively as follows:*

$\mathcal{O}_s$: *outputs samples of the form $(u_i, v_i) = \left(u_i, \; u_i^\top s + x_i\right) \in \mathbb{Z}_q^n \times \mathbb{Z}_q$, where, $s \in \mathbb{Z}_q^n$ is a uniformly distributed persistent value invariant across invocations, $x_i \in \mathbb{Z}_q$ is a fresh sample from $\chi$, and $u_i$ is uniform in $\mathbb{Z}_q^n$.*
$\mathcal{O}_\$$: *outputs truly uniform random samples from $\mathbb{Z}_q^n \times \mathbb{Z}_q$.*

*The $(\mathbb{Z}_q, n, \chi)$-LWE problem allows repeated queries to the challenge oracle $\mathcal{O}$. We say that an algorithm $\mathcal{A}$ decides the $(\mathbb{Z}_q, n, \chi)$-LWE problem if*

$$\textsf{LWE-adv}[\mathcal{A}] := \left| \Pr[\mathcal{A}^{\mathcal{O}_s} = 1] - \Pr[\mathcal{A}^{\mathcal{O}_\$} = 1] \right|$$

*is non-negligible for a random $s \in \mathbb{Z}_q^n$.*

Regev [Reg09] shows that for certain noise distributions $\chi$, denoted $\overline{\Psi}_\alpha$, the LWE problem is as hard as the worst-case SIVP and GapSVP under a quantum reduction (see also [Pei09]). Recall that for $x \in \mathbb{R}$ the symbol $\lfloor x \rceil$ denotes the closest integer to $x$.

**Definition 6.** *For an $\alpha \in (0, 1)$ and a prime $q$ let $\overline{\Psi}_\alpha$ denote the distribution over $\mathbb{Z}_q$ of the random variable $\lfloor q X \rceil \bmod q$ where $X$ is a normal random variable with mean 0 and standard deviation $\alpha/\sqrt{2\pi}$.*

**Theorem 2 ([Reg09]).** *If there exists an efficient, possibly quantum, algorithm for deciding the $(\mathbb{Z}_q, n, \overline{\Psi}_\alpha)$-LWE problem for $q > 2\sqrt{n}/\alpha$ then there exists an efficient quantum algorithm for approximating the SIVP and GapSVP problems, to within $\tilde{O}(n/\alpha)$ factors in the $\ell_2$ norm, in the worst case.*

The following lemma about the distribution $\overline{\Psi}_\alpha$ will be needed to show that decryption works correctly. The proof is implicit in [GPV08, Lemma 8.2].

**Lemma 3.** *Let $e$ be some vector in $\mathbb{Z}^m$ and let $y \overset{R}{\leftarrow} \overline{\Psi}_\alpha^m$. Then the quantity $|e^\top y|$ treated as an integer in $[0, q-1]$ satisfies*

$$|e^\top y| \leq \|e\| q \alpha \omega(\sqrt{\log m}) + \|e\| \sqrt{m}/2$$

*with all but negligible probability in $m$.*

As a special case, Lemma 3 shows that if $x \overset{R}{\leftarrow} \overline{\Psi}_\alpha$ is treated as an integer in $[0, q-1]$ then $|x| < q\alpha \, \omega(\sqrt{\log m}) + 1/2$ with all but negligible probability in $m$.

# 3    Basis Delegation without Dimension Increase

Let $A$ be a matrix in $\mathbb{Z}_q^{n \times m}$ and let $T_A$ be a "short" basis of $\Lambda_q^\perp(A)$, both given. We wish to "delegate" the basis $T_A$ in the following sense: we want to deterministically generate a matrix $B$ from $A$ and a random basis $T_B$ for $\Lambda_q^\perp(B)$ such that from $A, B$ and $T_B$ it is difficult to recover any short basis for $\Lambda_q^\perp(A)$. Basis delegation mechanisms were proposed by Cash et al [CHKP10] and Agrawal et al. [ABB10] where the dimension of the matrix $B$ was larger than the dimension of the given $A$. In the resulting HIBE systems ciphertext and private key sizes increase as the hierarchy deepens.

Here we consider a simple delegation mechanism that does not increase the dimension. To do so we use a public matrix $R$ in $\mathbb{Z}^{m \times m}$ where the columns of $R$ have "low" norm. We require that $R$ be invertible mod $q$. Now, define $B := AR^{-1}$ in $\mathbb{Z}_q^{n \times m}$ and observe that $B$ has the same dimension as $A$. We show how to build a "short" basis of $\Lambda_q^\perp(B)$ from which it is difficult to recover a short basis of $A$. In the next section we use this to build new HIBE systems.

We begin by defining distributions on matrices whose columns are low norm vectors. We then define the basis delegation mechanism.

**Distributions on low norm matrices.** We say that a matrix $R$ in $\mathbb{Z}^{m \times m}$ is $\mathbb{Z}_q$-**invertible** if $R \bmod q$ is invertible as a matrix in $\mathbb{Z}_q^{m \times m}$. Our construction makes use of $\mathbb{Z}_q$-invertible matrices $R$ in $\mathbb{Z}^{m \times m}$ where all the columns of $R$ are "low norm".

**Definition 7.** *Define* $\sigma_{\mathrm{R}} := \tilde{L}_{\mathrm{TG}} \, \omega(\sqrt{\log m}) = \sqrt{n \log q} \cdot \omega(\sqrt{\log m})$.

*We let* $\mathcal{D}_{m \times m}$ *denote the distribution on matrices in* $\mathbb{Z}^{m \times m}$ *defined as*

$$\left( \mathcal{D}_{\mathbb{Z}^m, \sigma_{\mathrm{R}}} \right)^m \quad \text{conditioned on the resulting matrix being } \mathbb{Z}_q\text{-invertible}$$

**Algorithm SampleR$(1^m)$.** The following simple algorithm samples matrices in $\mathbb{Z}^{m \times m}$ from a distribution that is statistically close to $\mathcal{D}_{m \times m}$.

1. Let $T$ be the canonical basis of the lattice $\mathbb{Z}^m$.
2. For $i = 1, \ldots, m$ do $r_i \xleftarrow{R}$ SampleGaussian$(\mathbb{Z}^m, T, \sigma_{\mathrm{R}}, 0)$.
3. If $R$ is $\mathbb{Z}_q$-invertible, output $R$; otherwise repeat step 2.

In the full version we show that step 2 will need to be repeated fewer than two times in expectation for prime $q$.

## 3.1    Basis Delegation: Algorithm BasisDel$(A, R, T_A, \sigma)$

We now describe a simple basis delegation algorithm that does not increase the dimension of the underlying matrices.

*Inputs:*
    a rank $n$ matrix $A$ in $\mathbb{Z}_q^{n \times m}$,
    a $\mathbb{Z}_q$-invertible matrix $R$ in $\mathbb{Z}^{m \times m}$ sampled from $\mathcal{D}_{m \times m}$
        (or a product of such),                                                    (1)
    a basis $T_A$ of $\Lambda_q^\perp(A)$,
    and a parameter $\sigma \in \mathbb{R}_{>0}$.

*Output:* Let $B := AR^{-1}$ in $\mathbb{Z}_q^{n \times m}$. The algorithm outputs a basis $T_B$ of $\Lambda_q^\perp(B)$.

**Algorithm BasisDel**$(A, R, T_A, \sigma)$works as follows:

1. Let $T_A = \{a_1, \ldots, a_m\} \subseteq \mathbb{Z}^m$. Calculate $T'_B := \{Ra_1, \ldots, Ra_m\} \subseteq \mathbb{Z}^m$.
   Observe that $T'_B$ is a set of independent vectors in $\Lambda_q^\perp(B)$.
2. Use Lemma 1 to convert $T'_B$ into a basis $T''_B$ of $\Lambda_q^\perp(B)$. The algorithm in the lemma takes as input $T'_B$ and an arbitrary basis of $\Lambda_q^\perp(B)$ and outputs a basis $T''_B$ whose Gram-Schmidt norm is no more than that of $T'_B$.
3. Call RandBasis$(T''_B, \sigma)$ and output the resulting basis $T_B$ of $\Lambda_q^\perp(B)$.

The following theorem shows that BasisDel produces a random basis of $\Lambda_q^\perp(B)$ whose Gram-Schmidt norm is bounded as a function of $\|\widetilde{T_A}\|$. The proof is given in the full version.

**Theorem 3.** *Using the notation in (1), suppose $R$ is sampled from $\mathcal{D}_{m \times m}$ and $\sigma$ satisfies*

$$\sigma > \|\widetilde{T_A}\| \cdot \sigma_{\mathrm{R}} \sqrt{m}\, \omega(\log^{3/2} m) \ .$$

*Let $T_B$ be the basis of $\Lambda_q^\perp(AR^{-1})$ output by BasisDel.*
*Then $T_B$ is distributed statistically close to the distribution RandBasis$(T, \sigma)$ where $T$ is an arbitrary basis of $\Lambda_q^\perp(AR^{-1})$ satisfying $\|\widetilde{T}\| < \sigma/\omega(\sqrt{\log m})$. If $R$ is a product of $\ell$ matrices sampled from $\mathcal{D}_{m \times m}$ then the bound on $\sigma$ degrades to $\sigma > \|\widetilde{T_A}\| \cdot \left(\sigma_{\mathrm{R}} \sqrt{m}\, \omega(\log^{1/2} m)\right)^\ell \cdot \omega(\log m)$ .*

When $R$ is a product for $\ell$ matrices sampled from $\mathcal{D}_{m \times m}$ then for the smallest possible $\sigma$ in Theorem 3 we obtain that w.h.p

$$\|\widetilde{T_B}\| / \|\widetilde{T_A}\| \leq \left(m\, \omega(\log m)\right)^\ell \sqrt{m}\, \omega(\log m) \ .$$

This quantity is the minimum degradation in basis quality as we delegate across $\ell$ levels of the HIBE hierarchy.

### 3.2   The Main Simulation Tool: Algorithm SampleRwithBasis$(A)$

All our proofs of security make heavy use of an algorithm SampleRwithBasis that given a random rank $n$ matrix $A$ in $\mathbb{Z}_q^{n \times m}$ as input generates a "low-norm" matrix $R$ (i.e., a matrix sampled from $\mathcal{D}_{m \times m}$) along with a short basis for $\Lambda_q^\perp(AR^{-1})$.

**Algorithm SampleRwithBasis$(A)$.** Let $a_1, \ldots, a_m \in \mathbb{Z}_q^n$ be the $m$ columns of the matrix $A \in \mathbb{Z}_q^{n \times m}$.

1. Run TrapGen$(q, n)$ to generate a random rank $n$ matrix $B \in \mathbb{Z}_q^{n \times m}$ and a basis $T_B$ of $\Lambda_q^\perp(B)$ such that $\|\widetilde{T_B}\| \leq \tilde{L}_{\mathrm{TG}} = \sigma_{\mathrm{R}}/\omega(\sqrt{\log m})$.
2. for $i = 1, \ldots, m$ do:
   (2a) sample $r_i \in \mathbb{Z}^m$ as the output of SamplePre$(B, T_B, a_i, \sigma_{\mathrm{R}})$,
        then $Br_i = a_i \bmod q$ and $r_i$ is sampled from a distribution statistically close to $\mathcal{D}_{\Lambda_q^{a_i}(B), \sigma_{\mathrm{R}}}$.
   (2b) repeat step (2a) until $r_i$ is $\mathbb{Z}_q$ linearly independent of $r_1, \ldots, r_{i-1}$.

3. Let $R \in \mathbb{Z}^{m \times m}$ be the matrix whose columns are $r_1, \ldots, r_m$.
   Then $R$ has rank $m$ over $\mathbb{Z}_q$. Output $R$ and $T_B$.

By construction $BR = A \bmod q$ and therefore $B = AR^{-1} \bmod q$. Hence, the basis $T_B$ is a short basis of $\Lambda_q^{\perp}(AR^{-1})$. It remains to show that $R$ is sampled from a distribution close to $\mathcal{D}_{m \times m}$.

**Theorem 4.** *Let $m > 2n \log q$ and $q > 2$ a prime. For all but at most a $q^{-n}$ fraction of rank $n$ matrices $A$ in $\mathbb{Z}_q^{n \times m}$ algorithm* SampleRwithBasis$(A)$ *outputs a matrix $R$ in $\mathbb{Z}^{m \times m}$ sampled from a distribution statistically close to $\mathcal{D}_{m \times m}$. The generated basis $T_B$ of $\Lambda_q^{\perp}(AR^{-1})$ satisfies $\|\widetilde{T_B}\| \leq \sigma_R / \omega(\sqrt{\log m})$ with overwhelming probability.*

The bound on $\|\widetilde{T_B}\|$ is from Theorem 1. The difficult part of the proof is arguing that $R$ is sampled from a distribution statistically close to $\mathcal{D}_{m \times m}$. The proof is based on a detailed analysis of the distribution from which $R$ is chosen and is given in the full version of the paper.

# 4   An HIBE in the Random-Oracle Model

Our first construction is a depth $d$ HIBE secure in the random oracle model. In the next section we describe an HIBE selectively secure in the standard model.

To encrypt a message $m$ for identity id, the encryptor builds a matrix $F_{id}$ and encrypts $m$ using the dual Regev public key system (described in [GPV08, sec. 7]) using $F_{id}$ as the public key. The matrix $F_{id}$ is built by multiplying a fixed matrix $A$, specified in the public parameters, by $\ell$ "low norm" square matrices generated by a random oracle $H$ described in (2) below.

At level $\ell$, let $id = (id_1, id_2, \ldots, id_\ell) \in (\{0,1\}^*)^\ell$, where $\ell \in [d]$. We assume the availability of a hash function $H$ that outputs matrices in $\mathbb{Z}_q^{m \times m}$:

$$H : (\{0,1\}^*)^{\leq d} \rightarrow \mathbb{Z}_q^{m \times m} : id \mapsto H(id) \sim \mathcal{D}_{m \times m} \qquad (2)$$

where the requirement is that, over the choice of the random oracle $H$, the output $H(id)$ is distributed as $\mathcal{D}_{m \times m}$ (as in Definition 7). In practice, the hash function $H$ can be built from a "standard" random function $h : (\{0,1\}^*)^{\leq d} \rightarrow \{0,1\}^t$ by using $h$ as a coin generator for the sampling process in Algorithm SampleR$(1^m)$. This method however is not indefferentiable in the sense of [CDMP05] and the analysis requires that $H$ itself be a random oracle.

## 4.1   Construction

The system uses a number of parameters that will be set in Section 4.2. The parameters $n, m$ and $q$ are fixed across the levels of the hierarchy. In addition, we have two level-dependent parameters: a guassian parameter $\bar{\sigma} = (\sigma_1, \ldots, \sigma_d)$ and a noise parameter $\bar{\alpha} = (\alpha_1, \ldots, \alpha_d)$.

For an identity $id = (id_1, \ldots, id_\ell)$ and $1 \leq k \leq \ell$ we use $id_{|k}$ to denote the vector $(id_1, \ldots, id_k)$. Now, for a hierarchy of maximum depth $d$ the scheme works as follows:

**Setup**$(1^n, 1^d)$ On input a security parameter $n$ and maximum depth $d$:

1. Invoke $\mathsf{TrapGen}(q, n)$ to generate a uniformly random matrix $A \in \mathbb{Z}_q^{n \times m}$ and a short basis $T_A = [a_1 | \ldots | a_m] \in \mathbb{Z}^{m \times m}$ for $\Lambda_q^\perp(A)$.
2. Generate a uniformly random vector $u_0 \in \mathbb{Z}_q^n$.
3. Output the public parameters $\mathsf{PP}$ and master key $\mathsf{MK}$ given by,

$$\mathsf{PP} = \left( A, \ u_0 \right) \qquad \mathsf{MK} = \left( T_A \right)$$

**Derive**$(\mathsf{PP}, \mathsf{SK}_{\mathsf{id}|\ell}, \mathsf{id})$: On input public parameters $\mathsf{PP}$, a secret key $\mathsf{SK}_{\mathsf{id}|\ell}$ corresponding to a "parent" identity $\mathsf{id}_{|\ell} = (\mathsf{id}_1, \ldots, \mathsf{id}_\ell)$, and a "child" identity $\mathsf{id} = (\mathsf{id}_1, \ldots, \mathsf{id}_\ell, \ldots \mathsf{id}_k)$ where $k \le d$ do:

1. Let $R_{\mathsf{id}|\ell} = H(\mathsf{id}_{|\ell}) \cdots H(\mathsf{id}_{|2}) H(\mathsf{id}_{|1}) \in \mathbb{Z}^{m \times m}$ and $F_{\mathsf{id}|\ell} = A R_{\mathsf{id}|\ell}^{-1}$ in $\mathbb{Z}_q^{n \times m}$. Then $\mathsf{SK}_{\mathsf{id}|\ell}$ is a short basis for $\Lambda_q^\perp(F_{\mathsf{id}|\ell})$.
2. Compute $R = H(\mathsf{id}_{|k}) \cdots H(\mathsf{id}_{|\ell+1}) \in \mathbb{Z}^{m \times m}$ and set $F_{\mathsf{id}} = F_{\mathsf{id}|\ell} R^{-1}$.
3. Evaluate $S' \leftarrow \mathsf{BasisDel}(F_{\mathsf{id}|\ell}, R, \mathsf{SK}_{\mathsf{id}|\ell}, \sigma_k)$ to obtain a short random basis for $\Lambda_q^\perp(F_{\mathsf{id}})$.
4. Output the delegated private key $\mathsf{SK}_{\mathsf{id}} = S'$.

Algorithm $\mathsf{Extract}(\mathsf{MK}, \mathsf{id})$ works the same way by running $\mathsf{Derive}(\mathsf{PP}, \mathsf{MK}, \mathsf{id})$ where $F_{\mathsf{id}|0} = A$ and $\mathsf{SK}_{\mathsf{id}|0} = \mathsf{MK}$.

**Encrypt**$(\mathsf{PP}, \mathsf{id}, b)$: On input public parameters $\mathsf{PP}$, a recipient identity $\mathsf{id}$ of depth $|\mathsf{id}| = \ell$, and a message bit $b \in \{0, 1\}$:

1. Compute $R_{\mathsf{id}} \leftarrow H(\mathsf{id}_{|\ell}) \ldots H(\mathsf{id}_{|2}) H(\mathsf{id}_{|1})$ in $\mathbb{Z}^{m \times m}$.
2. Compute the encryption matrix $F_{\mathsf{id}} \leftarrow A R_{\mathsf{id}}^{-1}$ in $\mathbb{Z}_q^{n \times m}$.
3. Now encrypt the message using Regev's dual public key encryption (as defined in [GPV08, sec. 7]) using $F_{\mathsf{id}}$ as the public key. To do so,

   (a) Pick a uniformly random vector $s \xleftarrow{R} \mathbb{Z}_q^n$.

   (b) Choose noise vectors $x \xleftarrow{\overline{\Psi}_{\alpha_\ell}} \mathbb{Z}_q$ and $y \xleftarrow{\overline{\Psi}_{\alpha_\ell}^m} \in \mathbb{Z}_q^m$. ($\overline{\Psi}_\alpha$ is as in def. 6)

   (c) Output the ciphertext,

$$\mathsf{CT} = \left( c_0 = u_0^\top s + x + b \lfloor \tfrac{q}{2} \rfloor, \ c_1 = F_{\mathsf{id}}^\top s + y \right) \quad \in \mathbb{Z}_q \times \mathbb{Z}_q^m$$

**Decrypt**$(\mathsf{PP}, \mathsf{SK}_{\mathsf{id}}, \mathsf{CT})$: On input public parameters $\mathsf{PP}$, a private key $\mathsf{SK}_{\mathsf{id}}$ for an identity $\mathsf{id}$ of length $|\mathsf{id}| = \ell$, and a ciphertext $\mathsf{CT}$:

1. Let $\tau_\ell = \sigma_\ell \sqrt{m} \, \omega(\sqrt{\log m}) \quad (\ge \|\widetilde{\mathsf{SK}_{\mathsf{id}}}\| \, \omega(\sqrt{\log m}) )$.
2. Construct the matrix $F_{\mathsf{id}} \in \mathbb{Z}_q^{n \times m}$ as in step (2) of $\mathsf{Encrypt}$.
3. Set $d_{\mathsf{id}} \leftarrow \mathsf{SamplePre}(F_{\mathsf{id}}, \mathsf{SK}_{\mathsf{id}}, u_0, \tau_\ell)$. Note that $F_{\mathsf{id}} \, d_{\mathsf{id}} = u_0$ in $\mathbb{Z}_q^n$.
4. Compute $w = c_0 - d_{\mathsf{id}}^\top c_1 \in \mathbb{Z}_q$.
5. Compare $w$ and $\lfloor \tfrac{q}{2} \rfloor$ treating them as integers in $[q] \subset \mathbb{Z}$:

   if they are close, i.e., if $\left| w - \lfloor \tfrac{q}{2} \rfloor \right| < \lfloor \tfrac{q}{4} \rfloor$ in $\mathbb{Z}$, output 1; otherwise output 0.

## 4.2   Parameters and Correctness

When the cryptosystem is operated as specified, during decryption of a cipher-text encrypted to an identity at level $\ell$ we have,

$$w = c_0 - d_{\mathsf{id}}^\top c_1 = b \lfloor \tfrac{q}{2} \rfloor + \underbrace{x - d_{\mathsf{id}}^\top y}_{\text{error term}}$$

Since $\|d_{\mathsf{id}}\| \le \tau_\ell \sqrt{m} = \sigma_\ell\, m\, \omega(\sqrt{\log m})$ w.h.p, we have by Lemma 3 that the norm of the error term is bounded w.h.p by

$$|x - d_{\mathsf{id}}^\top y| \le q\alpha_\ell\sigma_\ell m\, \omega(\log m) + \sigma_\ell m^{3/2}\, \omega(\sqrt{\log m}) \tag{3}$$

In addition, by properties of $\mathsf{RandBasis}(\cdot, \sigma_\ell)$ the Gram-Schmidt norm of a secret key $\mathsf{SK}_\ell$ at level $\ell$ satisfies w.h.p. $\|\widetilde{\mathsf{SK}_\ell}\| \le \sigma_\ell\sqrt{m}$. Therefore, with $\sigma_0 = \tilde{L}_{\mathrm{TG}}$, for the system to work correctly we need that:

- $\mathsf{TrapGen}$ can operate (i.e. $m > 6n \log q$),
- the error term in (3) is less than $q/5$ w.h.p
  (i.e. $\alpha_\ell < [\sigma_\ell m \omega(\log m)]^{-1}$ and $q > \sigma_\ell\, m^{3/2}\omega(\sqrt{\log m})$ ),
- $\mathsf{BasisDel}$ used in $\mathsf{Derive}$ can operate (i.e. $\sigma_\ell > \|\widetilde{\mathsf{SK}_{\ell-1}}\|\, \sigma_{\mathrm{R}} \sqrt{m}\, \omega(\log^{3/2} m)$
  which follows from $\sigma_\ell > \sigma_{\ell-1}\, m^{3/2}\, \omega(\log^2 m)$ ), and
- Regev's reduction applies (i.e. $q > 2\sqrt{n}/\alpha_\ell$ for all $\ell$).

To satisfy these requirements we set the parameters $(q, m, \bar{\sigma}, \bar{\alpha})$ as follows taking $n$ to be the security parameter (and letting $\ell = 1, \ldots, d$):

$$m = 6\,n^{1+\delta} = O(dn \log n) \qquad , \qquad q = m^{\frac{3}{2}d+2} \cdot \omega(\log^{2d+1} n)$$
$$\sigma_\ell = m^{\frac{3}{2}\ell+\frac{1}{2}} \cdot \omega(\log^{2\ell} n) \qquad , \qquad \alpha_\ell = [\sigma_\ell\, m\, \omega(\log n)]^{-1} \tag{4}$$

and round up $m$ to the nearest larger integer and $q$ to the nearest larger prime. Here we assume that $\delta$ is such that $n^\delta > \lceil \log q \rceil = O(d \log n)$.

Observe that since $\sigma_\ell$ is increasing with $\ell$ algorithm $\mathsf{Extract}$ generates the same distribution on private keys as algorithm $\mathsf{Derive}$ for all identities at depth greater than one, as required from our definition of HIBE.

Overall, the ciphertext size for all identities is $\tilde{O}(d^2 n)$. Security depends on the assumption that worst-case SVP cannot be solved to within a factor $q\sqrt{n} = \tilde{O}((dn)^{1.5d})$.

## 4.3   Security

We state the system's security against both a selective and an adaptive adversary. Selective security implies adaptive security in the random oracle model via a simple generic transformation from [BB04]. However, proving adaptive security directly gives a slightly simpler system. Recall that selective security in the random oracle model means that the attacker must commit to the target identity before issuing any type of query.

**Theorem 5.** *Let $\mathcal{A}$ be a PPT adversary that attacks the scheme of Section 4.1 when $H$ is modeled as a random oracle. Let $Q_H$ is the number of $H$ queries made by $\mathcal{A}$ and $d$ be the max hierarchy depth. Then there is a PPT algorithm $\mathcal{B}$ that decides LWE such that*

1. *If $\mathcal{A}$ is a selective adversary (INDr–sID-CPA) with advantage $\epsilon$ then $\epsilon \leq$ LWE-adv$[\mathcal{B}]$.*
2. *If $\mathcal{A}$ is an adaptive adversary (INDr–ID-CPA) with advantage $\epsilon$ then $\epsilon \leq$ LWE-adv$[\mathcal{B}] \cdot (dQ_H^d) + \mathrm{negl}(n)$.*

*where LWE-adv$[\mathcal{B}]$ is with respect to the parameters $(\mathbb{Z}_q, n, \overline{\Psi}_\alpha)$ from Section 4.2.*

*Proof.* We prove part (2) of the theorem. The proof of part (1) is similar and a little simpler. Recall that LWE is about recognizing an oracle $\mathcal{O}$ defined in Section 2.6. We use $\mathcal{A}$ to construct an LWE algorithm $\mathcal{B}$ with advantage about $\epsilon/dQ_H^d$.

**Instance.** $\mathcal{B}$ requests from $\mathcal{O}$ and receives, for each $i = 0, \ldots, m$, a fresh pair $(u_i, v_i) \in \mathbb{Z}_q^n \times \mathbb{Z}_q$.

As the number of oracle calls is known *a priori*, the samples can be supplied non-interactively at the beginning, e.g., here in the form of an instance with $(m + 1)(n + 1)$ elements of $\mathbb{Z}_q$.

**Setup.** $\mathcal{B}$ prepares a simulated attack environment for $\mathcal{A}$ as follows.

1. Select $d$ uniform random integer $Q_1^*, \ldots, Q_d^* \in [Q_H]$. where $Q_H$ is the maximum number of queries to $H$ that $\mathcal{A}$ can make.
2. Sample $d$ random matrices $R_1^*, \ldots, R_d^* \sim \mathcal{D}_{m \times m}$ by running $R_i^* \leftarrow$ SampleR$(1^m)$ for $i = 1, \ldots, d$.
3. Assemble the random matrix $A_0 \in \mathbb{Z}_q^{n \times m}$ from $m$ of the given LWE samples, by letting the $i$-th column of $A_0$ be the $n$-vector $u_i$ for all $i = 1, \ldots, m$.
4. Choose a random $w \in [d]$ and set $A \leftarrow A_0 R_w^* \cdots R_1^*$. The matrix $A$ is uniform in $\mathbb{Z}_q^{n \times m}$ since all the $R_i^*$ are invertible mod $q$ and $A_0$ is uniform in $\mathbb{Z}_q^{n \times m}$.
5. Publish the public parameters PP $= (A, u_0)$.

**Random-oracle hash queries.** $\mathcal{A}$ may query the random oracle $H$ on any identity id $= (\mathrm{id}_1, \ldots, \mathrm{id}_i)$ of its choice, adaptively, and at any time. $\mathcal{B}$ answers the $Q$-th such query as follows. (We assume w.l.o.g. that the queries are unique; otherwise the simulator simply returns the same output on the same input without incrementing the query counter $Q$.)

Let $i = |\mathrm{id}|$ be the depth of id. If this is query number $Q_i^*$ (i.e. $Q = Q_i^*$), define $H(\mathrm{id}) \leftarrow R_i^*$ and return $H(\mathrm{id})$.

Otherwise, if $Q \neq Q_i^*$:

1. Compute $A_i = A \cdot \left(R_{i-1}^* \cdots R_2^* R_1^*\right)^{-1} \in \mathbb{Z}_q^{m \times m}$ (where $A_1 = A$).
2. Run SampleRwithBasis$(A_i)$ to obtain a random $R \sim \mathcal{D}_{m \times m}$ and a short basis $T_B$ for $B = A_i R^{-1} \bmod q$.
3. Save the tuple $(i, \mathrm{id}, R, B, T_B)$ for future use, and return $H(\mathrm{id}) \leftarrow R$.

**Secret key queries.** $\mathcal{A}$ makes interactive key-extraction queries on arbitrary identities id, chosen adaptively. $\mathcal{B}$ answers a query on id $= (\mathsf{id}_1, \mathsf{id}_2, \ldots, \mathsf{id}_k)$ of length $|\mathsf{id}| = k \in [d]$ as follows.

1. Let $j \in [k]$ be the shallowest level at which $H(\mathsf{id}_{|j}) \neq R_j^*$. In the unlikely event that $H(\mathsf{id}_{|j}) = R_j^*$ for all $j = 1, \ldots, k$ the simulator aborts and fails.
2. Retrieve the saved tuple $(j, \mathsf{id}_{|j}, R, B, T_B)$ from the hash oracle query history. This tuple was created when responding to a query for $H(\mathsf{id}_{|j})$ (w.l.o.g., we can assume that an extraction query on id is preceded by a hash query on all prefixes of id). By construction

$$B = A \cdot (R_1^*)^{-1} \cdots (R_{j-1}^*)^{-1} \cdot H(\mathsf{id}_{|j})^{-1} \bmod q$$

and $T_B$ is a short basis for $\Lambda_q^{\perp}(B)$.
Notice that $B$ is exactly the encryption matrix $F_{\mathsf{id}_{|j}}$ (as defined in the Encrypt algorithm) for the ancestor identity $\mathsf{id}_{|j} = (\mathsf{id}_1, \mathsf{id}_2, \ldots, \mathsf{id}_j)$ and therefore $T_B$ is a trapdoor for $\Lambda_q^{\perp}(F_{\mathsf{id}_{|j}})$.
3. Run $\mathsf{Derive}(\mathsf{PP}, T_B, \mathsf{id})$ to generate a secret key for id from the private key $T_B$ for the identity $\mathsf{id}_{|j}$. Send the resulting secret key to the adversary.

**Challenge.** $\mathcal{A}$ announces to $\mathcal{B}$ the identity $\mathsf{id}^*$ on which it wishes to be challenged and a message $b^* \in \{0,1\}$ to be encrypted. We require that $\mathsf{id}^*$ not be equal to, or a descendant of, any identity id for which a private key has been or will be requested in any preceding and subsequent key extraction query.

Let $\ell = |\mathsf{id}^*|$. If there is an $i \in [\ell]$ such that $H(\mathsf{id}_{|i}^*) \neq R_i^*$, then the simulator must abort. (Indeed, when this is the case, $\mathcal{B}$ is able extract a private key for $\mathsf{id}^*$ and thus answer by itself the challenge that it intended to ask.)

Recall that $A = A_0 R_w^* \cdots R_1^*$. If $w \neq \ell$ then the simulator aborts and fails.

Now, suppose $w = \ell$ and $\mathsf{id}^*$ is such that $H(\mathsf{id}_{|i}^*) = R_i^*$ for all $i \in [\ell]$. Then by definition

$$F_{\mathsf{id}^*} = A \ (R_1^*)^{-1} \cdots (R_{\ell}^*)^{-1} = A_0 \quad \in \mathbb{Z}_q^{n \times m}$$

and $\mathcal{B}$ proceeds as follows:

1. Retrieve $v_0, \ldots, v_m \in \mathbb{Z}_q$ from the LWE instance and set $v^* = \begin{bmatrix} v_1 \\ \vdots \\ v_m \end{bmatrix} \in \mathbb{Z}_q^m$.
2. Blind the message bit by letting $c_0^* = v_0 + b^* \lfloor \frac{q}{2} \rceil \in \mathbb{Z}_q$.
3. Set $c_1^* = v^* \in \mathbb{Z}_q^m$.
4. Set $\mathsf{CT}^* = (c_0^*, c_1^*)$ and send it to the adversary.

When $\mathcal{O}$ is a pseudo-random LWE oracle then $c_0 = u_0^\top s + x + b \lfloor \frac{q}{2} \rceil$ and $c_1 = F_{\mathsf{id}^*}^\top s + y$ for some random $s \in \mathbb{Z}_q^n$ and noise values $x$ and $y$. In this case $(c_0, c_1)$ is a valid encryption of $b$ for $\mathsf{id}^*$.

When $\mathcal{O}$ is a random oracle then $(v_0, v^*)$ are uniform in $(\mathbb{Z}_q \times \mathbb{Z}_q^m)$ and therefore $(c_0, c_1)$ is uniform in $(\mathbb{Z}_q \times \mathbb{Z}_q^m)$.

Now, $\mathcal{A}$ makes more secret key queries, answered by $\mathcal{B}$ in the same manner as before. Finally, $\mathcal{A}$ guesses whether $\mathsf{CT}^*$ was an encryption of $b^*$ for $\mathsf{id}^*$. $\mathcal{B}$ outputs $\mathcal{A}$'s guess and ends the simulation.

The distribution of the public parameters is identical to its distribution in the real system as are responses to private key queries. By Theorem 3, responses to $H$ oracle queries are as in the real system. Finally, if $\mathcal{B}$ does not abort then the challenge ciphertext is distributed either as in the real system or is independently random in $(\mathbb{Z}_q, \mathbb{Z}_q^m)$. Hence, if $\mathcal{B}$ does not abort then its advantage in solving LWE is the same as $\mathcal{A}$'s advantage in attacking the system.

Since $\mathcal{A}$ is PPT it only finds collisions on $H$ with negligible probability. A standard argument shows that the simulator can proceed without aborting with probability $\Pr[\neg\text{abort}] \geq Q_H^{-\ell}/d - \text{negl}(n) \geq Q_H^{-d}/d - \text{negl}(n)$ for some constant $c > 0$. Then if $\mathcal{A}$ has advantage $\epsilon \geq 0$, $\mathcal{B}$ has advantage at least $[\epsilon/(dQ_H^d)] - \text{negl}(n)$ in deciding the LWE problem instance.

## 5  Selectively Secure HIBE in the Standard Model

We briefly describe an HIBE of depth $d$ that is selectively secure without random oracles. The details are in the full version of the paper. The construction is a binary tree encryption (BTE) which means that identities at each level are binary (i.e. 0 or 1). To build an HIBE with $k$-bit identities at each level we assign $k$ levels of the BTE hierarchy to each level of the HIBE. The parameters used by this system are shown in Table 1.

**Setup:** For a BTE of depth $d$ the setup algorithm runs $\mathsf{TrapGen}(q, n)$ to generate a random $n \times m$ matrix $A \in \mathbb{Z}_q^{n \times m}$ with a short basis $T_A \in \mathbb{Z}^{m \times m}$ for $\Lambda_q^\perp(A)$ and samples $2d$ matrices $R_{1,0}, R_{1,1}, \ldots, R_{d,0}, R_{d,1} \in \mathbb{Z}^{m \times m}$ from the distribution $\mathcal{D}_{m \times m}$ using $\mathsf{SampleR}(1^m)$. With $u_0$ random in $\mathbb{Z}_q^n$ the public params and master key are

$$\mathsf{PP} = \left( A, \; u_0, \; R_{1,0}, R_{1,1}, \; R_{2,0}, R_{2,1}, \; \ldots, \; R_{d,0}, R_{d,1} \right) \quad, \quad \mathsf{MK} = \left( T_A \right)$$

**Extract:** the secret key for an identity $\mathsf{id} = (\mathsf{id}_1, \ldots, \mathsf{id}_\ell) \in \{0,1\}^{\ell \leq d}$ is a short random basis for the lattice $\Lambda_q^\perp(F_{\mathsf{id}})$ where

$$F_{\mathsf{id}} = A \, (R_{1,\mathsf{id}_1})^{-1} \, (R_{2,\mathsf{id}_2})^{-1} \cdots (R_{\ell,\mathsf{id}_\ell})^{-1} \in \mathbb{Z}_q^{n \times m} \tag{5}$$

Encryption and decryption are as in the system of Section 4.1 using the matrix $F_{\mathsf{id}}$ from (5) in a dual-Regev encryption.

**Security.** The simulator is given an identity $\mathsf{id} = (\mathsf{id}_1, \ldots, \mathsf{id}_\ell) \in \{0,1\}^\ell$ where the attacker will be challenged. To simplify the description assume $\mathsf{id}$ is at maximum depth, namely $\ell = d$. The case $\ell < d$ is just as easy, but complicates the notation.

The simulator first constructs a matrix $A_0 \in \mathbb{Z}_q^{n \times m}$ from the given LWE challenge. It then samples random matrices

$$R_{1,\mathsf{id}_1}, R_{2,\mathsf{id}_2}, \ldots, R_{\ell,\mathsf{id}_\ell} \in \mathbb{Z}^{m \times m}$$

from the distribution $\mathcal{D}_{m \times m}$ and sets $A = A_0 \, R_{\ell,\mathsf{id}_\ell} \cdots R_{2,\mathsf{id}_2} \, R_{1,\mathsf{id}_1} \in \mathbb{Z}_q^{n \times m}$. Now, consider the $d$ matrices

$$F_i = A \, (R_{1,\mathsf{id}_1})^{-1} \cdots (R_{i,\mathsf{id}_i})^{-1} \quad \text{for } i = 0, \ldots, d-1.$$

For each matrix $F_i$ the simulator invokes $\mathsf{SampleRwithBasis}(F_i)$ to obtain a matrix $R_{i,1-\mathsf{id}_i} \in \mathbb{Z}^{m \times m}$ and a short basis $T_i$ for $\Lambda_q^{\perp}(F_i \cdot (R_{i,1-\mathsf{id}_i})^{-1})$. Finally, it sends to the adversary the public parameters

$$\mathsf{PP} = \left( A , \ u_0 , \ R_{1,0}, R_{1,1} , \ R_{2,0}, R_{2,1} , \ \ldots , \ R_{d,0}, R_{d,1} \right)$$

where $u_0$ is a random vector in $\mathbb{Z}_q^n$ from the LWE challenge.

It is not difficult to see that the simulator can use $T_1, \ldots, T_d$ to generate private keys for every node in the hierarchy except for the challenge identity id. Moreover, for the challenge identity it can generate a ciphertext that will help it solve the given LWE challenge as in Section 4.3, as required.

## 6   Conclusions

We presented a new lattice basis delegation mechanism and used it to construct two HIBE systems, one secure in the random oracle model and one secure without random oracles. The random oracle construction provides a lattice HIBE with short ciphertexts and private keys. The standard model system is not as short.

This work raises a number of interesting open problems. First, our standard model system processes bits of the identity one at a time. It would be interesting to apply the techniques of [ABB10, Boy10] to obtain a selective HIBE that processes many bits at a time so that the encryption matrix $F_{\mathsf{id}}$ is a product of only $\ell$ low-norm matrices for identities at depth $\ell$.

Another interesting problem is an adaptively secure HIBE in the standard model where performance does not degrade exponentially in the hierarchy depth. Using the lattice basis delegation method from this paper or from [CHKP10] in Waters' dual encryption system [Wat09] is a promising direction.

**Acknowledgments.** We thank David Freeman, Daniele Micciancio and Brent Waters for helpful comments about this work.

## References

[ABB10]   Agrawal, S., Boneh, D., Boyen, X.: Efficient lattice (H)IBE in the standard model. In: Gilbert, H. (ed.) EUROCRYPT 2010. LNCS, vol. 6110, pp. 553–572. Springer, Heidelberg (2010)

[Ajt99]   Ajtai, M.: Generating hard instances of the short basis problem. In: Wiedermann, J., Van Emde Boas, P., Nielsen, M. (eds.) ICALP 1999. LNCS, vol. 1644, pp. 1–9. Springer, Heidelberg (1999)

[AP09]   Alwen, J., Peikert, C.: Generating shorter bases for hard random lattices. In: STACS, pp. 75–86 (2009)

[BB04]   Boneh, D., Boyen, X.: Efficient selective-id secure identity-based encryption without random oracles. In: Cachin, C., Camenisch, J.L. (eds.) EUROCRYPT 2004. LNCS, vol. 3027, pp. 223–238. Springer, Heidelberg (2004)

[BBG05]   Boneh, D., Boyen, X., Goh, E.-J.: Hierarchical identity based encryption with constant size ciphertext. In: Cramer, R. (ed.) EUROCRYPT 2005. LNCS, vol. 3494, pp. 440–456. Springer, Heidelberg (2005)

[BF01]     Boneh, D., Franklin, M.: Identity-based encryption from the Weil pairing. In: Kilian, J. (ed.) CRYPTO 2001. LNCS, vol. 2139, pp. 213–229. Springer, Heidelberg (2001)

[Boy10]    Boyen, X.: Lattices mixing and vanishing trapdoors: A framework for fully secure short signatures and more. In: Nguyen, P.Q., Pointcheval, D. (eds.) PKC 2010. LNCS, vol. 6056, pp. 499–517. Springer, Heidelberg (2010)

[BW06]     Boyen, X., Waters, B.: Anonymous hierarchical identity-based encryption (without random oracles). In: Dwork, C. (ed.) CRYPTO 2006. LNCS, vol. 4117, pp. 290–307. Springer, Heidelberg (2006)

[CDMP05]   Coron, J.-S., Dodis, Y., Malinaud, C., Puniya, P.: Merkle-damgard revisited: how to construct a hash function. In: Shoup, V. (ed.) CRYPTO 2005. LNCS, vol. 3621, pp. 430–448. Springer, Heidelberg (2005)

[CHK07]    Canetti, R., Halevi, S., Katz, J.: A forward-secure public-key encryption scheme. J. Crypto 20(3), 265–294 (2007); Abstract in Eurocrypt 2003 (2003)

[CHKP10]   Cash, D., Hofheinz, D., Kiltz, E., Peikert, C.: Bonsai trees, or how to delegate a lattice basis. In: Gilbert, H. (ed.) EUROCRYPT 2010. LNCS, vol. 6110, pp. 523–552. Springer, Heidelberg (2010)

[GH09]     Gentry, C., Halevi, S.: Hierarchical identity based encryption with polynomially many levels. In: Reingold, O. (ed.) TCC 2009. LNCS, vol. 5444, pp. 437–456. Springer, Heidelberg (2009)

[GN08]     Gama, N., Nguyen, P.: Predicting lattice reduction. In: Smart, N.P. (ed.) EUROCRYPT 2008. LNCS, vol. 4965, pp. 31–51. Springer, Heidelberg (2008)

[GPV08]    Gentry, C., Peikert, C., Vaikuntanathan, V.: Trapdoors for hard lattices and new cryptographic constructions. In: STOC (2008)

[GS02]     Gentry, C., Silverberg, A.: Hierarchical id-based cryptography. In: Zheng, Y. (ed.) ASIACRYPT 2002. LNCS, vol. 2501, pp. 548–566. Springer, Heidelberg (2002)

[HL02]     Horwitz, J., Lynn, B.: Toward hierarchical identity-based encryption. In: Knudsen, L.R. (ed.) EUROCRYPT 2002. LNCS, vol. 2332, pp. 466–481. Springer, Heidelberg (2002)

[MG02]     Micciancio, D., Goldwasser, S.: Complexity of Lattice Problems: a cryptographic perspective, vol. 671. Kluwer Academic Publishers, Boston (March 2002)

[MR07]     Micciancio, D., Regev, O.: Worst-case to average-case reductions based on gaussian measures. SIAM Journal on Computing (SICOMP) 37(1), 267–302 (2007); Extended abstract in FOCS 2004 (2004)

[Pei]      Peikert, C.: Bonsai trees (or, arboriculture in lattice-based cryptography). Cryptology ePrint Archive, Report (2009), /359, http://eprint.iacr.org/

[Pei09]    Peikert, C.: Public-key cryptosystems from the worst-case shortest vector problem. In: STOC, pp. 333–342 (2009)

[Reg09]    Regev, O.: On lattices, learning with errors, random linear codes, and cryptography. J. ACM 56(6) (2009); Extended abstract in STOC 2005 (2005)

[Sha85]    Shamir, A.: Identity-based cryptosystems and signature schemes. In: Blakely, G.R., Chaum, D. (eds.) CRYPTO 1984. LNCS, vol. 196, pp. 47–53. Springer, Heidelberg (1985)

[Wat05]    Waters, B.: Efficient identity-based encryption without random oracles. In: Cramer, R. (ed.) EUROCRYPT 2005. LNCS, vol. 3494, pp. 114–127. Springer, Heidelberg (2005)

[Wat09]    Waters, B.: Dual key encryption: Realizing fully secure IBE and HIBE under simple assumption. In: Halevi, S. (ed.) CRYPTO 2009. LNCS, vol. 5677, pp. 619–636. Springer, Heidelberg (2009)

# Toward Basing Fully Homomorphic Encryption on Worst-Case Hardness

Craig Gentry

IBM T.J Watson Research Center
`cbgentry@us.ibm.com`

**Abstract.** Gentry proposed a fully homomorphic public key encryption scheme that uses ideal lattices. He based the security of his scheme on the hardness of two problems: an average-case decision problem over ideal lattices, and the sparse (or "low-weight") subset sum problem (SSSP).

We provide a key generation algorithm for Gentry's scheme that generates ideal lattices according to a "nice" average-case distribution. Then, we prove a worst-case / average-case connection that bases Gentry's scheme (in part) on the quantum hardness of the shortest independent vector problem (SIVP) over ideal lattices in the *worst-case*. (We cannot remove the need to assume that the SSSP is hard.) Our worst-case / average-case connection is the first where the average-case lattice is an ideal lattice, which seems to be necessary to support the security of Gentry's scheme.

## 1 Introduction

Recently, Gentry [10] presented a somewhat homomorphic encryption scheme that uses ideal lattices, and proved its security based on an average-case decision problem. In this paper, we focus on this somewhat homomorphic scheme and its security. Our main results are:

- Algorithms for his scheme – most importantly, a KeyGen algorithm for generating secret and public bases of an ideal lattice – that permit the scheme's semantic security to be based on a *search* problem over ideal lattices having a *nice* average-case distribution.
- A quantum worst-case / average-case reduction, which ultimately bases the security of Gentry's somewhat homomorphic scheme on the worst-case quantum hardness of the shortest independent vector problem (SIVP) over ideal lattices.

Gentry also showed that his somewhat homomorphic scheme, after some modifications, becomes "bootstrappable" and therefore can be used to construct a fully homomorphic encryption (FHE) scheme [31,10]. He proved that the FHE scheme is semantically secure if the original somewhat homomorphic scheme is semantically secure and the sparse (or "low-weight") subset sum problem (SSSP) [11,25] is hard. Those results are generic enough to work with our instantiation

T. Rabin (Ed.): CRYPTO 2010, LNCS 6223, pp. 116–137, 2010.

of KeyGen and the other algorithms. That is, we immediately obtain a FHE scheme whose security is based on two problems: the SSSP and worst-case quantum SIVP over ideal lattices.[1] Since the SSSP is an average-case problem, it remains an open problem to base FHE *entirely* on worst-case hardness. However, the more "troubling" of Gentry's two assumptions (in our opinion) is that the average-case decision problem over ideal lattices is hard. At least we can replace this assumption with one involving worst-case hardness.

## 1.1   Related Work

In 1996, Ajtai [1] found a surprising reduction of worst-case lattice problems to average-case ones. Unlike the random self-reduction of Diffie-Hellman, where the worst-case and average-case instances are over the same group $G$, Ajtai's worst-case problem is a completely general problem (over lattices) that is unconstrained by any parameters in the average-case problem. The average-case lattices in Ajtai's reduction are of a certain type: those generated by random parity-check matrices modulo an integer $q$.

Following Ajtai, improved worst-case / average-case connections were described in [8,22,23,28,24,26,20]. Also, various primitives have been based on worst-case hardness, including collision-resistant hash functions [1,8,22,17,24,27], public-key encryption [3,29,30,12,26,19], signatures [18,12], and (hierarchical) identity-based encryption [12,9,7]. Ajtai [2] showed how to generate his average-case lattices together with a short secret basis for the lattice that can be used as a decryption key in an encryption scheme [12]; Alwen and Peikert [4] tightened this result.

However, as far as we know, previous worst-case / average-case reductions cannot be used to base Gentry's somewhat homomorphic scheme on worst-case hardness. The essential problem is that Gentry's scheme [10] uses *ideal lattices* and relies heavily on the structure of these lattices as algebraic ideals in a ring to obtain homomorphism. However, in none of the previous reductions is the *average-case* lattice an ideal lattice.

Some previous work describes worst-case / average-case reductions where the worst-case lattice is an ideal lattice, and the average-case instances are *derived* from ideal lattices, in a fashion somewhat similar to how Ajtai's average-case lattices are derived from a worst-case instance. For example, for the ring $R = \mathbb{Z}[x]/(x^n - 1)$ and fixed $\mathbf{a}_1, \dots, \mathbf{a}_m \in R^m$, Micciancio [22,23] considered the lattice formed by solutions $\mathbf{v}_1, \dots, \mathbf{v}_m \in R^m$ to $\sum_i \mathbf{a}_i \times \mathbf{v}_i = \mathbf{0}$, and showed that solving the bounded distance decoding problem (BDDP) or SIVP for such "quasi-cyclic" lattices in the average-case allows one to solve the BDDP or SIVP for "cyclic lattices" (ideal lattices in $R$) in the worst-case. While Micciancio's worst-case lattices are ideal lattices, the average-case lattices are not; they correspond to modules, rather than ideals. Peikert and Rosen [28] demonstrated a

---

[1] Technically, both in [10] and here, a "circular-security" assumption is also needed to obtain an FHE scheme whose public key size is independent of the circuit depth of the functions being homomorphically evaluated.

very tight worst-case / average-case reduction where the worst-case lattices are ideal lattices, and where the average-case lattices are derived from ideal lattices in a way similar to that used by Micciancio. Some other results in this line of work include [27,17,20].

However, again, previous work does not provide a worst-case / average-case "random self-reduction" where both average-case and worst-case lattices are ideal lattices of the same dimension in the same ring, which seems to be necessary to preserve the algebraic structure used by Gentry's scheme, and thus necessary to support the security of Gentry's somewhat homomorphic scheme. This suggests that we need an approach fundamentally different from Ajtai's and other previous work. We also need a KeyGen algorithm for Gentry's scheme that generates an ideal lattice, together with a secret basis of the lattice, according to the appropriate average-case distribution.

## 1.2   Our Worst-Case / Average-Case Self-reduction

We provide the first worst-case / average-case self-reduction where the average-case lattice is an ideal lattice. We focus on the reduction for BDDP over ideal lattices, but this reduction can be extended to other ideal lattice problems. Combining with other results presented here and in prior work, this reduction bases the security of Gentry's somewhat homomorphic scheme on worst-case hardness.

Our reduction makes heavy use of the algebraic properties of ideals. Interestingly, and quite unlike other worst-case / average-case reductions, our reduction uses an integer factoring oracle to factor ideals in the ring. This integer factoring oracle can be instantiated efficiently with quantum computation [32], and hence we get an efficient quantum reduction. The reduction is also meaningful in the classical setting, since there are known sub-exponential factoring algorithms for factoring (e.g., the number field sieve). If solving average-case problems over ideal lattices is easy, our reduction implies that there are surprising sub-exponential algorithms for solving worst-case problems over ideal lattices.

Since our worst-case and average-case instances involve ideal lattices of the same dimension within the same ring $R$, one may prefer to think of our reduction as a "random self-reduction". It is an "imperfect" self-reduction in that the approximation factor is larger in the worst-case problem than in the average-case problem by a $\text{poly}(n)$ factor (for the rings $R$ that we use). However, as far as we know, the BDDP is hard even for sub-exponential approximation factors – i.e., for factors much larger than our reduction's $\text{poly}(n)$ lossiness.

Roughly speaking, the reduction works as follows. We are given the basis $\mathbf{B}_M$ of a worst-case ideal lattice $M$ that corresponds to an ideal in the ring $R$, together with a vector $\mathbf{t} \in \mathbb{R}^n$ that is close to some vector $\mathbf{u} \in M$; the BDDP is to output $\mathbf{u}$. To generate an average-case instance, we first sample a "random" vector $\mathbf{v}$ from the inverse ideal $M^{-1}$ according to a particular distribution. We multiply (in the ring $R$) each of the basis elements of $\mathbf{B}_M$ by $\mathbf{v}$ to obtain a basis $\mathbf{B}_L$ of the lattice for the ideal $L = M \cdot (\mathbf{v})$, and set $\mathbf{u}' \leftarrow \mathbf{v} \times \mathbf{u}$. $L$ will be an ideal in $R$ that is not divisible by $M$, since $\mathbf{v} \in M^{-1}$ and thus "cancels" $M$.

However, due to $\mathbf{v}$'s distribution, $L$'s *geometry* will be very closely related to $M$; in particular, solving BDDP for $(L, \mathbf{u}')$ will help solve BDDP for $(M, \mathbf{u})$. Toward solving BDDP for $(L, \mathbf{u}')$, we use our factoring oracle to find a "suitable" ideal $J$ that divides $L$ (restarting if no suitable one exists), and output the instance $(J, \mathbf{u}')$ to our average-case BDDP solver. Note that $L$ is a subset of $J$. As long as $L$ is not an overly sparse subset, and for suitable parameters, the closest vector in $J$ to $\mathbf{u}'$ will also be in $L$. Hence, a BDDP solution to average-case instance $(J, \mathbf{u}')$ leads to a BDDP solution to the worst-case instance $(M, \mathbf{u})$. We show that $J$ comes from our desired average-case distribution – i.e., that it is uniformly random among regular prime ideals in $R$ whose norms are in a prescribed interval. Of course, the target vector $\mathbf{u}'$'s distribution is not random – i.e., is not independent of the worst-case instance – but we also show how to randomize the target vector's distribution. See Section 3 for details and proofs.

### 1.3   How to Generate an Average Ideal Lattice, and Other Results

In [10], Gentry mentions some *ad hoc* ways of generating an ideal lattice, together with a secret basis for it. Here, we show how to generate ideal lattices (together with a secret basis) according to the average-case distribution used in our worst-case / average-case connection. Generating an ideal lattice according to this distribution is easy, but generating it together with a "good" secret basis is surprisingly difficult. Our solution to this problem is provided in Section 4.

Although the worst-case / average-case connection for BDDP over ideal lattices (Section 3) and the key generation algorithm (Section 4) are our main results, several other reductions are necessary to base our version of Gentry's somewhat homomorphic scheme on worst-case SIVP over ideal lattices. We summarize these reductions in Section 5.

## 2   Preliminaries

### 2.1   Ideal Lattices

By an *ideal lattice*, we mean an *ideal* in the *ring of integers* $R = \mathcal{O}_F$, where $f(x)$ is a monic, irreducible polynomial of degree $n$, and $F$ is the field $\mathbb{Q}[x]/(f(x))$. A good example to keep in mind is $f(x) = x^n + 1$, where $n$ is a power of 2. Then, the ring of integers is simply $\mathbb{Z}[x]/(f(x))$, integer polynomials modulo $f(x)$. In the full version, we address the general case $\mathbb{Z}[x]/(f(x)) \subseteq R \subseteq \mathcal{O}_F$.

Each element of $R$ is associated to a coefficient vector in $\mathbb{Q}^n$ (in $\mathbb{Z}^n$ in our example). Since an ideal $I \subset R$ is additively closed, the coefficient vectors associated to elements of $I$ form a *lattice*. The term "ideal lattice" emphasizes this object's dual nature as an algebraic ideal and a lattice.[2]

Ideals have additive structure as lattices, but they also have multiplicative structure. The *product* of two ideals $I$ and $J$ is $IJ = \{\sum \mathbf{v} \times \mathbf{w} : \mathbf{v} \in I, \mathbf{w} \in J\}$, where '$\times$' is ring multiplication. Let $F = \mathbb{Q}[x]/(f(x))$ be the field containing $R$.

---

[2] Alternative representations of an ideal lattice are possible – e.g., see [28,20].

The *inverse* of a ideal $I$ is $I^{-1} = \{\mathbf{w} \in F : \forall \mathbf{v} \in I, \mathbf{v} \times \mathbf{w} \in R\}$. For example, the inverse of (2) is $(1/2) = \{\mathbf{r}/2 : \mathbf{r} \in R\}$. (The inverse of any *principal* ideal $(\mathbf{v})$ is given by $(\mathbf{v}^{-1})$, where the inverse $\mathbf{v}^{-1}$ is taken in $F$, but for a non-principal ideal the inverse is not always so simple.) We say that ideal $I$ *divides* ideal $J$ if $JI^{-1} \subset R$. $I$ is a prime ideal if $I$ dividing $A \cdot B$ implies $I$ divides $A$ or $B$. The ideal $I^{-1}$ or $JI^{-1}$ is sometimes called a fractional ideal, particularly when it is not a subset of $R$.

Ideals in $R$ have many of the nice properties of integers, especially when $R$ is the ring of integers. For example, in this case, ideals in $R$ factor uniquely as a product of prime ideals. Also, all ideals in $R$ are *invertible* – i.e., $I \cdot I^{-1} = R$. Furthermore, one can define the norm of a fractional ideal $\mathrm{Nm}(I)$ as the index $[R : I]$, and this map is multiplicative: $\mathrm{Nm}(IJ) = \mathrm{Nm}(I) \cdot \mathrm{Nm}(J)$.

Just as the prime number theorem states that the number of primes less than $x$ is approximately $x/\ln x$, we have Landau's prime ideal theorem [15]:

**Theorem 1 (Theorem 8.7.2 from [5]).** *Let $F$ be an algebraic number field of degree $n$. Let $\pi_F(x)$ denote the number of prime ideals in $\mathcal{O}_F$ whose norm is $\leq x$. Let $\lambda(x) = (\ln x)^{3/5}(\ln \ln x)^{-1/5}$. There is a $c > 0$ (depending on $F$) such that*

$$\pi_F(x) = x/\ln x + O(xe^{-c\lambda(x)})$$

With the Generalized Riemann Hypothesis, one can make a stronger statement.

**Theorem 2 (Theorem 8.7.4 from [5]).** *Assume GRH. Let $F$ be an algebraic number field of degree $n$ and discriminant $\Delta_F$. For $x \geq 2$, we have*

$$|\pi_F(x) - x/\ln x| = O(\sqrt{x}(n \ln x + \ln |\Delta_F|))$$

*The constant implied by the "O" symbol is absolute.*

Regarding Theorem 2, $\Delta_F$ is upper-bounded by $\Delta(f)$, the discriminant of the polynomial $f$. Since $\Delta(f)$ is the determinant of the Sylvester matrix formed by $f(x)$ and its derivative $f'(x)$, it is upper bounded by $n^n \|f\|^{2n}$, where $\|f\|$ is the Euclidean length of the coefficient vector of $f(x)$ [33]. As in [10], we will always use $f(x)$ such that $\|f\| = \mathrm{poly}(n)$, which implies that $\ln |\Delta_F| = \mathrm{poly}(n)$.

We let $\gamma_f$ denote the minimal value such that $\|\mathbf{u} \times \mathbf{v}\| \leq \gamma_f \cdot \|\mathbf{u}\| \cdot \|\mathbf{v}\|$ for all $\mathbf{u}, \mathbf{v} \in \mathbb{Q}[x]/(f(x))$. For the values of irreducible $f(x)$ recommended in [10], we have $\gamma_f = \mathrm{poly}(n)$. A nice property of ideal lattices in such rings is that they are never too "oblong." In particular, trivially, $\lambda_n(I)/\lambda_1(I) \leq \gamma_f$, where $\lambda_k(I)$ is the $k$-th minimum of the ideal lattice $I$.

Again, a good choice for $f(x)$ is $x^n + 1$, where $n$ is a power of 2. This polynomial has the virtues of being irreducible, satisfying $R = \mathcal{O}_F = \mathbb{Z}[x]/(f(x))$, and having small values of $\Delta(f)$, $\|f\|$, and $\gamma_f$.

## 2.2    Gaussian Distributions and Other Preliminaries

For any real $s > 0$, define the Gaussian function on $\mathbb{R}^n$ centered at $\mathbf{c}$ with parameter $s$ as $\rho_{s,\mathbf{c}}(\mathbf{x}) = \exp(-\pi \|\mathbf{x} - \mathbf{c}\|^2/s^2)$ for all $\mathbf{x} \in \mathbb{R}^n$. The associated *discrete* Gaussian distribution over lattice $L$ is

$$\forall \; \mathbf{x} \in L, D_{L,s,\mathbf{c}}(\mathbf{x}) = \frac{\rho_{s,\mathbf{c}}(\mathbf{x})}{\rho_{s,\mathbf{c}}(L)} \; ,$$

where $\rho_{s,\mathbf{c}}(A)$ for set $A$ denotes $\sum_{\mathbf{x} \in A} \rho_{s,\mathbf{c}}(\mathbf{x})$. In other words, the probability $D_{L,s,\mathbf{c}}(\mathbf{x})$ is simply proportional to $\rho_{s,\mathbf{c}}(\mathbf{x})$, the denominator being a normalization factor.

As in [24], for lattice $L$ and real $\epsilon > 0$, we define the *smoothing parameter* $\eta_\epsilon(L)$ to be the smallest $s$ such that $\rho_{1/s}(L^* \setminus \{\mathbf{0}\}) \leq \epsilon$. We say that $s$ "exceeds the smoothing parameter" of $L$ if $s \geq \eta_\epsilon(L)$ for negligible $\epsilon$. In particular, this is true when $s \geq \lambda_n(L) \cdot \omega(\sqrt{\log n})$. Some useful lemmas are the following.

**Lemma 1 (Lemma 4.4 of [24]).** *For any $n$-dimensional lattice $L$, vector $\mathbf{c} \in \mathbb{R}^n$, and reals $0 < \epsilon < 1$, $s \geq \eta_\epsilon(L)$, we have*

$$\Pr_{\mathbf{x} \leftarrow D_{L,s,\mathbf{c}}} \{\|\mathbf{x} - \mathbf{c}\| > s\sqrt{n}\} \leq \frac{1+\epsilon}{1-\epsilon} \cdot 2^{-n}$$

**Lemma 2.** *Let $I, J$ be ideal lattices in $R$. Then for any $\epsilon \in (0, 1/2)$, and $s \geq \max\{\eta_\epsilon(I), \eta_\epsilon(J)\}$, and any $\mathbf{c} \in \mathbb{R}^n$, $\rho_{s,\mathbf{c}}(I)/\rho_{s,\mathbf{c}}(J)$ equals $\mathrm{Nm}(J)/\mathrm{Nm}(I)$, up to a multiplicative factor of between $(1 + \epsilon)^2/(1 - \epsilon)$ and its inverse.*

*Proof.* See full version.

We use $\mathbf{e}_i$ to refer to the vector $(0, \ldots, 0, 1, 0, \ldots, 0)$ with '1' in the $i$th position. We say that an equality $a \approx b$ holds "up to negligible error" if $a = (1 \pm \epsilon) \cdot b$ for some negligible $\epsilon$.

# 3   Random Self-reduction of Ideal Lattice Problems

In this section, we present our worst-case / average-case "random self-reduction" for problems over ideal lattices, focusing on the bounded distance decoding problem (BDDP) [19,30]. We describe our average-case distribution, and specify our average-case and worst-case versions of BDDP. Then we show how to "randomize" worst-case ideal lattices into ideal lattices from our average-case distribution. In Section 4, we establish that the average-case distribution is suitable for KeyGen – i.e., we can efficiently (classically) sample an ideal lattice and a good basis for it according to this distribution.

## 3.1   Our Average-Case Distribution and Hard Problem

Our average-case distribution is simple: uniform over prime (non-fractional) ideals in $R$ that have norms in some specified interval $[a, b]$.

Our average-case problem is really a "hybrid" of worst-case and average-case.

**Definition 1 (Hybrid Bounded Distance Decoding Problem (HBDDP)).**
*Fix ring $R$ and algorithm IdealGen that samples ideals in $R$, outputting the Hermite normal form basis of the sampled ideal lattice. Fix a positive real*

$s_{HBDDP}$. *The challenger sets* $\mathbf{B}_J \overset{R}{\leftarrow} \mathsf{IdealGen}(R)$. *The challenger sets* $\mathbf{x}$ *subject to the constraint that* $\|\mathbf{x}\| < s_{HBDDP}$ *and sets* $\mathbf{t} \leftarrow \mathbf{x} \bmod \mathbf{B}_J$. *The problem is: given* $(\mathbf{B}_J, \mathbf{t})$ *(and the fixed values), output* $\mathbf{x}$.

The ideal lattice is generated according to an average-case distribution induced by an algorithm $\mathsf{IdealGen}$. However, the vector $\mathbf{t}$ is "worst-case", in that $\mathbf{t}$ is only required to be within a certain distance of the lattice; it need not be chosen according to any known (or even samplable) distribution.

The worst-case BDDP (WBDDP) is identical, except the ideal lattice is not necessarily chosen from an efficiently samplable distribution. For both of the BDDPs, we assume that the $s$ parameter is chosen so that the solution is unique.

We base the security of our version of Gentry's scheme on HBDDP in the full version (and sketch this result in Section 5). As part of this result, we reduce HBDDP to a "purely" average-case BDDP where $\mathbf{t}$ is sampled according to a Gaussian distribution. In the full version, we also provide more reductions that (quantumly) reduce worst-case SIVP to WBDDP. We choose to focus on our techniques for randomizing the lattice since they are more interesting.

### 3.2   Statement of the Reduction

Our reduction is stated in the following theorem. It uses parameters that must satisfy certain conditions that we will specify momentarily.

**Theorem 3.** *Let $R$ be the ring of integers for field $F = \mathbb{Q}(x)/(f(x))$. Let $M$, $N$, $s_{WBDDP}$, $t$, $a$, and $b$ satisfy the conditions. Suppose that there is an algorithm $\mathcal{A}$ that solves $s_{HBDDP}$-HBDDP with overwhelming probability (over the random coins chosen by $\mathcal{A}$) for a $\epsilon$ fraction of prime ideals $J$ of $R$ having norm in $[a, b]$. Then, there is an algorithm $\mathcal{B}$, which given access to a factoring oracle, solves with overwhelming probability the $s_{WBDDP}$-WBDDP for any (worst-case) ideal $M$ of $R$ with norm in $[N, 2N]$ when $2t \cdot s_{WBDDP} \le s_{HBDDP}$. Regarding running times, $time(\mathcal{B}) = time(\mathcal{A}) \cdot \mathrm{poly}(n)/\epsilon$.*

The conditions are as follows ($s$ refers to $s_{WBDDP}$):

- $\log N$ and $\log b$ are only polynomial in the lattice dimension $n$
- $s = \omega(\sqrt{\log n})$,
- $s = \gamma_f \cdot (b/N)^{1/n} \cdot \omega(\sqrt{\log n})$,
- $t \ge \gamma_f \cdot n^{1.5} \cdot s$,
- $|\mathcal{I}_{a,b}|/b$ is non-negligible, where $\mathcal{I}_{a,b}$ is the set of prime ideals with norm in $[a, b]$,
- $a/b$ is non-negligible,
- $a^2 > 2N \cdot e t_0^n$ where $e$ is Euler's constant and $t_0 = t + s \cdot \sqrt{n}$.

*Remark 1.* Asymptotically, the requirement that $|\mathcal{I}_{a,b}|/b$ be non-negligible will be satisfied if $(b - a)/b$ is non-negligible. See Theorems 1 and 2.

To make the conditions more comprehensible, let us consider a concrete choice of parameters. Set $N = b = 2a$. Then, for any $g(n) = \omega(\sqrt{\log n})$, we can set

$s = \gamma_f \cdot g(n)$ and $t = \gamma_f{}^2 \cdot n^{1.5} \cdot g(n)$. The condition $a^2 = N^2/4 > 2N \cdot et_0^n$ is met when $N/8 > et_0^n \approx et^n = e \cdot \gamma_f{}^{2n} \cdot n^{1.5n} \cdot g(n)^n$. This is a very mild lower bound for $N$, considering that $N$ is related to the norm of $M$. In particular, the condition $a^2 > 2N \cdot et_0^n$ can be met even when $\lambda_n(M)$ is small – e.g., polynomial in $n$.

A "deficiency" of the reduction is that, according to the conditions, the norm of the output average-case ideal is lower-bounded in terms of the norm of the worst-case ideal. It would be preferable to remove this constraint. In a reduction described in the full version, we show that ideals with "small" norms are the "hard case" when one is given access to a factoring oracle, and therefore our reductions ultimately apply even to average-case ideals with fairly small norms.

### 3.3    The Randomizeldeal Algorithm

Toward proving Theorem 3, we present an algorithm Randomizeldeal that, assuming the conditions are met, "randomizes" a worst-case lattice into our average-case distribution. In Section 3.4, we show that one can solve WBDDP by using Randomizeldeal in combination with a HBDDP-solver.

Randomizeldeal$(R, M, N, s, t, a, b)$:

---

1. Outputs $\perp$ if the parameters do not satisfy the conditions.
2. Generates a vector $\mathbf{v}$ per the distribution $D_{M^{-1}, s, t \cdot \mathbf{e}_1}$; sets $L \leftarrow M \cdot (\mathbf{v})$.
3. Uses a factoring oracle to compute lattice bases of the prime ideal divisors $\{\mathfrak{p}_i\}$ of $L$.
4. Sets $J$ to be an ideal in $\{\mathfrak{p}_i\}$ with norm in $[a, b]$; if none exists, it aborts.
5. With probability $\mathrm{Nm}(J)/b$, outputs a basis $\mathbf{B}_J$ of $J$, along with the vector $\mathbf{v}$; otherwise, it aborts.

---

Regarding Step 2, one can sample from $D_{M^{-1}, s, t \cdot \mathbf{e}_1}$ by using the GPV algorithm [12] with the independent set $\{\mathbf{e}_i\}$ in $M^{-1}$.

Regarding Step 3, let $R' = \mathbb{Z}[x]/(f(x))$ and consider the following theorem.

**Theorem 4 (Kummer-Dedekind, as given in [33]).** *Consider the factorization $f(x) = \prod_i g_i(x)^{e_i} \bmod p$ for prime integer $p$. The prime ideals $\mathfrak{p}_i \in \mathbb{Z}[x]/(f(x))$ of $R'$ whose norms are powers of $p$ are precisely*

$$\mathfrak{p}_i = (p, g_i(x))$$

There are polynomial time algorithms for factoring polynomials in $\mathbb{Z}_p[x]$ – e.g., by Kaltofen and Shoup [14]. Therefore, in $R'$, if we have an integer factoring algorithm to factor $\mathrm{Nm}(L)$, we can efficiently discover all of the prime ideals that divide $L$. See [33] for details on how to extend this approach to rings $R \supset R'$. Note that since $R = \mathcal{O}_F$, the factorization in Step 3 is unique.

Regarding Step 4, there will be at most one ideal in $\{\mathfrak{p}_i\}$ with norm in $[a, b]$. If there were two such ideals $\mathfrak{p}_i, \mathfrak{p}_j$, the norm of their product would be at least $a^2 > 2N \cdot et_0^n$, where we will show the latter term exceeds the norm of $L$, a contradiction.

Before proving the reduction, we must establish that Randomizeldeal outputs $J$ according to our desired average-case distribution. We prove this in Lemma 6. Lemmas 3, 4 and 5 establish some preliminary facts.

**Lemma 3.** *Suppose the conditions are met. The probability that the ideal $L$ has a divisor in $\mathcal{I}_{a,b}$ is non-negligible.*

*Proof.* See full version.

**Lemma 4.** *Suppose $\mathbf{v} = \mathbf{e}_1 + \mathbf{u}$ for $\|\mathbf{u}\| \leq 1/(2\gamma_f)$. Then, $e^{-2n \cdot \gamma_f \cdot \|\mathbf{u}\|} \leq \mathrm{Nm}((\mathbf{v}))$ $\leq e^{n \cdot \gamma_f \cdot \|\mathbf{u}\|}$. In particular, when $\mathbf{v} \in t \cdot \mathbf{e}_1 + \mathcal{B}(s\sqrt{n})$, $\mathrm{Nm}((\mathbf{v})) \leq e \cdot t_0^n$.*

*Proof.* (Lemma 4) See full version.

**Lemma 5.** *Suppose the conditions are met. Randomizeldeal$(R, M, N, s, t, a, b)$ aborts with non-overwhelming probability.*

*Proof.* (Lemma 5) For Step 5, the probability of aborting is non-overwhelming, since $a/b$ is non-negligible and $\mathrm{Nm}(J) \geq a$. Regarding Step 4, we use Lemma 3, which establishes that, for our choice of parameters, there is a non-negligible probability that $M \cdot (\mathbf{v})$ has a prime ideal divisor with norm in $[a, b]$ when $\mathbf{v}$ is sampled according to the above distribution. □

**Lemma 6.** *Suppose the conditions are met. Then, Randomizeldeal samples $J$ as a statistically uniform prime ideal (independent of $M$) subject to the constraint that $\mathrm{Nm}(J) \in [a, b]$.*

*Proof.* (Lemma 6) Consider the probability that a particular prime ideal $J_0$ with norm in $[a, b]$ is chosen as the ideal $J$ in Step 4 in a single trial if there is no abort. (By Lemma 5, the probability of abort is non-overwhelming.) Assuming $\mathbf{v} \in t \cdot \mathbf{e}_1 + \mathcal{B}(s \cdot \sqrt{n})$ (which is indeed the case with overwhelming probability by Lemma 1), we claim that $J_0$ is chosen iff $\mathbf{v} \in J_0 M^{-1}$.

For the 'if' direction of our claim, if $\mathbf{v} \in J_0 M^{-1}$, then $J_0$ divides (is a super-lattice of) $L \leftarrow M \cdot (\mathbf{v})$. Since $\mathrm{Nm}((\mathbf{v})) \leq e t_0^n$ when $\mathbf{v} \in t \cdot \mathbf{e}_1 + \mathcal{B}(s \cdot \sqrt{n})$ by Lemma 4, we have that $\mathrm{Nm}(L) = \mathrm{Nm}(M) \cdot \mathrm{Nm}((\mathbf{v})) \leq 2N \cdot e t_0^n < a^2 \leq \mathrm{Nm}(J_0)^2$. Consequently, besides $J_0$, $L$ cannot have any other prime ideal divisors with norm in $[a, b]$, and $J_0$ is chosen. For the 'only if' direction, that $J_0$ is chosen implies that $J_0$ divides (is a super-lattice of) $L = M \cdot (\mathbf{v})$. But then $J_0 M^{-1}$ is a super-lattice of $M^{-1} M \cdot (\mathbf{v}) = (\mathbf{v})$. Therefore, $(\mathbf{v})$ is contained in $J_0 M^{-1}$; in particular, $\mathbf{v} \in J_0 M^{-1}$.

Given our claim, for fixed $M$, the probability that $J_0$ is chosen in Step 4 is:

$$\Pr[J_0] \approx \frac{\sum_{\mathbf{v} \in J_0 M^{-1}} \Pr[\mathbf{v}]}{\sum_{\mathbf{v} \in M^{-1}} \Pr[\mathbf{v}]} = \frac{\rho_{s, t \cdot \mathbf{e}_1}(J_0 M^{-1})}{\rho_{s, t \cdot \mathbf{e}_1}(M^{-1})}$$

(The approximate equality holds up to negligible error, since it relies on $\mathbf{v} \in t \cdot \mathbf{e}_1 + \mathcal{B}(s \cdot \sqrt{n})$.)

We claim that $s$ exceeds the smoothing parameters of $J_0 M^{-1}$ and $M^{-1}$. Assuming this claim, Lemma 2 implies that

$$\frac{\rho_{s, t \cdot \mathbf{e}_1}(J_0 M^{-1})}{\rho_{s, t \cdot \mathbf{e}_1}(M^{-1})} \approx \mathrm{Nm}(M^{-1})/\mathrm{Nm}(J_0 M^{-1}) = 1/\mathrm{Nm}(J_0)$$

up to negligible error. Step 5 uses rejection to adjust this probability from $1/\mathrm{Nm}(J_0)$ to $1/b$, making the distribution statistically uniform (and statistically independent of $M$) over all prime ideals with norms in $[a, b]$.

It remains to prove our claim that $s$ exceeds the smoothing parameters of $J_0 M^{-1}$ and $M^{-1}$. This is clearly true for $M^{-1}$, which contains $\mathbb{Z}^n$ as a sublattice. Regarding $J_0 M^{-1}$, we have

$$
\begin{aligned}
s &= \gamma_f \cdot (b/N)^{1/n} \cdot \omega(\sqrt{\log n}) \\
&\geq \gamma_f \cdot \mathrm{Nm}(J_0)^{1/n}/\mathrm{Nm}(M)^{1/n} \cdot \omega(\sqrt{\log n}) \\
&\geq \gamma_f \cdot \mathrm{Nm}(J_0)^{1/n} \cdot \mathrm{Nm}(M^{-1})^{1/n} \cdot \omega(\sqrt{\log n}) \\
&\geq \gamma_f \cdot \mathrm{Nm}(J_0 M^{-1})^{1/n} \cdot \omega(\sqrt{\log n}) \\
&\geq \gamma_f \cdot \lambda_1(J_0 M^{-1}) \cdot \omega(\sqrt{\log n}) \\
&\geq \lambda_n(J_0 M^{-1}) \cdot \omega(\sqrt{\log n})
\end{aligned}
$$

and the claim follows.                                                          $\square$

## 3.4  Proof of the Reduction

Finally, we prove Theorem 3, showing how to use the procedure Randomizeldeal to reduce WBDDP to HBDDP.

Intuitively, Randomizeldeal samples a vector $\mathbf{v}$ that is "nearly parallel" to $\mathbf{e}_1$ (since $t \gg s$), so that multiplying the basis vectors in $\mathbf{B}_M$ by $\mathbf{v}$ is similar (from a geometric perspective) to multiplying by $t$. Thus, $L$ is geometrically similar to a simple scaling of $M$, and it is easy to see how a solution to a lattice problem over $L$ (e.g., to BDDP or SIVP) yields a solution to a lattice problem over $M$. As long as $L$ is not an overly sparse subset of $J$ – e.g., suppose that $(\mathrm{Nm}(L)/\mathrm{Nm}(J))^{1/n}$ is poly($n$) – then $\lambda_1(J)$ will be only poly($n$) less than $\lambda_1(L)$, and the BDDP solution to $(L, \mathbf{u}')$ will be the same as to $(J, \mathbf{u}')$ as long as $\mathbf{u}'$ is sufficiently close to $L$.

*Proof.* (Theorem 3) $\mathcal{B}$ wants to solve the WBDDP instance $(M, \mathbf{u})$. It does the following:

1. Runs $(\mathbf{B}_J, \mathbf{v}) \xleftarrow{\mathrm{R}} \mathsf{Randomizeldeal}(R, M, N, s, t, a, b)$.
2. Sets $\mathbf{u}' \leftarrow (\mathbf{u} \times \mathbf{v}) \bmod \mathbf{B}_J$.
3. Runs $\mathcal{A}$ on the instance $(J, \mathbf{u}')$, receiving back a vector $\mathbf{y}$ such that $\mathbf{u}' - \mathbf{y} \in J$. (If $\mathcal{A}$ does not solve this instance, restart.)
4. Outputs $\mathbf{x} \leftarrow \mathbf{y}/\mathbf{v}$.

First, we verify that $(J, \mathbf{u}')$ is a valid HBDDP instance that should be solvable by $\mathcal{A}$. By Lemma 6, RandomizeIdeal outputs the basis of an ideal $J$ that is statistically uniform among invertible prime ideals with norm in $[a, b]$.

Now let us check that $\mathbf{u}'$ is also valid. By assumption, there exist $\mathbf{m} \in M$ and $\mathbf{z}$ with $\|\mathbf{z}\| \leq s_{\text{WBDDP}}$ such that $\mathbf{u} = \mathbf{m} + \mathbf{z}$. So, $\mathbf{u}' = \mathbf{m}' + \mathbf{z}'$, where $\mathbf{m}' \in M \cdot (\mathbf{v})$ and $\mathbf{z}' = \mathbf{z} \times \mathbf{v}$. Assuming $\mathbf{v} \in t \cdot \mathbf{e}_1 + \mathcal{B}(s \cdot \sqrt{n})$, which occurs with overwhelming probability, we have

$$\|\mathbf{z}'\| = \|\mathbf{z} \times \mathbf{v}\| \leq t \cdot \|\mathbf{z}\| + \gamma_f \cdot s \cdot \sqrt{n} \cdot \|\mathbf{z}\| \leq 2t \cdot s_{\text{WBDDP}} \leq s_{\text{HBDDP}}$$

Since $M \cdot (\mathbf{v})$ is a sub-lattice of $J$, we have that $\mathbf{u}' = \mathbf{j} + \mathbf{z}'$ for some $\mathbf{j} \in J$.

By the analysis above, $\mathcal{A}$ should solve the instance $(J, \mathbf{u}')$ with probability at least $\epsilon$. If $\mathcal{A}$ solves this instance – i.e., $\mathcal{B}$ receives from $\mathcal{A}$ the unique vector $\mathbf{y}$ with $\|\mathbf{y}\| < s_{\text{HBDDP}}$ such that $\mathbf{u}' - \mathbf{y} \in J$. It must be that $\mathbf{y} = \mathbf{z}'$. Thus $\mathbf{x} = \mathbf{z}'/\mathbf{v} = \mathbf{z}$, and $\mathcal{B}$ solves its WBDDP instance.

The probability that RandomizeIdeal does not abort and $\mathcal{A}$ succeeds is at least $\epsilon/\text{poly}(n)$. These probabilities are independent over trials, and the claimed running time of $\mathcal{B}$ follows.    $\square$

## 4    KeyGen According to the Average-Case Distribution

### 4.1    Our Approach at a High Level

For KeyGen, we want an algorithm IdealGen that generates a random ideal $J$ together with a short vector in $\mathbf{w} \in J^{-1}$ to be used as the secret key. Recall how decryption works in Gentry's somewhat homomorphic scheme, and suppose that $R = \mathbb{Z}[x]/(f(x))$ in this subsection for simplicity. A ciphertext is an integer vector of the form $\mathbf{c} = \mathbf{j} + \mathbf{e}$, where $\mathbf{j} \in J$ and $\mathbf{e}$ is a short noise vector containing the message. Decryption involves computing the fractional part $[\mathbf{w} \times \mathbf{c}]$, which equals $[\mathbf{w} \times \mathbf{e}]$ since $\mathbf{w} \times \mathbf{j}$ is in $R$ and thus an integer vector. If $\mathbf{w}$ and $\mathbf{e}$ are short enough – in particular, if we have the guarantee that all of the coefficients of $\mathbf{w} \times \mathbf{e}$ have magnitude less than $1/2$ – then $[\mathbf{w} \times \mathbf{e}]$ equals $\mathbf{w} \times \mathbf{e}$ exactly. From $\mathbf{w} \times \mathbf{e}$, the decrypter can recover $\mathbf{e}$ and the message.

How short should $\mathbf{w}$ be? Since $\lambda_n(J^{-1})$ is at least $\text{Nm}(J)^{-1/n}$, we cannot expect $\mathbf{w}$ to be much shorter than this. (Recall that we choose $R$ such that $\lambda_n(I)/\lambda_1(I)$ is polynomial in $n$.) So, we will consider $\mathbf{w}$ to be a "good" secret key with respect to ideal $J$ if $\|\mathbf{w}\| \leq g(n) \cdot \text{Nm}(J)^{-1/n}$ for some small polynomial $g(n)$. Now, how do we generate a random ideal $J$ together with a "good" $\mathbf{w} \in J^{-1}$?

Our first step is to generate a "small" random ideal $K$ – "small" in the sense that its norm is in $[n^{cn}, 2n^{cn}]$ for some small constant $c$, which guarantees that $\lambda_n(K) = \text{poly}(n)$. Since the norm of $K$ is so small, $\mathbf{e}_1 \in K^{-1}$ is trivially a good secret key for $K$ according to our definition. $K$ is not useful as the ideal in Gentry's scheme, since even very small errors $\mathbf{e}$ make ciphertexts indecipherable.

But suppose, as a thought experiment, that we simply set $J = K \cdot (\mathbf{v})$ where $\mathbf{v} = T \cdot \mathbf{e}_1$ for some large integer $T$. That is, $J$ is simply a scaling of $K$. Then, $\mathbf{w} \leftarrow \mathbf{e}_1/T$ is a vector in $J^{-1}$ that satisfies our definition of a good secret key. And $J$ is "large" enough to handle larger error vectors.

However, the simple scaling approach is obviously unsatisfactory for a few reasons. First, it does not generate $J$ according to our desired average-case distribution. Also, it may not even be secure: all of the coefficients of $J$'s vectors are divisible by $T$, and thus a ciphertext $\mathbf{c}$ leaks the value of $\mathbf{e}$ mod $T$. Obviously, we want to avoid these deficiencies.

Instead, as our second step, we sample $\mathbf{v} \leftarrow D_{K^{-1}, S, T \cdot \mathbf{e}_1}$ where $T/S = \mathrm{poly}(n)$. Then, as before, we set $J = K \cdot (\mathbf{v})$, and $\mathbf{w} \leftarrow \mathbf{e}_1 / \mathbf{v}$. That is, we do the same thing as in the simple scaling approach, except that we sample $\mathbf{v}$ from $K^{-1}$ rather than from $R$, and we choose it to be very close to $T \cdot \mathbf{e}_1$ rather than being exactly equal. It turns out that, if $\mathbf{v}$ is very close to $T \cdot \mathbf{e}_1$, then $1/\mathbf{v}$ is very close to $\mathbf{e}_1/T$. In particular, $\mathbf{w}$ will be a good secret key for $J$. Fortunately, this approach avoids the deficiencies of simple scaling. We can prove that, by including a couple of rejection steps – to output $J$ only if it is prime, to fine-tune the output distribution, etc. – the $J$ sampled using this approach has the correct average-case distribution.

Intuitively, why does this approach induce a random distribution on $J$? At a very high level, we can ask: is $J$ random *geometrically* (e.g., when one considers the "shape" of the parallelepiped formed by $J$'s shortest independent set), and is $J$ random *algebraically* (e.g., when one considers $J$'s norm)? Geometrically, $J$ inherits $K$'s shape, since (up to some perturbation in the sampling of $\mathbf{v}$) it is a simple scaling of $K$. We choose $K$ from a large enough space so that its shape, and hence $J$'s shape, is quite "random". Algebraically, the fact that $\mathbf{v}$ is sampled from $K^{-1}$ "randomizes" $J$ algebraically – in particular, $J$ is not divisible by $K$. But these are only intuitions. Before providing a more precise explanation, we need to describe our IdealGen algorithm in more detail.

## 4.2   IdealGen: The Details

IdealGen uses parameters $s = \omega(\sqrt{\log n})$, $t$ such that $t \geq 42 \cdot \gamma_f \cdot s \cdot n^{1.5}$ and $t > 8 \cdot \gamma_f \cdot s \cdot n^{1.5} \cdot \|f\|^2$, and $\alpha \geq 1$; let $S = s \cdot \alpha$ and $T = t \cdot \alpha$. It invokes an algorithm TempIdeal$(R, i, j)$, described in Section 4.3, that outputs a uniformly random ideal $K$ with norm in $[i, j]$ (but not a nontrivial "good" key for $K$). IdealGen ultimately outputs a uniformly random prime ideal $J$ with norm in $[2, 3] \cdot t^{2n} T^n$.

IdealGen:

1. Runs $\mathbf{B}_K \xleftarrow{\mathrm{R}} \mathsf{TempIdeal}(R, t^{2n}, 4t^{2n})$.
2. Samples $\mathbf{v} \xleftarrow{\mathrm{R}} D_{K^{-1}, S, T \cdot \mathbf{e}_1}$ and sets $\mathbf{w} \leftarrow 1/\mathbf{v}$; aborts if $\mathbf{v} \notin T \cdot \mathbf{e}_1 + \mathcal{B}(2S\sqrt{n})$.
3. Sets $J \leftarrow K \cdot (\mathbf{v})$; aborts if $J$ is not prime or $\mathrm{Nm}(J) \notin [2, 3] \cdot t^{2n} T^n$.
4. Continues to Step 5 with probability $\mathrm{Nm}(K)/4t^{2n}$; otherwise, aborts.
5. Continues to Step 6 with probability $\beta \cdot \frac{\rho_{S/T^2, (1/T) \cdot \mathbf{e}_1}(\mathbf{w})}{\rho_{S, T \cdot \mathbf{e}_1}(\mathbf{v})}$, where $\beta$ will be defined later; otherwise, aborts.
6. With probability $2t^{2n} T^n / \mathrm{Nm}(J)$, outputs $\mathbf{w}$ and the Hermite normal form of $J$; otherwise, aborts.

*Remark 2.* IdealGen is precisely what we outlined above, aside from the probability of aborting in Steps 2-6. We will show that the probability of aborting is non-overwhelming, and that these steps fine-tune the distribution so that $J$ is a uniformly random prime ideal with norm in the prescribed interval. The algorithm can be re-run until it completes successfully.

*Remark 3.* In Step 2, one can sample from the distribution $D_{K^{-1},S,T\cdot\mathbf{e}_1}$ by using the GPV algorithm [12] with the independent set $\{\mathbf{e}_i\}$ in $K^{-1}$.

*Remark 4.* By Lemma 1, the vector $\mathbf{v}$ is in $T \cdot \mathbf{e}_1 + \mathcal{B}(S\sqrt{n})$ with overwhelming probability. Note that we only abort in Step 2 if $\mathbf{v} \notin T \cdot \mathbf{e}_1 + \mathcal{B}(2S\sqrt{n})$. We use a ball of radius $2S\sqrt{n}$ instead of $S\sqrt{n}$ in Step 2 for technical reasons – specifically, Corollary 2 below and its use in the proof of Theorem 7.

*Remark 5.* Regarding Step 5, we must ensure that the "probability" is a number in $[0,1]$. We show that $\rho_{S/T^2,(1/T)\cdot\mathbf{e}_1}(\mathbf{w})/\rho_{S,T\cdot\mathbf{e}_1}(\mathbf{v}) \in [e^{-6\pi\sqrt{1/n}}, e^{6\pi\sqrt{1/n}}]$. (See Lemma 10.) Therefore, we can take $\beta \leftarrow e^{-6\pi\sqrt{1/n}}$.

To begin analyzing our IdealGen algorithm, we state some useful lemmas about the vector $\mathbf{v}$ sampled in Step 2. Omitted proofs can be found in the full version. The theme of these lemmas is that since $\mathbf{v}$ is very close to $T \cdot \mathbf{e}_1$, it behaves in many respects like $T \cdot \mathbf{e}_1$.

**Lemma 7.** *If $\mathbf{v} \in T \cdot \mathbf{e}_1 + \mathcal{B}(2S\sqrt{n})$, then $\mathrm{Nm}((\mathbf{v})) \in [T^n/1.1, 1.1 \cdot T^n]$.*

**Lemma 8.** *If $\mathbf{v} \in T \cdot \mathbf{e}_1 + \mathcal{B}(2S\sqrt{n})$, then it is the only vector in $(\mathbf{v})$ inside that ball.*

**Lemma 9.** *If $\|\mathbf{u}\| < 1/\gamma_f$, then*

$$\mathbf{e}_1/(\mathbf{e}_1 - \mathbf{u}) = \mathbf{e}_1 + \mathbf{u} + \mathbf{x} \quad \text{for} \quad \|\mathbf{x}\| \le \frac{\gamma_f \cdot \|\mathbf{u}\|^2}{1 - \gamma_f \cdot \|\mathbf{u}\|}$$

**Corollary 1.** *If $\mathbf{v} \in T \cdot \mathbf{e}_1 + \mathcal{B}(2S\sqrt{n})$, then $\mathbf{w} \in \mathbf{e}_1/T + \mathcal{B}(4S\sqrt{n}/T^2)$.*

**Corollary 2.** *If $\mathbf{w} \in \mathbf{e}_1/T + \mathcal{B}(S\sqrt{n}/T^2)$, then $\mathbf{v} \in T \cdot \mathbf{e}_1 + \mathcal{B}(2S\sqrt{n})$.*

**Lemma 10.** *If $\mathbf{v} \in T \cdot \mathbf{e}_1 + \mathcal{B}(2S\sqrt{n})$, then*

$$\rho_{S,T\cdot\mathbf{e}_1}(\mathbf{v})/\rho_{S/T^2,(1/T)\cdot\mathbf{e}_1}(\mathbf{w}) \in [e^{-6\pi\sqrt{1/n}}, e^{6\pi\sqrt{1/n}}]$$

Our main results about IdealGen are captured in Theorems 5, 6, and 7 – namely, that it outputs a good secret key for $J$, it does not abort very often (and therefore can be efficiently re-run until it outputs a result), and it outputs $J$ according to the desired average-case distribution.

**Theorem 5.** *The vector $\mathbf{w}$ output by IdealGen is a "good" key for $J$. Specifically, $\|\mathbf{w}\| < 6t^2 \cdot \mathrm{Nm}(J)^{-1/n}$.*

*Proof.* (Theorem 5) By Corollary 1, $\mathbf{w} \in \mathbf{e}_1/T + \mathcal{B}(4S\sqrt{n}/T^2)$. So, clearly, $\|\mathbf{w}\| < 2/T$. On the other hand, $\mathrm{Nm}(J)^{-1/n} \geq 1/(3^{1/n}t^2T)$. The result follows.    □

**Theorem 6.** *The probability of aborting in Steps 2-6 is non-overwhelming.*

*Proof.* (Theorem 6) For Steps 4 and 6, the claim is clearly true. For Step 2, it follows from Lemma 1.

For Step 5, we invoke Lemma 10, which implies we can set $\beta \leftarrow e^{-6\pi\sqrt{1/n}}$, and the algorithm will continue to Step 6 with at least (non-negligible) probability $e^{-12\pi\sqrt{1/n}}$.

For Step 3, an abort occurs if $J$ is not prime or $\mathrm{Nm}(J) \notin [2,3] \cdot t^{2n}T^n$. Asymptotically, Theorems 1 and 2 imply that, for an interval $[cx, x]$ with constant $c < 1$, prime ideals are a $O(1/\log x)$ fraction of ideals. Given that $\mathrm{Nm}(J) = \mathrm{Nm}(K) \cdot \mathrm{Nm}((\mathbf{v}))$ and $\mathrm{Nm}((\mathbf{v})) \in [T^n/1.1, 1.1 \cdot T^n]$ (by Lemma 7), $\mathrm{Nm}(J)$ falls outside the interval only if $\mathrm{Nm}(K)$ falls outside of $[2 \cdot 1.1, 3/1.1] \cdot t^{2n}$. By the distribution of ideals (see Theorems 1 and 2) and the claimed distribution of TempIdeal, this occurs only with only constant probability, in which case the probability of aborting in Step 2 is a constant.    □

Before getting to the last theorem, we state one more lemma.

**Lemma 11.** *Let $J$ be an ideal such that $\mathrm{Nm}(J) \in [2,3] \cdot t^{2n}T^n$. Then $S/T^2$ exceeds the smoothing parameter of $J^{-1}$.*

*Proof.* (Lemma 11) We have

$$\frac{S}{T^2} = \frac{s}{tT} \geq \frac{s \cdot \gamma_f}{2^{1/n}t^2T} \geq \frac{s \cdot \gamma_f}{\mathrm{Nm}(J)^{1/n}} \geq s \cdot \gamma_f \cdot \lambda_1(J^{-1}) \geq s \cdot \lambda_n(J^{-1}) \ .$$

Since $s = \omega(\sqrt{\log n})$, the result follows.    □

**Theorem 7.** *For any $\alpha \geq 1$, IdealGen with parameter $\alpha$ efficiently outputs a prime ideal $J$ that is statistically uniform subject to the constraint that $\mathrm{Nm}(J) \in [2,3] \cdot t^{3n}\alpha^n$.*

*Proof.* (Theorem 7) Let $\mathcal{K}$ be the sets of ideals with norms in $[1,4] \cdot t^{2n}$, and let $\mathcal{J}$ be the sets of prime ideals with norms in $[2,3] \cdot t^{2n}T^n$. For convenience, we define some sets of ideals associated to $J \in \mathcal{J}$. Let

$$\mathcal{S}_J = \{K \in \mathcal{K} : \exists \mathbf{v} \text{ s.t. } J = K \cdot (\mathbf{v}) \text{ and } \mathbf{v} \in T \cdot \mathbf{e}_1 + \mathcal{B}(2S\sqrt{n})\}$$
$$\mathcal{V}_J = \{\mathbf{w} : J \cdot (\mathbf{w}) \in \mathcal{K} \text{ and } 1/\mathbf{w} \in T \cdot \mathbf{e}_1 + \mathcal{B}(2S\sqrt{n})\}$$
$$\mathcal{W}_J = \{\mathbf{w} : J \cdot (\mathbf{w}) \in \mathcal{K} \text{ and } \mathbf{w} \in (1/T) \cdot \mathbf{e}_1 + \mathcal{B}(S\sqrt{n}/T^2)\}$$

Define $\mathcal{S}'_J$ identically to $\mathcal{S}_J$, except they include only those $K$ for which there is *exactly* one such $\mathbf{v}$. Lemma 8 implies that $\mathcal{S}_J = \mathcal{S}'_J$.

Consider the probability $\Pr[J_0]$ that a particular ideal $J_0$ is chosen as $J$ in Step 3. We have

$$\Pr[J_0] = \sum_{K \in \mathcal{S}_{J_0}} \Pr[J_0 \wedge K] = c_1 \cdot \sum_{K \in \mathcal{S}_{J_0}} \Pr[J_0|K] = c_1 \cdot \sum_{K \in \mathcal{S}'_{J_0}} \Pr[J_0|K],$$

for some universal constant $c_1$, where the second inequality follows from the fact that $K$ is chosen uniformly by TempIdeal.

For a particular candidate pair $(K_0, J_0)$ with $K_0 \in S'_{J_0}$, let $\mathbf{v}_0$ be the unique vector in $J_0 K_0^{-1} \cap (T \cdot \mathbf{e}_1 + \mathcal{B}(2S\sqrt{n}))$. We claim that, at Step 3,

$$\Pr[J_0 | K_0] = \rho_{S, T \cdot \mathbf{e}_1}(\mathbf{v}_0) / \rho_{S, T \cdot \mathbf{e}_1}(K_0^{-1})$$

This follows because the latter quantity is $\Pr[\mathbf{v}_0 | K_0]$, and from the fact that $J_0$ and $\mathbf{v}_0$ determine each other once $K_0$ is fixed.

Now, consider the denominator $\rho_{S, T \cdot \mathbf{e}_1}(K_0^{-1})$; we claim that, for fixed $(S, T)$, this sum is proportional to $\mathrm{Nm}(K_0)$, up to negligible error. This follows from Lemma 2, and the fact that $S$ exceeds the smoothing parameter of $K_0^{-1}$ (since $\mathbb{Z}^n$ is a sub-lattice of $K_0^{-1}$). So, after Step 3, we have

$$\Pr[J_0 | K_0] = c_2 \cdot \rho_{S, T \cdot \mathbf{e}_1}(\mathbf{v}_0) / \mathrm{Nm}(K_0)$$

up to negligible error for some universal constant $c_2$. After Steps 4 and 5, we have

$$\Pr[J_0 | K_0] = c_3 \cdot \rho_{S/T^2, (1/T) \cdot \mathbf{e}_1}(\mathbf{w}_0)$$

up to negligible error for some universal constant $c_3$, where $\mathbf{w}_0 = 1/\mathbf{v}_0$ and thus

$$\Pr[J_0] = c_4 \cdot \sum_{K_0 \in S'_{J_0}} \rho_{S/T^2, (1/T) \cdot \mathbf{e}_1}(\mathbf{w}_0)$$

We claim that

$$\sum_{K_0 \in S'_{J_0}} \rho_{\frac{S}{T^2}, \frac{\mathbf{e}_1}{T}}(\mathbf{w}_0) = \sum_{\mathbf{w}_0 \in \mathcal{V}_{J_0}} \rho_{\frac{S}{T^2}, \frac{\mathbf{e}_1}{T}}(\mathbf{w}_0) = \rho_{\frac{S}{T^2}, \frac{\mathbf{e}_1}{T}}(J_0^{-1}) = c_5 \cdot \mathrm{Nm}(J_0) \quad (1)$$

up to negligible error for some universal constant $c_5$. This claim lets us complete the proof. The abort in Step 6 adjusts this probability so that it becomes $c_5 \cdot 2t^{2n}T^n$, independent of $J_0$, and thus makes $\Pr[J_0]$ statistically uniform across all $J_0 \in \mathcal{J}$.

In Equation 1, the second sum is just a syntactic rewriting of the first sum.

To prove the second equality in Equation 1, first note that $\mathcal{W}_{J_0} \subset \mathcal{V}_{J_0} \subset J_0^{-1}$. The first inclusion follows from the fact that, by Lemma 2, for every $\mathbf{w}_0 \in (1/T) \cdot \mathbf{e}_1 + \mathcal{B}(S\sqrt{n}/T^2)$, it is the case that $1/\mathbf{w}_0 \in T \cdot \mathbf{e}_1 + \mathcal{B}(2S\sqrt{n})$. The second inclusion follow from the fact that each $\mathbf{w}_0$ satisfies $(\mathbf{w}_0) = J^{-1}K$ for some $K \in \mathcal{K}$; in particular, $\mathbf{w}_0 \in J_0^{-1}$. Now, we claim that

$$\sum_{\mathbf{w}_0 \in \mathcal{W}_{J_0}} \rho_{\frac{S}{T^2}, \frac{\mathbf{e}_1}{T}}(\mathbf{w}_0) = \rho_{\frac{S}{T^2}, \frac{\mathbf{e}_1}{T}}(J_0^{-1})$$

up to negligible error, which would establish the second equality (up to negligible error). This equality holds because $\mathcal{W}_{J_0}$ contains all of the $\mathbf{w}_0$'s in $J_0^{-1}$ that contribute substantially to the sum. Specifically, since $S/T^2$ exceeds the smoothing parameter of $J_0^{-1}$ (by Lemma 11), the sum $\rho_{\frac{S}{T^2},\frac{e_1}{T}}(J_0^{-1})$ is only negligibly affected when restricted to the set $J_0^{-1} \cap ((1/T) \cdot e_1 + \mathcal{B}(S\sqrt{n}/T^2))$ (Lemma 1). However, this set is contained in $\mathcal{W}_J$, since if we set $K_0 \leftarrow J_0 \cdot (w_0)$, then $K_0$ is indeed in $\mathcal{K}$, since $\mathrm{Nm}(K_0) = \mathrm{Nm}(J_0) \cdot \mathrm{Nm}((w_0))$, which is in the interval $[2/1.1, 3 \cdot 1.1] \cdot t^{2n} \subset [1,4] \cdot t^{2n}$.

The third equality in Equation 1 follows from Lemma 11 and Lemma 2.    □

One aspect of the proof may seem a bit mysterious. Why did we use Step 5 to convert $\Pr[J_0]$ from a sum of $\rho(v)$'s to a sum of $\rho(1/v)$'s? Note that $v \in J_0 K^{-1}$ for some $K$, and $w = 1/v \in J_0^{-1} K$. Summing over $\rho(w)$'s is more natural, since all of the points are in a single ideal – namely, $J_0^{-1}$. In contrast, summing over vectors in $J_0 K^{-1}$ for different $K$'s is not a sum we know how to evaluate.

### 4.3    The TempIdeal Algorithm

Here, we construct an efficient algorithm $\mathsf{TempIdeal}(R, i, j)$ that outputs a uniformly random ideal $K \subset R$ with norm in $[i, j]$. $\mathsf{TempIdeal}$ only needs to output *some* basis of $K$, not necessarily a "good" basis. Let us begin at a high level by considering some possible approaches.

Suppose we sample random $v$ from $R$, and set $K \leftarrow (v)$, re-sampling if $\mathrm{Nm}(K) \notin [i, j]$. Then, $K$ is a principal ideal, and unfortunately the probability that a "random" ideal from $R$ is principal is typically negligible in $n$. (More accurately, the field $F = \mathbb{Q}(x)/(f(x))$ has an associated *class group*, where each member of the group consists of an equivalence class of ideals. The set of principal ideals is only one class, whereas the class group size is typically exponential in $n$.) Clearly, this approach does not sample a "random" ideal.

A more promising approach is to use Kummer-Dedekind (Theorem 4), which *can* actually be used to sample a uniformly random *prime* ideal, as follows. Sample a uniform prime power $p^e \in [i, j]$, and use Kaltofen and Shoup [14] to (efficiently) compute the factorization $f(x) = \prod_i g_i(x)^{e_i} \bmod p$. Kummer-Dedekind tells us that all prime ideals of $\mathbb{Z}[x]/(f(x))$ having norm $p^e$ are of the form $(p, g_i(x))$, where $g_i(x)$ is an irreducible degree-$e$ factor of $f(x)$ modulo $p$. There can be at most $n$ ideals of norm $p^e$. If there are $r \leq n$ such factors $g_i(x)$, restart with probability $1 - r/n$. Otherwise, sample one of these $g_i(x)$'s uniformly and output $K \leftarrow (p, g_i(x))$. (It is straightforward to extend this method recover all prime ideals with norm $p^e$ in rings $\mathbb{Z}[x]/(f(x)) \subset R \subseteq \mathcal{O}_F$ [33].) This works, but unfortunately we require $\mathsf{TempIdeal}$ to sample $K$ from all ideals with norm in $[i, j]$, not just from prime ideals.

Consider the following modification to the above approach: sample a uniform (possibly composite) integer $N \in [i, j]$, and compute the factorization $f(x) = \prod_i g_i(x)^{e_i} \bmod N$, etc. But computing this factorization is hard in general when $N$ is composite. In fact, we do not see a way to generate a random ideal $K$ without knowing the factorization of its norm.

These considerations lead us to construct an algorithm for generating a random *factored* ideal whose norm is in the prescribed interval, even though, in principle, we do not need the factorization. For this task, a good place to start is to look at existing algorithms for generating a random factored integer – especially Kalai's elegantly simple algorithm [13].

### Kalai's Algorithm for Generating a Random Factored Number:

**Input:** Integer $b > 0$.

**Output:** A uniformly random number $1 \leq N \leq b$, with its factorization.

1. Generate a sequence $b \geq s_1 > s_2 > \cdots > s_\ell = 1$ by uniformly choosing $s_{i+1} \in \{1, \ldots, s_i - 1\}$. (Use $b$ as $s_0$.) Put all prime $s_i$'s in a list $L$.
2. For each $s_i \in L$, put $s_i$ into $L$ at least $k$ additional times with probability $1/s_i^k$.
3. Let $N$ be the product of the numbers in $L$ (with repetition).
4. If $N > b$, restart.
5. Output $N$ and the prime $s_i$'s with probability $N/b$; otherwise, restart.

*Remark 6.* Kalai presents his algorithm somewhat differently.

As Kalai highlights, the reason this algorithm works is because a prime $p \leq b$ is in the sequence independently with probability exactly $1/p$, since it occurs iff it is chosen before any number in $\{1, \ldots, p-1\}$. That is, we could replace the first step of Kalai's algorithm with this alternative step without affecting the output distribution:

1. For each prime number $s_i \in [1, b]$, put $s_i$ in a list $L$ with probability $1/s_i$.

Of course, the algorithm with this alternative step is grossly inefficient; Kalai's main insight is a way to obtain the same output efficiently. After this insight, the remainder of the analysis is relatively straightforward. The prime $p$ appears at least $e$ times in $L$ independently with probability $1/p^e$ through Step 2, and thus the probability that a $b$-smooth number $N$ is selected in Step 3 is proportional to $1/N$. The final two rejection steps ensure uniformity across numbers in $[1, b]$. By Mertens' theorem, the algorithm will *not* restart in Step 4 with probability $\theta(1/\log b)$. See Kalai's one page paper for more details.

Our TempIdeal algorithm is a modification of Kalai's algorithm that accounts for the fact that there could be up to $n$ prime ideals that are "tied" with the same norm. To each integer $s$, we associate $n$ ideals $\{I_{s,j}\}$. Specifically, if there are $r \leq n$ distinct prime ideals of norm $s$, we let $I_{s,1}, \ldots, I_{s,r}$ be these ideals, and set $I_{s,r+1} = \cdots = I_{s,n} = 1$.

### TempIdeal$(R, a, b)$:

1. Generate a sequence $b \geq s_1 > s_2 > \cdots > s_\ell = 1$ by uniformly choosing $s_{i+1,j} \in \{1, \ldots, s_i - 1\}$ for all $j \in \{1, \ldots, n\}$ and setting $s_{i+1} \leftarrow \max_j\{s_{i+1,j}\}$. (Use $b$ as $s_0$.) Put each $s_i$ that is a norm of a prime ideal in a list $L$.
2. For each $s_i \in L$, do the following. First, generate $j \in [1, n]$ uniformly and put the ideal $I_{s_i,j}$ into multiset $M$. Then, for each $j$, insert at least $k$ additional instances of $I_{s_i,j}$ into $M$ with probability $1/s_i^k$.

3. Remove those ideals in $M$ that are equal to 1.
4. Let $K$ be the product of the ideals remaining in $M$ (with repetition).
5. If $\mathrm{Nm}(K) \notin [a,b]$, restart.
6. Output a basis for $K$ with probability $\mathrm{Nm}(K)/b$; otherwise, restart.

*Remark 7.* Obviously, in Step 2, we could have avoided putting any ideals that equal 1 in to $M$ in the first place, since we remove them in Step 3. But we leave this in, since it will make the analysis a bit simpler.

**Theorem 8.** TempIdeal *uniformly samples an ideal* $K \subset R$ *with norm in* $[a,b]$. *The algorithm takes time* $b/(a-b) \cdot \mathrm{poly}(n, \log b)$.

To simplify the proof of Theorem 8, we define a "slow" version of the above algorithm – SlowTempIdeal – which is analogous to the "slow" version of Kalai's algorithm with the alternative first step.

SlowTempIdeal$(R, a, b)$:

1. For each $s_i \in [1,b]$ that is the norm of a prime ideal, for each $j \in [1,n]$, put at least $k$ instances of $I_{s_i,j}$ into multiset $M'$ with probability $1/s_i^k$. If there is some ideal $I_{s_i,j}$ in $M'$, put $s_i$ into $L$.
2. Run Steps 2-6 of TempIdeal$(R, a, b)$.

Now, Theorem 8 follows from Lemmas 12, 14, and 15.

**Lemma 12.** *The distribution of $L$ is the same in* TempIdeal *and* SlowTempIdeal, *and hence the two algorithms have the same output distribution.*

*Proof.* (Lemma 12) Consider the probability that a fixed $s$ is in $L$. For TempIdeal, this equals the probability that $s$ is in the sequence. If $s_i > s$, the probability that $s_{i+1} \in [1,s]$ is $s^n/(s_i-1)^n$, whereas the probability that $s_{i+1} \in [1, s-1]$ is $(s-1)^n/(s_i-1)^n$. Thus, when sampling $s_i$, the probability that $s_{i+1}$ is in $[1, s-1]$ given that it is in $[1,s]$ is $(s-1)^n/s^n$. Consequently, since $s_{i+1}$ must eventually be in $[1,s]$ for some $i$, the probability that $s$ is in the sequence is $1-(s-1)^n/s^n$. This probability is independent of whether or not other values $s'$ are in $L$. For SlowTempIdeal, the probability that none of the $n$ ideals $I_{s,j}$ is in $M'$ is $(s-1)^n/s^n$. So, the probability that some ideal $I_{s,j}$ is in $M'$, and hence $s \in L$, is the same as in TempIdeal: $1-(s-1)^n/s^n$. $\square$

**Lemma 13.** *Through Step 4 of* SlowTempIdeal, *the probability that a fixed ideal $K_0$ with prime ideal factors in $[1,b]$ is selected is*

$$\frac{1}{\mathrm{Nm}(K_0)} \cdot \prod_{\mathrm{Nm}(\mathfrak{p}) \leq b} \frac{\mathrm{Nm}(\mathfrak{p})-1}{\mathrm{Nm}(\mathfrak{p})}$$

*where the product is over prime ideals.*

*Proof.* (Lemma 13) It is clear that the multisets $M$ and $M'$ have exactly the same distribution conditioned on the list $L$. That is, if $s_i \notin L$, neither multiset contains an ideal $I_{s_i,j}$. If $s_i \in L$, then both $M$ and $M'$ contain a random non-empty multiset $S$ with elements from $\{I_{s_i,1}, \ldots, I_{s_i,n}\}$, where $\Pr[S]$ is proportional to $1/s_i^{|S|}$. Therefore, we could have used $M'$ instead of $M$ beginning in Step 3 of SlowTempIdeal without affecting the output distribution.

Remove the primes that equal 1 from $M'$. A (nontrivial) ideal $I_{s_i,j}$ is in $M'$ at least $k$ times independently with probability $1/s_i^k = 1/\mathrm{Nm}(I_{s_i,j})^k$, and therefore *exactly* $k$ times independently with probability $(\mathrm{Nm}(I_{s_i,j})-1)/\mathrm{Nm}(I_{s_i,j})^{k+1}$. By the independence of these probabilities, and by multiplicativity of the norm map over ideals, the result follows.                                                       □

**Lemma 14.** SlowTempIdeal *uniformly samples an ideal* $K \subset R$ *with norm in* $[a,b]$.

*Proof.* (Lemma 14) Given Lemma 13 – i.e., the fact that through Step 4 the probability that some $K_0$ is chosen equals $1/\mathrm{Nm}(K_0)$ times some universal constant that is independent of $K_0$ – it is clear that the final two rejection sampling steps ensure that $K$ is uniform among ideals with norm in $[a,b]$.          □

**Lemma 15.** TempIdeal *takes time* $b/(a-b) \cdot \mathrm{poly}(n, \log b)$.

*Proof.* (Lemma 15) Let us consider the probability that a restart occurs.

Regarding Step 5, by Merten's theorem for number fields, we have

$$\prod_{\mathrm{Nm}(\mathfrak{p}) \leq b} (1 - 1/\mathrm{Nm}(\mathfrak{p})) = \frac{e^{-\gamma}}{a_K} \frac{1}{\log b} + O\left(\frac{1}{\log^2 b}\right)$$

where $a_K$ is the residue of $\zeta_K(s)$, the Dedekind zeta-function, at $s = 1$, and $\gamma$ denotes Euler's constant 0.577.... Denote the above term by $\alpha$. By Lemma 13, the probability that some $K$ with norm at most $b$ is selected in Step 4 is

$$\alpha \cdot \sum_{\mathrm{Nm}(K) \leq b} 1/\mathrm{Nm}(K)$$

There are $\theta(b)$ ideals of norm at most $b$ (this follows from Theorems 1 and 2), and thus the above sum is $\Omega(1/\log(b))$.

Regarding Step 6, among $K$'s with norm at most $b$, approximately a $(b-a)/b$ fraction of them have norm at least $a$. (Again this follows from Theorems 1 and 2.) The result follows.                                                                                      □

## 5    Basing Gentry's Somewhat Homomorphic Scheme on SIVP over Ideal Lattices

We showed how to reduce WBDDP to HBDDP for our average-case distribution. It remains to base our variant of Gentry's scheme on HBDDP, and to reduce SIVP to WBDDP. We sketch these results here. Details are in the full version.

First, we specify some details of our variant. As in [10], the public key includes ideals $I$ and $J$, and a short independent set $\mathbf{B}_I$ of $I$ – e.g., where $\|\mathbf{B}_I\| = \text{poly}(n)$. $J$ is output by our new IdealGen algorithm. The cosets of $I$ form the plaintext space. Regarding $I$, we have a new requirement: that $\text{Nm}(I)$ is prime and very small – i.e., $\text{poly}(n)$. To find such an $I$, we can either construct $f(x)$ to ensure that the associated ring of integers has an ideal of small prime norm, or we can apply Kummer-Dedekind (Theorem 4) to primes of size $\text{poly}(n)$. By Theorem 8.7.7. of [5], for appropriate values of $f(x)$ and assuming GRH, applying Kummer-Dedeking will eventually give us the basis a prime ideal $\mathfrak{p}$ in $R$ having $\text{poly}(n)$-norm. From this basis, we can compute an independent set of $\mathfrak{p}$ of length at most $\text{Nm}(\mathfrak{p})$. We set $I \leftarrow \mathfrak{p}$ and $\mathbf{B}_I$ to be this independent set. We sample ciphertexts per a Gaussian distribution: $c \leftarrow c' \bmod \mathbf{B}_J$ where $c' \leftarrow D_{m+I,s,0}$ for some $s$.

To reduce HBDDP to the semantic security of this scheme, we first reduce HB-DDP to a decision problem that we call the inner ideal membership problem (IIMP): (roughly) given $(\mathbf{B}_J, \mathbf{t})$ where $\mathbf{B}_J \xleftarrow{R} \text{IdealGen}(R)$ and $\mathbf{t} \leftarrow \mathbf{x} \bmod \mathbf{B}_J$ for some $\mathbf{x} \in R$ with $\|\mathbf{x}\| < s_{\text{IIMP}}$, decide whether or not $\mathbf{x} \in I$. Essentially, a HBBDP-solver can use a IIMP-solver to find out which coset of $I$ that $\mathbf{x}$ is in. (For this search to be efficient, $\text{Nm}(I)$ must be $\text{poly}(n)$.) Using "Hensel lifting", the HBDDP-solver can recover $\mathbf{x}$ modulo $I^k$ for large $k$ – large enough that $\mathbf{x}$ becomes the shortest vector in $\mathbf{x}+I^k$ by such a large margin that is efficient to recover $\mathbf{x}$ using Babai's nearest plane algorithm. To reduce the IIMP to the semantic security of the scheme, we sample a uniform coset of $I$, set $\mathbf{u} \in R$ to be a short vector in that coset, and set the challenge ciphertext as follows: $c^* \leftarrow c' \bmod \mathbf{B}_J$ where $c' \leftarrow m_b + \mathbf{t} \times \mathbf{u} + D_{I,s,0}$. When $\mathbf{x} \in I$, $c' \in m_b + I$, and the ciphertext has the correct distribution. (This is not quite true: but we can smooth out the discrepancy by choosing $s$ large enough – in particular, so that $s/s_{\text{IIMP}} = \text{poly}(n)/\epsilon$.) When $\mathbf{x} \notin I$, $c'$ is in a random coset of $I$ that conveys no information about $m_b$. Overall, for some polynomial $g(n)$, if there is an algorithm $\mathcal{A}$ that breaks the semantic security of the scheme in time $t$ with probability $\epsilon$ for parameter $s$, then there is an algorithm that, for a $O(\epsilon)$ fraction of bases output by IdealGen, solves HBDDP for parameter $s_{\text{HBDDP}} \leq s \cdot \epsilon/g(n)$ with overwhelming probability in time $O(t \cdot \text{Nm}(I)/\epsilon)$. This reduction is entirely classical (non-quantum).

To reduce SIVP to WBDDP (quantumly), the heavy lifting has already been done by Regev [30]. He provided a quantum reduction of SIVP over the dual lattice $L^*$ to BDDP over the lattice $L$. A bit more work is necessary to turn his result into a quantum reduction of SIVP over an inverse ideal lattice $I^{-1}$ to BDDP over the ideal lattice $I$ (the inverse of an ideal lattice is not the same as its dual), and then to extend this result to SIVP over (non-inverse) ideals of $R$.

# 6   Conclusions and Open Problems

We showed that ideal lattice problems within some fixed rings are, in a sense, random self-reducible. However, the reduction uses a factoring oracle. One open problem is to find a random self-reduction that is efficient in the classical setting – in particular, to find a reduction that does not use factorization.

We presented a KeyGen algorithm that generates ideals according to our average-case distribution, together with a secret key. However, this algorithm is rather complicated, and one wonders whether there is a simpler approach.

While we are able to base Gentry's *somewhat* homomorphic encryption scheme on worst-case hardness, his FHE scheme requires an additional computational assumption – namely, that the (average-case) SSSP is hard. Currently, we do not have a worst-case / average-case reduction for the SSSP that would allow his FHE scheme to be based entirely on worst-case hardness.

ACKNOWLEDGMENTS. We thank Dan Boneh, Shai Halevi, Vadim Lyubashevsky, Chris Peikert, Oded Regev, Vinod Vaikuntanathan, and the anonymous reviewers for helpful comments and discussions.

# References

1. Ajtai, M.: Generating hard instances of lattice problems (extended abstract). In: STOC 1996, pp. 99–108 (1996)
2. Ajtai, M.: Generating hard instances of the short basis problem. In: Wiedermann, J., Van Emde Boas, P., Nielsen, M. (eds.) ICALP 1999. LNCS, vol. 1644, pp. 1–9. Springer, Heidelberg (1999)
3. Ajtai, M., Dwork, C.: A public key cryptosystem with worst-case / average-case equivalence. In: STOC 1997, pp. 284–293 (1997)
4. Alwen, J., Peikert, C.: Generating Shorter Bases for Hard Random Lattices. In: STACS 2009, pp. 75–86 (2009)
5. Bach, E., Shallit, J.: Algorithmic Number Theory, vol. 1 (1996)
6. Banaszczyk, W.: New bounds in some transference theorems in the geometry of numbers. Mathematische Annalen 296(4), 625–635 (1993)
7. Boyen, X.: Of Lettuces of Lattices: a Framework for Short Signatures and IBE with Full Security. PKC 2010 (to appear 2010)
8. Cai, J.-Y., Nerurkar, A.P.: An Improved Worst-Case to Average-Case Connection for Lattice Problems (extended abstract). In: FOCS 1997, pp. 468–477. IEEE, Los Alamitos (1997)
9. Cash, D., Hofheinz, D., Kiltz, E., Peikert, C.: Bonsai Trees, or How to Delegate a Lattice Basis. In: Gilbert, H. (ed.) EUROCRYPT 2010. LNCS, vol. 6110, pp. 523–552. Springer, Heidelberg (2010)
10. Gentry, C.: Fully Homomorphic Encryption Using Ideal Lattices. In: STOC 2009, pp. 169–178 (2009)
11. Gentry, C.: A Fully Homomorphic Encryption Scheme. Ph.D. thesis, Stanford University (2009), http://crypto.stanford.edu/craig
12. Gentry, C., Peikert, C., Vaikuntanathan, V.: Trapdoors for Hard Lattices and New Cryptographic Constructions. In: STOC 2008, pp. 197–206 (2008)
13. Kalai, A.: Generating Random Factored Numbers. Easily. J. Cryptology 16(4), 287–289 (2003); Preliminary version in SODA 2002 (2002)
14. Kaltofen, E., Shoup, V.: Subquadratic-time factoring of polynomials over finite fields. In: STOC 1995, pp. 398–406. ACM, New York (1995)
15. Landau, E.: Neuer Beweis des Primzahlsatzes und Beweis des Primidealsatzes. Mathematische Annalen 56, 645–670
16. Lenstra, A.K., Lenstra, H.W., Lovász, L.: Factoring polynomials with rational coefficients. Math. Ann. 261(4), 515–534 (1982)

17. Lyubashevsky, V., Micciancio, D.: Generalized compact knapsacks are collision resistant. In: Bugliesi, M., Preneel, B., Sassone, V., Wegener, I. (eds.) ICALP 2006. LNCS, vol. 4052, pp. 144–155. Springer, Heidelberg (2006)
18. Lyubashevky, V., Micciancio, D.: Asymptotically efficient lattice-based digital signatures. In: Canetti, R. (ed.) TCC 2008. LNCS, vol. 4948, pp. 37–54. Springer, Heidelberg (2008)
19. Lyubashevky, V., Micciancio, D.: On Bounded Distance Decoding, Unique Shortest Vectors, and the Minimum Distance Problem. In: Halevi, S. (ed.) CRYPTO 2009. LNCS, vol. 5677, pp. 577–594. Springer, Heidelberg (2009)
20. Lyubashevky, V., Peikert, C., Regev, O.: On ideal lattices and learning with errors over rings. In: Gilbert, H. (ed.) EUROCRYPT 2010. LNCS, vol. 6110, pp. 1–23. Springer, Heidelberg (2010)
21. Micciancio, D.: Improving Lattice Based Cryptosystems Using the Hermite Normal Form. In: Silverman, J.H. (ed.) CaLC 2001. LNCS, vol. 2146, pp. 126–145. Springer, Heidelberg (2001)
22. Micciancio, D.: Improved cryptographic hash functions with worst-case / average-case connection. In: STOC 2002, pp. 609–618 (2002); Full version: Almost perfect lattices, the covering radius problem, and applications to Ajtai's connection factor. SIAM Journal on Computing, 34(1):118–169 (2004)
23. Micciancio, D.: Generalized compact knapsacks, cyclic lattices, and efficient one-way functions from worst-case complexity assumptions. In: FOCS 2002, pp. 356–365 (2002)
24. Micciancio, D., Regev, O.: Worst-Case to Average-Case Reductions Based on Gaussian Measures. In: FOCS 2004, pp. 372–381 (2004); Full version: SIAM J. Comput., 37(1), 267–302 (2007)
25. Nguyen, P.Q., Stern, J.: Adapting Density Attacks to Low-Weight Knapsacks. In: Roy, B. (ed.) ASIACRYPT 2005. LNCS, vol. 3788, pp. 41–58. Springer, Heidelberg (2005)
26. Peikert, C.: Public-key cryptosystems from the worst-case shortest vector problem: extended abstract. In: STOC 2009, pp. 333–342. ACM, New York (2009)
27. Peikert, C., Rosen, A.: Efficient collision-resistant hashing from worst-case assumptions on cyclic lattices. In: Halevi, S., Rabin, T. (eds.) TCC 2006. LNCS, vol. 3876, pp. 145–166. Springer, Heidelberg (2006)
28. Peikert, C., Rosen, A.: Lattices that Admit Logarithmic Worst-Case to Average-Case Connection Factors. In: Proc. of STOC 2007, pp. 478–487 (2007)
29. Regev, O.: New lattice-based cryptographic constructions. Journal of the ACM 51(6), 899–942 (2004); Extended abstract in STOC 2003 (2003)
30. Regev, O.: On Lattices, Learning with Errors, Random Linear Codes, and Cryptography. In: Proc. of STOC 2005, pp. 84–93 (2005)
31. Rivest, R., Adleman, L., Dertouzos, M.: On data banks and privacy homomorphisms. In: Foundations of Secure Computation, pp. 169–180 (1978)
32. Shor, P.W.: Polynomial-time algorithms for prime factorization and discrete logarithms on a quantum computer. SIAM Journal on Computing 26(5), 1484–1509 (1997); Extended abstract in FOCS 1994 (1994)
33. Stevenhagen, P.: The Arithmetic of Number Rings. In: Algorithmic Number Theory, vol. 44. MSRI Publications (2008); See also Stevenhagen's course notes Number Rings

# Additively Homomorphic Encryption with $d$-Operand Multiplications

Carlos Aguilar Melchor[1], Philippe Gaborit[1], and Javier Herranz[2]

[1] XLIM-DMI, Université de Limoges,
123, av. Albert Thomas
87060 Limoges Cedex, France
{carlos.aguilar,philippe.gaborit}@xlim.fr
[2] Dept. Matemàtica Aplicada IV,
Universitat Politècnica de Catalunya,
C/ Jordi Girona, 1-3, 08034 Barcelona, Spain
jherranz@ma4.upc.edu

**Abstract.** The search for encryption schemes that allow to evaluate functions (or circuits) over encrypted data has attracted a lot of attention since the seminal work on this subject by Rivest, Adleman and Dertouzos in 1978.

In this work we define a theoretical object, chained encryption schemes, which allow an efficient evaluation of polynomials of degree $d$ over encrypted data. Chained encryption schemes are generically constructed by concatenating cryptosystems with the appropriate homomorphic properties; such schemes are common in lattice-based cryptography. As a particular instantiation we propose a chained encryption scheme whose IND-CPA security is based on a worst-case/average-case reduction from uSVP.

**Keywords:** homomorphic encryption, secure function evaluation, lattices.

## 1 Introduction

*Secure function evaluation* (SFE) is an essential ingredient to design protocols where different users interact in order to obtain some information from the others, at the same time that each user keeps private some of his information. In (a simplified version of) SFE, a user Alice has a function $f$ and a user Bob has some data $x$. Depending on the setting, one of the two users, or both of them, must obtain $f(x)$ without learning each other's input.

One solution for this problem uses the concept of *garbled circuit*, introduced by Yao in [36]. Alice receives from Bob a garbled version of $x$, and sends back a garbled version of $f$ as well as some cryptographic material allowing Bob to evaluate this function on $x$. After the end of the protocol, Bob learns $f(x)$ and nothing else about $f$, and Alice learns nothing about $x$. This solution is based on the usage of encrypted truth tables for the garbled function and oblivious transfer for the garbled data. The main drawback is that the size of the evaluated

T. Rabin (Ed.): CRYPTO 2010, LNCS 6223, pp. 138–154, 2010.

ciphertext is at least linear in $|f|$. Alternative solutions were therefore proposed, following a different paradigm (denoted as *computing over encrypted data*), to get size sublinear in $|f|$. Here Bob sends to Alice some information related to $x$ (e.g. an encryption of $x$), Alice combines $f$ and the data received from Bob, and sends a reply to Bob. From this reply, Bob is able to learn $f(x)$ and Alice learns nothing about $x$ (not even $f(x)$). If moreover, Bob does not learn anything about $f$ (besides $f(x)$) we say the protocol provides function privacy.

The garbled circuit approach provides generic protocols that work for virtually any function $f$, which may not be the case in the computing over encrypted data setting. On the other hand, the computing over encrypted data setting (on which this paper is focused) can lead to protocols with a much lower communication cost. Indeed, in the garbled circuit approach the communication includes an encrypted description of $f$ and an encrypted description of $x$. In the computing over encrypted data setting only an encrypted description of $x$ is sent and the reply sent to Bob by Alice can be very compact, perhaps independent of the size of $f$. More precisely, we will say that a secure evaluation is *efficient* for a family of functions $\mathcal{F}$ if for $f \in \mathcal{F}$ the size of the information exchanged by Alice and Bob is at most sublinear in the function size (and thus less than the size of the information exchanged in the garbled circuit approach).

A family of functions that are specially interesting is the one of multivariate polynomials with $m$ monomials and degree $d$; that is $P(X_1, \ldots, X_v) = \sum_{\ell=1}^{m} P_\ell(X_1, \ldots, X_v)$, where $P_\ell(X_1, \ldots, X_v)$ are monomials of degree at most $d$. Many applications such as private information retrieval [21], or private searching on streaming data [25] are based on the secure evaluation of low-degree multivariate polynomials with a large number of monomials (varying in real world scenarios from thousands to billions and above). The only approach to obtain efficient (and secure) evaluations of multivariate polynomials has been until now the usage of homomorphic encryption schemes.

In order to provide such evaluations for degree $d$ polynomials, these encryption schemes must allow to compute products of $d$ plaintexts over encrypted data (possibly with a large expansion factor), and to sum a very large number $m$ of these encrypted products with a small expansion factor (sublinear or logarithmic in $m$). In this paper we propose a generic construction to obtain such properties and we instantiate this construction with a well-known lattice-based cryptosystem. The security of this particular instance is based on a worst-case/average-case reduction from uSVP (see [24] for more details on hard problems related to lattices), which has been proved as hard as other standard problems like GapSVP or the Bounded Distance Decoding (BDD) problem in [22]. Other instantiations can be found in [15] and [2], using respectively a cryptosystem with security based in the worst-case hardness of LWE, and a cryptosytem with security based in the average-case hardness of particular instances of BDD [1].

**Related Work.** Since the introduction of the concept of homomorphic encryption, by Rivest, Adleman and Dertouzos in [30], many schemes with homomorphic properties have been proposed. Most of them allow only to compute over encrypted data one of the operations, either the product (RSA [31], El

Gamal [11]) or the sum of the plaintexts (Goldwasser-Micali [17] modulo 2 and Paillier [26] modulo a hard-to-factor composite integer).

These schemes lead to an efficient evaluation of multivariate monomials of any degree or multivariate polynomials of degree 1, but obtaining a scheme that provides efficient evaluation of multivariate polynomials of arbitrary degree is a much more complex problem. In order to evaluate a larger span of functions, some protocols have tried to use the homomorphic encryption schemes that allow to compute just one operation (sum or product) in a less direct way than just using the provided plaintext-ciphertext map. In particular, Sander, Young and Yung proposed a solution in [32], which allows to evaluate any constant fan-in boolean circuit in $NC^1$. The major drawback of their approach is that communication complexity is exponential in the depth of the circuit, which restricts their protocol to circuits of logarithmic depth. Ishai and Paskin show in [19] how to evaluate any branching program $P$ through ciphertexts whose size depends polynomially on the length of $P$. Such branching programs include, by a result of Barrington [4], the circuits in $NC^1$. Unfortunately, in order to evaluate a multivariate polynomial with $m$ monomials, we need an $NC^1$ circuit of depth in $O(\log m)$ or a branching program of length in $O(m)$ (see [23]). Thus, neither of these protocols are able to provide efficient evaluation of polynomials.

Finding an encryption scheme allowing an efficient direct computation over encrypted data of degree $d$ multivariate polynomials for $d > 1$ has been an open issue for a long term. The first attempts that tried to provide a *fully homomorphic* encryption scheme (i.e. a scheme allowing to compute over encrypted data *both* sums and multiplications arbitrarily), failed to resist to the research community attacks: Fellows and Koblitz proposed Polly Cracker [12] which was broken in [34], Grigoriev and Ponomarenko proposed another public-key scheme [18] which was broken in [7]. For the case of symmetric cryptography, Domingo-Ferrer proposed two schemes [9,10] which were broken in [6,35]. Fortunately, as we already noted, in order to have efficient evaluations of degree $d$ multivariate polynomials we just need the encryption scheme to compute products of $d$ plaintexts over encrypted data (possibly with a large expansion factor), and to sum a very large number $m$ of these encrypted products with a small expansion factor (sublinear or logarithmic in $m$). We will say that such a scheme is *d-multiplicative fully homomorphic*. If for any $d$ a $d$-multiplicative fully homomorphic instance of the scheme can be produced (possibly with an exponential cost in $d$), we will say that the scheme is *constant-bounded fully homomorphic*. If moreover the computational costs of the different functions of the encryption scheme are at most polynomial in $d$, we will say it is *leveled fully homomorphic*.

Finding a non-trivial (i.e. for $d > 1$) $d$-multiplicative fully homomorphic encryption scheme has also been a long standing open problem. The first step forward was given in 2005, by Boneh, Goh and Nissim, who proposed [5] the first efficient 2-multiplicative fully homomorphic encryption scheme. Their scheme allows the SFE of polynomials of degree $d = 2$, as long as the output $P(a_1, \ldots, a_t)$ is a small number (the computational cost of decryption is polynomial on this number). The size of the ciphertexts is independent of the number $m$ of monomials in the polynomial, and the secure function evaluation protocol provides function privacy. In 2008, the authors proposed, in a preliminary version of this paper [2], a way to obtain efficient constant-bounded fully homomorphic encryption schemes (without function privacy).

In STOC 2009, Gentry proposed an elegant solution [13] for the (efficient) leveled fully homomorphic encryption problem, in two steps. First, he proposed an efficient constant-bounded fully homomorphic encryption scheme based on the hardness of a new problem, the Ideal Coset Problem, which is close to a decisional Closest Vector Problem (which is in turn an instance of the Bounded Distance Decoding problem, see [24]). Second, he proposed an efficient leveled fully homomorphic variant of this scheme, based on the Ideal Coset Problem and a second new problem, the SplitKey Distinguishing Problem which seems to be related to the Sparse Subset Sum Problem (in fact the scheme can be modified to be fully homomorphic if circular security is assumed, see [13] for details). In lattice-based encryption schemes the randomness distribution usually evolves as homomorphic operations are done until the ciphertext becomes impossible to decrypt, which places a limit on the number of operations that can be done. The groundbreaking idea of Gentry is the proposal of a scheme that can "refresh" this randomness to its initial state (more exactly close to the initial state), by the homomorphic evaluation of its own decryption circuit, without revealing the plaintext. In his PhD dissertation [14], Gentry recently presented a *quantum* reduction from the security of his leveled fully homomorphic scheme to the worst case of the Shortest Independent Vector Problem (SIVP, see [24]) on *ideal lattices in a given ring R*. Improvements and variations of Gentry's schemes have appeared very recently [33,8].

Finally, in [15], Gentry, Halevi and Vaikuntanathan have proposed a new efficient 2-multiplicative encryption scheme (GHV for short) which improves the proposal of Boneh, Goh, and Nissim in various ways. First, it is based on a worst-case/average-case *classical* reduction from LWE (again, see [24]). Moreover, it does not have restrictions in the size of the output. And finally, it can also be used with our construction to obtain a constant-bounded homomorphic encryption scheme.

**Our Contribution.** This paper is a major write up of [2]. With respect to the related work as a whole, our main contribution is to provide a *generic* construction of efficient constant-bounded fully homomorphic encryption schemes. This construction can be instantiated using different encryption schemes as a base. In particular, the encryption schemes of the fruitful field of lattice-based cryptography seem specially well adapted, but other fields such as code-based cryptography are promising too. The recent instantiation with GHV (proposed in [15]), highlights the generic aspect of our contribution.

With respect to the proposal of Gentry [13], the main contribution comes from the fact that the construction relies on the same security assumptions as encryption schemes with simple-to-achieve homomorphic properties. Thus, we benefit from the strong reductions available in simple lattice-based encryption schemes, instead of the assumptions needed to get (somewhat) fully homomorphic encryption schemes. In particular, this leads to assumptions on the *classical* (by opposition to *quantum*) worst-case hardness of standard problems, which moreover are done over pretty general lattices, namely integer lattices, instead of ideal lattices over a given ring.

Finally, with respect to the generic proposals of Yao [36] and Sander et al. [32], considering the secure evaluation of polynomials, the main advantage comes from the bandwidth efficiency for low-degree polynomials with a large number of monomials. Indeed, our construction can be used for the secure evaluation of a polynomial with $v$ variables, $M$ monomials and degree $d$. In Table 1, we compare the bandwidth required by these generic proposals with different instantiations of our construction. The number of monomials is supposed to be bounded by a polynomial in the security parameter $\kappa^{r/2}$ for a given $r$, and poly($\kappa$) generically denotes a polynomial function of the security parameter $\kappa$.

**Table 1.** Comparison of the bandwidth requirements of different solutions

| Approach | Required Bandwidth |
|---|---|
| Yao's garbled circuits [36] | $M \cdot d \cdot \text{poly}(\kappa)$ |
| Sander-Young-Yung [32] | $(M \cdot d)^2 \cdot \text{poly}(\kappa)$ |
| $\tilde{O}(\kappa^{1.5+r})$-uSVP instantiation | $\tilde{O}(M^{4d/r}) \cdot \text{poly}(\kappa)^d$ |
| $\tilde{O}(\kappa^{3.5+3r})$-LWE instantiation [15] | $\text{poly}(\log M)^d \cdot \text{poly}(\kappa)^d$ |
| Optimal bound (for our construction) | $\log M \cdot \text{poly}(\kappa)^d$ |

Note that in the uSVP instantiation $r$ can be chosen arbitrarily large to reduce the bandwidth usage, but at the cost of a stronger security assumption. In the LWE instantiation, this is pointless as bandwidth usage does not depend on $r$ and it is enough to set $r$ such that $\kappa^{r/2} = O(M)$.

Last row of Table 1 would be ideally achieved by combining our construction with an additively homomorphic encryption scheme, supporting $M$ additions, where the expansion factor between plaintexts and ciphertexts does not depend on $M$. The scheme by Gentry [13] could be a candidate for such an encryption scheme but, to the best of the authors knowledge, there is no such scheme with a classical reduction (nor based on integer lattices).

For a given degree $d$ and a growing number of monomials our construction beats asymptotically the other approaches. For variable $d$, the comparison depends on whether the polynomials are sparse or not, and on the number $v$ of variables. If the polynomials are very sparse, our solution will not be efficient. On the other hand if the polynomials are dense (i.e. we have $M \simeq v^d$), our construction will beat the other approaches if and only if the number of variables is larger than the number of bits in a ciphertext of the encryption scheme used to instantiate our construction. Classical applications of secure polynomial evaluation, such as private information retrieval [21] or private searching on streaming data [25], result in dense polynomials with a large number of variables and should thus benefit from this construction.

As opposed to the other solutions in this table, the construction that we introduce in this work does not provide function privacy. An alternative construction providing such a property is possible, but due to space restrictions it is left to the long version of this paper. Some details about this alternative construction are provided in Section 5.2.

## 2  Basic Idea

Let $\mathsf{PKC} = (\mathsf{KeyGen}, \mathsf{Enc}, \mathsf{Dec})$ be an encryption scheme such that the addition of ciphertexts over the integers maps to an addition on the plaintext space and such that 0 decrypts to 0. Let $a, b \in \{0, 1\}$ and $\alpha \in \mathsf{Enc}(pk, a), \beta \in \mathsf{Enc}(pk, b)$, with $(\alpha^{(1)}, \ldots, \alpha^{(t)})$ the bit-representation of $\alpha$. We define the reconstruction function $R((\alpha^{(1)}, \ldots, \alpha^{(t)})) = \sum_i 2^{i-1} \alpha^{(i)} = \alpha$.

The basic idea is to build the compound ciphertext $\alpha \otimes \beta \stackrel{\text{def}}{=} (\alpha^{(1)}\beta, \ldots, \alpha^{(t)}\beta)$ which encrypts redundantly $a$ and $b$. Consider the following decryption algorithm: first, decrypt each coordinate with $\mathsf{Dec}$; then reconstruct the inner ciphertext with $R$, and decrypt it again with $\mathsf{Dec}$.

What is interesting is that each coordinate of the compound ciphertext is either 0 (which decrypts to 0) or $\beta$. If $b = 0$, all the coordinates will decrypt to 0 and the resulting null-vector will also decrypt to $ab = 0$ (as $b = 0$) whatever the value of $a$ is. On the other hand, if $b = 1$ all the coordinates in which we have $\beta$ will decrypt to 1, and we will thus get back $(\alpha^{(1)}, \ldots, \alpha^{(t)})$ which decrypts to $ab = a$ (as $b = 1$).

---

Toy Example

---

$\alpha = 110 \quad \beta = 101 \quad \gamma = \alpha \otimes \beta = (101, 101, 000)$
If $\beta \in \mathsf{Enc}(pk, 0)$

$$\gamma \xrightarrow{\text{1}^{\text{st}}\ \text{decryption}} (0, 0, 0) \xrightarrow{\text{reconstruction}} 000 \xrightarrow{\text{2}^{\text{nd}}\ \text{decryption}} 0$$

If $\beta \in \mathsf{Enc}(pk, 1)$

$$\gamma \xrightarrow{\text{1}^{\text{st}}\ \text{decryption}} (1, 1, 0) \xrightarrow{\text{reconstruction}} 110 \xrightarrow{\text{2}^{\text{nd}}\ \text{decryption}} a$$

---

Thus, these compound ciphertexts encrypt a product, but not very efficiently (the data is redundant). However, as $\mathsf{PKC}$ provides an homomorphic operation, we can add many of these compound ciphertexts, and the result will decrypt to the sum of the products. This allows us to evaluate degree 2 polynomials over encrypted data using a single vector of $t$ coordinates, which will save bandwidth if the number of added monomials is over $t$.

Finally, we can note that this construction can be easily generalized if $\alpha$ is a vector of integers (we split each integer in bits and reconstruct them separately on the decryption phase), which allows us to iterate the construction and evaluate polynomials of degree $d$, at the price of an expansion factor for the length of the ciphertexts which is exponential in $d$.

## 3     Chaining Encryption Schemes

### 3.1     $(n, t)$-Chainable Schemes

In this subsection we provide a definition of chainable encryption schemes. This definition allows us to present the properties needed to chain schemes (or to compute with them) as well as to have a short naming convention that highlights a given scheme's performance parameters. For integer values $a < b$, we denote as $[a, b]$ the set $\{a, a + 1, \ldots, b - 1, b\}$.

**Definition 1.** *A scheme* PKC $=$ (KeyGen, Enc, Dec) *is said* $(n, t)$-*chainable if the key generation algorithm* KeyGen *takes as input a security parameter* $\kappa$ *and a positive integer* $m$, *and for any value of these parameters, there are two positive integers* $n, t$ *(which may be functions of* $\kappa$ *and* $m$*), such that for any keypair* $(pk, sk) \in$ KeyGen$(1^\kappa, m)$ *the following holds:*

- *The plaintext space* $\mathcal{P}$ *is a subset of* $\mathbb{Z}$, *and includes* $[0, m]$;
- *The ciphertext space* $\mathcal{C}$ *is a subset of* $\mathbb{Z}^n$ *and includes* $0^n$, *moreover* $0^n$ *is in the support of the output of* Enc$(pk, 0)$;
- *Bounded size: for any plaintext* $x \in \mathcal{P}$ *and any ciphertext* $c \in$ Enc$(pk, x)$, *all the entries of* $c$ *are smaller than* $2^t$ *(i.e.,* Enc$(pk, x) \subset [0, 2^t - 1]^n$*);*
- $m$-*limited homomorphism via integer addition: for any* $\ell \leq m$, $a_1, \ldots, a_\ell \in \{0, 1\}$ *and any* $c_1, \ldots, c_\ell$ *with* $c_i \in$ Enc$(pk, a_i)$, *the integer vector* $c = \sum_i c_i$ *is decrypted via* Dec *to the integer* $a = \sum_i a_i$ *(which is in* $[0, m]$*).*

Lattice-based schemes with homomorphic properties are usually suitable (sometimes with a small transformation) for this definition. Note, however, that we do not set any constraint on the ciphertext size $n \times t$ or its relation to $m$ and thus, that not all the schemes that fit into this definition will be able to provide efficient (sub-linear in $m$) evaluations of polynomials. These issues will be dealt with in Sections 4 and 5.

### 3.2     Chaining Schemes

In this subsection we present an algorithm that chains two encryption schemes PKC$_1$, PKC$_2$ that are respectively $(n_1, t_1)$-chainable and $(n_2, t_2)$-chainable, into a scheme PKC $=$ chain(PKC$_1$, PKC$_2$), that is $(n_2 n_1 t_1, t_2)$-chainable. This scheme has a worse ciphertext/plaintext expansion ratio than the two chained schemes, but is interesting because given $\alpha \in$ Enc$_1(pk_1, a_1)$ and $\beta \in$ Enc$_2(pk_2, a_2)$ we are able to generate an element of Enc$(pk, a_1 a_2)$ (see Section 4).

---

Chaining Algorithm: $\mathsf{PKC} = \mathsf{chain}(\mathsf{PKC}_1, \mathsf{PKC}_2)$

---

*Input*:
- An $(n_1, t_1)$-chainable scheme $\mathsf{PKC}_1 = (\mathsf{KeyGen}_1, \mathsf{Enc}_1, \mathsf{Dec}_1)$
- An $(n_2, t_2)$-chainable scheme $\mathsf{PKC}_2 = (\mathsf{KeyGen}_2, \mathsf{Enc}_2, \mathsf{Dec}_2)$

*Output*:
- An $(n_1 t_1 n_2, t_2)$-chainable scheme $\mathsf{PKC} = (\mathsf{KeyGen}, \mathsf{Enc}, \mathsf{Dec})$

Consider the intermediate encryption scheme $\mathsf{PKC}_1'$:

---

$\mathsf{KeyGen}_1'(1^\kappa, m)$:
1  Return $(pk_1, sk_1) \leftarrow \mathsf{KeyGen}_1(1^\kappa, m)$

$\mathsf{Enc}_1'(pk_1, a)$ :
1  Sample $\alpha = (\alpha^{(1)}, \ldots, \alpha^{(n_1)})$ from $\mathsf{Enc}_1(pk_1, a)$
2  Return $\alpha' = (\underbrace{\alpha'^{(1)}, \ldots, \alpha'^{(t_1)}}_{\text{bits of } \alpha^{(1)}}, \ldots, \underbrace{\alpha'^{((n_1-1)t_1+1)}, \ldots, \alpha'^{(n_1 t_1)}}_{\text{bits of } \alpha^{(n_1)}})$

$\mathsf{Dec}_1'(sk_1, \alpha')$:
1  Compute $\alpha = \mathcal{R}_1(\alpha') \stackrel{\text{def}}{=} (\sum_{j=1}^{t_1} 2^{j-1} \alpha'^{(j)}, \ldots, \sum_{j=1}^{t_1} 2^{j-1} \alpha'^{((n_1-1)t_1+j)})$
2  Return $a \leftarrow \mathsf{Dec}_1(sk_1, \alpha)$

---

Return a description of the final encryption scheme $\mathsf{PKC}$:

---

$\mathsf{KeyGen}(1^\kappa, m)$:
1  Set $(pk_1, sk_1) \leftarrow \mathsf{KeyGen}_1(1^\kappa, m)$, $(pk_2, sk_2) \leftarrow \mathsf{KeyGen}_2(1^\kappa, m)$
2  Return $((pk_1, pk_2), (sk_1, sk_2))$

$\mathsf{Enc}((pk_1, pk_2), a)$ :
1  Set $\alpha' = (\alpha'^{(1)}, \ldots, \alpha'^{(n_1 t_1)}) \leftarrow \mathsf{Enc}_1'(pk_1, a)$
2  For each $j \in [1, n_1 t_1]$ set $\beta_j \leftarrow \mathsf{Enc}_2(pk_2, \alpha'^{(j)})$
3  Return $\gamma = (\beta_1, \ldots, \beta_{n_1 t_1})$

$\mathsf{Dec}((sk_1, sk_2), \gamma)$:
1  For each $j \in [1, n_1 t_1]$, set $\alpha'^{(j)} \leftarrow \mathsf{Dec}_2(sk_2, \gamma^{(j)})$
2  Return $a \leftarrow \mathsf{Dec}_1'(sk_1, (\alpha'^{(1)}, \ldots, \alpha'^{(n_1 t_1)}))$

---

**Proposition 1.** PKC *is* $(n_1 t_1 n_2, t_2)$-*chainable. Moreover, if the instance of* $\mathsf{PKC}_1$ *associated to* $(pk_1, sk_1)$ *and the instance of* $\mathsf{PKC}_2$ *associated to* $(pk_2, sk_2)$ *are $m$-limited homomorphisms, the instance of* $\mathsf{PKC}$ *associated to* $((pk_1, pk_2), (sk_1, sk_2))$ *is also an $m$-limited homomorphism.*

*Proof.* Clearly, $\mathcal{R}_1$ is linear and therefore the instance of $\mathsf{PKC}_1'$ associated to $(pk_1, sk_1)$ is an $m$-limited homomorphism via integer addition, as the instance of $\mathsf{PKC}_1$ associated to the same keypair. For $i \in [1, m]$, let $a_i \in \{0, 1\}$ and $\gamma_i \leftarrow \mathsf{Enc}((pk_1, pk_2), a_i)$. We have

$$\sum_{i=1}^{m} \gamma_i = \left( \sum_{i=1}^{m} \beta_{i,1}, \ldots, \sum_{i=1}^{m} \beta_{i,n_1 t_1} \right)$$

with $\beta_{i,j} \leftarrow \mathsf{Enc}_2(pk_2, \alpha_i'^{(j)})$ and $\alpha_i' \leftarrow \mathsf{Enc}_1'(pk_1, a_i)$. Since the used instance of $\mathsf{PKC}_2$ is an $m$-limited homomorphism via integer addition and each $\alpha_i'^{(j)}$ is in $\{0, 1\}$, applying $\mathsf{Dec}_2$ to each coordinate, with secret key $sk_2$, we obtain

$$\left( \sum_{i=1}^{m} \alpha_i'^{(1)}, \ldots, \sum_{i=1}^{m} \alpha_i'^{(n_1 t_1)} \right) = \sum_{i=1}^{m} \alpha_i'$$

As the instance of $\mathsf{PKC}_1'$ associated to $(pk_1, sk_1)$ is also an $m$-limited homomorphism via integer addition, decrypting this vector with $\mathsf{Dec}_1'$ and the secret key $sk_1$ we obtain $\sum_{i=1}^{m} a_i$, and thus the instance of $\mathsf{PKC}$ associated to $((pk_1, pk_2), (sk_1, sk_2))$ is an $m$-limited homomorphism via integer addition.

Finally, as $(0, \ldots, 0) \in \mathsf{Enc}((pk_1, pk_2), 0)$, and the ciphertexts are clearly vectors of $n_1 \cdot t_1 \cdot n_2$ scalars of $t_2$ bits each, we therefore have that $\mathsf{PKC}$ is $(n_1 t_1 n_2, t_2)$-chainable. □

Let us prove now that the chained scheme $\mathsf{PKC}$ resulting from $\mathsf{PKC}_1$ and $\mathsf{PKC}_2$ is IND-CPA secure if either of $\mathsf{PKC}_1$, $\mathsf{PKC}_2$ is IND-CPA secure. We recall first the standard notion of indistinguishability under chosen-plaintext attacks (IND-CPA security), for an encryption scheme $\mathsf{PKC} = (\mathsf{KeyGen}, \mathsf{Enc}, \mathsf{Dec})$. We use the following game that an attacker $\mathcal{A}$ plays against a challenger:

$$(pk, sk) \leftarrow \mathsf{KeyGen}(1^\kappa)$$
$$(St, a_0, a_1) \leftarrow \mathcal{A}(\mathrm{find}, pk)$$
$$b \leftarrow \{0, 1\} \text{ at random}$$
$$c^* \leftarrow \mathsf{Enc}(pk, a_b)$$
$$b' \leftarrow \mathcal{A}(\mathrm{guess}, c^*, St).$$

The advantage of such an adversary $\mathcal{A}$ is defined as

$$\mathrm{Adv}(\mathcal{A}) = \left| \Pr[b' = b] - \frac{1}{2} \right|.$$

A public key encryption scheme enjoys IND-CPA security if $\mathrm{Adv}(\mathcal{A})$ is a negligible function of the security parameter $\kappa$, for any attacker $\mathcal{A}$ running in polynomial time (in $\kappa$).

**Proposition 2 (IND-CPA Security).** $\mathsf{PKC} = \mathsf{chain}(\mathsf{PKC}_1, \mathsf{PKC}_2)$ *is IND-CPA secure if either of* $\mathsf{PKC}_1$, $\mathsf{PKC}_2$ *is IND-CPA secure.*

*Proof (Sketch.).* Let us assume that there exists a CPA attacker $\mathcal{A}$ against $\mathsf{PKC}$ and let us prove, then, that neither of $\mathsf{PKC}_1$, $\mathsf{PKC}_2$ can be IND-CPA. Specifically, we can construct CPA attackers $\mathcal{A}_1, \mathcal{A}_2$ against the schemes $\mathsf{PKC}_1$ and $\mathsf{PKC}_2$.

For $\mathsf{PKC}_1$, the attacker $\mathcal{A}_1$ is trivial as a random keypair of $\mathsf{PKC}_1$ can be transformed in a random keypair of $\mathsf{PKC}$ by adding a random keypair of $\mathsf{PKC}_2$. Moreover, the choice of the two plaintexts by $\mathcal{A}$ is maintained by $\mathcal{A}_1$, and the challenges from $\mathsf{Enc}_1$ can be transformed into challenges following the distribution of $\mathsf{Enc}$ by splitting them into bits and encrypting them through $\mathsf{Enc}_2$. Finally, as the plaintexts are the same, Attacker $\mathcal{A}_1$ will output the same guess as $\mathcal{A}$ will, and the success probability of both attackers will be exactly the same.

For $\mathsf{PKC}_2$, the idea is similar, but we proceed in two steps. First, we define an attacker $\mathcal{A}_2'$ able to distinguish between the distributions associated to $n_1 t_1$ plaintexts. Namely, if $\mathcal{A}$ chooses two plaintexts $a_0, a_1$, $\mathcal{A}_2'$ chooses two sets of plaintexts $(\alpha_0^{(1)}, \ldots, \alpha_0^{(n_1 t_1)})$, $(\alpha_1^{(1)}, \ldots, \alpha_1^{(n_1 t_1)})$, for $\alpha_0 \leftarrow \mathsf{Enc}_1(pk_1, a_0)$ and $\alpha_1 \leftarrow \mathsf{Enc}_1(pk_1, a_1)$ which ensures that $\mathcal{A}$, and therefore $\mathcal{A}_2'$, is able to distinguish the challenges with an non-negligible advantage. Then, we use a standard hybrid argument to derive from $\mathcal{A}_2'$ an attacker $\mathcal{A}_2$ against $\mathsf{PKC}_2$.

$\square$

The output of the chaining algorithm being itself chainable we can iteratively construct a chain of $d$ encryption schemes $\mathsf{PKC}_1, \ldots, \mathsf{PKC}_d$, if for any $i \in [1, d]$ $\mathsf{PKC}_i$ is $(n_i, t_i)$-chainable, and obtain an $(n_d \prod_{j=1}^{d-1} n_j t_j, t_d)$-chainable encryption scheme $\mathsf{PKC}$, with $(pk_1, \ldots, pk_d)$ and $(sk_1, \ldots, sk_d)$ as public and secret keys. Note that $\mathsf{PKC}_1, \ldots, \mathsf{PKC}_d$ need not to be different schemes and that we can chain $d$ times an $(n, t)$-chainable scheme $\mathsf{PKC}$ to itself. In this case we get an $(n(nt)^{d-1}, t)$-chainable scheme, which has the same public/secret keypair $(pk, sk)$ as $\mathsf{PKC}$.

## 4    Computing with Chained Schemes

### 4.1    Product and Polynomial Evaluation

Chained schemes being themselves chainable they provide a limited homomorphism via integer addition (by Definition 1). Thus, in order to compute sums of plaintexts over encrypted data with them we just need to add up the corresponding ciphertexts. Computing products of plaintexts over encrypted data is not as straightforward and requires to use ciphertexts of the multiplication operands under the encryption schemes that form the chain. The following algorithm shows how to proceed.

---

Product Computation Algorithm: $\gamma = \mathsf{product}(\alpha, \beta)$

---

*Input*:
- $\alpha \in \mathsf{Enc}_1(pk_1, a_1)$ for $a_1 \in \{0, 1\}$ and $\mathsf{PKC}_1$ $(n_1, t_1)$-chainable
- $\beta \in \mathsf{Enc}_2(pk_2, a_2)$ for $a_2 \in \{0, 1\}$ and $\mathsf{PKC}_2$ $(n_2, t_2)$-chainable

*Output:*
- $\gamma \in \mathsf{Enc}((pk_1, pk_2), a_1 a_2)$ for $\mathsf{PKC} = (\mathsf{KeyGen}, \mathsf{Enc}, \mathsf{Dec}) = \mathsf{chain}(\mathsf{PKC}_1, \mathsf{PKC}_2)$

1  Split $\alpha$ into the bit vector $\alpha' = (\alpha'^{(1)}, \ldots, \alpha'^{(n_1 t_1)}) \in \mathsf{Enc}'_1(pk_1, a_1)$
2  Multiply each one-bit scalar of this vector by $\beta$ and output the result.

---

**Proposition 3.** *The output of the above-described protocol* product *belongs to* $\mathsf{Enc}((pk_1, pk_2), a_1 a_2)$.

*Proof.* We have $\mathsf{product}(\alpha, \beta) = (\alpha'^{(1)} \beta, \ldots, \alpha'^{(n_1 t_1)} \beta)$. We want to prove that there is $\gamma \in \mathsf{Enc}((pk_1, pk_2), a_1 a_2)$ such that for all $j \in [1, n_1 t_1]$ we have $\alpha'^{(j)} \beta = \gamma^{(j)}$. By the construction of a chained scheme, this is equivalent to: there is $\alpha'_{1,2} \in \mathsf{Enc}'_1(pk_1, a_1 a_2)$ such that $\alpha'^{(j)} \beta \in \mathsf{Enc}_2(pk_2, \alpha'^{(j)}_{1,2})$, for all $j \in [1, n_1 t_1]$.

**If** $a_2 = 1$ set $\alpha'_{1,2} = \alpha' \in \mathsf{Enc}'_1(pk_1, a_1) = \mathsf{Enc}'_1(pk_1, a_1 a_2)$. For each $j \in [1, n_1 t_1]$

- if $\alpha'^{(j)} = 1$,

$$\alpha'^{(j)} \beta = \beta \in \mathsf{Enc}_2(pk_2, a_2) = \mathsf{Enc}_2(pk_2, \alpha'^{(j)}) \Rightarrow \alpha'^{(j)} \beta \in \mathsf{Enc}_2(pk_2, \alpha'^{(j)}_{1,2})$$

- if $\alpha'^{(j)} = 0$,

$$\alpha'^{(j)} \beta = (0, \ldots, 0) \in \mathsf{Enc}_2(pk_2, 0) = \mathsf{Enc}_2(pk_2, \alpha'^{(j)}) \Rightarrow \alpha'^{(j)} \beta \in \mathsf{Enc}_2(pk_2, \alpha'^{(j)}_{1,2})$$

$\Rightarrow$ if $a_2 = 1$ the output of the algorithm is in $\mathsf{Enc}((pk_1, pk_2), a_1 a_2)$.

**If** $a_2 = 0$, set $\alpha'_{1,2} = (0, \ldots, 0) \in \mathsf{Enc}'_1(pk_1, a_1 a_2)$. For each $j \in [1, n_1 t_1]$

- if $\alpha'^{(j)} = 1$,

$$\alpha'^{(j)} \beta = \beta \in \mathsf{Enc}_2(pk_2, a_2) = \mathsf{Enc}_2(pk_2, 0) \Rightarrow \alpha'^{(j)} \beta \in \mathsf{Enc}_2(pk_2, \alpha'^{(j)}_{1,2})$$

- if $\alpha'^{(j)} = 0$,

$$\alpha'^{(j)} \beta = (0, \ldots, 0) \in \mathsf{Enc}_2(pk_2, 0) \Rightarrow \alpha'^{(j)} \beta \in \mathsf{Enc}_2(pk_2, \alpha'^{(j)}_{1,2})$$

$\Rightarrow$ if $a_2 = 0$ the output of the algorithm is also in $\mathsf{Enc}((pk_1, pk_2), a_1 a_2)$. $\qquad \square$

This algorithm can be used iteratively to obtain encrypted products of $d$ plaintexts. As these products are ciphertexts of a chained (and thus chainable) encryption scheme, we can add them and the result will decrypt to the evaluation of a degree $d$ binary polynomial (if the homomorphic parameter $m$ of the scheme is larger than the number of monomials $M$). The following algorithms provide a complete protocol for degree $d$ polynomial evaluation over encrypted data. We want to stress that the algorithms can be easily modified (to get more efficient and simple), in case the input polynomial $P$ has a more compact representation, e.g. $P = (X_1 + 1)^d$.

---

Polynomial Evaluation Algorithms: $P = \sum_{\ell=1}^{M} X_{\ell_1} \ldots X_{\ell_d} \in \mathbb{Z}_2[X_1, \ldots, X_v]^1$

---

**Setup** $\mathsf{KeyGen}_P(1^\kappa, M)$:

*Input*:

- A security parameter $1^\kappa$
- A maximum number of monomials $M$

*Output*: A keypair $(pk, sk) \in \mathsf{KeyGen}(1^\kappa, M)$ for $\mathsf{PKC} = (\mathsf{KeyGen}, \mathsf{Enc}, \mathsf{Dec})$ an $(n, t)$-chainable scheme $\mathsf{PKC}$

**Encryption** $\mathsf{Enc}_P(pk, (a_1, \ldots, a_v))$:

*Input*:

- A public key $pk$ of the afore-mentioned $(n, t)$-chainable scheme $\mathsf{PKC}$
- A point $(a_1, \ldots, a_v)$ in $\{0, 1\}^v$ in which the polynomial should be evaluated

*Output*: A set of ciphertexts $\alpha_i \in \mathsf{Enc}(pk, a_i)$ for $i \in [1, v]$

1      Set $\alpha_i \leftarrow \mathsf{Enc}(pk, a_i)$ for $i \in [1, v]$
2      Return $\alpha_1, \ldots, \alpha_v$

**Evaluation** $\mathsf{Eval}_P((\alpha_1, \ldots, \alpha_v), P)$:

*Input*:

- An encryption $(\alpha_1, \ldots, \alpha_v)$ of a point in $\{0, 1\}^v$, through $\mathsf{PKC}$
- The description of a polynomial $P = \sum_{\ell=1}^{M} X_{\ell_1} \ldots X_{\ell_d} \in \mathbb{Z}_2[X_1, \ldots, X_v]$

*Output*: A sum of ciphertexts, $\alpha$, that decrypts to $P(a_1, \ldots, a_v)$

1      For $\ell = 1, \ldots, M$
2          $\alpha_{\ell,1} \stackrel{\text{def}}{=} \alpha_{\ell_1}$
3          For $j = 2, \ldots, d$
4              $\alpha_{\ell,j} = \mathsf{product}(\alpha_{\ell,j-1}, \alpha_{\ell_j})$
5      Return $\alpha = \sum_{\ell=1}^{M} \alpha_{\ell,d}$

**Decryption** $\mathsf{Dec}_P(sk, \alpha)$

*Input*:

- A secret key $sk$ of the afore-mentioned $(n, t)$-chainable scheme $\mathsf{PKC}$
- The output, $\alpha$, of the evaluation algorithm

*Output*: $P(a_1, \ldots, a_v)$

1      $\mathsf{PKC}_{1,2} = \mathsf{chain}(\mathsf{PKC}, \mathsf{PKC})$
2      For $j = 3, \ldots, d$: $\mathsf{PKC}_{1,j} = \mathsf{chain}(\mathsf{PKC}_{1,j-1}, \mathsf{PKC})$
4      Return $(a \bmod 2)$ for $a \leftarrow \mathsf{Dec}_{1,d}(sk, \alpha)$

---

Note that if the polynomial has a monomial of degree $d' < d$ it is enough to add the following computation: For $j \in [d', d-1]$: $\alpha_{\ell,j+1} = \mathsf{product}(\alpha_{\ell,j}, \alpha_0)$, where $\alpha_0 \in \mathsf{Enc}(pk, 1)$. This step ensures that the protocol processes the polynomial correctly.

---

[1] Note that we do not use a standard indexing such as $\sum_{\ell=1}^{M} X_{i_{1,\ell}} \ldots X_{i_{d,\ell}}$ and rather implicitly associate to each $\ell \in [1, M]$ a tuple $(\ell_1, \ldots, \ell_d) \in [1, v]^d$ to reduce index notations.

**Proposition 4.** *The Polynomial Evaluation Algorithm is correct, produces an output of $(nt)^d \log m$ bits and if PKC is IND-CPA, the choice of the evaluation point is private.*

*Proof (Sketch.).* The correctness of the Product Algorithm guarantees that $\alpha_{\ell,j} \in \text{Enc}_{1,j}(pk, a_{\ell_1} \cdots a_{\ell_j})$ for any $j \in [1, d]$ and any $\ell \in [1, M]$ (denoting $\text{PKC}_{1,1} = \text{PKC}$). Indeed, by induction, for $j = 1$ we have $\alpha_{\ell,1} \in \text{Enc}_{1,1}(pk, a_{\ell_1})$. Suppose that we have $\alpha_{\ell,j} \in \text{Enc}_{1,j}(pk, a_{\ell_1} \cdots a_{\ell_j})$. By the product algorithm correctness we know that $\alpha_{\ell,j+1} = \text{product}(\alpha_{\ell,j}, \alpha_{\ell,j+1})$ is an encryption of $a_{\ell_1} \cdots a_{\ell_{j+1}}$ using the encryption scheme $\text{chain}(\text{PKC}_{1,j}, \text{PKC}) \stackrel{\text{def}}{=} \text{PKC}_{1,j+1}$. In other words, $\alpha_{\ell,j+1} \in \text{Enc}_{1,j+1}(pk, a_{\ell_1} \cdots a_{\ell_{j+1}})$, which completes the induction proof.

As each monomial computed in the main loop of the evaluation algorithm is a ciphertext of $\text{PKC}_{1,d}$, and $\text{PKC}_{1,d}$ is $(n(nt)^{d-1}, t)$-chainable, the result of the final step has $(nt)^d \log m$ bits. Moreover, as the instance of $\text{PKC}_{1,d}$ associated to $(pk, sk)$ is an $M$-limited homomorphism the result decrypts (mod 2) to $P(a_1, \ldots, a_v)$.

If PKC is IND-CPA, the indistinguishability of the evaluation points is straightforward using a standard hybrid argument.    $\square$

In order to have an efficient evaluation of a polynomial through chained schemes, $(nt)^d \log m$ must be sub-linear in $m$. As $nt$ is the ciphertext size of PKC, we must use an encryption scheme such that ciphertext size grows as $o((m/\log m)^{1/d})$. Such instantiations are presented in the next section.

## 4.2   Higher Moduli

The definitions, algorithms, and propositions, in this and the previous section need only to be slightly changed in order to produce chained schemes that allow to compute sums, products, and more generally evaluate polynomials, over $\mathbb{Z}_r$ for $r > 2$. Namely,

- In Definition 1, the homomorphic property must hold $\forall a_1, \ldots, a_m \in [0, r-1]$
- In the product algorithm the output is a sum of up to $r$ ciphertexts;
- In the algorithm $\text{Eval}_P$ for the evaluation of a polynomial, we need an extra final step in each monomial computation in which the associated ciphertext is added to itself a given number of times (the coefficient in front of the monomial).

In order to remain correct, the product algorithm and the polynomial $\text{Eval}_P$ algorithm, require respectively $m > r$ and $m > Mr^d$ ($M$ being the number of monomials of the polynomial). The rest of the definitions, algorithms and propositions remain unchanged, but the proofs get harder to read, and we have thus preferred to provide them only in the long version of this paper.

## 5   Specific Realizations

In this section we describe some encryption schemes that satisfy the conditions given in Definition 1. These schemes are all based on lattices, and can be used at any point of a chain.[2] On the other hand, it is obvious that the last encryption scheme in a chain does not need to have a ciphertext space which is an additive group. Therefore, we can use for the last scheme $\mathsf{PKC}_d$ other homomorphic encryption schemes, not necessarily based on lattices, as long as their plaintext spaces are additive groups. This includes schemes like Paillier's [26] or Boneh-Goh-Nissim's [5] (BGN for short). The advantage of using the BGN scheme is that it provides an additional level of multiplications *for free*. That is, if we have a $d$-chained encryption scheme where $\mathsf{PKC}_d$ is the BGN cryptosystem, then we could use the global scheme to evaluate multivariate polynomials of degree up to $d+1$ (as long as the result of the evaluation is relatively small, which is the drawback of BGN). Such a hybrid lattice-based and number-theory encryption scheme allowing an efficient evaluation of degree $d > 2$ polynomials over encrypted data is a surprising consequence of our approach.

### 5.1   A Scheme Based on uSVP

In [20], Kawachi, Tanaka and Xagawa propose a set of lattice-based encryption schemes, derived from [16,28,29,3], that present some homomorphic properties. In particular, we are interested in the variant of [28], whose IND-CPA security is based on a worst-case/average-case reduction from $\tilde{O}(\kappa^{1.5+r}) - uSVP$ for given security parameters $\kappa, r$ (related to the underlying lattice). In this scheme, the plaintext space is $(\mathbb{Z}_p, +)$, for an arbitrary parameter $p$, and the ciphertext space is $(\mathbb{Z}_N, +)$, with $N = 2^{8\kappa^2}$. As it is proved in [20], the scheme is an $m$-limited additive homomorphism via addition modulo $N$, when $m \cdot p < \kappa^r$. As $\mathbb{Z}_N$ is a $\mathbb{Z}$-module, adding up the ciphertext as integers, and applying the mod $N$ operation just before the decryption gives the same result, and thus we have an $m$-limited additive homomorphism via integer addition. Moreover, for any keypair 0 is an encryption of 0 and thus, as long as $m < p$, the scheme is $(1, t)$-chainable, with $t = \log N = 8\kappa^2$.

The output of the secure evaluation of a degree $d$ polynomial with this scheme has a size $t^d \log m = 8^d \kappa^{2d} \log m$. Using $m < p$ and $m \cdot p < \kappa^r$ we get that the output of the evaluation has roughly a size of $8^d m^{4d/r} \log m$ bits, and therefore we must have $r > 4d$ in order to have an efficient evaluation. In terms of security, this implies that this instantiation relies on the worst case hardness of $\tilde{O}(\kappa^{1.5+4d}) - uSVP$.

### 5.2   Other Schemes

As noted in the related work section, GHV [15] is another lattice-based encryption scheme which can be used with our construction. The security of their scheme is based on the worst-case hardness of LWE (for a given approximation factor), which is equivalent to the worst-case hardness of several standard lattice problems (see [27]).

---

[2] Code-based schemes seem also an interesting alternative to be explored.

Gentry et al. note that their scheme has the same "multiplication-for-free" property (described at the beginning of this section) as the cryptosystem of Boneh et al. [5], allowing thus the evaluation of degree $d + 1$ polynomials with a chain of just $d$ schemes. In fact, it is possible to do much better, as even after the multiplication for free GHV's ciphertexts can undergo $m$ additive operations, and thus it is possible to alternate. First, we do a multiplication for free with their scheme and then a multiplication with our construction. As the result of our multiplication is a set of GHV's ciphertexts, they can again undergo a multiplication for free, and so on. With this improvement it is possible to evaluate polynomials of degree $2d$ with just a chain of $d$ schemes.

A second advantage of GHV is that ciphertext size grows only logarithmically in $m$ (whereas with the uSVP instantiation we present, it grows polynomially). In order to use the full potential of this fact we must change our construction and split the ciphertexts in groups of bits, instead of bits, just as it is presented in [2] for the instantiation of our construction with the lattice-based scheme of [1]. With such a construction it is possible to get a very small expansion factor at each iteration of the chain, asymptotically close to 1, tweaking slightly GHV. This is a major step forward, as it allows us to obtain an expansion factor linear, instead of exponential, in $d$. Indeed, by chaining instances with shrinking expansion factors, $(2, 3/2, ..., (d - 1)/(d - 2), d/(d - 1))$, the product of the expansion factors of $d$ chained schemes with the alternative construction is $d$. Moreover, using the scheme's blinding properties the instantiation also ensures formula privacy. The full details of this alternative construction and its instantiation with GHV are left to the long version of this paper.

## Acknowledgments

We want to warmly thank Shai Halevi for his support and very valuable comments on different versions of this work. We also want to thank Daniele Micciancio for his encouraging and useful recommendations, as well as the reviewers of Crypto'10 for their detailed comments.

The work of Javier Herranz is supported by Spanish MICINN Ministry, under project MTM2009-07694; and also by a *Ramón y Cajal* grant, partially funded by the European Social Fund (ESF) of the Spanish MICINN Ministry.

## References

1. Aguilar Melchor, C., Castagnos, G., Gaborit, P.: Lattice-based homomorphic encryption of vector spaces. In: The 2008 IEEE International Symposium on Information Theory (ISIT 2008), Toronto, Ontario, Canada, pp. 1858–1862. IEEE Computer Society Press, Los Alamitos (2008)
2. Aguilar Melchor, C., Gaborit, P., Herranz, J.: Additively homomorphic encryption with d-operand multiplications. Cryptology ePrint Archive, Report 2008/378 (2008), http://eprint.iacr.org/

3. Ajtai, M.: Representing hard lattices with O(n log n) bits. In: Gabow, H.N., Fagin, R. (eds.) Proceedings of the 37th Annual ACM Symposium on Theory of Computing, Baltimore, MD, USA, May 22-24, pp. 94–103. ACM, New York (2005)
4. Barrington, D.A.M.: Bounded-width polynomial-size branching programs recognize exactly those languages in $NC^1$. J. Comput. Syst. Sci. 38(1), 150–164 (1989)
5. Boneh, D., Goh, E.J., Nissim, K.: Evaluating 2-DNF formulas on ciphertexts. In: Kilian, J. (ed.) TCC 2005. LNCS, vol. 3378, pp. 325–341. Springer, Heidelberg (2005)
6. Cheon, J.H., Kim, W.H., Nam, H.S.: Known-plaintext cryptanalysis of the Domingo-Ferrer algebraic privacy homomorphism scheme. Inf. Process. Lett. 97(3), 118–123 (2006)
7. Choi, S.J., Blackburn, S.R., Wild, P.R.: Cryptanalysis of a homomorphic public-key cryptosystem over a finite group. J. Math. Cryptography 1, 351–358 (2007)
8. van Dijk, M., Gentry, C., Halevi, S., Vaikuntanathan, V.: Fully homomorphic encryption over the integers. In: Gilbert, H. (ed.) EUROCRYPT 2010, French Riviera. LNCS, vol. 6110, pp. 24–43. Springer, Heidelberg (2010)
9. Domingo-Ferrer, J.: A new privacy homomorphism and applications. Information Processing Letters 60(5), 277–282 (1996)
10. Domingo-Ferrer, J.: A provably secure additive and multiplicative privacy homomorphism. In: Chan, A.H., Gligor, V.D. (eds.) ISC 2002. LNCS, vol. 2433, pp. 471–483. Springer, Heidelberg (2002)
11. ElGamal, T.: A public key cryptosystem and a signature scheme based on discrete logarithms. IEEE Transactions on Information Theory 31(4), 469–472 (1985)
12. Fellows, M., Koblitz, N.: Combinatorial cryptosystems galore! In: Finite Fields: Theory, Applications, and Algorithms, Las Vegas, NV (1993). Contemp. Math., Amer. Math. Soc, vol. 168, pp. 51–61 (1994)
13. Gentry, C.: Fully homomorphic encryption using ideal lattices. In: Proceedings of STOC 2009, pp. 169–178. ACM Press, New York (2009)
14. Gentry, C.: A fully homomorphic encryption scheme. PhD thesis, Stanford University (2009), http://crypto.stanford.edu/craig
15. Gentry, C., Halevi, S., Vaikuntanathan, V.: A simple BGN-type cryptosystem from LWE. In: Gilbert, H. (ed.) EUROCRYPT 2010. LNCS, vol. 6110, pp. 506–522. Springer, Heidelberg (2010)
16. Goldreich, O., Goldwasser, S., Halevi, S.: Eliminating decryption errors in the Ajtai-Dwork cryptosystem. In: Kaliski Jr., B.S. (ed.) CRYPTO 1997. LNCS, vol. 1294, pp. 105–111. Springer, Heidelberg (1997)
17. Goldwasser, S., Micali, S.: Probabilistic encryption. Journal of Computer and System Sciences 28(2), 270–299 (1984)
18. Grigoriev, D., Ponomarenko, I.: Homomorphic public-key cryptosystems and encrypting boolean circuits. Applicable Algebra in Engineering, Communication and Computing 17(3), 239–255 (2006)
19. Ishai, Y., Paskin, A.: Evaluating branching programs on encrypted data. In: Vadhan, S.P. (ed.) TCC 2007. LNCS, vol. 4392, pp. 575–594. Springer, Heidelberg (2007)
20. Kawachi, A., Tanaka, K., Xagawa, K.: Multi-bit cryptosystems based on lattice problems. In: Okamoto, T., Wang, X. (eds.) PKC 2007. LNCS, vol. 4450, pp. 315–329. Springer, Heidelberg (2007)
21. Kushilevitz, E., Ostrovsky, R.: Replication is not needed: Single database, computationally-private information retrieval (extended abstract). In: FOCS: IEEE Symposium on Foundations of Computer Science (FOCS), pp. 364–373 (1997)

22. Lyubashevsky, V., Micciancio, D.: On bounded distance decoding, unique shortest vectors, and the minimum distance problem. In: Halevi, S. (ed.) CRYPTO 2009. LNCS, vol. 5677, pp. 577–594. Springer, Heidelberg (2009)
23. Mahajan, M.: Polynomial size log depth circuits: between $NC^1$ and $AC^1$. BEATCS: Bulletin of the European Association for Theoretical Computer Science 91 (2007)
24. Micciancio, D., Regev, O.: Lattice-Based Cryptography. In: Post Quantum Cryptography, pp. 147–191. Springer, Heidelberg (2009)
25. Ostrovsky, R., Skeith III, W.E.: Private searching on streaming data. J. Cryptology 20(4), 397–430 (2007)
26. Paillier, P.: Public-key cryptosystems based on composite degree residuosity classes. In: Stern, J. (ed.) EUROCRYPT 1999. LNCS, vol. 1592, pp. 223–238. Springer, Heidelberg (1999)
27. Peikert, C.: Public-key cryptosystems from the worst-case shortest vector problem. In: Proceedings of STOC 2009, pp. 333–342. ACM Press, New York (2009)
28. Regev, O.: New lattice based cryptographic constructions. Journal of the ACM 51(6), 899–942 (2004)
29. Regev, O.: On lattices, learning with errors, random linear codes, and cryptography. Journal of the ACM 56(6), 34 (2009)
30. Rivest, R.L., Adleman, L., Dertouzos, M.L.: On data banks and privacy homomorphisms. In: Foundations of Secure Computation, pp. 169–180. Academic Press, London (1978)
31. Rivest, R., Shamir, A., Adleman, L.: A method for obtaining digital signatures and public key cryptosystems. Communications of the ACM 21(2), 120–126 (1978)
32. Sander, T., Young, A., Yung, M.: Non-interactive CryptoComputing for $NC^1$. In: Proceedings of the 40th Symposium on Foundations of Computer Science (FOCS), pp. 554–567. IEEE Computer Society Press, New York (1999)
33. Smart, N., Vercauteren, F.: Fully homomorphic encryption with relatively small key and ciphertext sizes. In: Nguyen, P., Pointcheval, D. (eds.) PKC 2010. LNCS, vol. 6056, pp. 420–443. Springer, Heidelberg (2010)
34. Steinwandt, R., Geiselmann, W.: Cryptanalysis of Polly Cracker. IEEE Transactions on Information Theory 48(11), 2990–2991 (2002)
35. Wagner, D.: Cryptanalysis of an algebraic privacy homomorphism. In: Boyd, C., Mao, W. (eds.) ISC 2003. LNCS, vol. 2851, pp. 234–239. Springer, Heidelberg (2003)
36. Yao, A.C.: How to generate and exchange secrets (extended abstract). In: 27th Annual Symposium on Foundations of Computer Science, Toronto, Ontario, Canada, pp. 162–167. IEEE, Los Alamitos (1986)

# $i$-Hop Homomorphic Encryption and Rerandomizable Yao Circuits

Craig Gentry, Shai Halevi, and Vinod Vaikuntanathan

IBM T.J. Watson Research Center

**Abstract.** Homomorphic encryption (HE) schemes enable computing functions on encrypted data, by means of a public Eval procedure that can be applied to ciphertexts. But the evaluated ciphertexts so generated may differ from freshly encrypted ones. This brings up the question of whether one can keep computing on evaluated ciphertexts. An *i-hop* homomorphic encryption scheme is one where Eval can be called on its own output up to $i$ times, while still being able to decrypt the result. A *multi-hop* homomorphic encryption is a scheme which is $i$-hop for all $i$. In this work we study $i$-hop and multi-hop schemes in conjunction with the properties of function-privacy (i.e., Eval's output hides the function) and compactness (i.e., the output of Eval is short). We provide formal definitions and describe several constructions.

First, we observe that "bootstrapping" techniques can be used to convert any (1-hop) homomorphic encryption scheme into an $i$-hop scheme for any $i$, and the result inherits the function-privacy and/or compactness of the underlying scheme. However, if the underlying scheme is not compact (such as schemes derived from Yao circuits) then the complexity of the resulting $i$-hop scheme can be as high as $n^{O(i)}$.

We then describe a specific DDH-based multi-hop homomorphic encryption scheme that does not suffer from this exponential blowup. Although not compact, this scheme has complexity linear in the size of the composed function, independently of the number of hops. The main technical ingredient in this solution is a *re-randomizable* variant of the Yao circuits. Namely, given a garbled circuit, anyone can re-garble it in such a way that even the party that generated the original garbled circuit cannot recognize it. This construction may be of independent interest.

## 1   Introduction

Computing on encrypted data epitomizes the conflict between privacy and functionality, and has been receiving a great deal of attention lately. In the canonical setting of this problem there are two parties – a client that holds an input $x$, and a server that holds a function $f$. The client wishes to learn $f(x)$ using minimal interaction with the server and without giving away information about its input. Similarly, the server may want to hide information about the function $f$ from the client (except, of course, the value $f(x)$). This problem arises in a wide variety of practical applications such as secure cloud computing, searching encrypted e-mail and so on.

T. Rabin (Ed.): CRYPTO 2010, LNCS 6223, pp. 155–172, 2010.

One way to achieve this goal is via the paradigm of "computing with encrypted data" [15]: namely, the client encrypts its input $x$ and sends the ciphertext to the server, and the server "evaluates the function $f$ on the encrypted input". The server returns the evaluated ciphertext to the client, who decrypts it and recovers the result. An encryption scheme that supports computation on encrypted data is called a *homomorphic* encryption (HE) scheme. Namely, in addition to the usual encryption and decryption procedure, it has an *evaluation procedure*, that takes a ciphertext and a function and returns an "evaluated ciphertext", which can then be decrypted to obtain the value $f(x)$. Over the years there were many proposals for encryption schemes that support computations of *some* functions on encrypted data. In this work, however, we are only interested in schemes that allow computation of *any* function on encrypted data.

A trivial implementation of the evaluation procedure is for the evaluated ciphertext to include both the original ciphertext and the function $f$, and for the client to decrypt the original ciphertext and then evaluate $f$ on the result. The problem with this trivial solution is that it does not hide the server's function from the client, and that it does not offload any of the client's work to the server. We are therefore interested also in the properties of *function privacy* (meaning that the evaluated ciphertext hides the function) and *compactness* (meaning roughly that the work involved in decrypting the evaluated ciphertext is less than in computing the function "from scratch").

## 1.1   Homomorphic Encryption vs. Secure Function Evaluation

Cachin, Camenisch, Kilian, and Müller [5] observed that the paradigm of "computing with encrypted data" with function privacy can be instantiated using any two-message protocol for two-party secure function evaluation (SFE). Indeed, the specifications of these two primitives are very similar: we can think of the first message in a 2-message SFE protocol as "encrypting" the first party's input, and the second message is the evaluation of a function held by the second party on that encryption.

Following the observation of Cachin et al., there is a simple folklore construction of public-key homomorphic encryption scheme from any two-message SFE protocol and an auxiliary CPA-secure public key encryption (e.g., [10,3], see also Section 1.3 below). In particular, this construction can be used to convert a protocol based on Yao's garbled circuits [19] into a public-key homomorphic encryption scheme. The resulting scheme is function private but not compact: the client complexity is linear in the circuit size of the evaluated function $f$.

Many other schemes for "computing with encrypted data" can be found in the literature, with client complexity that depends in various forms on the complexity of the evaluated function $f$ (e.g., its truth-table size [11], circuit depth [16], branching-program length [10], polynomial degree [1], etc.) The new scheme of Gentry [7] and its variants [18,17] are the first schemes where the client complexity is independent of the complexity of $f$.

A REMARK ABOUT "FULLY HOMOMORPHIC" ENCRYPTION. We note that the schemes in [7,18,17] are unique in that evaluated ciphertexts can be made statistically close to freshly encrypted ones. We refer to schemes with this property as "fully homomorphic" (as opposed to just "homomorphic" for schemes without this property). It is easy to see that fully homomorphic schemes are both compact and function private. Also, all the issues with multi-hop evaluation that we consider in this work are trivialized for such schemes. For that reason, fully homomorphic schemes are not the focus of the current work.

## 1.2   Multi-hop Homomorphic Encryption

Beyond the simple client-server setting from above, computing with encrypted data is useful also in settings where several functions are computed on the same encrypted data. For example, consider an email message encrypted under the public-key of Alice, which is sent to `alice@yahoo.com` and promptly forwarded to `alice@gmail.com`. Both Yahoo and Google have their own spam-tagging algorithms that they want to apply to incoming emails, hence we may want to use a homomorphic encryption scheme so that they can apply these algorithms to the encrypted email. In this example, Yahoo can apply its spam-tagging algorithm to the encrypted email and produce an (encrypted and) tagged email, and then Google needs to apply its own spam-tagging algorithm to the result.

Another application with similar requirements is the setting of "autonomous mobile agents" that was considered by Cachin et al. [5]. In this application, a software agent is originated in some node in the network, and includes within it an encryption of data from that node. The agent then roams the network, visiting one node after another, and at each visited node it computes a function that depends on its current state and on the data from the visited node. Finally, the agent returns to its originator, and the originator learns the result of the composed function from all the visited nodes, as applied to the original data.

What we need in these applications is a *multi-hop* homomorphic encryption scheme, where the homomorphic function evaluation can be applied not only to a fresh ciphertext, but also a ciphertext that was already subjected to another homomorphic evaluation. We stress that evaluated ciphertexts may be very different from fresh ciphertexts, and it is not clear that the evaluation procedure of the scheme can process this modified form. (Indeed, homomorphic encryption schemes that are derived from generic secure computation protocols tend to have this problem; see below.) Cachin et al. [5] described a solution to the multi-hop setting based on Yao circuits, and our second construction in this work is an extension of that solution.

The multi-hop setting implies a new function-privacy requirement, namely *multi-hop function privacy*. For example, in the mail-forwarding example above, Google may worry that Yahoo! will try to collude with the sender and receiver of the email, in order to learn something about Google's spam-tagging techniques. Indeed, the solution of Cachin et al., which is described in Section 1.3 below, suffers from exactly this problem. Ensuring multi-hop function privacy is the main focus of our work.

### 1.3  Homomorphic Encryption from Yao Circuits

For the sake of concreteness, we now describe the folklore construction of (1-hop) homomorphic encryption from any two-message SFE protocol, and the extension of Cachin et al. to the multi-hop setting based on Yao circuits. Consider the structure of a two-message SFE protocol where a client holds an input $x$, a server holds a function $f$, and the client wishes to receive $f(x)$.

• The client sends to the server a message that "encodes" its input $x$, and yet does not reveal $x$ to a computationally bounded server. In other words, the client's message acts as an *encryption* of $x$.

• The server's response encodes the result of the computation (namely $f(x)$), and yet, reveals no more information to the client about the function $f$. In other words, the server essentially performs a function-private *evaluation* of the function $f$ on an encrypted input.

• The client recovers the result $f(x)$ from the server's message, using her secret randomness. This is the *decryption* procedure.

The above is still not quite a public-key encryption scheme: in particular, there is no public key involved, and the same party (the client) is doing both the encryption and the decryption. In contrast, a public key homomorphic encryption should be thought of as a three-player game: first a recipient publishes a public key, then a sender (client) encrypts the data $x$ under that public key, next an evaluator (server) computes a function $f$ on the encrypted data, and finally the recipient decrypts the result and recovers $f(x)$.

Fortunately, we can get a public key HE scheme from a two-message SFE protocol by using an auxiliary standard public-key encryption scheme: The recipient chooses a public/secret key pair for some semantically secure encryption scheme, the sender sends the first-message SFE message and in addition also the encryption of the SFE randomness under the public key, and the evaluator forwards the encrypted randomness to the recipient together with the second-message SFE message. The recipient uses its secret key to decrypt and recover the SFE randomness, and then uses the SFE procedure with this randomness to recover $f(x)$.

EXTENDING TO MORE THAN ONE HOP. Consider next the setting where there is a sender who holds an input $x$, two evaluators $E_1$ and $E_2$ who hold functions $f_1$ and $f_2$ respectively, and the recipient wishes to receive $f_2(f_1(x))$. To achieve this, the client would like to compute an encryption of $x$ and send it to the first evaluator, who computes an encryption of $f_1(x)$ and passes it to the second evaluator. The question we ask is: Can $E_2$ now compute on the output of $E_1$? For generic 1-hop homomorphic encryption (such as the construction above from a generic 2-message SFE protocol), we only offer a partial answer to this question: In Theorem 1 we show that "bootstrapping" techniques [7] can be used to transform a 1-hop HE scheme into an $i$-Hop scheme for any $i$, but the size of the ciphertext could grow by a polynomial factor for every hop (and hence we can only carry out this procedure for a constant number of hops).

On the other hand, a scheme based on Yao's garbled circuits [19] is easy to extend to many hops without the exponential blowup in complexity. Recall that in Yao's garbled circuit construction, the server (who has a function) chooses two random labels for every wire in the circuit that computes that function, and for every gate it computes a "gate gadget" that allows the client to learn one of the output labels if it knows one label on each input wire. The collection of all these gate gadgets is called the "garbled circuit." The server sends the garbled circuit to the client, and engages in an *oblivious transfer protocol* where it reveals to the client exactly one of the two labels on every input wire (without learning which was revealed). The client uses the gadgets to learn one label on each wire, all the way to the output wires of the circuit. The server also provides the client with a mapping between the output labels and zero/one, hence allowing the client to learn the output.

Cachin et al. [5] noted that this construction is extendable to more than one hop: the second evaluator $E_2$ receives the garbled circuit from the first evaluator $E_1$, and it can now just use $E_1$'s output labels for its own input labels, thus "connecting" these two circuits and proceeding with the protocol. Moreover this extension offers a weak form of function privacy: if only the client is corrupted, then the composed garbled circuit looks as if it was generated "from scratch" on the compositions of the two circuits, and thus it hides them from the recipient.

However, privacy breaks down completely when $E_1$ colludes with the recipient. Now, $E_1$ knows both the labels for each input wire of the garbled circuit that $E_2$ prepares. Thus, from the point of view of $E_1$, the output of $E_2$ is not garbled at all, in fact $E_1$ can completely recover $f_2$.

*Our main technical contribution is a re-randomizable variant of Yao circuits, allowing $E_2$ to re-randomize the labels of $E_1$'s garbled circuit, thus obtaining privacy even against a collusion of $E_1$ and the recipient.*

## 1.4   Summary of Our Results

DEFINITION OF MULTI-HOP HOMOMORPHIC ENCRYPTION. Informally, in an *i*-hop HE scheme, a sequence of *i* functions $f_1, \ldots, f_i$ can be homomorphically evaluated one by one on a ciphertext $c$ produced by encrypting a message $x$. This is pictorially depicted as follows. (Here $E_1, \ldots, E_i$ denote the *i* players – evaluators – that hold the functions $f_1, \ldots, f_i$).

$$\text{Encryptor}(x) \xrightarrow{c_0 = Enc(x)} E_1(f_1, c_0) \xrightarrow{c_1} \ldots \rightarrow \boxed{E_j(f_j, c_{j-1})} \xrightarrow{c_j} \ldots \xrightarrow{c_i} \text{Decryptor}$$

A multi-hop HE scheme is simply an *i*-hop scheme that works for any (polynomial) *i*.

The definition of multi-hop function privacy requires that for every $j \in [d]$, even if all the evaluators except $E_j$ combine their information, they still learn no information about $f_j$ (other than its input and output). The formal definition is simulation-based, extending the (1-hop) definition of Ishai and Paskin [10]. In this work we only deal with the honest-but-curious setting, and only consider the case where all but one of the evaluators are corrupted (as opposed to an arbitrary subset of them). Treatment of the more general cases is left for future work.

CONSTRUCTION I: 1-HOP → $i$-HOP. In Section 3, we show how to convert a 1-hop HE scheme into an $i$-hop HE scheme for any $i$. This construction uses a bootstrapping technique, similar to [7]: given a function $f$ and an evaluated ciphertext $c$ that decrypts to some value $x$, we can express the value $f(x)$ as a function of the secret key, $F_{f,c}(\text{SK}) \overset{\text{def}}{=} f(\text{Dec}(\text{SK}, c)) = f(x)$. If we add to the public key a fresh encryption of the secret key, we can then use the evaluation procedure of the scheme to evaluate $F_{f,c}$ on this fresh encryption, thus obtaining a ciphertext that decrypts to $f(x)$. As described, this construction relies on circular security of the underlying scheme (since we publish an encryption of the secret key). Just as in [7], we can avoid relying on circular security and still support up to $i$ hops, by having $i$ public/secret key pairs and encrypting the $j$'th secret key under the $j + 1$'st public key.

We note, however, that for non-compact HE schemes, the size of the evaluated ciphertext can be polynomially larger than the size of the evaluated function. Hence the ciphertext in the resulting $i$-hop scheme could grow by a factor of up to $k^{O(i)}$ after $i$ hops, where $k$ is the security parameter. Thus, this construction is viable only for a constant number of hops. Since by the folklore construction (described in section 1.3), the existence of 1-hop HE schemes is equivalent to the existence of two-message SFE protocols, we get:

**Theorem 1 (Informal).** *If two-message secure function evaluation protocols exist, then for any constant $i$ there is a public key encryption scheme $\mathcal{H}^{(i)}$ which is $i$-hop homomorphic and $i$-hop function-private. There is a fixed polynomial $q(k)$ in the security parameter $k$ such that on evaluating functions $f_1, \ldots, f_i$ on a fresh ciphertext of $\mathcal{H}^{(i)}$, the resulting evaluated ciphertext has size at most $\left(\sum_{j=1}^{i} |f_j|\right) \cdot q(k)^i$.*

We also note that if the underlying 1-hop HE scheme is compact, then the construction above can be carried out without the exponential blowup, hence we can extend it to an $i$-hop scheme for any polynomial $i$. Moreover, similar bootstrapping techniques can be used to combine two 1-hop HE schemes – one compact but not private and the other private but not compact – into a single 1-hop scheme which is both private and compact. Using the construction above we can then extend it to a *compact and private* $i$-hop scheme for any polynomial $i$.

**Theorem 2 (Informal).** *Assume that there exist a 1-hop compact HE scheme, and a (possibly different) 1-hop function-private HE scheme. Then, for every polynomial $p(k)$ there is an encryption scheme $\mathcal{H}^{(p)}$, which is $p(k)$-hop homomorphic and $p(k)$-hop private. There is a fixed polynomial $q(k)$ such that on evaluating functions $f_1, \ldots, f_{p(k)}$ on a fresh ciphertext of $\mathcal{H}^{(p)}$, the resulting ciphertext has size $q(k)$ (independent of the size of the functions $f_j$).*

CONSTRUCTION II: RE-RANDOMIZABLE YAO → MULTI-HOP. In Section 5, we describe a scheme that can handle any polynomial number of hops, and is semantically secure and function private under the decisional Diffie Hellman assumption. The size of the ciphertext in this scheme grows linearly with the size of the functions that are evaluated on the ciphertext, but independently of the number of hops.

This encryption scheme essentially amends the Yao-garbled-circuit construction from the previous section, which only offered a weak form of function privacy. The problem there was that the garbled circuit produced by the second evaluator $E_2$ contains (as a sub-circuit) the garbled circuit produced by $E_1$; this reveals non-trivial information about the function $f_2$ to the first evaluator. The solution to this problem is to come up with a way to *re-randomize Yao garbled circuits*. Roughly speaking, this is a procedure that takes a garbled circuit and constructs a *random garbled circuit* for the same function.

We describe a variant of the garbled circuit construction that allows such re-randomization. For the construction, we rely on the encryption scheme of Boneh-Halevi-Hamburg-Ostrovsky [4], and on the oblivious-transfer protocol of Naor-Pinkas and Aiello-Ishai-Reingold [13,2] (both of which are based on the decisional Diffie-Hellman assumption, and have "nice" additive homomorphic properties).

**Theorem 3 (Informal).** *Under the decisional Diffie-Hellman assumption, there is a public-key multi-hop homomorphic encryption scheme $\mathcal{H}^*$ which is function-private for any number of hops. Moreover, there is a fixed polynomial $q(k)$ in the security parameter such that on evaluating functions $f_1, \ldots, f_d$ on a fresh ciphertext, the resulting ciphertext has size $\left( \sum_{i=1}^{d} |f_i| \right) \cdot q(k)$.*

## 2   Definitions of Homomorphic Encryption

Nearly all our definitions rely on a security parameter, which is usually implicit. By $x \leftarrow X$ and $x \in_R S$ we denote drawing from a distribution and choosing uniformly from a set. We call a procedure efficient if it runs in time polynomial in the length of its input. We say that two distributions are computationally indistinguishable if any *efficient* distinguisher has only a negligible advantage in distinguishing them. Throughout the writeup, adversarial algorithms are always nonuniform.

A homomorphic encryption scheme consists of four procedures, $\mathcal{E} = (\mathsf{KeyGen}, \mathsf{Enc}, \mathsf{Dec}, \mathsf{Eval})$. $\mathsf{KeyGen}$ takes as input the security parameter and outputs a public/secret key-pair, $\mathsf{Enc}$ takes the public key and a plaintext and outputs a ciphertext, and $\mathsf{Dec}$ takes the secret key and a ciphertext and outputs a plaintext. The $\mathsf{Eval}$ procedure takes a description of a function, the public key, and a ciphertext, and outputs another ciphertext.

MULTI-HOP EVALUATION. We extend the $\mathsf{Eval}$ procedure from a single function to a sequence of functions in the natural way. Below we say that an ordered sequence of functions $\boldsymbol{f} = \langle f_1, \ldots, f_t \rangle$ is *compatible* if the output length of $f_j$ is the same as the input length of $f_{j+1}$ for all $j$. If $\boldsymbol{f}$ is a compatible sequence of $t$ functions, we denote its $j^{th}$ prefix by $\boldsymbol{f}_j = \langle f_1, \ldots, f_j \rangle$. The composed function $f_t(\cdots f_2(f_1(\cdot)) \cdots)$ is denoted $(f_t \circ \cdots \circ f_1)$.

We define an extended procedure $\mathsf{Eval}^*$ that takes as input the public key, a compatible sequence $\boldsymbol{f} = \langle f_1, \ldots, f_t \rangle$, and a ciphertext $c_0$. For $i = 1, 2, \ldots, t$ it sets $c_i \leftarrow \mathsf{Eval}(\mathrm{PK}, f_i, c_{i-1})$, outputting the last ciphertext $c_t$.

**Definition 1 ($i$-Hop Homomorphic Encryption).** *Let $i = i(k)$ be a function of the security parameter. A scheme $\mathcal{E} = (\mathsf{KeyGen}, \mathsf{Enc}, \mathsf{Dec}, \mathsf{Eval})$ is an $i$-hop homomorphic encryption scheme if for every compatible sequence $\boldsymbol{f} = \langle f_1, \ldots, f_t \rangle$ with $t \leq i$ functions, every input $x$ to $f_1$, every $(\mathrm{PK}, \mathrm{SK})$ in the support of $\mathsf{KeyGen}$, and every $c$ in the support of $\mathsf{Enc}(\mathrm{PK}; x)$,*

$$\mathsf{Dec}\big(\mathrm{SK}, \mathsf{Eval}^*(\mathrm{PK}, \boldsymbol{f}, c)\big) = (f_t \circ \cdots \circ f_1)(x)$$

*We say that $\mathcal{E}$ is a* multi-hop *homomorphic encryption scheme if it is $i$-hop for any polynomial $i$.*

We note that 1-hop homomorphic encryption is just the usual notion of homomorphic encryption, as formalized, e.g., in [10, Def 5].

FUNCTION PRIVACY AND COMPACTNESS. Semantic security [9] is defined exactly as for regular public-key encryption schemes (without regard to $\mathsf{Eval}$). We omit this definition due to space limitations.

To define function privacy, we view the operation of $\mathsf{Eval}^*$ as a multi-party protocol with one party per function, and formalize function-privacy as the usual input-privacy property for these parties: roughly speaking, we require that even if the recipient who holds the secret key colludes with all the parties but one, the function of that one party still remains hidden, except perhaps (its size and) the value that this function assumes on a single input.

**Definition 2 (function privacy - honest-but-curious).** *An $i$-hop homomorphic encryption scheme $\mathcal{E} = (\mathsf{KeyGen}, \mathsf{Enc}, \mathsf{Dec}, \mathsf{Eval})$ is function-private if there exists an efficient simulator $\mathsf{Sim}$ such that for every compatible sequence of functions $\boldsymbol{f} = \langle f_1, \ldots, f_t \rangle$ with $t \leq i$, every $j \leq t$, every input $x$ for $f_1$, every $(\mathrm{PK}, \mathrm{SK})$ in the support of $\mathsf{KeyGen}$, and every ciphertext $c_{j-1}$ in the support of $\mathsf{Eval}^*\big(\mathrm{PK}, \boldsymbol{f}_{j-1}, \mathsf{Enc}(\mathrm{PK}; x)\big)$, the following two distributions are indistinguishable (even given $x$, $\boldsymbol{f}_j$ and $\mathrm{SK}$):*

$$\mathsf{Eval}(\mathrm{PK}, f_j, c_{j-1}) \quad and \quad \mathsf{Sim}\big(\mathrm{PK}, c_{j-1}, 1^{|f_j|}, (f_1 \circ \cdots \circ f_j)(x)\big)$$

We remark that Definition 2 can be extended in several different ways. An obvious extension would be to consider the malicious case (with or without assuming that the public key and the initial ciphertext were created honestly). A second possible extension is to consider a more general adversarial structure, where the attacker can corrupt an arbitrary subset of the players (the encryptor, the evaluators, and the decryptor), and we still want to ensure the privacy of the non-corrupted ones. Yet another extension to Definitions 1 and 2 is to consider an arbitrary network of functions (and not just a single chain). Finally, one could strengthen the privacy guarantee, requiring that $\mathsf{Eval}^*$ hides not only the functions that the nodes compute but also the structure of the network itself (e.g., the number of functions in the chain). We leave all of these extensions to future work.

**Definition 3 (Compactness).** *A scheme* $\mathcal{E} = (\mathsf{KeyGen}, \mathsf{Enc}, \mathsf{Dec}, \mathsf{Eval})$ *is* $i$-*hop* compact *homomorphic if there exists a polynomial* $p(\cdot)$ *in (only) the security parameter* $k$, *such that decryption of any ciphertext (even one that is the output of* $\mathsf{Eval}^*$) *w.r.t. the security parameter* $k$ *can be implemented by a circuit of size at most* $p(k)$.

*Namely, for every value of* $k$, *there exists a circuit* $\mathsf{Dec}^{(k)}$ *of size at most* $p(k)$, *such that the* $i$-*Hop property from Definition 1 holds for that decryption circuit.*

The name "compactness" is justified by the fact that the length of the evaluated ciphertexts cannot grow beyond $p(k)$ (regardless of $f$), if they are to be decrypted by a $p(k)$-size circuit. We comment that compactness and function privacy together are still formally weaker than the Ishai-Paskin notion of "privacy with size hiding" [10, Def 8].

## 3    From 1-Hop to *i*-Hop Homomorphic Encryption

Below we show how to transform a 1-hop HE scheme to an $i$-hop scheme for any constant $i > 0$. The price that we pay, however, is that the complexity of the $i$-hop scheme (and in particular, the length of the evaluated ciphertexts) may grow as large as $k^{O(i)}$ (for security parameter $k$).

The idea is that each evaluator (with function $\tilde{f}$) in the chain, upon receiving the "evaluated ciphertext" $c$ from its predecessor, applies again the evaluation procedure, but not to its original function $f$. Rather, it applies the evaluation procedure to the concatenation of $f$ with the decryption function, namely to the function $F_{f,c}(\mathsf{SK}) \overset{\text{def}}{=} f\big(\mathsf{Dec}(\mathsf{SK}, c)\big)$. This technique, which is reminiscent of Gentry's "bootstrapping" technique [7], works because (by induction) applying $\mathsf{Dec}(\mathsf{SK}, c)$ on the previous evaluated ciphertext outputs the value $(f_{j-1} \circ \cdots \circ f_1)(x)$.

THE CONSTRUCTION. Let $\mathcal{H} = (\mathsf{KeyGen}, \mathsf{Enc}, \mathsf{Eval}, \mathsf{Dec})$ be a function-private homomorphic 1-hop encryption scheme (that need not be compact). Let $i$ be a constant parameter of the system (that represents the number of hops that we are shooting for). We construct a function-private $i$-hop homomorphic encryption scheme $\mathcal{H}^{(i)} = (\mathsf{KeyGen}^{(i)}, \mathsf{Enc}^{(i)}, \mathsf{Eval}^{(i)}, \mathsf{Dec}^{(i)})$ as follows.

$\mathsf{KeyGen}^{(i)}$: Run $\mathsf{KeyGen}$ for $i + 1$ times, to get for $j = 0, 1, \ldots, i$:

$$(\mathrm{PK}_j, \mathrm{SK}_j) \leftarrow \mathsf{KeyGen}, \text{ and for } j < i \text{ also: } \alpha_j \leftarrow \mathsf{Enc}\big(\underbrace{\mathrm{PK}_{j+1}}_{\text{key}}; \underbrace{\mathrm{SK}_j}_{\text{ptxt}}\big)$$

Defining $\alpha_i = \bot$, the public key is the set $\mathrm{PK}^{(i)} = \{(\mathrm{PK}_j, \alpha_j) : j = 0, 1, \ldots, i\}$, and the secret key is $\mathrm{SK}^{(i)} = (\mathrm{SK}_0, \mathrm{SK}_1, \ldots, \mathrm{SK}_i)$.

$\mathsf{Enc}^{(i)}(\mathrm{PK}^{(i)}; x)$: Set $c_0 \leftarrow \mathsf{Enc}(\mathrm{PK}_0; x)$ and output $\big[\text{level-0}, c_0\big]$.

$\mathsf{Eval}^{(i)}(\mathrm{PK}^{(i)}, \tilde{c}, f_{j+1})$: Parse the ciphertext as $\tilde{c} = \big[\text{level-}j, c_j\big]$. Compute the description of the function $F_{f_{j+1},c_j}(s) \overset{\text{def}}{=} f_{j+1}(\mathsf{Dec}(s; c_j))$, and set $c_{j+1} \leftarrow \mathsf{Eval}(\mathrm{PK}_{j+1}; F_{f_{j+1},c_j}, \alpha_j)$. Output $\big[\text{level-}(j + 1), c_{j+1}\big]$.

$\mathsf{Dec}^{(i)}(\mathrm{SK}^{(i)};\ \tilde{c})$: Parse the ciphertext as $\tilde{c} = \big[\text{level-}j, c_j\big]$. Compute and output $y \leftarrow \mathsf{Dec}(\mathrm{SK}_j;\ c_j)$.

**Theorem 4.** *The scheme $\mathcal{H}^{(i)}$ above is an $i$-hop function private homomorphic encryption scheme.*

*Proof.* (sketch) Correctness is easy to establish by induction. The correctness of the underlying 1-hop homomorphic encryption scheme $\mathcal{H}$ implies that for all $j \leq i$ we have

$$\mathsf{Dec}(\mathrm{SK}_j, c_j) = \mathsf{Dec}(\mathrm{SK}_j, \mathsf{Eval}(\mathrm{PK}_j;\ F_{f_j, c_{j-1}},\ \alpha_{j-1}))$$

$$\overset{(a)}{=} F_{f_j, c_{j-1}}(\mathrm{SK}_{j-1}) \overset{(b)}{=} f_j(\mathsf{Dec}(\mathrm{SK}_{j-1}, c_{j-1})) \overset{(c)}{=} (f_j \circ \ldots \circ f_1)(x),$$

where $f_j$ is the function that was used in the $j$'th hop, Equality $(a)$ holds by correctness of the underlying 1-hop scheme, Equality $(b)$ holds by definition of $F_{f_j, c_{j-1}}$, and Equality $(c)$ holds by the induction hypothesis.

Semantic security of $\mathcal{H}^{(i)}$ follows trivially from that of the underlying (1-hop) encryption scheme. Similarly, $i$-hop function privacy follows easily from the 1-hop privacy of the underlying scheme (and the fact that the size of $F_{f_j, c_{j-1}}$ that the $\mathcal{H}$ simulator needs can be computed easily from the size of $f_j$ and the size of $c_{j-1}$ both of which the simulator for $\mathcal{H}^{(i)}$ knows).

COMPLEXITY. For "generic" 1-hop encryption schemes (such as the one that we can obtain from two-message SFE using the folklore construction described in Section 1.3), the size of the ciphertext resulting from $\mathsf{Eval}(f, c)$ is larger than the input length $|c| + |f|$ by some factor $K$ which is polynomial in the security parameter $k$. Hence the size of the circuit for $F_{f_j, c_{j-1}}$ in our construction is at least

$$K(\cdots K(K(|c_0|+|f_1|)+|f_2|)\cdots)+|f_j| = |c_0|K^{j-1} + \sum_{t=1}^{j} |f_t| K^{j-t} = \Big(\sum_{t=1}^{j} |f_j|\Big) \cdot k^{O(j)}$$

which means that after $i$ hops the ciphertext grows as $k^{O(i)}$.

## 3.1   Compact and Function-Private Homomorphic Encryption

Recall that the exponential blowup in the construction above is due to the fact that the ciphertext that results from $\mathsf{Eval}$ is larger than the function size (by a multiplicative factor). On the other hand, if the underlying 1-hop scheme is compact (and function-private), then the construction above would yield a compact (and function-private) $i$-hop scheme.

Below we show that given a 1-hop scheme which is compact but not private, and another 1-hop scheme which is private but not compact, we can combine them to get a 1-hop scheme which is *both compact and private* (and thus also $i$-hop compact and private scheme for all $i$, by the observation above).

The idea is to iterate the two schemes at every hop. First we apply the private scheme to the function $f$ that we want to evaluate, thus getting a "private ciphertext" which is large but does not reveal information about $f$. Then we apply the compact scheme to the decryption function of the private scheme, in essence "compressing" the large ciphertext into a compact one which is still decrypted to the same value. The result is clearly compact (since it outputs the "compact ciphertext"). It is also function-private since the only dependence of the compact ciphertext on the function $f$ is via the value of the intermediate "private ciphertext", and even if we were to give the adversary the "private ciphertext" itself, it would still not violate the function-privacy of $f$.[1]

We note that when using this technique, we again get a "hard-wired" parameter $i$ that limits the number of hops that we can handle: to get an $i$-hop scheme, the public key must have size linear in $i$. Thus, the resulting scheme is not multi-hop, according to Definition 1. This limitation can be circumvented by relying on the circular security of the resulting 1-hop schemes; the details are deferred to the full version.

## 4    Extendable and Re-randomizable Secure Computation

Below we define the tool of "extendable and re-randomizable SFE", and show how it is used for multi-hop homomorphic encryption. In the next section we show that this tool can be implemented under the decisional Diffie-Hellman assumption. We begin with definitions (which are similar to Ishai et al. [10]).

We fix a particular "universal circuit evaluator" $U(\cdot, \cdot)$, taking as input a description of a function $f$ and an argument $x$, and returning $f(x)$. Using $U$ we can view every bit-string $f$ as describing a function (where $f(x)$ is just a shorthand for $U(f, x)$).

A two-message protocol for secure two-party computation to be run by Alice (the client) and Bob (the server), is implemented by three polynomial-time procedures $\Pi = (\mathsf{SFE1}, \mathsf{SFE2}, \mathsf{SFE\text{-}Out})$, where:

**1.** The procedure $\mathsf{SFE1}(x)$ is run by the client with input $x$ and randomness $r_1$ to get the "first message" $m_1$. $m_1$ is then sent to the server and $r_1$ is kept for later. We assume that $r_1$ includes in particular all the randomness that the client uses.

**2.** The procedure $\mathsf{SFE2}(f, m_1)$ is run by the server with input a function $f$ and randomness $r_2$. The output of this procedure $m_2$ is then sent to the client.

**3.** Finally, the client runs the procedure $\mathsf{SFE\text{-}Out}(r_1, m_2)$ to recover an output $y$. Correctness of the SFE protocol demands that the value $y$ is equal to $f(x)$.

By $\mathsf{SFE1}(x)$ (resp. $\mathsf{SFE2}(m_1, f)$), we mean the distribution generated by the respective algorithms (over the choice of their randomness). We also say that

---

[1] We comment that iterating the two systems in the opposite order also works: we can apply the compact scheme to the function $f$ and the private scheme to the decryption of the compact one.

$(m_1, r_1) \in \mathsf{SFE1}(x)$ (resp. $(m_2, r_2) \in \mathsf{SFE2}(m_1, f)$) to denote a particular element in the support of the distribution (together with the randomness involved).

**Definition 4 (Client and (honest-but-curious) Server privacy).** *A protocol $\Pi = (\mathsf{SFE1}, \mathsf{SFE2}, \mathsf{SFE\text{-}Out})$ is said to be:*

- Client-private, *if for any two inputs $x, x'$ of the same length, the distributions $\mathsf{SFE1}(x)$ and $\mathsf{SFE1}(x')$ are indistinguishable (even given $x, x'$).*
- Server-private *in the honest-but-curious model, if there exists a polynomial time simulator $\mathsf{Sim}$ such that for every input $x$ and function $f$, and every $(m_1, r_1) \in \mathsf{SFE1}(x)$, the distributions $\mathsf{SFE2}(f, m_1)$ and $\mathsf{Sim}(m_1, 1^{|f|}, f(x))$ are indistinguishable (even given $f, x, m_1$ and $r_1$).*

We now define the notion of an extendable SFE protocol.

**Definition 5 (Extendable SFE, honest-but-curious).** *A two-message SFE protocol $\Pi = (\mathsf{SFE1}, \mathsf{SFE2}, \mathsf{SFE\text{-}Out})$ is extendable, if there exists an efficient procedure $\mathsf{Extend}$ such that for any two compatible functions $f$ and $f'$, any input $x$ to $f$, and for every $(m_1, r_1) \in \mathsf{SFE1}(x)$, the distributions $\mathsf{Extend}(\mathsf{SFE2}(m_1, f), f')$ and $\mathsf{SFE2}(m_1, f' \circ f)$ are indistinguishable (even given $x, f, f', m_1$ and $r_1$).*

EXTENDABLE SFE FROM YAO CIRCUITS. The construction of Cachin et al. [5, Sec. 5] can be cast in our language as describing an extendable SFE protocol based on Yao's garbled circuit construction [19]. As described in the introduction, the idea is that since the garbled circuit for $f$ includes both the 0-label and the 1-label on any *output wire*, it can be extended by treating these labels as the input labels for $f'$.

We comment that garbling the gates hides only the type of these gates and not the topology of a circuit. To hide the function we must also use some form of canonicalization of circuits, so that all circuits of a given size will have the same topology. Moreover, to meet our definition of extendibility, it must be the case that canonicalizing $f$, then extending it with $f'$ and canonicalizing the whole thing yields the same topology as canonicalizing the composed function $f' \circ f$.

We note that such canonicalization is possible, and the size of the canonicalized circuits does not grow much. For example, a circuit of maximum width $w$ can be canonicalized to a leveled circuit with width $w$ at every level, and a big "multiplexer gate" between every two successive levels that determines what output from the lower level goes to what input in the upper one. To get the additional property that we need (where the order of canonicalization does not matter) we would also have $w$ output wires in the circuit, where the redundant output wires have both labels set to 0. (We may also need to supply some dummy gates that take as input the input wires and have both output labels set to 0, to be able to pad the circuit if the maximum width of $f'$ is larger than that of $f$.)

FROM EXTENDABLE TO RE-RANDOMIZABLE. Note that extendable SFE by itself already yields multi-hop homomorphic encryption with a weak form of function-privacy: to a recipient that does not know the intermediate values (namely, the output of $\mathsf{SFE2}(m_1, f)$), the output of $\mathsf{Extend}$ looks just as if it was generated "from scratch" by running SFE2 with input $f' \circ f$, so Extend hides the function if SFE2 does. This means that when the protocol $\Pi$ is used for many hops, then as long as all the intermediate hops are "trusted" not to reveal their intermediate results (and only the sender and the recipient are honest-but-curious), using Extend would maintain the privacy of everyone's functions.

However, this solution still falls short of our function-privacy goal, since a collusion between the recipient and the node that computed $SFE2(m_1, f)$ can reveal the function $f'$. In other words, the output of Extend may not be distributed like $SFE2(m_1, f' \circ f)$ *given also the intermediate results from* $SFE2(m_1, f)$. To overcome this problem, we introduce the notion of a *re-randomizable SFE*: In a nutshell, we want to transform the second message $m_2 \leftarrow SFE2(m_1, f)$ into $m_2'$ such that even if the recipient and the party that computed $m_2$, they cannot distinguish $m_2'$ from random. Then, a node can re-randomize the message from its predecessor, thus rendering the intermediate results held by the predecessor irrelevant.

**Definition 6 (Re-randomizable SFE, honest-but-curious).** *A two-message SFE protocol $\Pi$ is re-randomizable if there exists an efficient procedure* reRand *such that for every input $x$ and function $f$ and every $(m_1, r_1) \in$ SFE1$(x)$ and $(m_2, r_2) \in$ SFE2$(m_1, f)$, the distributions* reRand$(m_1, m_2)$ *and* SFE2$(m_1, f)$ *are indistinguishable, even given $x, f, m_1, r_1, m_2, r_2$.*

FROM EXTENDABLE AND RE-RANDOMIZABLE SFE TO MULTI-HOP HE. Let $\Pi = (\mathsf{SFE1}, \mathsf{SFE2}, \mathsf{SFE\text{-}Out})$ be an extendable and re-randomizable two message SFE protocol with client and server privacy, and let $\mathcal{E} = (\mathsf{KeyGen}, \mathsf{Enc}, \mathsf{Dec})$ be a semantically secure public-key encryption scheme. We now describe the construction of the multi-hop homomorphic scheme $\mathcal{H}^* = (\mathsf{KeyGen}^*, \mathsf{Enc}^*, \mathsf{Dec}^*, \mathsf{Eval}^*)$.

The key generation $\mathsf{KeyGen}^*$ is the same as $\mathsf{KeyGen}$ for the underlying encryption. The encryption procedure $\mathsf{Enc}^*(\mathrm{PK}; x)$ first runs $(m_1, r_1) \leftarrow \mathsf{SFE1}(x)$, then encrypts $r_1$ using $\mathcal{E}$ to get $c \leftarrow \mathsf{Enc}(\mathrm{PK}; r_1)$, and finally, computes $m_2 \leftarrow \mathsf{SFE2}(m_1, f_{ID})$ (where $f_{ID}$ is the identity function). The ciphertext is $(c, m_1, m_2)$.

To evaluate a function $f_j$ on an $\mathcal{H}^*$-ciphertext $c_{j-1}$, first parse $c_{j-1}$ as a tuple $(c, m_1, m_2^{(j-1)})$, then set $m_2' \leftarrow \mathsf{Extend}(m_2^{(j-1)}, f_j)$ and $m_2^{(j)} \leftarrow \mathsf{reRand}(m_1, m_2')$. The evaluated ciphertext is $(c, m_1, m_2^{(j)})$. Decrypting $c_j = (c, m_1, m_2^{(j)})$ consists of using the decryption of $\mathcal{E}$ to get $r_1 \leftarrow \mathsf{Dec}(\mathrm{SK}, c)$, then outputting $y \leftarrow \mathsf{SFE\text{-}Out}(r_1, m_2^{(j)})$.

**Theorem 5 (Extendable+Re-randomizable $\Rightarrow$ Multi-hop).** *Assume that the encryption scheme $\mathcal{E}$ is semantically secure, the SFE protocol $\Pi$ is extendable and re-randomizable with client and server privacy, and in addition that the size of any function $f$ can be efficiently determined from the output of* SFE2$(m_1, f)$.

*Then the scheme $\mathcal{H}^*$ above is a multi-hop function-private homomorphic encryption scheme. Moreover, the size of an evaluated ciphertext in $\mathcal{H}^*$ does not depend on the number of hops, but only on the size of the composed function.*

*Proof.* (sketch) Correctness of $\mathcal{H}^*$ follows from the the correctness of $\Pi$, and its extendability and re-randomizability: we know that SFE-Out would recover the right $y$ when given the second message from SFE2, and by extendability the output of Extend is the same as that of SFE2, no matter how many hops were used. Semantic security follows from semantic security of the underlying encryption and from the client-privacy of $\Pi$.

To show function privacy, we need to describe a simulator $\mathsf{Sim}_{\mathcal{H}^*}$ that on input $c_{j-1} = (c, m_1, m_2^{(j-1)})$, $|f_j|$, and $y_j = (f_1 \circ \cdots \circ f_j)(x)$, generates a distribution indistinguishable from $c_j = (c, m_1, m_2^{(j)})$. The simulator recovers from $m_2^{(j-1)}$ the size $|f_1 \circ \cdots \circ f_{j-1}|$ and adds it to $|f_j|$ to get $\gamma = |f_1 \circ \cdots \circ f_j|$. Then $\mathsf{Sim}_{\mathcal{H}^*}$ uses the simulator for $\Pi$ to get $m_2^{(j)} \leftarrow \mathsf{Sim}_\Pi(m_1, \gamma, y_j)$ and outputs $c_j = (c, m_1, m_2^{(j)})$.

By the server-privacy of $\Pi$, the distribution of $m_2^{(j)}$ so generated is indistinguishable from $\mathsf{SFE2}(m_1, f_1 \circ \cdots \circ f_j)$. On the other hand, by the extendability and re-randomizability properties of $\Pi$, the distribution of $m_2^{(j)}$ in $\mathcal{H}^*$ is also indistinguishable from the same $\mathsf{SFE2}(m_1, f_1 \circ \cdots \circ f_j)$. Hence these two distributions are indistinguishable from each other.                                        □

## 5   Extendable and Re-randomizable SFE from DDH

Given Theorem 5, we now focus on building an extendable and re-randomizable SFE protocol. Our starting point is Yao's garbled circuit construction [19], which is extendable, but not re-randomizable. We seek a re-randomizable implementation of this scheme by using building blocks that are "sufficiently homomorphic" to support the randomization that we need. Specifically, we rely on the oblivious-transfer protocol of Naor-Pinkas/Aiello-Ishai-Reingold [13,2], and on the encryption scheme of Boneh-Halevi-Hamburg-Ostrovsky [4], the security of both of which is equivalent to the decisional Diffie-Hellman assumption. Below we briefly summarize some properties of these building blocks; a slightly longer description (and the definitions of OT) can be found in the full version of this paper [8].

RE-RANDOMIZABLE OBLIVIOUS TRANSFER. The protocol in [13,2] is a two-message protocol. The receiver that has a choice bit $\sigma \in \{0, 1\}$ sends the first message $m_1 \leftarrow OT1(\sigma)$, the sender that has two bits $\gamma_0, \gamma_1 \in \{0, 1\}$ replies with $m_2 \leftarrow OT2(m_1, \gamma_0, \gamma_1)$, and the receiver can recover the bit $\gamma_\sigma$ from $m_2$ and the state that it keeps. Receiver security means that $OT1(0)$, $OT1(1)$ are indistinguishable, and sender security means that $OT2(m_1, \gamma_0, \gamma_1)$ can be simulated knowing only $m_1$ and $\gamma_\sigma$. We note that if the sender has two strings $\gamma_0, \gamma_1$, (rather than just two bits) then it can use the same $m_1$ from the receiver and send many $m_2$'s in reply, one for every bit position in the input vectors.

Another property we use is that the protocol from [13,2] is re-randomizable: given $m_1, m_2$, anyone can re-randomize the reply, computing another random $m_2'$ from the distribution $OT2(m_1, \gamma_0, \gamma_1)$ (even without knowing $\gamma_0, \gamma_1$).

KEY AND PLAINTEXT ADDITIVELY HOMOMORPHIC ENCRYPTION. The BHHO scheme [4] is a semantically secure public key encryption scheme where the secret key is a string $s \in \{0, 1\}^\ell$ and the plaintext is also a string $x \in \{0, 1\}^n$.

(In our application we use $n = 2\ell$.) The public key and ciphertexts are vectors of elements over a group of some prime order $q$.

The BHHO scheme has the following "additively homomorphic" property: Let $T, T'$ be two known affine transformations on vectors over $Z_q$ *that map 0-1 vectors to 0-1 vectors of the same length*. Then, given a public key PK corresponding to some secret key $s$ and a ciphertext $c \in \mathsf{Enc}(\text{PK}; \boldsymbol{x})$, anyone can generate a random public key PK$'$ corresponding to $T(\boldsymbol{s})$ and a random ciphertext $c' \in \mathsf{Enc}(\text{PK}'; T'(\boldsymbol{x}))$. In particular, this means that anyone can XOR known strings $\Delta, \Delta'$ into $\boldsymbol{s}$ and $\boldsymbol{x}$, and also anyone can permute the bits in either $\boldsymbol{s}$ or $\boldsymbol{x}$ (or both) according to known permutations.

## 5.1   Our Construction

Our construction closely follows Yao's original garbled circuit construction [19]. The client (Alice) on input $\boldsymbol{x} = \langle x_1, \ldots, x_n \rangle$, sends $n$ first messages of the OT protocol from above, using her input bit $x_i$ as the choice bit for the $i$'th message, $m_1[i] \leftarrow OT1(x_i)$.

The server (Bob) has a boolean circuit with fan-in-2 gates. Bob's circuit has $n$ input ports, some number of output ports, and some number of internal gates. Each wire in the circuit is therefore either an input wire (connecting an input port to some internal gates and/or output ports), or a gate-output wire (connecting the output of one internal gate to some other internal gates and/or output ports). We stress that we allow the same wire to be used as input to several internal gates or output ports.[2]

Bob receives from Alice the $n$ OT first messages, $m_1[1], \ldots, m_1[n]$. He begins by choosing at random two $\ell$-bit labels $L_{w,0}, L_{w,1}$ for every wire $w$, each having exactly $\lceil \ell/2 \rceil$ 1's. (Here $\ell$ is the length of the BHHO secret key.) For each input wire $w_i$, corresponding to Alice's first message $m_1[i]$, Bob computes the OT second message for the two labels on the corresponding input wire, $m_2[i] \leftarrow OT2(m_1[i]; L_{w_i,0}, L_{w_i,1})$.

Then, for an internal fan-in-2 gate (computing the binary operation $\star$), Bob computes four pairs of ciphertexts as follows: Let $w_1, w_2$ be the two input wires for this gate and $w_3$ be the output wire. Bob chooses four fresh random $2\ell$-bit masks $\delta_{i,j}$ for $i, j \in \{0, 1\}$ and computes the four pairs:

$$\left\{ \left( \mathsf{Enc}_{L_{w_1,i}}(\delta_{i,j}), \ \mathsf{Enc}_{L_{w_2,j}}((L_{w_3,k}|0^\ell) \oplus \delta_{i,j}) \right) \ : \ i, j \in \{0, 1\}, \ k = i \star j \right\} \quad (1)$$

Namely, Bob uses the secret key $L_{w_1,i}$ to encrypt the mask $\delta_{i,j}$ itself, and the other secret key $L_{w_2,j}$ to encrypt the masked label (concatenated with $\ell$ zeros). The "gadget" for this gate consists of the four pairs of ciphertexts from Eq. (1) in random order. The garbled circuit that Bob sends back to Alice consists of the $n$ OT second messages $m_2[1], \ldots, m_2[n]$, and the gadgets for all the gates in the circuit (which we assume include an indication of which wire connects what

---

[2] We assume that the two input wires at each gate are always distinct. This can be enforced, e.g., by implementing a fan-in-1 gate (i.e., NOT) via a fan-in-2 XOR-with-one gate.

gates). In addition, for each output wire $w$ with labels $L_{w,0}$ and $L_{w,1}$, Bob sends an ordered pair of public keys, the first corresponding to $L_{w,0}$ and the second to $L_{w,1}$. (We chose this particular mapping to enable re-randomization.)

Upon receiving this garbled circuit, Alice first uses the recovery procedure of the OT protocol to recover one of the labels for each input wire. Then she goes over the garbled circuit gate by gate as follows: For a fan-in-2 gate where she knows the labels $L_1, L_2$ for the two inputs, she uses the key $L_1$ to decrypt the first component in each of the four pairs and uses the key $L_2$ to decrypt the second component of the four pairs. Then she XORs the two decrypted strings from each pair, and if any of the resulting strings is of the form $L^*|0^\ell$ then she takes $L^*$ to be the label of the output wire. (If more than one string has the form $L^*|0$ then Alice takes the first one, and if none has this form then she sets $L^* = 0^\ell$.) Upon recovering a label on an output port, she checks if this label corresponds to the first or the second public keys that were provided for this port, outputting zero or one accordingly. (Or $\perp$ if it does not correspond to any of them.) The proof of the following theorem is very similar to [12], and is given in the full version.

**Theorem 6.** *The protocol from above, using the BHHO encryption scheme, enjoys both client and server privacy, under the DDH assumption.*

*Remark: balanced secret keys.* We note that the BHHO scheme is used here with secret keys that have exactly $\ell/2$ 1's in them, rather than with completely uniform secret keys. This is used for the purpose of re-randomization, as described in Section 5.2. We note that this variant of BHHO is also semantically secure: In fact, Naor and Segev proved that under DDH, the BHHO scheme is semantically-secure for *every secret-key distribution with sufficient min-entropy* (cf. [14, Sec 5.2]). We use this stronger result in our proof of the re-randomization property in Section 5.2.

## 5.2   Re-randomizing Garbled Circuits

We proceed to show how garbled circuits from above can be re-randomized. We begin by observing that a simple re-randomization method that only XORs random masks into the labels does not work: Observe that the re-randomizer does not know which of the two labels on a wire was used as key (or input) in what ciphertext, so it cannot use two different masks to randomize the two different labels on a wire. Rather, it can only apply the *same mask* $\Delta_w$ to both labels on a wire. But this is clearly not sufficient for randomization, since it leaves the XOR of the two labels on each wire as it was before.

Moreover, such "partial randomization" is clearly insecure in our application: Note that the predecessor of a node knows the two "old labels" for every wire in its circuit, including the labels for the output wires (which are the current node's input wires). Also, the receiver (Alice) would learn one of the "new labels" on these wire upon evaluation. Hence between the predecessor and Alice, they will be able to reconstruct both new labels for every input, thus un-garbling the circuit of the current node.

To overcome this problem, we rely on stronger homomorphic properties of BHHO: Namely, viewing keys and plaintexts as vectors, it is homomorphic with respect to any affine function over $Z_q$. This means, in particular, that it is homomorphic with respect to permutations (i.e., multiplications by permutation matrices). Namely, we can transform a ciphertext $\mathsf{Enc}_L(L')$ into $\mathsf{Enc}_{\pi(L)}(\pi'(L'))$ for any two permutations $\pi, \pi'$ of the bits. We therefore work with balanced secret keys that have exactly $\ell/2$ 1's, and use permutations to randomize them.

Note that in the attack scenario from above, where a predecessor colludes with the recipient, they will now know the old labels $L, L'$, and also one new label, computed as $\pi(L)$. In Lemma 1 we show that given these three values, the other new label $\pi(L')$ still has a lot of min-entropy, provided that the Hamming distance between $L, L'$ is not too small. In the honest-but-curious model, $L$ and $L'$ will be about $\ell/2$ apart, hence $\pi(L')$ will have min-entropy close to $\ell$ (see Lemma 1 below). The Naor-Segev result [14] then implies that it is safe to use $\pi(L')$ as a secret key, which is indeed the way that it is used in the re-garbled circuit. Putting all these arguments together, we have the following theorem:

**Theorem 7.** *Under the DDH assumption, the BHHO-based protocol from above is computationally re-randomizable.*

THE PERMUTATIONS LEMMA. Let $HW_{\ell,k} \subseteq \{0,1\}^\ell$ denote the set of all $\ell$-bit strings with Hamming weight exactly $k$, and also let $S_\ell$ denote the set of all permutations over $\ell$ elements. Assume that $\ell$ is even from now on. The lemma below shows that for two strings $L_1$ and $L_2$, chosen uniformly at random from $HW_{\ell,\ell/2}$, and a random permutation $\pi : [\ell] \rightarrow [\ell]$, the string $\pi(L_2)$ has large residual min-entropy *even given* $L_1, L_2$ *and* $\pi(L_1)$. For the lemma below, let $\tilde{H}_\infty(X|Y)$ be the average min-entropy of $X$ given $Y$ (cf. [6]), that is

$$\tilde{H}_\infty(X|Y) \stackrel{\text{def}}{=} - \log \mathop{\mathbb{E}}_{y \leftarrow Y} \left( \max_x \Pr[X = x | Y = y] \right) = - \log \mathop{\mathbb{E}}_{y \leftarrow Y} \left( 2^{-H_\infty(X|Y=y)} \right)$$

**Lemma 1.** *Let* $L_1, L_2 \in_R HW_{\ell,\ell/2}$, *and* $\pi \in_R S_\ell$ *be uniformly random. Then:*

$$\tilde{H}_\infty\big(\pi(L_2) \mid L_1, L_2, \pi(L_1)\big) \geq \ell - \frac{3}{2} \log \ell$$

The proof is in the full version. It follows easily from the observation that given $L_1, L_2$ and $\pi(L_1)$, the string $\pi(L_2)$ is distributed uniformly from among all strings in $HW_{\ell,\ell/2}$ whose Hamming distance from $\pi(L_1)$ equals the Hamming distance between $L_1$ and $L_2$. $\square$

**Acknowledgments.** We thank Yuval Ishai for several inspiring discussions.

# References

1. Aguilar-Melchor, C., Gaborit, P., Herranz, J.: Additively Homomorphic Encryption with d-Operand Multiplications. In: Rabin, T. (ed.) CRYPTO 2010. LNCS, vol. 6223, pp. 138–154. Springer, Heidelberg (2010)

2. Aiello, W., Ishai, Y., Reingold, O.: Priced oblivious transfer: How to sell digital goods. In: Pfitzmann, B. (ed.) EUROCRYPT 2001. LNCS, vol. 2045, pp. 119–135. Springer, Heidelberg (2001)

3. Barak, B., Haitner, I., Hofheinz, D., Ishai, Y.: Bounded key-dependent message security. In: Gilbert, H. (ed.) EUROCRYPT 2010. LNCS, vol. 6110, pp. 423–444. Springer, Heidelberg (2010)

4. Boneh, D., Halevi, S., Hamburg, M., Ostrovsky, R.: Circular-secure encryption from decision diffie-hellman. In: Wagner, D. (ed.) CRYPTO 2008. LNCS, vol. 5157, pp. 108–125. Springer, Heidelberg (2008)

5. Cachin, C., Camenisch, J., Kilian, J., Müller, J.: One-round secure computation and secure autonomous mobile agents. In: Welzl, E., Montanari, U., Rolim, J.D.P. (eds.) ICALP 2000. LNCS, vol. 1853, pp. 512–523. Springer, Heidelberg (2000)

6. Dodis, Y., Reyzin, L., Smith, A.: Fuzzy extractors: How to generate strong keys from biometrics and other noisy data. In: Cachin, C., Camenisch, J.L. (eds.) EUROCRYPT 2004. LNCS, vol. 3027, pp. 523–540. Springer, Heidelberg (2004)

7. Gentry, C.: Fully homomorphic encryption using ideal lattices. In: STOC 2009, pp. 169–178. ACM, New York (2009)

8. Gentry, C., Halevi, S., Vaikuntanathan, V.: i-hop homomorphic encryption schemes. Cryptology ePrint Archive, Report 2010/145 (2010)

9. Goldwasser, S., Micali, S.: Probabilistic encryption. J. Comput. Syst. Sci. 28(2), 270–299 (1984)

10. Ishai, Y., Paskin, A.: Evaluating branching programs on encrypted data. In: Vadhan, S.P. (ed.) TCC 2007. LNCS, vol. 4392, pp. 575–594. Springer, Heidelberg (2007)

11. Kushilevitz, E., Ostrovsky, R.: Replication is NOT Needed: SINGLE Database, Computationally-Private Information Retrieval. In: FOCS 1997, pp. 364–373. IEEE, Los Alamitos (1997)

12. Lindell, Y., Pinkas, B.: A Proof of Security of Yao's Protocol for Two-Party Computation. J. Cryptology 22(2), 161–188 (2009)

13. Naor, M., Pinkas, B.: Efficient oblivious transfer protocols. In: ACM-SIAM Symposium on Discrete Algorithms - SODA 2001, pp. 448–457. ACM, New York (2001)

14. Naor, M., Segev, G.: Public-key cryptosystems resilient to key leakage. In: Halevi, S. (ed.) CRYPTO 2009. LNCS, vol. 5677, pp. 18–35. Springer, Heidelberg (2009)

15. Rivest, R., Adleman, L., Dertouzos, M.: On data banks and privacy homomorphisms. In: Foundations of Secure Computation, pp. 169–177. Academic Press, London (1978)

16. Sander, T., Young, A., Yung, M.: Non-interactive CryptoComputing for NC1. In: 40th Annual Symposium on Foundations of Computer Science - FOCS 1999, pp. 554–567. IEEE, Los Alamitos (1999)

17. Smart, N., Vercauteren, F.: Fully homomorphic encryption with relatively small key and ciphertext sizes. In: Nguyen, P.Q., Pointcheval, D. (eds.) PKC 2010. LNCS, vol. 6056, pp. 420–443. Springer, Heidelberg (2010)

18. van Dijk, M., Gentry, C., Halevi, S., Vaikuntanathan, V.: Fully homomorphic encryption over the integers. In: Gilbert, H. (ed.) EUROCRYPT 2010. LNCS, vol. 6110, pp. 24–43. Springer, Heidelberg (2010)

19. Yao, A.C.: Protocols for secure computations (extended abstract). In: 23rd Annual Symposium on Foundations of Computer Science – FOCS 1982, pp. 160–164. IEEE, Los Alamitos (1982)

# Interactive Locking, Zero-Knowledge PCPs, and Unconditional Cryptography[*]

Vipul Goyal[1,**], Yuval Ishai[2,***], Mohammad Mahmoody[3,†], and Amit Sahai[4,‡]

[1] Microsoft Research, India
vipul@microsoft.com
[2] Technion and UCLA
yuvali@cs.technion.ac.il
[3] Princeton University
mohammad@cs.princeton.edu
[4] UCLA
sahai@cs.ucla.edu

**Abstract.** Motivated by the question of basing cryptographic protocols on stateless tamper-proof hardware tokens, we revisit the question of unconditional two-prover zero-knowledge proofs for **NP**. We show that such protocols exist in the *interactive PCP* model of Kalai and Raz (ICALP '08), where one of the provers is replaced by a PCP oracle. This strengthens the feasibility result of Ben-Or, Goldwasser, Kilian, and Wigderson (STOC '88) which requires two stateful provers. In contrast to previous zero-knowledge PCPs of Kilian, Petrank, and Tardos (STOC '97), in our protocol both the prover and the PCP oracle are efficient given an **NP** witness.

Our main technical tool is a new primitive that we call *interactive locking*, an efficient realization of an unconditionally secure commitment scheme in the interactive PCP model. We implement interactive locking by adapting previous constructions of *interactive hashing* protocols to our setting, and also provide a direct construction which uses a minimal amount of interaction and improves over our interactive hashing based constructions.

Finally, we apply the above results towards showing the feasibility of basing unconditional cryptography on *stateless* tamper-proof hardware tokens, and obtain the following results. **(1)** We show that if tokens can be used to encapsulate other tokens, then there exist unconditional and statistically secure (in fact, UC secure) protocols for general secure

---

[*] The full version of the paper is available at: http://eprint.iacr.org/2010/089
[**] This work was done mostly while this author was at UCLA, supported in part from NSF grants listed below.
[***] Supported in part by ISF grant 1310/06, BSF grants 2004361, 2008411, and NSF grants 0716835, 0716389, 0830803, 0916574.
[†] This work done partially while this author was visiting UCLA. Supported by NSF grants 0627526, 0426582 and 0832797.
[‡] Supported in part by BSF grants 2004361, 2008411, and NSF grants 0916574, 0716389, 0627781, 0830803.

computation. (**2**) Even if token encapsulation is not possible, there are unconditional and statistically secure commitment protocols and zero-knowledge proofs for **NP**. (**3**) Finally, if token encapsulation is not possible, then no protocol can realize statistically secure oblivious transfer.

# 1 Introduction

What is the minimal amount of trust required for unconditionally secure cryptography? Unconditional cryptography can be based on trusted two-party functionalities such as oblivious transfer [1,2] or noisy channels [3], on bounded storage assumptions [4], on the presence of an honest majority [5,6,7], or even on the presence of a dishonest majority of *non-communicating* parties [8]. More recently, there has been a considerable amount of work on cryptographic protocols in which parties can generate and exchange tamper-proof hardware tokens. In this model it was shown that unconditionally secure commitments [9] or even general secure two-party computation [10] are possible, provided that the tokens can be *stateful*. In particular, stateful tokens can erase their secrets after being invoked. The present work is motivated by the goal of establishing unconditional feasibility results for cryptography using *stateless* hardware tokens. This question turns out to be related to the classical question of unconditional multi-prover zero-knowledge proofs, which we revisit in this work. We start with some relevant background.

**Multi-Prover Zero-Knowledge.** Since the introduction of zero-knowledge proofs in the seminal work of Goldwasser, Micali, and Rackoff [11], a large body of work has been devoted to understanding the capabilities and limitations of such proofs. A particularly successful line of research studied the power of *statistical zero-knowledge* (SZK) proofs — ones which guarantee that even computationally unbounded verifiers can learn nothing from the interaction with the prover. In contrast to computational zero-knowledge proofs [12], a major limitation of SZK proofs which restricts their usefulness in cryptography is that they seem unlikely to cover the entire class of **NP** [13,14]. The related goal of obtaining any kind of *unconditional* zero-knowledge proofs for **NP**, which do not rely on unproven intractability assumptions, seems as unlikely to be achieved (cf. [15]) at least until the elusive **P** vs. **NP** question is resolved.

Motivated by the above goals, Ben-Or, Goldwasser, Kilian, and Wigderson [8] introduced in 1988 the model of *multi-prover interactive proofs* (MIPs), a natural extension of the standard model of interactive proofs which allows the verifier to interact with two or more non-communicating provers. The main result of [8] is an unconditional two-prover SZK proof for any language in **NP** (see [16,17,18] for subsequent improvements). A direct cryptographic application suggested in [8] is that of proving one's identity using a pair of bank cards. We will further discuss these types of applications later.

In a very surprising turn of events, the initial work on zero-knowledge in the MIP model led to a rapid sequence of developments that have literally transformed the theory of computer science. This line of research culminated in the first proof of the PCP Theorem [19,20].

The notion of probabilistically checkable proofs (PCPs) is very relevant to our work. In 1988, Fortnow, Rompel, and Sipser [21] suggested an alternative model for MIPs in which multiple provers are replaced by a single oracle, subsequently called a *PCP oracle* or just a PCP. The difference between an oracle and a prover is that an oracle, like a classical proof, cannot keep an internal state. When a prover is asked multiple queries, the answer to each query can depend on all previous queries, whereas the answer of an oracle to each query must depend on that query alone. The latter difference makes soundness against PCP oracles easier to achieve than soundness against provers, which explains the extra power of PCPs over traditional interactive proofs. However, as already observed in [8], the *zero-knowledge* property becomes harder to achieve when converting provers into oracles because oracles have no control over the number of queries made by a dishonest verifier. In particular, if the verifier may query the entire domain of the oracle (as in the case of traditional polynomial-length PCPs) then the oracle can no longer hide any secrets.

The question of replacing zero-knowledge provers by stateless oracles is motivated by practical scenarios in which verifiers can "reset" provers to their initial state, say by cutting off their power supply. (Note that similarly to zero-knowledge provers, zero-knowledge PCP oracles should be *randomized* in the sense that their answer depends both on the query and on a secret source of randomness which is picked once and for all when the oracle is initialized.) This motivation led to a recent line of work on *resettable zero-knowledge*, initiated by Canetti, Goldreich, Goldwasser, and Micali [22]. The main results from [22] show that, under standard cryptographic assumptions, there exist resettable (computational) zero-knowledge proofs for **NP**. However, results along this line do not seem relevant to the case of *unconditional* (and statistical) zero-knowledge proofs, which are the focus of the present work.

*Zero-knowledge PCPs.* The question of *unconditional* zero-knowledge PCPs was studied by Kilian, Petrank and Tardos [23] (improving over previous results implicit in [18]). Specifically, it is shown in [23] that any language in **NEXP** admits a proof system with a *single* PCP which is statistical zero-knowledge against verifiers that can make any polynomial number of PCP queries (but are otherwise computationally unbounded). However, as expected from proof systems for **NEXP**, the answers of the PCP oracle cannot be computed in polynomial time. This still leaves hope for scaling down the result to **NP** and making the PCP oracle efficient given an **NP** witness. Unfortunately, such a scaled down version presented in [23] has the undesirable side effect of scaling down the zero-knowledge property as well, effectively restricting the number of queries made by a cheating verifier to be much smaller than the (fixed polynomial) entropy of the oracle. Thus, compared to typical feasibility results in cryptography, the results of [23] for **NP** require us to either make an unreasonable assumption about the computational capability of the (stateless) prover, or to make an unreasonable assumption about the limitations of a cheating verifier.

*Interactive PCPs.* The above state of affairs motivates us to consider the *Interactive PCP* (IPCP) model, which was recently put forward by Kalai and Raz [24] and further studied in [25]. This model can be seen as lying in between the pure PCP model and the pure MIP model, thus aiding us in our quest for a "minimal" model for efficient unconditional zero-knowledge proofs for **NP**. In the IPCP model there is one interactive prover as in the MIP model and one PCP as in the PCP model. The study of IPCPs in [24] was motivated by the efficiency goal of allowing *shorter* PCPs for certain **NP** languages than in the traditional PCP model, at the price of a small amount of interaction with a prover. In contrast, our use of the IPCP model is motivated by the *feasibility* goal of obtaining unconditional zero-knowledge proofs for **NP** with polynomial-time prover and PCP oracle. Another difference is that while in the context of [24] a PCP is at least as helpful as a prover, the zero-knowledge property we consider is harder to satisfy with a PCP oracle than with a prover (as discussed above). The IPCP model can be made strictly stronger than the MIP model by requiring soundness to hold also with respect to *stateful* PCP oracles. We tackle this stronger variant as well, but we stick to the basic IPCP model by default.

To meaningfully capture zero-knowledge proofs with polynomial-time provers in the IPCP model, we extend the original IPCP model from [24] in two natural ways. First, we allow the PCP to be randomized. Concretely, we assume that both the prover and the PCP are implemented by polynomial-time algorithms with three common inputs: an instance $x$, a witness $w$, and a random input $r$. (This is analogous to earlier models for efficient multi-prover zero-knowledge proofs for **NP**.) The length of both $w$ and $r$ is polynomial in $|x|$. Second, as discussed above, in order to allow the PCP oracle to hide secrets from the verifier we need to use PCP oracles with a super-polynomial query domain, and we restrict cheating verifiers to make (an arbitrary) polynomial number of queries to the oracle, but otherwise allow them to be computationally unbounded. Note, however, that in contrast to the solutions from [23] we cannot use PCP oracles with a super-polynomial entropy since we want our PCP to be efficiently implementable.

This gives rise to the following feasibility question:

> *Are there (efficient-prover) statistical zero-knowledge proofs for* **NP** *in the interactive PCP model?*

**Our Results.** We answer the above question affirmatively, presenting an *unconditional* SZK proof for **NP** in the interactive PCP model with efficient prover and PCP oracle. Zero-knowledge holds against cheating verifiers which can make any polynomial (in fact, even sub-exponential) number of PCP queries, but are otherwise computationally unbounded. Our protocol can be implemented in a constant number of rounds. We also show how to get a similar protocol (with a non-constant number of rounds) in the stronger variant of the IPCP model in which a cheating PCP oracle may be stateful, thus strengthening the previous feasibility result from [8].

*Interactive locking.* The main technical tool we use to obtain the above results (as well as additional applications discussed below) is a new primitive which we call an *interactive locking scheme* (ILS). This primitive extends in a natural way the notion of non-interactive locking schemes which were defined and implemented in [23]. The original locking primitive can be viewed as a PCP-based implementation of a non-interactive commitment with statistical hiding and binding. Roughly speaking, a locking scheme is an oracle which hides a secret that can later be revealed to the receiver by sending it a decommitment key. Given access to the oracle alone, it is hard for the receiver to learn anything about the secret. However, it is easy for the receiver to become convinced that at most one secret can be successfully decommitted even when the oracle is badly formed.

The locking scheme from [23] requires the oracle to have bigger entropy than the number of queries against which the hiding property should hold. We prove the intuitive fact that such a limitation is inherent, and therefore there is no efficient-oracle non-interactive locking scheme which resists an arbitrary polynomial number of queries. This is because intuitively if the entropy of the oracle is bounded, then either: (1) the receiver is able to learn all the entropy by making a polynomial number of queries, and therefore break the hiding property; or (2) if some entropy is hidden no matter what queries the receiver makes, then a cheating sender is able to create a "fake" oracle that can cheat on this entropy and therefore be opened to any value, breaking the binding property.

This motivates our notion of an *interactive* locking scheme. An ILS is a locking scheme in the IPCP model: the commitment phase can involve, in addition to oracle queries by the receiver, interaction with the sender from whom the secret originated. Here the sender and the oracle play the roles of the prover and PCP oracle in the IPCP model, respectively. Decommitment still involves a single message from the sender to the receiver. Somewhat surprisingly (and counter to our own initial intuition), we show that interaction can be used to disrupt the intuitive argument above.

We present several constructions of efficient interactive locking schemes. We show how to obtain such schemes from *interactive hashing* — a primitive which was introduced by Naor, Ostrovsky, Venkatesan, and Yung [26] for the purpose of constructing statistically hiding and *computationally* binding commitment schemes from any one-way permutation (see also [27,28,29]). The high level idea of the transformation from interactive hashing to ILS is to "implement" a one-way permutation by an oracle which contains a random *point function* (i.e., a function that outputs 0 on all but one random point). To ensure the binding property even when the oracle is badly formed, the receiver should query the oracle on a small number random points to verify that it is not "too far" from a point function. The (black-box) proof of security of the interactive hashing protocol implies (unconditional) proof of security for the ILS.

The above connection allows us to use interactive hashing protocols from the literature for obtaining interactive locking schemes, but leaves open the question of minimizing the amount of interaction with the sender. We resolve this question by presenting a novel direct construction of ILS which requires only a single round of interaction with the sender.

The high level idea behind our single round ILS is as follows. The oracle $\pi$ constructed by the sender will be the zero function over $\{0,1\}^n$ except for an "interval" of size $2^{cn}$. That is, $\pi(x) = 1$ for $a \leq x \leq a + 2^{cn}$ and $\pi(x) = 0$ elsewhere. Depending on whether the sender commits to zero or one, the interval will be planted in the first or second half of the oracle $\pi$. The position $a$ of the interval will be revealed to the receiver in the decommitment phase. When $c < 1$, the interval size $2^{cn}$ will be small enough to prevent the receiver from finding the committed bit during the commitment phase. But now the sender is able to cheat by planting intervals in both the first and second half of $\pi$. To guarantee binding, we let the receiver ask a "challenge" question about the interval in such a way that the sender cannot find a *pair* of planted intervals in the first and second half of $\pi$ with the same challenge answer. A natural idea is to use a pairwise independent function $h\colon \{0,1\}^n \to \{0,1\}^{dn}$ and ask the sender to reveal $h(a)$. The sender is able to plant at most $2^{(1-c)n}$ *separate* intervals in each half of $\pi$. Each of the intervals in the first and second half of $\pi$ will have the same hashes with probability $2^{-dn}$. Therefore if $2(1 - c) < d$, then with high probability over the choice of $h$ the sender is *not* able to find two intervals with the same hash value $h(a)$ and thus gets committed to a fixed bit. But now the information revealed by $h(a)$ might help the receiver find a non-zero point in $\pi$ and break the hiding property. We show how to modify the a known construction of pairwise independent hash functions to get another function which is still almost pairwise independent but has the additional property that the preimages of any hash value are "scattered" in the domain of the hash function. The latter property prevents the receiver from taking advantage of the knowledge of $h(a)$ to find where the interval is planted. Using this approach we simultaneously guarantee binding and hiding.

*Cryptography using hardware tokens.* The above study of zero-knowledge interactive PCPs and interactive locking schemes is motivated by a recent line of research on the capabilities of cryptographic protocols in which parties can generate tamper-proof hardware tokens and send them to each other. Katz [30] shows that, under computational assumptions, general *universally composable* (UC) secure two-party computation [31] is possible in this model if the tokens are allowed to be *stateful*, and in particular can erase their secrets after being invoked. It was subsequently shown that even *unconditional* security is possible in this model, first for the case of commitment [9] and then for general tasks [10]. See [32,33,34] and references therein for other applications of stateful tokens in cryptography.

Obtaining similar results using *stateless* tokens turns out to be more challenging. Part of the difficulty stems from the fact that there is no guarantee on the functionality of tokens generated by malicious parties — they may compute arbitrary functions of their inputs and may even carry state information from one invocation to another. It was recently shown in [10], improving on [35], that any *one-way function* can be used for basing (computationally) UC-secure two-party computation on stateless tokens. More practical protocols which satisfy weaker notions of security were given in [36]. These works leave open the question of obtaining a similar result *unconditionally*, and with *statistical* security. (To get around impossibility results in the plain model, the number of queries to a token should be polynomially bounded, but otherwise malicious parties may be computationally unbounded.) In fact, the constructions from [35,10,36] may lead to a natural conjecture that achieving statistical security in this setting is impossible, since in these constructions all the "useful information" contained in tokens can be learned by a computationally unbounded adversary using a polynomial number of queries.

However, similar to the case of ILS discussed above, the combination of stateless tokens and interaction turns out to be surprisingly powerful. As already alluded to in [8], MIP protocols can naturally give rise to protocols in the hardware token model. In our case, we implement the ILS (or IPCP) by having a single sender (prover) create a stateless tamper-proof hardware token which implements the PCP oracle and send it to the receiver (verifier). Applying this to our results, this directly gives rise to the first unconditionally secure commitment protocols and SZK proofs for **NP** using stateless tokens.

We show how this can be extended to general unconditionally secure (in fact, UC-secure) two-party computation if parties are allowed to build tokens which encapsulate other tokens: namely, the receiver of a token $A$ is allowed to build another token $B$ which internally invokes $A$. The high level idea is the following. By the completeness of oblivious transfer (OT) [2,37], it suffices to realize OT using stateless tokens. This is done as follows. The OT sender's input is a pair of strings $(s_0, s_1)$ and the OT receiver's input is a selection bit $b$. The OT receiver commits $b$ using an ILS. Applying our best construction, this involves sending a token $A$ to the OT sender and responding to a random challenge message received from the OT sender. The OT sender now prepares and sends to the receiver a token $B$ with the following functionality. Token $B$ accepts a selection bit $b$ along with a corresponding decommitment message. It checks that the decommitment is valid (this involves invocations of the token $A$, which token $B$ encapsulates) and then returns the string $s_b$ if decommitment was successful. The binding property of the ILS guarantees that the OT receiver can learn at most one string $s_b$. The hiding property of the ILS guarantees that the sender cannot learn $b$.

Interestingly, we also show a matching negative result: if token encapsulation is not allowed, then statistically secure OT is impossible. This holds even if both parties are guaranteed to follow the protocol except for making additional queries to tokens in order to learn information about the other party's input. The proof of this negative result employs a variant of the recent notion of accessible entropy from [38] and has the following high level intuition: In the standard model *without* tokens, one way to explain why statistical OT is not possible is to consider the randomness $r_R$ of the receiver conditioned on the transcript $\tau$ of the protocol. If this conditional distribution reveals information about the receiver's choice $b$, then an unbounded sender can cheat by sampling from this distribution. But if not, then an unbounded receiver can cheat by sampling from this distribution for both values of $b$, and using the result to obtain both strings $s_0$ and $s_1$ of the sender.

In the token model, however, this situation is not symmetric, since the sender might not know what queries the receiver has asked from the tokens it holds (or vice versa). Informally, we define a protocol $(A, B)$ to have *accessible entropy* if the parties can nevertheless (information theoretically) sample their randomness conditioned on the *other* party's view. If an OT protocol did have accessible entropy, then essentially the above impossibility argument would apply. (In contrast, the original definition of accessible entropy of [38] required that the parties could *efficiently* sample, since the focus in that work was on analyzing protocols secure against computationally bounded parties.)

The technical core of our impossibility result is the following technical lemma: For any protocol $(A, B)$ in the stateless token model, there is another protocol $(A', B')$ that differs from $(A, B)$ only in that the parties ask (a polynomial number) more queries to the tokens that they hold. Furthermore, almost all the entropy of the new protocol $(A', B')$ is accessible. This lemma allows us to carry out the intuition above and rule out statistically secure OT in the stateless token model.

*Organization.* In Section 2, we define the notions of zero-knowledge IPCPs and ILS, and show how to use ILS to build unconditional zero-knowledge IPCPs for **NP**. We also show that interaction is required for efficient ILS. In Section 3, we show how to construct ILS. In Section 4, we show the implications of our work on (unconditionally secure) cryptography with tamper-proof hardware tokens.

## 2   Statistically Zero-Knowledge IPCP for NP

Interactive PCPs (Definition 1 below) were first introduced in [24] and combine the notion of oracle algorithms with interactive algorithms. Here we define IPCPs in a general way, not only for the purpose of a proof system, but rather as a model of interaction consisting of two interactive algorithms and a prover. (This way we can define our notion of interactive locking schemes as a protocol in the IPCP model implementing the commitment functionality.)

**Definition 1.** *(Adapted from [24]) An* interactive probabilistically checkable proof *(IPCP)* $\Gamma = (P, \pi, V)$ *consists of an interactive algorithm $P$ (the* prover*), an oracle $\pi$ (the* PCP oracle*), and an interactive algorithm $V$ (the* verifier*) such that:*

- *$P$ and $\pi$ share common randomness $r_P$, and $V$ is given the randomness $r_V$.*
- *$P$, $\pi$, and $V$ will be given an input $x$ of length $|x| = n$. $P$ and $\pi$ may also receive a common private input $w$.[1]*
- *The PCP oracle $\pi$ is a function of $(r_P, x, w, q)$ where $q$ is a query of the verifier $V$. Since $(r_P, x, w)$ is fixed at the beginning of the protocol, we might simply use $\pi(q)$ to denote the answer to the query $q$.*

---

[1] For example when $(P, \pi)$ are efficient and $L \in$ **NP**, $w$ could be a witness for $x \in L$.

– $P$ and $V^\pi$ engage in an interactive protocol during which $V$ can query the PCP oracle $\pi$ and at the end $V$ accepts or rejects.

By an efficient IPCP we mean one in which the prover $P$, the PCP oracle $\pi$, and the verifier $V$ run in polynomial time over the input length $|x| = n$.

By the *round complexity* of an IPCP we mean the number of rounds of interaction between the verifier and the *prover* (and not the PCP oracle) where each round consists of a message from the verifier followed by a message from the prover. (See the full version of the paper for more discussion on this definition and a comprehensive elaboration on the IPCP model.)

Now we define the notion of a proof system in the IPCP model which directly incorporates the statistical zero-knowledge feature. We use a quantitative definition allowing us to speak about exponential zero-knowledge (rather than just super-polynomial security).

**Definition 2 (SZK-IPCP for languages).** *We say that $\Gamma = (P, \pi, V)$ is an SZK-IPCP for the language $L$ with SZK $(u(n), \epsilon(n))$ and soundness $1 - \delta(n)$ if the following holds:*

- **Completeness:** *If $x \in L$, then $\Pr[\langle P, V^\pi \rangle(x) = 1] = 1$.*
- **Soundness:** *$\Gamma$ has soundness $1 - \delta$ if for all $x \notin L$ and for any arbitrary prover $\widehat{P}$ and oracles $\widehat{\pi}$ it holds that $\Pr[\langle \widehat{P}, V^{\widehat{\pi}} \rangle(x) = 1] \leq \delta(n)$.*
- **Statistical zero-knowledge (SZK):** *We say that the IPCP $\Gamma$ is $(u, \epsilon)$-SZK for $L$ with a straight-line[2] simulator if there is a simulator Sim as follows. The (straight-line) simulator Sim interacts with a (potentially malicious) verifier $\widehat{V}$, while the simulator Sim receives all the queries of the the verifier (including both the queries asked from the prover and from the oracle) and responds to them. Since an unbounded verifier can ask arbitrary number of queries from its oracle, here we put a bound $u$ on the number of oracle queries asked by $\widehat{V}$ and demand the following to hold: For any $v \leq u$, if $\widehat{V}$ asks at most $v$ oracle queries, then Sim runs in time $\mathrm{poly}(n, v)$ and produces a view for $\widehat{V}$ which is $\epsilon$-close to the view of $\widehat{V}$ when interacting with $(P, \pi)$.*

*We simply call $\Gamma$ an SZK-IPCP for $L$ with security $u$, if $\Gamma$ is $(1 - 1/u)$-(adaptively)-sound and $(u, 1/u)$-SZK.*

Note that when $u(n)$ is super-polynomial, Definition 2 implies zero-knowledge against polynomial-time verifiers.

We prove that $2^{\Omega(n)}$-secure constant-round SZK-IPCPs exist for any language $L \in \mathbf{NP}$ where both the prover and the PCP oracle in our construction can be implemented efficiently given a witness $w$ for $x \in L$.

**Theorem 3 (Constant-round SZK-IPCP for NP).** *For any language $L \in \mathbf{NP}$ there exists a 2-round efficient SZK-IPCP $\Gamma_{2R}$ for $L$ with security $2^{\Omega(n)}$.*

---

[2] Since all of our simulators in this paper are straight-line, for sake of simplicity here we only describe how to define SZK for IPCPs with straight-line simulators.

The simulator of $\Gamma_{2R}$ in Theorem 3 is straight-line and therefore by a result of [39], for a small enough constant $c$, a $2^{cn}$-fold concurrent composition of $\Gamma_{2R}$ remains $(2^{\Omega(n)}, 2^{-\Omega(n)})$-SZK if the inputs to the instances of $\Gamma_{2R}$ are fixed in the beginning.

*Ideas of the proof of Theorem 3.* Our main step to prove Theorems 3 is to construct an "interactive locking scheme" (ILS) (Definition 5), a primitive corresponding to commitment schemes in the IPCP model. In Theorem 6 we present an ILS with optimal round complexity (i.e. one round). Then we feed our ILS (as a commitment scheme) into the well-known construction of [12] to achieve zero-knowledge for **NP** with non-negligible soundness. A classical way to amplify the soundness of proof systems (while keeping the round-complexity) in the standard model of interaction is to use parallel composition. Firstly we define parallel composition of IPCPs (see the full version) in a careful way and prove an optimal bound on how the soundness amplifies in such a parallel composition. The latter result is interesting on its own since the IPCP model lies in between the single-prover and the multi-prover models and it is known [21] that the parallel repetition does *not* amplify the soundness in a simple exponential form (as one would wish). Secondly, we show that although the parallel composition might hurt the zero-knowledge in general, by crucially using a special feature of our ILS called "equivocability" (see Definition 5) one can prove that SZK is preserved under parallel composition. Roughly speaking, an ILS is equivocable, if a malicious sender can efficiently decommit to any desired value by changing the content of the oracle *after* the commitment phase. See the full version for the full proof of Theorem.

We also show how to achieve a $2^{\Omega(n)}$-secure SZK-IPCP for any $L \in$ **NP** where the security holds even against *stateful* oracles. A stateful oracle can save a state and behave as maliciously as an interactive algorithm. Namely, the answers returned by a (malicious) stateful oracle can depend on the previous queries asked to the oracle as well as the other queries asked in the same "round" of queries. We call such IPCPs (secure against stateful oracles) *adaptively*-sound.

**Theorem 4 (Adaptively-secure SZK-IPCP for NP).** *There exists a* (poly($n$)-*round) efficient SZK-IPCP $\Gamma_{\mathsf{adap}}$ for any $L \in$ **NP** *with adaptive-security* $2^{\Omega(n)}$.

*Ideas of the proof of Theorem 4.* To prove Theorem 4, we use ideas from [40] about converting multi-prover proof systems into an equivalent two-prover one (with non-negligible soundness) where the second prover is asked only one query. When a prover is asked only one query, it can be considered as an oracle. In our transformation to achieve adaptive security in the IPCP model, we use a similar compiler to that of [40] over the IPCP $\Gamma_{2R}$ of Theorem 3 and crucially use the fact that $\Gamma_{2R}$ is "public-coin" (i.e. the soundness holds even if the prover gets to see which oracle queries are asked). A public-coin IPCP is one which is sound even if the prover gets to see the oracle queries asked by the verifier. Finally we use sequential composition to amplify the soundness. See the full version of the paper for the full proof of Theorem.

# 3   Interactive Locking Schemes

An *Interactive locking scheme* is a commitment scheme implemented in the IPCP model. A similar definition appeared in [23] without the interaction (i.e. only with an oracle), but as we will see in Theorem 6 *non-interactive* locking schemes are inherently inefficient and therefore not as applicable in cryptographic settings.

**Definition 5 (Interactive locking scheme).** *Let $\Lambda = (S, \sigma, R)$ be an efficient IPCP (where we call $S$ the sender, $\sigma$ the locking oracle and $R$ the receiver). $\Lambda$ is called an* interactive locking scheme *(ILS) for the message space $W_n$ if it of the following form:*

*The common input is $1^n$ where $n$ is the security parameter. $(S, \sigma)$ receive a private input $w \in W_n$ which is called the committed message as well as the private randomness $r_S$. The receiver $R$ gets the randomness $r_R$. The receiver $R$ gets oracle access to the locking oracle $\sigma$ and $R^\sigma$ interacts with $S$ in two phases: (1) commitment phase and (2) decommitment phase. The decommitment phase consists of only one message from the sender $S$ to the receiver $R$ which includes the committed message $w$ and the private randomness $r_S$ used by $S$. Following this message the receiver $R$ (perhaps after asking more queries from the oracle $\sigma$) accepts or rejects. We demand the following properties to hold:*

- **Completeness:** *For any $w \in W_n$ if all parties are honest the receiver always accepts.*
- **Binding:** *We define $\Lambda$ to be $(1-\delta)$-binding if for any sender $\widehat{S}$ and any oracle $\widehat{\sigma}$, with probability at least $1 - \delta$ over the interaction of the commitment phase there is at most one possible $w$ such that $\widehat{S}$ can decommit to successfully.*
- **Hiding:** *Let $\widehat{R}$ be any malicious receiver who asks at most $u$ oracle queries from $\sigma$, and let $\tau_w$ be the random variable which consists of the transcript of the interaction of $R$ with $(S, \sigma)$ till the end of the commitment phase when the committed message is $w \in W$. $\Lambda$ is $(u, \epsilon)$-hiding if for every such malicious receiver $\widehat{R}$ and every $\{w_1, w_2\} \subseteq W$ it holds that $\mathsf{SD}(\tau_{w_1}, \tau_{w_2}) \leq \epsilon$.*
- **Equivocability:** *$\Lambda$ is equivocable if there is an efficient sampling algorithm $\mathsf{Sam}$ that given $(\tau, w)$ where $\tau$ is the transcript (including the oracle queries) of the commitment phase of $\langle S, \widehat{R}^\sigma \rangle$ (for an arbitrary receiver $\widehat{R}$) and any $w \in W$, $\mathsf{Sam}(\tau, w)$ outputs $r$ according to the distribution $(r_S \mid \tau, w)$. Namely $r$ is sampled according to the distribution of the private randomness $r_S$ of $(S, \sigma)$ conditioned on $w$ being the committed message and $\tau$ being the transcript of the commitment phase.*

*We simply call the ILS $\Lambda$ $u$-secure if it is $(1 - 1/u)$-binding and $(u, 1/u)$-hiding. If $W = \{0, 1\}$, we call $\Lambda$ a bit-ILS.*

The following theorem presents an ILS with optimal round complexity.

**Theorem 6.** *(A round-optimal ILS) Let $\ell(n) = \mathrm{poly}(n)$, then*

1. *There exist an efficient ILS $\Lambda_{1R} = (S, \sigma, R)$ for the message space $\{0, 1\}^\ell$ with security $2^{\Omega(n)}$ which has a commitment phase of only one round.*

2. *Any ILS with a noninteractive commitment phase needs an inefficient oracle*
   *σ and thus Λ has optimal round-complexity (as an efficient ILS).*

In the full version of the paper we give a general construction of ILS from any
interactive hashing scheme with some minimal properties. Unfortunately non-
trivial interactive hashing needs at least two rounds of interaction and thus this
approach is incapable of giving us a round-optimal ILS. Due to space limit we
refer the reader for this connection to the full version and here will only present
the optimal construction.

Before proving Theorem 6 we need the following lemma whose proof is immediate.

**Lemma 7.** *For $n > m$ let $\mathcal{A}$ be the family of $n \times m$ Boolean matrices as follows.
To get a uniform member of $A$, choose the first $n-m$ rows all at random, and take
the last $m$ rows to be an independently chosen at random conditioned on having
full rank $m$. Then for any $0 \neq x \in \{0,1\}^n$, it holds that $\Pr_{A \leftarrow \mathcal{A}}[xA = 0] \leq 2^{-m}$
(and equivalently for any $x_1 \neq x_2 \in \{0,1\}^n$ and $y \in \{0,1\}^m$, it holds that
$\Pr_{A \leftarrow \mathcal{A}}[x_1 A = x_2 A] \leq 2^{-m}$).*

**Construction 8 (A 1-round ILS)** *Suppose $b \in \{0,1\}$ is the private message
given to sender and the oracle $(S, \sigma)$, and suppose $R$ is the receiver. Let $m =
3n/4$. Below we associate $\{0,1\}^n$ with the integers $[0, 2^n)$ and all additions and
subtractions below are modulo $2^n$.*

*The commitment phase of $\Lambda_{1R}$:*

1. *Sender $S$ chooses $a \leftarrow \{0,1\}^n$ at random. Let $f_b$ be the function: $f_b(x) = 1$
   iff $a \leq x < a + 2^m$, and let $f_{1-b}$ be the zero function over $\{0,1\}^n$. The locking
   oracle will be the combination of the two functions $\sigma = (f_0|f_1)$ (indexed by
   the first bit of the query to $\sigma$).*
2. *Receiver $R$ samples $A \leftarrow \mathcal{A}$ from the family of matrices of Lemma 7 conditioned on the last $m$ rows of $A$ being independent[3] and sends $A$ to $S$.*
3. *Sender $S$ checks that the last $m$ rows of $A$ are independent, and if so he
   sends $h = aA$ to the receiver $R$.*

*The decommitment phase of $\Lambda_{1R}$:*

1. *Sender $S$ sends $(b, a)$ to the receiver $R$.*
2. *Receiver $R$ does the following checks and rejects if any of them does not hold.*
   (a) *Check that $aA = h$.*
   (b) *Check that $f_{1-b}(a) = 0$, and $f_b(a) = 1$.*
   (c) *For each $i \in [0, m]$, sample $10n$ random points from $[a, a + 2^i)$ and check
        that $f_b(x) = 1$ for all of them, and also sample $10n$ random points from
        $(a - 2^i, a - 1]$ and check that $f_b(x) = 0$ for all of them*

*Proof (of Theorem 6).*
   Now we study the properties of the ILS $\Lambda_{1R}$.
   Completeness and Equivocability are immediate.

---

[3] Note that the last rows of $A$ are independent with probability $1 - 2^{-m} = 1 - 2^{-n}$.

*Binding.* As a mental experiment we pretend that the randomness used during the decommitment phase by $R$ is chosen in the decommitment phase (rather than in the beginning of the commitment phase).

For a fixed locking oracle $\sigma$, Let $X_0$ (resp. $X_1$) be the set of possible values of $a$ that sender $S$ can send to the receiver $R$ as the decommitment of $b = 0$ (resp. $b = 1$) and get accepted in the decommitment phase with probability at least $2^{-2n}$. We prove that by the end of the commitment phase, with probability at least $1 - 2^{-n/8}$, it holds that $|X_0| = 0$ or $|X_1| = 0$ which means that the sender has only one way to decommit the value $b$ and get accepted with probability more than $2^{-2n}$. But now if we choose the receiver's randomness in the commitment phase, since there are at most $2^{n+1}$ possible values for $(b, a)$, it follows by a simple average argument that with probability at least $1 - 2^{2n-n-1}$ over the commitment phase, the prover gets committed to only one possible value for $(b, a)$ which he can use to pass the decommitment phase successfully.

*Claim.* $X_0 \cap X_1 = \varnothing$.

*Proof.* If $a \in X_0 \cap X_1$. Then when $a$ is used as the decommitment of $0$, in Step 2b of the decommitment phase the receiver $R$ checks that $f_0(a) = 1, f_1(a) = 0$. On the other hand in the case of decommitting to $1$, receiver $R$ checks that $f_b(a) = 0, f_{1-b}(a) = 1$, but they can't both hold at the same time.

*Claim.* It holds that $|X_0| \le 2^{n-m}$ and $|X_1| \le 2^{n-m}$.

*Proof.* We show that if $\{a, a'\} \subset X_0$ then $|a - a'| \ge 2^m$ (and this would show that $X_0 \le 2^n/2^m$). Assume on the contrary that $a' < a$ and $a - a' < 2^m$. Let $i \in [1, m]$ be such that $2^{i-1} \le a - a' < 2^i$. Then by the pigeonhole principle ether at least half of $\sigma([a', a])$ are zero or at least half of the values $\sigma([a', a])$ are one. Without loss of generality let assume that at least half of $\sigma([a', a])$ is zero. In this case at least $1/4$ of the values $\sigma([[a', a' + 2^i)])$ are zero. But then by Step 2c of the decommitment phase $(0, a')$ will be accepted with probability at most $(3/4)^{10n} < 2^{-2n}$, and therefore $a' \notin X_0$ which is a contradiction.

*Claim.* With probability at least $1 - 2^{\Omega(n)}$ over the choice of $A$, it holds that $|X_0| = 0$ or $|X_1| = 0$.

*Proof.* Fix any pair $a_0 \in X_0$ and $a_1 \in X_1$, we know that $a_0 \ne a_1$. Therefore, $\Pr_A[a_0 A = a_1 A] = \Pr_A[(a_0 - a_1)A = 0] \le 2^{-m}$. Claim 3 yields that there are at most $2^{n-m}2^{n-m}$ such pairs, so by using a union bound, with probability at least $1 - 2^{-m}2^{2n-2m} = 1 - 2^{2n-3m}$ over the choice of $A$, it holds that $X_0 A \cap X_1 A = \varnothing$ which implies that if the sender sends any hash value $h$, the consistency check of Step 2a of the decommitment phase either makes $|X_0| = 0$ or $|X_1| = 0$.

As we said before Claim 3 implies that with probability $1 - \text{poly}(n) \cdot 2^{2n-3m} = 1 - \text{poly}(n) \cdot 2^{-n/4} \ge 1 - 2^{-n/8}$ over the interaction in the commitment phase the sender gets bound to a fixed $b \in \{0, 1\}$ to which he can decommit successfully.

*Hiding.* Suppose receiver $R$ can ask at most $u \leq 2^{n/8}$ queries from the locking oracle $\sigma$. We claim that before sending the matrix $A$, all of receiver $R$'s queries to $\sigma$ are answered zero with probability at least $1 - 2^{-n/4}$. To see why, think of $Z_{2^n}$ as being divided into $2^{n-m} = 2^{n/4}$ equal intervals such that $a$ is the beginning of one of them. Since receiver $R$ asks up to $2^{n/8}$ queries, before sending the matrix $Z$, he will ask a query from the interval beginning with $a$ with probability at most $2^{n/8}/2^{n/4} = 2^{-n/8}$. Therefore (up to $2^{-n/8}$ statistical distance in the experiment) we can assume that the matrix $A$ is chosen by receiver $R$ independently of $a$.

After receiving $h$, the information that the receiver $R$ knows about $a$ is that it satisfies the equation $aA = h$. If we choose and fix the first $n - m$ bits of (a potential) $a$, then the remaining bits are determined uniquely because the last $m$ rows of $A$ are full rank. It means that for every $y \in [0, 2^{n-m})$ there is a unique solution for $a$ in the interval $[y2^m, y2^m + 2^m)$, and they are all equally probable to be the true answer from the receiver's point of view.

Now again we claim that (although there are $2^m$ nonzero points in $f_b$) all the queries that the receiver $R$ asks from $f_b$ are answered $0$ with probability at least $1 - 2^{-n/8}$. Let $Z = \{z \mid zA = h\}$ be the set of possible values for $a$. For $z \in Z$, let $I(z) = [z, z + 2^m)$. We claim that no $x \in \{0,1\}^n$ can be in $I(z)$ for three different $z$'s from $Z$. To see why, let $z_1 < z_2 < z_3$ and that $x \in I(z_1) \cap I(z_2) \cap I(z_3)$. But now the interval $[y2^m, y2^m + 2^m)$, containing $z_2$ separates $z_1$ and $z_3$, and so $z_3 - z_1 > 2^m$. Therefore $I(z_1) \cap I(z_3) = \varnothing$ which is a contradiction. So, if the receiver $R$ asks $u$ queries from $f_b$, he can ask queries from $I(z)$'s for at most $2u$ different $z$'s (out of $2^{n-m}$ many of them). As a mental experiment assume that $a$ is chosen from $Z$ after the receiver $R$ asked his queries, it holds that $I(a)$ will be an interval that the receiver $R$ never asked any query from with probability at least $1 - u/2^{n-m} \geq 1 - \cdot 2^{-n/8}$. Therefore with probability at least $1 - 2^{-n/9}$ all of receiver $R$'s queries during the commitment phase will be answered zero. But putting the oracle queries aside, the hash value $h$ does not carry any information about the bit-message $b$ and therefore the scheme is $(1 - 2^{n/8})$-hiding.

Now we turn to proving Part 2 of Theorem 6.

By a *noninteractive locking scheme* (NLS), we mean an ILS where the commitment phase is noninteractive and sender $S$ only participates in the decommitment phase. Note that an efficient locking scheme by definition uses poly$(n)$-sized circuits to implement the locking oracle $\sigma$, and therefore $\sigma$ can have at most poly$(n)$ entropy. In this section we show that there exist no efficient NLS with super-polynomial security.

Since we are going to prove that NLS's cannot be efficient, we need to deal with unbounded senders. Thus we can no longer assume that the decommitment phase is only a message $(b, r_S)$ sent to the receiver, because the randomness $r_S$ used by the sender can be exponentially long. Therefore to prove the strongest possible *negative* result, we allow the decommitment phase of a NLS to be interactive.

The following theorem clearly implies Part 2 of Theorem 6.

**Theorem 9.** *Let $\Lambda = (S, \sigma, R)$ be any NLS for message space $\{0, 1\}$ in which the function $\sigma$ of the locking oracle has Shannon entropy at most $H(\sigma) \leq \frac{uq}{1000}$ when the committed bit $b$ is chosen at random $b \leftarrow \{0, 1\}$. Let $u$ be an upper bound on* *the number of oracle queries to $\sigma$ asked by the receiver $R$ in the decommitment phase. Then either of the following holds:*

- **Violation of binding:** *There is a fixed locking oracle $\widehat{\sigma}$, and a sender strategy $\widehat{S}$ such that when $\widehat{\sigma}$ is used as the locking orale, for both $b = 0$ and $b = 1$, $\widehat{S}$ can decommit successfully with probability at least $4/5$.*
- **Violation of hiding:** *There exists an unbounded receiver $\widehat{R}$ who can guess the random bit $b \leftarrow \{0,1\}$ used by $(S, \sigma)$ with probability at least $4/5$ by asking at most $u$ queries to the locking oracle $\sigma$.*

*Ideas of the proof of Theorem 9.* Our main tool in proving Theorem 9 is the notion of "canonical entropy learner" (EL). Roughly speaking, EL is an efficient-query (computationally unbounded) algorithm which learns a randomized function $f$ (with an oracle access to $f$) under the uniform distribution assuming that $f$ has a bounded amount of entropy. EL proceeds by choosing to ask one of the "unbiased" queries of $f$ at any step and stop if such queries do not exist. An unbiased query $x$ is one whose answer $f(x)$ is not highly predictable with the current knowledge gathered about $f$ by EL. Whenever EL chooses to ask a query it learns non-negligible entropy of $f$, and thus the process will stop after $\text{poly}(n)$ steps. On the other hand, when EL stops, all the remaining queries are biased and thus will have a predictable answer *over the randomness of $f$*. We prove that either the receiver is able to find out the secret message of the sender (in an NLS) by running the EL algorithm, or otherwise if by the end of the learning phase still part of the entropy left in the locking oracle is hiding the secret message, then a malicious prover can plant at least two different messages in the locking oracle in such a way that it can decommit to successfully.

# 4    On Oblivious Transfer from Stateless Hardware Tokens

In this section we prove that in the stateless hardware token model, there is no statistically secure protocol for oblivious transfer (OT), when the only limitation on malicious parties is being bounded to make polynomially many queries to the tokens.

*The stateless token model.* In the stateless (tamper-proof hardware) token model, two (computationally unbounded) interactive algorithms $A$ and $B$ will interact with the following extra feature to the standard model. Each party at any time during the protocol can construct a circuit $T$ and put it inside a "token" and send the token $T$ to the other party. The party receiving the token $T$ will have *oracle access* to $T$ and is limited to ask $\text{poly}(n)$ number of queries to the token. The parties can exchange $\text{poly}(n)$ number of tokens during the interaction. The stateless token model clearly extends the IPCP model in which there is only one token sent from the prover to the verifier in the beginning of the game. Therefore proving any *impossibility* result in the stateless token model clearly implies the same result for the the IPCP model. It is easy to see that without loss of generality the parties can avoid sending "explicit messages" to each other and can only use tokens (with messages planted inside the tokens) to simulate all the classical communication with the tokens.

*Oblivious transfer by semi-honest parties.* If one of the parties is semi-honest (i.e. runs the protocol honestly, and only remember's its view for further off-line investigation), then in fact unconditionally secure OT *is* possible in the stateless token model. If the receiver is honest, then the protocol is simply a token $T$ sent from the sender which encodes $T(0) = x_0, T(1) = x_1$. The receiver will read $T(i)$ to learn $x_i$. Moreover it is well known that secure OT in one direction implies the existence of secure OT in the other direction, so if the sender is semi-honest unconditionally secure OT is possible in the stateless token model.

We prove that unconditionally secure OT is impossible in the stateless token model, if both parties are *slightly* more malicious than just being semi-honest. Roughly speaking, we define the notion of "curious" parties who run the original protocol (honestly), but will ask more queries from the tokens along the way.[4] We will prove that for any protocol $(A, B)$ aiming to implement OT, there are curious extensions of the original parties $(A_{cur}, B_{cur})$ who break the security of the protocol. We prove the following theorem.

**Theorem 10 (No unconditional OT from stateless tokens).** *Let $(S, R)$ be any protocol for the oblivious transfer in the stateless token model. Then there are curious extensions $(S_{cur}, R_{cur})$ to the original algorithms where $(S_{cur}, R_{cur})$ (and thus $(S, R)$) is not a secure protocol for oblivious transfer even when the inputs are random. More formally either of the following holds:*

- **Violation of sender's security:** *When the sender $S$ chooses $x_0$ and $x_1$ at random from $\{0, 1\}$ and interacts with $R_{cur}$, then $R_{cur}$ can find out both of $x_0$ and $x_1$ with probability at least $51/100$.*
- **Violation of receiver's security:** *When the receiver $R$ chooses $i \leftarrow \{0, 1\}$ at random and interacts with $S_{cur}$, then $S_{cur}$ can guess $i$ correctly with probability at least $51/100$.*

For a high level description of the ideas behind Theorem 10 we refer the reader to the discussion in the Introduction.

Perhaps surprisingly we show that if the parties are allowed to build tokens *around* the tokens received from the other party, then unconditional (UC) secure computation is possible by using *stateless* tokens.

*UC secure OT by encapsulation.* For a discussion on ideas behind our UC secure OT by token encapsulation we refer the reader to the Introduction and for more details to the full version of the paper.

---

[4] The term "honest but curious" is sometimes used equivalent to "semi-honest". Our notion is different from both of them because a curious party deviates from the protocol slightly by learning more but emulates the original protocol honestly.

# References

1. Rabin, M.O.: How to exchange secrets by oblivious transfer. TR-81, Harvard (1981)
2. Kilian, J.: Founding cryptography on oblivious transfer. In: Proceedings of the 20th Annual ACM Symposium on Theory of Computing, STOC (1988)
3. Crépeau, C., Kilian, J.: Achieving oblivious transfer using weakened security assumptions (extended abstract). In: FOCS, pp. 42–52 (1988)
4. Maurer, U.M.: Conditionally-perfect secrecy and a provably-secure randomized cipher. J. Cryptology 5(1), 53–66 (1992)
5. Ben-Or, M., Goldwasser, S., Wigderson, A.: Completeness theorems for non-cryptographic fault-tolerant distributed computation (extended abstract). In: STOC, pp. 1–10 (1988)
6. Chaum, D., Crépeau, C., Damgård, I.: Multiparty unconditionally secure protocols (extended abstract). In: STOC, pp. 11–19 (1988)
7. Rabin, T., Ben-Or, M.: Verifiable secret sharing and multiparty protocols with honest majority (extended abstract). In: STOC, pp. 73–85 (1989)
8. Ben-Or, M., Goldwasser, S., Kilian, J., Wigderson, A.: Multi-prover interactive proofs: How to remove intractability assumptions. In: STOC, pp. 113–131 (1988)
9. Moran, T., Segev, G.: David and Goliath commitments: UC computation for asymmetric parties using tamper-proof hardware. In: Smart, N.P. (ed.) EUROCRYPT 2008. LNCS, vol. 4965, pp. 527–544. Springer, Heidelberg (2008)
10. Goyal, V., Ishai, Y., Sahai, A., Venkatesan, R., Wadia, A.: Founding cryptography on tamper-proof hardware tokens. In: Micciancio, D. (ed.) TCC 2010. LNCS, vol. 5978, pp. 308–326. Springer, Heidelberg (2010)
11. Goldwasser, S., Micali, S., Rackoff, C.: The knowledge complexity of interactive proof systems. SIAM Journal on Computing 18(1), 186–208 (1989); Preliminary version in STOC 1985 (1985)
12. Goldreich, O., Micali, S., Wigderson, A.: Proofs that yield nothing but their validity or all languages in NP have zero-knowledge proof systems. Journal of the ACM 38(1), 691–729 (1991); Preliminary version in FOCS 1986 (1986)
13. Fortnow, L.: The complexity of perfect zero-knowledge. Advances in Computing Research: Randomness and Computation 5, 327–343 (1989)
14. Aiello, W., Håstad, J.: Statistical zero-knowledge languages can be recognized in two rounds. J. Comput. Syst. Sci. 42(3), 327–345 (1991)
15. Ostrovsky, R., Wigderson, A.: One-way fuctions are essential for non-trivial zero-knowledge. In: ISTCS, pp. 3–17 (1993)
16. Lapidot, D., Shamir, A.: A one-round, two-prover, zero-knowledge protocol for np. Combinatorica 15(2), 204–214 (1995)
17. Babai, L., Fortnow, L., Lund, C.: Non-deterministic exponential time has two-prover interactive protocols. In: FOCS, pp. 16–25 (1990)
18. Dwork, C., Feige, U., Kilian, J., Naor, M., Safra, S.: Low communication 2-prover zero-knowledge proofs for np. In: Brickell, E.F. (ed.) CRYPTO 1992. LNCS, vol. 740, pp. 215–227. Springer, Heidelberg (1993)
19. Arora, S., Safra, S.: Probabilistic checking of proofs: A new characterization of np. J. ACM 45(1), 70–122 (1998)
20. Arora, S., Lund, C., Motwani, R., Sudan, M., Szegedy, M.: Proof verification and the hardness of approximation problems. J. ACM 45(3), 501–555 (1998)
21. Fortnow, L., Rompel, J., Sipser, M.: On the power of multi-prover interactive protocols. In: Theoretical Computer Science, pp. 156–161 (1988)

22. Canetti, R., Goldreich, O., Goldwasser, S., Micali, S.: Resettable zero-knowledge (extended abstract). In: STOC, pp. 235–244 (2000)
23. Kilian, J., Petrank, E., Tardos, G.: Probabilistically checkable proofs with zero knowledge. In: STOC: ACM Symposium on Theory of Computing, STOC (1997)
24. Kalai, Y.T., Raz, R.: Interactive PCP. In: Aceto, L., Damgård, I., Goldberg, L.A., Halldórsson, M.M., Ingólfsdóttir, A., Walukiewicz, I. (eds.) ICALP 2008, Part II. LNCS, vol. 5126, pp. 536–547. Springer, Heidelberg (2008)
25. Goldwasser, S., Kalai, Y.T., Rothblum, G.N.: Delegating computation: interactive proofs for muggles. In: STOC, pp. 113–122 (2008)
26. Naor, M., Ostrovsky, R., Venkatesan, R., Yung, M.: Perfect zero-knowledge arguments for NP using any one-way permutation. Journal of Cryptology 11(2), 87–108 (1998); Preliminary version in CRYPTO 1992 (1992)
27. Ostrovsky, R., Venkatesan, R., Yung, M.: Fair games against an all-powerful adversary. In: AMS DIMACS Series in Discrete Mathematics and Theoretical Computer Science, pp. 155–169 (1993); Preliminary version in SEQUENCES 1991 (1991)
28. Ding, Y.Z., Harnik, D., Rosen, A., Shaltiel, R.: Constant-round oblivious transfer in the bounded storage model. In: Naor, M. (ed.) TCC 2004. LNCS, vol. 2951, pp. 446–472. Springer, Heidelberg (2004)
29. Haitner, I., Reingold, O.: A new interactive hashing theorem. In: IEEE Conference on Computational Complexity, pp. 319–332 (2007); See also preliminary draft of full version at the first author's home page
30. Katz, J.: Universally composable multi-party computation using tamper-proof hardware. In: Naor, M. (ed.) EUROCRYPT 2007. LNCS, vol. 4515, pp. 115–128. Springer, Heidelberg (2007)
31. Canetti, R.: Universally composable security: A new paradigm for cryptographic protocols. In: FOCS, pp. 136–145 (2001)
32. Goldreich, O., Ostrovsky, R.: Software protection and simulation on oblivious rams. J. ACM 43(3), 431–473 (1996)
33. Goldwasser, S., Kalai, Y.T., Rothblum, G.: One-time programs. In: Wagner, D. (ed.) CRYPTO 2008. LNCS, vol. 5157, pp. 39–56. Springer, Heidelberg (2008)
34. Hazay, C., Lindell, Y.: Constructions of truly practical secure protocols using standardsmartcards. In: ACM Conference on Computer and Communications Security, pp. 491–500 (2008)
35. Chandran, N., Goyal, V., Sahai, A.: New constructions for UC secure computation using tamper-proof hardware. In: Smart, N.P. (ed.) EUROCRYPT 2008. LNCS, vol. 4965, pp. 545–562. Springer, Heidelberg (2008)
36. Kolesnikov, V.: Truly efficient string oblivious transfer using resettable tamper-proof tokens. In: Micciancio, D. (ed.) TCC 2010. LNCS, vol. 5978, pp. 327–342. Springer, Heidelberg (2010)
37. Ishai, Y., Prabhakaran, M., Sahai, A.: Founding cryptography on oblivious transfer - efficiently. In: Wagner, D. (ed.) CRYPTO 2008. LNCS, vol. 5157, pp. 572–591. Springer, Heidelberg (2008)
38. Haitner, I., Reingold, O., Vadhan, S.P., Wee, H.: Inaccessible entropy. In: STOC, pp. 611–620 (2009)
39. Kushilevitz, E., Lindell, Y., Rabin, T.: Information-theoretically secure protocols and security under composition. In: STOC: ACM Symposium on Theory of Computing, STOC (2006)
40. Ben-Or, M., Goldreich, O., Goldwasser, S., Håstad, J., Kilian, J., Micali, S., Rogaway, P.: Everything provable is provable in zero-knowledge. In: Goldwasser, S. (ed.) CRYPTO 1988. LNCS, vol. 403, pp. 37–56. Springer, Heidelberg (1990)

# Fully Secure Functional Encryption with General Relations from the Decisional Linear Assumption

Tatsuaki Okamoto[1] and Katsuyuki Takashima[2]

[1] NTT, 3-9-11 Midori-cho, Musashino-shi, Tokyo 180-8585, Japan
okamoto.tatsuaki@lab.ntt.co.jp
[2] Mitsubishi Electric, 5-1-1 Ofuna, Kamakura, Kanagawa 247-8501, Japan
Takashima.Katsuyuki@aj.MitsubishiElectric.co.jp

**Abstract.** This paper presents a fully secure functional encryption scheme for a wide class of relations, that are specified by non-monotone access structures combined with inner-product relations. The security is proven under a well-established assumption, the decisional linear (DLIN) assumption, in the standard model. The proposed functional encryption scheme covers, as special cases, (1) key-policy and ciphertext-policy attribute-based encryption with non-monotone access structures, and (2) (hierarchical) predicate encryption with inner-product relations and functional encryption with non-zero inner-product relations.

## 1 Introduction

### 1.1 Background

Although numerous encryption systems have been developed over several thousand years, any traditional encryption system before the 1970's had a great restriction on the relation between a ciphertext encrypted by an encryption-key (ek) and the decryption-key (dk) such that ek and dk should be equivalent. The innovative notion of public-key cryptosystems in the 1970's relaxed this restriction, where ek and dk differ and ek can be published.

Recently, a new innovative class of encryption systems, *functional encryption* (FE), has been extensively studied. FE provides more sophisticated and flexible relations between the ek and dk where the ek and dk are parameterized by $x$ and $v$, respectively, and $dk_v$ can decrypt a ciphertext encrypted with $ek_x :=$ $(ek, x)$ iff $R(x, v)$ holds for some relation $R$. FE has various applications in the areas of access control for databases, mail services, and contents distribution [2,7,9,16,17,22,23,24,25,27].

When $R$ is the simplest relation or equality relation, i.e., $R(x, v)$ holds iff $x = v$, it is *identity-based encryption* (IBE) [3,4,5,6,10,12,13,15].

As a more general class of FE, *attribute-based encryption* (ABE) schemes have been proposed [2,7,9,16,17,22,23,24,25,27], where either one of the parameters for ek and dk is a tuple of attributes and the other is a access structure or (monotone) span program $\hat{M}$ along with a tuple of attributes, e.g.,

T. Rabin (Ed.): CRYPTO 2010, LNCS 6223, pp. 191–208, 2010.
© International Association for Cryptologic Research 2010

$x := (x_1, \ldots, x_d)$ for ek and $v := (\hat{M}, (v_1, \ldots, v_d))$ for dk, or $v := (v_1, \ldots, v_d)$ for dk and $x := (\hat{M}, (x_1, \ldots, x_d))$ for ek. Here, some elements of the tuple may be empty. The component-wise equality relations for (non-empty) attribute components, e.g., $\{x_t = v_t\}_{t \in \{1, \ldots, d\}}$, are input to (monotone) span program $\hat{M}$, and $R(x, v)$ holds iff the truth-value vector of $(\mathsf{T}(x_1 = v_1), \ldots, \mathsf{T}(x_d = v_d))$ is accepted by $\hat{M}$, where $\mathsf{T}(\psi) := 1$ if $\psi$ is true, and $\mathsf{T}(\psi) := 0$ if $\psi$ is false (For example, $\mathsf{T}(x = v) := 1$ if $x = v$, and $\mathsf{T}(x = v) := 0$ if $x \neq v$). If $\hat{M}$ is embedded into decryption-key $\mathsf{dk}_v$ (e.g., $v := (\hat{M}, (v_1, \ldots, v_d))$ for dk and $x := (x_1, \ldots, x_d)$ for ek), it is called key-policy ABE (KP-ABE). If $\hat{M}$ is embedded into a ciphertext (e.g., $x := (\hat{M}, (x_1, \ldots, x_d))$ for ek and $v := (v_1, \ldots, v_d)$ for dk), it is ciphertext-policy ABE (CP-ABE).

*Inner-product encryption* (IPE) [17] is also a class of FE, where each parameter for ek and dk is a vector over a field or ring (e.g., $\overrightarrow{x} := (x_1, \ldots, x_n) \in \mathbb{F}_q^n$ and $\overrightarrow{v} := (v_1, \ldots, v_n) \in \mathbb{F}_q^n$ for ek and dk, respectively), and $R(\overrightarrow{x}, \overrightarrow{v})$ holds iff $\overrightarrow{x} \cdot \overrightarrow{v} = 0$, where $\overrightarrow{x} \cdot \overrightarrow{v}$ is the inner-product of $\overrightarrow{x}$ and $\overrightarrow{v}$. The inner-product relation represents a wide class of relations including equality, conjunction and disjunction (more generally, CNF and DNF) of equality relations and polynomial relations.

There are two types of secrecy in FE, *attribute-hiding* and *payload-hiding* [17]. Roughly speaking, attribute-hiding requires that a ciphertext conceal the associated attribute as well as the plaintext, while payload-hiding only requires that a ciphertext conceal the plaintext. Attribute-hiding FE is called *predicate encryption* (PE) [17]. *Anonymous* IBE and *hidden-vector encryption* (HVE) [9] are a class of PE and covered by predicate IPE, or PE with inner-product relations.

Although many ABE and IPE schemes have been presented over the last several years, no adaptively-secure (or fully-secure) scheme has been proposed in the standard model except [18]. The ABE scheme in [18] supports monotone access structures with equality relations and is secure under non-standard assumptions over composite order pairing groups. The IPE scheme in [18] supports inner-product relations and is secure under a non-standard assumption, whose size depends on some parameter that is not the security parameter.

No adaptively-secure (or fully-secure) ABE (even for monotone access structures) or IPE scheme has been proposed under a *well-established* assumption in the standard model, and no adaptively-secure (or fully-secure) ABE scheme with *non-monotone* access structures has been proposed (even under non-standard assumptions) in the standard model. In addition, to the best of our knowledge, no FE scheme (even with selective security) has been presented that supports more general relations than those for ABE, i.e., access structures with equality relations, and those for IPE, i.e., inner-product relations.

## 1.2 Our Result

– This paper proposes an adaptively secure functional encryption (FE) scheme for a wide class of relations, that are specified by non-monotone access structures combined with inner-product relations. More precisely, either one of the parameters for ek and dk is a tuple of attribute vectors and the other is a non-monotone access structure or span program $\hat{M} := (M, \rho)$ along with a tuple of attribute vectors, e.g., $x := (\overrightarrow{x}_1, \ldots, \overrightarrow{x}_d) \in \mathbb{F}_q^{n_1 + \cdots + n_d}$ for ek and $v := (\hat{M}, (\overrightarrow{v}_1, \ldots, \overrightarrow{v}_d) \in \mathbb{F}_q^{n_1 + \cdots + n_d})$ for dk. The component-wise inner-product relations for attribute vector components, e.g., $\{\overrightarrow{x}_t \cdot \overrightarrow{v}_t = 0$ or not $\}_{t \in \{1, \ldots, d\}}$, are input to span program $\hat{M}$, and $R(x, v)$ holds iff the truth-value vector of $(\mathsf{T}(\overrightarrow{x}_1 \cdot \overrightarrow{v}_1 = 0), \ldots, \mathsf{T}(\overrightarrow{x}_t \cdot \overrightarrow{v}_t = 0))$ is accepted by span program $\hat{M}$.

Similarly to ABE, we propose two types of FE schemes, the KP-FE and CP-FE schemes. Although this paper focuses on the KP-FE scheme, similar results are obtained for the CP-FE scheme (see the full version of this paper). Note that in Section 5, parameter $x$ for encryption is expressed by $\Gamma := \{(t, \overrightarrow{x}_t) \mid 1 \leq t \leq d\}$ in place of a tuple of vectors $(\overrightarrow{x}_1, \ldots, \overrightarrow{x}_d)$, where $1 \leq t \leq d$ means that $t$ is an element of some subset of $\{1, \ldots, d\}$, and parameter $v$ for the decryption key is expressed by $\mathbb{S} := (M, \rho)$ (not by $\hat{M} := (M, \rho)$ along with $(\overrightarrow{v}_1, \ldots, \overrightarrow{v}_d)$ as described above), where $\rho$ in $\mathbb{S}$ is abused as $\rho$ in $\hat{M}$ combined with $(\overrightarrow{v}_1, \ldots, \overrightarrow{v}_d)$ (see Definition 4).

Since the class of relations supported by the proposed FE scheme is more general than that for ABE and IPE, the proposed FE scheme includes the following schemes as special cases:

1. The (KP and CP)-ABE schemes for non-monotone access structures with equality relations. Here, the underlying attribute vectors of the FE scheme, $\{\overrightarrow{x}_t\}_{t \in \{1, \ldots, d\}}$ and $\{\overrightarrow{v}_t\}_{t \in \{1, \ldots, d\}}$, are specialized to two-dimensional vectors for the equality relation, e.g., $\overrightarrow{x}_t := (1, x_t)$ and $\overrightarrow{v}_t := (v_t, -1)$, where $\overrightarrow{x}_t \cdot \overrightarrow{v}_t = 0$ iff $x_t = v_t$.

2. The IPE and non-zero-IPE schemes, where a non-zero-IPE scheme is a class of FE with $R(\overrightarrow{x}, \overrightarrow{v})$ iff $\overrightarrow{x} \cdot \overrightarrow{v} \neq 0$. Here, the underlying access structure $\mathbb{S}$ of the FE scheme is specialized to the 1-out-of-1 secret sharing. The IPE scheme is 'attribute-hiding,' i.e., it is the PE scheme for the inner-product relations (see the full version for the proof).

   In addition, if the underlying access structure is specialized to the $d$-out-of-$d$ secret sharing, our FE scheme can be specialized to a *hierarchical* zero/non-zero IPE scheme by adding delegation and rerandomization mechanisms (see the full version for the construction and proof).

- The proposed FE scheme with such a wide class of relations is proven to be *adaptively secure* (adaptively payload-hiding against CPA) under a well-established assumption, the *decisional linear (DLIN)* assumption (over prime order pairing groups), in the standard model.

Note that even for FE with the simplest relations or the equality relations, i.e., IBE, only a few IBE schemes are known to be adaptively secure under well-established assumptions; the Waters IBE scheme [26] under the DBDH assumption, and the Waters IBE scheme [28] under the DBDH and DLIN assumptions.

The DLIN assumption is considered to be the simplest decisional assumption regarding pairing group $\mathbb{G}$, since the DLIN assumption is defined only over $\mathbb{G}$, the DDH assumption does not hold in $\mathbb{G}$, and the DBDH assumption is defined over two groups $\mathbb{G}$ and $\mathbb{G}_T$.

- To prove the security, this paper elaborately combines the dual system encryption methodology proposed by Waters [28] and the concept of dual pairing vector spaces (DPVS) proposed by Okamoto and Takashima [20,21], in a manner similar to that in [18]. See Section 2 (and the full version of this paper) for the concept and actual construction of DPVS.
  This paper also develops a new technique to prove the security based on the DLIN assumption. This provides a new methodology of employing a simple assumption defined on primitive groups to prove a complicated scheme that is designed on a higher level concept, DPVS.
  In our methodology, the top level of the security proof (based on the dual system encryption methodology) directly employs only top level assumptions (assumptions by Problems 1 and 2), that are defined on DPVS. The methodology bridges the top level assumptions and the primitive one, the DLIN assumption, in a hierarchical manner, where several levels of assumptions are constructed hierarchically. Such a modular way of proof greatly clarifies the logic of a complicated security proof.
- The efficiency of the proposed FE scheme is comparable to that of the existing ABE and IPE schemes. For example, if the proposed FE scheme is specialized to the IPE scheme, the key and ciphertext sizes are $(4n+5) \cdot |G|$, while they are $(2n+3) \cdot |G|$ for the IPE scheme in [18], where $n$ is the dimension of the attribute vectors, and $|G|$ denotes the size of an element of pairing group $G$, e.g., 256 bits.
- It is easy to convert the (CPA-secure) proposed FE scheme to a CCA-secure FE scheme by employing an existing general conversion such as that by Canetti, Halevi and Katz [11] or that by Boneh and Katz [8] (using additional 8-dimensional dual spaces $(\mathbb{B}_{d+1}, \mathbb{B}_{d+1}^*)$ with $n_{d+1} := 2$ on the proposed FE scheme, and a strongly unforgeable one-time signature scheme or message authentication code with encapsulation). That is, we can present a *fully secure* (adaptively payload-hiding against CCA) FE scheme for the same class of relations in the *standard model* under the DLIN assumption as well as a strongly unforgeable one-time signature scheme or message authentication code with encapsulation (see the full version of this paper for the construction and security proof).

## 1.3   Notations

When $A$ is a random variable or distribution, $y \xleftarrow{\mathsf{R}} A$ denotes that $y$ is randomly selected from $A$ according to its distribution. When $A$ is a set, $y \xleftarrow{\mathsf{U}} A$ denotes that $y$ is uniformly selected from $A$. $y := z$ denotes that $y$ is set, defined or substituted by $z$. When $a$ is a fixed value, $A(x) \to a$ (e.g., $A(x) \to 1$) denotes the event that machine (algorithm) $A$ outputs $a$ on input $x$. A function $f : \mathbb{N} \to \mathbb{R}$ is *negligible* in $\lambda$, if for every constant $c > 0$, there exists an integer $n$ such that $f(\lambda) < \lambda^{-c}$ for all $\lambda > n$.

We denote the finite field of order $q$ by $\mathbb{F}_q$, and $\mathbb{F}_q \setminus \{0\}$ by $\mathbb{F}_q^\times$. A vector symbol denotes a vector representation over $\mathbb{F}_q$, e.g., $\overrightarrow{x}$ denotes $(x_1, \ldots, x_n) \in \mathbb{F}_q^n$. For two vectors $\overrightarrow{x} = (x_1, \ldots, x_n)$ and $\overrightarrow{v} = (v_1, \ldots, v_n)$, $\overrightarrow{x} \cdot \overrightarrow{v}$ denotes the inner-product $\sum_{i=1}^{n} x_i v_i$. The vector $\overrightarrow{0}$ is abused as the zero vector in $\mathbb{F}_q^n$ for any $n$. $X^{\mathrm{T}}$ denotes the transpose of matrix $X$. $I_\ell$ and $0_\ell$ denote the $\ell \times \ell$ identity matrix and the $\ell \times \ell$ zero matrix, respectively. A bold face letter denotes an element of vector space $\mathbb{V}$, e.g., $\boldsymbol{x} \in \mathbb{V}$. When $\boldsymbol{b}_i \in \mathbb{V}$ $(i = 1, \ldots, n)$, $\mathsf{span}\langle \boldsymbol{b}_1, \ldots, \boldsymbol{b}_n \rangle \subseteq \mathbb{V}$ (resp. $\mathsf{span}\langle \overrightarrow{x}_1, \ldots, \overrightarrow{x}_n \rangle$) denotes the subspace generated by $\boldsymbol{b}_1, \ldots, \boldsymbol{b}_n$ (resp. $\overrightarrow{x}_1, \ldots, \overrightarrow{x}_n$). For bases $\mathbb{B} := (\boldsymbol{b}_1, \ldots, \boldsymbol{b}_N)$ and $\mathbb{B}^* := (\boldsymbol{b}_1^*, \ldots, \boldsymbol{b}_N^*)$, $(x_1, \ldots, x_N)_{\mathbb{B}} := \sum_{i=1}^{N} x_i \boldsymbol{b}_i$ and $(y_1, \ldots, y_N)_{\mathbb{B}^*} := \sum_{i=1}^{N} y_i \boldsymbol{b}_i^*$.

## 2 Dual Pairing Vector Spaces by Direct Product of Symmetric Pairing Groups

**Definition 1.** *"Symmetric bilinear pairing groups" $(q, \mathbb{G}, \mathbb{G}_T, G, e)$ are a tuple of a prime $q$, cyclic additive group $\mathbb{G}$ and multiplicative group $\mathbb{G}_T$ of order $q$, $G \neq 0 \in \mathbb{G}$, and a polynomial-time computable nondegenerate bilinear pairing $e : \mathbb{G} \times \mathbb{G} \to \mathbb{G}_T$ i.e., $e(sG, tG) = e(G, G)^{st}$ and $e(G, G) \neq 1$.*

*Let $\mathcal{G}_{\mathsf{bpg}}$ be an algorithm that takes input $1^\lambda$ and outputs a description of bilinear pairing groups $(q, \mathbb{G}, \mathbb{G}_T, G, e)$ with security parameter $\lambda$.*

In this paper, we concentrate on the symmetric version of dual pairing vector spaces [20,21] constructed by using symmetric bilinear pairing groups given in Definition 1.

**Definition 2.** *"Dual pairing vector spaces (DPVS)" $(q, \mathbb{V}, \mathbb{G}_T, \mathbb{A}, e)$ by a direct product of symmetric pairing groups $(q, \mathbb{G}, \mathbb{G}_T, G, e)$ are a tuple of prime $q$, $N$-dimensional vector space $\mathbb{V} := \overbrace{\mathbb{G} \times \cdots \times \mathbb{G}}^{N}$ over $\mathbb{F}_q$, cyclic group $\mathbb{G}_T$ of order $q$, canonical basis $\mathbb{A} := (\boldsymbol{a}_1, \ldots, \boldsymbol{a}_N)$ of $\mathbb{V}$, where $\boldsymbol{a}_i := (\overbrace{0, \ldots, 0}^{i-1}, G, \overbrace{0, \ldots, 0}^{N-i})$, and pairing $e : \mathbb{V} \times \mathbb{V} \to \mathbb{G}_T$.*

*The pairing is defined by $e(\boldsymbol{x}, \boldsymbol{y}) := \prod_{i=1}^{N} e(G_i, H_i) \in \mathbb{G}_T$ where $\boldsymbol{x} := (G_1, \ldots, G_N) \in \mathbb{V}$ and $\boldsymbol{y} := (H_1, \ldots, H_N) \in \mathbb{V}$. This is nondegenerate bilinear i.e., $e(s\boldsymbol{x}, t\boldsymbol{y}) = e(\boldsymbol{x}, \boldsymbol{y})^{st}$ and if $e(\boldsymbol{x}, \boldsymbol{y}) = 1$ for all $\boldsymbol{y} \in \mathbb{V}$, then $\boldsymbol{x} = \boldsymbol{0}$. For all $i$ and $j$, $e(\boldsymbol{a}_i, \boldsymbol{a}_j) = e(G, G)^{\delta_{i,j}}$ where $\delta_{i,j} = 1$ if $i = j$, and $0$ otherwise, and $e(G, G) \neq 1 \in \mathbb{G}_T$.*

*DPVS also has linear transformations $\phi_{i,j}$ on $\mathbb{V}$ s.t. $\phi_{i,j}(\boldsymbol{a}_j) = \boldsymbol{a}_i$ and $\phi_{i,j}(\boldsymbol{a}_k) = \boldsymbol{0}$ if $k \neq j$, which can be easily achieved by $\phi_{i,j}(\boldsymbol{x}) := (\overbrace{0, \ldots, 0}^{i-1}, G_j, \overbrace{0, \ldots, 0}^{N-i})$ where $\boldsymbol{x} := (G_1, \ldots, G_N)$. We call $\phi_{i,j}$ "distortion maps".*

*DPVS generation algorithm $\mathcal{G}_{\mathsf{dpvs}}$ takes input $1^\lambda$ $(\lambda \in \mathbb{N})$ and $N \in \mathbb{N}$, and outputs a description of $\mathsf{param}_{\mathbb{V}} := (q, \mathbb{V}, \mathbb{G}_T, \mathbb{A}, e)$ with security parameter $\lambda$ and $N$-dimensional $\mathbb{V}$. It can be constructed by using $\mathcal{G}_{\mathsf{bpg}}$.*

For the asymmetric version of DPVS, $(q, \mathbb{V}, \mathbb{V}^*, \mathbb{G}_T, \mathbb{A}, \mathbb{A}^*, e)$, see the full version of this paper. The above symmetric version is obtained by identifying $\mathbb{V} = \mathbb{V}^*$ and $\mathbb{A} = \mathbb{A}^*$ in the asymmetric version.

We describe random dual orthonormal bases generator $\mathcal{G}_{ob}$ below, which is used as a subroutine in the proposed FE scheme.

$$\mathcal{G}_{ob}(1^\lambda, \overrightarrow{n} := (d; n_1, \ldots, n_d)) : \mathsf{param}_{\mathbb{G}} := (q, \mathbb{G}, \mathbb{G}_T, G, e) \overset{\mathsf{R}}{\leftarrow} \mathcal{G}_{bpg}(1^\lambda), \quad \psi \overset{\mathsf{U}}{\leftarrow} \mathbb{F}_q^\times,$$

$$N_0 := 5, \quad N_t := 4n_t \ \text{ for } t = 1, \ldots, d,$$

$$\text{for } t = 0, \ldots, d, \quad \mathsf{param}_{\mathbb{V}_t} := (q, \mathbb{V}_t, \mathbb{G}_T, \mathbb{A}_t, e) := \mathcal{G}_{dpvs}(1^\lambda, N_t, \mathsf{param}_{\mathbb{G}}),$$

$$X_t := (\chi_{t,i,j})_{i,j} \overset{\mathsf{U}}{\leftarrow} GL(N_t, \mathbb{F}_q), \quad (\vartheta_{t,i,j})_{i,j} := \psi \cdot (X_t^{\mathsf{T}})^{-1},$$

$$\boldsymbol{b}_{t,i} := (\chi_{t,i,1}, \ldots, \chi_{t,i,N_t})_{\mathbb{A}_t} = \sum_{j=1}^{N_t} \chi_{t,i,j} \boldsymbol{a}_{t,j}, \quad \mathbb{B}_t := (\boldsymbol{b}_{t,1}, \ldots, \boldsymbol{b}_{t,N_t}),$$

$$\boldsymbol{b}_{t,i}^* := (\vartheta_{t,i,1}, \ldots, \vartheta_{t,i,N_t})_{\mathbb{A}_t} = \sum_{j=1}^{N_t} \vartheta_{t,i,j} \boldsymbol{a}_{t,j}, \quad \mathbb{B}_t^* := (\boldsymbol{b}_{t,1}^*, \ldots, \boldsymbol{b}_{t,N_t}^*),$$

$$g_T := e(G, G)^\psi, \quad \mathsf{param}_{\overrightarrow{n}} := (\{\mathsf{param}_{\mathbb{V}_t}\}_{t=0,\ldots,d}, \ g_T)$$

$$\text{return } (\mathsf{param}_{\overrightarrow{n}}, \{\mathbb{B}_t, \mathbb{B}_t^*\}_{t=0,\ldots,d}).$$

We note that $g_T = e(\boldsymbol{b}_{t,i}, \boldsymbol{b}_{t,i}^*)$ for $t = 0, \ldots, d; i = 1, \ldots, N_t$.

# 3   Functional Encryption with General Relations

## 3.1   Span Programs and Non-monotone Access Structures

**Definition 3 (Span Programs [1]).** *Let $\{p_1, \ldots, p_n\}$ be a set of variables. A span program over $\mathbb{F}_q$ is a labeled matrix $\hat{M} := (M, \rho)$ where $M$ is a ($\ell \times r$) matrix over $\mathbb{F}_q$ and $\rho$ is a labeling of the rows of $M$ by literals from $\{p_1, \ldots, p_n, \neg p_1, \ldots, \neg p_n\}$ (every row is labeled by one literal), i.e., $\rho : \{1, \ldots, \ell\} \to \{p_1, \ldots, p_n, \neg p_1, \ldots, \neg p_n\}$.*

*A span program accepts or rejects an input by the following criterion. For every input sequence $\delta \in \{0, 1\}^n$ define the submatrix $M_\delta$ of $M$ consisting of those rows whose labels are set to 1 by the input $\delta$, i.e., either rows labeled by some $p_i$ such that $\delta_i = 1$ or rows labeled by some $\neg p_i$ such that $\delta_i = 0$. (i.e., $\gamma : \{1, \ldots, \ell\} \to \{0, 1\}$ is defined by $\gamma(j) = 1$ if $[\rho(j) = p_i] \wedge [\delta_i = 1]$ or $[\rho(j) = \neg p_i] \wedge [\delta_i = 0]$, and $\gamma(j) = 0$ otherwise. $M_\delta := (M_j)_{\gamma(j)=1}$, where $M_j$ is the $j$-th row of $M$.)*

*The span program $\hat{M}$ accepts $\delta$ if and only if $\overrightarrow{1} \in \text{span}\langle M_\delta \rangle$, i.e., some linear combination of the rows of $M_\delta$ gives the all one vector $\overrightarrow{1}$. (The row vector has the value 1 in each coordinate.) A span program computes a Boolean function $f$ if it accepts exactly those inputs $\delta$ where $f(\delta) = 1$.*

*A span program is called monotone if the labels of the rows are only the positive literals $\{p_1, \ldots, p_n\}$. Monotone span programs compute monotone functions. (So, a span program in general is "non"-monotone.)*

We assume that the matrix $M$ satisfies the condition: $M_i \neq \overrightarrow{0}$ for $i = 1, \ldots, \ell$.

We now introduce a non-monotone access structure with evaluating map $\gamma$ by using the inner-product of attribute vectors, that is employed in the proposed functional encryption schemes.

**Definition 4 (Inner-Products of Attribute Vectors and Access Structures).** $\mathcal{U}_t$ $(t = 1, \ldots, d$ and $\mathcal{U}_t \subset \{0,1\}^*)$ is a sub-universe, a set of attributes, each of which is expressed by a pair of sub-universe id and $n_t$-dimensional vector, i.e., $(t, \overrightarrow{v})$, where $t \in \{1, \ldots, d\}$ and $\overrightarrow{v} \in \mathbb{F}_q^{n_t} \setminus \{\overrightarrow{0}\}$.

We now define such an attribute to be a variable $p$ of a span program $\hat{M} := (M, \rho)$, i.e., $p := (t, \overrightarrow{v})$. An access structure $\mathbb{S}$ is span program $\hat{M} := (M, \rho)$ along with variables $p := (t, \overrightarrow{v}), p' := (t', \overrightarrow{v}'), \ldots$, i.e., $\mathbb{S} := (M, \rho)$ such that $\rho : \{1, \ldots, \ell\} \to \{(t, \overrightarrow{v}), (t', \overrightarrow{v}'), \ldots, \neg(t, \overrightarrow{v}), \neg(t', \overrightarrow{v}'), \ldots\}$.

Let $\Gamma$ be a set of attributes, i.e., $\Gamma := \{(t, \overrightarrow{x}_t) \mid \overrightarrow{x}_t \in \mathbb{F}_q^{n_t} \setminus \{\overrightarrow{0}\}, 1 \leq t \leq d\}$, where $1 \leq t \leq d$ means that $t$ is an element of some subset of $\{1, \ldots, d\}$.

When $\Gamma$ is given to access structure $\mathbb{S}$, map $\gamma : \{1, \ldots, \ell\} \to \{0, 1\}$ for span program $\hat{M} := (M, \rho)$ is defined as follows: For $i = 1, \ldots, \ell$, set $\gamma(i) = 1$ if $[\rho(i) = (t, \overrightarrow{v}_i)] \wedge [(t, \overrightarrow{x}_t) \in \Gamma] \wedge [\overrightarrow{v}_i \cdot \overrightarrow{x}_t = 0]$ or $[\rho(i) = \neg(t, \overrightarrow{v}_i)] \wedge [(t, \overrightarrow{x}_t) \in \Gamma]$ $\wedge [\overrightarrow{v}_i \cdot \overrightarrow{x}_t \neq 0]$. Set $\gamma(i) = 0$ otherwise.

Access structure $\mathbb{S} := (M, \rho)$ accepts $\Gamma$ iff $\overrightarrow{1} \in \mathsf{span}\langle (M_i)_{\gamma(i)=1} \rangle$.

We now construct a secret-sharing scheme for a non-monotone access structure or span program.

**Definition 5.** A secret-sharing scheme for span program $\hat{M} := (M, \rho)$ is:

1. Let $M$ be $\ell \times r$ matrix. Let column vector $\overrightarrow{f}^{\mathrm{T}} := (f_1, \ldots, f_r)^{\mathrm{T}} \xleftarrow{\mathsf{U}} \mathbb{F}_q^r$. Then, $s_0 := \overrightarrow{1} \cdot \overrightarrow{f}^{\mathrm{T}} = \sum_{k=1}^r f_k$ is the secret to be shared, and $\overrightarrow{s}^{\mathrm{T}} := (s_1, \ldots, s_\ell)^{\mathrm{T}} := M \cdot \overrightarrow{f}^{\mathrm{T}}$ is the vector of $\ell$ shares of the secret $s_0$ and the share $s_i$ belongs to $\rho(i)$.

2. If span program $\hat{M} := (M, \rho)$ accept $\delta$, or access structure $\mathbb{S} := (M, \rho)$ accepts $\Gamma$, i.e., $\overrightarrow{1} \in \mathsf{span}\langle (M_i)_{\gamma(i)=1} \rangle$ with $\gamma : \{1, \ldots, \ell\} \to \{0, 1\}$, then there exist constants $\{\alpha_i \in \mathbb{F}_q \mid i \in I\}$ such that $I \subseteq \{i \in \{1, \ldots, \ell\} \mid \gamma(i) = 1\}$ and $\sum_{i \in I} \alpha_i s_i = s_0$. Furthermore, these constants $\{\alpha_i\}$ can be computed in time polynomial in the size of matrix $M$.

### 3.2 Key-Policy Functional Encryption with General Relations

**Definition 6 (Key-Policy Functional Encryption : KP-FE).** A key-policy functional encryption scheme consists of four algorithms.

Setup. This is a randomized algorithm that takes as input security parameter and format $\overrightarrow{n} := (d; n_1, \ldots, n_d)$ of attributes. It outputs public parameters pk and master secret key sk.

KeyGen. This is a randomized algorithm that takes as input access structure $\mathbb{S} := (M, \rho)$, pk and sk. It outputs a decryption key $\mathsf{sk}_{\mathbb{S}}$.

Enc. This is a randomized algorithm that takes as input message $m$, a set of attributes, $\Gamma := \{(t, \overrightarrow{x}_t) \mid \overrightarrow{x}_t \in \mathbb{F}_q^{n_t} \setminus \{\overrightarrow{0}\}, 1 \leq t \leq d\}$, and public parameters pk. It outputs a ciphertext $\mathsf{ct}_\Gamma$.

Dec. *This takes as input ciphertext* $\mathsf{ct}_\Gamma$ *that was encrypted under a set of attributes* $\Gamma$, *decryption key* $\mathsf{sk}_\mathbb{S}$ *for access structure* $\mathbb{S}$, *and public parameters* $\mathsf{pk}$. *It outputs either plaintext* $m$ *or the distinguished symbol* $\perp$.

A KP-FE scheme should have the following correctness property: for all $(\mathsf{pk}, \mathsf{sk}) \xleftarrow{\mathsf{R}} \mathsf{Setup}(1^\lambda, \overrightarrow{n})$, all access structures $\mathbb{S}$, all decryption keys $\mathsf{sk}_\mathbb{S} \xleftarrow{\mathsf{R}} \mathsf{KeyGen}(\mathsf{pk}, \mathsf{sk}, \mathbb{S})$, all messages $m$, all attribute sets $\Gamma$, all ciphertexts $\mathsf{ct}_\Gamma \xleftarrow{\mathsf{R}} \mathsf{Enc}(\mathsf{pk}, m, \Gamma)$, it holds that $m = \mathsf{Dec}(\mathsf{pk}, \mathsf{sk}_\mathbb{S}, \mathsf{ct}_\Gamma)$ with overwhelming probability, if $\mathbb{S}$ accepts $\Gamma$.

**Definition 7.** *The model for proving the adaptively payload-hiding security of KP-FE under chosen plaintext attack is:*

**Setup.** *The challenger runs the setup algorithm,* $(\mathsf{pk}, \mathsf{sk}) \xleftarrow{\mathsf{R}} \mathsf{Setup}(1^\lambda, \overrightarrow{n})$, *and gives public parameters* $\mathsf{pk}$ *to the adversary.*
**Phase 1.** *The adversary is allowed to adaptively issue a polynomial number of queries,* $\mathbb{S}$, *to the challenger or oracle* $\mathsf{KeyGen}(\mathsf{pk}, \mathsf{sk}, \cdot)$ *for private keys,* $\mathsf{sk}_\mathbb{S}$ *associated with* $\mathbb{S}$.
**Challenge.** *The adversary submits two messages* $m^{(0)}, m^{(1)}$ *and a set of attributes,* $\Gamma$, *provided that no* $\mathbb{S}$ *queried to the challenger in Phase 1 accepts* $\Gamma$. *The challenger flips a coin* $b \xleftarrow{\mathsf{U}} \{0, 1\}$, *and computes* $\mathsf{ct}_\Gamma^{(b)} \xleftarrow{\mathsf{R}} \mathsf{Enc}(\mathsf{pk}, m^{(b)}, \Gamma)$. *It gives* $\mathsf{ct}_\Gamma^{(b)}$ *to the adversary.*
**Phase 2.** *The adversary is allowed to adaptively issue a polynomial number of queries,* $\mathbb{S}$, *to the challenger or oracle* $\mathsf{KeyGen}(\mathsf{pk}, \mathsf{sk}, \cdot)$ *for private keys,* $\mathsf{sk}_\mathbb{S}$ *associated with* $\mathbb{S}$, *provided that* $\mathbb{S}$ *does not accept* $\Gamma$.
**Guess.** *The adversary outputs a guess* $b'$ *of* $b$.

*We note that the model can easily be extended to handle chosen-ciphertext attacks by allowing for decryption queries in Phases 1 and 2.*

*The advantage of adversary* $\mathcal{A}$ *in the above game is defined as* $\mathsf{Adv}_\mathcal{A}^{\mathsf{KP\text{-}FE,PH}}(\lambda)$ $:= \Pr[b' = b] - 1/2$ *for any security parameter* $\lambda$. *A KP-FE scheme is secure if all polynomial time adversaries have at most a negligible advantage in the above game.*

Similarly we can define a *ciphertext-policy* FE (CP-FE) scheme (see the full version of this paper).

## 4    Assumption

**Definition 8 (DLIN: Decisional Linear Assumption).** *The DLIN problem is to guess* $\beta \in \{0, 1\}$, *given* $(\mathsf{param}_\mathbb{G}, G, \xi G, \kappa G, \omega \xi G, \gamma \kappa G, Y_\beta) \xleftarrow{\mathsf{R}} \mathcal{G}_\beta^{\mathsf{DLIN}}(1^\lambda)$, *where*

$$\mathcal{G}_\beta^{\mathsf{DLIN}}(1^\lambda) : \mathsf{param}_\mathbb{G} := (q, \mathbb{G}, \mathbb{G}_T, G, e) \xleftarrow{\mathsf{R}} \mathcal{G}_{\mathsf{bpg}}(1^\lambda),$$

$$\kappa, \omega, \xi, \gamma \xleftarrow{\mathsf{U}} \mathbb{F}_q, \quad Y_0 := (\omega + \gamma)G, \quad Y_1 \xleftarrow{\mathsf{U}} \mathbb{G},$$

$$\text{return } (\mathsf{param}_\mathbb{G}, G, \xi G, \kappa G, \omega \xi G, \gamma \kappa G, Y_\beta),$$

for $\beta \xleftarrow{\mathsf{U}} \{0,1\}$. *For a probabilistic machine $\mathcal{E}$, we define the advantage of $\mathcal{E}$ for the DLIN problem as:*

$$\mathsf{Adv}_{\mathcal{E}}^{\mathsf{DLIN}}(\lambda) := \left| \Pr\left[\mathcal{E}(1^\lambda, \varrho) \to 1 \,\middle|\, \varrho \xleftarrow{\mathsf{R}} \mathcal{G}_0^{\mathsf{DLIN}}(1^\lambda)\right] - \Pr\left[\mathcal{E}(1^\lambda, \varrho) \to 1 \,\middle|\, \varrho \xleftarrow{\mathsf{R}} \mathcal{G}_1^{\mathsf{DLIN}}(1^\lambda)\right] \right|.$$

*The DLIN assumption is: For any probabilistic polynomial-time adversary $\mathcal{E}$, the advantage $\mathsf{Adv}_{\mathcal{E}}^{\mathsf{DLIN}}(\lambda)$ is negligible in $\lambda$.*

## 5   Proposed KP-FE Scheme

We define function $\widetilde{\rho} : \{1, \ldots, \ell\} \to \{1, \ldots, d\}$ by $\widetilde{\rho}(i) := t$ if $\rho(i) = (t, \overrightarrow{v})$ or $\rho(i) = \neg(t, \overrightarrow{v})$, where $\rho$ is given in access structure $\mathbb{S} := (M, \rho)$. In the proposed scheme, we assume that $\widetilde{\rho}$ is injective for $\mathbb{S} := (M, \rho)$ with decryption key $\mathsf{sk}_{\mathbb{S}}$. We will show how to relax the restriction in the full version of this paper.

In the description of the scheme, we assume that input vector, $\overrightarrow{x}_t := (x_{t,1}, \ldots, x_{t,n_t})$, is normalized such that $x_{t,1} := 1$. (If $\overrightarrow{x}_t$ is not normalized, change it to a normalized one by $(1/x_{t,1}) \cdot \overrightarrow{x}_t$, assuming that $x_{t,1}$ is non-zero).

$\mathsf{Setup}(1^\lambda, \overrightarrow{n} := (d; n_1, \ldots, n_d)):$   $(\mathsf{param}_{\overrightarrow{n}}, \{\mathbb{B}_t, \mathbb{B}_t^*\}_{t=0,\ldots,d}) \xleftarrow{\mathsf{R}} \mathcal{G}_{\mathsf{ob}}(1^\lambda, \overrightarrow{n}),$

$\widehat{\mathbb{B}}_0 := (\boldsymbol{b}_{0,1}, \boldsymbol{b}_{0,3}, \boldsymbol{b}_{0,5}),$   $\widehat{\mathbb{B}}_t := (\boldsymbol{b}_{t,1}, .., \boldsymbol{b}_{t,n_t}, \boldsymbol{b}_{t,3n_t+1}, .., \boldsymbol{b}_{t,4n_t})$   for $t = 1, .., d,$

$\widehat{\mathbb{B}}_0^* := (\boldsymbol{b}_{0,1}^*, \boldsymbol{b}_{0,3}^*, \boldsymbol{b}_{0,4}^*),$   $\widehat{\mathbb{B}}_t^* := (\boldsymbol{b}_{t,1}^*, .., \boldsymbol{b}_{t,n_t}^*, \boldsymbol{b}_{t,2n_t+1}^*, .., \boldsymbol{b}_{t,3n_t}^*)$   for $t = 1, .., d,$

$\mathsf{pk} := (1^\lambda, \mathsf{param}_{\overrightarrow{n}}, \{\widehat{\mathbb{B}}_t\}_{t=0,\ldots,d}),$   $\mathsf{sk} := \{\widehat{\mathbb{B}}_t^*\}_{t=0,\ldots,d},$

return $\mathsf{pk}, \mathsf{sk}.$

$\mathsf{KeyGen}(\mathsf{pk}, \mathsf{sk}, \mathbb{S} := (M, \rho)):$

$\overrightarrow{f} \xleftarrow{\mathsf{U}} \mathbb{F}_q^r,$   $\overrightarrow{s}^{\mathrm{T}} := (s_1, \ldots, s_\ell)^{\mathrm{T}} := M \cdot \overrightarrow{f}^{\mathrm{T}},$   $s_0 := \overrightarrow{1} \cdot \overrightarrow{f}^{\mathrm{T}},$   $\eta_0 \xleftarrow{\mathsf{U}} \mathbb{F}_q,$

$\boldsymbol{k}_0^* := (-s_0, 0, 1, \eta_0, 0)_{\mathbb{B}_0^*},$

for $i = 1, \ldots, \ell,$

if $\rho(i) = (t, \overrightarrow{v}_i := (v_{i,1}, \ldots, v_{i,n_t}) \in \mathbb{F}_q^{n_t} \setminus \{\overrightarrow{0}\}),$   $\theta_i, \eta_{i,1}, .., \eta_{i,n_t} \xleftarrow{\mathsf{U}} \mathbb{F}_q,$

$$\boldsymbol{k}_i^* := (\ \overbrace{s_i + \theta_i v_{i,1}, \theta_i v_{t,2}, .., \theta_i v_{i,n_t}}^{n_t},\ \overbrace{0^{n_t}}^{n_t},\ \overbrace{\eta_{i,1}, .., \eta_{i,n_t}}^{n_t},\ \overbrace{0^{n_t}}^{n_t}\ )_{\mathbb{B}_t^*},$$

if $\rho(i) = \neg(t, \overrightarrow{v}_i),$   $\eta_{i,1}, \ldots, \eta_{i,n_t} \xleftarrow{\mathsf{U}} \mathbb{F}_q,$

$$\boldsymbol{k}_i^* := (\ \overbrace{s_i(v_{i,1}, .., v_{i,n_t})}^{n_t},\ \overbrace{0^{n_t}}^{n_t},\ \overbrace{\eta_{i,1}, .., \eta_{i,n_t}}^{n_t},\ \overbrace{0^{n_t}}^{n_t}\ )_{\mathbb{B}_t^*},$$

return $\mathsf{sk}_{\mathbb{S}} := (\mathbb{S}, \boldsymbol{k}_0^*, \boldsymbol{k}_1^*, \ldots, \boldsymbol{k}_\ell^*).$

$\mathsf{Enc}(\mathsf{pk}, m, \Gamma := \{(t, \overrightarrow{x}_t := (x_{t,1}, .., x_{t,n_t}) \in \mathbb{F}_q^{n_t} \setminus \{\overrightarrow{0}\}) \mid 1 \le t \le d, x_{t,1} := 1\}):$

$\delta, \varphi_0, \varphi_{t,1}, \ldots, \varphi_{t,n_t}, \zeta \xleftarrow{\mathsf{U}} \mathbb{F}_q$ for $(t, \overrightarrow{x}_t) \in \Gamma,$

$\boldsymbol{c}_0 := (\delta, 0, \zeta, 0, \varphi_0)_{\mathbb{B}_0},$

$$\overbrace{\qquad}^{n_t}\quad\overbrace{\qquad}^{n_t}\quad\overbrace{\qquad}^{n_t}\quad\overbrace{\qquad}^{n_t}$$

$$\boldsymbol{c}_t := (\ \delta(x_{t,1},..,x_{t,n_t}),\quad 0^{n_t},\quad 0^{n_t},\quad \varphi_{t,1},..,\varphi_{t,n_t}\ )_{\mathbb{B}_t}\ \text{ for } (t,\overrightarrow{x}_t) \in \Gamma,$$

$$c_{d+1} := g_T^\zeta m, \quad \mathsf{ct}_\Gamma := (\Gamma, \boldsymbol{c}_0, \{\boldsymbol{c}_t\}_{(t,\overrightarrow{x}_t)\in\Gamma}, c_{d+1}).$$

return $\mathsf{ct}_\Gamma$.

$\mathsf{Dec}(\mathsf{pk}, \mathsf{sk}_{\mathbb{S}} := (\mathbb{S}, \boldsymbol{k}_0^*, \boldsymbol{k}_1^*, \dots, \boldsymbol{k}_\ell^*),\ \mathsf{ct}_\Gamma := (\Gamma, \boldsymbol{c}_0, \{\boldsymbol{c}_t\}_{(t,\overrightarrow{x}_t)\in\Gamma}, c_{d+1}))$ :

If $\mathbb{S} := (M, \rho)$ accepts $\Gamma := \{(t, \overrightarrow{x}_t)\}$, then compute $I$ and $\{\alpha_i\}_{i \in I}$ such that

$$s_0 = \sum_{i \in I} \alpha_i s_i, \text{ and}$$

$$I \subseteq \{ i \in \{1, \dots, \ell\} \mid [\rho(i) = (t, \overrightarrow{v}_i) \ \wedge \ (t, \overrightarrow{x}_t) \in \Gamma \ \wedge \ \overrightarrow{v}_i \cdot \overrightarrow{x}_t = 0]$$

$$\vee\ [\rho(i) = \neg(t, \overrightarrow{v}_i) \ \wedge \ (t, \overrightarrow{x}_t) \in \Gamma \ \wedge \ \overrightarrow{v}_i \cdot \overrightarrow{x}_t \neq 0]\ \}.$$

$$K := e(\boldsymbol{c}_0, \boldsymbol{k}_0^*) \prod_{i \in I\ \wedge\ \rho(i)=(t,\overrightarrow{v}_i)} e(\boldsymbol{c}_t, \boldsymbol{k}_i^*)^{\alpha_i} \prod_{i \in I\ \wedge\ \rho(i)=\neg(t,\overrightarrow{v}_i)} e(\boldsymbol{c}_t, \boldsymbol{k}_i^*)^{\alpha_i/(\overrightarrow{v}_i\cdot\overrightarrow{x}_t)}$$

return $m' := c_{d+1}/K$.

**[Correctness]**

$$e(\boldsymbol{c}_0, \boldsymbol{k}_0^*) \prod_{i \in I\ \wedge\ \rho(i)=(t,\overrightarrow{v}_i)} e(\boldsymbol{c}_t, \boldsymbol{k}_i^*)^{\alpha_i} \cdot \prod_{i \in I\ \wedge\ \rho(i)=\neg(t,\overrightarrow{v}_i)} e(\boldsymbol{c}_t, \boldsymbol{k}_i^*)^{\alpha_i/(\overrightarrow{v}_i\cdot\overrightarrow{x}_t)}$$

$$= g_T^{-\delta s_0 + \zeta} \prod_{i \in I\ \wedge\ \rho(i)=(t,\overrightarrow{v}_i)} g_T^{\delta\alpha_i s_i} \prod_{i \in I\ \wedge\ \rho(i)=\neg(t,\overrightarrow{v}_i)} g_T^{\delta\alpha_i s_i(\overrightarrow{v}_i\cdot\overrightarrow{x}_t)/(\overrightarrow{v}_i\cdot\overrightarrow{x}_t)}$$

$$= g_T^{\delta(-s_0 + \sum_{i \in I}\alpha_i s_i) + \zeta} = g_T^\zeta.$$

# 6   Security

The proofs of Lemmas 1–4 and 6–8, and Claim 1 are given in the full version of this paper.

## 6.1   Theorem

**Theorem 1.** *The proposed KP-FE scheme is adaptively payload-hiding against chosen plaintext attacks under the DLIN assumption.*

*For any adversary $\mathcal{A}$, there exist probabilistic machines $\mathcal{E}_0, \mathcal{E}_h^+, \mathcal{E}_{h+1}$ ($h = 0, \dots, \nu - 1$), whose running times are essentially the same as that of $\mathcal{A}$, such that for any security parameter $\lambda$,*

$$\mathsf{Adv}_{\mathcal{A}}^{\mathsf{KP\text{-}FE,PH}}(\lambda) \leq \mathsf{Adv}_{\mathcal{E}_0}^{\mathsf{DLIN}}(\lambda) + \sum_{h=0}^{\nu-1}\left(\mathsf{Adv}_{\mathcal{E}_h^+}^{\mathsf{DLIN}}(\lambda) + \mathsf{Adv}_{\mathcal{E}_{h+1}}^{\mathsf{DLIN}}(\lambda)\right) + \epsilon,$$

*where $\nu$ is the maximum number of $\mathcal{A}$'s key queries and $\epsilon := (2d\nu + 12\nu + d + 7)/q$.*

## 6.2   Lemmas

We will show three lemmas for the proof of Theorem 1.

**Definition 9 (Problem 1).** *Problem 1 is to guess $\beta$, given* $(\mathsf{param}_{\overrightarrow{n}}, \widehat{\mathbb{B}}_0, \widehat{\mathbb{B}}_0^*,$
$e_{\beta,0}, \{\widehat{\mathbb{B}}_t, \widehat{\mathbb{B}}_t^*, e_{\beta,t,i}\}_{t=1,\ldots,d;i=1,\ldots,n_t}) \xleftarrow{\mathsf{R}} \mathcal{G}_\beta^{\mathsf{P1}}(1^\lambda, \overrightarrow{n})$, *where*

$$\mathcal{G}_\beta^{\mathsf{P1}}(1^\lambda, \overrightarrow{n}) : \quad (\mathsf{param}_{\overrightarrow{n}}, \{\mathbb{B}_t, \mathbb{B}_t^*\}_{t=0,\ldots,d}) \xleftarrow{\mathsf{R}} \mathcal{G}_{\mathsf{ob}}(1^\lambda, \overrightarrow{n}),$$

$\widehat{\mathbb{B}}_0 := (\boldsymbol{b}_{0,1}, \boldsymbol{b}_{0,3}, \boldsymbol{b}_{0,5}), \quad \widehat{\mathbb{B}}_t := (\boldsymbol{b}_{t,1}, .., \boldsymbol{b}_{t,n_t}, \boldsymbol{b}_{t,3n_t+1}, .., \boldsymbol{b}_{t,4n_t})$ for $t = 1, .., d,$

$\widehat{\mathbb{B}}_0^* := (\boldsymbol{b}_{0,1}^*, \boldsymbol{b}_{0,3}^*, \boldsymbol{b}_{0,4}^*), \quad \widehat{\mathbb{B}}_t^* := (\boldsymbol{b}_{t,1}^*, .., \boldsymbol{b}_{t,n_t}^*, \boldsymbol{b}_{t,2n_t+1}^*, .., \boldsymbol{b}_{t,3n_t}^*)$ for $t = 1, .., d,$

$u_0 \xleftarrow{\mathsf{U}} \mathbb{F}_q^\times, \quad \delta, \delta_0 \xleftarrow{\mathsf{U}} \mathbb{F}_q, \quad (u_{t,i,j})_{i,j=1,\ldots,n_t} \xleftarrow{\mathsf{U}} GL(n_t, \mathbb{F}_q)$ for $t = 1, .., d,$

$\boldsymbol{e}_{0,0} := (\delta, 0, 0, 0, \delta_0)_{\mathbb{B}_0}, \quad \boldsymbol{e}_{1,0} := (\delta, u_0, 0, 0, \delta_0)_{\mathbb{B}_0},$

for $t = 1, \ldots, d; \quad i = 1, \ldots, n_t;$

$\delta_{t,i,j} \xleftarrow{\mathsf{U}} \mathbb{F}_q$ for $j = 1, \ldots, n_t,$

$$
\begin{array}{llll}
\overbrace{\phantom{0^{i-1}, \delta, 0^{n_t-i}}}^{n_t} & \overbrace{\phantom{0^{n_t}}}^{n_t} & \overbrace{\phantom{0^{n_t}}}^{n_t} & \overbrace{\phantom{\delta_{t,i,1}, .., \delta_{t,i,n_t}}}^{n_t}
\end{array}
$$

$\boldsymbol{e}_{0,t,i} := (\ 0^{i-1}, \delta, 0^{n_t-i}, \quad 0^{n_t}, \quad 0^{n_t}, \quad \delta_{t,i,1}, .., \delta_{t,i,n_t}\ )_{\mathbb{B}_t},$

$\boldsymbol{e}_{1,t,i} := (\ 0^{i-1}, \delta, 0^{n_t-i}, \ u_{t,i,1}, .., u_{t,i,n_t}, \quad 0^{n_t}, \quad \delta_{t,i,1}, .., \delta_{t,i,n_t}\ )_{\mathbb{B}_t},$

return $(\mathsf{param}_{\overrightarrow{n}}, \widehat{\mathbb{B}}_0, \widehat{\mathbb{B}}_0^*, \boldsymbol{e}_{\beta,0}, \{\widehat{\mathbb{B}}_t, \widehat{\mathbb{B}}_t^*, \boldsymbol{e}_{\beta,t,i}\}_{t=1,\ldots,d;i=1,\ldots,n_t}),$

*for* $\beta \xleftarrow{\mathsf{U}} \{0,1\}$. *For a probabilistic machine $\mathcal{B}$, we define the advantage of $\mathcal{B}$ as the quantity*

$$\mathsf{Adv}_{\mathcal{B}}^{\mathsf{P1}}(\lambda) := \left| \Pr\left[ \mathcal{B}(1^\lambda, \varrho) \to 1 \ \middle| \ \varrho \xleftarrow{\mathsf{R}} \mathcal{G}_0^{\mathsf{P1}}(1^\lambda, \overrightarrow{n}) \right] - \Pr\left[ \mathcal{B}(1^\lambda, \varrho) \to 1 \ \middle| \ \varrho \xleftarrow{\mathsf{R}} \mathcal{G}_1^{\mathsf{P1}}(1^\lambda, \overrightarrow{n}) \right] \right|.$$

**Lemma 1.** *For any adversary $\mathcal{B}$, there exists a probabilistic machine $\mathcal{E}$, whose running time is essentially the same as that of $\mathcal{B}$, such that for any security parameter $\lambda$, $\mathsf{Adv}_{\mathcal{B}}^{\mathsf{P1}}(\lambda) \leq \mathsf{Adv}_{\mathcal{E}}^{\mathsf{DLIN}}(\lambda) + 5/q$.*

**Definition 10 (Problem 2).** *Problem 2 is to guess $\beta$, given* $(\mathsf{param}_{\overrightarrow{n}}, \widehat{\mathbb{B}}_0, \widehat{\mathbb{B}}_0^*,$
$\boldsymbol{h}_{\beta,0}^*, \boldsymbol{e}_0, \{\widehat{\mathbb{B}}_t, \widehat{\mathbb{B}}_t^*, \boldsymbol{h}_{\beta,t,i}^*, \boldsymbol{e}_{t,i}\}_{t=1,\ldots,d;i=1,\ldots,n_t}) \xleftarrow{\mathsf{R}} \mathcal{G}_\beta^{\mathsf{P2}}(1^\lambda, \overrightarrow{n})$, *where*

$$\mathcal{G}_\beta^{\mathsf{P2}}(1^\lambda, \overrightarrow{n}) : \quad (\mathsf{param}_{\overrightarrow{n}}, \{\mathbb{B}_t, \mathbb{B}_t^*\}_{t=0,\ldots,d}) \xleftarrow{\mathsf{R}} \mathcal{G}_{\mathsf{ob}}(1^\lambda, \overrightarrow{n}),$$

$\widehat{\mathbb{B}}_0 := (\boldsymbol{b}_{0,1}, \boldsymbol{b}_{0,3}, \boldsymbol{b}_{0,5}), \quad \widehat{\mathbb{B}}_t := (\boldsymbol{b}_{t,1}, .., \boldsymbol{b}_{t,n_t}, \boldsymbol{b}_{t,3n_t+1}, .., \boldsymbol{b}_{t,4n_t})$ for $t = 1, .., d,$

$\widehat{\mathbb{B}}_0^* := (\boldsymbol{b}_{0,1}^*, .., \boldsymbol{b}_{0,4}^*), \quad \widehat{\mathbb{B}}_t^* := (\boldsymbol{b}_{t,1}^*, .., \boldsymbol{b}_{t,n_t}^*, \boldsymbol{b}_{t,2n_t+1}^*, .., \boldsymbol{b}_{t,3n_t}^*)$ for $t = 1, .., d,$

$\tau, u_0 \xleftarrow{\mathsf{U}} \mathbb{F}_q^\times, \quad \omega, \delta, \gamma_0 \xleftarrow{\mathsf{U}} \mathbb{F}_q, \quad w_0 := \tau/u_0,$

$(z_{t,i,j})_{i,j=1,\ldots,n_t} := Z_t \xleftarrow{\mathsf{U}} GL(n_t, \mathbb{F}_q), \quad (u_{t,i,j})_{i,j=1,\ldots,n_t} := (Z_t^{-1})^\mathrm{T}$ for $t = 1, .., d,$

$\boldsymbol{h}_{0,0}^* := (\omega, 0, 0, \gamma_0, 0)_{\mathbb{B}_0^*}, \quad \boldsymbol{h}_{1,0}^* := (\omega, w_0, 0, \gamma_0, 0)_{\mathbb{B}_0^*}, \quad \boldsymbol{e}_0 := (\delta, u_0, 0, 0, 0)_{\mathbb{B}_0},$

for $t = 1, \ldots, d; \quad i = 1, \ldots, n_t;$

$\left(w_{t,i,j}\right)_{i,j=1,\ldots,n_t} := \tau \cdot Z_t, \quad \gamma_{t,i,j} \xleftarrow{\mathsf{U}} \mathbb{F}_q$ for $j = 1, .., n_t,$

$$\overbrace{\qquad}^{n_t}\quad\overbrace{\qquad}^{n_t}\quad\overbrace{\qquad}^{n_t}\quad\overbrace{\qquad}^{n_t}$$

$$
\begin{aligned}
\boldsymbol{h}_{0,t,i}^* &:= (\ 0^{i-1}, \omega, 0^{n_t-i}, & 0^{n_t}, & & \gamma_{t,i,1},..,\gamma_{t,i,n_t}, & & 0^{n_t} & )_{\mathbb{B}_t^*}\\
\boldsymbol{h}_{1,t,i}^* &:= (\ 0^{i-1}, \omega, 0^{n_t-i}, & w_{t,i,1},..,w_{t,i,n_t}, & & \gamma_{t,i,1},..,\gamma_{t,i,n_t}, & & 0^{n_t} & )_{\mathbb{B}_t^*}\\
\boldsymbol{e}_{t,i} &:= (\ 0^{i-1}, \delta, 0^{n_t-i}, & u_{t,i,1},..,u_{t,i,n_t}, & & 0^{n_t}, & & 0^{n_t} & )_{\mathbb{B}_t},
\end{aligned}
$$

$$\text{return } (\mathsf{param}_{\overrightarrow{n}}, \widehat{\mathbb{B}}_0, \widehat{\mathbb{B}}_0^*, \boldsymbol{h}_{\beta,0}^*, \boldsymbol{e}_0, \{\widehat{\mathbb{B}}_t, \widehat{\mathbb{B}}_t^*, \boldsymbol{h}_{\beta,t,i}^*, \boldsymbol{e}_{t,i}\}_{t=1,..,d;i=1,..,n_t}),$$

for $\beta \xleftarrow{\mathsf{U}} \{0,1\}$. For a probabilistic adversary $\mathcal{B}$, the advantage of $\mathcal{B}$ for Problem 2, $\mathsf{Adv}_{\mathcal{B}}^{\mathsf{P2}}(\lambda)$, is similarly defined as in Definition 9.

**Lemma 2.** *For any adversary $\mathcal{B}$, there exists a probabilistic machine $\mathcal{E}$, whose running time is essentially the same as that of $\mathcal{B}$, such that for any security parameter $\lambda$, $\mathsf{Adv}_{\mathcal{B}}^{\mathsf{P2}}(\lambda) \le \mathsf{Adv}_{\mathcal{E}}^{\mathsf{DLIN}}(\lambda) + 5/q$.*

**Lemma 3.** *For $p \in \mathbb{F}_q$, let $C_p := \{(\overrightarrow{x}, \overrightarrow{v}) \mid \overrightarrow{x} \cdot \overrightarrow{v} = p\} \subset V \times V^*$ where $V$ is $n$-dimensional vector space $\mathbb{F}_q^n$, and $V^*$ its dual. For all $(\overrightarrow{x}, \overrightarrow{v}) \in C_p$, for all $(\overrightarrow{r}, \overrightarrow{w}) \in C_p$, $\Pr[\overrightarrow{x} U = \overrightarrow{r} \wedge \overrightarrow{v} Z = \overrightarrow{w}] = 1/\sharp C_p$, where $Z \xleftarrow{\mathsf{U}} GL(n, \mathbb{F}_q), U := (Z^{-1})^{\mathrm{T}}$.*

## 6.3  Proof of Theorem 1

**Proof Outline :** At the top level of strategy of the security proof, we follow the dual system encryption methodology proposed by Waters [28]. In the methodology, ciphertexts and secret keys have two forms, *normal* and *semi-functional*. In the proof herein, we also introduce another form called *pre-semi-functional*. The real system uses only normal ciphertexts and normal secret keys, and semi-functional/pre-semi-functional ciphertexts and keys are used only in a sequence of security games for the security proof.

To prove this theorem, we employ Game 0 (original adaptive-security game) through Game 3. In Game 1, the target ciphertext is changed to semi-functional. When at most $\nu$ secret key queries are issued by an adversary, there are $2\nu$ game changes from Game 1 (Game 2-0), Game 2-0$^+$, Game 2-1 through Game 2-$(\nu - 1)^+$ and Game 2-$\nu$. In Game 2-$h$, the first $h$ keys are semi-functional while the remaining keys are normal, and the target ciphertext is semi-functional. In Game 2-$h^+$, the first $h$ keys are semi-functional and the $(h + 1)$-th key is *pre-semi-functional* while the remaining keys are normal, and the target ciphertext is *pre-semi-functional*. The final game with advantage 0 is changed from Game 2-$\nu$. As usual, we prove that the advantage gaps between neighboring games are negligible.

For $\mathsf{sk}_{\mathbb{S}} := (\mathbb{S}, \boldsymbol{k}_0^*, \boldsymbol{k}_1^*, \ldots, \boldsymbol{k}_\ell^*)$ and $\mathsf{ct}_{\varGamma} := (\varGamma, \boldsymbol{c}_0, \{\boldsymbol{c}_t\}_{(t,\overrightarrow{x}_t)\in\varGamma}, c_{d+1})$, we focus on $\overrightarrow{\boldsymbol{k}}_{\mathbb{S}}^* := (\boldsymbol{k}_0^*, \boldsymbol{k}_1^*, \ldots, \boldsymbol{k}_\ell^*)$ and $\overrightarrow{\boldsymbol{c}}_{\varGamma} := (\boldsymbol{c}_0, \{\boldsymbol{c}_t\}_{(t,\overrightarrow{x}_t)\in\varGamma})$, and ignore the other part of $\mathsf{sk}_{\mathbb{S}}$ and $\mathsf{ct}_{\varGamma}$ (and call them secret key and ciphertext, respectively) in this proof outline. In addition, we ignore a negligible factor in the (informal) descriptions of this proof outline. For example, we say "$A$ is bounded by $B$" when $A \le B + \epsilon(\lambda)$ where $\epsilon(\lambda)$ is negligible in security parameter $\lambda$.

A *normal* secret key, $\overrightarrow{k}^*_{\mathbb{S}}{}^{\text{norm}}$ (with access structure $\mathbb{S}$), is the correct form of the secret key of the proposed FE scheme, and is expressed by Eq. (1). Similarly, a *normal* ciphertext (with attribute set $\Gamma$), $\overrightarrow{c}_\Gamma^{\text{norm}}$, is expressed by Eq. (2). A *semi-functional* secret key, $\overrightarrow{k}^*_{\mathbb{S}}{}^{\text{semi}}$, is expressed by Eq. (8), and a *semi-functional* ciphertext, $\overrightarrow{c}_\Gamma^{\text{semi}}$, is expressed by Eqs. (3)-(5). A *pre-semi-functional* secret key, $\overrightarrow{k}^*_{\mathbb{S}}{}^{\text{pre-semi}}$, and *pre-semi-functional* ciphertext, $\overrightarrow{c}_\Gamma^{\text{pre-semi}}$, are expressed by Eq. (6) and Eqs. (3), (7) and (5), respectively.

To prove that the advantage gap between Games 0 and 1 is bounded by the advantage of Problem 1 (to guess $\beta \in \{0, 1\}$), we construct a simulator of the challenger of Game 0 (or 1) (against an adversary $\mathcal{A}$) by using an instance with $\beta \xleftarrow{\mathsf{U}} \{0, 1\}$ of Problem 1. We then show that the distribution of the secret keys and target ciphertext replied by the simulator is equivalent to those of Game 0 when $\beta = 0$ and Game 1 when $\beta = 1$. That is, the advantage of Problem 1 is equivalent to the advantage gap between Games 0 and 1 (Lemma 4). The advantage of Problem 1 is proven to be equivalent to that of the DLIN assumption (Lemma 1).

The advantage gap between Games 2-$h$ and 2-$h^+$ is similarly shown to be bounded by the advantage of Problem 2 (i.e., advantage of the DLIN assumption) (Lemmas 5 and 2). Here, we introduce *special forms of pre-semi-functional* keys and ciphertexts, $\overrightarrow{k}^*_{\mathbb{S}}{}^{\text{spec.pre-semi}}$ and $\overrightarrow{c}_\Gamma^{\text{spec.pre-semi}}$, respectively, such that they are equivalent to pre-semi-functional keys and ciphertexts, $\overrightarrow{k}^*_{\mathbb{S}}{}^{\text{pre-semi}}$ and $\overrightarrow{c}_\Gamma^{\text{pre-semi}}$, respectively, except that $w_0 r_0 = a_0 := \sum_{k=1}^r g_k$ and $r_0 \xleftarrow{\mathsf{U}} \mathbb{F}_q$ (note that $r_0, w_0 \xleftarrow{\mathsf{U}} \mathbb{F}_q$ for $\overrightarrow{k}^*_{\mathbb{S}}{}^{\text{pre-semi}}$ and $\overrightarrow{c}_\Gamma^{\text{pre-semi}}$). These forms of keys and ciphertexts, $\overrightarrow{k}^*_{\mathbb{S}}{}^{\text{spec.pre-semi}}$ and $\overrightarrow{c}_\Gamma^{\text{spec.pre-semi}}$, are simulated by using Problem 2 with $\beta = 1$. From the definition of these forms, $\overrightarrow{k}^*_{\mathbb{S}}{}^{\text{spec.pre-semi}}$ can decrypt $\overrightarrow{c}_\Gamma^{\text{spec.pre-semi}}$ for any $\Gamma$ when $\mathbb{S}$ accepts $\Gamma$, i.e., it is hard for simulator $\mathcal{B}^+_h$ to tell $(\overrightarrow{k}^*_{\mathbb{S}}{}^{\text{spec.pre-semi}}, \overrightarrow{c}_\Gamma^{\text{spec.pre-semi}})$ for Game 2-$h^+$ from $(\overrightarrow{k}^*_{\mathbb{S}}{}^{\text{norm}}, \overrightarrow{c}_\Gamma^{\text{semi}})$ for Game 2-$h$ under the assumption of Problem 2. On the other hand, $a_0 (= w_0 r_0)$ is independently distributed from the other variables when $\mathbb{S}$ does not accept $\Gamma$ (shown in Proof of Claim 1 by using Lemma 3). That is, the joint distribution of $\overrightarrow{k}^*_{\mathbb{S}}{}^{\text{pre-semi}}$ and $\overrightarrow{c}_\Gamma^{\text{pre-semi}}$ is equivalent to that of $\overrightarrow{k}^*_{\mathbb{S}}{}^{\text{spec.pre-semi}}$ and $\overrightarrow{c}_\Gamma^{\text{spec.pre-semi}}$, when $\mathbb{S}$ does not accept $\Gamma$ (i.e., $\mathcal{B}^+_h$'s simulation using Problem 2 with $\beta = 1$ is the same distribution as that of Game 2-$h^+$ from the adversary's view). In other words, $w_0$ and $r_0$ in $\overrightarrow{k}^*_{\mathbb{S}}{}^{\text{spec.pre-semi}}$ and $\overrightarrow{c}_\Gamma^{\text{spec.pre-semi}}$ (given by $\mathcal{B}^+_h$'s simulation using Problem 2 with $\beta = 1$) are correlated for the case that $\mathbb{S}$ accepts $\Gamma$ or for simulator $\mathcal{B}^+_h$'s view, but adversary $\mathcal{A}$ cannot notice the correlation since $\mathcal{A}$'s queries should satisfy the condition that $\mathbb{S}$ does not accept $\Gamma$.

The advantage gap between Games 2-$h^+$ and 2-$(h+1)$ is similarly shown to be bounded by the advantage of Problem 2, i.e., advantage of the DLIN assumption (Lemmas 6 and 2).

Finally we show that Game 2-$\nu$ can be conceptually changed to Game 3 (Lemma 7).

**Proof of Theorem 1 :** To prove Theorem 1, we consider the following $(2\nu + 3)$ games. In Game 0, a part framed by a box indicates coefficients to be changed in a subsequent game. In the other games, a part framed by a box indicates coefficients which were changed in a game from the previous game.

**Game 0 :** Original game. That is, the reply to a key query for $\mathbb{S} := (M, \rho)$ with $\ell \times r$ matrix $M$ is:

$$
\left.
\begin{aligned}
&\boldsymbol{k}_0^* := (-s_0, \boxed{0}, 1, \eta_0, 0)_{\mathbb{B}_0^*}, \\
&\text{for } i = 1, \ldots, \ell, \\
&\quad \text{if } \rho(i) = (t, \overrightarrow{v}_i), \boldsymbol{k}_i^* := (s_i + \theta_i v_{i,1}, \theta_i v_{i,2}, .., \theta_i v_{i,n_t}, \boxed{0^{n_t}}, \eta_{i,1}, .., \eta_{i,n_t}, 0^{n_t})_{\mathbb{B}_t^*}, \\
&\quad \text{if } \rho(i) = \neg(t, \overrightarrow{v}_i), \boldsymbol{k}_i^* := (s_i(v_{i,1}, .., v_{i,n_t}), \boxed{0^{n_t}}, \eta_{i,1}, .., \eta_{i,n_t}, 0^{n_t})_{\mathbb{B}_t^*},
\end{aligned}
\right\} \quad (1)
$$

where $\overrightarrow{f} \xleftarrow{\mathsf{U}} \mathbb{F}_q^r$, $\overrightarrow{s}^{\mathrm{T}} := (s_1, \ldots, s_\ell)^{\mathrm{T}} := M \cdot \overrightarrow{f}^{\mathrm{T}}$, $s_0 := \overrightarrow{1} \cdot \overrightarrow{f}^{\mathrm{T}}$, $\theta_i, \eta_0, \eta_{i,1}, \ldots,$ $\eta_{i,n_t} \xleftarrow{\mathsf{U}} \mathbb{F}_q$, and $\overrightarrow{v}_i := (v_{i,1}, \ldots, v_{i,n_t}) \in \mathbb{F}_q^{n_t} \setminus \{\overrightarrow{0}\}$. The target ciphertext for challenge plaintexts $(m^{(0)}, m^{(1)})$ and $\Gamma := \{(t, \overrightarrow{x}_t) \mid 1 \le t \le d\}$ is:

$$
\left.
\begin{aligned}
&\boldsymbol{c}_0 := (\delta, \boxed{0}, \zeta, 0, \varphi_0)_{\mathbb{B}_0}, \\
&\boldsymbol{c}_t := (\delta(x_{t,1}, \ldots, x_{t,n_t}), \boxed{0^{n_t}}, 0^{n_t}, \varphi_{t,1}, \ldots, \varphi_{t,n_t})_{\mathbb{B}_t} \text{ for } (t, \overrightarrow{x}_t) \in \Gamma, \\
&\boldsymbol{c}_{d+1} := g_T^\zeta m^{(b)},
\end{aligned}
\right\} \quad (2)
$$

where $b \xleftarrow{\mathsf{U}} \{0, 1\}$; $\delta, \zeta, \varphi_0, \varphi_{t,1}, \ldots, \varphi_{t,n_t} \xleftarrow{\mathsf{U}} \mathbb{F}_q$, and $\overrightarrow{x}_t := (x_{t,1}, \ldots, x_{t,n_t}) \in \mathbb{F}_q^{n_t} \setminus \{\overrightarrow{0}\}$.

**Game 1 :** Same as Game 0 except that the target ciphertext is:

$$
\boldsymbol{c}_0 := (\delta, \boxed{r_0}, \zeta, 0, \varphi_0)_{\mathbb{B}_0}, \tag{3}
$$

$$
\boldsymbol{c}_t := (\delta(x_{t,1}, .., x_{t,n_t}), \boxed{r_{t,1}, .., r_{t,n_t}}, 0^{n_t}, \varphi_{t,1}, .., \varphi_{t,n_t})_{\mathbb{B}_t} \text{ for } (t, \overrightarrow{x}_t) \in \Gamma, \tag{4}
$$

$$
\boldsymbol{c}_{d+1} := g_T^\zeta m^{(b)}, \tag{5}
$$

where $r_0, r_{t,1}, \ldots, r_{t,n_t} \xleftarrow{\mathsf{U}} \mathbb{F}_q$.

**Game 2-$h^+$ ($h = 0, \ldots, \nu - 1$) :** Game 2-0 is Game 1. Game 2-$h^+$ is the same as Game 2-$h$ except the reply to the $(h+1)$-th key query for $\mathbb{S} := (M, \rho)$ with $\ell \times r$ matrix $M$, and $\boldsymbol{c}_t$ of the target ciphertext are:

$$
\left.
\begin{aligned}
&\boldsymbol{k}_0^* := (-s_0, \boxed{w_0}, 1, \eta_0, 0)_{\mathbb{B}_0^*}, \\
&\text{for } i = 1, \ldots, \ell, \\
&\quad \text{if } \rho(i) = (t, \overrightarrow{v}_i) \\
&\quad\quad \boldsymbol{k}_i^* := (s_i + \theta_i v_{i,1}, \theta_i v_{i,2}, .., \theta_i v_{i,n_t}, \boxed{w_{i,1}, .., w_{i,n_t}}, \eta_{i,1}, .., \eta_{i,n_t}, 0^{n_t})_{\mathbb{B}_t^*}, \\
&\quad \text{if } \rho(i) = \neg(t, \overrightarrow{v}_i) \\
&\quad\quad \boldsymbol{k}_i^* := (s_i(v_{i,1}, .., v_{i,n_t}), \boxed{\overline{w}_{i,1}, .., \overline{w}_{i,n_t}}, \eta_{i,1}, .., \eta_{i,n_t}, 0^{n_t})_{\mathbb{B}_t^*},
\end{aligned}
\right\} \quad (6)
$$

$$
\boldsymbol{c}_t := (\delta(x_{t,1}, .., x_{t,n_t}), \boxed{r_{t,1}, .., r_{t,n_t}}, 0^{n_t}, \varphi_{t,1}, , , , \varphi_{t,n_t})_{\mathbb{B}_t} \text{ for } (t, \overrightarrow{x}_t) \in \Gamma, \tag{7}
$$

where $w_0 \xleftarrow{\mathsf{U}} \mathbb{F}_q$, $\overrightarrow{g} \xleftarrow{\mathsf{U}} \mathbb{F}_q^r$, $\overrightarrow{a}^{\mathrm{T}} := (a_1, \ldots, a_\ell)^{\mathrm{T}} := M \cdot \overrightarrow{g}^{\mathrm{T}}$, $\tau_i \xleftarrow{\mathsf{U}} \mathbb{F}_q$ ($i = 1, \ldots, \ell$), $Z_t \xleftarrow{\mathsf{U}} GL(n_t, \mathbb{F}_q)$, $U_t := (Z_t^{-1})^{\mathrm{T}}$ for $t = 1, \ldots, d$,

$$(w_{i,1}, \ldots, w_{i,n_t}) := (a_i + \tau_i v_{i,1}, \tau_i v_{i,2}, \ldots, \tau_i v_{i,n_t}) \cdot Z_t,$$
$$(\overline{w}_{i,1}, \ldots, \overline{w}_{i,n_t}) := a_i(v_{i,1}, \ldots, v_{i,n_t}) \cdot Z_t,$$
$$(r_{t,1}, \ldots, r_{t,n_t}) := (x_{t,1}, \ldots, x_{t,n_t}) \cdot U_t.$$

**Game 2-$(h+1)$ ($h = 0, \ldots, \nu - 1$) :** Game 2-$(h+1)$ is the same as Game 2-$h^+$ except the reply to the $(h+1)$-th key query for $\mathbb{S} := (M, \rho)$ with $\ell \times r$ matrix $M$, and $c_t$ of the target ciphertext are:

$$\left.\begin{aligned}
&\boldsymbol{k}_0^* := (-s_0, w_0, 1, \eta_0, 0)_{\mathbb{B}_0^*}, \\
&\text{for } i = 1, \ldots, \ell, \\
&\text{if } \rho(i) = (t, \overrightarrow{v}_i), \boldsymbol{k}_i^* := (s_i + \theta_i v_{i,1}, \theta_i v_{i,2}, .., \theta_i v_{i,n_t}, \boxed{0^{n_t}}, \eta_{i,1}, .., \eta_{i,n_t}, 0^{n_t})_{\mathbb{B}_t^*}, \\
&\text{if } \rho(i) = \neg(t, \overrightarrow{v}_i), \boldsymbol{k}_i^* := (s_i(v_{i,1}, .., v_{i,n_t}), \boxed{0^{n_t}}, \eta_{i,1}, .., \eta_{i,n_t}, 0^{n_t})_{\mathbb{B}_t^*}, \\
&\boldsymbol{c}_t := (\delta(x_{t,1}, \ldots, x_{t,n_t}), \boxed{r_{t,1}, \ldots, r_{t,n_t}}, 0^{n_t}, \varphi_{t,1}, \ldots, \varphi_{t,n_t})_{\mathbb{B}_t} \text{ for } (t, \overrightarrow{x}_t) \in \Gamma,
\end{aligned}\right\} \quad (8)$$

where $r_{t,1}, \ldots, r_{t,n_t} \xleftarrow{\mathsf{U}} \mathbb{F}_q$.

**Game 3 :** Same as Game 2-$\nu$ except that $c_0$ and $c_{d+1}$ of the target ciphertext are

$$c_0 := (\delta, r_0, \boxed{\zeta'}, 0, \varphi_0)_{\mathbb{B}_0}, \quad c_{d+1} := g_T^\zeta m^{(b)},$$

where $\zeta' \xleftarrow{\mathsf{U}} \mathbb{F}_q$ (i.e., independent from $\zeta \xleftarrow{\mathsf{U}} \mathbb{F}_q$).

Let $\mathsf{Adv}_{\mathcal{A}}^{(0)}(\lambda)$, $\mathsf{Adv}_{\mathcal{A}}^{(1)}(\lambda)$, $\mathsf{Adv}_{\mathcal{A}}^{(2\text{-}h)}(\lambda)$, $\mathsf{Adv}_{\mathcal{A}}^{(2\text{-}h^+)}(\lambda)$ and $\mathsf{Adv}_{\mathcal{A}}^{(3)}(\lambda)$ be the advantage of $\mathcal{A}$ in Game $0, 1, 2\text{-}h, 2\text{-}h^+$ and $3$, respectively. $\mathsf{Adv}_{\mathcal{A}}^{(0)}(\lambda)$ is equivalent to $\mathsf{Adv}_{\mathcal{A}}^{\mathsf{KP\text{-}FE,PH}}(\lambda)$ and it is clear that $\mathsf{Adv}_{\mathcal{A}}^{(3)}(\lambda) = 0$ by Lemma 8.

We will show four lemmas (Lemmas 4-7) that evaluate the gaps between pairs of $\mathsf{Adv}_{\mathcal{A}}^{(0)}(\lambda), \mathsf{Adv}_{\mathcal{A}}^{(1)}(\lambda)$, $\mathsf{Adv}_{\mathcal{A}}^{(2\text{-}h)}(\lambda), \mathsf{Adv}_{\mathcal{A}}^{(2\text{-}h^+)}(\lambda), \mathsf{Adv}_{\mathcal{A}}^{(2\text{-}(h+1))}(\lambda)$ for $h = 0, \ldots, \nu - 1$ and $\mathsf{Adv}_{\mathcal{A}}^{(3)}(\lambda)$. From these lemmas and Lemmas 1 and 2, we obtain $\mathsf{Adv}_{\mathcal{A}}^{\mathsf{KP\text{-}FE,PH}}(\lambda) = \mathsf{Adv}_{\mathcal{A}}^{(0)}(\lambda) \leq \left|\mathsf{Adv}_{\mathcal{A}}^{(0)}(\lambda) - \mathsf{Adv}_{\mathcal{A}}^{(1)}(\lambda)\right| + \sum_{h=0}^{\nu-1}\left|\mathsf{Adv}_{\mathcal{A}}^{(2\text{-}h)}(\lambda) - \mathsf{Adv}_{\mathcal{A}}^{(2\text{-}h^+)}(\lambda)\right| + \sum_{h=0}^{\nu-1}\left|\mathsf{Adv}_{\mathcal{A}}^{(2\text{-}h^+)}(\lambda) - \mathsf{Adv}_{\mathcal{A}}^{(2\text{-}(h+1))}(\lambda)\right| + \left|\mathsf{Adv}_{\mathcal{A}}^{(2\text{-}\nu)}(\lambda) - \mathsf{Adv}_{\mathcal{A}}^{(3)}(\lambda)\right| + \mathsf{Adv}_{\mathcal{A}}^{(3)}(\lambda) \leq \mathsf{Adv}_{\mathcal{B}_0}^{\mathsf{P1}}(\lambda) + \sum_{h=0}^{\nu-1}\mathsf{Adv}_{\mathcal{B}_h^+}^{\mathsf{P2}}(\lambda) + \sum_{h=0}^{\nu-1}\mathsf{Adv}_{\mathcal{B}_{h+1}}^{\mathsf{P2}}(\lambda) + (2d\nu + 2\nu + d + 2)/q \leq \mathsf{Adv}_{\mathcal{E}_0}^{\mathsf{DLIN}}(\lambda) + \sum_{h=0}^{\nu-1}\left(\mathsf{Adv}_{\mathcal{E}_h^+}^{\mathsf{DLIN}}(\lambda) + \mathsf{Adv}_{\mathcal{E}_{h+1}}^{\mathsf{DLIN}}(\lambda)\right) + (2d\nu + 12\nu + d + 7)/q.$ This completes the proof of Theorem 1. $\qquad\square$

**Lemma 4.** *For any adversary $\mathcal{A}$, there exists a probabilistic machine $\mathcal{B}_0$, whose running time is essentially the same as that of $\mathcal{A}$, such that for any security parameter $\lambda$, $|\mathsf{Adv}_{\mathcal{A}}^{(0)}(\lambda) - \mathsf{Adv}_{\mathcal{A}}^{(1)}(\lambda)| \leq \mathsf{Adv}_{\mathcal{B}_0}^{\mathsf{P1}}(\lambda) + (d+1)/q$.*

**Lemma 5.** *For any adversary $\mathcal{A}$, there exists a probabilistic machine $\mathcal{B}_h^+$, whose running time is essentially the same as that of $\mathcal{A}$, such that for any security parameter $\lambda$, $|\mathsf{Adv}_{\mathcal{A}}^{(2\text{-}h)}(\lambda) - \mathsf{Adv}_{\mathcal{A}}^{(2\text{-}h^+)}(\lambda)| \le \mathsf{Adv}_{\mathcal{B}_h^+}^{\mathsf{P2}}(\lambda) + (d+1)/q$.*

*Proof.* In order to prove Lemma 5, we construct a probabilistic machine $\mathcal{B}_h^+$ against Problem 2 by using an adversary $\mathcal{A}$ in a security game (Game 2-$h$ or 2-$h^+$) as a black box as follows:

1. $\mathcal{B}_h^+$ is given a Problem 2 instance, $(\mathsf{param}_{\overrightarrow{n}}, \widehat{\mathbb{B}}_0, \widehat{\mathbb{B}}_0^*, \boldsymbol{h}_{\beta,0}^*, \boldsymbol{e}_0, \{\widehat{\mathbb{B}}_t, \widehat{\mathbb{B}}_t^*, \boldsymbol{h}_{\beta,t,j}^*, \boldsymbol{e}_{t,j}\}_{t=1,\dots,d;j=1,\dots,n_t})$.
2. $\mathcal{B}_h^+$ plays a role of the challenger in the security game against adversary $\mathcal{A}$.
3. At the first step of the game, $\mathcal{B}_h^+$ provides $\mathcal{A}$ a public key $\mathsf{pk} := (1^\lambda, \mathsf{param}_{\overrightarrow{n}}, \{\widehat{\mathbb{B}}_t\}_{t=0,\dots,d})$ of Game 2-$h$ (and 2-$h^+$), that is a part of the Problem 2 instance.
4. When the $\iota$-th key query is issued for access structure $\mathbb{S} := (M, \rho)$, $\mathcal{B}_h^+$ answers as follows:
   (a) When $1 \le \iota \le h$, $\mathcal{B}_h^+$ answers semi-functional key $(\boldsymbol{k}_0^*, \dots, \boldsymbol{k}_\ell^*)$ with Eq. (8), that is computed by using $\{\widehat{\mathbb{B}}_t^*\}_{t=0,\dots,d}$ of the Problem 2 instance.
   (b) When $\iota = h + 1$, $\mathcal{B}_h^+$ calculates $(\boldsymbol{k}_0^*, \dots, \boldsymbol{k}_\ell^*)$ by using $(\boldsymbol{h}_{\beta,0}^*, \{\boldsymbol{h}_{\beta,t,j}^*\}_{t=1,\dots,d;j=1,\dots,n_t})$ of the Problem 2 instance as follows:

$$\mu_{t,l}, \widetilde{\mu}_{k,l} \overset{\mathsf{U}}{\leftarrow} \mathbb{F}_q \quad \text{for } t = 1, \dots, d; \ k = 1, \dots, r; \ l = 1, 2,$$

$$\boldsymbol{p}_{\beta,0}^* := \sum_{k=1}^r \left( \widetilde{\mu}_{k,1} \boldsymbol{h}_{\beta,0}^* + \widetilde{\mu}_{k,2} \boldsymbol{b}_{0,1}^* \right),$$

$$\text{for } t = 1, \dots, d; \ k = 1, \dots, r; \ j = 1, \dots, n_t;$$

$$\boldsymbol{p}_{\beta,t,j}^* := \mu_{t,1} \boldsymbol{h}_{\beta,t,j}^* + \mu_{t,2} \boldsymbol{b}_{t,j}^*, \qquad \widetilde{\boldsymbol{p}}_{\beta,t,k,j}^* := \widetilde{\mu}_{k,1} \boldsymbol{h}_{\beta,t,j}^* + \widetilde{\mu}_{k,2} \boldsymbol{b}_{t,j}^*,$$

$$\boldsymbol{k}_0^* := -\boldsymbol{p}_{\beta,0}^* + \boldsymbol{b}_{0,3}^*,$$

$$\text{for } i = 1, \dots, \ell,$$

$$\text{if } \rho(i) = (t, \overrightarrow{v}_i), \quad \boldsymbol{k}_i^* := \sum_{j=1}^{n_t} v_{i,j} \boldsymbol{p}_{\beta,t,j}^* + \sum_{k=1}^r M_{i,k} \widetilde{\boldsymbol{p}}_{\beta,t,k,n_t}^*,$$

$$\text{if } \rho(i) = \neg(t, \overrightarrow{v}_i), \quad \boldsymbol{k}_i^* := \sum_{j=1}^{n_t} v_{i,j} (\sum_{k=1}^r M_{i,k} \widetilde{\boldsymbol{p}}_{\beta,t,k,j}^*),$$

where $(M_{i,k})_{i=1,\dots,\ell;k=1,\dots,r} := M$.
   (c) When $\iota \ge h + 2$, $\mathcal{B}_h^+$ answers normal key $(\boldsymbol{k}_0^*, \dots, \boldsymbol{k}_\ell^*)$ with Eq. (1), that is computed by using $\{\widehat{\mathbb{B}}_t^*\}_{t=0,\dots,d}$ of the Problem 2 instance.
5. When $\mathcal{B}_h^+$ receives an encryption query with challenge plaintexts $(m^{(0)}, m^{(1)})$ and $\Gamma := \{(t, \overrightarrow{x}_t) \mid 1 \le t \le d\}$ from $\mathcal{A}$, $\mathcal{B}_h^+$ computes the challenge ciphertext $(c_0, \{c_t\}_{(t,\overrightarrow{x}_t) \in \Gamma}, c_{d+1})$ such that for $(t, \overrightarrow{x}_t) \in \Gamma$,

$$c_0 := \boldsymbol{e}_0 + \zeta \boldsymbol{b}_{0,3} + \boldsymbol{q}_0, \qquad c_t := \sum_{j=1}^{n_t} x_{t,j} \boldsymbol{e}_{t,j} + \boldsymbol{q}_t, \qquad c_{d+1} := g_T^\zeta m^{(b)},$$

where $\zeta \overset{\mathsf{U}}{\leftarrow} \mathbb{F}_q$, $b \overset{\mathsf{U}}{\leftarrow} \{0, 1\}$, $\boldsymbol{q}_0 \overset{\mathsf{U}}{\leftarrow} \mathsf{span}\langle \boldsymbol{b}_{0,5} \rangle$, $\boldsymbol{q}_t \overset{\mathsf{U}}{\leftarrow} \mathsf{span}\langle \boldsymbol{b}_{t,3n_t+1}, \dots, \boldsymbol{b}_{t,4n_t} \rangle$, and $(\boldsymbol{b}_{0,3}, \boldsymbol{e}_0, \{\boldsymbol{e}_{t,j}\}_{t=1,\dots,d;j=1,\dots,n_t})$ is a part of the Problem 2 instance.
6. When a key query is issued by $\mathcal{A}$ after the encryption query, $\mathcal{B}_h^+$ executes the same procedure as that of step 4.
7. $\mathcal{A}$ finally outputs bit $b'$. If $b = b'$, $\mathcal{B}_h^+$ outputs $\beta' := 1$. Otherwise, $\mathcal{B}_h^+$ outputs $\beta' := 0$.

**Claim 1.** *The distribution of the view of adversary $\mathcal{A}$ in the above-mentioned game simulated by $\mathcal{B}_h^+$ given a Problem 2 instance with $\beta \in \{0,1\}$ is the same as that in Game 2-h (resp. Game 2-$h^+$) if $\beta = 0$ (resp. $\beta = 1$).*

The proof of Claim 1 is given in the full version of this paper. This completes the proof of Lemma 5.                                                              □

**Lemma 6.** *For any adversary $\mathcal{A}$, there exists a probabilistic machine $\mathcal{B}_{h+1}$, whose running time is essentially the same as that of $\mathcal{A}$, such that for any security parameter $\lambda$, $|\mathsf{Adv}_{\mathcal{A}}^{(2\text{-}h^+)}(\lambda) - \mathsf{Adv}_{\mathcal{A}}^{(2\text{-}(h+1))}(\lambda)| \leq \mathsf{Adv}_{\mathcal{B}_{h+1}}^{P2}(\lambda) + (d+1)/q.$*

**Lemma 7.** *For any adversary $\mathcal{A}$, $\mathsf{Adv}_{\mathcal{A}}^{(3)}(\lambda) \leq \mathsf{Adv}_{\mathcal{A}}^{(2\text{-}\nu)}(\lambda) + 1/q.$*

**Lemma 8.** *For any adversary $\mathcal{A}$, $\mathsf{Adv}_{\mathcal{A}}^{(3)}(\lambda) = 0.$*

# References

1. Beimel, A.: Secure schemes for secret sharing and key distribution. PhD Thesis, Israel Institute of Technology, Technion, Haifa, Israel (1996)
2. Bethencourt, J., Sahai, A., Waters, B.: Ciphertext-policy attribute-based encryption. In: 2007 IEEE Symposium on Security and Privacy, pp. 321–334. IEEE Press, Los Alamitos (2007)
3. Boneh, D., Boyen, X.: Efficient selective-ID secure identity based encryption without random oracles. In: Cachin, C., Camenisch, J.L. (eds.) EUROCRYPT 2004. LNCS, vol. 3027, pp. 223–238. Springer, Heidelberg (2004)
4. Boneh, D., Boyen, X.: Secure identity based encryption without random oracles. In: Franklin, M.K. (ed.) CRYPTO 2004. LNCS, vol. 3152, pp. 443–459. Springer, Heidelberg (2004)
5. Boneh, D., Boyen, X., Goh, E.: Hierarchical identity based encryption with constant size ciphertext. In: Cramer, R. (ed.) EUROCRYPT 2005. LNCS, vol. 3494, pp. 440–456. Springer, Heidelberg (2005)
6. Boneh, D., Franklin, M.: Identity-based encryption from the Weil pairing. In: Kilian, J. (ed.) CRYPTO 2001. LNCS, vol. 2139, pp. 213–229. Springer, Heidelberg (2001)
7. Boneh, D., Hamburg, M.: Generalized identity based and broadcast encryption scheme. In: Pieprzyk, J. (ed.) ASIACRYPT 2008. LNCS, vol. 5350, pp. 455–470. Springer, Heidelberg (2008)
8. Boneh, D., Katz, J.: Improved efficiency for CCA-secure cryptosystems built using identity based encryption. In: Menezes, A. (ed.) CT-RSA 2005. LNCS, vol. 3376, pp. 87–103. Springer, Heidelberg (2005)
9. Boneh, D., Waters, B.: Conjunctive, subset, and range queries on encrypted data. In: Vadhan, S.P. (ed.) TCC 2007. LNCS, vol. 4392, pp. 535–554. Springer, Heidelberg (2007)
10. Boyen, X., Waters, B.: Anonymous hierarchical identity-based encryption (without random oracles). In: Dwork, C. (ed.) CRYPTO 2006. LNCS, vol. 4117, pp. 290–307. Springer, Heidelberg (2006)
11. Canetti, R., Halevi, S., Katz, J.: Chosen-ciphertext security from identity-based encryption. In: Cachin, C., Camenisch, J.L. (eds.) EUROCRYPT 2004. LNCS, vol. 3027, pp. 207–222. Springer, Heidelberg (2004)

12. Cocks, C.: An identity based encryption scheme based on quadratic residues. In: Honary, B. (ed.) Cryptography and Coding 2001. LNCS, vol. 2260, pp. 360–363. Springer, Heidelberg (2001)
13. Gentry, C.: Practical identity-based encryption without random oracles. In: Vaudenay, S. (ed.) EUROCRYPT 2006. LNCS, vol. 4004, pp. 445–464. Springer, Heidelberg (2006)
14. Gentry, C., Halevi, S.: Hierarchical identity-based encryption with polynomially many levels. In: Reingold, O. (ed.) TCC 2009. LNCS, vol. 5444, pp. 437–456. Springer, Heidelberg (2009)
15. Gentry, C., Silverberg, A.: Hierarchical ID-based cryptography. In: Zheng, Y. (ed.) ASIACRYPT 2002. LNCS, vol. 2501, pp. 548–566. Springer, Heidelberg (2002)
16. Goyal, V., Pandey, O., Sahai, A., Waters, B.: Attribute-based encryption for fine-grained access control of encrypted data. In: ACM Conference on Computer and Communication Security 2006, pp. 89–98. ACM, New York (2006)
17. Katz, J., Sahai, A., Waters, B.: Predicate encryption supporting disjunctions, polynomial equations, and inner products. In: Smart, N.P. (ed.) EUROCRYPT 2008. LNCS, vol. 4965, pp. 146–162. Springer, Heidelberg (2008)
18. Lewko, A., Okamoto, T., Sahai, A., Takashima, K., Waters, B.: Fully secure functional encryption: Attribute-based encryption and (hierarchical) inner product encryption. In: Gilbert, H. (ed.) EUROCRYPT 2010. LNCS, vol. 6110, pp. 62–91. Springer, Heidelberg (2010)
19. Lewko, A.B., Waters, B.: New techniques for dual system encryption and fully secure HIBE with short ciphertexts. In: Micciancio, D. (ed.) TCC 2010. LNCS, vol. 5978, pp. 455–479. Springer, Heidelberg (2010)
20. Okamoto, T., Takashima, K.: Homomorphic encryption and signatures from vector decomposition. In: Galbraith, S.D., Paterson, K.G. (eds.) Pairing 2008. LNCS, vol. 5209, pp. 57–74. Springer, Heidelberg (2008)
21. Okamoto, T., Takashima, K.: Hierarchical predicate encryption for inner-products. In: Matsui, M. (ed.) ASIACRYPT 2009. LNCS, vol. 5912, pp. 114–231. Springer, Heidelberg (2009)
22. Ostrovsky, R., Sahai, A., Waters, B.: Attribute-based encryption with non-monotonic access structures. In: ACM Conference on Computer and Communication Security 2007, pp. 195–203. ACM, New York (2007)
23. Pirretti, M., Traynor, P., McDaniel, P., Waters, B.: Secure attribute-based systems. In: ACM Conference on Computer and Communication Security 2006, pp. 99–112. ACM, New York (2006)
24. Sahai, A., Waters, B.: Fuzzy identity-based encryption. In: Cramer, R. (ed.) EUROCRYPT 2005. LNCS, vol. 3494, pp. 457–473. Springer, Heidelberg (2005)
25. Shi, E., Waters, B.: Delegating capability in predicate encryption systems. In: Aceto, L., Damgård, I., Goldberg, L.A., Halldórsson, M.M., Ingólfsdóttir, A., Walukiewicz, I. (eds.) ICALP 2008, Part II. LNCS, vol. 5126, pp. 560–578. Springer, Heidelberg (2008)
26. Waters, B.: Efficient identity based encryption without random oracles. In: Cramer, R. (ed.) EUROCRYPT 2005. LNCS, vol. 3494, pp. 114–127. Springer, Heidelberg (2005)
27. Waters, B.: Ciphertext-policy attribute-based encryption: an expressive, efficient, and provably secure realization. ePrint, IACR, http://eprint.iacr.org/2008/290
28. Waters, B.: Dual system encryption: realizing fully secure IBE and HIBE under simple assumptions. In: Halevi, S. (ed.) CRYPTO 2009. LNCS, vol. 5677, pp. 619–636. Springer, Heidelberg (2009)

# Structure-Preserving Signatures and Commitments to Group Elements

Masayuki Abe[1], Georg Fuchsbauer[2], Jens Groth[3], Kristiyan Haralambiev[4,*], and Miyako Ohkubo[5,*]

[1] Information Sharing Platform Laboratories, NTT Corporation, Japan
abe.masyuki@lab.ntt.co.jp
[2] École normale supérieure, CNRS-INRIA, Paris, France
http://www.di.ens.fr/~fuchsbau
[3] University College London, UK
j.groth@ucl.ac.uk
[4] Computer Science Department, New York University, USA
kkh@cs.nyu.edu
[5] National Institute of Information and Communications Technology, Japan
m.ohkubo@nict.go.jp

**Abstract.** A modular approach for cryptographic protocols leads to a simple design but often inefficient constructions. On the other hand, ad hoc constructions may yield efficient protocols at the cost of losing conceptual simplicity. We suggest *structure-preserving* commitments and signatures to overcome this dilemma and provide a way to construct modular protocols with reasonable efficiency, while retaining conceptual simplicity.

We focus on schemes in bilinear groups that preserve parts of the group structure, which makes it easy to combine them with other primitives such as non-interactive zero-knowledge proofs for bilinear groups.

We say that a signature scheme is *structure-preserving* if its verification keys, signatures, and messages are elements in a bilinear group, and the verification equation is a conjunction of pairing-product equations. If moreover the verification keys lie in the message space, we call them *automorphic*. We present several efficient instantiations of automorphic and structure-preserving signatures, enjoying various other additional properties, such as *simulatability*. Among many applications, we give three examples: adaptively secure round-optimal blind signature schemes, a group signature scheme with efficient concurrent join, and an efficient instantiation of anonymous proxy signatures.

A further contribution is *homomorphic trapdoor commitments to group elements* which are also length reducing. In contrast, the messages of previous homomorphic trapdoor commitment schemes are exponents.

## 1 Introduction

The designer of cryptographic protocols faces a tension between choosing a modular approach using generic primitives that lead to a simple design but inefficient

---

* Work done while at NTT Information Sharing Platform Laboratories.

T. Rabin (Ed.): CRYPTO 2010, LNCS 6223, pp. 209–236, 2010.

protocols or using ad hoc constructions that sometimes yield efficient protocols at the cost of losing conceptual simplicity. Cryptographic protocols often combine general building blocks such as commitments, encryption, signatures, and zero-knowledge proofs. While modular design is useful to show feasibility of cryptographic tasks and also to illustrate a comprehensible framework, efficient instantiations are sometimes left as a next challenge. Some cryptographic tasks find "cleverly crafted" efficient solutions dedicated to their specific purposes. Nevertheless, modular construction makes implementing more complex primitives easier when the building blocks have reasonable instantiations. We suggest *structure-preserving* commitments and signatures to provide a way to construct modular protocols that retain conceptual simplicity and at the same time yield reasonable efficiency.

A classical way of realizing efficient instantiations is to rely on the random-oracle heuristic [BR93] for non-interactive zero-knowledge (NIZK) proofs—or to directly use *interactive* assumptions (like the LRSW assumption [LRSW00] and its variants, or "one-more" assumptions [BNPS03]). Due to a series of criticisms starting with [CGH98] more and more practical schemes are being proposed and proved secure in the *standard model* (i.e., without random oracles) and under *falsifiable* (and thus non-interactive) assumptions [Nao03]. All schemes given in this work satisfy these criteria.

STRUCTURE-PRESERVING SIGNATURES. The combination of NIZK proofs of knowledge and signatures appears frequently in privacy-protecting cryptographic protocols such as group signatures [BMW03, KY05, BSZ05, Gro07], blind signatures [Fis06, AO09], anonymous credentials [BCKL08, BCC+09], verifiably encrypted signatures [BGLS03, RS09], non-interactive group encryption [CLY09] and many more.

An efficient non-interactive proof system in the standard model, however, has been absent until recently. In [GS08], Groth and Sahai presented the first (and currently the only) efficient non-interactive proof system for a large class of statements over bilinear groups. The most interesting and widely used type is a conjunction of pairing-product equations (PPE) whose variables are elements of the bilinear group (cf. Section 2.4). A PPE consists of products of pairings applied to the variables and constants from the group. For this type of equations, the proofs are fully extractable which actually makes them proofs of knowledge. This renders GS proofs particularly interesting for modular protocol design.

Research on signature schemes that are compatible with GS proofs was initiated in [Gro06]. While the design goal is clear and simple, giving an efficient instantiation has proved hard for years. There are efficient signature schemes, e.g., [BB04, CL04, BCKL08, CKS09], whose verification predicates are pairing-product equations, but none of them have signatures *and* messages that exclusively consist of group elements. Since only group elements can be extracted from GS commitments, this entailed limited applicability of each scheme or stronger security notions such as *F-unforgeability* [BCKL08].

The desirable properties of a signature scheme enabling modular design together with GS proofs are the following:

1. the scheme is unforgeable against chosen-message attacks;
2. the verification keys, messages, and signatures are elements of a bilinear group; and
3. the verification predicate is a conjunction of pairing-product equations over the key, the message and the signature.

Note that this proscribes the use of hash functions, which usually play a central role in making signature schemes unforgeable against adaptive chosen-message attacks. We therefore call such a scheme *structure preserving*. If in addition its verification keys lie in the message space, we call it an *automorphic signature* (since it signs its *own* keys besides preserving *structure*).

Combined with GS proofs, structure-preserving signatures allow to prove knowledge of messages, signatures and/or verification keys without actually revealing them. Proving knowledge of signatures has been used in many construction of group signatures, anonymous proxy signatures, anonymous credentials, blind signatures, and others. Clearly, structure-preserving signatures combined with the GS proof system will allow to instantiate those constructions without resorting to interactive assumptions nor to the random-oracle model while maintaining a modular design.

For example, Fischlin [Fis06] presented the following framework for round-optimal blind signatures in the common reference string model. To obtain a signature from the signer, the user commits to a message and sends the commitment to the signer. Then, the signer signs the commitment and sends back the signature. The user produces a NIZK proof of knowledge of a commitment, an opening of the commitment to that message, and a signature on the commitment. This proof constitutes a blind signature for the message. Despite its simplicity, the scheme has not been instantiated efficiently in the standard model because it requires a signature scheme which signs trapdoor commitments and whose verification equations should mesh well with the GS proof system.

An application that also requires signing verification keys are *anonymous proxy signatures* [FP08]. They enable users to delegate (and redelegate) their signing rights to other users. A signature on behalf of another user (proxy signature) hides the identity of the proxy signer and possible intermediate delegators. Instantiating anonymous proxy signatures requires a signature scheme that is both GS compatible and enables users to sign other user's verification keys to delegate. Automorphic signatures can thus be used create a *delegation chain* of which the proxy signer proves knowledge using GS proofs.

TRAPDOOR COMMITMENTS TO GROUP ELEMENTS. A non-interactive commitment scheme allows to create a *commitment c* to a message $m$. The commitment *hides* the message, but we may later disclose $m$ and demonstrate that $c$ was a commitment to $m$ by revealing the randomness $r$ used when creating it. This is called *opening* the commitment. It is essential that once a commitment is made, it is *binding*, meaning that it is infeasible to find two openings of the same commitment to two different messages.

In this paper, we consider public-key trapdoor commitments [GQ88, Ped92] which are also *homomorphic* and *length reducing*. The former means that messages and commitments belong to abelian groups and if we multiply two commitments, we get a new commitment that contains the product of the two messages, whereas the latter requires that the commitment is shorter than the message.

An example would be a generalization of Pedersen commitments whose $n$ message components are in $\mathbb{Z}_p$. The public key consists of $n+1$ group elements $G_1, \ldots, G_n, H$ and a commitment to $(m_1, \ldots, m_n)$ is $C = H^r \prod_{i=1}^{n} G_i^{m_i}$. This scheme is length-reducing since a commitment to $n$ messages consists of only one group element, a feature that has been found useful in contexts such as mix-nets/voting, digital credentials, blind signatures, leakage-resilient one-way functions, and zero-knowledge proofs [FS01, Nef01, Bra99, KZ06, ADW09, Lip03].

Common to all the homomorphic trapdoor commitment schemes is that they are homomorphic with respect to *addition* in a ring or a field. However, in public-key cryptography we often work over groups that are not rings or fields and it is useful to commit to elements from such groups. Of course, if we know the discrete logarithms of the group elements we want to commit to, we can commit to them using Pedersen commitments. In general, we cannot expect to know the discrete logarithms of the messages though, leaving us with the open problem of constructing homomorphic trapdoor commitments to group elements.

Furthermore, such schemes could be combined with Pedersen commitments since commitments of the latter scheme are single group element. So, if we have a homomorphic trapdoor commitment scheme whose commitments to $O(n)$ group elements are of size $O(1)$, we can commit to $m \cdot n$ elements in $\mathbb{Z}_p$ using commitment schemes with public keys of total size $O(m+n)$. In comparison, when using only Pedersen commitments the public key would be of size $O(m \cdot n)$.

Finally, note that similarly to structure-preserving signatures, "GS compatibility" of a homomorphic trapdoor commitment scheme makes it a useful component in constructing more advanced zero-knowledge arguments or giving an efficient proof of knowledge of a message and/or an opening of a commitment.

### 1.1   Our Contribution

The paper presents three main results, all of them based on groups with a bilinear map. We focus on constructions in asymmetric bilinear groups whereas those in the symmetric setting are given in the full versions.

Firstly, we present a *homomorphic trapdoor commitment to group elements*. The commitments are perfectly hiding, computationally binding, and *length reducing*. An advantage of our commitment scheme is that the construction is very simple. The public key consists of $n+1$ group elements $(G_R, G_1, \ldots, G_n)$ from $\mathbb{G}_1$ and we commit to $M_1, \ldots, M_n \in \mathbb{G}_2$ by choosing $R \in \mathbb{G}_2$ at random and computing the commitment

$$C = e(G_R, R) \prod_{i=1}^{n} e(G_i, M_i) \ .$$

The commitment scheme is computationally binding under the *double pairing assumption*, which we show to be implied by decisional Diffie-Hellman assumption in $\mathbb{G}_1$. We extend our construction to commit to commitments as mentioned above and present an honest verifier zero-knowledge argument of knowledge of the contents of such commitments.

Next, we present the first instantiation of structure-preserving signatures on group elements. The messages consist of 2 group elements from an asymmetric bilinear group and signatures of 5 elements. Since the verification keys lie in the message space, the scheme is actually an *automorphic* signature. The scheme is proved secure under a variant of the *strong Diffie-Hellman* assumption [BB04], a "q-type" assumption which holds in the generic-group model. We combine the scheme with the GS proof system to construct the first efficient round-optimal blind signature scheme, which also remains automorphic. Moreover, we give a generic transformation from any automorphic signature scheme to one that signs message vectors of arbitrary length that leaves the keys unchanged.

Lastly, we present a structure-preserving signature scheme which signs vectors of general group elements. It has a *constant signature size* regardless of the message length. Our scheme does not rely on setup assumptions nor the messages having a specific structure, e.g. Diffie-Hellman pairs, like in the previous construction. While its verification key grows linearly in the maximum message length, it is possible to extend the scheme to sign unbounded-length messages at the cost of signatures growing proportionally to the length. This way, it is possible to make the signature automorphic albeit less efficient than the scheme above. The security is based on a novel strong, "q-type", assumption which is fairly complex. However, it has an *optimal quadratic security bound* in generic bilinear groups unlike the popular strong Diffie-Hellman assumption and its variations. Finally, we define the notion of *simulatable signatures* and give an efficient instantiation. It is defined in the common reference string (CRS) model and allows to create valid signatures using the trapdoor associated with the CRS.

APPLICATIONS: We illustrate the advantages of structure-preserving signature schemes by presenting several useful applications. The round-optimal blind signature scheme of Fischlin described before, which is secure in the universal-composability framework [Can01], is easily instantiated with such a building block in hand. The only extra tool we need is a trapdoor commitment on messages in $\mathbb{Z}_p$ whose commitments and openings are group elements. Such scheme is easily derived from the Pedersen commitment scheme when working in bilinear groups.

We then present a practical group signature scheme in the strongest security model [BSZ05] which moreover supports concurrent join. The construction follows a commonly used approach, based on the technique of proving knowledge of a signature.

Finally, we present the first efficient instantiation of anonymous proxy signatures (APS) in the standard model. Since automorphic signatures allow certifying public keys, delegation can be done by signing the delegatee's public key. An anonymous proxy signature is a GS proof of knowledge of a certification chain that starts at the original delegator and ends at the message. We also discuss how to strengthen the anonymity guarantees of APS. Using blind automorphic signatures, we give a protocol that hides the identity of the delegatee from the delegator. Moreover, using randomizability of GS proofs, we show how to maintain anonymity of the intermediate delegators w.r.t. the delegatee.

We note that since the announcement of our work, automorphic signatures have been used to construct the first fair blind signatures without random oracles [FV10] and non-interactively delegatable anonymous credentials [Fuc10]. The commitment schemes and the related assumptions have been used to construct efficient leakage-resilient signatures and one-way relations [DHLAW10]. Moreover, one can use the commitment schemes to reduce the communication complexity of Groth's [Gro09b] sub-linear size zero-knowledge argument for circuit satisfiability from $O(|C|^{\frac{1}{2}})$ group elements to $O(|C|^{\frac{1}{3}})$ group elements.

## 1.2   Related Work

There are many examples of homomorphic commitments. Homomorphic cryptosystems such as [ElG86, OU98, Pai99, BGN05] or Linear Encryption [BBS04] can be seen as homomorphic commitment schemes that are perfectly binding and computationally hiding. Commitments based on homomorphic encryption can be converted into computationally binding and perfectly hiding homomorphic commitments, see for instance the mixed commitments of Damgård and Nielsen [DN02] and the commitment schemes used by Groth, Ostrovsky and Sahai [GOS06], Boyen and Waters [BW06], Groth [Gro06] and Groth and Sahai [GS08]. Even in the perfectly hiding versions of these schemes the size of a commitment is larger than the size of a message though. This length increase follows from the fact that the underlying building block is a cryptosystem whose ciphertexts must be large enough to include the message.

There are also direct constructions of homomorphic trapdoor commitment schemes such as Guillou and Quisquater commitments [GQ88] and Pedersen commitments [Ped92]. The latter are one of the most used commitment schemes in the field of cryptography. They are perfectly hiding with a trapdoor and if the discrete-logarithm problem is hard they are computationally binding. There are many variants of the Pedersen commitment scheme. Fujisaki and Okamoto [FO97] and Damgård and Fujisaki [DF02] for instance suggest a variant where the messages can be arbitrary integers. However, none of the previous trapdoor commitment schemes has messages from a *group*.

Feasibility of structure-preserving signatures on group elements was first shown by Groth [Gro06], who presents a construction based on the decision linear assumption (DLIN) [BBS04]. While it is remarkable that the security can be based on a simple standard assumption, the scheme is not practical as signatures consist of hundreds of thousands of group elements. Based on the q-Hidden LRSW assumption, Green and Hohenberger [GH08] presented an efficient scheme that provides security against random-message attacks. An extension to chosen-message security is not known.

Independently of our work, Cathalo, Libert and Yung [CLY09] gave a practical scheme based on a combination of the *hidden strong Diffie-Hellman assumption*, the *flexible Diffie-Hellman assumption*, and the DLIN assumption. It was the first structure-preserving signature scheme to sign single group elements. However, it cannot sign its own verification keys and signatures on vectors grow linearly in their length.

An instantiation, though not practical, of anonymous proxy signatures was given in [FP09]. Moreover, they are similar to the *delegatable anonymous credentials* from [BCC+09] in that they provide mechanisms enabling users to prove possession of certain rights while remaining anonymous; and they consider re-delegation of received rights. The interactive delegation protocol for anonymous credentials provides even *mutual* anonymity of the delegator and the delegatee. The two instantiations rely on similar assumptions.

### 1.3 Merging Our Results

This paper combines the results of three different lines of research. In [Gro09a] Groth presented the first homomorphic trapdoor commitments to group elements which are moreover length-reducing (Section 3). Fuchsbauer [Fuc09] gave the first structure-preserving signatures on group elements and used it to efficiently implement round-optimal blind signatures in the standard model (Section 4). Abe, Haralambiev and Ohkubo [AHO10] gave the first constant-size signature scheme on vectors of general group elements. They also explicitly defined the notion of simulatable signatures, gave an efficient construction, and used it to implement UC-secure round-optimal blind signatures (Sections 5 and 6.1).

## 2 Preliminaries

### 2.1 Bilinear Groups

We will work in bilinear groups of the form $\Lambda = (p, \mathbb{G}_1, \mathbb{G}_2, \mathbb{G}_T, e, G, H)$ where

- $p$ is a $\lambda$-bit prime, where $\lambda$ is a security parameter
- $\mathbb{G}_1, \mathbb{G}_2, \mathbb{G}_T$ are order $p$ groups with efficiently computable group operations, membership tests and map $e : \mathbb{G}_1 \times \mathbb{G}_2 \to \mathbb{G}_T$
- $G$ generates $\mathbb{G}_1$, $H$ generates $\mathbb{G}_2$ and $e(G, H)$ generates $\mathbb{G}_T$
- The map $e$ is bilinear $\forall A \in \mathbb{G}_1 \forall B \in \mathbb{G}_2 \forall x, y \in \mathbb{Z}_p : e(A^x, B^y) = e(A, B)^{xy}$

To simplify notation, we define $\mathbb{G}_1^* = \mathbb{G}_1 \setminus \{1\}$, $\mathbb{G}_2^* = \mathbb{G}_2 \setminus \{1\}$ and $\mathbb{G}_T^* = \mathbb{G}_T \setminus \{1\}$.

### 2.2 Assumptions

We will work with bilinear groups generated by a probabilistic polynomial-time algorithm $\mathcal{G}$ that takes the security parameter as input. The schemes we present will rely on one or more of the following computational assumptions about the bilinear groups generated by $\mathcal{G}$. We note right away that the assumptions imply $\mathbb{G}_1 \neq \mathbb{G}_2$ and furthermore some of them imply that we are working in so called type III bilinear groups [GPS08] where there are no efficiently computable non-trivial homomorphisms between the two base groups $\mathbb{G}_1$ and $\mathbb{G}_2$. We refer to the full papers for schemes that work in type I and type II bilinear groups.

**Variants of DDH and CDH.** The *decisional Diffie-Hellman (DDH) problem* in a group $\mathbb{G}$ is, given $(G, G^a, G^b, G^c)$, to decide whether $c = ab$. The *symmetric external Diffie-Hellman (SXDH) assumption* in a bilinear group states that DDH is hard in both groups.

**Assumption 1 (SXDH).** *For* $\Lambda = (p, \mathbb{G}_1, \mathbb{G}_2, \mathbb{G}_T, e, G, H) \leftarrow \mathcal{G}(1^\lambda)$, *the decisional Diffie-Hellman assumption holds in both* $\mathbb{G}_1$ *and* $\mathbb{G}_2$.

The *2-out-of-3 CDH* assumption [KP06] states that given $(G, G^a, H)$, it is hard to output $(G^r, H^{ar})$ for an arbitrary $r \neq 0$. To break the *Flexible CDH* assumption [LV08, CLY09], an adversary must additionally compute $G^{ar}$. We further weaken the assumption by defining a solution as $(G^r, G^{ar}, H^r, H^{ar})$, and generalize it to asymmetric groups by letting $G \in \mathbb{G}_1$ and $H \in \mathbb{G}_2$. The *asymmetric weak flexible CDH* is defined as follows:

**Assumption 2 (AWF-CDH).** *Let* $G \in \mathbb{G}_1$, $H \in \mathbb{G}_2$ *and* $a \in \mathbb{Z}_p$ *be random. Given* $(G, A = G^a, H)$, *it is hard to output* $(G^r, G^{ar}, H^r, H^{ar})$ *with* $r \neq 0$, *i.e., a tuple* $(R, M, S, N)$ *that satisfies*

$$e(A, S) = e(M, H) \qquad e(M, H) = e(G, N) \qquad e(R, H) = e(G, S) \quad (1)$$

Given a DDH instance $(G, G^a, G^b, G^c)$, solving AWF-CDH for $(G, G^a, H)$ yields $(G^r, G^{ar}, H^r, H^{ar})$; thus $G^c = G^{ab}$ can be checked by $e(G^{ab}, H^r) = e(G^b, H^{ar})$. We have thus

**Lemma 1.** *The AWF-CDH assumption holds if the decisional Diffie-Hellman assumption is hard in* $\mathbb{G}_1$.

**The Double Pairing Assumption.** The double pairing problem is given random $G_R, G_T \in \mathbb{G}_1$ to find non-trivial $R, S \in \mathbb{G}_2$ satisfying $e(G_R, R)e(G_T, T) = 1$.

**Assumption 3 (DBP).** *For all nonuniform polynomial-time adversaries* $\mathcal{A}$

$$\Pr\left[\Lambda \leftarrow \mathcal{G}(1^\lambda);\ G_R, G_T \leftarrow \mathbb{G}_1;\ (R, T) \leftarrow \mathcal{A}(\Lambda, G_R, G_T) : \right.$$
$$\left. (R, T) \in \mathbb{G}_2^* \times \mathbb{G}_2^* \ \wedge \ e(G_R, R)e(G_T, T) = 1\right] = \mathrm{negl}(\lambda).$$

We show in the full papers the following lemma:

**Lemma 2.** *The double pairing assumption holds if the decisional Diffie-Hellman assumption is hard in* $\mathbb{G}_1$.

The reverse double pairing problem, where the base groups are interchanged and the challenge is to find a non-trivial pair $(R, S) \in \mathbb{G}_1^2$ is defined analogously.

Next, observe that given an answer to an instance of the DBP problem, one can easily yield more answers. We eliminate such possibility by multiplying random pairings to both sides of the equation. As one of those stays the same in all instances, whereas the other, $e(V, W)$, changes in each instance, the intuition is that it would be hard to combine $e(V, W)$ and $e(V', W')$ into one equivalent pairing $e(V'', W'')$ — we call such a pairing *flexible* as it can be easily randomized and, when relations with respect to the same base is known, combined with another. Also, to make the assumption valid, we make a system of two such equations and require that their solutions share a common element, $Z$.

**Assumption 4 (Simultaneous Flexible Pairing Assumption ($q$-SFP)).**
*Let $\Lambda$ be a bilinear groups setup and let $G_Z$, $F_Z$, $G_R$, and $F_U$ be random generators of $\mathbb{G}_1$. Let $(A, \tilde{A}), (B, \tilde{B})$ be random pairs in $\mathbb{G}_1 \times \mathbb{G}_2$. For $j = 1, \ldots, q$, let $R_j = (Z, R, S, T, U, V, W)$ that satisfies*

$$e(A, \tilde{A}) = e(G_Z, Z)\, e(G_R, R)\, e(S, T) \quad and \tag{2}$$

$$e(B, \tilde{B}) = e(F_Z, Z)\, e(F_U, U)\, e(V, W). \tag{3}$$

*Given $(\Lambda, G_Z, F_Z, G_R, F_U, A, \tilde{A}, B, \tilde{B})$ and uniformly chosen $R_1, \ldots, R_q$, it is hard to find $(Z^\star, R^\star, S^\star, T^\star, U^\star, V^\star, W^\star)$ that fulfill relations (2) and (3) under the restriction that $Z^\star \neq 1$ and $Z^\star \neq Z \in R_j$ for every $R_j$.*

We also show that the SFP assumption can be justified and has an optimal bound in the generic bilinear group model.

**Lemma 3.** *For any generic algorithm $\mathcal{A}$, the probability that $\mathcal{A}$ breaks SFP with $\ell$ group operations and pairings is bound by $\mathcal{O}(q^2 + \ell^2)/p$.*

**A variant of the $q$-strong Diffie Helmman assumption.** The *$q$-strong Diffie-Hellman* (SDH) assumption [BB04] implies hardness of the following two problems in bilinear groups [FPV09]:

1. Given $G, G^x$ and $q - 1$ pairs $(G^{\frac{1}{x+c_i}}, c_i)$, output a new pair $(G^{\frac{1}{x+c}}, c)$.
2. Given $G, K, G^x, \left((K \cdot G^{v_i})^{\frac{1}{x+c_i}}, c_i, v_i\right)_{i=1}^{q-1}$, output a new $((K \cdot G^v)^{\frac{1}{x+c}}, c, v)$.

Boyen and Waters [BW07] define the *hidden* SDH assumption which states that the first problem is hard when the pairs are substituted with triples of the form $(G^{1/(x+c_i)}, G^{c_i}, H^{c_i})$, for a fixed $H$. Analogously, Fuchsbauer et al. [FPV09] define the *double hidden* SDH (DHSDH) by giving the scalars in the second problem as exponentiations of two group elements. We adapt DHSDH to asymmetric groups by giving generators $G, F, K \in \mathbb{G}_1$ and $H \in \mathbb{G}_2$; the elements $c_i$ and $v_i$ are given as $(F^{c_i}, H^{c_i})$ and $(G^{v_i}, H^{v_i})$. Due to the pairing, a tuple can thus be effectively verified. The assumption holds in the generic-group model [Sho97] for both asymmetric and symmetric groups [Fuc09] and falls in the generalized "Uber-Assumption" family [Boy08].

**Assumption 5 ($q$-ADH-SDH).** *Let $G, F, K \in \mathbb{G}_1$, $H \in \mathbb{G}_2$ and $x, c_i, v_i \in \mathbb{Z}_p$ be random. Given $(G, F, K, X = G^x; H, Y = H^x)$ and*

$$\left(A_i = (K \cdot G^{v_i})^{\frac{1}{x+c_i}}, \; C_i = F^{c_i}, \; D_i = H^{c_i}, \; V_i = G^{v_i}, \; W_i = H^{v_i}\right),$$

*for $1 \leq i \leq q - 1$, it is hard to output a new tuple $((K \cdot G^v)^{\frac{1}{x+c}}, F^c, H^c, G^v, H^v)$ with $(c, v) \neq (c_i, v_i)$ for all $i$.*

Note that a tuple $(A, C, D, V, W)$ of this form satisfies the following equations:

$$e(A, Y \cdot D) = e(K \cdot V, H) \qquad e(C, H) = e(F, D) \qquad e(V, H) = e(G, W) \qquad (4)$$

### 2.3  Digital Signatures

A digital signature scheme **Sig** = (Setup, KeyGen, Sign, Verify) consists of the following algorithms: Setup outputs system parameters; KeyGen outputs a pair $(vk, sk)$ of verification and signing keys; and Sign$(sk, M)$ outputs a signature $\sigma$, which is verified by Verify$(vk, M, \sigma)$. Signatures are *existentially unforgeable under chosen-message attack* (EUF-CMA) [GMR88] if no adversary, given $vk$ and a signing oracle for messages of its choice, can output a pair $(M, \sigma)$ s.t. $M$ was never queried and Verify$(vk, M, \sigma) = 1$.

Signatures are *strongly* EUF-CMA (sEUF-CMA) if no adversary can output a valid pair $(M, \sigma)$ such that $(M, \sigma) \neq (M_i, \sigma_i)$ for all $i$, with $M_i$ being the $i$-th oracle query and $\sigma_i$ the response.

### 2.4  SXDH Groth-Sahai Proofs for Pairing-Product Equations

One of the main motivations of structure-preserving signatures is to combine them with Groth-Sahai (GS) proofs [GS08], in particular witness-indistinguishable (WI) proofs of satisfiability of *pairing-product equations* (PPE). A PPE over variables $X_1, \ldots, X_m \in \mathbb{G}_1, Y_1, \ldots, Y_n \in \mathbb{G}_2$ is an equation of the form

$$\prod_{i=1}^{n} e(A_i, Y_i) \prod_{i=1}^{m} e(X_i, B_i) \prod_{i=1}^{m} \prod_{j=1}^{n} e(X_i, Y_j)^{\gamma_{i,j}} = \mathbf{t}_T , \qquad (E)$$

determined by $A_j \in \mathbb{G}_1, B_i \in \mathbb{G}_2, \gamma_{i,j} \in \mathbb{Z}_p$, and $\mathbf{t}_T \in \mathbb{G}_T$.

Groth and Sahai define an extractable commitment scheme for group elements. The setup algorithm is given a bilinear group and outputs a commitment key $ck \in \mathbb{G}_1^4 \times \mathbb{G}_2^4$. A commitment Com$(ck, X, \rho)$ to $X \in \mathbb{G}_i$ using randomness $\rho \in \mathbb{Z}_p^2$ is in $\mathbb{G}_i^2$ (for $i = 1, 2$). These commitments are perfectly binding and given an *extraction key*, the committed values can be recovered.

A *proof of satisfiability* of a PPE is constructed as follows. First, make commitments to the satisfying witness $(X_1, \ldots, X_m, Y_1, \ldots, Y_n)$. Then make a *proof* $\phi$ that the committed values satisfy the equation, using the values and the randomness of the commitments. The proofs, which are in $\mathbb{G}_1^4 \times \mathbb{G}_2^4$, are perfectly sound: if a proof passes verification for a set of commitments then the committed (and extractable) values satisfy the equation.

There is an alternative setup that outputs keys $ck^*$ which lead to commitments and proofs that are equally distributed for all witnesses. Under SXDH, these keys are indistinguishable from original keys; witness indistinguishability of GS proofs follows thus from SXDH.

Note that due to extractability, a proof of satisfiability is actually a non-interactive *proof of knowledge* of a witness; we will write thus

$$\mathsf{NIPK}\{(X_1, .., X_m, Y_1, .., Y_n) : \prod e(A_i, Y_i) \prod e(X_i, B_i) \prod \prod e(X_i, Y_j)^{\gamma_{i,j}} = \mathbf{t}_T\}$$

and PKVrf for the verification algorithm.

If for a signature scheme, public keys, messages and signatures are group elements that are verified by checking PPEs, we can commit to (encrypt) keys, messages and/or signatures and prove validity of the committed values using GS proofs.

**Randomization.** Groth-Sahai commitments can be randomized, in particular, given $\mathbf{c} = \mathsf{Com}(ck, X, \rho)$, one can compute $\mathsf{Com}(ck, X, \rho + \rho')$ for any $\rho'$ without knowledge of $X$ or $\rho$. Moreover, given commitments and a proof $\phi$ that the committed values satisfy a PPE, we can randomize the commitments and adapt $\phi$ to the randomized commitments [BCC+09]. WI implies that a randomized proof is indistinguishable from a proof computed with a different witness.

## 3    Commitments

A non-interactive commitment scheme consists of three polynomial-time algorithms $(\mathcal{G}, \mathcal{G}_{com}, com)$. $\mathcal{G}$ is a probabilistic polynomial-time setup algorithm that takes as input the security parameter $\lambda$ and outputs some setup information $\Lambda$; in our commitment scheme $\mathcal{G}$ will be a bilinear group generator. $\mathcal{G}_{com}$ is a probabilistic polynomial-time algorithm that takes as input the setup $\Lambda$ and and generates a public commitment key $ck$ and a trapdoor key $tk$. The commitment key $ck$ specifies a message space $\mathcal{M}_{ck}$, a randomizer space $\mathcal{R}_{ck}$ and a commitment space $\mathcal{C}_{ck}$. We assume it is easy to verify membership of the message space, randomizer space and the commitment space and it is possible to sample randomizers uniformly at random from $\mathcal{R}_{ck}$. The algorithm $\mathsf{Com}$ takes as input the commitment key $ck$, a message $m$ from the message space, a randomizer $r$ from the randomizer space and outputs a commitment $c$ in the commitment space. We call a message-randomizer pair an opening. Anybody with an opening and a commitment can check whether the commitment is a commitment to the message specified in the opening.

A commitment scheme should be binding, which means it is infeasible to find two openings of the same commitment to two different messages. A commitment scheme should also be hiding such that the commitment does not disclose anything about the message. Our commitment scheme is a trapdoor commitment scheme, which makes it hiding in a very strong sense. The commitment has a trapdoor opening algorithm $\mathsf{Topen}$ that takes the trapdoor key, an opening of a commitment and a message and outputs a randomizer such that the message and the randomizer constitute a new opening of the commitment.

We will now describe our commitment scheme. The commitment scheme will have message space $\mathcal{M}_{ck} = \mathbb{G}_2^n$, randomizer space $\mathcal{R}_{ck} = \mathbb{G}_2$ and commitment space $\mathcal{C}_{ck} = \mathbb{G}_T$. In other words, we can commit to an $n$-tuple of base group elements with a commitment that consists of a single target group element.

**Setup:** On input $1^\lambda$ return $\Lambda = (p, \mathbb{G}_1, \mathbb{G}_2, \mathbb{G}_T, e, G, H) \leftarrow \mathcal{G}(1^\lambda)$.

**Key generation:** On input $\Lambda$ pick $G_R \leftarrow \mathbb{G}_1^*$ and $x_1, \ldots, x_n \leftarrow \mathbb{Z}_p$ and set $G_1 = G_R^{x_1}, \cdots, G_n = G_R^{x_n}$. The commitment and trapdoor keys are

$$ck = (\Lambda, G_R, G_1, \ldots, G_n) \quad \text{and} \quad tk = (ck, x_1, \ldots, x_n).$$

**Commitment:** Using commitment key $ck$ on input message $(M_1, \ldots, M_n) \in \mathbb{G}_2^n$ pick randomizer $R \leftarrow \mathbb{G}_2$. The commitment is given by

$$C = e(G_R, R) \prod_{i=1}^{n} e(G_i, M_i) \ .$$

**Trapdoor opening:** On a commitment $C \in \mathbb{G}_T$ with opening $(M_1, \ldots, M_n, R) \in \mathbb{G}_2^n \times \mathbb{G}_2$ and another message $(M_1', \ldots, M_n') \in \mathbb{G}_2^n$ use the trapdoor key $tk$ to compute the trapdoor randomizer $R' = R \prod_{i=1}^{n} (M_i/M_i')^{x_i}$. This gives us a trapdoor opening $(M_1', \ldots, M_n', R')$ satisfying

$$C = e(G_R, R) \prod_{i=1}^{n} e(G_i, M_i) = e(G_R, R') \prod_{i=1}^{n} e(G_i, M_i') \ .$$

The commitment scheme has several useful properties. The commitment is length-reducing, since a commitment to a tuple of messages yields a commitment consisting of a single target group element. The commitment scheme is homomorphic since multiplying two commitments yields a commitment to the entry-wise product of the messages, i.e.,

$$e(G_R, R) \prod_{i=1}^{n} e(G_i, M_i) \cdot e(G_R, R') \prod_{i=1}^{n} e(G_i, M_i') = e(G_R, RR') \prod_{i=1}^{n} e(G_i, M_i M_i').$$

The commitment scheme is perfectly hiding since for all messages $(M_1, \ldots, M_n) \in \mathbb{G}_2^n$ the commitment procedure returns a uniformly random commitment $C \in \mathbb{G}_T$ and therefore no information is leaked about the commitment. Indeed, with the trapdoor key we can even take a commitment and its opening and create an opening to any other message. Finally, we prove in the full papers that the commitment scheme is computationally binding if the double pairing assumption holds for the bilinear group generator $\mathcal{G}$. We summarize these properties in the theorem below, which we prove in the full papers.

**Theorem 1.** $(\mathcal{G}, \mathcal{G}_{\mathrm{com}}, \mathsf{Com}, \mathsf{Topen})$ *described above is a homomorphic, perfectly hiding trapdoor commitment scheme; and assuming the double pairing assumption holds for $\mathcal{G}$ the commitment scheme is computationally binding.*

It is straightforward to construct a similar type of commitment scheme for tuples in $\mathbb{G}_1^n$ using the reverse double pairing assumption.

**Committing to commitments.** The defining characteristic of our commitment scheme is that we commit to base group elements as opposed to field elements. This opens up new applications for commitment schemes. As a simple example, we can for instance construct commitments to commitments. Recall that Pedersen commitments to tuples $(m_1, \ldots, m_n) \in \mathbb{Z}_p^n$ are of the form $C = H^r \prod_{j=1}^{n} H_j^{m_j}$. Each Pedersen commitment is a group element, and we can commit to many Pedersen commitments using our commitment scheme. Combining the two commitment schemes we can commit to $n^2$ field elements from

$\mathbb{Z}_p$. Since both Pedersen commitments and our commitments are homomorphic, the combined commitment scheme is also homomorphic. It also preserves the trapdoor opening property and is perfectly hiding. A commitment consists of a single group element in $\mathbb{G}_T$ and the commitment key consists of approximately $2n$ group elements, so unlike the Pedersen commitment we have a commitment key that is much smaller than the messages.

# 4  Automorphic Signatures

For elaborate applications, Groth-Sahai compatibility of a signature scheme is not sufficient; in addition, the verification keys have to lie in the message space. This enables constructions of *certification chains* (sequences of public keys linked by certificates from one key on the next one), which can be anonymized by GS proofs, as required by anonymous proxy signatures (see Section 6.3) and delegatable anonymous credentials. We call such a scheme an *automorphic signature*, as it is able to sign its *own* keys and it is *structure* preserving.

**Definition 1.** *An* automorphic signature *over* $\Lambda = (p, \mathbb{G}_1, \mathbb{G}_2, \mathbb{G}_T, e, G, H)$ *is an EUF-CMA secure signature whose verification keys lie in the message space. Moreover, the messages and signatures consist of elements from* $\mathbb{G}_1$ *and* $\mathbb{G}_2$, *and the verification predicate is a conjunction of pairing-product equations.*

The trick that enables an efficient instantiation of automorphic signatures is to define a message (and thus a verification key) as a *pair* of group elements of the form $(G^v, H^v)$. Hence, the message space is the set of *Diffie-Hellman pairs* $\mathcal{DH} = \{(G^v, H^v) \,|\, v \in \mathbb{Z}_p\}$. In Assumption 5, we could interpret $G, F, K, H$ as parameters, $(X, Y)$ as the public key, $(V, W)$ as the message and $(A, C, D)$ as the signature—since a signer holding the secret key $x$ can choose $c$ and produce $(A, C, D)$ without knowing $v$. ADH-SDH states that these signatures are unforgeable when the adversary gets $q - 1$ signatures on *random* messages.

   To make the scheme secure against chosen-message attacks, we interpret $G^v$ in the definition of $A$ as a *trapdoor commitment* to the message $(M, N)$. The key is an element $T := G^t \in \mathbb{G}_1$, where $t$ is the trapdoor, and a commitment to $(M, N)$ is defined as $V := T^r \cdot M$ with opening $(G^r, H^r)$. AWF-CDH implies that the commitments are computationally binding. Trapdoor opening requires knowledge of $W$ such that $(V, W) \in \mathcal{DH}$: for any $(V, W), (M, N) \in \mathcal{DH}$, a valid opening is $((V \cdot M)^{-t}, (W \cdot N)^{-t})$.

   The final signature will be $(A, C, D)$ together with the opening of the commitment $(R, S)$; a signature is thus in $\mathbb{G}_1^3 \times \mathbb{G}_2^2$.

## 4.1  Instantiation

Our automorphic signature scheme **Sig** = (Setup, KeyGen, Sign, Verify) is defined as follows.

**Setup:** On input $1^\lambda$ run $\Lambda = (p, \mathbb{G}_1, \mathbb{G}_2, \mathbb{G}_T, e, G, H) \leftarrow \mathcal{G}(1^\lambda)$, choose random elements $F, K, T \in \mathbb{G}_1$ and output the parameters $pp := (\Lambda, F, K, T)$. The message space is $\mathcal{DH} := \{(G^m, H^m) \,|\, m \in \mathbb{Z}_p\}$.

**Key generation:** On input $pp$ choose $x \leftarrow \mathbb{Z}_p$ and return the verification key $vk := (G^x, H^x)$ and the signing key $sk := x$.

**Signing:** On input the parameters $pp$, a secret key $x$ and a message $(M, N) \in \mathcal{DH}$, choose $c, r \leftarrow \mathbb{Z}_p$ and return

$$A := (K \cdot T^r \cdot M)^{\frac{1}{x+c}} \quad C := F^c \quad D := H^c \quad R := G^r \quad S := H^r$$

**Verification:** On input $pp$, a public key $(X, Y)$ and a message $(M, N)$, both in $\mathcal{DH}$, and a signature $(A, C, D, R, S)$, return 1 if

$$e(A, Y \cdot D) = e(K \cdot M, H) \, e(T, S) \qquad \begin{aligned} e(C, H) &= e(F, D) \\ e(R, H) &= e(G, S) \end{aligned} \qquad (5)$$

**Theorem 2.** *Under ADH-SDH and AWF-CDH,* **Sig** *is strongly unforgeable against chosen-message attacks.*

We refer to the full version [Fuc09] for a proof. Note that the scheme can also be instantiated for $\mathbb{G}_1 = \mathbb{G}_2$. Our scheme (and the blind signature scheme in the next section) can also be used to sign bit strings if we assume a collision-resistant hash function $\mathsf{Hash} \colon \{0, 1\}^* \to \mathbb{Z}_p$: before signing a message or verifying a signature, we map $m \in \{0, 1\}^*$ to $(M, N) := (G^{\mathsf{Hash}(m)}, H^{\mathsf{Hash}(m)}) \in \mathcal{DH}$.

### 4.2 Automorphic Blind Signatures

We now show how to combine automorphic signatures with the Groth-Sahai proof system to construct the first round-optimal blind signature scheme, satisfying standard security requirements as in [Oka06] (see Section 6.1 for a universally composable scheme). Similarly to Fischlin's generic construction, our blind signatures are defined as a proof of knowledge of a signature from an underlying scheme, which perfectly hides the signature. We thus only have to ensure that the signer does not learn the message while signing. In our scheme the user sends a *randomization* of the message, on which the signer makes a "pre-signature". By adapting the randomness, the user can retrieve a signature *on the message* (rather than on a commitment for which the user has to prove knowledge of the opening, as in Fischlin's construction). This increases useability of our blind signatures for applications (cf. Section 6.3) and also makes them shorter.

To obtain a blind signature on $(M, N)$, the user randomly picks $\rho \leftarrow \mathbb{Z}_p$ and *blinds* $M$ by the factor $T^\rho$. In addition to $U := T^\rho \cdot M$, she sends a GS proof of knowledge of $(M, N, G^\rho, H^\rho)$. The signer now formally produces a signature[1] on $U$, for which we have $A = (K \cdot T^r \cdot U)^{1/(x+c)} = (K \cdot T^{r+\rho} \cdot M)^{1/(x+c)}$; thus $A$ is the first component of a signature on $(M, N)$ with randomness $r + \rho$. The user can complete the signature by adapting randomness $r$ to $r + \rho$ in the other components. The blind signature is a GS proof of knowledge of this signature.

---

[1] Note that the user does *not* obtain a signature on $U$ (unless $U = M$), since it is not an element of the message space; to produce $(U, H^{\log_G U}) \in \mathcal{DH}$, the user would have to break AWF-CDH.

---

**Obtain**$\big((pp', ck), vk, (M, N)\big)$. Choose $\rho \leftarrow \mathbb{Z}_p$, set $P := G^\rho$, $Q := H^\rho$, and send:

- $U := T^\rho \cdot M$
- $\phi := \mathsf{NIPK}\{(M, N, P, Q) : e(M, H) = e(G, N)$
$$\wedge\ e(P, H) = e(G, Q) \wedge\ e(T, Q)\, e(M, H) = e(U, H)\}$$

**Issue**$\big((pp', ck), x\big)$. If $\phi$ is valid, choose $c, r \leftarrow \mathbb{Z}_p$ and send:

$$A := (K \cdot T^r \cdot U)^{\frac{1}{x+c}} \qquad C := F^c \qquad D := H^c \qquad R' := G^r \qquad S' := H^r$$

**Obtain** sets $R := R' \cdot P$, $S := S' \cdot Q$. If $(A, C, D, R, S)$ is valid on $(M, N)$ under $vk$, it outputs

$$\sigma := \mathsf{NIPK}\big\{(A, C, D, R, S) : \mathsf{Verify}_{\mathsf{Sig}}\big(pp, vk, (M, N), (A, C, D, R, S)\big)\big\} \ .$$

---

**Fig. 1.** Two-move blind signing protocol

Our blind signature scheme **BSig** $=$ (Setup, KeyGen, Obtain, Issue, Verify) is defined as follows.

**Setup:** On input $1^\lambda$ run the setup algorithms for **Sig** and for Groth-Sahai proofs; return the respective outputs $pp'$ and $ck$ as parameters $pp$.

**Key generation:** The message space and key generation are defined as for **Sig**.

**Signature issuing:** The protocol consists of interactive algorithms Obtain, run by the user, and Issue, run by the signer. Obtain has inputs $pp$, the signer's verification key $vk$ and a message $(M, N) \in \mathcal{DH}$. Issue has inputs $pp$ and the signing key $x$. The protocol is given in Figure 1.

**Verification:** On input $pp$, a verification key $vk$, a message $(M, N) \in \mathcal{DH}$ and a signature $\sigma$, return 1 if $\sigma$ is a valid Groth-Sahai proof, i.e.,

$$\mathsf{PKVrf}\{\sigma : \mathsf{Verify}_A(vk, (M, N), \cdot)\} = 1 \ .$$

**Theorem 3.** *Under ADH-SDH and SXDH, scheme* **BSig** *is an unforgeable blind-signature scheme.*

Using soundness of Groth-Sahai proofs, unforgeability is shown by reduction to the unforgeability of **Sig**, which holds under ADH-SDH and SXDH (the latter implies AWF-CDH). Under SXDH, the user's message $(U, \phi)$ computationally hides $(M, N)$ and the blind signature hides what the signer sends in the issuing; together this can be shown to imply blindness. See [Fuc09] for a formal proof of Theorem 3.

The round complexity of the scheme is optimal. A blind signature consists of commitments to $(A, C, D, R, S)$ in $\mathbb{G}_1^6 \times \mathbb{G}_2^4$ and GS proofs, which are in $\mathbb{G}_1^4 \times \mathbb{G}_2^4$, for 3 equations. A blind signature is thus in $\mathbb{G}_1^{18} \times \mathbb{G}_2^{16}$, the two messages sent during issuing are in $\mathbb{G}_1^{17} \times \mathbb{G}_2^{16}$ and $\mathbb{G}_1^3 \times \mathbb{G}_2^2$, respectively. Note that the scheme remains automorphic since GS proofs consists of group elements and are verified by checking pairing-product equations.

### 4.3  Automorphic Signatures on Message Vectors

In order to sign vectors of messages of arbitrary length, we proceed as follows. We first show how to transform any signature scheme whose message space forms an algebraic group (and contains the public-key space) into one that signs 2 messages at once—if we exclude the neutral element from the message space of the transform. A signature on a message pair will contain 3 signatures (of the original scheme) on different *products* of the components. Note that $\mathcal{DH}$, the message space of **Sig**, is a group when the group operation is defined as component-wise multiplication.

We then give a straightforward generic transformation from any scheme signing 2 messages (and whose verification keys lie in the message space) to one signing message vectors of arbitrary length (Definition 3). Both transformations do not modify setup and key generation and they are invariant w.r.t. the structure of verification; in particular, if the verification predicate of the original scheme is a conjunction of PPEs then so is that of the transform.

**Definition 2.** *Let* **Sig** $=$ (Setup, KeyGen, Sign, Verify) *be a signature scheme whose message space* $(\mathcal{M}, \cdot)$ *is an algebraic group that contains the verification keys. The* pair transform *of* **Sig** *with message space* $\mathcal{M}^* \times \mathcal{M}^*$ *is defined as* **Sig**$'$ $=$ (Setup, KeyGen, Sign$'$, Verify$'$) *with*

Sign$'(sk, (M_1, M_2))$: *Set* $(vk_0, sk_0) \leftarrow$ KeyGen *and return*

$$\sigma := \big( vk_0,\ \mathsf{Sign}(sk, vk_0),$$
$$\mathsf{Sign}(sk_0, M_1),\ \mathsf{Sign}(sk_0, M_1 \cdot M_2),\ \mathsf{Sign}(sk_0, M_1 \cdot M_2^3) \big)\ .$$

Verify$'\big(vk, (M_1, M_2), (vk_0, \sigma_0, \sigma_1, \sigma_2, \sigma_3)\big)$: *Return 1 if all of the following are 1:*

Verify$(vk, vk_0, \sigma_0)$

Verify$(vk_0, M_1, \sigma_1)$     Verify$(vk_0, M_1 \cdot M_2, \sigma_2)$     Verify$(vk_0, M_1 \cdot M_2^3, \sigma_3)$

**Theorem 4.** *If* **Sig** *is EUF-CMA secure then so is* **Sig**$'$.

**Definition 3.** *Let* **Sig** $=$ (Setup, KeyGen, Sign, Verify) *be a signature scheme with message space* $\mathcal{M} \times \mathcal{M}$, *such that* $\mathcal{M}$ *contains the verification keys. Assume an efficiently computable injection* $I \colon \{1, \ldots, |\mathcal{M}|\} \to \mathcal{M}$. *The* vector transform *of* **Sig** *is defined as* **Sig**$''$ $=$ (Setup, KeyGen, Sign$''$, Verify$''$) *with*

Sign$''(sk, (M_1, \ldots, M_n))$: *Set* $(vk_0, sk_0) \leftarrow$ KeyGen *and return*

$$\sigma := \big( vk_0,\ \mathsf{Sign}(sk, vk_0, I(n)),$$
$$\mathsf{Sign}(sk_0, M_1, I(1)), \ldots, \mathsf{Sign}(sk_0, M_n, I(n))) \big)\ .$$

Verify$''\big(vk, (M_1, \ldots, M_n), (vk_0, \sigma_0, \sigma_1, \ldots, \sigma_n)\big)$: *Return 1 if the following are 1:*

Verify$\big(vk, (vk_0, I(n)), \sigma_0\big)$     Verify$\big(vk_0, (M_i, I(i)), \sigma_i\big)$ *(for all* $1 \leq i \leq n$)

**Theorem 5.** *If* **Sig** *is EUF-CMA secure then so is* **Sig**$''$.

We refer to [Fuc09] for proofs of Theorems 4 and 5 where we also discuss why the construction in Definition 2 is optimal and why it seems somehow hard to construct a generic vector transform directly.

# 5    Signatures on Vectors of Group Elements

In this section, we present the first *constant-size* structure-preserving signature scheme for messages of general bilinear groups elements. We start by describing useful randomization techniques, followed by the scheme description and various extensions. Full details, as well as the byproduct of several trapdoor commitment schemes, can be found in [AHO10].

## 5.1    Randomization Techniques

We introduce techniques that randomize elements in a pairing or a pairing product without changing their value in $\mathbb{G}_T$. Let $(p, \mathbb{G}_1, \mathbb{G}_2, \mathbb{G}_T, e, G, H) \leftarrow \mathcal{G}(1^\lambda)$.

**Inner Randomization** $(X', Y') \leftarrow \mathsf{Rand}(X, Y)$: A pairing $A = e(X, Y) \neq 1$ is randomized as follows. Choose $\gamma \leftarrow \mathbb{Z}_p^*$ and let $(X', Y') = (X^\gamma, Y^{1/\gamma})$. It then holds that $(X', Y')$ distributes uniformly over $\mathbb{G}_1 \times \mathbb{G}_2$ under the condition of $A = e(X', Y')$. If $A = 1$, then first flip a coin and pick $e(1, 1)$ with probability $1/(2p - 1)$. If it is not selected, flip a coin and pick either $e(1, X)$ or $e(X, 1)$ with probability $1/2$. Then select $X$ uniformly from the corresponding group except for 1.

**Sequential Randomization** $\{X_i', Y_i'\}_{i=1}^k \leftarrow \mathsf{RandSeq}(\{X_i, Y_i\}_{i=1}^k)$: A pairing product $A = e(X_1, Y_1)\, e(X_2, Y_2) \ldots e(X_k, Y_k)$ is randomized into $A = e(X_1', Y_1')$ $e(X_2', Y_2') \ldots e(X_k', Y_k')$ as follows: Let $(\gamma_1, \ldots, \gamma_{k-1}) \leftarrow \mathbb{Z}_p^{k-1}$. We begin with randomizing the first pairing by using the second pairing as follows. First verify that $Y_1 \neq 1$ and $X_2 \neq 1$. If $Y_1 = 1$, replace the first pairing $e(X_1, 1)$ with $e(1, Y_1)$ with a new random $Y_1(\neq 1)$. The case of $X_2 = 1$ is handled in the same manner. Then multiply $1 = e(X_2^{-\gamma_1}, Y_1)\, e(X_2, Y_1^{\gamma_1})$ to both sides of the formula. We thus obtain

$$A = e(X_1 X_2^{-\gamma_1}, Y_1)\, e(X_2, Y_1^{\gamma_1} Y_2)\, e(X_3, Y_3) \cdots e(X_k, Y_k).$$

Next we randomize the second pairing by using the third one. As before, if $Y_1^{\gamma_1} Y_2 = 1$ or $X_3 = 1$, replace them to random values. Then multiply $1 = e(X_3^{-\gamma_2}, Y_1^{\gamma_1} Y_2)\, e(X_3, (Y_1^{\gamma_1} Y_2)^{\gamma_2})$. We thus have

$$A = e(X_1 X_2^{-\gamma_1}, Y_1)\, e(X_2 X_3^{-\gamma_2}, Y_1^{\gamma_1} Y_2)\, e(X_3, (Y_1^{\gamma_1} Y_2)^{\gamma_2} Y_3) \cdots e(X_k, Y_k).$$

This continues up to the $(k-1)$-st pairing. When done, the value of the $i$-th pairing distributes uniformly in $\mathbb{G}_T$ due to the uniform choice of $\gamma_i$. The $k$-th pairing follows the distribution determined by $A$ and preceding $k - 1$ pairings. To complete the randomization, every pairing is processed by the inner randomization.

The sequential randomization can be used to **extend** a product of $k$ pairings to a product of arbitrary $k'$ pairings, $k' \geq k$, by appending $e(1, 1)$ before randomization. By $\{X_i', Y_i'\}_{i=1}^{k'} \leftarrow \mathsf{RandExtend}(\{X_i, Y_i\}_{i=1}^k)$ we denote the sequential randomization with extension. Parameters $k$ and $k'$, $k' \geq k$, should be clear from the input and the output.

Note that the algorithms yield uniform elements and thus may include pairings that evaluate to $1_{\mathbb{G}_T}$. If it is not preferable, it can be avoided by repeating that particular step once again excluding the bad randomness.

## 5.2 Basic Signature Scheme

We define the signature scheme $\textbf{Sig} = (\mathcal{G}, \textsf{KeyGen}, \textsf{Sign}, \textsf{Verify})$ below. In addition to the common parameters outputted by the $\mathcal{G}$ algorithm, the key generation algorithm $\textsf{KeyGen}$ also takes a parameters $k$ which determines the message space $\mathbb{G}_2^k$; messages of shorter length are implicitly padded with $1_{\mathbb{G}_2}$-s. We do not use any trusted setup, but only the bilinear group generation.

**Setup:** On input $1^\lambda$ return $\Lambda = (p, \mathbb{G}_1, \mathbb{G}_2, \mathbb{G}_T, e, G, H) \leftarrow \mathcal{G}(1^\lambda)$.

**Key generation:** On input $\Lambda$ and $k$, choose random generators $G_R, F_U \leftarrow \mathbb{G}_1^*$. For $i = 1, \ldots, k$, choose $\gamma_i, \delta_i \leftarrow \mathbb{Z}_p^{*2}$ and compute $G_i = G_R^{\gamma_i}$ and $F_i = F_U^{\delta_i}$. Choose $\gamma_Z, \delta_Z \leftarrow \mathbb{Z}_p^{*2}$ and compute $G_Z = G_R^{\gamma_Z}$ and $F_Z = F_U^{\delta_Z}$. Also choose $\alpha, \beta \leftarrow \mathbb{Z}_p^{*2}$ and compute $\{A_i, \tilde{A}_i\}_{i=0}^1 \leftarrow \textsf{RandExtend}(G_R, H^\alpha)$ and $\{B_i, \tilde{B}_i\}_{i=0}^1 \leftarrow \textsf{RandExtend}(F_U, H^\beta)$. Set $sk = (vk, \alpha, \beta, \gamma_Z, \delta_Z, \{\gamma_i, \delta_i\}_{i=1}^k)$ and $vk = (\Lambda, G_Z, F_Z, G_R, F_U, \{G_i, F_i\}_{i=1}^k, \{A_i, \tilde{A}_i, B_i, \tilde{B}_i\}_{i=0}^1)$. Output $(vk, sk)$.

**Signature issuing:** On input $sk$ and $\vec{M}$, choose $\zeta, \rho, \tau, \varphi, \omega$ randomly from $\mathbb{Z}_p^*$ and set:

$$Z = H^\zeta,\, R = H^{\rho - \gamma_Z \zeta} \prod_{i=1}^k M_i^{-\gamma_i},\, S = G_R^\tau,\, T = H^{(\alpha - \rho)/\tau},$$

$$U = H^{\varphi - \delta_Z \zeta} \prod_{i=1}^k M_i^{-\delta_i},\, V = F_U^\omega,\, W = H^{(\beta - \varphi)/\omega}.$$

Output $\sigma = (Z, R, S, T, U, V, W)$ as a signature.

**Verification:** On input $vk, \vec{M}$, and $\sigma$, parse the signature $\sigma$ as $(Z, R, S, T, U, V, W)$. Output 1 if the following equations:

$$A = e(G_Z, Z)\, e(G_R, R)\, e(S, T) \prod_{i=1}^k e(G_i, M_i) \text{ and} \tag{6}$$

$$B = e(F_Z, Z)\, e(F_U, U)\, e(V, W) \prod_{i=1}^k e(F_i, M_i) \tag{7}$$

hold for $A = e(A_0, \tilde{A}_0)\, e(A_1, \tilde{A}_1)$ and $B = e(B_0, \tilde{B}_0)\, e(B_1, \tilde{B}_1)$. Output 0, otherwise.

The following theorem is proved in [AHO10]:

**Theorem 6.** $(\mathcal{G}, \textsf{KeyGen}, \textsf{Sign}, \textsf{Verify})$ *described above provides perfect correctness. It is existentially unforgeable against adaptive chosen-message attack if the SFP assumption holds for* $\mathcal{G}$.

Next, we describe some notable properties of the signature scheme:

**Partial Perfect Randomizability.** Given a signature $(Z, R, S, T, U, V, W)$ one can randomize every element except for $Z$ by applying the sequential randomization technique with small tweak as follows. Define the function $(R', S', T', U', V', W') \leftarrow \textsf{SigRand}(R, S, T, U, V, W)$, as:

- Randomize $(R, S, T)$ into $(R', S', T')$ as follows.
  - First, if $T = 1$, set $S = 1$ and choose $T \leftarrow \mathbb{G}_2^*$.
  - Then, choose $\varrho \leftarrow \mathbb{Z}_p$ and compute

$$R' = RT^\varrho, \quad (S', T') \leftarrow \mathsf{Rand}(SG_R^{-\varrho}, T)$$

- Randomize $(U, S, T)$ into $(U', S', T')$ analogously.

**Lemma 4.** *The above $(R', S', T', U', V', W')$ distributes uniformly over $(\mathbb{G}_2 \times \mathbb{G}_1 \times \mathbb{G}_2)^2$ under constraint that $e(G_R, R)\, e(S, T) = e(G_R, R')\, e(S', T')$ and $e(F_U, U)\, e(V, W) = e(F_U, U')\, e(V', W')$.*

The claim implies that $(S', T', V', W')$ is information theoretically independent of $Z$, the message, and the verification key. (In general, the same is true for publishing any two elements from $(R', S', T')$ and $(U', V', W')$ respectively.)

**Signature Binding Property.** Roughly, it claims that no one but the signer can obtain two signatures which have the same $S$ and $V$. In the following formal statement, the adversary is allowed to submit both $\vec{M}$ and $\vec{M}^\dagger$ to the signing oracle. That is way the property is not implied by EUF-CMA in general.

**Lemma 5.** *Under adaptive chosen message attacks, no adversary can output $(\vec{M}, \sigma)$ and $(\vec{M}^\dagger, \sigma^\dagger)$ such that $1 = \mathsf{Verify}(vk, \vec{M}, \sigma) = \mathsf{Verify}(vk, \vec{M}^\dagger, \sigma^\dagger)$, $\vec{M} \neq \vec{M}^\dagger$, and $(S, V)$ are shared in $\sigma$ and $\sigma^\dagger$.*

Hence, in a way, publishing $(S, V)$ together with the verification key works as a commitment on the signature and the message without revealing any information (recall that $(S, V)$ are chosen uniformly in the signing algorithm).

## 5.3   Variations and Extensions

In this section we describe various extensions and modifications of the above scheme. Due to the space limitations, the ideas are only described briefly and the full description is presented in the full version.

**Messages $\in \mathbb{G}_1^k$.** When working with asymmetric pairings, it is possible to define a "dual scheme" with a message space $\mathbb{G}_1^k$ (by essentially swapping $\mathbb{G}_1$ and $\mathbb{G}_2$ in the above description).

**Messages $\in \mathbb{G}_1^{k_1} \times \mathbb{G}_2^{k_2}$.** It is possible to combine the signature schemes with message spaces $\mathbb{G}_1^{k_1}$ and $\mathbb{G}_2^{k_2}$ to obtain a signature scheme whose message space is $\mathbb{G}_1^{k_1} \times \mathbb{G}_2^{k_2}$. Note that this is not trivial, as there is no efficient mappings between $\mathbb{G}_1$ and $\mathbb{G}_2$, and straightforward independent signing allows a forgery. The transformation is applicable to (or required by) the extensions below.

**Short Variable-Length Messages.** Let $\langle n \rangle$ denote a deterministic encoding of non-negative integer $n$ $(< p)$ to an element of $\mathbb{G}_2^*$. By limiting the maximum message length to be $k-1$, for a signature with message space $\mathbb{G}_2^k$, and appending $\langle |\vec{M}| \rangle$ to the input message $\vec{M}$, messages with length less than $k$ can be treated.

**Unbounded-Length Messages.** For a signature scheme with message space $\mathbb{G}_2^k$, it is possible to sign messages from the space $\mathbb{G}_2^n$, $n > k$, by using a "chaining" technique. The basic idea is to split the message vector into (almost) equal chunks and sign each chunk along with the signature of the previous chunk (or part of it using the signature binding property described above). This is useful when the signer does not know a priori the maximum length of the messages or has to sign her own verification key (e.g. *automorphic signatures*).

**Strong One-time Signatures.** Dropping the flexible part $e(S, T)$ and $e(V, W)$ from the construction results in a strongly unforgeable one-time signature scheme based on a (weaker) static assumption which is implied by the DBP.

**Strongly Unforgeable Signatures.** We construct a structure-preserving signature scheme with constant-size signatures that is sEUF-CMA secure. The generic construction, combining a EUF-CMA and a one-time sEUF-CMA signature schemes, is optimized by sharing some parts of the verification keys.

$vk$ **Variations.** We can replace $\{A_i, \tilde{A}_i, B_i, \tilde{B}_i\}_{i=0}^1$ with $A = e(G_R, H^\alpha)$ and $B = e(F_U, H^\beta)$ in a verification key, and use $A$ and $B$ directly in the verification equations (6) and (7). The reason we include a representation of $A$ (and $B$) in $\mathbb{G}_1$ and $\mathbb{G}_2$ is to address the needs to put the verification key into the base groups. The GS proof system provides zero-knowledge property for statements that do not include elements from $\mathbb{G}_T$ except for $1_{\mathbb{G}_T}$. When WI is of only concern, we do such replacement.

**Symmetric Pairings.** The signature scheme is also secure when working with symmetric pairings ($\mathbb{G}_1 = \mathbb{G}_2$). The above extensions apply in that case as well.

### 5.4   Simulatable Signatures

A *simulatable signature scheme* **SSig** $= (\mathcal{G}, \mathsf{CrsGen}, \mathsf{KeyGen}, \mathsf{Check}, \mathsf{Sign}, \mathsf{Verify}, \mathsf{Sim})$ consists of algorithms for which **Sig** $= ((\mathcal{G} + \mathsf{CrsGen}), \mathsf{KeyGen}, \mathsf{Sign}, \mathsf{Verify})$ constitutes a regular signature scheme. It is defined in the common reference string (CRS) model and allows to create valid signatures using the trapdoor associated with the CRS. The three algorithms not defined for regular signatures ($\mathsf{CrsGen}$, $\mathsf{Check}$, $\mathsf{Sim}$) are, respectively, for generating a CRS and the associated trapdoor, for checking that a verification key produced by a user is valid, and for simulating a signature on any valid message on behalf of any user using the trapdoor key rather than the corresponding signing key. A simulatable signature is a useful tool in combination with a witness indistinguishable (WI) proof system. Unlike zero-knowledge (ZK) proofs, WI proof system does not accompany a simulator. So when a signature is part of the witness and the signer is corrupt and useless, simulatable signature can provide a correct witness to the entity having the trapdoor.

The notion is introduced in [AO09] but in an informal way dedicated for their purposes. We present a formal treatment and give an efficient construction, but due to the space limitation, we can only sketch the intuition, the security definitions, and the construction details. Full details are presented in [AHO10].

The security properties we require from a simulatable signature scheme are correctness, simulatability, and unforgeability, extended to a multi-user setting where the adversary has access to a signing oracle for all correctly generated verification keys in addition to a proof oracle for simulated signatures on any valid verification key and message. Our construction shares a lot with our basic signature scheme. The main difference is that to sign messages of length $k$, we need $k$ flexible pairings rather than 1, so the signature is of size $4k + 3$ group elements. The Verify algorithm is defined similarly, with the verification equations being:

$$A = e(G_Z, Z) \, e(G_R, R) \prod_{i=1}^{k} e(G_i, M_i) \, e(S_i, T_i) \quad \text{and} \tag{8}$$

$$B = e(F_Z, Z) \, e(F_U, U) \prod_{i=1}^{k} e(F_i, M_i) \, e(V_i, W_i) \ . \tag{9}$$

So, for $k = 1$, the two schemes have the same signature distribution and verification algorithms. The key generation algorithm of the basic scheme is divided into two parts: CrsGen generating the elements on the right side of equations (8)-(9) and KeyGen computing those on the left as well as a signature on the default message (e.g. the all-$1_{\mathbb{G}_2}$ vector). The CRS is, in fact, a commitment key for a trapdoor commitment scheme similar to the one presented in Section 3, whereas any $vk$ is a commitment to the default message. The signing algorithm is quite intricate as it opens the commitment, the signer's $vk$, to any given message without using the commitment trapdoor. That is why we need $k$ flexible pairings to achieve perfectly random distribution for a signature under the condition that the verification equations are satisfied.

**Theorem 7.** *The* SSig *described above is a perfectly correct signature scheme and signature-simulatable. It is EUF-CMA with WI-simulation in the multi-user setting for* $k = 1$ *if the SFP assumption holds for* $\mathcal{G}$*.*

The security for the case of $k > 1$ is shown under a generalization of the SFP assumption and also presented in the full version.

# 6   Applications of Signatures on Group Elements

This section highlights the benefits of combining structure-preserving signatures on group elements with the GS proof system when building applications. We present the first efficient round-optimal non-committing blind signature scheme which is adaptively secure in the universal-composability framework, efficient group signatures with concurrent join under the strongest security definitions, and efficient anonymous proxy signatures with enhanced anonymity properties.

### 6.1  UC-Secure Blind Signatures

It has been an open problem to efficiently instantiate Fischlin's [Fis06] framework for UC-secure round-optimal blind signatures. We do so using our signature scheme from Section 5 and a variant of Pedersen commitments [Ped92]. In fact, we use the modification of [HKKL07, AO09] for which the generic construction uses a NIWI proof system and a simulatable signature scheme as it achieves *adaptive* security.

We instantiate the framework as follows: a user commits to a message $m \in \mathbb{Z}_p$, with opening $D = G^r$, as $C = H^m Y^r$ and sends $C$ to the signer. Note that the verification equation for $(D, m)$ being a valid opening is $e(G, C) \, e(D, Y^{-1}) = e(G, H^m)$ which could be viewed as a "pairing-based variant" of Pedersen commitment. The signer signs the commitment $c$ using the simulatable signature scheme from Section 5 and returns the signature to the user. Then, the user computes a NIWI proof of knowledge $\pi$ of a commitment $C$ to the message $m$, an opening $D$ of the commitment for that message, and a valid signature on $C$ with respect to the signer's verification key. The user outputs that proof as a blind signature on the message $m$.

Details of the instantiation can be found in [AHO10]. The signature size is 28 group elements when working with symmetric pairings and 28 group elements with asymmetric, while the total communication complexity is only 8 group elements in both cases.

### 6.2  Group Signatures

Group signatures have enjoyed much interest since they were introduced by Chaum and van Heyst [Cv91] almost twenty years ago. Most previous constructions, [CS97, ACJT00, BBS04, CL04, BSZ05, BW06, BW07, Gro06] among others, could be viewed as unsatisfactory in some aspect: relying on the random-oracle model, satisfying weaker security definitions, or not being efficient. The scheme by Groth [Gro07] both is practical and satisfies the strengthened security definitions of [BSZ05]. However, it does not support concurrent join of new users. Using our signature schemes in combination with the GS proof system and an appropriate encryption scheme [Kil06, Sha07], we overcome this shortcoming and construct a group signature scheme under the strongest security definitions which supports concurrent join while achieving comparable efficiency.

Our construction follows a common approach used, e.g., in [CS97, Gro07]. The dynamic join protocol between a group member and the issuer simply consists in the issuer signing the member's verification key. To sign a message $m$, the member signs the message using her secret key and gives a NIWI proof of knowledge of a verification key, a signature on that key by the issuer, and a signature on the message under that key. For the details of our constructions and further discussions, we refer to the full versions of our papers.

## 6.3   Anonymous Proxy Signatures

Combined with Groth-Sahai proofs, automorphic signatures enable the first efficient instantiation of anonymous proxy signatures [FP08]. This primitive generalizes (multi-level) proxy signatures [MUO96, BPW03] and group signatures. Consider a setting where users publish signature verification keys, which they have previously registered with an authority. Proxy signatures enable users to delegate others to sign on their behalf; moreover, received rights can be redelegated. Anonymity of proxy signatures guarantees that they neither reveal who signed nor who redelegated. As for group signatures, an *opening authority* can revoke anonymity to deter from misuse. Every valid signature can be opened to registered users (*traceability*) and no coalition even comprising the authorities can produce a signature that wrongfully accuses an honest user (*non-frameability*).

Automorphic signatures allow a straightforward instantiation of the generic construction. To delegate to Bob, Alice signs his public key (and possibly some public attributes). To redelegate to Carol, Bob forwards her the received signature and signs her public key. Carol makes a proxy signature by signing the message and then making a proof of knowledge of the following: Bob's key, Alice's signature on it, her own key, Bob's signature on it, and her signature on the message.[2] Since all of them consist of elements of a bilinear group and validity is expressed as pairing-product equations, Groth-Sahai (GS) proofs apply perfectly. The extraction key is given to the opener who can thus revoke anonymity of a signature by retrieving the public keys of the intermediate delegators and the proxy signer. A signature is verified by checking validity of the GS proof with respect to Alice's public key.

**Enhanced Anonymity Guarantees.** In the model of [FP08], anonymity holds only w.r.t. the verifier. We show how to protect the privacy of the delegatee and the delegators even during delegation. The delegatee remains anonymous if we use the issuing protocol of the blind signature from Section 4.2 for delegation. In the end, the delegatee holds an actual signature on her public key, as in the original scheme, but without the delegator having learned her identity.

The previous delegators can remain anonymous w.r.t. the delegatee as well, as due to the modularity of Groth-Sahai proofs, the "anonymization" of a signature need not be delayed until the proxy signing: instead of forwarding the received delegation chain, a delegator forwards a proof of knowledge of it. The delegatee can then extend the proof by one delegation step, or make a proxy signature; before doing so, she *randomizes* the proof, which prevents linkability of delegations and signatures. By additionally proving knowledge of his public key and signature, the delegator can also hide *his own identity*. Unfortunately, this is not compatible with blind delegation, while hiding the previous delegators is. We refer to [Fuc09] for the details.

---

[2] To guarantee traceability, Carol additionally proves knowledge of certificates from the authority on the public keys. Moreover, to delegate, a user actually signs (a hash value of) an identifier set by the original delegator and his position in the chain in addition to the public key to achieve non-frameability.

## Acknowledgments

The second author is supported by EADS, the French ANR-07-TCOM-013-04 PACE Project and the European Commission through the ICT Program under Contract ICT-2007-216676 ECRYPT II.

## References

[ACJT00]   Ateniese, G., Camenisch, J.L., Joye, M., Tsudik, G.: A practical and provably secure coalition-resistant group signature scheme. In: Bellare, M. (ed.) CRYPTO 2000. LNCS, vol. 1880, pp. 255–270. Springer, Heidelberg (2000)

[ADW09]   Alwen, J., Dodis, Y., Wichs, D.: Survey: Leakage-resilience and the bounded-retrieval model. Invited Paper to International Conference on Information Theoretic Security (2009), http://cs.nyu.edu/~dodis/surveys.html

[AHO10]   Abe, M., Haralambiev, K., Ohkubo, M.: Signing on elements in bilinear groups for modular protocol design. Cryptology ePrint Archive, Report 2010/133 (2010), http://eprint.iacr.org/

[AO09]   Abe, M., Ohkubo, M.: A framework for universally composable non-committing blind signatures. In: Matsui, M. (ed.) ASIACRYPT 2009. LNCS, vol. 5912, pp. 435–450. Springer, Heidelberg (2009)

[BB04]   Boneh, D., Boyen, X.: Short signatures without random oracles. In: Cachin, C., Camenisch, J.L. (eds.) EUROCRYPT 2004. LNCS, vol. 3027, pp. 56–73. Springer, Heidelberg (2004)

[BBS04]   Boneh, D., Boyen, X., Shacham, H.: Short group signatures. In: Franklin, M. (ed.) CRYPTO 2004. LNCS, vol. 3152, pp. 41–55. Springer, Heidelberg (2004)

[BCC+09]   Belenkiy, M., Camenisch, J., Chase, M., Kohlweiss, M., Lysyanskaya, A., Shacham, H.: Randomizable proofs and delegatable anonymous credentials. In: Halevi, S. (ed.) CRYPTO 2009. LNCS, vol. 5677, pp. 108–125. Springer, Heidelberg (2009)

[BCKL08]   Belenkiy, M., Chase, M., Kohlweiss, M., Lysyanskaya, A.: P-signatures and noninteractive anonymous credentials. In: Canetti, R. (ed.) TCC 2008. LNCS, vol. 4948, pp. 356–374. Springer, Heidelberg (2008)

[BGLS03]   Boneh, D., Gentry, C., Lynn, B., Shacham, H.: Aggregate and verifiably encrypted signatures from bilinear maps. In: Biham, E. (ed.) EUROCRYPT 2003. LNCS, vol. 2656, pp. 416–432. Springer, Heidelberg (2003)

[BGN05]   Boneh, D., Goh, E.-J., Nissim, K.: Evaluating 2-DNF formulas on ciphertexts. In: Kilian, J. (ed.) TCC 2005. LNCS, vol. 3378, pp. 325–341. Springer, Heidelberg (2005)

[BMW03]   Bellare, M., Micciancio, D., Warinschi, B.: Foundations of group signatures: Formal definitions, simplified requirements, and a construction based on general assumptions. In: Biham, E. (ed.) EUROCRYPT 2003. LNCS, vol. 2656, pp. 614–629. Springer, Heidelberg (2003)

[BNPS03]   Bellare, M., Namprempre, C., Pointcheval, D., Semanko, M.: The one-more-RSA-inversion problems and the security of Chaum's blind signature scheme. Journal of Cryptology 16(3), 185–215 (2003)

[Boy08]      Boyen, X.: The uber-assumption family (invited talk). In: Galbraith,
             S.D., Paterson, K.G. (eds.) Pairing 2008. LNCS, vol. 5209, pp. 39–56.
             Springer, Heidelberg (2008)

[BPW03]      Boldyreva, A., Palacio, A., Warinschi, B.: Secure proxy signature
             schemes for delegation of signing rights. Cryptology ePrint Archive, Re-
             port 2003/096 (2003), http://eprint.iacr.org/

[BR93]       Bellare, M., Rogaway, P.: Random oracles are practical: A paradigm for
             designing efficient protocols. In: Ashby, V. (ed.) ACM CCS 1993, pp.
             62–73. ACM Press, New York (1993)

[Bra99]      Brands, S.: Rethinking public key infrastructure and digital certificates–
             building privacy. PhD thesis, Eindhoven Inst. of Tech., The Netherlands
             (1999)

[BSZ05]      Bellare, M., Shi, H., Zhang, C.: Foundations of group signatures: The
             case of dynamic groups. In: Menezes, A. (ed.) CT-RSA 2005. LNCS,
             vol. 3376, pp. 136–153. Springer, Heidelberg (2005)

[BW06]       Boyen, X., Waters, B.: Compact group signatures without random ora-
             cles. In: Vaudenay, S. (ed.) EUROCRYPT 2006. LNCS, vol. 4004, pp.
             427–444. Springer, Heidelberg (2006)

[BW07]       Boyen, X., Waters, B.: Full-domain subgroup hiding and constant-size
             group signatures. In: Okamoto, T., Wang, X. (eds.) PKC 2007. LNCS,
             vol. 4450, pp. 1–15. Springer, Heidelberg (2007)

[Can01]      Canetti, R.: Universally composable security: A new paradigm for cryp-
             tographic protocols. In: 42nd FOCS, pp. 136–145. IEEE Computer So-
             ciety Press, Los Alamitos (2001)

[CGH98]      Canetti, R., Goldreich, O., Halevi, S.: The random oracle methodology,
             revisited (preliminary version). In: 30th ACM STOC, pp. 209–218. ACM
             Press, New York (1998)

[CKS09]      Camenisch, J., Kohlweiss, M., Soriente, C.: An accumulator based on
             bilinear maps and efficient revocation for anonymous credentials. In:
             Jarecki, S., Tsudik, G. (eds.) PKC 2009. LNCS, vol. 5443, pp. 481–500.
             Springer, Heidelberg (2009)

[CL04]       Camenisch, J., Lysyanskaya, A.: Signature schemes and anonymous cre-
             dentials from bilinear maps. In: Franklin, M. (ed.) CRYPTO 2004.
             LNCS, vol. 3152, pp. 56–72. Springer, Heidelberg (2004)

[CLY09]      Cathalo, J., Libert, B., Yung, M.: Group encryption: Non-interactive
             realization in the standard model. In: Matsui, M. (ed.) ASIACRYPT
             2009. LNCS, vol. 5912, pp. 179–196. Springer, Heidelberg (2009)

[CS97]       Camenisch, J., Stadler, M.: Efficient group signature schemes for large
             groups (extended abstract). In: Kaliski Jr., B.S. (ed.) CRYPTO 1997.
             LNCS, vol. 1294, pp. 410–424. Springer, Heidelberg (1997)

[Cv91]       Chaum, D., van Heyst, E.: Group signatures. In: Davies, D.W. (ed.)
             EUROCRYPT 1991. LNCS, vol. 547, pp. 257–265. Springer, Heidelberg
             (1991)

[DF02]       Damgård, I., Fujisaki, E.: A statistically-hiding integer commitment
             scheme based on groups with hidden order. In: Zheng, Y. (ed.) ASI-
             ACRYPT 2002. LNCS, vol. 2501, pp. 125–142. Springer, Heidelberg
             (2002)

[DHLAW10]    Dodis, Y., Haralambiev, K., Lopez-Alt, A., Wichs, D.: Efficient public-
             key cryptography in the presence of key leakage. Cryptology ePrint
             Archive, Report 2010/154 (2010), http://eprint.iacr.org/

234    M. Abe et al.

[DN02]     Damgård, I., Nielsen, J.B.: Perfect hiding and perfect binding universally composable commitment schemes with constant expansion factor. In: Yung, M. (ed.) CRYPTO 2002. LNCS, vol. 2442, pp. 581–596. Springer, Heidelberg (2002)

[ElG86]    El Gamal, T.: On computing logarithms over finite fields. In: Williams, H.C. (ed.) CRYPTO 1985. LNCS, vol. 218, pp. 396–402. Springer, Heidelberg (1986)

[Fis06]    Fischlin, M.: Round-optimal composable blind signatures in the common reference string model. In: Dwork, C. (ed.) CRYPTO 2006. LNCS, vol. 4117, pp. 60–77. Springer, Heidelberg (2006)

[FO97]     Fujisaki, E., Okamoto, T.: Statistical zero knowledge protocols to prove modular polynomial relations. In: Kaliski Jr., B.S. (ed.) CRYPTO 1997. LNCS, vol. 1294, pp. 16–30. Springer, Heidelberg (1997)

[FP08]     Fuchsbauer, G., Pointcheval, D.: Anonymous proxy signatures. In: Ostrovsky, R., De Prisco, R., Visconti, I. (eds.) SCN 2008. LNCS, vol. 5229, pp. 201–217. Springer, Heidelberg (2008)

[FP09]     Fuchsbauer, G., Pointcheval, D.: Proofs on encrypted values in bilinear groups and an application to anonymity of signatures. In: Shacham, H., Waters, B. (eds.) Pairing 2009. LNCS, vol. 5671, pp. 132–149. Springer, Heidelberg (2009)

[FPV09]    Fuchsbauer, G., Pointcheval, D., Vergnaud, D.: Transferable constant-size fair E-cash. In: Garay, J.A., Miyaji, A., Otsuka, A. (eds.) CANS 2009. LNCS, vol. 5888, pp. 226–247. Springer, Heidelberg (2009)

[FS01]     Furukawa, J., Sako, K.: An efficient scheme for proving a shuffle. In: Kilian, J. (ed.) CRYPTO 2001. LNCS, vol. 2139, pp. 368–387. Springer, Heidelberg (2001)

[Fuc09]    Fuchsbauer, G.: Automorphic signatures in bilinear groups and an application to round-optimal blind signatures. Cryptology ePrint Archive, Report 2009/320 (2009), http://eprint.iacr.org/

[Fuc10]    Fuchsbauer, G.: Commuting signatures and verifiable encryption and an application to non-interactively delegatable credentials. Cryptology ePrint Archive, Report 2010/233 (2010), http://eprint.iacr.org/

[FV10]     Fuchsbauer, G., Vergnaud, D.: Fair blind signatures without random oracles. In: Bernstein, D.J., Lange, T. (eds.) AFRICACRYPT 2010. LNCS, vol. 6055, pp. 16–33. Springer, Heidelberg (2010)

[GH08]     Green, M., Hohenberger, S.: Universally composable adaptive oblivious transfer. In: Pieprzyk, J. (ed.) ASIACRYPT 2008. LNCS, vol. 5350, pp. 179–197. Springer, Heidelberg (2008)

[GMR88]    Goldwasser, S., Micali, S., Rivest, R.L.: A digital signature scheme secure against adaptive chosen-message attacks. SIAM Journal on Computing 17(2), 281–308 (1988)

[GOS06]    Groth, J., Ostrovsky, R., Sahai, A.: Non-interactive zaps and new techniques for NIZK. In: Dwork, C. (ed.) CRYPTO 2006. LNCS, vol. 4117, pp. 97–111. Springer, Heidelberg (2006)

[GPS08]    Galbraith, S.D., Paterson, K.G., Smart, N.P.: Pairings for cryptographers. Discrete Applied Mathematics 156(16), 3113–3121 (2008)

[GQ88]     Guillou, L.C., Quisquater, J.-J.: A practical zero-knowledge protocol fitted to security microprocessor minimizing both trasmission and memory. In: Günther, C.G. (ed.) EUROCRYPT 1988. LNCS, vol. 330, pp. 123–128. Springer, Heidelberg (1988)

[Gro06]    Groth, J.: Simulation-sound NIZK proofs for a practical language and constant size group signatures. In: Lai, X., Chen, K. (eds.) ASIACRYPT 2006. LNCS, vol. 4284, pp. 444–459. Springer, Heidelberg (2006)

[Gro07]    Groth, J.: Fully anonymous group signatures without random oracles. In: Kurosawa, K. (ed.) ASIACRYPT 2007. LNCS, vol. 4833, pp. 164–180. Springer, Heidelberg (2007)

[Gro09a]   Groth, J.: Homomorphic trapdoor commitments to group elements. Cryptology ePrint Archive, Report 2009/007 (2009), http://eprint.iacr.org/

[Gro09b]   Groth, J.: Linear algebra with sub-linear zero-knowledge arguments. In: Halevi, S. (ed.) CRYPTO 2009. LNCS, vol. 5677, pp. 192–208. Springer, Heidelberg (2009)

[GS08]     Groth, J., Sahai, A.: Efficient non-interactive proof systems for bilinear groups. In: Smart, N.P. (ed.) EUROCRYPT 2008. LNCS, vol. 4965, pp. 415–432. Springer, Heidelberg (2008)

[HKKL07]   Hazay, C., Katz, J., Koo, C.-Y., Lindell, Y.: Concurrently-secure blind signatures without random oracles or setup assumptions. In: Vadhan, S.P. (ed.) TCC 2007. LNCS, vol. 4392, pp. 323–341. Springer, Heidelberg (2007)

[Kil06]    Kiltz, E.: Chosen-ciphertext security from tag-based encryption. In: Halevi, S., Rabin, T. (eds.) TCC 2006. LNCS, vol. 3876, pp. 581–600. Springer, Heidelberg (2006)

[KP06]     Kunz-Jacques, S., Pointcheval, D.: About the security of MTI/C0 and MQV. In: De Prisco, R., Yung, M. (eds.) SCN 2006. LNCS, vol. 4116, pp. 156–172. Springer, Heidelberg (2006)

[KY05]     Kiayias, A., Yung, M.: Group signatures with efficient concurrent join. In: Cramer, R. (ed.) EUROCRYPT 2005. LNCS, vol. 3494, pp. 198–214. Springer, Heidelberg (2005)

[KZ06]     Kiayias, A., Zhou, H.-S.: Concurrent blind signatures without random oracles. In: De Prisco, R., Yung, M. (eds.) SCN 2006. LNCS, vol. 4116, pp. 49–62. Springer, Heidelberg (2006)

[Lip03]    Lipmaa, H.: Verifiable homomorphic oblivious transfer and private equality test. In: Laih, C.-S. (ed.) ASIACRYPT 2003. LNCS, vol. 2894, pp. 416–433. Springer, Heidelberg (2003)

[LRSW00]   Lysyanskaya, A., Rivest, R.L., Sahai, A., Wolf, S.: Pseudonym systems. In: Heys, H.M., Adams, C.M. (eds.) SAC 1999. LNCS, vol. 1758, pp. 184–199. Springer, Heidelberg (2000)

[LV08]     Libert, B., Vergnaud, D.: Multi-use unidirectional proxy re-signatures. In: Ning, P., Syverson, P.F., Jha, S. (eds.) ACM CCS 2008, pp. 511–520. ACM Press, New York (2008)

[MUO96]    Mambo, M., Usuda, K., Okamoto, E.: Proxy signatures for delegating signing operation. In: ACM CCS 1996, pp. 48–57. ACM Press, New York (1996)

[Nao03]    Naor, M.: On cryptographic assumptions and challenges (invited talk). In: Boneh, D. (ed.) CRYPTO 2003. LNCS, vol. 2729, pp. 96–109. Springer, Heidelberg (2003)

[Nef01]    Andrew Neff, C.: A verifiable secret shuffle and its application to e-voting. In: ACM CCS 2001, pp. 116–125. ACM Press, New York (2001)

[Oka06]    Okamoto, T.: Efficient blind and partially blind signatures without random oracles. In: Halevi, S., Rabin, T. (eds.) TCC 2006. LNCS, vol. 3876, pp. 80–99. Springer, Heidelberg (2006)

[OU98]     Okamoto, T., Uchiyama, S.: A new public-key cryptosystem as secure
           as factoring. In: Nyberg, K. (ed.) EUROCRYPT 1998. LNCS, vol. 1403,
           pp. 308–318. Springer, Heidelberg (1998)
[Pai99]    Paillier, P.: Public-key cryptosystems based on composite degree resid-
           uosity classes. In: Stern, J. (ed.) EUROCRYPT 1999. LNCS, vol. 1592,
           pp. 223–238. Springer, Heidelberg (1999)
[Ped92]    Pedersen, T.P.: Non-interactive and information-theoretic secure veri-
           fiable secret sharing. In: Feigenbaum, J. (ed.) CRYPTO 1991. LNCS,
           vol. 576, pp. 129–140. Springer, Heidelberg (1992)
[RS09]     Rückert, M., Schröder, D.: Security of verifiably encrypted signatures
           and a construction without random oracles. In: Shacham, H. (ed.) Pair-
           ing 2009. LNCS, vol. 5671, pp. 19–35. Springer, Heidelberg (2009)
[Sha07]    Shacham, H.: A cramer-shoup encryption scheme from the linear as-
           sumption and from progressively weaker linear variants. Cryptology
           ePrint Archive, Report 2007/074 (2007), http://eprint.iacr.org/
[Sho97]    Shoup, V.: Lower bounds for discrete logarithms and related problems.
           In: Fumy, W. (ed.) EUROCRYPT 1997. LNCS, vol. 1233, pp. 256–266.
           Springer, Heidelberg (1997)

# Efficient Indifferentiable Hashing into Ordinary Elliptic Curves

Eric Brier[1], Jean-Sébastien Coron[2], Thomas Icart[2,*], David Madore[3],
Hugues Randriam[3], and Mehdi Tibouchi[2,4]

[1] Ingenico
eric.brier@ingenico.com
[2] Université du Luxembourg
jean-sebastien.coron@uni.lu, thomas.icart@m4x.org
[3] TELECOM-ParisTech
{david.madore,randriam}@enst.fr
[4] École normale supérieure
mehdi.tibouchi@ens.fr

**Abstract.** We provide the first construction of a hash function into ordinary elliptic curves that is indifferentiable from a random oracle, based on Icart's deterministic encoding from Crypto 2009. While almost as efficient as Icart's encoding, this hash function can be plugged into any cryptosystem that requires hashing into elliptic curves, while not compromising proofs of security in the random oracle model.

We also describe a more general (but less efficient) construction that works for a large class of encodings into elliptic curves, for example the Shallue-Woestijne-Ulas (SWU) algorithm. Finally we describe the first deterministic encoding algorithm into elliptic curves in characteristic 3.

## 1 Introduction

**Hashing into Elliptic Curves.** Many elliptic curve cryptosystems require to hash into an elliptic curve. For example in the Boneh-Franklin IBE scheme [4], the public-key for identity $id \in \{0,1\}^*$ is a point $Q_{id} = H_1(id)$ on the curve. This is also the case in many other pairing-based cryptosystems including IBE and HIBE schemes [1,17,18], signature and identity-based signature schemes [3,5,6,12,27] and identity-based signcryption schemes [8,21].

Hashing into elliptic curves is also required for some passwords based authentication protocols, for instance the SPEKE (Simple Password Exponential Key Exchange) [20] and the PAK (Password Authenticated Key exchange) [9], and also for discrete-log based signature schemes such as [13] when instantiated over an elliptic curve. In all those previous cryptosystems, security is proven when the hash function is seen as a random oracle into the curve. However, it remains to

---
* Work done while working for SAGEM company.

T. Rabin (Ed.): CRYPTO 2010, LNCS 6223, pp. 237–254, 2010.

determine which hashing algorithm should be used, and whether it is reasonable to see it as a random oracle.

In [4], Boneh and Franklin use a particular supersingular elliptic curve $E$ for which, in addition to the pairing operation, there exists a one-to-one mapping $f$ from the base field $\mathbb{F}_p$ to $E(\mathbb{F}_p)$. This enables to hash using $H_1(m) = f(h(m))$ where $h$ is a classical hash function from $\{0,1\}^*$ to $\mathbb{F}_p$. The authors show that their IBE scheme remains secure when $h$ is seen as a random oracle into $\mathbb{F}_p$ (instead of $H_1$ being seen as a random oracle into $E(\mathbb{F}_p)$). However, when no pairing operation is required (as in [9,13,20]), it is more efficient to use ordinary elliptic curves, since supersingular curves require much larger security parameters (due to the MOV attack [23]).

For hashing into an ordinary elliptic curve, the classical approach is inherently probabilistic: one can first compute an integer hash value $x = h(m)$ and then determine whether $x$ is the abscissa of a point on the elliptic curve:

$$y^2 = x^3 + ax + b$$

otherwise one can try $x + 1$ and so on. Using this approach the number of operations required to hash a message $m$ depends on $m$, which can lead to a timing attack (see [7]). To avoid this attack, one can determine whether $x + i$ is the abscissa of a point, for all $i$ between $0 \leq i < k$, and use for example the smallest such $i$; here $k$ is a security parameter that gives an error probability of roughly $2^{-k}$. However, this leads to a very lengthy hash computation.

The first algorithm to generate elliptic curve points in *deterministic* polynomial time was published in ANTS 2006 by Shallue and Woestijne [25]. The algorithm has running time $\mathcal{O}(\log^4 p)$ for any $p$, and $\mathcal{O}(\log^3 p)$ when $p \equiv 3 \pmod 4$. The rational maps in [25] were later simplified and generalized to hyper-elliptic curves by Ulas in [26]; we refer to this algorithm as the Shallue-Woestijne-Ulas (SWU) algorithm. Letting $f : \mathbb{F}_p \to E(\mathbb{F}_p)$ be the function defined by SWU, one can then hash in deterministic polynomial time using $H(m) = f(h(m))$ where $h$ is any hash function into $\mathbb{F}_p$.

Another deterministic hash algorithm for ordinary elliptic curves was recently published by Icart in [19]. The algorithm works for $p \equiv 2 \pmod 3$, with complexity $\mathcal{O}(\log^3 p)$. Given any elliptic curve $E$ defined over $\mathbb{F}_p$, Icart defines a function $f$ that is an algebraic function from $\mathbb{F}_p$ into the curve. As previously given any hash function $h$ into $\mathbb{F}_p$, one can use $H(m) = f(h(m))$ to hash into $E(\mathbb{F}_p)$. As shown in [19], $H$ is one-way if $h$ is one-way.

**The Random Oracle Model (ROM).** Many cryptosystems based on elliptic curves have been proven secure in the random oracle model, see for example [1,3,4,5,6,8,9,12,17,18,20,21,27]. In the random oracle model [2], the hash function is replaced by a publicly accessible random function (the random oracle); the adversary cannot compute the hash function by himself but instead he must query the random oracle. Obviously, a proof in the random oracle model is not fully satisfactory, because such a proof does not imply that the scheme will remain secure when the random oracle is replaced by a concrete hash function. Numerous papers have shown artificial schemes that are provably secure in the ROM but completely insecure when the RO is instantiated with any function family (see [11]). Despite these separation results, a proof in the ROM is believed to indicate that there are no structural flaws in the design of the system, and that no flaw will suddenly appear when a "well designed" hash function is used instead.

For a cryptosystem that requires a hash function $H$ into an ordinary elliptic curve (such as [9,20]), one possibility could be to use $H(m) = f(h(m))$ where $f$ is either Icart or SWU's function and $h$ is a hash function into $\mathbb{F}_p$. However we know that neither Icart nor SWU's function generate all the points of $E$; for example, Icart's function covers only $\simeq 5/8$ of the points [15,16]; moreover it is easy to see that the distribution of $f(h(m))$ is not uniform in $\mathsf{Im} f$. Therefore the current proofs in the random oracle model for $H$ do not guarantee the security of the resulting scheme when $H(m) = f(h(m))$ is used instead (even if $h$ is assumed to be ideal). In other words, even if a proof in the random oracle for $H$ can indicate that there are no structural flaws in the design of the cryptosystem, using $H(m) = f(h(m))$ could introduce a flaw that would make the resulting cryptosystem completely insecure (we give an example in Section 5.1).

**Our Results.** We provide the first construction of a hash function $H$ into ordinary elliptic curves with the property that *any cryptosystem* proven secure assuming $H$ is a random oracle remains secure when our construction is plugged instead (still assuming that the underlying $h$ is a random oracle). For this we use the indifferentiability framework of Maurer *et al.* [22]. As shown in [14], when a construction $H$ is indifferentiable from a random oracle, such a construction can then replace a random oracle in any cryptosystem, and the resulting scheme remains secure in the random oracle model for $h$.

Since the output of Icart and SWU functions only covers a fraction of the elliptic curve points, we cannot use the construction $H(m) = f(h(m))$ for indifferentiable hashing. Our main result is to show that for Icart's function $f$, we can use the following alternative construction which is almost as efficient:

$$H(m) := f(h_1(m)) + f(h_2(m))$$

where $h_1, h_2$ are two hash functions into $\mathbb{F}_p$, and $+$ denotes elliptic curve addition. Therefore $H(m)$ can be used in any cryptosystem provably secure with random oracle into elliptic curves, and the resulting cryptosystem remains secure in the random oracle model for $h_1$ and $h_2$.

However the proof involves somewhat technical tools from algebraic geometry, and it is not so simple to adapt to other encodings such as the SWU algorithm. Therefore we describe a more general (but less efficient) construction that applies to a large class of encoding functions satisfying a few simple axioms. Those encodings include Icart's function, the SWU algorithm, new deterministic encodings in characteristic 3, etc. More precisely, given an elliptic curve $E$ defined over $\mathbb{F}_p$ whose group of points is cyclic of order $N$ with generator $G$, our general construction is as follows:

$$H(m) := f(h_1(m)) + h_2(m)G$$

where $h_1 : \{0,1\}^* \to \mathbb{F}_p$ and $h_2 : \{0,1\}^* \to \mathbb{Z}_N$ are two hash functions, and $f$ is SWU or Icart's function. We show that $H(m)$ is indifferentiable from a random oracle when $h_1$ and $h_2$ are seen as random oracles. Intuitively, the term $h_2(m)G$ plays the role of a one-time pad; this ensures that $H(m)$ can behave as a random oracle even though $f(h_1(m))$ does not reach all the points in $E$. Note that one could not use $H(m) = h_2(m)G$ only since in this case the discrete logarithm of $H(m)$ would be known, which would make most protocols insecure.[1]

We also show how to extend the two previous constructions to hashing into the subgroup of an elliptic curve (with cyclic or non-cyclic group) and to hash-functions into strings (rather than $\mathbb{F}_p$). We also describe a slightly more efficient variant of the SWU algorithm when $p \equiv 3 \pmod 4$. Finally, we describe the first deterministic encoding algorithm into elliptic curves in characteristic 3. We summarize in Table 1 the known hashing algorithms into ordinary elliptic curves.

## 2    Preliminaries

### 2.1    Icart's Function

Consider an elliptic curve $E$ over a finite field $\mathbb{F}_q$, with $q$ odd and congruent to $2 \bmod 3$, with equation:

$$Y^2 = X^3 + aX + b$$

Icart's function is defined in [19] as the map $f_{a,b} : \mathbb{F}_q \to E(\mathbb{F}_q)$ such that $f_{a,b}(u) = (x, y)$ where:

$$x = \left(v^2 - b - \frac{u^6}{27}\right)^{1/3} + \frac{u^2}{3} \qquad y = ux + v \qquad v = \frac{3a - u^4}{6u}$$

for $u \neq 0$, and $f_{a,b}(0) = O$, the neutral element of the elliptic curve. When $q \equiv 2 \pmod 3$ we have that $x \mapsto x^3$ is a bijection in $\mathbb{F}_q$ so cube roots are uniquely defined with $x^{1/3} = x^{(2q-1)/3}$. We recall the following properties of $f_{a,b}$:

**Lemma 1 (Icart).** *The function $f_{a,b}$ is computable in deterministic polynomial time. For any point $\varpi \in f_{a,b}(\mathbb{F}_q)$, the set $f_{a,b}^{-1}(\varpi)$ is computable in polynomial time and $\#f_{a,b}^{-1}(\varpi) \leq 4$. Moreover $q/4 < \#f_{a,b}(\mathbb{F}_q) < q$.*

---

[1] For example in Boneh-Franklin IBE one could then decrypt any ciphertext.

**Table 1.** Known deterministic hashing algorithms into ordinary elliptic curves with discriminant $\Delta \neq 0$. We denote by $Q$ the set of quadratic residues. In char 2 we denote by $n$ the extension degree.

| char($K$) | normal form | discriminant $\Delta$ | encoding | condition |
|-----------|-------------|-----------------------|----------|-----------|
| $\neq 2, 3$ | $y^2 = x^3 + ax + b$ | $-16(4a^3 + 27b^2)$ | Icart [19] | $p \equiv 2 \pmod 3$ |
| | | | SW [25] | $-$ |
| | | | SWU [26] | $-$ |
| | | | SWU, Sec. 7 | $p \equiv 3 \pmod 4$ |
| 2 | $y^2 + xy = x^3 + ax^2 + b$ | $b$ | Icart [19] | odd $n$ |
| | | | SW [25] | $-$ |
| 3 | $y^2 = x^3 + ax^2 + b$ | $-a^3 b$ | Sec. 8.1 | $\Delta \in Q$ |
| | | | Sec. 8.2 | $\Delta \notin Q$ |
| | | | Sec. 8.3 | $-$ |

## 2.2 Indifferentiability

We recall the notion of indifferentiability introduced by Maurer *et al.* in [22].

**Definition 1 (Indifferentiability [22]).** *A Turing machine $C$ with oracle access to an ideal primitive $h$ is said to be $(t_D, t_S, q_D, \varepsilon)$-indifferentiable from an ideal primitive $H$ if there exists a simulator $S$ with oracle access to $H$ and running in time at most $t_S$, such that for any distinguisher $\mathcal{D}$ running in time at most $t_D$ and making at most $q_D$ queries, it holds that:*

$$\left| \Pr\left[ \mathcal{D}^{C^h, h} = 1 \right] - \Pr\left[ \mathcal{D}^{H, S^H} = 1 \right] \right| < \varepsilon$$

*$C^h$ is said to be indifferentiable from $H$ if $\varepsilon$ is a negligible function of the security parameter $k$, for polynomially bounded $q_D$, $t_D$ and $t_S$.*

It is shown in [22] that the indifferentiability notion is the "right" notion for substituting one ideal primitive by a construction based on another ideal primitive. That is, if the construction $C^h$ is indifferentiable from an ideal primitive $H$, then $C^h$ can replace $H$ in any cryptosystem, and the resulting cryptosystem is at least as secure in the $h$ model as in the $H$ model; see [22] or [14] for a proof.

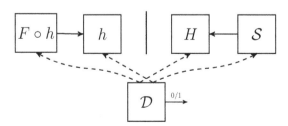

**Fig. 1.** The indifferentiability notion, illustrated with construction $C^h = F \circ h$ for some function $F$, and random oracles $h$ and $H$

# 3  Admissible Encodings and Indifferentiability

Our goal is to construct a hash function into elliptic curves that is indifferentiable from a random oracle. First, we introduce our new notion of *admissible encoding*. It can be seen as a generalization of the definition used in [4].

**Definition 2 (Admissible Encoding).** *A function* $F : S \to R$ *between finite sets is an $\varepsilon$-admissible encoding if it satisfies the following properties:*
1. *Computable: F is computable in deterministic polynomial time.*
2. *Regular: for s uniformly distributed in S, the distribution of $F(s)$ is $\varepsilon$-statistically indistinguishable from the uniform distribution in R.*
3. *Samplable: there is an efficient randomized algorithm $\mathcal{I}$ such that for any $r \in R$, $\mathcal{I}(r)$ induces a distribution that is $\varepsilon$-statistically indistinguishable from the uniform distribution in $F^{-1}(r)$.*

*F is an admissible encoding if $\varepsilon$ is a negligible function of the security parameter.*

The following theorem shows that if $F : S \to R$ is an admissible encoding, then the hash function $H : \{0,1\}^* \to R$ with:

$$H(m) := F(h(m))$$

is indifferentiable from a random oracle into $R$ when $h : \{0,1\}^* \to S$ is seen as a random oracle. This shows that the construction $H(m) = F(h(m))$ can replace a random oracle into $R$, and the resulting scheme remains secure in the random oracle model for $h$.

**Theorem 1.** *Let $F : S \to R$ be an $\varepsilon$-admissible encoding. The construction $H(m) = F(h(m))$ is $(t_D, t_S, q_D, \varepsilon')$-indifferentiable from a random oracle, in the random oracle model for $h : \{0,1\}^* \to S$, with $\varepsilon' = 4q_D\varepsilon$ and $t_S = 2q_D \cdot t_I$, where $t_I$ is the maximum running time of $F$'s sampling algorithm.*

*Proof.* We first describe our simulator; then we prove the indistinguishability property. As illustrated in Figure 1, the simulator must simulate random oracle $h$ to the distinguisher $\mathcal{D}$, and the simulator has oracle access to random oracle $H$. It maintains a list $L$ of previously answered queries. Our simulator is based on sampling algorithm $\mathcal{I}$ from $F$.

Simulator $\mathcal{S}$:
Input: $m \in \{0,1\}^*$
Output: $s \in S$
1. If $(m, s) \in L$, then return $s$
2. Query $H(m) = r$ and let $s \leftarrow \mathcal{I}(r)$
3. Append $(m, s)$ to $L$ and return $s$.

We must show that the systems $(C^h, h)$ and $(H, \mathcal{S}^H)$ are indistinguishable. We consider a distinguisher making at most $q_D$ queries. Without loss of generality, we can assume that the distinguisher makes all queries to $h(m)$ (or $\mathcal{S}^H$) for which there was a query to $C^h(m)$ (or $H(m)$), and conversely; this gives a total of at most $2q_D$ queries. We can then describe the full interaction between the distinguisher and the system as a sequence of triples:

$$\mathsf{View} = (m_i, s_i, r_i)_{1 \leq i \leq 2q}$$

where $s_i = h(m_i)$ (or $\mathcal{S}^H(m_i)$) and $r_i = C^h(m_i)$ (or $H(m_i)$). Without loss of generality we assume that the $m_i$'s are distinct.

In system $(C^h, h)$ we have that $s_i = h(m_i)$. Therefore the $s_i$'s are uniformly and independently distributed in $S$. Moreover we have $r_i = C^h(m_i) = F(s_i)$ for all $i$.

In system $(H, \mathcal{S}^H)$ we have that $r_i = H(m_i)$. Therefore the $r_i$'s are uniformly and independently distributed in $R$. Moreover we have $s_i = \mathcal{I}(r_i)$ for all $i$. The proof of the following Lemma is given in the full version of the paper [10]:

**Lemma 2.** *For $r$ uniformly distributed in $R$, the distribution of $s = \mathcal{I}(r)$ is $2\varepsilon$-statistically indistinguishable from the uniform distribution in $S$.*

This implies that in system $(H, \mathcal{S}^H)$ the distribution of $s_i = \mathcal{I}(r_i)$ is $2\varepsilon$-indistinguishable from the uniform distribution in $S$. Moreover from the definition of algorithm $\mathcal{I}$ we have that $r_i = F(s_i)$ except if $s_i = \bot$. Therefore, the statistical distance between $\mathsf{View}$ in system $(C^h, h)$ and $\mathsf{View}$ in system $(H, \mathcal{S}^H)$ is at most $4q_D\varepsilon$. This concludes the proof of Theorem 1. □

## 4    Our Main Construction

Let $E$ be an elliptic curve over a finite field $\mathbb{F}_q$ with $q \equiv 2 \pmod 3$. Let $f : \mathbb{F}_q \to E(\mathbb{F}_q)$ denote Icart's function to $E$. It is easy to see that Icart's function $f$ is *not* an admissible encoding into $E$ since as mentioned previously, the image of $f$ comprises only a fraction of the elliptic curve points. Therefore we cannot use the construction $H(m) = f(h(m))$ for indifferentiable hashing (not even on $\mathsf{Im}f$ since the distribution of $f(u)$ is not uniform in $\mathsf{Im}f$ for uniform $u \in \mathbb{F}_q$).

In this section, we describe a different construction which is almost as efficient. Namely we prove that if $h_1, h_2 : \{0, 1\}^* \to \mathbb{F}_q$ are two hash functions in the random oracle model, then the hash function $H : \{0, 1\}^* \to E(\mathbb{F}_q)$ defined by

$$H(m) := f(h_1(m)) + f(h_2(m))$$

is indifferentiable from a random oracle into the elliptic curve.

**Theorem 2.** *If $q > 2^{13}$ is any $2k$-bit prime power congruent to 2 mod 3 (even or odd), and if the $j$-invariant of $E$ is not in $\{0; 2592\}$, then the function*

$$H(m) := f(h_1(m)) + f(h_2(m))$$

*is $(t_D, t_S, q_D, \varepsilon')$-indifferentiable from a random oracle, where $\varepsilon' = 2^{10} \cdot q_D \cdot 2^{-k}$, in the random oracle model for $h_1, h_2 : \{0, 1\}^* \to \mathbb{F}_q$.*

Theorem 2 implies that this construction $H(m)$ can be used in any cryptosystem provably secure with random oracles into elliptic curves, and the resulting cryptosystem remains secure in the random oracle model for $h_1$ and $h_2$. We note that to prevent timing attacks (as in [7]), our construction $H$ can easily be implemented in constant time since Icart's function can be implemented in constant time.

To prove this result, it is enough, in view of Theorem 1, to show that the function $F : (\mathbb{F}_q)^2 \to E(\mathbb{F}_q)$ given by:

$$F(u, v) = f(u) + f(v)$$

is an $\varepsilon$-admissible encoding with $\varepsilon = 2^8 \cdot q^{-1/2}$.

$F$ is clearly computable in deterministic polynomial time, so Criterion 1 of admissible encodings is satisfied. To prove Criterion 2, we denote for any $\varpi \in E(\mathbb{F}_q)$:

$$N(\varpi) = \#\{(u, v) \in (\mathbb{F}_q)^2 \mid f(u) + f(v) = \varpi\} = \#F^{-1}(\varpi)$$

**Proposition 1.** *If $q$ is an odd prime power congruent to $2 \bmod 3$, and if the $j$-invariant of $E$ is not in $\{0; 2592\}$, then for every point $\varpi \in E(\mathbb{F}_q)$ except at most 144, we have*

$$|q - N(\varpi)| \leq 2^7 \cdot \sqrt{q}$$

*and all the remaining points $\varpi$ satisfy $N(\varpi) \leq 2^5 \cdot q$.*

Sections A.1 and A.2 are devoted to the proof of this proposition. Intuitively, the idea of the proof is to show that, for all points $\varpi \in E(\mathbb{F}_q)$ except a few exceptional ones, $F^{-1}(\varpi)$ is an irreducible algebraic curve of bounded genus in the affine plane $\mathbb{A}^2$ over $\mathbb{F}_q$. The estimate for the number of points then follows from the Hasse-Weil bound.

In the full version of this paper, we show that Proposition 1 directly implies Criterion 2, and that Criterion 3 easily follows from the point counting of [15,16]. Additionally, we prove that $F$ is also an admissible encoding when using Icart's function $f$ in characteristic 2.

# 5 A More General Construction

Our construction of Section 4 has the advantage of being simple and efficient as it only requires two evaluations of Icart's function. However, the proof involves somewhat technical tools from algebraic geometry, and it is not so simple to adapt to other encoding functions, such as the SWU algorithm.

At the cost of a small performance penalty, however, we describe a more general construction that applies to a large class of encoding functions satisfying a few simple axioms. Those encoding functions include Icart's function, a simpler variant of the SWU function, new deterministic encodings in characteristic 3, etc. We call them *weak encodings*. They are defined as follows.

**Definition 3 (Weak Encoding).** *A function $f : S \to R$ between finite sets is said to be an $\alpha$-weak encoding if it satisfies the following properties:*

1. *Computable: f is computable in deterministic polynomial time.*
2. *$\alpha$-bounded: for s uniformly distributed in S, the distribution of $f(s)$ is $\alpha$-bounded in R, i.e. the inequality $\Pr_s[f(s) = r] \leq \alpha/\#R$ holds for any $r \in R$.*
3. *Samplable: there is an efficient randomized algorithm $\mathcal{I}$ such that $\mathcal{I}(r)$ induces the uniform distribution in $f^{-1}(r)$ for any $r \in R$. Additionally $\mathcal{I}(r)$ returns $N_r = \#f^{-1}(r)$ for all $r \in R$.*

*The function f is a weak encoding if $\alpha$ is a polynomial function of the security parameter.*

The main difference with an admissible encoding is that in Criterion 2, the distribution of $f(s)$ is only required to be $\alpha$-bounded instead of being $\varepsilon$-indistinguishable from the uniform distribution. More precisely Criterion 2 for a weak encoding requires:

$$\forall r \in R, \quad \Pr_s[f(s) = r] = \frac{\#f^{-1}(r)}{\#S} \leq \frac{\alpha}{\#R} \tag{1}$$

From inequality (1) we have that any invertible function with bounded pre-image and bounded $\#R/\#S$ is a weak encoding; in particular, this is the case for Icart's function (the proof is given in the full version of the paper [10]).

**Lemma 3.** *Icart's function $f_{a,b}$ is an $\alpha$-weak encoding from $\mathbb{F}_q$ to $E_{a,b}(\mathbb{F}_q)$, with $\alpha = 4N/q$, where $N$ is the order of $E_{a,b}(\mathbb{F}_q)$.*

When the output set is a group (such as the group of points on an elliptic curve), we demonstrate how to construct an admissible encoding from any weak encoding.

**Theorem 3 (Weak → Admissible Encoding).** *Let $\mathbb{G}$ be cyclic group of order N noted additively, and let G be a generator of $\mathbb{G}$. Let $f : S \to \mathbb{G}$ be an $\alpha$-weak encoding. Then the function $F : S \times \mathbb{Z}_N \to \mathbb{G}$ with $F(s, x) := f(s) + xG$ is an $\varepsilon$-admissible encoding into $\mathbb{G}$, with $\varepsilon = (1 - 1/\alpha)^t$ for any t polynomial in the security parameter k, and $\varepsilon = 2^{-k}$ for $t = \alpha \cdot k$.*

We prove this theorem in the full version of this paper [10]. As a consequence, we get that if $f : S \to \mathbb{G}$ is any weak encoding to a cyclic group with generator G, then the hash function $H : \{0,1\}^* \to \mathbb{G}$ defined by:

$$H(m) := f(h_1(m)) + h_2(m)G$$

where $h_1 : \{0,1\}^* \to \mathbb{F}_p$ and $h_2 : \{0,1\}^* \to \mathbb{Z}_N$ are two hash functions, is indifferentiable from a random oracle in the random oracle model for $h_1$ and $h_2$. In particular, this is the case when $f$ is Icart's function. We note that for elliptic curves with non-cyclic group, we can easily adapt the previous construction with $H(m) = f(h_1(m)) + h_2(m)G_1 + h_3(m)G_2$ where $(G_1, G_2)$ are the generators of the group.

## 5.1   Discussion

We see that the construction $H(m) = f_{a,b}(h_1(m)) + f_{a,b}(h_2(m))$ of Section 4 requires two evaluations of Icart's function $f_{a,b}$ but no scalar multiplication. Since $f_{a,b}$ is essentially a field exponentiation, and in practice field exponentiation is roughly 10 times faster than scalar multiplication, the construction of Section 4 is approximately 5 times faster than the general construction of this section.

We note that for a number of existing schemes that are proven secure in the random oracle model into an elliptic curve, it would actually be sufficient to use $H(m) = f_{a,b}(h(m))$ only. This is because for many existing schemes the underlying complexity assumption (such as CDH or DDH) has the random self-reducibility property. So in the security proof one "programs" the RO using a random instance generated from the original problem instance. Then instead of letting $H(m) = P$ where $P$ is from the random instance, one can adapt the proof by letting $f(h(m)) = P$. To make sure that $h(m)$ is uniformly distributed, one can "replay" the random instance generation depending on the number of solutions to the equation $f(u) = P$, as we do in the proof of Theorem 3.

However it is easy to construct a cryptosystem that is secure in the ROM but insecure with $H(m) = f(h(m))$. Consider for example the following symmetric-key encryption scheme: to encrypt with symmetric key $k$, generate a random $r$ and compute $c = m + H(k, r)$ where the message $m$ is a point on the curve and $H$ hashes into the curve; the ciphertext is $(c, r)$. This scheme is semantically secure in the ROM for $H$, since this is a one-time pad. But the scheme is insecure with $H(k, r) = f(h(k, r))$ because in this case $H(k, r)$ is not uniformly distributed, and for two messages $m_0$ and $m_1$ the attacker has a good advantage in distinguishing between the encryption of $m_0$ and $m_1$.

## 6   Extensions

### 6.1   Extension to a Prime Order Subgroup

In many applications only a prime order subgroup of $E$ is used, so we show how to adapt the constructions of Sections 4 and 5 into a subgroup. Let $E$ be an elliptic curve over $\mathbb{F}_q$ with $N$ points, and let $\mathbb{G}$ be a subgroup of prime order $N'$ and generator $G$. Let $\ell$ be the co-factor, i.e. $N = \ell \cdot N'$. We require that $N'$ does not divide $\ell$ (i.e. that $(N')^2$ does not divide $N$), which is satisfied in practice for key size and efficiency reasons.

We show that it suffices to scalar multiply by co-factor $\ell$ the constructions of Sections 4 and 5 and the resulting constructions are still indifferentiable hash functions. More precisely, we consider the construction $H : \{0,1\}^* \to \mathbb{G}$ with:

$$H(m) := \ell\big(f_{a,b}(h_1(m)) + f_{a,b}(h_2(m))\big) \tag{2}$$

with $h_1, h_2 : \{0,1\}^* \to \mathbb{F}_q$ and $f_{a,b}$ is Icart's function.

**Proposition 2.** $H$ is $(t_D, t_S, q_D, \varepsilon)$-indifferentiable from a random oracle, in the random oracle model for $h_1$ and $h_2$, with $\varepsilon = 2^{10} \cdot q_D \cdot 2^{-k}$.

Informally, we show that the composition of two admissible encodings remains an (almost) admissible encoding, and that multiplication by a co-factor is an $\varepsilon$-admissible encoding, with $\varepsilon = 0$. This proves that $H$ is an indifferentiable hash function. See the full version of the paper [10] for the proof.

The same result holds for the construction of Section 5. In this case for both cyclic and non-cyclic elliptic curves we simply use $H(m) = \ell f(h_1(m)) + h_2(m)G$ where $G$ is a generator of the subgroup.

### 6.2    Extension to Random Oracles into Strings

The constructions in the previous sections are based on hash functions into $\mathbb{F}_{p^n}$ or $\mathbb{Z}_N$. However in practice a hash function outputs a fixed length string in $\{0,1\}^\ell$. We can modify our construction as follows. We consider an elliptic curve $E_{a,b}$ over $\mathbb{F}_p$, with $p$ a $2k$-bit prime. We define the hash function $H : \{0,1\}^* \to E_{a,b}(\mathbb{F}_p)$ with:

$$H(m) := f_{a,b}\big(h_1(m) \bmod p\big) + f_{a,b}\big(h_2(m) \bmod p\big)$$

where $h_1$ and $h_2$ are two hash functions from $\{0,1\}^*$ to $\{0,1\}^{3k}$ and $f_{a,b}$ is Icart's function.

**Proposition 3.** *The previous hash function $H$ is $(t_D, t_S, q_D, \varepsilon)$-indifferentiable from a random oracle, in the random oracle model for $h_1$ and $h_2$, with $\varepsilon = 2^{11} \cdot q_D \cdot 2^{-k}$.*

Informally, we first show that reduction modulo $p$ is an admissible encoding from $\{0,1\}^\ell$ to $\mathbb{F}_p$ if $2^\ell \gg p$. Since the composition of two admissible encodings remains an (almost) admissible encoding, this shows that $F(u, v) = f(u \bmod p) + f(v \bmod p)$ is also an admissible encoding into $E(\mathbb{F}_p)$ and therefore $H$ is an indifferentiable hash function. The same result holds for the general construction of Section 5. See the full version of the paper [10] for the proof.

## 7    A Simpler Variant of the SWU Algorithm

In this section, we describe a slightly simpler variant of the Shallue-Woestijne-Ulas (SWU) algorithm over $\mathbb{F}_q$, for $q \equiv 3 \pmod 4$. Note that this condition is usually satisfied in practice, since it enables to compute square roots efficiently.

**Proposition 4 (Simplified Ulas maps).** *Let $\mathbb{F}_q$ be a field and let $g(x) := x^3 + ax + b$, where $a, b \neq 0$. Let:*

$$X_2(t) = \frac{-b}{a}\left(1 + \frac{1}{t^4 - t^2}\right), \qquad X_3(t) = -t^2 X_2(t), \qquad U(t) = t^3 g(X_2(t))$$

*Then $U(t)^2 = -g\big(X_2(t)\big) \cdot g\big(X_3(t)\big)$.*

*Proof.* Let $g(x) = x^3 + ax + b$. Let $u$ be a non-quadratic residue and consider the equation in $x$:[2]

$$g(u \cdot x) = u^3 \cdot g(x) \qquad (3)$$

The first observation is that we can solve this equation for $x$ because the terms of degree 3 cancel:

$$g(u \cdot x) = u^3 \cdot g(x) \Leftrightarrow (ux)^3 + a(ux) + b = u^3(x^3 + ax + b)$$
$$\Leftrightarrow aux + b = u^3 ax + u^3 b$$
$$\Leftrightarrow x = \frac{b(u^3 - 1)}{a(u - u^3)} = \frac{-b}{a} \cdot \left(1 + \frac{1}{u + u^2}\right)$$

The second observation is that since $u$ is not a square, either $g(u \cdot x)$ or $g(x)$ must be a square. Therefore either $x$ or $u \cdot x$ must be the abscissa of a point on the curve. Moreover when $q \equiv 3 \pmod 4$ we have that $-1$ is a quadratic non-residue and we can take $u = -t^2$. Finally from (3) we get:

$$g(u \cdot x) \cdot g(x) = u^3 \cdot g^2(x) = -t^6 \cdot g^2(x) = -(t^3 \cdot g(x))^2$$

which gives the maps of Proposition 4.                                            □

Simplified SWU algorithm:
Input: $\mathbb{F}_q$ such that $q \equiv 3 \pmod 4$, parameters $a, b$ and input $t \in \mathbb{F}_q$
Output: $(x, y) \in E_{a,b}(\mathbb{F}_q)$ where $E_{a,b} : y^2 = x^3 + ax + b$

1. $\alpha \leftarrow -t^2$
2. $X_2 \leftarrow \frac{-b}{a} \left(1 + \frac{1}{\alpha^2 + \alpha}\right)$
3. $X_3 \leftarrow \alpha \cdot X_2$
4. $h_2 \leftarrow (X_2)^3 + a \cdot X_2 + b$;   $h_3 \leftarrow (X_3)^3 + a \cdot X_3 + b$
5. If $h_2$ is a square, return $(X_2, h_2^{(q+1)/4})$, otherwise return $(X_3, h_3^{(q+1)/4})$

In the full version of the paper [10] we show that our simplified SWU algorithm is a weak encoding into the curve. Therefore it can be used with the general construction from Section 5. An implementation is also provided in the full version of the paper [10].

## 8    Hashing in Characteristic 3

In characteristic 3 the normal form of an elliptic curve with $j$-invariant $j \neq 0$ and discriminant $\Delta \neq 0$ is:

$$Y^2 = X^3 + aX^2 + b$$

with $\Delta = -a^3 b$. It is easy to see that Icart's technique cannot work in characteristic 3, and the SWU algorithm does not work in characteristic 3 because the

---

[2]  A similar equation was used in [24] to show that there exists infinitely many elliptic-curves with $j$-invariant equal to given $j \neq 0, 1728$ and with Mordell-Weil rank $\geq 2$.

elliptic curve has a different equation. In this section we show the first deterministic[3] encoding algorithms for elliptic curves in characteristic 3. We denote by $Q$ the set of quadratic residues in the field. An implementation of the three algorithms is provided in the full version of the paper [10].

## 8.1 Algorithm for $\Delta \in Q$

**Proposition 5.** *Let $\mathbb{F}$ be a field of characteristic 3 and $g(x) = x^3 + ax^2 + b$ with $a \neq 0$ and $\Delta = -a^3 b \in Q$. Let $\eta \notin Q$ and let $c$ such that $c^2 = -b/a$. Let*

$$X(t) = c \cdot \left(1 - \frac{1}{\eta \cdot t^2}\right)$$

*Then either $g(X(t))$ or $g(\eta \cdot t^2 \cdot X(t))$ is a quadratic residue.*

*Proof.* As previously we choose $u \notin Q$ and we consider the equation in $x$:

$$g(u \cdot x) = u^3 \cdot g(x) \tag{4}$$

As previously the terms of degree 3 cancel, and using $u^3 - 1 = (u - 1)^3$ in char 3, we get:

$$g(u \cdot x) = u^3 \cdot g(x) \Leftrightarrow au^2 x^2 + b = au^3 x^2 + bu^3$$

$$\Leftrightarrow x^2 = \frac{b(u^3 - 1)}{a(u^2 - u^3)} = \frac{b(u-1)^3}{au^2(1-u)} = \frac{-b}{a} \cdot \left(\frac{u-1}{u}\right)^2$$

Since $\Delta = -a^3 b \in Q$, we have $-b/a \in Q$ so we can compute $c$ such that $c^2 = -b/a$. Therefore we can take the following solution for equation (4):

$$x = c \cdot \left(1 - \frac{1}{u}\right)$$

For $u$ we can take $u = \eta \cdot t^2$ where $\eta \notin Q$ is pre-computed. We recover the map $X(t)$ of Proposition 5. Moreover from equation (4) since $u^3 \notin Q$ either $g(x)$ or $g(u \cdot x)$ must be a quadratic residue.                                        □

From Proposition 5 we easily deduce a deterministic encoding algorithm.

## 8.2 Algorithm for $\Delta \notin Q$

**Proposition 6.** *Let $\mathbb{F}$ be a field of characteristic 3 and $g(x) = x^3 + ax^2 + b$ with $\Delta = -a^3 b \notin Q$. Let $x_0 \in \mathbb{F}$ such that $g(x_0) = 0$. Let $\eta \notin Q$. Let :*

$$X(t) = -2 \cdot x_0 \cdot \left(1 + \frac{1}{\eta \cdot t^2}\right)$$

*Let $X_1(t) = X(t) + x_0$ and $X_2(t) = \eta \cdot t^2 \cdot X(t) + x_0$. Then either $g(X_1(t))$ or $g(X_2(t))$ is a quadratic residue.*

*Proof.* When $\Delta \notin Q$ we have that $g(x) = x^3 + ax^2 + b$ has a (unique) root $x_0 \in \mathbb{F}$. Therefore we can let:

---

[3] We allow for a probabilistic pre-computation phase given the elliptic curve parameters.

$$f(x) = g(x + x_0) = x^3 + ax^2 + b'x$$

where $b' = 2 \cdot a \cdot x_0$. A deterministic encoding for elliptic curves of equation $y^2 = x^3 + ax^2 + b'x$ is already described in [26]. Given $u \notin Q$ one considers the equation in $x$:

$$
\begin{aligned}
f(u \cdot x) = u^3 \cdot f(x) &\Leftrightarrow au^2x^2 + b'ux = au^3x^2 + b'u^3x \\
&\Leftrightarrow ax(u^2 - u^3) = b'(u^3 - u) \\
&\Leftrightarrow axu^2(1 - u) = b'u(u - 1)(u + 1) \\
&\Leftrightarrow x = \frac{-b'}{a} \cdot \left(\frac{u+1}{u}\right) = -2 \cdot x_0 \cdot \left(1 + \frac{1}{u}\right)
\end{aligned}
$$

Then either $f(x)$ or $f(u \cdot x)$ is a square, which implies that either $g(x + x_0)$ or $g(u \cdot x + x_0)$ is a square. Letting $u = \eta \cdot t^2$ where $\eta \notin Q$ one recovers the maps $X(t)$, $X_1(t)$ and $X_2(t)$.     □

### 8.3    Algorithm for Any $\Delta$

In this section we describe a different encoding algorithm that works for any discriminant $\Delta$. We pre-compute $\eta \notin Q$ and $z_0, y_0$ such that $a\eta \cdot z_0^2 - y_0^2 + b = 0$.

Deterministic Encoding Algorithm in char 3:
Input: $t \in \mathbb{F}$
Output: $(x, y) \in E(\mathbb{F})$
  1. Let $z = (-z_0t^2 + 2y_0t - a\eta z_0)/(a\eta - t^2)$
  2. Let $y = y_0 + t \cdot (z - z_0)$
  3. Let $k = a/(b - y^2)$
  4. Find the unique solution $\alpha$ of the linear system $\alpha^3 + k \cdot \alpha = -k/a$
  5. Let $x = 1/\alpha$ and output $(x, y)$

We show in Appendix B that this also defines a deterministic encoding into elliptic curves.

## Acknowledgments

We would like to thank Pierre-Alain Fouque and the anonymous referees for useful comments on this paper.

## References

1. Baek, J., Zheng, Y.: Identity-based threshold decryption. In: Bao, F., Deng, R., Zhou, J. (eds.) PKC 2004. LNCS, vol. 2947, pp. 262–276. Springer, Heidelberg (2004)
2. Bellare, M., Rogaway, P.: Random oracles are practical: A paradigm for designing efficient protocols. In: ACM Conference on Computer and Communications Security, pp. 62–73 (1993)

3. Boldyreva, A.: Threshold signatures, multisignatures and blind signatures based on the gap-diffie-hellman-group signature scheme. In: Desmedt, Y.G. (ed.) PKC 2003. LNCS, vol. 2567, pp. 31–46. Springer, Heidelberg (2002)

4. Boneh, D., Franklin, M.K.: Identity-based encryption from the weil pairing. In: Kilian, J. (ed.) CRYPTO 2001. LNCS, vol. 2139, pp. 213–229. Springer, Heidelberg (2001)

5. Boneh, D., Gentry, C., Lynn, B., Shacham, H.: Aggregate and verifiably encrypted signatures from bilinear maps. In: Biham, E. (ed.) EUROCRYPT 2003. LNCS, vol. 2656, pp. 416–432. Springer, Heidelberg (2003)

6. Boneh, D., Lynn, B., Shacham, H.: Short signatures from the weil pairing. In: Boyd, C. (ed.) ASIACRYPT 2001. LNCS, vol. 2248, pp. 514–532. Springer, Heidelberg (2001)

7. Boyd, C., Montague, P., Nguyen, K.Q.: Elliptic curve based password authenticated key exchange protocols. In: Varadharajan, V., Mu, Y. (eds.) ACISP 2001. LNCS, vol. 2119, pp. 487–501. Springer, Heidelberg (2001)

8. Boyen, X.: Multipurpose identity-based signcryption (a swiss army knife for identity-based cryptography). In: Boneh, D. (ed.) CRYPTO 2003. LNCS, vol. 2729, pp. 383–399. Springer, Heidelberg (2003)

9. Boyko, V., MacKenzie, P.D., Patel, S.: Provably secure password-authenticated key exchange using diffie-hellman. In: Preneel, B. (ed.) EUROCRYPT 2000. LNCS, vol. 1807, pp. 156–171. Springer, Heidelberg (2000)

10. Brier, E., Coron, J.-S., Icart, T., Madore, D., Randriam, H., Tibouchi, M.: Efficient indifferentiable hashing into ordinary elliptic curves. Cryptology ePrint Archive, Report 2009/340 (2009) (full version of this paper), http://eprint.iacr.org/

11. Canetti, R., Goldreich, O., Halevi, S.: The random oracle methodology, revisited. J. ACM 51(4), 557–594 (2004)

12. Cha, J.C., Cheon, J.H.: An identity-based signature from gap diffie-hellman groups. In: Desmedt, Y.G. (ed.) PKC 2003. LNCS, vol. 2567, pp. 18–30. Springer, Heidelberg (2002)

13. Chevallier-Mames, B.: An efficient cdh-based signature scheme with a tight security reduction. In: Shoup, V. (ed.) CRYPTO 2005. LNCS, vol. 3621, pp. 511–526. Springer, Heidelberg (2005)

14. Coron, J.-S., Dodis, Y., Malinaud, C., Puniya, P.: Merkle-damgård revisited: How to construct a hash function. In: Shoup, V. (ed.) CRYPTO 2005. LNCS, vol. 3621, pp. 430–448. Springer, Heidelberg (2005)

15. Farashahi, R.R., Shparlinski, I.E., Voloch, J.F.: On hashing into elliptic curves (2010) (preprint), http://www.ma.utexas.edu/users/voloch/preprint.html

16. Fouque, P.-A., Tibouchi, M.: Estimating the size of the image of deterministic hash functions to elliptic curves. Cryptology ePrint Archive, Report 2010/037 (2010), http://eprint.iacr.org/

17. Gentry, C., Silverberg, A.: Hierarchical id-based cryptography. In: Zheng, Y. (ed.) ASIACRYPT 2002. LNCS, vol. 2501, pp. 548–566. Springer, Heidelberg (2002)

18. Horwitz, J., Lynn, B.: Toward hierarchical identity-based encryption. In: Knudsen, L.R. (ed.) EUROCRYPT 2002. LNCS, vol. 2332, pp. 466–481. Springer, Heidelberg (2002)

19. Icart, T.: How to hash into elliptic curves. In: Halevi, S. (ed.) CRYPTO 2009. LNCS, vol. 5677, pp. 303–316. Springer, Heidelberg (2009)

20. Jablon, D.P.: Strong password-only authenticated key exchange. SIGCOMM Comput. Commun. Rev. 26(5), 5–26 (1996)

21. Libert, B., Quisquater, J.-J.: Efficient signcryption with key privacy from gap diffie-hellman groups. In: Bao, F., Deng, R., Zhou, J. (eds.) PKC 2004. LNCS, vol. 2947, pp. 187–200. Springer, Heidelberg (2004)
22. Maurer, U.M., Renner, R., Holenstein, C.: Indifferentiability, impossibility results on reductions, and applications to the random oracle methodology. In: Naor, M. (ed.) TCC 2004. LNCS, vol. 2951, pp. 21–39. Springer, Heidelberg (2004)
23. Menezes, A., Okamoto, T., Vanstone, S.A.: Reducing elliptic curve logarithms to logarithms in a finite field. IEEE Transactions on Information Theory 39(5), 1639–1646 (1993)
24. Mestre, J.-F.: Rang de courbe elliptiques d'invariant donné. Comptes rendus de l'Académie des sciences. Série 1, Mathématique 314(12), 297–319 (1992)
25. Shallue, A., van de Woestijne, C.E.: Construction of rational points on elliptic curves over finite fields. In: Hess, F., Pauli, S., Pohst, M. (eds.) ANTS 2006. LNCS, vol. 4076, pp. 510–524. Springer, Heidelberg (2006)
26. Ulas, M.: Rational points on certain hyperelliptic curves over finite fields. Bull. Polish Acad. Sci. Math. 55(2), 97–104 (2007)
27. Zhang, F., Kim, K.: Id-based blind signature and ring signature from pairings. In: Zheng, Y. (ed.) ASIACRYPT 2002. LNCS, vol. 2501, pp. 533–547. Springer, Heidelberg (2002)

# A    Proof of Proposition 1

This appendix gives a proof of Proposition 1. For the sake of brevity, the proofs of some technical lemmas are omitted in this extended abstract, and can be found in the full version [10].

## A.1    Geometric Interpretation of Icart's Function

Icart's function $f$ admits a natural extension to the projective line over $\mathbb{F}_q$ by setting $f(\infty) = O$, the neutral element of the elliptic curve. Then, consider the graph of $f$:
$$C = \{(u, \varpi) \in \mathbb{P}^1 \times E \mid f(u) = \varpi\}$$
As shown in [19, Lemma 3], $C$ is the closed subscheme of $\mathbb{P}^1 \times E$ defined by

$$u^4 - 6xu^2 + 6yu - 3a = 0 \tag{5}$$

In other words, Icart's function is the algebraic correspondence between $\mathbb{P}^1$ and $E$ given by (5).

Let $j$ be the $j$-invariant of $E$:

$$j = 1728 \cdot \frac{4a^3}{4a^3 + 27b^2} \in \mathbb{F}_q$$

Save for a few exceptional values of $j$, we can precisely describe the geometry of $C$.

**Lemma 4.** *If $j \notin \{0; 2592\}$, the subscheme $C$ is a geometrically integral curve on $\mathbb{P}^1 \times E$ with one triple point at infinity and no other singularity. Its normalization $\widetilde{C}$ is a smooth, geometrically integral curve of genus 7. The natural map $h\colon \widetilde{C} \to E$ is a morphism of degree 4 ramified at 12 distinct finite points of $E(\overline{\mathbb{F}}_q)$, with ramification index 2.*

## A.2   The Square Correspondence

In this context, the function $(u, v) \mapsto f(u) + f(v)$ occurring in our hash function construction admits the following description. A point $(u, v)$ in the affine plane $\mathbb{A}^2$, or more generally in $\mathbb{P}^1 \times \mathbb{P}^1$, corresponds to $\varpi$ on the elliptic curve $E$ if and only if there is some point $(\alpha, \beta) \in \widetilde{C} \times \widetilde{C}$ over $(u, v)$ such that $h(\alpha) + h(\beta) = \varpi$.

Consider the surface $S = \widetilde{C} \times \widetilde{C}$, and define the following two morphisms. The map $p : S \to \mathbb{P}^1 \times \mathbb{P}^1$ is the square of the first projection, and $s : S \to E$ is obtained by composing $h \times h : S \to E \times E$ with the group law $E \times E \to E$. Then the set of points $(u, v) \in \mathbb{P}^1 \times \mathbb{P}^1$ corresponding to a given $\varpi \in E$ is exactly $p(s^{-1}(\varpi))$ (and we can take the intersection with $\mathbb{A}^2$ if we are only interested in affine points). This allows us to give a geometric proof of Proposition 1.

Let us first describe the geometry of the fibers $s^{-1}(\varpi)$. Denote by $\rho_1, \ldots, \rho_{12}$ the 12 geometric points of $E$ over which $h$ is ramified, and let $R = \{\rho_i + \rho_j\}_{1 \leq i,j \leq 12} \subset E$. The map $s$ is of rank 1 at $(\alpha, \beta)$ if and only if $h$ is of rank 1 at at least one of $\alpha$ or $\beta$, which is certainly the case when $h(\alpha)$ or $h(\beta)$ is not one the $\rho_i$. Therefore, $s$ is smooth of relative dimension 1 over the open subscheme $E_0 = E - R$, and all points in $E_0$ have smooth curves on $S$ as fibers. The following lemma makes this more precise.

**Lemma 5.** *The fibers of $s$ at all geometric points of $E_0$ are smooth connected curves on $S_{\overline{\mathbb{F}}_q}$ of genus 49.*

Consider now a fiber $Z$ of $s$ at some $\mathbb{F}_q$-point $\varpi$ of $E$ not in $R$. The previous description says that $Z$ is a smooth geometrically integral curve of genus 49 on $S$. This gives a precise estimate of the number of $\mathbb{F}_q$-points on $Z$ in view of the Hasse-Weil bound:

$$\left| q + 1 - \#Z(\mathbb{F}_q) \right| \leq 98\sqrt{q}$$

What we are interested in, however, is the number of points in $p(Z)$, or more precisely even, in $p(Z) \cap \mathbb{A}^2$. But those numbers are related in a simple way when Icart's function is well-defined, i.e. $q \equiv 2 \pmod 3$.

**Lemma 6.** *Suppose that $q \equiv 2 \pmod 3$, and let $N$ be the number of $\mathbb{F}_q$-points in $p(Z) \cap \mathbb{A}^2$. Then we have*

$$q - 98\sqrt{q} - 23 \leq N \leq q + 98\sqrt{q} + 1$$

The first part of Proposition 1 now follows from the previous propositions: under the hypotheses of that theorem, if $\varpi \in E(\mathbb{F}_q)$ does not belong to $R$, then $N(\varpi) = \#\{(u, v) \in (\mathbb{F}_q)^2 \mid f(u) + f(v) = \varpi\}$ satisfies

$$\left| q - N(\varpi) \right| \leq 98\sqrt{q} + 23 \leq 2^7 \cdot \sqrt{q}$$

as required. And obviously, there are at most $12^2 = 144$ points in $R$.

It remains to bound $N(\varpi)$ for an $\mathbb{F}_q$-point $\varpi \in R \cap E(\mathbb{F}_q)$. To do so, consider again $Z = s^{-1}(\varpi)$ the fiber at such a point, and $E' \subset E \times E$ the image of $Z$

under $h \times h$ (or equivalently, the fiber of the group law of $E$ at $\varpi$). The morphism $Z \to E'$ is of degree 16, so each point has at most 16 pre-images. Hence

$$N(\varpi) \leq 16 \cdot \#E'(\mathbb{F}_q) \leq 16(q + 1 + 2\sqrt{q}) \leq 2^5 \cdot q$$

since $q \geq 5$. This concludes the proof.

# B     Analysis of the Algorithm from Section 8.3

We consider the elliptic curve equation $y^2 = x^3 + ax^2 + b$ which we rewrite $x^3 + ax^2 + (b - y^2) = 0$. Letting $\alpha = 1/x$, we get:

$$\frac{1}{\alpha^3} + \frac{a}{\alpha^2} + (b - y^2) = 0$$

Multiplying by $\alpha^3/(b - y^2)$, this gives:

$$\alpha^3 + \frac{a}{b - y^2} \cdot \alpha = -1/(b - y^2) \tag{6}$$

Given $k \in \mathbb{F}$ we consider the function $f(\alpha) = \alpha^3 + k \cdot \alpha$. In char 3 this is a linear function. We have:

$$f(\alpha) = 0 \Leftrightarrow \alpha = 0 \text{ or } \alpha^2 = -k$$

Therefore $f$ is bijective if and only if $-k \notin Q$. When $f$ is bijective its inverse can be computed in deterministic polynomial time by solving a linear system.

Since $k = a/(b - y^2)$ in equation (6), we must have $-a/(b - y^2) \notin Q$ so that equation (6) has a unique solution. This is equivalent to $-(b - y^2)/a \notin Q$ or $-(b - y^2)/a = \eta \cdot z^2$ for some fixed $\eta \notin Q$. This gives:

$$a\eta z^2 - y^2 + b = 0$$

which is the equation of a conic which is easy to parameterize. Such parameterization is computed at steps 1 and 2 of the algorithm in Section 8.3.

# Credential Authenticated Identification and Key Exchange

Jan Camenisch[1], Nathalie Casati[1], Thomas Gross[1], and Victor Shoup[2]

[1] IBM Research, work funded by the European Community's Seventh Framework Programme (FP7/2007-2013) under grant agreement no. 216483
[2] NYU, work done while visiting IBM Research, supported by NSF grant CNS-0716690

**Abstract.** This paper initiates a study of two-party identification and key-exchange protocols in which users authenticate themselves by proving possession of credentials satisfying arbitrary policies, instead of using the more traditional mechanism of a public-key infrastructure. Definitions in the universal composability framework are given, and practical protocols satisfying these definitions, for policies of practical interest, are presented. All protocols are analyzed in the common reference string model, assuming adaptive corruptions with erasures, and no random oracles. The new security notion includes password-authenticated key exchange as a special case, and new, practical protocols for this problem are presented as well, including the first such protocol that provides resilience against server compromise (without random oracles).

## 1 Introduction

Secure two-party authentication and key exchange are fundamental problems. Traditionally, the parties authenticate each other by means of their identities, using a public-key infrastructure (PKI). However, this is not always feasible or desirable: an appropriate PKI may not be available, or the parties may want to remain anonymous, and not reveal their identities.

To address these needs, we introduce the notion of credential-authenticated identification (CAID) and key exchange key exchange (CAKE), where the compatibility of the parties' *credentials* is the criteria for authentication, rather than the parties' *identities* relative to some PKI.

We assume that prior to the protocol, the parties agree upon a *policy*, which specifies the types of credentials they each should hold, along with additional constraints that each credential should satisfy, and (possibly) relationships that should hold *between* the two credentials. The protocol should then determine whether or not the two parties have credentials that satisfy the policy, and in the CAKE case, should generate a session key, which could then be used to implement a secure communication session between the two parties. In any case, neither party should learn anything else about the other party's credentials, other than whether or not they satisfied the policy.

T. Rabin (Ed.): CRYPTO 2010, LNCS 6223, pp. 255–276, 2010.

For example, Alice and Bob may agree on a policy that says that Alice should hold an electronic ID card that says her age is at least 18, and that Bob should hold a valid electronic library card. If Alice then inputs an appropriate ID card and Bob inputs an appropriate library card, the protocol should succeed, and, in the CAKE case, both parties should obtain a session key. However, if, say, Alice tries to run the protocol without an appropriate ID card, the protocol should fail; moreover, Alice should not learn anything at all about Bob's input; in particular, Alice should not even be able to tell whether Bob had a library card or not.

As mentioned above, we may even consider policies that require that certain relationships hold between the two credentials. For example, Alice and Bob may agree upon a policy that says that they both should have national ID cards, and that they should live in the same state.

Both of the two previous examples illustrate that the CAKE problem is closely related to the "secret handshake" problem. In the latter problem, two parties wish to determine if they belong to the same group, so that neither party's status as a group member is revealed to the other, unless both parties belong to the same group. There are many papers on secret handshakes (see [12] for a recent paper, and the references therein). The system setup assumptions and security requirements vary significantly among the papers in the secret handshakes literature, and so we do not attempt a formal comparison of CAKE with secret handshakes. Nevertheless, the two problems share a common motivation, and to the extent that one can view owning a credential as belonging to a group, the two problems are very similar.

We also observe that the CAKE problem essentially includes the PAKE (password-authenticated key exchange) problem as a special case: the credentials are just passwords, and the policy says that the two passwords must be equal.

**Our Contributions.** So that our results are as general as possible, we work in the Universal Composability (UC) framework of Canetti [7]. We give natural ideal functionalities for CAID and CAKE, and give efficient, modularly designed protocols that realize these functionalities. If the underlying credential system is practical and comes equipped with practical proof-of-ownership protocols (such as the IDEMIX system, based on Camenisch and Lysyanskaya [5]), and if the policies are not too complex, the resulting CAKE protocols are fairly practical. In addition, if the credential system provides extra features such as traceability or revocability, or other mechanisms that mitigate against unwanted "credential sharing", then our protocols inherit these features as well.

All of our protocols are proved UC-secure in the adaptive corruption model, assuming parties can effectively erase internal data. Our protocols require a common reference string (CRS), but otherwise make use of standard cryptographic assumptions, and do not rely on random oracles.

As mentioned above, CAKE includes PAKE, and we also obtain two new practical PAKE protocols. The first is a practical PAKE protocol that is secure in the adaptive corruption model (with erasures); this is not the first such protocol (this was achieved recently by Abdalla, Chevalier, and Pointcheval [1], using completely different techniques). The second PAKE protocol is a simple variant of the first, but provides security against server compromise: the protocol is an asymmetric protocol run between a client, who knows the password, and a server, who only stores a function of the password; if the server is compromised, it is still hard to recover the password. Our new protocol is the first fairly practical PAKE protocol (UC-secure or otherwise) that is secure against server compromise and that does not rely on random oracles. Previous practical PAKE protocols that provide security against server compromise (such as Gentry, MacKenzie, and Ramzan [11]) all relied on random oracles (and also were analyzed only in the static corruption model).

**Outline of the paper.** In §2, we provide some background on the UC framework; in addition, we provide some recommendations for improving some of the low-level mechanics of the UC framework, to address some minor problems with the existing formulation in [7] that were uncovered in the course of this work. In any case, our results can be understood independently of these recommendations.

In §3, we introduce ideal functionalities for *strong* CAID and CAKE. These ideal functionalities are stronger than we want, as they can only be realized by protocols that use authenticated channels. Nevertheless, they serve as a useful building block. We also discuss there the types of policies that will be of interest to us here, as we want to restrict our attention to policies that are useful and that admit practical protocols.

In §4, we show how a protocol that realizes the strong CAID or CAKE functionalities can be easily and efficiently transformed into a protocol that realizes the CAID and CAKE functionality. The resulting protocol does not rely on authenticated channels. To this end, we utilize the idea of "split functionalities", introduced in [2]. Although the idea of using split functionalities for nonstandard authentication mechanisms was briefly mentioned in [2], it was not pursued there, and no new types of authentication protocols were presented. In this section, we review the basic notions introduced in [2], adjusting the definitions and results slightly to better meet our needs. We also give some new constructions, which are simpler and more efficient in the two-party setting.

In §5 we review definitions of UC zero knowledge (UCZK), and provide some new definitions that will be useful to us. UCZK will be a critical building block in the design of our CAID/CAKE protocols. In this section, we discuss a general language of statements we will want to be able to prove, as well as practical implementations of UCZK protocols for proving such statements. In a companion paper, we plan on fleshing out the details of this general framework, but it should be clear, based on these discussions, that there are, in fact, practical UCZK protocols for all the statements we need to prove in our CAID/CAKE protocols.

In §6, we present practical strong CAID/CAKE protocols for some fairly general policies of interest, and prove their security in the UC-framework, assuming secure channels. Using the split functionalities ideal in §4, these protocols can be transformed into practical CAID/CAKE protocols, which do not assume secure channels.

In §7, we present practical strong CAID/CAKE protocols for the equality relation and an interesting relation related to discrete logarithms. The former gives rise to our first new PAKE protocol, while the latter gives rise to our second new PAKE protocol (which provides resilience against server compromise).

Due to space limitations, many details, and all proofs, are left to the full paper [4].

## 2    Some UC Background

Our corruption model is always *adaptive corruptions with erasures*. We believe that allowing adaptive corruptions is important — there are known examples of protocols that are secure with respect to static corruptions, but trivially insecure if adaptive corruptions are allowed. Allowing erasures is a bit of a compromise: on the one hand, properly implementing secure erasures is difficult — but not impossible; on the other hand, if erasures are not allowed, then it becomes very difficult to obtain truly practical protocols, leading to results that are of theoretical interest only.

To streamline the descriptions of ideal functionalities, we assume the following convention in any two-party ideal functionality: *the adversary may at any time tell the ideal functionality to abort the protocol for one of the parties — the ideal functionality sends the special message* abort *to that party, and does not communicate any further with that party.*

In an actual protocol, an abort output would be generated when a "time out" or "error" condition was detected; the aborting party will also erase all internal data, and all future incoming messages will be ignored. While not essential for modeling security, it does allow us to distinguish between detectable and undetectable unfairness in protocols.

We clarify here a number of issues regarding terminology and notation in the UC framework. By a **party** we always mean an *interactive Turing machine (ITM)*. A party $P$ is addressed by **party ID (PID)** and **session ID (SID)**. So if $P$ has PID $P_{\text{pid}}$ and SID $P_{\text{sid}}$, then the PID/SID pair $(P_{\text{pid}}, P_{\text{sid}})$ uniquely identifies the party: no two parties in the system may have the same PID/SID pair. The convention is that the participants of any single protocol instance share the same SID, and conversely, if two parties share the same SID, then they are regarded as participants in the same protocol.

In [7] there are no semantics associated with PIDs, other than their role to distinguish participants in a protocol instance. Some authors (sometimes implicitly) tend to use the term "party" to refer to all ITMs that share a PID. We shall not do this: **a party is just a single ITM** (but see §2.2 below).

In describing protocols and ideal functionalities, we generally omit SIDs in messages — these can always be assumed to be implicitly defined.

### 2.1    Notions of Security

We recall some basic security notions from [7], with some extensions in [10] and [2]. We will not be too formal here.

We say that a protocol $\Pi$ **realizes** a protocol $\Pi^*$, if for every adversary $A$, there exists an adversary (i.e., simulator) $A^*$, such that for every environment $Z$, $Z$ cannot distinguish an attack of $A$ on $\Pi$ from an attack of $A^*$ on $\Pi^*$. Here, $Z$ is allowed to interact directly with the adversary and (via subroutine input/output) with parties (running the code for $\Pi$ or $\Pi^*$) that share the same SID.

If we like, we can remove the restriction that parties must share the same SID, which effectively allows $Z$ to interact with multiple, concurrently running instances of a single protocol in the above experiment. With this relaxation, we say that $\Pi$ **multi-realizes** $\Pi^*$. If these multiple instances of $\Pi$ access a *common* instance of a setup functionality $\mathcal{G}$, then we say that $\Pi$ multi-realizes $\Pi^*$ **with joint access to** $\mathcal{G}$. In applications, $\mathcal{G}$ is typically a common reference string (CRS).

The UC Theorem [7] implies that if $\Pi$ realizes $\Pi^*$, then $\Pi$ multi-realizes $\Pi^*$. However, if $\Pi$ makes use of a setup functionality $\mathcal{G}$, then it does not necessarily follow that $\Pi$ multi-realizes $\Pi^*$ with joint access to $\mathcal{G}$: one typically has to analyze the multiple-instance experiment directly.

In the above definitions, if $\Pi^*$ is the ideal protocol associated with an ideal functionality $\mathcal{F}$, then we simply say that $\Pi$ (multi-)realizes $\mathcal{F}$ (with joint access to $\mathcal{G}$). We also have some simple transitivity properties: if $\Pi_1$ realizes $\Pi_2$, and $\Pi_2$ realizes $\Pi_3$, then $\Pi_1$ realizes $\Pi_3$; also, if $\Pi_1$ multi-realizes $\Pi_2$ with joint access to $\mathcal{G}$, and $\Pi_2$ realizes $\Pi_3$, then $\Pi_1$ multi-realizes $\Pi_3$ with joint access to $\mathcal{G}$.

A protocol $\Pi$ may itself make use of an ideal functionality $\mathcal{F}'$ as a subroutine (where an instance of $\Pi$ may make use of multiple, independent instances of $\mathcal{F}'$). In this case, we call $\Pi$ an $\mathcal{F}'$**-hybrid protocol**. We may modify $\Pi$ by instantiating each instance of $\mathcal{F}'$ with an instance of a protocol $\Pi'$, and we denote the modified version of $\Pi$ by $\Pi[\mathcal{F}'/\Pi']$. The UC Theorem implies that if $\Pi'$ realizes $\mathcal{F}'$, then $\Pi[\mathcal{F}'/\Pi']$ realizes $\Pi$. Also, if $\Pi'$ multi-realizes $\mathcal{F}'$ with joint access to $\mathcal{G}$, then $\Pi[\mathcal{F}'/\Pi']$ multi-realizes $\Pi$ with joint access to $\mathcal{G}$.

This last statement is essentially a reformulation of a special case of the JUC Theorem [10], but in a form that is more convenient to apply. The notion of multi-realization (introduced, somewhat informally, in [2], and which can be easily expressed in the Generalized UC (GUC) framework [8]) seems a more elegant and direct way of modeling joint access to a CRS or similar setup functionality.

## 2.2    Conventions Regarding SIDs

We shall assume that an SID is structured as a **pathname**: $name_0/name_1/\cdots/name_k$. These pathnames reflect the subroutine call stack: when an honest party invokes an instance of subprotocol as a separate party, the new party has the same PID of the invoking party, and the SID is extended on the right by one element. Furthermore, we shall assume for two-party protocols, the rightmost element $name_k$, called the **basename**, has the form $ext : P_{\mathrm{pid}} : Q_{\mathrm{pid}} : data$, where $ext$ is a "local name" used to ensure unique basenames, $P_{\mathrm{pid}}$ and $Q_{\mathrm{pid}}$ are the PIDs of the participants $P$ and $Q$, and *data* represents shared public parameters. The ordering of these PIDs can be important in protocols where the two participants play different roles.

These conventions streamline and clarify a number of things. In application of the UC Theorem, we will be interested exclusively in protocols that act as subroutines: they are explicitly invoked by a single caller, who provides all inputs, and who receives all outputs.

The main points here are: (i) a subroutine is explicitly invoked by the caller, and (ii) the callee implicitly knows where to write its output. We can (and will) design protocols that deviate from this simple subroutine structure, although the UC Theorem will not directly apply in these cases.

With these restrictions, it also is convenient to make some restrictions on ideal functionalities: *we shall assume that an ideal functionality only delivers an output to a party that has previously supplied the ideal functionality with an input.*

These conventions are simply self-imposed restrictions, and do not represent a modification of the UC framework itself. However, in the full paper, we discuss some modifications to the UC framework that strictly impose these restrictions, along with a few other rules. Our rules guarantee that if $P$ is a party with PID *pid* and SID *sid*, and if $P'$ is a party with PID *pid* and SID *sid/basename*, then $P'$ is a subroutine of $P$ that was created by $P$, and moreover, so long as $P$ remains honest, then so does $P'$. As discussed in the full paper, we believe that without some type of restrictions such as these, there are some fundamental problems with the UC framework itself.

## 2.3    System Parameters

A common reference string, or CRS, is sometimes very useful. Sometimes, however, a different, but related notion is useful: a **system parameter**. Like a CRS, a system parameter is assumed to be generated by a trusted party, but unlike a system parameter, a CRS is visible to *all* parties, including the environment. A nice way to model this is using some elements of the GUC framework (although we do not attempt to design any protocols that achieve full GUC security here).

In designing a protocol that realizes some ideal functionality, a system parameter is a much better type of setup functionality than a CRS, as the security properties of protocols that use a CRS are not always so clear (e.g., "deniability" — see discussion in [8]). These problems do not arise with system parameters. Moreover, if a protocol $\Pi$ realizes an ideal functionality $\mathcal{F}$ using a system parameter, then it is easy to see that $\Pi$ multi-realizes $\mathcal{F}$ as well — there is no need to separately analyze a multi-instance experiment. A system parameter can also be used to parameterize an ideal functionality — a CRS cannot be used for this purpose, as that would conflate specification and implementation.

We can distinguish between two types of system parameters: **public coin** and **private coin**. In a public-coin system parameter, even the random bits used to generate the system parameter are visible to the environment (but no one else). In a private-coin system parameter, the random bits used to generate the system parameter remain hidden from all parties.

## 2.4   Authenticated Channels

We present here an ideal functionality for an authenticated channel. We have tuned this functionality to adhere to our conventions. We call this *ideal functionality $\mathcal{F}_{ach}$*.

For an SID is of the form $sid := parent/ext : P_{pid} : Q_{pid} :$, where $P$ is the sender and $Q$ is the receiver, and for an adversary $A$, the ideal functionality $\mathcal{F}_{ach}$ runs as follows:

1. *Wait for both: (a) an input message (send, $x$) from $P$, then send (send, $x$) to $A$;  (b) an input message* ready *from $Q$, then send* ready *to $A$.*
2. *Wait for the message* deliver *from $A$, then send the output message (deliver, $x$) to $Q$.*

**Corruption rule:** *If $P$ is corrupted between Steps 1a and 2, then $A$ is allowed to change the value of $x$ (at any time before Step 2).*

NOTES: *(i)* Like the corresponding functionality in [7], this one allows delivery of a single message per session. Multiple sessions should be used to send multiple messages. Alternatively, one could also define a multi-message functionality. *(ii)* Unlike the corresponding functionality in [7], the receiver here must explicitly initialize the channel before receiving a message. This design conforms to our conventions stated above, and is further discussed in the full paper.

## 2.5   Secure Channels

Secure channels provide both authentication and secrecy. We present a ideal functionality that is tuned to adhere to our conventions, and to our adaptive corruptions with erasures assumption.

One way to define secure channels is to modify $\mathcal{F}_{ach}$ as follows: in Step 1, send (send, len($x$)) to $A$, and in the corruption rule, $A$ is given $x$ and allowed to modify it $x$ as well. Here, len($x$) is the length of $x$. However, it turns out that a different functionality can be implemented more efficiently:

1. *Wait for both: (a) an input message (send, $x$) from $P$, then send the message (send, len($x$)) to $A$;  (b) an input message (ready, $maxlen$) from $Q$, then send the message (ready, $maxlen$) to $A$.*
2. *Wait for the message* lock *from $A$; verify that len($x$) $\leq maxlen$; if not, halt.*
3. *Wait for both: (a) a message* done *from $A$, then send the output message* done *to $P$; (b) a message* deliver *from $A$, then send the output message (deliver, $x$) to $Q$.*

**Corruption rule:** *If $P$ is corrupted between Steps 1a and 2, then $A$ is given $x$ and is allowed to change the value of $x$ (at any time before Step 2).*

We call this *ideal functionality $\mathcal{F}_{sch}$*. Here, the receiver specifies the maximum length message he is prepared to accept. This functionality reflects the fact that most of the time, the receiver knows the general "size and shape" of the message it is expecting, and so no additional interaction is required. In the cases where this information is not known in advance, the sender can transmit the length information to the receiver ahead of time on an authenticated channel.

# 3   Ideal Functionalities for Strong CAID and CAKE

In this section, we present ideal functionalities for *strong* CAID and CAKE. These ideal functionalities are stronger than we want, as they can only be realized by protocols that use authenticated channels. However, in the next section, we discuss how to we can very easily modify such protocols to obtain protocols that realize the desired CAID/CAKE functionalities (which will be defined in terms of strong CAID/CAKE).

We start with strong CAID. At a high level, the ideal functionality for strong CAID, denoted $\mathcal{F}^*_{caid}$, works as follows. We have two parties, $P$ and $Q$. $P$ and $Q$ agree (somehow) on a binary relation $R$, which consists of a set of pairs $(s,t)$. Then $P$ and $Q$ submit values to the ideal functionality: $P$ submits a value $s$ and $Q$ a value $t$. The ideal functionality then checks if $(s,t) \in R$; if so, it sends $P$ and $Q$ the value 1, and otherwise the value 0. The relation $R$ represents the "policy", discussed in §1.

The above description is lacking in details: some essential, and others not. We now describe some detailed variants of the above general idea. We assume that party $P$ has PID $P_{pid}$ and SID $P_{sid}$. Likewise, we assume that party $Q$ has PID $Q_{pid}$ and SID $Q_{sid}$.

## 3.1   Ideal Functionality $\mathcal{F}^*_{caid}$

We assume that the SIDs of the two parties are of the form $parent/ext : P_{pid} : Q_{pid} : \langle R \rangle$, where $\langle R \rangle$ is a description of the relation $R$. In principle, any efficiently computable family of relations is allowable, but specific realizations may implement only relations from some specific family of relations. It will convenient to assume that the special symbol $\perp$ has the following semantics: for all $s, t$, neither $(\perp, t)$ nor $(s, \perp)$ are in $R$.

An instance of $\mathcal{F}^*_{caid}$ with SID $parent/ext : P_{pid} : Q_{pid} : \langle R \rangle$ runs as follows.

1. *Wait for both:* (a) *an input message* (left-input, $s$) *from $P$, then send* left-input *to $A$;*   (b) *an input message* (right-input, $t$) *from $Q$, then send* right-input *to $A$.*
2. *Wait for a message* lock *from $A$; set res to 1 if $(s,t) \in R$, and 0 otherwise.*
3. *Wait for both:* (a) *a message* deliver-left *from $A$, then send the output message* (return, *res*) *to $P$;*   (b) *a message* deliver-right *from $A$, then send the output message* (return, *res*) *to $Q$.*

**Corruption rules:** (i) *If $P$ (resp., $Q$) is corrupted between Steps 1a (resp., 1b) and 2, then $A$ is given $s$ (resp., $t$), and is allowed to change the value of $s$ (resp., $t$) at any time before Step 2.*   (ii) *If $P$ (resp., $Q$) is corrupted between Steps 2 and Steps 3a (resp., 3b), then $A$ is given $s$ (resp., $t$).*

Note that the inclusion of $P_{pid}, Q_{pid}$ in the SID serves to break symmetry, and establish $P$ as the "left" party and $Q$ as the "right" party. The above ideal functionality captures the inherent "unfairness" in any such protocol: if one party is corrupt, they may learn that the relation holds, while the other may not.

However, such unfairness is at least detectable: since we do not conflate `abort` with a result of 0, if any party is being treated unfairly, this will at least be detected by an `abort` message. One could consider a weaker notion of security, in which 0 and `abort` were represented by the same value. While this may allow for more efficient protocols, such protocols may allow "undetectable unfairness". With our present formulation, a result of `abort` may indicate an unfair run of the protocol (or it may just indicate that there are network problems). The functionality $\mathcal{F}^*_{\text{caid}}$ does not provide as much privacy as one might like; in particular, if $P$ and $Q$ are honest, then $A$ still learns the relation $R$. In the full version of the paper, we discuss variations that prevent this.

## 3.2   From Authentication to Key Exchange

Functionality $\mathcal{F}^*_{\text{caid}}$ may be extended to provide key exchange in addition to authentication modifying Step 2 as follows:

2. *Wait for a message* $(\texttt{lock}, K_{\text{adv}})$ *from* $A$; *then set* res *to* $(1, K)$ *if* $(s, t) \in R$, *and* 0 *otherwise, where the key* $K$ *is determined as follows: if either* $P$ *or* $Q$ *are currently corrupted, set* $K := K_{\text{adv}}$; *otherwise, generate* $K$ *at random (according to some prescribed distribution).*

Corruption rules are unchanged. We call this *ideal functionality* $\mathcal{F}^*_{\text{cake}}$.

## 3.3   Some Relations of Interest

One type of relation that is of particular interest is a simple ***product relation***, where $R = S \times T$. For example, we may have $S = \{s : (x, s) \in E\}$, for a given $x$ and a fixed relation $E$. Here, $s$ might be an "anonymous credential" issued by some authority whose public key is $x$; the relation $E$ would assert that $s$ is a valid credential relative to $x$, possibly satisfying some other constraints as well.

A well-known example of an anonymous credential system of this type is the IDEMIX system [5]. This system comes with efficient zero-knowledge protocols for proofs of possession of credentials that we will be able to exploit. IDEMIX may also be equipped with mechanisms for identity escrow, revocation, etc., which automatically enhances the functionality of any strong CAID/CAKE protocol.

Similarly, we may have $T = \{t : (y, t) \in F\}$, for a given $y$ and fixed relation $F$. In this case the description $\langle R \rangle$ of $R$ is the pair $(x, y)$.

Two generalizations of potential interest are as follows. First, suppose we have binary relations $R_1, \ldots, R_k$. We can define their ***vectored union*** as the binary relation $R = \{((s_1, \ldots, s_k), (t_1, \ldots, t_k)) : (s_i, t_i) \in R_i \text{ for some } i = 1 .. k\}$. For example, each relation $R_i$ may represent a pair of "compatible" credentials, and the protocol should succeed if the two parties hold one such pair between them. Or more simply, the two parties may agree on a list of "clubs", and then determine if there is any one club to which they both belong.

Second, we might consider the intersection of a product relation with a ***partial equality relation***: $\{(s, t) : \sigma(s) = \tau(t)\}$, where $\sigma$ and $\tau$ are appropriate functions. Such relations can usefully model the "secret handshake" scenario, where $\sigma(s)$ and $\tau(t)$ perhaps represent "group names". A special case of this, of course, is the equality relation. A CAKE protocol for equality is essentially a PAKE protocol — this is discussed in §7.

One might even combine the above, considering vectored unions of such intersections. The reason for singling out these types of relations is that they are of potential practical interest, and admit efficient protocols.

# 4    Bootstrapping an Authentication Protocol

We shall presently give efficient protocols that realize strong CAID/CAKE functionalities for various relations of interest. All of these protocols work assuming secure channels. Of course, this is not interesting by itself, since we really want to use these protocols to establish secure channels in a setting without any existing authentication mechanism.

Without at least authenticated channels, it is impossible to realize strong CAID/CAKE. The solution is to weaken the notion of security, using the idea of "split functionalities", introduced in [2]. Our definitions of the CAID/CAKE functionalities are simply the split versions of the strong CAID/CAKE functionalities.

Although the idea of using split functionalities for nonstandard authentication mechanisms was briefly mentioned in [2], it was not pursued there, and no new types of authentication protocols were presented. In this section, we review the basic notions introduced in [2], adjusting the definitions and results slightly to better meet our needs, and give some new constructions, as well.

## 4.1    Details: Split Functionalities

We give a slight reformulation of the definitions and results in [2]: we focus on the two-party case, and we also make a few small syntactic changes that will allow us to apply the results in a more convenient way.

The basic idea is the same as in [2]. If $\mathcal{F}$ is a two-party ideal functionality involving two parties, $P$ and $Q$, then the split functionality $s\mathcal{F}$ works roughly as follows. Before any computation begins, the adversary partitions the set $\{P, Q\}$ into **authentication sets**: in the two-party case, the authentication sets are either $\{P\}$ and $\{Q\}$, or the single authentication set $\{P, Q\}$. The parties within an authentication set access a common instance of $\mathcal{F}$, while parties in different authentication sets access independent instances of $\mathcal{F}$. This is achieved by "mangling" SIDs appropriately: each authentication set is assigned a unique "channel ID" *chid*, which is used to "mangle" the SIDs of the instances of $\mathcal{F}$. Thus, the most damage an adversary can do is to make $P$ and $Q$ run two *independent* instances of $\mathcal{F}$.

As we shall see, one can transform any protocol $\Pi$ that realizes $\mathcal{F}$, where $\Pi$ relies on authenticated and/or secure channels, into a protocol $s\Pi$ that realizes $s\mathcal{F}$, where $s\Pi$ relies on neither authenticated nor secure channels. Moreover, $s\Pi$

is almost as efficient as $\Pi$. This result was first proved in [2]; however, we give a more efficient transformation — based on Diffie-Hellman key exchange — that is better suited to the two-party case.

Our CAID/CAKE functionalities are simply defined as the split versions of the strong CAID/CAKE functionalities: $\mathsf{s}\mathcal{F}^*_{\mathrm{caid}}$ and $\mathsf{s}\mathcal{F}^*_{\mathrm{cake}}$. Protocols for these functionalities may be obtained by applying the split transformation to the protocols for the corresponding strong functionalities.

### 4.2 General Split Functionalities

Now we give the general split functionality in more detail. Let $\mathcal{F}$ be an ideal functionality for a two party protocol. As in §2.2, we assume that the SID for $\mathcal{F}$ is of the form $parent/ext : P_{\mathrm{pid}} : Q_{\mathrm{pid}} : data$, and that $\mathcal{F}$ never generates an output for a party before receiving an input from that party.

The **split functionality** $\mathsf{s}\mathcal{F}$ has an SID $s := parent/ext : P_{\mathrm{pid}} : Q_{\mathrm{pid}} : data$ of the same form as $\mathcal{F}$, and for an adversary $A$ runs as follows.

- *Upon receiving a message* init *from a party* $X \in \{P, Q\}$: *record* $(\mathtt{init}, X_{\mathrm{pid}})$, *send* $(\mathtt{init}, X_{\mathrm{pid}})$ *to* $A$.
- *Upon receiving a message* $(\mathtt{authorize}, X_{\mathrm{pid}}, \mathcal{H}, chid)$ *from* $A$, *such that*
    (1) $X_{\mathrm{pid}}$ *is the PID of some* $X \in \{P, Q\}$; (2) $\{X_{\mathrm{pid}}\} \subseteq \mathcal{H} \subseteq \{P_{\mathrm{pid}}, Q_{\mathrm{pid}}\}$;
    (3) $(\mathtt{init}, X_{\mathrm{pid}})$ *has been recorded*; (4) *no tuple* $(\mathtt{authorize}, X_{\mathrm{pid}}, \ldots)$ *has been recorded; and*  (5) *if a tuple* $(\mathtt{authorize}, X'_{\mathrm{pid}}, \mathcal{H}', chid')$ *has been recorded, then either (a)* $\mathcal{H}' = \mathcal{H}$ *and* $chid' = chid$ *or (b)* $\mathcal{H}' \cap \mathcal{H} = \emptyset$ *and* $chid' \neq chid$
  *do the following:*
    (1) *if no tuple of the form* $(\mathtt{authorize}, \cdot, \mathcal{H}, chid)$ *has already been recorded, then initialize a "virtual" instance of* $\mathcal{F}$ *with SID* $sid_{\mathcal{H}} := chid/sid$; *we denote this instance* $\mathcal{F}_{\mathcal{H}}$ *and define* $chid_{\mathcal{H}} := chid$; *in addition, for each* $Y \in \{P, Q\}$, *if* $Y_{\mathrm{pid}} \notin \mathcal{H}$ *or* $Y$ *is corrupt, then notify* $\mathcal{F}_{\mathcal{H}}$ *that the party with PID* $Y_{\mathrm{pid}}$ *and SID* $sid_{\mathcal{H}}$ *is corrupt, and forward to* $A$ *the response of* $\mathcal{F}_{\mathcal{H}}$ *to this notification;*  (2) *record the tuple* $(\mathtt{authorize}, X_{\mathrm{pid}}, \mathcal{H}, chid)$;  (3) *send the output message* $(\mathtt{authorize}, chid)$ *to* $X$.
- *Upon receiving a message* $(\mathtt{input}, v)$ *from* $X \in \{P, Q\}$, *such that a tuple* $(\mathtt{authorize}, X_{\mathrm{pid}}, \mathcal{H}, chid)$ *has been recorded: send the message* $v$ *to* $\mathcal{F}_{\mathcal{H}}$, *as if coming as an input from the party with PID* $X_{\mathrm{pid}}$ *and SID* $sid_{\mathcal{H}}$.
- *Upon receiving a message* $(\mathtt{input}, X_{\mathrm{pid}}, \mathcal{H}, v)$ *from* $A$, *such that*
    (1) $X_{\mathrm{pid}}$ *is the PID of some* $X \in \{P, Q\}$,  (2) *a (uniquely determined) instance* $\mathcal{F}_{\mathcal{H}}$ *with* $X_{\mathrm{pid}} \in \mathcal{H}$ *has been initialized; and,*  (3) $X_{\mathrm{pid}} \notin \mathcal{H}$
  *send the message* $v$ *to* $\mathcal{F}_{\mathcal{H}}$, *as if coming as an input from the party with PID* $X_{\mathrm{pid}}$ *and SID* $sid_{\mathcal{H}}$.
- *Whenever an instance* $\mathcal{F}_{\mathcal{H}}$ *delivers an output* $v$ *to a party with PID* $X_{\mathrm{pid}}$, *where* $X_{\mathrm{pid}}$ *is the PID of some* $X \in \{P, Q\}$, *do the following: if* $X_{\mathrm{pid}} \in \mathcal{H}$, *then send the output message* $(\mathtt{output}, v)$ *to* $X$, *else send the output message* $(\mathtt{output}, X_{\mathrm{pid}}, v)$ *to* $A$.

- Upon receiving notification that a party $X \in \{P, Q\}$ is corrupted, such that a (uniquely determined) instance $\mathcal{F}_{\mathcal{H}}$ with $X_{\text{pid}} \in \mathcal{H}$ has been initialized: notify $\mathcal{F}_{\mathcal{H}}$ that the party with PID $X_{\text{pid}}$ and SID $sid_{\mathcal{H}}$ is corrupted, and forward to $A$ the response of $\mathcal{F}_{\mathcal{H}}$ to this notification.

We have a slightly different formulation of split functionalities than in [2], but the differences are mainly syntactic — our method of mangling the SIDs fits nicely in to our set of conventions on SIDs. In addition, in [2], a party is allowed to send an input as long as its authentication set is defined, whereas we require that a party wait for its explicit authorization notification before proceeding. This seems to avoid some potential confusion.

### 4.3  A Multi-session Secure Channels Functionality

We need a "multi-session extension" of our ideal functionality for secure channels. One approach would be to use the definition in [10]. However, a direct application of that definition would be unworkable, for two reasons: first, it would require that any implementation keep track of all subsession IDs that were ever used; second, the multi-session extension applies to all possible parties, whereas, we can really only deal with the same two parties in all subsessions. So for these reasons, we present our own multi-session extension, which we denote $\mathcal{F}_{\text{msc}}$. Note that in addition to secure channels (corresponding to the functionality $\mathcal{F}_{\text{sch}}$), it also provides for channels that only provide authentication (corresponding to the functionality $\mathcal{F}_{\text{ach}}$). It is quite tedious, and not very enlightening. The details are in the full paper.

### 4.4  Split Key Exchange

We now discuss a simple, low-level primitive: split key exchange. Let $\mathcal{K}$ be a key set. The **ideal functionality $\mathcal{F}_{\text{ske}}$** (parameterized by $\mathcal{K}$) has an SID of the form $parent/ext : P_{\text{pid}} : Q_{\text{pid}} :$, and for an adversary $A$, runs as follows:

- Upon receiving a message init from a party $X \in \{P, Q\}$: record $(\text{init}, X_{\text{pid}})$, send $(\text{init}, X_{\text{pid}})$ to $A$.
- Upon receiving a message $(\text{authorize}, X_{\text{pid}}, \mathcal{H}, chid, K)$ from $A$, such that
  (1) $X_{\text{pid}}$ is the PID of some $X \in \{P, Q\}$;  (2) $\{X_{\text{pid}}\} \subseteq \mathcal{H} \subseteq \{P_{\text{pid}}, Q_{\text{pid}}\}$;  (3) $K \in \mathcal{K}$;  (4) $(\text{init}, X_{\text{pid}})$ has been recorded;  (5) no tuple $(\text{authorize}, X_{\text{pid}}, \ldots)$ has been recorded; and,  (6) if a tuple $(\text{authorize}, X'_{\text{pid}}, \mathcal{H}', chid', K')$ has been recorded, then either (a) $\mathcal{H}' = \mathcal{H}$ and $chid' = chid$ or (b) $\mathcal{H}' \cap \mathcal{H} = \emptyset$ and $chid' \neq chid$
  do the following:
  (1) record the tuple $(\text{authorize}, X_{\text{pid}}, \mathcal{H}, chid, K)$;  (2) if $K_{\mathcal{H}}$ is not yet defined, then define it as follows: if $\mathcal{H} = \{X_{\text{pid}}\}$, then $K_{\mathcal{H}} \leftarrow K$, else $K_{\mathcal{H}} \leftarrow_{\text{R}} \mathcal{K}$;  (3) send the output message $(\text{key}, chid, K_{\mathcal{H}})$ to $X$.

We now present a simple protocol, $\Pi_{\mathrm{ske}}$, that realizes the functionality $\mathcal{F}_{\mathrm{ske}}$, under the decisional Diffie-Hellman (DDH) assumption. Assume a group $\mathbb{G}$ of prime order $q$ generated by $g \in \mathbb{G}$ where the DDH holds. The description of $\mathbb{G}$, $q$, and $g$ is viewed here as a system parameter. We also assume a PRG that maps a random $w \in \mathbb{G}$ to a pair of keys $(K, K_{\mathrm{auth}}) \in \mathcal{K} \times \mathcal{K}_{\mathrm{auth}}$, where $\mathcal{K}_{\mathrm{auth}}$ is some large set.

For two parties $P$ and $Q$ with SID $sid := parent/ext : P_{\mathrm{pid}} : Q_{\mathrm{pid}} :$, **protocol** $\Pi_{\mathrm{ske}}$ runs as follows. The roles played by $P$ and $Q$ are asymmetric. The protocol for $P$ runs as follows:

1. $P$ waits for an input init; then it computes $x \leftarrow_{\mathrm{R}} \mathbb{Z}_q$, $u \leftarrow g^x$, and sends $u$ to $Q$.
2. $P$ waits for $v \in \mathbb{G}$ from $Q$; then it computes $w \leftarrow v^x$, derives keys $K, K_{\mathrm{auth}}$ from $w$ using the PRG, sets $chid \leftarrow \langle u, v \rangle$, sends the key $K_{\mathrm{auth}}$ to $Q$ (after erasing all internal state other than $chid$ and $K$).
3. $P$ waits for a continuation signal, and then outputs and outputs (key, $chid$, $K$) (after erasing all internal state).

Note that in the UC framework, a party is allowed to only send one message at a time; therefore, $P$ first sends a message to $Q$ (via the adversary, of course), and then waits for a continuation signal (provided by the adversary) before delivering its own output.

The protocol for $Q$ runs as follows:

1. $Q$ waits for an input init; then it then does nothing, except to notify the network (i.e., adversary) that it is ready.
2. $Q$ waits for $u \in \mathbb{G}$ from $P$; then it computes $y \leftarrow_{\mathrm{R}} \mathbb{Z}_q$, $v \leftarrow g^y$, $w \leftarrow u^y$, derives keys $K, K_{\mathrm{auth}}$ from $w$, erases $y, w$, sets $chid \leftarrow \langle u, v \rangle$, and sends $v$ to $P$.
3. $Q$ waits for $K'_{\mathrm{auth}} \in \mathcal{K}_{\mathrm{auth}}$ from $P$; then it tests if $K_{\mathrm{auth}} = K'_{\mathrm{auth}}$; if so, it outputs (key, $chid$, $K$) (after erasing all internal state).

**Theorem 1.** *Assuming the DDH for $\mathbb{G}$, an appropriate PRG, and assuming the set $\mathcal{K}_{\mathrm{auth}}$ is large, protocol $\Pi_{\mathrm{ske}}$ realizes the ideal functionality $\mathcal{F}_{\mathrm{ske}}$.*

### 4.5   Realizing Split Multi-session Secure Channels

Our goal now is to realize the split version $\mathrm{s}\mathcal{F}_{\mathrm{msc}}$ of the multi-session secure channels functionality $\mathcal{F}_{\mathrm{msc}}$ presented in §4.3. This will be done with an $\mathcal{F}_{\mathrm{ske}}$-hybrid protocol $\Pi_{\mathrm{smsc}}$, where $\mathcal{F}_{\mathrm{ske}}$ is the split key exchange functionality discussed in §4.4. At a high-level, **protocol $\Pi_{\mathrm{smsc}}$** works as follows:

1. Wait for an input message init, then send the message init to $\mathcal{F}_{\mathrm{ske}}$.
2. Wait for a message (key, $chid$, $K$) from $\mathcal{F}_{\mathrm{ske}}$; then do the following:
   (a) derive subkeys required to implement bidirectional secure channels, erasing the key $K$; these channels will be implemented using a variant of Beaver and Haber's technique [3] (see full paper).  (b) generate the output message (authorize, $chid$).

3. *Now use the keys derived in the previous step to process the secure channels logic.*

**Theorem 2.** *The $\mathcal{F}_{ske}$-hybrid protocol $\Pi_{smsc}$ realizes the ideal functionality $s\mathcal{F}_{msc}$, assuming a secure PRG and secure MAC.*

### 4.6  Realizing General Split Functionalities

Let $\mathcal{F}$ be an arbitrary two-party ideal functionality. Let $\mathcal{G}$ be a setup functionality, such as a CRS. Let $\Pi$ be an $(\mathcal{F}_{ach}, \mathcal{F}_{sch})$-hybrid protocol that multi-realizes $\mathcal{F}$ with joint access to $\mathcal{G}$ (where $\mathcal{F}_{ach}$ is defined in §2.4 and $\mathcal{F}_{sch}$ is defined in §2.5).

Our goal is to use $\Pi$ to design an $s\mathcal{F}_{msc}$-hybrid protocol $s\Pi$ that multi-realizes $s\mathcal{F}$ with joint access to $\mathcal{G}$. The point is, $s\Pi$ does not require secure channels. Moreover, instantiating $s\mathcal{F}_{msc}$ with $\Pi_{smsc}$, we obtain the a protocol $s\Pi[s\mathcal{F}_{msc}/\Pi_{smsc}]$ that multi-realizes $s\mathcal{F}$ with joint access to $\mathcal{G}$.

At a high level, *protocol $s\Pi$* works as follows:

1. *Wait for an input message* init, *then send the message* init *to* $s\mathcal{F}_{msc}$.
2. *Wait for a message* (authorze, *chid*) *from* $s\mathcal{F}_{msc}$; *then do the following:* (a) *initialize a "virtual" instance of $\Pi$, assigning it a PID and SID that are the same as that of this protocol instance, except that the SID pathname is prefixed* chid; (b) *generate the output message* (authorize, *chid*).
3. *Proceed as follows:* (a) *process input requests by passing them to the virtual instance of $\Pi$;* (b) *pass along outputs of the virtual instance of $\Pi$ as outputs of this protocol instance;* (c) *use* $s\mathcal{F}_{msc}$ *to implement the secure channels used by the virtual instance of $\Pi$.*

**Theorem 3.** *If $\Pi$ is an $(\mathcal{F}_{ach}, \mathcal{F}_{sch})$-hybrid protocol that multi-realizes $\mathcal{F}$ with joint access to $\mathcal{G}$, then $s\Pi$ is an $s\mathcal{F}_{msc}$-hybrid protocol that multi-realizes $s\mathcal{F}$ with joint access to $\mathcal{G}$.*

This is essentially the same as the main technical result (Lemma 4.1) of [2], but there are some technical differences — see full paper for more discussion.

## 5  Practical UC Zero Knowledge

Before getting into strong CAID/CAKE protocols, we need to discuss an essential building block: practical protocols for UC ZK (zero knowledge). We will need a slightly stronger version of ZK, which we call "enhanced ZK". In the adaptive corruptions with erasures model, this is no more difficult to realize than ordinary ZK.

Let $R$ be a binary relation, consisting of pairs $(x, w)$: for such a pair, $x$ is called the "statement" and $w$ is called the "witness".

Let $\ell : \{0,1\}^* \to \{0,1\}^*$ be an "information leakage" function. The SID for an enhanced ZK protocol is of the form $parent/ext : P_{pid} : Q_{pid} :$, where $P$ is the prover and $Q$ the verifier. For an adversary $A$, an instance of *ideal functionality* $\mathcal{F}_{ezk}$ with SID $sid := parent/ext : P_{pid} : Q_{pid} :$ runs as follows:

1. *Wait for both:* (a) *an input message* (send, $x, w$) *from* $P$ *such that* $(x, w) \in R$, *then send the message* (send, $\ell(x)$) *to* $A$;  (b) *an input message* ready *from* $Q$, *then send* ready *to* $A$.
2. *Wait for the message* lock *from* $A$.
3. *Wait for both:* (a) *a message* done *from* $A$, *then send the output message* done *to* $P$;  (b) *a message* deliver *from* $A$, *then send the output message* (deliver, $x$) *to* $Q$.

**Corruption rule:** *If* $P$ *is corrupted between Steps 1a and 2, then* $A$ *is given* $(x, w)$ *and is allowed to change the value of* $(x, w)$ *to any value* $(x', w') \in R$ *(at any time before Step 2).*

Note the similarity with our secure channels functionality. Here, the functionality is parameterized by the information leakage function $\ell$, which is used to model the fact that some information about $x$ may be leaked to an eavesdropping adversary. Typically, this information will be some rough information about the "size and shape" of $x$ that ultimately determines the lengths of the ciphertexts that must be sent in an implementation.

**Parameterized relations.** In the above discussion, the relation $R$ was considered to be a fixed relation. However, for many applications, it is convenient to let $R$ be parameterized by a some system parameter (see §2.3). To realize the ZK (or extended ZK) functionality, it may be necessary to assume that the system parameter was generated in a certain way.

For example, a ZK protocol might require that the system parameter contains an RSA modulus $N$ that is the product of two primes. To realize the ZK ideal functionality, it might not even be necessary that the factorization of $N$ remain hidden. In such a case, the system parameter might be profitably viewed as a public-coin system parameter. This means that the environment may know the factorization of $N$, which may be useful to model situations where the factorization of $N$ is used, say, to sign messages in higher-level protocols that use a ZK protocol as a subprotocol.

## 5.1   Practical Protocols

Practical ZK protocols exist for the types of relations that we will be needed in our strong CAID/CAKE protocols — indeed, our protocols were designed with such protocols specifically in mind. In a companion paper, we give a detailed account of the current state of the art for such protocols. Here, we give a very brief sketch — see the full paper for more details.

We will be proving statements of the form

$$\exists\mathllap{\text{/}}\; w_1 \in \mathcal{D}_1, \ldots, w_n \in \mathcal{D}_n : \phi(w_1, \ldots, w_n). \tag{1}$$

Here, we use the symbol "$\exists\mathllap{\text{/}}$" instead of "$\exists$" to indicate that we are proving "knowledge" of a witness, rather than just its existence. The $\mathcal{D}_i$'s are domains which are finite intervals of integers centered around 0. $\phi$ is a predicate — we will presently place restrictions on the form of the domains and the predicate. A witness for a statement of the form (1) is a tuple $(w_1, \ldots, w_n)$ of integers such that $w_i \in \mathcal{D}_i$ for $i = 1 .. n$ and $\phi(w_1, \ldots, w_n)$. In cases where only the residue class of $w_i$ modulo $m$ is important, we may write the corresponding domain as $\mathbb{Z}_m$.

The predicate $\phi(w_1, \ldots, w_n)$ is given by a formula that is built up from "atoms" using arbitrary combinations of ANDs and ORs. An atom may express several types of relations among the $w_i$'s: (i) integer relations, such as $F = 0$, $F \geq 0$, $F \equiv 0 \pmod{m}$, or $\gcd(F, m) = 1$, where $F$ is an integer polynomial in the variables $w_1, \ldots, w_n$, and $m$ is a positive integer; (ii) group relations, such as $\prod_{j=1}^{k} g_j^{F_j} = 1$, where the $g_j$'s are elements of an abelian group, and the $F_j$'s are integer polynomials in the variables $w_1, \ldots, w_n$; the descriptions of the groups appearing in such atoms will in general be given as system parameters (see 2.3); the group order need not be known, but certain technical restrictions apply.

It is known how to construct efficient protocols for these types of statements that, under reasonable assumptions, multi-realize $\mathcal{F}_{\text{ezk}}$ with joint access to a CRS. (As discussed in the full paper, we actually allow corrupt provers to submit witnesses lying in somewhat larger intervals; the ideal functionality has to be modified to allow for this.) The computational complexity of these proof systems can be easily related to the arithmetic circuit complexity of the polynomials that appear in the description of $\phi$: the number of exponentiations is proportional to the sum of the circuit complexities; a more precise running time estimate depends on the types of groups and domains.

In some cases, we will write statements that quantify over certain variables using $\exists$ rather than $⋊$. Roughly speaking, witnesses quantified under $\exists$ are asserted just to exist, rather than to be explicitly "known" by the prover. Making sense of this formally requires some effort; however, the effort pays off in that the resulting ZK protocols may be substantially more efficient.

# 6    Strong CAID/CAKE Protocols

## 6.1    A Protocol for Vectored Unions of Product Relations

We present here a **protocol $\varPi_0$** for $\mathcal{F}_{\text{caid}}^*$ that works for a vectored union of product relations (see §3.3).

We assume the relation is described by values $x_1, \ldots, x_k$ and $y_1, \ldots, y_k$. Party $P$ has inputs $s_1 \in S_1^*, \ldots, s_k \in S_k^*$, and $Q$ has inputs $t_1 \in T_1^*, \ldots, t_k \in T_k^*$. They are trying to determine if $\bigvee_{i=1}^{k} \big[(x_i, s_i) \in E_i \wedge (y_i, t_i) \in F_i\big]$, for fixed relations $E_1, \ldots, E_k$ and $F_1, \ldots, F_k$.

We assume that as system parameters, we have a group $\mathbb{G}$ of prime order $q$, and random generator $g$. We will need to assume that the computational Diffie-Hellman (CDH) assumption holds in this group. This protocol also requires some extra machinery, described below.

1a. *P computes $h_{\mathrm{L}} \leftarrow_{\mathrm{R}} \mathbb{G}$ and sends $h_{\mathrm{L}}$ to $Q$ over a secure channel.*
1b. *Q computes $h_{\mathrm{R}} \leftarrow_{\mathrm{R}} \mathbb{G}$ and sends $h_{\mathrm{R}}$ to $P$ over a secure channel.*
2a. *P waits for $h_{\mathrm{R}}$, and then computes:*

$$\text{for } i = 1 .. k: \begin{cases} \alpha_i \leftarrow_{\mathrm{R}} \mathbb{Z}_q, \ \alpha_i' \leftarrow_{\mathrm{R}} \mathbb{Z}_q \\ \text{if } (x_i, s_i) \in E_i \text{ then } e_i \leftarrow g^{\alpha_i} \text{ else } e_i \leftarrow h_{\mathrm{R}}/g^{\alpha_i} \end{cases}$$

*Using $\mathcal{F}_{\mathrm{ezk}}$, $P$ proves to $Q$:*

$$\rtimes \{s_i \in S_i^*, \alpha_i' \in \mathbb{Z}_q\}_{i=1}^k : \left[ \bigwedge_{i=1}^k \Big( (x_i, s_i) \in E_i \vee g^{\alpha_i'} = h_{\mathrm{R}}/e_i \Big) \right].$$

*Note that $e_1, \ldots, e_k$ are delivered to $Q$ via the $\mathcal{F}_{\mathrm{ezk}}$ functionality after $P$ erases $\alpha_1', \ldots, \alpha_k'$.*
2b. *Q waits for $h_L$, and then computes:*

$$\text{for } i = 1 .. k: \begin{cases} \beta_i \leftarrow_{\mathrm{R}} \mathbb{Z}_q, \ \beta_i' \leftarrow_{\mathrm{R}} \mathbb{Z}_q \\ \text{if } (y_i, t_i) \in F_i \text{ then } f_i \leftarrow g^{\beta_i} \text{ else } f_i \leftarrow h_{\mathrm{L}}/g^{\beta_i'} \end{cases}$$

*Using $\mathcal{F}_{\mathrm{ezk}}$, $Q$ proves to $P$:*

$$\rtimes \{t_i \in T_i^*, \beta_i' \in \mathbb{Z}_q\}_{i=1}^k : \left[ \bigwedge_{i=1}^k \Big( (y_i, t_i) \in F_i \vee g^{\beta_i'} = h_{\mathrm{L}}/f_i \Big) \right].$$

*Note that $f_1, \ldots, f_k$ are delivered to $P$ via the $\mathcal{F}_{\mathrm{ezk}}$ functionality after $Q$ erases $\beta_1', \ldots, \beta_k'$.*
3a. *P computes: for $i = 1 .. k$: if $(x_i, s_i) \in E_i$ then $u_i \leftarrow f_i^{\alpha_i}$ else $u_i \leftarrow_{\mathrm{R}} \mathbb{G}$*
3b. *Q computes: for $i = 1 .. k$: if $(y_i, t_i) \in F_i$ then $v_i \leftarrow e_i^{\beta_i}$ else $v_i \leftarrow_{\mathrm{R}} \mathbb{G}$*
4. *P and Q run a strong CAID subprotocol to evaluate the predicate $\bigvee_{i=1}^k (u_i = v_i)$, and output the result of this computation after erasing all local data.*

NOTES: *(i)* We have reduced our original strong CAID problem to a simpler strong CAID problem in Step 4. We discuss implementations of Step 4 below. *(ii)* The intuition for the main idea of the protocol runs as follows. Suppose, for example, that $P$ is honest and $Q$ is corrupt. In Step 2b, $Q$ intuitively proves for each $i = 1 .. k$, either that it knows $t_i$ such that $(y_i, t_i) \in E_i$ or that it *does not know* $\beta_i$; in the latter case, $Q$ will not be able to predict the value $g^{\alpha_i \beta_i}$ when it comes to Step 4. *(iii)* Assuming the $E_i$'s and $F_i$'s are relations based on an anonymous credential system like IDEMIX, then all of the ZK protocols have relatively efficient implementations (see §5.1).

## 6.2  Security Analysis

Our goal now is to show that protocol $\Pi_0$ realizes $\mathcal{F}_{\mathrm{caid}}^*$. Note that $\Pi_0$ is a hybrid protocol that uses the following ideal functionalities as subroutines: secure channels (i.e., $\mathcal{F}_{\mathrm{sch}}$), enhanced ZK (i.e., $\mathcal{F}_{\mathrm{ezk}}$) for relations of the form appearing in Steps 2a and 2b of the protocol, and $\mathcal{F}_{\mathrm{caid}}^*$ for relations of the form appearing in Step 4 of the protocol.

**Theorem 4.** *Under the CDH assumption for $\mathbb{G}$, protocol $\Pi_0$ realizes $\mathcal{F}_{\mathrm{caid}}^*$.*

### 6.3   Implementing Step 4

In the case where $k = 1$, one can use the equality test protocol in §7. As an alternative to protocol $\Pi_0$, in the case where $k = 1$ one can use a different protocol altogether, described in the full paper.

In the general case where $k \geq 1$, we suggest the following method. Assume we have a UC protocol for evaluating an arithmetic circuit mod $N$, where $N$ is a system parameter that is the product of two large primes. Then to evaluate the boolean expression $\bigvee_{i=1}^{k}(u_i = v_i)$, $P$ chooses $a_0 \in \mathbb{Z}_N$ at random, and for $i = 1 .. k$, encodes $u_i$ as an element $a_i$ of $\mathbb{Z}_N$; similarly, $Q$ chooses $b_0 \in \mathbb{Z}_N$ at random, and for $i = 1 .. k$, encodes $v_i$ as an element $b_i$ of $\mathbb{Z}_N$. Then $P$ and $Q$ jointly evaluate in the expression $\prod_{i=0}^{k}(a_i - b_i)$ over $\mathbb{Z}_N$. If the boolean expression is true, then the expression over $\mathbb{Z}_N$ is zero; otherwise, the expression over $\mathbb{Z}_N$ evaluates to a random element of $\mathbb{Z}_N$.

Thus, we reduce the original strong CAID problem to a strong CAID problem for a simpler predicate, namely, boolean expressions of the form $\bigvee_{i=1}^{k}(u_i = v_i)$, and the latter is easily reduced to a simple circuit evaluation problem for expressions of the form $\prod_{i=0}^{k}(a_i - b_i)$ over $\mathbb{Z}_N$. There are quite practical protocols for circuit evaluation, which we discuss in detail in a companion paper. The basic idea is to use known techniques for circuit evaluation based on homomorphic encryption, making use of a semantically secure variant of Camenisch and Shoup's encryption scheme [6], which has the advantage that generating public keys is very inexpensive (making security with adaptive corruptions and erasures more practical) and proofs about plaintexts fit very nicely into the framework for ZK proofs discussed in §5.1. These protocols (and hence the resulting strong CAID protocols) require $O(k)$ exponentiations, and $O(\log k)$ (the circuit depth) rounds of communication, and $O(k)$ total communication complexity.

### 6.4   Adding Key Exchange

Adding key exchange is simple, especially since we are already assuming secure channels. We simply modify the protocol so that $P$ generates a random key, and sends it to $Q$ over a secure channel at the beginning of the protocol. In addition, whenever either party would output 1, it instead outputs $(1, K)$. This is a generic transformation that converts any $\mathcal{F}_{\text{caid}}^*$ protocol into an $\mathcal{F}_{\text{cake}}^*$ protocol. Some other variations of protocol $\Pi_0$, including one that deals with partial equality relations, are discussed in the full paper.

### 6.5   From Strong CAID/CAKE to CAID/CAKE

We can instantiate protocol $\Pi_0$ to get a practical $\mathcal{F}_{\text{sch}}$-hybrid protocol $\Pi_0'$ that multi-realizes $\mathcal{F}_{\text{caid}}^*$ (or any of the variations discussed above) with joint access to a CRS — the crucial building block is $\mathcal{F}_{\text{ezk}}$, discussed in §5. Then using the split functionalities techniques in §4, we can turn $\Pi_0'$ into a protocol $s\Pi_0'$ that multi-realizes $s\mathcal{F}_{\text{caid}}^*$ with joint access to a CRS. The resulting protocol is a CAID/CAKE protocol that works without secure channels.

Typically, the purpose of running a CAKE protocol is to use the session key to implement a secure session. If, in fact, this is the goal, a more straightforward way of achieving it is as follows. Simply design a $\mathcal{F}_{\text{sch}}$-hybrid protocol that works as follows: first, it runs a strong CAID protocol, and if that succeeds, the parties continue to communicate, using the secure channels provided by the $\mathcal{F}_{\text{sch}}$ functionality. Now apply the split functionalities techniques in §4 to this protocol, obtaining a protocol that essentially provides a "credential authenticated secure channel".

# 7   A Protocol for Equality Testing and a Related Problem

Here is a simple protocol for equality testing, called ***protocol $\Pi_{\text{eq}}$***. We assume that a group $\mathbb{G}$ of prime order $q$, along with a generator $g \in \mathbb{G}$, are given as system parameters. We will need to assume the DDH for $\mathbb{G}$. We assume the inputs to the two parties are encoded as elements of $\mathbb{Z}_q$. Again, we use $\mathcal{F}_{\text{ezk}}$ as a subprotocol. The protocol runs as follows, where $P$ has input $a \in \mathbb{Z}_q$, and $Q$ has input $b \in \mathbb{Z}_q$:

1. $P$ computes: $h \leftarrow_R \mathbb{G}$, $x_1, x_2, r \leftarrow_R \mathbb{Z}_q$, $c \leftarrow g^{x_1} h^{x_2}$, $u_1 \leftarrow g^r$, $u_2 \leftarrow h^r$, $e \leftarrow g^a c^r$, and using $\mathcal{F}_{\text{ezk}}$ proves to $Q$: $⋏\, a \in \mathbb{Z}_q\, \exists\, r \in \mathbb{Z}_q\,:\, g^r = u_1 \wedge h^r = u_2 \wedge g^a c^r = e$; note that $h, c, u_1, u_2, e$ are delivered to $Q$ via the $\mathcal{F}_{\text{ezk}}$ functionality after erasing $r$.
2. $Q$ computes: $s \leftarrow_R \mathbb{Z}_q^*$, $t \leftarrow_R \mathbb{Z}_q$, $\tilde{u}_1 \leftarrow u_1^s g^t$, $\tilde{u}_2 \leftarrow u_2^s h^t$, $\tilde{e} \leftarrow e^s g^{-bs} c^t$, and using $\mathcal{F}_{\text{ezk}}$ proves to $P$: $⋏\, b \in \mathbb{Z}_q\, \exists\, s, t \in \mathbb{Z}_q\,:\, u_1^s g^t = \tilde{u}_1 \wedge u_2^s h^t = \tilde{u}_2 \wedge e^s g^{-bs} c^t = \tilde{e} \wedge \gcd(s, q) = 1$; note that $\tilde{u}_1, \tilde{u}_2, \tilde{e}$ are delivered to $P$ via the $\mathcal{F}_{\text{ezk}}$ functionality after erasing $s, t$.
3. $P$ computes: $z \leftarrow_R \mathbb{Z}_q^*$, $d \leftarrow \tilde{e}^z (\tilde{u}_1)^{-zx_1} (\tilde{u}_2)^{-zx_2}$, and using $\mathcal{F}_{\text{ezk}}$ proves to $Q$: $\exists\, x_1, x_2, z \in \mathbb{Z}_q\,:\, g^{x_1} h^{x_2} = c \wedge \tilde{e}^z (\tilde{u}_1)^{-zx_1} (\tilde{u}_2)^{-zx_2} = d \wedge \gcd(z, q) = 1$; here, $d$ is delivered to $Q$ via the $\mathcal{F}_{\text{ezk}}$ functionality after erasing $x_1, x_2, z$.
4. After erasing all local data, both parties output 1 if $d = 1$, and output 0 otherwise.

NOTES: *(i)* We are using $\exists$ as well as $⋏$ quantifiers here. This allows for certain optimizations, since values quantified under $\exists$ are never explicitly needed in the simulator in the security proof below, other than to verify the that the corresponding relation holds. *(ii)* In Step 1, $(h, c)$ is the public key and $(x_1, x_2)$ the private key for "Cramer-Shoup Ultra-Lite" — the semantically secure version of Cramer-Shoup encryption. $(u_1, u_2, e)$ is an encryption of $g^a$. We will exploit the fact that this scheme is "receiver non-committing", as was demonstrated by Jarecki and Lysyanskaya [13]. This property will allow us to simulate adaptive corruptions. *(iii)* In Step 2, assuming $(u_1, u_2, e)$ encrypts $g^a$, then $(\tilde{u}_1, \tilde{u}_2, \tilde{e})$ is a random encryption of $g^{s(a-b)}$. *(iv)* In Step 3, $P$ is decrypting $(\tilde{u}_1, \tilde{u}_2, \tilde{e})$, and raising it to the power $z$, so that $d = g^{zs(a-b)}$ *(v)* All of the ZK protocols have practical implementations, as discussed in §5.1.

**Theorem 5.** *Assuming the DDH for $\mathbb{G}$, protocol $\Pi_{\text{eq}}$ realizes functionality $\mathcal{F}_{\text{caid}}^*$ for the equality relation.*

**Applications.** One application of protocol $\Pi_{eq}$ is in the implementation of Step 4 of protocol $\Pi_0$ (see §6.3). However, in this situation, a specialized protocol in the full paper is more efficient.

Another application is to PAKE protocols. We can efficiently implement $\mathcal{F}_{ezk}$ for the necessary relations using secure channels and a common reference string, and augment the protocol to share a random key over a secure channel. This gives us a fairly efficient strong CAKE protocol for the equality relation that uses secure channels. We then derive the split version of the protocol, using a simple Diffie-Hellman key exchange as in §4, which realizes the CAKE functionality (more precisely, it multi-realizes the CAKE functionality with joint access to a CRS). As observed in [2], a protocol that realizes this functionality in fact realizes the PAKE functionality (as defined in [9]). Our particular protocol is probably a bit less efficient than the one in [9]; however, our protocol has the advantage of being secure against adaptive corruptions (assuming erasures). A very different PAKE protocol, with a structure similar to that in [9], that is secure against adaptive corruptions was recently presented in [1].

### 7.1    A Variation

A variation on the above protocol gives a strong CAID protocol for the relation $DL := \{(a, g^a) : a \in \mathbb{Z}_q\}$. That is, it tests if $g^a = v$, where $a$ is the input to $P$ and $v$ is the input to $Q$. The idea is to have $Q$ "verifiably encrypt" $v$. The protocol, which we call ***protocol $\Pi_{dl}$***, runs as follows:

1. $P$ computes: $h \leftarrow_R \mathbb{G}$, $x_1, x_2, r \leftarrow_R \mathbb{Z}_q$, $c \leftarrow g^{x_1}h^{x_2}$, $u_1 \leftarrow g^r$, $u_2 \leftarrow h^r$, $e \leftarrow g^a c^r$, and using $\mathcal{F}_{ezk}$ proves to $Q$: $\lambda a \in \mathbb{Z}_q \exists r \in \mathbb{Z}_q : g^r = u_1 \wedge h^r = u_2 \wedge g^a c^r = e$; note that $h, c, u_1, u_2, e$ are delivered to $Q$ via the $\mathcal{F}_{ezk}$ functionality after erasing $r$.
2. $Q$ computes: $s \leftarrow_R \mathbb{Z}_q^*$, $t \leftarrow_R \mathbb{Z}_q$, $y \leftarrow_R \mathbb{Z}_q$, $\tilde{v} \leftarrow g^y v$, $\tilde{u}_1 \leftarrow u_1^s g^t$, $\tilde{u}_2 \leftarrow u_2^s h^t$, $\tilde{e} \leftarrow e^s v^{-s} c^t$ and using $\mathcal{F}_{ezk}$ proves to $P$: $\lambda y \in \mathbb{Z}_q \exists s, t \in \mathbb{Z}_q : u_1^s g^t = \tilde{u}_1 \wedge u_2^s h^t = \tilde{u}_2 \wedge e^s \tilde{v}^{-s} g^{ys} c^t = \tilde{e} \wedge \gcd(s, q) = 1$; note that $\tilde{v}, \tilde{u}_1, \tilde{u}_2, \tilde{e}$ are delivered to $P$ via the $\mathcal{F}_{ezk}$ functionality after erasing $y, s, t$.
3. $P$ computes: $z \leftarrow_R \mathbb{Z}_q^*$, $d \leftarrow \tilde{e}^z (\tilde{u}_1)^{-zx_1} (\tilde{u}_2)^{-zx_2}$, and using $\mathcal{F}_{ezk}$ proves to $Q$: $\exists x_1, x_2, z \in \mathbb{Z}_q : g^{x_1}h^{x_2} = c \wedge \tilde{e}^z (\tilde{u}_1)^{-zx_1} (\tilde{u}_2)^{-zx_2} = d \wedge \gcd(z, q) = 1$; here, $d$ is delivered to $Q$ via the $\mathcal{F}_{ezk}$ functionality after erasing $x_1, x_2, z$.
4. Both parties output 1 if $d = 1$, and output 0 otherwise.

NOTES: *(i)* Step 1 is exactly the same as before. *(ii)* In Step 2, $Q$ is generating a random encryption of $(g^a/v)^s$. Moreover, by giving $\tilde{v}$ and $y$ to $\mathcal{F}_{ezk}$, $Q$ is effectively giving $v$ to $\mathcal{F}_{ezk}$. *(iii)* Step 3 is the same as before, but now $d = (g^a/v)^{sz}$.

**Theorem 6.** *Assuming the DDH for $\mathbb{G}$, protocol $\Pi_{dl}$ realizes functionality $\mathcal{F}_{caid}^*$ for the relation DL.*

**Applications.** This protocol, when augmented with a key sharing step over a secure channel, and "split" as in §4, gives us a practical PAKE protocol that is secure against adaptive corruptions *and server compromise*. That is, the client stores the password $a$, while the server stores $g^a$. If the password file on the server is compromised, then it will not be easy to an attacker to login to the server as the client.

Unlike previous protocols, such as in [11], our protocol does not rely on random oracles. To be fair, the definition of security in [11] is so strong that it probably cannot be achieved without random oracles: the security definition in [11] requires that in the event of a server compromise, an attacker must carry out an offline dictionary attack in order to guess the password. Also, note that the protocol in [11] is proved secure only in the static corruption model.

In a complete PAKE protocol, one would likely set $a :=$ $H(pw, clientID, serverID)$, where $H$ is a cryptographic hash, $pw$ is the actual password, and *clientID* and *serverID* are the names of the client and server, respectively. If $H$ is entropy preserving, $pw$ is a high-entropy password, and the discrete logarithm problem in $\mathbb{G}$ is hard, then it will be infeasible to login as the client, even if the server is compromised. Moreover, if $H$ is modeled as a random oracle, and the discrete logarithm problem in $\mathbb{G}$ is hard, then even in the event of a server compromise, an attacker must still carry out an offline dictionary attack in order to login as the client. Thus, our new protocol is the first fairly practical PAKE protocol (UC-secure or otherwise) that is secure against server compromise and does not rely on random oracles; as a bonus, it is also secure against adaptive corruptions.

# References

1. Abdalla, M., Chevalier, C., Pointcheval, D.: Smooth projective hashing for conditionally extractable commitments. In: Halevi, S. (ed.) CRYPTO 2009. LNCS, vol. 5677, pp. 671–689. Springer, Heidelberg (2009)
2. Barak, B., Canetti, R., Lindell, Y., Pass, R., Rabin, T.: Secure computation without authentication. In: Shoup, V. (ed.) CRYPTO 2005. LNCS, vol. 3621, pp. 361–377. Springer, Heidelberg (2005), http://eprint.iacr.org/2007/464
3. Beaver, D., Haber, S.: Cryptographic protocols provably secure against dynamic adversaries. In: Rueppel, R.A. (ed.) EUROCRYPT 1992. LNCS, vol. 658, pp. 307–323. Springer, Heidelberg (1993)
4. Camenisch, J., Casati, N., Gross, T., Shoup, V.: Credential authenticated identification and key exchange. Cryptology ePrint Archive, Report 2010/055 (2010), http://eprint.iacr.org/
5. Camenisch, J., Lysyanskaya, L.: Efficient non-transferable anonymous multi-show credential system with optional anonymity revocation. In: Crypto 2001, pp. 93–118 (2001)
6. Camenisch, J., Shoup, V.: Practical verifiable encryption and decryption of discrete logarithms. In: Boneh, D. (ed.) CRYPTO 2003. LNCS, vol. 2729, pp. 126–144. Springer, Heidelberg (2003), http://eprint.iacr.org/2002/161
7. Canetti, R.: Universally composable security: a new paradigm for cryptographic protocols. Cryptology ePrint Archive, Report 2000/067 (December 14, 2005 version) (2005), http://eprint.iacr.org

8. Canetti, R., Dodis, Y., Pass, R., Walfish, S.: Universally composable security with global setup. In: Theory of Cryptography 2007, pp. 61–85 (2007), Full version at http://eprint.iacr.org/2006/432

9. Canetti, R., Halevi, S., Katz, J., Lindell, Y., MacKenzie, P.: Universally composable password-based key exchange. In: Cramer, R. (ed.) EUROCRYPT 2005. LNCS, vol. 3494, pp. 404–421. Springer, Heidelberg (2005)

10. Canetti, R., Rabin, T.: Universal composition with joint state. In: Boneh, D. (ed.) CRYPTO 2003. LNCS, vol. 2729, pp. 265–281. Springer, Heidelberg (2003), http://eprint.iacr.org/2002/047

11. Gentry, C., MacKenzie, P., Ramzan, Z.: A method for making password-based key exchange resilient to server compromise. In: Dwork, C. (ed.) CRYPTO 2006. LNCS, vol. 4117, pp. 142–159. Springer, Heidelberg (2006)

12. Jarecki, S., Kim, J., Tsudik, G.: Beyond secret handshakes: affiliation-hiding authenticated key agreement. In: Malkin, T.G. (ed.) CT-RSA 2008. LNCS, vol. 4964, pp. 352–369. Springer, Heidelberg (2008)

13. Jarecki, S., Lysyanskaya, A.: Adaptively secure threshold cryptography: introducing concurrency, removing erasures. In: Preneel, B. (ed.) EUROCRYPT 2000. LNCS, vol. 1807, pp. 221–242. Springer, Heidelberg (2000)

# Password-Authenticated Session-Key Generation on the Internet in the Plain Model

Vipul Goyal[1], Abhishek Jain[2,*], and Rafail Ostrovsky[3,**]

[1] Microsoft Research, India
vipul@microsoft.com
[2] UCLA
abhishek@cs.ucla.edu
[3] UCLA
rafail@cs.ucla.edu

**Abstract.** The problem of password-authenticated key exchange (PAKE) has been extensively studied for the last two decades. Despite extensive studies, no construction was known for a PAKE protocol that is secure in the plain model in the setting of *concurrent self-composition*, where polynomially many protocol sessions with the same password may be executed on the distributed network (such as the Internet) in an arbitrarily interleaved manner, and where the adversary may corrupt any number of participating parties.

In this paper, we resolve this long-standing open problem. In particular, we give the first construction of a PAKE protocol that is secure (with respect to the standard definition of Goldreich and Lindell) in the fully concurrent setting and without requiring any trusted setup assumptions. We stress that we allow polynomially-many concurrent sessions, where polynomial is not fixed in advance and can be determined by an adversary an an adaptive manner. Interestingly, our proof, among other things, requires important ideas from *Precise* Zero Knowledge theory recently developed by Micali and Pass in their STOC'06 paper.

## 1 Introduction

The problem of password authenticated key exchange (PAKE) has been studied since early 1990's. PAKE involves a pair of parties who wish to establish a high entropy session key in an authenticated manner when their *a priori* shared secret information only consists of a (possibly low entropy) password. More formally, the problem of PAKE can be modeled as a two-party functionality $\mathcal{F}$ involving a pair of parties $P_1$ and $P_2$; if the inputs (passwords) of the parties match, then $\mathcal{F}$ outputs a uniformly distributed session key, else it outputs $\perp$. Hence the goal of PAKE is to design a protocol that securely realizes the functionality $\mathcal{F}$. Unfortunately, positive results for secure multi-party computation (MPC) [1,2] do not immediately translate to this setting; the reason being that known solutions

---

* Supported in Part by NSF grants 0830803, 0916574.
** Supported in part by IBM Faculty Award, Xerox Innovation Group Award, the Okawa Foundation Award, Intel, Lockheed Martin, Teradata, NSF grants 0716835, 0716389, 0830803, 0916574 and U.C. MICRO grant..

T. Rabin (Ed.): CRYPTO 2010, LNCS 6223, pp. 277–294, 2010.

for secure MPC require the existence of authenticated channels – which is in fact the end goal of PAKE. Therefore, very informally speaking, secure multi-party computation and PAKE can be viewed as complementary problems.

The problem of password authenticated key exchange was first studied by Bellovin and Meritt [3]. This was followed by several additional works proposing protocols with only heuristic security arguments (see [4] for a survey). Subsequently, PAKE was formally studied in various models, including the random oracle/ideal cipher model, common reference string (CRS) model, and the plain model (which is the focus of this work). We briefly survey the state of the art on this problem. The works of Bellare et al [5] and Boyko et al [6] deal with defining and constructing PAKE protocols in the ideal cipher model and random oracle model respectively. In the CRS model, Katz, Ostrovsky and Yung [7] gave the first construction for PAKE without random oracles based on the DDH assumption. Their result were subsequently improved by Gennaro and Lindell [8], and Genarro [9]. Again in the CRS model, Canetti, Halevi, Katz, Lindell and MacKenzie [10] proposed new definitions and constructions for a PAKE protocol in the framework of Universal Composability [11]. They further proved the *impossibility* of a Universally Composable PAKE construction in the plain model.

Goldreich and Lindell [12] formulated a new simulation-based definition for PAKE and gave the first construction for a PAKE protocol in the *plain model*. Their construction was further simplified (albeit at the cost of a weaker security guarantee) by Nguyen and Vadhan [13]. Recently, Barak et al [14] gave a very general construction for a PAKE protocol that is secure in the bounded-concurrent setting (see below) in the plain model.

To date, [12,13] and [14] remain the only known solutions for PAKE in the plain model. However, an important limitation of Goldreich and Lindell [12] (as well as Nguyen and Vadhan [13]) is that their solution is only relevant to the stand-alone setting where security holds only if a single protocol session is executed on the network. A more natural and demanding setting is where several protocol sessions may be executed concurrently (a typical example being protocols executed over the Internet). In such a setting, an adversary who controls parties across different sessions may be able to mount a coordinated attack; as such, stand-alone security does not immediately translate to concurrent security [15]. In the context of PAKE, this problem was was fully resolved assuming CRS trusted setup (see below) and only partially addressed in the plain model by Barak, Canetti, Lindell, Pass and Rabin [14] who gave a construction that maintains security in the setting of *bounded-concurrency*. In this setting, an *a priori* bound is known over the number of sessions that may be executed concurrently at any time; this bound is crucially used in the design of the protocol. It is natural to consider the more general setting of full concurrent self-composition, where any polynomially many protocol sessions (with no a priori bound) with the same password may be executed in an arbitrary interleaved manner by an adversary who may corrupt any number of parties. We stress that although the works of [7,16,8,10,4] solve this problem (where [7,8] are secure under self-composition, and [16] also enjoy forward secrecy, while [10] is secure under general-composition), they all require a trusted setup in the form of a common reference string. Indeed, to date, no constructions are known for a PAKE protocol that is secure in the plain model in the setting of concurrent self-composition.

*Our Contribution.* In this paper, we resolve this open problem. In particular, we give the first construction of a PAKE protocol in the plain model that allows for concurrent executions of the protocol between parties with the same password. Our techniques rely on several previous works, most notably the works of Barak, Prabhakaran and Sahai [17] and Micali and Pass [18].

Our construction is proven secure as per the definition of Goldreich and Lindell [12] in the concurrent setting. We stress that Lindell's impossibility result [19] for concurrent self-composition is *not* applicable here since (a) Goldreich and Lindell used a *specific* definition that is different from the standard paradigm for defining secure computation[1], and (b) further, they only consider the scenario where the honest parties hold fixed inputs (while Lindell's impossibility result crucially requires adaptive inputs).

In fact, our security definition is somewhat stronger than the one by Goldreich and Lindell [12]. The definition in [12], for example, does not consider the case where the adversary may have some a priori information on the password of the honest parties in a protocol execution. We consider an improved simulation-based security model similar to that proposed by [6]. More specifically, in our model, the simulator in the ideal world is empowered to make a *constant* number of queries per (real world) session to the ideal functionality (as opposed to just one). Our security definition then requires computational indistinguishability of the output distributions of real and ideal world executions in keeping with the standard paradigm for secure computation. As noted in [20], this improved definition implies the original definition of Goldreich and Lindell (see full version for a proof).

In our main construction, we consider the setting where the honest parties across the (polynomially-many) concurrent executions hold the same password or independently chosen passwords[2]. An example of the same password case is

---

[1] Note that in the standard simulation paradigm, the output distributions of the "real" and "ideal" worlds must be computationally indistinguishable; in contrast, the definition of Goldreich and Lindell [12] allows these distributions to be $\mathcal{O}(1/|D|)$ apart (where $D$ is the password dictionary).

[2] A more general question is to consider the setting where the passwords of honest parties in different sessions might be *correlated in any arbitrary way*. Towards this end, we note that our construction can be easily extended to this setting. However, in this case we require the ideal simulator to be able to query the ideal functionality an *expected constant* number of times per session. Jumping ahead, in case the honest parties were using the same password or fully independent passwords, the simulator is able to "trade" ideal functionality calls in one session for another. Hence, the simulator is able to even out the number of calls to a fixed constant in each session. This in turn means that for the setting of correlated passwords, our construction will satisfy a security definition which is slightly weaker (in that the number of ideal functionality calls are constant only in expectation). Obtaining a construction for correlated (in an arbitrary way) passwords where the number of calls are not just constant in expectation but always bounded by a constant is left as an interesting open question.

when a server expects a specific password for authentication and several parties are trying to authenticate simultaneously.

We note that our techniques and constructions are quite general. Our construction can be instantiated with a basic semi-honest secure computation protocol for any PPT computable functionality. This would lead to a concurrently secure protocol for that functionality as per the security definition where we allow the simulator to make an expected constant number of calls to the ideal function per (real world) session. The meaningfulness of such a definition is shown in the case of password based key exchange which is the focus of this work (more precisely, by comparing it with the definition of [20]). However we anticipate that the above general construction with such security guarantees might be acceptable in many other settings as well.

A related model is that of *resettably secure computation* proposed by Goyal and Sahai [21]. In resettably secure computation, the ideal simulator is given the power to reset and query the trusted party any (polynomial) number of times. However there are important differences. Goyal and Sahai [21] consider only the "fixed role" setting and only one of the parties can be thought of as accepting concurrent sessions. This means that the key technical problems we face in the current work (arising out of the possibility of mauling attacks in the concurrent setting) do not arise in [21]. Secondly, [21] do not try to optimize (or even bound) the number of queries the ideal simulator makes to the trusted party per session.

*Overview of Main Ideas.* Note that in the setting of concurrent self-composition, an adversary may corrupt different parties across the various sessions. Consider for instance two different sessions where one of the parties is corrupted in each session. We can view one of these sessions as a "left" session and the other as a "right session", while the corrupted parties can be jointly viewed as an adversarial man-in-the-middle. An immediate side-effect of this setting is that it allows an adversary to possibly "maul" a "left" session in order to successfully establish a session key with an honest party (say) $P$ in a "right" session *without* the knowledge of $P$'s secret password. Clearly, in order to provide any security guarantee in such a setting, it is imperative to achieve independence between various protocol sessions executing on the network. Note that this is akin to guaranteeing non-malleability across various sessions in the concurrent setting. Then, as a first step towards solving this problem, we borrow techniques from the construction of concurrent non-malleable zero knowledge argument due to Barak, Prabhakaran and Sahai [17] (BPS-CNMZK). In fact, at a first glance, it might seem that compiling a semi-honest two-party computation protocol (that emulates the PAKE functionality in the stand-alone setting) with the BPS-CNMZK argument or some similar approach might fully resolve this problem. However, such an approach fails on account of several reasons. We highlight some important problems in such an approach.

We first note that the simulation of BPS-CNMZK is based on a rewinding strategy. In a concurrent setting, the adversary is allowed to control the scheduling of the messages of different sessions. Then for a given adversarial scheduling, it is possible that the simulator of BPS-CNMZK may rewind past the beginning of a session (say) $s$ when "simulating" another session. Now, every time session $s$ is re-executed, an adversary may be able to change his input (i.e., make a new password guess possibly based on the auxiliary information it has). In such a case, the simulator would have to query the ideal functionality for that session more than once; therefore, we need to allow the simulator to make extra (i.e., more than one) queries per session to ideal functionality. In order to satisfy our definition, we would need to limit the number of queries to a *constant* per session. However, the simulator for BPS-CNMZK, if used naively, may require large polynomially many queries per session to the ideal functionality, and therefore, fail to satisfy our definition.

In order to overcome this problem, we build on the techniques of precise simulation, introduced by Micali and Pass [18] in the context of (stand-alone) zero knowledge and later extended to the setting of concurrent zero knowledge by Pandey, Pass, Sahai, Tseng, and Venkitasubramaniam [22]. Specifically, Pandey et. al. [22] use a time-oblivious rewinding schedule that (with a careful choice of system parameters) ensures that the the time spent by the simulator in the "look-ahead" threads[3] is only within a *constant* factor of the time spent by the simulator in the "main" thread. We remark that we do not require this precision in simulation time; instead we require that the number of queries made by the simulator in the look-ahead threads is only within a constant factor of the number of queries made in the main thread. For this purpose, we employ the precise Zero-Knowedlge paradigm of Micali and Pass and consider an imaginary experiment in which our adversary takes a disproportionately large amount of time in generating the message after which the simulator has to query the trusted party. Our rewinding strategy is determined by running the PPSTV [22] simulator using the next message generation timings of such an (imaginary) adversary (even though our simulator is fully black-box and does not even measure the timings for the real adversary) in order to bound the number of queries.

We further note that in the security proof of the above approach, the simulator must be able to extract the inputs of the adversary in *all* the sessions in order to simulate its view. However, the extractor of [17] is unsuitable for this task since it can extract adversary's inputs (in the setting of BPS-CNMZK) only on a *session-by-session* basis. To further elaborate, let us first recall the setting of BPS-CNMZK, where an adversary is interacting with some honest provers as well as some honest verifiers. Now, in order to extract the input of an adversarial prover in a particular session $s$, the extractor in [17] honestly runs all the uncorrupted verifiers except the verifier in session $s$. We stress that the extractor is able to run the honest verifiers by itself since they do not possess any secret inputs; clearly, such an extraction technique would fail in our setting since the simulator does not know the inputs of the honest parties.

---

[3] Very roughly speaking, a "thread of execution" between the simulator and the adversary is a simulation of a prefix of an actual execution. The simulator may run multiple threads of execution, and finally output a single thread, called the *main thread*. Any other thread is referred to as a *look-ahead thread*.

To address this problem, we require each party in our protocol to commit to its input and randomness inside a separate preamble [22,23] that allows extraction of the committed values in a concurrent setting. However, we note that such a preamble requires a complicated rewinding strategy for extraction of committed value, and so is the case for simulating the BPS-CNMZK argument. Indeed, it seems that we might need to *compose* the (possibly conflicting) individual rewinding strategies of BPS-CNMZK and the additional preamble into a new uniform rewinding strategy. Fortunately, by ensuring that we use the same kind of preamble (for committing to the input of a party) as the one used inside BPS-CNMZK, we are able to avoid such a scenario, and crucially, we are able to use the BPS-CNMZK strategy as a single coherent rewinding strategy. The above idea also gives us a *new construction of a concurrent non-malleable zero-knowledge* protocol where the extraction can be *automatically done in-line* along with the simulation. We believe this implication to be of independent interest.

Finally, the construction in [17] is only analyzed for the setting where the theorems to be proven by the honest parties are fixed in advance before any session starts (in keeping with the impossibility results of Lindell [19]). Towards that end, our protocol only makes use of BPS-CNMZK in the very beginning of the protocol to prove a statement which could be generated by the honest parties before the start of any session.

## 2    Definitions and Preliminaries

### 2.1    Our Model

We first summarize the main differences in our model with respect to [12]. We first note that even in the stand-alone setting, if an adversary $\mathcal{A}$ controls the communication link between two honest parties, then $\mathcal{A}$ can execute separate "left" and "right" executions with the honest parties. Therefore, these executions can be viewed as two concurrent executions where $\mathcal{A}$ is the common party. In keeping with this observation, in our model, the adversary $\mathcal{A}$ is cast as a party participating in the protocol instead of being a separate entity who controls the communication link (as in [12], see full version for more details). We stress that this modeling allows us to assume that the communication between protocol participants takes place over authenticated channels. Furthermore, in contrast to [12], we allow the adversary to have a-priori information on the password. More details follow.

**Description of $\mathcal{F}$.** We model the problem of password-authenticated key exchange as a two-party functionality $\mathcal{F}$ involving parties $P_1$ and $P_2$ (where either party may be adversarial). If the inputs (password from a dictionary $D$) of $P_1$ and $P_2$ match, then $\mathcal{F}$ sends them a uniformly distributed session key (whose length is determined by the security parameter), else it sends $\perp$.

Further, in contrast to the stand-alone setting of [12] (where security holds only if a single protocol session is executed on the network), we consider the more general setting of *concurrent self-composition*, where polynomially many (in the security parameter) protocols with the same password may be executed on the network in an arbitrarily interleaved manner. In this setting, an adversary $\mathcal{A}$ may corrupt several parties across all the different sessions.

To formalize the above requirements and define security, we extend the standard simulation paradigm for defining secure computation. In particular, we allow the adversary in the ideal world to make a constant number of (output) queries to the trusted party for each protocol session. In the definition below, we focus only on the case where the honest parties hold the same password $p$. However it can be extended to the case of arbitrarily correlated passwords (or, in fact, general secure computation) in a natural way where the simulator in the ideal world might make an expected constant number of calls to the ideal functionality for every session in the real world.

We consider the static corruption model and probabilistic polynomial time (PPT) adversaries only. We denote computational indistinguishability by $\stackrel{c}{\equiv}$, and the security parameter by $\kappa$. Let $D$ be the dictionary of passwords.

IDEAL MODEL. In the ideal model, there is a trusted party that computes the password functionality $\mathcal{F}$ (described above) based on the inputs handed to it by the players. Let there be $n$ parties $P_1, \ldots, P_n$ where different pairs of parties are involved in one or more sessions, such that the total number of sessions is polynomial in the security parameter $\kappa$. Let $M \subset [n]$ denote the subset of corrupted parties controlled by an adversary. An execution in the ideal model with an adversary who controls the parties $M$ proceeds as follows:

**I. Inputs:** The honest parties hold a fixed input which is a password $p$ chosen from a dictionary $D$. The input of a corrupted party is not fixed in advance.

**II. Session initiation:** If a party $P_i$ wishes to initiate a session with another party $P_j$, it sends a (start-session, $i, j$) message to the trusted party. On receiving a message of the form (start-session, $i, j$), the trusted party sends (new-session, $i, j, k$) to both $P_i$ and $P_j$, where $k$ is the index of the new session.

**III. Honest parties send inputs to trusted party:** Upon receiving (new-session, $i, j, k$) from the trusted party, an honest party $P_i$ sends its real input along with the session identifier. More specifically, $P_i$ sets its session $k$ input $x_{i,k}$ to be the password $p$ and sends $(k, x_{i,k})$ to the trusted party.

**IV. Corrupted parties send inputs to trusted party:** A corrupted party $P_i$ sends a message $(k, x_{i,k})$ to the trusted party, for any $x_{i,k} \in D$ of its choice.

**V. Trusted party sends results to adversary:** For a session $k$ involving parties $P_i$ and $P_j$, when the trusted party has received messages $(k, x_{i,k})$ and $(k, x_{j,k})$, it computes the output $\mathcal{F}(x_{i,k}, x_{j,k})$. If at least one of the parties is corrupted, then the trusted party sends $(k, \mathcal{F}(x_{i,k}, x_{j,k}))$ to the adversary[4]. On the other hand, if both $P_i$ and $P_j$ are honest, then the trusted party sends the output message $(k, \mathcal{F}(x_{i,k}, x_{j,k}))$ to them.

---

[4] Note that here, the ideal functionality does not restrict the adversary to a *fixed* constant number of queries per session. However, in our security definition, we will require that the ideal adversary only makes a constant number of queries per session.

**VI. Adversary instructs the trusted party to answer honest players:** For a session $k$ involving parties $P_i$ and $P_j$ where exactly one party is honest, the adversary, depending on its view up to this point, may send the (output, $k$) message in which case the trusted party sends the most recently computed session $k$ output $(k, \mathcal{F}(x_{i,k}, x_{j,k}))$ to the honest party. (Intuitively, for each session $k$ where exactly one party is honest, we allow the adversary to choose which one of the $\lambda$ output values would be received by the honest party.)

**VII. Adversary makes more queries for a session:** The corrupted party $P_i$, depending upon its view up to this point, can send the message (new-query, $k$) to the trusted party. In this case, execution of session $k$ in the ideal world comes back to stage IV. $P_i$ can then choose its next input *adaptively* (i.e., based on previous outputs).

**VIII. Outputs:** An honest party always outputs the value that it received from the trusted party. The adversary outputs an arbitrary (PPT computable) function of its entire view (including the view of all corrupted parties) throughout the execution of the protocol.

Let $\mathcal{S}$ be a probabilistic polynomial-time ideal-model adversary that controls the subset of corrupted parties $M \subset [n]$. Then the ideal execution of $\mathcal{F}$ (or the ideal distribution) with security parameter $\kappa$, password $p \in D$ and auxiliary input $z$ to $\mathcal{S}$ is defined as the output of the honest parties along with the output of the adversary $\mathcal{S}$ resulting from the ideal process described above. It is denoted by $\text{IDEAL}_{M,\mathcal{S}}^{\mathcal{F}}(\kappa, p, z)$.

REAL MODEL. We now consider the real model in which a real two-party password-based key exchange protocol is executed.

Let $\mathcal{F}, P_1, \ldots, P_n, M$ be as above. Let $\Sigma$ be the password-based key exchange protocol in question. Let $\mathcal{A}$ be probabilistic polynomial-time (PPT) machine such that for every $i \in M$, the adversary $\mathcal{A}$ controls the party $P_i$.

In the real model, a polynomial number (in the security parameter $\kappa$) of sessions of $\Sigma$ may be executed concurrently, where the scheduling of all messages throughout the executions is controlled by the adversary. We *do not* assume that all the sessions have a unique session index. We assume that the communication between the parties takes place over authenticated channels[5]. An honest party follows all instructions of the prescribed protocol, while an adversarial party may behave arbitrarily. At the conclusion of the protocol, an honest party computes its output as prescribed by the protocol. Without loss of generality, we assume the adversary outputs exactly its entire view of the execution of the protocol.

The real concurrent execution of $\Sigma$ (or the real distribution) with security parameter $\kappa$, password $p \in D$ and auxiliary input $z$ to $\mathcal{A}$ is defined as the output of all the honest parties along with the output of the adversary resulting from the above process. It is denoted as $\text{REAL}_{M,\mathcal{A}}^{\Sigma}(\kappa, p, z)$.

We now give our definition of concurrently-secure password-authenticated key exchange protocol.

---

[5] As mentioned earlier, this is a reasonable assumption since in our model, the adversary is a protocol participant instead of being a separate entity that controls the communication links (as in [12]).

**Definition 1.** *Let $\mathcal{F}$ and $\Sigma$ be as above. Let $D$ be the dictionary of passwords. Then protocol $\Sigma$ for computing $\mathcal{F}$ is a concurrently secure password authenticated key exchange protocol if for every probabilistic polynomial-time adversary $\mathcal{A}$ in the real model, there exists a probabilistic expected polynomial-time adversary $\mathcal{S}$ such that $\mathcal{S}$ makes a constant number of queries to the ideal functionality per session, and, for every $z \in \{0,1\}^{*}$, $p \in D$, $M \subset [n]$,*

$$\left\{ \text{IDEAL}_{M,\mathcal{S}}^{\mathcal{F}}(\kappa, p, z) \right\}_{\kappa \in N} \stackrel{c}{\equiv} \left\{ \text{REAL}_{M,\mathcal{A}}^{\Sigma}(\kappa, p, z) \right\}_{\kappa \in N}$$

We note that our security definition implies the original definition of Goldreich and Lindell [12] (adapted to the concurrent setting). We refer the reader to the full version for a formal proof. We now state our main result.

**Theorem 1.** (Main Result) *Assume the existence of 1-out-of-2 oblivious transfer protocol secure against honest but curious adversaries[6]. Let $\mathcal{F}$ be the two-party PAKE functionality as described above. Then, there exists a protocol $\Sigma$ that securely realizes $\mathcal{F}$ as per Definition 1.*

We prove the above theorem by constructing such a protocol $\Sigma$ in section 3. If the underlying primitives are uniform (resp., non-uniform), then the protocol $\Sigma$ is uniform (resp., non-uniform) as well. A polynomial time adversary against $\Sigma$ translates to a polynomial time adversary against one of the underlying primitives.

## 2.2   Building Blocks

We now briefly mention some of the main cryptographic primitives that we use in our construction. We refer the the reader to the full version of the paper for more details.

*Statistically Binding Commitments.* In our protocol, we shall use the 2-round statistically binding commitment scheme of Naor [25] based on one-way functions. Given a random string $z$ from the receiver, let $\text{COM}_{z}(\cdot)$ denote the commitment function of the scheme.

*Preamble from PPSTV [22].* A PPSTV preamble is a protocol between a committer and a receiver that consists of two main phases, namely, (a) the commitment phase, and (b) the challenge-response phase. Let $k$ be a parameter that determines the round-complexity of the protocol. Then, in the commit phase, very roughly speaking, the committer commits to a secret string $\sigma$ and $k^2$ pairs of its 2-out-of-2 secret shares. The challenge-response phase consists of $k$ iterations, where in each iteration, very roughly speaking, the committer "opens" $k$ shares, one each from $k$ different pairs of secret shares as chosen by the receiver.

The goal of this protocol is to enable the simulator to be able to rewind and extract the "preamble secret" $\sigma$ with high probability. In the concurrent setting, rewinding can be difficult since one may rewind past the start of some other protocol [26]. However, as it has been demonstrated in [22] (see also [23,27])

---

[6] Note that 1-out-of-2 oblivious transfer (OT) secure against honest but curious adversaries implies 1-out-of-2 OT secure against malicious adversaries [24].

there is a "time-oblivious" rewinding strategy that the simulator can use to extract the preamble secrets from every concurrent cheating committer, with high probability. In the sequel, we will refer to the preamble simulator as *CEC-Sim*. For our purpose, we will use PPSTV preambles with linear (in the security parameter $\kappa$) number of rounds. Then, the simulation strategy in [22] guarantees a *linear precision* in the running time of the simulator. Specifically, the running time of the simulator is only a constant multiple of the running time of the adversarial committer in the real execution.

*Concurrent Non-Malleable Zero Knowledge Argument.* We shall use a concurrent non-malleable zero knowledge (CNMZK) argument for every language in **NP** with perfect completeness and negligible soundness error. In particular, we will use a slightly modified version of the CNMZK protocol of Barak, Prabhakaran and Sahai [17], henceforth referred to as $m$BPS-CNMZK. In the modified version, we replace the PRS [23] preamble used in the original construction with a PPSTV preamble with linear (in the security parameter) number of rounds. We will also require that the non-malleable commitment scheme used in the protocol is public-coin [28].

*Statistically Witness Indistinguishable Arguments.* In our construction, we shall use a statistically witness indistinguishable argument (sWI) for proving membership in any **NP** language with perfect completeness and negligible soundness error. Such a scheme can be constructed by using $\omega(\log n)$ copies of Blum's Hamiltonicity protocol [29] in parallel, with the modification that the prover's commitments in the Hamiltonicity protocol are made using a statistically hiding commitment scheme. Statistically hiding commitments were constructed by Naor, Ostrovsky, Venkatesan and Yung [30] in $O(k/\log(k))$ rounds using a one way permutation ([30] in turn builds on the *interactive hashing* technique introduced in [31]). Constructions based on one way functions were given in [32,33].

*Semi-Honest Two Party Computation.* We will also use a semi-honest two party computation protocol $\Pi_{\text{SH-PAKE}}$ that emulates the PAKE functionality $\mathcal{F}$ (as described in section 2.1) in the stand-alone setting as per the standard definition of secure computation. The existence of such a protocol $\Pi_{\text{SH-PAKE}}$ follows from [1,34].

# 3    Our Construction

In this section, we describe our two-party protocol $\Sigma$ that securely realizes the password functionality $\mathcal{F}$ in the setting of concurrent self-composition as per Definition 1.

*Notation.* Let $P_1$ and $P_2$ be two parties with private inputs (password from dictionary $D$) $x_1$ and $x_2$ respectively. Given a random string $z$ (from the receiver), let $\text{COM}_z(\cdot)$ denote the commitment function of Naor's commitment scheme [25]. By $m$BPS-CNMZK, we will refer to the modified version of the CNMZK protocol of [17] described in section 2.2. Let $\Pi_{m\text{BPS},P_i \to P_j}$ denote an instance of the $m$BPS-CNMZK protocol where $P_i$ and $P_j$ play the roles of prover and verifier respectively. Let $\Pi_{\text{SH-PAKE}}$ be any *semi-honest* two party computation protocol that emulates the functionality $\mathcal{F}$ in the stand-alone setting. Let $U_\eta$ denote the uniform distribution over $\{0,1\}^\eta$, where $\eta$ is a function of the security parameter.

The protocol $\Sigma$ proceeds as follows.

## I. Trapdoor Creation Phase

1. $P_1 \leftrightarrow P_2$ : $P_1$ sends a random string $z_2$ (of appropriate length) to $P_2$ as the first message of Naor's commitment scheme. Similarly, $P_2$ sends a random string $z_1$ to $P_1$.
2. $P_1 \rightarrow P_2$ : $P_1$ creates a commitment $com_1 = \text{COM}_{z_1}(0)$ to bit 0 and sends it to $P_2$. $P_1$ and $P_2$ now engage in the execution of a $m$BPS-CNMZK argument $\Pi_{m\text{BPS},P_1 \rightarrow P_2}$ where $P_1$ proves that $com_1$ is a commitment to 0.
3. $P_2 \rightarrow P_1$ : $P_2$ now acts symmetrically. Specifically, it creates a commitment $com_2 = \text{COM}_{z_2}(0)$ to bit 0 and sends it to $P_1$. $P_2$ and $P_1$ now engage in the execution of a $m$BPS-CNMZK argument $\Pi_{m\text{BPS},P_2 \rightarrow P_1}$ where $P_2$ proves that $com_2$ is a commitment to 0.

## II. $m$PPSTV Preamble Phase.
In this phase, each party $P_i$ engages in the execution of a *modified* PPSTV preamble (henceforth referred to as $m$PPSTV) with $P_j$ where it commits to its input and randomness. In our modified version of the PPSTV preamble, for a given receiver challenge, the committer does not "open" the commitments, but instead simply reveals the committed value (without the randomness) and proves its correctness by using a sWI. Let $\Pi_{m\text{PPSTV},P_i \rightarrow P_j}$ denote an instance of the $m$PPSTV protocol where $P_i$ plays the role of the committer. We now describe the steps in this phase.

1. $P_1 \leftrightarrow P_2$ : Generate a string $r_1 \xleftarrow{\$} U_\eta$ and let $\beta_1 = \{x_1, r_1\}$. Here $r_1$ is the randomness to be used (after coin-flipping with $P_2$) by $P_1$ in the execution of the protocol $\Pi_{\text{SH-PAKE}}$ in Phase III. We assume that $|r_1| = \eta$ is sufficiently long for that purpose. Now $P_1$ and $P_2$ engage in the execution of a $m$PPSTV preamble $\Pi_{m\text{PPSTV},P_1 \rightarrow P_2}$ in the following manner.
   Let $k$ be a polynomial in the security parameter $\kappa$. $P_1$ first prepares $2k^2$ secret shares $\{\alpha_{i,j}^0\}_{i,j=1}^k$, $\{\alpha_{i,j}^1\}_{i,j=1}^k$ such that $\alpha_{i,j}^0 \oplus \alpha_{i,j}^1 = \beta_1 (= \{x_1, r_1\})$ for all $i, j$. Using the commitment function $\text{COM}_{z_1}(\cdot)$, $P_1$ commits to $\beta_1$ and all its secret shares. Denote these commitments by $B_1, \{A_{i,j}^0\}_{i,j=1}^k, \{A_{i,j}^1\}_{i,j=1}^k$. $P_1$ now engages in the execution of a sWI with $\mathcal{A}$ in order to prove the following statement: either
   (a) the above commit phase is "valid", i.e., there exist values $\hat{\beta}_1, \{\hat{\alpha}_{i,j}^0, \hat{\alpha}_{i,j}^1\}_{i,j=1}^k$ such that (a) $\hat{\alpha}_{i,j}^0 \oplus \hat{\alpha}_{i,j}^1 = \hat{\beta}_1$ for all $i, j$, and, (b) commitments $B_1$, $\{A_{i,j}^0\}_{i,j=1}^k, \{A_{i,j}^1\}_{i,j=1}^k$ can be decommitted to $\hat{\beta}_1, \{\hat{\alpha}_{i,j}^0, \hat{\alpha}_{i,j}^1\}_{i,j=1}^k$, or,
   (b) $com_1$ in phase I is a commitment to bit 1.
   It uses the witness corresponding to the first part of the statement. $P_1$ and $P_2$ now execute a challenge-response phase. For $j = 1, \ldots, k$:
   (a) $P_2 \rightarrow P_1$ : Send challenge bits $z_{1,j}, \ldots, z_{k,j} \xleftarrow{\$} \{0,1\}^k$.
   (b) $P_1 \rightarrow P_2$ : Send $\alpha_{1,j}^{z_{1,j}}, \ldots, \alpha_{k,j}^{z_{k,j}}$. Now, $P_1$ and $P_2$ engage in the execution of a sWI, where $P_1$ proves the following statement: either
   (a) commitments $A_{1,j}^{z_{1,j}}, \ldots, A_{k,j}^{z_{k,j}}$ can be decommitted to $\alpha_{1,j}^{z_{1,j}}, \ldots, \alpha_{k,j}^{z_{k,j}}$ respectively, or (b) $com_1$ in Phase I is a commitment to bit 1. It uses the witness corresponding to the first part of the statement.

2. $P_2 \leftrightarrow P_1 : P_2$ now acts symmetrically.

**III. Secure Computation Phase.** In this phase, the parties run an execution of the semi-honest two party protocol $\Pi_{\text{SH-PAKE}}$ "compiled" with sWI.

*Coin Flipping.* $P_1$ (resp., $P_2$) first engage in a coin-flipping protocol. More specifically, $P_1$ (resp., $P_2$) generates $r_2' \overset{\$}{\leftarrow} U_\eta$ (resp., $r_1' \overset{\$}{\leftarrow} U_\eta$) and sends it to $P_2$ (resp., $P_1$). Define $r_1'' = r_1 \oplus r_1'$ and $r_2'' = r_2 \oplus r_2'$. Now $r_1''$ and $r_2''$ respectively are the random coins that $P_1$ and $P_2$ will use in the execution of protocol $\Pi_{\text{SH-PAKE}}$.

*Protocol $\Pi_{\text{SH-PAKE}}$.* Let the protocol $\Pi_{\text{SH-PAKE}}$ have $t$ rounds where one round is defined to have a message from $P_1$ to $P_2$ followed by a reply from $P_2$ to $P_1$. Let transcript $T_{1,j}$ (resp., $T_{2,j}$) be defined to contain all the messages exchanged between $P_1$ and $P_2$ before the point party $P_1$ (resp., $P_2$) is supposed to send a message in round $j$. Now, each message sent by either party in protocol $\Pi_{\text{SH-PAKE}}$ is compiled into a message block in $\Sigma$. For $j = 1, \ldots, t$:

1. $P_1 \rightarrow P_2 : P_1$ sends the next message $\Delta_{1,j} (= \Pi_{\text{SH-PAKE}}(T_{1,j}, x_1, r_1''))$ as per protocol $\Pi_{\text{SH-PAKE}}$. Now, $P_1$ and $P_2$ engage in the execution of a sWI where $P_1$ proves the following statement: either
   (a) there exists a value $\hat{\beta}_1 = \{\hat{x}_1, \hat{r}_1\}$ such that (a) the commitment $B_1$ in phase II.1 can be decommitted to $\hat{\beta}_1 = \{\hat{x}_1, \hat{r}_1\}$, and (b) the sent message $\Delta_{1,j}$ is consistent with input $\hat{x}_1$ and randomness $\hat{r}_1 \oplus r_1'$ (i.e., $\Delta_{1,j} (= \Pi_{\text{SH-PAKE}}(T_{1,j}, \hat{x}_1, \hat{r}_1 \oplus r_1'))$, or
   (b) $com_1$ in Phase I is a commitment to bit 1.
   It uses the witness corresponding to the first part of the statement.
2. $P_2 \rightarrow P_1 : P_2$ now acts symmetrically.

This completes the description of the protocol $\Sigma$. Note that $\Sigma$ consists of several instances of sWI where the proof statement in each instance consists of two parts. Specifically, the second part of the statement states that prover committed to bit 1 in the trapdoor creation phase. In the sequel, we will refer to the second part of the proof statement as the *trapdoor* condition. Further, we will call the witness corresponding to the first part of the statement as *real* witness and that corresponding to the second part of the statement as the *trapdoor* witness.

## 4    Proof of Security

**Theorem 2.** *The proposed protocol $\Sigma$ is a concurrently secure PAKE protocol as per Definition 1.*

Let there be $n$ parties in the system where different pairs of parties are involved in one or more sessions of $\Sigma$, such that the total number of sessions $m$ is polynomial in the security parameter $\kappa$. Let $\mathcal{A}$ be an adversary who controls an arbitrary number of parties. In order to prove theorem 2, we will first construct a simulator $\mathcal{S}$ that will simulate the view of $\mathcal{A}$ in the ideal world. We will then show that $\mathcal{S}$ makes only a constant number of queries per session while simulating the view of $\mathcal{A}$. Finally, we will argue that the output distributions of the real and ideal world executions are computationally indistinguishable. For simplicity of exposition, we will assume that exactly one party is corrupted in each session. We note that if the real and ideal distributions are indistinguishable for this case, then by using standard techniques we can easily remove this assumption. Due to lack of space, in this version, we only give the description of the simulator and bound its output queries. We refer the reader to the full version of the paper for a complete proof.

*Notation.* In the sequel, for any session $\ell \in [m]$, we will use the notation $H$ to denote the honest party and $\mathcal{A}$ to denote the corrupted party. Let $\Pi_{m\text{BPS},H \to \mathcal{A}}$ (resp., $\Pi_{m\text{BPS},\mathcal{A} \to H}$) denote an instance of $m$BPS-CNMZK where $H$ (resp., $\mathcal{A}$) plays the role of the prover and $\mathcal{A}$ (resp., $H$) plays the verifier. Similarly, let $\Pi_{m\text{PPSTV},H \to \mathcal{A}}$ (resp., $\Pi_{m\text{PPSTV},\mathcal{A} \to H}$) denote an instance of $m$PPSTV where $H$ (resp., $\mathcal{A}$) plays the role of the committer and $\mathcal{A}$ (resp., $H$) plays the receiver. Wherever necessary, we shall augment our notations with a super-script that denotes the session number.

Consider any session between $H$ and $\mathcal{A}$. Consider the last message from $\mathcal{A}$ before $H$ sends a message to $\mathcal{A}$ during the coin-flipping sub-phase in the secure computation phase. Note that this message could either be the first message of the coin-flipping phase or the last message of the $m$PPSTV phase, depending upon whether $\mathcal{A}$ or $H$ sends the first message in the coin-flipping phase. In the sequel, we will refer to this message from $\mathcal{A}$ as the special message. Intuitively, this message is important because our simulator will need to query the ideal functionality every time it receives such a message from $\mathcal{A}$. Looking ahead, in order to bound the number of queries made by our simulator, we will be counting the number of special messages sent by $\mathcal{A}$ during the simulation.

### 4.1   Description of Simulator $\mathcal{S}$

The simulator $\mathcal{S}$ consists of two parts, $S_{\text{CEC}}$ and $S_{\text{CORE}}$. Informally speaking, $S_{\text{CEC}}$ is essentially the simulator CEC-Sim (see section 2.2) whose goal is to extract the *preamble secret* in each instance of the PPSTV preamble where $\mathcal{A}$ acts as the committer. These extracted values are passed on to $S_{\text{CORE}}$, who uses them crucially to simulate the view of $\mathcal{A}$. We now give more details.

**Description of $S_{\text{CEC}}$.** $S_{\text{CEC}}$ is essentially the main simulator in that it handles all communication with $\mathcal{A}$. However, for each session $\ell \in [m]$, $S_{\text{CEC}}$ by itself only answers $\mathcal{A}$'s messages in those instances of the PPSTV preamble where $\mathcal{A}$ plays the role of the committer; $S_{\text{CEC}}$ in turn communicates with the core simulator $S_{\text{CORE}}$ to answer all other messages from $\mathcal{A}$.

Specifically, recall that our protocol consists of two instances of the PPSTV preamble where $\mathcal{A}$ plays the role of the committer. Consider any session $\ell \in [m]$. The first instance is inside the $m$BPS-CNMZK instance $\Pi^\ell_{m\text{BPS},H \to \mathcal{A}}$ in the trapdoor creation phase, while the second instance is in fact the $m$PPSTV preamble $\Pi^\ell_{m\text{PPSTV},\mathcal{A} \to H}$ in the second phase. Then, $S_{\text{CEC}}$ is essentially the simulator CEC-Sim that interacts with $\mathcal{A}$ in order to extract the preamble secret in each of the above instances of the PPSTV preamble. Specifically, in order to perform these extractions, $S_{\text{CEC}}$ employs the time-oblivious rewinding strategy of CEC-Sim for an imaginary adversary (see next paragraph). During the simulation, whenever $S_{\text{CEC}}$ receives a message from $\mathcal{A}$ in any of the above instance of the PPSTV preamble, then it answers it on its own in the same manner as CEC-Sim does (i.e., by sending a random challenge string). However, on receiving any other message, it simply passes it to the core simulator $S_{\text{CORE}}$ (described below), and transfers its response to $\mathcal{A}$. Whenever $S_{\text{CEC}}$ extracts a preamble secret from $\mathcal{A}$ at any point during the simulation, it immediately passes it to $S_{\text{CORE}}$. If $S_{\text{CEC}}$ fails to extract any of the preamble secrets from $\mathcal{A}$, then it outputs the abort symbol $\perp$.

*Message generation timings of $\mathcal{A}$.* We note that in order to employ the time-oblivious rewinding strategy of CEC-Sim, $S_{\text{CEC}}$ needs to know the amount of time that $\mathcal{A}$ takes to send each message in the protocol (see [22]). We remark that we do not seek precision in simulation time (guaranteed by the rewinding strategy of CEC-Sim); instead we only require that the number of queries made by the simulator in the look-ahead threads is only within a constant factor of the number of the number of sessions. To this end, we consider an imaginary experiment in which $\mathcal{A}$ takes a disproportionately large amount of time in generating the message after which our simulator has to query the trusted party. Then the rewinding strategy of $S_{\text{CEC}}$ is determined by running CEC-Sim using the next message generation timings of such an (imaginary) adversary, explained as follows.

Consider all the messages sent by $\mathcal{A}$ during a protocol execution. We will assign $q$ time units to the special message, where $q$ is the round complexity (linear in the security parameter) of our protocol; any other message from $\mathcal{A}$ is simply assigned one time unit. Intuitively, by assigning more weight to the special message, we ensure that if the running time of our simulator is only within a constant factor of the running time of $\mathcal{A}$ in the real execution, then the number of special messages sent by $\mathcal{A}$ during the simulation must be a constant as well. Looking ahead, this in turn will allow us to prove that the number of queries made by the simulator are only a constant.

**Description of $S_{\text{CORE}}$.** We describe the strategy of $S_{\text{CORE}}$ in each phase of the protocol, for each session $\ell \in [m]$. We stress that $S_{\text{CORE}}$ uses the same strategy in the main-thread as well as all look-ahead threads (unless mentioned otherwise).

*Trapdoor Creation Phase.* $S_{\text{CORE}}$ first sends a commitment to bit 1, instead of committing to bit 0. Now, recall that $S_{\text{CEC}}$ interacts with $\mathcal{A}$ during the preamble phase in $\Pi^{\ell}_{m\text{BPS}, H \to \mathcal{A}}$ and extracts the preamble secret $\sigma^{\ell}_{m\text{BPS}, H \to \mathcal{A}}$ from $\mathcal{A}$ at the conclusion of the preamble. Then, on receiving $\sigma^{\ell}_{m\text{BPS}, H \to \mathcal{A}}$ from $S_{\text{CEC}}$, $S_{\text{CORE}}$ simulates the *post-preamble* phase of $\Pi^{\ell}_{m\text{BPS}, H \to \mathcal{A}}$ (see [17] for protocol description) in a "straight-line" fashion, as described below.

Let $y^{\ell}$ be the proof statement in $\Pi^{\ell}_{m\text{BPS}, H \to \mathcal{A}}$. Then, in phase II of $\Pi^{\ell}_{m\text{BPS}, H \to \mathcal{A}}$, $S_{\text{CORE}}$ creates a statistically hiding commitment (sCOM) to $\sigma^{\ell}_{m\text{BPS}, H \to \mathcal{A}}$ (instead of a string of all zeros) and follows it up with an honest execution of statistical zero knowledge argument of knowledge (sZKAOK) to prove knowledge of the decommitment. In phase IV of $\Pi^{\ell}_{mBPS, H \to \mathcal{A}}$, $S_{\text{CORE}}$ creates a non-malleable commitment (NMCOM) to an all zeros string (instead of a valid witness to $y^{\ell}$). Finally, in phase V, $S_{\text{CORE}}$ proves the following statement using sZKAOK: (a) the value committed to in phase IV is a valid witness to $y^{\ell}$, or (b) the value committed to in phase II is $\sigma^{\ell}_{mBPS, H \to \mathcal{A}}$. Here it uses the witness corresponding to the *second* part of the statement.

Now, consider the $m$BPS-CNMZK instance $\Pi^{\ell}_{mBPS, \mathcal{A} \to H}$, where $H$ plays the role of the verifier. Here, $S_{\text{CORE}}$ simply uses the honest verifier strategy to interact with $\mathcal{A}$.

$m$PPSTV *Preamble Phase.* Consider the execution of the $m$PPSTV instance $\Pi^{\ell}_{mPPSTV, H \to \mathcal{A}}$. Here, $S_{\text{CORE}}$ commits to a random string and answers $\mathcal{A}$'s challenges with random strings. Note that the trapdoor condition is true for each instance of $s$WI in $\Pi^{\ell}_{mPPSTV, H \to \mathcal{A}}$ since $S_{\text{CORE}}$ committed to bit 1 (instead of 0) in the trapdoor creation phase. Therefore, $S_{\text{CORE}}$ uses the trapdoor witness in order to successfully simulate each instance of $s$WI in $\Pi^{\ell}_{mPPSTV, H \to \mathcal{A}}$.

Now consider the $m$PPSTV instance $\Pi^{\ell}_{mPPSTV, \mathcal{A} \to H}$. Note that in this preamble, $S_{\text{CEC}}$ interacts with $\mathcal{A}$ without the help of $S_{\text{CORE}}$. As explained earlier, $S_{\text{CEC}}$ extracts the preamble secret (that contains the input and randomness of $\mathcal{A}$ in session $\ell$) and passes it to $S_{\text{CORE}}$.

*Secure Computation Phase.* Let $S_{\Pi_{\text{SH-PAKE}}}$ denote the simulator for the semi-honest two-party protocol $\Pi_{\text{SH-PAKE}}$ used in our construction. $S_{\text{CORE}}$ internally runs the simulator $S_{\Pi_{\text{SH-PAKE}}}$ on adversary's input in session $\ell$. $S_{\Pi_{\text{SH-PAKE}}}$ starts executing, and, at some point, it makes a call to the trusted party in the ideal world with some input (say) $x$. $S_{\text{CORE}}$ uses the following strategy to manage queries to the trusted party.

$S_{\text{CORE}}$ maintains a counter $c$ to count the total number of queries (including all sessions) made to the trusted party on the look-ahead threads so far in the simulation (note that there will be exactly $m$ queries on the main thread). Now, when $S_{\Pi_{\text{SH-PAKE}}}$ makes a call to the trusted party, $S_{\text{CORE}}$ computes a session index $s$ in the following manner. If the query corresponds to the main thread, then $S_{\text{CORE}}$ sets $s = \ell$, else it computes $s = c \mod m$. Now, if $S_{\text{CORE}}$ has already queried the trusted party at least once for session $s$, then it first sends the (new-query, $s$) message to the trusted party. Otherwise, it simply sends the message $(s, x)$ to the trusted party.[7,8] The response from the trusted party is passed on to $S_{\Pi_{\text{SH-PAKE}}}$. If

---

[7] We stress that the simulator is able to "trade" the ideal functionality calls in one session for another since the inputs of the honest parties are the same across all sessions.

[8] Note that by choosing the session index for the output query in the above fashion, $S_{\text{CORE}}$ is able to equally distribute the queries across all the sessions. Looking ahead, in the next subsection, we will argue that the total number of queries across all the sessions are only within a constant factor of the number of sessions. Then, this strategy of distributing the queries will ensure that the queries per session are also a constant.

the query corresponds to the main thread, $S_{\text{CORE}}$ sends the message (output, $s$) to the trusted party, indicating it to send the output to the honest party in session $s$.[9]

Having received the trusted party's response from $S_{\text{CORE}}$, $S_{\Pi_{\text{SH-PAKE}}}$ runs further, and finally halts and outputs a transcript $\Delta^\ell_{1,1}, \Delta^\ell_{2,1}, \ldots, \Delta^\ell_{1,t}, \Delta^\ell_{2,t}$ of the execution of $\Pi_{\text{SH-PAKE}}$, and an associated randomness $r^\ell_{\mathcal{A}}$. Let $\hat{r}^\ell_{\mathcal{A}}$ be the randomness that $S$ extracted from $\mathcal{A}$ in phase II. Now, $S_{\text{CORE}}$ computes a random string $\tilde{r}^\ell_{\mathcal{A}}$ such that $r^\ell_{\mathcal{A}} = \tilde{r}^\ell_{\mathcal{A}} \oplus \hat{r}^\ell_{\mathcal{A}}$.

Now, in order to force $\mathcal{A}$ to use randomness $r^\ell_{\mathcal{A}}$ during the execution of $\Pi_{\text{SH-PAKE}}$, $S_{\text{CORE}}$ sends $\tilde{r}^\ell_{\mathcal{A}}$ to $\mathcal{A}$ during the coin-flipping phase prior to the execution of $\Pi_{\text{SH-PAKE}}$. Finally, $S_{\text{CORE}}$ forces the transcript $\Delta^\ell_{1,1}, \Delta^\ell_{2,1}, \ldots, \Delta^\ell_{1,t}, \Delta^\ell_{2,t}$ onto $\mathcal{A}$ during the execution of $\Pi_{\text{SH-PAKE}}$. This is done as follows. Without loss of generality, let us assume that the honest party sends the first message in this instance of $\Pi_{\text{SH-PAKE}}$. Then, in round $j$, $1 \leq j \leq t$, $S_{\text{CORE}}$ sends $\Delta^\ell_{1,j}$ to $\mathcal{A}$ (instead of sending a message as per the input and randomness committed to in the preamble in Phase II). $S_{\text{CORE}}$ uses the trapdoor witness to complete the associated sWI. If the reply of $\mathcal{A}$ is different from the (expected) message $\Delta^\ell_{2,j}$, then $S_{\text{CORE}}$ outputs the abort symbol $\perp$.

This completes the description of our simulator $S = \{S_{\text{CEC}}, S_{\text{CORE}}\}$.

## 4.2   Total Queries by $S$

**Lemma 1.** *Let $m$ be the total number of sessions of $\Sigma$ being executed concurrently. Then, the total number of queries made by $S$ to the trusted party is within a constant factor of $m$.*

*Proof.* Let $T$ be the total running time of the adversary in the real execution, as per the time assignment strategy described in section 4.1. Now, since $S$ employs the time-oblivious rewinding strategy of CEC-Sim (see section 2.2), it follows that the total running time of $S$ is within a constant factor of $T$. Let us now assume that our claim is false, i.e., the total number of queries made by $S$ is a super-constant multiple of $m$. We will show that in this case, the running time of $S$ must be super-constant multiple of $T$, which is a contradiction. We now give more details.

Let $q$ be the round complexity of $\Sigma$. Then, as per the time assignment strategy given in section 4.1, $T = (q - 1 + q) \cdot m$ (recall that the special message is assigned a weight of $q$ time units, while each of the remaining $q - 1$ messages is assigned one time unit). Now, let $\lambda$ be a value that is super-constant in the security parameter such that $S$ makes $\lambda \cdot m$ total queries during the simulation. Note

---

[9] Note that $s = \ell$ in this case. We stress that by setting $s = \ell$ for a query on the main thread, $S_{\text{CORE}}$ ensures that the honest party in session $\ell$ receives the correct output. (Note that an honest party does not receive any output for an output query on a look-ahead thread.)

that each output query corresponds to a unique special message. Let $T'$ be the total running time of $S$. We calculate $T'$ as follows:

$$T' \geq q \cdot (\lambda \cdot m) + (q - 1) \cdot m$$
$$> q \cdot (\lambda \cdot m)$$
$$> \frac{\lambda \cdot q}{(q - 1 + q)} \cdot T$$

Since $\frac{\lambda \cdot q}{(q-1+q)}$ is a super-constant in the security parameter, we have that $T'$ is a super-constant multiple of $T$, which is a contradiction. Hence the claim follows.

The corollary below immediately follows from lemma 1 and the description of $S$ in section 4.1.

**Corollary 1.** *$S$ makes a constant number of queries per session to the trusted party.*

## Acknowledgements

We thank Rafael Pass for pointing out that one of the arguments in an earlier draft of this paper was insufficient. We also thank Omkant Pandey, Rafael Pass and Akshay Wadia for useful discussions.

## References

1. Yao, A.C.C.: How to generate and exchange secrets (extended abstract). In: FOCS (1986)
2. Goldreich, O., Micali, S., Wigderson, A.: How to play any mental game. In: STOC (1987)
3. Bellovin, S.M., Merritt, M.: Encrypted key exchange: Password-based protocols secure against dictionary attacks. In: IEEE Symposium on Security and Privacy (1992)
4. Katz, J., Ostrovsky, R., Yung, M.: Efficient and secure authenticated key exchange using weak passwords. J. ACM 57(1) (2009)
5. Bellare, M., Pointcheval, D., Rogaway, P.: Authenticated key exchange secure against dictionary attacks. In: Preneel, B. (ed.) EUROCRYPT 2000. LNCS, vol. 1807, p. 139. Springer, Heidelberg (2000)
6. Boyko, V., MacKenzie, P.D., Patel, S.: Provably secure password-authenticated key exchange using diffie-hellman. In: Preneel, B. (ed.) EUROCRYPT 2000. LNCS, vol. 1807, p. 156. Springer, Heidelberg (2000)
7. Katz, J., Ostrovsky, R., Yung, M.: Efficient password-authenticated key exchange using human-memorable passwords. In: Pfitzmann, B. (ed.) EUROCRYPT 2001. LNCS, vol. 2045, p. 475. Springer, Heidelberg (2001)
8. Gennaro, R., Lindell, Y.: A framework for password-based authenticated key exchange. In: Biham, E. (ed.) EUROCRYPT 2003. LNCS, vol. 2656, pp. 524–543. Springer, Heidelberg (2003)
9. Genarro, R.: Faster and shorter password-authenticated key exchange. In: ACM Conference on Computer and Communications Security (2008)
10. Canetti, R., Halevi, S., Katz, J., Lindell, Y., MacKenzie, P.D.: Universally composable password-based key exchange. In: Cramer, R. (ed.) EUROCRYPT 2005. LNCS, vol. 3494, pp. 404–421. Springer, Heidelberg (2005)

11. Canetti, R.: Universally composable security: A new paradigm for cryptographic protocols. In: FOCS (2001)
12. Goldreich, O., Lindell, Y.: Session-key generation using human passwords only. In: Kilian, J. (ed.) CRYPTO 2001. LNCS, vol. 2139, p. 408. Springer, Heidelberg (2001)
13. Nguyen, M.H., Vadhan, S.P.: Simpler session-key generation from short random passwords. In: Naor, M. (ed.) TCC 2004. LNCS, vol. 2951, pp. 428–445. Springer, Heidelberg (2004)
14. Barak, B., Canetti, R., Lindell, Y., Pass, R., Rabin, T.: Secure computation without authentication. In: Shoup, V. (ed.) CRYPTO 2005. LNCS, vol. 3621, pp. 361–377. Springer, Heidelberg (2005)
15. Feige, U., Shamir, A.: Witness indistinguishable and witness hiding protocols. In: STOC (1990)
16. Katz, J., Ostrovsky, R., Yung, M.: Forward secrecy in password-only key exchange protocols. In: Cimato, S., Galdi, C., Persiano, G. (eds.) SCN 2002. LNCS, vol. 2576, pp. 29–44. Springer, Heidelberg (2003)
17. Barak, B., Prabhakaran, M., Sahai, A.: Concurrent non-malleable zero knowledge. In: FOCS (2006)
18. Micali, S., Pass, R.: Local zero knowledge. In: STOC (2006)
19. Lindell, Y.: Lower bounds for concurrent self composition. In: Naor, M. (ed.) TCC 2004. LNCS, vol. 2951, pp. 203–222. Springer, Heidelberg (2004)
20. Goldreich, O., Lindell, Y.: Session-key generation using human passwords only. J. Cryptology 19(3) (2006)
21. Goyal, V., Sahai, A.: Resettably secure computation. In: Joux, A. (ed.) EUROCRYPT 2009. LNCS, vol. 5479, pp. 54–71. Springer, Heidelberg (2010)
22. Pandey, O., Pass, R., Sahai, A., Tseng, W.L.D., Venkitasubramaniam, M.: Precise concurrent zero knowledge. In: Smart, N.P. (ed.) EUROCRYPT 2008. LNCS, vol. 4965, pp. 397–414. Springer, Heidelberg (2008)
23. Prabhakaran, M., Rosen, A., Sahai, A.: Concurrent zero knowledge with logarithmic round-complexity. In: FOCS (2002)
24. Haitner, I.: Semi-honest to malicious oblivious transfer - the black-box way. In: Canetti, R. (ed.) TCC 2008. LNCS, vol. 4948, pp. 412–426. Springer, Heidelberg (2008)
25. Naor, M.: Bit commitment using pseudorandomness. J. Cryptology (1991)
26. Dwork, C., Naor, M., Sahai, A.: Concurrent zero-knowledge. In: STOC (1998)
27. Kilian, J., Petrank, E.: Concurrent and resettable zero-knowledge in polyloalgorithm rounds. In: STOC (2001)
28. Dolev, D., Dwork, C., Naor, M.: Nonmalleable cryptography. SIAM J. Comput. 30(2) (2000)
29. Blum, M.: How to prove a theorem so no one else can claim it. In: International Congress of Mathematicians (1987)
30. Naor, M., Ostrovsky, R., Venkatesan, R., Yung, M.: Perfect zero-knowledge arguments for np can be based on general complexity assumptions. In: Brickell, E.F. (ed.) CRYPTO 1992. LNCS, vol. 740, pp. 196–214. Springer, Heidelberg (1993)
31. Ostrovsky, R., Venkatesan, R., Yung, M.: Fair games against an all-powerful adversary. DIMACS workshop presentation (1990); Extended abstract, In: Capocelli, R.M., De-Santis, A., Vaccaro, U. (eds.) Proceedings of Sequences II, Positano, Italy. Springer, Heidelberg (June 1991); Journal version in AMS DIMACS Series in Discrete Mathematics and Theoretical Computer Science 13 (1991)
32. Haitner, I., Nguyen, M.H., Ong, S.J., Reingold, O., Vadhan, S.P.: Statistically hiding commitments and statistical zero-knowledge arguments from any one-way function. SIAM J. Comput (2009)
33. Haitner, I., Reingold, O., Vadhan, S.P., Wee, H.: Inaccessible entropy. In: STOC (2009)
34. Kilian, J.: Founding cryptography on oblivious transfer. In: STOC (1988)

# Instantiability of RSA-OAEP under Chosen-Plaintext Attack

Eike Kiltz[1], Adam O'Neill[2], and Adam Smith[3]

[1] Centrum voor Wiskunde en Informatica, Amsterdam, Netherlands
kiltz@cwi.nl
[2] Georgia Institute of Technology, Atlanta, GA, USA
amoneill@cc.gatech.edu
[3] Pennsylvania State University, University Park, PA, USA
asmith@cse.psu.edu

**Abstract.** We show that the widely deployed RSA-OAEP encryption scheme of Bellare and Rogaway (Eurocrypt 1994), which combines RSA with two rounds of an underlying Feistel network whose hash (*i.e.*, round) functions are modeled as random oracles, meets indistinguishability under chosen-plaintext attack (IND-CPA) in the *standard model* based on simple, non-interactive, and non-interdependent assumptions on RSA and the hash functions. To prove this, we first give a result on a more general notion called "padding-based" encryption, saying that such a scheme is IND-CPA if (1) its underlying padding transform satisfies a "fooling" condition against small-range distinguishers on a class of high-entropy input distributions, and (2) its trapdoor permutation is sufficiently *lossy* as defined by Peikert and Waters (STOC 2008). We then show that the first round of OAEP satifies condition (1) if its hash function is $t$-wise independent for appopriate $t$ and that RSA satisfies condition (2) under the $\Phi$-Hiding Assumption of Cachin *et al.* (Eurocrypt 1999).

This appears to be the first non-trivial *positive* result about the instantiability of RSA-OAEP. In particular, it increases our confidence that chosen-plaintext attacks are unlikely to be found against the scheme. In contrast, RSA-OAEP's predecessor in PKCS #1 v1.5 was shown to be vulnerable to such attacks by Coron *et al.* (Eurocrypt 2000).

## 1 Introduction

The RSA-OAEP encryption scheme was designed by Bellare and Rogaway [5] as a drop-in replacement for RSA PKCS #1 v1.5 [37] with provable security guarantees. In particular, it follows the same paradigm as RSA PKCS #1 v1.5 in that it encrypts a message of less than $k$ bits to a $k$-bit ciphertext (where $k$ is the modulus size) by first applying a fast, randomized, and invertible "padding transform" to the message before applying RSA. In the case of RSA-OAEP, the underlying padding transform (which is itself called 'OAEP'[1]) embeds a

---

[1] We often use the same terminology for '$f$-OAEP,' which refers to OAEP using an abstract TDP $f$, with the meaning hopefully clear from context.

T. Rabin (Ed.): CRYPTO 2010, LNCS 6223, pp. 295–313, 2010.

message $m$ and random coins $r$ as $s\|(H(s) \oplus r)$ where '$\|$' denotes concatenation, $s = (m\|0^{k_1}) \oplus G(r)$ for some parameter $k_1$, and $G$ and $H$ are hash functions (see Figure 1 on p. 305). In contrast, PKCS #1 v1.5 essentially just concatenates $m$ with $r$.

RSA-OAEP was designed using the random oracle (RO) methodology [4]. This means that, for the security analysis, its hash functions are modeled as independent truly random functions, available as oracles to all parties. When the scheme is implemented in practice, they are heuristically "instantiated" in certain ways using a cryptographic hash function like SHA1. A cryptographic hash function is certainly not random (it has a short public description), but schemes designed using this methodology are hoped to be secure. Unfortunately, a series of works, starting with the seminal paper of Canetti *et al.* [16] showed that there are schemes secure in the RO model that are insecure under *every* instantiation of the oracle; such RO model schemes are called *uninstantiable*. Thus, to gain confidence in an RO model scheme, we should show that it is not uninstantiable, *i.e.*, that it admits a secure instantiation by an efficiently computable function under well-defined assumptions. Then, when we instantiate the scheme, we know that our goal is at least plausible. This is especially important for a scheme such as RSA-OAEP, which is by now widely standardized and deployed.

Yet, while RO model schemes continue to be proposed, few have been shown to be instantiable. In particular, we are not aware of *any* result showing instantiability of RSA-OAEP, even under a relatively modest security model. In fact, the scheme has come under criticism lately due to several works (discussed in Section 1.2) showing the impossibility of certain types of instantiations under chosen-ciphertext attack (IND-CCA). Fortunately, we bring some good news: We give reasonable assumptions under which RSA-OAEP is secure against *chosen-plaintext attack* (IND-CPA). We believe this is an important step towards a better understanding of the scheme's security.

## 1.1   Our Contributions

Our result on the instantiability of RSA-OAEP is obtained via three steps or other results. (These other results may also be of independent interest.) First, we show a general result on the instantiability of "padding-based encryption," of which $f$-OAEP is a special case, under the assumption that the underlying padding transform is what we call a *fooling extractor* and the trapdoor permutation is sufficiently *lossy* [36]. We then show that OAEP and RSA satisfy the respective conditions.

PADDING-BASED ENCRYPTION WITHOUT ROS. Our first result is a general theorem about *padding-based encryption* (PBE), a notion formalized recently by Kiltz and Pietrzak [29].[2] PBE generalizes the design methology of PKCS #1 and RSA-OAEP we already mentioned. Namely, we start with a $k$-bit to $k$-bit trapdoor permutation (TDP) that satisfies a weak security notion like one-wayness.

---

[2] Such schemes were called "simple embedding schemes" by Bellare and Rogaway [5], who discussed them only on an intuitive level.

To "upgrade" the TDP to an encryption scheme satisfying a strong security notion like IND-CPA, we design an invertible "padding transform" which embeds a plaintext and random coins into a $k$-bit string, to which we then apply the TDP. This methodology is quite natural and has long been prevalent in practice, motivating the design of OAEP and later schemes such as SAEP [9] and PSS-E [20]. The latter were all designed and analyzed in the RO model.

We show that the RO model is *unnecessary* in the design and analysis of IND-CPA secure PBE. To do so, we formulate an interesting connection between PBE and a new notion we call "fooling extractor for small-range distinguishers" or just "fooling extractor." Intuitively, a fooling extractor transforms a high-entropy source into something that "looks random" to any function (or distinguisher) with a *small range*.[3] Our result says that if the underlying padding transform of a PBE scheme is a fooling extractor for all sources of the form $(m, R)$ where $m$ is a plaintext and $R$ is the random coins (which we call "encryption sources") and its TDP is *lossy* as defined by Peikert and Waters [36] then the PBE scheme is IND-CPA. We call such padding transforms "encryption-compatible."

OAEP FOOLS SMALL-RANGE DISTINGUISHERS. Our second result says that the OAEP padding transform is encryption-compatible as we defined it above if the hash function $G$ is $t$-wise independent for appropriate $t$ (essentially, proportional to the allowed message length, where the latter is determined by how large an output range of the distinguisher should be tolerated in the definition of encryption-compatibility). Note that no restriction is put on hash function $H$; in particular, neither hash function is modeled as a RO.

The inspiration for our proof comes from the "Crooked" Leftover Hash Lemma (LHL) of Dodis and Smith [22] (see [6] for a simpler proof of the latter). Qualitatively, the Crooked LHL says that $K, f(\Pi(K, X))$ looks like $K, f(U)$ for any small-range function $f$, pairwise-independent function $\Pi$ keyed by $K$, and high-entropy source $X$; in our terminology, this says that a pairwise-independent function is a fooling extractor for such $X$. In our application, we might naïvely view $\Pi$ as the OAEP. There are two problems with this. First, OAEP is *not* pairwise independent, even in the RO model. Second, showing that OAEP is encryption-compatible entails showing it fools $f$ on *all* encryption sources simultaneously, whereas the lemma pertains to a *fixed* source. To solve the first problem, we show that the lemma can be strengthened to say that $K, f(X, \Pi(K, X))$ looks like $K, f(X, U)$; *i.e.*, that $\Pi(K, X)$ looks random to $f$ *even given* $X$. Then, we view $X$ as the random coins in OAEP and $\Pi$ as the hash function $G$; we can conclude that OAEP is a fooling extractor for a *fixed* encryption source $(m, R)$ (note that our analysis does not use any properties of $H$—the only fact we use about the second Feistel round is that it is invertible). To solve the second problem, we extend an idea of Trevisan and Vadhan [42] to our setting and show that if $G$ is in fact $t$-*wise* independent for large enough $t$, the error probability for a particular encryption source is so small that we can take a union bound and conclude that OAEP is a fooling extractor on all of them, as required.

---

[3] In the formal defintion there is also an "outer" distinguisher who gets the extractor seed; see Section 3 for details.

LOSSINESS OF RSA. To instantiate RSA-OAEP, it remains to show lossiness of RSA. Our final result is that RSA is indeed lossy under reasonable assumptions. Intuitively, lossiness [36] means that there is an alternative, "lossy" key generation algorithm that outputs a public key indistinguishable from a normal one, but which induces a small-range (uninvertible) function. We first show lossiness of RSA under the $\Phi$-Hiding Assumption ($\Phi$A) of Cachin, Micali, and Stadler [13]. $\Phi$A has been used as the basis for a number of efficient protocols, e.g., [13, 12, 24, 25]. $\Phi$A states roughly that given an RSA modulus $N = pq$, it is hard to distinguish primes that divide $\phi(N) = (p-1)(q-1)$ from those that do not. Normal RSA parameters $(N, e)$ are such that $\gcd(e, \phi(N) = 1$. Under $\Phi$A, we may alternatively choose $(N', e)$ such that $e$ *divides* $p-1$. The range of the RSA function is then reduced by a factor $1/e$. To resist known attacks, we can take the bit-length of $e$ up to almost $1/4$ that of $N$, giving RSA lossiness of almost $k/4$ bits, where $k$ is the modulus length.[4]

We then observe that for small $e$ lossiness may be amplified for a fixed modulus length by considering *multi-prime* RSA where $N = p_1 \cdots p_m$ for $m \geq 2$, and in the lossy case choosing $(N', e)$ such that $e$ divides $p_i$ for *all* $1 \leq i \leq m-1$; the range of the RSA function is then reduced by a factor $1/e^{m-1}$. (The maximum bit-length of $e$ in this case to avoid known attacks is roughly $k(1/m - 2/m^2)$ where $k$ is the modulus length, so for a fixed modulus size we gain in lossiness only for small $e$.) If we assume such multi-prime RSA moduli are indistinguishable from two-prime ones, we can achieve such lossiness in the case of standard (two-prime) RSA as well.

IMPLICATIONS FOR RSA-OAEP. Combining the above implies that RSA-OAEP is IND-CPA in the standard model under (rather surprisingly) simple, non-interactive, and non-interdependent assumptions on RSA and the hash functions. The parameters for RSA-OAEP supported by our proofs are discussed in Section 6. While they are considerably worse than what is expected in practice, we view the upshot of our results not as the concrete parameters they support, but rather that they increase the theoretical backing for the scheme's security at a more qualitative level, showing it can be instantiated at least for larger parameters. In particular, our results give us greater confidence that chosen-plaintext attacks are unlikely to be found against the scheme; such attacks are known against the predecessor of RSA-OAEP in PKCS #1 v1.5 [19]. That said, we strongly encourage further research to try to improve the concrete parameters.

Moreover, our analysis brings to light to some simple modifications that may increase the scheme's security. The first is to key the hash function $G$. Although our results have some interpretation in the case that $G$ is a fixed function (see below), it may be preferable for $G$ to have an explicit, randomly selected key. It is in an interesting open question whether our proof can be extended to function families that use shorter keys. The second possible modification is to increase the length of the randomness versus that of the redundancy in the message

---

[4] We remark that the recent attacks on $\Phi$A [40] are for moduli of a special form that does not include RSA.

when encrypting short messages under RSA-OAEP. Of course, we suggest these modifications only in cases where they do not impact efficiency too severely.

USING UNKEYED HASH FUNCTIONS. Formally, our results assume $G$ is randomly chosen from a large family (*i.e.*, it is a keyed hash function). However, our analysis actually shows that *almost every* function (*i.e.*, all but a negligible fraction) from the family yields a secure instantiation; we just do not know an explicit member that works. In other words, it is not strictly necessary that $G$ be randomly chosen. When $G$ is instantiated in practice using a cryptographic hash function, it is plausible that the resulting instantiation is secure.

CHOSEN-CIPHERTEXT SECURITY. Any extension of our results to CCA security must get around the recent negative results of Kiltz and Pietrzak [29] (which we discuss in more detail below). We outline some possible approaches in the full version [27].

## 1.2   Related Work

SECURITY OF OAEP IN THE RO MODEL. In their original paper [5], Bellare and Rogaway showed that OAEP is IND-CPA assuming the TDP is one-way. They further showed it achieves a notion they called "plaintext awareness." Subsequently, Shoup [41] observed that the latter notion is too weak to imply security against chosen-ciphertext attacks, and in fact there is no black-box proof of IND-CCA security of OAEP based on one-wayness of the TDP. Fortunately, Fujisaki *et al.* [23] proved that OAEP is nevertheless IND-CCA assuming so-called "partial-domain" one-wayness, and that partial-domain one-wayness and (standard) one-wayness of RSA are equivalent.

SECURITY OF OAEP WITHOUT ROs. Results on instantiability of OAEP have so far mainly been negative. Boldyreva and Fischlin [7] showed that (contrary to a conjecture of Canetti [14]) one cannot securely instantiate even *one* of the two hash functions (while still modeling the other as a RO) of OAEP under IND-CCA by a "perfectly one-way" hash function [14, 17] if one assumes only that $f$ is partial-domain one-way. Brown [10] and Paillier and Villar [34] later showed that there are no "key-preserving" black-box proofs of IND-CCA security of RSA-OAEP based on one-wayness of RSA. Recently, Kiltz and Pietrzak [29] (building on the earlier work of Dodis *et al.* [21] in the signature context) generalized these results and showed that there is no black-box proof of IND-CCA (or even NM-CPA) security of OAEP based on any property of the TDP satisfied by an *ideal* (truly random) permutation.[5] In fact, their result can be extended to rule out a black-box proof of NM-CPA security of OAEP assuming the TDP is lossy [30], so our results are in some sense *optimal* given our assumptions.

INSTANTIATIONS OF RELATED SCHEMES. A positive instantiation result about a variant of OAEP called OAEP++ [26] (where part of the transform is output

---

[5] Note, however, that their result does not rule out such a proof based on other properties of the TDP, non-black-box assumptions on the hash functions, or in the case of a specific TDP like RSA.

in the clear) was obtained by Boldyreva and Fischlin in [8]. They showed an instantiation that achieves (some weak form of) non-malleability under chosen-plaintext attacks (NM-CPA) for random messages, assuming the existence of non-malleable pseudorandom generators (NM-PRGs).[6] We note that the approach of trying to obtain positive results for instantiations under security notions weaker than IND-CCA originates from their work, and the authors explicitly ask whether OAEP can be shown IND-CPA in the standard model based on reasonable assumptions on the TDP and hash functions.

Another line of work has looked at instantiating other RO model schemes related at least in spirit to OAEP. Canetti [14] showed that the IND-CPA scheme in [4] can be instantiated using (a strong form of) perfectly-one way probabilistic hash functions. More recently, the works of Canetti and Dakdouk [15], Pandey *et al.* [35], and Boldyreva *et al.* [11] obtained (partial) instantiations of the earlier IND-CCA scheme of [4]. Hofheinz and Kiltz [28] recently showed an IND-CCA secure instantiation of a variant the DHIES scheme of [1].

## 2 Preliminaries

NOTATION AND CONVENTIONS. For a probabilistic algorithm $A$, by $y \xleftarrow{\$} A(x)$ we mean that $A$ is executed on input $x$ and the output is assigned to $y$, whereas if $S$ is a finite set then by $s \xleftarrow{\$} S$ we mean that $s$ is assigned a uniformly random element of $S$. We sometimes use $y \leftarrow A(x; \text{Coins})$ to make $A$'s random coins explicit. We denote by $\Pr[A(x) \Rightarrow y : \ldots]$ the probability that $A$ outputs $y$ on input $x$ when $x$ is sampled according to the elided experiment. Unless otherwise specified, an algorithm may be probabilistic and its running-time includes that of any overlying experiment. We denote by $1^k$ the unary encoding of the security parameter $k$. We sometimes surpress dependence on $k$ for readability. For $i \in \mathbb{N}$ we denote by $\{0, 1\}^i$ the set of all binary strings of length $i$. If $s$ is a string then $|s|$ denotes its length in bits, whereas if $S$ is a set then $|S|$ denotes its cardinality. By '$\|$' we denote string concatenation. All logarithms are base 2.

BASIC DEFINITIONS. Writing $P_X(x)$ for the probability that a random variable $X$ puts on $x$, the *statistical distance* between random variables $X$ and $Y$ with the same range is given by $\Delta(X, Y) = \frac{1}{2} \sum_x |P_X(x) - P_Y(x)|$. If $\Delta(X, Y)$ is at most $\varepsilon$ then we say $X, Y$ are $\varepsilon$-*close* and write $X \approx_\varepsilon Y$. The *min-entropy* of $X$ is $H_\infty(X) = -\log(\max_x P_X(x))$. A random variable $X$ over $\{0, 1\}^n$ is called a $(n, \ell)$-*source* if $H_\infty(X) \geq \ell$. Let $f : A \to B$ be a function. We denote by $R(f)$ the *range* of $f$, *i.e.*, $\{b \in B \mid \exists a \in A, f(a) = b\}$. We call $|R(f)|$ the *range-size* of $f$. We call $f$ *regular* if each pre-image set is the same size, *i.e.*, $|\{x \in D \mid f(x) = y\}|$ is the same for all $y \in R$.

PUBLIC-KEY ENCRYPTION AND ITS SECURITY. A *public-key encryption scheme* with message-space MsgSp is a triple of algorithms $\mathcal{AE} = (\mathcal{K}, \mathcal{E}, \mathcal{D})$. The

---

[6] In particular, their security notion does *not* imply IND-CPA since they consider random messages. We also point out that it remains an open question whether NM-PRGs can be constructed.

key-generation algorithm $\mathcal{K}$ returns a public key $pk$ and matching secret key $sk$. The encryption algorithm $\mathcal{E}$ takes $pk$ and a plaintext $m$ to return a cipher-text. The deterministic decryption algorithm $\mathcal{D}$ takes $sk$ and a ciphertext $c$ to return a plaintext. We require that for all messages $m \in \mathrm{MsgSp}$

$$\Pr\left[\, \mathcal{D}(sk, \mathcal{E}(pk, m)) \neq m \; : \; (pk, sk) \xleftarrow{\$} \mathcal{K} \,\right]$$

is negligible.

To an encryption scheme $\Pi = (\mathcal{K}, \mathcal{E}, \mathcal{D})$ and an adversary $A = (A_1, A_2)$ we associate a chosen-plaintext attack experiment,

> **Experiment $\mathbf{Exp}_{\Pi,A}^{\text{ind-cpa}}(k)$**
> $b \xleftarrow{\$} \{0, 1\}$ ; $(pk, sk) \xleftarrow{\$} \mathcal{K}(1^k)$
> $(m_0, m_1, \text{state}) \xleftarrow{\$} A_1(pk)$
> $c \xleftarrow{\$} \mathcal{E}(pk, m_b)$
> $d \xleftarrow{\$} B_2(pk, c, \text{state})$
> If $d = b$ then return 1 else return 0

where we require $A$'s output to satisfy $|m_0| = |m_1|$. Define the *ind-cpa advantage* of $A$ against $\Pi$ as

$$\mathbf{Adv}_{\Pi,A}^{\text{ind-cpa}}(k) = 2 \cdot \Pr\left[\, \mathbf{Exp}_{\Pi,A}^{\text{ind-cpa}}(k) \Rightarrow 1 \,\right] - 1 \,.$$

LOSSY TRAPDOOR PERMUTATIONS. A *lossy trapdoor permutation (LTDP) gen-erator* [36][7] is a pair $\mathsf{LTDP} = (\mathcal{F}, \mathcal{F}')$ of algorithms. Algorithm $\mathcal{F}$ is a usual trapdoor permutation (TDP) generator, namely it outputs a pair $(f, f^{-1})$ where $f$ is a (description of a) permutation on $\{0, 1\}^k$ and $f^{-1}$ its inverse. Algorithm $\mathcal{F}'$ outputs a (description of a) function $f'$ on $\{0, 1\}^k$. We call $\mathcal{F}$ the "injective mode" and $\mathcal{F}'$ the "lossy mode" of $\mathsf{LTDP}$ respectively, and we call $\mathcal{F}$ "lossy" if it is the first component of some lossy TDP. For a distinguisher $D$, define its *ltdp-advantage* against $\mathsf{LTDP}$ as

$$\mathbf{Adv}_{\mathsf{LTDP},D}^{\text{ltdp}}(k) = \Pr\left[\, D(f) \Rightarrow 1 \; : \; (f, f^{-1}) \xleftarrow{\$} \mathcal{F} \,\right] - \Pr\left[\, D(f') \Rightarrow 1 \; : \; f' \xleftarrow{\$} \mathcal{F}' \,\right] \,.$$

We say $\mathsf{LTDP}$ has *residual leakage* $s$ if for all $f'$ output by $\mathcal{F}'$ we have $|R(f')| \leq 2^s$. The *lossiness* of $\mathsf{LTDP}$ is $\ell = k - s$.

$t$-WISE INDEPENDENT HASHING. Let $H \colon \mathcal{K} \times D \to R$ be a hash function. We say that $H$ is *$t$-wise independent* if for all distinct $x_1, \ldots, x_t \in D$ and all $y_1, \ldots, y_t \in R$

$$\Pr\left[\, H(K, x_1) = y_1 \wedge \; \ldots \; \wedge \; H(K, x_t) = y_t \; : \; K \xleftarrow{\$} \mathcal{K} \,\right] = \frac{1}{|R|^t} \,.$$

In other words, $H(K, x_1), \ldots, H(K, x_t)$ are all uniformly and independently random.

---

[7] We note that [36] actually defines lossy trapdoor *functions*, but the extension to permutations is straightforward.

# 3   Padding-Based Encryption from Lossy TDP + Fooling Extractor

In this section, we show a general result on how to build IND-CPA secure padding-based encryption (PBE) without using random oracles, by combining a lossy TDP with a "fooling extractor" for small-range distinguishers.

## 3.1   Background and Tools

We first provide the definitions relevant to our result.

PADDING-BASED ENCRYPTION. The idea behind padding-based encryption (PBE) is as follows: We start with a $k$-bit to $k$-bit trapdoor permutation (*e.g.*, RSA) and wish to build a secure encryption scheme. As in [5], we are interested in encrypting messages of less than $k$ bits to ciphertexts of length $k$. It is well-known that we cannot simply encrypt messages under the TDP directly to achieve strong security. So, in a PBE scheme we "upgrade" the TDP by first applying a randomized and invertible "padding transform" to a message prior to encryption.

Our definition of PBE largely follows the recent formalization in [29]. Let $k, \mu, \rho$ be three integers such that $\mu + \rho \leq k$. A *padding transform* $(\pi, \hat{\pi})$ consists of two mappings $\pi : \{0,1\}^{\mu+\rho} \to \{0,1\}^k$ and $\hat{\pi} : \{0,1\}^k \to \{0,1\}^\mu \cup \{\perp\}$ such that $\pi$ is injective and the following consistency requirement is fulfilled:

$$\forall m \in \{0,1\}^\mu, r \in \{0,1\}^\rho : \quad \hat{\pi}(\pi(m \,\|\, r)) = m .$$

A *padding transform generator* is an algorithm $\Pi$ that on input $1^k$ outputs a (description of a) padding transform $(\pi, \hat{\pi})$. Let $\mathcal{F}$ be a $k$-bit trapdoor permutation generator and $\Pi$ be a padding transform generator. Define the associated *padding-based encryption scheme* $\mathcal{AE}_\Pi[\mathcal{F}] = (\mathcal{K}, \mathcal{E}, \mathcal{D})$ with message-space $\{0,1\}^\mu$ by

| **Alg** $\mathcal{K}(1^k)$ | **Alg** $\mathcal{E}((\pi, f), m)$ | **Alg** $\mathcal{D}((\pi, f^{-1}), y)$ |
|---|---|---|
| $(\pi, \hat{\pi}) \xleftarrow{\$} \Pi(1^k)$ | $r \xleftarrow{\$} \{0,1\}^\rho$ ; $x \leftarrow \pi(m\|r)$ | $x \leftarrow f^{-1}(y)$ |
| $\boldsymbol{\pi} \leftarrow (\pi, \hat{\pi})$ | $y \leftarrow f(x)$ | $m \leftarrow \hat{\pi}(x)$ |
| $(f, f^{-1}) \xleftarrow{\$} \mathcal{F}(1^k)$ | Return $y$ | Return $m$ |
| Return $((\boldsymbol{\pi}, f), (\boldsymbol{\pi}, f^{-1}))$ | | |

Padding-based encryption schemes have long been prevalent in practice, for example PKCS #1 [37]. While OAEP [5] is the best-known, the notion also captures later schemes such as SAEP [9] and PSS-E [20].

FOOLING EXTRACTORS. We define a new notion that we call "fooling extractor for small-range distinguishers" or just "fooling extractor." Intuitively, fooling extractors are a type of randomness extractor that "fools" distinguishers with small-range output. We give some more intuition after the formal definition.

**Definition 1.** *Let* FExt: $\{0,1\}^c \times \{0,1\}^n \to \{0,1\}^k$ *be a function and let* $\mathcal{X} = \{X_1, \ldots, X_q\}$ *be a class of $n$-bit sources. We say that* FExt *fools range-$2^s$*

distinguishers on $\mathcal{X}$ with probability $1 - \varepsilon$ *(or is an $(s, \varepsilon)$-fooling extractor for*
$\mathcal{X}$*) if for all functions $f$ on $\{0,1\}^k$ with range-size at most $2^s$ and all $1 \le i \le q$:*

$$(K, f(\mathsf{FExt}(K, X_i))) \approx_\varepsilon (K, f(U)) \,,$$

*where $K$ is uniform on $\{0,1\}^c$ and $U$ is uniform and independent on $\{0,1\}^n$.*
*(Here $K$ is the key or seed of $\mathsf{FExt}$.) For example, one is often interested in the*
*class $\mathcal{X}_{n,\ell}$ consisting of all $(n, \ell)$-sources $X$. As a strengthening of the above, we*
*say that $\mathsf{FExt}$ simultaneously fools range-$2^s$ distinguishers on $\mathcal{X}$ with probability*
$1 - \varepsilon$ *(or is a simultaneous $(s, \varepsilon)$-fooling extractor for $\mathcal{X}$) if for all functions $f$*
*on $\{0,1\}^k$ with range-size at most $2^s$:*

$$\mathop{\mathbf{E}}_{k \xleftarrow{\$} \{0,1\}^c} \left[ \max_{1 \le i \le q} \Delta\Big( f(\mathsf{FExt}(k, X_i)) \,,\, f(U) \Big) \right] \le \varepsilon \,.$$

*As a useful special case, we say that $\mathsf{FExt}$ fools regular range-$2^s$ distinguishers*
*on $\mathcal{X}$ with probability $1 - \varepsilon$ (or is a regular $(s, \varepsilon)$-fooling extractor for $\mathcal{X}$) if we*
*quantify only over regular $f$ in the definition. A simultaneous regular $(s, \varepsilon)$-*
*fooling extractor for $\mathcal{X}$ is defined analogously.*

Intuitively, one can think of the definition of a fooling extractor as involving a
two-stage distinguisher. The first stage is represented by the function $f$, which
takes as input $\mathsf{FExt}(K, X_i)$. The second stage is represented only implicitly, and
takes as input $f(\mathsf{FExt}(K, X_i))$ and $K$. While the intuition given prior to the
definition captures only the first stage, the second stage is crucial for the definition to be meaningful. Indeed, just asking that $f(\mathsf{FExt}(K, X_i))$ be indistinguishable from $f(U)$ for all small-range functions $f$ is equivalent to asking only that
$\mathsf{FExt}(K, X_i)$ be indistinguishable from $U$. This latter requirement is trivial to
achieve–for example, by using $K$ as a one-time pad.

We note that the concept of fooling extractors was implicit in the work of
Dodis and Smith [22] on error-correction without leaking partial information,
whose "Crooked" Leftover Hash Lemma establishes in our language that a
pairwise-independent function is a $(s, \varepsilon)$-fooling extractor for every singleton
$(n, \ell)$-source $X$ where $s \le \ell - 2\log(1/\varepsilon) + 2$.

### 3.2    The Result

To state our result, we first formalize the concept of *encryption-compatible*
padding transforms.

**Definition 2.** *Let $\Pi$ be a padding transform generator whose coins are drawn*
*from $\mathsf{Coins}$. Define the function $h_\Pi : \mathsf{Coins} \times \{0,1\}^{\mu+\rho} \to \{0,1\}^k$ by $h(c, m\|r) =$*
$\pi(m\|r)$ *for all $c \in \mathsf{Coins}, m \in \{0,1\}^\mu, r \in \{0,1\}^\rho$, where $(\pi, \hat{\pi}) \leftarrow \Pi(1^k; \mathsf{Coins})$.*
*We say that $\Pi$ is $(s, \varepsilon)$-encryption-compatible if $h_\Pi$ as above is a simultaneous*
$(s, \varepsilon)$*-fooling extractor for the class $\mathcal{X}_\Pi$ of sources of the form $(m, R)$, where*
$m \in \{0,1\}^\mu$ *is fixed and $R \in \{0,1\}^\rho$ is uniformly random. (Note that the class*
$\mathcal{X}_\Pi$ *contains $2^\mu$ distinct $(\mu + \rho)$-bit sources.) We call $\mathcal{X}_\Pi$ the class of encryption*
*sources associated to $\Pi$. A regular $(s, \varepsilon)$-encryption-compatible padding trans-*
*form generator is defined analogously.*

**Theorem 1.** *Let* LTDP $= (\mathcal{F}, \mathcal{F}')$ *be an LTDP with residual leakage* $s$*, and let* $\Pi$ *be an* $(s, \varepsilon)$*-encryption-compatible padding transform generator. Then for any IND-CPA adversary* $A$ *against* $\mathcal{AE}_\Pi[\mathcal{F}]$ *there is a adversary* $D$ *against* LTDP *such that for all* $k \in \mathbb{N}$

$$\mathbf{Adv}_{\mathcal{AE},A}^{\text{ind-cpa}}(k) \leq \mathbf{Adv}_{\text{LTDP},D}^{\text{ltdp}}(k) + \varepsilon .$$

*Furthermore, the running-time of* $D$ *is the time to run* $A$.

*Remark 1.* The analogous result to the above holds for regular LTDPs and regular encryption-compatible padding transforms. That is, if the LTDP is *regular* (meaning $\mathcal{F}'$ is) then it suffices to use a regular encryption-compatible padding transform to obtain the same conclusion. The latter may be easier to design or more efficient than in the general case; indeed, we get better parameters for OAEP in the regular case in Section 4. Furthermore, known examples of LTDPs (including RSA, as shown in Section 5) are regular, although some technical issues make it difficult to exploit this for RSA-OAEP; cf. Section 6.

## 4   OAEP as a Fooling Extractor

In this section, we show that the OAEP padding transform of Bellare and Rogaway [5] is encryption-compatible as defined in Section 3 if its initial hash function is $t$-wise independent for appropriate $t$.

### 4.1   OAEP

We recall the OAEP padding transform of Bellare and Rogaway [5], lifted to the "instantiated" setting where hash functions may be keyed. Let $G\colon \mathcal{K}_G \times \{0,1\}^\rho \to \{0,1\}^\mu$ and $H\colon \mathcal{K}_H \times \{0,1\}^\mu \to \{0,1\}^\rho$ be hash functions. The associated padding transform generator OAEP$[G,H]$ on input $1^k$ returns $(\pi_{K_G,K_H}, \hat{\pi}_{K_G,K_G})$, where $K_G \xleftarrow{\$} \mathcal{K}_G(1^k)$ and $K_H \xleftarrow{\$} \mathcal{K}_H(1^k)$, defined via

| **Algorithm** $\pi_{K_G,K_H}(m\|r)$ | **Algorithm** $\hat{\pi}_{K_G,K_H}(x)$ |
|---|---|
| $s \leftarrow m \oplus G(K_G, r)$ | $s\|t \leftarrow x$ |
| $t \leftarrow r \oplus H(K_H, s)$ | $r \leftarrow t \oplus H(K_H, s)$ |
| $x \leftarrow s\|t$ | $m \leftarrow s \oplus G(K_G, r)$ |
| Return $x$ | Return $m$ |

See Figure 1 for a graphical illustration.

*Remark 2.* Since we mainly study IND-CPA security, for simplicity we define above the "no-redundancy" version of the OAEP, *i.e.*, corresponding to the "basic scheme" in [5]. However, our results also hold for the redundant version. Additionally, as is typical in the literature we have defined OAEP to apply the $G$-function to the least-significant bits of the input; in standards and implementations it is typically the most significant bits (where the order of $m$ and $r$ are switched). Again, we stress that our results hold in either case.

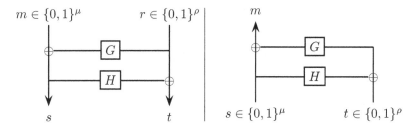

**Fig. 1.** Algorithms $\pi_{K_G,K_H}(m,r)$ and $\hat{\pi}_{K_G,K_H}(s,t)$ for OAEP$[G,H]$

## 4.2   Analysis

The following establishes that OAEP is encryption-compatible if the hash function $G$ is $t$-wise independent for appropriate $t$. No restriction is put on the other hash function $H$. Indeed, our result also applies to SAEP [9] (although the latter is neither standardized nor known to provide CCA security in the RO model, except in certain cases).

**Theorem 2.** *Let $G\colon \mathcal{K}_G \times \{0,1\}^\rho \to \{0,1\}^\mu$ and $H\colon \mathcal{K}_H \times \{0,1\}^\mu \to \{0,1\}^\rho$ be hash functions, and suppose $G$ is $t$-wise independent. Let $\mathsf{OAEP} = \mathsf{OAEP}[G,H]$. Then*

(1)   OAEP *is $(s,\varepsilon)$-encryption-compatible where $\varepsilon = 2^{-u}$ for $u = \frac{t}{3t+2}(\rho - s - \log t + 2) - \frac{2(\mu+s)}{3t+2} - 1$.*

(2)   OAEP *is regular $(s,\varepsilon)$-encryption-compatible where $\varepsilon = 2^{-u}$ for $u = \frac{t}{2t+2}(\rho - s - \log t + 2) - \frac{\mu+s+2}{t+1} - 1$.*

(3)   *When $t = 2$, OAEP $(s,\varepsilon)$-encryption-compatible where $\varepsilon = 2^{-u}$ for $u = (\rho - s - 2\mu)/4 - 1$.*

Note that parts (2) and (3) capture special cases of (1) in which we get better bounds. We give a high-level idea of the proof; details are deferred to the full version [27].

The high-level idea for all three parts of the theorem is the same. Fix a lossy function $f$ with range-size at most $2^s$. We first show that for every *fixed* message $m \in \{0,1\}^\mu$, with high probability (say $1-\delta$) over the choice of the hash function $G$, the statistical distance between $(K_G, f(\mathsf{OAEP}(m,R)))$ and $((K_G, f(U))$ is small (say $\hat{\varepsilon}$). Namely, we first compute the *expected* statistical distance over the choice of $G$ and then apply tail bounds. This aspect of the proof changes from part to part. For part (3) we use a strengthened version of the Crooked Leftover Hash Lemma (LHL) of [22] and Markov's inequality. For parts (1) and (2) we adapt the techniques of [42] (see also [2]) developed in the context of the standard LHL and use the tail inequality for $t$-wise independent random variables due to Bellare and Rompel [3]. (For part (2) this is relatively easy, but for part (1) we first apply a "balancing" lemma saying that for any non-regular $f$ we can find a "almost-regular" function $g$ that agrees with $f$ on a large fraction of its domain.) In all three parts, we can then take a union bound to show that OAEP is good for *all* messages with probability at least $1 - 2^\mu \delta$. This means that the statistical distance between the pair $(K_G, f(\text{OAEP}(m, R)))$ and $(K_G, f(\text{OAEP}(U)))$ is at most $\varepsilon = \hat\varepsilon + 2^\mu \delta$. Finally, we express $\delta$ as a function of $\hat\varepsilon$, and select $\hat\varepsilon$ to minimize this sum. Note that the entire argument works for any choice of $H$.

In order to get a more qualitative "feel" for the bounds in the theorem, we give the following simplification as a corollary:

**Corollary 1.** *Let $G \colon \mathcal{K}_G \times \{0,1\}^\rho \to \{0,1\}^\mu$ and $H \colon \mathcal{K}_H \times \{0,1\}^\mu \to \{0,1\}^\rho$ be hash functions and suppose that $G$ is $t$-wise independent for $t \geq 3\frac{\mu+s}{\rho-s}$. Then $\text{OAEP}[G, H]$ is $(s, \varepsilon)$-encryption-compatible where $\varepsilon = \exp(-c(\rho - s - \log t))$ for a constant $c > 0$.*

In particular, $c \approx 1/2$ for regular functions. For such a function, if $\rho - s$ is at least 180 then $\varepsilon$ is roughly $2^{-80}$ for $t = 10$ and message lengths $\mu \leq 2^{15}$ (which for practical purposes does not restrict the message-space). Applying Theorem 1, we see that if $G$ is 10-wise independent and the number of random bits used in OAEP is at least 180 bits larger than the residual lossiness of the TDP, then the security of OAEP is tightly related to that of the lossy TDP.

*Remark 3.* To show security of OAEP against what we call *key-independent* chosen-plaintext attack, it suffices to argue that $\text{OAEP}[G, H]$ is a fooling extractor for any *fixed* encryption source $X = (m, R)$ where $m \in \{0,1\}^\mu$. The latter holds for any $\varepsilon > 0$ and $s \leq \rho - 2\log(1/\varepsilon) + 2$ assuming $G$ is only pairwise-independent (*i.e.*, $t = 2$). See the full version [27] for details.

# 5   Lossiness of RSA

In this section, we show that the RSA trapdoor permutation is lossy under reasonable assumptions. In particular, we show that, for large enough encryption exponent $e$, RSA is considerably lossy under the $\Phi$-Hiding Assumption of [13]. We then show that by generalizing this assumption to multi-prime RSA we can get even more lossiness. Finally, we propose a "Two-Or-$m$-Primes" Assumption that, when combined with the former, amplifies the lossiness of standard (two-prime) RSA for small $e$.

## 5.1   Background on RSA and Notation

We denote by $\mathcal{RSA}_k$ the set of all tuples $(N, p, q)$ such that $N = pq$ is the product of two distinct $k/2$-bit primes. Such an $N$ is called an *RSA modulus*. By $(N, p, q) \xleftarrow{\$} \mathcal{RSA}_k$ we mean that $(N, p, q)$ is sampled according to the uniform distribution on $\mathcal{RSA}_k$. An *RSA TDP generator* [38] is an algorithm $\mathcal{F}$ that returns $(N, e), (N, d)$, where $N$ is an RSA modulus and $ed \equiv 1 \pmod{\phi(N)}$. (Here $\phi(\cdot)$ denotes Euler's totient function, so in particular $\phi(N) = (p-1)(q-1)$.) The tuple $(N, e)$ defines the permutation on $\mathbb{Z}_N^*$ given by $f(x) = x^e \bmod N$,

and similarly $(N, d)$ defines its inverse. We say that a lossy TDP generator
LTDP $= (\mathcal{F}, \mathcal{F}')$ is an RSA LTDP if $\mathcal{F}$ is an RSA TDP generator.

To define the $\Phi$-Hiding Assumption and later some extensions of it, the follow-
ing notation is also useful. For $i \in \mathbb{N}$ we denote by $\mathcal{P}_i$ the set of all $i$-bit primes.
Let $R$ be a relation on $p$ and $q$. By $\mathcal{RSA}_k[R]$ we denote the subset of $\mathcal{RSA}_k$
for that the relation $R$ holds on $p$ and $q$. For example, let $e$ be a prime. Then
$\mathcal{RSA}_k[p = 1 \bmod e]$ is the set of all $(N, p, q)$, where where $N = pq$ is the product
of two distinct $k/2$-bit primes $p, q$ and $p = 1 \bmod e$. That is, the relation $R(p, q)$
is true if $p = 1 \bmod e$ and $q$ is arbitrary. By $(N, p, q) \xleftarrow{\$} \mathcal{RSA}_k[R]$ we mean that
$(N, p, q)$ is sampled according to the uniform distribution on $\mathcal{RSA}_k[R]$.

## 5.2    RSA Lossy TDP from $\Phi$-Hiding

$\Phi$-HIDING ASSUMPTION ($\Phi$A). We recall the $\Phi$-Hiding Assumption of [13]. For
an RSA modulus $N$, we say that $N$ $\phi$-hides a prime $e$ if $e \mid \phi(N)$. Intuitively,
the assumption is that, given RSA modulus $N$, it is hard to distinguish primes
which are $\phi$-hidden by $N$ from those that are not. Formally, let $0 < c < 1/2$ be
a (public) constant determined later. Consider the following two distributions:

$$\mathcal{R}_1 = \{(e, N) : e, e' \xleftarrow{\$} \mathcal{P}_{ck} ; (N, p, q) \xleftarrow{\$} \mathcal{RSA}_k[p = 1 \bmod e']\}$$
$$\mathcal{L}_1 = \{(e, N) : e \xleftarrow{\$} \mathcal{P}_{ck} ; (N, p, q) \xleftarrow{\$} \mathcal{RSA}_k[p = 1 \bmod e])\} .$$

To a distinguisher $D$ we associate its $\Phi$A advantage defined as
$$\mathbf{Adv}_{c,D}^{\Phi A}(k) = \Pr[D(\mathcal{R}_1) \Rightarrow 1] - \Pr[D(\mathcal{L}_1) \Rightarrow 1] .$$

As shown in [13], distributions $\mathcal{R}_1, \mathcal{L}_1$ can be sampled efficiently assuming the
widely-accepted Extended Riemann Hypothesis.[8]

RSA LTDP FROM $\Phi$A. We construct an RSA LTDP based on $\Phi$A. In injective
mode the public key is $(N, e)$ where $e$ is not $\phi$-hidden by $N$, whereas in lossy
mode it is. Namely, define LTDP$_1 = (\mathcal{F}_1, \mathcal{F}_1')$ as follows:

| Algorithm $\mathcal{F}_1$ | Algorithm $\mathcal{F}_1'$ |
|---|---|
| $e, e' \xleftarrow{\$} \mathcal{P}_{ck}$ | $e \xleftarrow{\$} \mathcal{P}_{ck}$ |
| $(N, p, q) \xleftarrow{\$} \mathcal{RSA}_k[p = 1 \bmod e', p]$ | $(N, p, q) \xleftarrow{\$} \mathcal{RSA}_k[p = 1 \bmod e]$ |
| If $\gcd(e, \phi(N)) \neq 1$ then return $\bot$ | Return $(N, e)$ |
| $d \leftarrow e^{-1} \bmod \phi(N)$ | |
| Return $((N, e), (N, d))$ | |

The fact that algorithm $\mathcal{F}_1$ has only a negligible probability of failure (returning
$\bot$) follows from the fact that $\phi(N)$ can have only a constant number of prime
factors of length $ck$ and Bertrand's Postulate.

---

[8] This is done by choosing a uniform $(1/2 - c)k$-bit number $x$ until $p = xe + 1$ is a
prime.

**Proposition 1.** *Suppose there is a distinguisher $D$ against $\mathsf{LTDP}_1$. Then there is a distinguisher $D'$ such that for all $k \in \mathbb{N}$*

$$\mathbf{Adv}^{\mathrm{ltdp}}_{\mathsf{LTDP}_1,D}(k) \leq 2 \cdot \mathbf{Adv}^{\Phi\mathrm{A}}_{c,D'}(k) .$$

*Furthermore, the running-time of $D'$ is that of $D$. $\mathsf{LTDP}_1$ has lossiness $ck$.*

*Remark 4.* From a practical perspective, a drawback of $\mathsf{LTDP}_1$ is that $\mathcal{F}_1$ chooses $N = pq$ in a non-standard way, so that it hides a prime of the same length as $e$. Moreover, for small values of $e$ it returns $\perp$ with high probability. This is done for consistency with how [13] formulated $\Phi\mathrm{A}$. But, to address this, we also propose what we call the *Enhanced $\Phi\mathrm{A}$* (E$\Phi\mathrm{A}$), which says that $N$ generated in the non-standard way (*i.e.*, by $\mathcal{F}_1$) is indistinguishable from one chosen at random subject to $\gcd(e, \phi(N)) = 1$.[9] We conjecture that E$\Phi\mathrm{A}$ holds for all values of $c$ that $\Phi\mathrm{A}$ does. Details are given in the full version [27]. An analogous enhancement pertains to later extensions of $\Phi\mathrm{A}$.

PARAMETERS FOR $\mathsf{LTDP}_1$. When $e$ is too large, $\Phi\mathrm{A}$ can be broken by using Coppersmith's method for finding small roots of a univariate modulo an unknown divisor of $N$ [18, 32]. (No other attack on $\Phi\mathrm{A}$ here is known.) Namely, consider the polynomial $r(x) = ex + 1 \bmod p$. Coppersmith's method allows us to find all roots of $r$ smaller than $N^{1/4}$, and thus factor $N$, in lossy mode in polynomial time if $c \geq 1/4$. (This is essentially the "factoring with high bits known" attack.) More specifically, applying [32, Theorem 1], $N$ can be factored in time $O(N^\varepsilon)$ if $c = 1/4 - \varepsilon$ (*i.e.*, $\log e \geq k(1/4 - \varepsilon)$). For example, with modulus size $k = 2048$, for about 80-bit security in lossy mode we set $\varepsilon = .04$ (to enforce $k\varepsilon \geq 80$). The lossiness of $\mathsf{LTDP}_1$ is then 432 bits according to Proposition 1. A similar calculation shows that for a modulus of size 1024 (resp., 3072) the lossiness of $\mathsf{LTDP}_1$ we get is 176 (resp., 688) bits.

### 5.3   RSA Lossy TDP from Multi-prime $\Phi$-Hiding

Multi-prime RSA (according to [31] the earliest reference is [39]) is a generalization of RSA to moduli $N = p_1 \cdots p_m$ of length $k$ with $m \geq 2$ prime factors of equal bit-length. Multi-prime RSA is of interest to practitioners since it allows to speed up decryption and is included in RSA PKCS #1 v2.1. We are interested in it here because for it we can show greater lossiness and even with smaller encryption exponent $e$.

NOTATION AND TERMINOLOGY. Let $m \geq 2$ be fixed. We denote by $\mathcal{MRSA}_k$ the set of all tuples $(N, p_1, \ldots, p_m)$, where $N = p_1 \cdots p_m$ is the product of distinct $k/m$-bit primes. Such an $N$ is called an $m$-*prime RSA modulus*. By $(N, p_1, \ldots, p_m) \xleftarrow{\$} \mathcal{MRSA}_k$ we mean that $(N, p_1, \ldots, p_m)$ is sampled according

---

[9] Additionally, in practice the encryption exponent $e$ is usually fixed. This can be addressed by parameterizing E$\Phi\mathrm{A}$ by a fixed $e$ instead of choosing it at random. Note that for $e = 3$ one should make both $e \mid p - 1$ and $e \mid q - 1$ in the lossy case (otherwise the assumption is false; cf. [13, Remark 2, p. 6]).

to the uniform distribution on $\mathcal{MRSA}_k$. The rest of the notation and terminology of Section 5 is extended to the multi-prime setting in the obvious way.

MULTI $\Phi$-HIDING ASSUMPTION. For an $m$-prime RSA modulus $N$, let us say that $N$ $m\phi$-hides a prime $e$ if $e \mid p_i - 1$ for all $1 \leq i \leq m - 1$.Intuitively, the assumption is that, given such $N$, it is hard to distinguish primes which are $m\phi$-hidden by $N$ from those that do not divide $p_i - 1$ for any $1 \leq i \leq m$. Formally, let $m = m(k) \geq 2$ be a polynomial and let $c = c(k)$ be an inverse polynomial determined later. Consider the following two distributions:

$$\mathcal{R}_2 = \{(e, N) : e, e' \overset{\$}{\leftarrow} \mathcal{P}_{ck} ; (N, p_1, \ldots, p_t) \overset{\$}{\leftarrow} \mathcal{MRSA}_k[p_{i \leq m-1} = 1 \bmod e']\}$$
$$\mathcal{L}_2 = \{(e, N) : e \overset{\$}{\leftarrow} \mathcal{P}_{ck} ; (N, p_1, \ldots, p_t) \overset{\$}{\leftarrow} \mathcal{MRSA}_k[p_{i \leq m-1} = 1 \bmod e]\} .$$

Above and in what follows, by $p_{i \leq m-1} = 1 \bmod e$ we mean that $p_i = 1 \bmod e$ for all $1 \leq i \leq m - 1$. To a distinguisher $D$ we associate its $M\Phi A$ advantage defined as

$$\mathbf{Adv}^{M\Phi A}_{m,c,D}(k) = \Pr[D(\mathcal{R}_2) \Rightarrow 1] - \Pr[D(\mathcal{L}_2) \Rightarrow 1] .$$

As before, distributions $\mathcal{R}_2, \mathcal{L}_2$ can be sampled efficiently assuming the widely-accepted Extended Riemann Hypothesis.

Note that if we had required that in the lossy case $N = p_1 \cdots p_m$ is such that $e \mid p_i$ for all $1 \leq i \leq m$, then in this case we would always have $N = 1 \bmod e$. But in the injective case $N \bmod e$ is random, which would lead to a trivial distinguishing algorithm. This explains why we do not impose $e \mid p_m$ in the lossy case above.

MULTI-PRIME RSA LTDP FROM $M\Phi A$. We construct a multi-prime RSA LTDP based on $M\Phi A$ having lossiness $(m - 1) \log e$, where in lossy mode $N$ $m\phi$-hides $e$. Namely, define $\mathsf{LTDP}_2 = (\mathcal{F}_2, \mathcal{F}_2')$ as follows:

| Algorithm $\mathcal{F}_2$ | Algorithm $\mathcal{F}_2'$ |
|---|---|
| $e, e' \overset{\$}{\leftarrow} \mathcal{P}_{ck}$ | $e \overset{\$}{\leftarrow} \mathcal{P}_{ck}$ |
| $(N, p_1, \ldots, p_m)$ | $(N, p_1, \ldots, p_m)$ |
| $\overset{\$}{\leftarrow} \mathcal{MRSA}_k[p_{i \leq m-1} = 1 \bmod e']$ | $\overset{\$}{\leftarrow} \mathcal{MRSA}_k[p_{i \leq m-1} = 1 \bmod e]$ |
| If $\gcd(e, \phi(N)) \neq 1$ then Return $\bot$ | Return $(N, e)$ |
| $d \leftarrow e^{-1} \bmod \phi(N)$ | |
| Else return $(N, e), (N, d)$ | |

**Proposition 2.** *Suppose there is a distinguisher $D$ against* $\mathsf{LTDP}_2$. *Then there is a distinguisher $D'$ such that for all $k \in \mathbb{N}$*

$$\mathbf{Adv}^{\mathrm{ltdp}}_{\mathsf{LTDP}_2, D}(k) \leq 2 \cdot \mathbf{Adv}^{M\Phi A}_{m,c,D'}(k) .$$

*Furthermore, the running-time of $D'$ is that of $D$.* $\mathsf{LTDP}_2$ *has lossiness $(m-1)ck$.*

PARAMETERS FOR $\mathsf{LTDP}_2$. As in the case of $\mathsf{LTDP}_1$, if $e$ is too large then Coppersmith's method [18] can be used to factor $N$ in the lossy case. But this time the attack is more involved than "factoring with high bits known." Let us first consider $m = 3$. Consider the polynomial $r(x_1', x_2') = (ex_1' + 1)(ex_2' + 1) \bmod p_1 p_2$.

Substituting $x_1 = x_1'x_2'$ and $x_2 = x_1' + x_2'$ gives $r(x_1, x_2) = e^2 x_1 + e x_2 + 1 \bmod p_1 p_2$. Applying [33, Theorem 3] with $\beta = 2/3$ and $\gamma = 2\delta$ tells us that we can find all roots smaller than $N^\delta$ for $\delta = (2(1 - 2/3)^{3/2})/3 \approx .12$ in polynomial time, so we require $c \leq 1/3 - .12 \approx .21$ to prevent this attack. (Note that is slightly smaller than what we would deduce from "factoring with high bits known" [32], which gives $c \leq .22$.) More specifically, for $m = 3$ we can factor $N$ in the lossy case in time $O(N^\varepsilon)$ if $c \geq 1/3 - \delta - \varepsilon$ (i.e., $\log e = k(1/3 - \delta - \varepsilon)$) with $\delta$ as above.

In the general case, we can apply [33, Theorem 4] to deduce we must require $c \leq 1/m - \delta$ where

$$\delta = \frac{2((1/m)^{(1/m)-1} - (1/m)^{m/(m-1)})}{m(m-1)} \leq \frac{2}{m(m-1)}.$$

Note that this is only smaller than the bound with $\delta = 1/m^2$ obtained from "factoring with high bits known" for $m \geq 5$, namely for $m = 5$ we have $\delta \approx 0.06$. (The reason we also had a better attack for $m = 3$ is that we used a specialized theorem.)

We note that this may not be the best attack possible based on Coppersmith's method (in particular the coefficients of the polynomial we use are highly correlated). It is an interesting open question whether there is a better attack. We also remark that for a *fixed* modulus length, $m$ cannot be too large since the Elliptic Curve Method for factoring can compute a factor $p_i$ of $N$ faster than the Number Field Sieve one if $p_i$ is significantly smaller than $N^{1/2}$ [31].

### 5.4   Small-Exponent RSA LTDP from 2-Or-$m$-Primes

For efficiency reasons, the public RSA exponent $e$ is typically not chosen to be too large in practice. (For example, researchers at UC San Diego [43] observed that 99.5% of the certificates in the campus's TLS corpus had $e = 2^{16} + 1$.) Therefore, we investigate the possibility of using an additional assumption to amplify the lossiness of RSA for small $e$.

The high-level idea is to assume that it is hard to distinguish $N = pq$ where $p, q$ are primes of length $k/2$ from $N = p_1 \cdots p_m$ for $m > 2$, where $p_1, \ldots, p_m$ are primes of length $k/m$ (which we call the "2-or-$m$ Primes" Assumption). Combined with the MΦA Assumption of Section 5.3, we obtain $(m - 1) \log e$ bits of lossiness from *standard* (two-prime) RSA. Due to space constraints, details are deferred to the full version [27].

## 6   Instantiating RSA-OAEP

By combining the results of Section 3, Section 4, and Section 5, we obtain standard model instantiations of RSA-OAEP under chosen-plaintext attack.

REGULARITY. In particular, we would like to apply part (2) of Theorem 2 in this case, as it is not hard to see that under all of the assumptions discussed in Section 5, RSA is a *regular* lossy TDP on the domain $\mathbb{Z}_N^*$. Unfortunately, this domain is different from $\{0,1\}^{\rho+\mu}$ (identified as integers), the range of OAEP. In RSA PKCS #1 v2.1, the mismatch is handled by selecting $\rho + \mu = \lfloor \log N \rfloor - 2$, and viewing OAEP's output as an integer less than $2^{\rho+\mu} < N/4$. The problem is that in the lossy case RSA may not be regular on the subdomain $\{0, ..., 2^{\rho+\mu}-1\}$.

We can prove, in some cases, that in the lossy case RSA is *approximately* regular on this subdomain, and in those cases we obtain the better parameters given by part (2) of Theorem 2. However, here use just use the weaker parameters given by part (1) of Theorem 2. We leave a detailed discussion of approximate regularity to future work. In particular, understanding the regularity of RSA on subintervals of the domain is a first step towards improving the concrete parameters we obtain.

CONCRETE PARAMETERS. Since the results in Section 5 have several cases and the parameter settings are rather involved, we avoid stating an explicit theorem about RSA-OAEP. From part (1) of Theorem 2 one can see that for $u = 80$ bits security and assuming RSA has $\ell$ bits of lossiness, messages of roughly $\mu \approx \ell - 3 \cdot 80$ bits can be encrypted (for sufficintly large $t$). For concreteness, we give two example parameter settings. Using the Multi $\Phi$-Hiding Assumption with $N = 1024$ bits and 3 primes, we obtain $\ell = k - s = 291$ bits of lossiness and hence can encrypt messages of length $\mu = 40$ bits (for $t \approx 400$); using the $\Phi$-Hiding Assumption with $N = 2048$, we obtain $\ell = k - s = 430$ bits of lossiness and hence can encrypt messages of length $\mu = 160$ bits (for $t \approx 150$). We stress that while we view our results as providing important theoretical backing for the scheme at a more qualitative level, we strongly encourage further research to try to improve the concrete parameters.

# Acknowledgements

We thank Mihir Bellare, Alexandra Boldyreva, Dan Brown, Yevgeniy Dodis, Jason Hinek, Arjen Lenstra, Alex May, Phil Rogaway, and the anonymous reviewers of Crypto 2010 for helpful comments. In particular, we thank Dan for reminding us of [13, Remark 2, p. 6], Alex for pointing out the improved attack in Section 5.3, and Phil for encouraging us to consider the case of small $e$ more closely. A.O. was supported in part by Alexandra Boldyreva's NSF CAREER award 0545659 and NSF Cyber Trust award 0831184 and thanks her for her support. A.S. was supported in part by NSF awards #0747294, 0729171.

# References

[1] Abdalla, M., Bellare, M., Rogaway, P.: The Oracle Diffie-Hellman Assumptions and an Analysis of DHIES. In: Naccache, D. (ed.) CT-RSA 2001. LNCS, vol. 2020, p. 143. Springer, Heidelberg (2001)

[2] Barak, B., Shaltiel, R., Tromer, E.: True Random Number Generators Secure in a Changing Environment. In: Walter, C.D., Koç, Ç.K., Paar, C. (eds.) CHES 2003. LNCS, vol. 2779, pp. 166–180. Springer, Heidelberg (2003)

[3] Bellare, M., Rompel, J.: Randomness-Efficient Oblivious Sampling. In: FOCS 1994. ACM, New York (1994)

[4] Bellare, M., Rogaway, P.: Random oracles are practical: A paradigm for designing efficient protocols. In: The Conference on Computer and Communications Security. ACM, New York (1993)

[5] Bellare, M., Rogaway, P.: Optimal asymmetric encryption: How to encrypt with RSA. In: De Santis, A. (ed.) EUROCRYPT 1994. LNCS, vol. 950, pp. 92–111. Springer, Heidelberg (1995)

[6] Boldyreva, A., Fehr, S., O'Neill, A.: On notions of security for deterministic encryption, and efficient constructions without random oracles. In: Wagner, D. (ed.) CRYPTO 2008. LNCS, vol. 5157, pp. 335–359. Springer, Heidelberg (2008)

[7] Boldyreva, A., Fischlin, M.: Analysis of random oracle instantiation scenarios for OAEP and other practical schemes. In: Shoup, V. (ed.) CRYPTO 2005. LNCS, vol. 3621, pp. 412–429. Springer, Heidelberg (2005)

[8] Boldyreva, A., Fischlin, M.: On the security of OAEP. In: Lai, X., Chen, K. (eds.) ASIACRYPT 2006. LNCS, vol. 4284, pp. 210–225. Springer, Heidelberg (2006)

[9] Boneh, D.: Simplified OAEP for the RSA and Rabin functions. In: Kilian, J. (ed.) CRYPTO 2001. LNCS, vol. 2139, p. 275. Springer, Heidelberg (2001)

[10] Brown, D.: What hashes make RSA-OAEP secure? In: Cryptology ePrint Archive, Report 2006/223 (2006)

[11] Boldyreva, A., Cash, C., Fischlin, M., Warinschi, B.: Efficient private bidding and auctions with an oblivious third party. In: ASIACRYPT 2009 (2009)

[12] Cachin, C.: Efficient private bidding and auctions with an oblivious third party. In: CCS 1999. ACM, New York (1999)

[13] Cachin, C., Micali, S., Stadler, M.: Computationally private information retrieval with polylogarithmic communication. In: Stern, J. (ed.) EUROCRYPT 1999. LNCS, vol. 1592, p. 402. Springer, Heidelberg (1999), http://www.zurich.ibm.com/~cca/papers/cpir.pdf

[14] Canetti, R.: Towards realizing random oracles: Hash functions that hide all partial information. In: Kaliski Jr., B.S. (ed.) CRYPTO 1997. LNCS, vol. 1294, pp. 455–469. Springer, Heidelberg (1997)

[15] Canetti, R., Dakdouk, R.: Extractable Perfectly One-Way Functions. In: Aceto, L., Damgård, I., Goldberg, L.A., Halldórsson, M.M., Ingólfsdóttir, A., Walukiewicz, I. (eds.) ICALP 2008, Part II. LNCS, vol. 5126, pp. 449–460. Springer, Heidelberg (2008)

[16] Canetti, R., Goldreich, O., Halevi, S.: The random oracle methodology, revisited. J. ACM 51(4), 557–594 (2004)

[17] Canetti, R., Micciancio, D., Reingold, O.: Perfectly one-way probabilistic hash functions. In: STOC 1998. ACM, New York (1998)

[18] Coppersmith, D.: Small solutions to polynomial equations, and low exponent RSA vulnerabilities. J. Cryptology 10 (1997)

[19] Coron, J.-S., Joye, M., Naccache, D., Paillier, P.: New Attacks on PKCS #1 v1.5 Encryption. In: Preneel, B. (ed.) EUROCRYPT 2000. LNCS, vol. 1807, p. 369. Springer, Heidelberg (2000)

[20] Coron, J.-S., Joye, M., Naccache, D., Paillier, P.: Universal Padding Schemes for RSA. In: Yung, M. (ed.) CRYPTO 2002. LNCS, vol. 2442, p. 226. Springer, Heidelberg (2002)

[21] Dodis, Y., Oliveira, R., Pietrzak, K.: On the Generic Insecurity of the Full Domain Hash. In: Shoup, V. (ed.) CRYPTO 2005. LNCS, vol. 3621, pp. 449–466. Springer, Heidelberg (2005)

[22] Dodis, Y., Smith, A.: Correcting errors without leaking partial information. In: STOC 2005. ACM Press, New York (2005)

[23] Fujisaki, E., Okamoto, T., Pointcheval, D., Stern, J.: RSA-OAEP is secure under the RSA assumption. J. Cryptology 17(2), 81–104 (2004)

[24] Gentry, C., Mackenzie, P., Ramzan, Z.: Password authenticated key exchange using hidden smooth subgroups. In: CCS 2005. ACM, New York (2005)

[25] Hemenway, B., Ostrovsky, R.: Public-key locally-decodable codes. In: Wagner, D. (ed.) CRYPTO 2008. LNCS, vol. 5157, pp. 126–143. Springer, Heidelberg (2008)

[26] Kazukuni, K., Imai, H.: OAEP++: A Very Simple Way to Apply OAEP to Deterministic OW-CPA Primitives. In: Cryptology ePrint Archive, Report 2002/130 (2002)

[27] Kiltz, E., O'Neill, A., Smith, A.: Instantiability of RSA-OAEP under Chosen-Plaintexts Attacks. Full version of this paper

[28] Kiltz, E., Pietrzak, K.: The Group of Signed Quadratic Residues and Applications. In: Halevi, S. (ed.) CRYPTO 2009. LNCS, vol. 5677, pp. 637–653. Springer, Heidelberg (2009)

[29] Kiltz, E., Pietrzak, K.: On the security of padding-based encryption schemes (or: Why we cannot prove OAEP secure in the standard model). In: Joux, A. (ed.) EUROCRYPT 2009. LNCS, vol. 5479, pp. 389–406. Springer, Heidelberg (2009)

[30] Kiltz, E., Pietrzak, K.: Personal Communication (2009)

[31] Lenstra, A.K.: Unbelievable security: Matching AES security using public key systems. In: Boyd, C. (ed.) ASIACRYPT 2001. LNCS, vol. 2248, p. 67. Springer, Heidelberg (2001)

[32] May, A.: Using LLL-Reduction for Solving RSA and Factorization Problems: A Survey. In: LLL+25 Conference in Honour of the 25th Birthday of the LLL Algorithm (2007)

[33] Herrmann, M., May, A.: Solving Linear Equations Modulo Divisors: On Factoring Given Any Bits. In: Pieprzyk, J. (ed.) ASIACRYPT 2008. LNCS, vol. 5350, pp. 406–424. Springer, Heidelberg (2008)

[34] Paillier, P., Villar, J.: Trading one-wayness against chosen-ciphertext security in factoring-based encryption. In: Lai, X., Chen, K. (eds.) ASIACRYPT 2006. LNCS, vol. 4284, pp. 252–266. Springer, Heidelberg (2006)

[35] Pandey, O., Pass, R., Vaikuntanathan, V.: Adaptive One-Way Functions and Applications. In: Wagner, D. (ed.) CRYPTO 2008. LNCS, vol. 5157, pp. 57–74. Springer, Heidelberg (2008)

[36] Peikert, C., Waters, B.: Lossy trapdoor functions and their applications. In: STOC 2008. ACM, New York (2008)

[37] RSA Laboratories Public-Key Cryptography Standards, http://www.rsa.com/rsalabs/pkcs/

[38] Rivest, R., Shamir, A., Adelman, L.: A method for obtaining public-key cryptosystems and digital signatures. Technical Report MIT/LCS/TM-82 (1977)

[39] Rivest, R., Shamir, A., Adelman, L.: Cryptographic communications system and method. U.S. Patent 4,405,829 (1983)

[40] Schridde, C., Freisleben, B.: On the validity of the $\Phi$-Hiding Assumption in cryptographic protocols. In: Pieprzyk, J. (ed.) ASIACRYPT 2008. LNCS, vol. 5350, pp. 344–354. Springer, Heidelberg (2008)

[41] Shoup, V.: OAEP Reconsidered. J. Cryptology 15(4), 223–249 (2002)

[42] Trevisan, L., Vadhan, S.: Extracting Randomness from Samplable Distributions. In: FOCS 2000. ACM, New York (2000)

[43] Yilek, S., Rescorla, E., Shacham, H., Enright, B., Savage, S.: When Private Keys are Public: Results from the 2008 Debian OpenSSL Debacle. In: IMC 2009 (2009)

# Efficient Chosen-Ciphertext Security via Extractable Hash Proofs

Hoeteck Wee*

Queens College, CUNY
hoeteck@cs.qc.cuny.edu

**Abstract.** We introduce the notion of an extractable hash proof system. Essentially, this is a special kind of non-interactive zero-knowledge proof of knowledge system where the secret keys may be generated in one of two modes to allow for either simulation or extraction.

- We show how to derive efficient CCA-secure encryption schemes via extractable hash proofs in a simple and modular fashion. Our construction clarifies and generalizes the recent factoring-based cryptosystem of Hofheinz and Kiltz (Eurocrypt '09), and is reminiscent of an approach proposed by Rackoff and Simon (Crypto '91). We show how to instantiate extractable hash proof system for hard search problems, notably factoring and computational Diffie-Hellman. Using our framework, we obtain the first CCA-secure encryption scheme based on CDH where the public key is a constant number of group elements and a more modular and conceptually simpler variant of the Hofheinz-Kiltz cryptosystem (though less efficient).

- We introduce adaptive trapdoor relations, a relaxation of the adaptive trapdoor functions considered by Kiltz, Mohassel and O'Neil (Eurocrypt '10), but nonetheless imply CCA-secure encryption schemes. We show how to construct such relations using extractable hash proofs, which in turn yields realizations from hardness of factoring and CDH.

## 1  Introduction

The most basic security guarantee we require of a public key encryption scheme (PKE) is that of semantic security against chosen-plaintext attacks (CPA) [21]: it is infeasible to learn anything about the plaintext from the ciphertext. On the other hand, there is a general consensus within the cryptographic research community that in virtually every practical application, we require semantic security against adaptive chosen-ciphertext attacks (CCA) [37, 15], wherein an adversary is given access to decryptions of ciphertexts of her choice. So far, there have been two largely separate lines of works addressing the construction of CCA-secure encryption schemes: the first examines constructions from general assumptions starting with the beautiful works of Dolev, Dwork, Naor and Yung [15, 34, 37, 39, 31, 18, 36, 38, 33, 29] and related questions pertaining to minimal assumptions; the second examines practical and efficient constructions from specific number-theoretic assumptions, starting from those

---

* Supported by NSF CAREER Award CNS-0953626 and PSC-CUNY Award # 6014939 40.

of Cramer and Shoup [11, 40, 12, 30, 2, 24, 9, 10, 25]. In recent years, two distinct trends have surfaced in each of these lines of works.

**Practical CCA from Search Problems.** Until very recently, all of the practical CCA-secure encryption schemes (namely the Cramer-Shoup encryption scheme and all its variants) inherently relied on decisional assumptions, e.g., the Decisional Diffie-Hellman (DDH) assumption or the quadratic residuosity assumption. In general, decisional assumptions are a much stronger class of assumptions than computational assumptions based on search problems, such as factoring, finding shortest vectors in lattices, or even the Computational Diffie-Hellman (CDH) problem. Indeed, there are groups, such as certain elliptic curve groups with bilinear pairing map, where the DDH assumption does not hold, but the Computational Diffie-Hellman (CDH) problem appears to be hard. As such, schemes based on search problems are generally preferred to those based on decisional assumptions. However, such schemes seem to be very hard to obtain.

Several years ago, Canetti, Halevi and Katz [9] proposed the first practical CCA-secure PKE based on a computational assumption, namely the Bilinear DH assumption in bilinear groups (BDH). Since then, a series of works have shown how to base CCA-secure encryption schemes on CDH [10, 22, 23] and on hardness of factoring [25]. However, there seems to be no overarching framework explaining these schemes. Partial progress towards a unifying approach was made recently by Cramer, Hofheinz and Kiltz [13]; their approach remains unsatisfactory in two ways: first, it does not encompass constructions from hardness of factoring (it does cover the RSA assumption, which is possibly a stronger assumption), and second, the ensuing schemes even with suitable algebraic optimizations, do not quite match the efficiencies obtained in preceding works (for instance, the public key in the RSA-based scheme contains a linear number of group elements, whereas that in the factoring-based scheme of Hofheinz and Kiltz [25] only requires a constant number of group elements).

**CCA from weaker general assumptions.** Since the breakthrough work of Peikert and Waters on lossy trapdoor functions [36], a series of works has identified successively weaker general assumptions from which we may realize CCA-secure encryption schemes [38, 29] (in a black-box way). The current state-of-the-art is the (tag-based) adaptive trapdoor functions of Kiltz, Mohassel and O'Neil [29]; roughly speaking, these are trapdoor functions that remain one-way even if the adversary is given access to a restricted inversion oracle that inverts the function on "most" inputs. In spite of the black-box separations indicating that adaptive trapdoor functions are strictly weaker than its predecessors [29, 41], all of the concrete (standard) assumptions from which we can realize adaptive trapdoor functions are not significantly different from those known to imply lossy trapdoor functions. Most notably, we do not know how to base adaptive trapdoor functions on hardness of factoring (or the standard RSA assumption, and more generally, any hard *search* problem not related to lattices). On the other hand, we do know how to derive CCA-secure encryption schemes from enhanced trapdoor permutations, which may in turn be based on hardness of factoring [15, 16, 19].

## 1.1 Our Contributions

We introduce the notion of an *extractable hash proof system*, inspired in part by the Cramer-Shoup universal hash proof systems [12]. Informally, extractable hash proofs are like universal hash proofs in that they are a special kind of non-interactive zero-knowledge proofs [4], except we replace the soundness requirement (corresponding to smoothness) with a "proof of knowledge property" [37, 14]. That is, the secret keys may be generated in one of two modes to allow for either simulation or extraction. Using extractable hash proofs, we obtain new insights into the construction of CCA-secure encryption schemes, and obtain new results for both lines of works described earlier. Before we describe our results, we present an overview of extractable hash proofs.

**Extractable Hash Proof Systems.** Fix R to be a relation corresponding to some hard search problem – namely, R is efficiently samplable, but given a random $u$, it is hard to find an $s$ such that $(u, s) \in R$. (For instance, $s$ is the pre-image of $u$ under a one-way permutation.) We consider a family of hash functions $\{H_{PK}\}$ indexed by a public key PK which maps an input $u$ to some value. (We clarify that the name is somewhat of a misnomer since the "hash function" will in fact be injective, and possibly even length-increasing.) Moreover, we require that the hash function be efficiently computable given PK and the coin tosses $r$ used to sample $(u, s) \in R$. We denote this public evaluation algorithm by $\mathsf{Pub}(PK, r)$ and the hash value by $H_{PK}(u)$.

   Associated with this family of functions is a set-up algorithm that generates the public key PK along with a secret key. The set-up algorithm operates in one of two modes. In both modes, the algorithm generates exactly the same distribution of public keys; however, the functionality afforded by the secret key depends on the mode:

- In the hashing mode, the secret key $SK^*$ allows us to compute the hash value $\mathsf{Pub}(PK, u)$ without knowing either $s$ or $r$. Specifically, there is a private evaluation algorithm $\mathsf{Priv}$ such that for all $(u, s) \in R$, $\mathsf{Priv}(SK^*, u) = H_{PK}(u)$.

- In the extraction mode, the secret key $SK$ allows us to verify whether a hash value is correctly computed and if so extract a witness $s$. More formally, there is an extraction algorithm $\mathsf{Ext}$, such that for all $u, \tau$: $\mathsf{Ext}(SK, u, \tau)$ outputs $s$ satisfying $(u, s) \in R$ iff $\tau = H_{PK}(u)$. This implies efficient verification of the hash value (given $SK$) whenever R is efficiently computable.

Looking ahead, we will rely on the extraction mode for decryption in a CCA-secure encryption scheme, and on the hashing mode for the proof of security. This is opposite to the use of universal hash proofs in the Cramer-Shoup framework, where the hashing mode is used for decryption and the smoothness property (corresponding to soundness and thus extraction) is used to establish security. Moreover, unlike Cramer-Shoup hash proofs, extractable hash proofs are designed in tandem with families of relations, and are particularly well-suited for use with computationally hard search problems.

**Practical CCA via Extractable Hash Proofs.** We provide a generic construction of CCA-secure encryption schemes from extractable hash proofs. We use as an intermediate building block a somewhat richer cryptographic abstraction called *all-but-one extractable hash proofs* (which can be constructed generically from extractable hash proofs). The overall construction follows a variant of the Rackoff-Simon paradigm [37] (as opposed to the Naor-Yung double-encryption paradigm [34], also used in [13]): encrypt (or commit to) a one-time symmetric key (which is in turn used to encrypt the message, following the hybrid encryption paradigm), and then provide a zero-knowledge proof of knowledge of the key using an extractable hash proof. Indeed, such an approach was used implicitly in the afore-mentioned cryptosystems based on computational assumptions; however, the connection to the Rackoff-Simon paradigm has never been made explicit. Our framework may be viewed as a clarification and unification of all these constructions. We present extractable hash proofs related to hardness of factoring and CDH; in addition, we obtain the following new cryptosystems:

- a variant of the Hofheinz-Kiltz CCA-secure encryption scheme based on hardness of factoring (Fig 3), which is more modular and both conceptually and mathematically simpler, albeit less efficient — there is a linear blow-up in both ciphertext overhead and public key size over the previous scheme;

- a CCA-secure encryption scheme based on CDH where the public key comprises a constant number of group elements (Fig 5) and a linear ciphertext overhead; previous works all require a linear number of group elements [10, 22, 23] in the public key. Our construction offers a trade-off between public key size and ciphertext overhead when compared with the schemes in [22, 23]; such a trade-off may be preferable when encrypting very long messages via the hybrid encryption paradigm.

Our framework also encompasses a series of CCA-secure encryption schemes [9, 7, 27, 28] derived from the identity-based encryption schemes in [5, 8] whose security are based on decisional assumptions.

**CCA from Adaptive Trapdoor Relations.** We also propose a relaxation of adaptive trapdoor functions, which we call *adaptive trapdoor relations*. The relaxation here lies in the functionality requirement for evaluation: we only require that there exists an efficient sampling algorithm that generates a random input to the trapdoor function along with its image; the function itself need not be efficiently computable. It follows immediately from [29] (with essentially the same construction as that in [36, 38]) that adaptive trapdoor relations imply CCA-secure encryption schemes. Interestingly, the ensuing construction unlike previous constructions, is not witness-recovering (that is, the decryption algorithm does not completely recover the randomness used for encryption, c.f. [36, Section 1.1]).

Next, we show how to derive adaptive trapdoor relations from hardness of factoring and CDH. This partially answers an open problem posed in [29] on realizing adaptive trapdoor functions from hard search problems not related to lattices. (A comparison with previous works is shown in Fig 1.) Our construction relies on the use of extractable hash proofs and is very similar to our CCA-secure encryption schemes. Moreover, our adaptive trapdoor relations are fairly efficient and achieve parameters similar to the state-of-the-art lossy trapdoor functions based on DCR and DDH respectively [17].

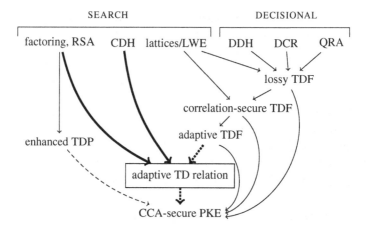

**Fig. 1.** Summary of CCA-secure PKEs from general assumptions, and how the latter relate to (standard) specific assumptions [15, 16, 19, 36, 38, 35, 29, 32, 17]. Here, lossy TDF and adaptive TDF refer to the respective all-but-one/tag-based variants. The bold lines denote our contributions (the dotted lines denote those that are straight-forward or follow readily from previous work). All of the constructions from general assumptions are black-box, except for the one marked with dashed lines. (Following current conventions, we do not regard hash proof systems [12] as a general assumption.)

## 2 Preliminaries and Definitions

### 2.1 Key Encapsulation Mechanisms

A *key encapsulation mechanism* (KEM) (Gen, Enc, Dec) with key-space $\{0,1\}^k$ consists three polynomial-time algorithms. Via $(\text{PK}, \text{SK}) \leftarrow \text{Gen}(1^k)$ the randomized key-generation algorithm produces public/secret keys for security parameter $1^k$; via $(C, K) \leftarrow \text{Enc}(\text{PK})$, the randomized encapsulation algorithm creates a uniformly distributed symmetric key $K \in \{0,1\}^k$, together with a ciphertext $C$; via $K \leftarrow \text{Dec}(\text{SK}, C)$, the possessor of secret key SK decrypts ciphertext $C$ to get back a key $K$ which is an element in $\{0,1\}^k$ or a special reject symbol $\perp$. For consistency, we require that for all $k$ and all $(C, K) \leftarrow \text{Enc}(\text{PK})$, we have $\Pr[\text{Dec}(\text{SK}, C) = K] = 1$, where the probability is taken over the choice $(\text{PK}, \text{SK}) \leftarrow \text{Gen}(1^k)$ and the coins of all the algorithms in the expression above.

*Chosen-Ciphertext Security.* The common requirement for a KEM is indistinguishability against chosen-ciphertext attacks (IND-CCA) [12] where an adversary is allowed to adaptively query a decapsulation oracle with ciphertexts to obtain the corresponding session key. More formally, for an adversary $\mathcal{A}$, we define the advantage function

$$\text{AdvCCA}^{\mathcal{A}}_{\text{KEM}}(k) := \Pr \left[ b = b' : \begin{array}{l} (\text{PK}, \text{SK}) \leftarrow \text{Gen}(1^k); \\ (C, K_0) \leftarrow \text{Enc}(\text{PK}); K_1 \leftarrow_{\text{R}} \{0,1\}^k; \\ b \leftarrow_{\text{R}} \{0,1\}; \\ b' \leftarrow \mathcal{A}^{\text{Dec}(\text{SK}, \cdot)}(\text{PK}, K_b, C) \end{array} \right]$$

with the restriction that $\mathcal{A}$ is only allowed to query $\mathsf{Dec}(\mathsf{SK}, \cdot)$ on ciphertexts different from the challenge ciphertext $C$. A KEM is said to be *indistinguishable against chosen ciphertext attacks* (IND-CCA) if for all PTA adversaries $\mathcal{A}$, the advantage $\mathsf{AdvCCA}_{\mathsf{KEM}}^{\mathcal{A}}(k)$ is a negligible function in $k$.

It was shown in [12] that an IND-CCA secure KEM with a CCA-secure symmetric encryption scheme yields an IND-CCA secure hybrid encryption scheme.

### 2.2 Binary Relations for Search Problems

Fix a family of (binary) relations $\mathsf{R}_{\mathsf{PP}}$ indexed by a public parameter PP. We require that PP be efficiently samplable given a security parameter $1^k$, and assume that all algorithms are given PP as part of its input. We omit PP henceforth whenever the context is clear. We will also require that $\mathsf{R}_{\mathsf{PP}}$ be efficiently verifiable (possibly given some trapdoor for PP) and efficiently samplable, where the sampling algorithm is denoted by SampR.

Intuitively, the relation $\mathsf{R}_{\mathsf{PP}}$ corresponds to a hard search problem, that is, given a random $u$, it is hard to find $s$ such $(u, s) \in \mathsf{R}_{\mathsf{PP}}$. More formally, we say that a binary relation $\mathsf{R}_{\mathsf{PP}}$ is *one-way* if:

- with overwhelming probability over PP, for all $u$, there exists at most one $s$ such that $(u, s) \in \mathsf{R}_{\mathsf{PP}}$; and
- there is an efficiently computable generator G such that $\mathsf{G}_{\mathsf{PP}}(s)$ is pseudorandom even against an adversary that gets PP, $u$ and oracle access to $\mathsf{R}_{\mathsf{PP}}$, where $(u, s) \leftarrow_{\mathsf{R}} \mathsf{SampR}(\mathsf{PP})$. (We will also refer to G as extracting hard-core bits from $s$.)

For relations where computing $s$ given $u$ is hard on average, we may derive a generator $\mathsf{G}_{\mathsf{PP}}$ with a one-bit output via the Goldreich-Levin hard-core bit $\mathsf{GL}(\cdot)$ [20] (with the randomness in PP). In many cases as we shall see shortly, we may derive a linear number of hard-core bits by either iterating a one-way permutation or relying on decisional assumptions. Next, we present one-way relations related to hardness of factoring and the Diffie-Hellman assumption.

**Iterated Squaring.** Fix a Blum integer $N = PQ$ for safe primes $P, Q \equiv 3 \pmod 4$ (such that $P = 2p + 1$ and $Q = 2q + 1$ for primes $p, q$). Following [26], we work over the cyclic group of signed quadratic residues, given by the quotient group $\mathbb{QR}_N^+ := \mathbb{QR}_N / \pm 1$. $\mathbb{QR}_N^+$ is a cyclic group of order $pq$ and is efficiently recognizable (by verifying that the Jacobi symbol is $+1$). In addition, the map $x \mapsto x^2$ is a permutation over $\mathbb{QR}_N^+$. Furthermore, assuming that factoring Blum integers are hard on average and that safe primes are dense, the family of permutations $x \mapsto x^2$ (indexed by $N$) acting on the groups $\mathbb{QR}_N^+$ is one-way.

In our constructions, the public parameter PP comprises $(N, g)$, where $N$ is a random $2k$-bit Blum integer and $g$ is chosen uniformly from $\mathbb{QR}_N^+$. We will henceforth assume that $g$ is a generator for $\mathbb{QR}_N^+$, which happens with probability $1 - O(1/\sqrt{N})$. We consider the relation:

$$\mathsf{R}_{\mathsf{PP}}^{\mathsf{isqr}} = \left\{ (u, s) \in \mathbb{QR}_N^+ \times \mathbb{QR}_N^+ : u = s^{2^k} \right\}$$

The associated sampling algorithm SampR picks a random $r \in [(N - 1)/4]$ and outputs $(g^{2^k r}, g^r)$. Note that the output distribution is statistically close to the uniform distribution over $\mathbb{QR}_N^+$ whenever $g$ is a generator. Using the Blum-Blum-Shub (BBS) pseudorandom generator [3], we may extract $k$ hard-core bits from $s$ that are pseudorandom even given $u$, that is:

$$G_{PP}^{bbs}(s) := (\mathsf{lsb}_N(s), \mathsf{lsb}_N(s^2), \ldots, \mathsf{lsb}_N(s^{2^{k-1}}))$$

**Diffie-Hellman Relation.** We consider a family of groups $\mathbb{G}$ of prime order $q$. The public parameter PP is given by $(g, g^\alpha)$ for a random $g \leftarrow_R \mathbb{G}$ and a random $\alpha \leftarrow_R \mathbb{Z}_q$. We consider the Diffie-Hellman relation

$$R_{PP}^{dh} = \left\{ (u, s) \in \mathbb{G} \times \mathbb{G} : s = u^\alpha \right\}$$

Note that $R_{PP}^{dh}$ is efficiently verifiable in bilinear groups (by computing a pairing) or if provided with $\alpha$ as a trapdoor. The associated sampling algorithm SampR picks a $r \leftarrow_R \mathbb{Z}_q$ and outputs $(g^r, g^{\alpha r})$. Next, we explain how to obtain hard-bit bits for $R_{PP}^{dh}$ under various assumptions.

- The Strong DH assumption assumption [1] asserts that computing $g^{ab}$ given $(g, g^a, g^b)$ is hard on average, even given oracle access to $R_{(g,g^a)}(\cdot, \cdot)$ (note that in bilinear groups, this is equivalent to CDH). Under Strong DH, we may extract a single hard-core bit from $s$ using $GL(s)$.
- The Bilinear DDH (BDDH) assumption [6] asserts that $e(g, g)^{abc}$ is pseudorandom given $g, g^a, g^b, g^c$ where $g, g^a, g^b, g^c$ are random elements of a bilinear group. Under BDDH, we may extract a linear number of hard-core bits from $s$ using:

$$G_{PP}^{bddh}(s) := e(s, g^\gamma) \quad \left( \Rightarrow G_{PP}^{bddh}(g^{\alpha r}) = e(g, g)^{\alpha \gamma r} \right)$$

where PP is now given by $(g, g^\alpha, g^\gamma)$. In addition, we may improve efficiency by pre-computing the pairing and setting PP to be $(g, g^\alpha, e(g, g^\gamma))$ and computing $G_{PP}^{bddh}(g^r) := e(g, g^\gamma)^r$. This construction extends naturally to the Gap Hashed DH assumption [28].

Following the twinning framework [10], we will also consider the twin Diffie-Hellman relation given by:

$$R_{PP}^{2dh} = \left\{ (u, (s_0, s_1)) \in \mathbb{G} \times \mathbb{G}^2 : (s_0, s_1) = (u^\alpha, u^\beta) \right\}$$

where PP is given by $(g, g^\alpha, g^\beta)$ for random $\alpha, \beta \leftarrow_R \mathbb{Z}_q$, and SampR picks $r \leftarrow_R \mathbb{Z}_q$ and outputs $(g^r, (g^{\alpha r}, g^{\beta r}))$. As shown in [10, Theorem 9], $GL(s_0)$ is a hard-core bit for the relation $R_{PP}^{2dh}$ under CDH.

## 2.3  Extractable Hash Proofs

We consider a family of hash functions $\{H_{PK}\}$ indexed by a public key PK. An *extractable hash proof system* associated with a one-way relation $R_{PP}$ is a tuple of algorithms (SetupExt, SetupHash, Pub, Ext, Priv) satisfying the following properties with overwhelming probability over PP:

(PUBLIC EVALUATION.) For all $(PK, SK) \leftarrow$ SetupExt(PP) and $(u, s) =$ SampR($r$):
   Pub(PK, $r$) = $H_{PK}(u)$.

(EXTRACTION MODE.) For all $(\text{PK}, \text{SK}) \leftarrow \text{SetupExt}(\text{PP})$ and all $(u, \tau)$:

$$\tau = \text{H}_{\text{PK}}(u) \quad \Longleftrightarrow \quad (u, \text{Ext}(\text{SK}, u, \tau)) \in \text{R}$$

(HASHING MODE.) For all $(\text{PK}, \text{SK}^*) \leftarrow \text{SetupHash}(\text{PP})$ and all $(u, s) \in \text{R}$,

$$\text{Priv}(\text{SK}^*, u) = \text{H}_{\text{PK}}(u)$$

(INDISTINGUISHABILITY.) The first outputs (namely PK) of $\text{SetupHash}(\text{PP})$ and $\text{SetupExt}(\text{PP})$ are statistically indistinguishable.

**All-But-One Extractable Hash Proofs.** For all of our applications, it is convenient to work with a richer abstraction, where as before, we consider a family of hash functions indexed by a public key PK, that takes a tag as an additional input. More formally, an *all-but-one (ABO) extractable hash proof system* is a tuple of algorithms $(\text{SetupExt}, \text{SetupABO}, \text{Pub}, \text{Ext}, \text{Ext}^*, \text{Priv})$ satisfying the following properties with overwhelming probability over PP:

(PUBLIC EVALUATION.) For all PK, TAG and $(u, s) = \text{SampR}(r)$: $\text{Pub}(\text{PK}, \text{TAG}, r) = \text{H}_{\text{PK}}(\text{TAG}, u)$.

(EXTRACTION MODE.) For all $(\text{PK}, \text{SK}) \leftarrow \text{SetupExt}(\text{PP})$ and all $(\text{TAG}, u, \tau)$:

$$\tau = \text{H}_{\text{PK}}(\text{TAG}, u) \quad \Longleftrightarrow \quad (u, \text{Ext}(\text{SK}, \text{TAG}, u, \tau)) \in \text{R}$$

(ALL-BUT-ONE MODE.) For all $\text{TAG}^*$ and all $(\text{PK}, \text{SK}^*) \leftarrow \text{SetupABO}(\text{PP}, \text{TAG}^*)$: for all $(u, s) \in \text{R}$,

$$\text{Priv}(\text{SK}^*, \text{TAG}^*, u) = \text{H}_{\text{PK}}(\text{TAG}^*, u)$$

In addition, for all $\text{TAG} \neq \text{TAG}^*$ and all $(u, \tau)$:

$$\tau = \text{H}_{\text{PK}}(\text{TAG}, u) \quad \Longleftrightarrow \quad (u, \text{Ext}^*(\text{SK}^*, \text{TAG}, u, \tau)) \in \text{R}$$

(INDISTINGUISHABILITY.) For all $\text{TAG}^*$, the first outputs (namely PK) of $\text{SetupABO}(\text{PP}, \text{TAG}^*)$ and $\text{SetupExt}(\text{PP})$ are statistically indistinguishable.

## 2.4  Trapdoor Functions

Informally, trapdoor functions are a family of functions $\{\text{F}_{\text{FID}}\}$ that are easy to sample, compute and invert with trapdoor, and hard to invert without the trapdoor (in this work, we always assume that the functions are injective). In the tag-based setting, the function takes an additional input, namely the tag; also, the trapdoor is independent of the tag. A family of *adaptive trapdoor functions* [29] is one that remains one-way even if the adversary is given access to a inversion oracle, except the adversary cannot query the oracle on the same tag as that in the challenge.

**Adaptive Trapdoor Relations.** In this work, we consider a relaxation of the functionality guarantee for adaptive trapdoor functions, that is, instead of requiring that $\text{F}_{\text{FID}}$ be efficiently computable, we only require that we can efficiently sample from the distribution $(s, \mathsf{F}_{\text{FID}}(\text{TAG}, s))$ for a random $s$ given FID, TAG. More precisely, a family of (tag-based) *adaptive trapdoor relations* is given by a family of injective functions $\{\mathsf{F}_{\text{FID}}\}$ that satisfies the following properties:

(TRAPDOOR GENERATION.) There is an efficient randomized algorithm TDG that outputs a random (FID, TID).

(PUBLIC SAMPLING.) There is an efficient randomized algorithm PSamp that on input (FID, TAG), outputs $(s, \mathsf{F}_{\text{FID}}(\text{TAG}, s))$ for a random $s$.[1]

(TRAPDOOR INVERSION.) There is an efficient algorithm TdInv such that for all (FID, TID) ← TDG and for all TAG, $y$, computes $\mathsf{TdInv}(\text{TID}, \text{TAG}, y) = \mathsf{F}_{\text{FID}}^{-1}(\text{TAG}, y)$.[2]

(ADAPTIVE ONE-WAYNESS.) For all efficient stateful adversaries $\mathcal{A}$, the following quantity is negligible:

$$\Pr\left[ s = s' : \begin{array}{l} \text{TAG}^* \leftarrow \mathcal{A}(1^k); \\ (\text{FID}, \text{TID}) \leftarrow_{\text{R}} \mathsf{TDG}(1^k); \\ (s, y) \leftarrow_{\text{R}} \mathsf{PSamp}(\text{FID}, \text{TAG}^*); \\ s' \leftarrow \mathcal{A}^{\mathsf{F}_{\text{FID}}^{-1}(\cdot, \cdot)}(\text{FID}, y) \end{array} \right]$$

where $\mathcal{A}$ is allowed to query $\mathsf{F}_{\text{FID}}^{-1}(\cdot, \cdot)$ on any tag different from $\text{TAG}^*$.

It follows immediately from [29, Theorem 2] that adaptive trapdoor relations imply IND-CCA secure encryption.

# 3    Generic Constructions from Extractable Hash Proofs

In this section, we show that starting from an extractable hash proof, we may derive (1) a IND-CPA secure encryption scheme (as a simple warm-up exercise); (2) an ABO-extractable hash proof; (3) an ABO-extractable hash proof with multiple hard-core bits; and finally, (4) a IND-CCA secure KEM.

## 3.1    CPA-Secure Encryption

Starting from an extractable hash proof (SetupExt, SetupHash, Pub, Ext, Ext*, Priv) for a one-way relation $\mathsf{R}_{\text{PP}}$ with an associated generator $\mathsf{G}_{\text{PP}}$, we may derive a IND-CPA secure bit encryption scheme as follows:

- Gen(PP): same as SetupExt(PP).
- Enc(PK, $b$): sample $(u, s) := \mathsf{SampR}(r)$ and output $(u, \mathsf{Pub}(\text{PK}, r), \mathsf{G}(s) \oplus b)$.
- Dec(SK, $(u, \tau, c)$): compute $s := \mathsf{Ext}(\text{SK}, u, \tau)$ and return $\mathsf{G}(s) \oplus c$.

---

[1] This is essentially the only distinction from the adaptive trapdoor functions in [29]; there, they require that $\mathsf{F}_{\text{FID}}$ be efficiently computable.

[2] Since $\mathsf{F}_{\text{FID}}$ is not necessarily efficiently computable given FID, it is crucial here that we quantify over all $y$ and that TdInv outputs $\perp$ if $y$ does not have a pre-image under $\mathsf{F}_{\text{FID}}(\text{TAG}, \cdot)$. In our constructions, it will be the case $\mathsf{F}_{\text{FID}}$ is efficiently computable given TID.

Observe that correctness of the encryption scheme follows readily from correctness of the extraction mode. To establish IND-CPA security, we consider an intermediate game where we generate $(\text{PK}, \text{SK}^*)$ using SetupHash(PP) and computes $\text{H}_{\text{PK}}(u)$ in the ciphertext using $\text{Priv}(\text{SK}^*, u)$. Any adversary that can distinguish between encryptions of 0 and 1 in this game yields a distinguisher that given PP, $u$ distinguishes $G(s)$ from random.

### 3.2 From Extractable to ABO-Extractable

Starting from an extractable hash proof for a relation $\text{R}_{\text{PP}}$, we may derive a ABO-extractable hash proof $(\text{SetupExt}', \text{SetupABO}', \text{Pub}', \text{Ext}', \text{Ext}'^*, \text{Priv}')$ for the same relation and tag space $\{0, 1\}^\ell$ via a construction analogous to those in [34, 15, 36, 38]:

- SetupExt'(PP): run SetupExt(PP) to obtain $(\text{PK}_{i,0}, \text{SK}_{i,0}), (\text{PK}_{i,1}, \text{SK}_{i,1})$, $i = 1, \ldots, \ell$; output $\widetilde{\text{PK}} = (\text{PK}_{i,0}, \text{PK}_{i,1})_{i \in [\ell]}$ and $\widetilde{\text{SK}} = (\text{SK}_{i,0}, \text{SK}_{i,1})_{i \in [\ell]}$.

- Pub'$(\widetilde{\text{PK}}, \text{TAG}, r)$: parse TAG=$(\text{TAG}_1, \ldots, \text{TAG}_\ell)$ and output $(\text{Pub}(\text{PK}_{i,\text{TAG}_i}, r))_{i \in [\ell]}$.

- Ext'$(\widetilde{\text{SK}}, \text{TAG}, u, (\tau_1, \ldots, \tau_\ell))$: compute $s_i := \text{Ext}(\text{SK}_{i,\text{TAG}_i}, u, \tau_i)$ for $i = 1, \ldots, \ell$, and output $s_1$ if all $\ell$ values agree, and $\perp$ otherwise.

- SetupABO'(PP, TAG$^*$): run SetupHash(PP) to generate $(\text{PK}_{i,\text{TAG}_i^*}, \text{SK}_{i,\text{TAG}_i^*})$ and SetupExt(PP) to generate $(\text{PK}_{i,1-\text{TAG}_i^*}, \text{SK}_{i,1-\text{TAG}_i^*})$, for $i = 1, \ldots, \ell$; output $\widetilde{\text{PK}} = (\text{PK}_{i,0}, \text{PK}_{i,1})_{i \in [\ell]}$ and $\widetilde{\text{SK}}^* = (\text{SK}_{i,0}, \text{SK}_{i,1})_{i \in [\ell]}$.

- Priv'$(\widetilde{\text{PK}}, \text{TAG}, u)$: output $(\text{Priv}(\text{SK}_{i,\text{TAG}_i}, u))_{i \in [\ell]}$.

- Ext'$^*(\widetilde{\text{SK}}^*, \text{TAG}, u, (\tau_1, \ldots, \tau_\ell))$: first, check that $\tau_i = \text{Priv}(\text{SK}_{i,\text{TAG}_i}, u)$ for all $i$ such that $\text{TAG}_i^* = \text{TAG}_i$ and if not, output $\perp$; next, compute $s_i := \text{Ext}(\text{SK}_{i,\text{TAG}_i}, u, \tau_i)$ for all $i$ such that $\text{TAG}_i^* \neq \text{TAG}_i$; output the common value if all these values agree and $\perp$ otherwise.

### 3.3 Obtaining Multiple Hard-Core Bits

Starting from an ABO-extractable hash proof for a relation $\text{R}_{\text{PP}}$, we may derive a ABO-extractable hash proof $(\text{SetupExt}', \text{SetupABO}', \text{Pub}', \text{Ext}', \text{Ext}'^*, \text{Priv}')$ for the $k$-wise direct product $\text{R}_{\text{PP}}^{\otimes k}$ of $\text{R}_{\text{PP}}$. This allows us to obtain more hard-core bits by using the $k$-wise direct product $G_{\text{PP}}^{\otimes k}$ of $G_{\text{PP}}$. The construction is as follows:

- SampG'$(r_1, \ldots, r_k) = (\text{SampG}(r_1), \ldots, \text{SampG}(r_k))$

- SetupExt' and SetupABO' are the same as SetupExt and SetupABO respectively.

- Pub'$(\text{PK}, \text{TAG}, (r_1, \ldots, r_k))$: output $(\text{Pub}(\widetilde{\text{PK}}, \text{TAG}, r_i))_{i \in [k]}$.

- Ext'$(\text{SK}, \text{TAG}, (u_1, \ldots, u_k), (\tau_1, \ldots, \tau_\ell))$: compute $s_i := \text{Ext}(\text{SK}, u_i, \tau_i)$ for $i = 1, \ldots, \ell$, and output $(s_1, \ldots, s_k)$.

- Priv'$(\widetilde{\text{PK}}, \text{TAG}, (u_1, \ldots, u_k))$: output $(\text{Priv}(\text{SK}, u_i))_{i \in [\ell]}$.

- Ext'$^*(\text{SK}, \text{TAG}, (u_1, \ldots, u_k), (\tau_1, \ldots, \tau_\ell))$: output $(\text{Ext}(\text{SK}, u_i, \tau_i))_{i \in [k]}$.

### 3.4 CCA-Secure Encryption

Starting from an ABO-extractable hash proof for a one-way relation $R_{PP}$ along with a target collision-resistant hash function TCR, we may derive a IND-CCA KEM (Gen, Enc, Dec) as follows:

- Gen(PP): same as SetupExt(PP).
- Enc(PK): sample $(u, s) := \mathsf{SampR}(r)$, compute TAG $:= \mathsf{TCR}(u), \tau := \mathsf{Pub}(\mathsf{PK}, \mathsf{TAG}, r)$, and return $(C, K) := ((u, \tau), \mathsf{G}(s))$.
- Dec(SK, $(u, \tau)$): compute TAG $:= \mathsf{TCR}(u)$ and $s := \mathsf{Ext}(\mathsf{SK}, \mathsf{TAG}, u, \tau)$; if $(u, s) \in R_{PP}$, return $\mathsf{G}(s)$, else return $\bot$.

We assume here that $\mathsf{G}_{PP}$ has linear output length; if not, we first apply the transformation in Section 3.3.

**Theorem 1.** *If $R_{PP}$ is a one-way relation, then the above* KEM (Gen, Enc, Dec) *is IND-CCA secure.*

*Proof.* Observe that correctness of the encryption scheme follows readily from correctness of the extraction mode. We proceed to establish IND-CCA security. In the following, we write $(u^*, s^*) = \mathsf{SampR}(r), C^* = (u^*, \tau^*), K_0^*, K_1^*$ to denote the challenge ciphertext and keys chosen by the IND-CCA experiment, and we set TAG$^*$ to denote the tag $\mathsf{TCR}(u^*)$ used in computing $C^*$. We proceed via a sequence of games. We start with Game 0, where the challenger proceeds like in the standard IND-CCA game (i.e, $K_0^*$ is a real key and $K_1^*$ is a random key) and end up with a game where both $K_0^*$ and $K_1^*$ are chosen uniformly at random. Then, we show that all games are indistinguishable under the assumption that $\mathsf{G}(s)$ is pseudorandom even given $u$.

GAME 1: ELIMINATING COLLISIONS. We replace the decapsulation mechanism Dec with Dec$'$ that outputs $\bot$ on inputs $(u, \tau)$ such that $\mathsf{TCR}(u) = $ TAG$^*$ but otherwise proceeds like Dec. We show that Games 0 and 1 are computationally indistinguishable, by arguing that Dec and Dec$'$ essentially agree on all inputs $(u, \tau)$. We consider three cases:

- case 1: $\mathsf{TCR}(u) \neq $ TAG$^*$. Here, Dec and Dec$'$ agree by definition.
- case 2: $u \neq u^*$ but $\mathsf{TCR}(u) = \mathsf{TCR}(u^*) = $ TAG$^*$. This only occurs with negligible probability, by target collision-resistance of TCR.
- case 3: $u = u^*$ but $\tau \neq \tau^*$. This means $\tau \neq \mathsf{H}_{PK}(\mathsf{TAG}^*, u)$ and therefore Dec returns $\bot$ and agrees with Dec$'$.

GAME 2: DECAPSULATION WITH SetupABO. We modify the IND-CCA experiment from Game 1, we generate the keys (PK, SK$^*$) using SetupABO instead of SetupExt and we answer decapsulation queries using SK$^*$ instead of SK. More precisely, the IND-CCA experiment proceeds as follows:

$(u^*, s^*) \leftarrow \mathsf{SampR}(r); \mathrm{TAG}^* := \mathsf{TCR}(u^*);$

$(\mathrm{PK}, \mathrm{SK}^*) \leftarrow \mathsf{SetupABO}(\mathrm{PP}, \mathrm{TAG}^*);$

$C^* := (u^*, \mathsf{Pub}(\mathrm{PK}, \mathrm{TAG}^*, r)); K_0^* := \mathsf{G}(s^*); K_1^* \leftarrow_\mathrm{R} \{0,1\}^k;$

$b \leftarrow_\mathrm{R} \{0,1\};$

$b' \leftarrow \mathcal{A}^{\mathsf{Dec}^*(\mathrm{SK}^*, \cdot)}(\mathrm{PK}, K_b^*, C^*)$

and where we replace $\mathsf{Dec}'(\mathrm{SK}, \cdot)$ from Game 1 with $\mathsf{Dec}^*(\mathrm{SK}^*, \cdot)$ which is defined as follows:

On input $(u, \tau)$: compute $\mathrm{TAG} = \mathsf{TCR}(u)$;
- if $\mathrm{TAG} = \mathrm{TAG}^*$ return $\bot$.
- if $\mathrm{TAG} \neq \mathrm{TAG}^*$, compute $s = \mathsf{Ext}^*(\mathrm{SK}^*, \mathrm{TAG}, u, \tau)$. If $(u, s) \in \mathrm{R_{PP}}$, return $\mathsf{G}(s)$, else return $\bot$.

We claim a stronger statement, namely that for all $r$, the outputs of Games 1 and 2 are statistically indistinguishable. First, indistinguishability of the two modes imply that the view $(\mathrm{PK}, K_b^*, C^*)$ in Games 1 and 2 are statistically indistinguishable. As such, it suffices to show that for all $\mathrm{PK}$, $\mathsf{Dec}'(\mathrm{SK}, \cdot)$ and $\mathsf{Dec}^*(\mathrm{SK}^*, \cdot)$ agree on all inputs $(u, \tau)$. Let $s$ denote the unique value such that $(u, s) \in \mathrm{R_{PP}}$ (if no such $s$ exists, then both $\mathsf{Dec}'$ and $\mathsf{Dec}^*$ return $\bot$) and let $\mathrm{TAG} = \mathsf{TCR}(u)$. We consider three cases:

- case 1: $\mathrm{TAG} = \mathrm{TAG}^*$. Both $\mathsf{Dec}'$ and $\mathsf{Dec}^*$ output $\bot$ by definition.

- case 2: $\mathrm{TAG} \neq \mathrm{TAG}^*$. Here, $\mathsf{Dec}'$ always agrees with $\mathsf{Dec}$ by definition. By correctness of the extraction mode, $\mathsf{Ext}(\mathrm{SK}, \mathrm{TAG}^*, \tau)$ returns $s$ iff $\tau = \mathsf{H_{PK}}(\mathrm{TAG}, u)$. Similarly, by correctness of the all-but-one mode, $\mathsf{Ext}^*(\mathrm{SK}^*, \mathrm{TAG}^*, \tau)$ returns $s$ iff $\tau = \mathsf{H_{PK}}(\mathrm{TAG}, u)$. It follows that both $\mathsf{Dec}$ (and thus $\mathsf{Dec}'$) and $\mathsf{Dec}^*$ return $\mathsf{G}(s)$ if $\tau = \mathsf{H_{PK}}(\mathrm{TAG}, u)$ and $\bot$ otherwise.

GAME 3: ENCAPSULATION WITH $\mathsf{Priv}$. We compute $\mathsf{H_{PK}}(\mathrm{TAG}^*, u^*)$ in $C^*$ using $\mathsf{Priv}$ instead of $\mathsf{Pub}$; that is, in the IND-CCA experiment from Game 2, we set

$$C^* := (u^*, \mathsf{Priv}(\mathrm{SK}^*, \mathrm{TAG}^*, u^*))$$

Games 2 and 3 are identically distributed by correctness of the all-but-one mode.

GAME 4: REPLACING $\mathsf{G}(s^*)$ WITH RANDOM. We generate $K_0^*$ at random from $\{0,1\}^k$ instead of using $\mathsf{G}(s^*)$ (recall here that $(u^*, s^*) = \mathsf{SampR}(r)$). Observe that in Game 3, we never use knowledge of the witness $s^*$ or randomness $r$ associated with $u^*$. It follows from the pseudorandomness of $\mathsf{G}$ that Games 3 and 4 are computationally indistinguishable. Specifically, we may transform any distinguisher for Games 3 and 4 into a distinguisher $K_0^*$ and $\mathsf{G}(s^*)$, given $\mathrm{PP}$, $u^*$ and oracle access to $\mathrm{R_{PP}}$ (the latter to simulate $\mathsf{Dec}^*$).

We conclude by observing that in Game 4, both $K_0^*$ and $K_1^*$ are identically distributed, so the probability that $b' = b$ is exactly $1/2$.                                                                              $\square$

## 4    Instantiations from Hardness of Factoring

We present a simple extractable hash proof for the iterated squaring relation from Section 2.2, namely $\mathsf{R}_{\mathrm{PP}}^{\mathsf{isqr}} := \{(u,s) \in \mathbb{QR}_N^+ \times \mathbb{QR}_N^+ : u = s^{2^k}\}$ where $N$ is a Blum integer. We also present an efficient ABO-extractable hash proof for iterated squaring that avoids the linear blow-up incurred by the transformation in Section 3.2. Both of these extractable hash proofs appear implicitly in the Hofheinz-Kiltz cryptosystem [25, 26].

Applying the generic transformations in Section 3 to the first hash proof, we obtain (i) a simple factoring-based IND-CPA encryption scheme shown in Fig 2 where decryption does not require knowing the factorization of the modulus; and (ii) a simple factoring-based IND-CCA encryption shown in Fig 3. Applying the transformation in Section 3.4 to the efficient ABO-extractable hash proof, we recover the original Hofheinz-Kiltz cryptosystem.

### 4.1    A Simple Extractable Hash Proof

SYSTEM PARAMETERS.   Here, $\mathrm{PP} = (N, g)$, $\mathrm{PK} \in \mathbb{QR}_N^+$. and $\mathsf{SampR}(r) := (g^{2^k r}, g^r)$, where $r \in [(N-1)/4]$. We define

$$\mathsf{H}_{\mathrm{PK}}(u) := (\mathrm{PK} \cdot g)^r \text{ where } u = g^{2^k r}.$$

PUBLIC EVALUATION / EXTRACTION.

- SetupExt: $\mathrm{PK} = g^{2^k \cdot \mathrm{SK}}$, $\mathrm{SK} \leftarrow_{\mathrm{R}} [(N-1)/4]$
- $\mathsf{Pub}(\mathrm{PK}, r) = (\mathrm{PK} \cdot g)^r$
- $\mathsf{Ext}(\mathrm{SK}, u, \tau)$: output $\tau \cdot u^{-\mathrm{SK}}$ if $u, \tau \in \mathbb{QR}_N^+$ and $\perp$ otherwise

Correctness of the extraction mode follows from the following simple calculation:

$$\tau = \mathsf{H}_{\mathrm{PK}}(u) = s^{2^k \cdot \mathrm{SK}+1} = u^{\mathrm{SK}} \cdot s \iff \tau \cdot u^{-\mathrm{SK}} = s$$

HASHING MODE.

- SetupHash: $\mathrm{PK} = g^{2^k \cdot \mathrm{SK}^* - 1}$, $\mathrm{SK}^* \leftarrow_{\mathrm{R}} [(N-1)/4]$
- $\mathsf{Priv}(\mathrm{SK}^*, u) = u^{\mathrm{SK}^*}$

Correctness of the hashing mode follows from the observation that $2^k \cdot \mathrm{SK}^* = 2^k \cdot \mathrm{SK} + 1 \pmod{\phi(N)/4}$ and thus

$$\mathsf{H}_{\mathrm{PK}}(u) = (g^{2^k \cdot \mathrm{SK}+1})^r = (g^{2^k \cdot \mathrm{SK}^*})^r = u^{\mathrm{SK}^*}$$

To establish indistinguishability, observe that the distributions of PK in both modes are identical if we sample SK and $\mathrm{SK}^*$ uniformly at random from $\mathbb{Z}_{\phi(N)/4}$ instead of $[(N-1)/4]$; moreover, sampling SK and $\mathrm{SK}^*$ this way only changes the distributions by a negligible quantity.

| Gen(PP), PP $= (N, g)$: | Enc(PK, $b$): | Dec(SK, $C$): |
|---|---|---|
| PK $:= g^{2\text{SK}}$, SK $\leftarrow_R [(N-1)/4]$ | $r \leftarrow_R [(N-1)/4]$ | parse $C$ as $(u, \tau, \psi)$ |
| return (PK, SK) | return $(g^{2r}, (\text{PK} \cdot g)^r, \text{lsb}(g^r) \oplus b)$ | return $\text{lsb}(\tau \cdot u^{-\text{SK}}) \cdot \psi$ |

**Fig. 2.** An IND-CPA bit encryption scheme based on hardness of factoring

### 4.2 Efficient ABO-Extractable Hash Proof

SYSTEM PARAMETERS. As before, PP $= (N, g)$ and PK $\in \mathbb{QR}_N^+$. The tag space is $\mathbb{Z}_{2^\ell}$ and $\text{SampR}(r) := (g^{2^{k+\ell}r}, g^{2^\ell r})$, where $r \in [(N-1)/4]$. We define

$$\mathsf{H}_{\text{PK}}(\text{TAG}, u) := (\text{PK} \cdot g^{\text{TAG}})^r \text{ where } u = g^{2^{k+\ell}r}.$$

PUBLIC EVALUATION / EXTRACTION.

- SetupExt: PK $= g^{2^{k+\ell} \cdot \text{SK}}$, SK $\leftarrow_R [(N-1)/4]$
- Pub(PK, TAG, $r$) $= (\text{PK} \cdot g^{\text{TAG}})^r$
- Ext(SK, TAG, $u, \tau$) : check that $u, \tau \in \mathbb{QR}_N^+$ and that $\tau^{2^{\ell+k}} = u^{\text{TAG}+2^{\ell+k} \cdot \text{SK}}$ and output $\bot$ otherwise. Compute $a, b, c \in \mathbb{Z}$ such that $2^c = \gcd(\text{TAG}, 2^{\ell+k}) = a \cdot \text{TAG} + b2^{\ell+k}$ and then output $(\tau^a \cdot u^{b-a \cdot \text{SK}})^{2^{\ell-c}}$.

Correctness of the extraction mode follows from the calculations: write $u = s^{2^k}$ and $s = g^{2^\ell \cdot r}$. Then,

$$\tau = \mathsf{H}_{\text{PK}}(\text{TAG}, s^{2^k}) = g^{r \cdot (\text{TAG}+2^{k+\ell} \cdot \text{SK})} \iff \tau^{2^{\ell+k}} = u^{\text{TAG}+2^{\ell+k} \cdot \text{SK}}$$

Moreover, if this holds, we have that $g^{r \cdot \text{TAG}} = \tau \cdot u^{-\text{SK}}$ and together with $u = g^{r2^{\ell+k}}$, we may compute $g^{r \cdot \gcd(\text{TAG}, 2^{\ell+k})} = g^{r2^c}$ from which we may compute $s = g^{r2^\ell}$ since $\gcd(\text{TAG}, 2^{\ell+k}) \leq 2^\ell$.

ABO-EXTRACTION MODE. We may write $2^{k+\ell} \cdot \text{SK}^* = 2^{k+\ell} \cdot \text{SK} + \text{TAG}^*$

- SetupABO: PK $= g^{2^{k+\ell} \cdot \text{SK}^* - \text{TAG}^*}$, SK$^* \leftarrow_R [(N-1)/4]$
- Priv(SK$^*$, $u$) $= u^{\text{SK}^*}$
- Ext$^*$(SK$^*$, TAG, $u, \tau$) :check that $u, \tau \in \mathbb{QR}_N^+$ and that $\tau^{2^{\ell+k}} \neq u^{\text{TAG}-\text{TAG}^*+2^{\ell+k} \cdot \text{SK}^*}$ and output $\bot$ otherwise. Compute $a, b, c \in \mathbb{Z}$ such that $2^c = \gcd(\text{TAG} - \text{TAG}^*, 2^{\ell+k}) = a(\text{TAG} - \text{TAG}^*) + b2^{\ell+k}$ and then output $(\tau^a \cdot u^{b-a \cdot \text{SK}^*})^{2^{\ell-c}}$.

Correctness of the ABO-extraction mode is similar to that for the extraction mode.

## 5 Instantiations from Diffie-Hellman Assumptions

We present an ABO-extractable hash proof for the Diffie-Hellman relation from Section 2.2, namely $\mathsf{R}_{\text{PP}}^{\text{dh}} = \{(u, s) \in \mathbb{G} \times \mathbb{G} : s = u^\alpha\}$ where $\mathbb{G}$ is a group of order $q$ and PP $= (g, g^\alpha)$. The construction is implicit in [5] and also [7, 27, 28, 23]. Applying the transformation in Section 3.4 to this hash proof system and the generator $\mathsf{G}_{\text{PP}}^{\text{bddh}}(\cdot)$, we obtain a variant of the BDDH-based IND-CCA KEM in [7, 27] (see Fig 4).

| Gen(PP), PP $= (N, g)$: | Enc(PK): | Dec(SK, $C$): |
|---|---|---|
| for $i = 1, \ldots, k$, for $b = 0, 1$: | $r \leftarrow_R [(N-1)/4]$ | parse $C$ as $(u, \tau_1, \ldots, \tau_k)$ |
| $\quad$ SK$_{i,b} \leftarrow_R [(N-1)/4]$ | $u := g^{2^k r}, t := \mathsf{TCR}(u)$ | check $u, \tau_1, \ldots, \tau_k \in \mathbb{QR}_N^+$ |
| $\quad$ PK$_{i,b} := g^{2^k \mathrm{SK}_{i,b}}$ | for $i = 1, \ldots, k$: | $t := \mathsf{TCR}(u)$ |
| PK $:= (\mathrm{PK}_{i,0}, \mathrm{PK}_{i,1})_{i \in [k]}$ | $\quad \tau_i := (\mathrm{PK}_{i,t_i} \cdot g)^r$ | for $i = 1, \ldots, k$: |
| SK $:= (\mathrm{SK}_{i,0}, \mathrm{SK}_{i,1})_{i \in [k]}$ | $C := (u, \tau_1, \ldots, \tau_k)$ | $\quad$ check $\tau_i^{2^k} = u^{2^k \mathrm{SK}_{i,t_i}+1}$ |
| return (PK, SK) | return $(C, \mathsf{G}_{\mathrm{PP}}^{\mathsf{bbs}}(g^r))$ | return $\mathsf{G}_{\mathrm{PP}}^{\mathsf{bbs}}(\tau_1 \cdot u^{-\mathrm{SK}_{1,t_1}})$ |

**Fig. 3.** An IND-CCA KEM based on hardness of factoring

## 5.1  ABO-Extractable Hash Proof for the Diffie-Hellman Relation

SYSTEM PARAMETERS. Here, PP $= (g, g^\alpha)$, SP $= \alpha$; the tag space is $\mathbb{Z}_q$; SampR$(r) :=$ $(g^r, g^{\alpha r})$ where $r \in \mathbb{Z}_q$. We define

$$H_{\mathrm{PK}}(u) := (g^{\alpha \cdot \mathrm{TAG}} \cdot \mathrm{PK})^r \text{ where } u = g^r.$$

PUBLIC EVALUATION / EXTRACTION.

- SetupExt: PK $= g^{\mathrm{SK}}$, SK $\leftarrow_R \mathbb{Z}_q$
- Pub(PK, TAG, $r$) $= (g^{\alpha \cdot \mathrm{TAG}} \cdot \mathrm{PK})^r$
- Ext(SK, TAG, $u, \tau$) $= (\tau \cdot u^{-\mathrm{SK}})^{\mathrm{TAG}^{-1}}$

Correctness of the extraction mode follows from the following simple calculation:

$$\tau = H_{\mathrm{PK}}(\mathrm{TAG}, u) = u^{\alpha \cdot \mathrm{TAG} + \mathrm{SK}} \iff (\tau \cdot u^{-\mathrm{SK}})^{\mathrm{TAG}^{-1}} = u^\alpha$$

ABO-EXTRACTION MODE.

- SetupABO: PK $= g^{\mathrm{SK}^*} \cdot (g^\alpha)^{-\mathrm{TAG}^*}$, SK$^* \leftarrow_R \mathbb{Z}_q$
- Priv(SK$^*, u$) $= u^{\mathrm{SK}^*}$
- Ext$^*$(SK$^*$, TAG, $u, \tau$) $= (\tau \cdot u^{-\mathrm{SK}^*})^{(\mathrm{TAG} - \mathrm{TAG}^*)^{-1}}$

Correctness of the ABO-extraction mode follows from the fact that SK$^* = \alpha \cdot$ TAG$^* +$ SK and thus

$$\tau = H_{\mathrm{PK}}(\mathrm{TAG}, u) = u^{\alpha(\mathrm{TAG} - \mathrm{TAG}^*)} \cdot u^{\mathrm{SK}^*} \iff (\tau \cdot u^{-\mathrm{SK}^*})^{(\mathrm{TAG} - \mathrm{TAG}^*)^{-1}} = u^\alpha$$

## 5.2  Constructions for the Twin Diffie-Hellman Relation

The construction in the previous section extends naturally to yield an ABO-extractable hash proof for the Twin Diffie-Hellman relation, by considering:

$$H_{\mathrm{PK}_0, \mathrm{PK}_1}(u) := ((g^{\alpha \cdot \mathrm{TAG}} \cdot \mathrm{PK}_0)^r, (g^{\beta \cdot \mathrm{TAG}} \cdot \mathrm{PK}_1)^r) \text{ where } u = g^r.$$

We may then apply the transformations from Sections 3.3 and 3.4 to obtain a CDH-based IND-CCA KEM, shown in Fig 5. The public key comprises 5 group elements and the ciphertext comprises $O(k)$ group elements.

| Gen(PP), $PP = (g, e(g, g^\gamma))$: | Enc(PK): | Dec(SK, $C$): |
|---|---|---|
| $SK := (\alpha, \widetilde{SK}, ) \leftarrow_R \mathbb{Z}_q^2$ | $u := g^r, r \leftarrow_R \mathbb{Z}_q$ | parse $C$ as $(u, \tau)$ |
| $(h, \widetilde{PK}) := (g^\alpha, g^{\widetilde{SK}})$ | $t := TCR(u), \tau := (\widetilde{PK} \cdot h^t)^r$ | $t := TCR(u)$ |
| $PK := (h, \widetilde{PK})$ | $C := (u, \tau), K := e(g, g^\gamma)^r$ | check $\tau = u^{\alpha t + \widetilde{SK}}$ |
| return $(PK, SK)$ | return $(C, K)$ | return $e(u^\alpha, g^\gamma)$ |

**Fig. 4.** An IND-CCA KEM based on BDDH (variant of [7, 27])

| Gen(PP), $PP = (g, R)$: | Enc(PK): | Dec(SK, $C$): |
|---|---|---|
| $(\alpha, \beta, SK_0, SK_1) \leftarrow_R \mathbb{Z}_q^4$ | for $i = 1, \ldots, k$: | parse $C$ as $(u_i, \tau_i^0, \tau_i^1)_{i \in [k]}$ |
| $(h_0, h_1) := (g^\alpha, g^\beta)$ | $u_i := g^{r_i}, r_i \leftarrow_R \mathbb{Z}_q$ | $t := TCR(u_1, \ldots, u_k)$ |
| $(PK_0, PK_1) := (g^{SK_0}, g^{SK_1})$ | $t := TCR(u_1, \ldots, u_k)$ | for $i = 1, \ldots, k$: |
| $PK := (h_0, h_1, PK_0, PK_1)$ | for $i = 1, \ldots, k$, for $b = 0, 1$: | check $\tau_i^0 = u_i^{\alpha t + SK_0}$ |
| $SK := (\alpha, \beta, SK_0, SK_1)$ | $\tau_i^b := (PK_b \cdot h_b^t)^{r_i}$ | check $\tau_i^1 = u_i^{\beta t + SK_1}$ |
| return $(PK, SK)$ | $C := (u_i, \tau_i^0, \tau_i^1)_{i \in [k]}$ | return $(GL_R(u_i^\alpha))_{i \in [k]}$ |
| | $K := (GL_R(h_0^{r_i}))_{i \in [k]}$ | |
| | return $(C, K)$ | |

**Fig. 5.** An IND-CCA KEM based on CDH

# 6 Adaptive Trapdoor Relations

Starting from an extractable hash proof (SetupExt, SetupABO, Pub, Ext, Ext*, Priv) for a one-way relation $R_{PP}$, we may derive an adaptive trapdoor relation as follows:

- FID is $(PP, PK)$ and for all $(u, s) \in R_{PP}$, $F_{FID}(TAG, s) := (u, H_{PK}(TAG, u))$.
- $TDG(1^k)$: computes $(PK, SK) \leftarrow SetupExt(PP)$ for a random $PP$ and returns $FID := (PP, PK)$ and $TID := SK$
- $PSamp(FID, TAG; r)$: computes $(u, s) := SampR(r), y := (u, Pub(PK, TAG, r))$ and return $(s, y)$.
- $TdInv(TID, TAG, (u, \tau))$: computes $s := Ext(SK, TAG, u)$ and returns $s$ if $(u, s) \in R_{PP}$ and $\perp$ otherwise.

From an adaptive trapdoor relation, we may derive a one-bit IND-CCA encryption scheme following the construction in [29, Theorem 2], or a more efficient $k$-bit IND-CCA scheme by using the construction with multiple hard-core bits from Section 3.3.

**Theorem 2.** *If $R_{PP}$ is a one-way relation, then the above construction yields an adaptive trapdoor relation.*

*Proof (sketch).* Trapdoor generation, public sampling and trapdoor inversion are straight-forward, so we only sketch the reduction for establishing adaptive one-wayness, which is very similar to that for our IND-CCA KEM in Section 3.4. Given an adversary $\mathcal{A}$ that breaks adaptive one-wayness with probability $\epsilon$, we may construct an adversary $\mathcal{B}$ given $(PP, u)$ and oracle access to $R_{PP}$, computes $s$ with probability roughly $\epsilon$:

| TDG(PP), PP $= (N, g)$: | PSamp(FID, TAG; $r$): | TdInv(SK, TAG, $(u, \tau)$): |
|---|---|---|
| TID $\leftarrow_{\text{R}} [(N-1)/4]$ | $(s, u) := (g^{2^{\ell}r}, g^{2^{k+\ell}r})$ | check $u, \tau \in \mathbb{QR}_N^+$ |
| FID $:= g^{2^{k+\ell} \cdot \text{TID}}$ | $\tau := (\text{FID} \cdot g^{\text{TAG}})^r$ | check $\tau^{2^{\ell+k}} = u^{\text{TAG} + 2^{\ell+k} \cdot \text{SK}}$ |
| return (FID, TID) | return $(s, (u, \tau))$ | find $a, b, c \in \mathbb{Z}: 2^c = a \cdot \text{TAG} + b 2^{\ell+k}$ |
| | | return $(\tau^a \cdot u^{b - a \cdot \text{SK}})^{2^{\ell-c}}$ |

**Fig. 6.** An adaptive trapdoor relation based on factoring

| TDG(PP), PP $= (g)$: | PSamp(FID, TAG; $r$): | TdInv(SK, TAG, $(u, \tau)$): |
|---|---|---|
| TID $:= (\alpha, \widetilde{\text{SK}}) \leftarrow_{\text{R}} \mathbb{Z}_q^2$ | return $(h^r, (g^r, (\widetilde{\text{PK}} \cdot h^{\text{TAG}})^r))$ | if $\tau = u^{\alpha \cdot \text{TAG} + \widetilde{\text{SK}}}$: |
| FID $:= (h, \widetilde{\text{PK}}) := (g^\alpha, g^{\widetilde{\text{SK}}})$ | | return $u^\alpha$, else $\perp$ |
| return (FID, TID) | | |

**Fig. 7.** An adaptive trapdoor relation based on Strong DH

- runs $\mathcal{A}(1^k)$ to get a tag $\text{TAG}^*$;
- computes $(\text{PK}, \text{SK}^*) \leftarrow_{\text{R}} \text{SetupABO}(\text{PP}, \text{TAG}^*)$;
- computes FID $:= (\text{PP}, \text{PK})$ and $\tau := \text{Priv}(\text{SK}^*, \text{TAG}^*, u)$
- computes and outputs $s' \leftarrow \mathcal{A}(\text{FID}, (u, \tau))$, by simulating $\text{F}_{\text{FID}}^{-1}(\cdot, \cdot)$ as follows:
  on input $(\text{TAG}, (u', \tau'))$ where $\text{TAG} \neq \text{TAG}^*$, compute $s' := \text{Ext}^*(\text{SK}^*, \text{TAG}, u')$; output $s'$ if $(u', s') \in \text{R}_{\text{PP}}$ and $\perp$ otherwise.

It is easy to check that $\Pr[\mathcal{B}^{\text{R}_{\text{PP}}(\cdot)}(\text{PP}, u) = s : (u, s) \leftarrow_{\text{R}} \text{SampR}(\text{PP})] \approx \epsilon$, which contradicts the pseudorandomness of $\text{G}_{\text{PP}}$. $\square$

Instantiating this construction with the ABO-extractable hash proofs in Sections 4.2 and 5.1, we derive the adaptive trapdoor relations shown in Fig 6 and 7, whose security are based on hardness of factoring and Strong DH respectively. By using the ABO-extractable hash proof in Section 5.2, we may also obtain an adaptive trapdoor relation based on CDH.

**Acknowledgments.** I thank the anonymous Crypto 2010 reviewers for pointing out that our framework applies to the constructions in [7, 27, 28], for suggesting the name "adaptive trapdoor relations", and for many helpful comments. I also thank Payman Mohassel for helpful discussions.

## References

[1] Abdalla, M., Bellare, M., Rogaway, P.: The oracle Diffie-Hellman assumptions and an analysis of DHIES. In: Naccache, D. (ed.) CT-RSA 2001. LNCS, vol. 2020, pp. 143–158. Springer, Heidelberg (2001)

[2] Abe, M., Gennaro, R., Kurosawa, K., Shoup, V.: Tag-KEM/DEM: A new framework for hybrid encryption and a new analysis of Kurosawa-Desmedt KEM. In: Cramer, R. (ed.) EUROCRYPT 2005. LNCS, vol. 3494, pp. 128–146. Springer, Heidelberg (2005)

[3] Blum, L., Blum, M., Shub, M.: Comparison of two pseudo-random number generators. In: CRYPTO, pp. 61–78 (1982)

[4] Blum, M., Feldman, P., Micali, S.: Non-interactive zero-knowledge and its applications. In: STOC, pp. 103–112 (1988)

[5] Boneh, D., Boyen, X.: Efficient selective-id secure identity-based encryption without random oracles. In: Cachin, C., Camenisch, J.L. (eds.) EUROCRYPT 2004. LNCS, vol. 3027, pp. 223–238. Springer, Heidelberg (2004)

[6] Boneh, D., Franklin, M.K.: Identity-based encryption from the Weil pairing. SIAM J. Comput. 32(3), 586–615 (2003)

[7] Boyen, X., Mei, Q., Waters, B.: Direct chosen ciphertext security from identity-based techniques. In: ACM CCS, pp. 320–329 (2005)

[8] Canetti, R., Halevi, S., Katz, J.: A forward-secure public-key encryption scheme. In: Biham, E. (ed.) EUROCRYPT 2003. LNCS, vol. 2656, pp. 255–271. Springer, Heidelberg (2003)

[9] Canetti, R., Halevi, S., Katz, J.: Chosen-ciphertext security from identity-based encryption. In: Cachin, C., Camenisch, J.L. (eds.) EUROCRYPT 2004. LNCS, vol. 3027, pp. 207–222. Springer, Heidelberg (2004)

[10] Cash, D., Kiltz, E., Shoup, V.: The Twin Diffie-Hellman problem and applications. J. Cryptology 22(4), 470–504 (2009)

[11] Cramer, R., Shoup, V.: A practical public key cryptosystem provably secure against adaptive chosen ciphertext attack. In: Krawczyk, H. (ed.) CRYPTO 1998. LNCS, vol. 1462, pp. 13–25. Springer, Heidelberg (1998)

[12] Cramer, R., Shoup, V.: Universal hash proofs and a paradigm for adaptive chosen ciphertext secure public-key encryption. In: Knudsen, L.R. (ed.) EUROCRYPT 2002. LNCS, vol. 2332, pp. 45–64. Springer, Heidelberg (2002)

[13] Cramer, R., Hofheinz, D., Kiltz, E.: A twist on the Naor-Yung paradigm and its application to efficient CCA-secure encryption from hard search problems. In: Micciancio, D. (ed.) TCC 2010. LNCS, vol. 5978, pp. 146–164. Springer, Heidelberg (2010)

[14] De Santis, A., Persiano, G.: Zero-knowledge proofs of knowledge without interaction. In: FOCS, pp. 427–436 (1992)

[15] Dolev, D., Dwork, C., Naor, M.: Nonmalleable cryptography. SIAM J. Comput. 30(2), 391–437 (2000)

[16] Feige, U., Lapidot, D., Shamir, A.: Multiple noninteractive zero knowledge proofs under general assumptions. SICOMP 29(1), 1–28 (1999)

[17] Freeman, D.M., Goldreich, O., Kiltz, E., Rosen, A., Segev, G.: More constructions of lossy and correlation-secure trapdoor functions. In: Nguyen, P.Q., Pointcheval, D. (eds.) PKC 2010. LNCS, vol. 6056, pp. 279–295. Springer, Heidelberg (2010)

[18] Gertner, Y., Malkin, T., Myers, S.: Towards a separation of semantic and CCA security for public key encryption. In: Vadhan, S.P. (ed.) TCC 2007. LNCS, vol. 4392, pp. 434–455. Springer, Heidelberg (2007)

[19] Goldreich, O.: Foundations of Cryptography: Volume II, Basic Applications. Cambridge University Press, Cambridge (2004)

[20] Goldreich, O., Levin, L.A.: A hard-core predicate for all one-way functions. In: STOC, pp. 25–32 (1989)

[21] Goldwasser, S., Micali, S.: Probabilistic encryption. J. Comput. Syst. Sci. 28(2), 270–299 (1984)

[22] Hanaoka, G., Kurosawa, K.: Efficient chosen ciphertext secure public key encryption under the computational Diffie-Hellman assumption. In: Pieprzyk, J. (ed.) ASIACRYPT 2008. LNCS, vol. 5350, pp. 308–325. Springer, Heidelberg (2008)

[23] Haralambiev, K., Jager, T., Kiltz, E., Shoup, V.: Simple and efficient public-key encryption from computational Diffie-Hellman in the standard model. In: Nguyen, P.Q., Pointcheval, D. (eds.) PKC 2010. LNCS, vol. 6056, pp. 1–18. Springer, Heidelberg (2010)

[24] Hofheinz, D., Kiltz, E.: Secure hybrid encryption from weakened key encapsulation. In: Menezes, A. (ed.) CRYPTO 2007. LNCS, vol. 4622, pp. 553–571. Springer, Heidelberg (2007)

[25] Hofheinz, D., Kiltz, E.: Practical chosen ciphertext secure encryption from factoring. In: Joux, A. (ed.) EUROCRYPT 2009. LNCS, vol. 5479, pp. 313–332. Springer, Heidelberg (2009)

[26] Hofheinz, D., Kiltz, E.: The group of signed quadratic residues and applications. In: Halevi, S. (ed.) CRYPTO 2009. LNCS, vol. 5677, pp. 637–653. Springer, Heidelberg (2009)

[27] Kiltz, E.: Chosen-ciphertext security from tag-based encryption. In: Halevi, S., Rabin, T. (eds.) TCC 2006. LNCS, vol. 3876, pp. 581–600. Springer, Heidelberg (2006)

[28] Kiltz, E.: Chosen-ciphertext secure key-encapsulation based on gap hashed diffie-hellman. In: Okamoto, T., Wang, X. (eds.) PKC 2007. LNCS, vol. 4450, pp. 282–297. Springer, Heidelberg (2007)

[29] Kiltz, E., Mohassel, P., O'Neil, A.: Adaptive trapdoor functions and chosen ciphertext security. In: Gilbert, H. (ed.) EUROCRYPT 2010. LNCS, vol. 6110, pp. 673–692. Springer, Heidelberg (2010)

[30] Kurosawa, K., Desmedt, Y.: A new paradigm of hybrid encryption scheme. In: Franklin, M. (ed.) CRYPTO 2004. LNCS, vol. 3152, pp. 426–442. Springer, Heidelberg (2004)

[31] Lindell, Y.: A simpler construction of CCA2-secure public-key encryption under general assumptions. J. Cryptology 19(3), 359–377 (2006)

[32] Mol, P., Yilek, S.: Chosen-ciphertext security from slightly lossy trapdoor functions. In: Nguyen, P.Q., Pointcheval, D. (eds.) PKC 2010. LNCS, vol. 6056, pp. 296–311. Springer, Heidelberg (2010)

[33] Myers, S., Shelat, A.: Bit encryption is complete. In: FOCS, pp. 607–616 (2009)

[34] Naor, M., Yung, M.: Public-key cryptosystems provably secure against chosen ciphertext attacks. In: STOC, pp. 427–437 (1990)

[35] Peikert, C.: Public-key cryptosystems from the worst-case shortest vector problem. In: STOC, pp. 333–342 (2009)

[36] Peikert, C., Waters, B.: Lossy trapdoor functions and their applications. In: STOC, pp. 187–196 (2008)

[37] Rackoff, C., Simon, D.R.: Non-interactive zero-knowledge proof of knowledge and chosen ciphertext attack. In: Feigenbaum, J. (ed.) CRYPTO 1991. LNCS, vol. 576, pp. 433–444. Springer, Heidelberg (1992)

[38] Rosen, A., Segev, G.: Chosen-ciphertext security via correlated products. In: Reingold, O. (ed.) TCC 2009. LNCS, vol. 5444, pp. 419–436. Springer, Heidelberg (2009)

[39] Sahai, A.: Non-malleable non-interactive zero knowledge and adaptive chosen-ciphertext security. In: FOCS, pp. 543–553 (1999)

[40] Shoup, V.: Using hash functions as a hedge against chosen ciphertext attack. In: Preneel, B. (ed.) EUROCRYPT 2000. LNCS, vol. 1807, pp. 275–288. Springer, Heidelberg (2000)

[41] Vahlis, Y.: Two is a crowd? a black-box separation of one-wayness and security under correlated inputs. In: Micciancio, D. (ed.) TCC 2010. LNCS, vol. 5978, pp. 165–182. Springer, Heidelberg (2010)

# Factorization of a 768-Bit RSA Modulus

Thorsten Kleinjung[1], Kazumaro Aoki[2], Jens Franke[3], Arjen K. Lenstra[1],
Emmanuel Thomé[4], Joppe W. Bos[1], Pierrick Gaudry[4], Alexander Kruppa[4],
Peter L. Montgomery[5,6], Dag Arne Osvik[1], Herman te Riele[6],
Andrey Timofeev[6], and Paul Zimmermann[4]

[1] EPFL IC LACAL, Station 14, CH-1015 Lausanne, Switzerland
[2] NTT, 3-9-11 Midori-cho, Musashino-shi, Tokyo, 180-8585 Japan
[3] University of Bonn, Dept. of Math., Beringstraße 1, D-53115 Bonn, Germany
[4] INRIA CNRS LORIA, Équipe CARAMEL - bâtiment A,
615 rue du jardin botanique, F-54602 Villers-lès-Nancy Cedex, France
[5] Microsoft Research, One Microsoft Way, Redmond, WA 98052, USA
[6] CWI, P.O. Box 94079, 1090 GB Amsterdam, The Netherlands

**Abstract.** This paper reports on the factorization of the 768-bit number RSA-768 by the number field sieve factoring method and discusses some implications for RSA.

**Keywords:** RSA, number field sieve.

## 1 Introduction

On December 12, 2009, we factored the 768-bit, 232-digit number RSA-768 by the number field sieve (NFS, [19]). RSA-768 is a representative 768-bit RSA modulus [34], taken from the RSA Challenge list [35]. Our result is a new record for factoring general integers. Factoring a 1024-bit RSA modulus would be about a thousand times harder and a 512-bit one was several thousands times easier. Because the factorization of a 512-bit RSA modulus [7] was first reported in 1999, it is not unreasonable to expect that 1024-bit RSA moduli can be factored well within the next decade by a similar academic effort. Thus, it would be prudent to phase out usage of 1024-bit RSA within the next three to four years.

The previous NFS record was the May 9, 2005, factorization of the 663-bit, 200-digit number RSA-200 [4]. NFS records must not be confused with *special* NFS (SNFS) records. The current SNSF record is the May 21, 2007, factorization of the 1039-bit number $2^{1039} - 1$ [2]. Although much bigger than RSA-768, its special form made $2^{1039} - 1$ an order of magnitude easier to factor.

The new NFS record required the following effort. We spent half a year on 80 processors on polynomial selection. This was about 3% of the main task, the *sieving*, which took almost two years on many hundreds of machines. On a single core 2.2 GHz AMD Opteron processor with 2 GB RAM, sieving would have taken about fifteen hundred years. We did about twice the sieving strictly necessary, to make the most cumbersome step, the *matrix step*, more manageable. Preparing the sieving data for the matrix step took a couple of weeks on a few processors. The final step after the matrix step took less than half a day of computing.

T. Rabin (Ed.): CRYPTO 2010, LNCS 6223, pp. 333–350, 2010.

Sieving is a laid back process that, once running, does not require much care beyond occasionally restarting a machine. The matrix step is more subtle. A slight disturbance easily causes major trouble, in particular if the problem stretches the available resources. *Oversieving* led to a matrix that could be handled relatively smoothly. More importantly, the extra sieving data allow experiments aimed at getting a better understanding of the relation between sieving and matrix efforts and the effect on NFS feasibility and overall performance. That work is in progress. All in all, the extra sieving time was well spent.

In [2] the block Wiedemann algorithm [9] was used, making it possible to process the matrix on disjoint clusters. Larger problems (such as 1024-bit moduli) require wider parallelization. Here we solve some of the challenges by dividing the workload in a more flexible manner. As a result a matrix nine times harder than in [2] was solved in less than four months, on clusters in three countries. Larger matrices are within reach and there can be little doubt about the feasibility by the year 2020 of the matrix for a 1024-bit modulus. We are studying if our current matrix can be handled by block Lanczos [8] on a single cluster.

The steps taken to factor RSA-768 are described in Section 2. The factors are given in Section 2.5. Implications for moduli larger than RSA-768 are briefly discussed in Section 3. Appendix A presents the sieving approach that we used, and Appendix B describes a new twist of the block Wiedemann algorithm that makes it easier to share large calculations among different parties.

## 2  Factoring RSA-768

### 2.1  Factoring Using the Morrison-Brillhart Approach

The *congruence of squares* method factors a composite $n$ by writing it as $\gcd(x - y, n) \cdot \gcd(x + y, n)$ for integers $x, y$ with $x^2 \equiv y^2 \bmod n$: for random such pairs $(x, y)$ the probability of success is at least $\frac{1}{2}$. We explain the Morrison-Brillhart approach [25] to solve $x^2 \equiv y^2 \bmod n$ and, roughly, how NFS works.

A non-zero integer $u$ is $b$-smooth if the prime factors of $|u|$ are at most $b$. Each $b$-smooth integer $u$ corresponds to the $(\pi(b)+1)$-dimensional integer vector $\mathfrak{v}(u)$ of exponents of the primes $\leq b$ in its factorization, where $\pi(b)$ is the number of primes $\leq b$ and the "+1" accounts for the sign. The *factor base* consists of the primes at most equal to the smoothness bound.

Let $n$ be a composite integer, $b$ a smoothness bound, and $t$ a positive integer. Let $V$ be a set of $\pi(b) + 1 + t$ integers $v$ for which the least absolute remainders $r(v) = v^2 \bmod n$ are $b$-smooth. Because the $(\pi(b)+1)$-dimensional vectors $\mathfrak{v}(r(v))$ are linearly dependent, at least $t$ independent subsets $T \subset V$ can be found using linear algebra such that $\sum_{v \in T} \mathfrak{v}(r(v))$ is an all-even vector. Thus, each $T$ leads to a solution $x = \prod_{v \in T} v$ and $y = \sqrt{\prod_{v \in T} r(v)}$ to $x^2 \equiv y^2 \bmod n$. Overall, this *combining of congruences* results in $t$ chances of at least $\frac{1}{2}$ to factor $n$.

In Dixon's random squares method [11] the set $V$ is generated by randomly selecting integers $v$ until enough have been found for which $r(v)$ is smooth. The expected runtime can be proved rigorously. With quadratic residues $r(v)$ of order $n$, however, the method is not practical: earlier, Morrison and Brillhart [25] had already shown how to use continued fractions to generate quadratic residues of order $n^{1/2}$. The much higher smoothness probabilities make their method much faster than Dixon's, despite the lack of a formal proof. Schroeppel with his *linear sieve* was the first, in about 1976, to combine similarly high smoothness probabilities with fast *sieving*-based smoothness detection [31, Section 6] and to analyze the resulting heuristic expected runtime [17, Section 4.2.6]. A variation led to Pomerance's *quadratic sieve* [31,32]. Factoring methods of this sort that rely on smoothness of residues of order $n^{\theta(1)}$ have expected runtimes of the form

$$e^{(c+o(1))(\ln n)^{1/2}(\ln \ln n)^{1/2}} \quad \text{(for } n \to \infty)$$

for positive constants $c$. The number field sieve [19] was the first, and so far the only, practical factoring method to break through the barrier of the $\ln n$-exponent of $\frac{1}{2}$. It uses more contrived congruences that involve smoothness of numbers of order $n^{o(1)}$, for $n \to \infty$, that can, as usual, be combined into a congruence of squares $x^2 \equiv y^2 \bmod n$. NFS factors a composite integer $n$ in heuristic expected time

$$e^{((64/9)^{1/3}+o(1))(\ln n)^{1/3}(\ln \ln n)^{2/3}} \quad \text{(for } n \to \infty).$$

It is currently the best algorithm to factor numbers without special properties, such as RSA-768, a 768-bit, 232-digit RSA modulus taken from [35]:

123018668453011775513049495838496272077285356959533479219732245215172640050726
365751874520219978646938995647494277406384592519255732630345373154826850791702
6122142913461670429214311602221240479274737794080665351419597459856902143413.

Similar to Schroeppel's linear sieve, the most important steps of NFS are *sieving* and the *matrix step*. In the former *relations* are collected, congruences involving smooth values similar to the smooth $r(v)$-values above. In the latter linear dependencies are found among the exponent vectors of the smooth values. NFS requires two non-trivial additional steps: a pre-processing *polynomial selection* step before the sieving can start, and a post-processing *square root step* to convert the linear dependencies into congruences of squares. A rough operational description of these steps as applied to RSA-768 is given below. For an explanation why these steps work, we refer to the many expositions on NFS [19,20,33].

## 2.2 Polynomial Selection

With $n$ the integer to be factored, let $f_1(X), f_2(X) \in \mathbf{Z}[X]$ be two irreducible integer polynomials of degrees $d_1$ and $d_2$, respectively, with a common root $m$ modulo $n$, i.e., $f_1(m) \equiv f_2(m) \equiv 0 \bmod n$. For simplicity we assume that $f_1$ and $f_2$ are monic, even though the actual $f_1$ and $f_2$ are not. With $v_k(a, b) = b^{d_k} f_k(a/b) \in \mathbf{Z}$ ($k = 1, 2$), relations are coprime pairs of integers $(a, b)$ with $b > 0$ such that $v_1(a, b)$ and $v_2(a, b)$ are simultaneously smooth, $v_1(a, b)$ with respect to some $b_1$ and $v_2(a, b)$ with respect to some $b_2$. Sufficiently many more than $\pi(b_1) + \pi(b_2) + 2$ relations lead to enough chances to factor $n$, as sketched below.

Let $\mathbf{Q}(\alpha_k) = \mathbf{Q}[X]/(f_k(X))$ for $k = 1, 2$ be two algebraic number fields. The elements $a - b\alpha_k \in \mathbf{Z}[\alpha_k]$ have *norm* $v_k(a, b)$ and belong to the first degree prime ideals in $\mathbf{Q}(\alpha_k)$ of (prime) norms equal to the prime factors of $v_k(a, b)$. These prime ideals in $\mathbf{Q}(\alpha_k)$ correspond bijectively to the pairs $(p, r \bmod p)$ where $p$ is prime and $f_k(r) \equiv 0 \bmod p$: excluding factors of $f_k$'s discriminant, such a first degree prime ideal has norm $p$ and is generated by $p$ and $r - \alpha_k$.

Because $f_k(m) \equiv 0 \bmod n$, the two natural ring homomorphisms $\phi_k : \mathbf{Z}[\alpha_k] \to \mathbf{Z}/n\mathbf{Z}$ for $k = 1, 2$ map $\sum_{i=0}^{d_k-1} a_i\alpha_k^i$ to $\sum_{i=0}^{d_k-1} a_im^i \bmod n$ and $\phi_1(a - b\alpha_1) \equiv \phi_2(a - b\alpha_2) \bmod n$. Linear dependencies modulo 2 among the exponent-vectors of the primes in the $b_1$-smooth $v_1(a, b)$, $b_2$-smooth $v_2(a, b)$ pairs lead to subsets $T$ such that $\prod_{(a,b)\in T}(a - b\alpha_k)$ is a square $\sigma_k$ in $\mathbf{Q}(\alpha_k)$, for $k = 1, 2$. With $\phi_1(\sigma_1) \equiv \phi_2(\sigma_2) \bmod n$ it then remains to compute square roots $\tau_k = \sqrt{\sigma_k} \in \mathbf{Q}(\alpha_k)$ for $k = 1, 2$ to find a solution $x = \phi_1(\tau_1)$ and $y = \phi_2(\tau_2)$ to $x^2 \equiv y^2 \bmod n$.

It is easy to find $f_1$ and $f_2$ so that numbers of order $n^{o(1)}$, for $n \to \infty$, must be smooth. Let $d_1 \in \mathbf{N}$ be of order $(\frac{3\ln n}{\ln\ln n})^{1/3}$, let $d_2 = 1$, let $m$ be an integer slightly smaller than $n^{1/d_1}$, and write $n$ in radix $m$ as $n = \sum_{i=0}^{d_1} n_im^i$ with $0 \le n_i < m$. Then $f_1(X) = \sum_{i=0}^{d_1} n_iX^i$ and $f_2(X) = X - m$ have common root $m$ modulo $n$, the coefficients are $n^{o(1)}$ for $n \to \infty$, and the values $a, b$ that suffice to generate enough relations are small enough to keep $b^{d_1}f_1(a/b)$ and $b^{d_2}f_2(a/b)$ of order $n^{o(1)}$ as well. Finally, if $f_1$ is not irreducible, it can be used to directly factor $n$ or, if that fails, one of its factors can be used instead of $f_1$. If $d_1 > 1$ and $d_2 = 1$ we refer to "$k = 1$" as the *algebraic side* and "$k = 2$" as the *rational side*. With $d_2 = 1$ the algebraic number field $\mathbf{Q}(\alpha_2)$ is simply $\mathbf{Q}$, the first degree prime ideals in $\mathbf{Q}$ are the regular primes and, with $f_2(X) = X - m$, the element $a - b\alpha_2$ of $\mathbf{Z}[\alpha_2]$ is $a - bm = v_2(a, b) \in \mathbf{Z}$ with $\phi_2(a - b\alpha_2) = a - bm \bmod n$.

Although with these polynomials NFS achieves its asymptotic runtime, there is a lot of freedom in the choices of $m$, $f_1$, and $f_2$. Exploiting this involves extensive searches, comparing choices based on smoothness probabilities, and thus with respect to coefficient size, number of real roots and roots modulo small primes, smoothness properties of leading coefficients, and sieving experiments. How the search is best conducted is the subject of active research; current approaches are guided by experience, helped by luck, and profit from patience.

One method is known that produces two good polynomials of degrees greater than one (namely, twice degree two [5]). Its results are not competitive with the current best $d_1 > 1$, $d_2 = 1$ methods which are all based on refinements [15] of the approach from [24,26] as summarized in [7, Section 3.1]. A search of three months on a cluster of 80 Opteron cores (i.e., $\frac{3}{12} \cdot 80 = 20$ core years), conducted at BSI in 2005 already and thus not including the idea from [16], produced three pairs of polynomials of comparable quality. We used

$$f_1(X) = 265482057982680X^6$$
$$+ 1276509360768321888X^5$$
$$- 5006815697800138351796828X^4$$
$$- 46477854471727854271772677450X^3$$
$$+ 6525437261935989397109667371894785X^2$$
$$- 18185779352088594356726018862434803054X$$
$$- 277565266791543881995216199713801103343120,$$
$$f_2(X) = 34661003550492501851445829X - 129118745658002122316354779157481 0881.$$

The leading coefficients factor as $2^3 \cdot 3^2 \cdot 5 \cdot 7^2 \cdot 11 \cdot 17 \cdot 23 \cdot 31 \cdot 112\,877$ and $13 \cdot 37 \cdot 79 \cdot 97 \cdot 103 \cdot 331 \cdot 601 \cdot 619 \cdot 769 \cdot 907 \cdot 1\,063$, respectively. The discriminant of $f_1$ equals $2^{12} \cdot 3^2 \cdot 5^2 \cdot 13 \cdot 17 \cdot 17\,722\,398\,737 \cdot c273$, for a 273-digit composite integer $c273$ that is most likely free of squares and of factors less than $10^{40}$. The discriminant of $f_2$ equals one. A renewed search at EPFL in the spring of 2007 (also not using the idea from [16]) produced a couple of candidates of similar quality, again after spending about 20 core years.

Following [15], during the search, the leading coefficient of $f_2$ allowed 11 (search at BSI) or 10 (search at EPFL) prime factors equal to 1 mod 6 and at most one other factor $< 2^{15.5}$. The leading coefficient of $f_1$ was a multiple of $258\,060 = 2^2 \cdot 3 \cdot 5 \cdot 11 \cdot 17 \cdot 23$. At least $2 \cdot 10^{18}$ pairs $(f_1, f_2)$ were considered.

## 2.3   Sieving

To be able to profit from near misses during the search for relations an integer $x$ is defined to be $(b_k, b_\ell)$-smooth if with the exception of, say, four prime factors between $b_k$ and $b_\ell$, all remaining prime factors of $|x|$ are at most $b_k$. We thus change the definition of a relation into a coprime pair of integers $(a, b)$ with $b > 0$ such that $b^{d_1} f_1(a/b)$ is $(b_1, b_\ell)$-smooth and $b^{d_2} f_2(a/b)$ is $(b_2, b_\ell)$-smooth. Although *large primes* speed up the sieving, they make it harder to decide whether enough relations have been found, as the criterion that more than $\pi(b_1) + \pi(b_2) + 2$ are needed is no longer adequate. The decision requires duplicate and singleton removal. It is briefly touched upon at the end of Section 2.3.

We used $b_1 = 11 \cdot 10^8$, $b_2 = 2 \cdot 10^8$ and $b_\ell = 2^{40}$ on cores with at least 2 GB RAM (the majority) and $b_1 = 4.5 \cdot 10^8$, $b_2 = 10^8$ on others (preferably with at least a GB RAM). Based on sieving experiments it was expected that it would suffice to use as sieving region the subset $S$ of $\mathbf{Z} \times \mathbf{Z}_{>0}$ of about $11 \cdot 10^{18}$ coprime pairs $(a, b)$ with $|a| \leq 3 \cdot 10^9 \cdot \kappa^{1/2} \approx 6.3 \cdot 10^{11}$ and $0 < b \leq 3 \cdot 10^9 / \kappa^{1/2} \approx 1.4 \cdot 10^7$. Here $\kappa = 44\,000$ approximates the *skewness* of $f_1$. It is used to approximately minimize the largest norm $v_1(a, b)$ encountered in the sieving region. Although prime ideal norms up to $2^{40}$ were accepted, the parameters were optimized for norms up to $2^{37}$. Most jobs attempted to factor after the sieving algebraic and rational cofactors up to $2^{140}$ and $2^{110}$, respectively, only considering the most promising candidates [14]. As far as we know, this was the first NFS factorization allowing more than three algebraic large primes.

Disregarding factors of $f_k$'s discriminant, a prime $p$ dividing $f_k(r)$ is equivalent to $(r \bmod p)$ being a root of $f_k$ modulo $p$. Because $d_2 = 1$, the polynomial $f_2$ has one root modulo $p$ for each prime $p$ not dividing its leading coefficient, and each such $p$ divides $f_2(j)$ once every $p$ consecutive $j$-values. For $f_1$ there may be between zero to $d_1$ roots modulo $p$: some primes $p$ do not divide $f_1(j)$ for any $j$, whereas other $p$ may divide $f_1(j)$ a total of $d_1$ times for every $p$ consecutive $j$-values. The $(p, r)$ pairs with $p \leq b_1$ for $f_1$ and $p \leq b_2$ for $f_2$ are precomputed.

Early implementations of NFS used *line sieving*: for some $b$-value and $k$, one marks for each precomputed $(p, r)$ pair for $f_k$ the $a$-values of the form $rb + ip$ for $i \in \mathbf{Z}$ with "$p$," since for those $a$-values $p$ divides $b^{d_k} f(a/b) = v_k(a, b)$. The locations hit by many different $p$'s are remembered, and the process is repeated for the other $k$. Relations may be found at locations that were hit twice. With many lines ($b$-values) to be processed, line sieving can easily be parallelized.

For RSA-768 we did not use line sieving but a more efficient approach that has gained popularity since the mid 1990s: the *lattice sieve* as described in [30]. For a (prime,root) pair $\mathfrak{q} = (q, s)$ define $L_{\mathfrak{q}}$ as the lattice of integer linear combinations of the 2-dimensional integer (row-)vectors $(q, 0), (s, 1) \in \mathbf{Z}^2$. Let $S_{\mathfrak{q}} = S \cap L_{\mathfrak{q}}$. Fix a (prime,root) pair $\mathfrak{q} = (q, s)$ for, say, $f_1$. The *special prime* $q$ (as it was referred to in [30]) is chosen smaller than $b_\ell$, and it divides $b^{d_1} f_1(a/b)$ for $(a, b) \in S_{\mathfrak{q}}$. Lattice sieving consists of marking, for each precomputed (prime,root) pair $\mathfrak{p}$ for $f_1$, the points in the intersection $L_{\mathfrak{p}} \cap S_{\mathfrak{q}}$. Locations that are hit often are remembered, and the process is repeated for the precomputed (prime,root) pairs for $f_2$. Relations may be found at locations that were hit twice. For each relation thus found, $q$ divides $v_1(a, b)$. The process is repeated for other $\mathfrak{q}$ until enough relations have been found. Because relations may be found for each special prime occurring in $v_1(a, b)$, duplicates will be found when lattice sieving.

In practice one fixes bounds $I$ and $J$ independent of $\mathfrak{q}$ and defines $S_{\mathfrak{q}} = \{iu + jv : i, j \in \mathbf{Z}, -I/2 \le i < I/2, 0 < j < J\}$, where $u, v$ form a basis for $L_{\mathfrak{q}}$ that minimizes the norms $v_1(a, b)$ for $(a, b) \in S_{\mathfrak{q}}$. Such a basis is found by partially reducing the basis $(q, 0), (s, 1)$ for $L_{\mathfrak{q}}$ such that the first coordinate is roughly $\kappa$ times bigger than the second, cf. skewness of $S$. Sieving is carried out over the set $\{(i, j) \in \mathbf{Z} \times \mathbf{Z}_{>0} : -I/2 \le i < I/2, 0 < j < J\}$, interpreted as $S_{\mathfrak{q}}$.

We used $I = 2^{16}$ and $J = 2^{15}$, i.e., a lattice sieving area of size roughly $2^{31} \approx 2 \cdot 10^9$. With $b_1 = 11 \cdot 10^8$ and $b_2 = 2 \cdot 10^8$, the majority of the sieving-primes can be expected to hit $S_{\mathfrak{q}}$ only a few times. Thus, for any sieving-$\mathfrak{p}$, only a few of the $j$-values (the lines) will be hit, unlike line sieving where each line will be hit several times by each prime. Therefore, when lattice sieving, a more sophisticated sieving method must be used that avoids looking at all lines $0 < j < J$ for each $\mathfrak{p}$. This *sieving by vectors* [30] was first implemented in [13] and used for many factorizations in the 1990s [10,7]. We used the implementation from [12], described in Appendix A. Most of the about 0.48 billion (prime,root) pairs $(q, s)$ for special primes $q$ between 0.45 and 11.1 billion (and some special primes below 0.45 billion, with a smaller $b_1$-value) were processed by eight contributing parties (cf. Table 1) during the period August 2007 until April 2009. Scaled to a 2.2 GHz Opteron core with 2 GB RAM, a single $L_{\mathfrak{q}}$ was processed in less than 100 seconds on average and produced about 134 relations, for an average of about four relations every three seconds. This average rate varies by a factor of about two between both ends of the special primes range that we used.

We collected 64 334 489 730 relations in total, each requiring about 150 bytes. Compressed they occupied about 5 terabytes of disk space, backed up at various locations. The 27.4% duplicates were removed using hashing. This was done mostly during the sieving, overall taking less than 10 days on a 2.66 GHz Core2 processor with ten 1TB hard disks. After including 57 223 462 *free relations* [19], we ended up with 47 762 243 404 relations involving 35 288 334 017 prime ideals.

Given the set of unique relations, those that involve a prime ideal that does not occur in any other relation, the singletons, cannot be part of a dependency.

**Table 1.** Percentages contributed

| contributor | relations contribution | matrix stages, % of matrix effort 1 (60%) | 2 (0%) | 3 (40%) | total |
|---|---|---|---|---|---|
| Bonn (University and BSI) | 8.14% | | | | |
| CWI | 3.44% | | | | |
| EPFL | 29.78% | 34.3% | 100% | 78.2% | 51.9% |
| INRIA LORIA (ALADDIN-G5K) | 37.97% | 46.8% | | 17.3% | 35.0% |
| NTT | 15.01% | 18.9% | | 4.5% | 13.1% |
| Scott Contini (Australia) | 0.43% | | | | |
| Paul Leyland (UK) | 0.69% | | | | |
| Heinz Stockinger (Enabling Grids for E-sciencE) | 4.54% | | | | |

Singletons were removed using hashing. Doing this once reduced the set of relations to 28 984 986 047 elements with 14 498 007 183 prime ideals. Removal of singletons usually creates new singletons, and the process must be repeated until no new singletons are created. After a few more singleton removals 24 615 168 385 relations involving at most 9 976 671 468 prime ideals were left.

Further singleton removal was combined with *clique removal* [6], i.e., search of combinations with matching first degree prime ideals of norms larger than $b_k$. Ultimately, this led to 2 458 248 361 relations with 1 697 618 199 prime ideals, still containing an almost insignificant number (604 423) of free relations. Since there are more relations than prime ideals (so that dependencies exist), we had done enough sieving and lots of flexibility to create a matrix. Singleton and clique removal took less than 10 days on the same platform as above.

## 2.4   The Matrix Step

Current best methods to find dependencies among the rows of a sparse matrix take time proportional to the product of the dimension and the weight (i.e., number of non-zero entries) of the matrix. Merging is a generic term for the set of strategies developed to build a matrix for which close to optimal dependency search can be expected. It is described in [6]. We ran about 10 separate merging jobs, aiming for various optimizations (low dimension, low weight, best-of-both, etc.), which each took a couple of days on a single core per node of a 37-node 2.66 GHz Core2 cluster with 16 GB RAM per node, and a not particularly fast interconnection network. The best alternative was a 192 796 550 × 192 795 550-matrix of total weight 27 797 115 920 (on average 144 non-zeros per row), requiring about 105 GB. It was generated in 5-days on two to three cores on the 37-node cluster, where the long duration was probably due to the large communication overhead. When we started the project, we expected dimension about a quarter billion and density per row of about 150, which would have been about $\frac{7}{4}$ times harder.

To find dependencies we used block Wiedemann [9,38] as described in [2, Section 5.1] . We give a high level description [18, Section 2.19]. Given a non-singular $d \times d$ matrix $M$ over the finite field $\mathbf{F}_2$ and $b \in \mathbf{F}_2^d$, we wish to solve the system $Mx = b$. The minimal polynomial $F$ of $M$ on the vector space spanned by $b$, $Mb$, $M^2b$, ... has degree at most $d$, so that $F(M)b = \sum_{i=0}^d F_i M^i b = 0$. From $F_0 = 1$ it follows that $x = \sum_{i=1}^d F_i M^{i-1} b$, so it suffices to find the $F_i$'s.

Denoting by $m_{i,j}$ the $j$th coordinate of the vector $M^i b$, it follows that for each $j$ with $1 \le j \le d$ the sequence $(m_{i,j})_{i=0}^{\infty}$ satisfies a linear recurrence relation of order at most $d$ defined by the coefficients $F_i$: for any $t \ge 0$ and $1 \le j \le d$ we have that $\sum_{i=0}^{d} F_i m_{i+t,j} = 0$. Given $2d + 1$ consecutive terms of an order $d$ linear recurrence, its coefficients can be computed using the Berlekamp-Massey method [22,38]. Each $j$ may lead to a polynomial of smaller degree than $F$, but taking, if necessary, the least common multiple of the polynomials found for a few different indices $j$, the correct minimal polynomial will be found.

Summarizing the above, there are three major stages: a first iteration consisting of $2d$ matrix×vector steps to generate $2d + 1$ terms of the linear recurrence, the Berlekamp-Massey stage to calculate the $F_i$'s, and a second iteration consisting of $d$ matrix×vector steps to calculate the solution using the $F_i$'s. For large matrices the first and the final stage are the most time consuming.

In practice it is common to use *blocking*, to take advantage of the fact that on 64-bit machines 64 different vectors $b$ over $\mathbf{F}_2$ can be processed simultaneously, at little or no extra cost compared to a single vector [9], while using the same three main stages. If the vector $\bar{b}$ is 64 bits wide and in the first stage the first 64 coordinates of each of the generated 64 bits wide vectors $M^i \bar{b}$ are kept, the number of matrix ($M$) times vector ($\bar{b}$) multiplications in both the first and the final stage is reduced by a factor of 64 compared to the number of $M$ times $b$ multiplications, while making the central Berlekamp-Massey stage a bit more cumbersome. It is less common to take the blocking a step further and run both iteration stages spread over a small number $n'$ of different sequences, possibly run on disjoint clusters; in [2] this was done with $n' = 4$ sequences run on three clusters. If for each sequence one keeps the first $64 \cdot n'$ coordinates of each of the 64 bits wide vectors they generate during the first stage, the number of steps to be carried out (per sequence) is further reduced by a factor of $n'$, while allowing independent and simultaneous execution on possibly $n'$ disjoint clusters. After the first stage the data generated for the $n'$ sequences have to be gathered at a central location where the Berlekamp-Massey stage will be carried out.

While keeping the first $64 \cdot n'$ coordinates per step for each sequence results in a reduction of the number of steps per sequence by a factor of $64 \cdot n'$, keeping a different number of coordinates while using $n'$ sequences results in another reduction in the number of steps for the first stage. Following [2, Section 5.1], if the first $64 \cdot m'$ coordinates are kept of the 64 bits wide vectors for $n'$ sequences, the numbers of steps become $\frac{d}{64 \cdot m'} + \frac{d}{64 \cdot n'} = (\frac{n'}{m'} + 1)\frac{d}{64 \cdot n'}$ and $\frac{d}{64 \cdot n'}$ for the first and third stage, respectively and for each of the $n'$ sequences. The choices of $m'$ and $n'$ should be weighed off against the cost of the Berlekamp-Massey step with time and space complexities proportional to $\frac{(m'+n')^3}{n'} d^{1+o(1)}$ and $\frac{(m'+n')^2}{n'} d$, respectively and for $d \to \infty$, and where the exponent "3" may be replaced by the matrix multiplication exponent (our implementation uses "3").

When running the first stage using $n'$ sequences, the effect of non-identical resources used for different sequences quickly becomes apparent: some locations finish their work faster than others (depicted in Fig. 1). To keep the fast contributors busy and to reduce the work of the slower ones (thereby reducing the wall-clock time), a quickly processed first stage sequence may continue for $s$ steps beyond $(\frac{n'}{m'}+1)\frac{d}{64\cdot n'}$ while reducing the number of steps in another first stage sequence by the same $s$. As described in Appendix B, this can be done in a very flexible way, as long as the overall number of steps over all first stage sequences adds up to $n' \cdot (\frac{n'}{m'}+1)\frac{d}{64\cdot n'}$. The termination points of the sequences in the third stage need to be adapted accordingly. This is easily arranged for, since the third stage allows much easier and much wider parallelization anyhow (assuming checkpoints from the first stage are kept). Another way to keep all parties busy is swapping jobs, thus requiring data exchanges, synchronization, and more human interaction, making it a less attractive option altogether.

For our matrix with $d \approx 193 \cdot 10^6$ we used, as in [2], $m' = 2n'$. But where $n' = 4$ was used in [2], we used $n' = 8$. This quadrupled the Berlekamp-Massey runtime and doubled its memory compared to the matrix from [2], on top of the increased runtime and memory demands caused by the larger dimension of the matrix. On the other hand, the compute intensive first and third stages could be split up into twice as many independent jobs as before. For the first stage on average $(\frac{8}{16}+1)\frac{193\cdot 10^6}{64\cdot 8} \approx 565\,000$ steps needed to be taken per sequence (for 8 sequences), for the third stage the average was about $\frac{193\cdot 10^6}{64\cdot 8} \approx 380\,000$ steps. The actual numbers of steps varied, approximately, between $490\,000$ and $650\,000$ for the first stage and between $300\,000$ and $430\,000$ for the third stage. The calculation of these stages was carried out on a wide variety of clusters accessed from three locations: a 56-node cluster of 2.2GHz dual hex-core AMD processors with Infiniband at EPFL (installed while the first stage was in progress), a variety of ALADDIN-G5K clusters in France accessed from INRIA LORIA, and a cluster of 110 3.0GHz Pentium-D processors on a Gb Ethernet at NTT.

On 12 nodes of a 12-cores-per-node cluster of 2.2 GHz AMD processors with 16 GB RAM per node and an Infiniband network, one multiplication step (of the matrix times a 64 bits wide vector) took between 4.3 and 4.5 seconds for the first stage and about 4.8 seconds for the slightly more involved third stage. Per-iteration timings for stage one on the Pentium cluster are 11.6 seconds per iteration when two sequences are run in parallel (thus, effectively, 5.8 seconds per sequence), and 6.4 seconds if one sequence is processed. For the third stage it was 7.8 seconds per iteration, for a single sequence. For the ALADDIN-G5K clusters the per-iteration timings for stages one and three varied between 2.3 and 4.1 seconds, and between 2.6 and 17.9 seconds, respectively. It follows that doing the entire first and third stage would have taken 98 days on 48 nodes (576 cores) of the 56-node EPFL cluster.

The first stage was split up into eight independent jobs run in parallel on those clusters, check-pointing once every $2^{14}$ steps. Running a first (or third) stage sequence required 180 GB RAM, a single 64 bits wide $\bar{b}$ took 1.5 GB, and a single $m_i$ matrix 8 KB, of which $565\,000$ were kept, on average, per first stage sequence. Each partial sum during the third stage evaluation required 12 GB.

The central Berlekamp-Massey stage was done in 17 hours and 20 minutes on the 56-node EPFL cluster (with 16 GB RAM per node), while using just 4 of the 12 available cores per node. Most of the time the available 896 GB RAM sufficed,

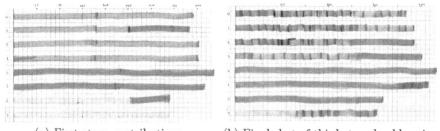

(a) First stage contributions.        (b) Final shot of third stage bookkeeping.

**Fig. 1.** Contributions to sequences 0-7: blue=INRIA, orange=EPFL, pink=NTT

but during a central part of the calculation more memory was needed (up to about 1 TB) and some swapping occurred. The third stage started right after completion of the second stage, running as many jobs in parallel as possible. The actual bookkeeping sheet used is pictured in Fig. 1b. Fig. 1a pictures the first stage contributions apocryphally but accurately. Calendar time for the entire block Wiedemann step was 119 days, finishing on December 8, 2009.

### 2.5  That's a Bingo[1]

As expected the matrix step resulted in $512 = 64 \cdot 8$ linear dependencies modulo 2 among the exponent vectors, more than enough to include the quadratic characters at this stage [1]. This reduced the solution space to 460 elements, giving us that many independent chances of about $\frac{1}{2}$ to factor RSA-768. In the $52 = 512 - 460$ difference, a dimension of 46 can be attributed to prime ideals not included in the matrix that divide the leading coefficients or the discriminant.

The square roots of the algebraic numbers were calculated by means of the method from [23] (see also [29]), which uses the known factorization of the algebraic numbers into small prime ideals of known norms. The implementation based on [3] turned out to have a bug when computing the valuations for the free relations of the prime ideals lying above the divisor $17\,722\,398\,737 > 2^{32}$ of the discriminant of $f_1$. Along with a bug in the quadratic character calculation, this delayed completion of the square root step by a few (harrowing) days.

Once the bugs were located and fixed, it took two hours using the hard disk and one core on each of twelve dual hex-core 2.2GHz AMD processors to compute the exponents of all prime ideals for eight solutions simultaneously. Computing a square root using the implementation from [3] took one hour and forty minutes on such a dual hex-core processor. The first one (and four of the other seven) led to the factorization $p \cdot q$, found at 20:16 GMT on December 12, 2009:

---

[1] "Is that the way you say it? "That's a bingo?""
"You just say "bingo"." [37]

$p = 3347807169895689878604416984821269081770479498371376856891$
$24313889828837938780022876147116525317430877378144679999489,$

$q = 3674604366679959042824463379962795263227915816434308764267$
$60322838157396665112792333734171433968102700927987363008917,$

where $p$ and $q$ are 384-bit, 116-digit primes. With "$pk$" a $k$-digit prime, we found:

$p - 1 = 2^8 \cdot 11^2 \cdot 13 \cdot 7193 \cdot 160378082551 \cdot 7721565388263419219 \cdot$
$111103163449484882484711393053 \cdot p47,$

$p + 1 = 2 \cdot 3 \cdot 5 \cdot 3193212274955337226200549186163034518\underline{3}416467 \cdot p71,$

$q - 1 = 2^2 \cdot 359 \cdot p113, \quad q + 1 = 2 \cdot 3 \cdot 23 \cdot 41 \cdot 47 \cdot 239875144072757917901 \cdot p90.$

# 3   Concluding Remarks

It is customary to conclude a paper reporting a new factoring record with a preview of coming attractions. Our main conclusion was summarized in the introduction and was already announced in [2, Section 7]: at this point factoring a 1024-bit RSA modulus looks more than five times easier than a 768-bit RSA modulus looked back in 1999, when we achieved the first public factorization of a 512-bit RSA modulus. Nevertheless, a 1024-bit RSA modulus is still about a thousand times harder to factor than a 768-bit one. It may be possible to factor a 1024-bit RSA modulus within the next decade by means of an academic effort on the same scale as the effort presented here. Recent standards recommend phasing out such moduli by the end of the year 2010 [28]. See also [21].

Another conclusion from our work is that we can confidently say that if we restrict ourselves to an open community, academic effort such as ours and unless something dramatic happens in factoring, we will not be able to factor a 1024-bit RSA modulus within the next five years [27]. After that, all bets are off.

The ratio between sieving and matrix time was almost 10. This is probably not optimal if one wants to minimize the overall runtime. But the latter may not be the most important criterion. Sieving is easy, and doing more of it may be a good investment if that leads to an easier matrix step. The relations collected for RSA-768 will give us a better insight in the trade-off between sieving and matrix efforts, where also the choice of the large prime bound $b_\ell$ may play a role. This is a subject for further study that may be expected to lead, ultimately, to a recommendation for close to optimal parameter choices – depending on what one wants to optimize – for NFS factorizations in the 700- to 800-bit range.

Our computation required more than $10^{20}$ operations. With the equivalent of almost 2000 years of computing on a single core 2.2GHz AMD Opteron, on the order of $2^{67}$ instructions were carried out. The overall effort is sufficiently low that even for short-term protection of data of little value, 768-bit RSA moduli can no longer be recommended. This conclusion is the opposite of the one on [36], which is based on a hypothetical factoring effort of six months on 100 000 workstations, i.e., about two orders of magnitude more than we spent.

**Acknowledgements.** This work was supported by the Swiss National Science Foundation under grant numbers 200021-119776 and 206021-128727 and by the Netherlands Organization for Scientific Research (NWO) as project 617.023.613. Experiments presented in this paper were carried out using the Grid'5000 experimental testbed, being developed under the INRIA ALADDIN development action with support from CNRS, RENATER and several universities as well as other funding bodies (see https://www.grid5000.fr). Condor middleware was used on EPFL's Greedy network. We acknowledge the help of Cyril Bouvier during filtering and merging experiments. We gratefully acknowledge sieving contributions by BSI, Scott Contini (using resources provided by AC3, the Australian Centre for Advanced Computing and Communications), Paul Leyland (using teaching lab machines at the Genetics Department of Cambridge University), and Heinz Stockinger (using EGEE, Enabling Grids for E-sciencE). Parts of this paper were inspired by Col. Hans Landa.

# References

1. Adleman, L.M.: Factoring numbers using singular integers. In: STOC, pp. 64–71. ACM, New York (1991)
2. Aoki, K., Franke, J., Kleinjung, T., Lenstra, A.K., Osvik, D.A.: A kilobit special number field sieve factorization. In: Kurosawa, K. (ed.) ASIACRYPT 2007. LNCS, vol. 4833, pp. 1–12. Springer, Heidelberg (2007)
3. Bahr, F.: Liniensieben und Quadratwurzelberechnung für das Zahlkörpersieb, Diplomarbeit, University of Bonn (2005)
4. Bahr, F., Böhm, M., Franke, J., Kleinjung, T.: Factorization of RSA-200 (May 2005), http://www.loria.fr/~zimmerma/records/rsa200
5. Buhler, J., Montgomery, P., Robson, R., Ruby, R.: Implementing the number field sieve. Technical report, Oregon State University (1994)
6. Cavallar, S.: Strategies in filtering in the number field sieve. In: Bosma, W. (ed.) ANTS 2000. LNCS, vol. 1838, pp. 209–232. Springer, Heidelberg (2000)
7. Cavallar, S., Dodson, B., Lenstra, A.K., Lioen, W.M., Montgomery, P.L., Murphy, B., te Riele, H.J.J., Aardal, K., Gilchrist, J., Guillerm, G., Leyland, P.C., Marchand, J., Morain, F., Muffett, A., Putnam, C., Putnam, C., Zimmermann, P.: Factorization of a 512-bit RSA modulus. In: Preneel, B. (ed.) EUROCRYPT 2000. LNCS, vol. 1807, pp. 1–18. Springer, Heidelberg (2000)
8. Coppersmith, D.: Solving linear equations over GF(2): block Lanczos algorithm. Linear Algebra and its Applications 192, 33–60 (1993)
9. Coppersmith, D.: Solving homogeneous linear equations over GF(2) via block Wiedemann algorithm. Math. Comput. 62(205), 333–350 (1994)
10. Cowie, J., Dodson, B., Elkenbracht-Huizing, R.M., Lenstra, A.K., Montgomery, P.L., Zayer, J.: A world wide number field sieve factoring record: On to 512 bits. In: Kim, K.-c., Matsumoto, T. (eds.) ASIACRYPT 1996. LNCS, vol. 1163, pp. 382–394. Springer, Heidelberg (1996)
11. Dixon, J.D.: Asymptotically fast factorization of integers. Math. Comp. 36, 255–260 (1981)
12. Franke, J., Kleinjung, T.: Continued fractions and lattice sieving. In: Workshop record of SHARCS (2005), http://www.ruhr-uni-bochum.de/itsc/tanja/SHARCS/talks/FrankeKleinjung.pdf

13. Golliver, R.A., Lenstra, A.K., McCurley, K.S.: Lattice sieving and trial division. In: Huang, M.-D.A., Adleman, L.M. (eds.) ANTS 1994. LNCS, vol. 877, pp. 18–27. Springer, Heidelberg (1994)
14. Kleinjung, T.: Cofactorisation strategies for the number field sieve and an estimate for the sieving step for factoring 1024 bit integers. In: Workshop record of SHARCS (2005), http://www.hyperelliptic.org/tanja/SHARCS/talks06/thorsten.pdf
15. Kleinjung, T.: On polynomial selection for the general number field sieve. Math. Comp. 75, 2037–2047 (2006)
16. Kleinjung, T.: Polynomial selection. Presented at the CADO Workshop on Integer Factorization (2008), http://cado.gforge.inria.fr/workshop/slides/kleinjung.pdf
17. Lenstra, A.K.: Computational methods in public key cryptology. In: Coding theory and cryptology. Lecture Notes Series, pp. 175–238. Institute for Mathematical Sciences, National University of Singapore (2002)
18. Lenstra, A.K., Lenstra Jr., H.W.: Algorithms in number theory. In: Handbook of Theoretical Computer Science, Volume A: Algorithms and Complexity, pp. 673–716. Elsevier, Amsterdam (1990)
19. Lenstra, A.K., Lenstra Jr., H.W.: The Development of the Number Field Sieve. LNM, vol. 1554. Springer, Heidelberg (1993)
20. Lenstra, A.K., Lenstra Jr., H.W., Manasse, M.S., Pollard, J.M.: The factorization of the ninth Fermat number. Math. of Comp. 61(203), 319–349 (1993)
21. Lenstra, A.K., Tromer, E., Shamir, A., Kortsmit, W., Dodson, B., Hughes, J., Leyland, P.C.: Factoring estimates for a 1024-bit RSA modulus. In: Laih, C.-S. (ed.) ASIACRYPT 2003. LNCS, vol. 2894, pp. 55–74. Springer, Heidelberg (2003)
22. Massey, J.: Shift-register synthesis and BCH decoding. IEEE Trans Information Theory 15, 122–127 (1969)
23. Montgomery, P.: Square roots of products of algebraic numbers, http://ftp.cwi.nl/pub/pmontgom/sqrt.ps.gz
24. Montgomery, P., Murphy, B.: Improved polynomial selection for the number field sieve. Technical report, the Fields institute, Toronto, Ontario, Canada (June 1999)
25. Morrison, M.A., Brillhart, J.: A method of factoring and the factorization of $F_7$. Math. of Comp. 29(129), 183–205 (1975)
26. Murphy, B.: Modelling the yield of number field sieve polynominals. In: Buhler, J.P. (ed.) ANTS 1998. LNCS, vol. 1423, pp. 137–150. Springer, Heidelberg (1998)
27. National Institute of Standards and Technology. Discussion paper: the transitioning of cryptographic algorithms and key sizes, http://csrc.nist.gov/groups/ST/key_mgmt/documents/Transitioning_CryptoAlgos_070209.pdf
28. National Institute of Standards and Technology. Special publication 800-57: Recommendation for key management part 1: General (revised), http://csrc.nist.gov/publications/nistpubs/800-57/sp800-57-Part1-revised2_Mar08-2007.pdf
29. Nguyen, P.Q.: A Montgomery-like square root for the number field sieve. In: Buhler, J.P. (ed.) ANTS 1998. LNCS, vol. 1423, pp. 151–168. Springer, Heidelberg (1998)
30. Pollard, J.M.: The lattice sieve. In: [19], pp. 43–49
31. Pomerance, C.: Analysis and comparison of some integer factoring algorithms. In: Lenstra Jr., H.W., Tijdeman, R. (eds.) Computational Methods in Number Theory, Math. Centrum Tract, Amsterdam, vol. 154, pp. 89–139 (1982)
32. Pomerance, C.: The quadratic sieve factoring algorithm. In: Beth, T., Cot, N., Ingemarsson, I. (eds.) EUROCRYPT 1984. LNCS, vol. 209, pp. 169–182. Springer, Heidelberg (1985)

33. Pomerance, C.: A tale of two sieves (1996),
    http://www.ams.org/notices/199612/pomerance.pdf
34. Rivest, R.L., Shamir, A., Adleman, L.: A method for obtaining digital signatures
    and public key cryptosystems. Communications of the ACM 21, 120–126 (1978)
35. RSA the security division of EMC. The RSA challenge numbers. formerly on
    http://www.rsa.com/rsalabs/node.asp?id=2093, now on
    http://en.wikipedia.org/wiki/RSA_numbers
36. RSA the security division of EMC. The RSA factoring challenge FAQ,
    http://www.rsa.com/rsalabs/node.asp?id=2094
37. Tarantino, Q.: That's a bingo! http://www.youtube.com/watch?v=WtHTc8wIo4Q,
    http://www.imdb.com/title/tt0361748/quotes
38. Thomé, E.: Subquadratic computation of vector generating polynomials and im-
    provement of the block Wiedemann algorithm. Journal of Symbolic Computa-
    tion 33(5), 757–775 (2002)

# A    Sieving by Vectors

We briefly describe the lattice sieve implementation from [12] which was used
for most NFS factorization records of the last decade.

Let $v_k(a, b) = b^{d_k} f_k(a/b)$. Lattice sieving, introduced by Pollard [30], increases
the smoothness probability of $v_1(a, b)$ by looking at $(a, b)$-pairs for which $v_1(a, b)$
is divisible by a large special prime $q$. Let $s \bmod q$ be a residue class such that this
is the case for $a \equiv sb \bmod q$. One constructs a reduced basis $(u, v)$ of the lattice
of all $(a, b) \in \mathbf{Z}^2$ with $a \equiv sb \bmod q$. A scalar product adapted to the skewness
of the polynomial pair is used for this reduction. The problem is then to find all
coprime pairs $(i, j)$, $-I/2 \le i < I/2, 0 < j < J$, such that $v_1(a, b)/q$ and $v_2(a, b)$
are smooth, with $(a, b) = iu + jv$. We assume $I$ to be even. As mentioned in
Section 2.3, for practical values of the parameters, $I$ is much smaller than the
smoothness bounds $b_1$ and $b_2$, and it is non-trivial to efficiently sieve such regions.

Pollard proposed to do this by using, for each (prime,root) pair $\mathfrak{p}$ with prime
$p$ bounded by the relevant smoothness bound $b_k$, a reduced base of the lattice
$\Gamma_\mathfrak{p}$ of pairs $(i, j)$ for which $v_k(a, b)$ for the corresponding $(a, b)$-pair is divisible
by $p$. In [13] that approach was used for $p$ larger than a small multiple of $I$,
while avoiding storage of "even, even" sieve locations (and using line sieving for
the other primes). Our approach uses a truncated continued fraction expansion
to determine a basis $B = ((\alpha, \beta), (\gamma, \delta))$ of $\Gamma_\mathfrak{p}$ with the following properties:

**a** The numbers $\beta$ and $\delta$ are positive.
**b** We have $-I < \alpha \le 0 \le \gamma < I$ and $\gamma - \alpha \ge I$.

Let us assume that $\Gamma_\mathfrak{p}$ consists of all $(i, j)$ for which $i \equiv \rho j \bmod p$, where $0 <
\rho < p$. The case $\rho = 0$ and the case where $\Gamma_\mathfrak{p}$ consists of all $(i, j)$ for which $p$
divides $j$ are not treated, because they produce just $(0, 1)$ and $(1, 0)$, respectively,
as only coprime pairs. We also assume $p \ge I$, as smaller primes are better
treated by line sieving. To construct a basis with the above properties, one takes
$(i_0, j_0) = (-p, 0)$, $(i_1, j_1) = (\rho, 1)$ and puts $(i_{\ell+1}, j_{\ell+1}) = (i_{\ell-1}, j_{\ell-1}) + r(i_\ell, j_\ell)$
with $r = \lfloor -\frac{i_{\ell-1}}{i_\ell} \rfloor$. Note that $(-1)^{\ell+1} i_\ell \ge 0$, that $r$ is positive and that the $j_\ell$ thus

form an increasing sequence of non-negative numbers. The process is stopped at the first $\ell$ with $\mid i_\ell \mid < I$. If $\ell$ is odd, we put $(\alpha, \beta) = (i_{\ell-1}, j_{\ell-1}) + r(i_\ell, j_\ell)$, where $r$ is the smallest integer for which $\alpha > -I$. If $\ell$ is even, we put $(\gamma, \delta) = (i_{\ell-1}, j_{\ell-1}) + r(i_\ell, j_\ell)$, where $r$ is the smallest integer such that $\gamma < I$. In both cases, the element of $B = ((\alpha, \beta), (\gamma, \delta))$ not yet described is given by $(i_\ell, j_\ell)$.

To explain how to efficiently sieve using a basis with these properties, let $(i, j) \in \Gamma_{\mathfrak{p}}$ such that $-I/2 \le i < I/2$. We want to find the (uniquely determined) $(i', j') \in \Gamma_{\mathfrak{p}}$ such that $-I/2 \le i' < I/2$, $j' > j$, and $j'$ is as small as possible. As $B$ is a basis of $\Gamma_{\mathfrak{p}}$, there are integers $d$ and $e$ with

$$(i', j') - (i, j) = d(\alpha, \beta) + e(\gamma, \delta).$$

If $d \cdot e < 0$, then condition **b** on $B$ would force the first component of the right hand side to have absolute value $\ge I$, whereas our constraints on $i$ and $i'$ force it to have absolute value $< I$. Since $j' - j$, $\beta$, and $\delta$ are all positive, we have $d \ge 0$ and $e \ge 0$. It is now easy to see that the solution to our problem is:

$$(d, e) = \begin{cases} (0, 1) & \text{if } i < I/2 - \gamma \\ (1, 1) & \text{if } I/2 - \gamma \le i < -I/2 - \alpha \\ (1, 0) & \text{if } i \ge -I/2 - \alpha. \end{cases}$$

The minimality of $j'$ follows because $d = 0$ leads to a violation of $i' < I/2$ unless $i < I/2 - \gamma$ (i.e., save for the first of the above cases) and $e = 0$ leads to $i' < -I/2$ unless $i \ge -I/2 - \alpha$ (i.e., save for the third of the above cases).

To implement this process on a modern CPU, it seems best to take $I = 2^\iota$ for some natural number $\iota$. It is possible to identify pairs $(i, j)$ of integers with $-I/2 \le i < I/2$ with integers $x$ by putting $x = j \cdot I + i + I/2$. If $x' = j' \cdot I + i' + I/2$ with $(i', j')$ as above, then $x' = x + C$, $x' = x + A + C$ and $x' = x + A$ in the three cases above, with $A = \alpha + I \cdot \beta$ and $C = \gamma + I \cdot \delta$. The first component of a pair $(i, j)$, $(\alpha, \beta)$ or $(\gamma, \delta)$ is extracted from these numbers by using a bitwise logical operation, and the selection of the appropriate one of the above three cases is best done using conditional move instructions.

For cache efficiency, the sieving region $S_{\mathfrak{q}}$ was split into areas $A_t$, $0 \le t < T$, of size equal to the L1-cache size. For primes $p$ larger than that size (or a small multiple thereof), sieving is not done directly. Instead, the numbers $x$ corresponding to elements of $S_{\mathfrak{q}} \cap \Gamma_{\mathfrak{p}}$ were calculated ahead of the sieving process, and their offsets into the appropriate region $A_t$ stored in the corresponding element of an array $\mathcal{S}$ of $T$ stacks. To implement the trial division sieve efficiently, the corresponding factor base index was also stored. Of course, this approach may also be used for line sieving, and in fact was used in [3]. A similar approach has been described by T. Oliveira e Silva in connection with his implementation of the Odlyzko-Lagarias-Lehmer-Meissel method.

Parallelization is possible in several different ways. A topology for splitting the sieving region among several nodes connected by a network is described in [12]. If one wants to split the task among several cores sharing their main memory, it seems best to distribute the regions $A_t$ and also the large factor base primes among them. Each core first calculates its part of $\mathcal{S}$, for its assigned part of the large factor base elements, and then uses the information generated by all cores to treat its share of regions $A_t$. A lattice siever parallelized that way was used for a small part of the RSA-576 sieving tasks, but the code fell out of use and was not used for the current project. The approach may be more useful today, with many cores per processor being a standard.

# B  Unbalanced Sequences in Block Wiedemann

Before describing the modification for unbalanced sequence lengths we give an overview of Coppersmith's block version of the Berlekamp-Massey algorithm. To avoid a too technical description we simplify the presentation of Coppersmith's algorithm and refer to [9] for details. The modification was also be applied to Thomé's subquadratic algorithm [38]. Below $m$ and $n$ are as in [9], and the terms "$+O(1)$" are constants depending on $m$ and $n$. We assume that $m$ and $n$, which play the role of $64 \cdot m'$ and $64 \cdot n'$ in Section 2.4, are much smaller than $d$.

Let $M$ be a $d \times d$ matrix over $\mathbf{F}_2$, $m \geq n$, $x_k \in \mathbf{F}_2^d$, $1 \leq k \leq m$ and $y_j \in \mathbf{F}_2^d$, $1 \leq j \leq n$ satisfying certain conditions. Set $a_{j,k}^{(i)} = x_k^T M^i y_j$ and

$$A = \sum_i (a_{j,k}^{(i)}) X^i \quad \in \mathrm{Mat}_{n,m}[X].$$

In the first step we calculate the coefficients of $A$ up to degree $\frac{d}{m} + \frac{d}{n} + O(1)$.

The goal of the Berlekamp-Massey step is to find polynomials of matrices $F \in \mathrm{Mat}_{n,n}[X]$, $G \in \mathrm{Mat}_{n,m}[X]$ with $\deg(F) \leq \frac{d}{n} + O(1)$, $\deg(G) \leq \frac{d}{n} + O(1)$ and

$$FA \equiv G \pmod{X^{\frac{d}{m} + \frac{d}{n} + O(1)}}.$$

Intuitively, we want to produce at least $d$ zero rows in the higher coefficients of $FA$ up to degree $\frac{d}{m} + \frac{d}{n} + O(1)$. Writing $F = \sum_{i=0}^{d_F} (f_{j,k}^{(i)}) X^i$, $d_F = \deg(F)$ the $j$th row of coefficient $d_F + b$ of $G$ being zero corresponds to

$$(M^b x_h)^T v_j = 0 \quad \text{for } 1 \leq h \leq m, 0 \leq b < \frac{d}{m} + O(1) \text{ where}$$

$$v_j = \sum_{k=1}^{n} \sum_{i=0}^{d_F} f_{j,k}^{(d_F - i)} \cdot M^i y_k.$$

Coppersmith's algorithm produces a sequence of matrices (of $m + n$ rows) $F_t \in \mathrm{Mat}_{m+n,n}[X]$ and $G_t \in \mathrm{Mat}_{m+n,m}[X]$ for $t = t_0, \ldots, \frac{d}{m} + \frac{d}{n} + O(1)$ (where $t_0 = O(1)$) such that

$$F_t A \equiv G_t \pmod{X^t}$$

and the degrees of $F_t$ and $G_t$ are roughly $\frac{m}{m+n} t$. In a first step $t_0$ and $F_{t_0}$ are chosen such that certain conditions are satisfied, in particular that $\deg(F_{t_0}) = O(1)$ and $\deg(G_{t_0}) = O(1)$. To go from $t$ to $t + 1$ a polynomial of degree 1 of matrices $P_t \in \mathrm{Mat}_{m+n,m+n}[X]$ is constructed and we set $F_{t+1} = P_t F_t$ and $G_{t+1} = P_t G_t$. This construction is done as follows. We have $F_t A \equiv G_t + E_t X^t$

(mod $X^{t+1}$) for some matrix $E_t$. Respecting a restriction involving the degrees of the rows of $G_t$ (essentially we avoid destroying previously constructed zero rows in the $G$'s) we perform a Gaussian elimination on $E_t$, i.e., we obtain $\tilde{P}_t$ such that

$$\tilde{P}_t E_t = \begin{pmatrix} 0 \\ 1_m \end{pmatrix}.$$

Then we set

$$P_t = \begin{pmatrix} 1_n & 0 \\ 0 & 1_m X \end{pmatrix} \cdot \tilde{P}_t.$$

In this way the degrees of at most $m$ rows are increased when passing from $G_t$ to $G_{t+1}$ (due to the restriction mentioned above $\tilde{P}_t$ does not increase the degrees), so the total number of zero rows in the coefficients is increased by $n$. Due to the restriction mentioned above the degrees of the rows of $F_t$ and of $G_t$ grow almost uniformly, i.e., they grow on average by $\frac{m}{m+n}$ when going from $t$ to $t+1$.

After $t = \frac{d}{m} + \frac{d}{n} + O(1)$ steps the total number of zero rows in the coefficients of $G_t$ is $\frac{m+n}{m}d + O(1)$ such that we can select $m$ rows that produce at least $d$ zero rows in the coefficients. These $m$ rows form $F$ and $G$.

We now consider unbalanced sequence lengths. Let $\ell_j$ be the length of sequence $j$, i.e., $a_{j,k}^{(i)}$ has been computed for all $k$ and $0 \leq i \leq \ell_j$. Without loss of generality we can assume $\ell_1 \leq \ell_2 \leq \cdots \leq \ell_n = \ell$. The sum of the lengths of all sequences has to satisfy again $\sum_j \ell_j \geq d \cdot (1 + \frac{n}{m}) + O(1)$. Moreover we can assume that $\ell_1 \geq \frac{d}{m}$, otherwise we could drop sequence 1 completely, thus facilitating our task. In this setting our goal is to achieve

$$FA \equiv G \pmod{X^{\ell+O(1)}}$$

with $d_F = \deg(F) \leq \ell - \frac{d}{m}$, $\deg(G) \leq \ell - \frac{d}{m}$ and

$$X^{\ell-\ell_k} \mid F_{\cdot,k} \qquad \text{(this denotes the } k\text{th column of } F\text{)}.$$

The latter condition will compensate our ignorance of some rows of the higher coefficients of $A$. Indeed, setting for simplicity $d_F = \ell - \frac{d}{m}$, the vectors

$$v_j = \sum_{k=1}^{n} \sum_{i=0}^{\ell_k - \frac{d}{m}} f_{j,k}^{(d_F-i)} \cdot M^i y_k$$

satisfy for $1 \leq h \leq m, 0 \leq b < \frac{d}{m}$

$$(M^b x_h)^T v_j = \sum_{k=1}^{n} \sum_{i=0}^{\ell_k - \frac{d}{m}} f_{j,k}^{(d_F-i)} a_{k,h}^{(i+b)} = g_{j,h}^{(d_F+b)} = 0.$$

If $i+b > \ell_k$ (thus $a_{k,h}^{(i+b)}$ not being computed), we have $d_F - i < d_F + b - \ell_k \leq \ell - \ell_k$, so $f_{j,k}^{(d_F-i)} = 0$ and the sum computes $g_{j,h}^{(d_F+b)}$.

Our new goal is achieved as before, but we will need $\ell$ steps and the construction of $P_t$ has to be modified as follows. In step $t$ we have $F_t A \equiv G_t + E_t X^t$ (mod $X^{t+1}$). Let $a \leq n$ be maximal such that

$$\sum_{i=1}^{a-1}(m+i)(\ell_{n-i+1}-\ell_{n-i}) \le mt$$

($a$ will increase during the computation). In the Gaussian elimination of $E_t$ we do not use the first $n-a$ rows for elimination. As a consequence, $\tilde{P}_t$ has the form

$$\tilde{P}_t = \begin{pmatrix} 1_{n-a} & * \\ 0 & * \end{pmatrix}.$$

Then we set

$$P_t = \begin{pmatrix} 1_{n-a}X & 0 & 0 \\ 0 & 1_a & 0 \\ 0 & 0 & 1_mX \end{pmatrix} \cdot \tilde{P}_t.$$

Therefore the sum of the degrees of $F_t$ will be increased by $m+n-a$ and the number of zero rows in $G_t$ will be increased by $a$ when passing from $t$ to $t+1$. For a fixed $a$, $\frac{(m+a)(\ell_{n-a+1}-\ell_{n-a})}{m}$ steps will increase the average degree of the last $m+a$ rows from $\ell - \ell_{n-a+1}$ to $\ell - \ell_{n-a}$. At this point $a$ will be increased.

To see why $X^{\ell-\ell_k} \mid F_{\cdot,k}$ holds we have to describe the choice of $F_{t_0}$ (and $t_0$). Let $c$ be the number of maximal $\ell_j$, i.e., $\ell_{n-c} < \ell_{n-c+1} = \ell_n$. Then $F_{t_0}$ will be of the form

$$F_{t_0} = \begin{pmatrix} 1_{n-c}X^{t_0} & 0 \\ 0 & * \end{pmatrix}.$$

The last $m+c$ rows will be chosen such that they are of degree at most $t_0 - 1$ and such that the conditions in Coppersmith's algorithm are satisfied. This construction will lead to a value of $t_0$ near $\frac{m}{c}$ instead of the lower value near $\frac{m}{n}$ in the original algorithm.

Let $k$ be such that $\ell_k < \ell$. As long as $n - a \ge k$ the $k$th column of $F_t$ will have the only non-zero entry at row $k$ and this will be $X^t$. Since $n - a \ge k$ holds for $t \le \ell - \ell_k$ this column will be divisible by $X^{\ell-\ell_k}$ for all $t \ge \ell - \ell_k$.

For RSA-768 we used the algorithm as described above in the subquadratic version of Thomé. The following variant might be useful in certain situations, e.g., if one of the sequences is much longer than the others.

If $\ell_{n-1} < \ell_n$, then for $t < \frac{(m+1)(\ell_n-\ell_{n-1})}{m}$ we have $a = 1$ and $P_t$ is of the form

$$P_t = \begin{pmatrix} 1_{n-1}X & * \\ 0 & * \end{pmatrix}.$$

A product of several of these $P_t$ will have a similar form, namely an $(n-1)\times(n-1)$ unit matrix times a power of $X$ in the upper left corner and zeros below it.

The basic operations in Thomé's subquadratic version are building a binary product tree of these $P_t$ and doing truncated multiplications of intermediate products with $F_{t_0}A$ or similar polynomials. If we split the computation into two stages, first computing the product of all $P_t$ for $t < \frac{(m+1)(\ell_n-\ell_{n-1})}{m}$ and then the remaining product, the matrix multiplications in the first stage become easier due to the special form of the $P_t$ and its products.

Obviously this can be done in as many stages as there are different $\ell_j$-values.

# Correcting Errors in RSA Private Keys

Wilko Henecka, Alexander May*, and Alexander Meurer**

Horst Görtz Institute for IT-Security
Ruhr-University Bochum, Germany
Faculty of Mathematics
wilko.henecka@rub.de, alex.may@rub.de, alexander.meurer@rub.de

**Abstract.** Let $\mathsf{pk} = (N, e)$ be an RSA public key with corresponding secret key $\mathsf{sk} = (p, q, d, d_p, d_q, q_p^{-1})$. Assume that we obtain partial error-free information of $\mathsf{sk}$, e.g., assume that we obtain half of the most significant bits of $p$. Then there are well-known algorithms to recover the full secret key. As opposed to these algorithms that allow for *correcting erasures* of the key $\mathsf{sk}$, we present for the first time a heuristic probabilistic algorithm that is capable of *correcting errors* in $\mathsf{sk}$ provided that $e$ is small. That is, on input of a full but error-prone secret key $\widetilde{\mathsf{sk}}$ we reconstruct the original $\mathsf{sk}$ by correcting the faults.

More precisely, consider an error rate of $\delta \in [0, \frac{1}{2})$, where we flip each bit in $\mathsf{sk}$ with probability $\delta$ resulting in an erroneous key $\widetilde{\mathsf{sk}}$. Our Las-Vegas type algorithm allows to recover $\mathsf{sk}$ from $\widetilde{\mathsf{sk}}$ in expected time polynomial in $\log N$ with success probability close to 1, provided that $\delta < 0.237$. We also obtain a polynomial time Las-Vegas factorization algorithm for recovering the factorization $(p, q)$ from an erroneous version with error rate $\delta < 0.084$.

**Keywords:** RSA, error correction, statistical cryptanalysis.

## 1   Introduction

RSA is the most widely deployed cryptosystem and has successfully withstood more than 30 years of cryptanalytic attacks [1]. An RSA modulus $N = pq$ is a product of two primes and the key-pair $e, d \in \mathbb{Z}_{\phi(N)}^*$ satisfies $ed = 1 \bmod \phi(N)$. Although theoretically, it would suffice to use $(N, d)$ as the RSA private key it is recommended in PKCS#1 standard [10] to use the highly redundant tuple $(N, e, d, p, q, d_p, d_q, q_p^{-1})$ in order to also allow for a fast Chinese Remainder type decryption process. Here, the last three components of $\mathsf{sk}$ are defined as usual by $d_p = d \bmod p - 1$, $d_q = d \bmod q - 1$ and $q_p = q^{-1} \bmod p$.

In the present work, we look at error-prone RSA keys, where we assume that the public information $(N, e)$ is never affected by errors. Thus, we only look

---

* This research was supported by the German Research Foundation (DFG) as part of the project MA 2536/3-1 and by the European Commission through the ICT programme under contract ICT-2007-216676 ECRYPT II.
** This work was supported by the Ruhr-University Research School funded by Germanys Excellence Initiative [DFG GSC 98/1].

T. Rabin (Ed.): CRYPTO 2010, LNCS 6223, pp. 351–369, 2010.

at erroneous tuples $\mathsf{sk} = (p, q, d, d_p, d_q, q_p^{-1})$. Our error-correction algorithm is motivated by side-channel attacks that are capable of extracting the complete private key but with some errors [5]. We assume that the errors are uniformly spread over the whole secret key, i.e., each secret key bit is flipped with some fixed error probability $\delta \in [0, \frac{1}{2})$. Notice that for $\delta = \frac{1}{2}$ we obtain a completely random string that does not provide any information about $\mathsf{sk}$.

Theoretically, our attack framework is modeled by oracle-assisted attacks on RSA. Oracle-assisted attacks were first introduced by Rivest and Shamir [9] at Eurocrypt 1985. Rivest and Shamir used an oracle that allowed for querying bits of $p$ in chosen positions. They showed that given such an oracle $\frac{3}{5} \log p$ queries are sufficient to factor $N$ in polynomial time. This was later improved by Coppersmith [3] to only $\frac{1}{2} \log p$ queries. In 1992, Maurer [8] showed that for stronger oracles, which allow for any type of oracle queries with YES/NO answers, $\epsilon \log p$ queries are sufficient for any $\epsilon > 0$.

In this oracle-based line of research, the goal is to minimize both the power of the oracle and the number of queries to the oracle. At Crypto 2009, Heninger and Shacham [6] presented a polynomial time attack on RSA that works whenever a 0.27-fraction of the key bits of $\mathsf{sk}$ is given, provided that the given bits are uniformly spread over the whole secret key. So as opposed to the oracle used by Rivest, Shamir and Coppersmith, the attacker has no control about the positions in which he receives some of the bits but he knows the positions and on expectation the erasure intervals between two known bits are never too large.

Notice that all these aforementioned attacks require a limited number of *fault-free* information provided by the oracles, mostly in the form of secret key bits. Since side-channel attacks are practical instantiations of oracles, in most scenarios it is questionable to put a limit on the number of bits that one can obtain. Why should an attacker stop at some point to extract bits? Why should he not proceed until he has the full bit information? In a more realistic scenario an attacker is capable of extracting the full $\mathsf{sk}$ bit string but subject to some errors that were caused by the physical measurements of his side-channel attack. This is the *error-prone* scenario that we address in our paper. Hence our work might motivate to look for weaker forms of side-channel attacks that produce only erroneous data.

**Our result and related work.** We present the first attack running in expected polynomial time that recovers a secret key $\mathsf{sk}$ from a disturbed secret key $\widetilde{\mathsf{sk}}$, where every bit is flipped with a fixed error rate of $\delta < 0.237$. That is, we allow for *error correction* of an RSA secret key, provided that the public RSA exponent is small. We also give results where an attackers obtains an erroneous version for a subset of the entries of $sk = (p, q, d, d_p, d_q, q_p^{-1})$. E.g., we obtain a polynomial time attack for erroneous versions of $(p, q, d)$ with error rates $\delta < 0.160$. Moreover, we obtain a polynomial time factorization algorithm that factors $N$ given a faulty version of $(p, q)$ with error rate $\delta < 0.084$. In this case, we do not need any restriction on the public exponent $e$.

Our work builds on the *erasure correction* algorithm of Heninger-Shacham [6] which allows for erasures of the secret key bits of sk with an erasure rate of $\delta < 0.73$. So as one might expect from coding theory, the correction of errors seems to be a much harder problem than the correction of erasures.

As our work builds on the Heninger-Shacham algorithm, let us briefly recall the idea of this construction. Heninger and Shacham recover the parameters $p, q, d, d_p, d_q$ bit by bit in a 2-adic fashion by growing a search tree. In their algorithm, the information from $q_p^{-1}$ is ignored. The nodes in depth $k$ of the search tree correspond to partial solutions of $p \bmod 2^k, q \bmod 2^k, \ldots, d_q \bmod 2^k$.

Since in the erasure correction scenario, one has fragmentary but correct key material, one can easily prune partial solutions that do not coincide with the known secret key bits. This process will never discard the correct solution, since the correct solution will always fully agree with the incomplete key material. Thus, such an algorithm will always succeed to recover sk.

The only remaining problem is to bound the algorithm's running time. Intuitively, whenever one has sufficiently many key bits to falsify incorrect partial solutions, one will obtain a bounded number of false partial solutions per iteration and so the total number of nodes in the search tree will stay small. Heninger and Shacham showed that with high probability the total number of partial solutions is quadratic in $\log(N)$ whenever the erasure rate is smaller than 0.73, i.e., we know at least a 0.27-fraction of the key bits. In order to show this result, Heninger and Shacham had to make the heuristic assumption that once a key candidate differs from the correct key, the subsequent candidate key bits are distributed uniformly at random.

Clearly, the Heninger-Shacham comparison of key candidates with the given key material cannot naively been transferred to the error correction scenario. The reason is that a disagreement of a key candidate may originate from an incorrect key candidate *or* from faulty bits of the key material. Thus, in our construction we do no longer compare bit by bit but we compare larger blocks of bits. More precisely, we grow subtrees of depth $t$ for each key candidate. This results in $2^t$ new candidates which we all compare with our faulty key material. If the bit agreement with our key material in these $t$ bits is above some threshold parameter $C$ we keep the candidate, otherwise we discard it.

Clearly, we obtain a trade-off for the choice $t$ of the depth of our subtrees. On the one hand, $t$ cannot be chosen too large since in each iteration wae grow our search tree by at least $2^t$ candidates. Thus, $t$ must be bounded by $\mathcal{O}(\log \log N)$ in order to guarantee a polynomial size of the search tree.

On the other hand, depending on the error rate $t$ has to be chosen sufficiently large to guarantee that the correct key candidate has large agreement with our key material $\widetilde{\mathsf{sk}}$ in each $t$ successive bit positions, such that the correct candidate will never be discarded during the execution of our algorithm. Moreover, $t$ has to be large enough such that the distribution corresponding to the correct candidate and the distribution derived from an incorrect candidate are separable by some threshold parameter $C$. If this property does not hold, we obtain too many faulty candidates for the next iteration.

We show that the above trade-off restrictions can be fulfilled whenever we have an error rate $\delta < 0.237 - \epsilon$ for some fixed $\epsilon > 0$. That is, if we choose $t$ of size polynomial in $\log \log N$ and $\frac{1}{\epsilon}$, we are able to define a threshold parameter $C$ such that the following holds.

1. With probability close to 1 the correct key candidate will never be discarded during the execution of the algorithm.
2. For fixed $\epsilon > 0$, our algorithm will consider no more than an expected total number of $\log^{\mathcal{O}(1)} N$ key candidates. E.g., our algorithm has expected running time polynomial in the bit-size of $N$.

We would like to point out that our proper choice of $t$ and $C$ assumes that we know a good upper bound for the error rate $\delta$. In practical side-channel attacks where $\delta$ might be unknown to an attacker, one can apply an additional search step which successively increases the value of $\delta$ until a solution is found. Alternatively, we provide a way to compute an equate upper bound for $\delta$ during the initialization phase of the algorithm.

Our algorithm is a probabilistic algorithm of Las Vegas type, i.e., whenever it outputs a solution the output is the correct secret key sk. Our error correction algorithm is elementary. The main work that has to be done is to carefully choose the subtree depth $t$ and the threshold parameter $C$, such that all trade-off restrictions hold. We achieve this goal by using a statistical analysis via Hoeffding bounds. Our analysis relies on a similar heuristic assumption as in [6], that is, as soon as a key candidate differs from the correct solution its subsequent bits are distributed uniformly at random.

Furthermore, we would like to stress that analogous to [6], our algorithm is restricted to the case of small public exponents $e$ – except for the case of correcting erroneous factorizations $(p, q)$. Small public exponent RSA appears to be the standard in practical applications [11].

We ran experiments to verify the predictions of our theoretical analysis and to validate the heuristic assumption. In practice, we achieved to correct error rates of up to $\delta = 0.2$ for 1024-bit RSA private keys with good success probabilities in a matter of seconds.

The paper is organized as follows. In Section 3, we briefly review the Heninger-Shacham algorithm. Section 4 introduces our new block-based threshold algorithm that grows subtrees of depth $t$. Section 5 is devoted to the theoretical analysis of the subtree depth $t$ and the choice of the threshold parameter for pruning nodes that correspond to incorrect candidates. Experimental results are given in Section 6.

## 2   Notation and Mathematical Background

For an $n$-bit string $\mathbf{x} = (x_{n-1}, \ldots, x_0) \in \{0, 1\}^n$ let $x[i] = x_i$ denote the $i$-th bit of $\mathbf{x}$ (where $x[0]$ is the least significant bit of $\mathbf{x}$) and let $\mathbf{x}[i..j] = (x_i, x_{i-1}, \ldots, x_j)$ for $i \geq j$. Throughout the paper we denote by $\ln(n)$ the natural logarithm of $n$ to base $e$ and we denote by $\log(n)$ the binary logarithm of $n$ to base 2.

The main technical tool used in our analysis is Hoeffding's bound [7], which upper bounds the absolute error of sums of independent random variables from their mean value.

**Theorem 1.** *Let* $X_1, \ldots, X_n$ *be a sequence of independent Bernoulli trials with identical success probability* $\mathbf{Pr}[X_i = 1] = p$ *for all* $i$. *Define* $X := \sum_{i=1}^n X_i$. *Then for every* $0 < \gamma < 1$ *we have*

i)  $\mathbf{Pr}[X \geq n(p + \gamma)] \leq e^{-2n\gamma^2}$,
ii) $\mathbf{Pr}[X \leq n(p - \gamma)] \leq e^{-2n\gamma^2}$.

A slightly more general version of Hoeffding's inequality allows for each random variable $X_i$ an individual expectation $\mathbb{E}[X_i]$ as well as a wider range, i.e. $X_i \in [a, b]$ for $a, b \in \mathbb{R}$. We define $\mathbb{E}[X] = \sum_{i=1}^n \mathbb{E}[X_i]$ and the above statement transforms to

$$\mathbf{Pr}\left[X \gtrless \mathbb{E}[X] \pm n\gamma\right] \leq e^{-\frac{2n\gamma^2}{(b-a)^2}}. \tag{1}$$

## 3 The Heninger-Shacham Algorithm

Let $(N, e)$ be an RSA public key with corresponding PKCS#1 standard secret key $\mathsf{sk} = (p, q, d, d_p, d_q, q_p^{-1})$, where

$$ed = 1 \bmod \phi(N), d_p = d \bmod p - 1, d_q = d \bmod q - 1 \text{ and } q_p^{-1} = q^{-1} \bmod p.$$

We will ignore the last parameter $q_p^{-1}$ as it is not used in the attack. Let $N$ be the product of two $\frac{n}{2}$-bit primes, i.e., all the secret key parameters except $d$ can be represented by $\frac{n}{2}$ bits. The Heninger-Shacham algorithm recovers these parameters bit by bit starting from the least significant bit until bit $\frac{n}{2} - 1$, where the factorization is revealed. Although by a result of Coppersmith [3] an amount of $\frac{n}{4}$ bits would suffice for factoring $N$ in polynomial time, going up to bit $\frac{n}{2} - 1$ instead does not significantly change the algorithm's analysis.

It is not hard to see that all parameters $p, q, d, d_p, d_q$ alone reveal the factorization of $N$, see [4]. Thus, the secret key is a highly redundant representation of the prime factorization. This redundancy in turn implies that the following four RSA identities simultaneously hold

$$N = pq \tag{2}$$
$$ed = 1 + k\phi(N) \tag{3}$$
$$ed_p = 1 + k_p(p - 1) \tag{4}$$
$$ed_q = 1 + k_q(q - 1), \tag{5}$$

for some parameters $k$, $k_p$ and $k_q$ that we are able to compute for small public exponents $e$.

We have $0 < k < e\frac{d}{\phi(N)} < e$, so there are at most $e - 1$ possible candidates for $k$. Therefore, we can brute-force search over all candidate values for $k$. Following an argument of Boneh, Durfee and Frankel [2], for each candidate value $k'$, we define

$$d(k') = \left\lfloor \frac{1 + k'(N+1)}{e} \right\rfloor, \tag{6}$$

which differs for the right choice $k' = k$ from $d$ by $\frac{k(p+q)}{e} < p + q$. Thus, for the right candidate choice of $k$ the values of $d(k)$ and $d$ agree roughly on half of their most significant bits.

In the *erasure correction* scenario, Heninger and Shacham simply compare each candidate $d(k')$ with the given fragmentary version of $d$ in order to determine $k$ uniquely with overwhelming probability.

We proceed similarly in the *error correction* szenario. Assume that we obtain some error-prone secret key

$$\widetilde{\mathsf{sk}} := (\tilde{p}, \tilde{q}, \tilde{d}, \tilde{d}_p, \tilde{d}_q),$$

which is derived from $\mathsf{sk}$ by flipping each bit individually with some fixed probability $\delta \in [0, \frac{1}{2})$. Intuitively, if $\delta$ is significantly below $\frac{1}{2}$, then among all $e - 1$ candidates $d(k'), k' = 1, \ldots, e - 1$, the Hamming distance between the upper half most significant bits of $d(k')$ and $\tilde{d}$ should be minimal for the correct choice $k' = k$. In Appendix A, we show that this is true with overwhelming probability for the error rates $\delta$ that we allow.

This means that we can learn the unknown $k$ in Eq. (3). Moreover, we can immediately correct almost half of the most significant bits of $d$. Notice that this information is not useful in the Heninger-Shacham algorithm as one stops to recover the secret key bits when reaching bit $\frac{n}{2} - 1$. However, we can use this information to compute a good approximation $\tilde{\delta}$ of the error rate $\delta$. Therefore, we simply compute the normalized Hamming distance of $d(k)$ and $\tilde{d}$ by

$$\tilde{\delta} := \frac{2}{n} \sum_{i=n/2}^{n-1} \tilde{d}[i] \oplus d(k)[i]. \tag{7}$$

For $n$ large enough and any fixed tolerance $\epsilon > 0$, we have $\delta \leq \tilde{\delta} + \epsilon$ with overwhelming probability. That is, in our asymptotic analysis it is reasonable to assume that the algorithm knows an upper bound of the error rate $\delta$. For practical values of $n$, one can easily show that

$$\mathbf{Pr}[\delta < \tilde{\delta} + \epsilon] \geq \frac{3}{4}$$

where $\epsilon = 0.037$ for $n = 1024$, see App. B for more details.

Now that we are able to compute $k$, Heninger and Shacham show that this directly allows us to compute candidates for $(k_p, k_q)$. If $e$ is prime then there are only two candidate values. In general, for $e$ with $m$ different prime factors there exist up to $2^m$ candidates. So one has to run $2^m$ copies of the Heninger-Shacham algorithm in parallel. Since $m = \mathcal{O}(\log e)$ and since we consider small public exponent RSA only, this factor can be neglected. We denote this whole precomputation process by $(k, k_p, k_q) \leftarrow \mathsf{Init}(N, e)$.

Now let us start to reconstruct a secret key in a bitwise manner. Since $p, q$ are odd primes, we have $p[0] = q[0] = 1$ and $2|p-1$ as well as $2|q-1$. Let $\tau(x)$ denote the largest exponent such that $2^{\tau(x)}$ divides $x$, i.e. $\tau(x) := \max\{k \in \mathbb{N} : 2^k|x\}$. From Eq. (4), we obtain

$$ed_p = 1 \bmod 2^{1+\tau(k_p)}.$$

Thus, we can immediately correct the least $1 + \tau(k_p)$ bits of $d_p$ from the knowledge of $e$ and $k_p$. Analogously, we can compute from Eq. (5) the $1 + \tau(k_q)$ least significant bits of $d_q$ and from Eq. (3) the $2 + \tau(k)$ least significant bits of $d$.

Moreover, if we fix all bits $p[i - 1..0]$ then changing bit $p[i]$ will change bit $d_p[i + \tau(k_p)]$. For odd $k_p$ this means that changing the $i$-th bit on the right hand side of Eq. (4) changes the corresponding $i$-th bit on the left hand side. Shifting by $\tau(k_p)$ on the right-hand side translates the change to position $i + \tau(k_p)$ on the left hand side.

Thus, Heninger and Shacham define for each bit index $i$ a so-called $i$-th bit slices, which we denote by

$$\mathsf{Slice}(i) := (p[i], q[i], d[i + \tau(k)], d_p[i + \tau(k_p)], d_q[i + \tau(k_q)]).$$

Let $\mathsf{Slice}(0) \leftarrow \mathsf{Mount}(e, k, k_p, k_q)$ be the computation of the initial first bit slice consisting of the steps described above, i.e., we set

$$\mathsf{Slice}(0) \leftarrow (1, 1, d[\tau(k)], d_p[\tau(k_p)], d_q[\tau(k_q)]),$$

where the last three entries can be easily computed once $k$, $k_p$ and $k_q$ are known. The running time of $\mathsf{Mount}(\cdot)$ can be neglected in our small public exponent RSA scenario.

**Lifting solutions.** Assume that we have computed a partial solution $\mathsf{sk}' = (p', q', d', d_p', d_q')$ up to $\mathsf{Slice}(i - 1)$. We would like to proceed by calculating all candidate solutions $(p, q, d, d_p, d_q)$ for the subsequent $\mathsf{Slice}(i)$. Heninger and Shacham show that by applying a multivariate version of Hensel's Lemma to Eq. (2)-(5) one obtains the following identities

$$p[i] + q[i] = (N - p'q')[i] \bmod 2 \tag{8}$$
$$d[i + \tau(k)] + p[i] + q[i] = (k(N + 1) + 1 - k(p' + q') - ed')[i + \tau(k)] \bmod 2 \tag{9}$$
$$d_p[i + \tau(k_p)] + p[i] = (k_p(p' - 1) + 1 - ed_p')[i + \tau(k_p)] \bmod 2 \tag{10}$$
$$d_q[i + \tau(k_q)] + q[i] = (k_q(q' - 1) + 1 - ed_q')[i + \tau(k_q)] \bmod 2. \tag{11}$$

This means we have four linearly independent equations in the five unknowns $p[i], q[i], d[i + \tau(k)], d_p[i + \tau(k_p)], d_q[i + \tau(k_q)]$ of $\mathsf{Slice}(i)$. Therefore, each Hensel lift yields exactly two candidate solutions. We denote this lifting process by $\mathsf{Expand}(p', q', d', d_p', d_q')$.

In the *erasure correction* scenario, Heninger and Shacham use their knowledge of the correct secret key bits to prune incorrect candidates produced by the lifting process. The analysis in [6] mainly shows that the number of candidates is sufficiently upper bounded as long as enough secret key bits are available.

Notice that in our *error correction* scenario, such a simple pruning is not possible, since a disagreement of $\mathsf{Slice}(i)$ with the corresponding bits of $\mathsf{sk}$ might be due to errors in our faulty secret key.

# 4     Blockwise Threshold-Based Vector Correction

## 4.1     Generic Description

In this section, we present our new algorithm for error correction. We would like to point out that our construction is a generic, elementary algorithm for reconstructing arbitrary *unknown* tuples of bit vectors $\mathbf{x}$ given only a corrupted version $\tilde{\mathbf{x}}$ and some public information on $\mathbf{x}$, which does not directly allow for extracting $\mathbf{x}$. For example, $\mathbf{x}$ may be the prime factorization of some public $N$.

In coding theory language, our construction resembles a maximum likelihood approach. In each iteration, we keep those vectors that are locally closest to $\tilde{\mathbf{x}}$ in the Hamming distance. Hopefully, we are also able to discard many incorrect partial solutions due to our public information.

Let $\mathbf{x} = (\mathbf{x_1}, \ldots, \mathbf{x_m})$. In a nutshell, our algorithm tries to reconstruct $\mathbf{x}$ iteratively by calculating a block of $t$ bits of each of $\mathbf{x_1}, \ldots, \mathbf{x_m}$ in each iteration. The algorithm proceeds in four phases, where the second and third phase are iterated until the candidates have the same bitlength as $\mathbf{x}$.

***Initialization phase:*** Use the public information to compute some initial partial solution to $\mathbf{x}$. This initialization is optional and may result in the empty string as the only partial solution.

***Expansion phase:*** Each partial solution is lifted for the next $t$ most significant bits, i.e., we compute the next $t$ bits of each of $\mathbf{x_1}, \ldots, \mathbf{x_m}$. Per partial solution this will result in up to $2^{mt}$ new partial solutions. By using our public information, we may hope to obtain considerably less than $2^{mt}$ candidates.

***Pruning phase:*** For every new partial solution we count the number of matches of the $mt$ expanded bits with the corresponding bits of $\tilde{\mathbf{x}}$. If this number is below some threshold parameter $C$ then we discard the partial solution.

***Finalization phase:*** We test with the help of our public information whether one of our candidate solutions is indeed equal to the desired $\mathbf{x}$.

Obviously, the choice of the blocksize $t$ is crucial for our algorithm. Since the number of partial solutions in the expansion phase grows exponentially in $t$, we cannot allow for large parameters $t$. On the other hand, we cannot choose $t$ too small, because we have to make sure that during the pruning phase the following two properties hold.

- The correct partial solution – the one that can be expanded to the desired **x** – is pruned only with small probability.
- Sufficiently many incorrect solutions are pruned such that the total number of candidates can be minimized.

## 4.2  Error Correction for RSA Keys

Let us now specialize the generic description from the previous section to our RSA error correction scenario. We want to compute some *unknown* RSA secret key $\mathsf{sk} = (p, q, d, d_p, d_q)$ from an erroneous version $\widetilde{\mathsf{sk}} = (\tilde{p}, \tilde{q}, \tilde{d}, \tilde{d}_p \tilde{d}_q)$ with the help of the public key $(N, e)$. For describing our algorithm, we use the notion introduced in Sect. 3.

**Algorithm** ERROR-CORRECTION
INPUT: $(N, e)$, erroneus $\widetilde{\mathsf{sk}} = (\tilde{p}, \tilde{q}, \tilde{d}, \tilde{d}_p, \tilde{d}_q)$ with error rate $\delta$

| |
|---|
| **Initialization phase:**<br>• $(k, k_p, k_q) \leftarrow \mathsf{Init}(N, e)$<br>• $\mathsf{Slice}(0) \leftarrow \mathsf{Mount}(e, k, k_p, k_q)$ |
| **For** $i = 1$ **to** $\left\lceil \frac{n/2-1}{t} \right\rceil$ |
|     **Expansion phase:** For every candidate $(p', q', d', d'_p, d'_q)$ with slices $0 \ldots (i-1)t$ expand the candidate $t$ times with the $\mathsf{Expand}(\cdot)$ procedure of Heninger-Shacham. This results in $2^t$ new candidates which differ in the slices $(i-1)t+1, \ldots, it$. |
|     **Pruning phase:** For every new candidate $(p', q', d', d'_p, d'_q)$ count the number of bits in the expanded slices $(i-1)t+1, \ldots, it$ that agree with the corresponding bits of $\widetilde{\mathsf{sk}}$. If this number is below some threshold parameter $C$, discard the solution. |
| **Finalization phase:** For every candidate $\mathsf{sk}' = (p', q', d', d'_p, d'_q)$ check all RSA identities (2)–(5). If all equations hold, output $\mathsf{sk}'$. |

OUTPUT: $\mathsf{sk} = (p, q, d, d_p, d_q)$

Notice that during the *Expansion phase* for every partial solution we only obtain $2^t$ new candidates for the $5t$ new bits instead of the naive $2^{5t}$ candidates. This is due to the clever usage of our public information in the $\mathsf{Expand}(\cdot)$ procedure of Heninger and Shacham.

In the subsequent section, we will analyze the probability that our algorithm succeeds in computing the secret key $\mathsf{sk}$. We will show that a choice of $t = \theta(\frac{\ln n}{\epsilon^2})$ will be sufficient for error rates $\delta < 0.237 - \epsilon$. The threshold parameter $C$ will be chosen such that the correct partial solution will survive each pruning phase with probability close to 1 and such that we expect that the number of candidates per iteration is bounded by $2^{t+1}$. For every fixed $\epsilon > 0$, this leads to an expected running time that is polynomial in $n$.

# 5    Choice of Parameters and Success / Runtime Analysis

We now give a detailed analysis for algorithm ERROR-CORRECTION from the previous section. Afterwards, we show that this analysis easily generalizes to settings where an attacker obtains instead of a faulty version of all five parameters in sk only faulty versions of e.g. $(p, q, d)$ or $(p, q)$.

## 5.1    Full Analysis for the RSA Case

Remember that in algorithm ERROR-CORRECTION, we count the number of matching bits between $5t$-bit blocks of $\widetilde{\text{sk}}$ and every partial candidate solution. Let us define a random variable $X_c$ for the number of matching bits between $\widetilde{\text{sk}}$ and a correct partial solution.

The distribution of $X_c$ is clearly the binomial distribution with parameters $5t$ and probability $(1 - \delta)$, denoted by $X_c \sim \text{Bin}(5t, 1 - \delta)$. That is, we have

$$\mathbf{Pr}[X_c = \gamma] = \binom{5t}{\gamma}(1 - \delta)^\gamma \delta^{5t-\gamma} \qquad (12)$$

for $\gamma = 0, \ldots, 5t$. The expected number of matches is thus $\mathbb{E}[X_c] = 5t(1 - \delta)$.

Assume that we expand some incorrect partial solution $(p', q', d', d'_p, d'_q)$ by $5t$ bits to $2^t$ new candidates. We let the random variable $X_b$ represent the number of matching bits of $\widetilde{\text{sk}}$ with the expanded $5t$ bits of these bad candidates.

In order to analyze the distribution of $X_b$, we make use of the following heuristic assumption which follows directly from the heuristic assumption of Heninger-Shacham [6] when applied to $t$-bit blocks.

**Heuristic 2.** *Every solution generated by applying the expansion phase to an incorrect partial solution is an ensemble of $t$ randomly chosen bit slices.*

That is under Heuristic 2, every expansion of an incorrect candidate in ERROR-CORRECTION results in an additional $5t$ uniformly random bits.

Heninger and Shacham verified the validity of this heuristic experimentally. Under Heuristic 2 we see that

$$\mathbf{Pr}[X_b = \gamma] = \binom{5t}{\gamma}2^{-5t}. \qquad (13)$$

Now, we basically have to choose our threshold $C$ such that the two distributions are sufficiently separated.

The remainder of this section is devoted to proof our main result.

**Main Theorem 3.** *Under Heuristic 2 for every fixed $\epsilon > 0$ the following holds. Let $(N, e)$ be an RSA public key with n-bit $N$ and fixed $e$. We choose*

$$t = \left\lceil \frac{\ln(n)}{10\epsilon^2} \right\rceil, \; \gamma_0 = \sqrt{(1 + \frac{1}{t}) \cdot \frac{\ln(2)}{10}} \; and \; C = 5t(\frac{1}{2} + \gamma_0).$$

*Further, let $\widetilde{sk} = (\tilde{p}, \tilde{q}, \tilde{d}, \tilde{d}_p, \tilde{d}_q)$ be an RSA secret key with noise rate*

$$\delta \leq \frac{1}{2} - \gamma_0 - \epsilon.$$

*Then algorithm* ERROR-CORRECTION *corrects $\widetilde{sk}$ in expected time $\mathcal{O}\left(n^{2 + \frac{\ln(2)}{5\epsilon^2}}\right)$ with success probability at least $1 - \left(\frac{5\epsilon^2}{\ln(n)} + \frac{1}{n}\right)$.*

**Remark.** Notice that for sufficiently large $n$, $t$ converges to infinity and thus $\gamma_0$ converges to $\sqrt{\frac{\ln(2)}{10}} \approx 0.263$. This means that ERROR-CORRECTION asymptotically allows for error rates $\frac{1}{2} - \gamma_0 - \epsilon \approx 0.237 - \epsilon$ and succeeds with probability close to 1.

*Proof.* The proof of our main theorem is organized as follows. First, we upper bound the expected number of bad solutions that arise in each iteration of the algorithm. Second, we show that our correct solution survives all pruning steps with probability close to 1. Third, we upper bound the total number of partial solutions that arise during the execution of ERROR-CORRECTION and conclude that ERROR-CORRECTION runs in polynomial time.

Let the random variables $Y_i$ represent the number of incorrect partial solutions that pass the threshold comparison in the pruning phase of the $i$-th iteration of ERROR-CORRECTION. Further, let the random variable $Y = \sum_{i=1}^{\tau(n)} Y_i$ denote the total number of incorrect solutions examined by ERROR-CORRECTION, where $\tau(n) := \left\lceil \frac{n/2 - 1}{t} \right\rceil$ denotes the total number of iterations.

**Lemma 4.** *The expected number of bad candidates that pass the i-th round's pruning phase is upper bounded by $\mathbb{E}[Y_i] < 2^{t+1}$.*

*Proof.* Define two random variables $Z_g$ and $Z_b$ as follows: $Z_g$ denotes the number of bad candidates arising from the unique correct solution, $Z_b$ counts the number of bad candidates generated from a single bad partial solution. It is not hard to see that

$$\mathbb{E}[Y_1] = \mathbb{E}[Z_g] \; and \; \mathbb{E}[Y_2] = \mathbb{E}[Z_g] + \mathbb{E}[Z_b] \cdot \mathbb{E}[Y_1] = \mathbb{E}[Z_g] \cdot (1 + \mathbb{E}[Z_b]).$$

More generally, we obtain

$$\mathbb{E}[Y_i] = \mathbb{E}[Z_g] + \mathbb{E}[Z_b] \cdot \mathbb{E}[Y_{i-1}] = \mathbb{E}[Z_g] + \mathbb{E}[Z_b] \cdot (\mathbb{E}[Z_g] + \mathbb{E}[Z_b] \cdot \mathbb{E}[Y_{i-2}])$$

$$= \ldots = \mathbb{E}[Z_g] \sum_{k=0}^{i-1} \mathbb{E}[Z_b]^k = \mathbb{E}[Z_g] \frac{1 - \mathbb{E}[Z_b]^i}{1 - \mathbb{E}[Z_b]}.$$

Now, we aim at upper bounding $\mathbb{E}[Z_b] < 1$ in order to upper bound

$$\mathbb{E}[Y_i] = \mathbb{E}[Z_g]\frac{1 - \mathbb{E}[Z_b]^i}{1 - \mathbb{E}[Z_b]} < \frac{\mathbb{E}[Z_g]}{1 - \mathbb{E}[Z_b]}. \tag{14}$$

Therefore, we define $2^t$ random variables $Z_b^i$ for $i = 1, \ldots, 2^t$ such that

$$Z_b^i = \begin{cases} 1 & i\text{-th bad candidate passes} \\ 0 & \text{otherwise} \end{cases}.$$

Write $Z_b = \sum_{i=1}^{2^t} Z_b^i$. Since all the $Z_b^i$ are identically distributed, we simplify this to $\mathbb{E}[Z_b] = 2^t \, \mathbb{E}[Z_b^i]$ and upper bound $\mathbb{E}[Z_b^i]$ for some fixed $i$. Note, that $Z_b^i = 1$ iff at least $C$ bits match, i.e.,

$$\mathbb{E}[Z_b^i] = \mathbf{Pr}[Z_b^i = 1] = \mathbf{Pr}[X_b \geq C],$$

where $X_b \sim \text{Bin}(5t, \frac{1}{2})$ is defined as in (13). Applying Hoeffding's bound (Theorem 1) directly yields

$$\mathbf{Pr}[X_b \geq C] = \mathbf{Pr}\left[X_b \geq 5t\left(\frac{1}{2} + \gamma_0\right)\right]$$
$$\leq \exp(-10t\gamma_0^2) = 2^{-(1+\frac{1}{t})t} \leq 2^{-(t+1)}.$$

This implies $\mathbb{E}[Z_b] \leq \frac{1}{2} < 1$ and we can simplify equation (14) to

$$\mathbb{E}[Y_i] < \frac{\mathbb{E}[Z_g]}{1 - \mathbb{E}[Z_b]} < 2^{t+1},$$

since we clearly have $\mathbb{E}[Z_g] \leq 2^t - 1$. $\qquad \square$

**Lemma 5.** ERROR-CORRECTION *succeeds with probability at least* $1 - \left(\frac{5\epsilon^2}{\ln(n)} + \frac{1}{n}\right)$.

*Proof.* The probability of pruning the correct solution at one single round is given by $\mathbf{Pr}[X_c < C]$, where $X_c \sim \text{Bin}(5t, 1 - \delta)$ as defined in (12). Using $\frac{1}{2} + \gamma_0 \leq 1 - \delta - \epsilon$ and applying Hoeffding's bound (Theorem 1) yields

$$\mathbf{Pr}[X_c < C] = \mathbf{Pr}\left[X_c < 5t\left(\frac{1}{2} + \gamma_0\right)\right] \leq \mathbf{Pr}\left[X_c < 5t(1 - \delta - \epsilon)\right]$$
$$\leq \exp(-10t\epsilon^2) \leq \frac{1}{n}.$$

Since algorithm ERROR-CORRECTION runs $\tau(n) \leq \frac{n}{2t} + 1$ rounds, the total success probability is given by

$$\mathbf{Pr}[\text{success}] = (1 - \mathbf{Pr}[X_c < C])^{\tau(n)} \geq \left(1 - \frac{1}{n}\right)^{\tau(n)} \geq 1 - \frac{\tau(n)}{n}$$
$$\geq 1 - \left(\frac{1}{2t} + \frac{1}{n}\right) = 1 - \left(\frac{5\epsilon^2}{\ln(n)} + \frac{1}{n}\right).$$

$\qquad \square$

**Lemma 6.** ERROR-CORRECTION *runs in expected time* $\mathcal{O}\left(n^{2+\frac{\ln(2)}{5\epsilon^2}}\right)$.

*Proof.* The total expected runtime $T$ of ERROR-CORRECTION is given by

$$T = T_{\mathsf{Init}} + \mathcal{O}(e) \cdot (T_{\mathsf{Mount}} + T_{\mathsf{main}})$$

where $T_{\mathsf{Init}}$, $T_{\mathsf{Mount}}$ and $T_{\mathsf{main}}$ represent the runtime of the procedures Init, Mount and the main loop of ERROR-CORRECTION, respectively. Recall that a factor of $\mathcal{O}(e)$ arises from the fact that $\mathsf{Init}(\cdot)$ possibly outputs up to $e$ candidate tuples $(k, k_p, k_q)$. Since we assume $e$ to be fixed, we can neglect $T_{\mathsf{Init}}$ as well as $T_{\mathsf{Mount}}$ and obtain $T = \mathcal{O}(T_{\mathsf{main}})$.

In order to upper bound $T_{\mathsf{main}}$, we upper bound the runtime needed by the expansion and pruning phase for one *single* partial solution:

- During the expanding phase, each partial solution implies the computation of $\sum_{i=0}^{t-1} 2^i < 2^t$ equation systems given by the equations (8)-(11). The right hand sides of equations (8)-(11) can be computed in time $\mathcal{O}(n)$ – when storing the results of the previous iteration. This yields a total computation time of $\mathcal{O}(n2^t)$ for the expanding phase.
- The pruning phase can be realized in time $\mathcal{O}(t)$ for each of the fresh $2^t$ partial solutions, summing up to $\mathcal{O}(t2^t)$.

We can upper bound $t \leq n$, which results in an overall runtime of $\mathcal{O}((n+t)\cdot 2^t) = \mathcal{O}(n2^t)$ per candidate.

An application of Lemma 4 yields an upper bound for the expected total number of partial solutions examined during the whole execution which is given by

$$\mathbb{E}[Y] = \sum_{i=1}^{\tau(n)} \mathbb{E}[Y_i] < \tau(n) \cdot 2^{t+1} \leq \left(\frac{n}{2t}+1\right) \cdot 2^{t+1} = \mathcal{O}\left(n2^t\right).$$

Putting both together finally yields

$$T_{\mathsf{main}} = \mathcal{O}\left(n2^t \cdot n2^t\right) = \mathcal{O}\left(n^2 2^{2t}\right) = \mathcal{O}\left(n^{2+\frac{\ln(2)}{5\epsilon^2}}\right).$$

$\square$

Combining Lemma 5 and 6 proves the Main Theorem.     $\square$

Although theoretically Lemma 6 gives us a polynomial running time for every fixed $\epsilon > 0$, our running time heavily depends on the parameter $t$ and thus on $\epsilon$. So one might expect that in practice one cannot achieve error rates close to the theoretical bound $\delta < 0.237$ since the running time already explodes for moderately small error terms $\epsilon$.

However, we give in Appendix C a more refined analysis of the parameter $t$ for moderately small $\epsilon$. This analysis shows that our choice of $t$ in Theorem 3 is quite conservative, since we insist on a success probability of ERROR CORRECTION close to 1. We obtain more flexibility if we also allow for smaller success rates. This in turn leads to a smaller choice of $t$, which allows to easily correct error rates up to $\delta = 0.2$ in practice. We will use this refined analysis in the experimental section (Section 6).

## 5.2 Generalization

We now formulate a slightly generalized version of our Main Theorem 3. Therefore, we parametrize algorithm ERROR-CORRECTION such that it allows for a secret key with $m$ components like in the generic description in Sect. 4.1.

So our RSA secret key $\mathsf{sk} = (p, q, d, d_p, d_q)$ resembles the parameter choice $m = 5$. We can apply the same analysis as in Section 5.1. The distributions of $X_c$ and $X_b$ are now given by $X_c \sim \mathrm{Bin}(mt, 1 - \delta)$ and $X_b \sim \mathrm{Bin}(mt, \frac{1}{2})$.

**Main Theorem 7.** *Under Heuristic 2 for every fixed $\epsilon > 0$ the following holds. Let $(N, e)$ be an RSA public key with $n$-bit $N$ and fixed $e$. We choose*

$$t = \left\lceil \frac{\ln(n)}{2m\epsilon^2} \right\rceil, \ \gamma_0 = \sqrt{(1 + \frac{1}{t}) \cdot \frac{\ln(2)}{2m}} \text{ and } C = mt(\frac{1}{2} + \gamma_0).$$

*Further, let $\widetilde{\mathsf{sk}} = (\widetilde{\mathsf{sk}}_1, \ldots, \widetilde{\mathsf{sk}}_m)$ be a generic RSA secret key with noise rate*

$$\delta \leq \frac{1}{2} - \gamma_0 - \epsilon.$$

*Then algorithm ERROR-CORRECTION corrects $\widetilde{\mathsf{sk}}$ in expected time $\mathcal{O}\left(n^{2 + \frac{\ln(2)}{m\epsilon^2}}\right)$ with success probability at least $1 - \left(\frac{m\epsilon^2}{\ln(n)} + \frac{1}{n}\right)$.*

As a consequence we obtain various results for scenarios where an attacker obtains an erroneous subset of the parameters in $\mathsf{sk} = (p, q, d, d_p, d_q)$. The resulting upper bounds for the error rates $\delta = \frac{1}{2} - \gamma_0$ are summarized in the following table. In the column "Equations" we indicate which of the Eqs. (8)-(11) are used.

**Table 1.** Parameters for varied RSA scenarios

| sk | m | Equations | $\delta$ |
|---|---|---|---|
| $(p, q)$ | 2 | (8) | 0.084 |
| $(p, q, d)$ | 3 | (8),(9) | 0.160 |
| $(p, q, d, d_p)$ | 4 | (8)-(10) | 0.206 |
| $(p, q, d, d_q)$ | 4 | (8),(9),(11) | 0.206 |
| $(p, q, d, d_p, d_q)$ | 5 | (8)-(11) | 0.237 |

The case $\mathsf{sk} = (p, q, d_p)$ can also be handled by our algorithm by using Eqs. (8),(10) with parameter $m = 3$. The only problem is that we cannot derive $k$ and therefore compute $k_p$ as described in Sect. 3, since we do not have information of $d$. Instead, we simply run $e - 1$ copies of the algorithm in parallel for each possible choice of $1 \leq k < e$.

## 6    Implementation and Experiments

We implemented our algorithm in Java and tested it on an Intel Xeon Quad-Core processor at 2.66 GHz with 8 GB of DDR2 SDRAM at 800 MHz. In all experiments we set the public exponent to $e = 2^{16} + 1$. For the case sk $=$ $(p, q, d, d_p, d_q)$ we ran a large number of experiments for a key size of 1024 bit and error rates $\delta \in [0.05, 0.2]$. We also carried out experiments for the scenarios sk $= (p, q)$ and sk $= (p, q, d)$ where we made experiments for different error rates up to the upper bounds presented in Table 1.

In each repetition, the RSA secret key was independently and randomly disturbed with error rate $\delta$. For simplicity, we omitted the mounting phase, i.e., the calculation of $k$ as well as $k_q$ and $k_p$. Thereby, we avoided to choose the wrong assignment for $k_q$ and $k_p$. Instead we just used the correct values for these parameters.

The choice of our tree depth $t$ roughly followed the refined analysis in Appendix C, where we made some manual adjustments for very small error rates and for $\delta \geq 0.18$. The threshold parameter $C$ was chosen as recommended in Theorem 3 with some rounding. All manual adjustments were made in order to obtain comparability of our experiments, i.e., we slightly tuned to achieve success probabilities in an interval between 20% and 50%. We point out that for small error rates it is easy to achieve much better success probabilities by a small increase of the parameter $t$.

For each experiment we generated 100 different RSA secret keys and disturbed each of these keys with 100 different error vectors resulting in a total sample size of 10.000 runs per error rate $\delta$.

The tables below summarize our results. We computed the success probability by calculating the term $\mathbf{Pr}[X_c < C]$ as defined in Eq. (12) exactly for the given parameters (row "$\mathbf{Pr}$ theoretical"). The experimental results perfectly match the exact calculations. In the last row, we give the average running time of algorithm ERROR-CORRECTION in order to reconstruct a *single key* successfully.

**Table 2.** Experimental results for $n = 1024$ and sk $= (p, q, d, d_p, d_q)$

| $\delta$ | 0.05 | 0.06 | 0.07 | 0.08 | 0.09 | 0.10 | 0.11 | 0.12 | 0.13 | 0.14 | 0.15 | 0.16 | 0.17 | 0.18 | 0.19 | 0.20 |
|---|---|---|---|---|---|---|---|---|---|---|---|---|---|---|---|---|
| $t$ | 3 | 4 | 5 | 6 | 7 | 9 | 9 | 10 | 10 | 11 | 12 | 12 | 13 | 16 | 16 | 20 |
| $C$ | 12 | 16 | 20 | 24 | 28 | 36 | 36 | 39 | 39 | 42 | 46 | 45 | 48 | 59 | 59 | 74 |
| Pr theoretical | 0.39 | 0.48 | 0.51 | 0.49 | 0.44 | 0.50 | 0.27 | 0.49 | 0.28 | 0.44 | 0.28 | 0.35 | 0.43 | 0.47 | 0.26 | 0.23 |
| experimental | 0.40 | 0.48 | 0.52 | 0.50 | 0.45 | 0.51 | 0.27 | 0.50 | 0.28 | 0.45 | 0.28 | 0.35 | 0.44 | 0.50 | 0.24 | 0.21 |
| time | < 1s | | | | ... | | | | | | | | < 1s | 3.7s | 23s 25s | 3m |

For error rates $\delta \leq 0.15$ we can easily achieve better success probabilities by using a slightly larger $t$, e.g., for $\delta = 0.15$ we experimentally achieved $\mathbf{Pr}[\text{success}] \approx 82\%$ with a modified choice $t = 15$ and $C = 56$.

For each run, we also recorded the total number of partial solutions examined by ERROR-CORRECTION. The following boxplot diagram represents the statistics of the total number of candidates. The thick horizontal line marks the median, the gray boxes describe the region bounded by the lower quartile Q1 and the upper quartile Q3, i.e., half of the candidate numbers fall in this intervall. The dashed lines mark the sample minimum and maximum, respectively.

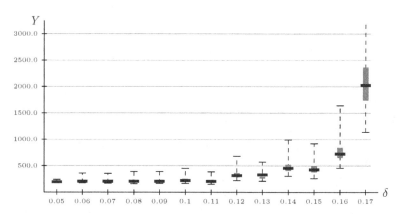

**Fig. 1.** Box plot diagram for 1024 bit key size and $\mathsf{sk} = (p, q, d, d_p, d_q)$

**Table 3.** Experimental results for $n = 1024$ and $\mathsf{sk} = (p, q, d)$

| $\delta$ | 0.05 | 0.06 | 0.07 | 0.08 | 0.09 | 0.10 | 0.11 | 0.12 | 0.13 | 0.14 |
|---|---|---|---|---|---|---|---|---|---|---|
| $t$ | 3 | 5 | 7 | 9 | 11 | 13 | 16 | 20 | 26 | 29 |
| $C$ | 7 | 12 | 17 | 22 | 27 | 32 | 39 | 49 | 64 | 71 |
| **Pr** theoretical | 0.24 | 0.34 | 0.38 | 0.36 | 0.30 | 0.23 | 0.33 | 0.26 | 0.21 | 0.17 |
| experimental | 0.24 | 0.34 | 0.38 | 0.36 | 0.30 | 0.25 | 0.34 | 0.24 | 0.21 | 0.15 |
| time | < 1s | | | | ... | | < 1s | 1.7s | 4.1s | 32.2s | 3m |

**Table 4.** Experimental results for $n = 1024$ and $\mathsf{sk} = (p, q)$

| $\delta$ | 0.01 | 0.02 | 0.03 | 0.04 | 0.05 | 0.06 | 0.07 | 0.08 |
|---|---|---|---|---|---|---|---|---|
| $t$ | 4 | 7 | 7 | 11 | 15 | 20 | 24 | 28 |
| $C$ | 7 | 12 | 12 | 19 | 26 | 35 | 42 | 49 |
| **Pr** theoretical | 0.70 | 0.83 | 0.56 | 0.61 | 0.58 | 0.45 | 0.34 | 0.24 |
| experimental | 0.71 | 0.84 | 0.57 | 0.62 | 0.58 | 0.47 | 0.35 | 0.22 |
| time | < 1s | | | ... | | < 1s | 2s | 12.7s |

In our experiments for error rates $\delta \leq 0.15$, we always examined around 300 candidates on the average and the maximum number of candidates never exceeded 1000 candidates. We omit the box plot for error rates $\delta \geq 0.18$ since the number of candidates increases rapidly beyond this bound. This is where the exponential dependence of our running time $\mathcal{O}\left(n^{2+\frac{\ln(2)}{5\epsilon^2}}\right)$ on the parameter $\epsilon$ comes into play.

**Acknowledgement.** The authors thank the anonymous CRYPTO reviewers for their comments, in particular for suggesting a way to approximate the error rate $\delta$.

# References

1. Boneh, D.: Twenty years of attacks on the rsa cryptosystem. Notices of the American Mathematical Society (AMS) 46(2), 203–213 (1999)
2. Boneh, D., Durfee, G., Frankel, Y.: An attack on RSA given a small fraction of the private key bits. In: Ohta, K., Pei, D. (eds.) ASIACRYPT 1998. LNCS, vol. 1514, pp. 25–34. Springer, Heidelberg (1998)
3. Coppersmith, D.: Small solutions to polynomial equations, and low exponent rsa vulnerabilities. J. Cryptology 10(4), 233–260 (1997)
4. Coron, J.-S., May, A.: Deterministic polynomial-time equivalence of computing the rsa secret key and factoring. J. Cryptology 20(1), 39–50 (2007)
5. Halderman, J.A., Schoen, S.D., Heninger, N., Clarkson, W., Paul, W., Calandrino, J.A., Feldman, A.J., Appelbaum, J., Felten, E.W.: Lest we remember: Cold boot attacks on encryption keys. In: van Oorschot, P.C. (ed.) USENIX Security Symposium, pp. 45–60. USENIX Association (2008)
6. Heninger, N., Shacham, H.: Reconstructing rsa private keys from random key bits. In: Halevi, S. (ed.) CRYPTO 2009. LNCS, vol. 5677, pp. 1–17. Springer, Heidelberg (2009)
7. Hoeffding, W.: Probability inequalities for sums of bounded random variables. Journal of the American Statistical Association 58(301), 13–30 (1963)
8. Maurer, U.M.: Factoring with an oracle. In: Rueppel, R.A. (ed.) EUROCRYPT 1992. LNCS, vol. 658, pp. 429–436. Springer, Heidelberg (1993)
9. Rivest, R.L., Shamir, A.: Efficient factoring based on partial information. In: Pichler, F. (ed.) EUROCRYPT 1985. LNCS, vol. 219, pp. 31–34. Springer, Heidelberg (1986)
10. RSA Laboratories. PKCS #1 v2.1: RSA Cryptography Standard (June 2002)
11. Yilek, S., Rescorla, E., Shacham, H., Enright, B., Savage, S.: When private keys are public: Results from the 2008 Debian OpenSSL vulnerability. In: Feldmann, A., Mathy, L. (eds.) Proceedings of IMC 2009, pp. 15–27. ACM Press, New York (November 2009)

# A    Mounting the Attack

Recall that for generating the initial $\mathsf{Slice}(0)$ one has to determine the correct $k$ in (3). We proposed in Sect. 3 to compute the $e - 1$ candidates $d(k')$ for $0 < k' < e$ as defined in (6) and chose the $k$ whose corresponding $d(k)$ has minimal Hamming distance to the error prone $\tilde{d}$.

We now give a formal justification for our claim that the Hamming distance between the error prone key $\tilde{d}$ and one of the candidates $d(k')$ is minimized for the correct $d(k)$ with probability close to 1. Therefore, we define random variables $X(k')$ counting the number of matching bits between $\tilde{d}$ and a fixed $d(k')$ on their $\alpha := \lfloor n/2 \rfloor - 2$ most significant bits. For every $k' \neq k$ let

$$D(k') := X(k) - X(k')$$

denote the gap of matching bits for the correct $d(k)$ and a fixed $d(k')$ in their window of $\alpha$ most significant bits. We aim to derive a lower bound for $\mathbf{Pr}[D(k') > 0]$ for arbitrary $k' \neq k$.

The main observation is that for the correct $k$ and balanced $p$ and $q$, we have $0 \leq d(k) < p+q < 3\sqrt{N}$. This implies that $d(k)$ agrees with the correct $d$ on at least $\alpha$ most significant bits. On the contrary, for $k' \neq k$ one obtains that $d(k')$ and $d$ agree on at most $\log(e)$ most significant bits. Notice that one can consider $D(k')$ as a sum of $\alpha$ random variables $D(k')_{n-i}$ where $i = 1, \ldots, \alpha$, each taking values in $\{-1, 0, 1\}$ representing the following three cases.

1. $D(k')_{n-i} = 1$ if $d(k)[n-i]$ and $\tilde{d}[n-i]$ do match *but* $d(k')[n-i]$ and $\tilde{d}[n-i]$ do not match.
2. $D(k')_{n-i} = 0$ if both $d(k)[n-i]$ and $d(k')[n-i]$ match with $\tilde{d}[n-i]$.
3. $D(k')_{n-i} = -1$ if $d(k)[n-i]$ and $\tilde{d}[n-i]$ do not match *but* $d(k')[n-i]$ and $\tilde{d}[n-i]$ do match.

Assuming that in the case $k' \neq k$ every bit of $d(k')$ and $\tilde{d}$ except for the $(n-\log(e))^{\text{th}}$ most significant bits matches with probability $\frac{1}{2}$, we obtain

$$\mathbb{E}[D(k')_{n-i}] = (1-\delta)\frac{1}{2} - \delta\frac{1}{2} = \frac{1}{2}(1-2\delta)$$

for $i = \log(e) + 1, \ldots, \alpha$. Summing over all $i$ yields

$$\mathbb{E}[D(k')] \geq \frac{(\alpha - \log(e))(1-2\delta)}{2}.$$

An application of the generalized Hoeffding inequality from (1) yields

$$\mathbf{Pr}[D(k') > 0] = 1 - \mathbf{Pr}[D(k') \leq 0] \geq 1 - \exp\left(-\frac{(\alpha - \log(e))^2(1-2\delta)^2}{8\alpha}\right)$$

for arbitrary $k' \neq k$. Hence, we can lower bound the probability of the event that $D(k') > 0$ for *every* $k' \neq k$ by taking this expression to the $(e-2)^{\text{th}}$ power. For fixed $e$ and $\delta \ll \frac{1}{2}$ we asymptotically achieve probability 1 since the exponent converges to $-\infty$. We calculated the probability for our experimental parameters $n = 1024$, $e = 2^{16} + 1\}$ and the theoretical upper bound $\delta = 0.237$. In this case the probability is very close to 1.

## B    Estimating the Error Rate

Recall the definition of $\tilde{\delta} := \frac{2}{n}\sum_{i=n/2}^{n-1} \tilde{d}[i] \oplus d(k)[i]$ from (7). We estimate the quality of $\tilde{\delta} + \epsilon$ as an upper bound for $\delta$, when we allow for an arbitrary small buffer $\epsilon > 0$. This can easily be done by regarding $\tilde{\delta}$ as a sum of $\frac{n}{2}$ random variables

$$D[i] := \tilde{d}[i] \oplus d(k)[i].$$

Notice that $\mathbf{Pr}[D[i] = 1] = \delta$ since $d(k)$ coincides with the correct secret key $d$ on its $\frac{n}{2}$ most significant bits. Applying Hoeffdings inequality yields

$$\mathbf{Pr}[\delta < \tilde{\delta} + \epsilon] = 1 - \mathbf{Pr}[\tilde{\delta} \leq \delta - \epsilon] \geq 1 - \exp\left(-2\epsilon^2\frac{n}{2}\right) = 1 - \exp(-n\epsilon^2),$$

i.e., for arbitrary fixed $\epsilon > 0$ and large enough $n$ we can use $\tilde{\delta} + \epsilon$ as a reasonable upper bound for $\delta$.

## C    Practical Choice of $t$

We give a slightly refined analysis of the parameter $t$ in order to obtain some flexibility in tuning the success probability $p$ of ERROR-CORRECTION. Therefore, we define a scaling parameter $\alpha := -5/\ln(p)$ and modify the choice of $t$ to

$$t := \frac{\ln\alpha + \ln n - \ln\ln n + 2\ln\epsilon}{10\epsilon^2}, \tag{15}$$

while keeping $C := 5t(\frac{1}{2} + \gamma_0)$ as proposed in Lemma 3. The following calculation which follows the proof of Lemma 5 derives a lower bound for the success probability depending on the additional parameter $\alpha$.

$$\mathbf{Pr}[\text{success}] \approx (1 - \mathbf{Pr}[X_c < C])^{n/2t} \geq \left(1 - e^{-10t\epsilon^2}\right)^{n/2t} = \left(1 - \frac{\ln n}{\alpha \cdot n \cdot \epsilon^2}\right)^{n/2t}$$

$$\approx e^{-\frac{5\ln n}{\alpha(\ln\alpha + \ln n - \ln\ln n + 2\ln\epsilon)}} \xrightarrow{n\to\infty} e^{-5/\alpha} = p$$

The above approximation should be taken with some care for very small $\epsilon$. Notice that our approximation gets tight for fixed $\epsilon$ and sufficiently large $n$. However, when we use it with realistic RSA values of $n \in [1024, .., 8192]$ the asymptotics of the Hoeffding bound do not yet apply for very small $\epsilon$.

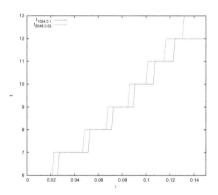

**Fig. 2.** Choice of $t_{n,p}$ for different $n$ and $p$ according to Eq. (15)

As one can see, the denominator in the exponent of $e$ yields a non-negativity restriction $\ln n + \ln\alpha - \ln\ln n + 2\ln\epsilon > 0$. This restriction simplifies to

$$\epsilon > \sqrt{\frac{\ln n}{n \cdot \alpha}},$$

e.g. for $n = 1024$ and $p = 0.1$ one obtains $\epsilon > 0.056$. Concerning our experiments, we modified our choice of $t$ manually when $\delta \geq 0.18$, i.e., when $\epsilon$ fell below 0.06.

# Improved Differential Attacks
# for ECHO and Grøstl

Thomas Peyrin

Ingenico, France
thomas.peyrin@ingenico.com

**Abstract.** We present improved cryptanalysis of two second-round SHA-3 candidates: the AES-based hash functions ECHO and Grøstl. We explain methods for building better differential trails for ECHO by increasing the granularity of the truncated differential paths previously considered. In the case of Grøstl, we describe a new technique, the *internal differential attack*, which shows that when using parallel computations designers should also consider the differential security between the parallel branches. Then, we exploit the recently introduced start-from-the-middle or Super-Sbox attacks, that proved to be very efficient when attacking AES-like permutations, to achieve a very efficient utilization of the available freedom degrees. Finally, we obtain the best known attacks so far for both ECHO and Grøstl. In particular, we are able to mount a distinguishing attack for the full Grøstl-256 compression function.

**Keywords:** hash function, cryptanalysis, ECHO, Grøstl, AES, internal differential attack.

## 1 Introduction

Cryptographic hash functions are very important tools in cryptography, used in many applications such as digital signatures, authentication schemes or message integrity. Informally, a hash function $H$ is a function that takes an arbitrarily long message as input and outputs a fixed-length hash value of size $n$ bits. The classical security requirements for such a function are collision resistance and (second)-preimage resistance. Namely, it should be impossible for an adversary to find a collision (two distinct messages that lead to the same hash value) in less than $2^{n/2}$ hash computations, or a (second)-preimage (a message hashing to a given challenge) in less than $2^n$ hash computations. Moreover, those primitives are traditionally used to simulate the behavior of a random oracle [2] and while the community is divided on such a requirement, in the ideal case an attacker should not be able to distinguish a hash function from a random oracle.

As many standardized hash functions [41, 31] have been broken a few years ago [45, 44], the NIST launched in 2008 the SHA-3 competition [33] that will lead to the future hash function standard. 14 candidates among 51 have been selected for the second round and many of them (like ECHO [3], Grøstl [14] or SHAvite-3 [5]) are actually using some parts of the standardized block cipher

T. Rabin (Ed.): CRYPTO 2010, LNCS 6223, pp. 370–392, 2010.

AES [32, 10] as internal primitives or mimicking the structure of this cipher. While AES-256 can no more be considered as secure in the related-key model [7], the cryptography community has made important progresses concerning the evaluation of AES-based hash functions security [35, 19, 27, 25, 23, 26, 15]. Those attacks make an extensive use of the freedom degrees that are available in a hash function and even provides the best known distinguishing attack against AES-128 [15] in the known-key model [21, 30]. Much recent analysis of AES-based hash functions has helped to identify the limits of current techniques, but as we show in this paper, it is possible to improve the differential path building methods used so far.

**Our contributions.** In this paper, we improve the best known cryptanalysis results [1, 18, 27, 26, 15] on two second round SHA-3 candidates: the hash functions ECHO [3] and Grøstl [14]. While we do not provide advances regarding the freedom degrees optimization, we use the recently introduced Super-Sbox techniques [15, 28] in order to find pairs of inputs verifying a given differential path. We then exploit some specific properties of ECHO and Grøstl to derive very good differential paths. More precisely, we improve the previously known truncated differential paths for ECHO by reducing the size of the truncated words considered. This allows us to broaden the differential trail search space, therefore increasing our probability to find a good path, but also augmenting the search complexity. We circumvent this constraint by giving a heuristic method to prune the potential candidates. Concerning Grøstl, we describe a novel yet simple cryptanalysis technique: the *internal differential attack*. It may be applied for functions using parallel branches that are not sufficiently made distinct. In that case, the attacker can find input instances (where a classical differential attack exhibits pairs of inputs) verifying non random properties on the output.

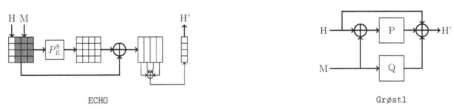

ECHO                                Grøstl

As a result, we improve the complexity for distinguishing the internal permutation of ECHO from a random 2048-bit permutation for a number of rounds corresponding to the full 256-bit version. Because of the folding phase after the permutation application at the end of the ECHO compression function, this attack does not translate into a distinguishing attack for the full ECHO compression function, nor the hash function itself. We provide also the first distinguishing attack on the full internal permutations for the 256-bit version of Grøstl, which can be directly derived into a distinguisher on the full Grøstl-256 compression function. Structural distinguishers (independent of the number of rounds) were already described in the original submission document [14]. For example, it was already identified that one can find fixed points or build a distinguisher for the

**Table 1.** Summary of results for ECHO, ECHO-SP and Grøstl compression functions. ECHO-256, ECHO-SP-256, ECHO-512 and ECHO-SP-512 compression functions have 8, 8, 10 and 10 rounds respectively, while Grøstl-256 and Grøstl-512 compression functions have 10 and 14 rounds respectively. All details of these attacks are given in the extended version of this article [36].

| target | rounds | computational complexity | memory requirements | type | section |
|--------|--------|--------------------------|---------------------|------|---------|
| ECHO-256 | 3 | $2^{64}$ | $2^{64}$ | semi-free-start collision[1] | this paper |
| comp. function | 4 | $2^{64}$ | $2^{64}$ | distinguisher | this paper |
| ECHO-512 | 3 | $2^{96}$ | $2^{64}$ | semi-free-start collision[1] | this paper |
| comp. function | 6 | $2^{96}$ | $2^{64}$ | distinguisher | this paper |
| ECHO-SP-256 | 3 | $2^{64}$ | $2^{64}$ | semi-free-start collision | this paper |
| comp. function | 3 | $2^{64}$ | $2^{64}$ | distinguisher | this paper |
| ECHO-SP-512 | 3 | $2^{64}$ | $2^{64}$ | semi-free-start collision[2] | this paper |
| comp. function | 4 | $2^{64}$ | $2^{64}$ | distinguisher | this paper |
| | 7 | $2^{56}$ | | distinguisher | see [26] |
| Grøstl-256 | 8 | $2^{112}$ | $2^{64}$ | distinguisher | see [15] |
| comp. function | 9 | $2^{80}$ | $2^{64}$ | distinguisher | this paper |
| | 10 | $2^{192}$ | $2^{64}$ | distinguisher | this paper |
| Grøstl-512 comp. function | 11 | $2^{640}$ | $2^{64}$ | distinguisher | this paper |

compression function with the generalized birthday paradox [43]. However, our results also allow to distinguish the Grøstl compression function from the same construction when assuming the two internal permutations $P$ and $Q$ as ideal. This is not the case for the known structural distinguishers since they already consider the two internal permutations as ideal. Our results are also interesting because they exploit the specificities of $P$ and $Q$ which is essential in order to really evaluate the security margin of this hash function in terms of number of rounds. All the results and the corresponding computation/memory complexities for ECHO, ECHO-SP (the simple-pipe version of ECHO) and Grøstl are summarized in Table 1 and available in the extended version of this article [36]. Note that none of the results described in this article seem to endanger the security of the ECHO compression function or the Grøstl hash function.

## 2   Previous Cryptanalysis

In this section, we recall the recent advances regarding cryptanalysis of AES-like permutations and their specificities. In the rest of the paper, we will use the start-from-the-middle and Super-Sbox attacks as basic tool for finding input pairs verifying a given differential path.

---

[1] A semi-free-start collision can be found for the 4-round reduced ECHO-256 or ECHO-512 compression functions with complexity $2^{224}$ computations and $2^{64}$ memory, if the salt value can be controlled by the attacker.

[2] A semi-free-start collision can be found for the 4-round reduced ECHO-SP-512 compression function with complexity $2^{224}$ computations and $2^{64}$ memory, if the salt value can be controlled by the attacker.

## 2.1  Building Differential Trails with Truncated Differences

Cryptanalysis of AES-based hash functions began with the hash family proposal Grindahl [20] for which collision attacks have been found [35, 19]. This showed that truncated differentials [22, 20] are very useful when cryptanalyzing a byte-oriented primitive such as the AES. Namely, instead of looking at the actual difference value of a byte, one only checks if a byte contains a difference (active byte) or not (inactive byte). In particular, this allows the attacker to handle the non-linear Sboxes quite nicely when building differential trails. On the other hand, the differential transitions through the linear MixColumns layer will now be verified probabilistically.

The matrix multiplication underlying the MixColumns transformation on a $r$-byte column for AES or Grøstl presents the interesting property of being a Maximum-Distance Separable (MDS) mapping: the number of active input and output bytes is always greater or equal to $r + 1$ (unless there is no active input and output byte at all). When picking random inputs, the probability of success of a differential transition that meets the MDS constraints through a MixColumns layer is determined by the number of active bytes in the output. More precisely, if such a differential transition contains $k$ active bytes in one column of the output, its probability of success will approximatively be equal to $2^{-8 \times (r-k)}$. For example, a $4 \mapsto 1$ transition for one column of the AES Mix-Columns layer has success probability of approximatively $2^{-24}$. Note that the same reasoning applies when dealing with the invert function of the MixColumns layer as well.

## 2.2  Rebound Attacks

The rebound attack [27] is a new technique for using efficiently the available freedom degrees. The authors utilize truncated differential paths in which most of the cost lies in the middle rounds. Then, by using a local meet-in-the-middle-like technique, the freedom degrees are consumed in the middle part of the differential path, right where they can improve at best the overall complexity. More precisely, some rounds in the middle (the *controlled rounds*) will be verified with only a few operations on average, while the rest of the path both in forward and backward direction (the *uncontrolled rounds*) is fulfilled probabilistically. This cryptanalysis provides good results [25, 23] and can work without any special constraint on the differential path. However, the controlled part is limited to two rounds.

## 2.3  Start-from-the-Middle Attacks

In [26], the start-from-the-middle attack for AES-like permutations is introduced. It can be seen as a generalization of the previous technique in the sense that the idea is simply to use the freedom degrees for AES-like permutations in the "most expensive" part of the differential trail, without setting any constraint in the way this is handled. The "cheaper" parts are then covered in an inside-out manner in both forward and backward directions. The authors describe a freedom degrees use example that can control 3 rounds in the middle part, without increasing the complexity (i.e. with only a few operations). However, the depicted technique only works for specific differential paths, in which the number of active bytes in the controlled rounds is not too important. We refer to the original publication [26] for more details.

### 2.4  The Super-Sbox Cryptanalysis Technique

Finally, another example of start-from-the-middle attacks is the Super-Sbox cryptanalysis ([15] and independently published in [28]). The idea is that one can view two rounds of an AES-like permutation as the parallel application of a layer of big Sboxes, named Super-Sboxes, preceded and followed by simple affine transformations. This technique can control 3-rounds in the middle of the differential trail with only a few operations on average, but works especially when the number of active bytes in the controlled rounds is important (this allowed to use longer differential paths which generally contain more active bytes). Because of some local precomputation steps, the drawback of this technique is its memory requirement when the size of the internal state of the scheme is too big. In the case of Grøstl this remains acceptable with a $2^{64}$ memory requirement, but in the case of ECHO as much as $2^{512}$ memory is required, making this tool unsuitable for this hash proposal. We refer to the original article [15] for more details.

## 3    Improved Differential Attack for ECHO

### 3.1   Description of ECHO

ECHO is a double-pipe hash function using HAIFA [4] as chaining iteration mode. The message to hash is first padded and divided into fixed-length blocks $M_i$ which are used to update iteratively the chaining variable $H_i$ (originally initialized with an initial vector $H_0 = IV$) thanks to the compression function $h$: $H_i = h(H_{i-1}, M_i)$. Finally, the hash output is obtained by truncating the last chaining variable. The compression function is built upon a 2048-bit AES-like permutation $P_E^R$ composed of $R$ rounds and its internal state can be viewed as a $4 \times 4$ matrix of 128-bit words (or cells). A cell will be denoted by $C_{i,j}$, where $i$ is its row position and $j$ its column position in the matrix, starting the counting from 0. One round of $P_E^R$ is composed of three layers: the "BIG SubBytes" layer (big Sbox or B.SB), the "BIG ShiftRows" layer (B.ShR) and the "BIG MixColumns" layer (B.MC).

The BIG SubBytes layer is a non-linear function defined by the application of a big Sbox $S$ on each 128-bit cell and this big Sbox is made of 2 AES rounds. The classical AddRoundKey part from the AES is not present in $P_E^R$ and in order to avoid trivial symmetric vulnerabilities that would occur, each big Sbox in ECHO is distinct thanks to different subkey additions in each of the 2-round AES uses. The first round subkey depends on the value of a 64-bit internal counter $K$ that is different at each use, while the second round subkey is set to the 128-bit salt value and thus always remains the same during the whole ECHO computation. So, for each cell $C_{i,j}$ of the internal state, we compute

$$C'_{i,j} = S[C_{i,j}] = AES_{salt}(AES_{0||K}(C_{i,j})).$$

where $AES_{sk}$ denotes the application of one AES round with the subkey $sk$. As for the AES, the BIG ShiftRows transformation permutes the position of each cell in its own row: for each cell $C_{i,j}$ of the internal state, we compute $C'_{i,j} = C_{i,Sub_i(j)}$ where $Sub_i(j) = (j - i) \bmod 4$. Finally, the BIG MixColumns function is a linear function that mixes all the columns of the internal state separately. More precisely, the 32-bit AES MixColumns function is reused: if $C_{i,j}^b$ denotes the $b$-th byte of the cell $C_{i,j}$, then we compute

$$(C'^b_{0,j}, C'^b_{1,j}, C'^b_{2,j}, C'^b_{3,j}) = AESMixColumns(C_{0,j}^b, C_{1,j}^b, C_{2,j}^b, C_{3,j}^b)$$

for all $0 \leq j \leq 3$ and $1 \leq b \leq 16$. The round function on an internal state $C$ can thus be defined as:

$$MixColumns \circ ShiftRows \circ SubBytes(C).$$

In the case of the ECHO-256 compression function, 8 rounds of the permutation are applied and a folding phase is processed after the final feedforward. Namely, the folding phase (denoted $\mathsf{fold}_{256}$) xors all the four 512-bit columns together. Finally, the compression function takes a 1536-bit message input $M$ (12 words) and a 512-bit chaining variable input $H$ (4 words) and outputs a new 512-bit chaining variable $H'$ with

$$H' = \mathsf{fold}_{256}(P_E^8(H||M) \oplus H||M)$$

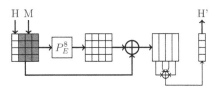

In the case of the ECHO-512 compression function, 10 rounds of the permutation are applied in order to turn a 1024-bit message input $M$ (8 words) and a 1024-bit chaining variable input $H$ (8 words) onto a new 1024-bit chaining variable $H'$. A different folding phase is processed after the final feedforward. Namely, the folding phase (denoted $\mathsf{fold}_{512}$) xors the two first and the two last 512-bit columns together.

$$H' = \mathsf{fold}_{512}(P_E^{10}(H||M) \oplus H||M)$$

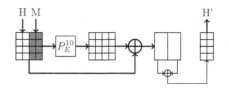

Since ECHO is a nested design of AES-like permutations, we will always use the prefix "BIG" when referring to one of the three layers of the 2048-bit permutation. When not using a prefix, we will refer to the layers of the 2-round AES permutation in the big Sboxes of ECHO.

In the following, $\mathsf{B.SB}_R^{in}$ (respectively $\mathsf{B.SB}_R^{out}$) will denote the whole internal state just before (respectively just after) application of the BIG SubBytes layer during round $R$ (starting the counting from 0). Similarly, $\mathsf{B.MC}_R^{in}$ and $\mathsf{B.MC}_R^{out}$ will stand for the input and output internal states of the BIG MixColumns layer during round $R$. Of course, we have $\mathsf{B.SB}_R^{in} = \mathsf{B.MC}_{R-1}^{out}$. We refer to [3] for the full specifications.

### 3.2 Generic Differential Paths

In order to fully use the power of recent freedom degrees optimization techniques, the core of the differential path we use will not differ from the ones described in [27, 26, 15]. The reason here is that this core characteristic is perfectly fit for using the available freedom degrees in the middle: it is computationally very costly in its middle part, but quite cheap on its side parts. This core truncated differential path is 7 rounds long and is depicted in Figure 1. Of course, when attacking a smaller number of rounds than 7, one can use this core to build a further reduced path by cutting off some of the first and/or last rounds.

The second advantage of this core characteristic is that the relatively low number of active cells in the controlled rounds makes it usable with the start-from-the-middle technique, as it is described in [26]: one can find a pair of internal states verifying the 128-bit truncated differential trail from the beginning of round 2 ($\mathsf{B.SB}_2^{in}$) up to the end of round 4 ($\mathsf{B.MC}_4^{out}$) with only one operation on average (and $2^{64}$ memory). Note that another view of the attack is to say that with one operation the attacker can find a pair of internal states such

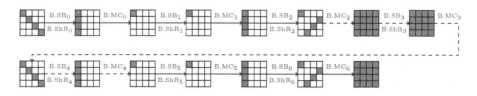

**Fig. 1.** Core of the truncated differential paths for 7-round reduced ECHO internal permutation. Each cell represents a 128-bit word and a gray cell stands for an active 128-bit word. The controlled rounds are depicted with dashed lines.

that the difference on $B.SB_2^{out}$ and on $B.MC_4^{out}$ are chosen (no more truncated differentials). Therefore, for ECHO we consider that the controlled rounds go from $B.SB_2^{out}$ up to $B.MC_4^{out}$.

One can easily check that the rest of the path (the uncontrolled rounds) is fulfilled with probability one, except round 1. Indeed, in round 1, a $4 \Rightarrow 1$ truncated differential transition is expected through the backward computation of the BIG MixColumns layer $B.MC_1$. When dealing with 128-bit truncated differentials, this will happen with approximate probability $2^{-24 \times 16} = 2^{-384}$ (i.e. a $4 \Rightarrow 1$ byte-wise truncated differential transition is expected through sixteen parallel $AESMixColumns$ functions) and this probability sets the overall $2^{384}$ complexity for finding a valid pair for the core path from Figure 1. We will see in the next section that by looking at byte-wise truncated differentials (instead of word-wise), one can sharpen the differential path and improve the success probability of this BIG MixColumns layer. On the other side, in order to be able to use the byte-wise truncated differentials at this stage and since he can control the difference only in $B.SB_2^{out}$ (and not in $B.SB_2^{in}$), the attacker will have to handle the backward computation of the BIG SubBytes layer of round 2 ($B.SB_2$) as well. He then hopes that controlling both $B.SB_2$ and $B.MC_1$ with byte-wise truncated differentials will cost less than $2^{384}$ operations. Not controlling $B.SB_2$ would lead us back to the 128-bit truncated differential cryptanalysis, as each active 128-bit word of $B.SB_2^{in}$ will very likely contain 16 active bytes (i.e. fully active word) since full diffusion is achieved with only two AES rounds.

### 3.3   Differential Transitions for 2 AES Rounds

Now that we introduced the core of the differential path, we need to study the word-wise differential transitions. That is, instead of looking for 128-bit truncated differentials, we will look at byte-wise truncated differentials. Of course, we still fully leverage the previous works on start-from-the-middle attacks [26]: the attacker can find a valid candidate pair verifying the controlled rounds and fully control the differences in $B.SB_2^{out}$ and $B.MC_4^{out}$ with one operation on average. Sharpening the differential path will improve the results since our scope is now wider, but it will also greatly increase the number of potential trails and complicate the analysis. For that reason, we need to heuristically filter them so that we place our search into a good subspace. First, we restrict ourselves to four types of byte-wise truncated differential words F, C, D and 1, all depicted in Figure 2. Due to symmetry and diffusion considerations, we believe that analyzing other differentials would not provide better results, while it would greatly increase the search space. Secondly, we add the constraint that all the active 128-bit words in an internal state will present the same byte-wise truncated differential (all words have the same truncated differential types F, C, D or 1). This seems a sound constraint as the processing of the BIG MixColumns layer on one word column of the internal state can be seen as the parallel application of sixteen $AESMixColumns$ functions (one for each byte position). Thus, for each word column, instead of analyzing the behavior of sixteen parallel $AESMixColumns$

**Fig. 2.** Byte-wise truncated differential states for one word of ECHO. Each cell represents a byte and a gray cell stands for an active byte.

functions one conceptually only handles a single function that will do for all the 16. Those two filters will really simplify the analysis.

Since the attacker will have to control the behavior of BIG SubBytes layer B.SB$_2$, we have to study the success probability for each possible transition for 2 AES rounds between the four bit-wise truncated differentials F, C, D and 1, especially in backward direction. First, we can compute the approximate probability of success for a one-round transition between those four 128-bit differential states and this is given in Table 2 for both forward and backward directions. Those probabilities are simply obtained by studying the $AESMixColumns$ transitions for one AES round (since we are dealing with byte-wise truncated differentials, all the probabilities comes only from the $AESMixColumns$ transitions, see [35]).

When computing backward through B.SB$_2$, the $AESMixColumns$ function from the second AES round is the first function to invert. But since this layer is fully linear, one can verify the expected backward transitions by carefully choosing the differences in B.SB$_2^{out}$ beforehand. Since the start-from-the-middle attack allows us such a liberty, the second AES round in B.SB$_2$ comes for free (one only has to check that the transition is not impossible, i.e. the probability in Table 2 is not null). Finally, having set all the constraints and the cost evaluation, we only have to pick the best backward differential transition through B.SB$_2$ in terms of probability and active byte weight: D $\Leftarrow$ 1 $\Leftarrow$ C. The transition D $\Leftarrow$ 1 is free as showed by Table 2, while the $2^{-24}$ probability for the transition 1 $\Leftarrow$ C is not taken in account since we can avoid it by carefully choosing the byte-wise truncated differences in B.SB$_2^{out}$ beforehand. Therefore, controlling B.SB$_2$ is now completely free for the attacker.

Now that we controlled the differential behavior of B.SB$_2$, what is the improvement obtained for the BIG MixColumns layer B.MC$_1$ ? Since we only have four active bytes in D, we can focus on controlling 4 parallel $AESMixColumns$ transitions instead of 16. We are looking for $4 \Rightarrow 1$ transitions, each happening with probability $2^{-24}$. Thus, for the whole BIG MixColumns layer, we get a probability of $2^{-24 \times 4} = 2^{-96}$ and this has to be compared to the previous $2^{-24 \times 16} = 2^{-384}$ probability.

Overall the whole 7-round differential path is depicted in Figure 3 and a valid candidate can be found with complexity $2^{96}$ operations and $2^{64}$ memory. Since the internal permutation of ECHO is much bigger than its hash output size, it should be easy to distinguish it from a random 2048-bit permutation. Note that our solution pair has four active 128-bit words in the input and four active 128-bit words in the output (the last BIG MixColumns call is not taken in account

**Table 2.** Byte-wise truncated differential transition approximated probabilities for one round of AES. The left table shows forward transitions, while the right one gives backward transitions.

| Forward | | | | | | Backward | | | | |
|---|---|---|---|---|---|---|---|---|---|---|
| in \ out | F | C | D | 1 | | in \ out | F | C | D | 1 |
| F | 1 | 0 | $2^{-96}$ | 0 | | F | 1 | $2^{-96}$ | 0 | 0 |
| C | 1 | 0 | 0 | 0 | | C | 0 | 0 | 1 | $2^{-24}$ |
| D | 0 | 1 | 0 | $2^{-24}$ | | D | 1 | 0 | 0 | 0 |
| 1 | 0 | 1 | 0 | 0 | | 1 | 0 | 0 | 1 | 0 |

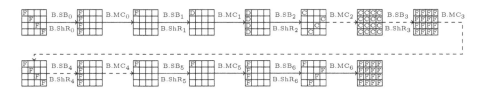

**Fig. 3.** 7-round differential path for the ECHO internal permutation. The controlled rounds are depicted with dashed lines.

since it is fully linear). A naive analysis would conclude that for a random 2048-bit permutation, finding such a pair with a birthday paradox technique should require at least $2^{(2048-512)/2} = 2^{768}$ operations. However, since the input and output amount of differences is low, the attacker can not fully leverage the power of the birthday paradox. We conclude by reusing the concept of *limited birthday distinguishers* [15] that for a random 2048-bit permutation, finding such a pair should require at least $2^{1024}$ operations.[1] Finally, 7 rounds of the internal permutation of ECHO can be distinguished from a random 2048-bit permutation with $2^{96}$ operations and $2^{64}$ memory. The amount of freedom degrees available during the attack is discussed in the Appendix A and a costly distinguisher for 8 rounds of the ECHO internal permutation is given in the extended version of this article [36].

## 4   Internal Differential Attack: Application to Grøstl

### 4.1   Description of Grøstl

We give in this section the description of Grøstl and refer to the submission document [14] for more details. Grøstl is a double-pipe hash function that uses a chaining mode similar to the Merkle-Damgård [29, 11] iteration. More

---

[1] The generic attack complexity for mapping through a permutation a fixed difference on $i$ bits on the input and $j$ bits on the output with $i \geq j$ is given by the formula $\max\{2^{j/2}, 2^{i+j-t}\}$, where $t$ is the size of the permutation.

precisely, after having initialized the internal state $H_0$ and padded the input message string, the iteration $i$ updates the $2n$-bit chaining variable $H_i$ with the $2n$-bit incoming message block $M_i$ by applying the compression function $h$: $H_i = h(H_{i-1}, M_i)$. After having processed all the $t$ message blocks, an output function is applied to the last chaining variable to obtain the $n$-bit hash result: $hash = trunc_n(P(H_t) \oplus H_t)$, where $trunc_n$ is the truncation function of the $n$ first bits and $P$ is an **AES**-based permutation. The double-pipe compression function $h$ is built upon two similar parallel **AES**-based permutations $P$ and $Q$ (that only differ by the constants addition layers) to update chaining variable $H$ with message block $M$:

$$H' = P(H \oplus M) \oplus Q(M) \oplus H$$

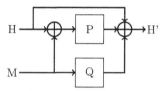

In the case of **Grøstl**-256, the 512-bit internal state of both permutations can be viewed as a $8 \times 8$ matrix of bytes. A byte for permutation $P$ is denoted by $CP_{i,j}$ (resp. $CQ_{i,j}$ for permutation $Q$), where $i$ is its row position and $j$ its column position in the matrix, starting the counting from 0. $P$ and $Q$ are both 10-round long and each round is composed of 4 layers. The first layer (AddConstant or AC) is a constant addition function. More precisely, for the round number $i$ (starting the counting from 0), in $P$ the byte $CP_{0,0}$ is xored with $i$ and in $Q$ the byte $CQ_{7,0}$ is xored with $i \oplus \texttt{0xff}$. **Note that this layer is the only difference between permutations $P$ and $Q$.** The second layer (SubBytes or SB) is a non-linear function defined by the application of the **AES** Sbox $S$ to each byte. The third layer (ShiftRows or ShR) cyclically rotates to the left the position of each byte in its own row with the following constants: $(0, 1, 2, 3, 4, 5, 6, 7)$. Finally, the last layer (MixColumns or MC) is a linear function that mixes all the columns of the internal state separately. As for $AESMixColumns$, the matrix multiplication underlying this transformation is a Maximum-Distance Separable mapping. In order to avoid overweighting the notations, we used the same notations for the **ECHO** and **Grøstl** subfunctions, but their meaning is implicit depending on which scheme we are dealing with. The round function on an internal state $C$ can thus be defined as $MixColumns \circ ShiftRows \circ SubBytes \circ AddConstant(C)$:

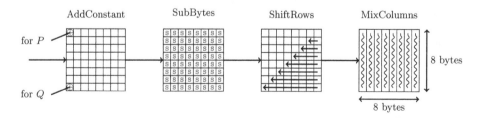

## 4.2   The Internal Differential Attack

In this section, we will show that very good differential trails can be found for Grøstl. Our new technique, *the internal differential attack*, may apply when a function is built upon parallel computation branches that are not distinct enough. The trick is **to devise a differential path representing the differences between the branches and not between two inputs of the function**. Usually this is avoided by a forcing strong separation between the two parallel branches. For example, for all steps of the hash function RIPEMD [39, 12], very distinct constants are used in the left and right branches. However, in the case of Grøstl, this separation is thin between permutations $P$ and $Q$, and we will describe in the next sections how to exploit this property in order to mount for example a distinguishing attack against the full Grøstl-256 compression function.

All the previous analysis of Grøstl studied the differential behavior of the permutations in a classic way. That is, they derived differential trails by dealing with two different inputs for each of the permutations $P$ and $Q$ (the two permutations were attacked separately). Those permutations mimicking the AES block cipher, the best usable differential paths naturally reached 8 rounds [15], but we argue that much more interesting trails can be built. We do not analyze the two permutations separately, but we build a differential path **between them**: we keep track of the differences ongoing between branch $P$ and branch $Q$ (see Figure 4). We compute two internal states $A$ and $B$, such that $A \oplus B = \Delta_{IN}$ and such that $P(A) \oplus Q(B) = \Delta_{OUT}$.

This idea comes naturally after having noticed that permutations $P$ and $Q$ are really similar, since their only distinction is the constant addition phase. Even in that step, the distinction is really thin: a different constant is added on only two different bytes. Thus, we can hope that the amount of differences will remain low when setting a differential trail.

Since using truncated differentials is very handy when attacking AES-like permutations, we will only keep track of active and inactive bytes through the path. Also, preparing for the utilization of Super-Sbox attacks, we aim for a differential path in which the costly part lies in the middle, and the cheap parts on the

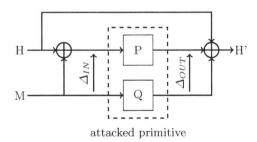

attacked primitive

**Fig. 4.** The differential path keeps track of the differences between permutations $P$ and $Q$

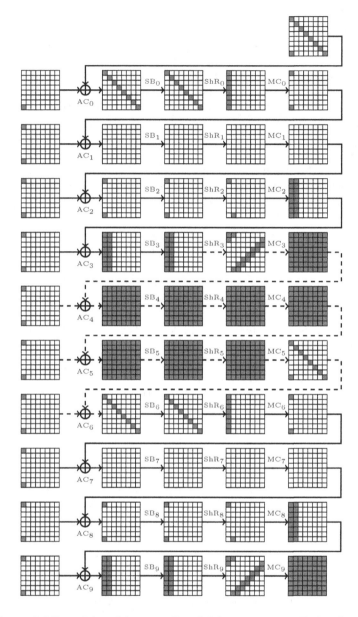

**Fig. 5.** 10-round differential path between $P$ and $Q$ for Grøstl-256. Each cell represents a byte and a gray cell stands for an active byte. The controlled rounds are depicted with dashed lines. The matrices on the left represent the differences incorporated during the AC layers.

sides. In Figure 5, we provide a differential path between the permutations $P$ and $Q$ of the Grøstl-256 compression function for the 10-round version. Note that only one difference is incorporated during $AC_0$ since the constant added in $P$ is 0.

### 4.3    Deriving a Distinguisher for Grøstl

In the following, our goal is to distinguish the Grøstl compression function from an ideal primitive on the same domain. As shown in Figure 4, once the differential path settled, we find a valid pair of internal states $(A, B)$ such that

$$A \oplus B = \Delta_{IN}$$
$$P(A) \oplus Q(B) = \Delta_{OUT}$$

where $\Delta_{IN}$ and $\Delta_{OUT}$ are respectively the input and output truncated differences. We then set $H = A \oplus B$ and $M = B$ and we obtain

$$h(H, M) = P(A) \oplus Q(B) \oplus A \oplus B = \Delta_{IN} \oplus \Delta_{OUT}.$$

We will show that $\Delta_{IN}$ and $\Delta_{OUT}$ are always maintained in a small subspace of $x$ and $y$ elements respectively. As a consequence, $\Delta_{IN} \oplus \Delta_{OUT}$ will also belong to a small subspace of the full output domain. Said in other words, we will be able to compute outputs of the $2n$-bit compression function that always belong to a predetermined set of at most $k = x \cdot y$ elements. In the ideal case, one such input/output property should not be obtained with substantially less than $2^{2n}/k$ compression function calls. Unlike the previously known distinguishers that find partially colliding outputs for AES-like permutations, the one we describe here is more "preimage" oriented.

One can go further and even try to distinguish the Grøstl compression function from its internal construction

$$h(H, M) = P(H \oplus M) \oplus Q(M) \oplus H = (P(A) \oplus A) \oplus (Q(B) \oplus B)$$

assuming $P$ and $Q$ as ideal permutations. We will compute pairs $(H, M)$ such that $H$ belong to a small subspace of $x$ elements and $H'$ to a small subspace of $k = x \cdot y$ elements. In the ideal case, one may think that the best attack can obtain such a property this with $\sqrt{2^{2n}/k}$ computations by performing a birthday method with the two branches. However, this is not the case here because a strong constraint on the input $H$ exists (see the limited birthday distinguishers [15]) and the best known complexity to obtain the input/output property with ideal permutations $P$ and $Q$ is $2^{2n}/(k \cdot x)$ computations. It is important to remark that this type of distinguisher is new since the already known ones are structural, i.e. they already consider $P$ and $Q$ as ideal permutations.

While formally defining a distinguisher for a keyless primitive is difficult [40], we argue that the property we exhibit here works for any choice of Sbox, Mix-Columns function or AddConstant positions for example. Note that such keyless primitive distinguishers have already been utilized in [26, 15].

Let Grøstl($a$) denote the Grøstl hash function for which the constant addition in $Q$ is $i \oplus a$ instead of $i \oplus \mathtt{0xff}$. Clearly, when choosing $a > \mathtt{0x1a}$, we ensure that the constant values added in $P$ and $Q$ are always distinct and each member of this family of Grøstl hash functions should have the same security as Grøstl($\mathtt{0xff}$). Overall, for each member of the family, the attacker can exhibit with good probability an output maintained in the set of $k$ elements, while the input $H$ belongs to the subspace of $x$ elements. Thus, if we are queried to distinguish the Grøstl compression function instantiated with permutations corresponding to Grøstl($a$) from the same construction with random permutations $P$ and $Q$, we have a very good probability to succeed. It shows a weakness in the Grøstl design philosophy.

# 5   Results

In this section we present some of our results on the compression functions of ECHO and Grøstl-256. For the complete results, and the differential paths concerning the internal permutation of ECHO, the single-pipe version ECHO-SP, or Grøstl-512 compression function, we refer to the extended version of this article [36]. Moreover, we also provide in the Appendix A a study of the amount of freedom degrees available during the attacks.

## 5.1   ECHO

**Compression function distinguishers.**   We provide here the first distinguishers for reduced ECHO compression functions. In the case of ECHO-256, we use the 4-round differential path from Figure 6 which is derived from the 7-round core path. One can find a solution with $2^{64}$ computations and memory ($2^{39}$ valid candidates can be generated by the attacker and $2^{167}$ if the salt is controlled as well). In the case of ECHO-512, we use the 6-round differential path from Figure 7 for which a solution can be found with $2^{96}$ computation and $2^{64}$ memory ($2^{71}$ valid candidates can be generated by the attacker and $2^{199}$ if the salt is controlled as well). In both cases, we obtain compression function outputs colliding on 2 predetermined words (i.e. 256 bits) and this should require $2^{128}$ computations in the ideal case.

**Collision attacks.**   We provide here the first collision attacks for reduced ECHO compression functions. In the case of ECHO-256, we use a special 3-round differential path depicted in Figure 8. In this trail, the start-from-the-middle technique can still be used and the only part uncontrolled is the first AES round of the very first BIG SubBytes layer. However, since we use the backward transition $\mathtt{D} \Leftarrow 1 \Leftarrow \mathtt{C}$, this layer will behave as expected with probability 1. Then, the feedforward is applied and only four 128-bit words will be active, each containing 4 active bytes at the exact same positions (truncated differential of type $\mathtt{D}$). Finally, since the four columns of 128-bit words are xored together to obtain the output chaining variable, a collision can occur if the truncated differences are

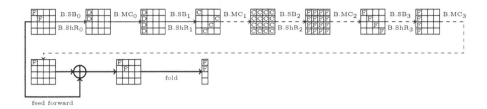

**Fig. 6.** 4-round differential path for the ECHO-256 compression function distinguisher. The controlled rounds are depicted with dashed lines.

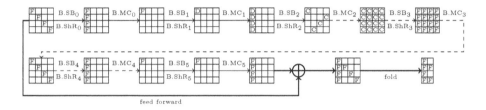

**Fig. 7.** 6-round differential path for the ECHO-512 compression function distinguisher. The controlled rounds are depicted with dashed lines.

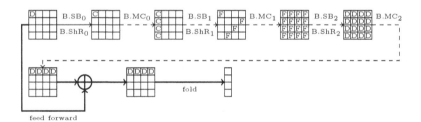

**Fig. 8.** 3-round differential path for the ECHO-256 compression function semi-free-start collision attack. The controlled rounds are depicted with dashed lines.

erased on the 4 byte positions. Thus, in order to get a semi-free-start collision, one should therefore test $2^{32}$ candidates (we have enough freedom degrees since one can generate $2^{143}$ valid candidates for the whole trail and $2^{271}$ if the salt is chosen by the attacker). However, the minimum cost for using the start-from-the-middle attack for ECHO is $2^{64}$ memory and precomputation. Thus, the overall cost is $2^{64}$ computations and memory in order to find one single semi-free-start collision for 3 rounds.

In the case of ECHO-512, we use the 4-round differential path from Figure 9 for which a solution can be found with $2^{96}$ computations and $2^{64}$ memory ($2^{71}$ valid candidates can be generated by the attacker and $2^{199}$ if the salt is controlled as well). Then, before the feedforward is applied, one active 128-bit word remains

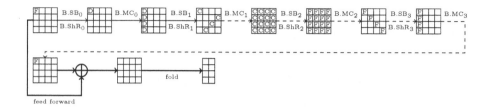

**Fig. 9.** 4-round differential path for the ECHO-512 compression function semi-free-start collision attack. The controlled rounds are depicted with dashed lines.

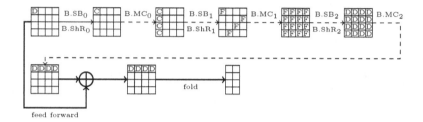

**Fig. 10.** 3-round differential path for the ECHO-512 compression function semi-free-start collision attack. The controlled rounds are depicted with dashed lines.

in the output of the permutation. In order to erase this difference and obtain a semi-free-start collision, this should be repeated $2^{128}$ times and the total cost of the attack is then $2^{224}$ computations and $2^{64}$ memory. Thus, this attack is valid only in the chosen-salt attacker model (otherwise the number of available freedom degrees is not sufficient). Since the collision happens before the final compression phase, this semi-free-start collision attack applies with the same complexity to ECHO-256 compression function.

When the attacker can not control the salt value, the 3-round attack from Figure 10 applies. Namely the reasoning is exactly the same as for the 256-bit case with Figure 8, except that we have 4 additional bytes to collide during the feed-forward phase. Finally, finding a semi-free-start collision for the 3-round reduced ECHO-512 compression function requires $2^{96}$ computations and $2^{64}$ memory.

## 5.2  Grøstl

We use the Super-Sbox technique to find two 512-bit internal states such that the 10-round differential path from Figure 5 between permutations $P$ and $Q$ is verified. Namely, one can find internal state values for $P$ and $Q$ verifying the truncated differential trail from the output of $SB_3$ up to the input of $SB_6$ with one computation on average. However, the two $8 \mapsto 1$ MixColumns transitions through $MC_2$ and the $8 \mapsto 2$ transition through $MC_6$ during the uncontrolled rounds happen with probability $2^{-2 \times 56} = 2^{-112}$ and $2^{-48}$ respectively. Also, 2 byte differences must be erased during both AddConstant functions $AC_2$ and $AC_7$ which adds another $2^{-4 \times 8} = 2^{-32}$ factor. Overall, one can find a valid candidate for the whole path with only $2^{112+48+32} = 2^{192}$ computations (an amount of $2^{64}$ memory is required by the utilization of the Super-Sbox technique).

The freedom degrees analysis from Appendix A shows that for the path from Figure 5, one can expect to obtain one solution with good probability. Indeed, when the success probability for a random input pair to verify the trail is $2^{-z}$, we have $2^{z-1}$ freedom degrees available. We argue in the Appendix that it is sufficient for the attack to be considered as valid.

**The distinguisher for Grøstl.**   In order to mount the distinguisher for Grøstl, one has to analyze the amount $k$ of reachable output difference values. In the differential path from Figure 5, we have 16 active bytes just before applying the very last MixColumns layer $MC_9$. Since the MixColumns layer is fully linear, the amounts of reachable difference values on its input and on its output are equal. Thus, we can deduce that at most $y = 2^{16 \times 8} = 2^{128}$ distinct output differences can be reached on the output of the differential trail. Regarding the input of the path, the same reasoning gives us that at most $x = 2^{8 \times 8} = 2^{64}$ distinct input differences can be reached. Note that the difference inserted during $AC_0$ can be ignored since it is the last layer when computing backward (the difference value on that byte will always be equal to the constant added, i.e. 0xff). Also, it is easy to verify that the differences on the output of $SB_0$ are always the same (since MixColumns is linear). Thus, since the inverse of the AES Sbox has the property that only $2^7$ distinct output differences can be reached when the input difference is fixed, we can conclude that $\Delta_{IN}$ can go through a maximum of $x = 2^{8 \times 7} = 2^{56}$ distinct values.

To summarize, the output chaining variable $H' = h(H, M) = \Delta_{IN} \oplus \Delta_{OUT}$ is limited to a set of at most $k = 2^{128+56} = 2^{184}$ values, with $H$ being limited to a set of at most $x = 2^{56}$ values. For an ideal 512-bit compression function, reaching any element of this set should require $2^{512-184} = 2^{328}$ operations. With $2^{80}$ and $2^{192}$ computations respectively (and $2^{64}$ memory), we finally conclude that our attack can distinguish 9-round reduced or the full 10-round compression function of Grøstl-256 from a random 512-bit compression function. One can even distinguish $h$ from the compression function construction with $P$ and $Q$ assumed ideal since the best known attack requires $2^{512-184-56} = 2^{272}$ computations.

Note that structural distinguishers (i.e. working for randomly chosen permutations $P$ and $Q$) already exist for Grøstl. For example, just like in the Davies-Meyer construction, one can very easily find fixed points for the compression function. Yet, as explained in Section 4.3, we believe that our distinguishers are very interesting because they exploit the real differential properties of the internal permutations $P$ and $Q$, which is essential in order to appropriately evaluate the security margin in terms of number of rounds. Moreover, such structural attacks can not distinguish $h$ from the compression function construction with $P$ and $Q$ assumed ideal.

# 6    Conclusion

In this article, based on recent advances on AES-like permutations studies, we provided a new cryptanalysis of ECHO and Grøstl, two second-round SHA-3 candidates. In particular, in the case of Grøstl, we introduce a new cryptanalysis technique: the internal differential attack. Overall, we obtain the best cryptanalysis results known so far for both ECHO and Grøstl. We are able to derive a distinguisher for the full (10 rounds) 256-bit version of the Grøstl compression function. This work also shows that designers must be careful when building a function with parallel branches computations as the internal differential paths may lead to unexpected attacks.

## Acknowledgments

The author would like to thank the Grøstl team, Henri Gilbert, Yannick Seurin and the CRYPTO 2010 committee for their helpful comments. Also, many thanks to Elmar Tischhauser, Jorge Nakahara and Kota Ideguchi for pointing me an omission in the complexity computation for the full ECHO internal permutation distinguisher.

## References

1. Barreto, P.S.L.M.: An observation on Grøstl. Comment submitted to the NIST hash function mailing list, hash-forum@nist.gov, http://www.larc.usp.br/~pbarreto/Grizzly.pdf
2. Bellare, M., Rogaway, P.: Random Oracles are Practical: A Paradigm for Designing Efficient Protocols. In: ACM Conference on Computer and Communications Security, pp. 62–73 (1993)
3. Benadjila, R., Billet, O., Gilbert, H., Macario-Rat, G., Peyrin, T., Robshaw, M., Seurin, Y.: SHA-3 Proposal: ECHO. Submission to NIST (2008), http://crypto.rd.francetelecom.com/echo/
4. Biham, E., Dunkelman, O.: A Framework for Iterative Hash Functions: HAIFA. In: Second NIST Cryptographic Hash Workshop (2006)
5. Biham, E., Dunkelman, O.: The SHAvite-3 Hash Function. Submission to NIST (2008)
6. Biryukov, A. (ed.): FSE 2007. LNCS, vol. 4593. Springer, Heidelberg (2007) (revised selected papers)
7. Biryukov, A., Khovratovich, D., Nikolic, I.: Distinguisher and Related-Key Attack on the Full AES-256. In: Halevi (ed.) [16], pp. 231–249
8. Brassard, G. (ed.): CRYPTO 1989. LNCS, vol. 435. Springer, Heidelberg (1990)
9. Cramer, R. (ed.): EUROCRYPT 2005. LNCS, vol. 3494. Springer, Heidelberg (2005)
10. Daemen, J., Rijmen, V.: The Design of Rijndael. In: Information Security and Cryptography. Springer, Heidelberg (2002), ISBN 3-540-42580-2
11. Damgård, I.: A Design Principle for Hash Functions. In: Brassard (ed.) [8], pp. 416–427

12. Dobbertin, H., Bosselaers, A., Preneel, B.: RIPEMD-160: A Strengthened Version of RIPEMD. In: Gollmann, D. (ed.) FSE 1996. LNCS, vol. 1039, pp. 71–82. Springer, Heidelberg (1996)
13. Dunkelman, O. (ed.): FSE 2009. LNCS, vol. 5665. Springer, Heidelberg (2009)
14. Gauravaram, P., Knudsen, L.R., Matusiewicz, K., Mendel, F., Rechberger, C., Schläffer, M., Thomsen, S.S.: Grøstl – a SHA-3 candidate. Submission to NIST (2008), http://www.groestl.info
15. Gilbert, H., Peyrin, T.: Super-Sbox Cryptanalysis: Improved Attacks for AES-like Permutations. In: FSE 2010. LNCS. Springer, Heidelberg (to appear 2010), http://eprint.iacr.org/2009/531
16. Halevi, S. (ed.): CRYPTO 2009. LNCS, vol. 5677. Springer, Heidelberg (2009)
17. Jacobson Jr., M.J., Rijmen, V., Safavi-Naini, R. (eds.): SAC 2009. LNCS, vol. 5867. Springer, Heidelberg (2009)
18. Kelsey, J.: Some notes on Grøstl. Comment submitted to the NIST hash function mailing list, hash-forum@nist.gov, http://ehash.iaik.tugraz.at/uploads/d/d0/Grostl-comment-april28.pdf
19. Khovratovich, D.: Cryptanalysis of Hash Functions with Structures. In: Jacobson Jr., M.J., et al. (eds.) [17], pp. 108–125
20. Knudsen, L.R., Rechberger, C., Thomsen, S.S.: The Grindahl Hash Functions. In: Biryukov (ed.) [6], pp. 39–57
21. Knudsen, L.R., Rijmen, V.: Known-Key Distinguishers for Some Block Ciphers. In: Kurosawa, K. (ed.) ASIACRYPT 2007. LNCS, vol. 4833, pp. 315–324. Springer, Heidelberg (2007)
22. Knudsen, L.R.: Truncated and Higher Order Differentials. In: Preneel, B. (ed.) FSE 1994. LNCS, vol. 1008, pp. 196–211. Springer, Heidelberg (1995)
23. Lamberger, M., Mendel, F., Rechberger, C., Rijmen, V., Schläffer, M.: Rebound Distinguishers: Results on the Full Whirlpool Compression Function. In: Matsui (ed.) [24], pp. 126–143
24. Matsui, M. (ed.): ASIACRYPT 2009. LNCS, vol. 5912. Springer, Heidelberg (2009)
25. Matusiewicz, K., Naya-Plasencia, M., Nikolic, I., Sasaki, Y., Schläffer, M.: Rebound Attack on the Full Lane Compression Function. In: Matsui (ed.) [24], pp. 106–125
26. Mendel, F., Peyrin, T., Rechberger, C., Schläffer, M.: Improved Cryptanalysis of the Reduced Grøstl Compression Function, ECHO Permutation and AES Block Cipher. In: Jacobson Jr., M.J., et al. (eds.) [17], pp. 16–35
27. Mendel, F., Rechberger, C., Schläffer, M., Thomsen, S.S.: The Rebound Attack: Cryptanalysis of Reduced Whirlpool and Grøstl. In: Dunkelman (ed.) [13], pp. 260–276
28. Mendel, F., Rechberger, C., Schläffer, M., Thomsen, S.S.: Rebound Attacks on the Reduced Grøstl Hash Function. In: Pieprzyk (ed.) [37], pp. 350–365
29. Merkle, R.C.: One Way Hash Functions and DES. In: Brassard (ed.) [8], pp. 428–446
30. Minier, M., Phan, R.C.-W., Pousse, B.: Distinguishers for Ciphers and Known Key Attack against Rijndael with Large Blocks. In: Preneel (ed.) [38], pp. 60–76
31. National Institute of Standards and Technology. FIPS 180-1: Secure Hash Standard (April 1995), http://csrc.nist.gov
32. National Institute of Standards and Technology. FIPS PUB 197, Advanced Encryption Standard (AES). Federal Information Processing Standards Publication 197, U.S. Department of Commerce (November 2001)

33. National Institute of Standards and Technology. Announcing Request for Candidate Algorithm Nominations for a New Cryptographic Hash Algorithm (SHA-3) Family. Federal Register 27(212), 62212–62220 (November 2007), http://csrc.nist.gov/groups/ST/hash/documents/FR_Notice_Nov07.pdf (2008/10/17)
34. Nguyên, P.Q. (ed.): VIETCRYPT 2006. LNCS, vol. 4341. Springer, Heidelberg (2006)
35. Peyrin, T.: Cryptanalysis of Grindahl. In: Kurosawa, K. (ed.) ASIACRYPT 2007. LNCS, vol. 4833, pp. 551–567. Springer, Heidelberg (2007)
36. Peyrin, T.: Improved Differential Attacks for ECHO and Grostl. Cryptology ePrint Archive, Report 2010/223 (2010), http://eprint.iacr.org/
37. Pieprzyk, J. (ed.): CT-RSA 2010. LNCS, vol. 5985. Springer, Heidelberg (2010)
38. Preneel, B. (ed.): AFRICACRYPT 2009. LNCS, vol. 5580. Springer, Heidelberg (2009)
39. RIPE. Integrity Primitives for Secure Information Systems. In: Bosselaers, A., Preneel, B. (eds.) RIPE 1992. LNCS, vol. 1007. Springer, Heidelberg (1995)
40. Rogaway, P.: Formalizing Human Ignorance. In: Nguyen (ed.) [34], pp. 211–228
41. Rivest, R.L.: RFC 1321: The MD5 Message-Digest Algorithm (April 1992), http://www.ietf.org/rfc/rfc1321.txt
42. Shoup, V. (ed.): CRYPTO 2005. LNCS, vol. 3621. Springer, Heidelberg (2005)
43. Wagner, D.: A Generalized Birthday Problem. In: Yung (ed.) [46], pp. 288–303
44. Wang, X., Yin, Y.L., Yu, H.: Finding Collisions in the Full SHA-1. In: Shoup (ed.) [42], pp. 17–36
45. Wang, X., Yu, H.: How to Break MD5 and Other Hash Functions. In: Cramer (ed.) [9], pp. 19–35
46. Yung, M. (ed.): CRYPTO 2002. LNCS, vol. 2442. Springer, Heidelberg (2002)

# Appendix A: The Amount of Freedom Degrees

Once a differential path settled, a point has to be clarified: the amount of freedom degrees available to the attacker. Indeed, one has to evaluate how much valid pairs can be found for the whole differential trail. We want to ensure that enough solutions for the controlled rounds exist so that we have a good probability that at least one of them will fulfill the entire differential characteristic. Moreover, when searching for semi-free-start collisions, we may even go further since we may require an important amount of valid candidates for the entire differential path.

### Freedom Degrees for ECHO

We use the same counting reasoning than in [15], except that we have to precisely evaluate what is the freedom degrees consumption for the various 2 AES-round differential transitions as well (in [15] it was implicitly assumed that all the BIG SubBytes transitions were F $\rightarrow$ F, thus happening with probability very close to 1 and consuming no freedom degrees). For example, let us take the D $\rightarrow$ 1 transition through the BIG SubBytes in the forward direction: we require one AES MixColumns transition 4 $\rightarrow$ 1 which happens with probability $2^{-24}$. Thus, if we have $k$ valid candidates on the input, we obtain $k \times 2^{-24}$ valid candidates on the output of this layer and we consumed $2^{24}$ freedom degrees. The amount of freedom degrees consumed during a transition is the invert of the probability of success of this transition. Thus, with Table 2, it is very easy to compute the freedom degrees consumption for all the AES round transitions considered so far.

We illustrate the counting method by applying it to the example of the 7-round path from Figure 3. First, note that the start-from-the-middle attack will find **all** the possible internal states such that the controlled rounds are verified. We start from state $\mathsf{B.MC}_3^{in}$ (located between $\mathsf{B.ShR}_3$ and $\mathsf{B.MC}_3$). This state is fully active which means that we can start with $2^{2048\times 2-1} = 2^{4095}$ distinct pairs. When going forward, the $\mathsf{B.MC}_3$ transition happens with probability $2^{-4\times 24\times 16} = 2^{-1536}$ and the transition through $\mathsf{B.MC}_4$ happens with probability $2^{-24\times 16} = 2^{-384}$. All the other layers are verified with probability one so the forward computation consumes $2^{1536+384} = 2^{1920}$ freedom degrees. Then, during the backward computation, the sixteen $\mathsf{C} \leftarrow \mathsf{F} \leftarrow \mathsf{F}$ transitions through $\mathsf{B.SB}_3$ happen with probability $2^{-16\times 96} = 2^{-1536}$ according to Table 2 ($\mathsf{C} \leftarrow \mathsf{F}$ with probability $2^{-96}$ and $\mathsf{F} \leftarrow \mathsf{F}$ with probability 1). Also, the four $\mathsf{D} \leftarrow \mathsf{1} \leftarrow \mathsf{C}$ transitions through $\mathsf{B.SB}_2$ happen with probability $2^{-4\times 24} = 2^{-96}$ ($\mathsf{D} \leftarrow \mathsf{1}$ with probability 1 and $\mathsf{1} \leftarrow \mathsf{C}$ with probability $2^{-24}$). Then, the BIG MixColumns transitions through $\mathsf{B.MC}_2$ are verified with probability $2^{-4\times 4\times 24} = 2^{-384}$ and through $\mathsf{B.MC}_1$ with probability $2^{-4\times 24} = 2^{-96}$. All the other layers in the backward direction are verified with probability one. Overall, the backward computation consumes $2^{1536+384+96+96} = 2^{2112}$ freedom degrees. We can finally conclude that we started with $2^{4095}$ pairs from which only a factor $2^{-1920-2112} = 2^{-4032}$ will be valid for the whole differential path. One can then generate $2^{63}$ distinct valid pairs for the 7-round path from Figure 3.

Note that the differential paths we use are just instances among a family of good differential trails. For example, in the case of the 7-round path from Figure 3, instead of placing the active word on the top left position of $\mathsf{B.MC}_0^{out}$ (between $\mathsf{B.MC}_0$ and $\mathsf{B.SB}_1$), one could place it in the 15 others locations. Those new paths present the same properties than the original one and this reasoning also applies to the active word located in $\mathsf{B.MC}_4^{out}$ (between $\mathsf{B.MC}_4$ and $\mathsf{B.SB}_5$). As a consequence, the attacker manages $16^2 = 2^8$ different core paths which provides him $2^8$ additional freedom degrees.

Finally, some additional freedom degrees can be obtained if one considers that the salt value can be fully controlled by the attacker. While this scenario is not very relevant in practice, it is interesting to see what the attacker is able to do in such a situation. In the case of ECHO, the salt value is 128-bit long and it then directly adds $2^{128}$ supplementary freedom degrees. To conclude, the attacker can generate $2^{71}$ distinct valid pairs for 7-round paths like the one depicted in Figure 3, and $2^{199}$ if he controls the salt. The same method is used for all the differential trails for ECHO considered in this article.

## Freedom Degrees for Grøstl

The case of Grøstl is easier to analyze since we don't have to handle word-wise and byte-wise truncated differentials at the same time. Yet, the same counting technique can be applied. Interestingly, for all the paths we provided concerning Grøstl, an attacker can expect only one solution for the whole trail with good probability. This explains why one can not really hope for a semi-free-start collision attack on reduced versions of Grøstl (such as 7 or 8-round versions) with the paths given. Or, said in other words, a semi-free-start collision attack may be mounted, but will only work with a low probability.

As an example, we provide here the freedom degrees analysis for the 10-round differential path from Figure 5. By starting from the fully active internal state located at the output of $MC_4$, we begin with about $2^{512 \times 2 - 1} = 2^{1023}$ distinct pairs of internal state values. When going forward, the first freedom degrees consuming operation is the $MC_5$ transition which happens with probability $2^{-7 \times 56 - 48} = 2^{-440}$. Then, one byte is erased during $AC_6$ while the transition through $MC_6$ happens with probability $2^{-48}$ and in total this round consumes $2^{8+48} = 2^{56}$ freedom degrees. Finally, the last consuming operation when computing forward is $AC_7$ for which two bytes have to be erased ($2^{16}$). When computing backward, the MixColumns functions $MC_3$ and $MC_2$ requires $2^{48 \times 8} = 2^{384}$ and $2^{2 \times 56} = 2^{112}$ freedom degrees respectively. Then, two bytes are erased through $AC_2$ and all the other differential transitions consume nothing since they are deterministic. Finally, we started with $2^{1023}$ freedom degrees from which only a fraction $2^{440+8+48+16+384+112+16} = 2^{1024}$ will verify the whole differential path. Thus, since this reasoning is done on average, an attacker has a good probability to obtain one single solution for the whole differential path.

Of course, one may argue that the attacker should have one more freedom degree to perform the attack. Yet, note that until really performed, most hash function attacks only have a certain success probability to actually find a solution. For example, in the case of SHA-1, even if very low, there is a probability that the known collision attacks eventually provide no solution. Therefore, with only a single freedom degree missing, we believe that the success probability is far sufficiently high in order to consider the attack as valid. Finally, if one really wants to increase this probability, additional freedom degrees could be found by defining a small family of Grøstl compression functions as explained in Section 4.3.

# A Practical-Time Related-Key Attack on the KASUMI Cryptosystem Used in GSM and 3G Telephony

Orr Dunkelman, Nathan Keller[*], and Adi Shamir

Faculty of Mathematics and Computer Science
Weizmann Institute of Science
P.O. Box 26, Rehovot 76100, Israel
{orr.dunkelman,nathan.keller,adi.shamir}@weizmann.ac.il

**Abstract.** The privacy of most GSM phone conversations is currently protected by the 20+ years old A5/1 and A5/2 stream ciphers, which were repeatedly shown to be cryptographically weak. They will soon be replaced by the new A5/3 (and the soon to be announced A5/4) algorithm based on the block cipher KASUMI, which is a modified version of MISTY. In this paper we describe a new type of attack called a *sandwich attack*, and use it to construct a simple distinguisher for 7 of the 8 rounds of KASUMI with an amazingly high probability of $2^{-14}$. By using this distinguisher and analyzing the single remaining round, we can derive the complete 128 bit key of the full KASUMI by using only 4 related keys, $2^{26}$ data, $2^{30}$ bytes of memory, and $2^{32}$ time. These complexities are so small that we have actually simulated the attack in less than two hours on a single PC, and experimentally verified its correctness and complexity. Interestingly, neither our technique nor any other published attack can break MISTY in less than the $2^{128}$ complexity of exhaustive search, which indicates that the changes made by ETSI's SAGE group in moving from MISTY to KASUMI resulted in a much weaker cipher.

## 1 Introduction

The privacy and security of GSM cellular telephony is protected by the A5 family of cryptosystems. The first two members of this family, A5/1 (developed primarily for European markets) and A5/2 (developed primarily for export markets) were designed in the late 1980's in an opaque process and were kept secret until they were reverse engineered in 1999 from actual handsets [14]. Once published, it became clear that A5/2 provided almost no security, and A5/1 could be attacked with practical complexity by a variety of techniques (e.g., [2,12,16]). The most recent attack was announced in December 2009, when a team of cryptographers led by Karsten Nohl [1] published a 2 terabyte rainbow table for A5/1, which makes it easy to derive the session key of any particular conversation with minimal hardware support.

---

[*] The second author was partially supported by the Koshland center for basic research.

T. Rabin (Ed.): CRYPTO 2010, LNCS 6223, pp. 393–410, 2010.

In response to these developments, the GSM Association had stated in [26] that they might speed up their transition to a new cryptosystem called A5/3, and they plan to discuss this matter in a meeting that was held in February 2010. This algorithm was developed for GSM telephony in 2002, and its specifications were published in 2003 [24]. It is already implemented in about 40% of the three billion available handsets, but very few of the 800 mobile carriers in more than 200 countries which currently use GSM cellular telephony have switched so far to the new standard. Once adopted, A5/3 will become one of the most widely used cryptosystems in the world, and its security will become one of the most important practical issues in cryptography.

The core of the A5/3 cryptosystem, as well as of the UAE1 cryptosystem (which replaces A5/3 in the third generation telephony networks), is the KA-SUMI block cipher, which is based on the MISTY block cipher which was published at FSE 1997 by Matsui [22]. MISTY has 64-bit blocks, 128-bit keys, and a complex recursive Feistel structure with 8 rounds, each one of which consists of 3 rounds, each one of which has 3 rounds of nonlinear SBox operations. MISTY has provable security properties against various types of attacks, and no attack is known on its full version. The best published attack can be applied to a 6-round reduced variant of the 8-round MISTY, and has a completely impractical time complexity of more than $2^{123}$ [15]. However, the designers of A5/3 decided to make MISTY faster and more hardware-friendly by simplifying its key schedule and modifying some of its components. In [25], the designers provide a rationale for each one of these changes, and in particular they analyze the resistance of KASUMI against related-key attacks by stating that "removing all the FI functions in the key scheduling part makes the hardware smaller and/or reduces the key set-up time. We expect that related key attacks do not work for this structure". The best attack found by the designers and external evaluators of KASUMI is described as follows:

> "There are chosen plaintext and/or related-key attacks against KASUMI reduced to 5 rounds. We believe that with further analysis it might be possible to extend some attacks to 6 rounds, but not to the full 8 round KASUMI."

The existence of better related-key attacks on the full KASUMI was already shown in [8,21]. Their attack had a data complexity of $2^{54.6}$ and time complexity of $2^{76.1}$, which are impractical but better than exhaustive search. In this paper we develop a new attack, which requires only 4 related keys, $2^{26}$ data, $2^{30}$ bytes of memory, and $2^{32}$ time. Since these complexities are so low, we could verify our attack experimentally, and our unoptimized implementation on a single PC recovered about 96 key bits in a few minutes, and the complete 128 bit key in less than two hours. Careful analysis of our attack technique indicates that it can not be applied against the original MISTY, since it exploits a sequence of coincidences and lucky strikes which were created when MISTY was changed to KASUMI by ETSI's SAGE group. This calls into question both the design of KASUMI and its security evaluation against related-key attacks. However, we would like to emphasize that even though our attack on the underlying cryptosystem has a *practical time complexity*, we do not claim that we can *practically apply* such a related key attack to the way KASUMI is used in the f8 and f9 modes of operation in cellular telephony.

We use a new type of attack which is an improved version of the boomerang attack introduced in [27]. We call it a "sandwich attack", since it uses a distinguisher which is divided into three parts: A thick slice ("bread") at the top, a thin slice ("meat") in the middle, and a thick slice ("bread") at the bottom. The top and bottom parts are assumed to have high probability differential characteristics, which can be combined into consistent quartet structures by the standard boomerang technique. However, in our case they are separated by the additional middle slice, which can significantly reduce the probability of the resulting boomerang structure. Nevertheless, as we show in this paper, careful analysis of the dependence between the top and bottom differentials allows us in some cases to combine the two properties above and below the middle slice with an enhanced probability. In particular, we show that in the case of KASUMI we can use top and bottom 3-round differential characteristics with an extremely high probability of $2^{-2}$ each, and combine them via a middle 1-round slice in such a way that the "price in probability" of the combination is $2^{-6}$, instead of the $2^{-32}$ we would expect from a naive analysis. This increases the probability of our 7-round distinguisher from $2^{-40}$ to $2^{-14}$, and has an even bigger impact on the amount of data and the time complexity of the attack due to the quadratic dependence of the number of cases we have to sample on the distinguishing probability. Such a three level structure was used in several previous attacks such as [10,11] (where it was called the "Feistel switch" or the "middle round S-box trick"), but to the best of our knowledge it was always used in the past in simpler situations in which the transition probability through the middle layer (in at least one direction) was 1 due to the structural properties of a single Feistel round, or due to the particular construction of a given SBox. Our sandwich attack is the first nontrivial application of such a structure, and the delicacy of the required probabilistic analysis is demonstrated by the fact that a tiny change in the key schedule of KASUMI (which has no effect on the differential probabilities of the top and bottom layers) can change the probability of the transition in the middle of the distinguisher from the surprisingly high value of $2^{-6}$ to 0.

This paper is organized as follows: Section 2 describes the new sandwich attack, and discusses the transition between the top and bottom parts of the cipher through the middle slice of the sandwich. Section 3 describes the KASUMI block cipher. Section 4 describes our new 7-round distinguisher for KASUMI which has a probability of $2^{-14}$. In Section 5 we use the new distinguisher to develop a practical-time key recovery attack on the full KASUMI cryptosystem.

# 2   Sandwich Attacks

In this section we describe the technique used in our attacks on KASUMI. We start with a description of the basic (related-key) boomerang attack, and then we describe a new framework, which we call a (related-key) *sandwich attack*, that exploits the dependence between the underlying differentials to obtain a more accurate estimation of the probability of the distinguisher. We note that the idea of using dependence between the differentials in order to improve the boomerang distinguisher was implicitly proposed by Wagner [27], and was also used in some simple scenarios in [10,11]. Therefore, our framework can be considered as a formal treatment and generalization of the ideas proposed in [10,11,27].

## 2.1   The Basic Related-Key Boomerang Attack

The related-key boomerang attack was introduced by Kim et al. [20,18], and independently by Biham et al. [7], as a combination of the boomerang attack [27] and the related-key differential attack [19]. In this attack, the cipher is treated as a cascade of two sub-ciphers $E = E_1 \circ E_0$, and related-key differentials of $E_0$ and $E_1$ are combined into an adaptive chosen plaintext and ciphertext distinguisher for $E$.

Let us assume that there exists a related-key differential $\alpha \to \beta$ for $E_0$ under key difference $\Delta K_{ab}$ with probability $p$. (i.e., $\mathbf{Pr}[E_{0(K)}(P) \oplus E_{0(K \oplus K_{ab})}(P \oplus \alpha) = \beta] = p$, where $E_{0(K)}$ denotes encryption through $E_0$ under the key $K$). Similarly, we assume that there exists a related-key differential $\gamma \to \delta$ for $E_1$ under key difference $\Delta K_{ac}$ with probability $q$. The related-key boomerang distinguisher requires encryption/decryption under the secret key $K_a$, and under the related-keys $K_b = K_a \oplus \Delta K_{ab}$, $K_c = K_a \oplus \Delta K_{ac}$, and $K_d = K_c \oplus \Delta K_{ab} = K_b \oplus \Delta K_{ac}$.

A boomerang quartet is generated by picking a plaintext $P_a$ at random, and asking for its encryption under $K_a$, namely, $C_a = E_{K_a}(P_a)$. Then, $P_b = P_a \oplus \alpha$ is encrypted under $K_b$ to obtain $C_b = E_{K_b}(P_b)$. Two new ciphertexts are computed, $C_c = C_a \oplus \delta$ and $C_d = C_b \oplus \delta$. Then, $C_c$ is decrypted under $K_c$, and $C_d$ is decrypted under $K_d$, i.e., $P_c = E_{K_c}^{-1}(C_c)$ and $P_d = E_{K_d}^{-1}(C_d)$. If $P_c \oplus P_d = \alpha$, a right boomerang quartet is found. The left side of Figure 1 describes such a right related-key boomerang quartet.

For a random permutation the probability that the last condition is satisfied is $2^{-n}$, where $n$ is the block size. For $E$, the probability that the pair $(P_a, P_b)$ is a right pair with respect to the first differential (i.e., the probability that the intermediate difference after $E_0$ equals $\beta$) is $p$. Assuming independence, the probability that both pairs $(C_a, C_c)$ and $(C_b, C_d)$ are right pairs with respect to the second differential is $q^2$. If all these are right pairs, then $E_1^{-1}(C_c) \oplus E_1^{-1}(C_d) = \beta = E_0(P_c) \oplus E_0(P_d)$. Thus, with probability $p$, $P_c \oplus P_d = \alpha$. Hence, the total probability of this quartet of plaintexts and ciphertexts to satisfy the condition $P_c \oplus P_d = \alpha$ is at least $(pq)^2$. Therefore, if $pq \gg 2^{-n/2}$, the algorithm above allows to distinguish $E$ from a random permutation given $O((pq)^{-2})$ adaptively chosen plaintexts and ciphertexts.

The distinguisher can be improved by considering multiple differentials of the form $\alpha \to \beta'$ and $\gamma' \to \delta$ (for the same $\alpha$ and $\delta$). We omit this improvement here since it is not used in our attack on KASUMI, and refer the reader to [7]. For a rigorous treatment of the related-key boomerang attack, including a discussion of the independence assumptions the attack relies upon, we refer the interested reader to [21,23].

## 2.2   Related-Key Sandwich Attacks

In this framework we consider the cipher as a cascade of three sub-ciphers: $E = E_1 \circ M \circ E_0$. Our assumptions are the same as in the basic attack: We assume that there exists a related-key differential $\alpha \to \beta$ for $E_0$ under key difference $\Delta K_{ab}$ with probability $p$, and a related-key differential $\gamma \to \delta$ for $E_1$ under key difference $\Delta K_{ac}$ with probability $q$. The attack algorithm is also exactly the same as in the basic attack (ignoring the middle sub-cipher $M$). However, the analysis is more delicate and requires great care in analyzing the dependence between the various distributions.

The main idea behind the sandwich attack is the transition in the middle. In the basic boomerang attack, if the pair $(P_a, P_b)$ is a right pair with respect to the first differential, and both pairs $(C_a, C_c)$ and $(C_b, C_d)$ are right pairs with respect to the second differential, then we have

$$(X_a \oplus X_b = \beta) \wedge (X_a \oplus X_c = \gamma) \wedge (X_b \oplus X_d = \gamma), \tag{1}$$

where $X_i$ is the intermediate encryption value of $P_i$, and thus

$$X_c \oplus X_d = (X_c \oplus X_a) \oplus (X_a \oplus X_b) \oplus (X_b \oplus X_d) = \beta \oplus \gamma \oplus \gamma = \beta, \tag{2}$$

resulting in $P_c \oplus P_d = \alpha$ with probability $p$ (see Figure 1).

In the new sandwich framework, instead of condition (1), we get

$$(X_a \oplus X_b = \beta) \wedge (Y_a \oplus Y_c = \gamma) \wedge (Y_b \oplus Y_d = \gamma). \tag{3}$$

Therefore, the probability of the three-layer related-key boomerang distinguisher is $p^2 q^2 r$, where

$$r = \mathbf{Pr}\left[(X_c \oplus X_d = \beta)\Big|(X_a \oplus X_b = \beta) \wedge (Y_a \oplus Y_c = \gamma) \wedge (Y_b \oplus Y_d = \gamma)\right]. \tag{4}$$

Without further assumptions on $M$, $r$ is expected to be very low (close to $2^{-n}$), and thus the distinguisher is expected to fail. However, as observed in [10,11,27], in some cases the differentials in $E_0$ and $E_1$ can be chosen such that the probability penalty $r$ in going through the middle sub-cipher (in at least one direction) is 1, which is much higher than expected.

An example of this phenomenon, introduced in [27] and described in [11] under the name "Feistel switch", is the following. Let $E$ be a Feistel cipher, decomposed as $E = E_1 \circ M \circ E_0$, where $M$ consists of one Feistel round (see Figure 2). Assume that the differentials $\alpha \to \beta$ (for $E_0$) and $\gamma \to \delta$ (for $E_1$) have no key difference (i.e., $\Delta K_{ab} = \Delta K_{ac} = 0$), and satisfy $\beta^R = \gamma^L$ (i.e., the right half of $\beta$ equals the left half of $\gamma$). We would like to compute the value of $r$.

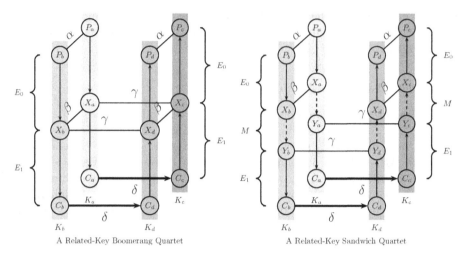

A Related-Key Boomerang Quartet          A Related-Key Sandwich Quartet

**Fig. 1.** Related-Key Boomerang and Sandwich Quartets

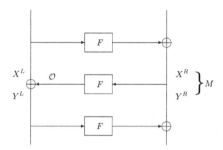

**Fig. 2.** A Feistel construction. $M$ is the second round

Assume that condition (3) holds. In this case, by the Feistel construction, $X_i^R = Y_i^L$ for all $i$, we have

$$X_a^R \oplus X_b^R = \beta^R = \gamma^L = X_a^R \oplus X_c^R = X_b^R \oplus X_d^R, \qquad (5)$$

and thus,

$$(X_a^R = X_d^R) \qquad \text{and} \qquad (X_b^R = X_c^R). \qquad (6)$$

Therefore, the output values of the F-function in the Feistel round represented by $M$, denoted by $(\mathcal{O}_a, \mathcal{O}_b, \mathcal{O}_c, \mathcal{O}_d)$, satisfy

$$(\mathcal{O}_a = \mathcal{O}_d) \qquad \text{and} \qquad (\mathcal{O}_b = \mathcal{O}_c).$$

Since by the Feistel construction, $X_i^L = Y_i^R \oplus \mathcal{O}_i$ and by condition (3), $Y_a \oplus Y_b \oplus Y_c \oplus Y_d = 0$ , it follows that

$$X_a \oplus X_b \oplus X_c \oplus X_d = 0,$$

which by condition (3) implies $X_c \oplus X_d = \beta$. Thus, in this case we get

$$r = \mathbf{Pr}\left[(X_c \oplus X_d = \beta)\big|(X_a \oplus X_b = \beta) \wedge (Y_a \oplus Y_c = \gamma) \wedge (Y_b \oplus Y_d = \gamma)\right] = 1,$$

independently of the choice of the F-function used.

Other examples of the same phenomenon are considered in [10] (under the name "middle round S-box trick"), and in [11] (under the names "ladder switch" and "S-box switch"). All these examples are methods for $r = 1$.

Our attack on KASUMI is the first non-trivial example of this phenomenon in which a careful analysis shows that $r$ is smaller than 1, but much larger than its expected value under the standard independence assumptions. In our attack, the cipher $E$ (7-round KASUMI) is a Feistel construction, $M$ consists of a single round, and $\beta = \gamma$. However, the argument presented above cannot be applied directly since there is a non-zero key difference in $M$, and thus a zero input difference to the F-function does not imply zero output difference. Instead, we analyze the F-function thoroughly and show that in this case, $r = 2^{-6}$ (instead of $2^{-32}$, which is the expected value for a random Feistel round in a 64-bit cipher).

*Remark 1.* We note that our treatment of the sandwich distinguisher allows us to specify the precise independence assumptions we rely upon. Since $r$ is defined as a conditional probability, the only independence assumptions we use are between the differentials of $E_0$ and $E_1$, and thus the formula $p^2q^2r$ relies on exactly the same assumptions as the ordinary boomerang attack. Moreover, in our case the assumptions seem more likely to hold since the insertion of $M$ in the middle decreases the potential dependencies between the differentials for $E_0$ and the differentials for $E_1$. In [10,11,27], this situation was treated as a "trick" allowing to increase the probability of the distinguisher, or in other words, as a failure of the formula $p^2q^2$ in favor of the attacker. This approach is problematic since once we claim that the entire formula does not hold due to dependencies, we cannot rely on independence assumptions in other places where such dependencies were not found yet.

## 3   The KASUMI Block Cipher

KASUMI [24] is a 64-bit block cipher with 128-bit keys. It has a recursive Feistel structure, following its ancestor MISTY. The cipher has eight Feistel rounds, where each round is composed of two functions: the $FO$ function which is in itself a 3-round 32-bit Feistel construction, and the $FL$ function that mixes a 32-bit subkey with the data in a linear way. The order of the two functions depends on the round number: in the even rounds the $FO$ function is applied first, and in the odd rounds the $FL$ function is applied first.

The $FO$ function also has a recursive structure: its $F$-function, called $FI$, is a four-round Feistel construction. The $FI$ function uses two non-linear S-boxes $S7$ and $S9$ (where $S7$ is a 7-bit to 7-bit permutation and $S9$ is a 9-bit to 9-bit permutation), and accepts an additional 16-bit subkey, which is mixed with the data. In total, a 96-bit subkey enters $FO$ in each round — 48 subkey bits are used in the $FI$ functions and 48 subkey bits are used in the key mixing stages.

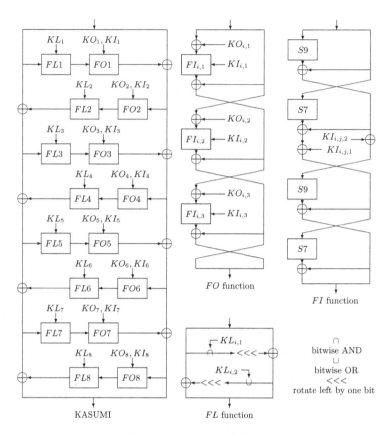

**Fig. 3.** Outline of KASUMI

**Table 1.** KASUMI's Key Schedule Algorithm

| Round | $KL_{i,1}$ | $KL_{i,2}$ | $KO_{i,1}$ | $KO_{i,2}$ | $KO_{i,3}$ | $KI_{i,1}$ | $KI_{i,2}$ | $KI_{i,3}$ |
|---|---|---|---|---|---|---|---|---|
| 1 | $K_1 \lll 1$ | $K_3'$ | $K_2 \lll 5$ | $K_6 \lll 8$ | $K_7 \lll 13$ | $K_5'$ | $K_4'$ | $K_8'$ |
| 2 | $K_2 \lll 1$ | $K_4'$ | $K_3 \lll 5$ | $K_7 \lll 8$ | $K_8 \lll 13$ | $K_6'$ | $K_5'$ | $K_1'$ |
| 3 | $K_3 \lll 1$ | $K_5'$ | $K_4 \lll 5$ | $K_8 \lll 8$ | $K_1 \lll 13$ | $K_7'$ | $K_6'$ | $K_2'$ |
| 4 | $K_4 \lll 1$ | $K_6'$ | $K_5 \lll 5$ | $K_1 \lll 8$ | $K_2 \lll 13$ | $K_8'$ | $K_7'$ | $K_3'$ |
| 5 | $K_5 \lll 1$ | $K_7'$ | $K_6 \lll 5$ | $K_2 \lll 8$ | $K_3 \lll 13$ | $K_1'$ | $K_8'$ | $K_4'$ |
| 6 | $K_6 \lll 1$ | $K_8'$ | $K_7 \lll 5$ | $K_3 \lll 8$ | $K_4 \lll 13$ | $K_2'$ | $K_1'$ | $K_5'$ |
| 7 | $K_7 \lll 1$ | $K_1'$ | $K_8 \lll 5$ | $K_4 \lll 8$ | $K_5 \lll 13$ | $K_3'$ | $K_2'$ | $K_6'$ |
| 8 | $K_8 \lll 1$ | $K_2'$ | $K_1 \lll 5$ | $K_5 \lll 8$ | $K_6 \lll 13$ | $K_4'$ | $K_3'$ | $K_7'$ |

$(X \lll i) - X$ rotated to the left by $i$ bits.

The $FL$ function accepts a 32-bit input and two 16-bit subkey words. One subkey word affects the data using the OR operation, while the second one affects the data using the AND operation. We outline the structure of KASUMI and its parts in Fig. 3.

The key schedule of KASUMI is much simpler than the original key schedule of MISTY, and the subkeys are linearly derived from the key. The 128-bit key $K$ is divided into eight 16-bit words: $K_1, K_2, \ldots, K_8$. Each $K_i$ is used to compute $K_i' = K_i \oplus C_i$, where the $C_i$'s are fixed constants (we omit these from the paper, and refer the intrigued reader to [24]). In each round, eight words are used as the round subkey (up to some in-word rotations). Hence, each 128-bit round subkey is a linearly modified version of the secret key. We summarize the details of the key schedule of KASUMI in Table 1.

# 4  A Related-Key Sandwich Distinguisher for 7-Round KASUMI

## 4.1  The New Distinguisher

In our distinguisher, we treat rounds 1–7 of KASUMI as a cascade $E = E_1 \circ M \circ E_0$, where $E_0$ consists of rounds 1–3, $M$ consists of round 4, and $E_1$ consists of rounds 5–7. The related-key differential we use for $E_0$ is a slight modification of the differential characteristic presented in [13], in which

$$\alpha = (0_x, 0010\ 0000_x) \rightarrow (0_x, 0010\ 0000_x) = \beta.$$

The corresponding key difference is $\Delta K_{ab} = (0, 0, 8000_x, 0, 0, 0, 0, 0)$, i.e., only the third key word has the single bit difference $\Delta K_3 = 8000_x$. This related-key differential is depicted in Figure 4. The related-key differential we use for $E_1$ is the same differential shifted by four rounds, in which the data difference is the same, but the key difference is $\Delta K_{ac} = (0, 0, 0, 0, 0, 0, 8000_x, 0)$ (to handle the different subkeys used in these rounds).

As shown in [13], the probability of each one of of these 3-round differential characteristics is $1/4$. In order to find the probability of the related-key sandwich distinguisher, we have to compute the probability

$$\mathbf{Pr}\left[(X_c \oplus X_d = \beta)\big|(X_a \oplus X_b = \beta) \wedge (Y_a \oplus Y_c = \gamma) \wedge (Y_b \oplus Y_d = \gamma)\right], \quad (7)$$

where $(X_a, X_b, X_c, X_d)$ and $(Y_a, Y_b, Y_c, Y_d)$ are the intermediate values before and after the middle slice of the sandwich during the encryption/decryption of the quartet $(P_a, P_b, P_c, P_d)$ (see the right side of Figure 1). This computation, which is a bit complicated, spans the rest of this subsection.

Consider a quartet $(P_a, P_b, P_c, P_d)$ for which the condition

$$(X_a \oplus X_b = \beta) \wedge (Y_a \oplus Y_c = \gamma) \wedge (Y_b \oplus Y_d = \gamma) \quad (8)$$

is satisfied. As explained in Section 2, since $M$ is a single Feistel round, this implies that

$$(X_a^R = X_d^R) \wedge (X_b^R = X_c^R), \quad (9)$$

where $X_i^R$ denotes the right half of $X_i$ that enters the function $FO4$. Moreover, as the right quarter of the differences $\beta = \gamma$ is zero, we have

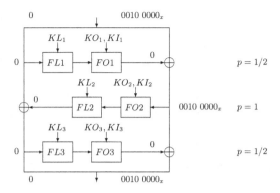

**Fig. 4.** 3-Round Related-Key Differential Characteristic of KASUMI

$$X_a^{RR} = X_b^{RR} = X_c^{RR} = X_d^{RR}, \tag{10}$$

where $X_i^{RR}$ denotes the right half (i.e., the 16 right bits) of $X_i^R$.

Consider now the computation depicted in Figure 5. The function $FO4$ is a 3-round Feistel construction whose 32-bit values after round $j$ are denoted by $(X_a^j, X_b^j, X_c^j, X_d^j)$, and the function $FI$ is a 4-round Feistel construction whose 16-bit values after round $j$ are denoted by $(I_a^j, I_b^j, I_c^j, I_d^j)$. Note that the key differences $\Delta K_{ab}$ and $\Delta K_{ac}$ affect in round 4 the subkeys $KI_{4,3}$ and $KI_{4,2}$, respectively, and in particular, there is no key difference in the first round of $FO4$. As a result, Equation (9) implies that

$$(X_a^1 = X_d^1) \wedge (X_b^1 = X_c^1). \tag{11}$$

Furthermore, there is no key difference in the pairs corresponding to $(P_a, P_b)$ and $(P_c, P_d)$ in the second round of $FO4$, and thus Equation (10) implies

$$(I_a^2 = I_b^2) \wedge (I_c^2 = I_d^2). \tag{12}$$

Combining equations (11) and (12), as depicted in Figure 5, we get the following relation in the right half of the intermediate values after round 3 of $FO4$:

$$X_a^{3R} \oplus X_b^{3R} \oplus X_c^{3R} \oplus X_d^{3R} = 0. \tag{13}$$

In the F-function of round 3 of $FO4$ we consider the pairs corresponding to $(P_a, P_d)$ and $(P_b, P_c)$. Since the key difference in these pairs (that equals $K_{ab} \oplus K_{ac}$) affects only the subkey $KI_{4,3,1}$, Equation (11) implies

$$I_a^{3R} \oplus I_b^{3R} \oplus I_c^{3R} \oplus I_d^{3R} = 0 \tag{14}$$

in the right hand side of the output. In the left hand side of the output, the XOR of the four values is not necessarily equal to zero, due to the subkey difference that affects the inputs to the second $S7$ in $FI_{4,3}$. However, if these 7-bit inputs, denoted by $(J_a, J_b, J_c, J_d)$, satisfy one of the conditions:

$$((J_a = J_b) \wedge (J_c = J_d)) \qquad \text{or} \qquad ((J_a = J_c) \wedge (J_b = J_d)), \tag{15}$$

then Equation (14) implies

$$I_a^{3L} \oplus I_b^{3L} \oplus I_c^{3L} \oplus I_d^{3L} = 0. \tag{16}$$

Since we have $J_a \oplus J_d = J_b \oplus J_c$ (both are equal to the subkey difference in $KI_{4,3,1}$), each one of the two conditions in Equation (15) is expected to hold[1] with probability $2^{-7}$. Therefore, combining Equations (13), (14), and (16) we get that the condition

$$X_a^3 \oplus X_b^3 \oplus X_c^3 \oplus X_d^3 = 0 \tag{17}$$

holds with probability $2^{-6}$.

Finally, since the $FL$ function is linear for a given key and there is no key difference in $FL4$, we can conclude that whenever Equation (17) holds, the outputs of the F-function in round 4 (denoted by $(O_a^4, O_b^4, O_c^4, O_d^4)$) satisfy

$$O_a^4 \oplus O_b^4 \oplus O_c^4 \oplus O_d^4 = 0 \tag{18}$$

with probability $2^{-6}$. Since by condition (8),

$$Y_a^L \oplus Y_b^L \oplus Y_c^L \oplus Y_d^L = 0,$$

it follows that

$$X_a^L \oplus X_b^L \oplus X_c^L \oplus X_d^L = 0 \tag{19}$$

also holds with probability $2^{-6}$. Combining it with Equation (9) yields

$$\mathbf{Pr}\left[(X_c \oplus X_d = \beta) \middle| (X_a \oplus X_b = \beta) \wedge (Y_a \oplus Y_c = \gamma) \wedge (Y_b \oplus Y_d = \gamma)\right] = 2^{-6}. \tag{20}$$

Therefore, the overall probability of the related-key sandwich distinguisher is

$$(1/4)^2 \cdot (1/4)^2 \cdot 2^{-6} = 2^{-14}, \tag{21}$$

which is much higher than the probability of $(1/4)^2 \cdot (1/4)^2 \cdot 2^{-32} = 2^{-40}$ which is expected by the naive analysis of the sandwich structure.

## 4.2   Experimental Verification

To verify the properties of the new distinguisher, we used the official code available as an appendix in [24]. The verification experiment was set up as follows: In each test we randomly chose a key quartet satisfying the required key differences. We then generated $2^{16}$ quartets by following the boomerang procedure described above. We utilized a slight improvement of the first differential suggested in [13] that increases its probability in the encryption direction by a factor of 2 by fixing the value of two plaintext bits. Hence, we expect the number of right quartets in each test to be distributed according to a Poisson distribution with a mean value of $2^{16} \cdot 2^{-14} \cdot 2 = 8$. We repeated the test 100,000 times, and obtained a distribution which is extremely close to the expected distribution. The full results are summarized in Table 3.

---

[1] This estimate is based on a randomness assumption that could be inaccurate in our case due to dependence between the differential characteristics. However, the experiments presented below verify that this probability is indeed as expected.

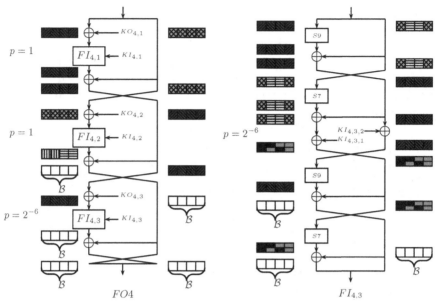

Values marked by the same color and style are equal. Values marked by $\mathcal{B}$ are balanced (i.e., the XOR of all four values is 0). The values in $FI_{4,3}$ which are either smooth gray or smooth black suggest one of two possible cases.

**Fig. 5.** The Development of Differences in $FO_4$ and in $FI_{4,3}$

## 5    Related-Key Sandwich Attack on the Full KASUMI

Our attack on the full KASUMI (depicted in Figure 6) applies the distinguisher presented in Section 4 to rounds 1–7, and retrieves subkey material in round 8. Let $\Delta K_{ab} = (0, 0, 8000_x, 0, 0, 0, 0, 0)$ and $\Delta K_{ac} = (0, 0, 0, 0, 0, 0, 8000_x, 0)$, and let $K_a$, $K_b = K_a \oplus \Delta K_{ab}$, $K_c = K_a \oplus \Delta K_{ac}$, and $K_d = K_c \oplus \Delta K_{ab}$ be the unknown related keys we wish to retrieve.

The attack algorithm is as follows:

1. **Data Collection Phase:**
   (a) Choose a structure of $2^{24}$ ciphertexts of the form $C_a = (X_a, A)$, where $A$ is fixed and $X_a$ assumes $2^{24}$ arbitrary different values. Ask for the decryption of all the ciphertexts under the key $K_a$ and denote the plaintext corresponding to $C_a$ by $P_a$. For each $P_a$, ask for the encryption of $P_b = P_a \oplus (0_x, 0010\ 0000_x)$ under the key $K_b$ and denote the resulting ciphertext by $C_b$. Store the pairs $(C_a, C_b)$ in a hash table indexed by the 32-bit value $C_b^R$ (i.e., the right half of $C_b$).
   (b) Choose a structure of $2^{24}$ ciphertexts of the form $C_c = (Y_c, A \oplus 0010\ 0000_x)$, where $A$ is the same constant as before, and $Y_c$ assumes $2^{24}$ arbitrary different values. Ask for the decryption of the ciphertexts under the key $K_c$ and denote the plaintext corresponding to $C_c$ by $P_c$. For each $P_c$, ask for the encryption of $P_d = P_c \oplus (0_x, 0010\ 0000_x)$ under the key $K_d$ and denote

the resulting ciphertext by $C_d$. Then, access the hash table in the entry corresponding to the value $C_d^R \oplus 0010\ 0000_x$, and for each pair $(C_a, C_b)$ found in this entry, apply Step 2 on the quartet $(C_a, C_b, C_c, C_d)$.

In the first step described above, the $(2^{24})^2 = 2^{48}$ possible quartets are filtered according to a condition on the 32 difference bits which are known (due to the output difference $\delta$ of the distinguisher), which leaves about $2^{16}$ quartets with the required differences.

In Step 2 we can identify the right quartets instantly using an extremely lucky property of the KASUMI structure. We note that a pair $(C_a, C_c)$ can be a right quartet if and only if

$$C_a^L \oplus FL8(FO8(C_a^R)) = C_c^L \oplus FL8(FO8(C_c^R)), \tag{22}$$

since by the Feistel structure, this is the only case of which the difference after round 7 is the output difference of the sandwich distinguisher (i.e., $\delta = (0_x, 0010\ 0000_x)$). However, the values $C_a^R$ and $C_c^R$ are fixed for all the considered ciphertexts, and hence Equation (22) yields

$$C_a^L \oplus C_c^L = FL8(FO8(A)) \oplus FL8(FO8(A \oplus 0010\ 0000_x)) = const. \tag{23}$$

Thus, the value $C_a^L \oplus C_c^L$ is equal for all the right quartets. This allows us to perform the following simple filtering:

2. **Identifying the Right Quartets:**
    (a) Insert the approximately $2^{16}$ remaining quartets $(C_a, C_b, C_c, C_d)$ into a hash table indexed by the 32-bit value $C_a^L \oplus C_c^L$, and apply Step 3 only to bins which contain at least three quartets.

Since the probability of a 3-collision in a list of $2^{16}$ random 32-bit values is lower than $\binom{2^{16}}{3} \cdot 2^{-64} \leq 2^{-18}$, with very high probability only the right quartets remain after this filtering.

In the following step, we treat all the remaining quartets as right quartets. Under this assumption, we know not only the actual inputs to round 8, but also the differences in the outputs of round 8.

3. **Analyzing Right Quartets:**
    (a) For each remaining quartet $(C_a, C_b, C_c, C_d)$, guess the 32-bit value of $KO_{8,1}$ and $KI_{8,1}$. For the two pairs $(C_a, C_c)$ and $(C_b, C_d)$ use the value of the guessed key to compute the input and output differences of the OR operation in the last round of both pairs. For each bit of this 16-bit OR operation of $FL8$, the possible values of the corresponding bit of $KL_{8,2}$ are given in Table 2. On average $(8/16)^{16} = 2^{-16}$ values of $KL_{8,2}$ are suggested by each quartet and guess of $KO_{8,1}$ and $KI_{8,1}$.[2] Since all the right quartets suggest the same key, all the wrong keys are discarded with overwhelming probability, and the attacker obtains the correct value of $(KO_{8,1}, KI_{8,1}, KL_{8,2})$.

---

[2] The simple proof of this claim is given in Section 4.3 of [8].

**Table 2.** Possible Values of $KL_{8,2}$ and $KL_{8,1}$

| OR — $KL_{8,2}$ | | | | | AND — $KL_{8,1}$ | | | | |
|---|---|---|---|---|---|---|---|---|---|
| | $(X'_{bd}, Y'_{bd})$ | | | | | $(X'_{bd}, Y'_{bd})$ | | | |
| $(X'_{ac}, Y'_{ac})$ | (0,0) | (0,1) | (1,0) | (1,1) | $(X'_{ac}, Y'_{ac})$ | (0,0) | (0,1) | (1,0) | (1,1) |
| (0,0) | {0,1} | — | 1 | 0 | (0,0) | {0,1} | — | 0 | 1 |
| (0,1) | — | — | — | — | (0,1) | — | — | — | — |
| (1,0) | 1 | — | 1 | — | (1,0) | 0 | — | 0 | — |
| (1,1) | 0 | — | — | 0 | (1,1) | 1 | — | — | 1 |

* The two bits of the differences are denoted by (input difference, output difference): $(X'_1, Y'_1)$ for one pair and $(X'_2, Y'_2)$ for the other pair.

**Table 3.** The Number of Right Quartets in 100,000 Experiments

| Right Quartets | 0 | 1 | 2 | 3 | 4 | 5 | 6 | 7 | 8 |
|---|---|---|---|---|---|---|---|---|---|
| Theory $(Poi(8))$ | 34 | 268 | 1,073 | 2,863 | 5,725 | 9,160 | 12,214 | 13,959 | 13,959 |
| Experiment | 32 | 259 | 1,094 | 2,861 | 5,773 | 9,166 | 12,407 | 13,960 | 13,956 |

| Right Quartets | 9 | 10 | 11 | 12 | 13 | 14 | 15 | 16 | 17 |
|---|---|---|---|---|---|---|---|---|---|
| Theory $(Poi(8))$ | 12,408 | 9,926 | 7,219 | 4,813 | 2,962 | 1,692 | 903 | 451 | 212 |
| Experiment | 12,230 | 9,839 | 7,218 | 4,804 | 3,023 | 1,672 | 859 | 472 | 219 |

| Right Quartets | 18 | 19 | 20 | 21 | 22 | 23 | 24 | 25 |
|---|---|---|---|---|---|---|---|---|
| Theory $(Poi(8))$ | 94 | 40 | 16 | 6 | 2 | 0.8 | 0.26 | 0.082 |
| Experiment | 89 | 39 | 13 | 12 | 2 | 0 | 0 | 1 |

(b) Guess the 32-bit value of $KO_{8,3}$ and $KI_{8,3}$, and use this information to compute the input and output differences of the AND operation in both pairs of each quartet. For each bit of the 16-bit AND operation of $FL8$, the possible values of the corresponding bit of $KL_{8,1}$ are given in Table 2. On average $(8/16)^{16} = 2^{-16}$ values of $KL_{8,1}$ are suggested by each quartet and guess of $KO_{8,3}$, $KI_{8,3}$, and thus the attacker obtains the correct value of $(KO_{8,3}, KI_{8,3}, KL_{8,1})$.

4. **Finding the Right Key:** For each value of the 96 bits of $(KO_{8,1}, KI_{8,1}, KO_{8,3}, KI_{8,3}, KL_{8,1}, KL_{8,2})$ suggested in Step 3, guess the remaining 32 bits of the key, and perform a trial encryption.

The data complexity of the attack is $2^{25}$ chosen ciphertexts and $2^{25}$ adaptively chosen plaintexts encrypted/decrypted under one of four keys. The time complexity is dominated by the trial encryptions performed in step 4 to find the last 32 bits of the key, and thus it is approximately equal to $2^{32}$ encryptions. The probability of success is approximately 76% (this is the probability of having at least three right pairs in the data pool).

The memory complexity of the attack is also very moderate. We just need to store $2^{26}$ plaintext/ciphertext pairs, where each pair takes 16 bytes. Hence, the total amount of memory used in the attack is $2^{30}$ bytes, i.e., 1 GByte of memory.

**Table 4.** The Number of Identified Right Quartets in 1,000 tests

| Right Quartets | 0/1/2 | 3 | 4 | 5 | 6 | 7 | 8 | 9 | 10 | 11 | 12 |
|---|---|---|---|---|---|---|---|---|---|---|---|
| Theory $(Poi(4))$ | 238 | 195 | 195 | 156 | 104 | 60 | 30 | 13 | 5 | 2 | 0.6 |
| Experiment | 247 | 197 | 180 | 167 | 112 | 52 | 30 | 7 | 4 | 3 | 1 |

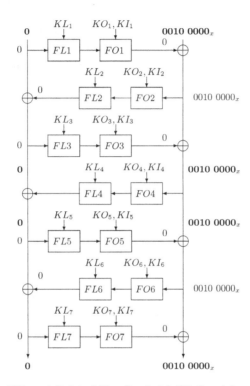

**Fig. 6.** The 7-Round Related-Key Sandwich Distinguisher of KASUMI

### 5.1   Experimental Verification

We performed two types of experiments to verify our attack. In the first experiment, we just generated the required data, and located the right quartets (thus verifying the correctness of our randomness assumptions). The second experiment was the application of the full attack (both with and without the final exhaustive search over the remaining 32 key bits). All our experiments were carried out on an Intel Core Duo 2 machine with a T7200 CPU (2 GHz, 4 MB L2 Cache, 2 GB RAM, Linux-2.6.27 kernel, with gcc 4.3.2 and standard optimization flags (-O3, -fomit-frame-pointers, -funroll-loops), single core, single thread).

The first experiment was conducted 1,000 times. In each test, we generated the data and found candidate quartets according to Steps 1 and 2 of the attack algorithm. Once these were found, we partially decrypted the quartets, and checked how many quartets were right ones. Table 4 details the outcome of these experiments, which follow the expected distribution.

The second experiment simulated the full attack. We repeated it 100 times, and counted in each case how many times the final exhaustive search over $2^{32}$ possible keys would have been evoked. In 78 out of these 100 experiments, the key was found when 3 or more quartets were identified to be right ones (the expected number was 76.1).

About 50% of the tests were able to identify the right key by invoking either 2 or 4 exhaustive searches. As the first part of the attack (which identifies candidate quartets) takes about 8 minutes, and each exhaustive search (using the official KASUMI source code) takes about 26 minutes, we could find the full 128 bit key in about 50% of our tests in less than 112 minutes (using a single core). It is important to note that by increasing the running time, one can increase the success rate of the attack without increasing its data requirements.

## 6   Summary

In this paper we develop a new sandwich attack on iterated block ciphers, and use it to reduce the time complexity of the best known attack on the full KASUMI from an impractical $2^{76}$ to the very practical $2^{32}$. However, the new attack uses both related keys and chosen messages, and thus it might not be applicable to the specific way in which KASUMI is used as the A5/3 encryption algorithm in third generation GSM telephony. Our main point was to show that contrary to the assurances of its designers, the transition from MISTY to KASUMI led to a much weaker cryptosystem, which should be avoided in any application in which related-key attacks can be mounted.

## References

1. A5/1 Security Project, Creating A5/1 Rainbow Tables (2009),
   http://reflextor.com/trac/a51
2. Barkan, E., Biham, E.: Conditional Estimators: an Effective Attack on A5/1. In: Preneel, B., Tavares, S. (eds.) SAC 2005. LNCS, vol. 3897, pp. 1–19. Springer, Heidelberg (2006)
3. Barkan, E., Biham, E., Keller, N.: Instant Ciphertext-Only Cryptanalysis of GSM Encrypted Communication. In: Boneh, D. (ed.) CRYPTO 2003. LNCS, vol. 2729, pp. 600–616. Springer, Heidelberg (2003)
4. Biham, E.: New Types of Cryptanalytic Attacks Using Related Keys. Journal of Cryptology 7(4), 229–246 (1994)
5. Biham, E., Dunkelman, O., Keller, N.: The Rectangle Attack — Rectangling the Serpent. In: Pfitzmann, B. (ed.) EUROCRYPT 2001. LNCS, vol. 2045, pp. 340–357. Springer, Heidelberg (2001)
6. Biham, E., Dunkelman, O., Keller, N.: New Results on Boomerang and Rectangle Attacks. In: Daemen, J., Rijmen, V. (eds.) FSE 2002. LNCS, vol. 2365, pp. 1–16. Springer, Heidelberg (2002)

7. Biham, E., Dunkelman, O., Keller, N.: Related-Key Boomerang and Rectangle Attacks. In: Cramer, R. (ed.) EUROCRYPT 2005. LNCS, vol. 3494, pp. 507–525. Springer, Heidelberg (2005)
8. Biham, E., Dunkelman, O., Keller, N.: A Related-Key Rectangle Attack on the Full KASUMI. In: Roy, B. (ed.) ASIACRYPT 2005. LNCS, vol. 3788, pp. 443–461. Springer, Heidelberg (2005)
9. Biryukov, A.: The Boomerang Attack on 5 and 6-Round Reduced AES. In: Dobbertin, H., Rijmen, V., Sowa, A. (eds.) AES 2005. LNCS, vol. 3373, pp. 11–15. Springer, Heidelberg (2005)
10. Biryukov, A., De Cannière, C., Dellkrantz, G.: Cryptanalysis of SAFER++. In: Boneh, D. (ed.) CRYPTO 2003. LNCS, vol. 2729, pp. 195–211. Springer, Heidelberg (2003)
11. Biryukov, A., Khovratovich, D.: Related-key Cryptanalysis of the Full AES-192 and AES-256. In: Matsui, M. (ed.) ASIACRYPT 2009. LNCS, vol. 5912, pp. 1–18. Springer, Heidelberg (2009)
12. Biryukov, A., Shamir, A., Wagner, D.: Real Time Cryptanalysis of A5/1 on a PC. In: Schneier, B. (ed.) FSE 2000. LNCS, vol. 1978, pp. 1–18. Springer, Heidelberg (2001)
13. Blunden, M., Escott, A.: Related Key Attacks on Reduced Round KASUMI. In: Matsui, M. (ed.) FSE 2001. LNCS, vol. 2355, pp. 277–285. Springer, Heidelberg (2002)
14. Briceno, M., Goldberg, I., Wagner, D.: A Pedagogical Implementation of the GSM A5/1 and A5/2 "voice privacy" encryption algorithms (1999), http://cryptome.org/gsm-a512.htm
15. Dunkelman, O., Keller, N.: An Improved Impossible Differential Attack on MISTY1. In: Pieprzyk, J. (ed.) ASIACRYPT 2008. LNCS, vol. 5350, pp. 441–454. Springer, Heidelberg (2008)
16. Ekdahl, P., Johansson, T.: Another Attack on A5/1. IEEE Transactions on Information Theory 49(1), 284–289 (2003)
17. Golic, J.D.: Cryptanalysis of Alleged A5 Stream Cipher. In: Fumy, W. (ed.) EUROCRYPT 1997. LNCS, vol. 1233, pp. 239–255. Springer, Heidelberg (1997)
18. Hong, S., Kim, J., Kim, G., Lee, S., Preneel, B.: Related-Key Rectangle Attacks on Reduced Versions of SHACAL-1 and AES-192. In: Gilbert, H., Handschuh, H. (eds.) FSE 2005. LNCS, vol. 3557, pp. 368–383. Springer, Heidelberg (2005)
19. Kelsey, J., Schneier, B., Wagner, D.: Key Schedule Cryptanalysis of IDEA, G-DES, GOST, SAFER, and Triple-DES. In: Koblitz, N. (ed.) CRYPTO 1996. LNCS, vol. 1109, pp. 237–251. Springer, Heidelberg (1996)
20. Kim, J., Kim, G., Hong, S., Hong, D.: The Related-Key Rectangle Attack — Application to SHACAL-1. In: Wang, H., Pieprzyk, J., Varadharajan, V. (eds.) ACISP 2004. LNCS, vol. 3108, pp. 123–136. Springer, Heidelberg (2004)
21. Kim, J., Hong, S., Preneel, B., Biham, E., Dunkelman, O., Keller, N.: Related-Key Boomerang and Rectangle Attacks, IACR ePrint report 2010/019
22. Matsui, M.: Block encryption algorithm MISTY. In: FSE 1997. LNCS, vol. 1267, pp. 64–74. Springer, Heidelberg (1997)
23. Murphy, S.: The Return of the Boomerang, technical report RHUL-MA-2009-20, Department of Mathematics, Royal Holloway, University of London (2009), http://www.rhul.ac.uk/mathematics/techreports

24. 3rd Generation Partnership Project, Technical Specification Group Services and System Aspects, 3G Security, Specification of the 3GPP Confidentiality and Integrity Algorithms; Document 2: KASUMI Specification, V3.1.1 (2001)

25. 3rd Generation Partnership Project, Technical Specification Group Services and System Aspects, 3G Security, Specification of the A5/3 Encryption Algorithms for GSM and ECSD, and the GEA3 Encryption Algorithm for GPRS; Document 4: Design and evaluation report, V6.1.0 (2002)

26. TECHNEWSWORLD, Hackers Jimmy GSM Cellphone Encryption (published 29/12/2009), `http://www.technewsworld.com/rsstory/68997.html`

27. Wagner, D.: The Boomerang Attack. In: Knudsen, L.R. (ed.) FSE 1999. LNCS, vol. 1636, pp. 156–170. Springer, Heidelberg (1999)

# Universally Composable Incoercibility*

Dominique Unruh[1] and Jörn Müller-Quade[2]

[1] Saarland University
[2] Karlsruhe Institute of Technology (KIT)

**Abstract.** We present the UC/c framework, a general definition for secure and incoercible multi-party protocols. Our framework allows to model arbitrary reactive protocol tasks (by specifying an ideal functionality) and comes with a universal composition theorem. We show that given natural setup assumptions, we can construct incoercible two-party protocols realising arbitrary functionalities (with respect to static adversaries).

**Keywords:** Incoercibility, universal composability, voting.

## 1   Introduction

Commonly, security of a cryptographic protocol encompasses (very roughly) two aspects: The protocol should guarantee that the private data of the parties stays secret (privacy), and it should ensure that all data transferred or computed is correct (integrity). Most security definitions ensure one or both of these requirements, and many protocols are known to satisfy these definitions (e.g., [16,1,11,8,9]).

There is, however, a requirement that does not fall into either category: coercion resistance (first noted by [17,2]). To illustrate this property, we use the example of a voting scheme. In a voting scheme, it might be possible for a voter to acquire a receipt that he cast a specific vote. This does not violate the anonymity of the voter since the voter is not required to reveal or even acquire such a receipt. Thus privacy is maintained. And getting a receipt does not allow to falsify the outcome of the election. Thus the integrity of the scheme is maintained. Yet the mere possibility of acquiring a receipt may make a party coercible. A coercive adversary may threaten certain reprisals if the party does not cast a specific vote and proves this by delivering a receipt to the adversary. Thus such an election protocol would not be coercion resistant (short: incoercible).

Incoercibility is an important property in any setting in which some malicious agent has the power to harm and thus threaten other protocol participants. Clearly, this is not restricted to the setting of voting but may be the case in other settings, too (e.g., when financial transactions are involved). Unfortunately, incoercibility turns out to be both difficult to define and to achieve.

* Partially funded by the Cluster of Excellence "Multimodal Computing and Interaction".

T. Rabin (Ed.): CRYPTO 2010, LNCS 6223, pp. 411–428, 2010.

Previous definitions of incoercibility are usually restricted to special domains such as voting (e.g., [2,19,13]). An exception are the models by Canetti and Gennaro [4] and by Moran and Naor [20] which give general definitions of incoercible multi-party computation. Their definitions are, however, restricted to the case of secure function evaluation. That is, they only consider protocols in which all parties need to first contribute their inputs, and then from these inputs the outputs for the parties are computed. Reactive protocols, protocols that have multiple phases and where the inputs in one phase can depend on the outputs of an earlier phase, are excluded. For example, the security of a commitment protocol could not be modelled in their settings.

Besides the problem of reactive protocols, the issue of composability arises. When building a complex protocol, it is often necessary to abstract from certain subprotocols in the analysis to make the analysis manageable. For example, one might first analyse the protocol assuming a perfectly secure mechanism for performing commitments (modelled by a trusted machine), and then later on prove the security of the subprotocol that is actually used for the commitments. To do so, and also to have a guarantee that the protocol does not become insecure when executed together with other protocols or instances of itself, one needs a security notion that comes with a composition theorem.

In the case of normal secure multi-party computation (i.e., without incoercibility) both the problem of modelling reactive protocols and of giving strong compositionality guarantees has been solved by Canetti's UC model [6]. In this model, we can define a protocol task by specifying a trusted machine, the ideal functionality, which by definition performs the required protocol task. Since this machine can interact with its environment in arbitrary ways, the security of very general reactive protocols can be modelled. Furthermore, the UC model guarantees that if a protocol is secure when using (as opposed to realising) an ideal functionality, then the protocol stays secure when instead of the ideal functionality, a subprotocol that securely realises the ideal functionality is used. The UC model, however, does not guarantee incoercibility.

**Our contribution.** We define the Composable Incoercibility framework (UC/c) which is an extension of the UC framework. Like UC, UC/c allows to model very general reactive protocol tasks and gives strong compositionality guarantees (universal composition). Additionally, protocols secure with respect to UC/c are incoercible. To illustrate the model, we show that a voting scheme that is UC/c secure is also incoercible with respect to a definition tailored specifically to voting. Finally, we show that in the restricted case of static coercions/deceptions (all corruptions and coercions happen at the beginning of the protocol), arbitrary UC/c secure two-party computation is possible assuming the availability of secure channels.

**Organisation.** In Section 1.1, we explain the intuition behind the UC/c framework. In Section 2 we define the UC/c framework and present the universal composition theorem. In Section 3 we illustrate our model by applying it to the setting of voting protocols. In Section 4 we present general feasibility results for two-party protocols. In Section 5 we give directions for further work.

## 1.1   The Intuition Behind UC/c

To understand the UC/c model, we first need to get an intuition of how inco-ercibility is achieved. The goal of an incoercible protocol is the following: When an adversary tries to coerce a party into performing a certain action (such as casting a particular vote $v^*$), the party should be able to perform the action it originally intended to perform (casting a vote $v$) without the adversary noticing. That is, the adversary should not be able to tell the difference between a party $P$ that follows the adversary's instructions (a *corrupted* party, casting the vote $v^*$) and a party $P$ that only tries to make the adversary believe that it follows the adversary's instructions (a *deceiving* party, casting the vote $v$ and giving fake evidence to the adversary that it cast the vote $v^*$).

The most natural way to define incoercibility would be to require that the adversary cannot distinguish between a coerced and a deceiving party. This, however, usually cannot be achieved. For example, in a voting protocol the adversary will eventually learn the tally. The distribution of the tally will, since there are only polynomially many voters, slightly but noticeably change when the vote of $P$ changes from $v$ to $v^*$. The adversary can hence distinguish coerced and deceiving parties by observing the tally.

Thus, we have to weaken the requirement. The adversary should not be able to distinguish a coerced and a deceiving party any better than he could do given only information that is "legally" available to him (the tally in our example). In general, however, it is not straightforward to define what information is "legally" available to the adversary in any particular situation. Neither is it straightforward to determine how much distinguishing advantage the adversary would get given only that information.

In order to circumvent this problem, we use a slightly different approach: We first define an ideal model in which the adversary has, by definition, only access to the "legally" available information. In the case of voting, such an ideal model would consist of a trusted machine (the ideal voting functionality $\mathcal{F}$) that collects the votes from all parties and gives only the tally to the adversary. In the ideal model, the distinguishing advantage between a coerced party (that gives $v^*$ to $\mathcal{F}$) and a deceiving party (that gives $v$ to $\mathcal{F}$) is, by definition, exactly the advantage the adversary gets from the "legally" available information (the tally).

To make this definition more formal, we introduce an additional entity, the deceiver [14]. The task of the deceiver is to instruct a deceiving party what it should do (i.e., how to deceive the adversary). More formally, a deceiving party will not run any program of its own, but instead follow the instructions of the deceiver. (In a sense, the deceiver models the party's free will.) In particular, the deceiver may instruct a party to cast a vote $v$ and to send to the adversary the fake notification that it cast vote $v^*$. (Since we are in the ideal model, no cryptographic receipts or similar need to be faked.) A corrupted party, on the other hand, will follow the adversaries instructions.

The combination of adversary and deceiver in the ideal model now allows to model any coercion situation that can occur in the ideal model. To define what it means that the real protocol is incoercible (or more precisely, as incoercible as the ideal model), we will use the concept of simulation that underlies many cryptographic definitions such as multi-party computation and zero-knowledge: We show that for any adversary in the real model that performs some coercion attack, there is another adversary in the ideal model (called the adversary-simulator) that performs a corresponding attack with as much success. In other words, we require that for any deceiver (specifying what a party would ideally want to do), and for any adversary in the real model (trying to coerce parties), there is an adversary-simulator in the ideal model such that the real and the ideal model are indistinguishable.

We are, however, missing one ingredient: We need to specify how the ideal deceptions (specified in terms of inputs to the ideal functionalities) translate into real deceptions (specified in terms of faked messages etc.). This is done by introducing a deceiver in the real model, too, called the deceiver-simulator. We then require that for any deceiver in the ideal model (representing a possible deception) there is a deceiver-simulator in the real model (that performs the corresponding real deceptions) such that for any adversary in the real model there is a adversary-simulator in the ideal model such that the two models are indistinguishable.

Finally, to model the indistinguishability of the two models, we follow the ideas from the UC framework and introduce a further machine, the environment, that either communicates with the machines in the real model or with the machines in the ideal model and that has to guess which model it is in. (For details on how this indistinguishability actually ensures that the adversary's advantage in distinguishing corrupted and deceiving parties carries over from the ideal to the real model we refer to the example in Section 3.)

## 1.2   Related Work

We are aware of only two works that tackle the problem of defining incoercibility or a similar property in a general fashion (i.e., not specialised to a particular protocol task such as voting).

**Incoercible secure function evaluation (Canetti-Gennaro, Moran-Naor).** Canetti and Gennaro [4] present a model for defining incoercible secure function evaluation which was subsequently refined by Moran and Naor [20]. The model by Moran and Naor is based on the so-called stand-alone model [5,15, Ch. 7]. In this model, one assumes that the inputs of all honest parties are fixed before the beginning of the protocol. This has several implications: First, reactive protocols where parties may decide on their inputs in later phases cannot be modelled. Second, when actually deploying the protocol, one would have to ensure very strong synchronisation: In order not to introduce possibilities for attacks not covered by the model, we have to ensure that no protocol message is sent until all honest parties have decided on their input. Third, the stand-alone model only guarantees sequential composability.[1] That is, we have no guarantee that the protocol stays secure when running concurrently with other protocols (which usually happens in real-life networks).

Since the model by Moran and Naor is based on the stand-alone model, in this model coerced parties only need to lie about their initial inputs. Because of this, Moran and Naor do not need to introduce an explicit deceiver; any deception a party might want to perform can be encoded by specifying a second input, the so-called "fake input". In contrast, the more complex deceptions that are possible in our setting necessitate the introduction of an explicit machine, the deceiver, to specify the deceptions.

Everything we said about the work by Moran and Naor also applies to the earlier work by Canetti and Gennaro [4]. Furthermore, the model by Canetti and Gennaro only models a very weak form of coercion-resistance; the adversary may instruct a coerced party to use a different input, but he may not instruct that party to deviate from the protocol. For a discussion of the difference between the models by Moran and Naor and by Canetti and Gennaro, we refer to [20].

**Externalized UC (deniability).** Another approach to define properties similar to incoercibility for general protocols is the Externalized UC (EUC) framework proposed by Canetti, Dodis, Pass, and Walfish [7] (also known as Generalized UC, UC with global setup, or, proposed independently by Hofheinz, Müller-Quade, and Unruh [18], UC with catalysts).

This framework is, like ours, an extension of the UC framework and inherits its support for reactive protocols and its universal composition theorem. The EUC framework differs from the UC framework by allowing the environment to directly access the ideal functionality used in the real protocol. As explained in [7], security in the EUC framework implies a property called deniability. This means that no (malicious) protocol party $P$ can collect any information during the protocol run that can later be used to prove to an outsider that some party $Q$ participated in the protocol. (An example for such incriminating information would be a message signed by $Q$.) In other words, $Q$ can plausibly claim that the whole protocol did not take place. Obviously, such a claim is only realistic with respect to an outsider who did not himself communicate with $Q$ during the protocol execution. In contrast, incoercibility as understood by this paper means that a party can lie about its actions towards an insider (e.g., a party could lie even towards another voter about the vote it has cast).

Thus the two models (EUC and UC/c) have very different aims. Technically they are, however, related: In the full version [21] we show that under certain conditions, EUC security implies UC/c security.

---

[1] Note that it has not been shown that the variant of the stand-alone model presented by Moran and Naor does compose sequentially. But it does not seem unlikely that this could be shown.

# 2   The Composable Incoercibility Framework (UC/c)

## 2.1   Review of the UC Framework

Our model is based on the Universal Composability (UC) framwork [6]. For self containment and to fix notation, we give a short overview over the UC framework. An interactive Turing machine (ITM) is a Turing machine that has additional tapes for incoming and for outgoing communication. An ITM may be activated by a message on an incoming communication tape. At the end of an activation, the ITM may send a message on an outgoing communication tape to another ITM. The recipient of a message is addressed by the unique identity of that ITM. The actions of an ITM may depend on a global parameter $k \in \mathbb{N}$, the so-called security parameter.

A network is modeled as a (possibly infinite) set of ITMs.[2] We call a network $S$ executable if it contains an ITM $\mathcal{Z}$ with distinguished input and output tape and with the special identity env. An execution of $S$ with input $z \in \{0,1\}^*$ and security parameter $k \in \mathbb{N}$ is the following random process: First, $\mathcal{Z}$ is activated with the message $z$ on its input tape. Whenever an ITM $M_1 \in S$ finishes an activation with an outgoing message $m$ addressed to another ITM $M_2 \in S$ on its outgoing communication tape, the other ITM $M_2$ is invoked with incoming message $m$ on its incoming communication tape (tagged with the identity of the sender $M_1$). If an ITM terminates its activation without an outgoing message or sends a message to a non-existing ITM, the ITM $\mathcal{Z}$ is activated. When the ITM $\mathcal{Z}$ sends a message on its output tape (not the communication tape!), the execution of $S$ terminates. The output of $\mathcal{Z}$ we denote by $\mathrm{EXEC}_S(k, z)$. An ITM $\mathcal{Z}$ with identity env we call an environment and an ITM $\mathcal{A}$ with identity adv we call an adversary. A protocol is a network that does not contain an environment or an adversary.

We call networks $S, S'$ indistinguishable if there is a negligible function $\mu$ such that for all $k \in \mathbb{N}$, $z \in \{0,1\}^*$, we have that $|\Pr[\mathrm{EXEC}_S(k,z) = 1] - \Pr[\mathrm{EXEC}_{S'}(k,z) = 1]| \leq \mu(k)$. We call $S, S'$ perfectly indistinguishable if $\mu = 0$.

Using the above network model, security is defined by comparison. We first define an ideal protocol $\rho$ that specifies the intended protocol behaviour. Then we define what it means that another protocol $\pi$ (securely) emulates $\rho$:

**Definition 1 (UC [6]).** *Let $\pi$ and $\rho$ be protocols. We say that $\pi$ UC emulates $\rho$ if for any polynomial-time adversary $\mathcal{A}$ there exists a polynomial-time adversary $\mathcal{S}$ (the adversary-simulator) such that for any polynomial-time environment $\mathcal{Z}$ the networks $\pi \cup \{\mathcal{A}, \mathcal{Z}\}$ (called the real model) and $\rho \cup \{\mathcal{S}, \mathcal{Z}\}$ (called the ideal model) are indistinguishable.*

In the UC framework, one can model secure channels (that do not even leak the length of the transmitted message) by direct communication between the ITMs; insecure channels can be modelled by sending messages to the adversary; secure

---

[2] In the case of infinite networks we require the network to be uniform in the sense that given the identity of an ITM, we can compute the code of that ITM in deterministic polynomial-time.

channels that leak the length of the message, as well as authenticated channels can be modelled as an ideal functionality.

Corruptions are modelled as follows: The environment $\mathcal{Z}$ can send special corruption requests to protocol parties (which are ITMs in $\pi$). If a protocol party receives such a request, it sends its current state to the adversary and from then on is controlled by the adversary (i.e., it forwards all incoming communication to the adversary and vice versa).

Usually, the ideal model will be described by a so-called ideal functionality. Such an ideal functionality is an incorruptible ITM that can be seen as a trusted third party accessible to the protocol parties. The ideal protocol corresponding to $\mathcal{F}$ consists of $\mathcal{F}$ itself and a so-called dummy-party $\tilde{P}$ for each party $P$ in the real model. The dummy-party $\tilde{P}$ simply forwards all messages received from the environment to $\mathcal{F}$ and vice versa. In slight abuse of notation, we write $\mathcal{F}$ for the ideal protocol corresponding to $\mathcal{F}$. Note that the dummy-parties can be corrupted, hence the inputs and outputs to $\mathcal{F}$ from corrupted parties can be influenced by the adversary-simulator. Using the concept of an ideal functionality, we can express many protocol tasks by first specifying an ideal functionality $\mathcal{F}$ that fulfils the protocol task by definition, and then requiring that the protocol $\pi$ UC emulates $\mathcal{F}$.

We can also consider real protocols $\pi$ which contain ideal functionalities $\mathcal{F}$ (e.g., a functionality modelling a CRS). These functionalities can then be accessed by all parties. We then say that $\pi$ is a protocol in the $\mathcal{F}$-hybrid model.

For more details, we refer the reader to the full version of [6].

## 2.2   The Composable Incoercibility Framework (UC/c)

In our framework (UC/c) the possibility of coercions is modelled by the presence of an additional adversarial entity, called the deceiver. Formally, a deceiver is an ITM $\mathcal{D}$ with the special identity dec. We further refine the notion of a protocol: A protocol is a network that does not contain an environment, adversary, or deceiver.

A typical network would consist of a protocol $\pi$, an adversary $\mathcal{A}$, a deceiver $\mathcal{D}$, and an environment $\mathcal{Z}$ (where the adversary and the deceiver may also be called adversary-simulator and deceiver-simulator for clarity depending on their role in the protocol). We put no restriction on the communication between machines, $\mathcal{A}, \mathcal{D}, \mathcal{Z}$ may all communicate with each other. Both the adversary and the deceiver may control parties. The exact mechanism of this is the following:

**Corruption model.** A protocol party may be in one of three corruption states: *Uncontrolled*, *corrupted*, and *deceiving*. We say a party is *controlled* if it is corrupted or deceiving. Initially, all machines are uncontrolled. Uncontrolled parties behave according to the protocol specification. If the environment $\mathcal{Z}$ sends a *corruption request* to an uncontrolled party, the party becomes corrupted. If the environment sends a *deception request* to an uncontrolled or a corrupted party, the party becomes deceiving. When a party becomes corrupted or deceiving, it sends its state to the adversary or the deceiver, respectively. From then on, it is controlled by the adversary or the deceiver, respectively (that is, it forwards all incoming communication to the controlling machine and sends messages as instructed by the controlling machine). The only exception is that if a corrupted machine receives a deception request, it will not forward that request to the adversary, because in that moment, it will become deceiving and hence be under the control of the deceiver. We stress that if a party is deceiving, the adversary cannot even observe that party's communication (unless the party communicates over an insecure channel or with a corrupted party).

We assume the existence of a globally readable register that contains the state of each party (whether it is uncontrolled, corrupted, or deceiving). However, when the adversary reads this register, the state of any deceiving machine will be reported as corrupted. (This reflects the fact that the adversary should not be able to know which machine is deceiving.) Protocol parties will not usually read this register; in some cases, however, it might be useful if the behaviour of an ideal functionality can depend on whether a machine is controlled or not.[3]

**Security definition.** We are now ready to specify the notion of UC/c security. In this notion, we do not only require the adversary-simulator (in the ideal model) to simulate the adversary's actions (in the real model), but simultaneously require that the deceiver-simulator (in the real model) simulates the actions of the deceiver (in the ideal model). The resulting notion is strictly stronger than UC.

**Definition 2 (UC/c).** *Let $\pi$ and $\rho$ be protocols. We say that $\pi$ UC/c emulates $\rho$ if for any polynomial-time deceiver $\mathcal{D}$ there exists a polynomial-time deceiver $\mathcal{D}_S$ (the deceiver-simulator) such that for any polynomial-time adversary $\mathcal{A}$ there exists a polynomial-time adversary $\mathcal{A}_S$ (the adversary-simulator) such that for any polynomial-time environment $\mathcal{Z}$ the following networks are indistinguishable: $\pi \cup \{\mathcal{A}, \mathcal{D}_S, \mathcal{Z}\}$ and $\rho \cup \{\mathcal{A}_S, \mathcal{D}, \mathcal{Z}\}$.*

**Where is the deception strategy?** The existence of a deception strategy that honest parties can follow when being coerced is an essential part of any notion of incoercibility. Such a deception strategy also exists in our model: if we consider the deceiver $\tilde{\mathcal{D}}$ that simply obeys any commands (such as "vote for Bob") sent to it by the environment (we call such a deceiver a dummy-deceiver $\tilde{\mathcal{D}}_S$, see Section 2.4), then the corresponding deceiver-simulator describes how a coerced party should behave in any situation. For an example of how to derive a special purpose deception strategy from $\tilde{\mathcal{D}}_S$, see the proof of Theorem 10.

**Why is the adversary not informed about deceiving parties?** The reader may notice an asymmetry in the definition: While the deceiver learns which party is corrupted and which party is deceiving, the adversary will be told that a party is corrupted even if it is deceiving. This is necessary because during a deception, the goal is to cheat the adversary into thinking that one behaves as he instructs

---

[3] A typical example is the key exchange functionality, which returns a random key for both parties [6, full version]. If one of the parties is corrupted, the key is instead chosen by the adversary. Thus the functionality needs to know which parties are corrupted.

(i.e., that one is corrupted). Therefore corrupted and deceiving parties should be indistinguishable from the point of view of the adversary.

**Why can deceiving party not become corrupted?** Another asymmetry is that a corrupted party can later become deceiving while the model does not allow to corrupt parties that are deceiving. Although formally both directions could be allowed, we have excluded the latter because we could not find an interpretation for such a scenario. For an interpretation of the former direction (bad-guy coercions), see the next section.

## 2.3   Corruption Schedules

The notion of UC/c (Theorem 2) allows the environment to corrupt or coerce any party at any point of time. This leads to a very strict definition. To get a definition that is more lenient but easier to fulfil, one can impose certain restrictions on the corruption and deception requests performed by the environment. We call such a restriction a corruption schedule.

**Bad-guy coercions.** There are no restrictions on the environment (except that the environment cannot corrupt a deceiving party, this is implicit in the modelling of the corruption mechanism).

We call this notion bad-guy coercions because the environment may first corrupt a party (make it a "bad-guy") and then later coerce it. It is very difficult to design protocols that are secure against bad-guy coercions because a corrupted party may be instructed by the adversary to actively deviate from the protocol to produce evidence against itself and thus thwart its own deniability. (In contrast, a deceiving party would, given the same instructions, only try to make the adversary believe that it follows these instructions.)

For example, in some protocol the ability to deceive the adversary (and thus the incoercibility of the protocol) might be based on the following fact: When the adversary requests a private secret $m$ of some party, that party may send a different secret $m'$ instead which contains a trapdoor. This trapdoor then is later essential for achieving incoercibility. In the setting of bad-guy coercions, a party might first be corrupted and then reveal the true secret $m$ to the adversary. This secret $m$ does not contain a trapdoor. Then later, if the party becomes deceiving, it will be unable to follow its deception strategy because it does not know any trapdoor for $m$. In a nutshell, while corrupted, a party may actively try to prevent its own incoercibility. Thus we expect that UC/c security with respect to bad-guy coercions is very hard to achieve.

In practise, bad-guy coercions are arguably a very rare event. A possible motivation for bad-guy coercions is the following thought experiment: A member (say, Bob) of a criminal organisation is required by the rules of that organisation to actively produce and deliver some evidence (e.g., certain keys) against himself to that organisation. While Bob still works for the organisation, he will not try to circumvent these rules and will deliver this evidence. But if Bob later decides to leave the criminal organisation and to cooperate with the police (undercover), Bob may have to convincingly act as if he was still following the criminal organisation's instructions. This is exactly the case that is modelled by bad-guy coercions.

In most cases, however, UC/c with bad-guy coercions will be much to strong a notion, and the notion of good-guy coercions (below) will be preferred.

**Good-guy coercions.** The environment may corrupt parties at any time and may send deception requests to uncontrolled parties at any time. The environment may not send deception requests to corrupted parties.

**Receipt-freeness.** The environment may corrupt parties at any time, and may send deception requests to uncontrolled parties after the end of the protocol (so that the adversary gets their state). The environment may not send deception requests to a corrupted party. Receipt-freeness implies that an honest party does not learn any data during the protocol that could later be used to prove after the protocol execution that the party performed a certain action. (Note that with erasing parties, receipt-freeness is probably easy to achieve: an honest party simply erases all intermediate protocol data.)

**Static corruptions/deceptions.** All corruption and deception requests must be sent at the very beginning of the protocol execution. In particular, this implies that the environment cannot choose which parties to corrupt depending on messages it observes during the protocol execution.

**Combinations.** The above corruptions schedules may be combined by requiring that the environment obeys a certain schedule with respect to some parties and another with respect to other parties. For example, one might have protocols that are UC/c secure with receipt-freeness for Alice and good-guy coercions for Bob.

### 2.4   Properties of UC/c Security

The proofs in this section are omitted for space reasons. They can be found in the full version [21].

**Dummy adversary and deceiver.** A dummy-adversary is an adversary that just follows the instructions of the environment. More precisely, it forwards all messages it receives to the environment, and sends only the messages the environment instructs it to send. It was shown by Canetti [6] in the UC setting that the dummy-adversary is complete, that is, without loss of generality we can consider only the dummy-adversary. Therefore we only have to specify the adversary-simulator for the dummy-adversary instead of having to specify the adversary-simulator for every possible adversary. This simplifies proofs.

In the setting of UC/c, we can additionally consider the dummy-deceiver that just follows the instructions of the environment. Below, we will show that both the dummy-adversary and the dummy-deceiver are complete. Besides strongly simplifying proofs, the completeness of the dummy-deceiver has an additional conceptual advantage. The deceiver-simulator corresponding to the dummy-deceiver encodes a universal deception strategy. That is, for any "ideal deception", it tells us how to perform this deception in the real protocol. The existence of such a universal deception strategy is very important in real life, protocol users need to have an explicit strategy how to lie in which situation; it is not sufficient that such a strategy exists for each situation.

**Definition 3 (Dummy-adversary, dummy-deceiver).** *The dummy-adversary $\tilde{\mathcal{A}}$ is an adversary that, when receiving a message $(id, m)$ from the environment, sends $m$ to the party with identity $id$, and that, when receiving $m$ from a party with identity $id$, sends $(id, m)$ to the environment. The dummy-deceiver $\tilde{\mathcal{D}}$ is defined analogously.*

**Lemma 4 (Completeness of dummy-adversary and dummy-deceiver).** *Let $\pi$ and $\rho$ be protocols. Then $\pi$ UC/c emulates $\rho$ iff $\pi$ UC/c emulates $\rho$ with respect to the dummy-adversary/deceiver (i.e., when only considering adversary $\tilde{\mathcal{A}}$ and deceiver $\tilde{\mathcal{D}}$ in Theorem 2).*

**Universal composition.** One of the main advantages of the UC framework is the universal composition theorem. This theorem guarantees that a UC secure protocol $\pi$ can be securely used as a subprotocol of arbitrary other protocols $\sigma$, even when $\sigma$ and polynomially many instances of $\pi$ run concurrently. The same compositionality result also holds for the UC/c security notion.

To formulate the composition theorem, we introduce some notation. Let $\pi$ and $\sigma$ be protocols. Then let $\sigma^\pi$ denote the protocol where $\sigma$ invokes a polynomial number of instances of the subprotocol $\pi$. That is, machines in $\sigma$ may give inputs to machines in $\pi$, these inputs are treated by $\pi$ as coming from the environment. When the machines in $\pi$ give output back to the environment, these are sent to the invoking machines in $\sigma$. Thus, in a sense, in $\sigma^\pi$, the protocol $\sigma$ plays the role of the environment for the instances of $\pi$. For example, if $\sigma^\mathcal{F}$ is a protocol using a commitment functionality $\mathcal{F}$ (i.e., $\sigma^\mathcal{F}$ is a protocol in the $\mathcal{F}$-hybrid model), then $\sigma^\pi$ would be the protocol that uses the subprotocol $\pi$ instead of using the commitment functionality $\mathcal{F}$. The following theorem guarantees that, if $\pi$ UC/c emulates some other protocol $\rho$ (e.g., $\rho = \mathcal{F}$), we do not loose security if we replace subprotocol invocations of $\rho$ by subprotocol invocations of $\pi$.

**Theorem 5 (Universal composition).** *Let $\pi$, $\rho$, and $\sigma$ be polynomial-time protocols. Assume that $\pi$ UC/c emulates $\rho$. Then $\sigma^\pi$ UC/c emulates $\sigma^\rho$.*

The most common use case of the composition theorem is given by the following corollary:

**Corollary 6.** *Let $\pi$ and $\sigma$ be polynomial-time protocols, and $\mathcal{F}$ and $\mathcal{G}$ be polynomial-time functionalities. Assume that $\pi$ UC/c emulates $\mathcal{F}$ and that $\sigma^\mathcal{F}$ UC/c emulates $\mathcal{G}$. Then $\sigma^\pi$ UC/c emulates $\mathcal{G}$.*

## 3  Voting Schemes

In this section we illustrate the UC/c security notion by applying it to the special case of voting schemes. We give a definition of incoercibility that is tailored to the specific case of voting protocols and show that this definition is implied by the UC/c security notion.

**Definition 7 (Voting scheme).** *Fix sets $\mathcal{V}$ (the set of votes), $\mathcal{T}$ (the set of tallies), $\mathcal{P}$ (the set of voters). A tally function is an efficiently computable function* tally : $(\mathcal{V} \cup \{\bot\})^{\mathcal{P}} \to \mathcal{T}$.

*A voting scheme for* tally *is a two-stage protocol. We call the stages voting phase and tallying phase. In such a protocol, each party $P_i \in \mathcal{P}$ gets an input $v_i \in \mathcal{V} \cup \{\bot\}$ (the vote of $P_i$). $v_i = \bot$ means that the $P_i$ does not participate in the protocol (abstention). In the end of the tallying phase a distinguished party $T$ outputs a value $t \in \mathcal{T}$.*

Typically, $\mathcal{V}$ would be the set of all candidates. In more complex schemes, elements of $\mathcal{V}$ might be, e.g., ordered lists of candidates in order of decreasing precedence. The set of tallies $\mathcal{T}$ usually is the set of all functions $\mathcal{V} \to \mathbb{N}_0$. Alternatively, in a voting scheme which only announces the winner, we would have have $\mathcal{T} = \mathcal{V}$. The tally function tally$(v_1, \ldots, v_n)$ specifies what the correct tally is for the votes $v_i \in \mathcal{V} \cup \{\bot\}$ where $v_i = \bot$ denotes abstention.

Note that we do not require that the parties $P_i \neq T$ are aware whether they are in the tallying or the voting phase. Such a requirement might be difficult to ensure in an asynchronous environment. In particular, votes cast during the tallying phase (but before the tally is announced) might or might not be counted.

An ideal voting scheme is given by the following functionality:

**Definition 8 (Voting functionality).** *The voting functionality $\mathcal{F}_{\mathrm{vote}} = \mathcal{F}_{\mathrm{vote}}^{\mathrm{tally}}$ expects (at most one) message $v_i \in \mathcal{V}$ from each party $P_i \in \mathcal{P}$. When receiving* tally *from $T$, $\mathcal{F}_{\mathrm{vote}}$ sets $v_i := \bot$ for all $P_i \in \mathcal{P}$ from which it did not receive a message $v_i \in \mathcal{V}$ yet. Then $\mathcal{F}_{\mathrm{vote}}$ computes $t := $ tally$(v_i, i \in \mathcal{P})$ (the tally) and sends $t$ to the adversary. Then, when $\mathcal{F}_{\mathrm{vote}}$ receives* deliver *from the adversary, it sends $t$ to the party $T$.*

This functionality models that the tally output by $T$ is correctly computed using the tally function (as long as $T$ itself is not corrupted) and that the individual votes are secret (even if $T$ is corrupted).

Natural properties of voting schemes are, e.g., correctness (the tally is correct even in the presence of an adversary) and anonymity (the adversary cannot tell who voted for whom, except as deducible from the tally itself). We will not formalise these properties here, but it is easy to see that a voting scheme that UC emulates the voting functionality $\mathcal{F}_{\mathrm{vote}}$ satisfies reasonable formalisations of these properties. Since the UC/c security notion is stronger than UC, this implies that these elementary properties are satisfied by UC/c secure voting scheme, too.

In our context, the most interesting property of a voting scheme is incoercibility. We will first formalise what incoercibility means for voting schemes (independently of our framework). Then we will show that incoercibility of voting schemes is implied by security in the UC/c framework. Assume some party $P$ that wants to cast a vote $v$. In an incoercible voting scheme, we expect that if the adversary $\mathcal{A}$ forces a party $P$ to deviate from the protocol, $\mathcal{A}$ should not be able to tell the difference between $P$ obeying the adversary $\mathcal{A}$, or the party $P$ casting the vote $v$ anyway (we say $P$ deceives the adversary). Of course, since the adversary learns the tally, this goal is unachievable – the tally always leaks a non-negligible amount of information about the vote of $P$ (at least if the number of voters is polynomial). We can only achieve the following: The adversary's advantage in distinguishing between $P$ obeying and $P$ deceiving is not greater than the advantage with which the adversary could distinguish these two cases given only the tally. To formulate this definition, we first introduce some notation:

Fix a voter $P \in \mathcal{P}$ and a vote $v \in \mathcal{V} \cup \{\bot\}$. Fix a distribution $\mathcal{B}$ on $(\mathcal{V} \cup \{\bot\})^{\mathcal{P} \setminus \{P\}}$. ($\mathcal{B}$ represents the distribution of the votes of the other voters.) Given a vote $v$, let $\mathcal{B}_v$ denote the distribution over $(\mathcal{V} \cup \{\bot\})^{\mathcal{P}}$ that chooses the votes for all $P_i \in \mathcal{P} \setminus \{P\}$ according to $\mathcal{B}$ and uses the vote $v$ for $P$. Accordingly, $\mathrm{tally}(\mathcal{B}_v)$ denotes the tally resulting from votes chosen according to $\mathcal{B}_v$. Let $\mathrm{Adv}_{ideal}(\mathcal{B}, v) := \max_{v^*} \Delta(\mathcal{B}_v, \mathcal{B}_{v^*})$ where $v^*$ ranges over $\mathcal{V} \cup \{\bot\}$ and $\Delta$ denotes the statistical distance. ($\mathrm{Adv}_{ideal}$ describes how well an adversary can distinguish between being obeyed and being deceived using only the tally.)

A voting adversary is an adversary that controls a party $P$ (however, depending on the setting, $P$ may choose to ignore the instructions given by the adversary) and that may decide when the tallying phase starts. We require that a voting adversary eventually starts the tallying phase. Furthermore, when the party $T$ outputs the tally, the tally is given to the voting adversary. In the end, the voting adversary outputs a bit $b$.

Given a voting adversary $\mathcal{A}$, let $\mathrm{Pr}_{obey}(\mathcal{A}, \mathcal{B})$ be the probability that $\mathcal{A}$ outputs 1 in the case that the party $P$ follows the instructions of the adversary (i.e., $P$ is corrupted) and all other parties honestly follow the protocol (with inputs chosen according to $\mathcal{B}$).

Given some program code $\mathfrak{d}$ (the deception strategy for $P$), let $\mathrm{Pr}_{deceive}(\mathcal{A}, \mathfrak{d}, \mathcal{B})$ denote the probability that the adversary $\mathcal{A}$ outputs 1 if $P$ follows the instructions in $\mathfrak{d}$ and all other parties honestly follow the protocol (with inputs chosen according to $\mathcal{B}$). (Intuitively, $\mathfrak{d}$ is a strategy that tells $P$ how to vote for $v$ and simultaneously make the adversary believe that $P$ obeys the adversary.) We assume that $\mathfrak{d}$ gets $v$ and the identity of $P$ as input. In the same setting, let $\mathrm{Tally}_{deceive}(\mathcal{A}, \mathfrak{d}, \mathcal{B})$ denote the tally output by $T$.

**Definition 9 (Incoercible voting schemes).** *A voting scheme is incoercible if there is a deception strategy $\mathfrak{d}$ such that for every polynomial-time voting adversary, every voter $P \in \mathcal{P}$, every vote $v \in \mathcal{V}$, and every efficiently sampleable distribution $\mathcal{B}$ the following holds:*

- *The deception strategy casts the right vote: The random variables* $\mathrm{Tally}_{deceive}(\mathcal{A}, \mathfrak{d}, \mathcal{B})$ *and* $\mathrm{tally}(\mathcal{B}_v)$ *are computationally indistinguishable.*
- *The adversary cannot distinguish between being obeyed and being deceived: For some negligible function $\mu$ we have that* $\big|\mathrm{Pr}_{obey}(\mathcal{A}, \mathcal{B}) - \mathrm{Pr}_{deceive}(\mathcal{A}, \mathfrak{d}, \mathcal{B})\big| \leq \mathrm{Adv}_{ideal}(\mathcal{B}, v) + \mu.$

Many variants of this definition are possible. For example, one could allow the voting adversary to corrupt additional parties from $\mathcal{P} \setminus \{P\}$. (In this case, one would have to adapt the definition of $\mathrm{Adv}_{ideal}$.) For the sake of simplicity, we do not strive to find the most general formulation of Theorem 9, especially in view of the fact that the UC/c framework already provides us with a very general definition of incoercibility.

We will now show that incoercibility in the sense of Theorem 9 is already implied by UC/c security. We find that the proof of the following theorem is very instructive because it gives some intuition for the UC/c framework, and because it illustrates how application-specific incoercibility definitions (not restricted to the application of voting) can be proven to be implied by UC/c security.

**Theorem 10.** *Let $\pi$ be a voting scheme for the tally function* tally. *Assume that $\pi$ UC/c emulates $\mathcal{F}_{\text{vote}}^{\text{tally}}$ with static corruption/deception. Then $\pi$ is an incoercible voting scheme.*

*Proof.* Fix a voting adversary $\mathcal{A}$. We define the UC/c adversary $\mathcal{A}'$ to behave like $\mathcal{A}$, except that when $\mathcal{A}$ starts the tallying phase, $\mathcal{A}'$ instead sends tally to the environment. When $\mathcal{A}$ would give an output $b$, $\mathcal{A}'$ sends $b$ to the environment.

We define an environment $\mathcal{Z}_{obey} := \mathcal{Z}_{obey}^{P,v,\mathcal{B}}$ as follows: Initially, $\mathcal{Z}_{obey}$ sends a corruption request to the party $P$. Then $\mathcal{Z}_{obey}$ chooses votes $v_1, \ldots, v_n$ according to the distribution $\mathcal{B}$ and gives these votes as input to the parties $P_i \in \mathcal{P} \setminus \{P\}$ (or, if $v_i = \bot$, sends no input to $P_i$). When the adversary sends tally to $\mathcal{Z}_{obey}$, $\mathcal{Z}_{obey}$ sends tally to the party $T$. When the adversary sends $b$ to $\mathcal{Z}_{obey}$, $\mathcal{Z}_{obey}$ terminates with output $b$.

Furthermore, we define $\mathcal{Z}_{deceive} := \mathcal{Z}_{deceive}^{P,v,\mathcal{B}}$ as follows: Initially, $\mathcal{Z}_{deceive}$ sends a deception request to the party $P$. Then $\mathcal{Z}_{deceive}$ chooses votes $v_1, \ldots, v_n$ according to the distribution $\mathcal{B}$ and gives these votes as input to the parties $P_i \in \mathcal{P} \setminus \{P\}$ (or, if $v_i = \bot$, sends no input to $P_i$). Then it sends $v$ to the deceiver. (This will make the deceiver $\mathcal{D}$ defined below instruct $P$ to cast vote $v$.) When the adversary sends tally to $\mathcal{Z}_{deceive}$, $\mathcal{Z}_{deceive}$ sends tally to the party $T$. When the adversary sends $b$ to $\mathcal{Z}_{deceive}$, $\mathcal{Z}_{deceive}$ terminates with output $b$.

We define the deceiver $\mathcal{D}$ as follows: When receiving a state from party $P$, $\mathcal{D}$ instructs $P$ to send this state to the adversary. (This is necessary only for formal reasons: since the adversary should believe that $P$ is corrupted, he expects a state from $P$. Since we are in the case of static corruptions/deceptions, the state is only sent before the start of the protocol and is thus empty.) When $\mathcal{D}$ receives $v$ from the environment, $\mathcal{D}$ instructs $P$ to send $v$ to the functionality $\mathcal{F}_{\text{vote}}$. (I.e., $P$ should cast the vote $v$.) Messages coming from the adversary are ignored. In particular, when the adversary instructs $P$ to cast some other vote, this is ignored.

Since $\pi$ UC/c emulates $\mathcal{F}_{\text{vote}} := \mathcal{F}_{\text{vote}}^{\text{tally}}$, there exist a polynomial-time deceiver-simulator $\mathcal{D}_S$ and a polynomial-time adversary-simulator $\mathcal{A}_S'$ such that for all polynomial-time environments $\mathcal{Z}$, the networks $\pi \cup \{\mathcal{A}', \mathcal{D}_S, \mathcal{Z}\}$ and $\mathcal{F}_{\text{vote}} \cup \{\mathcal{A}_S', \mathcal{D}, \mathcal{Z}\}$ are indistinguishable. (We write $\mathcal{F}_{\text{vote}}$ for the protocol containing $\mathcal{F}_{\text{vote}}$ and the dummy parties.)

By construction,

$$\Pr_{obey}(\mathcal{A}, \mathcal{B}) = \Pr[\text{EXEC}_{\pi \cup \{\mathcal{A}', \mathcal{D}_S, \mathcal{Z}_{obey}\}} = 1]. \tag{1}$$

(We omit the arguments $k, z$ from EXEC for brevity.) Note that since no party is deceiving, the deceiver-simulator $\mathcal{D}_S$ does nothing.

We define the deception strategy $\eth$ as follows: A party $P$ following $\eth$ and wishing to cast the vote $v$ internally simulates $\mathcal{D}_S$. Then $P$ sends the empty state to $\mathcal{D}_S$. (This is done for formal reasons: in the UC/c framework, $\mathcal{D}_S$ would get such an empty state when $P$ is deceiving from the start. Hence this message informs $\mathcal{D}_S$ that $P$ is deceiving.) Then $P$ sends $v$ to the internally simulated $\mathcal{D}_S$ as coming from the environment. Then $P$ follows the instructions that $\mathcal{D}_S$ gives to it. In the case that only $P$ is deceiving, $\mathcal{D}_S$ only sends instructions to $P$. Thus it is not necessary that $P$ simulates any other machines communicating with $\mathcal{D}_S$.

Then, by construction,

$$\text{Pr}_{deceive}(\mathcal{A}, \mathfrak{d}, \mathcal{B}) = \text{Pr}[\text{EXEC}_{\pi \cup \{\mathcal{A}', \mathcal{D}_S, \mathcal{Z}_{deceive}\}} = 1]. \tag{2}$$

Compare the networks $\mathcal{F}_{\text{vote}} \cup \{\mathcal{A}'_S, \mathcal{D}, \mathcal{Z}_{deceive}\}$ and $\mathcal{F}_{\text{vote}} \cup \{\mathcal{A}'_S, \mathcal{D}, \mathcal{Z}_{obey}\}$. In the first network, $\mathcal{Z}_{deceive}$ instructs the dummy-party $\tilde{P}$ (via the deceiver $\mathcal{D}$) to send the vote $v$ to $\mathcal{F}_{\text{vote}}$. In the second network, $\mathcal{A}'_S$ instructs $\tilde{P}$ to send some other vote $v^*$ to $\mathcal{F}_{\text{vote}}$ (where we write $v^* = \bot$ to indicate that $\mathcal{A}'_S$ does not instruct $\tilde{P}$ to vote before $\mathcal{A}'_S$ sends tally to the environment). In the ideal model, $\tilde{P}$ does not receive any incoming messages from other parties. Thus, in both networks, $\mathcal{A}'_S$ does not get any messages from $\tilde{P}$. Thus, $\mathcal{A}'_S$ can only use the tally to distinguish the networks. The distribution of the tally in the network $\mathcal{F}_{\text{vote}} \cup \{\mathcal{A}'_S, \mathcal{D}, \mathcal{Z}_{obey}\}$ is $\text{tally}(\mathcal{B}_{v^*})$, and the distribution of the tally in the network $\mathcal{F}_{\text{vote}} \cup \{\mathcal{A}'_S, \mathcal{D}, \mathcal{Z}_{deceive}\}$ is $\text{tally}(\mathcal{B}_v)$. Since $\mathcal{Z}_{obey}$ and $\mathcal{Z}_{deceive}$ output the bit $b$ received from $\mathcal{A}'_S$, it follows that

$$\Big| \text{Pr}[\text{EXEC}_{\mathcal{F}_{\text{vote}} \cup \{\mathcal{A}'_S, \mathcal{D}, \mathcal{Z}_{obey}\}} = 1] - \text{Pr}[\text{EXEC}_{\mathcal{F}_{\text{vote}} \cup \{\mathcal{A}'_S, \mathcal{D}, \mathcal{Z}_{deceive}\}} = 1] \Big|$$
$$\leq \max_{v^* \in \mathcal{V} \cup \{\bot\}} \Delta(\mathcal{B}_v, \mathcal{B}_{v^*}) = \text{Adv}_{ideal}(\mathcal{B}, v).$$

Since for all polynomial-time $\mathcal{Z}$, the networks $\pi \cup \{\mathcal{A}', \mathcal{D}_S, \mathcal{Z}\}$ and $\mathcal{F}_{\text{vote}} \cup \{\mathcal{A}'_S, \mathcal{D}, \mathcal{Z}\}$ are indistinguishable, it follows that

$$\Big| \text{Pr}[\text{EXEC}_{\pi \cup \{\mathcal{A}', \mathcal{D}_S, \mathcal{Z}_{obey}\}} = 1] - \text{Pr}[\text{EXEC}_{\pi \cup \{\mathcal{A}', \mathcal{D}_S, \mathcal{Z}_{deceive}\}} = 1] \Big| \leq \text{Adv}_{ideal}(\mathcal{B}, v) + \mu$$

for some negligible function $\mu$. Then with (1) and (2) we get that

$$\Big| \text{Pr}_{obey}(\mathcal{A}, \mathcal{B}) - \text{Pr}_{deceive}(\mathcal{A}, \mathfrak{d}, \mathcal{B}) \Big| \leq \text{Adv}_{ideal}(\mathcal{B}, v) + \mu.$$

This shows that the protocol $\pi$ satisfies the second condition in Theorem 9. (Notice that the construction of the deception strategy $\mathfrak{d}$ is independent of $\mathcal{A}$ and $\mathcal{B}$.)

We are left to show that $\text{Tally}_{deceive}(\mathcal{A}, \mathfrak{d}, \mathcal{B})$ and $\text{tally}(\mathcal{B}_v)$ are indistinguishable (first condition of Theorem 9).

Let $t$ denote the message received by $\mathcal{Z}_{deceive}$ from the party $T$ ($t$ is the tally). In the network $\mathcal{F}_{\text{vote}} \cup \{\mathcal{A}'_S, \mathcal{D}, \mathcal{Z}_{deceive}\}$, $t$ is the output of $\mathcal{F}_{\text{vote}}$. Thus the distribution of $t$ is $\text{tally}(\mathcal{B}_v)$: The party $P$ is instructed by $\mathcal{D}$ to send the vote $v$, all other parties cast votes chosen according to the distribution $\mathcal{B}$.

In the network $\pi \cup \{\mathcal{A}', \mathcal{D}_S, \mathcal{Z}_{deceive}\}$, by construction of $\mathcal{Z}_{deceive}$ and of $\mathfrak{d}$, the distribution of $t$ is $\text{Tally}_{deceive}(\mathcal{A}, \mathfrak{d}, \mathcal{B})$.

For contradiction, assume that $\text{Tally}_{deceive}(\mathcal{A}, \eth, \mathcal{B})$ and $\text{tally}(\mathcal{B}_v)$ were not computationally indistinguishable. Then there is an efficiently computable function $f : \{0,1\}^* \to \{0,1\}$ such that $|\Pr[f(\text{Tally}_{deceive}(\mathcal{A}, \eth, \mathcal{B})) = 1] - \Pr[f(\text{tally}(\mathcal{B}_v)) = 1]|$ is not negligible. Then we define $\mathcal{Z}^*_{deceive}$ like $\mathcal{Z}_{deceive}$, except that $\mathcal{Z}^*_{deceive}$ outputs $f(t)$. Then $|\Pr[\text{EXEC}_{\pi \cup \{\mathcal{A}', \mathcal{D}_S, \mathcal{Z}^*_{deceive}\}} = 1] - \Pr[\text{EXEC}_{\mathcal{F}_{vote} \cup \{\mathcal{A}'_S, \mathcal{D}, \mathcal{Z}^*_{deceive}\}} = 1]|$ is not negligible. This is a contradiction to the fact that for all polynomial-time $\mathcal{Z}$, the networks $\pi \cup \{\mathcal{A}', \mathcal{D}_S, \mathcal{Z}\}$ and $\mathcal{F}_{vote} \cup \{\mathcal{A}'_S, \mathcal{D}, \mathcal{Z}\}$ are indistinguishable. Thus $\text{Tally}_{deceive}(\mathcal{A}, \eth, \mathcal{B})$ and $\text{tally}(\mathcal{B}_v)$ are computationally indistinguishable and the first condition of Theorem 9 is satisfied by $\pi$. □

The design of voting protocols that are UC/c secure is, of course, an open problem. We believe designing UC/c secure remote voting schemes to be a challenging problem that may involve novel cryptographic techniques. In the case of non-remote voting (i.e., involving voting booths and other partially trusted setup such as in, e.g., [10,12,20,3]), realising UC/c security might be much easier. We therefore particularly propose UC/c as a security definition for that setting.

## 4    Incoercible Two-Party Protocols

In the previous section, we have seen that UC/c secure protocols are incoercible. We have not, however, shown that such protocols exist at all. Fortunately, the protocols that were presented in [18,7] for general multi-party computation in the externalized UC (EUC) model are also secure in our UC/c model *in the two-party case* and therefore enjoy incoercibility in addition to the properties guaranteed by the EUC model. The proof that their protocols work in our setting is quite technical; we defer it to the full version [21]. We only state the final result here. The protocols from [18,7] can be based on one of the following functionalities:

The *key registeration with knowledge (KRK) functionality* $\mathcal{F}_{krk}$ is a functionality where each party may register a public key/secret key pair and every party may request the public keys of all parties and the secret key of itself. The *augmented CRS (ACRS) functionality* $\mathcal{F}_{acrs}$ chooses a public key and a corresponding master secret key, and derives for each party a corresponding individual secret key. The public key is given to all parties, the secret key of each party is only given to that party. The *signature card functionality* $\mathcal{F}_{sc}$ with owner $P$ picks a signing/verification key pair and reveals the verification key to all parties. The party $P$ (the owner) may send arbitrary messages $m$ to $\mathcal{F}_{sc}$ and receives signatures of $m$ back. The signing key is never revealed.

**Theorem 11 (UC/c two-party computation).** *Let $\mathcal{F} \in \{\mathcal{F}_{krk}, \mathcal{F}_{acrs}, \mathcal{F}_{sc}\}$. Let $\mathcal{G}$ be a well-formed silent[4] functionality. Then there is a protocol $\pi$ in the $\mathcal{F}$-hybrid model such that $\pi$ UC/c emulates $\mathcal{G}$ with static corruptions/deceptions.*

---

[4] A well-formed functionality is one whose behaviour does not depend on which parties are corrupted or deceiving. We call $\mathcal{G}$ silent if it does not communicate with the adversary or deceiver.

# 5  Conclusions and Open Problems

We have presented the UC/c framework. This framework enables us to model the incoercibility of general multi-party protocols. The UC/c framework comes with a strong composition theorem (universal composition). We have shown that with respect to static coercions/deceptions, arbitrary two-party protocol tasks can be realised in the framework.

Directions for future work include:

- *Good-guy/bad-guy coercions.* Our feasibility results only hold for static coercions/deceptions. We believe that feasibility results similar to those presented in Section 4 can be shown for good-guy coercions. To achieve protocols that are secure with respect to bad-guy coercions, we believe that new cryptographic techniques will have to be developed.
- *Insecure channels.* We assumed perfectly secure channels, i.e., channels where the adversary does not even notice that a message is sent. Can the results from Section 4 be generalised to a setting with weaker assumptions on the channels?
- *Multi-party protocols.* Our feasibility results are restricted to two-party protocols. To capture important cases like voting protocols we need to extend this to multi-party protocols.
- *Impossibility results.* Since incoercibility is a strong requirement, we also expect that many protocol tasks cannot be fulfilled. For example, is it possible to realise a non-trivial protocol task using only a common reference string?
- *Not knowing who is coerced/corrupted.* In our setting, the deceiver-simulator's strategy may depend on who is corrupted/coerced. If we restrict every party's strategy to be independent of the other parties, can we still construct UC/c secure protocols?

**Acknowledgements.** We thank Yevgeniy Dodis and Daniel Wichs for extensive discussions. We also thank the anonymous reviewers for helpful comments.

# References

1. Ben-Or, M., Goldwasser, S., Wigderson, A.: Completeness theorems for non-cryptographic fault-tolerant distributed computation. In: STOC 1988, pp. 1–10. ACM, New York (1988)
2. Benaloh, J., Tuinstra, D.: Receipt-free secret-ballot elections (extended abstract). In: STOC 1994, pp. 544–553. ACM, New York (1994)
3. Bohli, J.M., Müller-Quade, J., Röhrich, S.: Bingo voting: Secure and coercion-free voting using a trusted random number generator. In: Alkassar, A., Volkamer, M. (eds.) VOTE-ID 2007. LNCS, vol. 4896, pp. 111–124. Springer, Heidelberg (2007)
4. Canetti, R., Gennaro, R.: Incoercible multiparty computation. In: FOCS 1996, p. 504. IEEE, Los Alamitos (1996)
5. Canetti, R.: Security and composition of multi-party cryptographic protocols. Journal of Cryptology 3(1), 143–202 (2000)

6. Canetti, R.: Universally composable security: A new paradigm for cryptographic protocols. In: FOCS 2001, pp. 136–145. IEEE, Los Alamitos (2001); Full version is IACR ePrint 2000/067
7. Canetti, R., Dodis, Y., Pass, R., Walfish, S.: Universally composable security with global setup. In: Vadhan, S.P. (ed.) TCC 2007. LNCS, vol. 4392, pp. 61–85. Springer, Heidelberg (2007)
8. Canetti, R., Feige, U., Goldreich, O., Naor, M.: Adaptively secure multi-party computation. In: STOC 1995, pp. 639–648. ACM Press, New York (1995)
9. Canetti, R., Lindell, Y., Ostrovsky, R., Sahai, A.: Universally composable two-party and multi-party secure computation. In: STOC 2002, pp. 494–503. ACM, New York (2002)
10. Chaum, D.: Secret-ballot receipts: True voter-verifiable elections. IEEE Security & Privacy 2(1), 38–47 (2004)
11. Chaum, D., Crépeau, C., Damgård, I.: Multiparty unconditionally secure protocols. In: STOC 1988, pp. 11–19. ACM Press, New York (1988)
12. Chaum, D., Ryan, P.Y.A., Schneider, S.A.: A practical voter-verifiable election scheme. In: di Vimercati, S.d.C., Syverson, P.F., Gollmann, D. (eds.) ESORICS 2005. LNCS, vol. 3679, pp. 118–139. Springer, Heidelberg (2005)
13. Delaune, S., Kremer, S., Ryan, M.D.: Verifying privacy-type properties of electronic voting protocols. Journal of Computer Security 17(4), 435–487 (2009)
14. Forsyth, F.: The Deceiver. Bantam Books (1991), http://tinyurl.com/ycvhuod
15. Goldreich, O.: Foundations of Cryptography (Basic Applications), vol. 2. Cambridge University Press, Cambridge (2004)
16. Goldreich, O., Micali, S., Wigderson, A.: How to play any mental game or a completeness theorem for protocols with honest majority. In: STOC 1987, pp. 218–229 (1987)
17. Herzberg, A.: Rumpsession. Crypto 1991 (1991)
18. Hofheinz, D., Unruh, D., Müller-Quade, J.: Universally composable zero-knowledge arguments and commitments from signature cards. In: Tatra. Mt. Math. Pub., pp. 93–103 (2007)
19. Juels, A., Catalano, D., Jakobsson, M.: Coercion-resistant electronic elections. In: 4nd ACM Workshop on Privacy in the Electronic Society (WPES), pp. 61–70. ACM Press, New York (2005)
20. Moran, T., Naor, M.: Receipt-free universally-verifiable voting with everlasting privacy. In: Dwork, C. (ed.) CRYPTO 2006. LNCS, vol. 4117, pp. 373–392. Springer, Heidelberg (2006)
21. Unruh, D., Müller-Quade, J.: Universally composable incoercibility. IACR ePrint 2009/520 (October 2009)

# Concurrent Non-Malleable Zero Knowledge Proofs

Huijia Lin*, Rafael Pass**, Wei-Lung Dustin Tseng***,
and Muthuramakrishnan Venkitasubramaniam

Cornell University
{huijia,rafael,wdtseng,vmuthu}@cs.cornell.edu

**Abstract.** Concurrent non-malleable zero-knowledge (NMZK) considers the concurrent execution of zero-knowledge protocols in a setting where the attacker can simultaneously corrupt multiple provers and verifiers. Barak, Prabhakaran and Sahai (FOCS'06) recently provided the first construction of a concurrent NMZK protocol without any set-up assumptions. Their protocol, however, is only computationally sound (a.k.a., a concurrent NMZK *argument*). In this work we present the first construction of a concurrent NMZK *proof* without any set-up assumptions. Our protocol requires poly($n$) rounds assuming one-way functions, or $\tilde{O}(\log n)$ rounds assuming collision-resistant hash functions.

As an additional contribution, we improve the round complexity of concurrent NMZK arguments based on one-way functions (from poly($n$) to $\tilde{O}(\log n)$), and achieve a near linear (instead of cubic) security reductions. Taken together, our results close the gap between concurrent ZK protocols and concurrent NMZK protocols (in terms of feasibility, round complexity, hardness assumptions, and tightness of the security reduction).

## 1   Introduction

Zero-knowledge ($\mathcal{ZK}$) interactive proofs [GMR89] are fundamental constructs that allow the Prover to convince the Verifier of the validity of a mathematical statement $x \in L$, while providing *zero additional knowledge* to the Verifier. *Concurrent $\mathcal{ZK}$*, first introduced and achieved by Dwork, Naor and Sahai [DNS04], considers the execution of zero-knowledge protocols in an asynchronous and concurrent setting. In this model, an adversary acts as verifiers in many concurrent executions of the zero-knowledge protocol, and launches a coordinated attack on multiple independent provers to gain knowledge. *Non-malleable $\mathcal{ZK}$*, first introduced and achieved by Dolev, Dwork and Naor [DDN00], also considers the concurrent execution of zero-knowledge protocols, but in a different manner. In this model, an adversary concurrently participates in only two executions,

---

* Supported in part by a Microsoft Research PhD Fellowship.
** Supported in part by a Microsoft New Faculty Fellowship, NSF CAREER Award CCF-0746990, AFOSR Award FA9550-08-1-0197 and BSF Grant 2006317.
*** Supported in part by a NSF Graduate Research Fellowship.

but plays different roles in the two executions; in the first execution (called the left execution), it acts as a verifier, whereas in the second execution (called the right execution) it acts as a prover. The notion of *Concurrent Non-malleable ZK* (*CNMZK*) considers both of the above attacks; the adversary may participate in an unbounded number of concurrent executions, playing the role of a prover in some, and the role of a verifier in others. Despite the generality of such an attacks scenario, this notion of security seems most appropriate for modeling the execution of cryptographic protocols in open networks, such as the internet.

Barak, Prabhakaran and Sahai (BPS) [BPS06] recently constructed the first *CNMZK* protocol for *NP* in the plain model (i.e., without any set-up assumptions).[1] They provide a poly($n$)-round construction based on one-way functions, and a $\tilde{O}(\log n)$-round construction based on collision-resistant hash-functions. Their constructions, however, are only computationally sound; that is, they only show the existence of *CNMZK* interactive *arguments* (as defined by [BCC88]). In contrast, for both concurrent *ZK* and non-malleable *ZK*, interactive *proofs* (as originally defined by [GMR89]) are known [RK99, KP01, PRS02, DDN00].

*Main result.* In this work, we provide the first construction of a *CNMZK* proof in the plain model.[2]

**Theorem 1.** *Assume the existence of one-way functions. Then there exists a* poly($n$)-*round concurrent non-malleable zero-knowledge proof (with a black-box simulator) for all of NP. Furthermore, assuming the existence of collision-resistant hash-functions, the round complexity is only* $\tilde{O}(\log n)$.

Due to the $\tilde{\Omega}(\log n)$-round lower bound for black-box concurrent *ZK* of [CKPR01], the round complexity of our construction based on collision-resistant hash-functions is essentially optimal (unless $NP \subseteq BPP$).

*Efficiency improvements.* As an additional contribution, we improve the round-complexity of *CNMZK* arguments based on one-way functions (recall that the BPS protocol requires poly($n$) rounds).

**Theorem 2.** *Assume the existence of one-way functions. Then there is a* $\tilde{O}(\log n)$-*round concurrent non-malleable zero-knowledge argument (with a black-box simulator) for all of NP.*

Combined with the black-box lower bounds of [CKPR01], this settles the round-complexity of *CNMZK* arguments based on minimal assumptions.

Finally, whereas the "knowledge security" [GMW91] of the BPS reduction (i.e., the overhead of the simulator w.r.t. to the adversary) is cubic, our analysis (for

---

[1] See also the more efficient construction of [OPV10].

[2] We mention that there are several works constructing *CNMZK* proofs in the Common Reference String (CRS) model (see e.g., [SCO+01, DN02]). A potential approach for getting *CNMZK* proofs in the plain model would thus be to try to implement the CRS in a way that prevents man-in-the-middle attacks. This task seems harder than constructing *CNMZK* proofs from scratch, so we have not pursued this approach.

both proofs and arguments) achieves a near linear security reduction; in fact, our protocols achieve the stronger notion of *precise zero-knowledge* [MP06] which bounds the overhead of the simulator in an execution-by-execution fashion (as opposed to only bounding the worst-case running time), and achieve the same level of security as the best concurrent $\mathcal{ZK}$ protocols [PPS+08].

*Techniques.* Our protocol attempts to combine previous techniques in concurrent and non-malleable $\mathcal{ZK}$ in a modular way. As a result, our $\mathcal{CNMZK}$ protocol largely consists of sub-protocols, more precisely commitments, that are developed in previous works.

To leverage existing techniques for concurrent $\mathcal{ZK}$, we follow the abstraction of *concurrently extractable commitments* (CECom) introduced by Micciancio, Ong, Sahai, and Vadhan [MOSV06]. Informally, values committed by CECom can be extracted by a rewinding simulator even in the concurrent setting. In our protocol (as in most concurrent $\mathcal{ZK}$ protocols), the verifier commits to a random trapdoor using CECom, so that our $\mathcal{ZK}$ simulator may extract this trapdoor to complete the simulation. Correspondingly, to leverage existing techniques for non-malleable $\mathcal{ZK}$, we employ *non-malleable commitments* as defined by Dolev, Dwork, and Naor [DDN00]. In our protocol (as in the work of [BPS06]), the prover commits to a witness of the proof statement using a non-malleable commitment, and next proves (using a stand-alone) $\mathcal{ZK}$ protocol that it either committed to a valid witness, or a valid trapdoor.

The crux of the proof is then to show that even during simulation, when the simulator commits to trapdoors (instead of real witnesses) in left interactions, the adversary still cannot commit to a trapdoor in right interactions. Intuitively this should follow from the security guarantees of the non-malleable commitments. The problem, however, is that even if the non-malleable commitments do not "leak" information about the simulator's trapdoors, other parts of the protocol, such as the zero-knowledge proof, might affect the values of the adversary commitments. On a high-level, BPS overcame this problem by relying on *statistical* zero-knowledge protocols for $\mathcal{NP}$; such protocols can only be computationally sound (unless the polynomial hierarchy collapses [AH91]), and known constructions based on one-way functions require $\text{poly}(n)$ rounds.

Instead, we overcome this obstacle by relying on the notion of *robust non-malleable commitments* introduced by [LP09];[3] informally, a robust non-malleable commitment is non-malleable with respect to any protocol that has small round complexity. As shown in [LP09], most known constructions of non-malleable commitment schemes are already robust, or can be made robust easily. Roughly speaking, by relying on this notion we can ensure that the witness used in the $\mathcal{ZK}$ protocol does not affect the witness committed by the adversary (using robust non-malleable commitments) in other executions; in particular, this is used to argue that the adversary essentially never commits to a trapdoor. The actual application of this technique, however, is not direct and requires a subtle treatment— in particular, for technical reasons, we require the prover to use two robust

---

[3] Robustness was originally referred to as *naturality.*

non-malleable commitments (the same technique is used in [LPV09] for constructing another primitive called strong non-malleable $\mathcal{WI}$ proofs). Furthermore, to make our simulation go through, we are unable to apply the original analysis of CECom as presented in [PRS02, MOSV06], but instead rely on the recent analysis of [PTV08]. Roughly speaking, the reason for this is that concurrently extractable commitments are traditionally used and analyzed in so-called *committed-verifier* protocols [MOSV06], where the verifier commits and *fixes* all of its messages at the start of the protocol. Our protocol does not fall into this category.

Finally, to improve the efficiency of the simulation we have the prover commit to its witness also using a CECom; doing this ensures that the concurrent non-malleability simulator becomes as efficient as the extractor of CECom. Our final result regarding precision is then obtained by relying on the precise $\mathcal{ZK}$ approach from [PPS+08] to implement CECom.

*Discussion and Perspectives.* Our work closes the "gap" between known constructions of concurrent $\mathcal{ZK}$ and $\mathcal{CNMZK}$ for the plain model (without set-up); that is, we have shown that all known results for concurrent $\mathcal{ZK}$ in the plain model extend to $\mathcal{CNMZK}$ (under the same assumptions, the same round complexity, and the same efficiency of security reductions). In essence, we reduce that task of constructing $\mathcal{CNMZK}$ protocols to constructing concurrently extractable commitments, and thus, concurrent non-malleability come for free. It seems promising that the same approach could be extended also to models with set-up. For instance, in the Bare Public Key model of [CGGM00], $O(1)$-round concurrent $\mathcal{ZK}$ with black-box simulation is known, whereas the only $O(1)$-round protocol for $\mathcal{CNMZK}$ of [OPV08] requires non-black-box simulation. Similar gaps exists for the Timing model [DNS04], and for the model of quasi-polynomial time security [Pas03]. We believe that, by providing appropriate implementations of concurrently extractable commitments (in line with the work on concurrent $\mathcal{ZK}$ in these models), our technique extends to close these gaps. We leave an exploration of these questions for future work.

*Overview.* Section 2 contains the basic notations and definitions of $\mathcal{CNMZK}$ and other primitives. In Section 3, we present our main result, a $\tilde{O}(\log n)$-round $\mathcal{CNMZK}$ proof system for all of $\mathcal{NP}$, from collision resistant hash functions, and provide the proof of security in Section 4. We also modify the protocol to obtain constructions of a $\text{poly}(n)$-round $\mathcal{CNMZK}$ proof, and a $\tilde{O}(\log n)$-round $\mathcal{CNMZK}$ argument system, from one-way functions at the end of Section 4. We defer our result on Precise $\mathcal{CNMZK}$ to the full version.

# 2   Preliminaries

Let $N$ denote the set of all positive integers. For any integer $n \in N$, let $[n]$ denote the set $\{1, 2, \ldots, n\}$, and let $\{0, 1\}^n$ denote the set of $n$-bit strings. We assume familiarity with interactive Turing machines, interactive protocols, statistical/computational indistinguishability, zero-knowledge, (strong) witness-indistinguishability (see [Gol01] for formal definitions).

## 2.1   Concurrent Non-Malleable Zero-Knowledge

We recall the definition of concurrent non-malleable zero-knowledge from [BPS06], which in turn closely follows the definition of simulation extractability of [PR05]. Let $\langle P, V \rangle$ be an interactive proof for a language $L \in \mathcal{NP}$ with witness relation $R_L$, and let $n$ be the security parameter. Consider a man-in-the-middle adversary $A$ that participates in many left and right interactions in which $m = m(n)$ proofs take place. In the left interactions, the adversary $A$ verifies the validity of statements $x_1, \ldots, x_m$ by interacting with an honest prover $P$, using identities $\mathsf{id}_1, \ldots, \mathsf{id}_m$. In the right interactions, $A$ proves the validity of statements $\tilde{x}_1, \ldots, \tilde{x}_m$ to an honest verifier $V$, using identities $\tilde{\mathsf{id}}_1, \ldots, \tilde{\mathsf{id}}_m$. Prior to the interactions, both $P$ and $A$ receives as common input the security parameter in unary $1^n$ and the statements $x_1, \ldots, x_m$. Additionally, $P$ receives as local input the witnesses $w_1, \ldots, w_m$, $w_i \in R_L(x_i)$, while $A$ receives as auxiliary input $z \in \{0,1\}^*$, which in particular might contain a-priori information about $x_1, \ldots, x_m$ and $w_1, \ldots, w_m$. On the other hand, the statements proved in the right interactions $\tilde{x}_1, \ldots, \tilde{x}_m$ and the identities in both the left and right interactions, $\mathsf{id}_1, \ldots, \mathsf{id}_m$ and $\tilde{\mathsf{id}}_1, \ldots, \tilde{\mathsf{id}}_m$, are chosen by $A$. Let $\mathsf{view}_A(n, x_1, \ldots, x_m, z)$ denote a random variable that describes the view of $A$ in the above experiment. Loosely speaking, an interactive proof is concurrent non-malleable zero-knowledge ($\mathcal{CNMZK}$) if for all man-in-the-middle adversary $A$, there exists a probabilistic polynomial time machine (called the simulator-extractor) that can simulate both the left and the right interactions for $A$, while outputting a witness for every statement proved by the adversary in the right interactions.

**Definition 1.** *An interactive proof $(P, V)$ for a language $L$ with witness relation $R_L$ is said to be* concurrent non-malleable zero-knowledge *if for every polynomial $m$, and every probabilistic polynomial-time man-in-the-middle adversary $A$ that participates in at most $m = m(n)$ concurrent executions, there exists a probabilistic polynomial time machine $S$ such that:*

1. *The following ensembles are computationally indistinguishable over $n \in N$*
   - $\{\mathsf{view}_A(n, x_1, \ldots, x_m, z)\}_{n \in N, x_1, \ldots, x_m \in L \cap \{0,1\}^n, z \in \{0,1\}^*}$
   - $\{S_1(1^n, x_1, \ldots, x_m, z)\}_{n \in N, x_1, \ldots, x_m \in L \cap \{0,1\}^n, z \in \{0,1\}^*}$
   *where $S_1(n, x_1, \ldots, x_m, z)$ denotes the first output of $S(1^n, x_1, \ldots, x_m, z)$.*
2. *Let $x_1, \ldots, x_m \in L \cap \{0,1\}^n$, $z \in \{0,1\}^*$, and let $(\mathsf{view}, \boldsymbol{w})$ denote the output of $S(1^n, x_1, \ldots, x_m, z)$. Let $\tilde{x}_1, \ldots, \tilde{x}_m$ be the statements of the right-interactions in view $\mathsf{view}$, and let $\mathsf{id}_1, \ldots, \mathsf{id}_m$ and $\tilde{\mathsf{id}}_1, \ldots, \tilde{\mathsf{id}}_m$ be the identities of the left-interaction and right-interactions, respectively, in view $\mathsf{view}$. Then for every $i \in [m]$, if the $i^{th}$ right-interaction is accepting and $\tilde{\mathsf{id}}_i \neq \mathsf{id}_j$ for all $j \in [m]$, $\boldsymbol{w}$ contains a witness $w_i$ such that $R_L(\tilde{x}_i, w_i) = 1$.*

## 2.2   Non-Malleable Commitment Schemes

We recall the definition of non-malleability from [LPV08] (which builds upon the definition of [DDN00, PR05]). Let $\langle C, R \rangle$ be a tag-based commitment scheme, and let $n \in N$ be a security parameter. Consider a man-in-the-middle adversary $A$ that, on auxiliary inputs $n$ and $z$, participates in one left and one right interaction simultaneously. In the left interaction, the man-in-the-middle adversary $A$ interacts with $C$, receiving a commitment to value $v$, using identity id of its choice. In the right interaction $A$ interacts with $R$ attempting to commit to a related value $\tilde{v}$, again using identity $\tilde{\text{id}}$ of its choice. If the right commitment is invalid, or undefined, its value is set to $\bot$. Furthermore, if $\tilde{\text{id}} = \text{id}$, $\tilde{v}$ is also set to $\bot$—i.e., a commitment where the adversary copies the identity of the left interaction is considered invalid. Let $\mathsf{nmc}^A_{\langle C,R \rangle} v_1, \ldots, v_m, z$ denote a random variable that describes the value $\tilde{v}$ and the view of $A$, in the above experiment.

**Definition 2.** *A commitment scheme $\langle C, R \rangle$ is said to be* non-malleable (with respect to itself) *if for every polynomial $p(\cdot)$, and every probabilistic polynomial-time man-in-the-middle adversary $A$, the following ensembles are computationally indistinguishable.*

$$\left\{ \mathsf{nmc}^A_{\langle C,R \rangle}(v, z) \right\}_{n \in N, v \in \{0,1\}^n, v' \in \{0,1\}^n, z \in \{0,1\}^*}$$

$$\left\{ \mathsf{nmc}^A_{\langle C,R \rangle}(v', z) \right\}_{n \in N, v \in \{0,1\}^n, v' \in \{0,1\}^n, z \in \{0,1\}^*}$$

*Remark 1.* The main difference of this definition compared to previous ones [PR03, DDN00] is that it considers not only the values the adversary commits to, but also the view of the adversary. This is particularly important in our analysis later. (See Hybrid $H_3$ and $H_4$ in case $j = 2$ in the proof of Lemma 7.)

**Non-Malleable Commitment Robust w.r.t. $k$-round Protocols** The notion of non-malleability w.r.t. arbitrary $k$-round protocols is introduced in [LP09]. Unlike traditional definitions of non-malleability, which only consider man-in-the middle adversaries that participate in two (or more) executions of the *same* protocol, non-malleability w.r.t. arbitrary protocols considers a class of adversaries that can participate in a left interaction of any arbitrary protocol. Below we recall the definition. Consider a one-many man-in-the-middle adversary $A$ that participates in one left interaction—communicating with a machine $B$—and one right interaction—acting as a committer using the commitment scheme $\langle C, R \rangle$. As in the standard definition of non-malleability, $A$ can adaptively choose the identity in the right interaction. We denote by $\mathsf{nmc}^{B,A}_{\langle C,R \rangle}(y, z)$ the random variable consisting of the view of $A(z)$ in a man-in-the-middle execution when communicating with $B(y)$ on the left and an honest receiver on the right, combined with the value $A(z)$ commits to on the right. Intuitively, we say that $\langle C, R \rangle$ is non-malleable w.r.t. $B$ if $\mathsf{nmc}^{B,A}_{\langle C,R \rangle}(y_1, z)$ and $\mathsf{nmc}^{B,A}_{\langle C,R \rangle}(y_2, z)$ are indistinguishable, whenever interactions with $B(y_1)$ and $B(y_2)$ cannot be distinguished. More formally, let $\mathsf{view}_A[\langle B(y), A(z) \rangle]$ denote the view of $A(z)$ in an interaction with $B(y)$.

**Definition 3.** *Let $\langle C, R \rangle$ be a commitment scheme, and $B$ a probabilistic polynomial-time machine. We say the commitment scheme $\langle C, R \rangle$ is* non-malleable w.r.t. $B$, *if for every probabilistic polynomial-time man-in-the-middle adversary $A$, and every two sequences $\{y_n^1\}_{n \in N}$ and $\{y_n^2\}_{n \in N}$ such that, for all probabilistic polynomial-time machine $\tilde{A}$, it holds that*

$$\left\{ \langle B(y_n^1), \tilde{A}(z) \rangle (1^n) \right\}_{n \in N, z \in \{0,1\}^*} \approx \left\{ \langle B(y_n^2), \tilde{A}(z) \rangle (1^n) \right\}_{n \in N, z \in \{0,1\}^*}$$

*where $\langle B(y), \tilde{A}(z) \rangle (1^n)$ denotes the view of $\tilde{A}$ in interaction with $B$ on common input $1^n$, and private inputs $z$ and $y$ respectively, then it holds that:*

$$\left\{ nmc_{\langle C,R \rangle}^{B,A} (y_n^1, z) \right\}_{n \in N, z \in \{0,1\}^*} \approx \left\{ nmc_{\langle C,R \rangle}^{B,A} (y_n^2, z) \right\}_{n \in N, z \in \{0,1\}^*}$$

We say that $\langle C, R \rangle$ is non-malleable w.r.t. $k$-round protocols if $\langle C, R \rangle$ is non-malleable w.r.t. any machine $B$ that interacts with the man-in-the-middle adversary in $k$ rounds. Below, we focus on commitment schemes that are non-malleable w.r.t. itself and arbitrary $\ell(n)$-round protocols, where $l$ is a super-logarithmic function. We say that such a commitment scheme is robust w.r.t. $\ell(n)$-round protocols

**Lemma 1.** *Let $\ell(n)$ be a super-logarithmic function. Then there exists a $O(\ell(n))$-round statistically binding commitment scheme that is robust w.r.t. $\ell(n)$-round protocols, assuming that one-way functions exist.*

The protocol is essentially identical to the $O(\log n)$-round protocol in [LPV08]. A formal proof of this lemma will appear in the full version.

### 2.3 Concurrently Extractable Commitment Schemes

Micciancio, Ong, Sahai and Vadhan introduce and construct *concurrently extractable commitment schemes*, CECom, in [MOSV06]. The commitment scheme is an abstraction of the preamble stage of the concurrent zero-knowledge protocol of [PRS02]. Informally, values committed by CECom can be extracted by a rewinding extractor (e.g., the zero-knowledge simulator of [KP01, PRS02, PTV08]), even in the concurrent setting. In this work, we use the same construction as in [PRS02, MOSV06], but are unable to employ their analysis.

## 3 A Concurrent Non-Malleable Zero-Knowledge Proof

In this section we construct a concurrent non-malleable zero-knowledge proof based on collision-resistant hash-functions. Let $\ell(n)$ be any super logarithmic function. Our concurrent non-malleable zero-knowledge protocol, CNMZKProof, employs several commitment protocols. Let $Com_{sh}$ be a 2-round statistically *hiding* commitment (based on collision-resistant hash-functions), $Com_{sb}$ be a 2-round statistically *binding* commitment (based on one-way functions), and NMCom be an $O(\ell(n))$-round statistically binding commitment scheme that is robust w.r.t. $\ell(n)$-round protocols (based on one-way functions).

Our protocol also employs $\ell(n)$-round, statistically hiding (respectively statistically binding) concurrently-extractable commitment schemes, $\mathsf{CECom}_{sh}$ (respectively $\mathsf{CECom}_{sb}$). These schemes are essentially instantiations of the PRS preamble [PRS02], and can be constructed given $\mathsf{Com}_{sh}$ and $\mathsf{Com}_{sb}$. We repeat their definitions below.

To commit a $n$-bit string $v$ under scheme $\mathsf{CECom}_{sh}$, the committer choses $n \times \ell(n)$ pairs of random $n$-bit strings $(\alpha_{i,j}^0, \alpha_{i,j}^1), i \in [n], j \in [\ell(n)]$, such that $\alpha_{i,j}^0 \oplus \alpha_{i,j}^1 = v$ for every $i$ and $j$. The sender then commits to $v$ and each of the $2n\ell(n)$ strings in parallel using $\mathsf{Com}_{sh}$. This is followed by $\ell(n)$ rounds of interactions. In the $j^{\text{th}}$ interaction, the receiver sends a random $n$-bit challenge $b_j = b_{1,j} \ldots b_{n,j}$, and the committer decommits the commitments of $\alpha_{1,j}^{b_{1,j}}, \ldots, \alpha_{n,j}^{b_{n,j}}$ according to the challenge.

A valid decommitment of $\mathsf{CECom}_{sh}$ requires the committer to decommit all initial commitments under scheme $\mathsf{Com}_{sh}$ (i.e., reveal the randomness of the commitments), and that the decommited values satisfy $\alpha_{i,j}^0 \oplus \alpha_{i,j}^1 = v$ for every $i$ and $j$.

$\mathsf{CECom}_{sb}$ is defined analogously as $\mathsf{CECom}_{sh}$ with the initial commitment $\mathsf{Com}_{sh}$ replaced by $\mathsf{Com}_{sb}$. Additionally, we say a transcript of $\mathsf{CECom}_{sb}$ is *valid* if there exists a valid decommitment. Formal definitions of $\mathsf{CECom}_{sh}$ and $\mathsf{CECom}_{sb}$ are shown in Fig. 1.

---

**Protocol $\mathsf{CECom}_{sh}$ (resp. $\mathsf{CECom}_{sb}$)**

**Inputs:** A security parameter $n$, and a value $v \in \{0,1\}^n$ given to the committer (to be committed).

**Protocol:**

The Committer selects $n\ell(n)$ pairs of random $n$-bit strings $(\alpha_{i,j}^0, \alpha_{i,j}^1)$, $i \in [n], j \in [\ell(n)]$ such that for all $i, j$ $\alpha_{i,j}^0 \oplus \alpha_{i,j}^1 = v$ and commits to $v$ and $\alpha_{i,j}^0$, $\alpha_{i,j}^1$ for every $i \in [n], j \in [\ell(n)]$ using scheme $\mathsf{Com}_{sh}$ (resp. $\mathsf{Com}_{sb}$) to the Receiver.

**For** $i = 1$ **to** $\ell(n)$:

The Receiver sends a $n$-bit string $b_j = b_{1,j} \ldots b_{n,j}$

The Committer decommits to $\alpha_{1,j}^{b_{1,j}}, \ldots, \alpha_{n,j}^{b_{n,j}}$.

**Decommitment:** The Committer decommits to all $n\ell(n) + 1$ commitments made under scheme $\mathsf{Com}_{sh}$ (resp. $\mathsf{Com}_{sb}$), and show that $\alpha_{i,j}^0 \oplus \alpha_{i,j}^1 = v$ for all $i$ and $j$.

---

**Fig. 1.** Concurrently extractable commitments [MOSV06, PRS02]

We now describe $\mathsf{CNMZKProof}$, our concurrent non-malleable zero-knowledge protocol. Protocol $\mathsf{CNMZKProof}$ for a language $L \in \mathcal{NP}$ proceeds in six stages given a security parameter $n$, a common input statement $x \in \{0,1\}^n$, an identity id of the Prover, and a private input $w \in R_L(x)$ to the Prover.

**Stage 1:** The Verifier choses a random string $r \in \{0,1\}^n$ and commits to $r$ using $\mathsf{CECom}_{sh}$; $r$ is called the "fake witness".

**Stage 2:** The Prover commits to the witness $w$ using $\mathsf{CECom}_{sb}$.

**Stage 3:** The Prover commits to the witness $w$ using $\mathsf{NMCom}$ under identity id.

**Stage 4:** The Prover commits to the witness $w$ using $\mathsf{NMCom}$ under identity id, again.

**Stage 5:** The Verifier decommits the Stage 1 commitment to value $v$.

**Stage 6:** The Prover, using a $\omega(1)$-round $\mathcal{ZK}$ proof (e.g., [Blu86]) proves that the commitments in Stages 2, 3 and 4 all commit to the same value $\tilde{w}$ (under identity id), and that either $\tilde{w} \in R_L(x)$, or $\tilde{w} = r$.

Protocol $\mathsf{CNMZKProof}$, in essence, is a modification of the Goldreich-Kahan protocol [GK96]. The protocol is trivially complete, and below we intuitively argue that the protocol is sound. To cheat in the protocol, because the Stage 2 commitment is statistically binding (and the Stage 6 protocol is a proof), the Prover must know the value $r$ committed by the Verifier in Stage 1, before the conclusion of Stage 2 (i.e., before the Verifier decommits to $r$). This violates that statistical hiding property of the commitment scheme $\mathsf{CECom}_{sh}$. A formal description of protocol $\mathsf{CNMZKProof}$ is shown in Figure 2.

# 4 Proof of Security

The definition of $\mathcal{CNMZK}$ requires a simulator-extractor $S$ that is able to simulate the view of a man-in-the-middle adversary $A$ (including both left and right interactions), while simultaneously extracting the witnesses to statements proved in the right interactions. We describe the construction of our simulator in the Sect. 4.1 and show its correctness in Sect. 4.2 and 4.3.

## 4.1 Our Simulator-Extractor

Our simulator-extractor, $S$, roughly follows this strategy:

**Simulating the view of the right interactions.** $S$ simply follows the honest verifier strategy.

**Simulating the view of the left interactions.** In each protocol execution, $S$ first extracts a "fake witness" $r$ from the $\mathsf{CECom}_{sh}$ committed by $A$ in Stage 1, then commits to $r$ in Stage 2, 3, and 4, and finally simulates the proof of knowledge using $r$ as a witness in Stage 6.

**Extracting the witnesses.** In each right interaction that completes successfully, $S$ extracts a witness $w$ from $\mathsf{CECom}_{sb}$ committed by $A$ in Stage 2 of the protocol.

Thus, the main task of $S$ is to extract the values committed by $A$, using $\mathsf{CECom}$, in Stage 1 and 2 of the protocol. This is done by rewinding $A$ during each $\mathsf{CECom}$. To that end, we employ the oblivious Killian-Petrank simulator [KP01] We also rely on the analysis of [PTV08], which is in turn based on the analysis of [PRS02].

---

Protocol CNMZKProof

**Common Input:** an instance $x$ of a language $L$ with witness relation $R_L$, an identifier id, and a security parameter $n$.

**Auxiliary Input for Prover:** a witness $w$, such that $(x, w) \in R_L(x)$.

**Stage 1:**

V uniformly chooses $r \in \{0, 1\}^n$ (the "fake witness").

V commits to $r$ using protocol $\mathsf{CECom}_{sh}$. Let $T_1$ be the commitment transcript.

**Stage 2:**

P commits to $w$ using protocol $\mathsf{CECom}_{sb}$. Let $T_2$ be the commitment transcript.

**Stage 3:**

P commits to $w$ using protocol NMCom and identity id. Let $T_3$ be the commitment transcript.

**Stage 4:**

P commits to $w$ using protocol NMCom and identity id. Let $T_4$ be the commitment transcript.

**Stage 5:**

V decommits $T_1$ to value $r$; P aborts if no valid decommitment is given.

**Stage 6:**

P $\leftrightarrow$ V: a $\omega(1)$-round $\mathcal{ZK}$ proof [Blu86] of the statement: There exists $\tilde{w}$ such that

  – $\tilde{w}$ is a valid decommitment of $T_2$,

  – *and* $\tilde{w}$ is a valid decommitment of $T_3$ and $T_4$ under identity id,

  – *and* $\tilde{w} \in R_L(x)$ or $\tilde{w} = r$.

---

**Fig. 2.** Concurrent Non-Malleable $\mathcal{ZK}$ argument for $\mathcal{NP}$

On a very high-level, $S$ attempts to simulate the view of $A$ (with "fake witnesses") in one continuously straight-line manner (so as to not skew the output distribution); this is aided by numerous auxiliary rewinds that allows $S$ to extract the "fake witnesses" in time. As implied by our simulation strategy, the view of $A$ generated by $S$ depends on the extracted "fake witnesses", but is otherwise independent of the interaction in auxiliary rewinds.

It is useful to know that $S$ may abort in two manners. At the end of a CECom, if $S$ is unable to extract the committed value (the rewinds were unhelpful), $S$ outputs $\perp_{ext}$. Or, in Stage 5 of a left interaction, if $A$ decommits its Stage 1 $\mathsf{CECom}_{sh}$ to a value that is different from the extracted value, $S$ outputs $\perp_{bind}$. The following claim bounds the abort probability of $S$.

**Claim 2.** *$S$ outputs $\perp_{ext}$ and $\perp_{bind}$ with negligible probability.*

*Proof.* This follows essentially from the analysis of [PTV08] in the setting of concurrent $\mathcal{ZK}$. We present the complete proof in the full version of the paper.

## 4.2   The View Generated by the Simulator

We next show that the view generated by $S$ is indistinguishable from the real view of $A$.

**Lemma 3.** *The following ensembles are computationally indistinguishable over* $n \in N$:

$$\{S(1^n, x_1, \ldots, x_m, z)\}_{n \in N, x_1, \ldots, x_m \in \{0,1\}^n \cap L, z \in \{0,1\}^*}$$
$$\{\mathsf{view}_A(1^n, x_1, \ldots, x_m, z)\}_{n \in N, x_1, \ldots, x_m \in \{0,1\}^n \cap L, z \in \{0,1\}^*}$$

To show Lemma 3, we introduce a series of hybrid simulators; the same hybrid simulators will also be helpful later in Sect. 4.3. Hybrids $\mathsf{hyb}^i$, $0 \le i \le m + 1$, receive the witnesses of the statements proved in any left interactions (i.e., "real witnesses"), and proceed in three steps. In the following description, we order the left interactions by the order in which Stage 1 is completed.

**Step 1:** Run the simulator $S$ with the adversary $A$ *in its entirety*. Output $\perp_{ext}$ or $\perp_{bind}$ if $S$ outputs $\perp_{ext}$ or $\perp_{bind}$. Otherwise, let $\mathcal{V}$ be the view of $A$ produced by $S$, and $r_j$ be the "fake witness" extracted by $S$ from the $j^{th}$ left interaction in $\mathcal{V}$.

**Step 2:** Let $\mathcal{V}_i$ be the prefix of $\mathcal{V}$ up until the $i^{th}$ left interaction has completed Stage 1 of the protocol. Simulate a new man-in-the-middle execution with $A$, continuing from $\mathcal{V}_i$, in a *straight-line manner*. In each of the following cases, we need to make sure that the view $\mathcal{V}_i$ can be completed in a consistent way. Note that we can continue any partial commitment or zero-knowledge proof contained in $\mathcal{V}_i$ as long as we don't change the committed value or proof witness.[4]

- Continue of the simulation of right interactions by following the honest verifier strategy (just like $S$).
- Continue the simulation of the first $i$ left interactions in the same manner as $S$: use the "fake witnesses" $r_j$'s for the commitments in Stage 2, 3 and 4, and the proof in Stage 6. This can be done in a straight line manner since the first $i$ extracted "fake witnesses" $(r_j, j \le i)$ are still useful; they correspond to the Stage 1 commitments of the first $i$ left interactions that are present in $\mathcal{V}_i$. Similar to $S$, if $A$ decommits the Stage 1 $\mathsf{CECom}_{sh}$ to a value different from the extracted "fake witness" $r$, $\mathsf{hyb}_i$ outputs $\perp_{bind}$.
- Continue the simulation of the $i + 1^{st}$ and later left interactions by following the honest prover strategy using the given "real witnesses". This does not conflict with the partial view $\mathcal{V}_i$, since Stage 2 of these left interactions have not yet started.

**Step 3:** Output the newly completed view of $A$ from step 2.

---

[4]   Recall that $S$ follows the honest committer and prover strategy in each stage of the protocol; it only cheats by using "fake witnesses". Formally, we can continue any partial commitment or zero-knowledge proof, for example, by requiring $S$ to output the state of every partial commitment and zero-knowledge proofs, for every prefix of the view $\mathcal{V}$.

We also define hybrids $\mathsf{hyb}^i_+$ that proceed identically as $\mathsf{hyb}^i$ except that in step 2, it simulates the $i^{\text{th}}$ left interaction following the honest prover strategy, using the given "real witness" (all other interactions are handled identically as before). Note that these hybrids are only concerned with producing a view of $A$, and do not extract the witnesses of the right interactions.

We start with a claim bounding the abort probability of the hybrids.

**Claim 4.** *For all $i$, $\mathsf{hyb}^i$ and $\mathsf{hyb}^i_+$ output $\bot$ with negligible probability.*

*Proof.* $\mathsf{hyb}^i$ and $\mathsf{hyb}^i_+$ abort when $S$ aborts, or if they output $\bot_{bind}$ during the second pass of the simulation (while mimicking $S$). The first event is bounded by Claim 2. The second event occurs with negligible probability due to the binding property of $\mathsf{CECom}$;

By Claim 4, the output of $\mathsf{hyb}^0$ is statistically close to the real view of $A$ (they only differ when $\mathsf{hyb}^0$ aborts, which occurs with negligible probability). The output of $\mathsf{hyb}^{m+1}$, on the other hand, is identical to the output of simulator $S$. Therefore Lemma 3 directly follows from the next two claims:

**Claim 5.** *The output of $\mathsf{hyb}^i$ and $\mathsf{hyb}^i_+$ are computationally indistinguishable.*

*Proof.* $\mathsf{hyb}^i$ and $\mathsf{hyb}^i_+$ differs only in how the $i^{\text{th}}$ left interaction is simulated (real or fake witness), which is done in a straight line fashion by both hybrids. Therefore they are computationally indistinguishable by the computational hiding property of the Stage 2, 3, and 4 commitments, and the strongly witness-indistinguishable property (implied by the $\mathcal{ZK}$ property) of the Stage 6 proof.

**Claim 6.** *The output of $\mathsf{hyb}^i_+$ and $\mathsf{hyb}^{i-1}$ are statistically close.*

*Proof.* Ignoring the fact that $\mathsf{hyb}^i_+$ and $\mathsf{hyb}^{i-1}$ may abort, their outputs are identical. This is because $\mathsf{hyb}^i_+$ differs from $\mathsf{hyb}^{i-1}$ only in that when generating the output view, from the end of the $i-1^{\text{st}}$ Stage 1 until the end of the $i^{\text{th}}$ Stage 1 of the left interactions, $\mathsf{hyb}^i_+$ employs *rewinds*. However, these rewinds do not extract any new "fake witnesses" for use in the output view, and do not skew the output distribution because the rewinding schedule (including which rewind determines the output view) is oblivious. Since both machines abort at most with negligible probability by Claim 4, their outputs are statistically close.

*Remark 2.* Note that Claim 4 is crucial to the analysis of the hybrids. The analysis of [PRS02, MOSV06] can only realize Claim 4 for *committed-verifier* protocols. Since $\mathsf{CNMZKProof}$ is not committed-verifier, we instead turn to the analysis of [PTV08]. Alternatively, it seems we can also utilize the analysis of [KP01], at the cost of $O(\log^2 n)$ round complexity.

## 4.3   The Witnesses Output by the Simulator

We now show that the extracted witnesses are indeed the $\mathcal{NP}$ witnesses of the statements proved in the right interactions; this is the main technical contribution of our work.

Observe that if $A$ commits to a valid witness using $\mathsf{CECom}_{sb}$ in Stage 2 of a right interaction, then by Claim 2, the simulator $S$ would extract this witness except with negligible probability. Therefore, the following lemma establishes the correctness of the output witnesses:

**Lemma 7.** *For every $\mathcal{PPT}$ adversary $A$, there exists a negligible function $\nu$, such that for every $n \in N$, $x_1, \ldots, x_m \in \{0,1\}^n \cap L$ and $z \in \{0,1\}^*$, the probability that $A$ fails to commit to a valid witness in Stage 2 of a right interaction that is accepting and uses a different identity from all left interactions, is less than $\nu(n)$.*

*Proof.* Assume for contradiction that there exists a man-in-the-middle adversary $A$ that participates in $m = m(n)$ left and right interactions, and a polynomial function $p$, such that for infinitely many $n \in N$, there exists $x_1, \ldots, x_m \in \{0,1\}^n \cap L$ and $z \in \{0,1\}^*$, such that $A$ *cheats* in an outcome of $S_1(n, x_1, \ldots, x_{m(n)}, z)$ with probability $1/p(n)$; by cheating, we mean that $A$ fails to commit to a valid witness in Stage 2 of any right interaction that is accepting and uses a different identity from all the left interactions. (Note that $A$ is not considered cheating if the simulator fails to output a view of $A$).

Consider the series of hybrids, $\mathsf{hyb}^i$ and $\mathsf{hyb}^i_+$, defined in section 4.2. Since $\mathsf{hyb}^{m+1}$ is identical to $S$, by our hypothesis, the probability that $A$ cheats in $\mathsf{hyb}^{m+1}$ is non-negligible. On the other hand, in $\mathsf{hyb}^0$, it follows from the soundness of Stage 6 that, except with negligible probability, in every accepting right interaction, $A$ commits (successfully) to either a real or a "fake witness"; it further follows from the statistically hiding property of Stage 1 and the (stand-alone) extractability of Stage 2 that, except with negligible probability, $A$ never commits to a "fake witness" in any accepting right interactions. Hence, by union bound, except with negligible probability, $A$ never cheats in $\mathsf{hyb}^0$. In addition, it follows from Claim 6 that the probabilities of $A$ cheating in $\mathsf{hyb}^i$ and $\mathsf{hyb}^{i+1}_+$ differ by at most a negligible amount. Therefore, for infinitely many $n$, there must exist an $i = i(n)$, such that, the probability of cheating differ by at least a polynomial amount in $\mathsf{hyb}^i_+$ and $\mathsf{hyb}^i$. Since the total number of right interactions is bounded by a polynomial, this implies that the probabilities that $A$ cheats in a *randomly chosen* right interaction in the two hybrids differ by a polynomial amount.

Notice that the hybrids $\mathsf{hyb}^i_+$ and $\mathsf{hyb}^i$ proceed identically up until the $i^{th}$ left interaction has completed Stage 1 of the protocol—we call it the *cutoff point*. After the cutoff point, the only difference between the two experiments lies in how the $i^{th}$ left interaction are simulated (using either the real or fake witness.) Recall that the adversary $A$ controls the message scheduling in the network; it can thus arrange messages in the $i^{th}$ left-proof and the randomly chosen right-proof in one of the following three ways; see figure 3. Below we omit specifying the $i^{th}$ left interaction and the randomly chosen right interaction, when it is clear in the context.

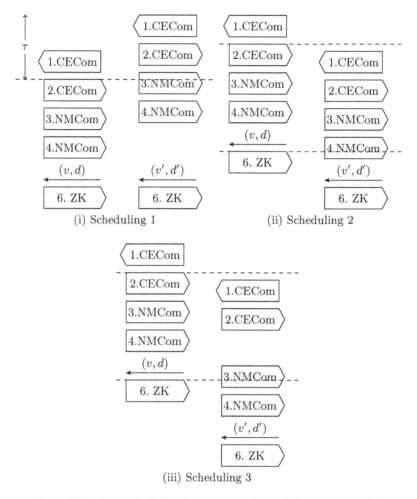

**Fig. 3.** The three scheduling in a man-in-the-middle execution of $A$

**Scheduling 1:** $A$ completes the Stage 2 commitment on the right before the cutoff point.

**Scheduling 2:** $A$ completes the Stage 2 commitment after the cutoff point, but completes the Stage 3 commitment before the Stage 6 proof starts on the left.

**Scheduling 3:** $A$ completes the Stage 2 commitment after the cutoff point, and completes the Stage 3 commitment after the Stage 6 proof starts on the left.

Now consider a variant of $\mathsf{hyb}^i$, $\mathsf{hyb}^{i,j}$ where $j \in \{1, 2, 3\}$, which proceeds identically to $\mathsf{hyb}^i$, except that it outputs $\perp$ if scheduling $j$ does not occur in the output view; define $\mathsf{hyb}_+^{i,j}$ correspondingly for $\mathsf{hyb}_+^i$. Since every man-in-the-middle execution must follow one of the three scheduling above, it holds that, there exists a $j \in \{1,2,3\}$, such that for infinitely many $n \in N$, the probabilities that $A$ cheats in a randomly chosen right interaction in $\mathsf{hyb}_+^{i,j}$ and $\mathsf{hyb}^{i,j}$ differ by a polynomial amount,

Towards reaching a contradiction, let $\mathsf{hyb}^{i,j}(n, x_1, \ldots, x_m, z)$ denote the combined view of $A$ and the value $v$ it commits to in Stage 2 of a randomly chosen right interaction in $\mathsf{hyb}^{i,j}$; $v$ is replaced with $\bot$ if any of the following three events happens: the hybrid experiment fails, or the right interaction $j$ fails, or the right interaction copies the identity of one of the left interactions. Define $\mathsf{hyb}_+^{i,j}(n, x_1, \ldots, x_m, z)$ correspondingly for $\mathsf{hyb}_+^{i,j}$. (For convenience, we refer to $v$ as the committed value of the right interaction.) Below we show that, for every $b$, and every function $i : N \rightarrow N$, $\{\mathsf{hyb}^{i,j}(n, x_1, \ldots, x_m, z)\}$ and $\left\{ \mathsf{hyb}_+^{i,j}(n, x_1, \ldots, x_m, z) \right\}$ are computationally indistinguishable, which implies that the probabilities that $A$ cheats in a *randomly chosen* right interaction differ by at most a negligible amount in the two hybrid experiments, which is a contradiction. The lemma thus follows.

**When $j = 1$,** $A$ completes the Stage 2 commitment on the right *before the cutoff point*, in hybrids $\mathsf{hyb}^{i,1}$ and $\mathsf{hyb}_+^{i,1}$. Since the two hybrid experiments proceed identically before the cutoff point, the values $A$ commits to in Stage 2 on the right are identical in the two experiments. It then follows using essentially the same argument as in Lemma 3 (by relying on the hiding property of Stage 2 to 4 and the strongly $\mathcal{WI}$ property of Stage 6) that the view and the committed value on the right are indistinguishable, i.e.,

**Claim 8.** *For every function $i : N \rightarrow N$, the following ensembles are computationally indistinguishable:*

- $\left\{ \mathsf{hyb}^{i(n),1}(n, x_1, \ldots, x_m, z) \right\}_{n \in N, x_1, \ldots, x_m \in (L \cap \{0,1\}^n)^m, z \in \{0,1\}^*}$
- $\left\{ \mathsf{hyb}_+^{i(n),1}(n, x_1, \ldots, x_m, z) \right\}_{n \in N, x_1, \ldots, x_m \in (L \cap \{0,1\}^n)^m, z \in \{0,1\}^*}$

**When $j = 2$,** Stage 3 to 6 of the right interaction are simulated completely after the cutoff point in a straight line fashion, in $\mathsf{hyb}^{i,2}$ and $\mathsf{hyb}_+^{i,2}$. It then follows from the soundness of Stage 6 that, except from negligible probability, $A$ always commits to the same value in Stage 2, 3 and 4 on the right, provided that the right interaction is *accepting*. Hence to show the indistinguishability of the view and the value $A$ commits to on the right, it suffices to show the indistinguishability of the view $\mathcal{V}$ and the value $v$ that $A$ commits to in Stage 3 (This is because the committed value on the right can be efficiently reconstructed from $\mathcal{V}$ and $v$, by replacing $v$ with $\bot$ appropriately according to $\mathcal{V}$). Then consider the following hybrids, $H_0 = \mathsf{hyb}_+^{i,2}$ to $H_5 = \mathsf{hyb}^{i,2}$.

**Hybrid $H_1$** proceeds identically to $H_0$, except that, in $H_1$, Stage 6 of the left interaction is simulated using the simulator of the $\mathcal{ZK}$ protocol $\langle P, V \rangle$. Since in Scheduling 2, the Stage 3 commitment on the right completes before the Stage 6 proof starts, the value $A$ commits to in Stage 3 is independent of the $\mathcal{ZK}$ proof. Therefore, the view and the value $A$ commits to in Stage 3 are indistinguishable in $H_0$ and $H_1$.

**Hybrid $H_2$** proceeds identically to $H_1$, except that the Stage 2 $\mathsf{CECom}_{sb}$ of the left interaction is now a commitment to the "fake witness" (whereas in $H_1$, it is a commitment to a valid witness). It then follows from the non-malleability w.r.t. $\ell(n)$-round protocols of $\mathsf{NMCom}$, (and the fact that Stage 2 of the protocol consists of $\ell(n)$ rounds) that, the view and the value $A$ commits to in Stage 3 are indistinguishable in $H_1$ and $H_2$.

**Hybrid $H_3$ (and $H_4$ resp.)** proceeds identically to $H_2$ (and $H_3$ resp.), except that, Stage 3 (and Stage 4 resp.) of the left interaction is now a commitment to the "fake witness". It follows using a similar argument as in $H_2$, but relying on the non-malleability w.r.t. itself of $\mathsf{NMCom}$ that the view and the value $A$ commits to in Stage 3 are indistinguishable in $H_2$ and $H_3$ (and in $H_3$ and $H_4$ resp.).

**Hybrid $H_5$** proceeds identically to $H_4$, except that Stage 6 of the left interaction is simulated by proving that Stage 2, 3 and 4 are valid commitments to the value revealed by $A$ in Stage 5 on the left. Note that, by defintion, $H_5$ proceeds identically to the experiment $\mathsf{hyb}^{i,2}$. Furthermore, it follows using the same argument as in $H_1$ that the view and the values $A$ commits to in Stage 3 are indistinguishable in $H_4$ and $H_5$.

Finally, it follows using a hybrid argument that the combined view and the value $A$ commits to in Stage 3 are indistinguishable in $\mathsf{hyb}^{i,2}$ and $\mathsf{hyb}_+^{i,2}$. Therefore,

**Claim 9.** *For every function $i : N \to N$, the following ensembles are computationally indistinguishable:*

- $\left\{\mathsf{hyb}^{i(n),2}(n, x_1, \ldots, x_m, z)\right\}_{n \in N, x_1, \ldots, x_m \in (L \cap \{0,1\}^n)^m, z \in \{0,1\}^*}$
- $\left\{\mathsf{hyb}_+^{i(n),2}(n, x_1, \ldots, x_m, z)\right\}_{n \in N, x_1, \ldots, x_m \in (L \cap \{0,1\}^n)^m, z \in \{0,1\}^*}$

**When $j = 3$,** by the same argument as in the case when $j = 2$, $A$ always commits to the same value in Stage 2, 3 and 4 of every accepting right interaction, and thus, it suffices to show that the view and the value $A$ commits to in Stage 4 are indistinguishable. In $\mathsf{hyb}^{i,3}$ and $\mathsf{hyb}_+^{i,3}$, (as $A$ completes the Stage 3 commitment on the right after the Stage 6 proof starts on the left), the Stage 4 commitment on the right starts completely after the Stage 6 proof on the left, which (by definition) consists of only $\omega(1)$ rounds. It thus follows from the non-malleability with respect to $\omega(1)$-round protocols of $\mathsf{NMCom}$ (along with the strongly $\mathcal{WI}$ property of Stage 6) that, the view and the value $A$ commits to in Stage 4 are indistinguishable. Therefore,

**Claim 10.** *For every function $i : N \to N$, the following ensembles are computationally indistinguishable:*

- $\left\{\mathsf{hyb}^{i(n),3}(n, x_1, \ldots, x_m, z)\right\}_{n \in N, x_1, \ldots, x_m \in (L \cap \{0,1\}^n)^m, z \in \{0,1\}^*}$
- $\left\{\mathsf{hyb}_+^{i(n),3}(n, x_1, \ldots, x_m, z)\right\}_{n \in N, x_1, \ldots, x_m \in (L \cap \{0,1\}^n)^m, z \in \{0,1\}^*}$

A formal proof of this claim will appear in the full version.

*Completing Theorem 1 and Theorem 2.* Above we constructed a $\tilde{O}(\log n)$-round $\mathcal{CNMZK}$ proof based on collision-resistant hash-functions. We obtain a $\tilde{O}(\log n)$-round $\mathcal{CNMZK}$ argument from one-way functions, simply by replacing the Stage 1 CECom$_{sh}$ commitment with protocol CECom$_{sb}$. Note that the resulting protocol is still sound since because the Stage 2 commitment by the prover (CECom$_{sb}$) is statistically binding and "extractable".[5]

Furthermore, to obtain a poly($n$)-round $\mathcal{CNMZK}$ proof based on one-way functions, we use the same protocol CNMZKProof, except that we construct the Stage 1 CECom$_{sh}$ using the public-coin statistically hiding commitment from one-way functions by Haitner et. al. [HNO$^+$09]. It follows using essentially the same security proof as for CNMZKProof that this protocol is $\mathcal{CNMZK}$; the difference lies in how to bound the "binding failure" However, as in the main proof, this can be bound using the analysis of [PTV08] since the commitment of [HNO$^+$09] is public-coin.

# References

[AH91]    Aiello, W., Håstad, J.: Statistical zero-knowledge languages can be recognized in two rounds. J. Comput. Syst. Sci. 42(3), 327–345 (1991)

[BCC88]   Brassard, G., Chaum, D., Crépeau, C.: Minimum disclosure proofs of knowledge. J. Comput. Syst. Sci. 37(2), 156–189 (1988)

[Blu86]   Blum, M.: How to prove a theorem so no one else can claim it. In: Proc. of the International Congress of Mathematicians, pp. 1444–1451 (1986)

[BPS06]   Barak, B., Prabhakaran, M., Sahai, A.: Concurrent non-malleable zero knowledge. In: FOCS, pp. 345–354 (2006)

[CGGM00]  Canetti, R., Goldreich, O., Goldwasser, S., Micali, S.: Resettable zero-knowledge (extended abstract). In: STOC 2000, pp. 235–244 (2000)

[CKPR01]  Canetti, R., Kilian, J., Petrank, E., Rosen, A.: Black-box concurrent zero-knowledge requires $\tilde{\omega}(\log n)$ rounds. In: STOC 2001, pp. 570–579 (2001)

[DDN00]   Dolev, D., Dwork, C., Naor, M.: Nonmalleable cryptography. SIAM Journal on Computing 30(2), 391–437 (2000)

[DN02]    Damgård, I., Nielsen, J.B.: Perfect hiding and perfect binding universally composable commitment schemes with constant expansion factor. In: Yung, M. (ed.) CRYPTO 2002. LNCS, vol. 2442, pp. 581–596. Springer, Heidelberg (2002)

[DNS04]   Dwork, C., Naor, M., Sahai, A.: Concurrent zero-knowledge. J. ACM 51(6), 851–898 (2004)

[GK96]    Goldreich, O., Kahan, A.: How to construct constant-round zero-knowledge proof systems for NP. Journal of Cryptology 9(3), 167–190 (1996)

[GMR89]   Goldwasser, S., Micali, S., Rackoff, C.: The knowledge complexity of interactive proof systems. SIAM Journal on Computing 18(1), 186–208 (1989)

[GMW91]   Goldreich, O., Micali, S., Wigderson, A.: Proofs that yield nothing but their validity or all languages in NP have zero-knowledge proof systems. J. ACM 38(3), 690–728 (1991)

---

[5] Given a prover that breaks soundness, we may break the computationally hiding property of the Stage 1 verifier CECom$_{sb}$ by rewinding the prover and extracting the committed value of the Stage 2 prover CECom$_{sb}$.

[Gol01]     Goldreich, O.: Foundations of Cryptography — Basic Tools. Cambridge University Press, Cambridge (2001)

[HNO+09]    Haitner, I., Nguyen, M.-H., Ong, S.J., Reingold, O., Vadhan, S.P.: Statistically hiding commitments and statistical zero-knowledge arguments from any one-way function. SIAM J. Comput. 39(3), 1153–1218 (2009)

[KP01]      Kilian, J., Petrank, E.: Concurrent and resettable zero-knowledge in polyloalgorithm rounds. In: STOC 2001, pp. 560–569 (2001)

[LP09]      Lin, H., Pass, R.: Non-malleability amplification. In: STOC 2009, pp. 189–198 (2009)

[LPV08]     Lin, H., Pass, R., Venkitasubramaniam, M.: Concurrent non-malleable commitments from any one-way function. In: Canetti, R. (ed.) TCC 2008. LNCS, vol. 4948, pp. 571–588. Springer, Heidelberg (2008)

[LPV09]     Lin, H., Pass, R., Venkitasubramaniam, M.: A unified framework for concurrent security: universal composability from stand-alone non-malleability. In: STOC 2009, pp. 179–188 (2009)

[MOSV06]    Micciancio, D., Ong, S.J.J., Sahai, A., Vadhan, S.: Concurrent zero knowledge without complexity assumptions. In: Halevi, S., Rabin, T. (eds.) TCC 2006. LNCS, vol. 3876, pp. 1–20. Springer, Heidelberg (2006)

[MP06]      Micali, S., Pass, R.: Local zero knowledge. In: STOC 2006, pp. 306–315 (2006)

[OPV08]     Ostrovsky, R., Persiano, G., Visconti, I.: Constant-round concurrent non-malleable zero knowledge in the bare public-key model. In: Aceto, L., Damgård, I., Goldberg, L.A., Halldórsson, M.M., Ingólfsdóttir, A., Walukiewicz, I. (eds.) ICALP 2008, Part II. LNCS, vol. 5126, pp. 548–559. Springer, Heidelberg (2008)

[OPV10]     Ostrovsky, R., Pandey, O., Visconti, I.: Efficiency preserving transformations for concurrent non-malleable zero knowledge. In: Micciancio, D. (ed.) TCC 2010. LNCS, vol. 5978, pp. 535–552. Springer, Heidelberg (2010)

[Pas03]     Pass, R.: Simulation in quasi-polynomial time, and its application to protocol composition. In: Biham, E. (ed.) EUROCRYPT 2003. LNCS, vol. 2656, pp. 160–176. Springer, Heidelberg (2003)

[PPS+08]    Pandey, O., Pass, R., Sahai, A., Tseng, W.-L.D., Venkitasubramaniam, M.: Precise concurrent zero knowledge. In: Smart, N.P. (ed.) EUROCRYPT 2008. LNCS, vol. 4965, pp. 397–414. Springer, Heidelberg (2008)

[PR03]      Pass, R., Rosen, A.: Bounded-concurrent secure two-party computation in a constant number of rounds. In: FOCS, p. 404 (2003)

[PR05]      Pass, R., Rosen, A.: New and improved constructions of non-malleable cryptographic protocols. In: STOC 2005, pp. 533–542 (2005)

[PRS02]     Prabhakaran, M., Rosen, A., Sahai, A.: Concurrent zero knowledge with logarithmic round-complexity. In: FOCS 2002, pp. 366–375 (2002)

[PTV08]     Pass, R., Tseng, W.-L.D., Venkitasubramaniam, M.: Concurrent zero knowledge: Simplifications and generalizations (2008) (manuscript), http://hdl.handle.net/1813/10772

[RK99]      Richardson, R., Kilian, J.: On the concurrent composition of zero-knowledge proofs. In: Stern, J. (ed.) EUROCRYPT 1999. LNCS, vol. 1592, pp. 415–432. Springer, Heidelberg (1999)

[SCO+01]    De Santis, A., Di Crescenzo, G., Ostrovsky, R., Persiano, G., Sahai, A.: Robust non-interactive zero knowledge. In: Kilian, J. (ed.) CRYPTO 2001. LNCS, vol. 2139, pp. 566–598. Springer, Heidelberg (2001)

# Equivalence of Uniform Key Agreement and Composition Insecurity*

Chongwon Cho[1,**], Chen-Kuei Lee[1], and Rafail Ostrovsky[2,***]

[1] Department of Computer Science, UCLA
[2] Department of Computer Science and Mathematics, UCLA
{ccho,jcklee,rafail}@cs.ucla.edu

**Abstract.** We prove that achieving adaptive security from composing two general non-adaptively secure pseudo-random functions is impossible if and only if a uniform-transcript key agreement protocol exists.

It is well known that proving the security of a key agreement protocol (even in a special case where the protocol transcript looks random to an outside observer) is at least as difficult as proving $P \neq NP$. Another (seemingly unrelated) statement in cryptography is the existence of two or more non-adaptively secure pseudo-random functions that do not become adaptively secure under sequential or parallel composition. In 2006, Pietrzak showed that *at least one* of these two seemingly unrelated statements is true. Pietrzak's result was significant since it showed a surprising connection between the worlds of public-key (i.e., "cryptomania") and private-key cryptography (i.e., "minicrypt"). In this paper we show that this duality is far stronger: we show that *at least one* of these two statements must also be false. In other words, we show their *equivalence*.

More specifically, Pietrzak's paper shows that if sequential composition of two non-adaptively secure pseudo-random functions is not adaptively secure, then there exists a key agreement protocol. However, Pietrzak's construction implies a slightly stronger fact: If sequential composition does not imply adaptive security (in the above sense), then a *uniform-transcript* key agreement protocol exists, where by uniform-transcript we mean a key agreement protocol where the transcript of the protocol execution is indistinguishable from uniform to eavesdroppers. In this paper, we complete the picture, and show the reverse direction as well as a strong equivalence between these two notions. More specifically, as our main result, we show that if there exists *any* uniform-transcript key agreement protocol, then composition does not imply adaptive security. Our result holds for both parallel and sequential composition. Our implication holds based on virtually all known key agreement protocols, and can also be based on general complexity assumptions of the existence of dense trapdoor permutations.

---

* Full version appeared on ECCC (Report No.: TR09-108, 31st October 2009).
** Supported in part by grants 0716835, 0716389, 0830803, 0916574.
*** Supported in part by IBM Faculty Award, Xerox Innovation Group Award, the Okawa Foundation Research Award, Intel, Teradata, Lockheed-Martin, NSF grants 0716835, 0716389, 0830803, 0916574, BSF grant, and U.C. MICRO grant.

T. Rabin (Ed.): CRYPTO 2010, LNCS 6223, pp. 447–464, 2010.

# 1    Introduction

One of the central questions in cryptography is the question of *composition*, which very broadly is the study of various ways to compose several basic primitives in a way that amplifies the hardness of the composed object. Naturally, this central question has received a lot of attention in various settings and we continue the study of this question here. More specifically, we investigate a question of whether a composition of pseudo-random functions, to be defined shortly, constitutes stronger security by utilizing the security of the component functions. We consider two very natural types of conventional compositions: a parallel composition with respect to Exclusive-Or (XOR) operation denoted by $\oplus$ and a sequential composition. Briefly, on input $x$ in the domain of F and G, the parallel XOR-composition of two functions F and G is defined as $F(x) \oplus G(x)$. The sequential composition of F and G is defined as $G(F(x))$ (or $F(G(x))$).

Seemingly unrelated to the notion of security amplification via composition, there is the question of designing Key Agreement protocol. Recall that Key Agreement (KA) is a protocol that enables two parties to generate a secret string (also called key) by communicating with each other over an insecure channel in the presence of a eavesdropping adversary. Uniform-transcript key agreement (UTKA) is a strengthened version of key agreement in which messages between two parties are indistinguishable from uniform distribution by all probabilistic polynomial-time (PPT) adversaries. The reason why key agreement seems unrelated to the security of composition is that key agreement belongs to the world of public-key cryptography (also known as "cryptomania") whereas the security of composed pseudo-random functions rather belongs to the world of private-key cryptography (also known as "minicrypt"). For further discussion on cryptomania and minicrypt, see [4].

Now, let us recall briefly recall the definition of Pseudo-Random Functions (PRF) [2]. There are two notions of security of PRF: adaptive security and non-adaptive security. Intuitively, a (pseudo-random) function is said to be non-adaptively secure if the function is indistinguishable from a random function against all PPT adversaries that evaluate the function on inputs chosen independently of the function outputs, that is, chosen prior to PPT adversary learning any of the outputs. Adaptive security is a far stronger notion of security than non-adaptive security: a PRF is said to be adaptively secure if the function remains indistinguishable from random function against all PPT adversaries preparing the current query based on the outputs of the function on all previous queries. Clearly, adaptive security implies non-adaptive security.

We show that the equivalence between the impossibility of achieving adaptive security by composing general non-adaptively secure pseudo-random functions and the existence of uniform transcript key-agreement protocol. We note that our impossibility result holds not only for the case in which the non-adaptively-secure component functions are drawn from the different function families (also known as the general composition) but also for the case where the component functions are drawn from the same function family (also known as self-composition).

## 1.1   Related Work

There has been extensive research on relationship between the security of component functions and the security of their parallel or sequential composition. In the information theoretic context, Vaudenay [11] proved that if F is a pseudo-random permutation with security $\epsilon$ against any distinguisher making $q$ (non-)adaptive queries, then the sequential composition of $k$ F's has improved security $2^{k-1}\epsilon^k$ against a (non-)adaptive distinguisher. F only needs to be a function instead of a permutation for the same security in parallel composition. Luby and Rackoff [5] show the similar security amplification result in the computational context.

In the information theoretic setting, Maurer and Pietrzak [6] proved that composition of non-adaptive secure functions amplifies its security $\epsilon$ to security $2\epsilon(1+\ln(\epsilon^{-1}))$ against an adaptive distinguisher. In 2007, Maurer et al. improved this bound to $2\epsilon$ [7].

Myers [8] showed that the existence of oracles relative to which there are non-adaptively secure permutations, but where the composition of such permutations fails to achieve adaptive security. Recently, Pietrzak [9] showed that the composition of non-adaptively secure functions does not imply adaptive security under the Decisional Diffie-Hellman (DDH) assumption. Pietrzak's more recent work [Pie06] showed that if sequential composition does not imply adaptive security, then there exists a key agreement protocol. Moreover, it turns out that Pietrzak's construction in [10] implies a slightly stronger result: that his key agreement protocol satisfies the property of uniform-transcript. Thus, we can restate the Pietrazak's result as follows:

**Theorem 1.** [10] *If sequential composition of pseudo-random functions is not adaptively secure, then there exists a UTKA.*

## 1.2   Our Results

Pietrzak's work left open the question of establishing the precise connection between the impossibility of adaptively secure composition and key agreement. Our main contribution is to establish sufficient and necessary conditions. In particular, we prove that the existence of UTKA implies the impossibility of obtaining an adaptively secure function from composing general non-adaptively secure functions. The main technique is the fully black-box construction of counterexample functions from UTKA. Therefore, our result holds with respect to any UTKA without relying on the actual code of the UTKA. We prove our result in both parallel and sequential compositions.

**Theorem 2.** *If there exists a UTKA, then parallel composition of non-adaptively secure pseudo-random functions does not imply a pseudo-random function with adaptive security.*

**Theorem 3.** *If there exists a UTKA, then sequential composition of non-adaptively secure pseudo-random functions does not imply a pseudo-random function with adaptive security.*

We also prove the analog of Pietrzak's Theorem 1 for parallel composition:

**Theorem 4.** *If a parallel composition of speudo-random functions is not adaptively secure, then there exists a UTKA.*

Putting all our results together with Theorem 1, we conclude the equivalence between the impossibility of adaptively secure composition and the existence of a uniform transcript key-agreement (both for parallel and sequential compositions). This is informally stated as follows.

**Theorem 5.** (MAIN) *The composition of two non-adaptively secure pseudo-random functions does not imply an adaptively secure pseudo-random function if and only if a UTKA exists.*

We emphasize that our main theorem holds regardless of whether PRFs being composed are taken from a single function family (called self-composition) or from two distinct function families (called general-composition). In particular, we show that the impossibility of secure general-compositions further implies the impossibility of secure self-compositions. The precise connection between the impossibility of adaptively secure composition and a UTKA protocol were not known prior to our work. We summarize these previously known results and our contributions in Fig. 1.

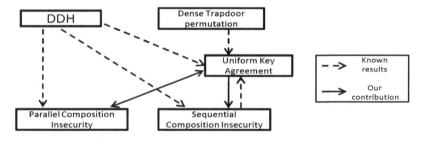

**Fig. 1.** Relationship between composition insecurity and other assumptions

### Organization of the rest of the paper

In Section 2, we review all basic cryptographic notions and definitions. To build the intuition of our main construction, we first show in section 3 a high level outline of somewhat weaker result. In particular, we outline the analogue of Theorem 2 and Theorem 3 not assuming UTKA, but rather assuming the existence of a family of enhanced trapdoor permutations. We note that even this weaker variant of our main result is a generalization from the result by [9], which relies on a specific assumption (i.e., DDH assumption). In section 4 we proceed to give the intuition of our main result assuming UTKA. In section 5, we extend our main results to the one in the context of self-composition. We provide the complete constructions of our functions and full proofs of all theorems in [1].

# 2    Preliminaries

We let $n \in \mathbb{N}$ be a security parameter. An algorithm is considered efficient if its computation can be carried out by a PPT machine whose running time is expected polynomial in the input length. We use the notation $x \leftarrow_\$ \{0,1\}^n$ when string $x$ is uniformly drawn from $\{0,1\}^n$. We omitted the rest of standard notations and (well-known) formal definitions. For those definitions, we refer the readers to [1].

# 3    Building Intuition: Composition Insecurity vs. Dense Trapdoor Permutation

For gentle introduction to our main result, we first present a special case of our main result as an example – The existence of dense trapdoor permutation (DTP) implies the impossibility of achieving the adaptive security by composing (in a black-box way) non-adaptively secure pseudo-random functions. The main idea behind showing this, is that a family of DTPs is well-known to provide a 2-pass (uniform-transcript) key agreement.

## 3.1    Parallel Composition Insecurity from Dense Trapdoor Permutation

We construct two counter-example pseudo-random functions F and G which are secure against any PPT adversary non-adaptively. Then, we prove that their parallel composition is not secure against a particular sequence of four adaptive queries.

**Intuitions of Parallel Composition of F and G.** We provide the high-level overview and intuition of our construction of pseudo-random functions F and G based on DTP, and show how to break the adaptive security of their parallel composition. The main technique of our constructions of counter-example functions is to design the functions to detect the adaptive query throughout the input and output behavior. In particular, F and G emulate a 2-pass key agreement protocol via adaptive inputs and outputs. Once F and G internally obtain a shared key, they generate outputs which hide a special relation with respect to the shared key. As we input these specially generated outputs to the parallel composition again, F and G retrieve the previously shared key and verify the special relation with respect to the shared key. Hence, function F and G are convinced that the queries must be indeed adaptively generated, and reveal their private keys through their outputs, which break their security.

Our counter-example functions F and G are both defined over $(\{0,1\}^n)^{2n+3}$. F and G hide the secret keys $k_F$ and $k_G$ respectively. P denotes an adaptively secure pseudo-random permutation. Let $(\mathsf{Gen}(\cdot), f, f^{-1})$ be a family of DTPs. $r_{ij}$ and $s_{ij}$ denote the $i$th pseudo-random string generated by F and G using their secret keys on $j$th input respectively. In addition, $\mathsf{Enc}_k(x)$ is defined to be a pseudo-random private-key encryption of $x$ with respect to key $k$. Hence, we have $x = \mathsf{Dec}_k(\mathsf{Enc}_k(x))$.

We first define $\mathsf{F}$ and $\mathsf{G}$ on the first *fixed* adaptive query $Q_1 = (0^n, 0^n, \cdots, 0^n)$:

- $\mathsf{F}$ generates $2n+3$ pseudo-random strings $r^*, r_{21}, r_{31}, \cdots, r_{(2n+3)1}$ computed by $\mathsf{P}_{k_\mathsf{F}}(Q_1)$.
- $\mathsf{G}$ on input $Q_1$ uses its secret key to first compute sufficiently long pseudo-random string which is then used to compute DTP pair $(k, t_k)$: a pair of a DTP key $k$ and its private trapdoor $t_k$ by $\mathsf{Gen}(1^n)$ of DTP. $\mathsf{G}$ generates $2n+2$ pseudo-random strings $s_{21}, s_{31}, \cdots, s_{(2n+3)1}$ by $\mathsf{P}_{k_\mathsf{G}}(Q_1)$, then it outputs $(k, s_{21}, \cdots, s_{(2n+3)1})$.

We describe the outputs of $\mathsf{F}$ and $\mathsf{G}$, and their parallel composition outputs below:

$$Q_1 \rightarrow \begin{bmatrix} \mathsf{F} \rightarrow (r^*, r_{21}, \cdots, r_{(2n+3)1}) \\ \mathsf{G} \rightarrow (k, s_{21}, \cdots, s_{(2n+3)1}) \end{bmatrix} \rightarrow (r^* \oplus k, r_{21} \oplus s_{21}, \cdots, r_{(2n+3)1} \oplus s_{(2n+3)1})$$

The second adaptive query is of the form $Q_2 = (u, 0^n, 0^n, \cdots, 0^n)$ where $u = r^* \oplus k$. We define $\mathsf{F}$ and $\mathsf{G}$ on $Q_2$ as follows.

- $\mathsf{F}$ first simulates the first-round of computation (by internally executing $\mathsf{P}_{k_\mathsf{F}}$ on the fixed query $Q_1$) to obtain $r^*$, then computes $u \oplus r^*$ which is equal to $k$; Now, $\mathsf{F}$ computes $2n+3$ pseudo-random strings $x_1, x_2, \cdots, x_n$ and $r_{(n+1)2}, r_{(n+2)2}, \cdots, r_{(2n+3)2}$ by $\mathsf{P}_{k_\mathsf{F}}(Q_2)$. $\mathsf{F}$ computes $y_i$ by $f_k(x_i)$ for $1 \leq i \leq n$, then outputs $(y_1, \cdots, y_n, r_{(n+1)2}, \cdots, r_{(2n+3)2})$.
- $\mathsf{G}$ generates fresh pseudo-random strings $(s_{12}, s_{22}, \cdots, s_{(2n+3)2})$ computed by $\mathsf{P}_{k_\mathsf{G}}(Q_2)$.

We describe what both $\mathsf{F}$ and $\mathsf{G}$ output individually and the output of their parallel composition:

$$Q_2 \rightarrow \begin{bmatrix} \mathsf{F} \rightarrow (y_1, \cdots, y_n, r_{(n+1)2}, \cdots, r_{(2n+3)2}) \\ \mathsf{G} \rightarrow (s_{12}, \cdots, s_{n2}, s_{(n+1)2} \cdots, s_{(2n+3)2}) \end{bmatrix}$$

$$\rightarrow (y_1 \oplus s_{12}, \cdots, y_n \oplus s_{n2}, r_{(n+1)2} \oplus s_{(n+1)2}, \cdots, r_{(2n+3)2} \oplus s_{(2n+3)2})$$

We define the third adaptive query $Q_3$ to consist of the selected coordinates in the previous outputs such that $Q_3 = (y_1 \oplus s_{12}, \cdots, y_n \oplus s_{n2}, r_{(n+1)2} \oplus s_{(n+1)2}, \cdots, r_{(2n)2} \oplus s_{(2n)2}, k \oplus r^*, 0^n, 0^n)$. On $Q_3$, we defined $\mathsf{F}$ and $\mathsf{G}$ as follows.

- $\mathsf{F}$ regenerates all the pseudo-random strings in the second round, $x_1, \cdots, x_n$, $r_{(n+1)2}, \cdots, r_{(2n+3)2}$ by $\mathsf{P}_{k_\mathsf{F}}(Q_2)$. Notice that $Q_2$ is $(k \oplus r^*, 0^n, \cdots, 0^n)$ where $\mathsf{F}$ can obtain $k \oplus r^*$ from $Q_3$. $\mathsf{F}$ can compute $b_i = <x_i, r_{(n+i)2}>$ for all $1 \leq i \leq n$ and retrieve a shared key $sk$ by letting $sk = b_1 b_2 \cdots b_n$. Now, $\mathsf{F}$ generates pseudo-random strings $r_{13}, r_{23}, \cdots, r_{(2n+3)3}$ by $\mathsf{P}_{k_\mathsf{F}}(Q_3)$ and encrypts $r_{13}$ with the shared key as $\mathsf{Enc}_{sk}(r_{13})$. Finally, $\mathsf{F}$ outputs $(\mathsf{Enc}_{sk}(r_{13}), r_{13}, r_{23}, \cdots, r_{(2n+2)3})$.

- G regenerates $s_{12}, s_{22}, \cdots, s_{(2n)2}$ by $\mathsf{P}_{k_\mathsf{G}}(Q_2)$. G can obtain $y_1, \cdots, y_n, r_{(n+1)2}$, $\cdots, r_{(2n)2}$ as it cancels $s_{12}, s_{22}, \cdots, s_{(2n)2}$ out of the first $2n$ coordinates in $Q_3$. By using the inverse permutation $f_{t_k}^{-1}$ with respect to the trapdoor $t_k$, G can obtain $x_i$ by computing $f_{t_k}^{-1}(y_i)$ for all i. Hence, G can compute $b_i = \langle x_i, r_i \rangle$ for all i and retrieve the shared key $sk$ by letting $sk = b_1 b_2 \cdots b_n$. Then, G generates pseudo-random strings $s_{13}, s_{23}, \cdots, s_{(2n+3)3}$ by $\mathsf{P}_{k_\mathsf{G}}(Q_3)$ and creates an encryption $\mathsf{Enc}_{sk}(s_{13})$. Finally, G outputs ($\mathsf{Enc}_{sk}(s_{13})$, $s_{13}, s_{23}, \cdots, s_{(2n+2)3}$).

Below we depict the individual outputs of F and G and the output of their parallel composition:

$$Q_3 \rightarrow \begin{bmatrix} \mathsf{F} \rightarrow (\mathsf{Enc}_{sk}(r_{13}), r_{13}, r_{23}, \cdots, r_{(2n+2)3}) \\ \mathsf{G} \rightarrow (\mathsf{Enc}_{sk}(s_{13}), s_{13}, s_{23}, \cdots, s_{(2n+2)3}) \end{bmatrix}$$

$$\rightarrow (\mathsf{Enc}_{sk}(r_{13}) \oplus \mathsf{Enc}_{sk}(s_{13}), r_{13} \oplus s_{13}, r_{23} \oplus s_{23}, \cdots, r_{(2n+2)3} \oplus s_{(2n+2)3})$$

Our fourth query $Q_4$ is a selective collection of the outputs in the previous round such that $Q_4 = (y_1 \oplus s_{12}, \cdots, y_n \oplus s_{n2}, r_{(n+1)2} \oplus s_{(n+1)2}, \cdots, r_{(2n)2} \oplus s_{(2n)2}, k \oplus r^*, \mathsf{Enc}_{sk}(r) \oplus \mathsf{Enc}_{sk}(s), r \oplus s)$. Notice that F and G can simulate all the computations of previous rounds upon $Q_4$. Hence, F and G can retrieve shared key $sk$. F computes $\mathsf{Enc}_{sk}(r_{13})$ and $r_{13}$ by the simulation of computations on $Q_3$. Then, F checks to see if equality $\mathsf{Dec}_{sk}(\mathsf{Enc}_{sk}(r_{13}) \oplus (\mathsf{Enc}_{sk}(r_{13}) \oplus \mathsf{Enc}_{sk}(s_{13})))$ $= r_{13} \oplus (r_{13} \oplus s_{13})$ holds where $(\mathsf{Enc}_{sk}(r_{13}) \oplus \mathsf{Enc}_{sk}(s_{13}))$ and $(r_{13} \oplus s_{13})$ are obtained from $Q_4$. Since the equality holds, F deduces that the input query is indeed an adaptive query. Hence, F outputs $(k_\mathsf{F}, 0^n, 0^n, \cdots, 0^n)$ containing its secret key $k_\mathsf{F}$. G does the same and outputs $(0^n, k_\mathsf{G}, 0^n, \cdots, 0^n)$. The individual outputs of F and G and the output of the parallel composition are described below.

$$Q_4 \rightarrow \begin{bmatrix} \mathsf{F} \rightarrow (k_\mathsf{F}, 0^n, 0^n, \cdots, 0^n) \\ \mathsf{G} \rightarrow (0^n, k_\mathsf{G}, 0^n, \cdots, 0^n) \end{bmatrix} \rightarrow (k_\mathsf{F}, k_\mathsf{G}, 0^n, \cdots, 0^n)$$

Now, it remains to prove that the above described functions are non-adaptively secure and their parallel composition is adaptively insecure. We prove the following claims that immediately substantiate Lemma 1. In this paper, a pseudo-random function is said to be *breakable by q adaptive queries* if there is a PPT adversary $\mathcal{A}$ such that $\mathcal{A}$ distinguishes the pseudo-random function from a uniform random function by asking $q$ adaptive queries to the pseudo-random function.

*Claim.* The function F and G described above are secure against any non-adaptive PPT adversary.

*Claim.* The parallel composition function F $\oplus$ G is breakable by four adaptive queries.

**Lemma 6.** *Suppose that a dense trapdoor permutation exists. Then, there exist non-adaptively secure pseudo-random functions F and G such that the parallel composition over XOR of F and G is breakable by four adaptive queries.*

## 3.2 Sequential Composition Insecurity from Dense Trapdoor Permutation

We now present a somewhat more interesting construction: namely a sequential composition of non-adaptively secure functions does not imply even *minimal* adaptive security. That is, we show that there exist non-adaptively secure pseudorandom functions F and G whose sequential composition is breakable by only two adaptive queries and yet it remains only non-adaptively secure.

**Intuitions of Sequential Composition of F and G.** We provide the high-level overview of their formal constructions of counter-example PRFs F and G. The standard notions and specifications of the underlying primitives are identical to the ones in the previous section. F (resp. G) contains two secret keys $k_F$ and $k'_F$ (resp. $k_G$ and $k'_G$).

We define the first adaptive query $Q_1$ to be a fixed query, $(0^n, 0^n, \cdots, 0^n)$. Then, F and G are defined on $Q_1$ as follows.

- F computes $(k, t_k)$ by $\mathsf{Gen}(1^n)$, a pair of a public key defining a one-way permutation and its corresponding trapdoor for the inverse permutation. F also computes pseudo-random strings $r_{21}, r_{31}, \cdots, r_{(2n+3)1}$ by $\mathsf{P}_{k_F}(Q_1)$. F outputs $(k, r_{21}, \cdots, r_{(2n+3)1})$.

- On $(k, r_{21}, \cdots, r_{(2n+3)1})$, function G is defined to generate $2n+3$ pseudo-random strings $x_1, \ldots, x_n, s_{(n+1)1}, \cdots, s_{(2n+3)1}$ by $\mathsf{P}_{k_G}(k, r_{21}, \cdots, r_{(2n+3)1})$ and computes the shared key $sk = b_1 b_2 \ldots b_i$, where $b_i = <x_i, s_{(n+1)1}>$ for all $1 \le i \le n$. In addition, G creates an encryption of $s_{(2n+1)1}$ with respect to the shared key, denoted by $\mathsf{Enc}_{sk}(s_{(2n+1)1})$. Also, G encrypts one of its own secrets $k'_G$ with respect to the shared key, resulting in $\mathsf{Enc}_{sk}(k'_G)$. Finally, G encrypts $x_i$s to $y_i$ by a one-way permutation defined by $k$ (i.e., $y_i = f_k(x_i)$ for all $1 \le i \le n$). Hence, G outputs $(y_1, \cdots, y_n, s_{(n+1)1}, \cdots, s_{(2n)1}, \mathsf{Enc}_{sk}(s_{(2n+1)1}), s_{(2n+1)1}, \mathsf{Enc}_{sk}(k'_G))$.

The computation of the sequential composition of F and G on $Q_1$ is described below:

$$Q_1 \xrightarrow{F} (k, r_{21}, \cdots, r_{(2n+3)1})$$
$$\xrightarrow{G} (y_1, \cdots, y_n, s_{(n+1)1}, \cdots, s_{(2n)1}, \mathsf{Enc}_{sk}(s_{(2n+1)1}), s_{(2n+1)1}, \mathsf{Enc}_{sk}(k'_G))$$

We define our second adaptive query $Q_2$ to be the output of the sequential composition on $Q_1$ such that $Q_2 = (y_1, \cdots, y_n, s_{(n+1)1}, \cdots, s_{(2n)1}, \mathsf{Enc}_{sk}(s_{(2n+1)1}), s_{(2n+1)1}, \mathsf{Enc}_{sk}(k'_G))$. On $Q_2$, we define F and G as follows.

- F obtains all $x_i$s by inverting $y_i$s with its private trapdoor information $t_k$ as $f_{t_k}^{-1}(y_i)$ for all $1 \le i \le n$. Now F can retrieve the shared key $sk$ by letting $sk = b_1 b_2 \cdots b_n$ where $b_i = <x_i, s_{(n+1)1}>$ for all $1 \le i \le n$. F takes $\mathsf{Enc}_{sk}(s_{(2n+1)1})$ from $Q_2$ and decrypts it to $s_{(2n+1)1}$ by $\mathsf{Dec}_{sk}(\mathsf{Enc}_{sk}(s_{(2n+1)1}))$. Finding the decrypted string equivalent to the $(2n+2)$th coordinate in $Q_2$ (i.e., $s_{(2n+1)1}$),

F is convinced that $Q_2$ is an adaptive query. Then, F inverts the final coordinate of $Q_2$ with the shared key $sk$, so F obtains $k'_G = \mathsf{Dec}_{sk}(\mathsf{Enc}_{sk}(k'_G))$. Finally, F outputs a vector $(k'_G, k_F, k'_F, 0^n, \cdots, 0^n)$ containing all the secrets of F.

– Upon the input $(k'_G, k_F, t_k, 0^n, \cdots, 0^n)$ from F, function G checks to see if the first coordinate of the input vector equals its own secret $k'_G$. Since the equality holds, G reveals all the secret keys of F and G by outputting $(k_G, k'_G, k_F, k'_F, 0^n, \cdots, 0^n)$.

All the individual outputs of F and G as a part of sequential composition is described as follows.

$$Q_2 \xrightarrow{\mathsf{F}} (k_G, k_F, k'_F, 0^n, \cdots, 0^n) \xrightarrow{\mathsf{G}} (k_G, k^{k'}_G, k_F, k'_F, 0^n, \cdots, 0^n)$$

We prove the following claims that constitute Lemma 7 below. Hence, by Lemma 6 and Lemma 7, we immediately obtain Theorem 8.

*Claim.* The functions F and G described above are secure against any non-adaptive PPT adversary.

*Claim.* The sequential composition $\mathsf{G}(\mathsf{F}(\cdot))$ is breakable by two adaptive queries.

**Lemma 7.** *Suppose that a dense trapdoor permutation exists. Then, there exist non-adaptively secure functions F and G whose sequential composition $\mathsf{G}(\mathsf{F}(\cdot))$ is breakable by two adaptive queries.*

**Theorem 8.** *If a dense trapdoor permutation exists, then the composition of non-adaptively secure functions does not imply the adaptive security.*

# 4 Composition Insecurity vs. Uniform Transcript Key Agreement

In this section, we prove our main result: the existence of UTKA protocol implies the impossibility of obtaining adaptive security by the general composition of non-adaptively secure functions. Moreover, Pietrzak showed that the insecurity of sequential composition implies the existence of key agreement protocol. In fact, the key agreement protocol satisfies the property of *uniform-transcript* even though Pietrzak did not mention it in [10]. For the whole equality between the impossibility of general adaptively secure composition and UTKA, we prove that the parallel composition insecurity also achieves a UTKA by employing a small trick to the technique given in [10].

## 4.1 Parallel Composition Insecurity vs. Uniform Transcript Key Agreement

**Constructing UTKA from the Adaptive Insecurity of $\mathsf{F} \oplus \mathsf{G}$.** We present the parallel version of the result by using the technique originally presented by

[10]. That is, if the parallel composition of two $k-1$ adaptively secure functions is not $k$-adaptively secure, then a $(2k-1)$-pass key agreement exists. For clarity, we rather present a special case where $k = 2$. Following the technique of [10], we construct a $(2k-1)$-pass bit agreement with $\epsilon$-correlation and $\delta$-security where $\epsilon$ is *non-negligible* and $\delta$ is *overwhelming*. It is known that $n$ parallel repetitions of bit agreement with $\epsilon$-correlation and $\delta$-security achieves a $n$-bit key agreement without increasing the round complexity when $\epsilon$ is *noticeable* and $\delta$ is *overwhelming* [3]. With non-negligible $\epsilon$, a bit agreement still realizes a key agreement which achieves correctness for (infinitely many) $n$ such that for any $c$, $\epsilon \geq 1/n^c$.

We present the pictorial description of a $(2k-1)$-pass UTKA from two adaptively pseudo-random functions whose parallel composition is not $k$-adaptively secure when $k = 2$ in Protocol 1. The 3-pass uniform-transcript bit agreement in Protocol 1 may be easily extended to the $(2k-1)$-pass bit agreement for arbitrary $k$.

**Protocol Bit-Agreement$(1^n)$**

| Alice | Transcript | Bob |
|---|---|---|
| $b_A \leftarrow_\$ \{0,1\}$ | | |
| $k_A \leftarrow \mathsf{Gen}_F(1^n)$ | | $k_B \leftarrow \mathsf{Gen}_G(1^n)$ |
| $x_1 \leftarrow \mathcal{D}(1^n)$ | | $x_1 \leftarrow \mathcal{D}(1^n)$ |
| If $b_A = 0$, | | |
| then $z_1 \leftarrow \mathsf{F}_{k_A}(x_1)$ | | |
| else $z_1 \leftarrow_\$ \{0,1\}^n$ | $\xrightarrow{z_1}$ | |
| | $\xleftarrow{y_1}$ | $y_1 \leftarrow z_1 \oplus \mathsf{G}_{k_B}(x_1)$ |
| $x_2 \leftarrow \mathcal{D}(y_1)$ | | $x_2 \leftarrow \mathcal{D}(y_1)$ |
| If $b_A = 0$, | | |
| then $z_2 \leftarrow \mathsf{F}_{k_A}(x_2)$ | | |
| else $z_2 \leftarrow_\$ \{0,1\}^n$ | $\xrightarrow{z_2}$ | $y_2 \leftarrow z_2 \oplus \mathsf{G}_{k_B}(x_2)$ |
| | | $b_B \leftarrow \mathcal{D}(y_1, y_2)$ |

**Protocol 1.** 3-pass uniform-transcript bit agreement based on 2-adaptive distinguisher $\mathcal{D}$

**Theorem 9.** *Let* $\mathsf{F}$ *and* $\mathsf{G}$ *be* $k$-*adaptively secure pseudo-random functions. If the parallel composition* $\mathsf{F} \oplus \mathsf{G}$ *is NOT* $k$-*adaptively secure, then a* $(2k-1)$-*pass UTKA exists.*

**Insecurity of Parallel Composition from UTKA.** A $\gamma$-round uniform-transcript key agreement protocol ($\gamma$-UTKA), denoted by $\Phi_u^\gamma = (\mathsf{A}, \mathsf{B})$, is a uniform-transcript key agreement protocol consisting of two sub-protocols $\mathsf{A}$ and $\mathsf{B}$, in which Alice (using $\mathsf{A}$) and Bob (using $\mathsf{B}$) exchange $2\gamma$ messages to each other ($\gamma$ messages from each party) in order to share a secret key $sk$.

In this section, we use the parallel version of $\gamma$-UTKA to construct counterexample functions. The parallel $\gamma$-UTKA is a $\gamma$-UTKA where Alice and Bob are *symmetric* to each other in Protocol. In particular, Bob's first message is completely independent of Alice's first message and is only dependent on his own private randomness. That is, $\alpha_1 \leftarrow A_1(r_A)$ while $\beta_1 \leftarrow B_1(r_B)$ where $r_A$ and $r_B$ are independent randomness of Alice and Bob. For $2 \leq i \leq \gamma$, $\alpha_i \leftarrow A_i(r_A, \beta_1, \cdots, \beta_{i-1})$ and $\beta_i \leftarrow B_i(r_B, \alpha_1, \cdots, \alpha_{i-1})$. Finally, $sk \leftarrow A_{\gamma+1}(r_A, \beta_1, \cdots, \beta_\gamma)$ and $sk \leftarrow B_{\gamma+1}(r_B, \alpha_1, \cdots, \alpha_\gamma)$ where $sk$ is the shared key.[1]

Now, we provide a high-level overview of our pseudo-random functions F and G from $\gamma$-UTKA and describe how to break the adaptive security of their parallel composition. For underlying primitives, we have a black-box access to $\Phi_u = (A, B)$, parallel $\gamma$-UTKA described above. $\alpha_i$ and $\beta_i$ denote the $i$th message computed by A and B respectively. We are given a pseudo-random private-key encryption scheme $(\mathsf{Enc}, \mathsf{Dec})$ such that $\mathsf{Dec}_k(\mathsf{Enc}_k(x)) = x$. Finally, let P be any given adaptively secure PRP.

Intuitively, F utilizes A as its subroutine as well as G utilizes B as its subroutine in order for them to share a secret key via input and outputs. Then, F and G create pseudo-random strings specially related with respect to the shared secret key. As we input the specially related pseudo-random strings to the composition, the functions retrieve the shared key, verify the special relation hidden in the input query, and reveal their secret keys in their outputs. F and G internally contain secret keys $k_F$ and $k_G$. F and G are defined over $(\{0,1\}^n)^{\gamma+2}$.

First, we define F and G upon the first adaptive (fixed) query $Q_1 = (0^n, \cdots, 0^n)$ as:

- F generates $\gamma + 2$ pseudo-random strings $r_F, r_{21}, \cdots, r_{(\gamma+2)1}$ by $P_{k_F}(Q_1)$. F creates Alice's first message $\alpha_1$ by $A_1(r_F)$ and then outputs $(\alpha_1, r_{21}, \cdots, r_{(\gamma+2)1})$.
- G does the same as it generates $s_G, s_{21}, \cdots, s_{(\gamma+2)1}$ by $P_{k_F}(Q_1)$, and then computes Bob's first message $\beta_1$ by $B_1(s_G)$, and outputs $(\beta_1, s_{21}, \cdots, s_{(\gamma+2)1})$.

Below we depict the individual outputs of F and G on $Q_1$ and their parallel composition:

$$Q_1 \rightarrow \begin{bmatrix} F \rightarrow (\alpha_1, r_{21}, \cdots, r_{(\gamma+2)1}) \\ G \rightarrow (\beta_1, s_{21}, \cdots, s_{(\gamma+2)1}) \end{bmatrix} \rightarrow (\alpha_1 \oplus \beta_1, r_{21} \oplus s_{21}, \cdots, r_{(\gamma+2)1} \oplus s_{(\gamma+2)1})$$

Inductively, for $2 \leq i \leq \gamma$, we define F and G to process the $i$-th adaptive query $Q_i = (\alpha_1 \oplus \beta_1, \alpha_2 \oplus \beta_2, \cdots, \alpha_{i-1} \oplus \beta_{i-1}, 0^n, \cdots, 0^n)$ as follows.

---

[1] We emphasize that we can construct the same counter-example functions to show the same impossibility of adaptively secure composition by using a (*sequential*) $\gamma - UTKA$ in which Bob's first message is *dependent* on Alice's first message. However, it requires more adaptive queries to break the parallel composition of such functions. The main reason for using this parallel version of $\gamma$-UTKA is that it is simpler to emulate the key agreement protocol in the context of parallel composition of our proposed counter-example pseudo-random functions F and G. Also, it provides us with a tighter bound on the number of adaptive queries required to break the adaptive security of the parallel composition.

– F first regenerates $r_F$ and $\alpha_1$ by simulating the first-round computation. That is, F first computes $P_{k_F}(Q_1)$ to obtain $r_F$ and then executes $A(r_F)$. Then, F processes the following *chain of computations* in the direction of left-to-right and top-to-bottom with $r_F$, $\alpha_1$ and $Q_i$,

$$\beta_1 \leftarrow (\alpha_1 \oplus u_1) \qquad\qquad \alpha_2 \leftarrow A_2(r_F, \beta_1)$$

$$\vdots \qquad\qquad\qquad\qquad \vdots$$

$$\beta_{i-1} \leftarrow (\alpha_{i-1} \oplus u_{i-1}) \qquad \alpha_i \leftarrow A_i(r_F, \beta_1, \beta_2, \ldots, \beta_{i-1})$$

Finally, F outputs $(\alpha_i, r_{2i}, \cdots, r_{(\gamma+2)i})$ where $r_{2i}, \cdots, r_{(\gamma+2)i}$ are fresh pseudo-random strings generated by $P_{k_F}(Q_i)$.

– G is symmetrically defined. Hence, G outputs $(\beta_i, s_{2i}, \cdots, s_{(\gamma+2)i})$ where $s_{2i}, \cdots, s_{(\gamma+2)i}$ are pseudo-random strings generated by $P_{k_G}(Q_i)$.

On $Q_i$ for $2 \le i \le \gamma$, we demonstrate the individual outputs of F and G and the output of their parallel composition below. Note that we obtain $\alpha_\gamma \oplus \beta_\gamma$ by feeding the parallel composition of F and G with $Q_\gamma$ to be $(\alpha_1 \oplus \beta_1, \alpha_2 \oplus \beta_2, \cdots, \alpha_{\gamma-1} \oplus \beta_{\gamma-1}, 0^n, 0^n)$.

$$Q_i \rightarrow \begin{bmatrix} F \rightarrow (\alpha_i, r_{2i}, \cdots, r_{(\gamma+2)i}) \\ G \rightarrow (\beta_i, s_{2i}, \cdots, s_{(\gamma+2)i}) \end{bmatrix} \rightarrow (\alpha_i \oplus \beta_i, r_{2i} \oplus s_{2i}, \cdots, r_{(\gamma+2)i} \oplus s_{(\gamma+2)i})$$

The $(\gamma+1)$th adaptive query is defined to be $Q_{\gamma+1} = (\alpha_1 \oplus \beta_1, \alpha_2 \oplus \beta_2, \cdots, \alpha_\gamma \oplus \beta_\gamma, 0^n, 0^n)$. Then, we define our functions F and G on $Q_{\gamma+1}$ as follows.

– F first regenerates $r_F$ and $\alpha_1$ by simulating the first-round computation as before. Then, F performs the chain of computations described above, and so obtains $\beta_1, \beta_2, \cdots, \beta_\gamma$. Hence, F can generate a shared key $sk$ by $A_{\gamma+1}(r_F, \beta_1, \beta_2, \ldots, \beta_\gamma)$. F generates pseudo-random strings $r_{1(\gamma+1)}, r_{2(\gamma+1)}, \cdots, r_{(\gamma+2)(\gamma+1)}$ by $P_{k_F}(Q_{\gamma+1})$. F creates an (pseudo-random) encryption $\mathsf{Enc}_{sk}(r_{1(\gamma+1)})$. F outputs $(\mathsf{Enc}_{sk}(r_{1(\gamma+1)}), r_{1(\gamma+1)}, r_{3(\gamma+1)}, \cdots, r_{(\gamma+2)(\gamma+1)})$.

– G is symmetrically defined. So, G outputs $(\mathsf{Enc}_{sk}(s_{1(\gamma+1)}), s_{1(\gamma+1)}, s_{3(\gamma+1)}, \cdots, s_{(\gamma+2)(\gamma+1)})$.

The following describes the each output of F and G, and that of parallel composition on $Q_{\gamma+1}$.

$$Q_{\gamma+1} \rightarrow \begin{bmatrix} F \rightarrow (\mathsf{Enc}_{sk}(r_{1(\gamma+1)}), r_{1(\gamma+1)}, r_{3(\gamma+1)}, \cdots, r_{(\gamma+2)(\gamma+1)}) \\ G \rightarrow (\mathsf{Enc}_{sk}(s_{1(\gamma+1)}), s_{1(\gamma+1)}, s_{3(\gamma+1)}, \cdots, s_{(\gamma+2)(\gamma+1)}) \end{bmatrix}$$
$$\rightarrow (\mathsf{Enc}_{sk}(r_{1(\gamma+1)}) \oplus \mathsf{Enc}_{sk}(s_{1(\gamma+1)}), r_{1(\gamma+1)} \oplus s_{1(\gamma+1)},$$
$$r_{3(\gamma+1)} \oplus s_{3(\gamma+1)}, \cdots, r_{(\gamma+2)(\gamma+1)} \oplus s_{(\gamma+2)(\gamma+1)})$$

The final $(\gamma+2)$th adaptive query is defined to be $Q_{\gamma+2} = (\alpha_1 \oplus \beta_1, \cdots, \alpha_\gamma \oplus \beta_\gamma, \mathsf{Enc}_{sk}(r_{1(\gamma+1)}) \oplus \mathsf{Enc}_{sk}(s_{1(\gamma+1)}), r_{1(\gamma+1)} \oplus s_{1(\gamma+1)})$ which is the combination of all the outputs of the parallel composition on the previous adaptive queries. Then, F and G are defined on $Q_{\gamma+2}$ as follows.

- F executes the chain of computations to retrieve $\beta_1, \beta_2, \cdots, \beta_\gamma$, then computes a shared key $sk$ by $A_{\gamma+1}(r_F, \beta_1, \beta_2, \ldots, \beta_\gamma)$. Since $Q_{\gamma+1} = (\alpha_1 \oplus \beta_1, \alpha_2 \oplus \beta_2, \cdots, \alpha_\gamma \oplus \beta_\gamma, 0^n, 0^n)$, F can obtain $\mathsf{Enc}_{sk}(r_{1(\gamma+1)})$ and $r_{1(\gamma+1)}$ generated by the internal *simulation* of $F(Q_{\gamma+1})$. F checks to see if equality $\mathsf{Dec}_{sk}(\mathsf{Enc}_{sk}(r_{1(\gamma+1)}) \oplus (\mathsf{Enc}_{sk}(r_{1(\gamma+1)}) \oplus \mathsf{Enc}_{sk}(s_{1(\gamma+1)}))) = r_{1(\gamma+1)} \oplus (r_{1(\gamma+1)} \oplus s_{1(\gamma+1)})$ holds where $(\mathsf{Enc}_{sk}(r_{1(\gamma+1)}) \oplus \mathsf{Enc}_{sk}(s_{1(\gamma+1)}))$ and $(r_{1(\gamma+1)} \oplus s_{1(\gamma+1)})$ are obtained from $Q_{\gamma+2}$. As the equality holds, F is convinced that $Q_{\gamma+2}$ is indeed an adaptively generated query. Hence, F outputs $(k_F, 0^n, 0^n, \cdots, 0^n)$.
- G is symmetrically defined. Hence, G similarly outputs $(0^n, k_G, 0^n, \cdots, 0^n)$.

Below we provide the overall picture of the individual computations of F and G and the output of their parallel composition.

$$Q_{\gamma+2} \rightarrow \begin{bmatrix} \mathsf{F} \rightarrow (k_F, 0^n, 0^n, \cdots, 0^n) \\ \mathsf{G} \rightarrow (0^n, k_G, 0^n, \cdots, 0^n) \end{bmatrix} \rightarrow (k_F, k_G, 0^n, \cdots, 0^n)$$

We prove the following claims that substantiate Theorem 10. Therefore, we immediately obtains Theorem 11 by Theorem 9 and 10.

*Claim.* The functions F and G described above are secure against any non-adaptive PPT adversary.

*Claim.* The parallel composition $F \oplus G$ is breakable by $\gamma+2$ adaptive queries.

**Theorem 10.** *If $\gamma$-UTKA $\Phi_u = (A, B)$ exists, then there exist non-adaptively secure pseudo-random functions F and G such that their parallel composition over XOR is $(\gamma+2)$-adaptive query breakable.*

**Theorem 11.** *The parallel composition of two pseudo-random functions does not imply adaptive security if and only if the uniform-transcript key agreement exists.*

### 4.2   Sequential Composition Insecurity vs. Uniform Transcript Key Agreement

We examine the equivalence between the insecurity of sequential composition and the existence of UTKA protocol. Pietrzak already showed that a key agreement protocol can be achieved from two functions whose sequential composition is not adaptively secure. His key agreement protocol satisfies the property of uniform-transcript. We prove this as a separate claim in [1] and formally restate Pietrzak's theorem below.

**Theorem 12 ([10]).** *Let F and G be $k$-adaptively secure pseudo-random functions. If the sequential composition $G(F(\cdot))$ is NOT $k$-adaptively secure, then a $(2k-1)$-pass UTKA exists.*

**Insecurity of Sequential Composition from UTKA.** In this section, we use the sequential version of $\gamma$-UTKA in which Bob's first message is *dependent* on Alice's first message to construct the counter-example PRFs. That is, $\beta_1 \leftarrow B_1(r_B, \alpha_1)$ where $\alpha_1 \leftarrow A_1(r_A)$ for $r_A$ and $r_B$, independent randomness of Alice and Bob. For $2 \leq i \leq \gamma$, $\alpha_i \leftarrow A_i(r_A, \beta_1, \cdots, \beta_{i-1})$ and $\beta_i \leftarrow B_i(r_B, \alpha_1, \cdots, \alpha_i)$. Consequently, $sk \leftarrow A_{\gamma+1}(r_A, \beta_1, \cdots, \beta_\gamma)$ and $sk \leftarrow B_{\gamma+1}(r_B, \alpha_1, \cdots, \alpha_\gamma)$ where $sk$ is the shared key. Notice that in this scenario Bob must wait for the first message $\alpha_1$ from Alice in order to compute his first message $\beta_1$.

In the following, we present the high-level overview on our constructions of counter-example functions F and G based on $\gamma$-UTKA described above. For the building blocks, we are given a sequential version of $\gamma$-UTKA, $\Phi_u = (A, B)$ and all the other primitives remain identical to the ones in Section 4.1. F (resp. G) is defined over $(\{0,1\}^n)^{\gamma+3}$ and internally possesses a secret key $k_F$ (resp. $k_G$).

Our first adaptive query is an arbitrary vector in $(\{0,1\}^n)^{\gamma+3}$ as $Q_1 = (u_1, u_2, \cdots, u_{\gamma+2}, u^*)$ for $u_1, u_2, \cdots, u_{\gamma+2}, u^* \leftarrow_\$ \{0,1\}^n$. On $Q_1$, we define F and G as follows.

- F computes a pseudo-random string $r_F$ by $P_{k_F}(u^*)$. Then, F generates the first message $\alpha_1$ by executing $A_1(r_F)$. F continues to compute $r_{21}, \cdots, r_{\gamma 1}$ by executing $A_2(r_F, u_1), \cdots, A_\gamma(r_F, u_1, \cdots, u_{\gamma-1})$. Notice that $Q_1$ is an arbitrarily chosen input so that running A (Alice) on $Q_1$ produces only pseudo-random strings except for the first message $\alpha_1$. F computes its first $n$-bit shared key $sk_F^1$ from $A_{\gamma+1}(r_F, u_1, \cdots, u_\gamma)$. F tests if $\text{Dec}_{sk_F^1}(u_{\gamma+1}) = u_{\gamma+2}$. The equality is satisfied only negligible probability since $u_{\gamma+1}$ and $u_{\gamma+2}$ are arbitrary chosen. Hence, with overwhelming probability, F concludes its computation by outputting $(\alpha_1, r_{21}, r_{31}, \cdots, \text{Enc}_{sk_F^1}(r_{(\gamma+1)1}), r_{(\gamma+1)1}, r_{(\gamma+3)1})$ where $r_{(\gamma+1)1}$, $r_{(\gamma+2)1}$ and $r_{(\gamma+3)1}$ are generated from $P_{k_F}(u_{\gamma+1}, u_{\gamma+2}, u_{\gamma+3})$.
- On $F(Q_1)$, G is defined to compute $\beta_1$ by $B_1(s_G, \alpha_1)$ where $s_G$ is generated by $P_{k_G}(u_1)$ and $\alpha_1$ is the first message validly generated by F. G continues to compute $s_{21}, \cdots, s_{\gamma 1}$ by executing $B_2(s_G, \alpha_1, r_{21}), \cdots, B_\gamma(s_G, \alpha_1, r_{21}, \cdots, r_{\gamma 1})$. Since $r_{21}, \cdots, r_{\gamma 1}$ are pseudo-random strings computed by F upon non-adaptive query $Q_1$, $s_{21}, \cdots, s_{\gamma 1}$ are pseudo-random strings. G computes $sk_G^1$ from $B_{\gamma+1}(r_G, u_1, \cdots, u_\gamma)$ and then tests if $\text{Dec}_{sk_G^1}(u_{\gamma+1}) = u_{\gamma+2}$ holds. This equality holds with only negligible probability. G computes pseudo-random strings $s_{(\gamma+1)1}, s_{(\gamma+2)1}$ and $s_{(\gamma+3)1}$ from $P_{k_G}(\pi_{sk_G^1}(r_{(\gamma+1)1}), r_{(\gamma+1)1}, r_{(\gamma+3)1})$. G outputs $(\beta_1, s_{21}, s_{31}, \cdots, \text{Enc}_{sk_G^1}(s_{(\gamma+1)1}), s_{(\gamma+1)1}, s_{(\gamma+3)1})$.

We describe the outputs of F and G in the computation of their sequential composition on $Q_1$:

$$Q_1 \xrightarrow{F} (\alpha_1, r_{21}, r_{31}, \cdots, \text{Enc}_{sk_F^1}(r_{(\gamma+1)1}), r_{(\gamma+1)1}, r_{(\gamma+3)1})$$

$$\xrightarrow{G} (\beta_1, s_{21}, s_{31}, \cdots, \text{Enc}_{sk_G^1}(s_{(\gamma+1)1}), s_{(\gamma+1)1}, s_{(\gamma+3)1}).$$

Inductively, for $2 \leq i \leq \gamma - 1$, the $i$th adaptive query $Q_i$ is in the form of $(\beta_1, \cdots, \beta_{i-1}, s_{i(i-1)}, \cdots, s_{\gamma(i-1)}, \text{Enc}_{sk_G^{i-1}}(s_{(\gamma+1)(i-1)}), s_{(\gamma+1)(i-1)}, u^*)$ where $u^*$ is the final coordinate of $Q_1$ and the rest of coordinates are the first $2\gamma + 2$

coordinates in the output of $G(F(Q_{i-1}))$. Then, F computes all the messages $\alpha_1$ to $\alpha_\gamma$ and shared key $sk_F^i$ based on $Q_i$ as described above. F tests if $Dec_{sk_F^i}(Enc_{sk_G^{i-1}}(s_{(\gamma+1)(i-1)})) = s_{(\gamma+1)(i-1)}$. Obviously, $sk_F^i \neq sk_G^{i-1}$ with overwhelming probability since the keys are computed based on insufficient number of valid messages. Hence, F outputs $(\alpha_1, \cdots, \alpha_i, r_{(i+1)i}, \cdots, r_{(\gamma)i}, Enc_{sk_F^i}(r_{(\gamma+1)i}), r_{(\gamma+1)i}, r_{(\gamma+3)i})$. Similarly, G undertakes the same course of computations: G computes messages and shared key, tests the equality and finally outputs $(\beta_1, \cdots, \beta_i, s_{(i+1)i}, \cdots, s_{(\gamma)i}, Enc_{sk_G^i}(s_{(\gamma+1)i}), s_{(\gamma+1)i}, s_{(\gamma+3)i})$. The individual output of F and the output of G in their sequential composition on $Q_i$ are described as follows:

$$Q_i \xrightarrow{F} (\alpha_1, \cdots, \alpha_i, r_{(i+1)i}, \cdots, r_{(\gamma)i}, Enc_{sk_F^i}(r_{(\gamma+1)i}), r_{(\gamma+1)i}, r_{(\gamma+3)i})$$
$$\xrightarrow{G} (\beta_1, \cdots, \beta_i, s_{(i+1)i}, \cdots, s_{(\gamma)i}, Enc_{sk_G^i}(s_{(\gamma+1)i}), s_{(\gamma+1)i}, s_{(\gamma+3)i}).$$

Hence, after the $(\gamma-1)$th adaptive query, our $\gamma$th adaptive query $Q_\gamma$ is $(\beta_1, \beta_2, \cdots, \beta_{\gamma-1}, s_{\gamma(\gamma-1)}, Enc_{sk_G^{\gamma-1}}(s_{(\gamma+1)(\gamma-1)}), s_{(\gamma+1)(\gamma-1)}, u^*)$. On $Q_\gamma$, we define F and G as follows.

- F computes $r_F$ from $P_{k_F}(u^*)$. Then, F internally regenerates all $\alpha_i$ by $A_i(r_F, \beta_1, \cdots, \beta_{i-1})$ for $1 \leq i \leq \gamma$ and shared key $sk_F^\gamma$ by $A_i(r_F, \beta_1, \cdots, \beta_{i-1}, s_{\gamma(\gamma-1)})$. $sk_F^\gamma$ is still a merely pseudo-random string since $s_{\gamma(\gamma-1)}$ is not a proper message. F performs the equality test $Dec_{sk_F^\gamma}(Enc_{sk_G^{\gamma-1}}(s_{(\gamma+1)(\gamma-1)})) = s_{(\gamma+1)(\gamma-1)}$ which fails with overwhelming probability. Hence, F outputs $(\alpha_1, \cdots, \alpha_\gamma, Enc_{sk_F^\gamma}(r_{(\gamma+1)\gamma}), r_{(\gamma+1)\gamma}, r_{(\gamma+3)\gamma})$ as $(r_{(\gamma+1)\gamma}, r_{(\gamma+2)\gamma}, r_{(\gamma+3)\gamma})$ is generated by $P_{k_F}(Enc_{sk_G^{\gamma-1}}(s_{(\gamma+1)(\gamma-1)}), s_{(\gamma+1)(\gamma-1)}, u^*)$.
- G obtains $r_G$ by $P_{k_G}(\alpha_1)$. Then, since G obtains its complete set of $\gamma$ messages $\alpha_i$'s from F, function G correctly generates all the messages $\beta_i$'s by executing $B_i(r_G, \alpha_1, \cdots, \alpha_i)$ for all $1 \leq i \leq \gamma$. In addition, G computes the shared key $sk_G^\gamma$ from executing $B_{\gamma+1}(r_G, \alpha_1, \cdots, \alpha_\gamma)$. Finally, G outputs $(\beta_1, \cdots, \beta_\gamma, Enc_{sk_G^\gamma}(s_{(\gamma+1)\gamma}), s_{(\gamma+1)\gamma}, s_{(\gamma+3)\gamma})$ since $Dec_{sk_G^\gamma}(Enc_{sk_F^\gamma}(r_{(\gamma+1)(\gamma)})) \neq r_{(\gamma+1)(\gamma)}$ with overwhelming probability, where $(s_{(\gamma+1)\gamma}, s_{(\gamma+2)\gamma}, s_{(\gamma+3)\gamma})$ is generated by $P_{k_G}(Enc_{sk_F^\gamma}(r_{(\gamma+1)\gamma}), r_{(\gamma+1)\gamma}, r_{(\gamma+3)\gamma})$.

We describe the overall picture of F and G in their sequential composition on input $Q_\gamma$ below:

$$Q_\gamma \xrightarrow{F} (\alpha_1, \cdots, \alpha_\gamma, Enc_{sk_F^\gamma}(r_{(\gamma+1)\gamma}), r_{(\gamma+1)\gamma}, r_{(\gamma+3)\gamma})$$
$$\xrightarrow{G} (\beta_1, \cdots, \beta_\gamma, Enc_{sk_G^\gamma}(s_{(\gamma+1)\gamma}), s_{(\gamma+1)\gamma}, s_{(\gamma+3)\gamma}).$$

The (final) $(\gamma+1)$th adaptive query $Q_{\gamma+1}$ is defined to be $(\beta_1, \cdots, \beta_\gamma, Enc_{sk_G^\gamma}(s_{(\gamma+1)\gamma}), s_{(\gamma+1)\gamma}, u^*)$. On $Q_{\gamma+1}$, we define functions F and G on $Q_{\gamma+1}$ as:

- F now obtains all the messages $\beta_i$'s from $Q_{\gamma+1}$ so that it can compute all the messages $\alpha_1, \cdots, \alpha_\gamma$ and the shared key $sk_F^{\gamma+1}$ by executing

$A_{\gamma+1}(r_F, \beta_1, \cdots, \beta_\gamma)$. F tests if the following equality is satisfied: $\mathsf{Dec}_{sk_F^{\gamma+1}}(\mathsf{Enc}_{sk_G^\gamma}(s_{(\gamma+1)(\gamma)})) = s_{(\gamma+1)(\gamma)}$. Notice that $sk_F^{\gamma+1} = sk_G^\gamma$ since both keys are computed on each complete set of messages. Hence, F verifies that the equality holds and is convinced that $Q_{\gamma+1}$ is adaptively generated. Finally, F outputs $(\alpha_1, \cdots, \alpha_\gamma, \pi_{sk_F^{\gamma+1}}(r_{(\gamma+1)(\gamma+1)}), r_{(\gamma+1)(\gamma+1)}, k_F)$ where $r_{(\gamma+1)(\gamma+1)}, r_{(\gamma+2)(\gamma+1)}$ and $r_{(\gamma+3)(\gamma+1)}$ are generated from $P_{k_F}(\mathsf{Enc}_{sk_G^\gamma}(s_{(\gamma+1)\gamma}), s_{(\gamma+1)\gamma}, u^*)$.

- On $(\alpha_1, \cdots, \alpha_\gamma, \pi_{sk_F^{\gamma+1}}(r_{(\gamma+1)(\gamma+1)}), r_{(\gamma+1)(\gamma+1)}, k_F)$, G also computes all of the messages and shared key $sk_G^{\gamma+1}$. Clearly, $sk_G^{\gamma+1} = sk_F^{\gamma+1}$ since both keys are computed based on the same set of messages $\alpha_1 \cdots \alpha_\gamma$. Then G tests if $\mathsf{Dec}_{sk_G^{\gamma+1}}(\mathsf{Enc}_{sk_F^{\gamma+1}}(r_{(\gamma+1)(\gamma+1)})) = r_{(\gamma+1)(\gamma+1)}$. Since both $sk_F^{\gamma+1}$ and $sk_G^{\gamma+1}$ are computed from the complete sets of messages, they must be equal. G is convinced that the query from F is adaptively generated. Therefore, G outputs $(k_G, k_F, 0^n, \cdots, 0^n)$ where $k_F$ can be obtained from the input (i.e., the final coordinate of the input vector).

The overall description of outputs of F and G on the final adaptive query is provided below:

$$Q_{\gamma+1} \xrightarrow{F} (\alpha_1, \cdots, \alpha_\gamma, \mathsf{Enc}_{sk_F^{\gamma+1}}(r_{(\gamma+1)(\gamma+1)}), r_{(\gamma+1)(\gamma+1)}, k_F) \xrightarrow{G} (k_G, k_F, 0^n, \cdots, 0^n).$$

We prove the following claims which substantiate Theorem 13. Putting Theorem 12 and 13 together, we immediately obtains Theorem 14.

*Claim.* The functions F and G described above are secure against any non-adaptive PPT adversary.

*Claim.* The sequential composition of functions F and G, defined by $S(\cdot) = G(F(\cdot))$, is breakable by $\gamma + 1$ adaptive queries.

**Theorem 13.** *If $\gamma$-UTKA $\Phi_u = (A, B)$ exists, then there exist non-adaptively secure functions F and G such that the sequential composition $G(F(\cdot))$ is $(\gamma+1)$-adaptive query breakable.*

**Theorem 14.** *The sequential composition of two pseudo-random functions does not imply adaptive security if and only if the uniform-transcript key agreement exists.*

## 5   Impossibility of Adaptively Secure Self-composition

Self-composition is a composition of two or more copies of a single function. For instance, we call $F(F(\cdot))$ the sequential self-composition of function F, and $F \oplus F$ the parallel self-composition of function F. Note that several copies of identical F's must contain independent secret seeds. That is, each copy of F's must be allowed to be independently drawn from its function family.

So far, we proved the equivalence relation between the insecurity of composition and UTKA protocols. In fact, when we mention the insecurity of composition in previous sections, the main argument is rather that, given a non-adaptively secure function, there might be another non-adaptively secure function such that their composition is adaptively insecure. We call this type of composition *general-composition*. Hence, we still have a lingering unanswered question of whether the self-composition of a non-adaptively secure function implies the unconditional adaptive security. We answered the question negatively as follows.

Suppose that we are given non-adaptively secure pseudo-random functions $F_k$ and $G_{k'}$, without loss of generality, both defined over $\{0,1\}^n$ such that their parallel (general-)composition $(F \oplus G)(\cdot)$ is adaptively insecure. Note that $k$ and $k'$ are independently chosen secret seeds for pseudo-random functions. That is, there exists a PPT adversary $\mathcal{A}$ with an adaptive adversarial strategy which succeeds in breaking the security of $(F \oplus G)(\cdot)$ with non-negligible probability $\delta$. Now, we define a function family $\mathcal{F}_{(b,s)} : \{0,1\}^n \to \{0,1\}^n$ on input string $u$ by

$$\mathcal{F}_{(b,s)}(u) = \begin{cases} F_s(u) & \text{if } b = 0 \\ G_s(u) & \text{if } b = 1 \end{cases} \tag{$*$}$$

where $b$ and $s$ are private seeds.

It is easy to see that function $\mathcal{F}(\cdot)$ is also non-adaptively secure due to the non-adaptive security of functions $F$ and $G$. This trivially leads to

$$\mathbf{Adv}_{\mathcal{A}}^{\mathcal{F}} \leq \mathbf{Adv}_{\mathcal{A}}^{F} + \mathbf{Adv}_{\mathcal{A}}^{G}.$$

To break the adaptive security of $(\mathcal{F} \oplus \mathcal{F})(\cdot)$, it suffices to draw two copies of functions from the family at random and then use the same adaptively adversarial strategy of $\mathcal{A}$ as follows: the first bit of seeds of $F$ and $G$ differ in their first bit with probability $1/2$. Therefore, if we draw two independent $\mathcal{F}$'s, then $\mathcal{F} \oplus \mathcal{F}$ is equivalent to $F \oplus G$ with probability $1/4$ which is adaptively insecure.

Informally, by the above construction of $\mathcal{F}$ from any two non-adaptively secure functions $F$ and $G$ such that their parallel composition is not adaptively secure, we actually show that the adaptive insecurity of the parallel general-composition implies the adaptive insecurity of the parallel self-composition. We formally state this as follows.

**Theorem 15.** *Suppose there are two non-adaptively secure functions $F$ and $G$ such that the parallel composition $(F \oplus G)(\cdot)$ is adaptively insecure. Then, there exists a non-adaptively secure function $\mathcal{F}$ such that the parallel self-composition is adaptively insecure.*

Combining the above theorem with the previous results of this paper in Sections 3.1 and 4.1 related to parallel composition insecurity from DTP and $\gamma$-UTKA, we obtain the following theorems.

**Theorem 16.** *If a family of dense trapdoor permutations or a UTKA exists, then the parallel self-composition of a non-adaptively secure function does not imply adaptive security.*

Furthermore, the above constructions of $\mathcal{F}$ defined in $(*)$ and its analysis of adaptive security can be easily extended to the context of sequential composition.

In particular, $\mathcal{F}$ is also non-adaptively secure while $\mathcal{F}(\mathcal{F}(\cdot))$ is equal to $G(F(\cdot))$ with probability $1/4$ when we draw two independent $\mathcal{F}$'s from its function family. Thus, $\mathcal{F}(\mathcal{F}(\cdot))$ is also adaptively insecure. Consequently, we obtain the following theorem.

**Theorem 17.** *Suppose there are two non-adaptively secure functions* F *and* G *such that the sequential composition* $G(F(\cdot))$ *is adaptively insecure. Then, there exists a non-adaptively secure function* $\mathcal{F}$ *such that the self-composition is adaptively insecure.*

Again, combining the above theorem with the previous results of this paper in Sections 3.2 and 4.2 relevant to sequential composition insecurity from DTP and $\gamma$-UTKA, we derive the following theorem.

**Theorem 18.** *If a family of dense trapdoor permutations or a UTKA exists, then the sequential self-composition of a non-adaptively secure function does not imply adaptive security.*

# References

1. Cho, C., Lee, C.K., Ostrovsky, R.: Equivalence of uniform key agreement and composition insecurity. Electronic Colloquium on Computational Complexity (ECCC), Report No. 108 (2009)
2. Goldreich, O., Goldwasser, S., Micali, S.: How to construct random functions. J. ACM 33(4), 792–807 (1986)
3. Holenstein, T.: Key agreement from weak bit agreement. In: STOC 2005: Proceedings of the 37th Annual ACM Symposium on Theory of Computing, pp. 664–673. ACM, New York (2005)
4. Impagliazzo, R.: A personal view of average-case complexity. In: SCT 1995: Proceedings of the 10th Annual Structure in Complexity Theory Conference, p. 134. IEEE Computer Society, Washington (1995)
5. Luby, M., Rackoff, C.: Pseudo-random permutation generators and cryptographic composition. In: STOC 1986: Proceedings of the 18th Annual ACM Symposium on Theory of Computing, pp. 356–363. ACM, New York (1986)
6. Maurer, U., Pietrzak, K.: Composition of random systems: When two weak make one strong. In: Naor, M. (ed.) TCC 2004. LNCS, vol. 2951, pp. 410–427. Springer, Heidelberg (2004)
7. Maurer, U., Pietrzak, K., Renner, R.: Indistinguishability amplification. In: Menezes, A. (ed.) CRYPTO 2007. LNCS, vol. 4622, pp. 130–149. Springer, Heidelberg (2007)
8. Myers, S.: Black-box composition does not imply adaptive security. In: Cachin, C., Camenisch, J.L. (eds.) EUROCRYPT 2004. LNCS, vol. 3027, pp. 189–206. Springer, Heidelberg (2004)
9. Pietrzak, K.: Composition does not imply adaptive security. In: Shoup, V. (ed.) CRYPTO 2005. LNCS, vol. 3621, pp. 55–65. Springer, Heidelberg (2005)
10. Pietrzak, K.: Composition implies adaptive security in minicrypt. In: Vaudenay, S. (ed.) EUROCRYPT 2006. LNCS, vol. 4004, pp. 328–338. Springer, Heidelberg (2006)
11. Vaudenay, S.: Decorrelation: A theory for block cipher security. J. Cryptology 16(4), 249–286 (2003)

# Non-interactive Verifiable Computing: Outsourcing Computation to Untrusted Workers

Rosario Gennaro[1], Craig Gentry[1], and Bryan Parno[2]

[1] IBM T.J.Watson Research Center
[2] CyLab, Carnegie Mellon University

**Abstract.** We introduce and formalize the notion of *Verifiable Computation*, which enables a computationally weak client to "outsource" the computation of a function $F$ on various dynamically-chosen inputs $x_1,...,x_k$ to one or more workers. The workers return the result of the function evaluation, e.g., $y_i = F(x_i)$, as well as a proof that the computation of $F$ was carried out correctly on the given value $x_i$. The primary constraint is that the verification of the proof should require substantially less computational effort than computing $F(x_i)$ from scratch.

We present a protocol that allows the worker to return a computationally-sound, non-interactive proof that can be verified in $O(m \cdot \texttt{poly}(\lambda))$ time, where $m$ is the bit-length of the output of $F$, and $\lambda$ is a security parameter. The protocol requires a one-time pre-processing stage by the client which takes $O(|C| \cdot \texttt{poly}(\lambda))$ time, where $C$ is the smallest known Boolean circuit computing $F$. Unlike previous work in this area, our scheme also provides (at no additional cost) input and output privacy for the client, meaning that the workers do not learn any information about the $x_i$ or $y_i$ values.

## 1 Introduction

Several trends are contributing to a growing desire to "outsource" computing from a (relatively) weak computational device to a more powerful computation service. For years, a variety of projects, including SETI@Home [5], Folding@Home [2], and the Mersenne prime search [4], have distributed computations to millions of Internet clients to take advantage of their idle cycles. A perennial problem is dishonest clients: end users who modify their client software to return plausible results without performing any actual work [23]. Users commit such fraud even when the only incentive is to increase their ranking on a website listing. Many projects cope with such fraud via redundancy: the same work unit is sent to several clients and the results are compared for consistency. Apart from wasting resources, this provides little defense against colluding users.

A related fear plagues cloud computing, where businesses buy computing time from a service, rather than purchasing, provisioning, and maintaining their own computing resources [1,3]. Sometimes the applications outsourced to the cloud are so critical that it is imperative to rule out accidental errors during the computation. Moreover, in such arrangements, the business providing the computing services may have a strong financial incentive to return incorrect answers, if such answers require less work and are unlikely to be detected by the client.

The proliferation of mobile devices, such as smart phones and netbooks, provides yet another venue in which a computationally weak device would like to be able to outsource a computation, e.g., a cryptographic operation or a photo manipulation, to a third party and yet obtain a strong assurance that the result returned is correct.

T. Rabin (Ed.): CRYPTO 2010, LNCS 6223, pp. 465–482, 2010.

In all of these scenarios, a key requirement is that the amount of work performed by the client to generate and verify work instances must be substantially cheaper than performing the computation on its own. It is also desirable to keep the work performed by the workers as close as possible to the amount of work needed to compute the original function. Otherwise, the worker may be unable to complete the task in a reasonable amount of time, or the cost to the client may become prohibitive.

PRIOR WORK: In the security community, research has focused on solutions based on audits and various forms of secure co-processors. Audit-based solutions [24,9] typically require the client (or randomly selected workers) to recalculate some portion of the work done by untrusted workers. This may be infeasible for resource-constrained clients and often relies on some fraction of the workers to be honest, or at least non-colluding.

Secure co-processors [28, 33] provide isolated execution environments, but their strong tamper-resistance typically makes them quite expensive (thousands of dollars each) and sparsely deployed. The requirements of tamper-resistance also lead to the use of weak CPUs to limit the amount of heat dissipation needed. The growing ubiquity of Trusted Platform Modules (TPMs) [29] in commodity machines promises to improve platform security, but TPMs have achieved widespread deployment in part due to reduced costs (one to five dollars) that result in little to no physical tamper resistance.

In the cryptographic community, there is a long history of outsourcing expensive cryptographic operations to a semi-trusted device. Chaum and Pedersen define the notion of *wallets with observers* [10], a piece of secure hardware installed by a third party, e.g. a bank, on the client's computer to "help" with expensive computations. The hardware is not trusted by the client, who retains assurance that the hardware is performing correctly by analyzing its communication with the bank. Hohenberger and Lysyanskaya formalize this model [17], and present protocols for the computation of modular exponentiations (arguably the most expensive step in public-key cryptography operations). Their protocol requires the client to interact with *two* non-colluding servers. Other work targets specific classes of functions, such as one-way function inversion [16].

The theoretical community has devoted considerable attention to the verifiable computation of arbitrary functions. *Interactive proofs* [6, 15] are a way for a powerful (e.g. super-polynomial) prover to (probabilistically) convince a weak (e.g. polynomial) verifier of the truth of statements that the verifier could not compute on its own. As it is well known, the work on interactive proofs lead to the concept of *probabilistically checkable proofs* (PCPs), where a prover can prepare a proof that the verifier can check in only very few places (in particular only a constant number of bits of the proofs needed for NP languages). Notice, however, that the PCP proof might be very long, potentially too long for the verifier to process. To avoid this complication, Kilian proposed the use of efficient arguments[1] [19, 20] in which the prover sends the verifier a short commitment to the entire proof using a Merkle tree. The prover can then interactively open the bits requested by the verifier (this requires the use of a collision-resistant hash function). A non-interactive solution can be obtained using Micali's CS Proofs [22], which remove interaction from the above argument by choosing the bits to open based on the

---

[1] We follow the standard terminology: an *argument* is a computationally sound proof, i.e. a protocol in which the prover is assumed to be computationally bounded. In an argument, an infinitely powerful prover can convince the verifier of a false statement, as opposed to a proof where this is information-theoretically impossible or extremely unlikely.

application of a random oracle to the commitment string. In more recent work, which still uses some PCP machinery, Goldwasser et al. [14] show how to build an interactive proof to verify arbitrary polynomial-time computations in almost linear time. They also extend the result to a non-interactive argument for a restricted class of functions.

Therefore, if we restrict our attention to non-interactive protocols, the state of the art offers either Micali's CS Proofs [22] which are arguments that can only be proven in the random oracle model, or the arguments from [14] that can only be used for a restricted class of functions. Our scheme overcomes these limitations, since it is non-interactive, works for any function, and is provable in the standard model. It also provides the client with input and output privacy, a property not considered in previous work.

OUR CONTRIBUTION. We slightly move away from the notions of proofs and arguments, to define the notion of a *Verifiable Computation Scheme*: this is a protocol between two polynomial-time parties, a *client* and a *worker*, to collaborate on the computation of a function $F : \{0,1\}^n \rightarrow \{0,1\}^m$. Our definition uses an amortized notion of complexity for the client: he can perform some expensive pre-processing, but after this stage, he is required to run very efficiently. Since the preprocessing stage happens only once, it is important to stress that it can be performed in a trusted environment where the weak client, who does not have the computational power to perform it, outsources it to a trusted party (think of a military application in which the client loads the result of the preprocessing stage performed inside the military base by a trusted server, and then goes off into the field where outsourcing servers may not be trusted anymore – or think of the preprocessing phase as being executed on the client's home machine and then used by his portable device in the field).

By introducing a one-time preprocessing stage (and the resulting amortized notion of complexity), we can circumvent the result of Rothblum and Vadhan [26], which indicated that efficient verifiable computation requires the use of PCP constructions. In other words, unless a substantial improvement in the efficiency of PCP constructions is achieved, our model potentially allows much simpler and more efficient constructions than those possible in previous models.

More specifically, a verifiable computation scheme consists of three phases:

**Preprocessing.** A one-time stage in which the client computes some auxiliary (public and private) information associated with $F$. This phase can take time comparable to computing the function from scratch, but it is performed only once, and its cost is amortized over all the future executions.

**Input Preparation.** When the client wants the worker to compute $F(x)$, it prepares some auxiliary (public and private) information about $x$. The public information is sent to the worker.

**Output Computation and Verification.** Once the worker has the public information associated with $F$ and $x$, it computes a string $\pi_x$ which encodes the value $F(x)$ and returns it to the client. From the value $\pi_x$, the client can compute the value $F(x)$ and verify its correctness.

Notice that this is a minimally interactive protocol: the client sends a single message to the worker and vice versa. The crucial efficiency requirement is that Input Preparation and Output Verification must take less time than computing $F$ from scratch (ideally linear time, $O(n + m)$). Also, the Output Computation stage should take roughly the same amount of computation as $F$.

After formally defining the notion of verifiable computation, we present a verifiable computation scheme for *any* function. Assume that the function $F$ is described by a Boolean circuit $C$. Then the Preprocessing stage of our protocol takes time $O(|C| \cdot \text{poly}(\lambda))$, i.e., time linear in the size of the circuit $C$ that the client would have used to compute the function on its own (and polynomial in the security parameter $\lambda$). Apart from that, the client runs in linear time, as Input Preparation takes $O(n \cdot \text{poly}(\lambda))$ time and Output Verification takes $O(m \cdot \text{poly}(\lambda))$ time. Finally the worker takes time $O(|C| \cdot \text{poly}(\lambda))$ to compute the function for the client.

The computational assumptions underlying the security of our scheme are the security of block ciphers (i.e., the existence of one-way functions) and the existence of a secure fully homomorphic encryption scheme [13, 12, 27, 30] (more details below).

*Dynamic and Adaptive Input Choice.* We note that in this amortized model of computation, Goldwasser et al.'s protocol [14] can be modified using Kalai and Raz's transformation [18] to achieve a non-interactive scheme (see [25]). However an important feature of our scheme, that is not enjoyed by Goldwasser et al.'s protocol [14], is that the inputs to the computation of $F$ can be chosen in a dynamic and adaptive fashion throughout the execution of the protocol (as opposed to [14] where they must be fixed and known in advance).

*Privacy.* We also note that our construction has the added benefit of providing input and output privacy for the client, meaning that the worker does not learn any information about $x$ or $F(x)$ (details below). This privacy feature is bundled into the protocol and comes at no additional cost. This is a critical feature for many real-life outsourcing scenarios in which a function is computed over highly sensitive data (e.g., medical records or trade secrets). Our work is the first to provide a weak client with the ability to efficiently and verifiably offload computation to an untrusted server in such a way that the input remains secret.

OUR SOLUTION IN A NUTSHELL. Our work is based on the crucial (and somewhat surprising) observation that Yao's Garbled Circuit Construction [31, 32], in addition to providing secure two-party computation, also provides a "one-time" verifiable computation. In other words, we can adapt Yao's construction to allow a client to outsource the computation of a function on a single input. More specifically, in the preprocessing stage the client garbles the circuit $C$ according to Yao's construction. Then in the "input preparation" stage, the client reveals the random labels associated with the input bits of $x$ in the garbling. This allows the worker to compute the random labels associated with the output bits, and from them the client will reconstruct $F(x)$. If the output bit labels are sufficiently long and random, the worker will not be able to guess the labels for an incorrect output, and therefore the client is assured that $F(x)$ is the correct output.

Unfortunately, reusing the circuit for a second input $x'$ is insecure, since once the output labels of $F(x)$ are revealed, nothing can stop the worker from presenting those labels as correct for $F(x')$. Creating a new garbled circuit requires as much work as if the client computed the function itself, so on its own, Yao's Circuits do not provide an efficient method for outsourcing computation.

The second crucial idea of the paper is to combine Yao's Garbled Circuit with a fully homomorphic encryption system[2] (e.g., Gentry's recent proposal [13]) to be able

---

[2] While homomorphic encryption already solves the problem of computing over private data, it does not address the main problem of this paper: to efficiently verify the result.

to safely reuse the garbled circuit for multiple inputs. More specifically, instead of revealing the labels associated with the bits of input $x$, the client will encrypt those labels under the public key of a fully homomorphic scheme. A new public key is generated for every input in order to prevent information from one execution from being useful for later executions. The worker can then use the homomorphic property to compute an encryption of the output labels and provide them to the client, who decrypts them and reconstructs $F(x)$.

While existing fully-homomorphic encryption schemes [13, 12, 27, 30] are expensive (leading to large constants in our protocol's performance), we anticipate that any performance improvements in future schemes will directly result in similar performance gains for our protocol as well, since we use the fully-homomorphic encryption scheme in a black-box fashion.

*One pre-processing step for many workers.* Note that the pre-processing stage is independent of the worker, since it simply produces a Yao-garbled version of the circuit $C$. Therefore, in addition to being reused many times, this garbled circuit can also be sent to many different workers, which is the usage scenario for applications like Folding@Home [2], which employ a multitude of workers across the Internet.

*Handling malicious workers.* In our scheme, if we assume that the worker learns whether or not the client accepts the proof $\pi_x$, then for every execution, a malicious worker potentially learns a bit of information about the labels of the Yao-garbled circuit. For example, the worker could try to guess one of the labels, encrypt it with the homomorphic encryption and see if the client accepts. In a sense, the output of the client at the end of the execution can be seen as a very restricted "decryption oracle" for the homomorphic encryption scheme (which is, by definition, not CCA secure). Because of this one-bit leakage, we are unable to prove security in this case.

There are two ways to deal with this. One is to assume that the verification output bit by the client remains private until all of the workers' results have been returned. The other is to repeat the pre-processing stage, i.e. the Yao garbling of the circuit, every time a verification fails. In this case, in order to preserve a good amortized complexity, we must assume that failures do not happen very often. This is indeed the case in the previous scenario, where the same garbled circuit is used with several workers, under the assumption that only a small fraction of workers will be malicious. See Section 5.

## 2   Background

YAO'S PROTOCOL FOR TWO-PARTY COMPUTATION. We summarize Yao's protocol for two-party private computation [31, 32]. For more details, we refer the interested reader to Lindell and Pinkas' excellent description [21].

We assume two parties, Alice and Bob, wish to compute a function $F$ over their private inputs $a$ and $b$. For simplicity, we focus on polynomial-time deterministic functions, but the generalization to stochastic functions is straightforward.

At a high-level, Alice converts $F$ into a boolean circuit $C$. She prepares a garbled version of the circuit, $G(C)$, and sends it to Bob, along with a garbled version, $G(a)$, of her input. Alice and Bob then engage in a series of oblivious transfers so that Bob obtains $G(b)$ without Alice learning anything about $b$. Bob then applies the garbled circuit to the two garbled outputs to derive a garbled version of the output: $G(F(a,b))$.

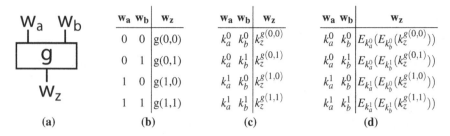

**Fig. 1. Yao's Garbled Circuits.** *The original binary gate* **(a)** *can be represented by a standard truth table* **(b)**. *We then replace the 0 and 1 values with the corresponding randomly chosen λ-bit values* **(c)**. *Finally, we use the values for $w_a$ and $w_b$ to encrypt the values for the output wire $w_z$* **(d)**. *The random permutation of these ciphertexts is the garbled representation of gate g.*

Alice can then translate this into the actual output and share the result with Bob. Note that the privacy of this protocol assumes an honest-but-curious adversary model.

In more detail, Alice constructs the garbled version of the circuit as follows. For each wire $w$ in the circuit, Alice chooses two random values $k_w^0, k_w^1 \overset{R}{\leftarrow} \{0,1\}^\lambda$ to represent the bit values of 0 or 1 on that wire. Once she has chosen wire values for every wire in the circuit, Alice constructs a garbled version of each gate $g$ (see Figure 1). Let $g$ be a gate with input wires $w_a$ and $w_b$, and output wire $w_z$. Then the garbled version $G(g)$ of $g$ is simply four ciphertexts:

$$\gamma_{ij} = E_{k_a^i}(E_{k_b^j}(k_z^{g(i,j)})), \text{ where } i \in \{0,1\}, j \in \{0,1\} \qquad (1)$$

where $E$ is an secure symmetric encryption scheme with an "elusive range" (more details below). The order of the ciphertexts is randomly permuted to hide the structure of the circuit (i.e., we shuffle the ciphertexts, so that the first ciphertext does not necessarily encode the output for $(0,0)$).

In Yao's protocol, Alice transfers all of the ciphertexts to Bob, along with the wire values corresponding to the bit-level representation of her input. In other words, she transfers either $k_a^0$ if her input bit is 0 or $k_a^1$ if her input bit is 1. Since these are randomly chosen values, Bob learns nothing about Alice's input. Alice and Bob then engage in an oblivious transfer so that Bob can obtain the wire values corresponding to his inputs (e.g., $k_b^0$ or $k_b^1$). Bob learns exactly one value for each wire, and Alice learns nothing about his input. Bob can then use the wire values to recursively decrypt the gate ciphertexts, until he arrives at the final output wire values. When he transmits these to Alice, she can map them back to 0 or 1 values and hence obtain the result of the function computation.

HOMOMORPHIC ENCRYPTION. A fully-homomorphic encryption scheme $\mathcal{E}$ is defined by four algorithms: the standard encryption functions **KeyGen**$_{\mathcal{E}}$, **Encrypt**$_{\mathcal{E}}$, and **Decrypt**$_{\mathcal{E}}$, as well as a fourth function **Evaluate**$_{\mathcal{E}}$. **Evaluate**$_{\mathcal{E}}$ takes in a circuit $C$ and a tuple of ciphertexts and outputs a ciphertext that decrypts to the result of applying $C$ to the plaintexts. A nontrivial scheme requires that **Encrypt**$_{\mathcal{E}}$ and **Decrypt**$_{\mathcal{E}}$ operate in time independent of $C$ [13, 12, 27, 30]. More precisely, the time needed to generate a ciphertext for an input wire of $C$, or decrypt a ciphertext for an output wire, is polynomial in the security parameter of the scheme (independent of $C$). Note that this implies that the length of the ciphertexts for the output wires is bounded by some polynomial in the security parameter (independent of $C$).

Gentry recently proposed a scheme, based on ideal lattices, that satisfies these requirements for arbitrary circuits [13, 12] (since Gentry's proposal, additional integer-based schemes have been proposed [27, 30]). The complexity of **KeyGen**$_{\mathcal{E}}$ in his initial *leveled* fully homomorphic encryption scheme grows linearly with the depth of $C$. However, under the assumption that his encryption scheme is *circular secure* – i.e., roughly, that it is "safe" to reveal an encryption of a secret key under its associated public key – the complexity of **KeyGen**$_{\mathcal{E}}$ is independent of $C$. See [13, 12, 8] for more discussion on circular-security (and, more generally, key-dependent-message security) as it relates to fully homomorphic encryption.

# 3 Problem Definition

At a high-level, a verifiable computation scheme is a two-party protocol in which a *client* chooses a function and then provides an encoding of the function and inputs to the function to a *worker*. The worker is expected to evaluate the function on the input and respond with the output. The client then verifies that the output provided by the worker is indeed the output of the function computed on the input provided.

## 3.1 Basic Requirements

A *verifiable computation scheme* $\mathcal{VC} = ($**KeyGen, ProbGen, Compute, Verify**$)$ consists of the four algorithms defined below.

1. **KeyGen**$(F, \lambda) \rightarrow (PK, SK)$: Based on the security parameter $\lambda$, the randomized *key generation* algorithm generates a public key that encodes the target function $F$, which is used by the worker to compute $F$. It also computes a matching secret key, which is kept private by the client.
2. **ProbGen**$_{SK}(x) \rightarrow (\sigma_x, \tau_x)$: The *problem generation* algorithm uses the secret key $SK$ to encode the function input $x$ as a public value $\sigma_x$ which is given to the worker to compute with, and a secret value $\tau_x$ which is kept private by the client.
3. **Compute**$_{PK}(\sigma_x) \rightarrow \sigma_y$: Using the client's public key and the encoded input, the worker *computes* an encoded version of the function's output $y = F(x)$.
4. **Verify**$_{SK}(\tau_x, \sigma_y) \rightarrow y \cup \bot$: Using the secret key $SK$ and the secret "decoding" $\tau_x$, the *verification* algorithm converts the worker's encoded output into the output of the function, e.g., $y = F(x)$ or outputs $\bot$ indicating that $\sigma_y$ does not represent the valid output of $F$ on $x$.

A verifiable computation scheme should be both correct and secure. A scheme is correct if the problem generation algorithm produces values that allows an honest worker to compute values that will verify successfully and correspond to the evaluation of $F$ on those inputs. More formally:

**Definition 1 (Correctness).** *A verifiable computation scheme $\mathcal{VC}$ is correct if for any function $F$, the key generation algorithm produces keys $(PK, SK) \leftarrow$ **KeyGen**$(F, \lambda)$ such that, $\forall x \in$ Domain$(F)$, if $(\sigma_x, \tau_x) \leftarrow$ **ProbGen**$_{SK}(x)$ and $\sigma_y \leftarrow$ **Compute**$_{PK}(\sigma_x)$ then $y = F(x) \leftarrow$ **Verify**$_{SK}(\tau_x, \sigma_y)$.*

Intuitively, a verifiable computation scheme is secure if a malicious worker cannot persuade the verification algorithm to accept an incorrect output. In other words, for a given function $F$ and input $x$, a malicious worker should not be able to convince the verification algorithm to output $\hat{y}$ such that $F(x) \neq \hat{y}$. Below, we formalize this intuition with an experiment, where $poly(\cdot)$ is a polynomial.

Experiment $\mathbf{Exp}_A^{Verif}[\mathcal{V}C, F, \lambda]$
$\quad (PK, SK) \xleftarrow{R} \mathbf{KeyGen}(F, \lambda);$
$\quad$ For $i = 1, \ldots, \ell = poly(\lambda);$
$\quad\quad x_i \leftarrow A(PK, x_1, \sigma_1, \ldots, x_{i-1}, \sigma_{i-1});$
$\quad\quad (\sigma_i, \tau_i) \leftarrow \mathbf{ProbGen}_{SK}(x_i);$
$\quad (i, \hat{\sigma}_y) \leftarrow A(PK, x_1, \sigma_1, \ldots, x_\ell, \sigma_\ell);$
$\quad \hat{y} \leftarrow \mathbf{Verify}_{SK}(\tau_i, \hat{\sigma}_y)$
$\quad$ If $\hat{y} \neq \perp$ and $\hat{y} \neq F(x_i)$, output '1', else '0';

Essentially, the adversary is given oracle access to generate the encoding of multiple problem instances. The adversary succeeds if it produces an output that convinces the verification algorithm to accept on the wrong output value for a given input value. Note that in this experiment, the adversary does not learn whether or not he succeeded; we consider the implications of providing the adversary with this information in Section 5. We can now define the security of the system based on the adversary's success in the above experiment.

**Definition 2 (Security).** *For a verifiable computation scheme $\mathcal{V}C$, we define the advantage of an adversary $A$ in the experiment above as:*

$$Adv_A^{Verif}(\mathcal{V}C, F, \lambda) = Prob[\mathbf{Exp}_A^{Verif}[\mathcal{V}C, F, \lambda] = 1] \qquad (2)$$

*A verifiable computation scheme $\mathcal{V}C$ is secure for a function $F$, if for any adversary $A$ running in probabilistic polynomial time,*

$$Adv_A^{Verif}(\mathcal{V}C, F, \lambda) \leq \mathtt{negli}(\lambda) \qquad (3)$$

*where $\mathtt{negli}()$ is a negligible function of its input.*

In the above definition, we could have also allowed the adversary to select the function $F$. However, our protocol is a verifiable computation scheme that is secure for *all $F$*, so the above definition suffices.

### 3.2 Input and Output Privacy

While the basic definition of a verifiable computation protects the integrity of the computation, it is also desirable that the scheme protect the secrecy of the input given to the worker(s). We define input privacy based on a typical indistinguishability argument that guarantees that *no* information about the inputs is leaked.

Intuitively, a verifiable computation scheme is *private* when the public outputs of the problem generation algorithm **ProbGen** over two different inputs are indistinguishable; i.e., an adversary cannot decide which encoding is the correct one for a given input. More formally consider the following experiment: the adversary is given the public key for the scheme and selects two inputs $x_0, x_1$. He is then given the encoding of a randomly selected one of the two inputs and must guess which one was encoded. During this process the adversary is allowed to request the encoding of any input he desires. The experiment is described below. The oracle $\textbf{PubProbGen}_{SK}(x)$ calls $\textbf{ProbGen}_{SK}(x)$ to obtain $(\sigma_x, \tau_x)$ and returns only the public part $\sigma_x$.

Experiment $\textbf{Exp}_A^{Priv}[\mathcal{VC}, F, \lambda]$

$\quad (PK, SK) \xleftarrow{R} \textbf{KeyGen}(F, \lambda);$

$\quad (x_0, x_1) \leftarrow A^{\textbf{PubProbGen}_{SK}(\cdot)}(PK)$

$\quad (\sigma_0, \tau_0) \leftarrow \textbf{ProbGen}_{SK}(x_0);$

$\quad (\sigma_1, \tau_1) \leftarrow \textbf{ProbGen}_{SK}(x_1);$

$\quad b \xleftarrow{R} \{0, 1\};$

$\quad \hat{b} \leftarrow A^{\textbf{PubProbGen}_{SK}(\cdot)}(PK, x_0, x_1, \sigma_b)$

$\quad$ If $\hat{b} = b$, output '1', else '0';

**Definition 3 (Privacy).** *For a verifiable computation scheme $\mathcal{VC}$, we define the advantage of an adversary A in the experiment above as:*

$$Adv_A^{Priv}(\mathcal{VC}, F, \lambda) = \left| Prob[\textbf{Exp}_A^{Priv}[\mathcal{VC}, F, \lambda] = 1] - \frac{1}{2} \right| \qquad (4)$$

*A verifiable computation scheme $\mathcal{VC}$ is* private *for a function F, if for any adversary A running in probabilistic polynomial time,*

$$Adv_A^{Priv}(\mathcal{VC}, F, \lambda) \leq \texttt{negli}(\lambda) \qquad (5)$$

*where* $\texttt{negli}()$ *is a negligible function of its input.*

An immediate consequence of the above definition is that in a private scheme, the encoding of the input must be probabilistic (since the adversary can always query $x_0, x_1$ to the **PubProbGen** oracle, and if the answer were deterministic, he could decide which input is encoded in $\sigma_b$).

A similar definition can be made for output privacy.

## 3.3 Efficiency

The final condition we require from a verifiable computation scheme is that the time to encode the input and verify the output must be smaller than the time to compute the function from scratch.

**Definition 4 (Outsourceable).** *A $\mathcal{VC}$ can be outsourced if it permits efficient generation and efficient verification. This implies that for any x and any $\sigma_y$, the time required for $\textbf{ProbGen}_{SK}(x)$ plus the time required for $\textbf{Verify}(\sigma_y)$ is $o(T)$, where T is the fastest known time required to compute $F(x)$.*

Notice that we are not including the time to compute the key generation algorithm (i.e., the encoding of the function itself). Therefore, the above definition captures the idea of an outsourceable verifiable computation scheme which is more efficient than computing the function in an *amortized* sense, since the cost of encoding the function can be amortized over many different input computations.

# 4   An Efficient Verifiable-Computation Scheme with Input and Output Privacy

We are now ready to describe our scheme. Informally, our protocol works as follows. The key generation algorithm consists of running Yao's garbling procedure over a Boolean circuit computing the function $F$: the public key is the collection of ciphertexts representing the garbled circuit, and the secret key consists of all the random wire labels. The input is encoded in two steps: first a fresh public/secret key pair for a homomorphic encryption scheme is generated, and then the labels of the correct input wires are encrypted with it. These ciphertexts constitute the public encoding of the input, while the secret key is kept private by the client. Using the homomorphic properties of the encryption scheme, the worker performs the computation steps of Yao's protocol, but working over ciphertexts (i.e., for every gate, given the encrypted labels for the correct input wires, obtain an encryption of the correct output wire, by applying the homomorphic encryption over the circuit that computes the "double decryption" in Yao's protocol). At the end, the worker will hold the encryption of the labels of the correct output wires. He returns these ciphertexts to the client who decrypts them and then computes the output from them. We give a detailed description below.

**Protocol $\mathcal{VC}$**

1. **KeyGen**$(F, \lambda) \rightarrow (PK, SK)$: Represent $F$ as a circuit $C$. Following Yao's Circuit Construction (see Section 2), choose two values, $w_i^0, w_i^1 \xleftarrow{R} \{0,1\}^\lambda$ for each wire $w_i$. For each gate $g$, compute the four ciphertexts $(\gamma_{00}^g, \gamma_{01}^g, \gamma_{10}^g, \gamma_{11}^g)$ described in Equation 1. The public key $PK$ will be the full set of ciphertexts, i.e., $PK \leftarrow \cup_g (\gamma_{00}^g, \gamma_{01}^g, \gamma_{10}^g, \gamma_{11}^g)$, while the secret key will be the wire values chosen: $SK \leftarrow \cup_i (w_i^0, w_i^1)$.

2. **ProbGen**$_{SK}(x) \rightarrow \sigma_x$: Run the fully-homomorphic encryption scheme's key generation algorithm to create a new key pair: $(PK_{\mathcal{E}}, SK_{\mathcal{E}}) \leftarrow \textbf{KeyGen}_{\mathcal{E}}(\lambda)$. Let $w_i \subset SK$ be the wire values representing the binary expression of $x$. Set the public value $\sigma_x \leftarrow (PK_{\mathcal{E}}, \textbf{Encrypt}_{\mathcal{E}}(PK_{\mathcal{E}}, w_i))$ and the private value $\tau_x \leftarrow SK_{\mathcal{E}}$.

3. **Compute**$_{PK}(\sigma_x) \rightarrow \sigma_y$: Calculate $\textbf{Encrypt}_{\mathcal{E}}(PK_{\mathcal{E}}, \gamma_i)$. Construct a circuit $\Delta$ that on input $w, w', \gamma$ outputs $D_w(D_{w'}(\gamma))$, where $D$ is the decryption algorithm corresponding to the encryption $E$ used in Yao's garbling (therefore $\Delta$ computes the appropriate decryption in Yao's construction). Calculate $\textbf{Evaluate}_{\mathcal{E}}(\Delta, \textbf{Encrypt}_{\mathcal{E}}(PK_{\mathcal{E}}, w_i),$ $\textbf{Encrypt}_{\mathcal{E}}(PK_{\mathcal{E}}, \gamma_i))$ repeatedly, to decrypt your way through the ciphertexts, just as in the evaluation of Yao's garbled circuit. The result is $\sigma_y \leftarrow \textbf{Encrypt}_{\mathcal{E}}(PK_{\mathcal{E}}, \bar{w}_i)$, where $\bar{w}_i$ are the wire values representing $y = F(x)$ in binary.

4. **Verify**$_{SK}(\sigma_y) \rightarrow y \cup \bot$: Use $SK_{\mathcal{E}}$ to decrypt $\textbf{Encrypt}_{\mathcal{E}}(PK_{\mathcal{E}}, \bar{w}_i)$, obtaining $\bar{w}_i$. Use $SK$ to map the wire values to an output $y$. If the decryption or mapping fails, then output $\bot$.

**Remark:** *On verifying ciphertext ranges in an encrypted form.* Recall that Yao's scheme requires the encryption scheme $E$ to have an *efficiently verifiable range* [21]: Given the key $k$, it is possible to decide efficiently if a given ciphertext falls into the range of encryptions under $k$. In other words, there exists an efficient machine $M$ such that $M(k, \gamma) = 1$ iff $\gamma \in \mathsf{Range}_\lambda(k)$. This is needed to "recognize" which ciphertext to pick among the four ciphertexts associated with each gate.

In our verifiable computation scheme $\mathcal{VC}$, we need to perform this check using an encrypted form of the key $c = \mathbf{Encrypt}_{\mathcal{E}}(PK_{\mathcal{E}}, k)$. When applying the homomorphic properties of $\mathcal{E}$ to the range testing machine $M$, the worker obtains an encryption of 1 for the correct ciphertext, and an encryption of 0 for the others. Of course he is not able to distinguish which one is the correct one.

The worker then proceeds as follows: for the four ciphertexts $\gamma_1, \gamma_2, \gamma_3, \gamma_4$ associated with a gate $g$, he first computes $c_i = \mathbf{Encrypt}_{\mathcal{E}}(PK_{\mathcal{E}}, M(k, \gamma_i))$ using the homomorphic properties of $\mathcal{E}$ over the circuit describing $M$. Note that only one of these ciphertexts encrypts a 1, exactly the one corresponding to the correct $\gamma_i$. Then the worker computes $d_i = \mathbf{Encrypt}_{\mathcal{E}}(PK_{\mathcal{E}}, D_k(\gamma_i))$ using the homomorphic properties of $\mathcal{E}$ over the decryption circuit $\Delta$. Note that $k' = \Sigma_i M(k, \gamma_i) D_k(\gamma_i)$ is the correct label for the output wire. Therefore, the worker can use the homomorphic properties of $\mathcal{E}$ to compute $c = \mathbf{Encrypt}_{\mathcal{E}}(PK_{\mathcal{E}}, k') = \mathbf{Encrypt}_{\mathcal{E}}(PK_{\mathcal{E}}, \Sigma_i M(k, \gamma_i) D_k(\gamma_i))$ from $c_i, d_i$, as desired.

The main result of our paper is the following.

**Theorem 1.** *Let E be a Yao-secure symmetric encryption scheme and $\mathcal{E}$ be a semantically secure homomorphic encryption scheme. Then protocol $\mathcal{VC}$ is a secure, outsourceable and private verifiable computation scheme.*

The proof of Theorem 1 requires two high-level steps. First, we show that Yao's garbled circuit scheme is a one-time secure verifiable computation scheme, i.e. a scheme that can be used to compute $F$ securely on one input. This is an almost immediate reduction to the security of Yao's protocol as a two-party computation scheme. Then, by using the semantic security of the homomorphic encryption scheme, we reduce the security of our scheme (with multiple executions) to the security of a single execution where we expect the adversary to cheat. The proof appears in Appendix A.

INPUT AND OUTPUT PRIVACY. Note that for each oracle query the input and the output are encrypted under the homomorphic encryption scheme $\mathcal{E}$. It is not hard to see that the proof of correctness above, easily implies the proof of input and output privacy. For the one-time case, it obviously follows from the security of Yao's two-party protocol. For the general case, it follows from the semantic security of $\mathcal{E}$, and the proof relies on the same style of hybrid arguments described above.

## 5   How to Handle Cheating Workers

Our definition of security (Definition 2) assumes that the adversary does not see the output of the **Verify** procedure run by the client on the value $\sigma$ returned by the adversary. Theorem 1 is proven under the same assumption. In practice this means that our protocol $\mathcal{VC}$ is secure if the client keeps the result of the computation private.

In practice, there might be circumstances where this is not feasible, as the behavior of the client will change depending on the result of the evaluation (e.g., the client might refuse to pay the worker). Intuitively, and we prove this formally below, seeing the result of **Verify** on proofs the adversary correctly **Compute**s using the output of **PubProbGen** does not help the adversary (since it already knows the result based on the inputs it supplied to **PubProbGen**). But what if the worker returns a malformed response – i.e., something for which **Verify** outputs $\bot$. How does the client respond, if at all? One option is for the client to ask the worker to perform the computation again. But this repeated request informs the worker that its response was malformed, which is an additional bit of information that a cheating worker might exploit in its effort to generate forgeries. Is our scheme secure in this setting? In this section, we prove that our scheme remains secure as long as the client terminates after detecting a malformed response. We also consider the interesting question of whether our scheme is secure if the client terminates only after detecting $k > 1$ malformed responses, but we are unable to provide a proof of security in this modified setting.

Note that there is a real attack on the scheme in this setting if the client does not terminate. Specifically, for concreteness, suppose that each ciphertext output by $\mathbf{Encrypt}_{\mathcal{E}}$ encrypts a single bit of a label for an input wire of the garbled circuit, and that the adversary wants to determine the first bit $w_{11}^{b_1}$ of the first label (where that label stands in for unknown input $b_1 \in \{0,1\}$). To do this, the adversary runs $\mathbf{Compute}$ as before, obtaining ciphertexts that encrypt the bits $\bar{w}_i$ of a label for the output wire. Using the homomorphism of the encryption scheme $\mathcal{E}$, it XORs $w_{11}^{b_1}$ with the first bit of $\bar{w}_i$ to obtain $\bar{w}_i'$, and it sends (the encryption of) $\bar{w}_i'$ as its response. If $\mathbf{Verify}$ outputs $\perp$, then $w_{11}^{b_1}$ must have been a 1; otherwise, it is a 0 with overwhelming probability. The adversary can thereby learn the labels of the garbled circuit one bit at a time – in particular, it can similarly learn the labels of the output wire, and thereafter generate a verifiable response without actually performing the computation.

Intuitively, one might think that if the client terminates after detecting $k$ malformed responses, then the adversary should only be able to obtain about $k$ bits of information about the garbled circuit before the client terminates (using standard entropy arguments), and therefore it should still be hard for the adversary to output the entire "wrong" label for the output wire as long as $\lambda$ is sufficiently larger than $k$. However, we are unable to make this argument go through. In particular, the difficulty is with the hybrid argument in the proof of Theorem 1, where we gradually transition to an experiment in which the simulator is encrypting the same Yao input labels in every round. This experiment must be indistinguishable from the real world experiment, which permits different inputs in different rounds. When we don't give the adversary information about whether or not its response was well-formed or not, the hybrid argument is straightforward – it simply depends on the semantic security of the FHE scheme.

However, if we do give the adversary that information, then the adversary can easily distinguish rounds with the same input from rounds with random inputs. To do so, it chooses some "random" predicate $P$ over the input labels, such that $P(w_{b_1}^1, w_{b_2}^2, \ldots) = P(w_{b_1'}^1, w_{b_2'}^2, \ldots)$ with probability $1/2$ if $(b_1, b_2, \ldots) \neq (b_1', b_2', \ldots)$. Given the encryptions of $w_{b_1}^1, w_{b_2}^2, \ldots$, the adversary runs $\mathbf{Compute}$ as in the scheme, obtaining ciphertexts that encrypt the bits $\bar{w}_i$ of a label for the output wire, XORs (using the homomorphism) $P(w_{b_1}^1, w_{b_2}^2, \ldots)$ with the first bit of $\bar{w}_i$, and sends (an encryption of) the result $\bar{w}_i'$ as its response. If the client is making the same query in every round – i.e., the Yao input labels are the same every time – then, the predicate always outputs the same bit, and thus the adversary gets the same response (well-formed or malformed) in every round. Otherwise, the responses will tend to vary.

One could try to make the adversary's distinguishing attack more difficult by (for example) trying to hide which ciphertexts encrypt the bits of which labels – i.e., via some form of obfuscation. However, the adversary may define its predicate in such a way that it "analyzes" this obfuscated circuit, determines whether two ostensibly different inputs in fact represent the same set of Yao input labels, and outputs the same bit if they do. (It performs this analysis on the encrypted inputs, using the homomorphism.) We do not know of any way to prevent this attack, and preventing it may be rather difficult in light of Barak et al.'s result that there is no general obfuscator [7].

*Security with Verification Access.* We say that a verifiable computation scheme is secure *with verification access* if the adversary is allowed to see the result of **Verify** over the queries $x_i$ he has made to the **ProbGen** oracle in $\mathbf{Exp}_A^{Verif}$ (see Definition 2).

Let $\mathcal{VC}^\dagger$ be like $\mathcal{VC}$, except that the client terminates if it receives a malformed response from the worker. Below, we show that $\mathcal{VC}^\dagger$ is secure with verification access. In other words, it is secure to provide the worker with verification access (indicating whether a response was well-formed or not), until the worker gives a malformed response. Let $\mathbf{Exp}_A^{Verif^\dagger}\left[\mathcal{VC}^\dagger, F, \lambda\right]$ denote the experiment described in Section 3.1, with the obvious modifications.

**Theorem 2.** *If $\mathcal{VC}$ is a secure outsourceable verifiable computation scheme, then $\mathcal{VC}^\dagger$ is a secure outsourceable verifiable computation scheme with verification access. If $\mathcal{VC}$ is private, so is $\mathcal{VC}^\dagger$.*

The proof appears in Appendix B.

In practice Theorem 2 implies that every time a malformed response is received, the client must re-garble the circuit (or, as we said above, make sure that the results of the verification procedure remain secret). Therefore the amortized efficiency of the client holds only if we assume that malformed responses do not happen very frequently.

In some settings, it is not necessary to inform the worker that its response is malformed, at least not immediately. For example, in the Folding@Home application [2], suppose the client generates a new garbled circuit each morning for its many workers. At the end of the day, the client stops accepting computations using this garbled circuit, and it (optionally) gives the workers information about the well-formedness of their responses. Indeed, the client may reveal all of its secrets for that day. In this setting, our previous proof clearly holds even if there are arbitrarily many malformed responses.

## 6    Conclusions and Future Directions

In this work, we introduced the notion of Verifiable Computation as a natural formulation for the increasingly common phenomenon of outsourcing computational tasks to untrusted workers. We describe a scheme that combines Yao's Garbled Circuits with a fully-homomorphic encryption scheme to provide extremely efficient outsourcing, even in the presence of an adaptive adversary. As an additional benefit, our scheme maintains the privacy of the client's inputs and outputs.

Our work leaves open several interesting problems. It would be desirable to devise a verifiable computation scheme that used a more efficient primitive than fully-homomorphic encryption. Similarly, it seems plausible that a verifiable scheme might sacrifice input privacy to increase its efficiency. Finally, while our scheme is resilient against a single malformed response from the worker, ideally we would like a scheme that tolerates $k > 1$ malformed responses.

## Acknowledgements

The authors are grateful to Virgil Gligor for a number of useful discussions and to the anonymous reviewers for their helpful suggestions.

This research was supported in part by the US Army Research Laboratory and the UK Ministry of Defence under Agreement Number W911NF-06-3-0001, as well as by the National Science Foundation (NSF), under award number CCF-0424422. Bryan Parno was supported in part by an NSF Graduate Research Fellowship. The views and conclusions contained in this document are those of the authors and should not be interpreted as representing the official policies, either expressed or implied, of the US Army Research Laboratory, U.S. Government, UK Ministry of Defense, UK Government, or NSF. The US and UK Governments are authorized to reproduce and distribute reprints for Government purposes notwithstanding any copyright notation hereon.

## References

1. Amazon Elastic Compute Cloud, http://aws.amazon.com/ec2
2. The Folding@home project. Stanford University,
   http://www.stanford.edu/group/pandegroup/cosm/
3. Sun Utility Computing, http://www.sun.com/service/sungrid/index.jsp
4. The Great Internet Mersenne Prime Search, http://www.mersenne.org/
5. Anderson, D.P., Cobb, J., Korpela, E., Lebofsky, M., Werthimer, D.: SETI@Home: An experiment in public-resource computing. Communications of the ACM 45(11), 56–61 (2002)
6. Babai, L.: Trading group theory for randomness. In: Proceedings of the ACM Symposium on Theory of Computing (STOC), pp. 421–429. ACM, New York (1985)
7. Barak, B., Goldreich, O., Impagliazzo, R., Rudich, S., Sahay, A., Vadhan, S., Yang, K.: On the (im)possibility of obfuscating programs. In: Kilian, J. (ed.) CRYPTO 2001. LNCS, vol. 2139, pp. 1–18. Springer, Heidelberg (2001)
8. Barak, B., Haitner, I., Hofheinz, D., Ishai, Y.: Bounded key-dependent message security. In: Proceedings of EuroCrypt (June 2010)
9. Belenkiy, M., Chase, M., Erway, C.C., Jannotti, J., Küpçü, A., Lysyanskaya, A.: Incentivizing outsourced computation. In: Proceedings of the Workshop on Economics of Networked Systems (NetEcon), pp. 85–90. ACM, New York (2008)
10. Chaum, D., Pedersen, T.: Wallet databases with observers. In: Brickell, E.F. (ed.) CRYPTO 1992. LNCS, vol. 740, pp. 89–105. Springer, Heidelberg (1993)
11. Gennaro, R., Gentry, C., Parno, B.: Non-Interactive Verifiable Computing: Outsourcing Computation to Untrusted Workers, http://eprint.iacr.org/2009/547
12. Gentry, C.: A fully homomorphic encryption scheme. PhD thesis, Stanford University (2009)
13. Gentry, C.: Fully homomorphic encryption using ideal lattices. In: Proceedings of the ACM Symposium on the Theory of Computing (STOC) (2009)
14. Goldwasser, S., Kalai, Y.T., Rothblum, G.N.: Delegating computation: interactive proofs for muggles. In: Proceedings of the ACM Symposium on the Theory of Computing (2008)
15. Goldwasser, S., Micali, S., Rackoff, C.: The knowledge complexity of interactive proof-systems. SIAM Journal on Computing 18(1), 186–208 (1989)
16. Golle, P., Mironov, I.: Uncheatable distributed computations. In: Proceedings of the RSA Conference (2001)
17. Hohenberger, S., Lysyanskaya, A.: How to securely outsource cryptographic computations. In: Kilian, J. (ed.) TCC 2005. LNCS, vol. 3378, pp. 264–282. Springer, Heidelberg (2005)
18. Kalai, Y.T., Raz, R.: Probabilistically checkable arguments. In: Halevi, S. (ed.) CRYPTO 2009. LNCS, vol. 5677, pp. 143–159. Springer, Heidelberg (2009)

19. Kilian, J.: A note on efficient zero-knowledge proofs and arguments (extended abstract). In: Proceedings of the ACM Symposium on Theory of Computing (STOC) (1992)
20. Kilian, J.: Improved efficient arguments (preliminary version). In: Coppersmith, D. (ed.) CRYPTO 1995. LNCS, vol. 963, pp. 311–324. Springer, Heidelberg (1995)
21. Lindell, Y., Pinkas, B.: A proof of Yao's protocol for secure two-party computation. Journal of Cryptology 22(2), 161–188 (2009)
22. Micali, S.: CS proofs (extended abstract). In: Proceedings of the IEEE Symposium on Foundations of Computer Science (1994)
23. Molnar, D.: The SETI@Home problem. ACM Crossroads, 7.1 (2000)
24. Monrose, F., Wyckoff, P., Rubin, A.: Distributed execution with remote audit. In: Proceedings of ISOC Network and Distributed System Security Symposium (NDSS) (February 1999)
25. Rothblum, G.: Delegating Computation Reliably: Paradigms and Constructions. PhD thesis, Massachusetts Institute of Technology (2009)
26. Rothblum, G., Vadhan, S.: Are PCPs inherent in efficient arguments? In: Proceedings of Computational Complexity (CCC) (2009)
27. Smart, N.P., Vercauteren, F.: Fully homomorphic encryption with relatively small key and ciphertext sizes. In: Nguyen, P.Q., Pointcheval, D. (eds.) PKC 2010. LNCS, vol. 6056, pp. 420–443. Springer, Heidelberg (2010)
28. Smith, S., Weingart, S.: Building a high-performance, programmable secure coprocessor. Computer Networks (Special Issue on Computer Network Security) 31, 831–960 (1999)
29. Trusted Computing Group. Trusted platform module main specification. Version 1.2, Revision 103 (July 2007)
30. van Dijk, M., Gentry, C., Halevi, S., Vaikuntanathan, V.: Fully homomorphic encryption over the integers. In: Proceedings of EuroCrypt (June 2010)
31. Yao, A.: Protocols for secure computations. In: Proceedings of the IEEE Symposium on Foundations of Computer Science (1982)
32. Yao, A.: How to generate and exchange secrets. In: Proceedings of the IEEE Symposium on Foundations of Computer Science (1986)
33. Yee, B.S.: Using Secure Coprocessors. PhD thesis, Carnegie Mellon University (1994)

# A   Proof of Theorem 1

The proof of Theorem 1 proceeds in two steps. First, we show that Yao's garbled circuit scheme is a one-time secure verifiable computation scheme, in other words, a scheme that can be used to compute $F$ securely on one input. This is an almost immediate reduction to the security of Yao's protocol as a two-party computation scheme. Then, by using the semantic security of the homomorphic encryption scheme, we reduce the security of our scheme (with multiple executions) to the security of a single execution in which we expect the adversary to cheat.

### A.1   Proof Sketch of Yao's Security for One Execution

Consider the verifiable computation scheme $\mathcal{VC}_{Yao}$ defined as follows:

**Protocol $\mathcal{VC}_{Yao}$.**

1. **KeyGen**$(F, \lambda) \rightarrow (PK, SK)$: Represent $F$ as a circuit $C$. Following Yao's Circuit Construction (see Section 2), choose two values, $w_i^0, w_i^1 \xleftarrow{R} \{0,1\}^\lambda$ for each wire $w_i$. For each gate $g$, compute the four ciphertexts $(\gamma_{00}^g, \gamma_{01}^g, \gamma_{10}^g, \gamma_{11}^g)$ from Equation 1. The public key $PK$ will be the full set of ciphertexts, i.e, $PK \leftarrow \cup_g (\gamma_{00}^g, \gamma_{01}^g, \gamma_{10}^g, \gamma_{11}^g)$, while the secret key will be the wire values chosen: $SK \leftarrow \cup_i (w_i^0, w_i^1)$.

2. **ProbGen$_{SK}(x) \to \sigma_x$**: Reveal the labels of the input wires associated with $x$. In other words, let $w_i \subset SK$ be the wire values representing the binary expression of $x$, and set $\sigma_x \leftarrow w_i$. $\tau_x$ is the empty string.

3. **Compute$_{PK}(\sigma_x) \to \sigma_y$**: Compute the decryptions in Yao's protocol to obtain the labels of the correct output wires. Set $\sigma_y$ to be these labels.

4. **Verify$_{SK}(\sigma_y) \to y \cup \bot$**: Use $SK$ to map the wire values in $\sigma_y$ to the binary representation of the output $y$. If the mapping fails, output $\bot$.

**Theorem 3.** $\mathcal{VC}_{Yao}$ is a correct verifiable computation scheme.

**Proof of Theorem 3:** The proof of correctness follows directly from the proof of correctness for Yao's garbled circuit construction [21]. Using $C$ and $\tilde{x}$ produces a $\tilde{y}$ that represents the correct evaluation of $F(x)$.    ∎

In the full version of the paper [11], we prove that $\mathcal{VC}_{Yao}$ is a *one-time* secure verifiable computation scheme. The definition of *one-time secure* is the same as Definition 2 except that in experiment $\mathbf{Exp}_A^{Verif}$, the adversary is allowed to query the oracle **ProbGen$_{SK}(\cdot)$** only once (i.e., $\ell = 1$) and must cheat on that input.

Intuitively, an adversary who violates the security of this scheme must either guess the "incorrect" random value $k_w^{1-y_i}$ for one of the output bit values representing $y$, or he must break the encryption scheme used to encode the "incorrect" wire values in the circuit. The former happens with probability $\leq \frac{1}{2^\lambda}$, i.e., negligible in $\lambda$. The latter violates our security assumptions about the encryption scheme. We formalize this intuition using a hybrid argument similar to the one used in [21].

**Theorem 4.** Let $E$ be a Yao-secure symmetric encryption scheme. Then $\mathcal{VC}_{Yao}$ is a one-time secure verifiable computation scheme.

## A.2    Completing the Proof of Theorem 1

The proof of Theorem 1 follows from Theorem 4 and the semantic security of the homomorphic encryption scheme. More precisely, we show that if the homomorphic encryption scheme is semantically secure, then we can transform (via a simulation) a successful adversary against the full verifiable computation scheme $\mathcal{VC}$ into an attacker for the one-time secure protocol $\mathcal{VC}_{Yao}$. The intuition is that for each query, the labels in the circuit are encrypted with a semantically-secure encryption scheme (the homomorphic scheme), so multiple queries do not help the adversary to learn about the labels, and hence if he cheats, he must be able to cheat in the one-time case as well.

**Proof of Theorem 1:** Let us assume for the sake of contradiction that there is an adversary $A$ such that $Adv_A^{Verif}(\mathcal{VC}, F, \lambda) \geq \varepsilon$, where $\varepsilon$ is non-negligible in $\lambda$. We use $A$ to build another adversary $A'$ which queries the **ProbGen** oracle only once, and for which $Adv_{A'}^{Verif}(\mathcal{VC}_{Yao}, F, \lambda) \geq \varepsilon'$, where $\varepsilon'$ is close to $\varepsilon$. The details of $A'$ follow.

$A'$ receives as input the garbled circuit $PK$. It activates $A$ with the same input. Let $\ell$ be an upper bound on the number of queries that $A$ makes to its **ProbGen** oracle. The adversary $A'$ chooses an index $i$ at random between 1 and $\ell$ and continues as follows. For the $j^{th}$ query by $A$, with $j \neq i$, $A'$ will respond by (i) choosing a random private/public key pair for the homomorphic encryption scheme $(PK_{\mathcal{E}}^j, SK_{\mathcal{E}}^j)$ and (ii) encrypting random $\lambda$-bit strings under $PK_{\mathcal{E}}^j$. For the $i^{th}$ query, $x$, the adversary $A'$ gives $x$ to its own

**ProbGen** oracle and receives $\sigma_x$, the collection of active input labels corresponding to $x$. It then generates a random private/public key pair for the homomorphic encryption scheme $(PK_\mathcal{E}^i, SK_\mathcal{E}^i)$, and it encrypts $\sigma_x$ (label by label) under $PK_\mathcal{E}^i$.

Once we prove the Lemma 1 below, we have our contradiction and the proof of Theorem 1 is complete. ∎

**Lemma 1.** $Adv_{A'}^{Verif}(\mathcal{VC}_{Yao}, F, \lambda) \geq \varepsilon'$ where $\varepsilon'$ is non-negligible in $\lambda$.

**Proof of Lemma 1:** This proof also proceeds by defining, for any adversary $A$, a set of hybrid experiments $\mathcal{H}_A^k(\mathcal{VC}, F, \lambda)$ for $k = 0, \ldots, \ell - 1$. We define the experiments below. Let $i$ be an index randomly selected between 1 and $\ell$ as in the proof above.

**Experiment** $\mathcal{H}_A^k(\mathcal{VC}, F, \lambda) = 1$]: In this experiment, we change the way the oracle **ProbGen** computes its answers. For the $j^{th}$ query:

- $j \leq k$ and $j \neq i$: The oracle will respond by (i) choosing a random private/public key pair for the homomorphic encryption scheme $(PK_\mathcal{E}^j, SK_\mathcal{E}^j)$ and (ii) encrypting random $\lambda$-bit strings under $PK_\mathcal{E}^j$.
- $j > k$ or $j = i$: The oracle will respond exactly as in $\mathcal{VC}$, i.e. by (i) choosing a random private/public key pair for the homomorphic encryption scheme $(PK_\mathcal{E}^j, SK_\mathcal{E}^j)$ and (ii) encrypting the correct input labels in Yao's garbled circuit under $PK_\mathcal{E}^j$.

In the end, the bit output by the experiment $\mathcal{H}_A^k$ is 1 if $A$ successfully cheats on the $i^{th}$ input and otherwise is 0. We denote with $Adv_A^k(\mathcal{VC}, F, \lambda) = Prob[\mathcal{H}_A^k(\mathcal{VC}, F, \lambda) = 1]$. Note that

- $\mathcal{H}_A^0(\mathcal{VC}, F, \lambda)$ is identical to the experiment $\mathbf{Exp}_A^{Verif}[\mathcal{VC}, F, \lambda]$, except for the way the bit is computed at the end. Since the index $i$ is selected at random between 1 and $\ell$, we have that

$$Adv_A^0(\mathcal{VC}, F, \lambda) = \frac{Adv_A^{Verif}(\mathcal{VC}, F, \lambda)}{\ell} \geq \frac{\varepsilon}{\ell}$$

- $\mathcal{H}_A^{\ell-1}(\mathcal{VC}, F, \lambda)$ is equal to the simulation conducted by $A'$ above, so

$$Adv_A^{\ell-1}(\mathcal{VC}, F, \lambda) = Adv_{A'}^{Verif}(\mathcal{VC}_{Yao}, F, \lambda)$$

If we prove for $k = 0, \ldots, \ell - 1$ that experiments $\mathcal{H}_A^k(\mathcal{VC}, F, \lambda)$ and $\mathcal{H}_A^{k-1}(\mathcal{VC}, F, \lambda)$ are computationally indistinguishable, that is for every $A$

$$|Adv_A^k(\mathcal{VC}, F, \lambda) - Adv_A^{k-1}(\mathcal{VC}, F, \lambda)| \leq \mathtt{negli}(\lambda) \tag{6}$$

we are done, since that implies that

$$Adv_{A'}^{Verif}(\mathcal{VC}_{Yao}, F, \lambda) \geq \frac{\varepsilon}{\ell} - \ell \cdot \mathtt{negli}(\lambda)$$

which is the desired non-negligible $\varepsilon'$.

But Eq. 6 easily follows from the semantic security of the homomorphic encryption scheme. Indeed assume that we could distinguish between $\mathcal{H}_A^k$ and $\mathcal{H}_A^{k-1}$, then we can decide the following problem, which is easily reducible to the semantic security of $\mathcal{E}$:

**Security of $\mathcal{E}$ with respect to Yao Garbled Circuits:** *Given a Yao-garbled circuit $PK_{Yao}$, an input $x$ for it, a random public key $PK_{\mathcal{E}}$ for the homomorphic encryption scheme, a set of ciphertexts $c_1, \ldots, c_n$ where $n$ is the size of $x$, decide if for all $i$, $c_i = $* **Encrypt**$_{\mathcal{E}}(PK_{\mathcal{E}}, w_i^{x_i})$, *where $w_i$ is the $i^{th}$ input wire and $x_i$ is the $i^{th}$ input bit of $x$, or $c_i$ is the encryption of a random value.*

Now run experiment $\mathcal{H}_A^{k-1}$ with the following modification: at the $k^{th}$ query, instead of choosing a fresh random key for $\mathcal{E}$ and encrypting random labels, answer with $PK_{\mathcal{E}}$ and the ciphertexts $c_1, \ldots, c_n$ defined by the problem above. If $c_i$ is the encryption of a random value, then we are still running experiment $\mathcal{H}_A^{k-1}$, but if $c_i = $ **Encrypt**$_{\mathcal{E}}(PK_{\mathcal{E}}, w_i^{x_i})$, then we are actually running experiment $\mathcal{H}_A^k$. Therefore we can decide the Security of $\mathcal{E}$ with respect to Yao Garbled Circuits with the same advantage with which we can distinguish between $\mathcal{H}_A^k$ and $\mathcal{H}_A^{k-1}$.

The reduction of the Security of $\mathcal{E}$ with respect to Yao Garbled Circuits to the basic semantic security of $\mathcal{E}$ is an easy exercise, and details will appear in the final version. ∎

# B   Proof of Theorem 2

**Proof of Theorem 2:**   Consider two games between a challenger and an adversary $A$. In the real world game for $\mathcal{VC}^{\dagger}$, Game 0, the interactions between the challenger and $A$ are exactly like those between the client and a worker in the real world – in particular, if $A$'s response was well-formed, the challenger tells $A$ so, but the challenger immediately aborts if $A$'s response is malformed. Game 1 is identical to Game 0, except that when $A$ queries **Verify**, the challenger always answers with the correct $y$, whether $A$'s response was well-formed or not, and the challenger never aborts. Let $\varepsilon_i$ be $A$'s success probability in Game $i$.

First, we show that if $\mathcal{VC}$ is secure, then $\varepsilon_1$ must be negligible. The intuition is simple: since the challenger always responds with the correct $y$, there is actually no information in these responses, since $A$ could have computed $y$ on its own. More formally, there is an algorithm $B$ that breaks $\mathcal{VC}$ with probability $\varepsilon_1$ by using $A$ as a sub-routine. $B$ simply forwards communications between the challenger (now a challenger for the $\mathcal{VC}$ game) and $A$, except that $B$ tells $A$ the correct $y$ w.r.t. all of $A$'s responses. $B$ forwards $A$'s forgery along to the challenger.

Now, we show that $\varepsilon_0 \leq \varepsilon_1$, from which the result follows. Let $E_{mal}$ be the event that $A$ makes a malformed response, and let $E_f$ be the event that $A$ successfully outputs a forgery – i.e., where $\mathbf{Exp}_A^{Verif^{\dagger}}[\mathcal{VC}^{\dagger}, F, \lambda]$ outputs '1'. $A$'s success probability, in either Game 0 or Game 1, is:

$$Prob[E_f] = Prob[E_f|E_{mal}] \cdot Prob[E_{mal}] + Prob[E_f|\neg E_{mal}] \cdot Prob[\neg E_{mal}] \qquad (7)$$

If $A$ does not make a malformed response, then Games 0 and 1 are indistinguishable to $A$; therefore, the second term above has the same value in Games 0 and 1. In Game 0, $Prob[E_f|E_{mal}] = 0$, since the challenger aborts. Therefore, $\varepsilon_0 \leq \varepsilon_1$. ∎

# Improved Delegation of Computation Using Fully Homomorphic Encryption[*]

Kai-Min Chung[1,**], Yael Kalai[2], and Salil Vadhan[1,***]

[1] School of Engineering & Applied Sciences, Harvard University,
Cambridge MA, USA
kmchung@fas.harvard.edu, salil@seas.harvard.edu
[2] Microsoft Research New England, Cambridge MA, USA
yael@microsoft.com

**Abstract.** Following Gennaro, Gentry, and Parno (Cryptology ePrint Archive 2009/547), we use fully homomorphic encryption to design improved schemes for delegating computation. In such schemes, a *delegator* outsources the computation of a function $F$ on many, dynamically chosen inputs $x_i$ to a *worker* in such a way that it is infeasible for the worker to make the delegator accept a result other than $F(x_i)$. The "online stage" of the Gennaro et al. scheme is very efficient: the parties exchange two messages, the delegator runs in time $\mathrm{poly}(\log T)$, and the worker runs in time $\mathrm{poly}(T)$, where $T$ is the time complexity of $F$. However, the "offline stage" (which depends on the function $F$ but not the inputs to be delegated) is inefficient: the delegator runs in time $\mathrm{poly}(T)$ and generates a public key of length $\mathrm{poly}(T)$ that needs to be accessed by the worker during the online stage.

Our first construction eliminates the large public key from the Gennaro et al. scheme. The delegator still invests $\mathrm{poly}(T)$ time in the offline stage, but does not need to communicate or publish anything. Our second construction reduces the work of the delegator in the offline stage to $\mathrm{poly}(\log T)$ at the price of a 4-message (offline) interaction with a $\mathrm{poly}(T)$-time worker (which need not be the same as the workers used in the online stage). Finally, we describe a "pipelined" implementation of the second construction that avoids the need to re-run the offline construction after errors are detected (assuming errors are not too frequent).

**Keywords:** verifiable computation, outsourcing computation, worst-case/average-case reductions, computationally sound proofs, universal argument systems.

## 1 Introduction

The problem of delegating computation considers a scenario where one party, the *delegator*, wishes to delegate the computation of a function $f$ to another

---

[*] A full version of this paper can be found on [CKV10].
[**] Supported by US-Israel BSF grant 2006060 and NSF grant CNS-0831289.
[***] Supported by NSF grant CNS-0831289.

T. Rabin (Ed.): CRYPTO 2010, LNCS 6223, pp. 483–501, 2010.

party, the *worker*. The challenge is that the delegator may not trust the worker, and thus it is desirable to have the worker "prove" that the computation was done correctly. Obviously, we want verifying this proof to be easier than doing the computation.

This concept of "outsourcing" computation is relevant in several real world scenarios, as illustrated by the following three examples (taken from [GGP09,GKR08]):

1. **Volunteer computing.** The idea of volunteer computing is for a server to split large computations into small units, send these units to volunteers for processing, and reassemble the results (via a much easier computation). The Berkeley Open Infrastructure for Network Computing (BOINC) [And03,And04] is an example of such a platform. Some famous projects using the BOINC platform are SETI@home, and the Great Internet Mersenne Prime Search [Mer07]. We refer the reader to [GKR08] for more details on these projects.
2. **Cloud computing.** In the setting of cloud computing, businesses buy computing time from a service, rather than purchasing their own computing resources.
3. **Weak mobile devices.** Mobile devices, such as cell-phones, security access-cards, music players, and sensors, are typically very weak computationally, and thus need the help of remote computers to run costly computations.

A natural question about such settings is: *what if the workers are dishonest?* For example, in the volunteer computing setting, an adversarial volunteer may introduce errors into the computation. In the cloud computing example, the cloud (i.e., the business providing the computing services) may have a financial incentive to return incorrect answers, if such answers require less work and are unlikely to be detected by the client. Moreover, in some cases, the applications outsourced to the cloud may be so critical that the delegator wishes to rule out accidental errors during the computation. As for weak mobile devices, the communication channel between the device and the remote computer may be corrupted by an adversary.

In practice, many projects cope with such fraud by redundancy; the same work unit is sent to several workers and the results are compared for consistency. However, this requires the use of several workers and provides little defense against colluding workers.

Instead, we would like the worker to *prove* to the delegator that the computation was performed correctly. Of course, it is essential that the time it takes to verify the proof is significantly smaller than the time needed to actually run the computation. At the same time, the running time of the worker carrying out the proof should also be reasonable — comparable to the time it takes to do the computation. For example, when delegating the computation of a function $f$ that takes time $T$ and has inputs and outputs of length $n$, we would like the delegator to run in time $\text{poly}(n, \log T)$ and the worker to run in time $\text{poly}(T)$.

## 1.1   Previous Work

The large body of work on probabilistic proof systems, starting with [Bab85,GMR89], is very relevant to secure delegation. Indeed, after computing the delegated function $f$ on input $x$ and sending the result $y$, the worker can use various types of proof systems to convince the delegator of the statement "$f(x) = y$".

*Interactive Proofs.* The IP=PSPACE Theorem [LFKN92,Sha92] yields interactive proofs for any function $f$ computable in polynomial space, with a verifier (delegator) running in polynomial time. However, the complexity of the prover (worker) is also only bounded by polynomial space (and hence exponential time). This theorem was refined and scaled down in [FL93] to give verifier complexity $\text{poly}(n,s)$ and prover complexity $2^{\text{poly}(s)}$ for functions $f$ computable in time $T$ and space $s$, on inputs of length $n$. Note that the prover complexity is still superpolynomial in $T$, even for computations that run in the smallest possible space, namely $s = O(\log T)$. However, the prover complexity was recently improved by Goldwasser et al. [GKR08] to $\text{poly}(T, 2^s)$, which is $\text{poly}(T)$ when $s = O(\log T)$. More generally, Goldwasser et al. [GKR08] give interactive proofs for computations of small *depth* $d$ (i.e. parallel time). For these, they achieve prover complexity $\text{poly}(T)$ and verifier complexity $\text{poly}(n, d, \log T)$. (This implies the result for space-bounded computation because an algorithm that runs in time $T$ and space $s$ can be converted into one that runs in time $\text{poly}(T, 2^s)$ and depth $d = O(s^2)$.) However, if we do not restrict to computations of small space or depth, then we cannot use interactive proofs. Indeed, any language that has an interactive proof with verifier running time (and hence communication) $T_V$ can be decided in space $\text{poly}(n, T_V)$.

*PCPs and MIPs.* The MIP=NEXP Theorem [BFL91] and its scaled-down version by Babai et al. [BFLS91] yield multiprover interactive proofs and probabilistically checkable proofs for time $T$ computations with a prover running in time $\text{poly}(T)$ and a verifier running in time $\text{poly}(n, \log T)$, exactly as we want. However, using these for delegation require specialized communication models — either 2 noncommunicating provers, or a mechanism for the prover to give the verifier random access to a long PCP (of length $\text{poly}(T)$) that cannot be changed by the prover during the verification.

*Interactive Arguments.* Instead of changing the communication model, interactive arguments [BCC88] (aka computationally sound proofs [Mic94]) relax the soundness condition to be computational. That is, instead of requiring that no prover strategy whatsoever can convince the verifier of a false statement, we instead require that no computationally feasible prover strategy can convince the verifier of a false statement. In this model, Kilian [Kil92] and Micali [Mic94] gave constant-round protocols with prover complexity $\text{poly}(T, k)$ and verifier complexity $\text{poly}(n, k, \log T)$ (where $k$ is the security parameter), assuming the existence of collision-resistant functions. Under a subexponential hardness assumption, the security parameter can be taken as small as $\text{polylog}(T)$; this also holds for the schemes described below.

*Towards Non-interactive Solutions.* In this work, we are interested in getting closer to non-interactive solutions (with computational soundness). Ideally, the worker/prover should be able to send a proof to the delegator/verifier in the same message that it sends the result of the computation.

This possibility of efficient non-interactive arguments was suggested by Micali [Mic94], who showed that non-interactive arguments with prover complexity $\text{poly}(T, k)$ and verifier complexity $\text{poly}(n, k, \log T)$ are possible in the Random Oracle Model (the oracle is used to eliminate interaction a la Fiat–Shamir [FS86]). Heuristically, one might hope that by instantiating the random oracle with an appropriate family of hash functions, we could obtain a non-interactive solution to delegating computation: in an offline stage, the verifier/delegator (or a trusted third party) chooses and publishes a random hash function from the family, and in the online stage, the proofs are completely non-interactive (just one message from the prover to the verifier). However, the Random Oracle Heuristic is known to be unsound in general [CGH04] and even in the context of Fiat–Shamir [Bar01,GK03]. Thus, despite extensive effort, the existence of efficient non-interactive arguments remains a significant open problem in complexity and cryptography.

There has been some recent progress in reducing the amount of interaction needed. Using a transformation of Kalai and Raz [KR09], Goldwasser, Kalai, and Rothblum [GKR08] showed how to convert their interactive proofs for small-depth computations into non-interactive arguments in a "public key" model (assuming the existence of single-server private-information retrieval (PIR) schemes): in an offline stage, the verifier/delegator generates a public/secret key pair, publishes the public key and stores the secret key. Then, in the online stage, the prover/ worker retrieves the public key and can construct a proof to send along with the result of the computation. However, like the interactive proofs of [GKR08], this solution applies only to small-depth computations, as the verifier's complexity grows linearly with the depth.

Very recently, Gennaro, Gentry, and Parno [GGP09] showed how to delegate arbitrary computations by increasing the verifier's offline complexity and public-key size, and using a fully homomorphic encryption (FHE) scheme (as recently constructed by Gentry [Gen09]). In their construction, the delegator invests $\text{poly}(T, k)$ work in the offline stage to construct a public key of size $\text{poly}(T, k)$ and a secret key of size $\text{poly}(k)$ (for delegating a function $f$ that is computable in time $T$). In the online stage, the delegator's running time is reduced to $\text{poly}(n, k, \log T)$ for an input of length $n$, and the worker's complexity is $\text{poly}(T, k)$. Thus, the delegator's large investment in the offline stage can be amortized over many executions of the online stage to delegate the computation of $f$ on many inputs. Their online stage is not completely non-interactive, but consists of two messages. However, in many applications, two messages will be necessary anyway, as the delegator may need to communicate the input $x$ to the worker.

We remark that in the schemes where the delegator has a secret key (namely [GKR08] and [GGP09], as well as two of our constructions below), soundness is only guaranteed as long as the adversarial worker does not learn that the delegator has rejected a proof. Thus, either the accept/reject decision should be kept secret, or the (possibly expensive) offline stage should be re-run after rejection.

## 1.2   Our Results

In this work, we provide the following protocols that improve over the work of Gennaro et al. [GGP09]:

- Our first protocol eliminates the large public key of the Gennaro et al. scheme. That is, the delegator still performs $poly(T, k)$ work in the offline stage, but the result of this computation is just a secret key of length $poly(n, k, \log T)$; there is no need for any interaction with the worker(s) in advance of the online stage (not even to transmit a public key).
- Our second protocol reduces the work of the delegator in the offline stage to $poly(n, k, \log T)$, at the price of a constant-round interaction with a worker that runs in time $poly(T, k)$. With this protocol, re-running the offline stage after a rejected proof becomes more reasonable, and thus there is no reason to keep the accept/reject decisions secret.
- Finally, we describe a "pipelined" implementation of our second protocol that avoids the latency of re-running the offline stage, while maintaining soundness even if the accept/reject decisions are revealed. This solution requires both parties to maintain state, and completeness holds provided that faults do not occur too often. Thus, this solution is most suitable for cases where the delegator is using a single worker many times and there are random faults (in communication or computation) that may cause the delegator to reject occasionally.

Like [GGP09], all of our protocols require the use of a fully homomorphic encryption scheme, and have a 2-message online stage. A full comparison of our model and results with previous work is given in Table 1.

*Organization.* Brief preliminaries on fully homomorphic encryption schemes are presented in Section 2. Then we present a formal definition of our model in Section 3. In Section 4 – 8, we start with a simple scheme $Del_1$ that achieves rather weak properties, and strengthen it through a series of steps leading to our main delegation schemes $Del_4$ and $Del_5$.

Due to space constraints, we skip all the proofs. Please refer to the full version of this paper [CKV10] for details.

# 2   Preliminaries on Fully Homomorphic Encryption

Inspired by the recent work of Gennaro, Gentry, and Parno [GGP09] on secure delegation, our constructions rely on the use of a fully homomorphic encryption scheme.

**Table 1.** Results on Delegating Computation. $D$ = delegator/verifier, $W$ = worker/prover, $PK$ = $D$'s public key, $SK$ = $D$'s secret key, $k$ = security parameter. Parameters of computation $f$ being delegated: $n$ = input length, $T$ = time, $d$ = depth/parallel time (we assume $n \leq T \leq 2^d$).

| Ref | Assumption | Soundness | offline | | keys | | online | | |
| --- | --- | --- | --- | --- | --- | --- | --- | --- | --- |
| | | | # msgs | $D$ complexity | $|PK|$ | $|SK|$ | # msgs | $D$ complexity | $W$ complexity |
| [GKR08] | none | stat | 0 | 0 | 0 | 0 | $\mathrm{poly}(d, \log T)$ | $\mathrm{poly}(n, d, \log T)$ | $\mathrm{poly}(T)$ |
| [BFL91,BFLS91] | none | MIP/PCP | 0 | 0 | 0 | 0 | 1 | $\mathrm{poly}(n, \log T)$ | $\mathrm{poly}(T)$ |
| [Kil92,Mic00] | CRH | comp | 0 | 0 | 0 | 0 | 4 | $\mathrm{poly}(k, n, \log T)$ | $\mathrm{poly}(k, T)$ |
| [Kil92,Mic00] | RO-Heur | comp | 1 | $\mathrm{poly}(k)$ | $\mathrm{poly}(k)$ | 0 | 1 | $\mathrm{poly}(k, n, \log T)$ | $\mathrm{poly}(k, T)$ |
| [GKR08,KR09] | PIR | comp | 1 | $\mathrm{poly}(k)$ | $\mathrm{poly}(k, d, \log T)$ | $\mathrm{poly}(k, d, \log T)$ | 1 | $\mathrm{poly}(k, n, d, \log T)$ | $\mathrm{poly}(k, T)$ |
| [GGP09] | FHE | comp | 1 | $\mathrm{poly}(k, T)$ | $\mathrm{poly}(k, T)$ | $\mathrm{poly}(k, n)$ | 2 | $\mathrm{poly}(k, n, \log T)$ | $\mathrm{poly}(k, T)$ |
| Thm. 1 | FHE | comp | 0 | $\mathrm{poly}(k, T)$ | 0 | $\mathrm{poly}(k, n)$ | 2 | $\mathrm{poly}(k, n, \log T)$ | $\mathrm{poly}(k, T)$ |
| Thm. 3 | FHE | comp | 4 | $\mathrm{poly}(k, n, \log T)$ | 0 | $\mathrm{poly}(k, n)$ | 2 | $\mathrm{poly}(k, n, \log T)$ | $\mathrm{poly}(k, T)$ |

*Fully Homomorphic Encryption.* A public-key encryption scheme $E =$ (KeyGen, Enc, Dec) is said to be *fully homomorphic* if it is associated with an additional polynomial-time algorithm Eval, that takes as input a public key pk, a ciphertext $\hat{x} = \text{Enc}(x)$ and a circuit $C$, and outputs, a new ciphertext $c = \text{Eval}_{pk}(\hat{x}, C)$, such that $\text{Dec}_{sk}(c) = C(x)$, where sk is the secret key corresponding to the public key pk. It is required that the size of $c = \text{Eval}_{pk}(\text{Enc}_{pk}(x), C)$ depends polynomially on the security parameter and the length of $C(x)$, but is otherwise independent of the size of the circuit $C$. We also require that Eval is deterministic, and the scheme has perfect correctness (i.e. it always holds that $\text{Dec}_{sk}(\text{Enc}_{pk}(x)) = x$ and that $\text{Dec}_{sk}(\text{Eval}_{pk}(\text{Enc}_{pk}(x), C)) = C(x)$). For security, we simply require that $E$ is semantically secure.

In a recent breakthrough, Gentry [Gen09] proposed a fully homomorphic encryption scheme based on ideal lattices. In his basic scheme, the complexity of the algorithms (KeyGen, Enc, Dec) depends linearly on the *depth* of the circuit $C$, where $d$ is an upper bound on the depth of the circuit $C$ that are allowed as inputs to Eval. However, under the additional assumption that his scheme is circular secure (i.e., it remains secure even given an encryption of the secret key), the complexity of these algorithms are independent of $C$. Furthermore, Gentry's construction satisfies the perfect correctness and the Eval of his scheme can be made deterministic. We refer the reader to [Gen09] for details.

An interesting aspect of the [GGP09] construction is how they use the *secrecy* property of fully homomorphic encryption schemes in order to achieve a *soundness* property in their delegation scheme; this phenomenon also recurs several times in our work.

## 3   The Model

In this section, we formally define a model that captures the delegating computation scenario we are interested in.

**Definition 1 (Delegation Scheme).** *A* delegation scheme *is an interactive protocol* Del $= \langle D, W \rangle$ *between a delegator* D *and a worker* W *with the following structure:*

1. *The scheme* Del *consists of two stages: an offline/preprocessing stage and an online stage. The offline stage is executed once before the online stage, whereas the online stage can be executed many times.*
2. *In the offline stage, both the delegator* D *and the worker* W *receive a security parameter* $k$ *and a function* $F : \{0,1\}^n \rightarrow \{0,1\}^m$, *represented by a Turing machine* $M$ *and a time bound* $T$ *for* $M$. *At the end of the interaction, the delegator* D *decides whether to accept or reject. If* D *accepts, then* D *outputs a* secret key $\sigma_D$ *and a* public key $\sigma_W$. *We will denote this by* $(\sigma_D, \sigma_W) = \langle D, W \rangle(F, 1^k)$. *We will use the notation* $M$, $n$, $m$, *and* $T$ *as the Turing machine and parameters associated with* $F$ *throughout the paper, and we will often omit the security parameter from the notation.*
3. *In the online stage, both parties receive* $F$, $1^k$, *and an input* $x \in \{0,1\}^n$, *and execute    a    one    round    communication    protocol.    Namely,    D    sends*

$q = D(F, x, \sigma_D)$ to W, and then W sends $a = W(F, x, \sigma_W, q)$ to D. Then the delegator D either accepts or rejects. If D accepts, then D also generates a private output $y = D(F, x, \sigma_D, q, a) \in \{0, 1\}^m$, which is supposed to be $F(x)$. For simplicity, we will omit the function $F$ and the security parameter from the input of the online stage.

We also define the following properties of delegation schemes.

- A delegation scheme Del has an **efficient** delegator in the online (resp., offline) stage if the computational complexity of D in the online (resp., offline) stage is $\text{poly}(k, n, m, |M|, \log T)$.
- A delegation scheme Del has an **efficient** worker if the computational complexity of W is $\text{poly}(k, |M|, T)$.
- A delegation scheme Del has a **non-interactive** offline stage if D and W do not interact at all during the offline stage, and only D does some computation. Note that if Del has a non-interactive offline stage, then we can assume w.l.o.g. that D always accepts in the offline stage.

For a delegation scheme to be meaningful, it needs to have completeness and soundness properties. Informally, the completeness property says that the delegator D always learns the desired value $F(x)$, assuming both parties follow the prescribed protocol. The soundness property says that the delegator D mistakenly accepts a wrong value $y \neq F(x)$ from a malicious worker with only negligible probability.

**Definition 2 (Completeness).** A delegation scheme Del $= \langle D, W \rangle$ has perfect completeness if for all parameters $n, m, T, k$, for every function $F$ and every $x \in \{0, 1\}^n$, the following holds with probability 1: When D and W run the offline stage protocol with input $F$, and then run the online stage protocol with input $x$, the delegator D accepts in both the offline and the online stage, and outputs $y = F(x)$ in the online stage.

In order to define the soundness, we introduce the following security game.

**Definition 3 (Security Game for Delegation Schemes).** Let Del $= \langle D, W \rangle$ be a delegation scheme and $k \in \mathbb{N}$ be the security parameter. The security game $\mathcal{G}(k)$ for Del is the following game played by a worker strategy $W^*$.

- The game starts with the offline stage of Del, and is followed by many rounds of the online stage.
- $W^*(1^k)$ first chooses the delegation function $F$ and then D and $W^*$ interact in the offline stage of Del with input $F$.
- At the beginning of each round of the online stage (indexed by $\ell$), $W^*$ can either terminate the game or choose an input $x_\ell \in \{0, 1\}^n$. If the game is not terminated, D and $W^*$ interact in the online stage of Del on input $x_\ell$.
- Whenever the delegator D rejects, the game terminates.

$W^*$ succeeds in the game $\mathcal{G}(k)$ if there exists a round $\ell$ of the online stage such that D accepts and outputs a wrong value $y_\ell \neq F(x_\ell)$, where $x_\ell$ is the delegated input chosen by $W^*$.

**Definition 4 (Soundness).** *Let* $\varepsilon : \mathbb{N} \to [0,1]$ *and* $t : \mathbb{N} \to \mathbb{N}$ *be efficiently computable functions. A delegation scheme* $\mathsf{Del} = \langle \mathsf{D}, \mathsf{W} \rangle$ *has* **soundness error** $\varepsilon$ *for delegation functions with runtime at most* $t$ *if for every worker strategy* $\mathsf{W}^*$, *which runs in time* $t(k)$ *and chooses a delegation function with runtime at most* $t(k)$,

$$\Pr[\mathsf{W}^* \text{ succeeds in } \mathcal{G}(k)] \leq \varepsilon(k),$$

*for sufficiently large* $k$, *where* $\mathcal{G}(k)$ *is the corresponding security game for* $\mathsf{Del}$. *We say that* $\mathsf{Del}$ *is* **sound** *if* $\mathsf{Del}$ *has soundness error* $1/k^c$ *for delegation functions with runtime at most* $k^c$ *for every constant* $c$.

Note that the above definition does not guarantee soundness for delegating functions of complexity superpolynomial in $k$. However, we have soundness for functions of complexity that is an arbitrarily large polynomial in $k$, whereas an efficient delegator would run in time that is a fixed polynomial in $k$; so the delegation is still quite useful. This quantitative relationship stems from the standard asymptotic formulation of security as being with respect to polynomial-time adversaries. If we use a fully homomorphic encryption scheme that is secure against adversaries running time subexponential in $k$, then we would obtain soundness for delegating functions of subexponential complexity (while the delegator still runs in fixed polynomial time).

In terms of concrete security, the security parameter $k$ should be chosen by the delegator so that breaking the encryption scheme requires an infeasible amount $R$ of resources for the worker, and thus the delegator should only be delegating functions that require significantly less resources than $R$.

Note that in the security game $\mathcal{G}$, the delegator $\mathsf{D}$ rejects and terminates the game, whenever he catches the worker cheating. Thus, the soundness is only guaranteed until the worker cheats. In other words, once the worker cheats, the delegator $\mathsf{D}$ can catch this mistake with overwhelming probability, but the delegation scheme no longer guarantees soundness for the next delegated inputs. Therefore, $\mathsf{D}$ should restart the delegation scheme from the offline stage to ensure the soundness of future delegated inputs.

The model of [GGP09] takes a different approach. Rather than halting the game after a rejection, they instead consider a game where the delegator's accept/reject decisions are kept secret from the worker. Our protocols also satisfy their definition; indeed, the two definitions are equivalent for schemes where the delegator has no state (other than the secret key).

## 4    $\mathsf{Del}_1 = \langle \mathsf{D}_1, \mathsf{W}_1 \rangle$: One-Time, Random-Input Delegation Scheme

In this section, we present our first warmup delegation scheme $\mathsf{Del}_1 = \langle \mathsf{D}_1, \mathsf{W}_1 \rangle$ for the following one-time and random-input scenario.

**Scenario:** Suppose the delegator $\mathsf{D}$ knows that at some point in the future, he will receive a *random* input $x \in \{0,1\}^n$ drawn from a certain (samplable) distribution $\mathcal{D}$ and he will want to learn the value $F(x)$ quickly. Thus, $\mathsf{D}$ decides to delegate the computation of $F(x)$ to an untrusted worker W (who does *not* know the random $x$), and D wants to be able to verify the answer from W very efficiently.

The idea is simple and similar to the idea underlying reCAPTCHAs [vAMM+08]: In the offline stage, the delegator $D_1$ samples a random input $r \leftarrow \mathcal{D}$ and precomputes $F(r)$. In the online stage, $D_1$ sends both $x$ and $r$ to $W_1$ in a random order, and asks $W_1$ to compute both $F(x)$ and $F(r)$. Upon receiving the answers from $W_1$, the delegator $D_1$ checks the correctness of the returned value $F(r)$; if it is correct then he accepts the returned $F(x)$, and otherwise he reject. Thus, a malicious worker $W^*$ can convince $D_1$ with a wrong answer iff $W^*$ can guess which input is the delegator's real input. Since $x$ and $r$ are independent and identically distributed, no malicious prover can guess the real input $x$ and cheat successfully with probability greater than $1/2$. A formal description of our random-input delegation scheme $Del_1 = \langle D_1, W_1 \rangle$ can be found in Figure 1. A formal analysis of $Del_1$ can be found in the full version of this paper [CKV10].

---

- **Inputs.** Security parameter $1^k$ and function $F : \{0,1\}^n \rightarrow \{0,1\}^m$, specified by a Turing machine $M$, and a time bound $T$.
- **Offline Stage.** Both $D_1$ and $W_1$ receive input $(F, \mathcal{D})$
    1. $D_1$ samples a random input $r \leftarrow \mathcal{D}$, computes $w = F(r)$, and stores the pair $(r, w)$ as his secret state.
- **Online Stage.** $D_1$ receives $x \in \{0,1\}^n$ (where $x$ is expected to distribute according to $\mathcal{D}$), and $W_1$ does not receive any input.
    1. $D_1$ sets $r_0 = r$ and $r_1 = x$. It then samples a random bit $b \in_R \{0,1\}$, and sends $(z_0, z_1) = (r_b, r_{1-b})$ to $W_1$.
    2. $W_1$ computes and sends $(y_0, y_1) = (F(z_0), F(z_1))$ to $D_1$.
    3. $D_1$ accepts and outputs the answer $y_{1-b}$ iff $w = y_b$.

---

**Fig. 1.** Delegation Scheme $Del_1 = \langle D_1, W_1 \rangle$

## 5   $Del_2 = \langle D_2, W_2 \rangle$: One-Time, Arbitrary-Input Delegation Scheme

Recall that in the random-input delegation scheme $Del_1 = \langle D_1, W_1 \rangle$, it was essential that the input $x$ is hidden from the worker in the online stage to guarantee the soundness. If the worker knew $x$, he could discriminate between $r$ and $x$, and cheat by answering correctly on $r$ and incorrectly on $x$.

We eliminate this strong limitation by using a fully-homomorphic encryption scheme to "*computationally randomize*" the input: Instead of sending $x$ in the clear, the delegator will encrypt the input $x$ to obtain $\hat{x} \stackrel{\text{def}}{=} Enc_{pk}(x)$. Then the delegator will ask the worker to compute the *deterministic* homomorphic evaluation $\hat{F}(\hat{x}) \stackrel{\text{def}}{=} Eval_{pk}(\hat{x}, F)$ of $F$ on the encrypted value $\hat{x}$, from which he can decrypt to obtain the desired answer $F(x)$.[1] Notice that even if $x$ is fixed, the distribution of $\hat{x} = Enc_{pk}(x)$ is computationally indistinguishable from the distribution of

---

[1] We note that in order to compute $Eval_{pk}(\hat{x}, F)$, the Turing machine $F$ needs to be turned into a circuit. This can be done via a standard simulation of Turing machines by circuits.

$\text{Enc}_{pk}(\bar{0})$, which is efficiently samplable and independent of $x$. Thus, the delegator can precompute an encryption $\hat{r} = \text{Enc}_{pk}(\bar{0})$ together with $\hat{F}(\hat{r}) = \text{Eval}_{pk}(\hat{r}, F)$, and use the pair $(\hat{r}, \hat{F}(\hat{r}))$ to verify the worker's answer as before.

We emphasize that the delegator checks the correctness of the *ciphertext* $\hat{F}(\hat{r}) = \text{Eval}_{pk}(\hat{r}, F)$ obtained from homomorphic evaluation of $F$ on $\hat{r} = \text{Enc}_{pk}(\bar{0})$, as opposed to the *value* $f(\bar{0})$ underlying the ciphertext. Indeed, it is insufficient for the delegator to only check the correctness of the value $f(\bar{0})$, since an adversarial worker $W^*$, who knows the input $x$, could easily cheat by applying $\hat{G}(\hat{r}) = \text{Eval}_{pk}(\hat{r}, G)$, where $G(y)$ equals $F(y)$ iff $y \neq x$.

The above computational randomization technique extends the random-input delegation scheme $\text{Del}_1$ to a (standard) delegation scheme $\text{Del}_2$ with one-time soundness error $1/2$. We formally describe the delegation scheme $\text{Del}_2 = \langle D_2, W_2 \rangle$ in Figure 2 below.

---

- **Inputs.** Security parameter $1^k$ and function $F : \{0,1\}^n \to \{0,1\}^m$, specified by a Turing machine $M$, and a time bound $T$.
- **Offline Stage.** Both $D_2$ and $W_2$ receive as input a function $F$.
    1. $D_2$ generates a pair of keys $(pk, sk) \leftarrow \text{KeyGen}(1^k)$, computes an encryption $\hat{r} = \text{Enc}_{pk}(\bar{0})$ and the (deterministic) homomorphic evaluation $\hat{w} = \hat{F}(\hat{r}) = \text{Eval}_{pk}(\hat{r}, F)$, and stores the tuple $(pk, sk, \hat{r}, \hat{w})$ as his secret key.
- **Online Stage.** Both $D_2$ and $W_2$ receive an input $x \in \{0,1\}^n$.
    1. $D_2$ computes an encryption $\hat{x} = \text{Enc}_{pk}(x)$, sets $\hat{r}_0 = \hat{r}$ and $\hat{r}_1 = \hat{x}$, samples a random bit $b \in_R \{0,1\}$, and sends the public key $pk$ and $(\hat{z}_0, \hat{z}_1) = (\hat{r}_b, \hat{r}_{1-b})$ to $W_2$.
    2. $W_2$ computes $\hat{y}_i = \hat{F}(\hat{z}_i) = \text{Eval}_{pk}(\hat{z}_i, F)$ for $i \in \{0,1\}$, and sends $(\hat{y}_0, \hat{y}_1) = (\hat{F}(\hat{z}_0), \hat{F}(\hat{z}_1))$ to $D_2$.
    3. $D_2$ accepts and outputs the answer $\text{Dec}_{sk}(\hat{y}_{1-b})$ iff $\hat{w} = \hat{y}_b$.

---

**Fig. 2.** Delegation Scheme $\text{Del}_2 = \langle D_2, W_2 \rangle$

It is straightforward to check that if the fully homomorphic encryption scheme has perfect correctness, then $\text{Del}_2$ has the perfect completeness. To argue the soundness of the scheme, we first give the definition of one-time soundness.

**Definition 5 (One-time Soundness for Delegation Schemes).** *Let* $\text{Del} = \langle D, W \rangle$ *be a delegation scheme and* $k \in \mathbb{N}$ *be a security parameter. The* **one-time security game** $\mathcal{G}(k)$ *for* $\text{Del}$ *is the same as security game for* $\text{Del}$ *defined in Definition 3 excepts that it only allows one round in the online stage. We say that* $\text{Del}$ *has* **one-time soundness error** $\varepsilon$ *if for every* PPT *worker strategy* $W^*$ *who chooses a polynomial time delegation function, and all sufficiently large* $k$, $\Pr[W^*$ *succeeds in* $\mathcal{G}(k)] \leq \varepsilon(k)$.

**Lemma 1.** *Assume that the fully homomorphic encryption is semantically secure. Then the delegation scheme* $\text{Del}_2$ *has one-time soundness error* $1/2 + \text{ngl}(k)$.

## 6   $\text{Del}_3 = \langle D_3, W_3 \rangle$: One-Time, Arbitrary-Input Delegation Scheme with Negligible Soundness

In this section, we exploit the above computational randomization technique to improve the soundness. The idea is the following: The delegator $D$ asks the worker to compute $\hat{F}$ on multiple independent rerandomized inputs $\hat{x}_i = \text{Enc}_{\text{pk}}(x)$ together with multiple $\hat{r}_i$'s (sent in a random order), as opposed to a single $\hat{x}$ and a single $\hat{r}$. Upon receiving the worker's answers, the delegator D checks whether (i) the returned value for $\hat{r}_i$ is equal to $\hat{F}(\hat{r}_i)$ for every $\hat{r}_i$, and (ii) the decryption of the returned values for $\hat{x}_i$ are consistent, and accepts the consistent value if the worker's answers pass these two tests. Observe that for a malicious worker to cheat, he needs to simultaneously cheat on all the $\hat{x}_i$'s while providing correct answers on all the $\hat{r}_i$'s. The formal description of the delegation scheme $\text{Del}_3 = \langle D_3, W_3 \rangle$ appears in Figure 3.

---

- **Inputs.** Security parameter $1^k$ and function $F : \{0,1\}^n \to \{0,1\}^m$, specified by a Turing machine $M$ and a time bound $T$, and an additional parameter $t$.
- **Offline Stage.** Both $D_3$ and $W_3$ receive as input a function $F$.
  1. $D_3$ generates a pair of keys $(\text{pk}, \text{sk}) \leftarrow \text{KeyGen}(1^k)$, computes $t$ independent encryptions $\hat{r}_i = \text{Enc}_{\text{pk}}(\bar{0})$ and the homomorphic evaluations $\hat{w}_i = \hat{F}(\hat{r}_i) = \text{Eval}_{\text{pk}}(\hat{r}_i, F)$ for $i \in [t]$, and stores $\text{pk}$, $\text{sk}$, and the pairs $(\hat{r}_1, \hat{w}_1), \ldots, (\hat{r}_t, \hat{w}_t)$ as his secret key.
- **Online Stage.** Both $D_3$ and $W_3$ receive an input $x \in \{0,1\}^n$.
  1. $D_3$ computes $t$ independent encryptions $\hat{r}_{i+t} = \text{Enc}_{\text{pk}}(x)$ for $i \in [t]$, samples a random permutation $\pi \in_R S_{2t}$, and sends the public key $\text{pk}$ and $(\hat{z}_{\pi(1)}, \ldots, \hat{z}_{\pi(2t)}) = (\hat{r}_1, \ldots, \hat{r}_{2t})$ to $W_3$.
  2. $W_3$ computes $\hat{y}_i = \hat{F}(\hat{z}_i) = \text{Eval}_{\text{pk}}(\hat{z}_i, F)$ for $i \in [2t]$, and sends to $D_3$ the tuple

$$(\hat{y}_1, \ldots, \hat{y}_{2t}) = (\hat{F}(\hat{z}_1), \ldots, \hat{F}(\hat{z}_{2t})).$$

  3. $D_3$ checks two things. First, $D_3$ checks if $\hat{w}_i = \hat{y}_{\pi(i)}$ for all $i \in [t]$. Second, $D_3$ decrypts $\hat{y}_{\pi(i+t)}$ for $i \in [t]$, and checks if the decrypted values are the same. $D_3$ accepts and outputs the consistent decrypted value if the returned values pass the two tests.

---

**Fig. 3.** Delegation Scheme $\text{Del}_3 = \langle D_3, W_3 \rangle$

In the following lemma, we argue that since the $\hat{x}_i$'s and the $\hat{r}_i$'s are computationally indistinguishable, the probability of cheating is exponentially small in $t$ (which is the number of $\hat{x}_i$'s). Thus, by setting $t = \omega(\log k)$, the protocol $\langle D_3, W_3 \rangle$ achieves negligible soundness error.

**Lemma 2.** *Assume that the fully homomorphic encryption is semantically secure. Then the delegation scheme $\text{Del}_3$ has one-time soundness error* $\binom{2t}{t}^{-1} + \text{ngl}(k)$.

# 7    The First Main Delegation Schemes $\text{Del}_4$

All the delegation schemes presented in Section 4 – 6 had only *one-time* soundness. Namely, the delegator could delegate the computation of only one input $x$. In this section, we present *reusable* delegation schemes, which satisfy the (standard) soundness property of Definition 4. To this end, we abstract the idea of Gennaro, Gentry, and Parno [GGP09] and present a generic transformation that converts any delegation scheme with one-time soundness to a reusable delegation scheme (i.e., one which satisfies the soundness property of Definition 4). Applying the transformation to the previous delegation scheme $\text{Del}_3$, we obtain our first main delegation scheme $\text{Del}_4$.

For intuition, let us take a closer look at why the previous delegation scheme $\mathsf{Del}_3 = \langle \mathsf{D}_3, \mathsf{W}_3 \rangle$ is not reusable. Recall that in that scheme it is essential for the worker to not know the $\hat{r}_i$'s: Once a malicious worker $\mathsf{W}^*$ learns the values of the $\hat{r}_i$'s, he can easily cheat by answering correctly only on those $\hat{r}_i$'s. Therefore, each precomputed pair $(\hat{r}_i, \hat{C}(\hat{r}_i))$ can be used only once. Phrased more abstractly, the security of the protocol $\langle \mathsf{D}_3, \mathsf{W}_3 \rangle$ relies on assumption that the secret key of the delegator $\mathsf{D}_3$ (i.e., the pairs $(\hat{r}_i, \hat{C}(\hat{r}_i))$), remains secret. However, in that protocol, this secret key is revealed after delegating one input.

To make the protocol reusable, we use the idea of [GGP09], of running the protocol under a fully-homomorphic encryption scheme. Namely, our transformation takes any delegation scheme $\mathsf{Del} = \langle \mathsf{D}, \mathsf{W} \rangle$ which has only one-time soundness, and converts it into a new delegation scheme $\widetilde{\mathsf{Del}} = \langle \widetilde{\mathsf{D}}, \widetilde{\mathsf{W}} \rangle$ with (standard) soundness, as follows: The delegator $\widetilde{\mathsf{D}}$, instead of sending the message of $\mathsf{D}$ in the clear (which may reveal information about his secret key), will send a public key $\mathsf{pk}$ corresponding to a fully homomorphic encryption scheme, and will send the message of $\mathsf{D}$ encrypted under the public key $\mathsf{pk}$. The worker $\widetilde{\mathsf{W}}$ will then use the homomorphic property of the encryption scheme, to compute an encrypted reply of $\mathsf{W}$. This enables the delegator $\widetilde{\mathsf{D}}$ to hide its message (which contains the information about the delegator's secret key) from the worker $\widetilde{\mathsf{W}}$, while still allowing the worker to do the computation for the delegator. A formal description of the transformation can be found in Figure 4.

---

*The transformation.* Let $\mathsf{Del} = \langle \mathsf{D}, \mathsf{W} \rangle$ be a *one-time* delegation scheme. We define a transformed delegation scheme $\widetilde{\mathsf{Del}} = \langle \widetilde{\mathsf{D}}, \widetilde{\mathsf{W}} \rangle$ from $\mathsf{Del}$ as follows.

- **Inputs.** Security parameter $1^k$ and function $F : \{0,1\}^n \to \{0,1\}^m$, specified by a Turing machine $M$ and a time bound $T$.
- **Offline Stage.** $\widetilde{\mathsf{Del}}$ has exactly the same offline stage as $\mathsf{Del}$. (Recall that in this stage both players receive a function $F$.)
- **Online Stage.** both $\widetilde{\mathsf{D}}$ and $\widetilde{\mathsf{W}}$ receive input $x \in \{0,1\}^n$.
  1. $\widetilde{\mathsf{D}}$ generates a fresh pair of keys $(\mathsf{pk}, \mathsf{sk}) \leftarrow \mathsf{KeyGen}(1^k)$ of a fully-homomorphic encryption scheme, computes $\mathsf{D}$'s message $q = \mathsf{D}(F, x, \sigma_\mathsf{D})$ and its encryption $\hat{q} = \mathsf{Enc}_\mathsf{pk}(q)$, and sends $\mathsf{pk}$ and $\hat{q}$ to $\widetilde{\mathsf{W}}$.
  2. $\widetilde{\mathsf{W}}$ homomorphically computes an encrypted version of $\mathsf{W}$'s message $\hat{a} = \mathsf{Eval}(\hat{q}, \mathsf{W}(F, x, \sigma_\mathsf{W}, \cdot))$, and sends $\hat{a}$ to $\widetilde{\mathsf{D}}$.
  3. $\widetilde{\mathsf{D}}$ decrypts $\hat{a}$ to obtain $a = \mathsf{W}(F, x, \sigma_\mathsf{W}, q)$, and computes his decision and his output according to $\mathsf{D}$.

**Fig. 4.** Transforming *one-time* delegation scheme $\mathsf{Del}$ into a *reusable* delegation scheme $\widetilde{\mathsf{Del}}$

We next analyze the properties of the resulting (reusable) delegation scheme $\widetilde{\mathsf{Del}}$.

- If the one-time delegation scheme $\mathsf{Del}$ has a non-interactive off-line stage, then so does $\widetilde{\mathsf{Del}}$, since the offline stage remains unchanged.
- If the one-time delegation scheme $\mathsf{Del}$ has an efficient worker $\mathsf{W}$, then the resulting (reusable) delegation scheme $\widetilde{\mathsf{Del}}$ also has an efficient worker $\widetilde{\mathsf{W}}$, since $\widetilde{\mathsf{W}}$ does the same computation as $\mathsf{W}$, but in an encrypted manner.

– The fact that the complexity of the algorithms (KeyGen, Enc, Dec) are independent of the runtime of $F$, implies that if the one-time delegation scheme Del has an efficient delegator D then the delegator $\tilde{D}$ in the resulting (reusable) delegation scheme $\tilde{\text{Del}}$ is also efficient.
– The completeness of the fully homomorphic encryption scheme implies that if the one-time delegation scheme Del is complete then the resulting (reusable) delegation scheme $\tilde{\text{Del}}$ is also complete.

Thus, it remains to analyze the soundness of the resulting delegation scheme $\tilde{\text{Del}}$. Intuitively, by using a fully homomorphic encryption scheme, the information about the delegator's secret is not leaked, and so the delegator can reuse the secret key to delegate the computation on multiple inputs. However, note that not only the delegator's message, but also the delegator's decision bit can leak information about the delegator's secret key , since the delegator's decision depends on his secret key . Hence, in the security game (see Definition 3), the delegator terminates the scheme once he rejects to ensure the delegator's secret key is not leaked. (As discussed in Section 3, an alternative option is to assume that the worker does not learn the decision of the delegator, and our scheme is also sound in this model.)

**Lemma 3.** *Assume that the fully homomorphic encryption is semantically secure. Let* Del $= \langle \text{D}, \text{W} \rangle$ *be a delegation scheme with negligible one-time soundness error, and let* $\tilde{\text{Del}} = \langle \tilde{\text{D}}, \tilde{\text{W}} \rangle$ *be the delegation scheme obtained by applying to* Del *the transformation described in Figure 4. Then* $\tilde{\text{Del}}$ *also has negligible soundness error.*

Applying the above transformation to the previous delegation scheme $\text{Del}_3$, we obtain our main delegation scheme $\text{Del}_4$. We summarize the properties of $\text{Del}_4$ in the following theorem.

**Theorem 1.** *Assume that the fully homomorphic encryption scheme is semantically secure. Then the delegation scheme* $\text{Del}_4 = \langle \text{D}_4, \text{W}_4 \rangle$ *has the following properties, for delegating the computation of a function* $F : \{0,1\}^n \to \{0,1\}^m$ *computable by a Turing machine* $M$ *that runs in time* $T \geq \max\{n, m\}$, *on security parameter* $k$:

– *Perfect completeness and negligible soundness error.*
– *Non-interactive offline stage, with* $\text{D}_4$ *running in time* $\text{poly}(T, |M|, k)$ *and generating a secret key of length* $\text{poly}(n, m, k)$, *but not creating any public key.*
– *2-message online stage, with* $\text{D}_4$ *running in time* $\text{poly}(n, m, k)$ *and* $\text{W}_4$ *running in time* $\text{poly}(T, |M|, k)$. *That is, both* $\text{D}_4$ *and* $\text{W}_4$ *are efficient in the online stage.*

# 8   The Second Main Delegation Scheme $\text{Del}_5$

We note that in all the delegation schemes presented in Section 4 – 7, the delegator needs to run heavy computations in the offline stage. For example, in the offline stage of $\text{Del}_4$, the delegator needs to compute pairs of the form $(\hat{r}_i, \hat{F}(\hat{r}_i))$, where each $\hat{r}_i \leftarrow \text{Enc}_{\text{pk}}(\bar{0})$, and therefore runs in time comparable to the runtime of $F$.

In this section, we show how to make the offline stage efficient by delegating its computation as well. However, since we do not know how to do non-interactive delegation (this is the problem we started with!), this will come at the price of making the offline stage interactive. In particular, we will use *universal arguments*, a notion developed by [Mic94,Kil92,BG02], and which yields a 4-message delegation scheme. However, we cannot apply universal arguments directly, as they allow the worker to learn the result of the computation, which in our case is supposed to be the secret state of the delegator. To solve this problem, we use yet another layer of fully homomorphic encryption, and use the universal argument to delegate an encrypted form of the computation.

## 8.1   Universal Arguments

Consider the language

$$L_{\text{uni}} \triangleq \{(M, x, y, t) : M \text{ is a Turing machine that on input } x \text{ outputs } y \text{ after at most } t \text{ steps}\}$$

**Definition 6 (Universal Arguments [BG02]).** *A* **universal argument system** *is a pair of interactive Turing machines, denoted by* $(P, V)$, *that satisfy the following properties.*

- **Efficient verification.** *There exists a polynomial $p$ such that for any $z = (M, x, y, t)$ the total runtime of $V$, on common input $z$, is at most $p(|z|)$. In particular, all the messages exchanged in the protocol have length smaller than $p(|z|)$.*
- **Completeness via a relatively-efficient prover.** *For every $(M, x, y, t) \in L_{\text{uni}}$, $\Pr[(P, V)(M, x, y, t) = 1] = 1$.*
  *Furthermore, there exists a polynomial $p$ such that for every $(M, x, y, t) \in L_{\text{uni}}$, the total runtime of $P$ on input $z = (M, x, y, t)$ is at most $p(|M|, t)$.*
- **Computational soundness.** *For every polynomial-size circuit family $P^* = \{P_n^*\}_{n \in \mathbb{N}}$ there exists a negligible function $\mu$ such that for every $(M, x, y, t) \in \{0, 1\}^n \setminus L_{\text{uni}}$, $\Pr[(P_n^*, V)(M, x, y, t) = 1] \leq \mu(n)$.*

*Remark.* We note that Barak and Goldreich [BG02] consider a more general language $L$, where they allow the Turing machine $M$ to be non-deterministic. Moreover, they require an additional proof-of-knowledge type property. In this work, we are only interested in deterministic Turing machines, and only focus on the properties that we need.

**Theorem 2 ([Kil92,Mic94,BG02]).** *Assuming the existence of collision-resistant hash functions, there exists a 4-message (2-round) universal argument system.*

We remark that the existence of fully homomorphic encryption schemes implies the existence of collision-resistant hash functions [IKO05].

## 8.2   Our New Delegation Scheme Del₅

We now show how to use a universal argument $(P, V)$, together with a fully homomorphic encryption scheme $E = (KeyGen, Enc, Dec)$, to convert any delegation scheme $Del = (D, W)$ with a non-interactive offline stage into a delegation scheme $\tilde{Del} = (\tilde{D}, \tilde{W})$, such that the online stage remains unchanged, but the offline stage of $\tilde{Del}$ is now interactive (consists of 4 messages) and the delegator $\tilde{D}$ is *efficient* in the offline stage.

Instead of having the delegator carry out its computations on its own in the offline stage, it will use a worker to do it for him. However, as previously noted, there is a subtle issue here: the result of the computation done by the delegator in the offline stage should remain secret for soundness to hold. Therefore, we cannot simply delegate this computation. Instead, will delegate this computation in a secret manner; namely, we will do a universal argument over encrypted data, as follows.

Suppose without loss of generality, that in the offline stage the delegator $D$ chooses some randomness $r \in \{0,1\}^\ell$ for $\ell = poly(k)^2$ computes a function $g(r)$, where $g$ may depend on both the delegated function $F$ and the security parameter $k$. The delegator $D$ can delegate this computation, in a secret manner, by giving the worker an encryption of $r$ (rather than $r$ in the clear); i.e., giving the worker a pair $(pk, Enc_{pk}(r))$, and delegating the computation of the function $Eval_{pk}(Enc_{pk}(r), g)$ to the worker, by running a universal argument protocol. Then all the delegator needs to do is to decrypt the message he gets from the worker. A formal description of this transformation can be found in Figure 5.

---

*The transformation.* Let $Del = \langle D, W \rangle$ be a delegation scheme with a non-interactive offline stage. We define a transformed delegation scheme $\tilde{Del} = \langle \tilde{D}, \tilde{W} \rangle$ from $Del$ as follows.

- **Inputs.** Security parameter $1^k$ and function $F : \{0,1\}^n \to \{0,1\}^m$, specified by a Turing machine $M$ and a time bound $T$.
- **Offline Stage.** Both $\tilde{D}$ and $\tilde{W}$ receive as input the functin $F$.
  Suppose that in the delegation scheme $Del$, the delegator $D$ chooses a random $r \leftarrow \{0,1\}^\ell$ (where $\ell = poly(k)$) and computes $\sigma_D = D(1^k, F; r)$. We denote by $g(\cdot) \stackrel{def}{=} D(1^k, F; \cdot)$. The offline stage of $\tilde{Del}$ proceeds as follows.
  1. The delegator $\tilde{D}$ chooses a random $r \leftarrow \{0,1\}^\ell$; chooses a random key pair $(pk, sk) \leftarrow KeyGen(1^k)$; computes $\hat{r} = Enc_{pk}(r)$; and sends the pair $(pk, \hat{r})$ to the worker $\tilde{W}$.
  2. The worker $\tilde{W}$ computes $c = Eval_{pk}(\hat{r}, g)$ and sends $c$ to the delegator.
  3. Then the worker $\tilde{W}$ and the delegator $\tilde{D}$ engage in a universal argument, where the worker proves to the delegator that indeed $c = Eval_{pk}(\hat{r}, g)$.
  4. If the delegator $\tilde{D}$ accepts the universal argument, then he decrypts the ciphertext $c$ and outputs $\sigma_D \leftarrow Dec_{sk}(c)$.
- **Online Stage.** The online stage of $\tilde{Del}$ is identical to the online stage of $Del$.

---

**Fig. 5.** Transforming delegation scheme $Del$ with non-interactive but inefficient offline stage into $\tilde{Del}$ with an *efficient* but interactive offline stage

---

[2] The randomness of $D$ can always be reduced to $poly(k)$ by use of a pseudorandom generator if needed.

Analysis of our transformation can be found in the full version of this paper [CKV10]. By applying the transformation above to the delegation scheme $\mathsf{Del}_4$, and relying on Theorem 1, we get the following theorem.

**Theorem 3.** *Assume that there exists a fully homomorphic encryption scheme. Then there is a secure delegation scheme* $\mathsf{Del} = \langle \mathsf{D}, \mathsf{W} \rangle$ *with the following properties, for delegating the computation of a function* $F : \{0,1\}^n \to \{0,1\}^m$ *computable by a Turing machine $M$ that runs in time $T \geq \max\{n, m\}$, on security parameter $k$:*

- $\mathsf{Del}$ *has perfect completeness and negligible soundness error.*
- *The offline stage consists of 4 messages, with* $\mathsf{D}$ *running in time* $\mathrm{poly}(n, m, k, |M|, \log T)$ *and* $\mathsf{W}$ *running in time* $\mathrm{poly}(T, |M|, k)$.
- *The offline stage produces a secret key of length* $\mathrm{poly}(n, m, k)$ *for* $\mathsf{D}$*, and no public key.*
- *In the (2-message) online stage,* $\mathsf{D}$ *runs in time* $\mathrm{poly}(n, m, k)$ *and* $\mathsf{W}$ *runs in time* $\mathrm{poly}(T, |M|, k)$.

*Thus,* $\mathsf{D}$ *and* $\mathsf{W}$ *are efficient in both stages.*

### 8.3   Pipelined Implementation of $\mathsf{Del}_5$

As mentioned in the introduction, the soundness of our main schemes $\mathsf{Del}_4$ and $\mathsf{Del}_5$ is only guaranteed as long as the adversarial worker does not learn that the delegator has rejected a proof, as this may leak information about the delegator's secret key. Hence, the delegator needs to re-run the offline stage after rejection.

Our "pipelined" scheme avoids this issue by having the delegator keep $c$ secret keys (for a constant $c$) and continually refresh them during the online stage. Recall that $\mathsf{Del}_5$ has an efficient but 4-message offline stage where the delegator delegates the computation of his secret key to a worker. The idea is that, in each execution of the 2-message online stage, the delegator and the worker shall simultaneously run $2c$ copies of offline stages in the background. These are run in a pipelined fashion so that with each online stage, $c$ copies of the offline stage are finished and can be used to refresh secret keys that are expired. We consider a secret key to be expired when it has been used in an online stage of $\mathsf{Del}_5$ in which the delegator has rejected. Thus, the delegator will always have a fresh secret key available provided that for every $c$ online stages in which there is an error (i.e. rejection), there are at least 2 consecutive errorless stages. We note that this implementation requires the worker and delegator to maintain state, and thus is most useful for settings in which the delegator is interacting with a single worker for many executions and wishes to avoid disruption from benign faults. (If the worker were truly cheating, then it seems prudent to halt the interaction and restart with a different worker...)

## Acknowledgments

We are grateful to Boaz Barak for collaboration at the start of this research, and for sharing his insights with us.

# References

[And03]     Anderson, D.P.: Public computing: Reconnecting people to science. In: Conference on Shared Knowledge and the Web (2003)

[And04]     Anderson, D.P.: Boinc: A system for public-resource computing and storage. In: GRID, pp. 4–20 (2004)

[Bab85]     Babai, L.: Trading group theory for randomness. In: STOC, pp. 421–429 (1985)

[Bar01]     Barak, B.: How to go beyond the black-box simulation barrier. In: FOCS, pp. 106–115 (2001)

[BCC88]     Brassard, G., Chaum, D., Crépeau, C.: Minimum disclosure proofs of knowledge. Journal of Computer and System Sciences 37(2), 156–189 (1988)

[BFL91]     Babai, L., Fortnow, L., Lund, C.: Non-deterministic exponential time has two-prover interactive protocols. Computational Complexity 1, 3–40 (1991)

[BFLS91]    Babai, L., Fortnow, L., Levin, L.A., Szegedy, M.: Checking computations in polylogarithmic time. In: STOC, pp. 21–31 (1991)

[BG02]      Barak, B., Goldreich, O.: Universal arguments and their applications. In: IEEE Conference on Computational Complexity, pp. 194–203 (2002)

[CGH04]     Canetti, R., Goldreich, O., Halevi, S.: The random oracle methodology, revisited. Journal of the ACM 51(4), 557–594 (2004)

[CKV10]     Chung, K.-M., Kalai, Y., Vadhan, S.: Improved delegation of computation using fully homomorphic encryption. Cryptology ePrint Archive, Report 2010/241 (2010), http://eprint.iacr.org/

[FL93]      Fortnow, L., Lund, C.: Interactive proof systems and alternating time-space complexity. Theoretical Computer Science 113(1), 55–73 (1993)

[FS86]      Fiat, A., Shamir, A.: How to prove yourself: Practical solutions to identification and signature problems. In: Odlyzko, A.M. (ed.) CRYPTO 1986. LNCS, vol. 263, pp. 186–194. Springer, Heidelberg (1987)

[Gen09]     Gentry, C.: Fully homomorphic encryption using ideal lattices. In: STOC, pp. 169–178 (2009)

[GGP09]     Gennaro, R., Gentry, C., Parno, B.: Non-interactive verifiable computing: Outsourcing computation to untrusted workers. Cryptology ePrint Archive, Report 2009/547 (2009), http://eprint.iacr.org/

[GK03]      Goldwasser, S., Kalai, Y.T.: On the (in)security of the fiat-shamir paradigm, pp. 102–113 (2003)

[GKR08]     Goldwasser, S., Kalai, Y.T., Rothblum, G.N.: Delegating computation: interactive proofs for muggles. In: STOC, pp. 113–122 (2008)

[GMR89]     Goldwasser, S., Micali, S., Rackoff, C.: The knowledge complexity of interactive proof-systems. SIAM Journal on Computing 18(1), 186–208 (1989)

[IKO05]     Ishai, Y., Kushilevitz, E., Ostrovsky, R.: Sufficient conditions for collision-resistant hashing. In: Kilian, J. (ed.) TCC 2005. LNCS, vol. 3378, pp. 445–456. Springer, Heidelberg (2005)

[Kil92]     Kilian, J.: A note on efficient zero-knowledge proofs and arguments (extended abstract). In: STOC, pp. 723–732 (1992)

[KR09]      Kalai, Y.T., Raz, R.: Probabilistically checkable arguments. In: Halevi, S. (ed.) CRYPTO 2009. LNCS, vol. 5677, pp. 143–159. Springer, Heidelberg (2009)

[LFKN92]    Lund, C., Fortnow, L., Karloff, H.J., Nisan, N.: Algebraic methods for interactive proof systems. J. ACM 39(4), 859–868 (1992)

[Mer07]     The great internet mersenne prime search, project webpag (2007), `http://www.mersenne.org/`

[Mic94]     Micali, S.: Cs proofs (extended abstracts). In: FOCS, pp. 436–453 (1994)

[Mic00]     Micali, S.: Computationally sound proofs. SIAM J. Comput. 30(4), 1253–1298 (2000)

[Sha92]     Shamir, A.: IP = PSPACE. Journal of the ACM 39(4), 869–877 (1992)

[vAMM$^+$08]  von Ahn, L., Maurer, B., McMillen, C., Abraham, D., Blum, M.: re-CAPTCHA: Human-Based Character Recognition via Web Security Measures. Science 321(5895), 1465–1468 (2008)

# Oblivious RAM Revisited

Benny Pinkas[1,*] and Tzachy Reinman[2]

[1] Dept. of Computer Science, University of Haifa, Mount Carmel, Haifa 31905, Israel
benny@pinkas.net
[2] School of Computer Science and Engineering, The Hebrew University of Jerusalem,
Jerusalem 91904, Israel
reinman@cs.huji.ac.il

**Abstract.** We reinvestigate the oblivious RAM concept introduced by
Goldreich and Ostrovsky, which enables a client, that can store locally
only a constant amount of data, to store remotely $n$ data items, and
access them while hiding the identities of the items which are being
accessed. Oblivious RAM is often cited as a powerful tool, but is also
commonly considered to be impractical due to its overhead, which is
asymptotically efficient but is quite high. We redesign the oblivious RAM
protocol using modern tools, namely Cuckoo hashing and a new oblivious sorting algorithm. The resulting protocol uses only $O(n)$ external
memory, and replaces each data request by only $O(\log^2 n)$ requests.

**Keywords:** Secure two-party computation, oblivious RAM.

## 1 Introduction

The need to enhance the security of data storage systems and to encrypt the content they store is obvious. Various encryption algorithms are in common use for
many years, so content-encryption may be considered, for the most part, as an
already-solved issue. Apparently, encryption alone does not suffice. A server, which
maintains a data storage system, can gain information about its users' habits and
interests, and violate their privacy, even without being able to decrypt the data
that they store. The server can monitor the queries made by the clients and perform different traffic analysis tasks. It can learn the usual pattern of accessing the
encrypted data, and try to relate it to other information it might have about the
clients. For example, if a sequence of queries $q1, q2, q3$ is always followed by a stock-
exchange action, a curious server can learn about the content of these queries, even
though they are encrypted, and predict the user action when the same (or similar) sequence of queries appears again. Moreover, it is possible to analyze the importance of different areas in the database, e.g., by counting the frequency of the
client accessing the same data items. If the server is an adversary with significant
but limited power, it can concentrate its resources in trying to decrypt only data

* This research was supported by the European Research Council as part of the ERC
project SFEROT, and by the Israel Science Foundation (grant No. 860/06).

T. Rabin (Ed.): CRYPTO 2010, LNCS 6223, pp. 502–519, 2010.

items which are often accessed by the target-user. Another ability of the server is to draw conclusions about relations between queries, and so on.

In order to protect against this kind of privacy violation, one must hide the access patterns of clients of the storage system. This problem is related to the classic result of Pippenger and Fischer on oblivious simulation of Turing machines [18]. In the context of RAM machines, this problem was investigated by Goldreich [8] and Ostrovsky [14] as a software protection problem (the goal there was to hide the pattern of access of a program to memory in order to prevent reverse engineering of the software). The best results of Goldreich and Ostrovsky appear in [9].

Hiding the access pattern, or making it *oblivious*, means that any equal-length sequence of clients' data requests to the server are equivalent from the point of view of the eavesdropper (who might be the server itself). The server must only know the number of queries in the sequence.

The cost of the best protocol of Goldreich-Ostrovsky was efficient asymptotically but clearly unfeasible for any reasonable application: Storing $n$ data items was replaced with storing $O(n \log n)$ items; furthermore, each access to a data item was replaced by $O(\log^3 n)$ data requests to the stored data (this $O(\log^3 n)$ overhead comes with a very large constant factor; it can be replaced with $O(\log^4 n)$ with a reasonable constant).

Due to the overwhelming overhead of the oblivious RAM protocol, it was often cited as a "theoretical" solution which could in principal solve many problems (such as cache attacks, or search on encrypted data; see discussion below), but is clearly impractical. Our goal was to design an improved protocol which will be feasible in practice. We describe in this work a new construction with a considerably improved overhead: it requires the client to store only $O(n)$ items, and replace each data request with $O(\log^2 n)$ accesses to the stored data, where the constants in the "$O$" notation are small. A detailed comparison with previous schemes appears in Sect. 2.

**Other applications of oblivious RAM.** We mentioned above that oblivious RAM can be used to hide access patterns to data stored on a remote and untrusted server, or to enable a CPU to operate securely with an untrusted memory. Another application of oblivious RAM is for the symmetric encryption variant of "search on encrypted data", where a client stores data (e.g. mail messages) remotely, and wishes to use the data while protecting its privacy (see, e.g. [19]). Oblivious RAM can also be used for protecting against cache attacks, which are software side-channel attacks run by monitoring the state of the CPU's memory cache. These attacks have been demonstrated to reveal AES keys in real systems [15]. As noted in [15], an oblivious RAM can hide these access patterns, but at a cost which is definitely unacceptable for basic CPU operations.

**The basic ideas behind our new construction.** We base our solution on the Goldreich-Ostrovsky hierarchical solution, which is described in [9] (and in the full version of our paper). We improve its overhead by using the following primitives instead of the original components of the construction.

– *Cuckoo Hashing.* In the Goldreich-Ostrovsky construction the client maps data items into bins using a random hash function that is kept secret from the server. The number of items mapped into each bin must be hidden from the server. It is well known that when $n$ items are randomly mapped to $n$ bins then (with high probability) the most populated bin contains $O(\log n)$ items. Therefore in the original construction the client sets each bin to have sufficient room for $O(\log n)$ items, and stores in a bin fake items if less than this number of items are mapped to it. This increases the overall storage required by the construction to $O(n \log n)$.

In comparison, our construction uses Cuckoo hashing [16,17], which is a hashing scheme mapping $n$ items to $2(1 + \epsilon)n$ bins with the guarantee that at most a single item is mapped into a bin. Consequently, the construction uses a total of only $O(n)$ server storage.

- *Pseudo-random permutation.* The server observes where items are inserted to the Cuckoo hash table, and might use this information to identify "dummy" items (a discussion of the usage of dummy items is given in Sect. 4). In order to prevent that, the client needs to apply a pseudo-random permutation to the order of the items before inserting them to the hash table.
- *Randomized Shell sort.* The storage system is built of hierarchical levels. Periodically, the items of two adjacent levels are reshuffled. The reshuffling process uses sorting, which is composed of many steps where the client retrieves a pair of encrypted items from the server, decrypts them and compares the results, and stores a re-encrypted version of the sorted pair. The sorting must be oblivious in the sense that the indices of the pair of items that are compared must not leak any information about the results of previous comparisons. The original Goldreich-Ostrovsky construction uses a sorting network for this purpose, but this solution has an overhead of $O(n \log^2 n)$ comparisons, with a very small constant, using Batcher's network [5], or $O(n \log n)$ comparisons, with a constant of about 6100, using the AKS network [2]. We perform sorting using the new randomized Shell sort algorithm of Goodrich [10]. This algorithm is oblivious, sorts with very high probability, and works in $O(n \log n)$ comparisons; where the "$O$" notation hides only a very small constant.

We stress that even given these improved primitive building blocks, a lot of care had to be taken in order to compose them to a secure, and efficient, oblivious RAM protocol. Additional effort was needed in order to reduce the constant factors of the overhead.

## 1.1   Basics of Oblivious RAMs

The problem of hiding access patterns is modeled in the following way: The setting includes a client which has a small secure memory, and a server with a large insecure storage. The client can use the server's storage to store and retrieve its data. The client stores internally a secret key of a symmetric encryption scheme, and uses it to encrypt the data before storing it, and decrypt it after retrieving it.

We assume here and throughout the paper that encryption is done with a *semantically* secure probabilistic encryption scheme and therefore two encrypted copies of the same data look different. The server cannot identify whether these two copies correspond to the same data of the client.

The client has $n$ data items denoted as $(v_i, x_i)$, where $i = 1, \ldots, n$ is an index, $v_i$ is the data identifier or location-index (e.g., a serial number), different for each data item, and $x_i$ is the data payload. It is assumed that all $x_i$ values are of the same length. To simplify the description we assume that the storage service of the server has slots of a size which is equal to the size of an encryption of a data item used by the client. Therefore each slot can be used to store a data item, where the client can ask to store a specific data item in a specific slot location $j$. All requests to the server are therefore of the form "GET $j$", which provides the client with the (encrypted) content of slot $j$, or "PUT data $j$", which stores at slot $j$ the encrypted data provided by the client.

The client has a small amount of secure internal memory. It includes space for $O(1)$ data items, for $O(1)$ secret keys for symmetric key cryptographic functions, and for a constant number of counters which count up to $n$ and therefore are of length $O(\log n)$ bits.

We assume that the server does not tamper and modify the stored data, because this issue can be easily solved by the client authenticating the stored data using a message authentication code (MAC) and a secret key known only to the client. However, the server does learn which location in its storage is being accessed by the client in each operation.

By default, the client cannot hide the fact that it accesses a specific location in the server's storage. The server can examine the contents of its storage and of the requests from the client, but the server obviously cannot learn the contents of the stored data, since it is encrypted. The goal of the client is to hide its access pattern to the stored data. This is expressed in the following definition.

**Definition 1.** *The input $y$ of the client is a sequence of data items, denoted by $y = ((v_1, x_1), \ldots, (v_n, x_n))$ and a corresponding sequence of operations, denoted by $(op_1, \ldots, op_m)$, where each operation is either a read operation, denoted* read$(v)$, *which retrieves the data of the item indexed by $v$, or a write operation, denoted* write$(v, x)$, *which sets the value of item $v$ to be equal to $x$.*

*The access pattern $\mathcal{A}(y)$ is the sequence of accesses to the remote storage system, which contain both the indices accessed in the system and the data items read or written. An oblivious RAM system is considered secure if for any two inputs $y, y'$ of the client, of equal length, the access patterns $\mathcal{A}(y)$ and $\mathcal{A}(y')$ are computationally indistinguishable for anyone but the client.*

Hiding the access patterns, or "unifying" them, must have a cost – each access is simulated by more than one access. First, we would like to make the different types of accesses look the same. For example, if we want that *read* and *write* would be indistinguishable, we would have each of them both implement *read* and *write*, i.e., read the value in the accessed location, decrypt it and then rewrite it with an encryption of the same value or a different one. Note that since we use a semantically secure probabilistic encryption, the server cannot identify whether the data was changed before it was written back. We note that this element of making different types of data access look the same, by always using a read-and-then-write operation, is common for all the following solutions. From here on, we treat all *read*, *write*, or other access-operation, as equal. Adding a write operation to each read operation already multiplies the computational overhead by a factor of two. In addition, we would like to prevent the adversary from distinguishing between accesses to locations $\{v_1, v_2, v_3\}$ and $\{v_2, v_1, v_2\}$, etc. A trivial solution is to read and rewrite the entire set of stored data for each access. Applying this solution is usually infeasible, since it multiplies the computational overhead by a factor equal to the number of stored items, which is normally huge. On the other hand, it is easy to see that this is the best possible deterministic scheme. A probabilistic scheme, where the operation of the client depends on a random bits, can do much better.

## 2  Related Work

Most oblivious RAM constructions are based on the client having access to a secret (pseudo-)random function, which is implemented using symmetric cryptographic functionalities, such as encryption, and can therefore be constructed assuming the existence of one-way functions. Very recent results of Ajtai [1] and of Damgård et al. [7] construct an oblivious RAM based on no cryptographic assumption (but rather, letting the client use the oblivious RAM itself for storing random coin tosses and accessing them obliviously). The client needs to store remotely (for each of its data items) an equivalent to a poly-logarithmic amount of items, rather than $O(1)$ items in our scheme, and each data request is replaced with a poly-logarithmic number of requests to the server. It is not clear how high is the exponent of this poly-logarithmic overhead. We therefore focus our description of related work on cryptographic oblivious RAM schemes.

The investigation of oblivious RAM techniques was initiated by Goldreich and Ostrovsky [9]. A major tool used in their constructions is a primitive which performs an oblivious sorting of the stored data. That is, it sorts the stored data items according to some index, while hiding from the server all information about the permutation that orders the input set of data items. Specifically, this primitive was implemented in [9] using a sorting network: either the sorting network of Batcher [5] which performs $O(n \log^2 n)$ read operations with a very small constant (approximately $1/2$), or the sorting network of AKS [2] which performs only $O(n \log n)$ read operations, but whose complexity has a much larger constant. (The actual overhead of the AKS sorting network is about $6100 n \log n$ comparisons, and therefore it is clear that for any feasible input, the performance of the Batcher network is preferable.)

Goldreich and Ostrovsky presented a basic "square-root" algorithm, whose overhead is $O(\sqrt{n})$ read/write operations for each original access to a data item. They also designed a more complex hierarchical solution, using a data structure composed of levels, where each level is twice the size of the former level, and whose overhead is $O(\log^4 n)$ (using Batcher sorting network). A detailed description of these solutions can be found in [9] or in the full version of our paper.

Williams and Sion [20] modified the hierarchical solution of Goldreich and Ostrovsky, assuming that the client can locally store $O(\sqrt{n})$ data items, rather than $O(1)$ items. This extended local storage enables to run an oblivious merge sort and improve the run time overhead to $O(\log^2 n)$. A solution based on Bloom filters [6] was presented in [21]. In that solution the client stores at the server, for every level, an encrypted Bloom filter and uses it to check whether an item appears in the level. The work in [21] claims to reduce the storage overhead at the server to $O(n)$, and to reduce the number of actual data requests per item requested by the client to only $O(\log n \log \log n)$. That analysis is based on the assumption that the size of the Bloom filter encoding $m$ items is $O(m)$. The overhead is actually larger, since the size of the Bloom filter must also be a function of the number of the hash functions used and of the allowed error probability (which in inevitable when a Bloom filter is used). As a result, the overhead of the Bloom filter based scheme is worse than that of our scheme for any reasonable choice of the number of items $n$ and of the error probability of the filter.

**Table 1.**  A comparison of the different access hiding schemes. (For the scheme of [21], we note that the original analysis is inaccurate. The second line is for an invocation using an optimal number of hash functions, with specific numbers for an error probability of $2^{-64}$.)

| | computational overhead | client memory (data items) | server storage (data items) |
|---|---|---|---|
| Goldreich-Ostrovsky [9] $\sqrt{n}$ | $O(\sqrt{n}\log n)$ | $O(1)$ | $O(n+\sqrt{n})$ |
| Goldreich-Ostrovsky [9] Batcher | $O(\log^4 n)$ | $O(1)$ | $O(n\log n)$ |
| Goldreich-Ostrovsky [9] AKS | $O(\log^3 n)$, const $\geq 6100$ | $O(1)$ | $O(n\log n)$ |
| Merge sort [20] | $O(\log^2 n)$ | $O(\sqrt{n})$ | $O(n\log n)$ |
| Bloom filter [21] | | | |
| original analysis (inaccurate) | $O(\log n \log \log n)$ | $O(\sqrt{n})$ | $O(n)$ |
| optimal # of hash functions | $O(1.44c \log n \log \log n)$ for $c=64$: const $> 92$ | $O(\sqrt{n})$ | $O(n)$ $+1.44cn$ bits |
| This paper | $O(\log^2 n)$ | $O(1)$ | $O(n)$ |

Table 1 compares the performance of all schemes described in this section.[1] The performance comparison can be summarized as follows: (1) The constructions of

---

[1] Note that for the Bloom filter based scheme [21] the first line of the table lists the performance according to the original analysis in [21], which is inaccurate. The second line lists the performance according to a more careful analysis (detailed in the full version of our paper), assuming an allowed error probability of $2^{-c}$. The $O(\log n \log \log n)$ overhead in the second line has a constant factor of at least $1.44c$ (greater than 92 for $c = 64$), *in addition* to other constant factors which are similar to those incurred by all schemes. Given this finer analysis, the performance of [21] is worse than the performance of our scheme when $\log n < 1.44c \log \log n$, which is clearly the case for any reasonable choices of $n$ and $c$. For example, for $n < 2^{80}$ this holds for any error parameter $c \geq 9$.

Goldreich-Ostrovsky and of our work are the only ones using local storage of only $O(1)$ data items; (2) The computational overhead of our construction is better or equal to that of all other constructions (except for the asymptotic overhead of the Bloom filter construction for unreasonably high values of $n$); (3) The amount of server storage in our construction is better than that of all other constructions (except for the Bloom filer construction, which stores a comparable number of data items and in addition $1.44cn$ bits, which are more than $92n$ bits for $c = 64$).

# 3   Building Blocks

## 3.1   Randomized Shell Sort (Oblivious Sorting Algorithm)

Goodrich's recent randomized Shell sort algorithm [10] is an efficient sorting algorithm, using only $O(n \log n)$ comparisons with a relatively small constant factor. Equally important is the fact that this algorithm is also data oblivious. This property means that if we assume that the operation of comparing two items and reordering them according to their value is a black-box (i.e., the result of the comparison is hidden from an external observer, which is the server in our case), then the algorithm performs no operations which depend on the relative order of the elements in the input array. In other words, an external observer who can only observe the items which the algorithm compares, but not the results of the comparisons, sees a list of pairs of items which are compared, where the choice of items for these pairs is independent of the results of previous comparisons.

We note that other sorting algorithms are not known to be both oblivious and efficient. For example, bubble sort is oblivious, but is not efficient; quick sort is efficient (in the average case) but is not oblivious; sorting networks are oblivious, but, as noted in Sect. 1, the only sorting network constructions of size $O(n \log n)$ are not efficient in the practical sense, due to large constants. See [2,10] for details.

We use randomized Shell sort in our scheme in order to reorder items in the server database, according to a new permutation, in a way that prevents the server from tracking the new ordering. The details of the randomized Shell sort construction appear in [10] or in the full version of our paper.

## 3.2   Cuckoo Hashing

Cuckoo hashing [16,17] is a relatively new hashing algorithm, which in its basic form maps each item to one of two potential entries of a hash table, while ensuring constant lookup and deletion time in the *worst case*, and amortized constant time for insertions.

The basic idea of Cuckoo hashing is to use two hash functions denoted $h_0$ and $h_1$ (or multiple hash functions in the general case). The size of a hash table used for storing $n$ items must be slightly larger than $2n$ (to simplify the discussion, we consider the size of the table to be exactly $2n$). When a new item $x$ is inserted to the hash table, it is inserted to location $h_0(x)$. If this location is already occupied by another item $y$, then that item is "kicked out" of its current location and is re-located to its other possible location. Namely, if $h_b(y) = h_b(x)$ (for $b \in \{0,1\}$, and initially $b = 0$) then item $y$ is moved to location $h_{1-b}(y)$. If location $h_{1-b}(y)$ is already occupied by another item $z$ (i.e., $h_{1-b}(z) = h_{1-b}(y)$), this item ($z$) is re-located to location $h_b(z)$, and so on. If this chain of relocations continues for too long, then the table is rehashed using two new hash functions $h_0', h_1'$. In this case the insertion time is longer, but analysis shows that this event is rare, and therefore the amortized insertion time is constant. Lookup and deletion are natural – one just has to check the two possible locations of the given item.

Most recent works (e.g., [11,12,13,3,4]) present variants of Cuckoo hashing with guaranteed constant worst-case performance for insertion (this is also referred to as de-amortizing the insertion time of Cuckoo hashing).

## 4   Our Scheme

We first describe the basic form of our oblivious RAM scheme, which has the desired asymptotic overhead. Appendix A then describes how to improve the constant factors of the overhead of the scheme.

The construction is based on combining a modified version of the hierarchical solution of Goldreich and Ostrovsky with Cuckoo hashing and randomized Shell sort. The server stores the data, which can potentially consist of $n$ items, in a hierarchical data structure of $N = \lceil \log_2 n \rceil + 1$ levels, each of which is twice larger than its previous. Additional levels may be allocated, as necessary (when a new level is allocated, its size is twice the size of the last allocated level).

In the original scheme of Goldreich-Ostrovsky, level $i$ consists of $2^i$ buckets, where each bucket contains $O(\log n)$ entries. In our scheme level $i$ consists of a table of $4 \cdot 2^i$ single item entries, which will be used to store up to $2^i$ data items of the client. Storing the items is done in the following way: Along with the $2^i$ items of the client, up to $2^i$ "dummy" items might be stored in the level, where the client might access a dummy item in order to hide the fact that it does not need to search for a real item in this level (since the real item was already found in a previous level). All $2 \cdot 2^i$ items of the level are stored in a Cuckoo hashing table of size $4 \cdot 2^i$. (We note that according to this description the first level is used to store only two items. Any actual implementation would probably set the first level to be much larger, say, to contain 128 data items. To simplify the analysis we assume, however, that the first level stores only two items.)

For each level we associate an *epoch*, which is defined for level $i$ as $2^{i-1}$ requests (the epoch ends when a reshuffle from level $i-1$ to level $i$ occurs). For each level $i$ and its $\ell^{th}$ epoch, the client randomly chooses two hash functions whose ranges are $\{1 \ldots 2^{i+2}\}$: $h_{k,0}^{i,\ell}$ and $h_{k,1}^{i,\ell}$, where $k$ is a secret key known to the client and used to define these functions. At the end of every epoch each level is re-hashed obliviously, using a new pair of hash functions. The following table summarizes the properties of level $i$.

| | real items | dummy items | size | epoch-length | "moved down" |
|---|---|---|---|---|---|
| level $i$ | $2^i$ | $2^i$ | $4 \cdot 2^i$ | $2^{i-1}$ | every $2^i$ requests |

Each data request includes both reading and writing to the data structure, such that the server cannot distinguish which operation occurred. In addition, for any request, the accessed item is re-encrypted by the client, using a probabilistic encryption scheme.

**Data requests.** Initially, the data structure is empty. For each request (of any type) of a location-index (virtual address) $v$, the following operations are performed.[2]

1. Scan through the entire first level in a sequential order to find the item whose location-index is $v$. This step includes reading all the items in the first level. If the requested item is found, it is stored in the client's secure memory, and the process continues as usual.

2. Go through all other levels, and for each level $i = 2 \ldots N$, do:
   - If $v$ has not been found yet, examine its two possible locations in the Cuckoo hashing table of the current level ($i$): $h_{k,0}^{i,\ell}(v)$ and $h_{k,1}^{i,\ell}(v)$. If the requested item is found in one of the two locations, it is stored in the client's secure memory, and the process continues as usual.
   - If $v$ has been found, examine two random locations $h_{k,0}^{i,\ell}(\text{"}dummy\text{"} \circ t)$ and $h_{k,1}^{i,\ell}(\text{"}dummy\text{"} \circ t)$, where $t$ is a counter which is increased with every data request. (These are locations allocated by the Cuckoo hashing for two fresh dummy items which were not searched for before.)

3. Scan again through the entire first level in a sequential order, and write back the updated (and re-encrypted) item of location-index $v$ in the next available location. If $v$ is already in the first level, overwrite it. This step includes reading and writing all items of the first level.

The first level functions as a cache, meaning that for any request, the updated value is written to the first level. Since the capacity of the level is final, after a certain number of requests it becomes full. In order to avoid this, the content of the first level is "moved down" to the second level before the first level becomes potentially exhausted. Now the second level may become full, so the same process is repeated. When the content of the last level has to be "moved down", a new level with twice the number of entries is allocated.[3] "Moving down" the content of level $i$ is done every $2^i$ requests. This makes sure that no level is overflowed, and that the first level is emptied and has enough available slots at the beginning of each epoch of any of the levels (since the beginning of an epoch of level $i$ is

---

[2] We assume here, as was implicitly assumed by all previous constructions [9,20,21], that the client does not perform a "read" operation for an item which does not exist in the remote storage.

[3] In fact, if we are willing to disclose an upper bound on the number of items that are stored, there is no need to allocate a new level when the last level has to be "moved down". The system may instead re-order the entire database.

also a beginning of an epoch of all the levels $j < i$). In fact, this process makes sure that at any time level $i$ contains no more than $2^i$ items, as is stated in Lemma 1 below. Whenever level $i$ is moved down to level $i + 1$, the latter level is reshuffled.

When the client moves the content of level $i$ to level $i + 1$, it obliviously hashes the content of both levels to level $i + 1$. This reshuffling must fulfill the following requirements: (1) If there is a duplicate item (the same location-index, and possibly different data content) in level $i$ and level $i + 1$, the newer item (from level $i$) must be kept, and the older one must be deleted; (2) The resulting buffer, namely level $i + 1$ after the reshuffling, must be ordered independently of any of the levels before the reshuffling; (3) Level $i$ must be cleaned, i.e., its content must be deleted. As we continue, we see that all these requirements are fulfilled.

Before describing the reshuffle process, we state Fact 1, which trivially follows from the reshuffling algorithm, and Lemma 1, which is proved in the full version of the paper.

**Fact 1.** *When a reshuffle from level $i$ to level $i + 1$ occurs, all levels $j \leq i - 1$ are empty. At the end of the reshuffle, all levels $j \leq i$ are empty.*

**Lemma 1.** *When a reshuffle from level $i$ to level $i + 1$ occurs, each of these two levels contains at most $2^i$ real items. At the end of the reshuffle, level $i$ is empty and level $i + 1$ has at most $2^{i+1}$ real items.*

### 4.1   Reshuffling Levels Using Cuckoo Hashing and Randomized Shell Sort

The reshuffle of levels $i$ and $i + 1$ into level $i + 1$ is a complex process, consisting of the steps enumerated below and based on two basic primitives: (1) *Scanning*, which is reading and (possibly) writing in a sequential order all the items in a given buffer; (2) *Oblivious Sorting* (O-Sort), which is done by randomized Shell sort (see Sect. 3.1). Note that whenever an item is written to a storage (whether it is one of the levels or a temporary buffer), it is re-encrypted. The reshuffle process is also depicted in Fig. 1.

1. Allocate a temporary buffer $C$, whose size is $2^{i+1}$. (Recall that jointly, both levels contain at most $2^{i+1}$ real items).
2. O-sort each of the levels (level $i$ and level $i + 1$): The sorting is according to an order which locates real items before dummy and empty items. At the end of this step, all the real items of level $i$ (at most $2^i$) are in its first locations, and all the real items of level $i+1$ (at most $2^i$) are in its first locations.

   In the following two steps (3–4), $2^{i+1}$ items that include all the real items of the two levels are copied into the temporary buffer.
3. Move the first $2^i$ items from level $i$ to the left side of $C$. Mark each item as "new". At the end of this step, *all* the real items of level $i$ (and possibly additional items) are in $C$. (This step is depicted in Step I of Fig. 1.)

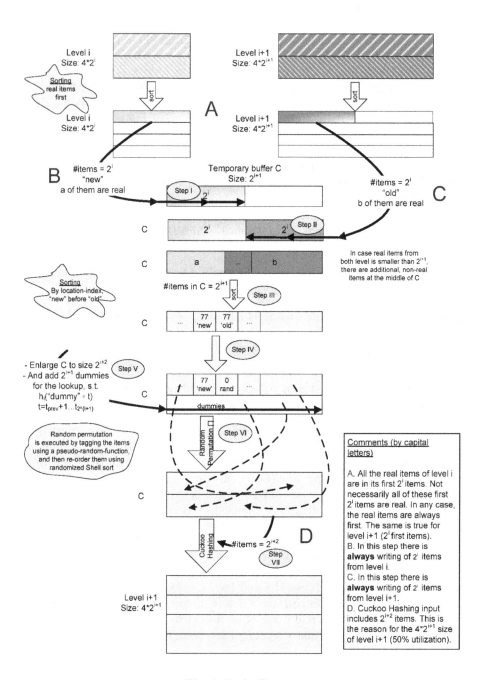

**Fig. 1.** Reshuffle steps

4. Move the first $2^i$ items from level $i+1$ to the right side of $C$. Mark each item as "old". At the end of this step there are $2^{i+1}$ items in $C$, that include *all* the real items of both levels and possibly additional items (if both levels together contained less than $2^{i+1}$ real items). (This step is depicted in Step II of Fig. 1.)

   The goal of the following two steps (5–6) is to erase duplications of items with the same location-indices.
5. O-sort $C$ according to these criterions (ordered): (a) smaller location-indices (virtual addresses) first, (b) items tagged "new" before items tagged "old". (This step is depicted in Step III of Fig. 1).
6. Removing duplicates: Sequentially scan $C$ and erase each old item preceded by a new item with the same location-index (the marks "old" and "new" of the remaining items can be ignored from now on). Replace each erased item and each dummy item with a random item (an item with a special location-index and random content). At the end of this step there are $2^{i+1}$ real and random items in $C$, without duplications. We will refer to all these items in the sequel as "real". (This step is depicted in Step IV of Fig. 1.)
7. Create $2^{i+1}$ dummy items with indices "dummy" $\circ (t+j)$, where $t$ is a counter of the number of requests so far and $j = 1\ldots 2^{i+1}$.[4] Add these items to $C$ (this requires increasing the size of $C$ from $2^{i+1}$ to $2^{i+2}$). (This step is depicted in Step V of Fig. 1.)

   The client then obliviously reorders the items in a pseudo-random order. This is done by (1) choosing a new keyed pseudo-random function $F_k$ and using it to tag each of the $2^{i+2}$ items with a new value which is the result of $F_k()$ applied to their location-index, and then (2) obliviously sorting the items by their tags using randomized Shell sort. The new order of the items is independent of their original order. (This step is depicted in Step VI of Fig. 1.)
8. Sequentially scan the buffer and use Cuckoo hashing with two new random hash functions that are kept secret from the server: $h_{0,k}^{i+1,\ell+1}$ and $h_{1,k}^{i+1,\ell+1}$, to map the $2^{i+2}$ items to the $4 \cdot 2^{i+1}$ entries of level $i + 1$ ($\ell$ is the index of the current epoch). The hash functions are applied to the location-index. At the end of this step there are $2^{i+2}$ real and dummy items in level $i + 1$, located according to the Cuckoo hashing functions. (This step is depicted in Step VII of Fig. 1.)
9. If the Cuckoo hashing fails (due to cycles, see [17]), choose new random secret hash functions $h_{0,k}^{i+1,\ell+1}$ and $h_{1,k}^{i+1,\ell+1}$ and repeat the previous step.

After step 4, all the real content of levels $i$ and $i + 1$ is in $C$ (possibly with additional items), Steps 5 and 6 handle possible duplications (of items with the

---

[4] These dummies are necessary in order to hide whether a data request in level $i + 1$ in the next epoch searches for an item which was found in a level prior to level $i+1$. If this event happens in the $j$th time slot of the epoch, then the client will look for item "$dummy'' \circ (t+j)$, which was inserted to the level in the reshuffle. As a result, every search in this level will be to an item which is stored in the Cuckoo hashing table and which was never searched before.

same location-index in the two original buffers), Step 7 reorders the real and dummy items (pseudo-) randomly, and Steps 8 and 9 insert the items to level $i+1$, according to the random secret Cuckoo hashing functions.

### 4.2    Analysis and Implementation

**Overhead.** The construction uses $\log n$ levels, where level $i$ contains $4 \cdot 2^i = O(2^i)$ items, yielding a server storage of $O(n)$ data items. The overall amortized computational overhead is $O(\log^2 n)$ data requests for each original request of the client: First, observe that accessing an item requires scanning through the first level (which is of constant size), and then accessing two locations in each other level. The reshuffling process uses randomized Shell sort that sorts $\ell$ elements in $O(\ell \log \ell)$ time, with a reasonable constant factor. Performing the oblivious sorting is the main time-consuming element of the reshuffle process. The size of the sorted array in level $i$ is $O(2^i)$, giving a sorting time of $O(2^i \log 2^i) = O(2^i \cdot i)$ for level $i$. Level $i$ is sorted every $2^i$ requests, giving an amortized cost of $\frac{O(2^i \cdot i)}{2^i} = O(i)$. Summing this for all the levels gives $\sum_{i=1}^{\log n} O(i) = O(\log^2 n)$.

Examining the performance more carefully, we note that level $i$ has room for $4 \cdot 2^i$ items (Appendix A shows how to reduce the storage by about 50%). The bulk of the computation overhead comes from the sorting operations. In particular, Step 2 sorts level $i+1$, which consists of $4 \cdot 2^{i+1}$ items (the other sorting operations are applied to smaller sets of items). However, we show in Appendix A how the sort operations in Step 2 can be changed to sort half as many items. This is estimated to reduce the overhead by 33%. Appendix A discusses an additional optimization which reduces the constant factors of the overhead of the construction by an additional 33%.

### Security

**Theorem 1.** *The oblivious RAM protocol described above is secure according to Definition 1.*

*Proof.* The security of the construction holds under the assumption of the existence of pseudo-random functions, or assuming that the client has access to random functions (e.g., an internal random number generator which always provides the same output when given the same input). The PRF assumption is probably more reasonable for most applications. A crucial ingredient of the protocol is that the hash functions $h_0$ and $h_1$, used to map items during the Cuckoo hashing, are randomly chosen by the client and are kept secret from the server. (When a PRF is used, these functions are defined by some function $F_k()$ where the key $k$ is chosen by the client and is unknown to the server.)

We will show that for any input sequence $y$ of the client, the access pattern $A(y)$ to the storage server is indistinguishable, by a polynomial-time server, from an access pattern $A'$ which can be simulated without any knowledge of $y$, except for the length of $y$.

The *contents* of the requests in the access pattern are encrypted with a semantically secure probabilistic encryption scheme, and therefore the server cannot distinguish between the contents of the requests in $A(y)$ and in $A'$. We therefore only need to show that the locations accessed by the two access patterns in the server's memory are indistinguishable.

Consider the reshuffle operation from level $i$ to level $i + 1$. The first steps of the reshuffle perform an oblivious sorting or a serial scan of data items, and are therefore independent of the actual data stored by the client and of the input sequence $y$. As such they can be easily simulated. Namely, the simulated access pattern $A'$ contains a serial scan in every step of the reshuffle where such a scan in performed (namely, Steps 1, 3, 4 and 6). In addition, whenever the reshuffle performs an oblivious sorting (in Steps 2, 5 and 7), $A'$ performs an oblivious sorting assuming that the values to be sorted are $1, 2, 3, \ldots$.

Let $M = 2^{i+1}$. In Step 7 of the protocol the client obliviously reorders in a pseudo-random order $M$ real values and $M$ dummy values. Step 8 maps these $2M$ values to a Cuckoo hash table of size $4M$, using two hash functions $h_0$ and $h_1$ which are chosen at random by the client and are unknown to the server. The client goes over the $2M$ items according to their new order, and attempts to insert each of them to the table according to the Cuckoo hashing algorithm. If $x$ is a certain item in the list, then the client probes locations $h_0(x)$ and $h_1(x)$ in the table and might perform some evictions of items to find a place for $x$. In this process the server might learn the $h_0$ and $h_1$ values of each of the $2M$ items. However, since the server does not know the hash functions used, these values are independent of the actual values of the items. Simulating this process is performed in the following way: Define random functions $h_0$, $h_1$, and apply the Cuckoo hashing algorithm to an *arbitrary set* of $2M$ values, say the values $1 \ldots 2M$, using these functions. This process results in exactly the same distribution, as in the real execution, for all the events observed by the server, including the locations probed in the hash table and the occurrences of evictions and cycles (which might cause a repeat of the Cuckoo hashing algorithm as defined by Step 9). Note that our security analysis does not have to analyze the exact probabilities with which evictions and cycles occur, but rather only observe that these probabilities are independent of the data items being hashed.

At the end of the hashing process the server knows, for each of the $2M$ items, the two locations to which this item is mapped by $h_0$ and $h_1$, and the exact location in this pair to which this item was eventually mapped. Recall, however, that the $2M$ items were randomly reordered, and that half of them are dummy items. In the epoch that follows, the server can observe which locations are probed in each request of this level. Namely, it might see that the $j^{th}$ request probes locations 10 and 17 to which, say, the first of the $2M$ items is mapped. However, the server does not know whether this is a real or a dummy item. Also, each item hashed into this level is probed at most once during the epoch, since each dummy value is probed at most once (due to the dummy counter being increased), and a real value that is accessed is immediately moved to the top level and is not accessed again in this level during the current epoch. Given these observations, the probes to the level in this epoch can be simulated in the following way: use the random functions $h_0, h_1$ that were used in the simulation of the Cuckoo hashing into this level; let $(a_1, \ldots, a_{2M})$ be a random permutation of the numbers $1, \ldots, 2M$; when performing the $j$th data request from level $i + 1$ in the current epoch, probe the locations to which item $a_j$ is mapped by $h_0$ and $h_1$.

We have described above how to simulate probes to a specific level during data requests. The entire sequence of probes during data request can therefore be simulated as follows: In Steps 1 and 3 the simulation scans the entire first level. In Step 2 the simulation goes through all levels, starting with the second one, and simulates a pair of probes to each level, as is described in the previous paragraph.                                                                                      □

**Implementation.** We implemented a basic prototype of our scheme, including the hierarchical data structures, the randomized Shellsort algorithm and the Cuckoo hashing algorithm. This allowed us to simulate the operation of the oblivious RAM construction, and to measure and estimate its performance. We chose Java as an initial platform and compiled using the Sun JDK 1.6.0_16. The testing environment was a standard PC. In our measurements we ignored network delays, and therefore we only provide measurements of the number of operations per data request, rather than of the amount of time each request takes.

We ran experiments on various databases, of sizes between $n = 2^{10}$ and $n = 2^{20}$. For a database of $n = 2^i$ potential items, we ran $k = 2^i - 10$ requests, and counted the number of read/write operations handled by the server. The results appear in Table 2. The constant of the $O(k \log^2 k)$ overhead seems to be about 160. We note that the two improvements described in Appendix A, (minimizing the amount of sorted items – either by not sorting empty items, or by using an advanced Cuckoo hashing algorithm; and reshuffling several levels together)

Table 2. Performance measurements

| $\log_2 n$ | $n$ | $k = n - 10$ (# of req.) | $k \log^2 k$ | #operations | ops per request | const of $O(k \log^2 k)$ |
|---|---|---|---|---|---|---|
| 10 | 1024 | 1014 | 101113 | 15445582 | 15232 | 152 |
| 11 | 2048 | 2038 | 246281 | 38081523 | 18685 | 154 |
| 12 | 4096 | 4086 | 588038 | 91975576 | 22509 | 156 |
| 13 | 8192 | 8182 | 1382383 | 218482493 | 26702 | 158 |
| 14 | 16384 | 16374 | 3208900 | 511882978 | 31261 | 159 |
| 15 | 32768 | 32758 | 7370117 | 1185355399 | 36185 | 160 |
| 16 | 65536 | 65526 | 16774194 | 2717439532 | 41471 | 162 |
| 17 | 131072 | 131062 | 37876427 | 6175479249 | 47118 | 163 |
| 18 | 262144 | 262134 | 84930896 | 13926487414 | 53127 | 163 |
| 19 | 524288 | 524278 | 189263809 | 31192732955 | 59496 | 164 |
| 20 | 1048576 | 1048566 | 419425822 | 69442426048 | 66226 | 165 |

which have not yet been implemented by us, are estimated to reduce the overhead by about 33% each. Applying both optimizations is likely to reduce the overhead by about 55%, and obtain a constant of about 72 in the "$O$" notation.

## 5    Open Questions

The efficiency analysis of our construction, as well as that of all other known constructions of oblivious RAM, is amortized. A data request which is followed by a reshuffle of level $i$ has a larger overhead than a request which requires a reshuffle of a level $j < i$, or one that does not require any reshuffling. A major open goal is, therefore, to reduce the worst case performance of oblivious RAM. Note that the recent result on deamortizing Cuckoo hash [3] does not help here, since it can be applied to the Cuckoo hashing part of the the reshuffling process, but not to the fact that the worst case overhead of reshuffling is high.

## Acknowledgements

The authors wish to thank Yuriy Arbitman for informing us of the randomized Shell sort result.

## References

1. Ajtai, M.: Oblivious RAMs without cryptographic assumptions. In: STOC 2010 (2010)
2. Ajtai, M., Kolmós, J., Szemerédi, E.: An $O(n \log n)$ sorting network. In: STOC, pp. 1–9 (1983)
3. Arbitman, Y., Naor, M., Segev, G.: De-amortized Cuckoo hashing: Provable worst-case performance and experimental results. In: ICALP (1), pp. 107–118 (2009)
4. Arbitman, Y., Naor, M., Segev, G.: Backyard Cuckoo hashing: Constant worst-case operations with a succinct representation (2010) (manuscript)
5. Batcher, K.: Sorting networks and their applications. In: AFIPS Spring Joint Computing Conference, vol. 32, pp. 307–314 (1968)
6. Bloom, B.H.: Space/time trade-offs in hash coding with allowable errors. Communications of the ACM 13(7), 422–426 (1970)
7. Damgård, I., Meldgaard, S., Nielsen, J.B.: Perfectly secure oblivious RAM without random oracles. Cryptology ePrint Archive, Report 2010/108 (2010),
   http://eprint.iacr.org/2010/108
8. Goldreich, O.: Towards a theory of software protection and simulation by oblivious RAMs. In: STOC, pp. 182–194. ACM, New York (1987)
9. Goldreich, O., Ostrovsky, R.: Software protection and simulation on oblivious RAMs. Journal of the ACM 43(3), 431–473 (1996)
10. Goodrich, M.T.: Randomized Shellsort: A simple oblivious sorting algorithm. In: Proceedings 21st ACM-SIAM Symposium on Discrete Algorithms, SODA (2010)
11. Kirsch, A., Mitzenmacher, M.: Using a queue to de-amortize Cuckoo hashing in hardware. In: Proceedings of the 45th Annual Allerton Conference on Communication, Control, and Computing, pp. 751–758 (2007)
12. Kirsch, A., Mitzenmacher, M.: Simple summaries for hashing with choices. IEEE/ACM Trans. Netw. 16(1), 218–231 (2008)

13. Kirsch, A., Mitzenmacher, M., Wieder, U.: More robust hashing: Cuckoo hashing with a stash. In: Halperin, D., Mehlhorn, K. (eds.) ESA 2008. LNCS, vol. 5193, pp. 611–622. Springer, Heidelberg (2008)
14. Ostrovsky, R.: Efficient computation on oblivious RAMs. In: STOC 1990, pp. 514–523. ACM Press, New York (1990)
15. Osvik, D.A., Shamir, A., Tromer, E.: Cache attacks and countermeasures: The case of AES. In: Pointcheval, D. (ed.) CT-RSA 2006. LNCS, vol. 3860, pp. 1–20. Springer, Heidelberg (2006)
16. Pagh, R., Rodler, F.F.: Cuckoo hashing. In: Meyer auf der Heide, F. (ed.) ESA 2001. LNCS, vol. 2161, pp. 121–133. Springer, Heidelberg (2001)
17. Pagh, R., Rodler, F.F.: Cuckoo hashing. J. Algorithms 51(2), 122–144 (2004)
18. Pippenger, N., Fischer, M.J.: Relations among complexity measures. J. ACM 26(2), 361–381 (1979)
19. Song, D., Wagner, D., Perrig, A.: Practical techniques for searches on encrypted data. In: Proceedings of 2000 IEEE Symposium on Security and Privacy, S&P 2000, pp. 44–55 (2000)
20. Williams, P., Sion, R.: Usable PIR. In: NDSS (2008)
21. Williams, P., Sion, R., Carbunar, B.: Building castles out of mud: Practical access pattern privacy and correctness on untrusted storage. In: ACM Conference on Computer and Communications Security, pp. 139–148 (2008)

# A   Optimizing the Construction

We present here several optimizations to the basic oblivious RAM construction presented in Sect. 4. The optimizations improve the constant factors of the overhead, but not its asymptotic performance. Still, they are beneficial for any implementation of the construction.

**Not storing empty items.** Recall that each level $i$ contains up to $2^i$ real items and $2^i$ dummy items which must be indistinguishable, from the server's point of view, from the real items. The remaining $2^{i+1}$ locations in this level are empty, and are needed for the Cuckoo hashing to succeed. The construction can be optimized by not storing in these locations encrypted "empty" data items, but rather using a flag signaling that the entry is empty. Since we can safely assume that a data item is much larger than this flag, this optimization saves about 50% of the storage required by the levels.

As for security, note that this change enables the server to identify empty locations, but it does not enable it to distinguish between real items and dummy items. Namely, this corresponds to revealing to the server the empty locations in a Cuckoo hashing table, but since the hash functions used are kept secret, no information is revealed about the items in the table.

**Implication to sorting.** Step 2 of the reshuffle algorithm sorts levels $i$ and $i + 1$, whose lengths are $4 \cdot 2^i$ and $8 \cdot 2^i$, respectively. These sorting operations are done in order to move the real items to the beginning of these buffers. If empty items are flagged, as suggested above, then there is no need to sort the corresponding entries in the level. Namely, the data to be sorted is half as long as in the basic protocol, and the overhead of sorting is reduced by more than 50%.

Let us therefore estimate how much is saved by this optimization. Note that Steps 5 and 7 sort $2 \cdot 2^i$ and $4 \cdot 2^i$ items, respectively. Assume that the overhead of sorting is linear (this is roughly the case when comparing the overhead of sorting adjacent levels, which are of similar sizes). Before the optimization, the algorithm sorts buffers of sizes $4 \cdot 2^i, 8 \cdot 2^i, 2 \cdot 2^i$ and $4 \cdot 2^i$, which are of total length $18 \cdot 2^i$. After the optimization, it sorts buffers of sizes $2 \cdot 2^i, 4 \cdot 2^i, 2 \cdot 2^i$ and $4 \cdot 2^i$, which are of total length $12 \cdot 2^i$. The overhead of sorting, which is the bulk of the overhead of the entire construction, is therefore reduced by about 33%.

**Using an advanced Cuckoo hashing scheme.** The basic Cuckoo hashing scheme used in our construction utilizes approximately only 50% of its storage to store real and dummy items, while the remaining storage is empty. The new Backyard Cuckoo hashing [4] algorithm has a much better space utilization – in order to store $n$ items, it requires only $(1 + \epsilon)n$ storage. Using this scheme has therefore the same effect as the optimization suggested above, of not storing empty entries in the hash table: it saves about 50% of the storage required by each level in the hierarchical structure. In addition, the overhead of each sorting operation is reduced by more than 50%, and the overhead of the entire construction is reduced by about 33%.

**Reshuffling several levels together.** In time $t$, where $t \bmod 2^i = 0$, and $t \bmod 2^{i+1} \neq 0$, the basic construction performs subsequent reshuffles of levels $1, 2, \ldots, i$. These reshuffles include many redundant steps. (For example, the first reshuffle inserts dummy items into the second level. Then, the reshuffle of the second level begins by (possibly) removing these items. Furthermore, the first reshuffle fills the second level, while the second reshuffle empties it.) Instead, it is possible to reshuffle together in a single step the contents of all these levels into level $i + 1$. According to our estimates this optimization saves an additional 33% of the total overhead.

# On Strong Simulation and Composable Point Obfuscation

Nir Bitansky and Ran Canetti

School of Computer Science, Tel Aviv University
{nirbitan,canetti}@tau.ac.il

**Abstract.** The Virtual Black Box (VBB) property for program obfuscators provides a strong guarantee: Anything computable by an efficient adversary given the obfuscated program can also be computed by an efficient simulator with only oracle access to the program. However, we know how to achieve this notion only for very restricted classes of programs.

This work studies a simple relaxation of VBB: Allow the simulator unbounded computation time, while still allowing only polynomially many queries to the oracle. We then demonstrate the viability of this relaxed notion, which we call Virtual Grey Box (VGB), in the context of fully composable obfuscators for point programs: It is known that, w.r.t. VBB, if such obfuscators exist then there exist multi-bit point obfuscators (aka "digital lockers") and subsequently also very strong variants of encryption that are resilient to various attacks, such as key leakage and key-dependent-messages. However, no composable VBB-obfuscators for point programs have been shown. We show fully composable *VGB*-obfuscators for point programs under a strong variant of the Decision Diffie Hellman assumption. We show they suffice for the above applications and even for extensions to the public key setting as well as for encryption schemes with resistance to certain related key attacks (RKA).

## 1 Introduction

Informally, an obfuscator is an algorithm which gets as input a program (e.g. a Turing machine or circuit) and outputs a new program which has the same functionality as the original one, but is otherwise "unintelligible". The rigorous study of obfuscation was initiated in the work of [3], who introduced the concept of *virtual black box* security (VBB in short). This concept requires the obfuscated program to behave like a "black box", in the sense that it should not leak any information about the program except its input-output behavior. More precisely, any efficient adversary with access to an obfuscated program can be simulated by an *efficient simulator* with only oracle access to the program. The same work presented the impossibility of *"universal VBB obfuscation"*, showing a family of programs which can not be VBB obfuscated.

In light of this negative result, subsequent work has included several research directions. One line of work extends the result of [3], ruling out obfuscation in various settings [16, 26]. Another line of work is aimed at constructing obfuscators

T. Rabin (Ed.): CRYPTO 2010, LNCS 6223, pp. 520–537, 2010.

for specific program families, which are not ruled out by the universal impossibility result. Here, if we stick to VBB obfuscators, our knowledge is limited essentially to obfuscating point programs and their extensions [8, 11, 23, 14, 26, 9, 13, 12]. A point program $P_v : \mathbb{D}_n \to \{0, 1\}$ holds a value $v \in \mathbb{D}_n$ in its code, and accepts its input $x$ iff $x = v$. We only know how to obfuscate point programs in which the point $v$ is explicitly obtainable from the code. Moreover, the known constructions depend on rather strong hardness assumptions, and somewhat inherently so [26].

A third line of work focuses on relaxations of VBB. In this context, [3] suggested the notion of *indistinguishability obfuscators* (INDO) , according to which obfuscations of two related size programs implementing the same functionality should be indistinguishable to any *efficient* adversary. Another relaxation, called *best possible obfuscation* (BPO) [17], requires that any information which the obfuscation leaks is efficiently learnable from any other program with the same functionality and related size (hence "best possible"). These two notions turn out to be equivalent, when restricted to efficient obfuscators.

Both notions are easier to satisfy than VBB. However, the security guarantee they provide is less clear. Unlike VBB, both seem to lose their meaning for a relatively wide range of program classes which are natural candidates for obfuscation. For instance, these notions become meaningless if we allow the obfuscator to work only when the program is given in some *"canonical"* representation in which case, no two programs have the same functionality. Another relaxation requires the obfuscation to be secure only when the program is sampled from some adequate distribution (rather than requiring security for any program in the family). This was done in the context of *perfect one-way hashing* [11] , *point proximity testing* [14], *re-encryption* [22] and more [1, 21, 18]. However, in some scenarios such a relaxation does not capture the security properties we would expect from an obfuscation.

A natural goal is thus to come up with a notion of secure obfuscation that is both meaningful and achievable. Here there is room to consider notions which might be meaningful only for certain program families but not for all.

## 1.1  This Work

We study a new relaxation of VBB security notion for obfuscators. The requirement is that an obfuscation leaks no information about the program, rather than what can also be learned by an *all-powerful learner* that witnesses only a limited number of input-output pairs (at his choice).

More formally, any *efficient* adversary with access to an obfuscated program can be simulated by an *all-powerful simulator* with poly many oracle queries to the program (in contrast to poly-time simulation which VBB requires). For lack of better name, we call this notion *virtual grey box* (or VGB in short). The extra power given to the simulator is intended to allow it to "reverse engineer" the adversary while avoiding technical difficulties that might be irrelevant to the overall goal. In certain cases (such as "highly unlearnable" programs), this could be done without losing too much of the meaningfulness of the guarantee.

*Relationship with existing notions.* VGB obfuscation is clearly weaker than VBB obfuscation. In particular, a VGB obfuscation is allowed to leak information which a VBB obfuscation can not. Formally, we show that VGB is strictly weaker, demonstrating a family of programs which can not be VBB obfuscated but is (trivially) VGB obfuscatable. On the other hand, we show that VGB is stronger than the INDO and BPO notions mentioned above. To do so, we observe that even if we further weaken the VGB security requirement by allowing the simulator an unlimited number of oracle queries, it still implies INDO.

For Turing machine obfuscators, the impossibility result of [3] extends to rule out *"universal VGB obfuscation"*. However, we could not rule out universal VGB *circuit* obfuscators (see more details within regarding this difference). We note that [17] show impossibility of strong universal BPO obfuscation that can handle even circuits that use *random oracle* gates. This impossibility applies to the stronger VGB notion.

*A setting where VGB is both meaningful and achievable.* Like INDO and BPO, VGB is not strong enough for some desirable obfuscation tasks. For example, its weakness might be revealed whenever the obfuscated program computes some kind of cryptographic functionality; indeed, in such cases an all-powerful simulator, even with limited oracle access to the program has a clear advantage over a bounded simulator. In general, it seems that VGB is mostly meaningful for program classes which are *unlearnable* with only poly many queries *even for learners with unbounded computation time*. We demonstrate concrete obfuscation tasks where VGB obfuscation is both meaningful and achievable (under appropriate hardness assumptions) while VBB is not known to be achievable. We hope that this notion will prove instrumental in other obfuscation settings as well.

The main task we consider is that of *composable obfuscation of point programs*. A point program obfuscator is $t$-composable if having access to $t$ obfuscated point programs, where the values hidden in the programs are related to each other in arbitrary ways, has the "expected effect". In other words, any adversary that has access to the obfuscated programs can be simulated given only oracle access to the programs. Ideally, $t$ could be any polynomial.

In the context of VBB obfuscation, composable point obfuscators were shown to suffice for obfuscating *multi-bit* point programs (MBPP). A MBPP has two hidden values $k, m$ in its code. It returns $m$ on input $k$, and $\bot$ on any other input. MBPP obfuscators (MBPO's) were in turn shown to imply strong symmetric encryption schemes that are simultaneously secure against weakly random keys (i.e., keys with any super-log entropy) and key dependent messages (KDM) [10]. However, as natural and fruitful as the composability property may seem, none of the known point program obfuscators were shown to be composable (w.r.t. VBB). In particular, existing MBPO's were only shown to be secure for the restricted case that the message $m$ is independent of the key $k$ [9, 10].

We show that, with respect to VGB obfuscation, composable point obfuscators do exist, under appropriate hardness assumptions. Specifically, we show that the point program obfuscator from [8] is VGB-composable for any polynomial number of instances. This is done under a strong variant of the Decision

Diffie Hellman assumption, which naturally extends the assumption used in [8] to demonstrate that this construction yields a VBB point obfuscator.

We then show that VGB composable point obfuscators suffice for constructing MBPO's which are VGB composable on their own. This yields very strong encryption schemes which are resilient to a variety of attacks. This includes the aforementioned KDM and weak keys resilience as well as resistance to certain *related key attacks* (RKA) [2]. The encryption schemes can also be extended to the public key setting (given an extra re-randomiztion property that the [8] obfuscator has). We remark that the result for KDM encryption should be contrasted with the fact that fully KDM-secure encryption schemes can not be proven secure using *fully black box reductions* to *standard cryptographic game* [19]. Our proof of security does not fit this characterization.

## 1.2   Our Techniques

Proving composability for point obfuscators encounters several difficulties. We sketch these difficulties as well as the ideas and techniques we use to overcome them. We also exhibit how the VGB relaxation comes to our aid.

*Simulation and distributional indistinguishability.* Ideally, we might try to require that for fixed sequence of points, the resulting obfuscated point programs would appear to an efficient adversary as a sequence of obfuscated *random* point programs (similarly to the *semantic security* requirement for encryption schemes). This would allow simple simulation, by running the adversary on obfuscations of random hidden values. However, in the context of obfuscation such a requirement is unachievable, since the adversary is able to run the program and verify any guesses it might have; in particular it can have some hardwired values which it can always recognize. Instead, we consider a weaker requirement which we call *Distributional Indistinguishability* (DI in short). We show that: (a) DI is necessary and sufficient for constructing VGB simulators, and (b) It is achievable under appropriate hardness assumptions.

DI is an extension of a notion used in [8] in the context of plain point obfuscators. The requirement refers to a specific type of distributions over tuples of points which we call *coordinatewise well spread* (CWS in short). $\mathcal{X} = \left\{ (X_n^{(1)} \ldots X_n^{(t)}) \right\}$ is a CWS distribution ensemble on $\{\mathbb{D}_n^t\}$ if for any $a \in \mathbb{D}_n$ and $i \in [t]$, $X_i \neq a$ except with negligible probability.

Essentially, $\mathcal{O}$ is a $t$-DI obfuscator if for any CWS distribution, $\mathcal{X}$, over $t$-tuples of elements in $\mathbb{D}_n$, no *efficient* adversary can distinguish obfuscations of $t$ uniform values from obfuscations of a tuple of values sampled from $\mathcal{X}$. We show:

**Theorem 1.1 (informal).** *If $\mathcal{O}$ is a $t$-DI point obfuscator then it is a $t$-composable VGB point obfuscator. Moreover, if $\mathcal{O}$ is $t$-DI for any polynomial $t$, then it is a composable VGB obfuscator for any polynomial number of point programs.*

The main technical difficulty in this work is in proving Theorem 1.1. We sketch the ideas used in the proof. Our starting point is a result of [8] showing that for point obfuscators (i.e. $t = 1$) the notions of DI and VBB obfuscation are equivalent and that DI obfuscation is achievable under certain number theoretic assumptions.

First we ask whether $t$-DI obfuscators imply $t$-composable VBB obfuscators for $t > 1$. We show that this is the case as long as $t = O(1)$. However, when $t = \omega(1)$, major (and seemingly inherent) difficulties rise. Specifically, recall that when constructing a simulator, we should deal with the fact that the adversary can run the obfuscated programs and might have some hardwired values which it can always recognize. When the adversary has access only to a single obfuscated point program, [8, 26, 10] show that in fact it cannot do much more than have a polynomial number of such hardwired test elements. We call these the *distinguishing elements*. This allows the simulator to check its oracle only on the polynomially many distinguishing elements.

However, in the case of multiple obfuscated points, this plan does not go through. The main difficulty is *adaptivity*. More specifically, while in the case of a single hidden point there is only a single secret, in the case of composable point obfuscators the adversary might first discover only some of the points, and then use this partial information to make his next choices. Fortunately, we can show that for any partial information already learnt there is a corresponding poly set of distinguishing elements. However, there still remains the question of how to compute these elements ahead of time.

We show that the total number of potentially queried elements is $n^{\Theta(t)}$. Here VGB comes to our aid when $t = \omega(1)$. That is, having limited oracle access to the point programs and sufficient power to compute the distinguishing elements allows performing the required simulation.

We remark that a converse statement is also true. That is, DI is necessary for VGB composable obfuscation (and thus also for the stronger VBB notion).

*A t-DI point obfuscator.* Finally, we reconsider the point program obfuscator constructed in [8]. Under a strong Decision Diffie Hellman (DDH) assumption, we show that this obfuscator is $t$-DI for any polynomial $t$ and hence is a $t$-composable VGB obfuscator. As evidence of plausibility, we show that our assumption holds in the *Generic Group Model* [25], where algorithms are only allowed to perform generic group operations and can not exploit the representation of group elements. We note that there exist well studied group ensembles (e.g. Quadratic Residues modulo a prime, and Elliptic Curves groups) where the best cryptanalytic techniques are in fact generic ones [6].

In addition to the above construction, Theorem 1.1 enables construction of composable point obfuscators, based on other hardness assumptions. One natural candidate is the *decisional learning with errors* assumption (LWE) [24] when considered with *weak* (non uniform) secrets. Indeed, under appropriate parameter settings, LWE with weak secrets can be reduced to LWE with uniform secrets [15]. This implies point obfuscators which are secure as long as the secret point is taken from a distribution with some poly-logarithmic entropy.

*Organization.* Section 2 is devoted to the definition and discussion of VGB obfuscation and its relations with the VBB notion and previous relaxations. Section 3 shows how to construct composable VGB obfuscators for point programs. Section 4 discusses the nature and plausibility of our hardness assumption. Section 5 demonstrates the applications of composable point obfuscators to multi-bit point programs, to set programs, and to strong encryption schemes. Most proofs and some of the secondary results appear in the full version [4].

## 2 Definitions

We formalize the notion of *virtual black box obfuscation with strong simulators,* and explore its relation to existing notions. In all following definitions, we consider the task of obfuscating an ensemble $C = \{C_n\}$, where each $C_n$ is a collection of circuits with input length $n$ and poly$(n)$ size.

### 2.1 VBB, IND and BP Obfuscation

We first recall the *virtual black box* definition and two of its previous relaxations.

**Definition 2.1 (*obfuscator* [3] ).** *A PPT $\mathcal{O}$ is a VBB obfuscator for $C$, if it satisfies:*

- *(Functionality) For any $n \in \mathbb{N}$, $C \in C_n$, $\mathcal{O}(C)$ is a circuit which computes the same function as $C$.*
- *(Polynomial Slowdown) There is a polynomial $q$ such that for any $n \in \mathbb{N}$, $C \in C_n$, $|\mathcal{O}(C) \leq q(|C|)$.*
- *(Virtual Black Box)*[1] *For any PPT adversary $\mathcal{A}$ and polynomial $p$ there is a PPT simulator $\mathcal{S}$ such that for all sufficiently large $n \in \mathbb{N}$ and $C \in C_n$:*

$$\left| \Pr_{\mathcal{A}, \mathcal{O}}[\mathcal{A}(\mathcal{O}(C)) = 1] - \Pr_{\mathcal{S}}[\mathcal{S}^C(1^{|C|}) = 1] \right| \leq 1/p(n)$$

**Definition 2.2 (*Indistinguishability Obfuscation* [3]).** $\mathcal{O}$ *is said to be an* indistinguishability obfuscator *(INDO in short) for $C$, if it satisfies the functionality and polynomial slowdown and for any ensemble of circuit pairs $C^{(1)} \times C^{(2)} = \{(C_n^{(1)}, C_n^{(2)}) \in C_n \times C_n\}$, where the two circuits in each pair are of the same size and functionality, it holds that: $\mathcal{O}(C^{(1)}) \approx_c \mathcal{O}(C^{(2)})$.*

Another relaxation of VBB is *Best Possible Obfuscation* (BPO in short) [17]. Here the requirement is that any information which the obfuscation leaks is efficiently learnable from any other circuit with the same functionality and related size (hence it is "best possible"). The two definitions are equivalent when the obfuscator is required to be a PPT [17].

Before presenting our definition we make the following preliminary observation regarding the nature of the above relaxations. The INDO (BPO) definition is

---

[1] As noted by [3] the above can be replaced with the equivalent requirement that $\left| \Pr[\mathcal{A}(\mathcal{O}(C) = \pi(C)] - \Pr[\mathcal{S}^C(1^{|C|}) = \pi(C)] \right| \leq \frac{1}{p(n)}$ for any predicate $\pi : C_n \rightarrow \{0, 1\}$. Also the size of the simulator can depend on $p(n)$, namely the required simulation quality.

in fact equivalent to a weak black-box definition, which allows an unbounded simulator with unlimited number of oracle queries (proof in [4]).

**Proposition 2.1.** $\mathcal{O}$ *is an indistinguishability obfuscator for an ensemble of circuits* $\mathcal{C} = \{\mathcal{C}_n\}$ *iff for any efficient distinguisher* $\mathcal{A}$ *and polynomial* $p$*, there is a (possibly inefficient) simulator* $\mathcal{S}$*, such that for all large enough* $n$ *and* $C \in \mathcal{C}_n$*:*

$$\left| \Pr_{\mathcal{A},\mathcal{O}}[\mathcal{A}(\mathcal{O}(C)) = 1] - \Pr_{\mathcal{S}}[\mathcal{S}^C(1^{|C|}) = 1] \right| \leq 1/p(n)$$

## 2.2 VGB Obfuscation

The new definition relaxes the VBB security requirement by allowing the simulator to have more computational power. However, we still restrict the number of oracle queries it is allowed to make. The *functionality* and *polynomial slowdown* requirements should be satisfied as in Definition 2.1. The VBB requirement is replaced by the following. Denote by $C[q]$ an oracle to the circuit (function) $C$ which allows at most $q$ queries.

**Definition 2.3 (*Virtual Grey Box - obfuscation with a strong simulator*).** *A PPT* $\mathcal{O}$ *has the VGB property if for any PPT adversary* $\mathcal{A}$ *and polynomial* $p$ *there is a (possibly* inefficient*) simulator* $\mathcal{S}$ *and a polynomial* $q$ *such that for all sufficiently large* $n \in \mathbb{N}$ *and any* $C \in \mathcal{C}_n$*:*

$$\left| \Pr_{\mathcal{A},\mathcal{O}}[\mathcal{A}(\mathcal{O}(C)) = 1] - \Pr_{\mathcal{S}}[\mathcal{S}^{C[q(n)]}(1^{|C|}) = 1] \right| \leq 1/p(n)$$

*Remark 2.1.* The definitions above concern obfuscators for circuits. That is, both the input program and the output of the obfuscator are given by circuits. One can naturally adjust these definitions to fit the case of *Turing Machine obfuscators* (both input and output are given by a description of a TM). In this work we shall focus on circuit obfuscators (see[4] for corresponding TM definition).

*When Is VGB Meaningful?* Like INDO and BPO, VGB obfuscation does not seem strong enough for some desirable obfuscation tasks. Examples include: transforming private key encryption schemes to public ones and constructing homomorphic encryption schemes[2]. Informally, the problem in these scenarios is that the obfuscated program computes some kind of cryptographic functionality, which does not remain secure in the presence of unbounded simulators. In general, it seems that VGB is mostly meaningful for program classes which are *unlearnable* with only poly many queries even for learners with unbounded computation time. For families of programs that are not *efficiently learnable*, but are *learnable* for unbounded algorithms with only polynomially many oracle queries, VGB might not guarantee the required security .

---

[2] See the section on applications in [3] for more details.

## 2.3   VGB vs. VBB and INDO

*VGB is strictly weaker than VBB.* The VGB definition is clearly implied by the VBB definition. We show that in fact it is strictly weaker. That is, we show a family which can not be obfuscated according to the VBB definition but is (trivially) obfuscatable under the weaker VGB definition. To do so, we use a slight variation of the family constructed in the [3] impossibility result.

**Proposition 2.2.** *Assuming the existence of one-way permutations, there exist a family of programs which is not VBB obfuscatable but is VGB obfuscatable.*

To prove the above we use the notion of TM obfuscation (in contrast to circuit obfuscation used in most of this work). The corresponding definitions and proof are given in in [4].

*VGB implies INDO (BPO).* The relation between VGB obfuscation and the INDO (BPO) follows from Proposition 2.1. That is, even when VGB is further weakened by allowing the (unbounded) simulator unlimited oracle access, it still implies INDO and (for *efficient* obfuscators) BPO.

## 2.4   Impossibility Results

We consider the possibility of *"universal VGB obfuscation"*. That is, could there exist a VGB obfuscator for the class of all programs? We observe that for TM's obfuscators the impossibility result of [3] extends and also applies for VGB obfuscation. However, for circuits obfuscators the [3] separation no longer holds. Essentially, the reason for this difference is that the *VBB unobfuscatable circuit family* constructed by [3] include cryptographic functionalities (such as *encryption schemes* and *pseudo random functions*) which fail to remain secure in the presence of unbounded simulators (even with limited oracle access). We could not rule out universal VGB obfuscation in the circuit case.

We note that [17] show impossibility of universal BPO obfuscation for circuits which are allowed to use *random oracle* gates. Their result also applies for the stronger VGB notion; however, the meaning of an impossibility result in such setup is somewhat less clear.

## 3   Composable Point Obfuscators

In this section we define and construct composable VGB point obfuscators. In next sections we show that such obfuscators suffice for meaningful applications.

### 3.1   Composition of Obfuscators

One central question in the context of obfuscation is the question of *composition*, which asks when and whether is it secure to obfuscate a sequence of programs by obfuscating each program on its own and combining the obfuscated programs. There are several forms of *composition* one could consider, in this work we consider one specific form, namely *composition by concatenation* [23].

**Definition 3.1** (*t-composable obfuscation* [23]). *A PPT $\mathcal{O}$ is a t*-composable obfuscator *for a circuit ensemble $\mathcal{C} = \{\mathcal{C}_n\}$ if it satisfies the* functionality *and* poly slow-down *as in Definition 2.1 and for any PPT binary adversary $\mathcal{A}$ and polynomial p, there is a simulator $\mathcal{S}$, such that for any sequence of circuits $C^1, \ldots, C^t \in \mathcal{C}_n$ (where $t = \mathrm{poly}(n)$) and any sufficiently large n:*

$$\left| \Pr[\mathcal{A}(\mathcal{O}(C^1), \ldots, \mathcal{O}(C^t)) = 1] - \Pr[\mathcal{S}^{C^1, \ldots, C^t}(1^{|C^1|}, \ldots, 1^{|C^t|}) = 1] \right| \leq 1/p(n)$$

*Where $C^1, \ldots, C^t$ gets as input $(x, i)$ and returns $C^i(x)$.*

Originally [23] naturally refer to VBB obfuscation, i.e. the simulator $\mathcal{S}$ is a polynomially bounded. We consider the definition also for VGB obfuscators, i.e. we allow the simulator to be unbounded with poly many oracle queries.

## 3.2    Point Obfuscators

*Point circuits.* For a security parameter $n \in \mathbb{N}$ and a domain $\mathbb{D}_n$, a point circuit $C_x : \mathbb{D}_n \to \{0, 1\}$ returns 1 on input $x$ and 0 on all other inputs. The point circuits we discuss are given in some "canonical" form where the point $x$ is explicit. As the size of the canonical circuits is determined by the parameter $n$, we simplify our notation by letting the simulator take input $1^n$ (instead of the circuit size). The natural choice for the domain is $\mathbb{D}_n = \{0, 1\}^n$. However, to avoid confusion when discussing tuples of points in $\mathbb{D}_n^t$, we shall stick to the more general notation. We refer to obfuscators for point circuits as **point obfuscators**.

*Is any point obfuscator composable?* Point obfuscators have been constructed, both in the plain model and in the random oracle model. A natural question is whether any VBB secure point obfuscator is also guaranteed to be composable (as in Definition 3.1). [23] conjectured that the answer is negative. To support their conjecture they give a point obfuscator in the *Random Oracle model* which is not even 2-composable. In the standard model, it can be shown that if point obfuscators exist, then there are also point obfuscators which are not $\Omega(n)$-composable [9]. In general, none of the constructions of point obfuscators were known to be composable.

## 3.3    Distributional Indistinguishability and Composable Point Obfuscation

To overcome the difficulties in achieving composable point obfuscators, we explore in this section an additional property of point obfuscators, called *Distributional Indistinguishability (or DI in short)*[3]. We will show that this additional property is necessary for composable obfuscation even under the weaker VGB notion. More importantly, we will show that in fact it suffices for VGB obfuscation. The definition we present generalizes the DI definition presented in [8].

---

[3] DI should not be confused with *Indistinguishability Obfuscators* of [3] which were presented in Definition 2.2

**Definition 3.2 (*Coordinatewise Well Spread*).** *Let* $\mathcal{X} = \{X_n\}$ *be an ensemble, where each* $X_n$ *is a distribution on* $\mathbb{D}_n^{t(n)}$ *for a domain ensemble* $\{\mathbb{D}_n\}$. *We say that* $\mathcal{X}$ *is CWS if:* $\max\limits_{a \in \mathbb{D}_n} \Pr_{\bar{x} \leftarrow X_n}[\exists i \in [t] : x_i = a] = n^{-\omega(1)}$.

That is any element has only a negligible chance of being picked within a vector sampled from the distribution. Equivalently, in a CWS ensemble the distributions $X_n^{(i)}$ all have super-log *min-entropy*, i.e. $\min_{i \in [t]} H_\infty(X_n^{(i)}) = \omega(\log n)$.

**Definition 3.3 (*Distributional Indistinguishability*).** $\mathcal{O}$ *is t-DI if for any CWS distribution ensemble,* $\mathcal{X} = \{X_n = \langle X_n^{(1)}, \ldots, X_n^{(t)} \rangle\}$, *it holds that:*

$$\mathcal{O}(C_{\mathcal{X}^{(1)}}), \ldots, \mathcal{O}(C_{\mathcal{X}^{(t)}}) \approx_c \mathcal{O}(C_{\mathcal{U}^{(1)}}), \ldots, \mathcal{O}(C_{\mathcal{U}^{(t)}})$$

*Where each* $\mathcal{O}(C_{\mathcal{X}^{(i)}})$ *is an ensemble of distributions on point obfuscations, and the hidden point is drawn from* $\mathcal{X}^{(i)}$ *and* $\mathcal{U}^{(1)}, \ldots, \mathcal{U}^{(t)}$ *are ensembles of independent uniform distributions over* $\{\mathbb{D}_n\}$.

We note that for the case $t = 1$ Definition 3.3 is equivalent to the DI definition in [8], where it is shown that for $t = 1$, DI and VBB are in fact equivalent. The proof there does not follow through for larger $t$. Nevertheless, we show:

**Theorem 3.1.** *Any t-DI point obfuscator is a t-composable VGB obfuscator. Moreover, for* $t = O(1)$ *it is VBB composable. Conversely, any t-composable VGB point obfuscator is t-DI.*

The proof of the second part of Theorem 3.1 (namely the necessity of DI for composable VGB obfuscation) is rather simple and brought in [4]. We focus on the first part of the theorem that is more involved. A high-level discussion of the proof, techniques and main ideas is given in the introduction. The proof itself is divided to several lemmas. We start with preliminary notations.

*Notations.* Given a vector of $t$ points $\bar{x} = \langle x_1, \ldots, x_t \rangle$ we abuse notation and denote by $C_{\bar{x}}$ the vector of point circuits $\langle C_{x_1}, \ldots, C_{x_t} \rangle$. We also denote by $\mathcal{O}(C_{\bar{x}})$ the composition $\mathcal{O}(C_{x_1}), \ldots, \mathcal{O}(C_{x_t})$. Speaking of vectors, we shall often be interested in the (unordered) set of their elements. Whenever we use set operators such as $\in, \cap, \cup$ on vectors, it should be interpreted as operating on the corresponding sets. For integers $s \leq t$ we denote by $\binom{[t]}{s}$ the family of subsets of $[t]$ of size $s$. For vectors $\bar{x}, \bar{z}$ of dimensions $s$ and $t - s$, and a set of indices $I \subseteq [t]$ of size $|I| = s$, we denote by $\text{CMB}_I(\bar{x}, \bar{z})$ the $t$-vector with the elements of $\bar{x}$ in coordinates $I$ and those of $\bar{z}$ in coordinates $[t] - I$ (the mapping is according to ascending order of indices)[4].

As explained in the introduction, we show that for any partial information the adversary learns, there is a relevant polynomial set of *distinguishing elements*. The first lemma deals with the case that no partial information is learnt, and can be viewed as a generalization of a similar claim in [8] to multiple points.

---

[4] For example $\text{CMB}_{\{2,5\}}((a,b),(c,d,e)) = (c,a,d,e,b)$.

**Lemma 3.1.** *Assume $\mathcal{O}$ is t-DI, then for any binary PPT $\mathcal{A}$ and $p = \text{poly}(n)$ there is a poly-size family $\mathcal{L} = \{L_n \subseteq \mathbb{D}_n\}$ such that any vector $\bar{x} \in \mathbb{D}_n^t$ which does not intersect $L_n$ (i.e. $\bar{x} \subseteq \mathbb{D}_n \setminus L_n$) satisfies:*

$$\left| \Pr_{\mathcal{A},\mathcal{O}}[\mathcal{A}(\mathcal{O}(C_{\bar{x}})) = 1] - \Pr_{\mathcal{A},\mathcal{O},\bar{u} \overset{U}{\leftarrow} \mathbb{D}_n^t}[\mathcal{A}(\mathcal{O}(C_{\bar{u}})) = 1] \right| \leq 1/p(n) \qquad (1)$$

*Proof.* Consider a binary PPT $\mathcal{A}$ and a polynomial $p$. We describe the corresponding family $\mathcal{L}$. Let $X_n$ be the set of all vectors which do not satisfy Equation (1). That is, $X_n = X_n^+ \cup X_n^-$, where:

$$X_n^+ = \left\{ \bar{x} \in \mathbb{D}_n^t : \Pr[\mathcal{A}(\mathcal{O}(C_{\bar{x}})) = 1] - \Pr[\mathcal{A}(\mathcal{O}(C_{\bar{u}})) = 1] \geq 1/p(n) \right\}$$
$$X_n^- = \left\{ \bar{x} \in \mathbb{D}_n^t : \Pr[\mathcal{A}(\mathcal{O}(C_{\bar{u}})) = 1] - \Pr[\mathcal{A}(\mathcal{O}(C_{\bar{x}})) = 1] \geq 1/p(n) \right\}$$

First reduce $X_n^+$ to a subset of vectors $Y_n^+ \subseteq X_n^+$ in which any element $a \in \mathbb{D}_n$ which appears in some vector $\bar{x} \in X_n^+$ appears in exactly one vector $\bar{y} \in Y_n^+$. Similarly reduce $X_n^-$ to $Y_n^-$. Let $Y_n = Y_n^+ \cup Y_n^-$ and define $L_n = \bigcup_{\bar{y} \in Y_n} \bar{y} = \{a \in \mathbb{D}_n : \exists \bar{y} \in Y_n, a \in \bar{y}\}$. By the construction of $L_n$, any $\bar{x} \subseteq \mathbb{D}_n \setminus L_n$ satisfies Equation (1). It remains to show that $|L_n| = \text{poly}(n)$. As $|L_n| \leq t|Y_n|$, it suffices to show that $|Y_n| = \text{poly}(n)$. Assume towards contradiction that the latter does not hold. We shall construct a CWS distribution ensemble $\mathcal{Z} = \{Z_n\}$ over $\mathbb{D}_n^t$, such that $\mathcal{A}$ distinguishes $\mathcal{O}(C_{\mathcal{Z}})$ from $\mathcal{O}(C_{\mathcal{U}(\mathcal{D}^t)})$ with advantage $1/p$ contradicting the DI property. By the assumption on the size of $|L_n|$ there exist a function $\ell(n) = n^{\omega(1)}$ such that for infinitely many $n$'s either $|Y_n^+| \geq \ell(n)$ or $|Y_n^-| \geq \ell(n)$. We assume WLOG the first case holds (the proof is similar for the second). For any $n \in \mathbb{N}$ such that $|L_n| \geq \ell(n)$, set $Z_n$ to be uniform on the set $Y_n^+$. For other $n$ let $Z_n$ be uniform on some arbitrary set of size $\ell(n)$ in which any element appears in at most one vector (we can take $\ell = o(|\mathbb{D}_n|)$ to assure such a choice is possible). The resulting ensemble $\mathcal{Z}$ is CWS since any single vector is drawn with probability at most $1/\ell$, and any single element appears in at most one vector. Moreover, for any $n$ such that $Z_n \triangleq U(Y_n^+)$, it holds that:

$$\Pr_{\bar{z} \leftarrow Z_n}[\mathcal{A}(\mathcal{O}(C_{\bar{z}})) = 1] - \Pr_{\bar{u} \leftarrow U(\mathbb{D}_n^t)}[\mathcal{A}(\mathcal{O}(C_{\bar{u}})) = 1] \geq$$
$$\min_{\bar{y} \in Y_n^+} \Pr[\mathcal{A}(\mathcal{O}(C_{\bar{y}})) = 1] - \Pr_{\bar{u} \leftarrow U(\mathbb{D}_n^t)}[\mathcal{A}(\mathcal{O}(C_{\bar{u}})) = 1] \geq 1/p(n) \qquad \square$$

While in [8] the above lemma suffices for constructing a simulator, in our setup it does not, since it does not cover the possibility that the adversary successfully learns only some of the points. The next lemma shows that for any partial information learnt by the adversary there is still a corresponding polynomial *distinguishing set*.

**Lemma 3.2.** *Assume $\mathcal{O}$ is t-DI. Let $s = s(n)$ be any length function such that $s \leq t$ and let $\mathcal{T} = \left\{ (\bar{x}_n, I_n) \in \mathbb{D}_n^s \times \binom{[t]}{s} \right\}_{n \in \mathbb{N}}$ be a family of vectors and index*

$sets^5$. Then for any binary PPT $\mathcal{A}$ and $p = \mathrm{poly}(n)$ there exists a poly-size family $\mathcal{L}^{\mathcal{T}} = \{L_n\}$ such that for any $\bar{y} \in \mathbb{D}_n^{t-s}$ that does not intersect $L_n$:

$$|\Pr[\mathcal{A}(\mathrm{CMB}_{I_n}(\mathcal{O}(C_{\bar{x}_n}), \mathcal{O}(C_{\bar{y}})) = 1] - \Pr[\mathcal{A}(\mathrm{CMB}_{I_n}(\mathcal{O}(C_{\bar{x}_n}), \mathcal{O}(C_{\bar{u}}))) = 1]| \le \frac{1}{p(n)}$$

Where $\bar{u} \xleftarrow{U} \mathbb{D}_n^{t-s}$ and the probabilities are over the coins of $\mathcal{A}, \mathcal{O}$ and $\bar{u}$.

To prove the lemma, we shall need the following intuitively correct claim.

*Claim.* If $\mathcal{O}$ is $t$-DI then it is also $s$-DI for any $s \le t$. (proof in [4]).

*Proof (of Lemma 3.2).* Consider the function $r = t - s$, then by Claim 3.3 $\mathcal{O}$ is $r$-DI. Consider an adversary $\mathcal{A}'$ (for $r$-compositions) which has $\mathcal{T}$ hardwired, and on input $\bar{w}$ (here $\bar{w} = \mathcal{O}(C_{\bar{y}})$ for some $y_1 \ldots y_r$), runs $\mathcal{A}$ on the valid obfuscation $\mathrm{CMB}_{I_n}(\mathcal{O}(C_{\bar{x}_n}), \mathcal{O}(C_{\bar{y}}))$ . By Lemma 3.2 this $\mathcal{A}'$ has a family $\mathcal{L}^{\mathcal{T}}$ which satisfies the required property with respect to the original adversary $\mathcal{A}$.    □

The next lemma shows there is a uniform polynomial bound on the size of all *distinguishing sets* (corresponding to any partial information), and hence there exists a *distinguishing function family*, which given any partial information outputs a poly-size set of all *distinguishing elements* (with respect to this information).

**Lemma 3.3.** Let $\mathcal{O}$ be a $t$-DI obfuscator. Then for any binary PPT $\mathcal{A}$ and $p = \mathrm{poly}(n)$, there exists a family of functions $\mathcal{F} = \{F_n\}$ and a $q = \mathrm{poly}(n)$ such that $F_n : \bigcup_{s \le t} \left( \mathbb{D}_n^s \times \binom{[t]}{s} \right) \longrightarrow \bigcup_{s \le q} \binom{\mathbb{D}_n}{s}$ and satisfies for any $(\bar{x}, I) \in \mathbb{D}_n^{|I|} \times \binom{[t]}{|I|}$ and any $\bar{y} \in \mathbb{D}_n^{t-|I|}$ which does not intersect the set $F_n(\bar{x}, I)$:

$$|\Pr[\mathcal{A}(\mathrm{CMB}_{I_n}(\mathcal{O}(C_{\bar{x}}), \mathcal{O}(C_{\bar{y}}))) = 1] - \Pr[\mathcal{A}(\mathrm{CMB}_{I_n}(\mathcal{O}(C_{\bar{x}}), \mathcal{O}(C_{\bar{u}}))) = 1]| \le \frac{1}{p(n)}$$

Where $\bar{u} \xleftarrow{U} \mathbb{D}_n^{t-|I|}$ and the probabilities are over the coins of $\mathcal{A}, \mathcal{O}$ and $\bar{u}$.

*Remark 3.1.* The function $F_n$ is defined for any "partial information". In particular the set of indices $I$ is allowed to be the empty set corresponding to no partial information as in Lemma 3.1.

*Proof.* For any $(\bar{x}, I) \in \mathbb{D}_n^{|I|} \times \binom{[t]}{|I|}$, let $F_n(\bar{x}, I) \subseteq \mathbb{D}_n$ be the minimal set which satisfies the above condition (note that such a set always exists as $\mathbb{D}_n$ trivially satisfies the requirement). We show that, there exists a $q = \mathrm{poly}(n)$, such that $|F_n| \le q(n)$ (i.e. $q$ is a uniform bound on all images). Let $(\bar{x}_n^*, I_n^*)$ be the pair which maximizes $F_n(\bar{x}, I)$, i.e. $|F_n(\bar{x}_n^*, I_n^*)| = \max_{I \subseteq [t], \bar{x} \in \mathbb{D}_n^{|I|}} |F_n(\bar{x}, I)|$. By Lemma 3.2 there exists a $q = \mathrm{poly}(n)$ for which $|F_n(\bar{x}_n^*, I_n^*)| \le q(n)$ (just by considering the family $\{(\bar{x}_n^*, I_n^*)\}_{n \in \mathbb{N}}$). The result follows.    □

---

[5] Any pair $(\bar{x}, I)$ should be thought of as partial information on a tuple of size $t$ with the elements of $\bar{x}$ in the indices $I$.

To complete the proof of the theorem, we construct a simulator using the family of *distinguishing functions* $\mathcal{F}$. However, as it might not be computable by a poly-size simulator, the result holds only for strong simulators as in the VGB definition.

*Proof (Any t-DI point obfuscator is also VGB t-composable (sketch)).* Let $\mathcal{A}$ be a binary PPT adversary and $p$ a polynomial. Let $\mathcal{F}$ be the corresponding family of functions given by Lemma 3.3 and let $q$ be the polynomial bound on the images of $\mathcal{F}$. We construct an unbounded simulator $\mathcal{S}$ which performs at most $q \cdot t$ oracle queries. Given oracle access to a tuple of circuits $C_{\bar{x}} = C_{x_1}, \ldots, C_{x_t}$, for some $\bar{x} \in \mathbb{D}_n^t$. $\mathcal{S}$ first runs $F_n$ (on the empty set), retrieves a set $L^{(0)}$ of all *distinguishing elements* with respect to no partial information, and queries its oracle on all the elements in $L^{(0)}$. In case it did not reveal any elements (i.e. $\bar{x} \cap L^{(0)} = \emptyset$), it chooses a uniform vector $\bar{u} \xleftarrow{U} \mathbb{D}_n^t$, computes obfuscations of the points in $\bar{u}$ and runs $\mathcal{A}$ on their composition. Otherwise, it revealed some elements given by a pair $(\bar{z}^{(0)}, I^{(0)})$. It then computes $L^{(1)} = F_n(\bar{z}^{(0)}, I^{(0)})$, and as in the first step, queries all the values in $L^{(1)}$. In case it did not reveal any new values, it chooses a uniform vector $\bar{u} \xleftarrow{U} \mathbb{D}_n^{t-|I^{(0)}|}$ and runs $\mathcal{A}$ on an obfuscation $\mathrm{CMB}_{I^{(0)}}(\mathcal{O}(C_{\bar{z}^{(0)}}), \mathcal{O}(C_{\bar{u}}))$. Otherwise it has updated partial information given by a pair $(\bar{z}^{(1)}, I^{(1)})$. It continues on in this manner. If at any point it revealed all the points in $\bar{x}$ it just runs $\mathcal{A}$ on a random composed obfuscation of the points in $\bar{x}$ performing a perfect simulation. Otherwise, it stops after at most $t$ iterations, performing a simulation of $1/p$ accuracy. This completes the main part of the proof of Theorem 3.1.    $\square$

A more careful analysis shows that we can somewhat "compress" the distinguishing function $\mathcal{F}$ to a set of distinguishing elements. This yields the following.

**Proposition 3.1.** *If $\mathcal{O}$ is a t-DI obfuscator, then any binary adversary given a sequence of $t$ obfuscations can be simulated by a simulator of size $n^{O(t)}$ and polynomially many queries. In particular, for $t = O(1)$ this yields a polynomially bounded simulator (VBB). (proof in [4])*

*On the possibility of bounded simulation (VBB).* We note that our result does not rule out the possibility of bounded simulation for any $t = \mathrm{poly}(n)$. It might be that there always exists a function family $\mathcal{F}$ such as the one required in Theorem 3.1 which is also efficiently computable, or even a "compressed" poly set of distinguishing elements as in Proposition 3.1. Alternatively, there might be other techniques which allow efficient simulation. In this context, we show an example of an adversary whose distinguishing function can not be compressed to a poly set. We also show that if bounded simulation exists then so does an efficiently computable function family $\mathcal{F}$ (i.e. simulation can be proven using the same technique we use above). The details are given in [4].

*Remark 3.2.* We note that the applications in Section 5.2 can be shown to hold using the DI obfuscation definition, the equivalence given by Theorem 3.1 allows considering a "simulation" definition that holds for *any input*, and provides a security guarantee even with keys (hidden points) from an arbitrary distribution.

## 3.4  A Composable Point Obfuscator

After establishing the proper framework in the previous, this section is devoted to a concrete construction for composable VGB point obfuscators. We consider the point obfuscator constructed in [8] and analyze its security under composition.

**Construction 3.2 (*The* $r, r^x$ *Point Obfuscator* [8]).** *Let $\mathcal{G} = \{\mathbb{G}_n\}_{n \in \mathbb{N}}$ be a group ensemble, where each $\mathbb{G}_n$ is a group of prime order $p_n \in (2^{n-1}, 2^n)$. We define an obfuscator, $\mathcal{O}$, for points in the domain $\mathbb{Z}_{p_n}^*$ as follows: $C_x \xmapsto{\mathcal{O}} \mathcal{C}(r, r^x)$ Where $r \xleftarrow{U} \mathbb{G}_n^*$ is a random generator of $\mathbb{G}_n$, and $\mathcal{C}(r, r^x)$ is a circuit which on input $z$, checks whether $r^x = r^z$.*

In [8] Construction 3.2 is shown to be secure under a strong variant of the Decision Diffie-Hellman assumption. We now present our assumption which is a generalization of the [8] assumption to tuples of points.

**Assumption 3.3 (*$t$-Strong Vector Decision Diffie Hellman*).** *Let $t = \text{poly}(n)$. There exist group ensemble $\mathcal{G} = \{\mathbb{G}_n : |\mathbb{G}_n| = p_n \text{ is prime}\}$ with efficient representation and operations, such that for any CWS distribution ensemble $\mathcal{X} = \{X_n\}$ over vectors in $(\mathbb{Z}_{p_n}^*)^t$ the following holds:*

$$\left\{ \begin{matrix} g_1, g_1^{a_1} \\ \vdots \\ g_t, g_t^{a_t} \end{matrix} \quad : \quad \begin{matrix} \bar{g} \xleftarrow{U} (\mathbb{G}_n^*)^t \\ \bar{a} \xleftarrow{X_n} (\mathbb{Z}_{p_n}^*)^t \end{matrix} \right\}_{n \in \mathbb{N}} \approx_c \left\{ \begin{matrix} g_1, g_1^{u_1} \\ \vdots \\ g_t, g_t^{u_t} \end{matrix} \quad : \quad \begin{matrix} \bar{g} \xleftarrow{U} (\mathbb{G}_n^*)^t \\ \bar{u} \xleftarrow{U} (\mathbb{Z}_{p_n}^*)^t \end{matrix} \right\}_{n \in \mathbb{N}}$$

We observe that Assumption 3.3 implies that the $r, r^x$ point obfuscator is $t$-DI with respect to the corresponding group ensemble $\mathcal{G}$, given by the construction. Hence, Theorem 3.1 yields:

**Theorem 3.4.** *Under Assumption 3.3, the $r, r^x$ point obfuscator is a $t$-composable VGB point obfuscator (w.r.t the group ensemble $\mathcal{G}$ given by the assumption). Assuming the existence of a "universal" group ensemble which satisfy Assumption 3.3 for any $t = \text{poly}(n)$ implies fully composable VGB obfuscators (i.e. $t$-composable for any $t = \text{poly}(n)$).*

# 4  On the Assumption

In this section we discuss Assumption 3.3 and its relation to previous Decision Diffie Hellman variants. We also show that it holds in the *Generic Group Model*.

*Relation to Previous DDH Assumptions.* We start by presenting another strong variant of DDH for tuples of points, which is in a sense a natural generalization to the standard and strong DDH assumptions [6, 8].

**Assumption 4.1 (*$t$-Strong Vector Decision Diffie Hellman II*).** *Let $t = \text{poly}(n)$. There exist group ensemble $\mathcal{G} = \{\mathbb{G}_n : |\mathbb{G}_n| = p_n \text{ is prime}\}$ with* *efficient representation and operations, such that for any CWS distribution ensemble* $\mathcal{X} = \{X_n\}$ *over vectors in* $(\mathbb{Z}_{p_n}^*)^t$ *the following holds:*

$$
\left\{
\begin{array}{ll}
g_1, g_1^{a_1}, g_1^{b_1}, g_1^{c_1} & \bar{g} \xleftarrow{U} (\mathbb{G}_n^*)^t \\
\quad\vdots & : \ \bar{a} \xleftarrow{X_n} (\mathbb{Z}_{p_n}^*)^t \\
g_t, g_t^{a_t}, g_t^{b_t}, g_t^{c_t} & \bar{b}, \bar{c} \xleftarrow{U} (\mathbb{Z}_{p_n}^*)^t
\end{array}
\right\}_{n \in \mathbb{N}}
\approx_c
\left\{
\begin{array}{ll}
g_1, g_1^{a_1}, g_1^{b_1}, g_1^{a_1 b_1} & \bar{g} \xleftarrow{U} (\mathbb{G}_n^*)^t \\
\quad\vdots & : \ \bar{a} \xleftarrow{X_n} (\mathbb{Z}_{p_n}^*)^t \\
g_t, g_t^{a_t}, g_t^{b_t}, g_t^{a_t b_t} & \bar{b} \xleftarrow{U} (\mathbb{Z}_{p_n}^*)^t
\end{array}
\right\}_{n \in \mathbb{N}}
$$

Restricting the assumption to $t = 1$ results in the strong DDH (SDDH) assumption in [8]. If in addition we restrict $\mathcal{X}$ to be the uniform distribution ensemble, we get the standard DDH assumption. Assumption 4.1 appears as a more familiar and natural generalization of SDDH and DDH than Assumption 3.3 does. However, 3.3 is somewhat simpler and is clearly weaker (the distributions induced by the last two elements of each foursome in 4.1 are identical to those in 3.3). It turns out that the assumptions are in fact equivalent (proof in [4]).

A natural question is whether assumptions 3.3 and 4.1 for $t = 1$ imply the corresponding assumptions for general polynomial $t$ (or even just larger constant $t$). For the case that the distribution ensemble $\mathcal{X}$ is the uniform distribution this is true (corresponds to showing DDH for any poly number of foursomes from DDH for a single foursome by an hybrid argument). However, when allowing any CWS distribution, such an argument fails to work for two main reasons: (a) dependence among coordinates. (b) the distribution ensemble might not even be efficiently samplable. In general we do not know whether SDDH implies SVDDH.

*SVDDH Holds in the Generic Group Model.* We show that Assumption 3.3 holds for any $t = \text{poly}(n)$ in the *generic group model* [25] where algorithms can not exploit the representation of the group elements, other than the fact that each element has a unique representation (formal model description and proof in [4]).

## 5   Applications

In this section, we show how composable VGB point obfuscators, can be used to construct composable VGB obfuscators for MBPC's. Then we discuss how these can be used to obtain strong encryption schemes that are simultaneously resilient to *key dependent messages* (KDM), *leakage* and *related key attacks* (RKA).

### 5.1   Obfuscation of Point Circuits with Multi-bit Output

A *multibit point circuit* (or MBPC in short), $C_{x \to y} : \mathbb{D}_n \to \{0, 1\}^m$, returns $y$ on input $x$ and $\perp$ on all other inputs (once again we assume $C_{x \to y}$ is given in some *canonical* form where $x, y$ are explicit). MBPC obfuscators were constructed by [9] assuming the existence of a composable VBB point obfuscators. However, as explained earlier no known obfuscator has been shown to be composable. We show that applying the [9] construction to composable VGB point obfuscators results in a strong VBB MBPC obfuscator which is also VGB composable. We remark that existing MBPO's were only shown to be secure for the restricted case that the message $m$ is independent of the key $k$ [9, 10]. Moreover, they were not shown to be composable. Both properties are essential for the encryption schemes discussed in the next subsection, in order to get resilience to *key-dependent-messages and related key attacks*.

**Construction 5.1 (*Multibit-bit Output Point Obfuscator* [9]).** *Let $\mathcal{O}$ be a point obfuscator. Define a PPT $\mathcal{O}^{(m)}$ for point circuits with $m$-bit output as follows. For a point $x \in \mathbb{D}_n$ and output $y = y_1 y_2 \ldots y_m \in \{0,1\}^m$, choose a random $s \in \mathbb{D}_n - \{x\}$ and define $\bar{a} = \langle a_0, a_1, \ldots, a_m \rangle$ as follows. $a_0 = x$, and for any $i \in [m]$ $a_i = x$ if $y_i = 1$ and $a_i = s$ otherwise. The output of the obfuscator is: $\mathcal{O}^{(m)}(C_{x \to y}) = \mathcal{C}(\mathcal{O}(C_{a_0}), \ldots, \mathcal{O}(C_{a_m}))$. Where $\mathcal{C}$ is a circuit which performs as follows. On input $z$, it first checks whether $a_0 = z$ (using the first point circuit). If it does not, it returns $\bot$. Otherwise, it finds all other coordinates such that $a_i = z$ and outputs $y_1 \ldots y_m$, where $y_i = 1$ if $a_i = z$ and 0 otherwise.*

**Proposition 5.1.** *if $\mathcal{O}$ is an $(m+1)$-composable VGB point obfuscator then $\mathcal{O}^{(m)}$ (given by Construction 5.1) is a VBB obfuscator for $m$-bit point circuits. Moreover, for any decomposition $m + 1 = t \times (m' + 1)$ $\mathcal{O}^{(m')}$ is a VGB $t$-composable MBPC obfuscator (proof in [4]).*

### 5.2 Strong Encryption Schemes

As noted in [9], obfuscation of MBPC's implies a very strong type of symmetric encryption (which they call a *digital locker*). This usage was further explored lately by [10] who showed tight relations between MBPC (VBB) obfuscation and the notions of *weak key encryption* and *key dependent messages encryption*. Informally, they show that the existence of MBPC VBB obfuscators imply the existence of strong symmetric encryption schemes which are secure for key dependent messages even with weak random keys. We extend their results by showing that using composable VGB MBPC obfuscators (as the ones described above), similar implications still hold, even for the scenario of multiple messages and keys which are correlated (KDM, RKA). We note that the implications of composable MBPC obfuscation to RKA encryption was not discussed prior to this work. We start by presenting the basic natural transformation between MBPC obfuscators and symmetric encryption schemes.

**Construction 5.2 (MBPCO to Symmetric Encryption).** *Let $\mathcal{O}$ be an MBPC obfuscator, define (probabilistic) encryption and decryption algorithms: $E_k^{\mathcal{O}}(m) \triangleq \mathcal{O}(C_{k \to m})$ and $D_k^{\mathcal{O}}(C) = C(k)$, where $C$ is interpreted as an MBPC and $k$ is a key taken from a domain of keys $\mathbb{D}_n$ (key sampling is addressed below).*

There are several definitions regarding KDM, RKA and leakage [5, 20, 7, 2]. We use a variant of the definition in [10] extended to the setup of multiple related keys. In this definition, $t$ keys are generated from a distribution $\mathcal{X} = \{X_n\}$ on key vectors in $\mathbb{D}_n^t$ and the adversary witnesses $t$ encryptions of predetermined functions of the keys. Any message might depend on any key, and the keys themselves might also be dependent, according to the joint distribution $X_n$. The definition considers the case where the distributions $X_n$ are not necessarily uniform but only have certain entropy guarantee.

**Definition 5.1 (encryption with multi keys-messages dependence).** *An encryption scheme $(E, D)$ is $(m, t)$-MKM secure if for any CWS distribution ensemble $\mathcal{X} = \{X_n\}$ on key vectors in $\mathbb{D}_n^t$, any PPT $\mathcal{A}$, and (predetermined) functions $f_1, \ldots, f_t : \mathbb{D}_n^t \to \{0, 1\}^m$ and all large enough $n$, the following difference is negligible.*

$$\left| \Pr_{\substack{\bar{k} \leftarrow X_n \\ E, \mathcal{A}}} [\mathcal{A}(E_{k_1}(f_1(\bar{k})), \ldots, E_{k_t}(f_t(\bar{k}))) = 1] - \Pr_{\substack{\bar{k} \xleftarrow{U} \mathbb{D}_n^t \\ E, \mathcal{A}}} [\mathcal{A}(E_{k_1}(\bar{0}), \ldots, E_{k_t}(\bar{0})) = 1] \right|$$

*Where $m(n), t(n)$ are polynomially bounded length functions and $\bar{0} = 0^m$.*

**Theorem 5.3.** *Let $\mathcal{O}$ be a $t$-composable VGB obfuscator for $m$-bit point circuits, then the encryption scheme $(E^{\mathcal{O}}, D^{\mathcal{O}})$ is $(m, t)$-MKM secure (proof in [4]).*

*Extension to asymmetric encryption.* In case the underlying point obfuscator used in Constructions 5.1, 5.2 can be re-randomized, we can in fact get a CPA-secure public key encryption scheme[6] with essentially the same strong properties described above. In particular, one can consider a CPA adaptive definition instead of the one given above. We note that the point obfuscator given by Construction 3.2 is indeed re-randomizable as required.

*Other extensions and remarks.* We note that the RKA resilience described above does not deal in general with adversaries which adaptively choose the key dependence. However, considering the instantiation of the scheme with the Construction 3.2 obfuscator, one gets RKA security for the family of affine functions of the key even against adaptive adversaries (this follows simply because the construction allows affine homomorphisms of the key). Another remark is that the KDM resilience the scheme is also restricted to a non-adaptive model in which the adversary has to choose in advance the functions of the key which it is interested in, this can be equivalently formulated as an adaptive definition where the family of correlation functions is polynomially bounded, nevertheless this is a meaningful setting which captures common KDM resilience such as the classical *circular dependence*.

# References

[1] Adida, B., Wikström, D.: How to shuffle in public. In: Vadhan, S.P. (ed.) TCC 2007. LNCS, vol. 4392, pp. 555–574. Springer, Heidelberg (2007)

[2] Applebaum, B., Harnik, D., Ishai, Y.: Semantic security under related-key attacks and applications (2010) (manusript)

[3] Barak, B., Goldreich, O., Impagliazzo, R., Rudich, S., Sahai, A., Vadhan, S.P., Yang, K.: On the (im)possibility of obfuscating programs. In: Kilian, J. (ed.) CRYPTO 2001. LNCS, vol. 2139, pp. 1–18. Springer, Heidelberg (2001)

[4] Bitansky, N., Canetti, R.: On strong simulation and composable point obfuscation (2010), http://eprint.iacr.org

---

[6] Given a secret key $k \in \mathbb{D}_n$ the public key is an obfuscation $\mathcal{O}(C_k)$ and encryption is done as in constructions 5.1, 5.2 using the re-randomization properties.

[5] Black, J., Rogaway, P., Shrimpton, T.: Encryption-scheme security in the presence of key-dependent messages. In: Nyberg, K., Heys, H.M. (eds.) SAC 2002. LNCS, vol. 2595, pp. 62–75. Springer, Heidelberg (2003)

[6] Boneh, D.: The decision diffie-hellman problem. In: Buhler, J.P. (ed.) ANTS 1998. LNCS, vol. 1423, pp. 48–63. Springer, Heidelberg (1998)

[7] Boneh, D., Halevi, S., Hamburg, M., Ostrovsky, R.: Circular-secure encryption from decision diffie-hellman. In: Wagner, D. (ed.) CRYPTO 2008. LNCS, vol. 5157, pp. 108–125. Springer, Heidelberg (2008)

[8] Canetti, R.: Towards realizing random oracles: Hash functions that hide all partial information. In: Kaliski Jr., B.S. (ed.) CRYPTO 1997. LNCS, vol. 1294, pp. 455–469. Springer, Heidelberg (1997)

[9] Canetti, R., Dakdouk, R.R.: Obfuscating point functions with multibit output. In: Smart, N.P. (ed.) EUROCRYPT 2008. LNCS, vol. 4965, pp. 489–508. Springer, Heidelberg (2008)

[10] Canetti, R., Kalai, Y.T., Varia, M., Wichs, D.: On symmetric encryption and point obfuscation. In: Micciancio, D. (ed.) TCC 2010. LNCS, vol. 5978, pp. 52–71. Springer, Heidelberg (2010)

[11] Canetti, R., Micciancio, D., Reingold, O.: Perfectly one-way probabilistic hash functions (preliminary version). In: STOC, pp. 131–140 (1998)

[12] Canetti, R., Rothblum, G.N., Varia, M.: Obfuscation of hyperplane membership. In: Micciancio, D. (ed.) TCC 2010. LNCS, vol. 5978, pp. 72–89. Springer, Heidelberg (2010)

[13] Canetti, R., Varia, M.: Non-malleable obfuscation. In: Reingold, O. (ed.) TCC 2009. LNCS, vol. 5444, pp. 73–90. Springer, Heidelberg (2009)

[14] Dodis, Y., Smith, A.: Correcting errors without leaking partial information. In: STOC, pp. 654–663 (2005)

[15] Goldwasser, S., Kalai, Y., Peikert, C., Vaikuntanathan, V.: Robustness of the learning with errors assumption. In: ICS (2010)

[16] Goldwasser, S., Kalai, Y.T.: On the impossibility of obfuscation with auxiliary input. In: FOCS, pp. 553–562 (2005)

[17] Goldwasser, S., Rothblum, G.N.: On best-possible obfuscation. In: Vadhan, S.P. (ed.) TCC 2007. LNCS, vol. 4392, pp. 194–213. Springer, Heidelberg (2007)

[18] Hada, S.: Secure obfuscation for encrypted signatures. In: Gilbert, H. (ed.) EUROCRYPT 2010. LNCS, vol. 6110, pp. 92–112. Springer, Heidelberg (2010)

[19] Haitner, I., Holenstein, T.: On the (im)possibility of key dependent encryption. In: Reingold, O. (ed.) TCC 2009. LNCS, vol. 5444, pp. 202–219. Springer, Heidelberg (2009)

[20] Halevi, S., Krawczyk, H.: Security under key-dependent inputs. In: ACM Conference on Computer and Communications Security, pp. 466–475 (2007)

[21] Hofheinz, D., Malone-Lee, J., Stam, M.: Obfuscation for cryptographic purposes. In: Vadhan, S.P. (ed.) TCC 2007. LNCS, vol. 4392, pp. 214–232. Springer, Heidelberg (2007)

[22] Hohenberger, S., Rothblum, G.N., Shelat, A., Vaikuntanathan, V.: Securely obfuscating re-encryption. In: Vadhan, S.P. (ed.) TCC 2007. LNCS, vol. 4392, pp. 233–252. Springer, Heidelberg (2007)

[23] Lynn, B., Prabhakaran, M., Sahai, A.: Positive results and techniques for obfuscation. In: Cachin, C., Camenisch, J.L. (eds.) EUROCRYPT 2004. LNCS, vol. 3027, pp. 20–39. Springer, Heidelberg (2004)

[24] Regev, O.: On lattices, learning with errors, random linear codes, and cryptography. In: STOC, pp. 84–93 (2005)

[25] Shoup, V.: Lower bounds for discrete logarithms and related problems. In: Fumy, W. (ed.) EUROCRYPT 1997. LNCS, vol. 1233, pp. 256–266. Springer, Heidelberg (1997)

[26] Wee, H.: On obfuscating point functions. In: STOC, pp. 523–532 (2005)

# Protocols for Multiparty Coin Toss with Dishonest Majority

Amos Beimel[1,*], Eran Omri[2,**], and Ilan Orlov[1,***]

[1] Dept. of Computer Science, Ben Gurion University, Be'er Sheva, Israel
[2] Dept. of Computer Science, Bar Ilan University, Ramat Gan, Israel

**Abstract.** Coin-tossing protocols are protocols that generate a random bit with uniform distribution. These protocols are used as a building block in many cryptographic protocols. Cleve [STOC 1986] has shown that if at least half of the parties can be malicious, then, in any $r$-round coin-tossing protocol, the malicious parties can cause a bias of $\Omega(1/r)$ to the bit that the honest parties output. However, for more than two decades the best known protocols had bias $\frac{t}{\sqrt{r}}$, where $t$ is the number of corrupted parties. Recently, in a surprising result, Moran, Naor, and Segev [TCC 2009] have shown that there is an $r$-round two-party coin-tossing protocol with the optimal bias of $O(1/r)$. We extend Moran et al. results to the *multiparty model* when less than $2/3$ of the parties are malicious. The bias of our protocol is proportional to $1/r$ and depends on the gap between the number of malicious parties and the number of honest parties in the protocol. Specifically, for a constant number of parties or when the number of malicious parties is somewhat larger than half, we present an $r$-round $m$-party coin-tossing protocol with optimal bias of $O(1/r)$.

## 1 Introduction

Secure multiparty computation in the malicious model allows distrustful parties to compute a function securely. Designing such secure protocols is a delicate task with a lot of subtleties. An interesting and basic functionality for secure computation is coin tossing – generating a random bit with uniform distribution. This is a simple task where the parties have no inputs. However, already this task raises questions of fairness and how malicious parties can bias the output. Understanding what can be achieved for coin tossing in various settings can be considered as a step towards understanding general secure and fair multiparty computation. Indeed, some of the early works on secure computation were on coin tossing, e.g., [3, 4, 6]. Furthermore, coin tossing is used as a basic tool in constructing many protocols that are secure in the malicious model. Secure

* Supported by ISF grant 938/09.
** This research was generously supported by the European Research Council as part of the ERC project "LAST". Part of the research was done while the author was a post-doctoral fellow at BGU supported by the ISF grant 860/06.
*** Supported by ISF grant 938/09 and by the Frankel Center for Computer Science.

T. Rabin (Ed.): CRYPTO 2010, LNCS 6223, pp. 538–557, 2010.

protocols for coin tossing are a digital analogue of physical coin tossing, which have been used throughout history to resolve disputes.

The main problem in designing coin-tossing protocols is the prevention of a bias of the output. The bias of a coin-tossing protocol measures the maximum influence of the adversary controlling a subset of malicious parties on the output of the honest parties, where the bias is 0 if the output is always uniformly distributed and the bias is $1/2$ if the adversary can force the output to be always (say) 1. To demonstrate the problems of designing a coin-tossing protocol, we describe Blum's two-party coin-tossing protocol [3].

*Example 1 (Blum's coin-tossing protocol [3]).* To toss a coin, Alice and Bob execute the following protocol.

- Alice chooses a random bit $a$ and sends a commitment $c = \text{commit}(a)$ to Bob.
- Bob chooses a random bit $b$ and sends it to Alice.
- Alice sends the bit $a$ to Bob together with de-commit$(c)$.
- If Bob does not abort during the protocol, Alice outputs $a \oplus b$, otherwise she outputs a random bit.
- If Alice does not abort during the protocol and $c$ is a commitment to $a$, then Bob outputs $a \oplus b$, otherwise he outputs a random bit.

If Alice is malicious, then she can bias the output toward (say) 1. If $a \oplus b = 1$, she opens the commitment and Bob outputs 1. However, if $a \oplus b = 0$, Alice aborts, and Bob outputs 1 with probability $1/2$. All together, the probability that the honest Bob outputs 1 is $3/4$. It can be proved that this is the best that Alice can do in this protocol, and hence, the bias of the protocol is $1/4$. This protocol demonstrates the problems caused by parties aborting the protocol and the need to define how the output of the other parties is computed after such aborts.

While the above protocol is a significant improvement over naive protocols whose bias is $1/2$, the protocol still has a constant bias. If *more than half* of the parties are honest, then, using general secure multiparty protocols there are constant-round protocols with negligible bias (assuming a broadcast channel), e.g., the protocol of [14]. Cleve [6] proved that when at least half of the parties can be malicious, the bias of every protocol with $r$ rounds is $\Omega(1/r)$. In particular, this proves that without honest majority no protocol with polynomial number of rounds (in the security parameter) can have negligible bias. On the positive side, it was shown in [2, 6] that there is a two-party protocol with bias $O(1/\sqrt{r})$. This can be generalized to an $m$-party protocol that tolerates any number of malicious parties and has bias $O(t/\sqrt{r})$. Cleve and Impagliazzo [7] have shown that, in a model where commitments are available only as black-box (and no other assumptions are made), the bias of every coin-tossing protocol is $\Omega(1/\sqrt{r})$.[1] The protocols of [3, 2, 6] are in this model.

---

[1] The lowerbound of [7] holds in a stronger model which we will not discuss in this paper.

The question if there is a coin-tossing protocol without an honest majority that has bias $O(1/r)$ was open for many years. Recently, in a surprising and elegant result, Moran, Naor, and Segev [12] have shown that there is an $r$-round two-party coin-tossing protocol with bias $O(1/r)$. Moran et al. ask the following question:

"An interesting problem is to identify the optimal trade-off between the number of parties, the round complexity, and the bias. Unfortunately, it seems that several natural variations of our approach fail to extend to the case of more than two parties. Informally, the main reason is that a coalition of malicious parties can guess the threshold round with a pretty good probability by simulating the protocol among themselves for any possible subset."

## 1.1   Our Results

Our main contribution is a multi-party coin-tossing protocol that has small bias when less than $2/3$ of the parties are malicious.

**Theorem 1 (Informal).** *Let $m, t$, and $r$ be integers such that $m/2 \leq t < 2m/3$. There exists an $r$-round $m$-party coin-tossing protocol tolerating $t$ malicious parties that has bias $O(2^{2^{k+1}}/r')$, where $k = 2t - m$ and $r' = r - O(k+1)$.*

The above protocol nearly has the desired dependency on $r$, i.e., the dependency implied by the lower bound of Cleve [6]. However, its dependency on $k$ has, in general, a prohibitive cost. Nevertheless, there are interesting cases where the bias is $O(1/r)$.

**Corollary 1 (Informal).** *Let $m, t$ be constants such that $m/2 \leq t < 2m/3$ and $r$ be an integer. There exists an $r$-round $m$-party coin-tossing protocol tolerating $t$ malicious parties that has bias $O(1/r)$.*

For example, we construct an $r$-round 5-party coin-tossing protocol tolerating 3 malicious parties that has bias $8/(r - O(1))$ (this follows from our general construction in Sections 4–6).

Notice that the protocol of Theorem 2 depends on $k$ and not on the number of malicious parties $t$. Thus, it is efficient when $k$ is small.

**Corollary 2 (Informal).** *Let $m, r$ be integers and $t = m/2 + O(1)$. There exists an $r$-round $m$-party coin-tossing protocol tolerating $t$ malicious parties that has bias $O(1/r)$.*

For example, for any even $m$ we construct an $r$-round $m$-party coin-tossing protocol tolerating $m/2$ malicious parties that has bias $1/(2r - O(1))$. Furthermore, even when $t = 0.5m + 0.5 \log \log m - 1$, the bias of our protocol is small, namely, $O(m/(r - O(\log \log m)))$.

We also reduce the bias compared to previous protocols when more than $2/3$ of the parties are malicious. The bias of the $m$-party protocol of [2, 6] is $O(t/\sqrt{r})$. We present a protocol whose bias is $O(1/\sqrt{r})$ when $t/m$ is constant, that is, when the fraction of malicious parties is constant we get rid of the factor $t$ in the bias.

*Communication Model.* We consider a communication model where all parties can only communicate through an authenticated broadcast channel. On one hand, if a party broadcasts a message, then all other parties see *the same* message. This ensures some consistency between the information the parties have. On the other hand, there are no private channels and all parties see all messages. We assume a synchronous model, however, the adversary is rushing.[2]

We note that our results can be translated to a model with authenticated point-to-point channels with a PKI infrastructure (in an $m$-party protocol, the translation will increase the number of rounds by a multiplicative factor of $O(m)$). Thus, our results hold in the two most common models for secure multiparty computation.

*The Idea of Our Protocol.* We generalize the two-party protocol of Moran et al. [12] to the multi-party setting. In the protocol of [12] in each round Alice and Bob get bits that are their output if the other party aborts: If a party aborts in round $i$, then the other party outputs the bit it got in round $i-1$. Furthermore, there is a special round $i^*$; prior to round $i^*$ the bits given to Alice and Bob are random independent bits and from round $i^*$ onward the bits given to Alice and Bob are the same fixed bit. The adversary can bias the output only if it guesses $i^*$. In our protocol, in each round there are many bits. We define a collection of subsets of the parties and each subset gets a bit. The bits are chosen similarly to [12]: prior to $i^*$ they are independent and from $i^*$ onward they are fixed. In our case we cannot give the bits themselves to the parties. We rather use a few layers of secret-sharing schemes to store these bits. For every subset in the collection, we use the first secret-sharing scheme to share the bit of the subset among the parties of the subset. We use an additional secret-sharing scheme to share the shares of the first secret-sharing scheme. The threshold in the latter secret sharing scheme is chosen such that the protocol can proceed until enough parties aborted. In the round when the number of aborted parties ensures that there is an honest majority, an appropriate subset in the collection is chosen, its bit is reconstructed, and this bit is the output of the honest parties. The description of how to implement these ideas appears in Sections 4–6.

The construction of Moran et al. [12] is presented in two phases. In the first phase they present a protocol with a trusted dealer, for which an adversary can only inflict bias $O(1/r)$. Then, they show how to implement this protocol in the real-world, using a constant round secure-with-abort multiparty protocol, as well as secret sharing and authentication schemes. This can be seen as a general transformation from any two-party coin-tossing protocol with a trusted dealer, into a real world two-party coin-tossing protocol. We observe that the transformation of Moran et al. to a real-world protocol requires some further care trying to generalize it for the multiparty case. We show how this can be achieved by adopting the definition of secure multiparty computation of [1], which requires the protocol to detect a cheating party, that is, at the end of

---

[2] If there is synchronous broadcast without a rushing adversary then coin tossing is trivial.

the protocol either the honest parties hold a correct output or all honest parties agree on a party that cheated during the protocol.

## 2  Preliminaries

A multi-party coin-tossing protocol with $m$ parties is defined using $m$ probabilistic polynomial-time Turing machines $p_1, \ldots, p_m$ having the security parameter $1^n$ as their only input. The coin-tossing computation proceeds in rounds, in each round, the parties broadcast and receive messages on a broadcast channel. The number of rounds in the protocol is typically expressed as some polynomially-bounded function $r$ in the security parameter. At the end of protocol, the (honest) parties should hold a common bit $w$. We denote by CoinToss() the ideal functionality that gives the honest parties the same uniformly distributed bit $w$, that is, $\Pr[w = 0] = \Pr[w = 1] = 1/2$. In our protocol, the output bit $w$ will have some bias.

In this work we consider a malicious static computationally-bounded (i.e., non-uniform probabilistic polynomial-time) adversary that is allowed to corrupt some subset of parties. That is, before the beginning of the protocol, the adversary corrupts a subset of the players that may deviate arbitrarily from the protocol, and thereafter the adversary controls the messages sent by the corrupted parties. The honest parties follow the instructions of the protocol.

The parties communicate in a synchronous network, using only a broadcast channel. The adversary is rushing, that is, in each round the adversary hears the messages sent by the honest parties before broadcasting the messages of the corrupted parties for this round (thus, the messages broadcast by corrupted parties can depend on the messages of the honest parties broadcast in this round).

The security of multiparty computation protocols is defined using the real vs. ideal paradigm. In this paradigm, we consider the real-world model, in which protocols are executed. We then formulate an ideal model for executing the task at hand. This ideal model involves a trusted party whose functionality captures the security requirements from the task. Finally, we show that the real-world protocol "emulates" the ideal-world protocol: For any real-life adversary $\mathcal{A}$ there should exist an ideal-model adversary $\mathcal{S}$ (also called simulator) such that the global output of an execution of the protocol with $\mathcal{A}$ in the real-world model is distributed similarly to the global output of running $\mathcal{S}$ in the ideal model.

*1/p-Secure Computation.* As explained in the introduction, the ideal functionality CoinToss() cannot be implemented when there is no honest majority. We use 1/p-secure computation, defined by [9, 10], to capture the divergence from the ideal worlds. This notion applies to general secure computation. We start with some notation.

**Definition 1 (1/p-indistinguishability).** *A function $\mu(\cdot)$ is* negligible *if for every positive polynomial $q(\cdot)$ and all sufficiently large $n$ it holds that $\mu(n) < 1/q(n)$. A distribution ensemble $X = \{X_n\}_{n \in \mathbb{N}}$ is an infinite sequence of random variables indexed by $n \in N$. For a fixed function $p$, two distribution ensembles $X = \{X_n\}_{n \in \mathbb{N}}$ and $Y = \{Y_n\}_{n \in \mathbb{N}}$ are* computationally 1/p-indistinguishable, denoted $X \overset{1/p}{\approx} Y$, if for every non-uniform polynomial-time algorithm $D$ there exists a negligible function $\mu(\cdot)$ such that for every $n$,

$$|\Pr[D(X_n) = 1] - \Pr[D(Y_n)) = 1]| \leq \frac{1}{p(n)} + \mu(\cdot).$$

We next define the notion of $1/p$-secure computation [9, 10]. The definition uses the standard real/ideal paradigm [8, 5], except that we consider a completely fair ideal model (as typically considered in the setting of honest majority), and require only $1/p$-indistinguishability rather than indistinguishability. In the coin-tossing protocol, the parties do not have inputs. Thus, to simplify the definitions, we define secure computation without inputs (except for the security parameters).

**Definition 2 (1/p-secure computation [9, 10]).** *Let $p = p(n)$ be a function. An $m$-party protocol $\Pi$ is said to $1/p$-securely compute a functionality $\mathcal{F}$ if for every non-uniform probabilistic polynomial-time adversary $\mathcal{A}$ in the real model, there exists a non-uniform probabilistic polynomial-time adversary $\mathcal{S}$ in the ideal model such that the following two distribution ensembles are computationally $1/p$-indistinguishable $\left\{ \text{IDEAL}_{\mathcal{F},\mathcal{S}(\text{aux})}(1^n) \right\}_{n \in \mathbb{N}} \overset{1/p}{\approx} \left\{ \text{REAL}_{\Pi,\mathcal{A}(\text{aux})}(1^n) \right\}_{n \in \mathbb{N}}$, where $\text{REAL}_{\Pi,\mathcal{A}(\text{aux})}(1^n)$ is a random variable consisting of the view of the adversary (i.e., its random input and the messages it got) and the output of the honest parties following an execution of $\Pi$, and $\text{IDEAL}_{\mathcal{F},\mathcal{S}(\text{aux})}(1^n)$ is a random variable consisting of the output of the adversary $\mathcal{S}$ in the ideal world execution and the output of the honest parties in that execution.*

**Definition 3.** *We say that a protocol is a coin-tossing protocol with bias $1/p$ if it is a $1/p$-secure protocol for the functionality* CoinToss().

## 2.1   The Two-Party Protocol of Moran et al.

Moran, Naor, and Segev [12] present a two-party coin-tossing protocol with optimal bias with respect to round complexity (i.e., meeting the lowerbound of Cleve [6]). We next briefly review their protocol, which later serves as the basis for our construction. Following the presentation of [12], we first describe a construction that uses an on-line trusted party acting as a dealer. Later, we describe how the trusted party can be eliminated.

The main underlying idea is that the dealer chooses a special round during which the parties actually unknowingly learn the output of the protocol. If the adversary guesses this round, it can bias the output by aborting. If the adversary aborts (or behaves maliciously) in any other time, then there is no bias. However, this special round is uniformly selected (out of $r$ possible rounds) and then concealed such that the adversary is unable to guess it with probability exceeding $1/r$. Therefore, the overall bias achievable by any adversary is $O(1/r)$.

More specifically, for two parties Alice and Bob to jointly toss a random coin, the protocol proceeds as follows. In a preprocessing phase, the trusted party selects a special round number $i^* \in \{1, \ldots, r\}$, uniformly at random, and selects bits $a_1, \ldots, a_{i^*-1}, b_1, \ldots, b_{i^*-1}$, independently, uniformly at random. It then uniformly selects a bit $w \in \{0, 1\}$ and sets $a_i = b_i = w$ for all $i^* \leq i \leq r$. Thereafter, the protocol proceeds in rounds: In round $i$, the dealer gives Alice the bit $a_i$ and Bob the bit $b_i$. If none of the parties abort, then at the end of the protocol both output $a_r = b_r = w$. If a party prematurely aborts in some round $i$, the honest party outputs the bit it received in the previous round (i.e., $a_{i-1}$ or $b_{i-1}$ respectively). If one party aborts before the other party received its first bit (i.e., $a_1$ or $b_1$), then the other party outputs a random bit.

The security of the protocol follows from the fact that, unless the adversary aborts in round $i^*$, it cannot bias the output of the protocol. The view of any of the parties up to round $i \leq i^*$ is independent of the value of $i^*$, hence, any adversary corrupting a single party can guess $i^*$ with probability at most $1/r$.

To eliminate the trusted party, Moran et al. first turn the trusted party from an on-line dealer into an off-line dealer, i.e., one that computes some values in a preprocessing phase, deals them to the parties, and halts. To achieve this, they use a 2-out-of-2 secret-sharing scheme and an authentication scheme. The trusted party selects $i^*$, bits $a_1, \ldots, a_{i^*-1}, b_1, \ldots, b_{i^*-1}$, and a bit $w \in \{0, 1\}$ as before. It then selects random shares for $a_i$ and $b_i$ for each $i \in \{1, \ldots, r\}$. That is, it computes shares $a_i^{(A)} \oplus a_i^{(B)} = a_i$ and $b_i^{(A)} \oplus b_i^{(B)} = b_i$. At the beginning of the protocol, the trusted party sends to Alice her shares of the $a_i$'s, that is $a_i^{(A)}$, and the shares $b_i^{(A)}$ together with an authentication of the $b_i^{(A)}$ (i.e., authenticated shares of the $b_i$'s), and sends to Bob his shares of the $b_i$'s and authenticated shares of the $a_i$'s. The protocol now proceeds in rounds. In each round $i$ Bob sends to Alice his authenticated share of $a_i$, so Alice can reconstruct the bit $a_i$. Alice then sends to Bob her authenticated share of $b_i$. An adversary cannot forge an authentication, and is, thus, essentially limited to aborting in deviating from the prescribed protocol.

The off-line dealer is then replaced by a (constant round) secure-with-abort two-party protocol [11] for computing the preprocessing functionality. That is, at the end of the initialization protocol, the parties get the authenticated shares of the $a_i$'s and the $b_i$'s, while the underlying $i^*$ and authentication keys stay secret. The security of the 2-party preprocessing protocol guarantees that a bounded adversary is essentially as powerful as in a computation with an off-line dealer.

## 3   Coin Tossing with Dishonest Majority – A Warm-Up

In this section we present two warm-up constructions for multiparty coin-tossing with bias $O(1/r)$ where $r$ is the number of rounds in the protocol. The first construction considers the case that at most half of the parties are malicious (however, there is no majority of honest parties). The second construction solves the problem of coin tossing with 5 parties, where at most 3 are malicious. These two protocols demonstrate the main difficulties in constructing multiparty coin-tossing protocols with dishonest majority, alongside the techniques we use to overcome these difficulties. In Sections 4–6, we present a construction for the general case that combines ideas from the two constructions presented in this section.

The main issue of any coin-tossing protocol is how to deal with premature aborts. The protocol must instruct any large enough subset of parties (i.e., at least as large as the set of honest parties) how to jointly reconstruct a bit if all other parties abort the protocol. Since there is no honest majority, an adversary controlling some set of parties can compute the output of this set assuming that the other parties abort. The problem in designing a protocol is how to ensure that this information does not enable the adversary to bias the output.

## 3.1 Multiparty Coin Tossing When Half of the Parties Can Be Malicious

In this section we present a protocol with optimal (up to a small constant) bias with respect to round complexity, when up to half the parties may be corrupt. We next give an informal description of the protocol with an on-line trusted party who acts as a dealer.

To construct a protocol for multiparty coin tossing for the case that up to half the parties may be malicious, the parties simulate the 2-party protocol of [12]. That is, we partition the parties into two sets $A$ and $B$, one will simulate Alice and the other will simulate Bob. The main difficulty is that the adversary is not restricted to corrupting parties only in one of these sets. To overcome this problem, in our partition $A$ contains a single party $p_1$, and the set $B$ consists of the parties $p_2, \ldots, p_m$. If the adversary corrupts $p_1$, it gains full access to the view of Alice in the 2-party protocol; however, in this case a strict majority of the parties simulating Bob is honest, and the adversary will gain no information about the bits of Bob, i.e., the $b_i$'s.

We next describe the protocol. In a preprocessing phase, the dealer uniformly selects $i^* \in \{1, \ldots, r\}$ and then uniformly and independently selects $a_1, \ldots, a_{i^*-1}, b_1, \ldots, b_{i^*-1}$. Finally, it uniformly selects $w \in \{0,1\}$ and sets $a_i = b_i = w$ for all $i^* \leq i \leq r$. In each round $i$, the dealer sends $a_i$ to $A$, selects random shares of $b_i$ in Shamir's $m/2$-out-of-$(m-1)$ secret-sharing scheme, and sends each share to the appropriate party in $B$. We stress that formally (to model a rushing adversary), the dealer first sends the malicious parties their messages, allows them to abort, and proceeds as described below.

During the execution some parties might abort; we say that a party is active if it has not aborted. If a party $p_j$ prematurely aborts, then the trusted party notifies all currently active parties that $p_j$ has aborted. We next describe the actions when a party aborts:

- If $p_1$ aborts in round $i$, then the parties in $B$ reconstruct $b_{i-1}$, output it, and halt. In this case, since $p_1$ is corrupt, at least $m/2$ honest parties exist in $B$ and, thus, they will be able to reconstruct the output.
- If in round $i$ parties from $B$ abort such that less than $m/2$ active parties remain in $B$, then $p_1$ broadcasts $a_{i-1}$ to the remaining $m/2 - 1$ parties in $B$ and all (honest) parties output $a_{i-1}$ and halt. In this case $p_1$ must be honest and hence can be trusted to broadcast $a_{i-1}$.

- While there are still at least $m/2$ active parties in $B$ (i.e., at most $m/2 - 1$ of them abort) and $p_1$ is active, the protocol proceeds without a change.

To prevent cheating, the dealer needs to sign the messages given to the parties. We omit these details in this section.

Recall that at most $m/2$ out of the $m$ parties are malicious. Thus, if $p_1$ is corrupted, then at most $(m/2) - 1$ parties in $B$ are corrupted, and they cannot reconstruct $b_i$. To see that the above protocol is $O(1/r)$-secure is now straightforward. An adversary wishing to bias the protocol must cause premature termination. To do so, it must either corrupt $p_1$ (and gain no information on the $b_i$'s) or otherwise corrupt $m/2$ parties in $B$ (hence, leaving $p_1$ uncorrupted). Thus, for any adversary in the multi-party protocol there is an adversary corrupting Alice or Bob in the on-line setting of the two party protocol of [12] that is essentially as powerful. An important feature that we exploit in our protocol is the fact that in the two-party protocol Bob does not need its bit $b_{i-1}$ if Alice does not abort. Thus, in our protocol the parties in $B$ do not reconstruct $b_{i-1}$ unless $p_1$ aborts in round $i$.

More work is required in order to eliminate the trusted dealer, however, the arguments justifying such a move are a special case of those described in Section 6.

## 3.2 A 5-Party Protocol That Tolerates up to 3 Malicious Parties

In this section we consider the case where $m = 5$ and $t = 3$, i.e., a 5-party protocol where up to 3 of the parties may be malicious. As in previous protocols, we first sketch our construction assuming there is a special on-line trusted dealer. This dealer interacts with the parties in rounds, sharing bits to subsets of parties, and proceeds with the normal execution as long as at least 4 of the parties are still active.

Denote the trusted dealer by $T$ and the parties by $p_1, \ldots, p_5$. Let $S_1, \ldots, S_{10}$ be all possible triplets of parties, i.e., $S_j \subset \{p_1, \ldots, p_5\}$ such that $|S_j| = 3$. Denote by $\sigma_{S_j}^i$ a bit to be recovered by $S_j$ if the protocol terminates in round $i + 1$. In a preprocessing phase, the dealer $T$ selects uniformly at random $i^* \in \{1, \ldots, r\}$, indicating the special round in this five-party protocol. Then, for every $0 \le i < i^*$ it selects $\sigma_{S_j}^i$ independently and uniformly at random for each $j \in \{1, \ldots, 10\}$. Finally, it independently and uniformly selects a random bit $w$ and sets $\sigma_{S_j}^i = w$, for every $i \in \{i^*, \ldots, r\}$ and for every $j \in \{1, \ldots, 10\}$.

The dealer $T$ interacts with $p_1, \ldots, p_5$ in rounds, where round $i$, for $1 \le i \le r$ consists of three phases:

**First phase.** The dealer sends to the adversary all the bits $\sigma_{S_j}^i$ such that there is a majority of corrupted parties in $S_j$, i.e., at least 2 parties in $S_j$ are controlled by the adversary.

**Second phase.** The adversary sends to $T$ a list of parties that abort in the current round. If there are less than 4 active parties (i.e., there are either 2

or 3 active parties),[3] $T$ sends $\sigma_{S_j}^{i-1}$ to the active parties, where $S_j$ is the lexicographically first triplet that contains all of the active parties. The honest
parties output this bit and halt.

**Third phase.** If at least 4 parties are active, $T$ notifies the active parties that
the protocol proceeds normally.

If after $r$ rounds, there are at least 4 active parties, $T$ simply sends $w$ to all
active parties and the honest parties output this bit.

As an example of a possible execution of the protocol, assume that $p_1$ aborts
in round 4 and $p_3$ and $p_4$ abort in round 26. In this case, $T$ sends $\sigma_{\{p_1,p_2,p_5\}}^{25}$ to
$p_2$ and $p_5$, which output this bit.

Recall that the adversary obtains the bit $S_j$ if at least two parties in $S_j$ are
malicious. If the adversary causes the dealer to halt, then, either there are two
active parties, both of them must be honest, or there are three active parties
and at most one of them is malicious. In either case, the adversary does not
know $\sigma_{S_j}^{i-1}$ in advance. Furthermore, the dealer reveals the appropriate bit $\sigma_{S_j}^{i-1}$
to the active parties. Jumping ahead, these properties are later preserved in a
real world protocol by using a 2-out-of-3 secret-sharing scheme.

We next argue that any adversary can bias the output of the above protocol
by at most $O(1/r)$. As in the protocol of Moran et al., the adversary can only
bias the output by causing the protocol to terminate in round $i^*$. In contrast to
the protocol of [12], in our protocol if in some round there are two bits $\sigma_S^i$ and
$\sigma_{S'}^i$ that the adversary can obtain such that $\sigma_S^i \neq \sigma_{S'}^i$, then the adversary can
deduce that $i \neq i^*$. However, there are at most 7 bits that the adversary can
obtain in each round (i.e., the bits of sets $S$ containing at least two malicious
parties). For a round $i$ such that $i < i^*$, the probability that all these bits are
equal to (say) 0 is $(1/2)^7$. Such rounds are indistinguishable to the adversary
from round $i^*$. Intuitively, the best an adversary can do is guess one of these
rounds, and therefore cannot succeed guessing with probability better than $1/2^7$.
Thus, the bias the adversary can cause is $2^7/r$.

**Eliminating the Dealer of the 5-Party Protocol.** We eliminate the trusted
on-line dealer in a few steps using a few layers of secret-sharing schemes. First,
we change the on-line dealer, so that in each round $i$ it shares the bit $\sigma_S^i$ of
each subset $S$ among the parties of $S$ using a 2-out-of-3 secret-sharing scheme
– called *inner* secret-sharing scheme. The same requirement on $\sigma_S^i$ as in the
above protocol are preserved using this inner secret-sharing scheme. That is, the
adversary is able to obtain information on $\sigma_S^i$ only if it controls at least two of
the parties in $S$. On the other hand, if the adversary does not control at least
two parties in $S$ (i.e., there is an honest majority in $S$), then, in round $i$, the
honest parties can reconstruct $\sigma_S^{i-1}$ (if so instructed by the protocol).

Next we turn the on-line dealer into an off-line dealer. That is, we show that
it is possible for the dealer to only interact with the parties once, sending each

---

[3] The reason for requiring that the dealer does not continue when at least two parties
abort will become clear when we transform the protocol to a protocol with an off-line
dealer.

party some input, so that thereafter, the parties interact in rounds (without the dealer) and in each round $i$ each party learns its shares in the $i$th inner secret-sharing scheme. That is, in each round $i$, each party $p$ learns a share of $\sigma_S^i$ in a 2-out-of-3 secret-sharing scheme, for every triplet $S$ such that $p \in S$. For this purpose, the dealer computes, in a preprocessing phase, the appropriate shares for the inner secret-sharing scheme. The shares of each round for each party $p$ are then shared in a 2-out-of-2 secret-sharing scheme, where $p$ gets one of the two shares (this serves as a mask, allowing only $p$ to later reconstruct its shares of the appropriate $\sigma_S^i$'s). All parties get shares in a 4-out-of-5 Shamir secret-sharing scheme of the other share of the 2-out-of-2 secret sharing. We call the resulting secret-sharing scheme the *outer* scheme.

The use of the 4-out-of-5 secret-sharing scheme plays a crucial role in eliminating the on-line dealer. On the one hand, it guarantees that an adversary, corrupting at most three parties, cannot reconstruct the shares of round $i$ before round $i$. On the other hand, at least two parties must not reveal their shares in order to prevent a reconstruction of the outer scheme (this is why we cannot proceed after 2 parties aborted). Hence, the protocol proceed normally as long as at least 4 parties are active. If, indeed, at least two parties abort (in round $i$), then the remaining parties use their shares of the inner scheme to reconstruct the bit $\sigma_S^{i-1}$ for the appropriate triplet $S$.

To prevent malicious parties from cheating, by say, sending false shares and causing reconstruction of wrong secrets, every message that a party should send during the execution of the protocol is signed in the preprocessing phase (together with the appropriate round number and with the party's index). In addition, the dealer sends a verification key to each of the parties. To conclude, the off-line dealer gives each party the signed shares for the outer secret sharing scheme together with the verification key.

The protocol with the off-line dealer proceeds in rounds. In round $i$ of the protocol all parties broadcast their (signed) shares in the outer (4-out-of-5) secret-sharing scheme. Thereafter, each party can unmask the message it receives (with its share in the appropriate 2-out-of-2 secret-sharing scheme) to obtain its shares in the 2-out-of-3 sharing of the bits $\sigma_S^i$ (for the appropriate sets $S$'s to which the party belongs). If a party stops broadcasting messages or broadcasts improperly signs messages, then all other parties consider it as aborted. If two or more parties abort, the remaining parties reconstruct the bit of the lexicographically first triplet that contains all of them, as described above. In the special case of premature termination already in the first round, the remaining parties engage in a fully secure protocol (with honest majority) to toss a completely random coin.

Finally, we replace the off-line dealer by using a secure with abort protocol with cheat detection computing the functionality computed by the dealer. The details of this final step are given in Section 6.

The above construction can be generalized in a straightforward manner to any number $m$ of parties and any number $t$ of malicious parties such that $t < 2m/3$. However, in the protocol described in Section 4 the bias on the output is substantially smaller; this is done using a better way for distributing bits to subsets.

# 4  Coin-Tossing with Dishonest Majority – Our Main Construction

In Sections 4–6 we present our main result – a coin-tossing protocol that has nearly optimal bias and can tolerate up to 2/3 fraction of malicious parties. More specifically, we prove the following theorem:

**Theorem 2.** *If enhanced trap-door permutations exist, then for any $m$, $t$, and $r$ such that $t < 2m/3$, there is an $r$-round $m$-party coin-tossing protocol tolerating up to $t$ malicious parties and has bias $O\left(2^{2^{k+1}}/r'\right)$, where $k = 2t - m$ and $r' = r - O(k + 1)$.*

In the above theorem $k$ is the difference between the number of malicious parties and the number of honest parties, i.e., $k = t - (m - t) = 2t - m$.

Following [12], we describe our protocol in two steps. In Section 5, we describe Protocol CoinTossWithDealer$_r$ that uses an online trusted party. In Section 6, we get rid of the on-line dealer. This simplifies the description and understanding of our protocols. More importantly, we can prove the security of our main protocol in a modular way. We first prove

**Theorem 3.** *Protocol* CoinTossWithDealer$_r$ *is an $r$-round $m$-party coin-tossing protocol with an on-line dealer tolerating up to $t$ malicious parties that has bias $O\left(2^{2^{k+1}}/r\right)$.*

We then consider the on-line dealer of Protocol CoinTossWithDealer$_r$ as an ideal functionality. In this protocol, the honest parties do not send any messages and in each round the dealer sends messages to the parties; we consider an interactive functionality sending the messages that the dealer sends. We prove

**Theorem 4.** *Let $t < 2m/3$. If enhanced trap-door permutations exist, the protocol presented in Section 6 is a computationally-secure implementation with $r' + O(k+1)$ rounds of the dealer functionality in Protocol* CoinTossWithDealer$_{r'}$.

The above theorem is proved using the hybrid model techniques of Canetti [5]. Theorem 2 follows from Theorem 3 and Theorem 4 by a composition argument.

We stress that constructing fair coin-tossing protocols assuming a trusted dealer is an easy task, e.g., the trusted party can choose a random bit and send it to each party. However, when considering a rushing adversary one cannot eliminate the trusted party in this protocol. The coin-tossing protocol we describe, Protocol CoinTossWithDealer$_r$, is designed such that it is possible to transform it to a protocol with no trusted party.

# 5  Coin-Tossing with Dishonest Majority and an On-Line Dealer

In this section we describe a protocol with a special trusted party $T$ who acts as an on-line dealer interacting with the parties in rounds. In the protocol we describe, in every round the trusted party $T$ chooses bits for some subsets of parties (the collection of subsets that receive a bit is part of the design of the protocol). Since, in the real world, the adversary can be rushing, the interaction between the parties and $T$ in each round has three phases. In the first phase, for each set $S$ that contains enough malicious parties, the trusted party sends the bit of the set to the malicious parties in the subset $S$. In the second phase, malicious parties may abort the computation (and by that prevent later reconstruction of some of the information). To do so, these parties send to $T$ an "abort" message. Finally, in the third phase, the actual (ideal) secret sharing takes place.

**Protocol CoinTossWithDealer$_r$**

**Inputs:** The input of each party $p_i$ is the security parameter $1^n$, a polynomial
  $r = r(n)$ specifying the number of rounds in the protocol, and an upper
  bound $t$ on the number of corrupted parties.
**Underlying Subsets:** Let $P_j = \{p_j\}$ for $1 \leq j \leq k+1$ and $P_{k+2} = \{p_{k+2}, \ldots, p_m\}$.
  Define $Q_J = \cup_{j \in J} P_j$ for each $J \subset \{1, \ldots, k+2\}$.
  For each subset $P_j$ define a reconstruction threshold value $o_j$: For $1 \leq j \leq$
  $k+1$ define $o_j = 1$ and define $o_{k+2} = m - t$. Finally, $o_J = \sum_{j \in J} o_j$ for each
  $J \subset \{1, \ldots, k+2\}$.
**Instructions for the (trusted) dealer:**
  **The preprocessing phase:** Select a bit $\sigma_J^i$ for every subset $Q_J$ and every
    round $i$ as follows:
      1. Select $i^* \in \{1, \ldots, r\}$ and $w \in \{0, 1\}$ independently with uniform
        distribution.
      2. For each $J \subset \{1, \ldots, k+2\}$, select $\sigma_J^0, \ldots, \sigma_J^{i^*-1}$ independently with
        uniform distribution.
      3. For each $J \subset \{1, \ldots, k+2\}$, set $\sigma_J^{i^*} = \ldots = \sigma_J^r = w$ for $i^* \leq i \leq r$.
  **Interaction rounds:** In each round $1 \leq i \leq r$ of the protocol, interact with
    the parties in three phases:
      – **The peeking phase:** For each $J \subset \{1, \ldots, k+2\}$, if $Q_J$ contains
        at least $o_J$ malicious parties, send the bit $\sigma_J^i$ to all malicious parties
        in $Q_J$.
      – **The abort phase:** Upon receiving an $abort_j$ message from party $p_j$,
        remove party $p_j$ from the list of active parties and notify all parties
        that party $p_j$ is inactive. (Ignore all other types of messages.)
        If at least $m - t$ parties have aborted so far, move to the premature
        termination process.
      – **The main phase:** Send "proceed" to all parties.
  **Premature termination process:** This round consists of two phases, af-
    ter which the protocol terminates and all honest parties hold the same
    output.
      – **The abort phase:** Upon receiving an $abort_j$ message from party
        $p_j$, remove party $p_j$ from the list of active parties.
      – **The default output phase:** Let $D$ be the set of indices of parties
        that aborted the protocol thus far, i.e., $D = \{j \mid p_j \text{ has aborted}\}$.

- If $|D \cap \{k+2, \ldots, m\}| \geq m - t$ then $J = \{1, \ldots, k+1\} \setminus D$.
- If $|D \cap \{k+2, \ldots, m\}| < m - t$ then $J = (\{1, \ldots, k+1\} \setminus D) \cup \{k+2\}$.
- Send $w' = \sigma_J^{i-1}$ to all parties.

**Normal termination:** This phase is executed if the last round of the protocol is completed.

Send $w$ to all parties.

**Instructions for honest parties:** Upon receiving output $y$ from the dealer, output $y$. (Honest parties do not send any message throughout the protocol.)

We next informally explain why the protocol has small bias, that is, we give a sketch of the proof of Theorem 3. First, we claim that the adversary can bias the output only if the premature termination occurs in round $i^*$:

1. If the premature termination round occurs after round $i^*$ (or does not occur at all), then the output is already fixed.
2. If the premature termination round occurs before round $i^*$, then the adversary does not know the random bit $\sigma_J^{i-1}$ that the honest parties output:
   (a) If $|D \cap \{k+2, \ldots, m\}| \geq m - t$, then $J = \{1, \ldots, k+1\} \setminus D$ and $o_J = |J|$. There are at most $t$ corrupt parties and at least $m - t$ of them are in $Q_{\{k+2\}}$, thus, at most $t - (m - t) = k$ in $\{p_1, \ldots, p_{k+1}\}$. In other words, there is at least one honest (and therefore active) party in $Q_J$, and the trusted party does not send the bit $\sigma_J^{i-1}$ to the parties in $Q_J$.
   (b) If $|D \cap \{k+2, \ldots, m\}| < m - t$, then $J = (\{1, \ldots, k+1\} \setminus D) \cup \{k+2\}$. Let $\alpha = |D \cap \{1, \ldots, k+1\}|$. In this case, $o_J = k+1-\alpha+m-t = t+1-\alpha$. The set $Q_J$ contains at most $t-\alpha < o_J$ corrupt parties, thus, these parties do not get the bit $\sigma_J^{i-1}$ from the trusted party.

Thus, the adversary can bias the output only if it guesses $i^*$. If $\sigma_J^i \neq \sigma_{J'}^i$ for two bits that the adversary gets from the trusted party, then it can learn that $i < i^*$. It can be shown that the adversary can get at most $2^{k+1}$ such bits (out of the $2^{k+2}$ bits). With probability $1/2^{2^{k+1}}$ all these bits are all equal in a round prior to $i^*$ and the adversary cannot distinguish such round from $i^*$. By Lemma 2 in [9], this implies that the adversary can guess $i^*$ with probability at most $2^{2^{k+1}}/r$. Therefore, the bias is $O(2^{2^{k+1}}/r)$.

Roughly speaking, transforming the above informal arguments into a formal proof, which uses the real vs. ideal paradigm, works as follows. We define a simulator $\mathcal{S}$ that for an adversary $\mathcal{A}$, first uses the ideal CoinToss() functionality to toss a completely fair coin $w_S$ (this coin is the output of the honest parties in the simulated execution). Then, in order to simulate the view of $\mathcal{A}$, the simulator $\mathcal{S}$ runs $\mathcal{A}$ internally and interacts with $\mathcal{A}$ playing the role of $T$ (with $w = w_S$); the only difference is that in the a premature termination, it always sends the parties $w_S$. The arguments of the above proof sketch show that view of $\mathcal{A}$ together with the output of the honest parties are identically distributed whenever premature termination does not occur in the special round $i^*$. The above bound on the ability of any adversary to correctly guess $i^*$ finalizes the proof. The formal proof will appear in the full version of this paper.

# 6    Omitting the On-Line Dealer

In this section we show how Protocol CoinTossWithDealer$_r$, presented in Section 5, can be transformed into a real-world protocol. That is, we present a fully secure $m$-party protocol implementing the ideal functionality described in Protocol CoinTossWithDealer$_r$. The resulting protocol has $r'$ rounds, where $r' = r + c(k + 1)$, for some constant $c$, and is executed in a network where the parties communicate via an authenticated broadcast channel. Before formally describing our construction, we outline its main components.

*The inner secret-sharing scheme.* To implement the ideal secret sharing functionality of the trusted party $T$ in the CoinTossWithDealer$_r$ protocol to share the bits $\sigma_j^i$, we use an $o_J$-out-of-$|Q_J|$ Shamir secret-sharing scheme. That is, in each round $i$, each party $p_j \in Q_J$ obtains a share $S_j^{i,J}$ in a $o_J$-out-of-$|Q_J|$ secret-sharing of $\sigma_j^i$. The same requirement on $\sigma_j^i$ as in the ideal protocol are preserved using this inner secret-sharing scheme. That is, the adversary is able to obtain information on $\sigma_j^i$ only if it controls at least $o_J$ of the parties in $Q_J$. On the other hand, if, in a premature termination in round $i$, at least $o_J$ parties in $Q_J$ work together, then they can reconstruct $\sigma_J^{i-1}$.

*The outer secret-sharing scheme.* In the ideal protocol, the adversary never learns anything about the bits $\sigma_j^i$ before round $i$ begins. To achieve this property in the real-world protocol, the shares of the inner secret-sharing schemes of all rounds are shared, in a preprocessing step, using a $(t+1)$-out-of-$m$ secret-sharing scheme. The $t + 1$ threshold guarantees that the adversary cannot see the shares of the inner secret-sharing scheme for a given round $i$ without the help of honest parties, which will not be given before round $i$.

In each round $i$ the parties send messages so that each party can reconstruct its shares in the inner secret-sharing schemes of round $i$. Since all messages are broadcast and all parties can see them, the shares that party $p_j$ receives in round $i$ are masked by using yet another layer of secret-sharing. Specifically, a share $S_j^{i,J}$ to be reconstructed by $p_j$ in round $i$ is signed and shared (already in the preprocessing phase) in a 2-out-of-2 secret sharing scheme, such that one share is given to $p_j$ and the other is shared among all parties in a $(t+1)$-out-of-$m$ secret-sharing scheme. We refer to the combination of these two layers of secret-sharing as the outer secret-sharing scheme.

*Premature Termination.* The $t+1$ threshold of the outer secret sharing scheme allows a successful reconstruction (of the shares of the inner scheme) as long as at least $t+1$ parties participate in the reconstruction. This allows the real-world protocol to proceed with normal interaction rounds as long as less than $m - t$ parties have aborted (as does the ideal-world protocol). This property is crucial to the success of the real world protocol, since in the complementary event that during round $i$ the number of parties that have aborted is at least $m - t$, then an honest majority is guaranteed (since $t < 2m/3$). Thus, in a premature termination in round $i$, the active parties can engage in a fully secure multiparty computation of the appropriate functionality, i.e., the CoinToss functionality in the special case that $i = 1$ and a reconstruction functionality otherwise.

*Signatures.* In order to confine adversarial strategies to premature aborts, the messages that the parties send are signed (together with the appropriate round number and the index of the sending party), and a verification key is given to all parties. Furthermore, all shares in the inner secret-sharing scheme are signed (as they are used as messages if reconstruction is required). Any message failing to comply with the prescribed protocol is considered an abort message. Since all messages are publicly broadcast, all parties can keep record of all aborts.

*The preliminary phase.* The goal of the preliminary phase is to compute the MultiShareGen$_r$ functionality, which computes the bits for the underlying sets and the signed shares for the inner and outer secret-sharing schemes. As an honest majority is not guaranteed, it is not possible to implement this functionality by a secure protocol with fairness. That is, we cannot implement an ideal functionality where a trusted party computes the MultiShareGen$_r$ functionality and sends the appropriate output to each party. However, since the outputs of the MultiShareGen$_r$ functionality do not reveal any information regarding the output of the protocol to any subset of size at most $t$, fairness is not essential for this part of the computation. We use a protocol with cheat detection, that is, if the protocols fails at least one corrupt party is identified by all honest parties. The computation is then repeated without the detected malicious parties.

More formally, we compute the MultiShareGen$_r$ functionality using a multiparty computation protocol that is secure-with-abort with cheat-detection. Informally, this means that we use a protocol that implements the following ideal model: the trusted party computes the MultiShareGen$_r$ functionality and gives the outputs of the corrupted parties to the adversary; the adversary either sends "proceed", in which case, the trusted party sends the appropriate output to each honest party; otherwise, the adversary sends "abort$_j$" (where, $p_j$ is in the set of corrupted parties) to the trusted party, which in turn notifies the honest parties that $p_j$ is malicious. Using methods from Pass [13], one can obtain a constant-round multiparty protocol secure-with-abort with cheat-detection. Since this protocol is repeated at most $k + 1$ times before an honest majority is guaranteed, the round complexity of the preliminary phase is $O(k)$.

We first present the initialization functionality of the protocol,

## Functionality MultiShareGen$_r$

### Computing default bits
1. Choose $w \in \{0, 1\}$ and $i^* \in \{1, \ldots, r\}$ uniformly at random.
2. For each $i \in \{1, \ldots, r\}$ and for each $J \subset \{1, \ldots, k + 2\}$,
   (a) if $i \in \{1, \ldots, i^* - 1\}$, then choose independently and uniformly at random $\sigma_J^i \in \{0, 1\}$.
   (b) if $i \in \{i^*, \ldots, r\}$, then set $\sigma_J^i = w$.

### Computing signed shares of the inner secret sharing scheme
3. Compute $(K_{\text{sign}}, K_{\text{ver}}) \leftarrow \text{Gen}(1^n)$.
4. For each $i \in \{1, \ldots, r\}$ and for each $J \subset \{1, \ldots, k + 2\}$
   (a) Choose random secret shares of $\sigma_J^i$ in an $o_J$-out-of-$|Q_J|$ Shamir's secret sharing scheme for the parties in $Q_J$.
   For each party $p_j \in Q_J$, let $S_j^{i,J}$ be its share of $\sigma_J^i$.

(b) For each share $S_j^{i,J}$, add the corresponding set index and the round number and sign:
$$R_j^{i,J} \leftarrow (S_j^{i,J}, J, i, \text{Sign}((S_j^{i,J}, J, i), K_{\text{sign}})).$$

**Computing shares of the outer secret sharing scheme**

5. For each $i \in \{1, \ldots, r\}$, for each $J \subset \{1, \ldots, k + 2\}$, and for each $p_j \in Q_J$, share $p_j$'s signed share $R_j^{i,J}$ using a 2-out-of-2 secret sharing scheme; one share is given to $p_j$ (a private mask only $p_j$ obtains) and the other share is shared among all parties in a $(t + 1)$-out-of-$m$ secret-sharing scheme.

**Signing the messages of all parties**

6. Compute the message $m_{(q,i)}$ that $p_q \in P$ broadcasts in round $i$ by concatenating (1) $p_q$'s identity, (2) the round number $i$, and (3) the shares of $R_j^{i,J}$ (for all $J$ and for all $j$ such that $p_j \in Q_J$) produced in Step 5 for $p_q$ (excluding $p_q$'s private masks).

7. Compute $M_{(q,i)} \leftarrow (m_{(q,i)}, \text{Sign}(m_{(q,i)}, K_{\text{sign}}))$.

**Outputs:** Each party $p_j$ receives
- The verification key $K_{\text{ver}}$.
- The messages $M_{(j,1)}, \ldots, M_{(j,r)}$ that $p_j$ broadcasts during the protocol.
- $p_j$'s private masks which were produced in Step 6 for each $J \subset \{1, \ldots, k+2\}$ such that $p_j \in Q_J$.

Next, we formally define the $m$-party coin-tossing protocol tolerating $t < 2m/3$ malicious parties without any dealer.

**Protocol MultiPartyCoinToss$_r$**

**Joint input:** Security parameter $1^n$.

**Preliminary phase:**
- The parties execute a secure with abort protocol with cheat detection computing Functionality MultiShareGen$_r$.
- If a party aborts, then this phase is repeated without the parties that were identified as cheaters so far.
- If the first phase was repeated $k + 1$ times (thus, an honest majority is guaranteed), the parties use a multiparty secure protocol (with fairness) to toss a fair coin, output this resulting bit, and halt.
- Denote the set of indices of inactive parties (i.e., parties that cheated or aborted so far) by $D$.

**In each round $i = 1, \ldots, r$ do:**
- Each party $p_j \in P$ broadcasts $M_{(j,i)}$ (containing its shares in the outer secret-sharing scheme).
- If $\text{Ver}(M_{(j,i)}, K_{\text{ver}}) = 0$ or if $p_j$ broadcasts an invalid or no message, then all parties mark $p_j$ as inactive, i.e., set $D \leftarrow D \cup \{j\}$. If $|D| \geq m - t$, then the premature termination step is executed.

**Premature termination step**
- If $i = 1$, then the active parties use a multiparty secure protocol (with fairness) to toss a fair coin, output this resulting bit, and halt.
- Otherwise,

1. Each party $p_j$ reconstructs $R_j^{i-1,J}$, the signed share of the "inner secret sharing scheme" produced in Step (4) of Functionality MultiShareGen$_r$, for every $J \subset \{1, \ldots, k+2\}$ such that $p_j \in Q_J$.
2. The active parties execute a secure multiparty protocol with an honest majority to compute Functionality Reconstruction, where the input of each party $p_j$ is $R_j^{i-1,J}$ for every $J \subset \{1, \ldots, k+2\}$ such that $p_j \in Q_J$.
3. The active parties output the output of this protocol, and halt.

**At the end of round** $r$:
- Each active party $p_j$ broadcasts the signed shares $R_j^{r,J}$ for each $J$ such that $p_j \in Q_J$.
- Each active party reconstructs the bit $\sigma_J^r$ for the lexicographically first set $J$ such that at least $o_J$ parties broadcast properly signed shares $R_j^{r,J}$, outputs $\sigma_J^r$, and halts.

## Functionality Reconstruction

**Joint Input:** The indices of inactive parties, $D$, and the verification key, $K_{\text{ver}}$.
**Private Input of** $p_j$**:** A set of signed shares $R_j^{i-1,J}$ for each $J \subset \{1, \ldots, k+2\}$ such that $p_j \in Q_J$.
**Computation:**
1. For each $p_j$, if $p_j$ sends a message that is not appropriately signed or malformed, then $D \leftarrow D \cup \{j\}$.
2. Define the set $J$:
   - If $|D \cap \{k+2, \ldots, m\}| \geq m - t$ then $J = \{1, \ldots, k+1\} \setminus D$.
   - If $|D \cap \{k+2, \ldots, m\}| < m - t$ then $J = (\{1, \ldots, k+1\} \setminus D) \cup \{k+2\}$.
3. Reconstruct $\sigma_J^{i-1}$ from the shares of the active parties in $Q_J$.

**Outputs:** Each honest party $p_j$ output the value $\sigma_J^{i-1}$.

We next claim that Functionality Reconstruction is well-defined, that is, if the functionality is computed (after premature termination in round $i > 1$), then, indeed, $\sigma_J^{i-1}$ can be reconstructed. To see this, observe that the number of parties in the appropriate set $Q_J$ that participate in the computation (i.e., not in $D$) is at least the reconstruction threshold $o_J$: If $|D \cap \{k+2, \ldots, m\}| \geq m - t$, then $Q_J$ contains only active parties and $|Q_J| = o_J$. Notice that we already proved that in this case $|Q_J| \geq 1$. If $|D \cap \{k+2, \ldots, m\}| \leq m - t - 1$, then $o_J = |J| - 1 + m - t$ and $|Q_J \setminus \{p_j : j \in D\}| = |J| - 1 + |\{k+2, \ldots, m\} \setminus D|$. To prove that $|Q_J \setminus \{p_j : j \in D\}| \geq o_J$, it suffices to show that $|\{k+2, \ldots, m\} \setminus D| \geq (m - k - 1) - (m - t - 1) = t - k = t - (2t - m) = m - t$.

## 7 Coin-Tossing Protocol for Any Constant Fraction of Corrupted Parties

In this section we describe an $r$-round $m$-party coin-tossing protocol that tolerates up to $t$ dishonest parties, where $t$ is some constant fraction of $m$, that is, $t = (1 - \varepsilon)m$, for some (constant) $0 < \varepsilon$. The bias of our protocol is $O\left(\varepsilon / \sqrt{r - t}\right)$.

Before our work, the best known protocol for this scenario is an extension of Blum's two-party coin-tossing protocol [3] to an $r$-round $m$-party protocol that has bias $O\left(t/\sqrt{r-t}\right)$ [2, 6]. In this protocol, in each round $i$ of the protocol, the parties jointly select a random bit $\sigma_i$ in two phases. In the first phase, each party commits to a "private" random bit, and in the second phase the private bits are all revealed and the output bit $\sigma_i$ is taken to be the XOR of all private bits. The output of the whole protocol is taken to be the value of the majority of the $\sigma_i$'s. When there is a premature abort in round $i$, the remaining parties repeat the computation of round $i$ and continue with the prescribed computation.

Intuitively, the best strategy for a rushing adversary to bias the output of the protocol, say toward 0, is in each round $i$ to instruct a corrupted party to abort before the completion of the revealing phase if $\sigma_i = 1$. This is possible, since the rushing adversary learns $\sigma_i$ before the completion of round $i$, specifically, a corrupted party can delay its message until all honest parties reveal their bit. This can go on at most $t$ times, adding a total bias of $O\left(t/\sqrt{r}\right)$, whenever $r = \Omega(t^2)$.

We use the notion of cheat detection to limit the adversary to abort in a constant number of rounds. Roughly speaking, we follow the general structure of the above protocol in computing the $\sigma_i$'s and taking the majority over them. However, we compute each $\sigma_i$ using a secure with abort with cheat-detection protocol, such that either the computation is completed or at least a constant fraction of the malicious parties abort (specifically, $m - t$ malicious parties abort). Next, we briefly describe the computation of each $\sigma_i$ in our protocol. That is, we show how to obtain a constant round secure-with-abort with cheat detection protocol to compute a random bit that identifies at least $m - t$ cheating parties. Let $m_i$ be the number of active parties at the beginning of round $i$ and $t_i$ be a bound on the number of active corrupted parties at the beginning of round $i$ (that is, if $t'$ parties have aborted in rounds $1, \ldots, i-1$, then $t_i = t - t'$). We assume the existence of a constant round secure-with-abort with cheat detection protocol.

In the first phase, a preprocessing phase, active parties execute a constant round secure-with-abort with cheat detection protocol to compute a $(t_i + 1)$-out-of-$m_i$ secret sharing of a random bit $\sigma_i$. That is, at the end of this phase, each party holds a share in a $(t_i + 1)$-out-of-$m_i$ Shamir secret sharing scheme of $\sigma_i$. To confine adversarial strategies to aborts, the share that each party receives is signed and a verification key is given to all parties. In a second phase, a revealing phase, all parties reveal their shares and reconstruct $\sigma_i$. Broadcasting anything other than a signed share is treated as abort. To see that the above protocol achieves the required properties, observe that after the first phase the adversary cannot reconstruct $\sigma_i$. Thus, by aborting the preprocessing round, malicious parties cannot bias the output. We stress that they are able to cause the preprocessing phase to fail, at the cost of at least one malicious party being detected by all honest parties. in such a case, the preprocessing stage is repeated without the detected party. This, however, can only happen at most $t$ times in total, throughout the whole protocol. In the revealing phase, a rushing adversary is already able to learn $\sigma_i$ before the corrupted parties broadcast their messages and thus can bias the output by not broadcasting these messages. However, by the properties of the secret sharing scheme, at least $m - t$ parties will have to not broadcast their message, and hence, effectively abort the computation. Hence, the adversary can do this at most $\frac{1-2\varepsilon}{\varepsilon}$ times throughout the protocol, before an honest majority among active parties is guaranteed. Thus, the majority function is applied to $\Omega(r - t)$ random bits, of which the adversary can bias $\frac{1-2\varepsilon}{\varepsilon}$. Thus, the total bias of the protocol is $O\left(\frac{1}{\varepsilon\sqrt{r-t}}\right)$.

*Acknowledgments.* We are grateful to Yehuda Lindell for many helpful discussions and great advice. We thank Oded Goldreich and Gil Segev for suggesting this problem and for useful conversations.

# References

[1] Aumann, Y., Lindell, Y.: Security against covert adversaries: Efficient protocols for realistic adversaries. In: Vadhan, S.P. (ed.) TCC 2007. LNCS, vol. 4392, pp. 137–156. Springer, Heidelberg (2007)

[2] Averbuch, B., Blum, M., Chor, B., Goldwasser, S., Micali, S.: How to implement Bracha's $O(\log n)$ Byzantine agreement algorithm (1985) (unpublished manuscript)

[3] Blum, M.: Coin flipping by telephone a protocol for solving impossible problems. SIGACT News 15(1), 23–27 (1983)

[4] Blum, M., Micali, S.: How to generate cryptographically strong sequences of pseudo-random bits. SIAM J. on Computing 13, 850–864 (1984)

[5] Canetti, R.: Security and composition of multiparty cryptographic protocols. J. of Cryptology 13(1), 143–202 (2000)

[6] Cleve, R.: Limits on the security of coin flips when half the processors are faulty. In: Proc. of the 18th STOC, pp. 364–369 (1986)

[7] Cleve, R., Impagliazzo, R.: Martingales, collective coin flipping and discrete control processes (1993) (manuscript)

[8] Goldreich, O.: Foundations of Cryptography, Voume II Basic Applications. Cambridge University Press, Cambridge (2004)

[9] Gordon, D., Katz, J.: Partial fairness in secure two-party computation. Cryptology ePrint Archive, Report 2008/206 (2008), http://eprint.iacr.org/

[10] Katz, J.: On achieving the "best of both worlds" in secure multiparty computation. In: Proc. of the 39th STOC, pp. 11–20. ACM Press, New York (2007)

[11] Lindell, Y.: Parallel coin-tossing and constant-round secure two-party computation. J. of Cryptology 16(3), 143–184 (2003)

[12] Moran, T., Naor, M., Segev, G.: An optimally fair coin toss. In: Reingold, O. (ed.) TCC 2009. LNCS, vol. 5444, pp. 1–18. Springer, Heidelberg (2009)

[13] Pass, R.: Bounded-concurrent secure multi-party computation with a dishonest majority. In: Proc. of the 36th STOC, pp. 232–241 (2004)

[14] Rabin, T., Ben-Or, M.: Verifiable secret sharing and multiparty protocols with honest majority. In: Proc. of the 21st STOC, pp. 73–85 (1989)

# Multiparty Computation for Dishonest Majority: From Passive to Active Security at Low Cost

Ivan Damgård and Claudio Orlandi

Department of Computer Science, Aarhus University
{ivan,claudio}@cs.au.dk

**Abstract.** Multiparty computation protocols have been known for more than twenty years now, but due to their lack of efficiency their use is still limited in real-world applications: the goal of this paper is the design of efficient two and multi party computation protocols aimed to fill the gap between theory and practice. We propose a new protocol to securely evaluate reactive arithmetic circuits, that offers security against an active adversary in the universally composable security framework. Instead of the "do-and-compile" approach (where the parties use zero-knowledge proofs to show that they are following the protocol) our key ingredient is an efficient version of the "cut-and-choose" technique, that allow us to achieve active security for just a (small) constant amount of work more than for passive security.

## 1 Introduction

In multi party computation (MPC) a set of parties $(P_1, P_2, \ldots, P_n)$ owns some private inputs $(x_1, x_2, \ldots, x_n)$ and wants to compute some function $f$ of these inputs in such a way that the output $z = f(x_1, x_2, \ldots, x_n)$ is correct and even if $n - 1$ parties are corrupted and cooperate, they cannot learn more information about the honest party's input than what they can learn from their inputs and the output of the computation.

The first solutions for this problem were given by Yao [Yao82] for the two party case and by Goldreich, Micali and Wigderson [GMW87] for the multi party case. Those solutions provide computational security: if we are willing to assume that a majority of the parties are honest, information-theoretical secure solutions were introduced by Ben-Or, Goldwasser and Widgerson [BGW88] and Chaum, Crepeau and Damgård: [CCD88]. An unexpected advantage of the latter kind of protocols with respect to the former, is that information-theoretical secure protocols are more efficient than the computational secure one, and therefore have been implemented and successfully used to solve real-world problems [BCD+09], while protocols that are secure against a dishonest majority – and therefore consider a more realistic threat model, and in particular can be used in the crucial two-party setting – are still too cumbersome to be used in real life.

The goal of this paper is to fill this gap and design an efficient protocol for arithmetic MPC secure against a dishonest majority.

T. Rabin (Ed.): CRYPTO 2010, LNCS 6223, pp. 558–576, 2010.

Another advantage of the protocols in [BGW88, CCD88] over the ones in [Yao82, GMW87], is to provide security also in concurrent settings: when we run an MPC protocol over an Internet-like network, we need to be sure that the protocol remains secure also when other protocols are running over the network: in particular, the adversary might use the information that he gets running one protocol in order to break the security of the other one. The universally composable (UC) security framework of [Can01] provides a strong definition of security, and if a protocol is UC secure then we know that it's going to be secure also when arbitrarily composed with itself or other protocols. The protocols of [BGW88, CCD88] are secure also in the UC sense, while the security of [Yao82, GMW87] does not hold in the concurrent case.

We achieve the best of both worlds and present a *truly efficient* protocol that can be implemented and used in real life, and that guarantees static UC security against any dishonest majority. An earlier version of this protocol, described in [Orl09], has already been implmenented and tested by Jakobsen, Makkes and Nielsen in [JMN10], where timings for different level of security and circuit sizes can be found.

The price to pay when designing protocols secure against any dishonest majority is high. First of all, it is clearly impossible to guarantee termination, meaning that even if one single party leaves the protocol, the protocol is going to abort. Also, it is not possible to guarantee fairness for general MPC [Cle86], meaning that the adversary can see the output and then decide whether to let the honest parties receive their output or not.

Our protocol requires a (small) constant amount of public key operations per gate of the circuit. The protocol has a preprocessing flavor with a first (heavier) preprocessing phase and a (lighter) on-line phase of actual computation. The preprocessing phase is independent of the function to be computed and the inputs.

**Informal Theorem 1.** *Assuming semi-honest multiplication protocols and homomorphic trapdoor commitment schemes, there exist a protocol for arithmetic multi party computation that is UC secure against any dishonest majority.*

- *If $n$ parties want to preprocess $M$ multiplication gates with security $1 - 2^{-s}$, every party calls the multiplication protocol $n(5M + 18s)$ times.*
- *In the on-line phase, 3 commitments are computed for each multiplication gate.*

*State of the art:* The first solution for MPC with dishonest majority in the UC framework was given by Canetti, Lindell, Ostrovsky and Sahai [CLOS02]: while their construction is an important feasibility result, the protocol is completely impractical due to the use of generic zero-knowledge proofs.

Efficient solutions for MPC over Boolean circuits have been extensively investigated in the past years [LP07, LPS08, NO09, PSSW09]. For the case of arithmetic computation, a step towards efficient solutions has been taken by Cramer, Damgård and Nielsen in [CDN01, DN03], based on threshold homomorphic encryption: however efficient protocols for the distributed key generation phase are still lacking and the use of homomorphic encryption during the on-line computation makes these protocol impractical.

In a recent work Ishai, Prabhakaran and Sahai [IPS09], following the "MPC in the head" approach of [IPS08], present a protocol for arithmetic computation with characteristics similar to ours, but where the constants involved are significantly bigger. On the other hand, the focus of [IPS09] is on optimizing the amortized asymptotic complexity, ignoring multiplicative constants and low-order additive terms, whereas our goal is to optimize practical efficiency.

## 1.1   Main Ideas

*Secret representation:* We call a *shared commitment* a secret-shared value in $\mathbb{Z}_p$ between the parties: the sharing of a value $a$ is represented by an additive secret sharing of the value $a$ and some randomness $r$, together with a public homomorphic trapdoor commitment to $\mathsf{Comm}(a; r)$.

*MPC with a trusted dealer:* Suppose there exists a trusted dealer that provides the parties with random triplets of multiplicative shared commitments $\mathsf{Comm}(a), \mathsf{Comm}(b), \mathsf{Comm}(c)$, with $c = a \cdot b$, and additive sharings of the openings. We will call these commitments to random multiplications together with the sharing of their openings *multiplicative triplets* or *triplets* from now on.

Given access to this trusted dealer, the parties can efficiently compute any arithmetic circuit over the field: given that shared commitments are linear (the commitments are homomorphic and the openings additively shared), it is possible to evaluate additions without any interaction. Using circuit derandomization from [Bea91], it is possible to evaluate a multiplication in the circuit using one of the preprocessed triplets.

The resulting protocol is extremely efficient as the interaction is limited to the opening of a triplets of commitments for every multiplication gate in the circuit, and some local computation. As for security, $n - 1$ corrupted parties have no information about the honest party's inputs, and cannot force the computation to output the wrong value without breaking the binding property of the commitment scheme.

*Implementing the trusted dealer:* The main challenge of this paper is to implement the trusted dealer i.e., to generate the triplets in an efficient way. We start from any two party multiplication protocol that satisfies *strong semi-honest security*. This could be done using homomorphic encryption, OT, or other cryptographic assumptions, see for instance [IPS09]. Intuitively, a protocol is *strongly secure against a semi-honest adversary if 1) the security is guaranteed for any choice of the corrupted parties' randomness and 2) the view of the protocol commits the adversary to his randomness and given the view and the randomness it is possible to verify whether any party deviated from the protocol.[1]

---

[1] Most "natural" multiplication protocols satisfy these requirement. If not, they can be easily modified to do that.

The main challenge now is to turn this semi-honest protocol into a protocol with security against a malicious adversary in the UC setting. In order to do so, we will employ a kind of cut-and-choose technique reminiscent of the one from [NO09], that works as follow:

1. First, many random triplets are created.
2. Then, a fraction (say half) of the triplets are checked to detect cheating attempts. The parties randomly select a subset of the generated triplets and disclose the randomness that they used during the multiplication protocol. If any cheating is detected the protocol aborts, otherwise the parties proceed to the next step.
3. If the test goes through, we know that with high probability the adversary didn't cheat in most of the executions of the multiplication protocol. Given that any triplet is checked with probability $1/2$, if the adversary cheats in the generation of $s$ triplets the cheating will be detected during the test except with probability $2^{-s}$. So the honest parties can reasonably assume that if the test goes through there are no more than, say 80, triplets that were generated maliciously among the untested ones. For this informal description let's call a triplet *good* if it was honestly generated, and *bad* if it was maliciously generated. Given that the protocol to generate the triplets is semi-honest secure, a good triplet will satisfy correctness ($c = a \cdot b$) and privacy ($a, b$ are uniformly random in the view of the adversary), while a bad triplet might nor be correct nor private.
4. The triplets are checked for correctness: they are paired two-by-two, and a sanity-check is performed. If any bad triplet is found, the protocol aborts, otherwise we know that all the triplets are correct i.e. for every triplets it holds that $c = a \cdot b$. Every check "burns" one of the two triplets.
5. At this point we know that the triplets are correct, but still the adversary might have some extra knowledge about some of the honest parties' shares: So we combine the remaining triplets in such a way that we can "distill" $M$ fully private triplets from a set of $O(M + s)$ triplets, where $s$ of them might not be private. The way the triplets are combined can be seen as a new and unexpected application of packed Shamir's secret sharing [FY92].
6. The last step to achieve UC security is, informally, to ask every party to prove knowledge of their shares — thus ensuring input independence. To do that, the parties generate some random homomorphic UC commitments, and open the differences of the triplets and those commitments. Opening the differences between those commitments can be seen as a very simple proof of knowledge.

*UC commitments:* For the last step of the protocol sketched above, we need some UC commitments that are compatible with the homomorphic commitments used during the MPC protocol.

A really easy way to construct UC commitments is to ask a party to provide a commitment $\mathsf{Comm}(a; r)$ together with an encryption of its opening. The encryption is relative to a public key in the common reference string (CRS).

Therefore, the simulator (by choosing the CRS) can "extract" the commitment by decrypting the ciphertext. Clearly a malicious committer can encrypt something different than the opening of the commitment. To force honest behavior, we use again a cut-and-choose technique. This protocol also has a preprocessing flavor, with a heavier preprocessing phase and a light on-line phase.

**Informal Theorem 2.** *Assuming semantic secure encryption and trapdoor homomorphic commitment schemes, it is possible to implement UC commitments in the CRS model.*

- *The protocol generates $M$ secure UC commitments with probability $1 - 2^{-s}$ using $4M + 4s$ invocations of both primitives.*
- *The actual commit phase uses no cryptographic primitives and in the open phase 1 trapdoor commitment is verified.*

*Higher level operations:* Our protocols are designed to be compatible with higher level protocols to perform complex operation such as exponentiation, bit decomposition and comparison in an efficient way — as in [DFK$^+$06] and related work

## 2   Preliminaries

*Homomorphic commitment schemes:* A *double-trapdoor homomorphic commitment scheme* is defined by four efficient algorithms $(\mathsf{Gen}, \mathsf{Comm}, \mathsf{TOpen}, \odot)$, where $(ck, \tau_1, \tau_2) \leftarrow \mathsf{Gen}(1^\kappa, p)$ generates a commitment key together with two trapdoors, $C = \mathsf{Comm}_{ck}(x; r)$ takes a message $x \in \mathbb{Z}_p$ and randomness $r$ in the commitment randomness space $\mathcal{RC}$ and produces a commitment $C$. Using one of the trapdoors it is possible to *trapdoor open* a commitment $C$ to any message $x' \neq x$. Finally the plain-text space defined by the commitment key $ck$ is the field $\mathbb{Z}_p$ of prime order $p$, with $|p| > \kappa$, and the commitments are homomorphic meaning that $\mathsf{Comm}(x; r) \odot \mathsf{Comm}(y; s) = \mathsf{Comm}(x + y \mod p; r + s)$.[2]

**Definition 1.** *We call a tuple of algorithms* $(\mathsf{Gen}, \mathsf{Comm}, \mathsf{TOpen}, \odot)$ *a double-trapdoor homomorphic commitment scheme if: let* $(ck, \tau_1, \tau_2) \leftarrow \mathsf{Gen}(1^\kappa, p)$*, then the following properties hold:*

**Trapdoor Security:** *There is no PPT $A$ s.t. $\tau_{3-i} \leftarrow A(1^\kappa, ck, \tau_i)$.*
**Computational Binding:** *There is an efficient PPT $E$ s.t. $\tau \leftarrow E(ck, x, r, x', r')$ if $\mathsf{Comm}_{ck}(x; r) = \mathsf{Comm}_{ck}(x'; r')$, $x \neq x'$, with $\tau \in \{\tau_1, \tau_2\}$.*
**Statistical Hiding:** *$\forall x, x' \in \mathbb{Z}_p$ and randomness $r$, let $r'_i = \mathsf{TOpen}(C, x, r, x', \tau_i)$ with $i = 1, 2$ then $\mathsf{Comm}_{ck}(x; r) = \mathsf{Comm}_{ck}(x'; r'_i)$; moreover $r'_1, r'_2$'s distributions are statistically close.*

Intuitively we need the commitments to have two trapdoors because we need to argue that even after the simulator opens some commitments towards the

---

[2] To ease the notation, we will write $\mathcal{RC}$ as an additive group.

adversary using one of the trapdoor, the adversary still cannot break the binding property of the commitment scheme.

In [CD98] it has been shown that trapdoor homomorphic commitment schemes can be instantiated using any $q$-*one-way group homomorphism*: this primitive can be built from the discrete logarithm assumption, RSA, and other standard assumptions.

*Semi-honest multiplication protocol:* The building block of our protocol is any strong-semi-honest multiplication protocol $(c_1, c_2) \leftarrow \pi_{\mathrm{MUL}}(a, b)$ where $a, b \in \mathbb{Z}_p$ are respectively the first and the second party's inputs, $c_1$ is random in $\mathbb{Z}_p$ and $c_2 = a \cdot b - c_1 \mod p$.

The two party multiplication protocol can be instantiated using a variety of assumption, like homomorphic encryption, OT, and more. The exact requirements for the multiplication protocol are slightly stronger than the standard definition of semi-honest security. Most "natural" semi-honest multiplication protocol would satisfy this stronger requirement, or can be easily modified in order to do so. Intuitively we need the protocol to be 1) secure also if the adversary chooses maliciously the randomness for the corrupted parties and 2) the adversary cannot cheat during the protocol and then pretend that he behaved honestly, if that instance of the protocol is checked during the cut-and-choose.

More in detail, consider any two party semi-honest secure protocol $view \leftarrow \pi(r_1, r_2)$ where $r_i$ is the randomness used by $P_i$. Without loss of generality assume that $P_1$ is honest and fix his randomness $r_1$.

**Definition 2.** *A protocol $\pi$ is strongly secure against a semi-honest adversary if $\pi$ is 1) secure for any adversary that follows the protocol but chooses its random $r_2$ maliciously and 2) if $P_2^*$ deviates from the protocol $\pi$ it holds that either a) $P_2^*$ does not break the security of $\pi$ or b) for all PPT $P_2^* : r_2^* \leftarrow P_2^*(view, r_2)$ then $view \neq \pi(r_1, r_2^*)$ with all but negligible probability.*

## 3    MPC Protocol

In Figure 1 the ideal functionality $\mathcal{F}_{\mathrm{AMPC}}$ is presented. This ideal functionality allows $n$ parties to input values in $\mathbb{Z}_p$, manipulate them (via additions and multiplications) and output the result to a given party.

In this description of the protocol[3] we assume that the parties already have secure and authenticated point to point channels, and a functionality for broadcast. Also, following the modular spirit of the UC framework we will implement the protocol in the presence of a "trusted dealer" that gives to the party a public key for the commitment scheme, together with random shared commitments. The ideal functionality describing the behavior of this trusted dealer is detailed in Figure 2.

---

[3] In the full version of this paper [DO10] an actual instantiation of the protocol, for a specific choice of trapdoor commitments and multiplication protocol is presented.

---

The functionality $\mathcal{F}_{\text{AMPC}}$ has the following commands:

**Initialize:** On input $(init, p)$ from all parties, activate and store the modulo $p$.

**Rand:** On input $(rand, P_i, varid)$ from all parties $P_i$, with $varid$ a fresh identifier, pick $r \leftarrow \mathbb{Z}_p$ and store $(varid, r)$.

**Input:** On input $(input, P_i, varid, x)$ from $P_i$ and $(input, P_i, varid, ?)$ from all other parties, with $varid$ a fresh identifier, store $(varid, x)$.

**Add:** On command $(add, varid_1, varid_2, varid_3)$ from all parties (if $varid_1, varid_2$ are present in memory and $varid_3$ is not), retrieve $(varid_1, x)$, $(varid_2, y)$ and store $(varid_3, x + y \mod p)$.

**Multiply:** On input $(multiply, varid_1, varid_2, varid_3)$ from all parties (if $varid_1, varid_2$ are present in memory and $varid_3$ is not), retrieve $(varid_1, x)$, $(varid_2, y)$ and store $(varid_3, x \cdot y \mod p)$.

**Output:** On input $(output, P_i, varid)$ from all parties (if $varid$ is present in memory), retrieve $(varid, x)$ and output it to $P_i$.

---

**Fig. 1.** The ideal functionality for arithmetic MPC

---

The functionality $\mathcal{F}_{\text{RAND}}$ has the following commands.

**Initialize:** On input $(init, p)$ from all parties, activate, generate a key for a double-trapdoor homomorphic commitment scheme $ck \leftarrow \text{Gen}(1^\kappa, p)$ with plain-text space $\mathbb{Z}_p$ and send $ck$ to the parties.

**Req. Share:** On input $(share, sid, a_i, r_i, P_i)$, with $sid$ a fresh identifier, create and output a shared commitment $\text{Comm}_{ck}(a, r)$ with $a = \sum a_i, r = \sum r_i$.

---

**Fig. 2.** The ideal functionality that models $\mathcal{F}_{\text{RAND}}$

### 3.1 Notation and Library

We will call a *shared commitment* of $x$ (and write $[x]$) the following configuration: $P_i$, $i = 1, \ldots, n$ owns $x_i \in \mathbb{Z}_p$, $r_i$ in the commitment scheme's randomness space $\mathcal{RC}$ and $\text{Comm}_{ck}(x; r)$, where it holds that $x = \sum_{i=1}^{n} x_i \mod p$ and $r = \sum_{i=1}^{n} r_i$.

For convenience we define a library of commands that the parties can perform on shared commitments. Call H, C respectively the sets of honest parties and the set of corrupted parties. $H \cap C = \emptyset$ and $H \cup C = \{1, \ldots, n\}$. Finally $|H| \geq 1$. In Figure 3 some basic commands are introduced and in Figure 4 some advanced commands are defined.

### 3.2 On-Line Phase

As mentioned our protocol has two phases: the preprocessing phase described in Figure 7 produces many random triplets, and in the on-line phase the triplets are used to implement the ideal functionality $\mathcal{F}_{\text{AMPC}}$: the on-line protocol, detailed in Figure 5, is quite simple. Parties provide inputs and compute multiplications by opening differences between random commitments generated during the preprocessing and the actual values of the computation. The security of the protocol

---

**Share Secret:** To share an element $x \in \mathbb{Z}_p$, choose random $x_1, \ldots, x_{n-1} \in_R \mathbb{Z}_p$, define $x_n = x - \sum_{i=1}^{n-1} x_i \mod p$. Choose random $\rho_{x,1}, \ldots, \rho_{x,n} \in \mathcal{RC}$, define $\rho_x = \sum_{i=1}^n \rho_{x,i}$ and $C_x = \mathsf{Comm}_{ck}(x, \rho_x)$. Send $[x]_i = (x_i, \rho_{x,i}, C_x)$ to party $P_i$. We denote this operation by $[x] = \mathsf{Share}(P_i, x, \rho_x)$.

**Open Secret:** every party $P_i$ broadcasts a share pair $(x_i', \rho_{x,i}')$. The parties compute the sums $x', \rho_x'$ and check $\mathsf{Comm}_{ck}(x', \rho_x') \stackrel{?}{=} C_x$. If yes, output $x = x'$, else output $x = \bot$. We denote this operation by $x = \mathsf{Open}([x])$. If just a party $P_i$ should learn the output, we modify the above protocol in the sense that all parties send their shares to $P_i$, that verifies the correctness and outputs the result in the same way. We denote this operation by $x = \mathsf{OpenTo}(P_i, [x])$.

**Random Share:** To generate a share of a random element $r \in_R \mathbb{Z}_p$, party $P_i$ chooses at random $(r_i, \rho_{r,i}) \in_R \mathbb{Z}_p \times \mathcal{RC}$ and broadcast $C_r^i = \mathsf{Comm}_{ck}(r_i, \rho_{r,i})$. Every party computes $C_r = \prod_{i=1}^n C_r^i = \mathsf{Comm}_{ck}(r, \rho_r)$, where $r = \sum_{i=1}^n r_i, \rho_r = \sum_{i=1}^n \rho_{r,i}$. Party $P_i$ sets $[r]_i = (r_i, \rho_{r,i}, C_r)$. We denote this operation by $[r] = \mathsf{Rand}()$.

**Addition:** We denote by $[z] = [x] + [y]$ the following: each $P_i$ computes $[z]_i = [x]_i + [y]_i = (x_i + y_i \mod p, \rho_{x,i} + \rho_{y,i}, C_x \odot C_y)$. From now on we will write commands like $[z] = 3[x] - [y] + 2$ with the obvious semantic. Any additive constant $c$ can be interpreted as $[c]_1 = (c, 0, \mathsf{Comm}_{ck}(c, 0))$, and $[c]_i = (0, 0, \mathsf{Comm}_{ck}(c, 0))$ for $i \neq 1$.

Note that no communication is involved in this command.

**Fig. 3.** Basic commands on shared commitments

---

**Shift:** Assume the parties have a shared commitment $[r]$. Then we denote by $[x] = \mathsf{Shift}(P_i, x, [r])$ the following protocol:
1. $r = \mathsf{OpenTo}(P_i, [r])$;
2. $P_i$ broadcast $\Delta = r - x \mod p$;
3. $[x] = [r] - \Delta$;

**Multiplication:** Assume the parties have a triplet of shared commitments $([a], [b], [c])$. Then we define the following command $[z] = \mathsf{Mul}([x], [y], [a], [b], [c])$ (the output $z$ is equal to $x \cdot y$ if $c = a \cdot b$). The command is implemented as:
1. $d = \mathsf{Open}([x] - [a]); e = \mathsf{Open}([y] - [b])$;
2. $[z] = e[x] + d[y] - de + [c]$;

**Fig. 4.** Advanced commands for shared commitments

intuitively follows from the fact that the random preprocessing material is used to mask the actual values of the computation. Also, when a value is opened, the presence of the commitment prevents cheating parties to force a wrong output value.

### 3.3 Preprocessing

The main contribution of this paper is in the way the random triplets are generated. The task is to start from a strong semi-honest multiplication protocol as defined in Definition 2 and a dealer that provides random shared commitments as

---

The protocol implements $\mathcal{F}_{\mathrm{AMPC}}$'s commands in the following way:

**Initialize:** The parties invoke $\mathcal{F}_{\mathrm{RAND}}(init, p)$ and store $ck$. Run the preprocessing as in Figure 7 to produce a big enough set of triplets.

**Rand:** The parties invoke $\mathcal{F}_{\mathrm{RAND}}(share, varid)$ and store the commitment $[a]$.

**Input:** The parties invoke $\mathcal{F}_{\mathrm{RAND}}(share, varid)$ and store the commitment $[a]$, then perform $[x] = \mathsf{Shift}(P_i, x, [a])$.

**Add:** To add $[x], [y]$ with identifiers $varid_1, varid_2$ the parties perform $[z] = [x] + [y]$ and assign $[z]$ the identifier $varid_3$.

**Multiply:** To multiply $[x], [y]$ with identifiers $varid_1, varid_2$ the parties take a triplet $([a], [b], [c])$ from the set of the available ones, perform $[z] = \mathsf{Mul}([x], [y], [a], [b], [c])$ and assign $[z]$ the identifier $varid_3$ and remove $([a], [b], [c])$ from the set of the available triplets.

**Output:** To output $[x]$ with identifier $varid$ to $P_i$ perform $x = \mathsf{OpenTo}(P_i, [x])$.

---

**Fig. 5.** The on-line protocol $\Pi_{\mathrm{AMPC}}$

---

Every party $P_i$ does the following:

1. Choose random shares $a_i, b_i \in \mathbb{Z}_p$.
2. For all $j \neq i$, run $(d_{ij}, e_{ji}) \leftarrow \pi_{\mathrm{MUL}}(a_i, b_j)$ as party 1.
3. For all $j \neq i$, run $(d_{ji}, e_{ij}) \leftarrow \pi_{\mathrm{MUL}}(a_j, b_i)$ as party 2.
4. Set $c_i = a_i \cdot b_i + \sum_j d_{ij} + \sum_j e_{ij} \mod p$.
5. Choose $r_i, s_i, t_i \in \mathcal{RC}$, compute $A_i = \mathsf{Comm}_{ck}(a_i, r_i), B_i = \mathsf{Comm}_{ck}(b_i, s_i)$, $C_i = \mathsf{Comm}_{ck}(c_i, t_i)$, and broadcast $A_i, B_i, C_i$.
6. Everyone computes $A = \odot_i A_i, B = \odot_i B_i, C = \odot_i C_i$

---

**Fig. 6.** The protocol to generate one triplet $\Pi_{\mathrm{TRI}}$

described in Figure 2, and finish with a fully secure protocol that outputs triplets of multiplicative shared commitments. The main technical tool is a somewhat new and surprising application of packed Shamir's secret sharing [FY92].

We start with a protocol to generate one triplets: the parties use $\pi_{\mathrm{MUL}}$ to compute cross products of their shares and broadcast commitments to their shares (details are given in Figure 6). This protocol is not secure against a malicious adversary (that could cheat in $\pi_{\mathrm{MUL}}$ or commit to inconsistent values): Intuitively to achieve full security we need the following: 1) the triplets are correct i.e. $c = a \cdot b$, 2) the triplets are private i.e. $a, b$ are uniformly random in the view of the adversary and 3) the adversary knows his shares of the shared commitments. The protocol, presented in Figure 7 will proceed in steps and ensure one property after the other.

Note that in the protocol of Figure 7 every "distilled" triplets is the product of every produced triplets. This give a quadratic blow-up in local computation. A solution that might be more efficient in practice is to change the step **Privacy** as follows: instead of creating just one big polynomial, randomly partition the

Start by running $(1 + \lambda)(4M + 4B - 2)$ times the protocol $\Pi_{\text{TRI}}$. Call $\mathcal{M} = \{([a_i], [b_i], [c_i])\}_{i=1,\ldots,(1+\lambda)(4M+4B-2)}$ the set of produced triplets.

**Test:** Using $\mathcal{F}_{\text{RAND}}$ sample a string $t$ that determines a subset $\mathcal{T} \subset \mathcal{M}$ of size $\lambda(4M + 4B - 2)$. For every triplet in $\mathcal{T}$, the parties reveal all the randomness used during $\Pi_{\text{TRI}}, \pi_{\text{MUL}}$. If any cheating is detected the protocol aborts.

**Proof of Knowledge:** for each of the untested triplets $([a], [b], [c])$, sample three random shared commitments $[r], [s], [u]$ using $\mathcal{F}_{\text{RAND}}$ and perform $\text{Open}([r - a]), \text{Open}([s - b]), \text{Open}([u - c])$.

**Correctness:** For every pair of triplets left $([a], [b], [c])$ and $([x], [y], [z])$ do: using $\mathcal{F}_{\text{RAND}}$ sample a random $r \in \mathbb{Z}_p$. Compute $[c'] = \text{Mul}([a], [b], r[x], r[y], r^2[z])$. Then if $\text{Open}([c - c']) \neq 0$ abort the protocol, otherwise store $[a], [b], [c]$ for future use and drop $[x], [y], [z]$.

**Privacy:** We are now left with $2M + 2B - 1$ triplets. Let $d = M + B - 1$.

1. The parties have a set of $2d + 1$ triplets $([a_i], [b_i], [c_i])$, $i = 1, \ldots, 2d + 1$
2. The parties generate $d + 1$ random commitments $[f_1], \ldots, [f_{d+1}]$
3. The parties generate $d + 1$ random commitments $[g_1], \ldots, [g_{d+1}]$
4. Those commitments define two *random shared polynomials* $[F(x)], [G(x)]$ of degree $d$, where $[F(x)] := \sum_{i=1}^{d+1} \delta_i^{(d)}(x)[f_i]$, $[G(x)] := \sum_{i=1}^{d+1} \delta_i^{(d)}(x)[g_i]$, where:

$$\delta_i^d(x) = \prod_{i \neq j=1}^{d+1} \frac{x - j}{i - j}$$

5. The parties locally evaluate $[F(d + 2)], \ldots, [F(2d + 1)]$ and $[G(d + 2)], \ldots, [G(2d + 1)]$
6. For all $i = 1, \ldots, 2d + 1$, the parties compute $[h_i] := [F(i) \cdot G(i)]$ using one of the triplets $([a_i], [b_i], [c_i])$
7. These new shared commitments $[h_i]$, $i = 1, \ldots, 2d + 1$ define a new shared polynomial $[H(x)] := \sum_{i=1}^{2d+1} \delta_i^{(2d)}(x)[h_i]$ of degree $2d$.
8. The parties locally compute $M$ new triplets $[a_i'], [b_i'], [c_i']$ where $[a_i'] = [F(-i)], [b_i'] = [G(-i)], [c_i'] = [H(-i)]$, with $i = 1, \ldots, M$.

**Fig. 7.** The preprocessing protocol $\Pi_{\text{PRE}}$

remaining triplets in subset of smaller size and use many polynomials of smaller degree. The analysis of this kind of approach can be found in [NO09].

**Theorem 1.** *Let $\pi_{\text{MUL}}$ be a strong semi-honest secure two-party multiplication protocol and $\text{Comm}$ a double trapdoor homomorphic commitment scheme, then the protocol $\Pi_{\text{AMPC}}$ $(\kappa, B \log_2(1 + \lambda))$-securely implements $\mathcal{F}_{\text{AMPC}}$ in the $\mathcal{F}_{\text{RAND}}$-hybrid model against any static, active adversary that corrupts any number of parties.*

*Remark:* The statistical security of the protocol depends on both parameters $B$ and $\lambda$. In practice one can set $\lambda = 1/4$ and $B = 3.6s$, so to get a protocol that is secure except with probability $2^{-3.6 \log_2(5/4)s} < 2^{-s}$, where the total number of invocation to $\Pi_{\text{TRI}}$ is now less than $5M + 18s$.

*Proof (sketch):* The simulator $\mathcal{S}_{\text{AMPC}}$ simulates every call to $\mathcal{F}_{\text{RAND}}$ and keeps a copy of what the internal state of the corrupted parties should look like if they had followed the protocol. The simulator can do so as this state is uniquely determined by the output of $\mathcal{F}_{\text{RAND}}$ and the protocol execution. A description of the simulator is provided in Figure 8.

---

The simulator $\mathcal{S}_{\text{AMPC}}$ maintains at any point a copy of the shares of all parties (honest and corrupted).

**Initialize:** The simulator runs $(ck, \tau_1, \tau_2) \leftarrow \text{Gen}(1^\kappa)$, gives $ck$ to the parties, flips a coin $b$ and stores $\tau = \tau_{1+b}$, and discards $\tau_{2-b}$. Call *init* on the ideal functionality $\mathcal{F}_{\text{AMPC}}$. The simulator simulates the preprocessing by following the protocol in Figure 7 as an honest party would do, except that it reads the corrupted parties shares from $\mathcal{F}_{\text{RAND}}$.

**Rand:** Simulate the call to $\mathcal{F}_{\text{RAND}}$ by reading the corrupted parties shares and choose random $a_i, r_i$ for the honest parties. Call *rand* on the ideal functionality $\mathcal{F}_{\text{AMPC}}$, and store internally the shares for all parties.

**Output:** To simulate an output of $[x]$ to $P_i$ where $i \in \mathsf{H}$, the simulator receives $(x'_i, r'_i)$ from all corrupted parties $P_i$, $i \in \mathsf{C}$. Let $(x_i, r_i)$ be the internal shares of the simulator corresponding to $P_i$. If $\sum_{i \in \mathsf{C}} x_i = \sum_{i \in \mathsf{C}} x'_i$ and $\sum_{i \in \mathsf{C}} r_i = \sum_{i \in \mathsf{C}} r'_i$ call *output* on the ideal functionality, otherwise abort the protocol.

To simulate an output of $[x]$ to $P_i$ where $i \in \mathsf{C}$, the simulator receives $x'$ from the ideal functionality, and the sum of the internal shares $x_i, r_i$ is $x, r$, the opening of $C_x$. The simulator picks the smallest $j \in \mathsf{H}$, executes $r'_j = \text{TOpen}(x_j, r_j, x_j + (x - x'), \tau)$ and sends $(x_j + (x - x'), r'_j)$, and $(x_i, r_i)$ for all $i \in \mathsf{H}, i \neq j$ to the adversary.

**Input:** To simulate the call for $P_i$, with $i \in \mathsf{C}$ simulate the call to $\mathcal{F}_{\text{RAND}}$ as described above, and perform $[x] = \text{Shift}(P_i, x, [a])$ as the honest parties would do (check for the abort condition in **Open** as described before). Internally update all the parties shares. Given $\Delta$ and $a$, compute $x' = \Delta + a \mod p$ and input it in the ideal functionality $\mathcal{F}_{\text{AMPC}}$.

To simulate the call for $P_i$, with $i \in \mathsf{H}$ simulate the call to $\mathcal{F}_{\text{RAND}}$ as described above, and perform $[x] = \text{Shift}(P_i, 0, [a])$ (check for the abort condition in **Open** as described before). Internally update all the parties shares.

**Add:** Run the protocol honestly and update all the internal shares and call *add* on the ideal functionality $\mathcal{F}_{\text{AMPC}}$.

**Multiply:** Run the protocol honestly, updating all the internal shares (check for the abort condition in **Open** as described before). Call *multiply* on the ideal functionality $\mathcal{F}_{\text{AMPC}}$.

---

**Fig. 8.** The simulator $\mathcal{S}_{\text{AMPC}}$

*On-line security:* Define an hybrid game where the adversary is restricted to following the protocol during the preprocessing phase $\Pi_{\text{PRE}}$, and then behaves arbitrarily during the on-line phase. The view of the protocol (excluding the preprocessing) contains statistically no information about the actual values of the computation: every value that is opened in **Input, Multiply** is masked with fresh randomness, and the commitments are statistically hiding.

Then the only way that the environment can distinguish between the real and the ideal execution is by forcing an output towards an honest party (or an input of a dishonest party) to be incorrect.

To do that, the adversary needs to send a set of shares $(x'_i, r'_i)$ with $i \in C$ with $\sum x_i \neq \sum x'_i$ and such that $\mathsf{Comm}(\sum x_i; \sum r_i) = \mathsf{Comm}(\sum x'_i; \sum r'_i)$, where $(x_i, r_i)$ are the simulator's internal shares for the corrupted parties. Using $E$ in Definition 1 we can extract a trapdoor $\tau_{b^*}$ from these values. Given that the view of the simulated protocol is statistically independent of the trapdoor used by the simulator $\tau_b$, then $\Pr[b = b^*] = 1/2$ and we can turn an adversary that distinguish the the real and the ideal world with probability $1/2 + q$, $q$ non negligible, into an adversary that break the security of the commitment scheme with non-negligible probability $q/2$, and we reach a contradiction.

*Preprocessing security:* For the sake of simplicity, let's assume $n = 2$, $P_1$ honest and $P_2$ corrupted[4]. The UC simulator runs the preprocessing protocol as the honest party would. If the corrupted party send values that would make a honest party abort, the simulator inputs abort to $\mathcal{F}_{\mathrm{AMPC}}$ on behalf of the corrupted party. If the simulator does not input abort to $\mathcal{F}_{\mathrm{AMPC}}$, the simulator stores the corrupted party's shares of $[a], [b], [c]$, namely $(a_2, r_2, b_2, s_2, c_2, t_2)$ that he learns during **Proof of Knowledge** (by simulating $\mathcal{F}_{\mathrm{RAND}}$) and proceed to the on-line phase. The simulation of the preprocessing phase is perfect, as the simulator behaves exactly as an honest party. What remains to argue is that if the protocol did not abort at the end of the preprocessing phase, then the triplets are correct and the honest parties' shares are uniformly random in the adversary's view, even if the adversary is corrupted.

Note that $\Pi_{\mathrm{TRI}}$ securely produces random multiplicative triplets against a strong semi-honest adversary. In fact: $c = \sum_i c_i = \sum_i a_i b_i \sum_{i \neq j} d_{ij} + \sum_{i \neq j} e_{ij} = \sum_i a_i b_i + \sum_{i \neq j} a_i b_j = ab \mod p$. If $\mathcal{A}$ can cheat during $\Pi_{\mathrm{TRI}}$ and then pretend he didn't during **Test** it can be used to break either the strong semi-honest security of $\pi_{\mathrm{MUL}}$ or the binding property of $\mathsf{Comm}$.

The step **Test** doesn't leak any information as it can be simulated as detailed in Lemma 2. We can use Lemma 1 to *define* a *good* triplet to be one where the adversary could open the triplet during **Test** and make an honest party accept, and a *bad* triplet otherwise. Note that the lemma uses rewinding techniques: this is fine, as we do not use the lemma to extract the adversary shares — we do this in **Proof of Knowledge** — but to prove that the simulation is correct. From the properties of $\pi_{\mathrm{MUL}}$ we know that for a good triplet $c = a \cdot b$ and $a, b$ are random in the adversary's view except with negligible probability. Therefore after **Test** we know that (except with negligible probability) the number of bad triplets is bounded by some constant $B$ except with probability $(1 + \lambda)^{-B}$.

After the **Correctness** step, if the protocol doesn't abort the triplets are correct except with probability $1/p$: let $z = x \cdot y + \Delta_z \mod p$ and $c = a \cdot b + \Delta_c$,

---

[4] If more malicious parties are present, one can just think of all of them as a new party whose shares are the sum of their shares. Clearly introducing more honest parties will not help the adversary.

then $c' - c = r^2 \Delta_z - \Delta_c$ that is $\neq 0$ if $(\Delta_c, \Delta_z) \neq (0,0)$ with probability $1 - 1/p$ over the choice of $r$. Then if the adversary doesn't break the binding property of Comm and $c' - c \neq 0$ for any pair of triplets the protocol aborts.

In **Privacy** after the triplets are randomly partitioned, we know that the probability that there are more than $B$ bad triplets left is less than $(1 + \lambda)^{-B}$. Therefore the adversary knows less than $B$ points on the polynomials $F, G$ of degree $d$, so from Lagrange interpolation theory those polynomials have still $M + 1$ degrees of freedom in the adversary's view. So the adversary gains statistically no information about the newly generated $M$ triplets $[a'], [b'], [c']$ and, even after $M - 1$ of those will be opened during the protocol, the last unopened triplet is still random in his view.    □

# 4    UC Commitment Scheme

In this section we show how to implement $\mathcal{F}_{\text{RAND}}$. For the sake of simplicity, we present a two party protocol for UC commitments[5]. In order to produce a random commitment between $n$ parties as required by $\mathcal{F}_{\text{RAND}}$ it will suffice to let every party publish a commitment and, using the homomorphic properties of the commitment, sum them up.

The protocol generates many commitments at once in a preprocessing flavor and it is *efficient* in the sense that to construct $M$ UC commitments with security $s$, one needs $O(M + s)$ call to the primitives — the efficiency of the protocol is roughly the efficiency of the primitives used.

*Protocol idea:* To let a semi-honest party UC commit to a message $m$ one can use the following protocol: the committer sends the pair $\text{Comm}(m, r), \text{Enc}(m||r, s)$ to the receiver, where the encryption and the commitments are relative two public keys in the CRS. To open, the committer sends $m, r$. The commitment scheme is UC secure as, intuitively, the simulator can choose the CRS together with the secret key for Enc and the trapdoor for Comm. So if the sender is corrupted the simulator can extract the message from Enc and if the receiver is corrupted the simulator can open Comm to any value using the trapdoor. Clearly if the committer is corrupted by an active adversary, he can send an inconsistent pair and break the security of the protocol. We solve this by using the cut-and-choose approach to force honest behavior.

First the committer selects at random two polynomials $f$ and $g$ of degree $d = 2M + s - 1$ over $\mathbb{Z}_p$. Then the committer sends to the receiver commitments to $2M + 2s$ points on both polynomials using the semi-honest protocol. Now a random challenge is coin-flipped, in order to determine a subset of $M + s$ commitments to be checked. The committer reveals the points and the randomness used in the semi-honest protocol to the receiver, who aborts if any opening is inconsistent. If the protocol doesn't abort we know that, with probability $1 - 2^{-s}$, at least $M$ out of the $M + s$ unopened commitments are well-formed. Therefore the simulator learns the required $2M + s$ points that uniquely determine $f$: the

---

[5] We refer to [CF01] for the definition of the ideal functionality $\mathcal{F}_{\text{MCOMM}}$.

first $M + s$ are disclosed during the cut-and-choose, while the last $M$ are extracted from the unopened (but well-formed) commitments. Also note that any $M$ out of the $M + s$ unopened points are still uniformly random in the view of the receiver.

In order for this to work, we need to ensure that $f$ is of the right degree $d$ (or the simulator will not have enough points to determine $f$): to do so the receiver will send a random challenge $w \in \mathbb{Z}_p$ and the committer will reveal $h(i) = w \cdot f(i) + g(i)$ for all $i$'s. Thanks to the homomorphic properties of the commitment Comm the receiver can verify that the committer is not lying about these points, and he can check that $h$ has degree most $d$. This implies, with probability $1 - 1/p$, that $f$ and $g$ have degree at most $d$. In the test $g$ is used to mask $f$, so that the points on $f$ are still random to the receiver.

The protocol actually implements a random commitment functionality. If one wants to commit to specific messages it is always possible to derandomize the commitments (the committer simply sends the difference between the random committed value and the actual messages).

## 4.1 UC Commitments with Preprocessing

In Figure 9 the protocol for UC commitments with preprocessing is presented.

We write (Gen, Enc, Dec) for a semantically secure encryption scheme where $(ek, dk) \leftarrow \mathsf{Gen}(1^\kappa)$ is the key generation algorithm, $C = \mathsf{Enc}_{ek}(x, r)$ is an encryption of $x$ using randomness $r$ and given the decryption key $dk$ is possible to recover the message $x = \mathsf{Dec}_{dk}(C)$. Security is defined in the standard way.

**Theorem 2.** *The protocol $\Pi_{\mathrm{COMM}}$ securely $(\kappa, s)$-implements $\mathcal{F}_{\mathrm{MCOMM}}$ in the $\mathcal{F}_{\mathrm{CRS}}$-hybrid model.*

*Proof (sketch):* To simulate against a corrupted receiver, just run the protocol honestly but simulate the test as in Lemma 2 i.e. commit to random values. In **Degree check** choose a random polynomial $h$ consistent with the revealed values and trapdoor open the remaining commitments. In **Commit:** send random $\Delta_j$'s. When opening, use the trapdoor to open the commitment to the value $\Delta_j + m_j$ where $m_j$ is the message that the simulator receives from the ideal functionality. If the environment can distinguish, then it can be turned into an adversary that breaks semantic security of Enc using standard techniques.

In the more interesting case where the committer is corrupted, the proof follows the one of Theorem 1: we use Lemma 1 to define which pairs are good and which bad. After **Cut-and-Choose** the number of openings that the simulator cannot extract is bounded by $s$ with probability $2^{-s}$. Therefore the simulator can reconstruct the unique polynomial $f'(x)$ defined by the $M + s$ point seen during **Cut-and-Choose** and the $M$ points it can extract from the consistent pairs. Once the simulator knows $f'$ it can compute the $a_j$'s for all $j$'s. Therefore it can extract the committed messages in **Commit** by just computing $m'_j = a_j - \Delta_j$ mod $p$. The only way for the environment can distinguish the real game from the simulated one is by forcing an opening to a message $m_j$ different from the

---

Parse the common reference string $CRS$ as $(ek, ck)$.

**Generation:**
1. $P_r$ chooses two random polynomials $f, g$ of degree at most $d = 2M + s - 1$;
2. For $i = 1, \ldots, 2(M + s)$, $P_c$ computes and sends
   $F_i = \mathsf{Comm}_{ck}(f(i); r_i)$, $U_i = \mathsf{Enc}_{ek}(f(i) \| r_i; u_i)$,
   $G_i = \mathsf{Comm}_{ck}(g(i); t_i)$, $V_i = \mathsf{Enc}_{ek}(g(i) \| t_i; v_i)$;

**Cut-and-Choose:**
1. $P_c$ computes and send $E_c = \mathsf{Comm}_{ck}(e_c, r_c)$;
2. $P_r$ sends a challenge $e_r$;
3. $P_c$ opens $E_c$;
4. Let $e = e_c \oplus e_r$ define a random subset $\mathcal{T} \subset \{1, \ldots, 2(M+s)\}$ of size $M+s$;
5. For $i \in \mathcal{T}$ the committer $P_c$ sends $(f(i), r_i, u_i)$ and $(g(i), t_i, v_i)$. The receiver $P_r$ checks for consistency and abort otherwise;

**Degree Check:**
1. $P_r$ sends a random challenge $w$;
2. For $i \in \{1, \ldots, 2(M+s)\} \setminus \mathcal{T}$ the committer $P_c$ sends $h(i) = w \cdot f(i) + g(i)$ and $t_i = w \cdot r_i + s_i$;
3. The receiver $P_r$ checks that $(h(i), t_i)$ is a valid opening of $F_i^w \cdot G_i$, and that $h$ is a polynomial of degree at most $d$. If not abort;
4. We renumber sequentially the unopened commitments: Let $C_j$ denote the $j$-th unopened commitment $F_i$, and $(a_j, z_j)$ its opening. The committer outputs $(C_j, a_j, z_j)$ and the sender outputs $C_j$ for all $j = 1, \ldots, M$.

**Commit:** To commit to the $j$-th message $m_j$, $P_c$ sends $\Delta_j = a_j - m_j \mod p$.

**Open:** To open a commitment $C_j$, $P_c$ sends $(m_j, z_i)$ to $P_r$ that accepts if $C_j = \mathsf{Comm}(m_j + \Delta_j, z_j)$.

**Fig. 9.** The $\Pi_{\mathrm{COMM}}$ protocol

one extracted by the simulator $m'_j$. Such an environment can be turned into one that break the binding property of Comm using standard techniques.  □

*Multi-party case:* It is possible to extend the protocol to the case of multi receivers by replacing the random choices of the receiver with a coin flip protocol. If one wants to allow multiple parties to play as committer, several modification to the protocol can be considered:

- Use a longer CRS that contains $n$ key pairs $(ck_1, ek_1, \ldots, ck_n, ek_n)$, and every party commits using his own keys.
- If one wants to keep the CRS short, 1) Comm needs to be a double-trapdoor commitment scheme and 2) either one uses semantic secure encryption scheme, and require the preprocessing to run sequentially (at any given point just one party is acting as $P_c$ before **Commit**) or one can replace Enc with a CCA secure encryption – in this case different parties can all encrypt using the same public key and non-malleability is still guaranteed. The proof for the multi party protocol is essentially the same as the two-party case.

# 5   Cut-and-Choose Toolkit

In both the protocols presented in this paper we achieve security against a malicious adversary by using a kind of cut-and-choose reminiscent of the one first used in [NO09]. To make this paper self contained, we restate two useful lemmas: Let's just define a component to be the output of a one-way function $f : \mathcal{X} \to \mathcal{Y}$: an image is *good* if the sender knows the preimage and *bad* if he doesn't. The structure of a cut-and-choose is shown in Figure 10: we will argue the cut-and-choose can be efficiently simulated and if the adversary passes the test then most of the images are good. The first observation is that if the test goes through then there are at most $B$ bad images between the unchecked ones, except with probability $(1 + \lambda)^{-B}$.

---

**Test:** Let $\mathcal{M} = \{1, \ldots, (1 + \lambda)M\}$.
1. $P_1$ computes $y_i = f(x_i)$ for $i \in \mathcal{M}$ for random $x_i$ and sends them to $P_2$;
2. $P_2$ sends $P_1$ a random challenge $r$ that defines a random $\mathcal{T} \subset \mathcal{M}$ of size $\lambda M$;
3. $P_1$ sends $\{x_i\}_{i \in \mathcal{T}}$ to $P_2$;
4. $P_2$ accepts if $y_i = f(x_i)$ for all $i \in \mathcal{T}$;

---

**Fig. 10.** A simple cut-and-choose

**Lemma 1 (Extraction).** *There exist a knowledge extractor $E$ s.t. for any $P_1^*$ in Figure 10 the following holds: consider an augmented execution of Figure 10 where if $P_2$ accepts we run $E$ on $P_1^*$. Then: 1) The augmented execution terminates in expected poly-time and 2) The probability that we start the extractor $E$, and the extractor outputs less than $(1 + \lambda)M - B$ preimages $x_i$ is negligible in $B$.*

*Proof:* Let accept be the event of $P_2$ accepting the test. Assume $\mu = \Pr[\text{accept}] \geq 2(1 + \lambda)^{-B}$ for some constant $B$. Then $\mathcal{B}$, the set of bad components for which $P_1^*$ doesn't know an opening is small.

Formally let $r_i = 1$ if $i \in \mathcal{T}$ and $r_i = 0$ otherwise. Then $\mathcal{B} = \{i | \Pr[\text{accept}|r_i = 1] < \mu/2\}$, then $|\mathcal{B}| \leq B$. If not:

$$\mu = \Pr[\exists i \in \mathcal{B} : r_i = 1] \Pr[\text{accept}|\exists i \in \mathcal{B} : r_i = 1] + \\ \Pr[\forall i \in \mathcal{B} : r_i = 0] \Pr[\text{accept}|\forall i \in \mathcal{B} : r_i = 0] \\ < 1 \cdot \mu/2 + (1 + \lambda)^{-|\mathcal{B}|} \cdot 1$$

But then $\mu/2 < (1 + \lambda)^{-|\mathcal{B}|}$ and we have a contradiction.

Now consider the following extractor $E$ that sets $\mathcal{W} = \emptyset$ and while $|\mathcal{W}| < (1 + \lambda)M - B$, runs the test with $P_1^*$ and stores the new preimages he gets, $\mathcal{W} = \mathcal{W} \cup \{(i, x_i)\}_{i \in \mathcal{T}}$. The extractor keeps also a counter $j$ of the number of runs and if it didn't stop before it stops when $j > S = (1 + \lambda)^B \text{poly}(s)$. When it stops it outputs $\mathcal{W}$.

For any $i \in \mathcal{M} \setminus \mathcal{B}$, consider the probability $\nu$ that $(i, x_i) \notin \mathcal{W}$ when $E$ terminates. Formally $\nu = \Pr[(i, x_i) \notin \mathcal{W} \leftarrow E^{P_1^*}(1^s) | i \notin \mathcal{B}]$. Remember that the challenges are uniformly random and independent. Then assuming $\mu/2 \geq (1 + \lambda)^{-B}$:

$$\nu = \prod_j \left(1 - \Pr[r_i^{(j)} = 1 \wedge \mathtt{accept}^{(j)}]\right) \leq \left(1 - \frac{\mu}{2} \frac{\lambda}{1 + \lambda}\right)^S < e^{-\frac{\lambda}{1+\lambda} \mathrm{poly}(s)}$$

The expected running time is given by the probability that we start rewinding $\mu$ times the time that we spend doing the extraction. If $\mu < 2(1 + \lambda)^{-B}$, then the running time is bounded by $\mu \cdot S = \mathrm{poly}(s)$. If $\mu \geq 2(1 + \lambda)^{-B}$, then the extractor stops with success after expected time $S' = \frac{1+\lambda}{\lambda} \frac{2}{\mu} M$, and therefore the total expected running is $\mu \cdot S' = O(M)$.    □

**Lemma 2 (Simulation).** *For any honest $P_2$ there exist an expected poly-time simulator $\mathcal{S}$ for the test in Figure 10 s.t. the view of $P_2$ when interacting with an honest $P_1$ and the output of $\mathcal{S}$ are indistinguishable.*

*Proof:* Consider the $\mathcal{S}$ that is given as input a set $\mathcal{B}$ of up to $\lambda M$ random images $y_i$. $\mathcal{S}$ chooses a random challenge $r$ and orders the $y_i$'s in such a way that $\mathcal{T} \cap \mathcal{B} = \emptyset$. Then $\mathcal{S}$ fills $\mathcal{M}$ with $M$ random fresh images $y_i = f(x_i)$ for random $x_i$. The produced view is distributed exactly as in the protocol.    □

*Remarks:* It is possible to simulate against malicious $P_2^*$, by replacing step 2 in Figure 10 with a coin flip protocol, and in particular an UC coin flip protocol leads to a UC simulator for the test. This means that running the test doesn't give $P_2^*$ any advantage when he tries to invert the one way function on $y_i$, $i \notin \mathcal{T}$.

**Acknowledgments.** The authors would like to thank Jesper Buus Nielsen for the essential suggestions in the protocol design, and Yuval Ishai, Yehuda Lindell for valuable comments.

# References

[BCD+09]  Bogetoft, P., Christensen, D.L., Damgård, I., Geisler, M., Jakobsen, T., Krøigaard, M., Nielsen, J.D., Nielsen, J.B., Nielsen, K., Pagter, J., Schwartzbach, M., Toft, T.: Secure multiparty computation goes live. In: Dingledine, R., Golle, P. (eds.) FC 2009. LNCS, vol. 5628, pp. 325–343. Springer, Heidelberg (2009)

[Bea91]  Beaver, D.: Efficient multiparty protocols using circuit randomization. In: Feigenbaum, J. (ed.) CRYPTO 1991. LNCS, vol. 576, pp. 420–432. Springer, Heidelberg (1992)

[BGW88]  Ben-Or, M., Goldwasser, S., Wigderson, A.: Completeness theorems for non-cryptographic fault-tolerant distributed computation (extended abstract). In: STOC, pp. 1–10 (1988)

[Can01]    Canetti, R.: Universally composable security: A new paradigm for crypto-
           graphic protocols. In: FOCS, pp. 136–145 (2001)
[CCD88]    Chaum, D., Crépeau, C., Damgård, I.: Multiparty unconditionally secure
           protocols. In: Proceedings of the Twentieth Annual ACM Symposium on
           Theory of Computing, pp. 11–19 (1988)
[CD98]     Cramer, R., Damgård, I.: Zero-knowledge proofs for finite field arithmetic
           or: Can zero-knowledge be for free? In: Krawczyk, H. (ed.) CRYPTO 1998.
           LNCS, vol. 1462, pp. 424–441. Springer, Heidelberg (1998)
[CDN01]    Cramer, R., Damgård, I., Nielsen, J.B.: Multiparty computation from
           threshold homomorphic encryption. In: Pfitzmann, B. (ed.) EUROCRYPT
           2001. LNCS, vol. 2045, pp. 280–299. Springer, Heidelberg (2001)
[CF01]     Canetti, R., Fischlin, M.: Universally composable commitments. In: Kilian,
           J. (ed.) CRYPTO 2001. LNCS, vol. 2139, pp. 19–40. Springer, Heidelberg
           (2001)
[Cle86]    Cleve, R.: Limits on the security of coin flips when half the processors are
           faulty (extended abstract). In: STOC, pp. 364–369 (1986)
[CLOS02]   Canetti, R., Lindell, Y., Ostrovsky, R., Sahai, A.: Universally composable
           two-party and multi-party secure computation. In: STOC, pp. 494–503
           (2002)
[DFK+06]   Damgård, I., Fitzi, M., Kiltz, E., Nielsen, J.B., Toft, T.: Unconditionally
           secure constant-rounds multi-party computation for equality, comparison,
           bits and exponentiation. In: Halevi, S., Rabin, T. (eds.) TCC 2006. LNCS,
           vol. 3876, pp. 285–304. Springer, Heidelberg (2006)
[DN03]     Damgård, I., Nielsen, J.B.: Universally composable efficient multiparty
           computation from threshold homomorphic encryption. In: Boneh, D. (ed.)
           CRYPTO 2003. LNCS, vol. 2729, pp. 247–264. Springer, Heidelberg (2003)
[DO10]     Damgård, I., Orlandi, C.: Multiparty computation for dishonest majority:
           from passive to active security at low cost (full version) (2010),
           http://eprint.iacr.org/2010/318
[FY92]     Franklin, M.K., Yung, M.: Communication complexity of secure computa-
           tion (extended abstract). In: STOC, pp. 699–710 (1992)
[GMW87]    Goldreich, O., Micali, S., Wigderson, A.: How to play any mental game or
           a completeness theorem for protocols with honest majority. In: STOC, pp.
           218–229 (1987)
[IPS08]    Ishai, Y., Prabhakaran, M., Sahai, A.: Founding cryptography on oblivious
           transfer - efficiently. In: Wagner, D. (ed.) CRYPTO 2008. LNCS, vol. 5157,
           pp. 572–591. Springer, Heidelberg (2008)
[IPS09]    Ishai, Y., Prabhakaran, M., Sahai, A.: Secure arithmetic computation with
           no honest majority. In: Reingold, O. (ed.) TCC 2009. LNCS, vol. 5444, pp.
           294–314. Springer, Heidelberg (2009)
[JMN10]    Jakobsen, T.P., Makkes, M.X., Nielsen, J.D.: Efficient implementation of
           the Orlandi protocol. In: ACNS, pp. 255–272 (2010)
[LP07]     Lindell, Y., Pinkas, B.: An efficient protocol for secure two-party compu-
           tation in the presence of malicious adversaries. In: Naor, M. (ed.) EURO-
           CRYPT 2007. LNCS, vol. 4515, pp. 52–78. Springer, Heidelberg (2007)

[LPS08]    Lindell, Y., Pinkas, B., Smart, N.P.: Implementing two-party computation efficiently with security against malicious adversaries. In: Ostrovsky, R., De Prisco, R., Visconti, I. (eds.) SCN 2008. LNCS, vol. 5229, pp. 2–20. Springer, Heidelberg (2008)

[NO09]    Nielsen, J.B., Orlandi, C.: LEGO for two-party secure computation. In: Reingold, O. (ed.) TCC 2009. LNCS, vol. 5444, pp. 368–386. Springer, Heidelberg (2009)

[Orl09]    Orlandi, C.: LEGO and other cryptographic constructions. Technical report, Aarhus University (2009)

[PSSW09]    Pinkas, B., Schneider, T., Smart, N.P., Williams, S.C.: Secure two-party computation is practical. In: Matsui, M. (ed.) ASIACRYPT 2009. LNCS, vol. 5912, pp. 250–267. Springer, Heidelberg (2009)

[Yao82]    Yao, A.C.: Protocols for secure computations. In: Proceedings of the 23rd Annual IEEE Symposium on Foundations of Computer Science, pp. 160–164 (1982)

# Secure Multiparty Computation with Minimal Interaction

Yuval Ishai[1,*], Eyal Kushilevitz[2,**], and Anat Paskin-Cherniavsky[2]

[1] Computer Science Department, Technion and UCLA
yuvali@cs.technion.ac.il
[2] Computer Science Department, Technion
{eyalk,anatp}@cs.technion.ac.il

**Abstract.** We revisit the question of secure multiparty computation (MPC) with two rounds of interaction. It was previously shown by Gennaro et al. (Crypto 2002) that 3 or more communication rounds are necessary for general MPC protocols with guaranteed output delivery, assuming that there may be $t \geq 2$ corrupted parties. This negative result holds regardless of the total number of parties, even if *broadcast* is allowed in each round, and even if only *fairness* is required. We complement this negative result by presenting matching positive results.

Our first main result is that if only *one* party may be corrupted, then $n \geq 5$ parties can securely compute any function of their inputs using only *two* rounds of interaction over secure point-to-point channels (without broadcast or any additional setup). The protocol makes a black-box use of a pseudorandom generator, or alternatively can offer unconditional security for functionalities in $NC^1$.

We also prove a similar result in a client-server setting, where there are $m \geq 2$ clients who hold inputs and should receive outputs, and $n$ additional servers with no inputs and outputs. For this setting, we obtain a general MPC protocol which requires a single message from each client to each server, followed by a single message from each server to each client. The protocol is secure against a single corrupted client and against coalitions of $t < n/3$ corrupted servers.

The above protocols guarantee output delivery and fairness. Our second main result shows that under a relaxed notion of security, allowing the adversary to selectively decide (after learning its own outputs) which honest parties will receive their (correct) output, there is a general 2-round MPC protocol which tolerates $t < n/3$ corrupted parties. This protocol relies on the existence of a pseudorandom generator in $NC^1$ (which is implied by standard cryptographic assumptions), or alternatively can offer unconditional security for functionalities in $NC^1$.

**Keywords:** Secure multiparty computation, round complexity.

## 1 Introduction

This work continues the study of the round complexity of secure multiparty computation (MPC) [53,29,9,13]. Consider the following motivating scenario. Two or more

* Supported in part by ISF grant 1310/06, BSF grant 2008411 and NSF grants 0830803, 0716835, 0627781.
** Work done in part while visiting UCLA. Supported in part by ISF grant 1310/06 and BSF grant 2008411.

T. Rabin (Ed.): CRYPTO 2010, LNCS 6223, pp. 577–594, 2010.

employees wish to take a vote on some sensitive issue and let their manager only learn whether a majority of the employees voted "yes". Given an external trusted server, we have the following minimalist protocol: each employee sends her vote to the server, who computes the result and sends it to the manager.

When no single server can be completely trusted, one can employ an MPC protocol involving the employees, the manager, and (possibly) additional servers. A practical disadvantage of MPC protocols from the literature that offer security against malicious parties is that they involve a substantial amount of interaction. This interaction includes 3 or more communication rounds, of which at least one requires broadcast messages.

The question we consider is whether it is possible to obtain protocols with only two rounds of interaction, which resemble the minimal interaction pattern of the centralized trusted server solution described above. That is, we would like to employ several *untrusted* servers instead of a single trusted server, but still require each employee to only send a single message to each server and each server to only send a single message to the manager. (All messages are sent over secure point-to-point channels, without relying on a broadcast channel or any further setup assumptions.)

In a more standard MPC setting, where there are $n$ parties who may contribute inputs and expect to receive outputs, the corresponding goal is to obtain MPC protocols which involve only two rounds of point-to-point communication between the parties.

The above goal may seem too ambitious. In particular:

– Broadcast is a special case of general MPC, and implementing broadcast over secure point-to-point channels generally requires more than two rounds [23].
– Even if a free use of broadcast messages is allowed in each round, it is known that three or more communication rounds are necessary for general MPC protocols which tolerate $t \geq 2$ corrupted parties and guarantee output delivery, regardless of the total number of parties [26].

However, neither of the above limitations rules out the possibility of realizing our goal in the case of a *single* corrupted party, even when the protocols should guarantee output delivery (and in particular fairness). This gives rise to the following question:

*Question 1.* Are there general MPC protocols (i.e., ones that apply to general functionalities with $n$ inputs and $n$ outputs) that resist a single malicious party, guarantee output delivery, and require only two rounds of communication over point-to-point channels?

The above question may be highly relevant to real world situations where the number of parties is small and the existence of two or more corrupted parties is unlikely.

Another possibility left open by the above negative results is to tolerate $t > 1$ malicious parties by settling for a weaker notion of security against malicious parties. A common relaxation is to allow the adversary who controls the malicious parties to *abort* the protocol. There are several flavors of "security with abort." The standard notion from the literature (cf. [28]) allows the adversary to first learn the output, and then decide whether to (1) have the correct outputs delivered to the uncorrupted parties, or (2) abort the protocol and have *all* uncorrupted parties output a special abort symbol "⊥".

Unfortunately, the latter notion of security is not liberal enough to get around the first negative result. But it turns out that a further relaxation of this notion, which we refer to as security with *selective abort*, is not ruled out by either of the above negative results. This notion, introduced in [30], differs from the standard notion of security with abort in that it allows the adversary (after learning its own outputs) to individually decide for each uncorrupted party whether this party will obtain its correct output or will output "⊥".[1] Indeed, it was shown in [30] that two rounds of communication over point-to-point channels are sufficient to realize *broadcast* under this notion, with an arbitrary number of corrupted parties. This gives rise to the following question:

*Question 2.* Are there general MPC protocols that require only two rounds of communication over point-to-point channels and provide security with *selective abort* against $t > 1$ malicious parties?

We note that both of the above questions are open even if broadcast messages are allowed in each of the two rounds.

## 1.1  Our Results

We answer both questions affirmatively, complementing the negative results in this area with matching positive results.

- Our first main result answers the first question by showing that if only *one* party can be corrupted, then $n \geq 5$ parties can securely compute any function of their inputs with guaranteed output delivery by using only *two* rounds of interaction over secure point-to-point channels (without broadcast or any additional setup). The protocol can provide computational security for general functionalities (assuming one-way functions exist) or statistical security for functionalities in $NC^1$.
- We also prove a similar result in the client-server setting (described in the initial motivating example), where there are $m \geq 2$ clients who hold inputs and/or should receive outputs, and $n$ additional servers with no inputs and outputs. For this setting, we obtain a general MPC protocol which requires a single message from each client to each server, followed by a single message from each server to each client. The protocol is secure against a single corrupted client and against coalitions of $t < n/3$ corrupted servers,[2] and guarantees output delivery to the clients. We note that the proofs of the negative results from [26] apply to this setting as well, ruling out protocols that resist a coalition of a client and a server.

As is typically the case for protocols in the setting of an honest majority, the above protocols are in fact UC-secure [12,43]. Moreover, similarly to the constant-round protocols from [21,46] (and in contrast to the protocol from [7]), the general version of the above protocols can provide computational security while making only a *black-box* use

---

[1] Our notions of "security with abort" and "security with selective abort" correspond to the notions of "security with unanimous abort and no fairness" and "security with abort and no fairness" from [30]. We note that the negative result from [26] can be extended to rule out the possibility of achieving fairness in our setting with $t > 1$.

[2] Achieving the latter threshold requires the complexity of the protocol to grow exponentially in the number of servers $n$. When $t = O(n^{1/2} \log n)$, the complexity of the protocol can be made polynomial in $n$.

of a pseudorandom generator (PRG). This suggests that the protocols may be suitable for practical implementations.

Our second main result answers the second question, showing that by settling for security with selective abort, one can tolerate a constant fraction of corrupted parties:

- There is a general 2-round MPC protocol over secure point-to-point channels which is secure with *selective abort* against $t < n/3$ malicious parties. The protocol can provide computational security for general functionalities (assuming there is a PRG in $NC^1$, which is implied by most standard cryptographic assumptions [2]) or statistical security for functionalities in $NC^1$.

We note that the bound $t < n/3$ matches the security threshold of the best known 2-round protocols in the *semi-honest* model [9,7,33]. Thus, the above result provides security against malicious parties without any loss in round complexity or resilience. In the case of security against malicious parties, previous constant-round MPC protocols (e.g., the ones from [7,25,39]) require at least 3 rounds using broadcast, or at least 4 rounds over point-to-point channels using a 2-round implementation of broadcast with selective abort [30].

Our results are motivated not only by the quantitative goal of minimizing the amount of interaction, but also by several *qualitative* advantages of 2-round protocols over protocols with three or more rounds. In a client-server setting, a 2-round protocol does not require servers to communicate with each other or even to know which other servers are employed. The minimal interaction pattern also allows to break the secure computation process into two *non-interactive* stages of input contribution and output delivery. These stages can be performed independently of each other in an asynchronous manner, allowing clients to go online only when their inputs change, and continue to (passively) receive periodic outputs while inputs of other parties may change. Finally, their minimal interaction pattern allows for a simpler and more direct *security analysis* than that of comparable protocols from the literature with security against malicious parties.

## 1.2  Related Work

The round complexity of secure computation has been the subject of intense study. In the 2-party setting, 2-round protocols (in different security models and under various setup assumptions) were given in [53,52,10,31,15]. Constant-round 2-party protocols with security against malicious parties were given in [45,41,46,37,35]. In [41] it was shown that the optimal round complexity for secure 2-party computation without setup is 5 (where the negative result is restricted to protocols with black-box simulation).

More relevant to our work is previous work on the round complexity of MPC with an honest majority and guaranteed output delivery. In this setting, constant-round protocols were given in [4,7,6,5,33,25,17,34,19,21,39,40,16]. In particular, it was shown in [25] that 3 rounds are sufficient for general secure computation with $t = \Omega(n)$ malicious parties, where one of the rounds requires broadcast. Since broadcast in the presence of a single malicious party can be easily done in two rounds, this yields 4-round protocols in our setting. The question of minimizing the exact round complexity of MPC over point-to-point networks was explicitly considered in [39,40]. In contrast to the present work, the focus of these works is on obtaining nearly optimal resilience.

Two-round protocols with guaranteed output delivery were given in [26] for specific functionalities, and for general functionalities in [19,16]. However, the protocols from [19,16] rely on broadcast as well as *setup* in the form of correlated randomness.

The round complexity of verifiable secret sharing (VSS) was studied in [25,24,40,50]. Most relevant to the present work is the existence of a 1-round VSS protocol which tolerates a single corrupted party [25]. However, it is not clear how to use this VSS protocol for the construction of two-round MPC protocols. The recent work on the round complexity of *statistical VSS* [50] is also of relevance to our work. In the case where $n = 4$ and $t = 1$, this work gives a VSS protocol in which both the sharing phase and the reconstruction phase require two rounds. Assuming that two rounds of reconstruction are indeed necessary (which is left open by [50]), the number of parties in the statistical variant of our first main result is optimal. (Indeed, 4-party VSS with a single round of reconstruction reduces to 4-party MPC of a linear function, which is in $NC^1$.)

Finally, a non-interactive model for secure computation, referred to as the *private simultaneous messages* (PSM) model, was suggested in [22] and further studied in [32]. In this model, two or more parties hold inputs as well as a shared secret random string. The parties privately communicate to an external referee some predetermined function of their inputs by simultaneously sending messages to the referee. Protocols for the PSM model serve as central building blocks in our constructions. However, the model of [22] falls short of our goal in that it requires setup in the form of shared private randomness, it cannot deliver outputs to some of the parties, and does not guarantee output delivery in the presence of malicious parties.

**Organization.** Following some preliminaries (Section 2), Section 3 presents a 2-round protocol in the client-server model. Our first main result (a fully secure protocol for $t = 1$ and $n \geq 5$) is presented in Section 4 and our second main result (security with selective abort for $t < n/3$) in Section 5. For lack of space, some of the definitions and protocols, as well as most of the proofs, are deferred to the full version.

## 2 Preliminaries

### 2.1 Secure Computation

We consider $n$-party protocols that involve two rounds of synchronous communication over secure point-to-point channels. All of our protocols are secure against rushing, adaptive adversaries, who may corrupt at most $t$ parties for some specified security threshold $t$. See [11,12,28] and the full version for more complete definitions.

In addition to the standard simulation-based notions of *full security* (with guaranteed output delivery) and *security with abort*, we consider several other relaxed notions of security. Security in the *semi-honest model* is defined similarly to the standard definition, except that the adversary cannot modify the behavior of corrupted parties (only observe their secrets). *Privacy* is also the same as in the standard definition, except that the environment can only obtain outputs from the adversary (or simulator) and not from the uncorrupted parties. Intuitively, this only ensures that the adversary does not learn anything about the inputs of uncorrupted parties beyond what it could have learned by submitting to the ideal functionality some (distribution over) valid inputs. Privacy, however, does not guarantee any form of correctness. *Privacy with knowledge of outputs* is similar to privacy except that the adversary is also required to "know" the (possibly incorrect) outputs of the honest parties. This notion is defined similarly to full security (in particular, the environment receives outputs from both the simulator and the honest parties), with the difference that the ideal functionality first delivers the corrupted parties' output to the simulator, and then receives from the simulator an output to deliver to *each* of the uncorrupted parties. Finally, *security with selective abort* is defined similarly to security with abort, except that the simulator can decide for each uncorrupted party whether this party will receive its output or $\perp$.

## 2.2   The PSM Model

A *private simultaneous messages* (PSM) protocol [22] is a non-interactive protocol involving $m$ parties $P_i$, who share a common random string $r$, and an external referee who has no access to $r$. In such a protocol, each party sends a single message to the referee based on its input $x_i$ and $r$. These $m$ messages should allow the referee to compute some function of the inputs without revealing any additional information about the inputs. Formally, a PSM protocol for a function $f : \{0,1\}^{\ell \times m} \to \{0,1\}^*$ is defined by a randomness length parameter $R(\ell)$, $m$ message algorithms $A_1, ..., A_m$ and a reconstruction algorithm Rec, such that the following requirements hold.

- Correctness: for every input length $\ell$, all $x_1, ..., x_m \in \{0,1\}^\ell$, and all $r \in \{0,1\}^{R(\ell)}$, we have $\text{Rec}(A_1(x_1, r), ..., A_m(x_m, r)) = f(x_1, ..., x_m)$.
- Privacy: there is a simulator $S$ such that, for all $x_1, ..., x_m$ of length $\ell$, the distribution $S(1^\ell, f(x_1, ..., x_m))$ is indistinguishable from $(A_1(x_1, r), ..., A_m(x_m, r))$.

We consider either perfect or computational privacy, depending on the notion of indistinguishability. (For simplicity, we use the input length $\ell$ also as security parameter, as in [28]; this is without loss of generality, by padding inputs to the required length.)

A *robust* PSM protocol should additionally guarantee that even if a subset of the $m$ parties is malicious, the protocol still satisfies a notion of "security with abort." That is, the effect of the messages sent by corrupted parties on the output can be simulated by either inputting to $f$ a valid set of inputs (independently of the honest parties' inputs) or by making the referee abort. This is formalized as follows.

- Statistical robustness: For any subset $T \subset [m]$, there is an efficient (black-box) simulator $S$ which, given access to the common $r$ and to the messages sent by (possibly malicious) parties $P_i^*, i \in T$, can generate a distribution $x_T^*$ over $x_i, i \in T$, such that the output of Rec on inputs $A_T(x_T^*, r), A_{\bar{T}}(x_{\bar{T}}, r)$ is statistically close to the "real-world" output of Rec when receiving messages from the $m$ parties on a randomly chosen $r$. The latter real-world output is defined by picking $r$ at random, letting party $P_i$ pick a message according to $A_i$, if $i \notin T$, and according to $P_i^*$ for $i \in T$, and applying Rec to the $m$ messages. In this definition, we allow $S$ to produce a special symbol $\perp$ (indicating "abort") on behalf of some party $P_i^*$, in which case Rec outputs $\perp$ as well.

The following theorem summarizes some known facts about PSM protocols that are relevant to our work.

**Theorem 1.** *[22] (i) For any $f \in \mathrm{NC}^1$, there is a polynomial-time, perfectly private and statistically robust PSM protocol. (ii) For any polynomial-time computable $f$, there is a polynomial-time, computationally private and statistically robust PSM protocol which uses any pseudorandom generator as a black box.*

For self-containment, the full version contains a full description and proof of the robust variants, which are only sketched in [22, Appendix C].

## 2.3   Secret Sharing

An $(n,t)$-*threshold* secret sharing scheme, also referred to as a $t$-*private* secret sharing scheme, is an $n$-party secret sharing scheme in which every $t$ parties learn nothing about the secret, and every $t + 1$ parties can jointly reconstruct it. In this work, we rely on variants of several standard secret sharing schemes, such as Shamir's scheme [49], a bivariate version of Shamir's scheme [9], and the CNF scheme [36].

Recall that in Shamir's scheme over a finite field $\mathbb{F}$ (where $|\mathbb{F}| > n$), a secret $s \in \mathbb{F}$ is shared by picking a random polynomial $p$ of degree (at most) $t$ over $\mathbb{F}$ such that $p(0) = s$, and distributing to each party $P_i$ the value of $p$ on a distinct field element associated with this party. In the bivariate version of Shamir's scheme, $P_i$ receives the $i$'th row and column (from an $n \times n$ matrix of evaluations) of a random bivariate polynomial $p(x, y)$ of degree at most $t$ in each variable such that $p(0, 0) = s$. In the CNF scheme over an Abelian group $G$, a secret $s \in G$ is shared by first additively breaking it into $\binom{n}{t}$ shares, a share per size-$t$ subset of $[n]$, and then distributing to $P_i$ all shares corresponding those subsets $T$ such that $i \notin T$.

We will refer to a few abstract properties of secret sharing schemes which will be useful for our protocols. In a $d$-*multiplicative* secret sharing scheme over $\mathbb{F}$, each party should be able to apply a local computation (denoted MULT) on its shares of $d$ secrets, such that the outcomes of the $n$ local computations always add up to the product of the $d$ secrets (where addition and multiplication are in $\mathbb{F}$). The standard notion of multiplicative secret sharing from [20] corresponds to the case $d = 2$. The three concrete schemes, mentioned above, are $d$-multiplicative when $n > dt$.

Another property we will rely on is *pairwise verifiability*. This property has been implicitly used in the context of verifiable secret sharing [9,20,25]. In a pairwise verifiable scheme, checking that the shares are globally consistent (with some sharing of some secret) reduces to pairwise equality tests between values that are locally computed by pairs of parties. More concretely, each pair $i, j$ defines an equality test between a value computed from the share of $P_i$ and a value computed from the share of $P_j$. These $\binom{n}{2}$ equality tests should have the property that for any subset $T$ of two or more parties, if all $\binom{|T|}{2}$ tests involving parties in $T$ pass then the shares given to $T$ are consistent with some valid sharing of a secret. The CNF and bivariate Shamir schemes are pairwise verifiable. For instance, in the CNF scheme, each pair of parties compares the $\binom{n-2}{t}$ additive shares they should have in common. See the full version for more details, including a construction of an efficient secret sharing scheme over the binary field $\mathbb{F}_2$ which is both $d$-multiplicative and pairwise verifiable.

One last property we will need is *efficient extendability*. A secret sharing scheme is efficiently extendable, if for any subset $T \subseteq [n]$, it is possible to efficiently check whether the (purported) shares to $T$ are consistent with a valid sharing of some secret $s$. Additionally, in case the shares are consistent, it is possible to efficiently sample a (full) sharing of some secret which is consistent with that partial sharing. This property is satisfied, in particular, by the schemes mentioned above, as well as any so-called "linear" secret sharing scheme.

# 3   A Protocol in the Client-Server Model

In this section, we present a two-round protocol which operates in a setting where the parties consist of $m$ *clients* and $n$ *servers*. The clients provide the inputs to the protocol (in its first round) and receive its output (in its second round) but the "computation" itself is performed by the servers alone. Our construction provides security against any adversary that corrupts either a single client or at most $t$ servers. We refer to this kind of security as $(1, t)$-security[3]. The protocol in this setting illustrates some of the techniques we use throughout the paper, and it can be viewed as a warmup towards our main results; hence, we do not present here the strongest statement (e.g., in terms of resilience) and defer various improvements to the full version. Specifically, for any functionality $f \in$ POLY, we present a 2-round $(1, t)$-secure MPC protocols (with guaranteed output delivery) for $m \geq 2$ clients and $n = \Theta(t^3)$ servers. The protocol makes a black-box use of a PRG, or alternatively can provide unconditional security for $f \in$ NC$^1$.

*Tools.* Our protocol relies on the following building blocks:

1. An $(n, t)$-*secret sharing scheme* for which it is possible to check in NC$^1$ whether a set of more than $t$ shares is consistent with some valid secret. For instance, Shamir's scheme satisfies this requirement. Unlike the typical use of secret sharing in the context of MPC, our constructions do not rely on linearity or multiplication property of the secret sharing scheme.
2. A set system $\mathcal{T} \subseteq 2^{[n]}$ of size $\ell$ such that (a) $\mathcal{T}$ is *t-resilient*, meaning that every $B \subseteq [n]$ of size $t$ avoids at least $\ell/2 + 1$ sets; and (b) $\mathcal{T}$ is $(t + 1)$-*pairwise intersecting*, meaning that for all $T_1, T_2 \subseteq \mathcal{T}$ we have $|T_1 \cap T_2| \geq t + 1$. See the full version for a construction with $n = \Theta(t^3), \ell = poly(n)$.
3. A PSM protocol, with the best possible privacy (according to Theorem 1, either perfect or computational) for some functions $f'$ depending on $f$ (see below).

*Perfect security with certified randomness.* We start with a protocol for $m \geq 2$ clients and $n = \Theta(t^3)$ servers, denoted $\Pi^R$, that works in a scenario where each set of servers $T \in \mathcal{T}$, shares a common random string $r_T$ (obtained in a trusted setup phase). We explain how to get rid of this assumption later.

– **Round 1:** Each Client $i$ secret-shares its input $x_i$ among the $n$ servers using the $t$-private secret sharing scheme.

---

[3] Recall that the impossibility results of [26] imply that general 2-round protocols in this setting tolerating a coalition of a client and a server are impossible.

- **Round 2:** For each $T \in \mathcal{T}$ and $i \in [m]$, the set $T$ runs a PSM protocol with the shares $s$ received from the clients in Round 1 as inputs, $r_T$ as the common randomness, and Client $i$ as the referee (i.e., one message is sent from each server in $T$ to Client $i$). This PSM protocol computes the following functionality $f_i'$:
  - If all shares are consistent with some input value $x$, then $f_i'(s) = f(x)$.
  - Else, if the shares of a single Client $i$ are inconsistent, let $f_i'(s) = \bot$.
  - Otherwise, let $j$ be the smallest such that the shares of Client $j$ are inconsistent. Then, $f_i'(s)$ is an "accusation" of Client $j$; i.e., a pair $(j, f(x'))$, where $x'$ is obtained from $x$ by replacing $x_j$ with 0.
- **Reconstruction:** Each Client $i$ computes its output as follows: If all sets $T$ blame some Client $j$, then output the (necessarily unanimous) "backup" output $f(x')$ given by the PSM protocols. Otherwise, output the majority of the outputs reported by non-blaming sets $T$.

*Proof idea.* If the adversary corrupts at most $t$ servers (and no client), then privacy follows from the use of a secret sharing scheme (with threshold $t$). By the $t$-resilience of the set system, a majority of the sets $T \in \mathcal{T}$ contain no corrupted server and thus will not blame any client and will output the correct value $f(x)$.

If the adversary corrupts Client $j$, then all servers are honest. Every set $T \in \mathcal{T}$ either does not blame any client or blames Client $j$. Consider two possible cases: (a) Client $j$ makes all sets $T$ observe inconsistency: in such a case, Client $j$ receives $\bot$ from all $T$ and hence does not learn any information; moreover, all honest clients will output the same backup output $f(x')$. (b) Client $j$ makes some subsets $T$ observe consistent shares: since the intersection of every two subsets in $\mathcal{T}$ is of size at least $t + 1$ then, using the $(t + 1)$ reconstruction threshold of the secret sharing scheme, every two non-blaming sets must agree on the same input $x$. This means that Client $j$ only learns $f(x)$. Moreover, all other (honest) clients will receive the actual output $f(x)$ from at least one non-blaming set $T$ and, as discussed above, all outputs from non-blaming sets must agree.

Observe that the fact that a set $T$ uses the same random string $r_T$ in all $m$ PSM instances it participates in does not compromise privacy. This is because in each of them the output goes to a different client and only a single client may be corrupted.[4]

**Lemma 1.** $\Pi^R$ *is a 2-round, $(1, t)$-secure MPC protocol for $m > 1$ clients and $n = \Theta(t^3)$ servers, assuming that the servers in each set $T \in \mathcal{T}$ have access to a common random string $r_T$ (unknown to the clients). The security can be made perfect for $f \in \mathrm{NC}^1$, and computational for $f \in POLY$ by making a black-box use of a PRG.*

Note that the claim about $f \in \mathrm{NC}^1$ holds since the functions $f'$ evaluated by the PSM sub-protocols are in $\mathrm{NC}^1$ whenever $f$ is.

*Removing the certified randomness assumption.* If we have at least 4 clients, we can let each Client $i$ generate its own candidate PSM randomness $r_T^i$ and send it to all servers in $T$. Each PSM protocol, corresponding to some set $T$, is executed using each of these

---

[4] Alternatively, $r_T$ can be made sufficiently long so that the set $T$ can use a distinct portion of $r_T$ in each invocation of a PSM sub-protocol.

strings, where in the $i$-th invocation (using randomness $r_T^i$) Client $i$ receives no message (otherwise, the privacy of the protocol could be compromised). The other clients receive the original messages as prescribed by the PSM protocol. Upon reconstruction, Client $i$ lets the PSM output for a set $T$ be the majority over the $m - 1$ PSM outputs it sees. We observe that this approach preserves perfect security. If statistical (or computational) security suffices, we can rely on 2 or 3 clients as well. Here, the high level idea is to use a transformation as described for the case $m \geq 4$, and let Client $i$ authenticate the consistency of the randomness $r_T^j$ used in the $j$'th PSM protocol executed by set $T$, using a random string it sent in Round 1. Upon reconstruction, each client only considers the PSM executions which passed the authentication. Combining the above discussion with Theorem 1, we obtain the following theorem.

**Theorem 2.** *There exists a statistically $(1, t)$-secure 2-round general MPC protocol in the client-server setting for $m \geq 2$ clients and $n = \Theta(t^3)$ servers. For $f \in \mathrm{NC}^1$, the protocol is perfectly secure if $m \geq 4$, and statistically secure otherwise. The protocol is computationally secure for $f \in \mathrm{POLY}$.*

## 4  Full Security for $t = 1$

In this section, we return to the standard model where all parties may contribute inputs and receive outputs. We present a 2-round protocol in this model for $n \geq 5$ parties and $t = 1$. This protocol uses some similar ideas to our basic client-server protocol above, but it is different in the types of secret sharing scheme and set system that it employs. Specifically, we use the following ingredients:

1. A 1-private *pairwise verifiable* secret sharing scheme (see Section 2.3). For simplicity, we use here the CNF scheme, though one could use the bivariate version of Shamir's scheme for better efficiency. Recall that in the 1-private CNF scheme the secret $s$ is shared by first randomly breaking it into $n$ additive parts $s = s_1 + \ldots + s_n$, and then distributing each $s_i$ to all parties *except* for party $i$. Here we can view a secret as an element of $\mathbb{F}_2^m$.
2. A robust $(n - 2)$-party PSM protocol (see Section 2.2). In particular, such a PSM protocol ensures that the effect of any single malicious party on the output can be simulated in the ideal model (allowing the simulator to send "abort" to the functionality).
3. A simple set system, consisting of the $\binom{n}{2}$ sets $T_{i,j} = [n] \setminus \{i, j\}$. (Note that, for $n \geq 5$, we have $|T_{i,j}| \geq 3$.)

Again, we assume for simplicity that members of each set $T_{i,j}$ share common randomness $r_{i,j}$. Similarly to the client-server setting, this assumption can be eliminated by letting 3 of the parties in $T_{i,j}$ pick their candidate for $r_{i,j}$ and distributing it to the parties in the set (in Round 1 of our protocol), and then letting $T_{i,j}$ execute the PSM sub-protocol (in Round 2) using each of the 3 candidates and sending the outputs to $P_i, P_j$ (which are not in the set); the final PSM output will be the majority of these three outputs. Finally, for a graph $G$, let $\mathrm{VC}(G)$ denote the size of the *minimal vertex cover* in $G$.

Our protocol proceeds as follows:

- **Round 1:** Each party $P_k$ shares its input $x_k$ among all *other* parties using a 1-private, $(n - 1)$-party CNF scheme (i.e., each party gets $n - 2$ out of the $n - 1$ additive shares of $x_k$). In addition, to set up the consistency checks, each pair $P_i, P_j$ $(i < j)$ generates a shared random pad $s_{i,j}$ by having $P_i$ pick such a pad and send it to $P_j$.
- **Round 2:** For each "dealer" $P_k$, each pair $P_i, P_j$ send the $n - 3$ additive shares from $P_k$ they should have in common, masked with the pad $s_{i,j}$, to all parties.[5] Following this stage, each party $P_i$ has an inconsistency graph $G_{i,k}$ corresponding to each dealer $P_k$ $(k \neq i)$, with node set $[n] \setminus \{k\}$ and edge $(j, l)$ if $P_j, P_l$ report inconsistent shares from $P_k$.

  In addition, each set $T_{i,j}$ invokes a robust PSM protocol whose inputs are all the shares received (in Round 1) by the $n - 2$ parties in this set, and whose outputs to $P_i, P_j$ (which are not in $T_{i,j}$) are as follows:

  - If all input shares are consistent with some input $x$, then both $P_i, P_j$ receive $v = f(x)$.
  - Else, if shares originating from exactly one $P_k$ are inconsistent, then $P_k$ gets $\bot$ (in case $k \in \{i, j\}$) and the other party(s) get an "accusation" of $P_k$; namely, a pair $(k, x^*)$ where $x^* = (x_1, \ldots, x_{k-1}, x'_k, x_{k+1}, \ldots, x_n)$. Here, each $x_j$ (for $j \neq k$) is the protocol input recovered from the (consistent) shares and $x'_k = x_k$ if the shares of any $n - 3$ out of the $n - 2$ parties in $T_{i,j}$ are consistent with each other and $x'_k = 0$ (a default value) otherwise.
  - Else, if shares originating from more than one party are inconsistent, output $\bot$.

- **Reconstruction:** Each party $P_i$ uses the $n - 1$ inconsistency graphs $G_{i,k}$ $(k \neq i)$, and the PSM outputs that it received, to compute its final output:

  (a) If some inconsistency graph $G_{i,k}$ has $\mathrm{VC}(G_{i,k}) \geq 2$ then the PSM output of $T_{i,k}$ is of the form $(k, x^*)$; substitute $x_k^*$ by 0, to obtain $x'$, and output $f(x')$.

  Else, (b) if some inconsistency graph $G_{i,k}$ has a vertex cover $\{j\}$ and at least 2 edges, consider the PSM outputs of $T_{i,j}, T_{i,k}$ (assume that $i \neq j$; if $i = j$ it is enough to consider the output of $T_{i,k}$). If any of them outputs $v$ of the form $f(x)$ then output $v$; otherwise, if the output is of the form $(k, x^*)$, output $f(x^*)$.

  Else, (c) if some inconsistency graph $G_{i,k}$ contains exactly one edge $(j, j')$, consider the outputs of $T_{i,j}, T_{i,j'}$ (again, assume $i \notin \{j, j'\}$), and use any of them which is non-$\bot$ to extract the output (either directly, if the output is of the form $f(x)$, or $f(x^*)$ from an output $(k, x^*)$).

  Finally, (d) if all $G_{i,k}$'s are empty, find some $T_{i,j}$ that outputs $f(x)$ (with no accusation), and output this value.

Intuitively, a dishonest party $P_d$ may deviate from the protocol in very limited ways: it may distribute inconsistent shares (in Round 1) which will be checked (in Round 2) and will either be caught (if the inconsistency graph has VC larger than 1) or will be "corrected" (either to a default value or to its original input, if the VC is of size at most 1). $P_d$ may report false masked shares, for the input of some parties, but this will result in very simple inconsistency graphs (with vertex cover of size 1) that can be

---

[5] This is similar to Round 2 of the 2-round VSS protocol of [25], except that we use point-to-point communication instead of broadcast; note that, in our case, if the dealer is dishonest, then all other parties are honest.

detected and fixed. And, finally, $P_d$ may misbehave in the robust PSM sub-protocols (in which it participates) but this has very limited influence on their output (recall that, for sets in which $P_d$ participates, it does not receive the output). A detailed analysis appears in the full version. This proves:

**Theorem 3.** *There exists a general, 2-round MPC protocol for $n \geq 5$ parties which is fully secure (with guaranteed output delivery) against a single malicious party. The protocol provides statistical security for functionalities in $\mathrm{NC}^1$ and computational security for general functionalities by making a black-box use of a pseudorandom generator.*

## 5   Security with Selective Abort

This section describes our second main result; namely, a 2-round protocol which achieves security with *selective abort* against $t < n/3$ corruptions. This means that the adversary, after learning its own outputs, can selectively decide which honest parties will receive their (correct) output and which will output "$\bot$". More precisely, we prove the following theorem:

**Theorem 4.** *There exists a general 2-round MPC protocol for $n > 3t$ parties which is $t$-secure, with* selective abort. *The protocol provides statistical security for functionalities in $\mathrm{NC}^1$ and computational security for functionalities in POLY, assuming the existence of a pseudorandom generator in $\mathrm{NC}^1$.*

Our high-level approach is to apply a sequence of reductions, where the end protocol we need to construct only satisfies the relaxed notion of "privacy with knowledge of outputs", described in Section 2, and only applies to vectors of degree-3 polynomials. In particular,

1. We concentrate, without loss of generality, on functionalities which are deterministic with a public output.
2. We reduce (using unconditional one-time MACs) the secure evaluation of a function $f \in$ POLY to a private evaluation, with knowledge of outputs, of a related functionality $f' \in$ POLY. The reduction is statistical, and if $f \in \mathrm{NC}^1$ then so is $f'$.
3. We reduce the private evaluation with knowledge of outputs of a function $f' \in$ POLY to a private evaluation with knowledge of outputs of a related functionality $f''$, where $f''$ is a vector of degree-3 polynomials. The reduction (using [34]) is perfect for functions in $\mathrm{NC}^1$, and only computationally secure (using [2]) for general functionalities in POLY.
4. We present a 2-round protocol that allows $dt + 1$ parties to evaluate a vector of degree-$d$ polynomials, for all $d \geq 1$, and provides privacy with knowledge of outputs. In particular, for $d = 3$ the protocol requires $n = 3t + 1$ parties.

In the following subsections, we describe steps 2–4 in detail.

### 5.1   A Private Protocol with Knowledge of Outputs

In this section, we present a 2-round protocol for degree-$d$ polynomials which is private with knowledge of outputs. Let $p(x_1, \ldots, x_m)$ be a multivariate polynomial over a finite field $\mathbb{F}$, of total degree $d$. Assume, without loss of generality, that the degree of each monomial in p is exactly $d$.[6] Hence, $p$ can be written as $p = \sum_{g_1 \leq \ldots \leq g_d} \alpha_g \prod_{l=1}^{d} x_{g_l}$. We start by describing a protocol for evaluating $p$ with security in the semi-honest model. (This protocol is similar to previous protocols from [9,33].) The protocol can rely on any $d$-multiplicative secret sharing scheme over $\mathbb{F}$. Recall that, in such a scheme, each party should be able to apply a local computation MULT to the shares it holds of some $d$ secrets, to obtain an additive share of their product.

- **Round 1:** Each party $P_i$, $i \in [n]$, shares every input $x_h$ it holds by computing shares $(s_1^h, \ldots, s_n^h)$, using the $d$-multiplicative scheme, and distributes them among the parties. $P_i$ also distributes random *additive* shares of 0; i.e., it sends to each $P_j$ a field element $z_i^j$ such that $z_1^j, \ldots, z_n^j$ are random subject to the restriction that they sum up to 0.
- **Round 2:** Each party $P_i$, $i \in [n]$, computes $y_i = p_i(s_i^1, \ldots, s_i^m) + \sum_{j=1}^{n} z_i^j$, where
  $p_i(s_i^1, \ldots, s_i^m) \triangleq \sum_{g_1 \leq \ldots \leq g_d} \alpha_g \text{MULT}(i, s_i^{g_1}, \ldots, s_i^{g_d})$. It sends $y_i$ to all parties.
- **Outputs:** Each party computes and outputs $\sum_{i=1}^{n} y_i$ which is equal to $p(s^1, \ldots, s^m)$, as required.

We will refer to the above protocol as the "basic protocol". The proof of correctness and privacy in the semi-honest case are standard, and are omitted. Interestingly, this basic protocol happens to be *private* with knowledge of outputs (but not secure) against malicious parties for $d \leq 2$, when using Shamir's scheme as its underlying secret sharing scheme. However, the following simple example demonstrates that the basic protocol is not private against malicious parties already for $d = 3$.[7]

*Example 1.* Consider 4 parties where only $P_1$ is corrupted and the parties want to compute the degree-3 polynomial $x_1 x_2 x_3$ (party $P_4$ has no input). We argue that, when $x_3 = 0$, party $P_1$ can compute $x_2$, contradicting the privacy requirement. Let $q_2(z) = r_2 z + x_2$ and $q_3(z) = r_3 z$ be the polynomials used by $P_2, P_3$ (respectively) to share their inputs. Their product is $q(z) = r_2 r_3 z^2 + x_2 r_3 z$. Note that the messages sent by $P_1$ to the other 3 parties in Round 1 can make $P_1$ learn (in Round 2) an arbitrary linear combination of the values of $q(z)$ at 3 distinct points. Since the degree of $p$ is at most 2, this means that $P_1$ can also learn an arbitrary linear combination of the coefficients of $q$. In particular, it can learn $x_2 r_3$. This alone suffices to violate the privacy of $x_2$, because it can be used to distinguish with high probability between, say, the case where $x_2 = 0$ and the case $x_2 = 1$.

To prevent badly-formed shares from compromising privacy, we use the following variant of *conditional disclosure of secrets (CDS)* [27] as a building block. This primitive will allow an honest player to reveal a secret $s$ subject to the condition that two secret values $a, b$ held by other two honest players are equal.

---

[6] Otherwise, replace each monomial $m(x)$ of degree $d' < d$ by $m(x) \cdot x_0^{d-d'}$, where $x_0$ is a dummy variable whose value is set to 1 (by using some fixed valid $n$-tuple of shares).

[7] Note that degree-3 polynomials are "complete", in the sense that they can be used to represent every function, whereas degree-2 polynomials are not [33].

**Definition 1.** *An MCDS (multiparty CDS) protocol is a protocol for $n$ parties, which include three distinct special parties $S, A, B$. The sender $S$ holds a secret $s$, and parties $A, B$ hold inputs $a, b$ (respectively). The protocol should satisfy the following properties (as usual, the adversary is rushing).*

1. *If $a = b$, and $A, B, S$ are honest, then all honest parties output $s$.*
2. *If $a = b$, and $A, B$ are honest, then the adversary's view is independent of $a$, even conditioned on $s$.*
3. *If $a \neq b$, and $A, B, S$ are honest, then the adversary's view is independent of $s$, even conditioned on $a, b$.*

Note that there is no requirement when $a \neq b$ and some of the special parties are corrupted (e.g., a corrupted $A$ may still learn $s$). To be useful for our purposes, an MCDS protocol needs to have only two rounds, and also needs to satisfy the technical requirement that the message sent by $A$ and $B$ in the first round do not depend on the values $a$ and $b$.

A simple MCDS protocol with the above properties may proceed as follows (see the full version for a proof): In Round 1, party $A$ picks random independent values $r, z \in \mathbb{F}$ and sends them to $B$, and party $S$ sends $s$ to $A$. In Round 2, $A$ sends to each of the parties $m_A = a \cdot r - z + s$ and $B$ sends $m_B = z - b \cdot r$. Each party outputs $m_A + m_B$.

An MCDS protocol as above will be used to compile the basic protocol for $n = dt + 1$ semi-honest parties into a protocol $\Pi_{priv}$ which is private against malicious parties. For this, we instantiate the basic protocol with a $d$-multiplicative secret sharing scheme which is also pairwise-verifiable and efficiently extendable (see Section 2.3). More precisely, the parties run the basic protocol, and each party $P_i$ masks its Round 2 message with a sum of random independent masks $s_{i,j,k,h}$, corresponding to a shared input $x_h$ and a pair of parties $P_j, P_k$ (not holding $x_h$). In parallel, the MCDS protocol is executed for revealing each pad $s_{i,j,k,h}$ under the condition that the shares of $x_h$ given to $P_j$ and $P_k$ are consistent, as required by the pairwise verifiable scheme (where $a, b$ in the MCDS are values locally computed by $P_j, P_k$ that should be equal by the corresponding local check). Intuitively, this addresses the problem in Example 1 by ensuring that, if a party sends inconsistent shares of one of its inputs to the honest parties, some consistency check would fail (by pairwise-verifiability), and thus at least one random mask is not "disclosed" to the adversary, and so the adversary learns nothing.

The resulting protocol $\Pi_{priv}$ proceeds as follows:

- **Round 1:**
    - Each party $P_i$, $i \in [n]$ shares every input $x_h$ it holds by computing shares $(s_1^h, \ldots, s_n^h)$ and distributing them among the parties. Each $P_i$ also sends to each $P_j$ a share $z_j^i$ where $z_1^i, \ldots, z_n^i$ form a random additive sharing of 0.
    - Each triple of distinct parties $P_i, P_j, P_k$ such that $j < k$ runs, for each $h \in [m]$ such that $x_h$ is not held by $\{P_i, P_j, P_k\}$, Round 1 of the MCDS protocol (playing the roles of $S, A, B$ respectively), where all $n$ parties receive the MCDS output), with secret $s = s_{i,j,k,h}$, selected independently at random by $P_i$.
- **Round 2:**
    - Each party $P_i$, $i \in [n]$, computes $y_i = p_i(s_i^1, \ldots, s_i^m) + \sum_{j=1}^{n} z_i^j$, where $p_i(s_i^1, \ldots, s_i^m) \triangleq \sum_{g_1 \leq \ldots \leq g_d \in [m]} \alpha_g \mathrm{MULT}(i, s_i^{g_1}, \ldots, s_i^{g_d})$. It sends $y_i' \triangleq y_i + \sum_{j,k,h} s_{i,j,k,h}$ to all parties.

- Each triple of parties $P_i, P_j, P_k$ runs Round 2 of their MCDS protocols for each (relevant) $x_h$, where $a, b$ are the outputs of the relevant local computations applied to shares of $x_h$ held by $P_j, P_k$ which should be equal. Denote by $s^u_{i,j,k,h}$ the output of $P_u$ in this MCDS protocol.
- **Outputs:** Each party $P_u$ computes $\sum_{i=1}^{n} y'_i - \sum_{i,j,k,h} s^u_{i,j,k,h}$.

See the full version, for a proof of the following lemma.

**Lemma 2.** *Suppose $n = dt + 1$. Then the protocol $\Pi_{priv}$, described above, computes the degree-d polynomial p and satisfies statistical t-privacy with knowledge of outputs.*

*Remark 1.* The above protocol can be easily generalized to support a larger number of parties $n > dt + 1$. This can be done by letting all parties share their inputs among only the first $dt + 1$ parties in the first round, and letting only these $dt + 1$ parties reply to all parties in the second round. A similar generalization applies to the other protocols in this section.

Our protocols were described as if we need to evaluate a single polynomial. To evaluate a vector of polynomials (which is actually required for our application), we make the following observation. Both the basic semi-honest protocol and $\Pi_{priv}$ can be directly generalized to this case by running one copy of Round 1, except for the additive shares od 0 that are distributed for each output, and then executing Round 2 separately for each polynomial (using the corresponding additive shares). The analysis of the extended protocols is essentially the same. Combining $\Pi_{priv}$, instantiated with bivariate Shamir, with the above discussion, we get the following lemma:

**Lemma 3.** *For any $d \geq 1$ and $t \geq 1$, there exists a 2-round protocol for $n = dt + 1$ parties which evaluates a vector of polynomials of total degree d over a finite field $\mathbb{F}$ of size $|\mathbb{F}| \geq n$, such that the protocol is statistically t-private with knowledge of outputs.*

The transition from degree-3 polynomials to general functions $f \in \text{POLY}$ is essentially done by adapting known representations of general functions by degree-3 polynomials [34,2]. That is, securely evaluating $f(x_1, \ldots, x_m) : \{0, 1\}^m \to \{0, 1\}^*$ is reduced to securely evaluating a vector of randomized polynomials $p(x_1, \ldots, x_m, r_1, \ldots, r_l)$ of degree $d = 3$, over (any) finite field $\mathbb{F}_p$. However, the reduction is not guaranteed to work if the adversary shares a value of $x_i$'s which is not in $\{0, 1\}$. If the secret domain of the underlying secret sharing is $\mathbb{F}_2$, then the adversary is unable to share non-binary values, and there is no problem. This is the case with the CNF scheme over $\mathbb{F}_2$, but using $(3t + 1, t)$-CNF would result in exponential (in $n$) complexity for the protocol. An alternative approach is to rely on (say) bivariate Shamir, but using a variant of the above reduction from [14], applied to a function $f'$ over $\mathbb{F}^m$ (rather than $\{0, 1\}^m$) related to $f$, which is always consistent with $f(x)$, for some $x \in \{0, 1\}^m$. In particular, $f' \in \text{NC}^1$ if $f \in \text{NC}^1$ and $f' \in \text{POLY}$ if $f \in \text{POLY}$. Another solution is to devise an efficient 3-multiplicative, pairwise-verifiable $(3t + 1)$-party scheme over $\mathbb{F}_2$. See the full version, for more details on both solutions. We obtain the following:

**Lemma 4.** *Suppose there exists a PRG in $\text{NC}^1$. Then, for any n-party functionality $f$, there exists a 2-round MPC protocol which is (computationally) t-private with knowledge of outputs, assuming that $n > 3t$. Alternatively, the protocol can provide statistical (and unconditional) privacy with knowledge of outputs for $f \in \text{NC}^1$.*

### 5.2   From Privacy with Knowledge of Outputs to Security with Selective Abort

The final step in our construction is a reduction from secure evaluation of functions with selective abort to private evaluation with knowledge of outputs. For this, we make use of unconditional MACs. Our transformation starts with a protocol $\Pi'$ for evaluating a single output function $f$, which is private with knowledge of outputs. We then use $\Pi'$ to evaluate an augmented (single output) functionality $f'$, which computes $f$ along with $n$ MACs on the output of $f$, where the $i$-th MAC uses a private key chosen by party $P_i$ at random. That is, $f'$ takes an input $x$, and $k_i \in \mathcal{K}$ from each party $P_i$, and returns $y = f(x)$ along with $\texttt{MAC}(y, k_1), \ldots, \texttt{MAC}(y, k_n)$. The protocol $\Pi$ is obtained by running $\Pi'$ on $f'$ and having each party $P_i$ locally verify that the output $y$ it gets is consistent with the corresponding MAC. If so, then $P_i$ outputs $y$; otherwise, it outputs $\perp$. Intuitively, this is enough for getting security with selective abort since to make an uncorrupted party output an inconsistent value, the adversary would have to find $y'$ with $\texttt{MAC}(y', k) = \texttt{MAC}(y, k)$ for a random unknown $k$ and a known $y$, which can only be done with negligible probability. A formal construction and a proof of Theorem 4 appear in the full version.

*Acknowledgements.* We thank the anonymous CRYPTO 2010 referees for helpful comments and suggestions. We also would like the thank the third author's husband, Beni, for a lot of help on technical issues and for proofreading the paper.

# References

1. Alon, N., Merritt, M., Reingold, O., Taubenfeld, G., Wright, R.N.: Tight bounds for shared memory systems accessed by Byzantine processes. Journal of Distributed Computing 18(2), 99–109 (2005)
2. Applebaum, B., Ishai, Y., Kushilevitz, E.: Computationally private randomizing polynomials and their applications. Computational Complexity 15(2), 115–162 (2006)
3. Barrington, D.A.: Bounded-width polynomial-size branching programs recognize exactly those languages in NC1. In: Proc. 18th STOC, pp. 150–164 (1986)
4. Bar-Ilan, J., Beaver, D.: Non-cryptographic fault-tolerant computing in a constant number of rounds. In: Proc. 8th ACM PODC, pp. 201–209 (1989)
5. Beaver, D.: Minimal-Latency Secure Function Evaluation. In: Preneel, B. (ed.) EURO-CRYPT 2000. LNCS, vol. 1807, pp. 335–350. Springer, Heidelberg (2000)
6. Beaver, D., Feigenbaum, J., Kilian, J., Rogaway, P.: Security with low communication overhead (extended abstract). In: Menezes, A., Vanstone, S.A. (eds.) CRYPTO 1990. LNCS, vol. 537, pp. 62–76. Springer, Heidelberg (1991)
7. Beaver, D., Micali, S., Rogaway, P.: The round complexity of secure protocols (extended abstract). In: Proc. 22nd STOC, pp. 503–513 (1990)
8. Beimel, A.: Secure Schemes for Secret Sharing and Key Distribution. Phd. thesis. Dept. of Computer Science (1996)
9. Ben-Or, M., Goldwasser, S., Wigderson, A.: Completeness Theorems for Noncryptographic Fault-Tolerant Distributed Computations. In: Proc. 20th STOC 1988, pp. 1–10 (1988)
10. Cachin, C., Camenisch, J., Kilian, J., Muller, J.: One-round secure computation and secure autonomous mobile agents. In: Welzl, E., Montanari, U., Rolim, J.D.P. (eds.) ICALP 2000. LNCS, vol. 1853, p. 512. Springer, Heidelberg (2000)
11. Canetti, R.: Security and composition of multiparty cryptographic protocols. Journal of Cryptology 13(1), 143–202 (2000)

12. Canetti, R.: Universally composable security: A new paradigm for cryptographic protocols.cfik03. In: FOCS, pp. 136–145 (2001)
13. Chaum, D., Crepeau, C., Damgard, I.: Multiparty Unconditionally Secure Protocols. In: Proc. 20th STOC 1988, pp. 11–19 (1988)
14. Cramer, R., Fehr, S., Ishai, Y., Kushilevitz, E.: Efficient Multi-party Computation over Rings. In: Biham, E. (ed.) EUROCRYPT 2003. LNCS, vol. 2656, pp. 596–613. Springer, Heidelberg (2003)
15. Choi, S.G., Elbaz, A., Juels, A., Malkin, T., Yung, M.: Two-Party Computing with Encrypted Data. In: Kurosawa, K. (ed.) ASIACRYPT 2007. LNCS, vol. 4833, pp. 298–314. Springer, Heidelberg (2007)
16. Choi, S.G., Elbaz, A., Malkin, T., Yung, M.: Secure Multi-party Computation Minimizing Online Rounds. In: Matsui, M. (ed.) ASIACRYPT 2009. LNCS, vol. 5912, pp. 268–286. Springer, Heidelberg (2009)
17. Cramer, R., Damgård, I.: Secure distributed linear algebra in a constant number of rounds. In: Kilian, J. (ed.) CRYPTO 2001. LNCS, vol. 2139, p. 119. Springer, Heidelberg (2001)
18. Cramer, R., Damgård, I., Dziembowski, S., Hirt, M., Rabin, T.: Efficient multiparty computations with dishonest minority. In: Stern, J. (ed.) EUROCRYPT 1999. LNCS, vol. 1592, pp. 311–326. Springer, Heidelberg (1999)
19. Cramer, R., Damgård, I., Ishai, Y.: Share conversion, pseudorandom secret-sharing and applications to secure computation. In: Kilian, J. (ed.) TCC 2005. LNCS, vol. 3378, pp. 342–362. Springer, Heidelberg (2005)
20. Cramer, R., Damgård, I., Maurer, U.M.: General Secure Multi-party Computation from any Linear Secret-Sharing Scheme. In: Preneel, B. (ed.) EUROCRYPT 2000. LNCS, vol. 1807, pp. 316–334. Springer, Heidelberg (2000)
21. Damgård, I., Ishai, Y.: Secure multiparty computation using a black-box pseudorandom generator. In: Proc. CRYPTO 2005 (2005)
22. Feige, U., Kilian, J., Naor, M.: A minimal model for secure computation. In: Proc. 26th STOC, pp. 554–563 (1994)
23. Fischer, M.J., Lynch, N.A.: A lower bound for the time to assure interactive consistency. Information Processing Letters 14(4), 183–186 (1982)
24. Fitzi, M., Garay, J.A., Gollakota, S., Rangan, C.P., Srinathan, K.: Round-Optimal and Efficient Verifiable Secret Sharing. In: Halevi, S., Rabin, T. (eds.) TCC 2006. LNCS, vol. 3876, pp. 329–342. Springer, Heidelberg (2006)
25. Gennaro, R., Ishai, Y., Kushilevitz, E., Rabin, T.: The Round Complexity of Verifiable Secret Sharing and Secure Multicast. In: Proc. 33th STOC (2001)
26. Gennaro, R., Ishai, Y., Kushilevitz, E., Rabin, T.: On 2-Round Secure Multiparty Computation. In: Yung, M. (ed.) CRYPTO 2002. LNCS, vol. 2442, pp. 178–193. Springer, Heidelberg (2002)
27. Gertner, Y., Ishai, Y., Kushilevitz, E., Malkin, T.: Protecting Data Privacy in Private Information Retrieval Schemes. In: STOC 1998, pp. 151–160 (1998)
28. Goldreich, O.: Foundations of Cryptography: Basic Applications. Cambridge University Press, Cambridge (2004)
29. Goldreich, O., Micali, S., Wigderson, A.: How to Play Any Mental Game. In: Proc. 19th STOC, pp. 218–229 (1987)
30. Goldwasser, S., Lindell, Y.: Secure Multi-Party Computation without Agreement. J. Cryptology 18(3), 247–287 (2005)
31. Horvitz, O., Katz, J.: Universally-Composable Two-Party Computation in Two Rounds. In: Menezes, A. (ed.) CRYPTO 2007. LNCS, vol. 4622, pp. 111–129. Springer, Heidelberg (2007)
32. Ishai, Y., Kushilevitz, E.: Private simultaneous messages protocols with applications. In: ISTCS 1997, pp. 174–184 (1997)

33. Ishai, Y., Kushilevitz, E.: Randomizing polynomials: A new representation with applications to round-efficient secure computation. In: Proc. 41st FOCS (2000)
34. Ishai, Y., Kushilevitz, E.: Perfect Constant-Round Secure Computation via Perfect Randomizing Polynomials. In: Widmayer, P., Triguero, F., Morales, R., Hennessy, M., Eidenbenz, S., Conejo, R. (eds.) ICALP 2002. LNCS, vol. 2380, p. 244. Springer, Heidelberg (2002)
35. Ishai, Y., Prabhakaran, M., Sahai, A.: Founding Cryptography on Oblivious Transfer - Efficiently. In: Wagner, D. (ed.) CRYPTO 2008. LNCS, vol. 5157, pp. 572–591. Springer, Heidelberg (2008)
36. Ito, M., Saito, A., Nishizeki, T.: Secret sharing scheme realizing general access structure. Electronics and Communications in Japan, Part III: Fundamental Electronic Science 72(9), 56–64
37. Jarecki, S., Shmatikov, V.: Efficient Two-Party Secure. Computation on Committed Inputs. In: Naor, M. (ed.) EUROCRYPT 2007. LNCS, vol. 4515, pp. 97–114. Springer, Heidelberg (2007)
38. Karchmer, M., Wigderson, A.: On Span Programs. In: Proceedings of the 8th Structures in Complexity conference, pp. 102–111 (1993)
39. Katz, J., Koo, C.-Y.: Round-Efficient Secure Computation in Point-to-Point Networks. In: Naor, M. (ed.) EUROCRYPT 2007. LNCS, vol. 4515, pp. 311–328. Springer, Heidelberg (2007)
40. Katz, J., Koo, C.-Y., Kumaresan, R.: Improving the Round Complexity of VSS in Point-to-Point Networks. In: Aceto, L., Damgård, I., Goldberg, L.A., Halldórsson, M.M., Ingólfsdóttir, A., Walukiewicz, I. (eds.) ICALP 2008, Part II. LNCS, vol. 5126, pp. 499–510. Springer, Heidelberg (2008)
41. Katz, J., Ostrovsky, R.: Round-Optimal Secure Two-Party Computation. In: Franklin, M. (ed.) CRYPTO 2004. LNCS, vol. 3152, pp. 335–354. Springer, Heidelberg (2004)
42. Katz, J., Ostrovsky, R., Smith, A.: Round Efficiency of Multi-party Computation with a Dishonest Majority. In: Biham, E. (ed.) EUROCRYPT 2003. LNCS, vol. 2656, pp. 578–595. Springer, Heidelberg (2003)
43. Kushilevitz, E., Lindell, Y., Rabin, T.: Information-theoretically secure protocols and security under composition. In: STOC 2006, pp. 109–118 (2006), Full version: Cryptology ePrint Archive, Report 2009/630 (2009)
44. Lamport, L., Shostack, R.E., Pease, M.: The Byzantine generals problem. ACM Trans. Prog. Lang. and Systems 4(3), 382–401 (1982)
45. Lindell, Y.: Parallel Coin-Tossing and Constant-Round Secure Two-Party Computation. In: Kilian, J. (ed.) CRYPTO 2001. LNCS, vol. 2139, pp. 171–189. Springer, Heidelberg (2001)
46. Lindell, Y., Pinkas, B.: An efficient protocol for secure two-party computation in the presence of malicious adversaries. In: Naor, M. (ed.) EUROCRYPT 2007. LNCS, vol. 4515, pp. 52–78. Springer, Heidelberg (2007)
47. Lynch, N.: Distributed Algorithms. Morgan Kaufmann, San Francisco (1996)
48. Pass, R.: Bounded-concurrent secure multi-party computation with a dishonest majority. In: Proc. STOC 2004, pp. 232–241 (2004)
49. Shamir, A.: How to share a secret. Communications of the ACM 22, 612–613
50. Patra, A., Choudhary, A., Rabin, T., Rangan, C.P.: The Round Complexity of Verifiable Secret Sharing Revisited. In: Halevi, S. (ed.) CRYPTO 2009. LNCS, vol. 5677, pp. 487–504. Springer, Heidelberg (2009)
51. Rabin, T., Ben-Or, M.: Verifiable Secret Sharing and Multiparty Protocols with Honest Majority. In: Proc. 21st STOC, pp. 73–85 (1989)
52. Sander, T., Young, A., Yung, M.: Non-Interactive CryptoComputing For NC1. In: Proc. 40th FOCS, pp. 554–567. IEEE, Los Alamitos (1999)
53. Yao, A.C.-C.: How to Generate and Exchange Secrets. In: Proc. 27th FOCS, pp. 162–167. IEEE, Los Alamitos (1986)

# A Zero-One Law for Cryptographic Complexity with Respect to Computational UC Security[*]

Hemanta K. Maji[1], Manoj Prabhakaran[1], and Mike Rosulek[2]

[1] Department of Computer Science, University of Illinois, Urbana-Champaign
{hmaji2,mmp}@uiuc.edu
[2] Department of Computer Science, University of Montana
mikero@cs.umt.edu

**Abstract.** It is well-known that most cryptographic tasks do not have universally composable (UC) secure protocols, if no trusted setup is available in the framework. On the other hand, if a task like fair coin-tossing is available as a trusted setup, then *all* cryptographic tasks have UC-secure protocols. What other trusted setups allow UC-secure protocols for all tasks? More generally, given a particular setup, what tasks have UC-secure protocols?

We show that, surprisingly, every trusted setup is either useless (equivalent to having no trusted setup) or all-powerful (allows UC-secure protocols for all tasks). There are no "intermediate" trusted setups in the UC framework. We prove this **zero-one law** under a natural intractability assumption, and consider the class of deterministic, finite, 2-party functionalities as candidate trusted setups.

One important technical contribution in this work is to initiate the comprehensive study of the cryptographic properties of reactive functionalities. We model these functionalities as finite automata and develop an automata-theoretic methodology for classifying and studying their cryptographic properties. Consequently, we completely characterize the reactive behaviors that lead to cryptographic non-triviality. Another contribution of independent interest is to optimize the hardness assumption used by Canetti et al. (*STOC* 2002) in showing that the common random string functionality is complete (a result independently obtained by Damgård et al. (*TCC* 2010)).

## 1 Introduction

Cryptographic tasks provide a fascinating arena to study the interplay of information, interaction and computation. Each cryptographic task has a fundamental "information-control fingerprint" that specifies how various parties involved in the task can learn and/or influence all the pieces of information in the system. This work forms part of a study that aims to systematically understand abstract cryptographic tasks, classifying them by how "cryptographically complex" their fingerprints are.

---

[*] Work supported by NSF grants CNS 07-16626 and CNS 07-47027.

T. Rabin (Ed.): CRYPTO 2010, LNCS 6223, pp. 595–612, 2010.

A crisp way to capture the information-control fingerprint of a cryptographic task is by modeling it as a program carried out by a third party. This program is simply called a **functionality**. Modeling tasks this way conveniently separates the security definition from the information-control fingerprint.[1] Understanding and classifying such functionalities has been the subject, implicitly or explicitly, of a wealth of literature in theoretical computer science.

In this work, we continue the study of cryptographic complexity of functionalities from [30,25,26], which explicitly define cryptographic complexity classes using various notions of reductions among functionalities. Our focus in this work is on cryptographic complexity defined using security in the universal composition (UC) framework.[2] We show that under standard computational intractability assumptions, the universe of functionalities collapse to just two distinct levels of cryptographic complexity.

For simplicity, first we describe our result in terms of standard UC security terminology, and then summarize its implications for the theory of cryptographic complexity developed in [30,25,26].

*Main Result.* The standard UC framework defines security in a *plain* model in which protocols are allowed access only to communication channels. However, the framework also allows protocols to use access to a trusted *setup functionality*, in what is called a hybrid model. It is known that in the plain model, very few tasks admit UC-secure protocols [6,30]. On the other hand, in certain hybrid models (say, those corresponding to oblivious transfer or fair coin-tossing functionalities being used as a trusted setup) *all* tasks admit UC-secure protocols [7,17].

In this work we classify the strength of a functionality in terms of how useful it is as a trusted setup in a UC hybrid model. We prove the following surprisingly simple classification that was conjectured in [30] — under a natural intractability assumption, every deterministic, finite, 2-party functionality is either:

**Trivial:** These functionalities already have secure protocols in the *plain* model. As such, they are useless as trusted setups; they admit no more tasks to have UC-secure protocols than in the plain model. Or,

**Complete:** When any of these functionalities is used as a trusted setup, *all tasks* have UC-secure protocols.[3]

---

[1] For instance, the commitment functionality is specified the same way for various security settings, independently of considerations in defining security; in contrast, traditional specification of commitment cannot be divorced from how binding and hiding are defined, and results in different notions (like statistically binding commitment, or statistically hiding commitment) which are not formally captured as the same functionality.

[2] While using UC security provides a fine resolution picture of cryptographic complexity, weaker security notions also yield meaningful complexity classes. For the computationally unbounded setting, semi-honest and standalone (as well as UC) security notions were considered in [25]. For the computationally bounded setting, results analogous to ours for these weaker security notions follow from the classical results in [33] and [12]. See related work.

[3] Only *well-formed* functionalities are considered here, without any fairness requirement.

We call this classification our **zero-one law**. In other words, every such functionality is at the *extremes* of usefulness: either trivial or complete. If a functionality is unrealizable in the plain model, then it is all-powerful if used as a trusted setup.

We consider UC security against probabilistic, polynomial-time adversaries who corrupt parties statically (non-adaptively). In sharp contrast, when considering the computationally unbounded setting, [25] shows that there are infinitely many setups that allow realization of infinitely many distinct classes of functionalities.

The intractability assumption we use is the existence of a protocol for oblivious transfer secure against standalone semi-honest PPT adversaries (sh-OT assumption). Interestingly, this intractability assumption is both necessary and sufficient. Then, our main result is formally stated as:

> **Main Theorem:** The sh-OT assumption is true if and only if every deterministic, finite, 2-party functionality is either trivial or complete.

The class of deterministic, finite, 2-party functionalities is defined formally in Section 3. Most notably, this class includes *reactive* functionalities, which receive inputs, give outputs, and keep internal state over many rounds of interaction with the parties. An important contribution in this work is to initiate an "automata-theoretic" study of reactive functionalities. Previous works on multi-party computation are almost exclusively restricted to SFE functionalities, except for the positive (i.e., triviality) results like in [12], which give secure realization of reactive functionalities.[4] In contrast, we develop techniques to *use* an arbitrary reactive functionality in a cryptographic protocol.

*Cryptographic Complexity and Intractability.* As alluded to above, one way to think about our main result is by taking a complexity-theoretic view of secure multi-party computation. Say that a functionality $\mathcal{F}$ *reduces to* another functionality $\mathcal{G}$ (written $\mathcal{F} \sqsubseteq_{\text{PPT}} \mathcal{G}$) if there is a UC-secure protocol for $\mathcal{F}$ in the $\mathcal{G}$-hybrid model. This reduction is reflexive and transitive (for standard notions of secure reductions), and is a natural complexity-theoretic reduction to compare the relative "cryptographic complexities" of cryptographic tasks. Throughout this work, we use this convenient $\sqsubseteq_{\text{PPT}}$ notation.

Under this interpretation, "completeness" (as defined above) indeed refers to $\sqsubseteq_{\text{PPT}}$-completeness in the complexity-theory sense. The zero-one law shows that there are only two *degrees* of the $\sqsubseteq_{\text{PPT}}$ reduction.[5]

While we show that the sh-OT assumption is both necessary and sufficient for the entire zero-one law, the sh-OT assumption may not be necessary for all individual reductions of the form $\mathcal{F} \sqsubseteq_{\text{PPT}} \mathcal{G}$. In a companion paper [26], we also classify which intractability assumptions are *necessary* for reductions of the form $\mathcal{F} \sqsubseteq_{\text{PPT}} \mathcal{G}$. Every reduction of the form $\mathcal{F} \sqsubseteq_{\text{PPT}} \mathcal{G}$ that we classify turns

---

[4] An important exception is [30] which gives a characterization of trivial functionalities which is applicable to reactive functionalities as well; however, [30] does not offer an explicit combinatorial or automata theoretic interpretation of their characterization.

[5] The degree of $\mathcal{G}$ under reduction $\sqsubseteq$ is the set $\{\mathcal{F} \mid \mathcal{F} \sqsubseteq \mathcal{G}\}$.

out to be unconditionally true or false, or else *exactly equivalent* to a well-known computational assumption (the sh-OT assumption or the existence of one-way functions). This suggests the possibility of *defining* intractability assumptions in terms of reductions of the form $\mathcal{F} \sqsubseteq_{\mathrm{PPT}} \mathcal{G}$. Such assumptions are of a fundamental nature for secure multi-party computation, since they are derived directly from the definitions of functionalities themselves.

Our results in this work imply that the sh-OT assumption is the *maximal* assumption that can emerge in this framework; we conjecture that the existence of one-way functions is the *minimal* assumption. A more intriguing question is whether there are other intermediate assumptions. Put differently, one likely outcome of this line of investigation is to discover new cryptographically interesting worlds in "Impagliazzo's multiverse" [15] between Cryptomania (which we interpret as a world where the sh-OT assumption is true) and Minicrypt (where only one-way functions exist), or to show there are none.

*Related Work.* There is a large body of work on complexity of 2-party functionalities in the computationally unbounded setting [18,8,24,2,19,20,21,23,25]. In the computationally bounded setting, the classical results of [33,12] imply that *all functionalities are trivial* (i.e., realizable without relying on any other functionality) *for the semi-honest and standalone security notions respectively*, under the sh-OT assumption. Our work could be considered a refinement of these early results, but for the UC security notion.

Beimel et al. [3], who showed (in the probabilistic polynomial-time setting, and for the special case of SFE functionalities in which only one party receives the output) that the sh-OT assumption is implied by the existence of a *semi-honest* secure protocol for any functionality that is not unconditionally trivial. [13] partially extends this result beyond finite functionalities, but is still restricted to the case of one-sided output. (In the full version, we show that, as in [13], there is a gap between triviality and completeness when our results are extended to unbounded-memory functionalities.)

The above results do not apply in a security setting with an arbitrary environment. Since Canetti introduced the Universal Composition (UC) framework [4], there have been several works on cryptographic complexity of functionalities in this setting. In particular, [5,6,30] characterize trivial functionalities. (For finite functionalities this class remains the same in computationally bounded and unbounded settings.)

Less was known about which functionalities were *complete* under UC-secure reductions. Results in [18,22,17] establish the completeness of oblivious transfer and many other non-reactive functionalities, for computationally unbounded adversaries. In the polynomial-time setting, the well-known CLOS construction [7] demonstrates the completness of the "coin-tossing" functionality, assuming enhanced trapdoor permutations and dense cryptosystems. Our result improves this by using the minimal sh-OT assumption, but more significantly by showing the completeness of *every* non-trivial deterministic functionality. (However, [7] proves completeness against adversaries which corrupt parties *adaptively*, whereas we consider only *static* security.) Independently of our work, Damgård et al. [9] also show the completeness of the coin-tossing functionality under the minimal sh-OT assumption, as we do. Their construction is similar in spirit to our protocol for the same task, though more complicated due to the use of an intermediate "public-key infrastructure" functionality. Our current protocol is the result of a simplification to a protocol in an earlier draft of this work, motivated by their recent result.

## 2   Overview of Our Techniques

In proving our main result, the more interesting direction is to show that sh-OT assumption implies the zero-one law. That is, we must construct secure protocols which demonstrate the completeness of every non-trivial functionality, proving security using only the sh-OT assumption. We do this in a series of steps, outlined in Figure 1.

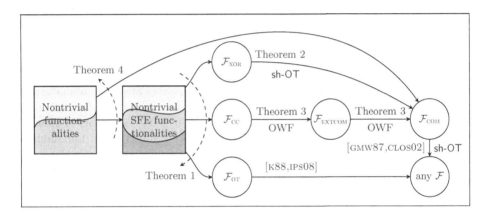

**Fig. 1.** Overview of protocol constructions. An arrow from functionality $\mathcal{G}$ to $\mathcal{F}$ denotes a secure protocol for $\mathcal{F}$ using ideal access to $\mathcal{G}$ (that is, $\mathcal{F} \sqsubseteq \mathcal{G}$). Arrows not labeled by a computational assumption indicate unconditionally secure protocols.

*Approach: Behavioral Components of Functionalities.* Our approach of proving the zero-one law centers around identifying four distinct behaviors of functionalities that lead to non-triviality. For each behavior we associate a familiar "canonical" functionality which is non-trivial for only that reason:

- Allowing simultaneous exchange of information, exemplified by the boolean XOR functionality $\mathcal{F}_{\text{XOR}}$. In this functionality, one party's output completely determines the other's input. Thus its cryptographic non-triviality stems not from *hiding* information, but ensuring that both party's inputs are chosen independent of the other's.
- Selectively hiding one party's inputs from the other, exemplified by a simple SFE functionality we introduce called *simple cut-and-choose* ($\mathcal{F}_{\text{CC}}$). In this

functionality, Alice gives a bit as input, and Bob gives an input indicating whether he wants to learn Alice's bit or not. Furthermore, Alice is told whether Bob learned her bit. Thus $\mathcal{F}_{CC}$ embodies selective hiding of Alice's input alone.

– Selectively hiding both party's inputs simultaneously, exemplified by the $\binom{2}{1}$-oblivious transfer functionality $\mathcal{F}_{OT}$. Recall that $\mathcal{F}_{OT}$ hides meaningful information about both parties from the other.

– Holding meaningful information hidden in internal memory between rounds, exemplified by the commitment functionality $\mathcal{F}_{COM}$. This functionality holds the sender's data in memory between the commit phase and reveal phase. This component can appear only in a reactive functionality.

To show the zero-one law, we do the following: First, we formally define what it means for a functionality to exhibit each of these four fundamental behaviors. Next, we show that these four behaviors are in fact an *exhaustive* characterization of non-triviality: in Theorems 1 and 4, we show that a reactive functionality $\mathcal{G}$ is non-trivial if and only if $\mathcal{F} \sqsubseteq \mathcal{G}$ *unconditionally* for some $\mathcal{F} \in \{\mathcal{F}_{XOR}, \mathcal{F}_{CC}, \mathcal{F}_{OT}, \mathcal{F}_{COM}\}$.[6] In other words, every non-trivial functionality must exhibit at least one of the above four behaviors. Finally, we show that each of the four canonical functionalities ($\mathcal{F}_{XOR}$, $\mathcal{F}_{CC}$, $\mathcal{F}_{OT}$, $\mathcal{F}_{COM}$) is complete under the sh-OT assumption.

Since our definitions of these four component behaviors are all combinatorial, we are able to give the first complete *combinatorial* characterization of non-triviality (and consequently completeness) for *reactive* functionalities. Further, this characterization holds even with respect to computationally unbounded adversaries.

*Non-Reactive Behaviors (Section 4).* Of the four behaviors enumerated above, only the $\mathcal{F}_{COM}$ behavior is exclusive to reactive functionalities. For the other three, which can apply to non-reactive functionalities, we give formal combinatorial definitions in terms of the input/output function table. Then it suffices to show that any non-reactive functionality not meeting one of these three criteria is in fact trivial (Theorem 1).

Next, we show that $\mathcal{F}_{XOR}$, $\mathcal{F}_{CC}$, and $\mathcal{F}_{OT}$ are each complete. It is well-known that $\mathcal{F}_{OT}$ is (unconditionally) complete, even under the strong notion of reduction that we consider [18,17]. For the other two cases, we use the fact that the commitment functionality $\mathcal{F}_{COM}$ is complete in the UC framework under the sh-OT assumption. This follows directly from the well-known CLOS construction [7]. Thus, to complete our claim, it suffices to show that the sh-OT assumption implies $\mathcal{F}_{COM} \sqsubseteq_{PPT} \mathcal{F}_{XOR}$ and $\mathcal{F}_{COM} \sqsubseteq_{PPT} \mathcal{F}_{CC}$.

We give new commitment protocols in the $\mathcal{F}_{XOR}$- and $\mathcal{F}_{CC}$-hybrid models (Theorems 2 and 3), secure under the sh-OT and OWF assumptions, respectively. We

---

[6] Indeed just $\mathcal{F}_{XOR}$ and $\mathcal{F}_{CC}$ by themselves are an exhaustive characterization of non-triviality, as they can both be unconditionally obtained from $\mathcal{F}_{OT}$ and $\mathcal{F}_{COM}$. However, we include all four functionalities in our list of fundamental behavioral components because we prove the complete of each one differently.

note that [7] shows (implicitly) that $\mathcal{F}_{\mathrm{XOR}}$ is complete;[7] however, their protocol focuses on achieving *adaptive* security and, as such, depends on a hardness assumption that is not known to be implied by sh-OT assumption. Our new protocol achieves *static* security using a new protocol and under the minimal sh-OT assumption.

*Reactive Behaviors (Section 5).* To complete the classification of reactive functionalities, we show that every reactive functionality is either trivial, contains a non-reactive behavioral component ($\mathcal{F}_{\mathrm{XOR}}$, $\mathcal{F}_{\mathrm{CC}}$, $\mathcal{F}_{\mathrm{OT}}$), or else can be used for a commitment ($\mathcal{F}_{\mathrm{COM}}$) protocol (Theorem 4). As mentioned above, $\mathcal{F}_{\mathrm{COM}}$ is complete under the sh-OT assumption, thus we establish the exhaustiveness of the four behavioral components, as well as the completeness of their respective canonical functionalities.

The bulk of our technical contributions for reactive functionalities involves formally defining this fourth behavioral component; namely, defining when an arbitrary functionality keeps meaningful information about a party's input hidden in memory between rounds. We model reactive functionalities as finite-state automata, and initiate an automata-theoretic analysis of their input/output behavior. This classification involves identifying states and transitions of an automaton which have specific cryptographic consequences, and then showing how such features can be leveraged to give a protocol for $\mathcal{F}_{\mathrm{COM}}$.

# 3   Preliminaries

*Model and Security Definition.* Our security definitions are grounded in the framework of Universal Composable (UC) security [4], with which we assume the reader has slight familiarity. For concreteness we consider the model used in [30], which in turn is based on that in [29]. However, we emphasize that very few specifics of the model (including ideal functionalities, an interactive environment and simulation based security) are important for the results.

The UC model allows a large class of MPC functionalities, not all of which are "natural." For instance, a functionality that announces the identities of the corrupt parties is not natural; a reactive functionality which introduces a race condition depending on the order in which it receives inputs from parties is also not natural. Following the convention in all previous works (to the best of our knowledge), we do not consider such functionalities. We note that the functionalities we consider do not offer a guarantee of output *fairness*; that is, they allow the adversary to control the delivery of outputs.

We write $\mathcal{F} \sqsubseteq \mathcal{G}$ if there is a protocol that securely realizes $\mathcal{F}$ in the "$\mathcal{G}$-hybrid model;" see [4] or [29] for a formal definition. In the $\mathcal{G}$-hybrid model, the parties in the protocol can interact with any number of (asynchronous) copies of $\mathcal{G}$, and can access $\mathcal{G}$ in any "role". This second convention is crucial to our results (see Section 7). We consider only efficient protocols, but make

---

[7] They show that the coin-tossing functionality, for which there is an elementary protocol using $\mathcal{F}_{\mathrm{XOR}}$, is complete.

a notational distinction between unconditionally (statistically) secure protocols (denoted by $\sqsubseteq_{STAT}$) and protocols whose security depends on a computational assumption (denoted by $\sqsubseteq_{PPT}$). As is standard, we require security against active (i.e., malicious) adversaries. However, as we point out in Section 7, our results extend to a stronger definition where security is required against both active and semi-honest adversaries.[8]

By default, we also allow protocols access to a communication channel. Following [30], we consider the natural model of a private communication channel, in which parties can send fixed-length messages (with the adversary controlling delivery). The choice of public vs. private channel is not crucial to our results (see Section 7).

All results in this work are restricted to static corruption (where the adversary has to corrupt any parties before the protocol begins). In fact, we leave open the possibility that our main theorem breaks down in the case of adaptive corruption.

*Classes of functionalities.* In this work we restrict our attention to **finite, deterministic, 2-party, reactive** functionalities. We formally model such functionalities as finite automata. Each state transition is labeled by a tuple in $X \times Y \times Z \times Z$, where $X$, $Y$, and $Z$ are finite sets. A transition from $q$ to $q'$ with label $(x, y, s, t)$ means that upon receiving input $x$ from Alice and $y$ from Bob in state $q$, the functionality will deliver output $s$ to Alice and $t$ to Bob, and change to state $q'$. We require the automaton to be deterministic; that is, for every state $q$ and every $(x, y) \in X \times Y$, there is at most one transition leaving $q$ whose label begins with $(x, y)$. We consider an asynchronous network setting in which the adversary has control over the timing of input/output delivery. In Figure 2 we give an example of how the (reactive) commitment functionality $\mathcal{F}_{COM}$ can be expressed in such a way.

We say that a functionality is a **secure function evaluation** (SFE; or non-reactive) functionality if it engages only one round of interaction; that is, all transitions leading from the start state lead to a dead state with no transitions.

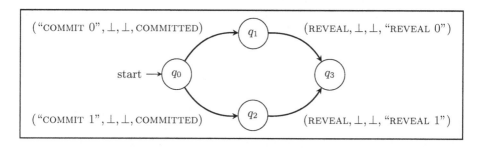

**Fig. 2.** Commitment functionality $\mathcal{F}_{COM}$ modeled as a deterministic finite functionality

---

[8] Note that when considering security against semi-honest adversaries, the simulator must also be semi-honest.

Alternatively, an SFE functionality is completely specified by a pair of functions $(f_A, f_B)$, where Alice's output is $f_A(x, y)$ and Bob's output is $f_B(x, y)$.

*The* sh-OT *assumption.* The primary intractability assumption we consider is the existence of a protocol for $\mathcal{F}_{\text{OT}}$ secure against semi-honest, PPT adversaries (sh-OT assumption, for short). It is possible to express this assumption using the definition of UC security restricted to semi-honest adversaries (in both the real and the ideal executions). However, we point out that the traditional (standalone) security definition is equivalent to the UC security definition, since the simulation required by semi-honest security does not, and need not, *extract* the inputs of the corrupt players; it simply uses the input given by the environment.

Some of our protocol constructions additionally rely on statistically binding (standalone secure) commitment schemes, pseudorandom generators, (standalone secure) witness-indistinguishable proofs or zero-knowledge proofs of knowledge for NP. All of these primitives have well-known constructions assuming the existence of one-way functions [27,14,10]. One-way functions are in turn implied by the sh-OT assumption [16].

# 4    Zero-One Law for Non-reactive Functionalities

*Three "Canonical" Non-Reactive Functionalities.* The following three SFE functionalities exemplify the three different behaviors that lead to cryptographic non-triviality for non-reactive functionalities:

$\mathcal{F}_{\text{XOR}}$ (exclusive-or): Alice gives input $x \in \{0, 1\}$ and Bob gives input $y \in \{0, 1\}$. Both parties receive output $x \oplus y$.

$\mathcal{F}_{\text{CC}}$ (simple cut-and-choose): Alice gives input $x \in \{0, 1\}$ and Bob gives input $y \in \{0, 1\}$. If $y = 0$, then both parties receive output $x$. If $y = 1$, then both parties receive output 2. Thus, Bob decides whether to learn Alice's bit, while Alice always learns Bob's choice.

$\mathcal{F}_{\text{OT}}$ (oblivious transfer): Alice gives inputs $x_0, x_1 \in \{0, 1\}$ and Bob gives input $y \in \{0, 1\}$. Bob receives output $x_y$ and Alice receives output $\perp$.

We show that these three fundamental behaviors completely characterize non-triviality (for non-reactive functionalities), as follows:

**Theorem 1.** *Let $\mathcal{F}$ be an SFE functionality. Then $\mathcal{F}$ is non-trivial if and only if $\mathcal{F}_{\text{XOR}} \sqsubseteq_{\text{STAT}} \mathcal{F}$ or $\mathcal{F}_{\text{CC}} \sqsubseteq_{\text{STAT}} \mathcal{F}$ or $\mathcal{F}_{\text{OT}} \sqsubseteq_{\text{STAT}} \mathcal{F}$.*

*Proof (Sketch).* ($\Leftarrow$) Each of $\mathcal{F}_{\text{XOR}}$, $\mathcal{F}_{\text{CC}}$, and $\mathcal{F}_{\text{OT}}$ is unconditionally non-trivial, from the characterization of trivial SFE functionalities in [30].

($\Rightarrow$) Kraschewski and Müller-Quade [22] identify a $2 \times 2$ minor within the function table of an SFE, which generalizes the (symmetric-output) boolean OR functionality $\begin{smallmatrix} 0 & 1 \\ 1 & 1 \end{smallmatrix}$ that is known to be complete. They show that an SFE $\mathcal{F}$ can be used to construct an unconditionally UC-secure protocol for $\mathcal{F}_{\text{OT}}$ if and only if $\mathcal{F}$ contains such a minor.

Similarly, we also identify another important $2 \times 2$ minor called a *generalized-CC minor*. A minor a generalized-CC if, when restricted to the minor, one party can choose whether to learn the other's input, and this choice is revealed to the other party in the function's output. We show that if $\mathcal{F}$ has such a minor, then the protocol in which the parties simply restrict their inputs to that minor while accessing $\mathcal{F}$ is a UC-secure protocol for $\mathcal{F}_{\mathrm{CC}}$.[9]

Finally, it is easy to see that if $\mathcal{F}$ does not have either kind of $2 \times 2$ minor mentioned above, then $\mathcal{F}$ must simply be (equivalent to) a function that takes inputs $x \in X$ from Alice and $y \in Y$ from Bob, then outputs $(x, y)$ to both parties. If $\max\{|X|, |Y|\} \geq 2$, then there is an elementary UC-secure protocol for $\mathcal{F}_{\mathrm{XOR}}$ in the $\mathcal{F}$-hybrid model. Otherwise, $\mathcal{F}$ is trivial: the protocol in which one party simply sends their input to the other party is a UC-secure protocol for $\mathcal{F}$ (without any set-up).

*Completeness of the Three Canonical Non-Reactive Functionalities.* Since $\mathcal{F}_{\mathrm{OT}}$ is unconditionally complete (even with respect to UC secure protocols) [18,17], and the commitment functionality $\mathcal{F}_{\mathrm{COM}}$ is complete under the sh-OT assumption [7], it suffices to prove the following two theorems:

**Theorem 2.** *If the* sh-OT *assumption is true, then* $\mathcal{F}_{\mathrm{COM}} \sqsubseteq_{\mathrm{PPT}} \mathcal{F}_{\mathrm{XOR}}$.

*Proof (Sketch).* We first observe that the coin-tossing functionality $\mathcal{F}_{\mathrm{COIN}}$[10] has an elementary, unconditionally secure protocol in the $\mathcal{F}_{\mathrm{XOR}}$-hybrid model. Thus it suffices to show that $\mathcal{F}_{\mathrm{COM}} \sqsubseteq_{\mathrm{PPT}} \mathcal{F}_{\mathrm{COIN}}$. The well-known CLOS construction [7] proves exactly this; however, their focus was on achieving *adaptive* security, and their protocol relied on a stronger computational assumption than the sh-OT assumption. Thus we must use an entirely different approach for achieving $\mathcal{F}_{\mathrm{COM}}$ (with static security) from $\mathcal{F}_{\mathrm{COIN}}$. We sketch an overview of our protocol below:

Suppose $\psi_{\mathsf{sh}}$ is the semi-honest protocol for $\mathcal{F}_{\mathrm{OT}}$ guaranteed by the sh-OT assumption. We suppose that the sender in $\psi_{\mathsf{sh}}$ provides two bits $(x_0, x_1)$, the receiver provides a bit $y$, and the receiver learns $x_y$.

Our commitment protocol is as follows, when Alice is committing to $b \in \{0, 1\}$. First, both parties use $\mathcal{F}_{\mathrm{COIN}}$ to generate a sequence of random coins $\sigma$. The sender Alice and receiver Bob interact in an instance of $\psi_{\mathsf{sh}}$, with Alice using inputs $(x_0 = 0, x_1 = b)$, and Bob using input $y = 0$. To ensure that both parties provide inputs of the required form, we "compile" the $\psi_{\mathsf{sh}}$ subprotocol using a variant of the standard GMW compiler [12]. Unlike the GMW compiler, at each step we make the parties give a witness-indistinguishable proof that *either* they are following the protocol honestly with the appropriate inputs, *or* the public coins $\sigma$ are from a pseudorandom distribution. In the reveal phase, Alice

---

[9] Note that, in general, restricting inputs to a minor of $\mathcal{F}$ does not give a secure protocol (against malicious adversaries) for the SFE corresponding to that minor, since a malicious adversary may send inputs to $\mathcal{F}$ outside of the prescribed minor.

[10] $\mathcal{F}_{\mathrm{COIN}}$ is a functionality which, upon activation, samples an unbiased coin $b \leftarrow \{0, 1\}$ and outputs it to both parties. It does not fall in our class of deterministic finite functionalities, but we use it as an intermediate step in our protocol construction.

gives a witness-indistinguishable proof that *either* $\sigma$ was from a pseudorandom distribution, *or* all her messages in the $\psi_{\text{sh}}$ subprotocol were consistent with her having input $x_1 = b$.

In the real interaction, $\sigma$ is generated honestly using $\mathcal{F}_{\text{COIN}}$ and is therefore in the pseudorandom distribution with negligible probability. Thus the GMW-style compilation ensures that both parties are executing the $\psi_{\text{sh}}$ subprotocol honestly as stated. Then applying the semi-honest security of $\psi_{\text{sh}}$, we see that Bob learns nothing about $b$ in the commit phase, and Alice can only open the commitment to the value of $b$ that she used in the commit phase.

However, when the simulator is corrupt it can choose $\sigma$ from a pseudorandom distribution. If Alice is corrupt, the simulator can play the role of Bob using input $y = 1$ to the $\psi_{\text{sh}}$ subprotocol, while still giving convincing GMW proofs. By the correctness of the $\psi_{\text{sh}}$ protocol, the simulated Bob obtains $x_1 = b$ from $\psi_{\text{sh}}$ (i.e., the simulator *extracts* $b$), and by the security of $\psi_{\text{sh}}$, the simulation is indistinguishable from the real interaction.

If Bob is corrupt, the simulator can give a commitment to 0 in the commit phase, but open it to any value in the reveal phase (using the clause in the witness-indistinguishable proof related to $\sigma$). Thus the simulator can successfully equivocate to a corrupt Bob.

To show that both of these simulations are sound, we must apply the semi-honest security of $\psi_{\text{sh}}$, which is the most delicate part of the proof, since the simulator exists in the UC setting. We construct a sequence of hybrid interactions between the real and ideal UC (straight-line) interactions, and show that if any adversary can distinguish between certain hybrids, then we can construct a corresponding adversary (possibly using rewinding) which violates the semi-honest security properties of $\psi_{\text{sh}}$. For technical reasons in this part, we require the interactive proofs to be proofs *of knowledge*.

**Theorem 3.** *If one-way functions exist, then* $\mathcal{F}_{\text{COM}} \sqsubseteq_{\text{PPT}} \mathcal{F}_{\text{CC}}$.

*Proof (Sketch).* The simulator for a UC-secure commitment protocol has two main tasks: (1) to extract the committed value from a corrupt sender during the commit phase, and (2) to give an equivocal commitment to a corrupt receiver that can then be convincingly opened to any value during the reveal phase. Our construction of a UC-secure commitment protocol is broken into two major conceptual steps, which tackle these two properties in a somewhat modular fashion.

We first define an intermediate "extractable commitment" functionality called $\mathcal{F}_{\text{EXTCOM}}$. The complete formulation of $\mathcal{F}_{\text{EXTCOM}}$ is highly non-trivial, and is deferred to the full version. $\mathcal{F}_{\text{EXTCOM}}$ succinctly expresses the requirements of a *statistically binding, computationally hiding* commitment scheme (in the traditional standalone-secure sense) which also admits a *straight-line extracting* simulator. We believe that this method of expressing a combination of standalone and universally composable security properties may be of independent interest. Using a technique similar in spirit to the $\binom{2}{1}$-commitments of Nguyen and Vadhan [28], we show that if one-way functions exist, then $\mathcal{F}_{\text{COM}} \sqsubseteq_{\text{PPT}} \mathcal{F}_{\text{EXTCOM}}$.

Thus it suffices to construct a commitment protocol which has a UC extraction property, but only a standalone-secure hiding property. This commitment protocol is as follows. To commit to a bit $b$, Alice first chooses a random bitstring $s$ and then applies a good linear error-correcting code to obtain a codeword $t$. She commits to $t$ using a statistically binding (standalone-secure) commitment protocol. For each bit $t_i$ of $t$, Alice gives $t_i$ as input to $\mathcal{F}_{\mathrm{CC}}$, and Bob chooses to learn it with some probability. Recall that in $\mathcal{F}_{\mathrm{CC}}$, Alice learns whether Bob choses to see her input. Alice ensures that Bob only learned sufficiently few bits of $t$ so that some uncertainty about $s$ remains. This remaining uncertainty can be deterministically extracted (as a linear function of $s$), and Alice uses it as a one-time pad to mask $b$. She sends the masked $b$ to Bob to complete the commitment phase. In the reveal phase, Alice opens the commitment to $t$, and Bob checks for consistency with the bits that he learned in the commit phase.

Intuitively, the protocol is computationally hiding and statistically binding because the deterministic extraction of the mask is perfect (using a simple linear function). The only information about the mask is given in a statistically-binding standalone-secure commitment to $t$.

However, the simulator provides the interface for $\mathcal{F}_{\mathrm{CC}}$ to a corrupt Alice. Consequently, the simulator can see *all* of Alice's inputs to $\mathcal{F}_{\mathrm{CC}}$, which are the (purported) bits of $t$. Because Bob has a certain probability of revealing each one of the bits of $t$ and he verifies them against Alice's statistically binding commitment to $t$, we argue that Alice could not supply too many incorrect values to $\mathcal{F}_{\mathrm{CC}}$. In particular, Alice cannot give more incorrect bits than can be corrected by the error correcting code, except with negligible probability. Thus the simulator can perform a noisy decoding to obtain $s$ and then easily extract $b$.

## 5   Classifying Reactive Functionalities

We show that a reactive functionality can be non-trivial only for two simple reasons: (1) behaving like a non-trivial SFE functionality during a single round, or (2) using its internal memory in a non-trivial way. Formally defining condition (2) requires a careful new automata-theoretic analysis of reactive behaviors. Intuitively, memory is used in a non-trivial way when some part of the memory is both *hidden* (has not yet affected its external behavior) and *meaningful* (may eventually influence its future external behavior). Such usage of internal memory is exemplified by the commitment functionality $\mathcal{F}_{\mathrm{COM}}$ (between the commit and reveal phases).

*Automata-theoretic Characterization.* We develop three new important properties of reactive functionalities, all defined combinatorially.

Say that an input $\hat{x}$ *dominates* another input $x$ if Alice can use $\hat{x}$ as her input to $\mathcal{F}$ in the first round of interaction, but then convince any environment that she had really used $x$. In other words, any behavior that can be induced by sending $x$ to $\mathcal{F}$ in the first round can also be induced by instead sending $\hat{x}$ and thereafter engaging in some local "translation" protocol. We emphasize that

Alice must perform this translation *online*, without knowledge of the inputs that the environment will provide in future rounds. When $\hat{x}$ dominates $x$, Alice can use $\hat{x}$ in place of $x$ in the first round without loss of generality. The condition of $\hat{x}$ dominating $x$ can be defined directly in terms of the UC security condition.

The input-output behavior of each state in the functionality naturally defines a corresponding SFE. Take any SFE and say that $x \sim x'$ if Alice inputs $x$ and $x'$ always induce the same output for Bob. In an SFE, Bob's output may reveal information about Alice's output, but up to $\sim$-equivalence at most. However, in a *trivial* SFE, Bob's output always reveals exactly the $\sim$-equivalence class of Alice's input. We say that the start state of $\mathcal{F}$ is *simple* if: (1) its associated SFE is a trivial SFE, and (2) each equivalence class of $\sim$ (for Alice inputs and Bob inputs) contains some input that dominates all other inputs in its class.

To understand this definition, suppose the start state of $\mathcal{F}$ is simple. Then just by looking at his own output from the first round, Bob can exactly determine the $\sim$-class of the input Alice used. There is some input, say, $\hat{x}$, which dominates all Alice inputs in this $\sim$-class. No matter how the environment instructs Alice to behave in the future, she could have achieved the same effect if she had used input $\hat{x}$ in the first round. Thus, Bob can assume without loss of generality that Alice in fact used $\hat{x}$. The same is true for Alice; she can determine, given her output, an input $\hat{y}$ for Bob, and assume without loss of generality that Bob supplied $\hat{y}$ in the first round.

Thus we can assume without loss of generality that Alice and Bob only use inputs $\hat{x}$ and $\hat{y}$ of this special kind (they dominate their respective $\sim$-equivalence classes). We call the transition from the start state on such inputs $(\hat{x}, \hat{y})$ a *safe transition*. Intuitively, only safe transitions are relevant; furthermore, after a safe transition, neither party has uncertainty about the functionality's resulting state.

We can now state our main automata-theoretic characterization:

**Theorem 4.** *Let $\mathcal{F}$ be a deterministic, finite (reactive) functionality. Then the following are equivalent:*

1. *$\mathcal{F}$ is non-trivial.*
2. *$\mathcal{F}_{\mathrm{COM}} \sqsubseteq_{\mathrm{STAT}} \mathcal{F}$ or $\mathcal{G} \sqsubseteq_{\mathrm{STAT}} \mathcal{F}$ for some non-trivial SFE functionality $\mathcal{G}$.*
3. *There is a non-simple state in $\mathcal{F}$ that is reachable from the start state via a sequence of safe transitions.*

The automata-theoretic properties defined above, and subsequently condition (3) of this theorem, can be expressed completely combinatorially, giving the first *combinatorial* characterization of triviality (and thus completeness) for any class of arbitrary reactive functionalities.

*Proof (Sketch).* $2 \Rightarrow 1$ follows from the non-triviality of $\mathcal{F}_{\mathrm{COM}}$.

$(1 \Rightarrow 3)$ Consider all the states of $\mathcal{F}$ reachable via a sequence of safe transitions; intuitively, these are the only states that matter. If all such states are simple, then $\mathcal{F}$ has the following trivial protocol: repeatedly evaluate the (trivial) SFE corresponding to $\mathcal{F}$'s current state, using that SFE's trivial protocol. Without loss of generality we can assume a safe transition was taken; thus, each party's output in the round determines the next state of $\mathcal{F}$, and the protocol can be repeated for each round.

$(3 \Rightarrow 2)$ Assume that one of the safely reachable states of $\mathcal{F}$ is non-simple; without loss of generality, we can take the start state to be non-simple. The definition of a simple state requires two conditions, so we consider two cases: (1) If the start state is non-simple because of its input-output behavior, then there is an elementary protocol which securely realizes that associated SFE in the $\mathcal{F}$-hybrid model (simply interact with $\mathcal{F}$ for one round only). (2) Otherwise, the start state is non-simple because there exist two inputs for (by symmetry) Alice, say $x_0$ and $x_1$, for which $x_0 \sim x_1$ (that is, these inputs always induce the same output for Bob in the first round), but no Alice input dominates *both* of $\{x_0, x_1\}$. In other words, Alice's first-round input "binds" her to the behaviors consistent with $x_0$ or to those consistent with $x_1$, but not both.

We formalize this natural connection to commitment by constructing a protocol for $\mathcal{F}_{\mathrm{COM}}$, as follows. Alice commits to $b$ by sending $x_b$ to $\mathcal{F}$ in the first round. The commitment is perfectly hiding since $x_0 \sim x_1$. To reveal, Alice must convince Bob that in the first round she used an input that dominates $x_b$, since no input can dominate both $x_0$ and $x_1$.

Suppose $x$ does not dominate $x'$. Then for every strategy for Alice which uses input $x$ in the first round, there is some environment that can distinguish between Alice's strategy and one which uses input $x'$ in the first round and thereafter runs the dummy protocol. Using an automata-theoretic characterization, we show that these quantifiers can be exchanged: there is a fixed environment such that for every $x$ not dominating $x'$ and every Alice strategy that uses input $x$ in the first round, the environment has a constant probability of "catching" Alice.[11] Our commitment protocol instructs Bob to play the role of such an environment in the reveal phase, sending a sequence inputs to $\mathcal{F}$ himself and a sequence of "challenge" inputs to Alice. Just like in the definition of domination, Alice must report back to Bob her own purported responses from $\mathcal{F}$, in an online manner. If Alice's first-round input did not dominate $x_b$, she is guaranteed to be caught with constant probability. By repeating this basic protocol in parallel an appropriate number of times, Bob can be assured of catching an equivocating Alice with overwhelming probability.

# 6   Necessity of the sh-OT Assumption

Finally, we show that the sh-OT assumption is not only sufficient but also necessary for the zero-one law to hold.

**Theorem 5.** *If the zero-one law is true, then the* sh-OT *assumption is true.*

---

[11] This environment results in a protocol for $\mathcal{F}_{\mathrm{COM}}$ whose worst-case $O(k)$ efficiency hides very large constants. However, it is usually possible to tailor such a distinguishing environment for a particular $\mathcal{F}$ to achieve much better efficiency bounds, resulting in a very practical commitment protocol.

*Proof.* If the zero-one law holds, then $\mathcal{F}_{\mathrm{XOR}}$ is complete, since it is unconditionally non-trivial. Thus $\mathcal{F}_{\mathrm{OT}} \sqsubseteq_{\mathrm{PPT}} \mathcal{F}_{\mathrm{XOR}}$. $\mathcal{F}_{\mathrm{OT}}$ has the property that any protocol that securely realizes $\mathcal{F}_{\mathrm{OT}}$ (against active adversaries) is also secure against semi-honest adversaries (see [30] for more details). Hence the given $\mathcal{F}_{\mathrm{OT}}$ protocol is secure against semi-honest adversaries, in the $\mathcal{F}_{\mathrm{XOR}}$-hybrid model. Since $\mathcal{F}_{\mathrm{XOR}}$ has an elementary plain protocol unconditionally secure against semi-honest adversaries, we can compose these two protocols to obtain a plain protocol that securely realizes $\mathcal{F}_{\mathrm{OT}}$ against semi-honest adversaries.

More generally, if $\mathcal{F}$ has an unconditionally secure protocol against semi-honest adversaries, then the $\sqsubseteq_{\mathrm{PPT}}$-completeness of $\mathcal{F}$ implies the sh-OT assumption.

## 7   Extensions, Limitations, and Open Problems

We discuss several natural extensions of our main theorem, discussed in greater detail in the full version:

*Strengthening the Reduction.* As the definition of a reduction is strengthened, fewer functionalities reduce to one another. In the extreme, the reduction could be made so restrictive that no functionality reduces to another. On the other hand, it is relatively easy to see that the zero-one law still applies as stated in this work if protocols are given only public channels instead of private channels, or if security is simultaneously required against both active and semi-honest adversaries.

If the reduction requires security against *computationally unbounded* adversaries, then the zero-one law breaks down. In fact, there exist infinitely many distinct intermediate (between trivial and complete) complexities with respect to this stronger reduction [25].

If the reduction requires parties to use the given ideal functionality with only *fixed roles* (i.e., Alice can access $\mathcal{F}$ only in the role of Alice), then $\mathcal{F}_{\mathrm{COM}} \not\sqsubseteq \mathcal{F}_{\mathrm{CC}}$ (note that the behavior of $\mathcal{F}_{\mathrm{CC}}$ is not symmetric with respect to the two parties), so the zero-one law does not hold under this strong reduction. This impossibility highlights the fact that $\mathcal{F}_{\mathrm{CC}}$ indeed has rather low complexity, and justifies our somewhat complicated protocol used to realize $\mathcal{F}_{\mathrm{COM}}$ using $\mathcal{F}_{\mathrm{CC}}$.

We leave open the question of whether the zero-one law holds if the reduction is strengthened to require security against *adaptive* corruption.

*Larger Classes of Functionalities.* We restricted our attention to a class of deterministic functionalities with finite memory and inputs. In fact, the zero-one law does not extend if we relax the restriction on finiteness. Let $\mathcal{F}$ be a channel which accepts an arbitrary-length string $x$ from Alice and sends $f(x)$ to Bob for a fixed function $f$. Assuming one-way functions exist, one can construct an $f$ so that the resulting functionality is neither trivial nor complete.[12] The construction

---

[12] Of course, if one-way functions do *not* exist, then the sh-OT assumption, and subsequently the zero-one law, is again false.

of this intermediate $\mathcal{F}$ is admittedly contrived, and we leave open the important problem of identifying the largest "natural" class of unbounded-memory functionalities that still satisfies the zero-one complexity law.

The other natural way to extend the scope of our results is to consider *randomized* functionalities. However, very little is known about randomized functionalities, even in the simplest case of SFE functionalities and considering perfect security against computationally unbounded, semi-honest adversaries; for comparison, the corresponding characterization for *deterministic* SFE has been known for 20 years [24,2].

*Optimizing Hardness Assumptions.* While our main theorem relies on the minimal sh-OT assumption, our use of the assumption itself is non-black-box. In Theorems 2 and 3 we use interactive proofs of statements regarding various cryptographic primitives (ultimately derived from the sh-OT assumption). We do not know whether such non-black-box usage of the assumption is necessary, although it seems that a fundamentally different approach is required to avoid the use of interactive proofs.

## Acknowledgments

We acknowledge helpful discussions with Ran Canetti, Yuval Ishai, Yehuda Lindell and Amit Sahai, as well as helpful suggestions from anonymous conference referees. The protocol in Theorem 2 was simplified from its original form in an earlier manuscript, partly motivated by the recent results of [9].

## References

1. Proc. 30th FOCS. IEEE, Los Alamitos (1989)
2. Beaver, D.: Perfect privacy for two-party protocols. In: Feigenbaum, J., Merritt, M. (eds.) Proceedings of DIMACS Workshop on Distributed Computing and Cryptography, vol. 2, pp. 65–77. American Mathematical Society, Providence (1989)
3. Beimel, A., Malkin, T., Micali, S.: The all-or-nothing nature of two-party secure computation. In: Wiener, M. (ed.) CRYPTO 1999. LNCS, vol. 1666, pp. 80–97. Springer, Heidelberg (1999)
4. Canetti, R.: Universally composable security: A new paradigm for cryptographic protocols. Electronic Colloquium on Computational Complexity (ECCC) TR01-016, 2001. Previous version. A unified framework for analyzing security of protocols" availabe at the ECCC archive TR01-016. Extended abstract in FOCS (2001)
5. Canetti, R., Fischlin, M.: Universally composable commitments. In: Kilian, J. (ed.) CRYPTO 2001. LNCS, vol. 2139, p. 19. Springer, Heidelberg (2001)
6. Canetti, R., Kushilevitz, E., Lindell, Y.: On the limitations of universally composable two-party computation without set-up assumptions. In: Biham, E. (ed.) EUROCRYPT 2003. LNCS, vol. 2656. Springer, Heidelberg (2003)

7. Canetti, R., Lindell, Y., Ostrovsky, R., Sahai, A.: Universally composable twoparty computation. In: Proc. 34th STOC, pp. 494–503. ACM, New York (2002)
8. Chor, B., Kushilevitz, E.: A zero-one law for boolean privacy (extended abstract). In: STOC, pp. 62–72. ACM, New York (1989)
9. Damgård, I., Nielsen, J.B., Orlandi, C.: On the necessary and sufficient assumptions for UC computation. In: Micciancio, D. (ed.) TCC 2010. LNCS, vol. 5978, pp. 109–127. Springer, Heidelberg (2010)
10. Goldreich, O.: Foundations of Cryptography: Basic Tools. Cambridge University Press, Cambridge (2001), Earlier version available on,
    `http://www.wisdom.weizmann.ac.il/~{}oded/frag.html`
11. Goldreich, O.: Foundations of Cryptography: Basic Applications. Cambridge University Press, Cambridge (2004)
12. Goldreich, O., Micali, S., Wigderson, A.: How to play ANY mental game. In: ACM (ed.) Proc. 19th STOC, pp. 218–229. ACM, New York (1987), See 11, Chap. 7 for more details
13. Harnik, D., Naor, M., Reingold, O., Rosen, A.: Completeness in two-party secure computation: A computational view. J. Cryptology 19(4), 521–552 (2006)
14. Håstad, J., Impagliazzo, R., Levin, L.A., Luby, M.: A pseudorandom generator from any one-way function. SIAM J. Comput. 28(4), 1364–1396 (1999); Preliminary versions appeared in STOC 1989 and STOC 1990
15. Impagliazzo, R.: A personal view of average-case complexity. In: Structure in Complexity Theory Conference, pp. 134–147 (1995)
16. Impagliazzo, R., Luby, M.: One-way functions are essential for complexity based cryptography (extended abstract). In: Proc. 30th FOCS [1], pp. 230–235
17. Ishai, Y., Prabhakaran, M., Sahai, A.: Founding cryptography on oblivious transfer - efficiently. In: Wagner (ed.) [32], pp. 572–591
18. Kilian, J.: Founding cryptography on oblivious transfer. In: STOC, pp. 20–31. ACM, New York (1988)
19. Kilian, J.: A general completeness theorem for two-party games. In: STOC, pp. 553–560. ACM, New York (1991)
20. Kilian, J.: More general completeness theorems for secure two-party computation. In: Proc. 32th STOC, pp. 316–324. ACM, New York (2000)
21. Kilian, J., Kushilevitz, E., Micali, S., Ostrovsky, R.: Reducibility and completeness in private computations. SIAM J. Comput. 29(4), 1189–1208 (2000)
22. Kraschewski, D., Müller-Quade, J.: Completeness theorems with constructive proofs for symmetric, asymmetric and general 2-party-functions, 2008 (2008) (Unpublished Manuscript), `http://iks.ira.uka.de/eiss/completeness`
23. Künzler, R., Müller-Quade, J., Raub, D.: Secure computability of functions in the it setting with dishonest majority and applications to long-term security (2009)
24. Kushilevitz, E.: Privacy and communication complexity. In: FOCS [1], pp. 416–421
25. Maji, H.K., Prabhakaran, M., Rosulek, M.: Complexity of multi-party computation problems: The case of 2-party symmetric secure function evaluation. In: Reingold (ed.) [31], pp. 256–273
26. Maji, H.K., Prabhakaran, M., Rosulek, M.: Cryptographic complexity classes and computational intractability assumptions. In: Yao, A.C.-C. (ed.) Innovations in Computer Science, pp. 266–289. Tsinghua University Press, Beijing (2010)
27. Naor, M.: Bit commitment using pseudorandomness 4(2), 151–158 (1991), Brassard, G. (ed.) CRYPTO 1989. LNCS, vol. 435, pp. 128–136. Springer, Heidelberg (1990)
28. Nguyen, M.-H., Vadhan, S.P.: Zero knowledge with efficient provers. In: STOC, pp. 287–295. ACM, New York (2006)

29. Prabhakaran, M.: New Notions of Security. PhD thesis, Department of Computer Science, Princeton University (2005)
30. Prabhakaran, M., Rosulek, M.: Cryptographic complexity of multi-party computation problems: Classifications and separations. In: Wagner (ed.) [32], pp. 262–279
31. Reingold, O. (ed.): TCC 2009. LNCS, vol. 5444. Springer, Heidelberg (2009)
32. Wagner, D. (ed.): CRYPTO 2008. LNCS, vol. 5157. Springer, Heidelberg (2008)
33. Yao, A.C.: How to generate and exchange secrets. In: Proc. 27th FOCS, pp. 162–167. IEEE, Los Alamitos (1986)

# On Generalized Feistel Networks

Viet Tung Hoang and Phillip Rogaway

Dept. of Computer Science, University of California, Davis, USA

**Abstract.** We prove beyond-birthday-bound security for most of the well-known types of generalized Feistel networks: (1) unbalanced Feistel networks, where the $n$-bit to $m$-bit round functions may have $n \neq m$; (2) alternating Feistel networks, where the round functions alternate between contracting and expanding; (3) type-1, type-2, and type-3 Feistel networks, where $n$-bit to $n$-bit round functions are used to encipher $kn$-bit strings for some $k \geq 2$; and (4) numeric variants of any of the above, where one enciphers numbers in some given range rather than strings of some given size. Using a unified analytic framework, we show that, in any of these settings, for any $\varepsilon > 0$, with enough rounds, the subject scheme can tolerate CCA attacks of up to $q \sim N^{1-\varepsilon}$ adversarial queries, where $N$ is the size of the round functions' domain (the larger domain for alternating Feistel). Prior analyses for most generalized Feistel networks established security to only $q \sim N^{0.5}$ queries.

**Keywords:** Block ciphers, coupling, Feistel networks, generalized Feistel networks, modes of operation, provable security, symmetric techniques.

## 1  Introduction

BACKGROUND. Feistel-like ciphers come in several flavors beyond the "classical" one used in DES [31,7]. In speaking of *generalized* Feistel networks we mean to encompass most all of them; see Fig. 1. In particular, we include: *unbalanced* Feistel networks with either expanding or contracting round functions, as described by Schneier and Kelsey [30]; *alternating* Feistel networks, where the rounds alternate between contracting and expanding steps, as described by Anderson and Biham [1] and by Lucks [11]; *type-1*, *type-2*, and *type-3* Feistel networks, as described by Zheng, Matsumoto, and Imai [35], each of which uses an $n$-bit to $n$-bit round function to create a $kn$-bit blockcipher for some $k \geq 2$; and *numeric* variants of any of the above, where one enciphers numbers in $\mathbb{Z}_N$, for some $N \in \mathbb{N}$, instead of enciphering binary strings. Well-known blockciphers that use generalized Feistel networks include Skipjack (an unbalanced Feistel network), BEAR/LION (alternating), CAST-256 (type-1), RC6 (type-2), and MARS (type-3).

The provable-security analysis of Feistel networks begins with the seminal work of Luby and Rackoff [10]. The $\imath$ round functions used are assumed to be selected uniformly and independently at random ($\imath = 3$ or $\imath = 4$ in [10]). One then considers how close to a random permutation the constructed cipher is. Subsequent

T. Rabin (Ed.): CRYPTO 2010, LNCS 6223, pp. 613–630, 2010.

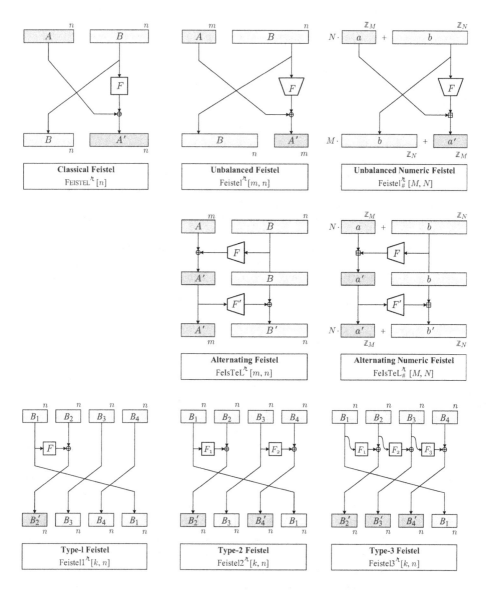

**Fig. 1. Generalized Feistel networks.** The superscript $\imath$ is the number of rounds. The illustrations show a single round $\imath = 1$ except for the alternating schemes, where $\imath = 2$ rounds are shown. Scheme FEISTEL is the classical balanced-Feistel scheme; all remaining schemes are generalizations of it. Schemes Feistel$_\sharp$ and FeIsTeL$_\sharp$ are numeric variants of Feistel (unbalanced Feistel) and FeIsTeL (alternating Feistel); they encipher a number $x = aN + b \in \mathbb{Z}_{MN}$ ($a \in \mathbb{Z}_M$, $b \in \mathbb{Z}_N$) instead of a string $X \in \{0, 1\}^{m+n}$. Schemes Feistel1, Feistel2, and Feistel3 are the so-called type-1, type-2, and type-3 Feistel networks. They are used in modern blockciphers like CAST-256, RC6, and MARS, respectively. Variable $k$ refers to the number of $n$-bit input blocks $B_1, \ldots, B_k$. The illustrations are for $k = 4$.

| scheme | $E =$ | $\mathbf{Adv}_E^{\mathrm{cca}}(q) \leq$ | where $r =$ |
|---|---|---|---|
| classical | $\mathrm{FEISTEL}^r[n]$ | $\frac{2q}{r+1}\left(4q \,/\, 2^n\right)^r$ | $6r - 1$ |
| unbalanced | $\mathrm{Feistel}^r[m, n]$ | | |
| with $n > m$ | | $\frac{2q}{r+1}\left((3\lceil n/m\rceil + 3)q \,/\, 2^n\right)^r$ | $r(4\lceil n/m\rceil + 4)$ |
| with $n \leq m$ | | $\frac{2q}{r+1}\left(4\lceil m/n\rceil q \,/\, 2^n\right)^r$ | $r(2\lceil m/n\rceil + 4)$ |
| unbalanced♯ | $\mathrm{Feistel}_\sharp^r[M, N]$ | | |
| with $N > M$ | | $\frac{2q}{r+1}\left((9\lceil\log_M N\rceil + 5)q \,/\, N\right)^r$ | $r(6\lceil\log_M N\rceil + 4)$ |
| with $N \leq M$ | | $\frac{2q}{r+1}\left((7\lceil\log_N M\rceil + 7)q \,/\, N\right)^r$ | $r(2\lceil\log_N M\rceil + 6)$ |
| alternating | $\mathrm{FeIsTeL}^r[m, n]$ | $\frac{2q}{r+1}\left((6\lceil n/m\rceil + 3)q \,/\, 2^n\right)^r$ | $r(12\lceil n/m\rceil + 8)$ |
| alternating♯ | $\mathrm{FeIsTeL}_\sharp^r[M, N]$ | $\frac{2q}{r+1}\left((6\lceil\log_M N\rceil + 3)q \,/\, N\right)^r$ | $r(12\lceil\log_M N\rceil + 8)$ |
| type-1 | $\mathrm{Feistel1}^r[k, n]$ | $\frac{2q}{r+1}\left(2k(k-1)q \,/\, 2^n\right)^r$ | $r(4k - 2)$ |
| type-2 | $\mathrm{Feistel2}^r[k, n]$ | $\frac{2q}{r+1}\left(2k(k-1)q \,/\, 2^n\right)^r$ | $r(2k + 2)$ |
| type-3 | $\mathrm{Feistel3}^r[k, n]$ | $\frac{2q}{r+1}\left(4(k-1)^2 q \,/\, 2^n\right)^r$ | $r(2k + 2)$ |

**Fig. 2. Summary of CCA bounds in this paper.** The rows correspond to the generalized Feistel networks pictured in Fig. 1. Unbalanced schemes are distinguished by their using contracting ($n > m$) or expanding ($n \leq m$) round functions. Parameters $k, m, n, M, N$ describe the scheme and $r \geq 1$ determines the number of rounds $r$. The specified results appear as Theorems 6–10.

work in this information-theoretic framework (still analyzing the classical Feistel construction) includes Maurer [12], Naor and Reingold [19], Vaudenay [33], Maurer and Pietrzak [13], and a sequence of papers by Patarin [26,21,23,24,22]. The last culminates with the claim that six rounds of (classical) Feistel on a $2n$-bit string is enough to defeat (meaning the advantage goes to 0 as $n \to \infty$) adaptive chosen-ciphertext attacks of $2^{n(1-\varepsilon)}$ queries, for any $\varepsilon > 0$.

Information-theoretic analysis of *generalized* Feistel schemes is less mature. We postpone describing the known results except to say that they are either absent (alternating Feistel with highly-imbalanced round functions), quantitatively weak (birthday bounds that generalize Luby and Rackoff's 25-year-old work), or highly specialized (unbalanced Feistel networks with maximally unbalanced contracting round functions).

CONTRIBUTIONS. Our CCA-security bounds for generalized Feistel networks are described in Fig. 2. Proofs omitted due to lack of space appear in the full version of this paper [8]. Let us briefly describe each result and how it compares with prior work.

For the classical Feistel network on $2n$ bits, our results are comparable to those of Maurer and Pietrzak (henceforth "MP") [13]. As with that work, the bounds get better as one increases the number of rounds $r$. Asymptotically, for any $\varepsilon > 0$, there is a corresponding number of rounds $r$ (about $6/\varepsilon$) such that any CCA-adversary has vanishing advantage if it asks at most $q = 2^{n(1-\varepsilon)}$ forwards or backwards queries. Our actual results are concrete, and are a little sharper than MP's bounds; see Fig. 3 for a graphical comparison. Our proof is much simpler than those of MP or Patarin. One reason for this is just that we employ the lovely result of Maurer, Pietrzak, and Renner for passing from NCPA-security to CCA-security [14]. The more important reason stems from our use of *coupling*, a well-known technique from the theory of Markov chains.

Next we look at unbalanced Feistel networks; the round functions are maps $F_i: \{0,1\}^n \to \{0,1\}^m$. For the contracting case $(n > m)$ we prove CCA-security to $2^{n(1-\varepsilon)}$ queries. Earlier work by Naor and Reingold provided bounds that topped out at $2^{n/2}$ adversarial queries. Interpreting our result, if one holds fixed the block length $\ell = m + n$, bounds improve with increasing imbalance, the best bounds at $m = 1$, the setting earlier studied by Morris, Rogaway, and Stegers ("MRS") [17]. In effect, we "connect up" MP's bounds on balanced Feistel with MRS's bounds on maximally unbalanced Feistel, demonstrating a smooth increase in security with increasing imbalance. This behavior is not an artifact of the analysis; corresponding information-theoretic attacks exist [22,27].

For unbalanced Feistel networks with expanding random round functions our concrete-security results (again see Fig. 2) can similarly be interpreted asymptotically to show CCA security to $2^{n(1-\varepsilon)}$ queries. But note that as imbalance increases in an expanding round functions the value of $n$ goes down, so provable security is effectively vanishing. Again this is no artifact; there are corresponding information-theoretic attacks [22,28].

We next treat unbalanced Feistel networks that acts on numbers instead of strings, the blockcipher we denote $\mathrm{Feistel}_{\sharp}^{\mathcal{N}}[M, N]$. This situation is seen in the card-shuffling technique of Thorp [32] (where $M = 2$) and is defined explicitly in the work of Bellare *et al.* [4]. While one might expect unbalanced Feistel schemes to behave similarly in the number-based and string-based settings, being able to show this is something else: the number-based setting is considerably more complex. We note that MRS only managed to deal with the case $M = 2$ and $N = 2^n$, leaving the generalization open. We show security to $q \sim N^{1-\varepsilon}$ queries.

Unbalanced Feistel networks are unpleasant in requiring a "repartitioning" of each round's output before it can be treated as the next round's input. An alternative is suggested by the "ladder" way of drawing DES (the way that avoids wire-crossings, as in our illustration of FeIsTeL). Information-theoretic security bounds for alternating Feistel networks [1,3,4,11] were weak in two ways: quantitatively, they top out at the birthday-bound; qualitatively, they depend on the domain size of the round function with *smaller* domain, leading to a non-result for the highly imbalanced setting. We overcome both issues. Our results cover the numeric as well as the string-based settings.

Finally, we consider type-1, type-2, and type-3 Feistel networks [35], as used in several modern blockciphers. We prove information-theoretically optimal bounds (as the number of rounds becomes large). The proofs here are straightforward compared to those for unbalanced and alternating Feistel, highlighting a strength of the coupling-based approach.

Unmentioned in all of the above is that our string-based results also work when the alphabet is non-binary. This turns out to be useful; for example, one could encipher a 16-digit credit card number (CCN) (the ciphertext again being a 16-digit number) using a scheme $\mathrm{FEISTEL}_{10}^{\curvearrowright}[8]$ just like $\mathrm{FEISTEL}^{\curvearrowright}[8]$ but over the decimal alphabet instead of the binary one [2] (re-interpret the xor operator as, say, modular addition). Our security bounds for schemes with non-binary alphabets are as given in Fig. 2 but with $2^n$ replaced by $d^n$, where $d$ is the radix of the alphabet.

In general, finding a unified framework with which to analyze Feistel-like schemes—one that gives concrete, asymptotically optimal, humanly-verifiable bounds—is a contribution we see as being at least as important as all the improved bounds.

ADDITIONAL RELATED WORK. In work just subsequent to our own, Patarin provides a concrete security bound for the classical Feistel construction $\mathrm{FEISTEL}^6[n]$ [25]. He goes on to claim beyond-birthday-bound security for the unbalanced scheme $\mathrm{Feistel}^8[n, 2n]$. Earlier versions of our paper confessed an inability to extract concrete security bounds from Patarin's body of work.

Nachef attacks a Feistel variant that she calls an alternating unbalanced Feistel scheme [18], but the scheme is different from the more classical one that we study here. The specific rotation operation used in Nachef's scheme makes this Feistel variant highly insecure.

The first use of a coupling argument in cryptography that we know is due to Mironov, who used the technique to gave a lovely (even if slightly heuristic) analysis of RC4 [15]. As mentioned earlier, Morris, Rogaway, and Stegers go on to use coupling to analyze the security of a maximally-unbalanced (contracting round function) Feistel network. Our work builds on theirs, but our use of coupling becomes considerably more complex.

Beyond their use in making conventional blockciphers, generalized Feistel networks have been proposed as blockcipher modes-of-operation for format-preserving encryption (FPE) [5,3,4]. Here one usually aims to encipher points within some arbitrary string-valued domain $\Sigma^n$, or within some arbitrary numeric domain $\mathbb{Z}_N$. Commercial interest in doing this has been spurred by PCI regulations [29] that require vendors to encipher CCNs they store; an architecturally clean way to do this is to encipher a column in a database without making any modification to the database's schema. There is now a NIST proposal for an FPE-providing mode of operation, FFX [2], that employs an unbalanced or alternating Feistel network over a possibly non-binary alphabet.

## 2   Preliminaries

NOTATION. For finite nonempty sets $A$ and $B$, let $\mathrm{Func}(A, B)$ be the set of all functions from $A$ to $B$ and let $\mathrm{Perm}(A)$ be the set of all permutations on $A$. For numbers $a, b \geq 1$, let $\mathrm{Func}(a, b)$ be the set of all functions from $\{0, 1\}^a$ to $\{0, 1\}^b$.

BLOCKCIPHERS. Let $E\colon \mathcal{K} \times \mathcal{M} \to \mathcal{M}$ be a blockcipher, meaning that each $E_K(\cdot) = E(K, \cdot)$ is a permutation on the finite nonempty set $\mathcal{M}$. We emphasize that $\mathcal{M}$ (and also $\mathcal{K}$) need not consist of binary strings of some particular length, as is often assumed to be the case. For any blockcipher $E$, we let $E^{-1}$ be its inverse blockcipher. For blockcipher $E\colon \mathcal{K} \times \mathcal{M} \to \mathcal{M}$ and adversary $A$ the *advantage* of $A$ in carrying out an (adaptive) chosen-ciphertext attack (CCA) on $E$ is
$$\mathbf{Adv}_E^{\mathrm{cca}}(A) = \Pr[K \xleftarrow{\$} \mathcal{K} \colon A^{E_K(\cdot), E_K^{-1}(\cdot)} \Rightarrow 1] - \Pr[\pi \xleftarrow{\$} \mathrm{Perm}(\mathcal{M}) \colon A^{\pi(\cdot), \pi^{-1}(\cdot)} \Rightarrow 1].$$
We say that $A$ carries out an (adaptive) chosen-plaintext attack (CPA) if it asks no queries to its second oracle. Adversary $A$ is *non-adaptive* if it asks the same queries on every run. Let $\mathbf{Adv}_E^{\mathrm{cca}}(q)$ be the maximum advantage of any (adaptive) CCA adversary against $E$ subject to the adversary asking at most $q$ total oracle queries. Similarly define $\mathbf{Adv}_E^{\mathrm{ncpa}}(q)$ for nonadaptive CPA attacks (NCPA).

For blockciphers $F, G\colon \mathcal{K} \times \mathcal{M} \to \mathcal{M}$ let $F \circ G$ denote their cascade, with $F$'s output fed into $G$'s input; formally, $F \circ G\colon \mathcal{K}^2 \times \mathcal{M} \to \mathcal{M}$ is defined by $(F \circ G)_{(K, K')} = G_{K'}(F_K(X))$. To be consistent with this left-to-right convention for composing blockciphers we define composition of permutations by $(f \circ g)(x) = g(f(x))$. (This won't be used often and should not cause confusion for those used to the opposite convention.)

COUPLING ARGUMENTS. The high-level idea for a coupling argument can be explained like this. We have a Markov chain $X_t$ that we want to analyze. For example, the Markov chain may consist of the image of the distinct, fixed strings $(x_1, \ldots, x_q) \in (\{0, 1\}^{2n})^q$ as each point is enciphered for $t$ rounds according to the classical Feistel network on $2n$ bits. We would like to show that, after $t = \sim$ rounds, the tuple of points $X_t$ is pretty close to being uniformly distributed. For this purpose, we introduce a *second* Markov chain $U_t$ that, after any number of rounds $t$, is indisputably uniform. We arrange so that $X_t$ and $U_t$ can be viewed as co-evolving on a common probability space; formally, we create a joint distribution that yields the correct marginal distributions. We try to arrange our joint distribution so that, usually, $X_t$ and $U_t$ quickly *couple*: for *most* random choices, it does not take long until $X_t = U_t$. After $X_t$ and $U_t$ come together, they should remain so. The basic observation underlying coupling is that the statistical distance between the distributions associated to $X_t$ and $U_t$ is upperbounded by the probability that $X_t \neq U_t$.

More formally, let $\mu$ and $\nu$ be probability distributions on a finite event space $\Omega$. The *total variation distance* between distributions $\mu$ and $\nu$ is defined as $\|\mu - \nu\| = \frac{1}{2} \sum_{x \in \Omega} |\mu(x) - \nu(x)| = \max_{S \subset \Omega} \{\mu(S) - \nu(S)\}$. A *coupling* of $\mu$ and $\nu$ is a pair of random variables $X, Y\colon \Omega \to R$ (the set $R$ is arbitrary) such that $X \sim \mu$ and $Y \sim \nu$, that is, variables $X$ and $Y$ have marginal distributions $\mu$ and $\nu$, respectively. The *coupling lemma* we will use is as follows.

**Lemma 1 (Coupling lemma).** *Let $\mu$ and $\nu$ be probability distributions on a finite event space $\Omega$ and let $(X, Y)$ be a coupling of $\mu$ and $\nu$. Then $\|\mu - \nu\| \leq \Pr[X \neq Y]$.*

FROM COUPLING TO NCPA-SECURITY. Suppose that an adversary asks some non-adaptive distinct queries. The adversary's NCPA advantage cannot exceed the total variation distance between the distribution of the outputs from her queries and the uniform distribution. The uniform distribution itself can be viewed as the distribution of outputs from a uniformly random choice of distinct queries. Think of a coupling argument as a computer program that accepts as its input either the actual adversarial queries or a pool of uniformly random, distinct queries. On each input, the program implements a Feistel network and gives a random output. The program tries to produce the same output on its two possible inputs. Hence the total variation distance between the distributions of the program's outputs is upperbounded by the program's probability of failure (that is, its failure to produce the same output in the two cases).

To ease the design of such a program, a hybrid argument is employed and a chain of inputs is created—the first being the adversarial queries and the last being the pool of uniformly random, distinct ones. The purpose of this hybrid argument is to reduce the difference between any pair of adjacent inputs in the chain. Given an arbitrary pair of adjacent inputs, our goal now is to design a coupling program that produces identical output on those two inputs with high probability. The program runs both inputs, one after another. When the program starts running the second input, it has finished the operations on the first input and now knows all the random choices of the first Feistel network. It then uses this knowledge in implementing the second Feistel network. For example, if at some step the second network needs a uniformly random string then the program may reuse the corresponding string from the first network. The random choices in the second network are geared toward the first output, but they are subject to the restriction that the round functions in the second network must be independent and uniformly random.

FROM NCPA TO CCA-SECURITY. We bound the CCA-security of a Feistel network from its NCPA-security by using the following result of Maurer, Pietrzak, and Renner [14, Corollary 5]. It is key to our approach, effectively letting us assume that our adversaries are of the simple, NCPA breed. Recall that in writing $F \circ G$, the blockciphers are, in effect, independently keyed.

**Lemma 2 (Maurer-Pietrzak-Renner).** *If $F$ and $G$ are blockciphers on the same message space then, for any $q$,* $\mathbf{Adv}_{F \circ G^{-1}}^{\mathrm{cca}}(q) \leq \mathbf{Adv}_F^{\mathrm{ncpa}}(q) + \mathbf{Adv}_G^{\mathrm{ncpa}}(q)$.

# 3   Classical Feistel

This section provides a strong, concrete security bound for conventional, balanced Feistel networks. It also serves as a pedagogical example for proving security of a Feistel network using coupling; some later examples get much more complex.

DEFINING THE SCHEME. Fix $n \geq 1$ and let $F \colon \{0,1\}^n \to \{0,1\}^n$ be a function. Define from $F$ the permutation $\Psi_F \colon \{0,1\}^{2n} \to \{0,1\}^{2n}$ by way of

$\Psi_F(A, B) = (B, A \oplus F(B))$ where $|A| = |B| = n$, and $\oplus$ denotes xor. Blockcipher FEISTEL$^{\varkappa}[n]\colon \mathcal{K} \times \{0,1\}^{2n} \to \{0,1\}^{2n}$ has key space $\mathcal{K} = (\text{Func}(n,n))^{\varkappa}$ and a key $(F_1, \ldots, F_{\varkappa}) \in \mathcal{K}$ names the permutation $\Psi_{F_1} \circ \cdots \circ \Psi_{F_{\varkappa}}$ on $\{0,1\}^{2n}$. Each $F_i$ is called the round function at round $i$. For an illustration, see Fig. 1.

INITIAL NOTATION. Given a query $X$ to $E = \text{FEISTEL}^{\varkappa}[n]$, define its round-0 output to be $X$ itself, while the round-$t$ output is $(\Psi_{F_1} \circ \cdots \circ \Psi_{F_t})(X)$. The *coin* of the query $X$ at round $t$ is the string $A \oplus F(B)$, where $F$ is the round function at round $t$ and $(A, B)$ is the round-$(t-1)$ output, with $|A| = |B| = n$. Two queries *collide* at time $t$ if their round-$t$ outputs have the same final $n$ bits.

NCPA-SECURITY. We will now prove the NCPA-security of $E$ by way of coupling, afterwards lifting this to show CCA-security using the result of [14] from Lemma 2. The lemma below will help us bound the probability that we *fail* to couple.

**Lemma 3.** *For the blockcipher $E = \text{FEISTEL}^{\varkappa}[n]$, the chance that two distinct non-adaptive queries collide at time $t \geq 1$ is at most $2^{-n}$.*

*Proof.* Suppose that the Feistel network receives distinct nonadaptive queries $X_1$ and $X_2$. For each $i \in \{1, 2\}$, let $(A_i, B_i)$ be the output at round $t-1$ of $X_i$, where $|A_i| = |B_i| = n$. The queries $X_1$ and $X_2$ collide at time $t$ if and only if $A_1 \oplus F(B_1) = A_2 \oplus F(B_2)$, with $F$ being the round function at round $t$. This occurs with probability $2^{-n}$ if $B_1$ and $B_2$ differ, because $F$ is uniformly random. If $B_1 = B_2$ then so are $A_1$ and $A_2$, which contradicts the hypothesis that $X_1$ and $X_2$ are distinct.                                                  $\square$

**Theorem 4.** *Let $E = \text{FEISTEL}^{\varkappa}[n]$, $\varkappa = 3r$. Then $\mathbf{Adv}_E^{\text{ncpa}}(q) \leq \frac{q}{r+1}(4q/2^n)^r$.*

*Proof.* Suppose that $E$ receives non-adaptive distinct queries $X_1, \ldots, X_q$. For each $\ell \leq q$, consider a vector of queries $(Z_1, \ldots, Z_q)$ such that $Z_i$ is $X_i$ if $i \leq \ell$ and $Z_i$ is chosen uniformly from $\{0,1\}^{2n} \backslash \{Z_1, \ldots, Z_{i-1}\}$ otherwise. Let $\mu_\ell$ be the distribution of the vector of $q$ outputs when $E$ receives queries $Z_1, \ldots, Z_q$. We will show in a moment that the total variation distance between $\mu_\ell$ and $\mu_{\ell+1}$ is at most $(4\ell / 2^n)^r$ for every $\ell \leq q - 1$. Assuming this, we have, by hybrid argument,

$$\mathbf{Adv}_E^{\text{ncpa}}(q) \leq \sum_{\ell=0}^{q-1} \|\mu_\ell - \mu_{\ell+1}\| \leq \sum_{\ell=0}^{q-1} (4\ell / 2^n)^r \leq 2^{r(2-n)} \int_0^q x^r dx,$$

which is $\frac{q}{r+1}(4q / 2^n)^r$. Now we show the claim. Fix a value $\ell \leq q - 1$. We must bound the total variation distance between $\mu_\ell$ and $\mu_{\ell+1}$, each of them is a distribution of a vector of $q$ outputs. However, only the first $\ell + 1$ components of the vector matter, because of the uniform sampling of the other. Consider a $3r$-round balanced Feistel network on $n$ bits that receives queries $X_1, \ldots, X_{\ell+1}$. Let $X_i(t)$ be the output at round $t$ from the query $X_i$.

THE COUPLING. We construct another $3r$-round balanced Feistel network on $n$ bits with its non-adaptive distinct queries $U_1, \ldots, U_{\ell+1}$. Let $U_i(t)$ be the output at round $t$ of the new Feistel network on input $U_i$. The construction of the new Feistel network will satisfy the following conditions:

- Query $U_j$ equals to $X_j$ for every $j \leq \ell$, and $U_{\ell+1}$ is uniformly chosen over $\{0,1\}^{2n} \backslash \{U_1, \ldots, U_\ell\}$.
- If for all $i \leq \ell + 1$, the outputs at round $t$ of $X_i$ and $U_i$ are identical then so are their outputs in any subsequent round.

Let $T$ be the smallest round for which $X_i$ and $U_i$ have identical outputs for every $i \leq \ell + 1$. From the second condition above and from Lemma 1, we have that

$$\|\mu_\ell - \mu_{\ell+1}\| \leq \Pr[X_i(3r) \neq U_i(3r) \text{ for some } i \leq \ell + 1] = \Pr[T > 3r] .$$

The first condition above describes how to initialize $U_1(0), \ldots, U_{\ell+1}(0)$. As the coin of $U_i$ at round $t+1$ dictates how to update $U_i(t+1)$ from $U_i(t)$, it suffices to show how to construct just that coin.

- If $U_i$ collides with some previous query $U_j$ at time $t$ then the coin at round $t+1$ of $U_i$ is defined so as to ensure consistency with the earlier query.
- Suppose that, in the new Feistel network, $U_i$ does not collide with any previous query at time $t$. If the query $X_i$ collides with some previous query $X_j$ at time $t$ then we choose a string uniformly from $\{0,1\}^n$ to be the coin of $U_i$ at round $t+1$. Otherwise, the coin of $X_i$ at round $t+1$ is uniformly distributed over $\{0,1\}^n$ and $U_i$ will use exactly the same coin at round $t+1$.

Note that $U_i$ and $X_i$ always have the same output at round $t$, for every $i \leq \ell$ and every $t$. Consider the event Coll that in either Feistel networks, the $(\ell+1)$-th query collides with some previous query at some time $t \in \{1, 2\}$. From Corollary 3, each such collision occurs with probability at most $2^{-n}$. Summing over the two Feistel networks, two rounds, and $\ell$ previous queries shows that the probability Coll occurs is at most $4\ell / 2^n$. Unless Coll occurs, $U_{\ell+1}$ and $X_{\ell+1}$ will share the coins at the second and third rounds, and then have identical outputs at the third round. Hence $\Pr[T > 3] \leq \Pr[\text{Coll}]$, which is at most $4\ell / 2^n$.

Now imagine that we run a sequence of trials. In each trial, we observe the outputs of $X_{\ell+1}$ and $U_{\ell+1}$ for an additional three rounds. The probability that $X_{\ell+1}$ and $U_{\ell+1}$ have different outputs after the first trial is at most $4\ell / 2^n$. Since the round functions of both Feistel networks in each trial are independent with those in previous trials, the conditional probability that $X_{\ell+1}$ and $U_{\ell+1}$ have different outputs after the $r$-th trial, given that their outputs remain different after the first $r - 1$ trials, is again at most $4\ell / 2^n$. Hence $\Pr[T > 3r] \leq (4\ell / 2^n)^r$. $\qquad\square$

CCA-SECURITY. Let Rev denote the permutation on $\{0,1\}^{2n}$ where $\mathsf{Rev}(A, B) = (B, A)$, for $|A| = |B| = n$. The following observation is standard; see [13] for proof.

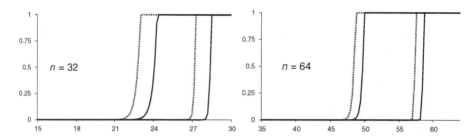

**Fig. 3. Proven CCA-security for the classical Feistel network: our own bounds and MP's.** The $x$-axis gives the log base-2 of the number of adversarial queries and the $y$-axis gives upper bounds on an adversary's CCA advantage. In the left-hand plot (64-bit inputs), the dashed lines depict MP's bounds for FEISTEL$^{24}$[32] (left) and FEISTEL$^{96}$[32] (right); the solid lines depict our own bounds. In the right-hand plot (128-bit inputs), the dashed lines likewise depict MP's bounds for FEISTEL$^{24}$[64] (left) and FEISTEL$^{96}$[64] (right); the solid lines depict our own bounds.

**Lemma 5.** *If $F$ and $G$ are the blockcipher* FEISTEL$^{\nu}$[n] *then* $F \circ G^{-1} \circ \mathrm{Rev}$ *is the blockcipher* FEISTEL$^{2\nu-1}$[n]. □

Employing Lemma 2 we conclude the following.

**Theorem 6.** *Let $E =$* FEISTEL$^{\nu}$[n], $\nu = 6r-1$. *Then* $\mathbf{Adv}_E^{\mathrm{cca}}(q) \le \frac{2q}{r+1}\left(4q/2^n\right)^r$.

ASYMPTOTIC INTERPRETATION.    For an asymptotic interpretation of Theorem 6, fix $r > 0$. Suppose that $q = 2^{n(1-1/r)}$. Let $E_n$ be the blockcipher FEISTEL$^{6r-1}$[n]. Then

$$\mathbf{Adv}_{E_n}^{\mathrm{cca}}(q) \le \frac{2q}{r+1}\left(4q/2^n\right)^r = \frac{2^{2r+1}}{r+1}/2^{n/r},$$

which goes to 0 as $n \to \infty$. Translating into English, CCA security is guaranteed to about $q = 2^{n(1-\varepsilon)}$ adversarial queries as long as one employs $\nu \ge 6/\varepsilon - 1$ rounds. At a higher level still, ignoring the $1-\varepsilon$ multiplier in the exponent, an appropriate number of rounds lets one tolerate nearly $q = 2^n$ adversarial queries.

COMPARISONS.    Maurer and Pietrzak's earlier work proves a security bound of $\mathbf{Adv}_E^{\mathrm{cca}}(q) \le 4q^2/2^{2n} + 2q\left(8q/2^n\right)^r$ for $E =$ FEISTEL$^{6r-1}$[n]. Our own bound is always tighter than this; see Fig. 3 for a comparison of Theorem 6 and MP's bound. Earlier versions of our paper explained that we were unable to plot Patarin's latest bounds [26] due to the absence of a concrete security statement. In very recent work [25] (subsequent to our own), Patarin bounds the security of $E =$ FEISTEL$^6$[n] by $\mathbf{Adv}_E^{\mathrm{cca}}(q) \le 8q/2^n + q^2/2^{2n+1}$ (assuming $q \le 2^n/128n$).

## 4    Unbalanced Feistel

DEFINING THE SCHEME.    Fix $n, m \ge 1$ and let $F\colon \{0,1\}^n \to \{0,1\}^m$ be a function. Define from $F$ the permutation $\Psi_F\colon \{0,1\}^{m+n} \to \{0,1\}^{m+n}$ by way of $\Psi_F(A, B) = (B, A \oplus F(B))$ where $|A| = m$ and $|B| = n$, and $\oplus$ denotes xor. We call $\Psi_F$ a Feistel $(m, n)$-permutation and $F$ its round function. Blockcipher Feistel$^{\mathcal{L}}[m, n]$: $\mathcal{K} \times \{0,1\}^{m+n} \to \{0,1\}^{m+n}$ has key space $\mathcal{K} = (\mathrm{Func}(m, n))^{\mathcal{L}}$ and a key $(F_1, \ldots, F_{\mathcal{L}}) \in \mathcal{K}$ names the permutation $\Psi_{F_1} \circ \cdots \circ \Psi_{F_{\mathcal{L}}}$ on $\{0,1\}^{m+n}$. For an illustration, see Fig. 1.

SECURITY OF UNBALANCED FEISTEL SCHEMES. The theorem below shows the CCA-security of Feistel$^{\mathcal{L}}[m, n]$. The proof can be found in Appendix A. Interpreted asymptotically, the result says that, with an adequate number of rounds, CCA security is guaranteed to about $2^n$ adversarial queries. Note that for *expanding* round functions this guarantee eventually becomes meaningless. This is as it should be; expanding round functions with small domains give rise to information-theoretically insecure schemes.

**Theorem 7.** *Fix integers $m, n, r \geq 1$.*

1) *Let $E = $ Feistel$^{\mathcal{L}}[m, n]$ where $n > m$ and $\mathcal{L} = r(4\lceil n/m \rceil + 4)$.*
   *Then $\mathbf{Adv}_E^{\mathrm{cca}}(q) \leq \frac{2q}{r+1}\left((3\lceil n/m \rceil + 3)q / 2^n\right)^r$.*

2) *Let $E = $ Feistel$^{\mathcal{L}}[m, n]$ where $n \leq m$ and $\mathcal{L} = r(2\lceil m/n \rceil + 4)$.*
   *Then $\mathbf{Adv}_E^{\mathrm{cca}}(q) \leq \frac{2q}{r+1}\left(4\lceil m/n \rceil q / 2^n\right)^r$.*

NON-BINARY ALPHABETS. We can replace the binary alphabet $\{0,1\}$ in an unbalanced Feistel scheme with an arbitrary alphabet $\Sigma$ where $d = |\Sigma| \geq 2$. Regard the characters as numbers $\{0, 1, \ldots, d-1\}$ and reinterpret $\oplus$ either as integer addition modulo $d^m$ or as characterwise addition modulo $d$. The analysis associated to Theorem 7 is trivially lifted to this setting; for example, if $E = $ Feistel$_d^{\mathcal{L}}[m, n]$, the radix of the alphabet indicated by the subscript, with $n > m$ and $\mathcal{L} = r(4\lceil n/m \rceil + 4)$, then $\mathbf{Adv}_E^{\mathrm{cca}}(q) \leq \frac{2q}{r+1}\left((3\lceil n/m \rceil + 3)q / d^n\right)^r$. We comment that our proof for part (1) of Theorem 7 works for any group operator on $\Sigma^m$, but our proof for part (2) does not.

GRAPHICAL ILLUSTRATION. Fig. 4 illustrates our CCA-security bounds for Feistel$^{\mathcal{L}}[32, 96]$ versus Feistel$^{\mathcal{L}}[64, 64]$. Given an adequate number of rounds, imbalance provably helps.

UNBALANCED NUMERIC FEISTEL. We now go on to show security for the numeric variant of the unbalanced Feistel scheme. We begin by defining this. Let $M \geq 2$ and $N \geq 2$ be numbers and let $F$ have signature $F: \mathbb{Z}_N \to \mathbb{Z}_M$. Let $\boxplus: \mathbb{Z}_M \times \mathbb{Z}_M \to \mathbb{Z}_M$ represent addition modulo $M$, that is, $a \boxplus b = (a+b) \bmod M$. Consider the permutation $\Psi_F: \mathbb{Z}_{MN} \to \mathbb{Z}_{MN}$ that maps $Na + b$ to $Mb + (a \boxplus F(b))$ for every $(a, b) \in \mathbb{Z}_M \times \mathbb{Z}_N$. We call $\Psi_F$ a numeric Feistel $(M, N)$-permutation and $F$ its round function. Blockcipher Feistel$_{\sharp}^{\mathcal{L}}[M, N]$: $\mathcal{K} \times \mathbb{Z}_{MN} \to \mathbb{Z}_{MN}$ has key space $(\mathrm{Func}(\mathbb{Z}_N, \mathbb{Z}_M))^{\mathcal{L}}$. A key $(F_1, \ldots, F_{\mathcal{L}}) \in \mathcal{K}$ names the permutation $\Psi_{F_1} \circ \cdots \circ \Psi_{F_{\mathcal{L}}}$ on $\mathbb{Z}_{MN}$, permutations composing from the left. For an illustration, see Fig. 1.

SECURITY OF NUMERIC FEISTEL SCHEMES. The following theorem establishes CCA-security for Feistel$_{\sharp}$. Interpreted asymptotically, the result implies that,

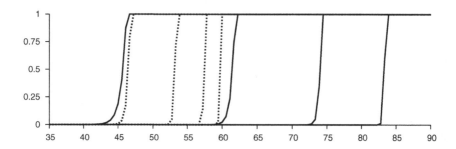

**Fig. 4. Unbalanced Feistel versus classical Feistel on a 128-bit string.** Proven CCA-security of Feistel$^{\wedge}$[32, 96] (bold lines) versus Feistel$^{\wedge}$[64, 64] = FEISTEL$^{\wedge}$[64] (dashed lines) when $\wedge$ is 18, 36, 72, and 144 (the curves from left to right). The $x$-axis gives the log base-2 of the number of queries; the $y$-axis gives an upper bound on an adversary's CCA advantage by Theorems 6 and 7.

with an adequate number of rounds, unbalanced numeric Feistel with a $\mathbb{Z}_N \to \mathbb{Z}_M$ round function withstands a chosen-ciphertext attack to nearly $N$ queries.

**Theorem 8.** *Fix $M, N \geq 2$, $r \geq 1$.*

1) *Let $E = \text{Feistel}_{\#}^{\wedge}[M, N]$ where $N > M$ and $\wedge = r(6 \lceil \log_M N \rceil + 4)$.*
   *Then $\mathbf{Adv}_E^{\text{cca}}(q) \leq \frac{2q}{r+1}\big((9 \lceil \log_M N \rceil + 5)q \,/\, N\big)^r$.*
2) *Let $E = \text{Feistel}_{\#}^{\wedge}[M, N]$ where $N \leq M$ and $\wedge = r(2 \lceil \log_N M \rceil + 6)$.*
   *Then $\mathbf{Adv}_E^{\text{cca}}(q) \leq \frac{2q}{r+1}\big((7 \lceil \log_N M \rceil + 7)q \,/\, N\big)^r$.*

PROOF IDEAS. Let us briefly give an overview of the proof; see the full version of this paper [8, Appendix B] for the complete proof. We begin by extending the concepts of *coin* and *collision* of Section 3. The coupling method in Section 3 requires that every pair of queries share coins at each round, if possible. But this does not work here because if $M$ and $N$ are relatively prime, we may find two deterministic queries that *never* yield the same output under such a coupling strategy. Instead, think of coupling as a computer program trying to produce the same output for two different inputs by manipulating the coins. The program first creates a rule for coin-renaming. For example, suppose that each Feistel network is programmed to create a sequence of uniformly random, independent coins. The rule will map each possible value of the random sequence in the first network to a *unique* value of the corresponding sequence in the second network. The program then runs the first input. Now, knowing the exact value of the sequence of coins in the first network, it runs the second input and uses the rule above to specify how the coins of the second network are created. The uniqueness property is to ensure that the round functions in the second network are independent and uniformly random.

# 5    Alternating Feistel

DEFINING THE SCHEMES. Let $m$ and $n$ be positive integers such that $m \leq n$. The blockcipher $\text{FeIsTeL}^{\nu}[m, n] \colon \mathcal{K} \times \{0, 1\}^{m+n} \to \{0, 1\}^{m+n}$ consists of $\nu$ rounds in which the odd rounds are Feistel $(m, n)$-permutations (contracting) and the even rounds are Feistel $(n, m)$-permutations (expanding). For simplicity, we assume that $\nu$ is even. The key space of $\text{FeIsTeL}^{\nu}[m, n]$ is then $\mathcal{K} = (\text{Func}(n, m) \times \text{Func}(m, n))^{\nu/2}$. Given integers $M$ and $N$ such that $2 \leq M \leq N$, we define the blockcipher $\text{FeIsTeL}_{\sharp}^{\nu}[M, N] \colon \mathcal{K} \times \mathbb{Z}_{MN} \to \mathbb{Z}_{MN}$, with numeric Feistel $(M, N)$ permutations at odd rounds and numeric Feistel $(N, M)$ permutations at even rounds. See Fig. 1 for illustration. We comment that it does not much matter whether one starts with a contracting or expanding round because a security bound with respect to one notion implies the same security bound with respect to the other after one additional round.

SECURITY OF ALTERNATING FEISTEL. The information-theoretic security of blockciphers FeIsTeL and FeIsTeL$_{\sharp}$ are established by the following results. Interpreted asymptotically, the result says that, with an adequate number of rounds, alternating Feistel can withstand a chosen-ciphertext attack to nearly $N$ adversarial queries.

**Theorem 9.** *Fix $r > 0$, $1 \leq m \leq n$, and $2 \leq M \leq N$.*

1) *Let $E = \text{FeIsTeL}^{\nu}[m, n]$ where $\nu = r(12\lceil n/m \rceil + 8)$.*
   *Then $\mathbf{Adv}_E^{\text{cca}}(q) \leq \frac{2q}{r+1}\big((6\lceil n/m \rceil + 3)q / 2^n\big)^r$.*

2) *Let $E = \text{FeIsTeL}_{\sharp}^{\nu}[M, N]$ where $\nu = r(12\lceil \log_M N \rceil + 8)$.*
   *Then $\mathbf{Adv}_E^{\text{cca}}(q) \leq \frac{2q}{r+1}\big((6\lceil \log_M N \rceil + 3)q / N\big)^r$.*

PROOF IDEAS. We give an overview; see the full version of this paper for all details [8, Appendix C]. We consider the generalization of FeIsTeL$_{\sharp}$ in which the operator $\boxplus$ is replaced by any two group operators on $\mathbb{Z}_M$ and $\mathbb{Z}_N$, regarding FeIsTeL as a special case. While we still follow the framework of Section 3, extending the concepts of *coin* and *collision* is tricky. Following the birthday-bound proof of Black and Rogaway [3] and using the simple coupling method for classical Feistel, one may be tempted to define two types of coins, one for odd rounds and one for even rounds; and, likewise, two types of collisions. This will indeed give rise to a bound, which however falls off with $\min(N, M)$ queries instead of $\max(N, M)$ queries; that is, the approach is only good in the nearly-balanced setting. Instead, we define coins only at odd rounds, and collisions only at even rounds.

We are left with the task of coupling two pools of queries. Coins alone cannot completely determine the outputs, because they dictate only the randomness at odd rounds. However, if we require that the two pools use the same expanding round functions (that control the randomness at even rounds), it suffices to specify how coins evolve. While some specific choice of expanding round functions may give us a poor chance of coupling, the expected value of the success probability is good when those functions are uniformly chosen.

## 6    Type-1, Type-2, and Type-3 Feistel

DEFINING THE SCHEMES.  For illustrations, refer again to Fig. 1.

1) Fix $k \geq 2$ and $n \geq 1$, and let $F\colon \{0,1\}^n \to \{0,1\}^n$ name a permutation $\Psi_F\colon \{0,1\}^{kn} \to \{0,1\}^{kn}$ by way of setting $\Psi_F(B_1, \cdots, B_k) = (B_2 \oplus F(B_1), B_3, \ldots, B_k, B_1)$, where $|B_i| = n$. Then Feistel1$^{\nu}[k,n]\colon \mathcal{K} \times \{0,1\}^{kn} \to \{0,1\}^{kn}$ is the blockcipher obtained by the $\nu$-fold composition of $\Psi_F$ permutations, the key space being $\mathcal{K} = (\mathrm{Func}(n,n))^{\nu}$.

2) Assume $k \geq 2$ is even, $n \geq 1$, and $f_i\colon \{0,1\}^n \to \{0,1\}^n$ for every $i \leq k/2$. Let $F = (f_1, \ldots, f_{k/2})$ name a permutation $\Psi_F\colon \{0,1\}^{kn} \to \{0,1\}^{kn}$ by $\Psi_F(B_1, \ldots, B_k) = (B_2 \oplus f_1(B_1), B_3, B_4 \oplus f_2(B_3), B_5, \ldots, B_k \oplus f_{k/2}(B_{k-1}), B_1)$ where $|B_i| = n$. Then the blockcipher Feistel2$^{\nu}[k,n]\colon \mathcal{K} \times \{0,1\}^{kn} \to \{0,1\}^{kn}$ is obtained by the $\nu$-fold composition of $\Psi_F$ permutations, the key space being $\mathcal{K} = (\mathrm{Func}(n,n))^{k\nu/2}$.

3) Finally, with $k \geq 2$ and $n \geq 1$, consider $f_i\colon \{0,1\}^n \to \{0,1\}^n$ for every $i \leq k-1$. Let $F = (f_1, \ldots, f_{k-1})$ name a permutation $\Psi_F\colon \{0,1\}^{kn} \to \{0,1\}^{kn}$ by $\Psi_F(B_1, \cdots, B_k) = (B_2 \oplus f_1(B_1), B_3 \oplus f_2(B_2), \ldots, B_k \oplus f_{k-1}(B_{k-1}), B_1)$, where $|B_i| = n$. Then Feistel3$^{\nu}[k,n]\colon \mathcal{K} \times \{0,1\}^{kn} \to \{0,1\}^{kn}$ is the blockcipher obtained by the $\nu$-fold composition of $\Psi_F$ permutations, the key space being $\mathcal{K} = (\mathrm{Func}(n,n))^{(k-1)\nu}$.

SECURITY RESULTS.  The following results show CCA-security of type-1, type-2, type-3 Feistel variants to $2^{n(1-\varepsilon)}$ queries. Of course this may be a disappointing bound when $n$ is small—and the type-$i$ Feistel variants are in part motivated by a desire to keep $n$ small despite a long block length. But the bound is the best possible, up to the asymptotic behavior, and substantially improves the prior bound in the literature [35].

**Theorem 10.** *Fix $k, r \geq 1$. Then:*

1) $E = \mathrm{Feistel1}^{\nu}[k,n],\ \nu = r(4k-2) \Rightarrow \mathbf{Adv}_E^{\mathrm{cca}}(q) \leq \frac{2q}{r+1}\left(2k(k-1)q/2^n\right)^r$.

2) $E = \mathrm{Feistel2}^{\nu}[k,n],\ \nu = r(2k+2) \Rightarrow \mathbf{Adv}_E^{\mathrm{cca}}(q) \leq \frac{2q}{r+1}\left(2k(k-1)q/2^n\right)^r$.

3) $E = \mathrm{Feistel3}^{\nu}[k,n],\ \nu = r(2k+2) \Rightarrow \mathbf{Adv}_E^{\mathrm{cca}}(q) \leq \frac{2q}{r+1}\left(4(k-1)^2q/2^n\right)^r$.

The proofs for the results above can be found in the full version of this paper [8, Appendix D].

## Acknowledgments

The authors gratefully acknowledge the support of NSF grant 0904380. Thanks particuarly to program directors Richard Beigel and Lenore Zuck.

# References

1. Anderson, R., Biham, E.: Two practical and provably secure block ciphers: BEAR and LION. In: Gollmann, D. (ed.) FSE 1996. LNCS, vol. 1039, pp. 113–120. Springer, Heidelberg (1996)
2. Bellare, M., Rogaway, P., Spies, T.: The FFX mode of operation for format-preserving encryption (draft 1.1). NIST submission (February 2010), http://csrc.nist.gov/groups/ST/toolkit/BCM/modes_development.html
3. Black, J., Rogaway, P.: Ciphers with arbitrary finite domains. In: Preneel, B. (ed.) CT-RSA 2002. LNCS, vol. 2271, pp. 114–130. Springer, Heidelberg (2002)
4. Bellare, M., Ristenpart, T., Rogaway, P., Stegers, T.: Format-preserving encryption. In: Jacobson Jr., M.J., Rijmen, V., Safavi-Naini, R. (eds.) SAC 2009. LNCS, vol. 5867, pp. 295–312. Springer, Heidelberg (2009)
5. Brightwell, M., Smith, H.: Using datatype-preserving encryption to enhance data warehouse security. In: 20th NISSC Proceedings, pp. 141–149 (1997), http://csrc.nist.gov/nissc/1997
6. Coppersmith, D.: Luby-Rackoff: four rounds is not enough. Technical Report RC 20674, IBM (December 1996)
7. Feistel, H., Notz, W., Smith, J.: Some cryptographic techniques for machine-to-machine data communications. Proc. of the IEEE 63, 1545–1554 (1975)
8. Hoang, V., Rogaway, P.: On generalized Feistel networks. Full version of this paper. Cryptology ePrint report 2010/301, May 26 (2010)
9. Jutla, C.: Generalized birthday attacks on unbalanced Feistel networks. In: Krawczyk, H. (ed.) CRYPTO 1998. LNCS, vol. 1462, pp. 186–199. Springer, Heidelberg (1998)
10. Luby, M., Rackoff, C.: How to construct pseudorandom permutations from pseudorandom functions. SIAM Journal on Computing 17(2), 373–386 (1988); Earlier version in CRYPTO 1985
11. Lucks, S.: Faster Luby-Rackoff ciphers. In: Gollmann, D. (ed.) FSE 1996. LNCS, vol. 1039, pp. 189–203. Springer, Heidelberg (1996)
12. Maurer, U.: A simplified and generalized treatment of Luby-Rackoff pseudorandom permutation generator. In: Rueppel, R.A. (ed.) EUROCRYPT 1992. LNCS, vol. 658, pp. 239–255. Springer, Heidelberg (1993)
13. Maurer, U., Pietrzak, K.: The security of many-round Luby-Rackoff pseudo-random permutations. In: Biham, E. (ed.) EUROCRYPT 2003. LNCS, vol. 2656, pp. 544–561. Springer, Heidelberg (2003)
14. Maurer, U., Pietrzak, K., Renner, R.: Indistinguishability amplification. In: Menezes, A. (ed.) CRYPTO 2007. LNCS, vol. 4622, pp. 130–149. Springer, Heidelberg (2007)
15. Mironov, I.: (Not so) random shuffles of RC4. In: Yung, M. (ed.) CRYPTO 2002. LNCS, vol. 2442, pp. 304–319. Springer, Heidelberg (2002)
16. Moriai, S., Vaudenay, S.: On the pseudorandomness of top-level schemes of block ciphers. In: Okamoto, T. (ed.) ASIACRYPT 2000. LNCS, vol. 1976, pp. 289–302. Springer, Heidelberg (2000)
17. Morris, B., Rogaway, P., Stegers, T.: How to encipher messages on a small domain: deterministic encryption and the Thorp shuffle. In: Halevi, S. (ed.) CRYPTO 2009. LNCS, vol. 5677, pp. 286–302. Springer, Heidelberg (2009)
18. Nachef, V.: Generic attacks on alternating unbalanced Feistel schemes. Cryptology ePrint report 2009/287, June 16 (2009)

19. Naor, M., Reingold, O.: On the construction of pseudo-random permutations: Luby-Rackoff revisited. Journal of Cryptology 12(1), 29–66 (1997)
20. Nyberg, K.: Generalized Feistel networks. In: Kim, K.-c., Matsumoto, T. (eds.) ASIACRYPT 1996. LNCS, vol. 1163, pp. 91–104. Springer, Heidelberg (1996)
21. Patarin, J.: About Feistel schemes with six (or more) rounds. In: Vaudenay, S. (ed.) FSE 1998. LNCS, vol. 1372, pp. 103–121. Springer, Heidelberg (1998)
22. Patarin, J.: Generic attacks on Feistel schemes. In: Boyd, C. (ed.) ASIACRYPT 2001. LNCS, vol. 2248, pp. 222–238. Springer, Heidelberg (2001)
23. Patarin, J.: Luby-Rackoff: 7 Rounds are enough for $2^{n-\varepsilon}$ security. In: Boneh, D. (ed.) CRYPTO 2003. LNCS, vol. 2729, pp. 513–529. Springer, Heidelberg (2003)
24. Patarin, J.: New results on pseudorandom permutation generators based on the DES scheme. In: Feigenbaum, J. (ed.) CRYPTO 1991. LNCS, vol. 576, pp. 301–312. Springer, Heidelberg (1992)
25. Patarin, J.: Security of balanced and unbalanced Feistel schemes with linear non equalities. Cryptology ePrint report 2010/293. May 17 (2010)
26. Patarin, J.: Security of random Feistel schemes with 5 or more rounds. In: Franklin, M. (ed.) CRYPTO 2004. LNCS, vol. 3152, pp. 106–122. Springer, Heidelberg (2004)
27. Patarin, J., Nachef, V., Berbain, C.: Generic attacks on unbalanced Feistel schemes with contracting functions. In: Lai, X., Chen, K. (eds.) ASIACRYPT 2006. LNCS, vol. 4284, pp. 396–411. Springer, Heidelberg (2006)
28. Patarin, J., Nachef, V., Berbain, C.: Generic attacks on unbalanced Feistel schemes with expanding functions. In: Kurosawa, K. (ed.) ASIACRYPT 2007. LNCS, vol. 4833, pp. 325–341. Springer, Heidelberg (2007)
29. PCI Security Standards Council. Payment Card Industry (PCI) Data Security Standard: Requirements and Security Assessment Procedures, version 1.2.1 (July 2009), www.pcisecuritystandards.org
30. Schneier, B., Kelsey, J.: Unbalanced Feistel networks and block cipher design. In: Gollmann, D. (ed.) FSE 1996. LNCS, vol. 1039, pp. 121–144. Springer, Heidelberg (1996)
31. Smith, J.: The design of Lucifer: a cryptographic device for data communications. IBM Research Report RC 3326. IBM T.J. Watson Research Center, Yorktown Heights, New York, USA (April 15, 1971)
32. Thorp, E.: Nonrandom shuffling with applications to the game of Faro. Journal of the American Statistical Association 68, 842–847 (1973)
33. Vaudenay, S.: Provable security for block ciphers by decorrelation. In: Meinel, C., Morvan, M. (eds.) STACS 1998. LNCS, vol. 1373, pp. 249–275. Springer, Heidelberg (1998)
34. Yun, A., Park, J., Lee, J.: On Lai-Massey and quasi-Feistel ciphers. In: Designs, Codes and Cryptography, Online First (2010)
35. Zheng, Y., Matsumoto, T., Imai, H.: On the construction of block ciphers provably secure and not relying on any unproved hypotheses. In: Brassard, G. (ed.) CRYPTO 1989. LNCS, vol. 435, pp. 461–480. Springer, Heidelberg (1990)

# A    Proof for Unbalanced Feistel — Theorem 7

Given a query $X$ to Feistel$^{\nu}[m,n]$, its *coin* at round $t$ is the string $A \oplus F(B)$, where $F$ is the round function at round $t$ and $(A, B)$ is the round-$(t-1)$ output, with $|A| = m$ and $|B| = n$. We say that two queries *collide* at time $t$ if their outputs at round $t$ have the same last $n$ bits. We begin with the following.

**Lemma 11.** *In the blockcipher* Feistel$^{\curlyvee}[m, n]$, *the chance that two distinct non-adaptive queries have the same coin at round $t \geq 1$ is at most $2^{-m}$.*

*Proof.* Suppose that the Feistel network receives distinct non-adaptive queries $X_1$ and $X_2$. For each $i \in \{1, 2\}$, let $(A_i, B_i)$ be the output at round $t - 1$ of $X_i$, where $|A_i| = m$ and $|B_i| = n$. The queries $X_1$ and $X_2$ collide at time $t$ if and only if $A_1 \oplus F(B_1) = A_2 \oplus F(B_2)$, with $F$ being the round function at round $t$. This occurs with probability $2^{-m}$ if $B_1$ and $B_2$ differ, because $F$ is uniformly random. If $B_1 = B_2$ then so are $A_1$ and $A_2$, which contradicts the hypothesis that the two queries are distinct. □

CONTRACTING ROUND FUNCTIONS. We first consider the security of the blockcipher Feistel$^{\curlyvee}[m, n]$ with $n > m$ (that is, the round functions are contracting). Later we show how to deal with expanding round functions.

**Lemma 12.** *In the blockcipher* Feistel$^{\curlyvee}[m, n]$ *with $n > m$, the chance that two distinct non-adaptive queries collide at time $t > \lceil n/m \rceil$ is at most $3/2^{n+1}$.*

*Proof.* Suppose that the Feistel network receives distinct non-adaptive queries $X_1$ and $X_2$. We shall prove by induction on $b$ that for any $b \leq n$, the probability that outputs at round $t > \lceil b/m \rceil$ of the two queries have the same last $b$ bits is at most $3/2^{b+1}$. The claim of this lemma corresponds to the special case $b = n$.

First consider the base case $b < m$. For each $i \in \{1, 2\}$, let $(A_i, B_i)$ be the output at round $t - 1$ of $X_i$, where $|A_i| = m$ and $|B_i| = n$. The last $m$-bit substring of the round-$t$ output of $X_i$ is $A_i \oplus F(B_i)$, with $F$ being the round function at round $t$. If $B_1$ and $B_2$ differ then the probability that outputs at round $t$ of the two queries have the same last $b$ bits is at most $2^{-b}$, because $F$ is uniformly random. If $B_1 = B_2$ then the two queries have the same coin at round $t - 1$, which by Lemma 11 occurs with probability at most $2^{-m}$. Hence, by union bound, the chance that the two queries have the same last $b$ bits is at most $2^{-b} + 2^{-m} \leq 3/2^{b+1}$.

Next consider $b \geq m$ and assume that the chance round-$(t - 1)$ outputs of the two queries have the same last $b - m$ bits is at most $3/2^{b-m+1}$. The outputs at round $t$ of the two queries have the same last $b$ bits if and only if (i) they have the same coin at round $t$, which by Lemma 11 occurs with probability at most $2^{-m}$, and (ii) their output at round $t - 1$ have the same lat $b - m$ bits, which occurs with probability at most $3/2^{b-m+1}$ by induction hypothesis. As the round functions in the network are independent, the chance that both (i) and (ii) occur is at most $2^{-m} \cdot 3 / 2^{b-m+1} = 3/2^{b+1}$. □

We now prove NCPA-security of Feistel$^{r(2\lceil n/m \rceil + 2)}[m, n]$. Employing Lemma 2 then yields the desired result. Let $b = \lceil n/m \rceil + 1$. Suppose that the network receives nonadaptive distinct queries $X_1, \ldots, X_q$. We shall use a similar strategy as in the proof of Theorem 4. Fix an integer $\ell \leq q - 1$. For every $i \leq \ell$, let $U_i = X_i$ and let $U_{\ell+1}$ be chosen uniformly from $\{0, 1\}^{n+m} \setminus \{U_1, \ldots, U_\ell\}$. We shall construct another Feistel$^{2rb}[m, n]$ for the queries $U_1, \ldots, U_\ell$. Let $X_i(t)$ and $U_i(t)$ be the outputs at round $t$ of $X_i$ and $U_i$ respectively. It suffices to define the coupling in the first $2b$ rounds, and then show that the probability that $X_i(2b) \neq U_i(2b)$ for some $i \leq \ell + 1$ is at most $3b\ell / 2^n$.

THE COUPLING. In the first $b$ rounds, for every $i \leq \ell$, we use the same coin to update $X_i(t)$ and $U_i(t)$, and couple $X_{\ell+1}(t)$ and $U_{\ell+1}(t)$ in an arbitrary way. In the next $b$ rounds, we couple as follows.

- If $U_i$ collides with some previous query $U_j$ at time $t$ then the coin at round $t + 1$ of $U_i$ is defined so as to ensure consistency with the earlier query.
- Suppose that, in the new Feistel network, $U_i$ does not collide with any previous query at time $t$. If the query $X_i$ collides with some previous query $X_j$ at time $t$ then we choose a string uniformly from $\{0,1\}^{n+m}$ to be the coin of $U_i$ at round $t + 1$. Otherwise, the coin of $X_i$ at round $t + 1$ is uniformly distributed over $\{0,1\}^{n+m}$ and $U_i$ will use exactly the same coin at round $t + 1$.

Note that $U_i$ and $X_i$ always have the same output at round $t$, for every $i \leq \ell$ and every $t$. Consider the event Coll that in either Feistel networks, the $(\ell+1)$-th query collides with some previous query at some time $t \in \{b, \ldots, 2b - 1\}$. From Lemma 12, each such collision occurs with probability at most $3/2^{n+1}$. Summing over the two Feistel networks, $b$ rounds, and $\ell$ previous queries shows that the probability Coll occurs is at most $3b\ell / 2^n$. Unless Coll occurs, $U_{\ell+1}$ and $X_{\ell+1}$ will share the coins at the rounds $b + 1, \ldots, 2b$, and then have identical outputs at the round $2b$. Hence the chance that we fail to couple at round $2b$ cannot exceed $3b\ell / 2^n$.

EXPANDING ROUND FUNCTIONS. We follow the same proof as before, but Lemma 12 is replaced by the following result.

**Lemma 13.** *In the blockcipher* Feistel$^{\curvearrowright}[m, n]$ *with $n \leq m$, the chance that two distinct non-adaptive queries collide at time $t \geq \lceil m/n \rceil$ is at most $\lceil m/n \rceil / 2^n$.*

*Proof.* Suppose that the Feistel network receives distinct non-adaptive queries $X_1$ and $X_2$. For each $i \in \{1,2\}$, let $(A_i, B_i)$ be the output at round $t - 1$ of $X_i$, where $|A_i| = m$ and $|B_i| = n$. The queries $X_1$ and $X_2$ collide at time $t$ if and only if the two strings $A_1 \oplus F(B_1)$ and $A_2 \oplus F(B_2)$ have the same last $n$ bits, with $F$ being the round function at round $t$. This occurs with probability $2^{-n}$ if $B_1$ and $B_2$ differ, because $F$ is uniformly random. If $B_1 = B_2$ then $A_1$ and $A_2$ must have the same last $n$ bits. In other words, the round-$(t-1)$ outputs of the two queries must agree at the last $2n$ bits. Repeating this argument leads us to examine the case that for every $j < \lceil m/n \rceil$ the round-$(t-j)$ outputs of the two queries must agree at the last $(j + 1)n$ bits. When this chain of reasoning stops at round $t - \lceil m/n \rceil + 1$, the outputs at that round must have the same last $m$ bits. In other words, the queries have the same coin at that round, which by Lemma 11 occurs with probability at most $2^{-m} \leq 2^{-n}$. Hence by union bound, the chance that the two queries collide at time $t$ is at most $\lceil m/n \rceil / 2^n$.  □

# Cryptographic Extraction and Key Derivation: The HKDF Scheme

Hugo Krawczyk

IBM T.J. Watson Research Center, Hawthorne, New York
hugo@ee.technion.ac.il
http://eprint.iacr.org/2010/264

**Abstract.** In spite of the central role of *key derivation functions (KDF)* in applied cryptography, there has been little formal work addressing the design and analysis of general multi-purpose KDFs. In practice, most KDFs (including those widely standardized) follow ad-hoc approaches that treat cryptographic hash functions as perfectly random functions. In this paper we close some gaps between theory and practice by contributing to the study and engineering of KDFs in several ways. We provide detailed rationale for the design of KDFs based on the *extract-then-expand* approach; we present the first general and rigorous definition of KDFs and their security that we base on the notion of *computational extractors*; we specify a concrete *fully practical* KDF based on the HMAC construction; and we provide an analysis of this construction based on the extraction and pseudorandom properties of HMAC. The resultant KDF design can support a large variety of KDF applications under suitable assumptions on the underlying hash function; particular attention and effort is devoted to minimizing these assumptions as much as possible for each usage scenario.

Beyond the theoretical interest in modeling KDFs, this work is intended to address two important and timely needs of cryptographic applications: (i) providing a single hash-based KDF design that can be standardized for use in multiple and diverse applications, and (ii) providing a conservative, yet efficient, design that exercises much care in the way it utilizes a cryptographic hash function.

(The HMAC-based scheme presented here, named HKDF, is being standardized by the IETF.)

## 1 Introduction

A Key derivation function (KDF) is a basic and essential component of cryptographic systems: Its goal is to take a *source of initial keying material,* usually containing some good amount of randomness, but not distributed uniformly or for which an attacker has some partial knowledge, and derive from it one or more *cryptographically strong* secret keys. We associate the notion of "cryptographically strong" keys with that of *pseudorandom* keys, namely, indistinguishable by feasible computation from a random uniform string of the same length. In particular, knowledge of part of the bits, or keys, output by the KDF should

T. Rabin (Ed.): CRYPTO 2010, LNCS 6223, pp. 631–648, 2010.
© International Association for Cryptologic Research 2010

not leak information on the other generated bits. Examples of initial keying material include the output of an imperfect physical random number generator, a bit sequence obtained by a statistical sampler (such as sampling system events or user keystrokes), system PRNGs that use renewable sources of randomness, and the less obvious case of a Diffie-Hellman value computed in a key exchange protocol.

The main difficulty in designing a KDF relates to the form of the initial keying material (which we refer to as source keying material). When this key material is given as a uniformly random or pseudorandom key $K$ then one can use $K$ to seed a pseudorandom function (PRF) or pseudorandom generator (PRG) to produce additional cryptographic keys. However, when the source keying material is not uniformly random or pseudorandom then the KDF needs to first "extract" from this "imperfect" source a first pseudorandom key from which further keys can be derived using a PRF. Thus, one identifies two logical modules in a KDF: a first module that takes the source keying material and extracts from it a fixed-length pseudorandom key $K$, and a second module that expands $K$ into several additional pseudorandom cryptographic keys.[1]

The expansion module is standard in cryptography and can be implemented on the basis of any secure PRF. The *extraction* functionality, in turn, is well modeled by the notion of *randomness extractors* [31,30] as studied in complexity theory and related areas (informally, an extractor maps input probability distributions with sufficient entropy into output distributions that are statistically close to uniform). However, in many cases the well-established extractors (e.g., via universal hashing) fall short of providing the security and/or functionality required in practice in the KDF context. Here we study randomness extraction from the cryptographic perspective and specifically in the context of KDFs (building upon and extending prior work [26,14,6,5]). A main objective is to develop a basis for designing and analyzing secure key derivation functions following the above natural extract-then-expand approach. We are interested in the *engineering* of practical designs that can serve a variety of applications and usage scenarios and hence can be standardized for wide use. In particular, we need to be able to design extractors that will be well-suited for a large variety of sources of keying material (see detailed examples in [28]) and work in liberal as well as constrained environments. For this we resort to the use of cryptographic functions, especially cryptographic hash functions, as the basis for such multi-purpose extraction.

We identify *computational extractors*, namely randomness extractors where the output is only required to be pseudorandom rather than statistically close to uniform, as the main component for extraction in cryptographic applications, and build the notion of a KDF and its implementations on the basis of such extractors. Computational extractors are well-suited for the crypto setting where

---

[1] KDF is sometimes used only with the meaning of expanding a given *strong* key into several additional keys (e.g., [33]); this ignores the extract functionality which is central to a general *multi-purpose* KDF.

attackers are computationally bounded and source entropy may only exist in a computational sense. In particular, one can build such extractors in more efficient and practical ways through the use of cryptographic functions *under suitable assumptions*. Advantages of such cryptographic extractors range from purely practical considerations, such as better performance and the operational advantage of re-using functions (e.g., cryptographic hash functions) that are already available in cryptographic applications, to their more essential use for bypassing some of the inherent limitations of statistical extractors. We use the ability of cryptographic hash functions to be keyed as a way to include a *salt* value (i.e., a random but non-secret key) which is essential to obtain *generic* extractors and KDFs that can extract randomness from arbitrary sources with sufficiently high entropy.

We then study the requirements from computational extractors in the cryptographic setting, ranging from applications where the source key material has large entropy (and a good source of public randomness is also available) to the much more constrained scenarios encountered in practice where these resources (entropy and randomness) are much more limited. On this basis, we offer a KDF design that accommodates these different scenarios under suitable assumptions from the underlying cryptographic functions. In some cases, well-defined combinatorial assumptions from the hash functions will suffice while in others one has to resort to idealized modeling and "random oracle" abstractions. Our goal is to *minimize such assumptions as much as possible for each usage scenario,* but for this we need to first develop a good understanding of the properties one can expect from cryptographic hash functions as well as an understanding of the extraction functionality and the intrinsic limitations of unconditional statistical extractors in practical settings. We provide a detailed account of these issues throughout the paper.

Based on the notion of computational extractors (and on a better understanding of the complexities and subtleties of the use of KDFs in practice), we present a formal definition of the key derivation functionality suitable for capturing multiple uses of KDFs and a basis for judging the quality of general KDF designs such as those considered here. Somewhat surprisingly, in spite of being one of the most central and widely used cryptographic functionalities (in particular, specified in numerous standards), there appears to be little formal work on the specific subject of multi-purpose key derivation functions. Ours seems to be the first general definition of the KDF functionality in the literature. Our definitions include a formalization of what is meant by a "source of keying material" and they spell the security requirements from the KDF taking into account realistic adversarial capabilities such as the possession by the attacker of side information on the input to the KDF. In our formulation, KDFs accept four inputs: a sample from the source of keying material from which the KDF needs to extract cryptographic keys, a parameter defining the number of key bits to be output, an (optional) randomizing salt value as mentioned before, and a fourth "contextual information" field. The latter is an important parameter for the KDF intended to include key-related information that needs to be uniquely and cryptographically bound to the produced key material (e.g., a protocol identifier, identities of principals, timestamps, etc.).

We then use the above theoretical background, including results from [14,12], to describe and analyze a concrete practical design of a multi-purpose KDF based on cryptographic hash functions. The scheme (denoted HKDF), that uses HMAC as the underlying mode of operation, supports multiple KDF scenarios and strives to minimize the required assumptions from the underlying hash function *for each such scenario*. For example, in some applications, assuming that the underlying hash function has simple combinatorial properties, e.g., universal hashing, will suffice while in the most constrained scenarios we will need to model the hash function as a random oracle. *The important point is that we will be able to use the same KDF scheme in all these cases as required for a standardized multi-purpose KDF.*[2]

We end by observing that most of today's standardized KDFs (e.g., [3,4,32,23]) do not differentiate between the extract and expand phases but rather combine the two in ad-hoc ways under a single cryptographic hash function (refer to [28] for a description and discussion of these KDF schemes and their shortcomings). This results in ad-hoc designs that are hard to justify with formal analysis and which tend to "abuse" the hash function, requiring it to behave in an "ideally random" way even when this is not strictly necessary in most KDF applications (these deficiencies are present even in the simple case where the source of keying material is fully random). In contrast, we formulate and analyze a fully practical KDF scheme based on current theoretical research as well as on sound engineering principles. The end result is a well-defined hash-based KDF scheme applicable to a wide variety of scenarios and which exercises much care in the way it utilizes cryptographic hash functions. Our view is that given the current (healthy) skepticism about the strength of our hash functions we must strive to design schemes that use the hash function as prudently as possible. Our work is intended to fulfill this principle in the context of key derivation functions (especially at a time that new standards based on hash functions are being developed, e.g., [33]).

**Related Work.** As already mentioned, in spite of their importance and wide use, there is little formal work on the specific subject of multi-purpose key derivation functions. The first work to analyze KDFs in the context of cryptographic hash functions and randomness extractors appears to be [14], which was followed-up in the context of random oracles by [12]. The former work laid the formal foundations for the HMAC-based KDF scheme presented here. This scheme, in turn, is based on the KDF originally designed by this author for the IKE protocols [19,24] and which put forth the extract-then-expand paradigm in the context of practical KDFs. A variant of the expansion stage of HKDF has also been adopted elsewhere, e.g. into TLS [13] (however, TLS does not use the extract approach; for example, keys from a DH exchange are used directly as PRF keys without any extraction operation). The extract-then-expand approach has subsequently been taken in [5] in the context of designing "system random number generators"; that work shares many elements with ours although the papers

---

[2] The proposed HKDF scheme is being standardized by the IETF as RFC 5869 [27].

differ significantly in emphasis and scope. Another related work is [6] which proposes the use of statistical extractors in the design of *physical* random-number generators and points out to the potential practicality of these extractors in this specific setting. Both [5,6] offer interesting perspectives on the use of randomness extractors in practice that complement our work; our HKDF design is well suited for use also in the settings studied by these works. A good discussion of extraction issues in the context of KDFs in the Diffie-Hellman setting can be found in [11] where a dedicated deterministic extractor for specific DH groups is presented. Another such extractor (very different in techniques and applicability) is presented in [16]. See more on related work in [28].

**Full version.** Due to space limitations we have omitted some important material that complements this presentation. Please refer to the full version [28] for expanded rationale, a treatment of the role of random oracles in the KDF setting, comparison with the most commonly used KDFs in practice, discussion of additional KDF applications, and more.

# 2  Statistical and Computational Extractors

This section is intended to introduce the basic notions behind the abstract randomness extraction functionality; in particular we define "computational extractors" that are central in our treatment.

The goal of the extract part of a KDF scheme is to transform the input source (seen as a probability distribution) into a close-to-uniform output. This corresponds to the functionality of *randomness extractors* which have been extensively studied in complexity theory and related areas [31]. Informally, a randomness extractor is a family of functions indexed by a public, i.e., non-secret, parameter (which we refer to as "salt") with the property that on any input distribution with sufficiently large entropy, if one chooses a salt value at random (and independently of the source distribution) the output of the extractor is statistically close to uniform (see below for a formal definition). Moreover, this statistical closeness holds even if conditioned on the salt value. Extractors with the latter property are called *strong* randomness extractors but since we only consider this type we often omit both the "strong" and "randomness" qualifiers. On the other hand, we often add the qualifier "statistical" to differentiate these extractors from computational ones (defined below) that are an essential part of our work.

Before presenting a formal definition of statistical extractors, we recall the notion of entropy considered in this context, called *min-entropy*, that captures a "worst case" notion of entropy different than the traditional average notion of Shannon's entropy (it is not hard to see that Shannon's notion is insufficient in the context of randomness extraction).

**Background definitions and notation.** Refer to Appendix A for some background definitions and notation used throughout the paper (e.g., the notion of $\delta$-close).

**Definition 1.** *A probability distribution* $\mathcal{X}$ *has* min-entropy (at least) $m$ *if for all* $a$ *in the support of* $\mathcal{X}$ *and for random variable* $X$ *drawn according to* $\mathcal{X}$, $Prob(X = a) \leq 2^{-m}$.

**Definition 2.** *Let* $\mathcal{X}$ *be a probability distribution over* $\{0,1\}^n$. *A function* ext : $\{0,1\}^t \times \{0,1\}^n \to \{0,1\}^{m'}$ *is called a $\delta$-statistical* extractor *with respect to* $\mathcal{X}$ *if the distribution of pairs* $(r, y)$, *where* $r$ *is chosen with uniform probability over* $\{0,1\}^t$ *and* $y = \text{ext}_r(x)$ *for* $x$ *chosen according to distribution* $\mathcal{X}$, *is $\delta$-close to the distribution of pairs* $(r, z)$ *where* $z$ *is chosen with uniform probability from* $\{0,1\}^{m'}$. *If* ext *is a $\delta$-statistical extractor with respect to all distributions over* $\{0,1\}^n$ *with min-entropy* $m$, *then we say that* ext *is a* $(m, \delta)$-statistical extractor.

This notion was first defined in [31]; see [30,38] for surveys.

Randomization of the extractor function via the parameter $r$ (the salt) is mandatory if the same extractor function is to be able to extract randomness from *any* high min-entropy distribution. Indeed, for any deterministic function one can construct a high min-entropy source on which the function will produce very non-uniform outputs. On the other hand, one may consider randomness extractors that are suited for a *specific* source (or family of sources). In the latter case, one can consider *deterministic extractors*. Examples of such source-specific extractors in the cryptographic setting include the well-known hard-core schemes for RSA [2,15] and for discrete-log based functions [21,34], and the recent elegant extraction functions specific to some Diffie-Hellman groups in [11,16]. For most of our study we focus on generic extractors, i.e., those that can extract randomness from *any* source with sufficient min-entropy, and hence require some non-secret salt.

A natural (and practical) question is whether common KDF applications may have a randomness source from which to obtain salt. After all, the whole purpose of extractors is to generate randomness, so if one already has such a random salt why not use it directly as a PRF key? The answer is that this randomness needs *not* be secret while in KDF applications we want the output of the extractor to be secret. Obtaining public randomness is much easier than producing secret bits, especially since in most applications the extractor key (or salt) can be used repeatedly with many (independent) samples from the same source (hence it can be chosen in an out-of-band or setup stage and be repeatedly used later). For example, a random number generator (RNG) that requires an extractor to "purify" its possibly imperfect output can simply have a random, non-secret, extractor key built-in; the same extractor key is used to purify each output from the RNG [6]. In other cases, such as key-exchange protocols, extraction keys can be generated as part of the protocol (e.g., by using random nonces exchanged in the clear [19,24]). See [28] for further elaboration on the issue of randomization in extractors, in particular as a means to enforce independence between the source distribution and the extractor.

Efficient constructions of generic (hence randomized) statistical extractors exist such as those built on the basis of universal hash functions [10]. However, in spite of their simplicity, combinatorial and algebraic constructions present significant limitations for their practical use in generic KDF applications. For example, statistical extractors require a significant difference (called the gap) between the min-entropy $m$ of the source and the required number $m'$ of extracted bits (in particular, no statistical extractor can achieve a statistical distance, on arbitrary sources, better than $2^{-\frac{m-m'}{2}}$ [35,38]). That is, one can use statistical extractors (with its provable properties) only when the min-entropy of the source is significantly higher than the length of output. These conditions are met by some applications, e.g., when sampling a physical random number generator or when gathering entropy from sources such as system events or human typing (where higher min-entropy can be achieved by repeated sampling). In other cases, very notably when extracting randomness from computational schemes such as the Diffie-Hellman key exchange, the available gap may not be sufficient (for example, when extracting 160 bits from a DH over a 192-bit group). In addition, depending on the implementation, statistical extractors may require from several hundred bits of randomness (or salt) to as many bits of salt as the number of input bits.

To obtain more practical instantiations of extractors we relax their requirements in several ways. Most significantly, we will not require that the output of the extractor be statistically close to uniform but just "computationally close", i.e., pseudorandom. The following notion is implicit in [17,14].

**Definition 3.** *A $(t, \varepsilon)$-computational* extractor *with respect to a probability distribution $\mathcal{X}$ is defined as in Definition 2 except that the requirement for statistical closeness between the distributions $(r, y)$ and $(r, z)$ is replaced with $(t, \varepsilon)$-computational indistinguishability[3]. An extractor that is $(t, \varepsilon)$-computational with respect to all distributions with min-entropy $m$ is called $(m, t, \varepsilon)$-computational.*

This relaxed notion will allow for more practical instantiations of extractors, particularly well-suited for the key derivation setting. Computational extractors fit the cryptographic settings where attackers are assumed to be computationally bounded, and they allow for constructions based on cryptographic hash functions. In addition, computational extraction is natural in settings such as the Diffie-Hellman protocol where the input $g^{xy}$ to the extractor is taken from a source that has zero statistical entropy (since an attacker that knows $g^x, g^y$ has full information to compute $g^{xy}$), yet may contain a significant amount of "computational min-entropy" [20] as defined next.

**Definition 4.** *A probability distribution $\mathcal{X}$ has $(t, \varepsilon)$-computational* min-entropy *$m$ if there exists a distribution $\mathcal{Y}$ with min-entropy $m$ such that $\mathcal{X}$ and $\mathcal{Y}$ are $(t, \varepsilon)$-computationally indistinguishable.*

*Note:* In our application of extraction to the KDF setting, where an attacker often has some a-priori information about the source (e.g., it knows the public DH values $g^x, g^y$ from which the source key material $g^{xy}$ is derived), we use a notion of min-entropy (statistical or computational) that is *conditioned* on such a-priori information (see following section).

---

[3] See Appendix A for the definition of computational indistinguishability.

# 3   Formalizing Key Derivation Functions

We present a formal definition of (secure) key derivation functions and a formalization of what is meant by a "source of keying material". To the best of our knowledge, no such general definitions have been given in the literature.[4] We start with a definition of KDF in terms of its inputs and outputs (consistent with the KDF description in Section 4). Later, after introducing the notion of sources of keying material, we define what it means for a KDF to be secure.

**Definition 5.** *A* key derivation function (KDF) *accepts as input four arguments: a value $\sigma$ sampled from a source of keying material (Def. 6), a length value $\ell$, and two additional arguments, a salt value $r$ defined over a set of possible salt values and a context variable $c$, both of which are optional, i.e., can be set to the null string or to a constant. The KDF output is a string of $\ell$ bits.*[5]

The security and quality of a KDF depends on the properties of the "source of keying material", defined next, from which the input $\sigma$ is chosen (see [28] for more examples of such sources.)

**Definition 6.** *A* source of keying material *(or simply* source*) $\Sigma$ is a two-valued probability distribution $(\sigma, \alpha)$ generated by an efficient probabilistic algorithm. (We will refer to both the probability distribution as well as the generating algorithm by $\Sigma$.)*

This definition does not specify the input to the $\Sigma$ algorithm (but see below for a discussion related to potential adversary-chosen inputs to such an algorithm). It does specify the form of the output: a pair $(\sigma, \alpha)$ where $\sigma$ (the "sample") represents the (secret) source key material to be input to a KDF, while $\alpha$ represents some *auxiliary knowledge* about $\sigma$ (or its distribution) that is available to the attacker. For example, in a Diffie-Hellman application the value $\sigma$ will consist of a value $g^{xy}$ while $\alpha$ could represent a quintuple $(p, q, g, g^x, g^y)$. In a different application, say a random number generator that works by hashing samples of system events in a computer system, the value $\alpha$ may include some of the sampled events used to generate $\sigma$. The importance of $\alpha$ in our formal treatment is that we will require a KDF to be secure on inputs $\sigma$ *even when the knowledge value $\alpha$ is given to the attacker.* The restriction to sources that can be generated efficiently represents our interest in sources that can arise (and be used/sampled) in practice.

Next, we define the security of a KDF with respect to a *specific* source $\Sigma$. See Definition 9 for the generic case.

**Definition 7.** *A* key derivation function KDF *is said to be $(t, q, \varepsilon)$*-secure with respect to a source of key material $\Sigma$ *if no attacker $\mathcal{A}$ running in time $t$ and making*

---

[4] Yao and Yin [40] provide a formal definition of KDFs specific to the password setting which is different from and inapplicable to the general setting treated here (see [28]).

[5] The values $\sigma, \ell, r, c$ correspond to the values $SKM, L, XTS, CTXinfo$ in the description of Section 4.

*at most $q$ queries can win the following distinguishing game with probability larger than $1/2 + \varepsilon$:*

1. *The algorithm $\Sigma$ is invoked to produce a pair $\sigma, \alpha$.*
2. *A salt value $r$ is chosen at random from the set of possible salt values defined by* KDF *($r$ may be set to a constant or a null value if so defined by* KDF*).*
3. *The attacker $\mathcal{A}$ is provided with $\alpha$ and $r$.*
4. *For $i = 1, \ldots, q' \leq q$: $\mathcal{A}$ chooses arbitrary values $c_i, \ell_i$ and receives the value* KDF$(\sigma, r, c_i, \ell_i)$ *(queries by $\mathcal{A}$ are adaptive, i.e., each query may depend on the responses to previous ones).*
5. *$\mathcal{A}$ chooses values $c$ and $\ell$ such that $c \notin \{c_1, \ldots, c_{q'}\}$.*
6. *A bit $b \in_R \{0, 1\}$ is chosen at random. If $b = 0$, $\mathcal{A}$ is provided with the output of* KDF$(\sigma, r, c, \ell)$*, else $\mathcal{A}$ is given a random string of $\ell$ bits.*
7. *Step 4 is repeated for up to $q - q'$ queries (subject to the restriction $c_i \neq c$).*
8. *$\mathcal{A}$ outputs a bit $b' \in \{0, 1\}$. It wins if $b' = b$.*

It is imperative for the applicability of this definition that the attacker is given access to both $\alpha$ and $r$. This models the requirement that the KDF needs to remain secure even when the side-information $\alpha$ and salt $r$ are known to the attacker (in particular, note that the choice of the $c$'s and $\ell$'s by the attacker may depend on $\alpha$ and $r$). Allowing for multiple values of $c_i$ to be chosen by the attacker under the same input $\sigma$ to KDF ensures that even if an attacker can force the use of the same input $\sigma$ to the KDF in two different contexts (represented by $c$), the outputs from the KDF in these cases are computationally independent (i.e., leak no useful information on each other).

The following definition extends the min-entropy definitions from Section 2 to the setting of keying material sources (for a detailed treatment of *conditional* (computational) entropy as used in the next definition see [36,37,22]).

**Definition 8.** *We say that $\Sigma$ is a* statistical $m$-entropy source *if for all $s$ and $a$ in the support of the distribution $\Sigma$, the conditional probability $Prob\,(\sigma = s \mid \alpha = a)$ induced by $\Sigma$ is at most $2^{-m}$.*
*We say that $\Sigma$ is a* computational $m$-entropy source *(or simply an $m$-entropy source) if there is a statistical $m$-entropy source $\Sigma'$ that is computationally indistinguishable from $\Sigma$.*

We note that in the above definition we can relax the "for all $a$" to "all but a negligible fraction of $a$". That is, we can define $\alpha = a$ to be "bad" (for a given value $m$) if there is $s$ such that $Prob\,(\sigma = s \mid \alpha = a) > 2^{-m}$ and require that the joint probability induced by $\Sigma$ on bad $a$'s be negligible.

**Definition 9.** *A KDF function is called $(t, q, \varepsilon)$ $m$-entropy secure if it is $(t, q, \varepsilon)$-secure with respect to all (computational) $m$-entropy sources.*

We note that for the most part of this paper the (implicit) notion of security of a KDF corresponds to this last definition, namely, we think of KDFs mainly as a *generic* function that can deal with different sources as long as the source has enough computational min-entropy. We stress that this notion of security can only be achieved for randomized KDFs where the salt value $r$ is chosen at random from a large enough set. Yet, this work also touches on deterministic KDFs (see more in [28]) that may be good for specific applications and sources and whose security is formalized in Definition 7.

**On adversarially-chosen inputs to $\Sigma$.** Please refer to [28] for a discussion on extending the above definitions to incorporate possiblly adversarial inputs to the generation of the source $\Sigma$.

# 4    Extract-Then-Expand KDF and an HMAC-Based Instantiation

In this section we first describe an abstract KDF that implements the *extract-then-expand* approach discussed throughout this paper, and then specify an instantiation solely based on HMAC [7].

An extract-then-expand key derivation function KDF comprises two modules: a randomness extractor XTR and a variable-length output pseudorandom function PRF$^*$ (the latter is usually built on the basis of a regular PRF with output extension via counter mode, feedback mode, etc.). The extractor XTR is assumed to produce "close-to-random", in the statistical or computational sense, outputs on inputs sampled from the source key material distribution (this should be the case also when the *SKM* value includes auxiliary knowledge $\alpha$, per Definition 6, that is provided to the distinguisher). XTR may be deterministic or keyed via an optional "salt value" (i.e., a non-secret random value) that we denote by *XTS* (for *extractor salt*). The key to PRF$^*$ is denoted by *PRK* (pseudorandom key) and in our scheme it is the output from XTR; thus, we are assuming that XTR produces outputs of the same length as the key to PRF$^*$. The function PRF$^*$ also gets a length parameter indicating the number of bits to be output by the function. In all, KDF receives four inputs: the source key material *SKM*, the extractor salt *XTS* (which may be null or constant), the number $L$ of key bits to be produced by KDF, and a "context information" string *CTXinfo* (which may be null). The latter string should include key-related information that needs to be uniquely (and cryptographically) bound to the produced key material. It may include, for example, information about the application or protocol calling the KDF, session-specific information (session identifiers, nonces, time, etc.), algorithm identifiers, parties identities, etc. The computation of the extract-then-expand KDF proceeds in two steps; the $L$-bit output is denoted *KM* (for "key material"):

1. $PRK$ = XTR$(XTS,\ SKM)$
2. $KM$ = PRF$^*(PRK,\ CTXinfo,\ L)$

The following theorem establishes the security of a KDF built using this extract-then-expand approach. The proof, presented in [28], follows from the definition of computational extractors (Definition 3), the security definition of variable-length-output pseudorandom functions (Definition 14 in Appendix A), and the definition of KDF security (Definition 7).

**Theorem 1.** *Let* XTR *be a* $(t_X, \varepsilon_X)$-*computational extractor w.r.t. a source* $\Sigma$ *and* PRF* *a* $(t_P, q_P, \varepsilon_P)$-*secure variable-length-output pseudorandom function family, then the above extract-then-expand* KDF *scheme is* $(\min\{t_X, t_P\}, q_P, \varepsilon_X + \varepsilon_P)$-*secure w.r.t. source* $\Sigma$.

**An HMAC-based instantiation.** For the sake of implementation in real applications we propose to instantiate the above general scheme with HMAC serving as the PRF underlying PRF* as well as the XTR function. We denote the resultant scheme by HKDF.

We use the following notational conventions: (i) the variable $k$ denotes the output (and key) length of the hash function used with HMAC; (ii) we represent HMAC as a two-argument function where the first argument always represents the HMAC key; (iii) the symbol $\|$ denotes string concatenation. Thus, when writing HMAC$(a,\ b\ \|\ c)$ we mean the HMAC function (using a given hash function) keyed with the value $a$ and applied to the concatenation of the strings $b$ and $c$.

The scheme HKDF is specified as:

$$\text{HKDF}(XTS,\ SKM,\ CTXinfo,\ L)\ =\ K(1)\ \|\ K(2)\ \|\ \dots\ \|\ K(t)$$

where the values $K(i)$ are defined as follows:

$$PRK\ =\ \text{HMAC}(XTS,\ SKM)$$
$$K(1) = \text{HMAC}(PRK, CTXinfo\ \|\ 0),$$
$$K(i+1) = \text{HMAC}(PRK,\ K(i)\ \|\ CTXinfo\ \|\ i),\quad 1 \le i < t,$$

where $t = \lceil L/k \rceil$ and the value $K(t)$ is truncated to its first $d = L \bmod k$ bits; the counter $i$ is non-wrapping and of a given fixed size, e.g., a single byte. Note that the length of the HMAC output is the same as its key length and therefore the scheme is well defined.

When the extractor salt $XTS$ is not provided (i.e., the extraction is deterministic) we set $XTS = 0$.

Example: Let HMAC-SHA256 be used to implement KDF. The salt $XTS$ will either be a provided 256-bit random (but not necessarily secret) value or, if not provided, $XTS$ will be set to 0. If the required key material consists of one AES key (128 bits) and one HMAC-SHA1 key (160 bits), then we have $L = 288$, $k = 256$, $t = 2$, $d = 32$ (i.e., we will apply HMAC-SHA256 with key $PRK$ twice to produce 512 bits but only 288 are output by truncating the second output from HMAC to its first 32 bits). Note that the values $K(i)$ do not necessarily correspond to individual keys but they are concatenated to produce as many key bits as required.

**Practical Notes.** Please refer to [28] for several notes on the use of the HKDF in practice, including some variants such as replacing HMAC with a block-cipher based construct or with other "multi-property preserving" hash schemes, and using hybrid schemes where the extract and expand modules are implemented with separate components. [28] also contains a discussion on the use of feedback mode in the expansion stage of HKDF.

## 5   The Security of HKDF

Theorem 1 from Section 4 allows us to argue the security of HKDF on the basis of the properties of the HMAC scheme both as extractor and PRF. In this section we review results concerning these properties of HMAC and use them to prove the security of HKDF. These results demonstrate that the structure of HMAC works well in achieving the basic functionalities that underline HKDF including PRF, extraction, and random-oracle domain extension. In particular, they exploit the versatility of HMAC that supports working with a secret key, a random non-secret key (salt), or deterministically (i.e., with a fixed-value key). We also note that the security analysis of HKDF uses in an essential way the structure of HMAC and *would not hold* if one simply replaces HMAC with a plain (Merkle-Damgard) hash function.

**Notation.** We use $H$ to denote a Merkle-Damgard hash function and $h$ the underlying compression function. We also consider these as keyed families, where $h_\kappa$ and $H_\kappa$ denote, respectively, the compression function and the Merkle-Damgard hash with their respective IVs set to $\kappa$; the key and output lengths of these functions is denoted by $k$. We will abuse notation and talk about "the family $h_\kappa$" instead of the more correct $\{h_\kappa\}_{\kappa \in \{0,1\}^k}$; same for $H_\kappa$. When we say that "the family $h_\kappa$ is random", we mean that each of the functions $h_\kappa$ is chosen at random (with the corresponding input/output lengths). When we talk about HMAC (or NMAC), we assume underlying functions $h$ and $H$ (or their keyed versions).

The properties of HMAC as a pseudorandom function family are well established [8,9] and are based on the assumed pseudorandomness of the underlying compression function family $h_\kappa$.[6] It is not hard to see that the use of HMAC in "feedback mode" in HKDF (for realizing PRF*) results in a secure variable-length-output pseudorandom function family. Indeed, the latter is a generic transformation from fixed-length output PRF into a variable-length output PRF* (see more on this transformation and the rationale for the use of feedback mode in [28]).

The suitability of HMAC as a computational extractor is more complex and is treated in detail below. These results show the extraction properties of HMAC for a wide variety of scenarios under suitable assumptions on the underlying hash function, ranging from purely combinatorial properties, such as universality, to the idealized modeling of compression functions as random oracles.

We first state the following general theorem.

**Theorem 2.** *(informal) Let $H$ be a Merkle-Damgard hash function built on a family of pseudorandom compression functions $\{h_\kappa\}_\kappa$. Let $S$ be a collection of probability distributions acting as sources of keying material. Assume that the instantiation of HMAC with the family $\{h_\kappa\}_\kappa$ is a secure computational extractor w.r.t. sources in $S$, then HKDF is a secure KDF w.r.t. sources in $S$.*

---

[6] Although the security of HMAC as PRF degrades quadratically with the number of queries, such attack would require the computation of the PRF (by the owner of the key) over inputs totaling $2^{k/2}$ blocks. This is not a concern in typical KDF applications where the number of applications of the PRF is relatively small.

The theorem follows from Theorem 1 applied to the collection of sources $\mathcal{S}$ and the fact, discussed above, that HMAC is a secure PRF when instantiated with a pseudorandom family of compression functions $h_\kappa$. *Each of the lemmas presented below provides a condition on HMAC extraction that can be plugged into this theorem to obtain a proof of security for* HKDF *for the appropriate sources of key material in a* well-defined and quantifiable *way.*

The results below involve the notion of "almost universal (AU)" hash functions [10,39]: A family $h_\kappa$ is $\delta$-AU if for any inputs $x \neq y$ and for random $\kappa$, $Prob(h_\kappa(x) = h_\kappa(y)) \leq \delta$. This is a natural (combinatorial) property of hash functions and also one that any (even mildly) collision resistant hash family must have and then a suitable assumption for cryptographic hash functions. Specifically, if the hash family $h_\kappa$ is $\delta$-collision-resistant against linear-size circuits (i.e., such an attacker finds collisions in $h_\kappa$ with probability at most $\delta$) then $h_\kappa$ is $\delta$-AU [14]. For results that apply to the most constrained scenarios (as those discussed in the Random Oracles section of [28]) we need to resort to stronger, idealized assumptions, in which we model functions as *random oracles (RO)*, namely, random functions which the attacker can only query on a limited number, $q$, of queries.

**NMAC as extractor.** The following lemmas are adapted from [14] and apply directly to the NMAC scheme underlying HMAC (recall that $\mathsf{NMAC}_{\kappa_1,\kappa_2}(x) = H_{\kappa_2}(H_{\kappa_1}(x))$, where $H_{\kappa_2}$ is called the "outer function" and $H_{\kappa_1}$ the "inner function"). The results extend to HMAC as explained below. They show that HMAC has a structure that supports its use as a generic extractor and, in particular, it offers a much better design for extraction than the plain hash function $H$ used in many of the existing KDFs.

**Lemma 1.** *If the outer function is modeled as a RO and the inner function is $\delta$-AU then NMAC applied to an $m$-entropy source produces an output that is $\sqrt{q(2^{-m} + \delta)}$-close to uniform where $q$ is a bound on the number of RO queries.*

The above modeling of the outer function as a random oracle applies to the case where the outer function is a single (fixed) random function (in which case the source distribution needs to be independent of this function) or when it is represented as a keyed family of random functions (in which case only the key, or salt, needs to be chosen independently of the source distribution).

One natural question is whether one can ensure good extraction properties for NMAC based on the extraction properties of the underlying compression functions and *without idealized assumptions*. The following result from [14] provides an affirmative answer for *m-blockwise sources*, namely, where each $k$-bit input block has min-entropy $m$ when conditioned on other blocks. Denote by $\hat{h}_\kappa$ a family identical to the compression function family $h_\kappa$ but where the roles of key and input are swapped relative to the definition of $h_\kappa$.

**Lemma 2.** *If $h_\kappa$ is a $(m,\delta)$-statistical extractor and $\hat{h}_\kappa$ is a $(t,q,\varepsilon)$-pseudorandom family for $q = 1$, then NMAC is a $(t, n\delta + \varepsilon)$-computational extractor for $m$-blockwise sources with $n$ input blocks.*

An example of a practical application where this non-idealized result can be used is the IKE protocol [19,24] where all the defined "mod $p$" DH groups have the required block-wise (computational) min-entropy. In particular, the output from HKDF is guaranteed to be pseudorandom in this case without having to model $h$ as a random oracle.

**Truncated NMAC.** Stronger results can be achieved if one truncates the output of NMAC by $c$ bits to obtain $k' = k - c$ bits of output (e.g., one computes NMAC with SHA-512 but only outputs 256 bits). In this case one can show that NMAC is a good *statistical* extractor (not just computational) under the following sets of assumptions:

**Lemma 3.** *If $h_\kappa$ is a family of random compression functions (with $k$ bits of output) then NMAC truncated by $c$ bits is a $(k, \sqrt{(n+2)2^{-c}})$-statistical extractor where $n$ is the number of input blocks.*

**Lemma 4.** *If the inner function is $\delta_1$-AU and the outer is $(2^{-k'} + \delta_2)$-AU then truncated NMAC (with $k'$ bits of output) is a $(m, \sqrt{2^{k'}(2^{-m} + \delta_1 + \delta_2)})$-statistical extractor.*

The latter lemma is particularly interesting as its guarantee is "unconditional": it does not depend on hardness assumptions or idealized assumptions, and it ensures statistical extraction. Moreover, it fits perfectly the HKDF setting if one implements the extract phase with, say, SHA2-512 (with output truncated to 256 bits) and the PRF part with SHA2-256 (as discussed in the practical notes section of [28]).

In particular, we have the following significant case:

**Corollary 1.** *If the family of compression functions $h_\kappa$ is strongly universal (or pairwise independent) and the family $H_\kappa$ is generically collision resistant against linear-size circuits, then NMAC truncated by $c$ bits is a $(k, (n+2)2^{-c/2})$-statistical extractor on $n$-block inputs.*

Indeed, the assumption that the family $h_\kappa$ is strongly universal means $\delta_2 = 0$; and it is not hard to see that if there is no trivial (linear size) algorithm to find collisions in the family $H_\kappa$ better than guessing then $\delta_1 \leq n2^{-k}$. Putting these values and $m = k$ into Lemma 4 the corollary follows.

Applying the corollary to the above example of SHA2-512 truncated to 256 bits we get a statistical distance of $(n+2)2^{-128}$. And there is plenty room to get good security even if the $\delta$ values deviate from the above numbers; e.g., for $\delta_1 = \delta_2 = 2^{100}2^{-k}$ we would get a statistical closeness of $(n+2)2^{-78}$. Finally, we note that a min-entropy of $m = k$ is a condition satisfied by many common distributions such as statistical samplers and Diffie-Hellman groups modulo safe primes.

**From NMAC to HMAC.** To use the above results with HMAC one needs to assume the computational independence of the values $h(\kappa \oplus \text{opad})$ and $h(\kappa \oplus \text{ipad})$ for random $\kappa$ (i.e., each of these values is indistinguishable from uniform even if the other is given). In the cases where $h$ is modeled as a random function this requires no additional assumption.

**HMAC as a random oracle.** In [28] (Random Oracles section) we point out to various scenarios that require the modeling of the extraction functionality through random oracles. This may be due to the stringent requirements of an application (e.g., when all of the min-entropy of the source is to be extracted), when extraction can be solely based on cryptographic hardness without assuming additional min-entropy (the "hard core" case), or when the application itself assumes the KDF to be a random oracle (as in certain key exchange protocols). In all these cases we are interested to model the extract part of HKDF as a random oracle. Fortunately, as shown in [12] (using the framework of indifferentiability from [29]), the HMAC structure preserves randomness in the sense that if the underlying compression function (computed on fixed length inputs) is modeled as a random oracle so is HMAC on variable length inputs ([12] claims the result for a variant of HMAC but it applies to HMAC itself).[7] This result together with the random-oracle-extraction lemma from [28] implies:

**Lemma 5.** *If the compression function $h$ is modeled as a RO, then the application of HMAC to an $m$-entropy source produces an output that is $q2^{-m}$-close to uniform where $q$ is a bound on the number of RO queries.*

As explained in [28], the above holds for source distributions that are (sufficiently) *independent* from the function $h$. To obtain a generic extractor one needs to randomize the scheme by keying HMAC with a salt value.

Finally, we point out that using random-oracle-hardcore lemma from [28], one obtains that if $h_\kappa$ is a RO family then HMAC over the family $H_\kappa$ is a generic hard-core family. This is needed when extraction is to be based on cryptographic hardness (or unpredictability) only without assuming additional min-entropy (e.g., in a Diffie-Hellman exchange where only CDH is assumed – see more in [28]).

# References

1. Adams, C., Kramer, G., Mister, S., Zuccherato, R.: On The Security of Key Derivation Functions. In: Zhang, K., Zheng, Y. (eds.) ISC 2004. LNCS, vol. 3225, pp. 134–145. Springer, Heidelberg (2004)
2. Alexi, W., Chor, B., Goldreich, O., Schnorr, C.-P.: RSA and Rabin Functions: Certain Parts are as Hard as the Whole. SIAM J. Comput. 17(2), 194–209 (1988)
3. ANSI X9.42-2001: Public Key Cryptography For The Financial Services Industry: Agreement of Symmetric Keys Using Discrete Logarithm Cryptography

---

[7] This "RO preserving" property does not hold for the plain Merkle-Damgard hash which is susceptible to extension attacks. Moreover, even if one considers fixed-length inputs (to avoid extension attacks), the Merkle-Damgard family $H_\kappa$ built on random compression functions is not a good statistical extractor (e.g., [14] show that the output of such family on any distribution for which the last block of input is fixed is statistically far from uniform).

4. ANSI X9.63-2002: Public Key Cryptography for the Financial Services Industry: Key Agreement and Key Transport
5. Barak, B., Halevi, S.: A model and architecture for pseudo-random generation with applications to /dev/random. In: ACM Conference on Computer and Communications Security (2005)
6. Barak, B., Shaltiel, R., Tromer, E.: True random number generators secure in a changing environment. In: Walter, C.D., Koç, Ç.K., Paar, C. (eds.) CHES 2003. LNCS, vol. 2779, pp. 166–180. Springer, Heidelberg (2003)
7. Bellare, M., Canetti, R., Krawczyk, H.: Keying hash functions for message authentication. In: Koblitz, N. (ed.) CRYPTO 1996. LNCS, vol. 1109, pp. 1–15. Springer, Heidelberg (1996)
8. Bellare, M., Canetti, R., Krawczyk, H.: Pseudorandom Functions Revisited: The Cascade Construction and Its Concrete Security. In: Proc. 37th FOCS, pp. 514–523. IEEE, Los Alamitos (1996)
9. Bellare, M.: New Proofs for NMAC and HMAC: Security Without Collision-Resistance. In: Dwork, C. (ed.) CRYPTO 2006. LNCS, vol. 4117, pp. 602–619. Springer, Heidelberg (2006)
10. Carter, L., Wegman, M.N.: Universal Classes of Hash Functions. JCSS 18(2) (1979)
11. Chevassut, O., Fouque, P.-A., Gaudry, P., Pointcheval, D.: The twist-aUgmented technique for key exchange. In: Yung, M., Dodis, Y., Kiayias, A., Malkin, T.G. (eds.) PKC 2006. LNCS, vol. 3958, pp. 410–426. Springer, Heidelberg (2006)
12. Coron, J.-S., Dodis, Y., Malinaud, C., Puniya, P.: Merkle-Damgard Revisited: How to Construct a Hash Function. In: Shoup, V. (ed.) CRYPTO 2005. LNCS, vol. 3621, pp. 430–448. Springer, Heidelberg (2005)
13. Dierks, T., Allen, C. (eds.): The TLS Protocol – Version 1. Request for Comments 2246 (1999)
14. Dodis, Y., Gennaro, R., Håstad, J., Krawczyk, H., Rabin, T.: Randomness Extraction and Key Derivation Using the CBC, Cascade and HMAC Modes. In: Franklin, M. (ed.) CRYPTO 2004. LNCS, vol. 3152, pp. 494–510. Springer, Heidelberg (2004)
15. Fischlin, R., Schnorr, C.-P.: Stronger Security Proofs for RSA and Rabin Bits. In: Fumy, W. (ed.) EUROCRYPT 1997. LNCS, vol. 1233, pp. 267–279. Springer, Heidelberg (1997)
16. Fouque, P.-A., Pointcheval, D., Stern, J., Zimmer, S.: Hardness of Distinguishing the MSB or LSB of Secret Keys in Diffie-Hellman Schemes. In: Bugliesi, M., Preneel, B., Sassone, V., Wegener, I. (eds.) ICALP 2006. LNCS, vol. 4052, pp. 240–251. Springer, Heidelberg (2006)
17. Gennaro, R., Krawczyk, H., Rabin, T.: Secure Hashed Diffie-Hellman over Non-DDH Groups. In: Cachin, C., Camenisch, J.L. (eds.) EUROCRYPT 2004. LNCS, vol. 3027, pp. 361–381. Springer, Heidelberg (2004)
18. Goldwasser, S., Micali, S.: Probabilistic Encryption. JCSS 28(2), 270–299 (1984)
19. Harkins, D., Carrel, D. (eds.): The Internet Key Exchange (IKE). RFC 2409 (November 1998)
20. Hastad, J., Impagliazzo, R., Levin, L., Luby, M.: Construction of a Pseudorandom Generator from any One-way Function. SIAM. J. Computing 28(4), 1364–1396 (1999)
21. Hastad, J., Schrift, A., Shamir, A.: The Discrete Logarithm Modulo a Composite Hides O(n) Bits. J. Comput. Syst. Sci. 47(3), 376–404 (1993)
22. Hsiao, C.-Y., Lu, C.-J., Reyzin, L.: Conditional Computational Entropy, or Toward Separating Pseudoentropy from Compressibility. In: Naor, M. (ed.) EUROCRYPT 2007. LNCS, vol. 4515, pp. 169–186. Springer, Heidelberg (2007)

23. IEEE P1363A: Standard Specifications for Public Key Cryptography: Additional Techniques, Institute of Electrical and Electronics Engineers
24. Kaufman, C. (ed.): Internet Key Exchange (IKEv2) Protocol. RFC 4306 (December 2005)
25. Krawczyk, H., Bellare, M., Canetti, R.: HMAC: Keyed-Hashing for Message Authentication. RFC 2104 (February 1997)
26. Krawczyk, H.: SIGMA: The 'SiGn-and-MAc' Approach to Authenticated Diffie-Hellman and Its Use in the IKE Protocols. In: Boneh, D. (ed.) CRYPTO 2003. LNCS, vol. 2729, pp. 400–425. Springer, Heidelberg (2003)
27. Krawczyk, H., Eronen, P.: HMAC-based Extract-and-Expand Key Derivation Function (HKDF), RFC 5869 (to appear)
28. Krawczyk, H.: Cryptographic Extraction and Key Derivation: The HKDF Scheme (full version of this paper), http://eprint.iacr.org/2010/264
29. Maurer, U.M., Renner, R., Holenstein, C.: Indifferentiability, Impossibility Results on Reductions, and Applications to the Random Oracle Methodology. In: Naor, M. (ed.) TCC 2004. LNCS, vol. 2951, pp. 21–39. Springer, Heidelberg (2004)
30. Nisan, N., Ta-Shma, A.: Extracting Randomness: A Survey and New Constructions. JCSS 58, 148–173 (1999)
31. Nisan, N., Zuckerman, D.: Randomness is linear in space. J. Comput. Syst. Sci. 52(1), 43–52 (1996)
32. NIST Special Publication (SP) 800-56A, Recommendation for Pair-Wise Key Establishment Schemes Using Discrete Logarithm Cryptography (March 2006)
33. NIST Special Publication (SP) 800-108, Recommendation for Key Derivation Using Pseudorandom Functions (October 2009)
34. Patel, S., Sundaram, G.: An Efficient Discrete Log Pseudo Random Generator. In: Krawczyk, H. (ed.) CRYPTO 1998. LNCS, vol. 1462, pp. 304–317. Springer, Heidelberg (1998)
35. Radhakrishnan, J., Ta-Shma, A.: Tight bounds for depth-two superconcentrators. SIAM J. Discrete Math. 13(1), 2–24 (2000)
36. Renner, R., Wolf, S.: Smooth Renyi entropy and applications. In: Proceedings of IEEE International Symposium on Information Theory (2004)
37. Renner, R., Wolf, S.: Simple and tight bounds for information reconciliation and privacy amplification. In: Roy, B. (ed.) ASIACRYPT 2005. LNCS, vol. 3788, pp. 199–216. Springer, Heidelberg (2005)
38. Shaltiel, R.: Recent developments in Extractors. Bulletin of the European Association for Theoretical Computer Science 77, 67–95 (2002), http://www.wisdom.weizmann.ac.il/~ronens/papers/survey.ps
39. Douglas, R.: Stinson: Universal Hashing and Authentication Codes. Des. Codes Cryptography 4(4), 369–380 (1994)
40. Yao, F.F., Yin, Y.L.: Design and Analysis of Password-Based Key Derivation Functions. In: Menezes, A. (ed.) CT-RSA 2005. LNCS, vol. 3376, pp. 245–261. Springer, Heidelberg (2005)

# A     Background Definitions

In this section we recall basic formal definitions for some of the notions used throughout this work.

**Notation.** In the sequel $\mathcal{X}$ and $\mathcal{Y}$ denote two (arbitrary) probability distributions over a common support set $A$; $X$ and $Y$ denote random variables drawn from $\mathcal{X}$ and $\mathcal{Y}$, respectively.

**Definition 10.** *We say that probability distributions* $\mathcal{X}$ *and* $\mathcal{Y}$ *have* statistical distance $\delta$ *(or are $\delta$-close) if* $\sum_{a \in A} |Prob(X = a) - Prob(Y = a)| \leq \delta$.

**Definition 11.** *An algorithm* $D$ *is an* $\varepsilon$-distinguisher *between distributions* $\mathcal{X}$ *and* $\mathcal{Y}$ *if* $|Prob(D(X) = 1) - Prob(D(Y) = 1| < \varepsilon$.

We note that two distributions $\mathcal{X}$ and $\mathcal{Y}$ are $\delta$-close iff there is no $\varepsilon$-distinguisher between $\mathcal{X}$ and $\mathcal{Y}$ for $\varepsilon > \delta$.

By restricting the computational power of the distinguisher in the above definition one obtains the following well-known definition of "computational indistinguishability" [18] (we formulate definitions using the "concrete security" $(t, \varepsilon)$ approach as a non-asymptotic alternative to the classical polynomial-time treatment; we also take the liberty of omitting the $(t, \varepsilon)$ notation when appropriate).

**Definition 12.** *Two probability distributions* $\mathcal{X}$, $\mathcal{Y}$ *are* $(t, \varepsilon)$-computationally indistinguishable *if there is no $\varepsilon$-distinguisher between $\mathcal{X}$ and $\mathcal{Y}$ that runs in time $t$.*

**Definition 13.** *A probability distribution* $\mathcal{X}$ *over the set* $\{0, 1\}^n$ *is called* $(t, \varepsilon)$-pseudorandom *if it is $(t, \varepsilon)$-computationally indistinguishable from the uniform distribution over* $\{0, 1\}^n$.

Next we recall the definition of security for a variable-length output pseudorandom function family. Such a family consists of a collection of keyed functions which on input a key $\kappa$, an input $c$ and a length parameter $\ell$, outputs $\ell$ bits.

**Definition 14.** *A variable-length output pseudorandom function family* PRF* *is* $(t, q, \varepsilon)$-secure *if no attacker* $\mathcal{A}$ *running in time $t$ and making at most $q$ queries can win the following distinguishing game with probability larger than $1/2 + \varepsilon$:*

1. *For $i = 1, \ldots, q' \leq q$: $\mathcal{A}$ chooses arbitrary values $c_i, \ell_i$ and receives the value* PRF*$(\kappa, c_i, \ell_i)$ *(queries by $\mathcal{A}$ are adaptive, i.e., each query may depend on the responses to previous ones).*
2. *$\mathcal{A}$ chooses values $c$ and $\ell$ such that $c \notin \{c_1, \ldots, c_q\}$.*
3. *A bit $b \in_R \{0, 1\}$ is chosen at random. If $b = 0$, $\mathcal{A}$ is provided with the output of* PRF*$(\kappa, c, \ell)$, *else $\mathcal{A}$ is given a random string of $\ell$ bits.*
4. *Step 4 is repeated for up to $q - q'$ queries.*
5. *$\mathcal{A}$ outputs a bit $b' \in \{0, 1\}$. It wins if $b' = b$.*

# Time Space Tradeoffs for Attacks against One-Way Functions and PRGs

Anindya De[1],[*], Luca Trevisan[2],[**], and Madhur Tulsiani[3],[* * *]

[1] University of California at Berkeley
anindya@cs.berkeley.edu
[2] University of California at Berkeley and Stanford University
luca@cs.berkeley.edu
[3] Institute for Advanced Study, Princeton
madhurt@math.ias.edu

**Abstract.** We study time space tradeoffs in the complexity of attacks against one-way functions and pseudorandom generators.

Fiat and Naor [7] show that for every function $f : [N] \to [N]$, there is an algorithm that inverts $f$ everywhere using (ignoring lower order factors) time, space and advice at most $N^{3/4}$.

We show that an algorithm using time, space and advice at most

$$\max\{\epsilon^{\frac{5}{4}} N^{\frac{3}{4}} , \ \sqrt{\epsilon N}\}$$

exists that inverts $f$ on at least an $\epsilon$ fraction of inputs. A lower bound of $\tilde{\Omega}(\sqrt{\epsilon N})$ also holds, making our result tight in the "low end" of $\epsilon \leq \sqrt[3]{\frac{1}{N}}$.

(Both the results of Fiat and Naor and ours are formulated as more general trade-offs between the time and the space and advice length of the algorithm. The results quoted above correspond to the interesting special case in which time equals space and advice length.)

We also show that for every length-increasing generator $G : [N] \to [2N]$ there is a algorithm that achieves distinguishing probability $\epsilon$ between the output of $G$ and the uniform distribution and that can be implemented in polynomial (in $\log N$) time and with advice and space $O(\epsilon^2 \cdot N \log N)$. We prove a lower bound of $S \cdot T \geq \Omega(\epsilon^2 N)$ where $T$ is the time used by the algorithm and $S$ is the amount of advice. This lower bound applies even when the distinguisher has oracle access to $G$.

We prove stronger lower bounds in the *common random string* model, for families of one-way permutations and of pseudorandom generators.

**Keywords:** One-way functions, pseudorandom generators, random permutations, time-space tradeoffs.

[*] Supported by the "Berkeley fellowship for Graduate Study" and by the BSF under grant 2006060.
[**] This material is based upon work supported by the National Science Foundation under grant No. CCF-0729137 and by the BSF under grant 2006060.
[* * *] This material is based upon work supported by the National Science Foundation under grant No. CCF-0832797 and IAS Sub-contract no. 00001583. Work done partly when the author was a graduate student at UC Berkeley.

# 1   Introduction

In the applied cryptography literature, a cryptographic primitive with a key of length $k$ is typically considered "broken" if the key can be recovered in time less than $2^k$, that is, faster than via an exhaustive brute force search. Implicit in this attitude is the belief in the existence of primitives for which a brute force attack is optimal. A time $t$ brute force attack against a one-way function $f : \{0,1\}^n \to \{0,1\}^n$, consisting in trying about $t$ random guesses for the inverse, only succeeds with probability about $t/2^n$, and a brute force attack that attempts to distinguish a length increasing generator $G : \{0,1\}^{n-1} \to \{0,1\}^n$ from the uniform distribution by attempting to guess the seed achieves distinguishing probability about $t/2^n$. Is it plausible that such trade-offs are optimal? Would it be plausible to assume that AES with 128 key bit cannot be distinguished from a random permutation with distinguishing probability more than $2^{-40}$ by adversaries running in time $2^{60}$?[1]

If we apply a non-uniform measure of complexity, that is, if we restrict ourselves to a fixed finite one-way function or pseudorandom generator, and allow our adversary to use precomputed information as advice, then it turns out that the above "brute force" bounds can always be improved upon.

In 1980, Hellman [12] proved that for every one-way *permutation* $f : [N] \to [N]$ (for this discussion, it will be convenient to set $N = 2^n$ and identify $\{0,1\}^n$ with $[N]$) and for every parameters $S, T$ satisfying $S \cdot T \geq N$, there is a data structure of size $\tilde{O}(S)$ and an algorithm that, with the help of the data structure, given $f(x)$ is always able to find $x$ in time $\tilde{O}(T)$. The notation $\tilde{O}(\cdot)$ hides lower order factors that are polynomial in $\log N$; we will ignore such factors from now on in the interest of readability. We shall refer to $S$, the size of the pre-computed data structure used by the algorithm, as the *space* used by the algorithm.

In particular, every one-way permutation can be inverted in time $\sqrt{N}$ using $\sqrt{N}$ bits of advice.[2]

Hellman's algorithm only requires oracle access to the permutation. Yao [18] proves that, in this oracle setting, Hellman's trade-off is tight for random permutations. (See also [10].)

---

[1] The answer to the last question is no. It follows from our results that there is a distinguisher that makes two queries, then performs a computation realizable as a circuit of size $2^{56}$, assuming a complete basis of fan-in two gates, and achieves distinguishing probability $\geq 2^{-40}$ between $AES_{128}$ and a random permutation $\{0,1\}^{128} \to \{0,1\}^{128}$. Otherwise, after the two oracle queries, the distinguisher can be implemented in a 64-bit architecture with two table look-ups and three unit-cost RAM operations, given access to a precomputed table of $2^{49}$ entries.

[2] This doesn't mean that there is a circuit of size $\tilde{O}(\sqrt{N})$; the running time of $\tilde{O}(\sqrt{N})$ is in the RAM model. The relationship between non-uniform time/space complexity measures and circuit complexity is the following: a circuit of size $C$ can be simulated using time at most $\tilde{O}(C)$ given a pre-computed data structure of size $\tilde{O}(C)$; and an algorithm that uses time $T$ and a pre-computed data structure of size $S$ can be simulated by a circuit of size $\tilde{O}((S + T)^2)$.

Hellman also considers the problem of inverting a random function $f : [N] \to [N]$ given oracle access to $f$. He provides a heuristic argument suggesting that for every $S, T$ satisfying $TS^2 \geq N^2$, and with high probability over the choice or a random function $f : [N] \to [N]$, there is a data structure of size $S$ and an algorithm of complexity $T$ that inverts $f$ everywhere using the data structure and given oracle access $f$. This trade-off yields the interesting special case $S = T = N^{2/3}$.

Fiat and Naor [7] prove Hellman's result rigorously, and are able to handle arbitrary functions, not just random functions. If the given function $f : [N] \to [N]$ has collision probability[3] $\lambda$, then the algorithm of Fiat and Naor requires the trade-off $TS^2 \geq \lambda \cdot N^3$. Note that with high probability a random function has collision probability about $1/N$ (recall that we ignore $(\log N)^{O(1)}$ terms), and so one recovers Hellman's tradeoff. For general functions, Fiat and Naor are able to prove the trade-off $TS^3 \geq N^3$, which has the special case $S = T = N^{3/4}$.

Barkan, Biham, and Shamir [4] prove that the $TS^2 = N^2$ trade-off of Fiat and Naor for random functions is optimal under certain assumptions on what is stored in the data structure and on the behavior of the algorithm.

The result of Fiat and Naor can also be applied to the task of distinguishing a given pseudorandom generator from the uniform distribution (and hence a given pseudorandom permutation from a random permutation or a given pseudorandom function from a random function) by recovering the seed. We are not aware of previous work that focused specifically on the complexity of distinguishers for pseudorandom generators. Two related results, however, should be mentioned. It has been known for a long time (going back to, as far we know, [2]) that every distribution that has constant statistical distance from the uniform distribution, and, in particular, the output of any length increasing generator, can be distinguished from the uniform distribution over $n$ bits using a parity function (of linear circuit complexity), and with distinguishing probability $\Omega(2^{-\frac{n}{2}})$. The other result is due to Andreev, Clementi and Rolim [3], who prove that for every boolean predicate $P : \{0,1\}^n \to \{0,1\}$ and every $\epsilon$ there is a circuit of size $O(\epsilon^2 2^n)$ that computes $P$ on at least a $1/2 + \epsilon$ fraction of inputs. This implies that for every pseudorandom generator of the form $x \to f(x)P(x)$, where $f$ is a permutation and $P$ is a hard-core predicate for $f$, and every $\epsilon > 0$, there is a circuit of size $O(\epsilon^2 2^n)$ that achieves distinguishing probability $\epsilon$.

## Our Results

**Upper Bounds for Inverting One-Way Functions.** We introduce a new way to analyze the Fiat-Naor construction. Instead of being limited by the collision probability, it is limited by the "irregularity" of the function. In particular, if $f$ is a regular function, then our bound is as good as that for a random function. While this approach yields no improvement for the worst-case complexity of inverting a function everywhere, it improves the complexity if we only seek

---

[3] Here by the collision probability of a function we mean the probability that after sampling two independent random inputs $x, y$ we have $f(x) = f(y)$.

to invert on an $\epsilon$ fraction of the inputs. In particular, we show that there is an algorithm such that for every $f : [N] \to [N]$ and every $\epsilon$, the algorithm inverts $f$ on an $\epsilon$ fraction of inputs and its time complexity, space complexity and advice length are bounded by

$$\tilde{O}\left(\max\left\{\sqrt{\epsilon N}, \epsilon^{\frac{5}{4}} N^{\frac{3}{4}}\right\}\right)$$

Here the $\tilde{O}$ hides factors of $2^{\text{poly} \log \log}$. It follows from known results, and we present a proof in the full version, that, in an oracle setting, it is not possible to do better than $\Omega(\sqrt{\epsilon N})$, so our result is best possible when $\epsilon < N^{-1/3}$.

Indeed, we establish the following more general trade-off: for every $T < 1/\epsilon$ and $S$ satisfying the trade-off $ST = \epsilon N$ we can construct an algorithm that has time $T$ and uses a data structure of size $S$ (up to lower order factors); for every $T > 1/\epsilon$, we can use time $T$ and space $S$ provided $TS^3 = \epsilon^5 N^3$. As we discuss below, a straight-forward application of the analysis of Fiat and Naor would have given a trade-off $TS^3 = \epsilon^3 N^3$, or a time and space complexity $\tilde{O}(\epsilon^{\frac{3}{4}} N^{\frac{3}{4}})$ in the $T = S$ case. For comparison, when $\epsilon = N^{-1/3}$, we can achieve (optimal) time and space $N^{1/3}$; the straight-forward use of the Fiat-Naor analysis would have given time and space $\sqrt{N}$. Given an upper bound $\lambda$ on the collision probability, we can achieve the optimal trade-off $TS = \epsilon N$ if $S \geq \epsilon^2 N^2 \lambda$, and the trade-off $TS = \epsilon^2 N^2 \lambda$ otherwise. For example, if we have a function with collision probability close to $1/N$, and we want to achieve inversion probability $\epsilon = N^{-1/4}$, then we can do so, using the latter construction, employing time, space and advice at most $N^{5/12} = N^{.416\cdots}$; using our generic construction (which applies to functions of arbitrary collision probability) would have given a complexity of $N^{7/16} = N^{.4375}$. We note however that all our tradeoffs (as well as the previous ones that we state) apply only for $S = \tilde{\Omega}(\sqrt{\epsilon N})$. The difference between our analysis and the one in [7] is explained in Section 2.

## Upper Bounds for Breaking pseudorandom Generators.

We also give non-uniform attacks to distinguish between distributions with significant statistical difference. For the sake of simplicity, in this version, we only consider the case of distinguishing the output of a pseudorandom generator from the uniform distribution. Given an arbitrary length-increasing generator $G : \{0,1\}^n \to \{0,1\}^m$, $m > n$, we show that, for every $\epsilon$, there is a distinguisher that runs in polynomial time, uses a data structure of size $O(\epsilon^2 2^n)$, and achieves distinguishing probability $\epsilon$. The distinguisher can also be implemented as a circuit of size $O(\epsilon^2 2^n)$. Notably, the distinguisher need not have oracle access to $G$, and so our result applies to generators constructed for applications in derandomization, in which the generator may have complexity $2^{O(n)}$, or even higher. In this setting, in which the complexity of the generator is not bounded, it is easy to see that advice $\Omega(\epsilon^2 2^n)$ is necessary. We also present a simpler construction that achieves the slightly worse circuit size $O(\epsilon^2 n 2^n)$.

## Lower Bounds.

We prove lower bounds for non-uniform attacks on one-way permutations and pseudorandom generators. Our lower bound for permutations is proven in the following model : Given any permutation $f$, the algorithm (call it $A$) is allowed to store a data structure of size $S$ which can be arbitrarily dependent on $f$. Further, on any input $x$, $A$ is allowed to make $T$ queries to $f$ along with any other computation it may perform. We say that there is a lower bound on time space tradeoff of time $T$ and space $S$ for inverting permutations on $\epsilon$ fraction of the inputs, if for any such algorithm $A$, there exists a permutation $f$ such that to invert $f$ on $\epsilon$ fraction of the inputs, if $A$ stores a data structure of size $S$, then it must make $T$ queries on some of its inputs. In this model, we prove that $S \cdot T = \tilde{\Omega}(\epsilon N)$. This in particular implies that the technique to invert a permutation described previously is optimal. While such lower bounds had previously been proven by Yao [18], Gennaro and Trevisan [8] and Wee [17], they were only applicable till $T = O(\sqrt{\epsilon N})$ while our proof shows the lower bound for the full range of $T$. Also, arguably our proof is simpler than the previous proofs.

Another problem we consider is that of showing lower bounds on time space tradeoffs for attacks on pseudorandom generators. The model is the same as that of permutations except that the algorithm is given access to the stretching function $G$ and it is required to distinguish between the output of the pseudorandom generator from the uniform by at least $\epsilon$. In this model, we get a lower bound of $S \cdot T = \tilde{\Omega}(\epsilon^2 N)$. From the previous discussion, this is tight even when restricted to distinguishers with no oracle access. To the best of our knowledge, this question has not been considered previously. Interestingly, the family $G$ that we use to prove the lower bound is a random permutation $f : [N] \rightarrow [N]$ followed by a random predicate $P : [N] \rightarrow \{0, 1\}$ i.e. $G(x) = f(x) \circ P(x)$.

**Common Random String Model.** Finally, we prove time space lower bounds for the problem of inverting a function (or breaking a pseudorandom generator) sampled from a family of functions (or a family of pseudorandom generators). This is the case when a common source of randomness is available to all the parties and this randomness is used to sample the one-way permutation or the pseudorandom generator as the case may be. We prove stronger lower bounds in this model. In particular, we show that if there is an algorithm which inverts any family of permutation $f : [N] \times [K] \rightarrow [N]$ (where $K$ denotes the common randomness), then for large $K$, the brute force attack in the best possible. Similarly, if there is an algorithm which for any family of pseudorandom generators, $G : [N] \times [K] \rightarrow [N] \times \{0, 1\}$ distinguishes the output of $G$ from uniform by more than $\epsilon$, then $S \cdot T = \tilde{\Omega}(\epsilon^2 K N)$ provided $K$ is large enough. Here $S$ and $T$ have their usual meanings. We specify the exact trade-offs with $K$ in the next section.

## Open Questions

It remains open to either improve the Fiat-Naor construction or to prove a stronger lower bound for the problem of inverting a random function or an arbitrary function everywhere. It is plausible that the optimal trade-off $ST = N$, while achievable for permutations, is impossible to achieve for general functions, maybe even impossible for random functions. Such a separation between the complexity of dealing with general or random functions versus permutations would be extremely interesting.

If one wants to invert a random permutation or function uniformly (that is, given no advice), then the lower bound $T \geq N$ (ignoring lower-order factors) holds. A quantum computer, however, can achieve $T = \sqrt{N}$ [11], which is optimal [5]. What is the complexity of inverting a random permutation, a random function, or an arbitrary function with a quantum computation that takes advice? It was pointed out to us by Scott Aaronson that extension of the techniques in [1] can prove that any pointer-jumping arguments (as the one in Hellman's scheme) cannot beat the $\sqrt{N}$ bound even with access to quantum advice. Hence, if at all quantum computation can beat the classical $\sqrt{N}$ bound, it will have to use significantly new techniques.

We do not have matching upper and lower bounds for the problem of constructing distinguishers for pseudorandom generators, except in the extremal case $T = O(1)$, $S = \epsilon^2 N$. Is $T = \epsilon^2 N$, $S = O(1)$ achievable? More generally, for what range of parameters is it possible to achieve distinguishability even though inversion of one-way permutations or functions is impossible?

## 2   Inverting One-Way Functions

How can one invert one-way functions, in general, faster than by brute force?

### 2.1   An Overview of the Ideas of Hellman and of Fiat and Naor

If we are given a one-way *permutation* $f : [N] \to [N]$, then it is easy to construct an inverter for $f()$ that uses time and space $\tilde{O}(\sqrt{N})$. Suppose for simplicity that $f()$ is a cyclic permutation and that $N = s^2$ is a perfect square: then pick $\sqrt{N}$ "equally spaced" points $x_1, \ldots, x_s$, such that $x_{i+1} = f^{(s)}(x_i)$, and create a data structure to store the pairs $(x_i, x_{i+1})$. Then given $y$, we compute $f(y)$, $f(f(y))$, and so on, until, for some $j$, we reach a point $f^{(j)}(x)$ which is one of the special points in the data structure. Then we can read from the data structure the value $f^{(j-s)}(y)$, and then by repeatedly computing $f$ again we will eventually reach $f^{(-1)}(y)$. Note that this takes $O(s)$ evaluations of $f$ and table look-ups, so both the time and space complexity are approximately $s = \sqrt{N}$. If $f()$ is not cyclic, we do a similar construction for each cycle of length less than $s$, and if $N$ is not a perfect square we can round $s$ to $\lceil \sqrt{N} \rceil$.

Abstractly, this construction works for the following reason. Consider the graph $G_f = ([N], E)$ that has $[N]$ as set of vertices and that for every $x$ has the directed edge $(x, f(x))$. Then, if $f$ is a permutation, it is possible to cover $G_f$ using $\sqrt{N}$ edge-disjoint paths, each of length $\sqrt{N}$ or, more generally, $S$ edge-disjoint paths of length $T$, provided $ST \geq N$. Furthermore, if $f$ is a function such that $G_f$ can be covered using $S$ edge-disjoint paths, each of length at most $T$, then we have an algorithm to invert $f$ using space $S$ and time $T$.

The problem is that, in general, no good collection of paths may exist. Suppose, for example, that $G_f$ looks like the graph on the left in Figure 1: a directed path of length $\frac{1}{3}N$ with a length-2 path joining in at each point. Then we see that there is a set $S$ (the vertices of indegree zero in the picture) of size $N/3$ such that no path can contain more than one vertex of $S$, and so no collection of $o(N)$ paths can cover the entire graph.

Hellman [12] considers the case in which $f()$ is a *random* function. Then, even though it's not clear how many edge-disjoint paths of what length can cover $G_f$ it is not hard to see that one can find $N^{\frac{1}{3}}$ paths of length $N^{\frac{1}{3}}$ having very few "collisions." This gives a construction that uses time and space $N^{\frac{1}{3}}$ and that inverts $f()$ at $N^{\frac{2}{3}}$ points. Hellman then suggests to modify $f()$ by composing it with a fixed permutation of the input bits, and to reason heuristically as if the new function behaved as an independently chosen new random function. Then one can repeat the construction, and have a new algorithm of time and space complexity $N^{\frac{1}{3}}$ that inverts $f()$ at $N^{\frac{2}{3}}$ points, which are assumed to be an independent random subset of size $N^{\frac{2}{3}}$. After iterating this process $N^{\frac{1}{3}}$ times one has $N^{\frac{1}{3}}$ candidate algorithms, each of time and space complexity $N^{\frac{1}{3}}$, such that, for every $x$, $f(x)$ is inverted by at least one of the algorithms. Overall, one gets an algorithm of complexity $N^{\frac{2}{3}}$ that inverts $f$ everywhere.

Fiat and Naor [7] make Hellman's argument rigorous. The idea of Fiat and Naor is to pick a good random hash function $g$, and then work with the new function $h(x) := g(f(x))$. (See Figure 1 for an example of the effect of this randomization.) If $g$ were a truly random function, and $f$ where a function such that every output has few pre-images, then one can repeat Hellman's calculation that $N^{\frac{1}{3}}$ nearly disjoint paths of length $N^{\frac{1}{3}}$ exist. Picking $N^{\frac{1}{3}}$ random functions $g_i$ then would give a rigorous version of the full argument, except for the dependency on several random oracles. For a more general trade-off, it is possible to pick $m$ nearly disjoint paths of length $t$ provided that $m \cdot t^2 < N$, and then iterate the construction $r$ times, where $r = N/mt$. Thus one gets a data structure of size $r \cdot m$, plus the space needed to store the descriptions of the hash functions, and an inversion procedure whose complexity is dominated by the complexity of evaluating the $r$ random hash functions at $t$ points each. Fiat and Naor then show that each $g_i$ only needs to be $k$-wise independent where $k$ is approximately $t$, the length of the paths. While one evaluation of a $t$-wise independent hash function would take time $t$, Fiat and Naor show that the overall time for the $rt$ evaluations can be made $t^2 + rt$ via a careful evaluation process and amortized analysis. The different $g_i$, in turn, only need to be pair-wise independent with respect to each other. Overall, the $r$ hash functions can be represented using only about $t$ bits, so that the space complexity is of the order of $r \cdot m + t$. Choosing the parameters $r, m, t$ optimally shows that the time-space tradeoff $TS^2 = N^2$ is achievable.

For general functions, the above ideas continue to work if the collision probability $\lambda$ of the distribution $f(U_{[N]})$ is small. In particular, one can have an algorithm of space $S = m \cdot r + t$ and time $T = t^2 + t \cdot r$ provided that $m \cdot t^2 \leq 1/\lambda$ and $m \cdot t \cdot r \geq N$. This optimizes to the time-space tradeoff $TS^2 = \lambda \cdot N^3$.

For functions having large collision probability, the idea is to create an additional look-up table $L$ (we also refer to it as a list), containing, for each of the $\ell$

**Fig. 1.** A graph $G_f$ that cannot be partitioned into few edge-disjoint paths and the graph $G_{f \circ g}$ where $g$ is a random permutation

elements $y$ such that $f^{(-1)}(y)$ is largest, the pair $(x, y)$ where $x$ is an arbitrary pre-image of $y$. Then, given $f(x) \in L$ we can immediately find an inverse by searching $L$, and the problem of inverting $f$ reduces to the problem of inverting the restriction of $f$ to $\{0,1\}^n - f^{(-1)}(L)$, which, intuitively, is the problem of inverting a function of low collision probability. More precisely, if we define the "effective" collision probability of $f$ relative to $L$ as the probability that, picking $x, x'$ uniformly at random we have $f(x) = f(x')$ conditioned on $f(x) \notin L$, then the effective collision probability is at most $1/\ell$. The $TS^2 = \lambda N^3$ trade-off can be extended to the case in which $\lambda$ is the effective collision probability, although at the additional cost of $\ell$ in the space. The optimal choice ends up being $\ell = S$, and so the trade-off becomes $TS^3 = N^3$. One additional difficulty that comes up in the analysis is that we need hash functions $g_i$ with the property that $g_i(f(x)) \notin f^{(-1)}(L)$ if $f(x) \notin L$. This is achieved by realizing $g_i$ by starting from a sequence of functions $g_i^1, \dots, g_i^k$, and then defining $g_i(y)$ to be $h_i^j(y)$ for the first $j$ such that $h_i^j(y) \notin f^{(-1)}(L)$.

## 2.2   Scaling Down the Fiat-Naor Construction

Consider now the issue of scaling down this construction in order to invert only $\epsilon N$ points.

If we fix parameters $r, m, t, \ell$ such that $r \cdot m \cdot t = \epsilon N$ and $m \cdot t^2 \leq \ell$, then we have an algorithm that inverts the function at $\epsilon N$ points and whose time complexity is $t^2 + rt$ and whose space complexity is $\ell + rm + t$. Some calculations show that this gives a time-space trade-off of $TS^3 = (\epsilon N)^3$.

Parameters

$\ell$ := Size of list $L$
$t$ := Length of each walk
$r$ := Number of independent functions $g \in \mathcal{F}$ used
$m$ := Number of walks according to each function $g_i$
$\mathcal{F}$ := Family of $k = 2t \cdot (\log N)^2$-wise independent functions.

Construction of Data Structure

1. Consider the $\ell$ elements in the range of $f$ with highest value of $I(y)$. For each such element $y$, store an entry $(y, x)$ in the list for some $x \in f^{-1}(y)$.
2. Choose functions $g_1, \ldots, g_r \in \mathcal{F}$ pairwise independently at random. For each function $g_i$, define the partial function $g_i^* : [N] \to [N]$ as

$$g_i^*(x) = \begin{cases} g_i(x,u) & \text{if } u \text{ is the least index such that } f(g_i(x,u)) \notin L \\ \text{undefined if } \forall u \in [(\log N)^2].\ f(g_i(x,u)) \in L \end{cases}$$

   For each $i$, define the partial function $h_i = g_i^* \circ f$.
3. For each $i \in [r]$ and $j \in [m]$, construct a walk $W_{ij}$ of length $t$; by starting at a random point $x_{ij}$ and computing the sequence $x_{ij}, h_i(x_{ij}), \ldots, h_i^t(x_{ij})$. Discard the walk if
   - for some $t_1 \le t$, $h_i^{t_1}(x_{ij})$ is undefined.
   - the walk cycles i.e. for $t_1, t_2 \le t$, $h_i^{t_1}(x_{ij}) = h_i^{t_2}(x_{ij})$.

   For walks $W_{ij}$ that are not discarded, store the pairs $(x_{ij}, h_i^t(x_{ij}))$.

**Fig. 2.** Description of data structure for inverting $f$

A first improvement comes by considering that if $|f^{(-1)}(L)| \ge \epsilon N$, then just by constructing $L$ we are done. This means that we may assume that the elements not in $L$ have each at most $\epsilon N/\ell$ pre-images, and there are $(1-\epsilon)N > N/2$ elements not in $f^{(-1)}(L)$, meaning that the collision probability of $f$ restricted to $\{0,1\}^n - f^{(-1)}(L)$ is at most $\epsilon/\ell$. This is a stronger bound than the "effective collision probability" bound $1/\ell$ in the Fiat-Naor analysis. This means that we can set the parameters so that $rmt = \epsilon N$, $mt^2 \le \ell/\epsilon$, and have $S = \ell + rm + t$ and $T = t^2 + rt$. This leads to the improved trade-off $TS^3 = \epsilon^4 N^3$, provided $\epsilon N > T > \epsilon^{-2}$.

A second improvement comes by using new constructions of $k$-wise independent hash functions (with $k = \tilde{O}(t)$) that can be evaluated in time negligible in $t$. We present such a construction in the full version of the paper. Using such a construction, the running time of the algorithm becomes just $rt$, rather than $t^2 + rt$. In the original Fiat-Naor construction, the two bounds are of the same order, because optimizing the parameters always leads to $r > t$. In the scaled-down construction we described above, however, $r > t$ is optimal only as long as $T > \epsilon^{-2}$, which is why we added such a constraint above. Hence, we require a family of hash functions with two properties:

---

Invert($y$)

1. If $(x, y) \in L$ for some $L$, return $x$.
2. For each $i \in [r]$
   (a) Construct the sequence $(g_i^*(y), h_i(g_i^*(y)), \ldots, h_i^{t-1}(g_i^*(y)))$.
   (b) If there are indices $j_0 \in [m]$ and $t_0 \leq t - 1$ such that $h_i^{t_0}(g_i^*(y)) = h_i^t(x_{ij_0})$, then compute $h_i^{t-t_0-1}(x_{ij_0})$. In case there are multiple choices for $j_0$, pick the smallest one.
   (c) If $f(h_i^{t-t_0-1}(x_{ij_0})) = y$, output $h_i^{t-t_0-1}(x_{ij_0})$ else output fail.

---

**Fig. 3.** Procedure for inverting a given element $y$

- *Small size:* it is sufficient for our purposes that each function be representable with $\Theta(t) + N^{o(1)}$ bits;
- *Efficient evaluation:* given the description of a function in the family and a point in the domain, we would like the evaluation of the function at that point to take time $t^{o(1)} \cdot N^{o(1)}$

We note that most known constructions with small size do not satisfy the efficient computation requirement. The construction that we use in this paper is based on an observation by Siegel [15] coupled with the lossless expander construction by Capalbo *et al.* [6]. The only other construction to us, which satisfies both the properties is the construction by Ostlin and Pagh [14] but their construction can differ from being uniform on a set of size $t$ by an inverse polynomial in $t$ which is too large an error for us. Using these hash functions would lead to the same trade-off $TS^3 = \epsilon^4 N^3$, but for the wider range of parameters $\epsilon N > T > \epsilon^{-1}$.

## 2.3   The Main New Idea

Our main improvement over the techniques of Fiat and Naor comes from the use of a more precise counting of the number of inputs $x$ such that $f(x)$ can be inverted using a given data structure.

We note that if we have the endpoints of a path of length $t$ in our data structure, then we are able to invert $f$ not just at $t$ inputs, but rather at as many inputs as the sum of the indegrees (in $G_f$) of the vertices of the path.[4] If, for example, the function $f$ is $k$-regular (meaning that, for every $x$, $f(x)$ has exactly $k$ pre-images), then a special case of the analysis that we provide shows that we can invert everywhere with trade-off $TS^2 = N^2/k^2$, while the Fiat-Naor analysis would give a trade-off $TS^2 = N^2 \cdot k$. They are the same when $k = \tilde{O}(1)$, but for larger $k$ the analysis of Fiat and Naor provides worse bounds, because the collision probability increases, while our analysis provides better bounds.

---

[4] Said differently, Fiat and Naor count the number of $y$ which are inverted, while one should count the number of $x$ such that $f(x)$ is inverted.

For functions that are not regular, providing a good bound on the number of elements that are inverted by the data structure is more challenging.

If the function has collision probability $\lambda$ (or "effective" collision probability $\lambda$ after discounting the elements in the high-indegree table), and we construct $r$ data structures, each having $m$ paths of length $t$, then the average sum of the indegrees of the vertices in the data structure is $m \cdot t \cdot r \cdot \lambda \cdot N$, which is potentially much more than $mtr$ if the collision probability is large. It seems, then, that we could fix parameters $m, t, r$ such that

$$m \cdot t^2 \leq \lambda^{-1}$$
$$m \cdot t \cdot r \cdot \lambda N \geq \epsilon N \qquad (1)$$

and be able to invert $\epsilon N$ elements using time $rt$ and space $rm+t$. This would optimize, in the interesting case in which space and time are equal, to having space and time $\max\{\sqrt{\epsilon N}, \epsilon N^{2/3}\}$, which would be great. In particular, it would improve the Fiat-Naor construction even when $\epsilon = 1$. Unfortunately, while $mrt\lambda N$ is the expectation of the sum of the indegrees of the vertices in all the paths of the data structure, it is not the expectation of the number of $x$ such that $f(x)$ is inverted: the problem is that, if the collision probability is very high, there might be elements $y$ with many pre-images that occur in multiple data structures, and which would then be counted multiple times.

We then proceed by considering three cases. If the collision probability is small, that is, less than $\epsilon^2/S$, where $S$ is the amount of space we plan to use, then we find parameters $m, t, r$ such that

$$mt^2 \leq S/\epsilon^2$$
$$mtr \geq \epsilon N$$

that is, we take advantage of the bound on collision probability but we do not attempt to improve the Fiat-Naor count on the number of inverted elements. This allows us to invert an $\epsilon$ fraction of elements using time and space at most $\max\{\sqrt{\epsilon N}, \epsilon^{5/4} N^{3/4}\}$.

If the collision probability is more than $\epsilon^2/S$, then we consider how much $mrt\lambda N$ is overcounting the real number of inverted elements. The overcounting is dominated by the elements $x$ such that, for a given choice of $r, m, t$, $f(x)$ has probability $\Omega(1)$, say, probability $\geq 1/100$, of belonging to one of the data structures. Call such a $y = f(x)$ a *heavy* image to invert.

If the number of pre-images of heavy elements is at least $100\epsilon N$, then we are done, because we expect to be able to invert at least a $1/100$ fraction of heavy elements.

The remaining case, then, is when the collision probability is more than $\epsilon^2/S$, but the total number of pre-images of heavy elements is less than $100\epsilon N$. This information, together with the fact that (thanks to the size-$S$ high-indegree table) we are only trying to invert elements with at most $\epsilon N/S$ pre-images, allows us to bound the total number of occurrences of heavy elements in the data structure, and to conclude that the total number of pre-images of non-heavy elements (which are inverted) is at least $\Omega(mrt\lambda N)$. This means that a choice of $m, r, t$ satisfying (1) leads us to invert an $\epsilon$ fraction of inputs, and to do so with time and space at most $\max\{\sqrt{\epsilon N}, \epsilon N^{2/3}\}$.

Applying these ideas to get full time-space trade-offs give us that, if $\epsilon < 1/N^{1/3}$ we can have the optimal trade-off $TS = \epsilon N$; otherwise we achieve the trade-off $TS^3 = \epsilon^5 N^3$. We now formally state the main theorem.

**Theorem 1.** *There is an oracle algorithm* Invert *such that given any* $f : \{0,1\}^n \to \{0,1\}^n$, *there is a data structure* DS *with parameters* $\ell$, $m$, $t$ *and* $r$ *such that*

$$\Pr_{x \in [N]} \left[ \text{Invert}^{f, \text{DS}}(f(x)) \in f^{-1}(f(x)) \right] \geq \epsilon$$

*where the total space required for* DS *is* $\tilde{O}(\ell + mr + t)$, *space required by* Invert *is* $\tilde{O}(t)$ *and the total time required by* Invert *is* $\tilde{O}(tr)$. *Hence, assuming that* $(\ell + mr + t) = O(S)$ *and* $tr = T$, *there is an algorithm (in the RAM model) which uses space* $\tilde{O}(S)$ *and time* $\tilde{O}(T)$ *and inverts* $f$ *on an* $\epsilon$ *fraction of inputs. Here* $\tilde{O}$ *hides factors of* $2^{\text{poly} \log \log N}$.

*Remark 1.* For most of the allowed range of the parameters we have $r < t$, and that in the "low-end" range $\epsilon < N^{-1/3}$ for which our result is optimal we have $r = 1$. For this reason there is a notable improvement in using our efficient hash functions instead of the amortized hash function evaluation of Fiat and Naor.

As described above, the algorithm by Hellman, Fiat and Naor, as well as ours involve a significant amount of search and hence it is interesting to ask if this search can be parallelized. There have been results in this direction by van Oorschot and Wiener [16] and more recently by Joux and Lucks [13]. Some of these results can be helpful in parallelizing even in the regime when $\epsilon$ is small (as in our case).

## 3   Attacks on Pseudorandom Generators

The starting point of our result for pseudorandom generators is the fact [2] that if two random variables ranging over $\{0,1\}^m$ have constant statistical distance, then there is a linear function (of $O(m)$ circuit complexity) that distinguishes the two random variables with advantage at least $2^{-m/2}$.

Suppose that we are given a length-increasing pseudorandom generator $G : \{0,1\}^{n-1} \to \{0,1\}^n$ and that we want to construct a distinguisher achieving distinguishing probability $\epsilon$.

Our idea is to partition $\{0,1\}^n$ into $\epsilon^2 2^n$ sets each of size $\epsilon^{-2}$, for example based on the value of the first $n - 2 \log 1/\epsilon$ bits, and then apply within each block the linear function that provides, within that block, the best distinguishing probability. Overall, this defines a function of circuit complexity $O(\epsilon^2 \cdot n \cdot 2^n)$. Then, intuitively, within each block we achieve distinguishing probability at least $\epsilon$, because each block is a set of size $\epsilon^{-2}$, and the distinguishing probability is at least the square root of the inverse of the block size.

The straightforward implementation of this intuition would be to use, in each block, the linear function that best distinguishes the uniform distribution within the block from the conditional distribution of the output of the generator conditioned on landing in the block. Unfortunately this approach would not work because the overall distinguishing probability is not a convex combination of the conditional distinguishing probabilities.[5]

Instead, in each block, we choose the linear function that most contributes to the overall distinguishing probability. In order to quantify this contribution we need a slight generalization of the result of [2].

We then present a more efficient distinguisher of circuit complexity $O(\epsilon^2 \cdot 2^n)$ which employs a hash function sampled from a 4-wise independent family, and whose analysis employs a more involved fourth-moment argument, inspired by [3]. As noted before, all our ideas apply to the case when we want to distinguish between two arbitrary distributions $D_1$ and $D_2$. In particular, given two distributions $D_1$ and $D_2$ with statistical distance $\delta$, we can construct a circuit of size $O(\epsilon^2 2^n)$ which distinguishes between $D_1$ and $D_2$ with probability $\epsilon\delta$. More, formally, we have the following result.

**Theorem 2.** *Given any two distributions $D_1$ and $D_2$ over $\{0,1\}^n$ such that their statistical distance is $\delta$ and $\epsilon \leq 2^{n/2}$, there is a circuit $C$ of size $O(\epsilon^2 \cdot 2^n)$ such that*

$$\mathbf{P}[C(D_1) = 1] - \mathbf{P}[C(D_2) = 1] \geq 2\epsilon\delta$$

## 4    Lower Bounds

Using techniques of Yao [18], Gennaro and Trevisan [8], and Wee [17], it is possible to show that, in the generic oracle setting that we consider in this paper, there are permutations for which the amount of advice $S$ and the oracle query complexity $T$ must satisfy

$$S \cdot T \geq \tilde{\Omega}(\epsilon N)$$

for any algorithm that inverts an $\epsilon$ fraction of inputs. More precisely, we prove the following theorem.

**Theorem 3.** *If $A$ is an oracle algorithm that runs in time at most $T$ and such that for every permutation $f : [N] \to [N]$ there is a data structure adv of size $\leq S$ such that*

$$\mathbf{P}_x[A^f_{adv}(f(x)) = x] \geq \epsilon$$

*Then*

$$S \cdot T = \tilde{\Omega}(\epsilon N)$$

---

[5] This is a subtle issue related to the fact that the condition of landing in a given block might have different probabilities in the uniform distribution versus the output of the generator. If so, then the respective conditional probabilities are normalized differently, and the use of a distinguisher for the conditional distributions in a block does not necessarily contribute to the task of distinguishing the original distributions.

Such lower bound proofs are based on the idea that an algorithm with better performance could be used to encode every permutation $f : [N] \rightarrow [N]$ using strictly less than $\log N!$ bits, which is impossible. Here, we simplify such proofs by using randomized encodings. (Even a randomized encodings cannot represent every permutation using less than $\log N!$, and showing that such an encoding would be possible if the lower bound were wrong is easier by using randomization.) In fact, while previous proofs gave a lower bound on the trade-off only when $T = \tilde{O}(\sqrt{\epsilon N})$, our lower bound works for the full range of parameters.

We then consider the question of the security of pseudorandom generators in the oracle setting. By using the aforementioned results for permutations and applying efficient hard-core predicates [9], it is possible to show the existence of generators for which $S \cdot T \geq \epsilon^7 N$. By instead applying the ideas of randomized encodings to a pair $f, p$ where $f$ is a random permutation (modeling a one-way permutation) and $p$ is a random predicate (modeling a "hard-core predicate" for $p$), we prove the existence of length-increasing generators such that for every distinguisher that makes $T$ oracle queries to the generator, and which has advice $S$ and distinguishing probability $\epsilon$, we have

$$S \cdot T \geq \tilde{\Omega}(\epsilon^2 N)$$

where $N$ is the number of seeds. Formally, we have the following theorem.

**Theorem 4.** *Suppose that $A$ is an oracle algorithm that makes $T$ queries, uses a $S$-bit advice string, and is such that for every length-increasing function $G : [N] \rightarrow [N] \times \{0, 1\}$ there is an advice string adv such that*

$$|\mathbf{P}[A_{adv}^G(G(x)) = 1] - \mathbf{P}[A_{adv}^G(y) = 1]| \geq \epsilon$$

*Then $S \cdot T \geq \tilde{\Omega}(\epsilon^2 N)$*

A key intermediate result used here is a lower bound on the following kind of computation. Given any predicate $P : \{0, 1\}^n \rightarrow \{0, 1\}$ we are required to compute $P(x)$ by querying the oracle $P$ on any point but $x$. For this kind of computation, Yao [18] had established an (optimal) lower bound on the trade-off between length of advice and the number of queries to the oracle $P$ when one is required to compute $P$ correctly at all places. We extend this to the case when we are only required to compute $P$ correctly at $1/2 + \epsilon$ fraction of the places, to get the following result.

**Theorem 5.** *Suppose that $A$ is an oracle algorithm that makes $T$ queries while never querying its input, uses a $S$ bit advice string, and is such that for every predicate $P : [N] \rightarrow \{0, 1\}$, there is an advice string adv such that*

$$\mathbf{P}[A_{adv}^P(x) = P(x)] \geq \frac{1}{2} + \epsilon$$

*Then $S \cdot T \geq \Omega(\epsilon^2 N)$.*

We note that the optimal lower bound in this case seems to be $S \cdot T \geq \Omega(\epsilon N)$ and we do not how to close this gap. This gap is reflected in the gap between our lower bound for pseudorandom generators of the form $x \to f(x)p(x)$ where $f$ is a permutation, and known constructions of distinguishers for such generators. In particular, the best known algorithm is one of the following (depending on $\epsilon$, $S$, $T$) : Use the algorithm for inverting functions which can be at best $S \cdot T \leq \tilde{\Omega}(\epsilon N)$ or use the circuit which we described in the previous subsection. That in particular uses $S = \tilde{\Omega}(\epsilon^2 N)$ and $T = \tilde{\Omega}(1)$.

Finally, we look at the *common random string* model, in which all parties share a common random string $k$, which they can use to select a permutation $f_k(\cdot)$ from a family of permutations, or a generator $G_k(\cdot)$ from a family of generators. In such a setting, the trivial brute force attack that achieves inverting (and distinguishing) probability $\epsilon = T/N$ with no advice remains possible. Alternatively, one can think of a family of permutations as a single permutation $(k, x) \to (k, f_k(x))$. We show that, for families of permutations, either the trivial uniform algorithm or Hellman's construction applied to the mapping $(k, x) \to (k, f_k(x))$ are best possible, depending on whether the available advice is shorter or longer than the number of keys. More precisely, we prove the following theorem.

**Theorem 6.** *Suppose that $A$ is an oracle algorithm that makes $T$ queries, uses an $S$-bit advice string, and is such that for every family of permutations $f :$ $[K] \times [N] \to [N]$ there is an advice string adv such that*

$$\mathbf{P}_{k\in[K],\ x\in[N]}[A^f_{adv}(k, f(k, x)) = x] \geq \epsilon$$

*Then $S \cdot T \geq \tilde{\Omega}(\epsilon KN) - \tilde{O}(KT)$*

We also prove strong lower bounds for distinguishers for pseudorandom generators in the common random string model. We show the following theorem.

**Theorem 7.** *Suppose that $A$ is an oracle algorithm that makes $T$ queries, uses an $S$-bit advice string, and is such that for every family of length-increasing functions $G : [K] \times [N] \to [N] \times \{0,1\}$ there is an advice string adv such that*

$$|\mathbf{P}_{x\in[N]}[A^G_{adv}(k, G(k, x)) = 1] - \mathbf{P}_{y\in[2N]}[A^G_{adv}(k, y) = 1]| \geq \epsilon$$

*Then $S \cdot T \geq \tilde{\Omega}(\epsilon^2 KN) - \tilde{O}(KT)$.*

This translates to saying that for generators in the common random string model, either $T \geq \tilde{\Omega}(\epsilon^2 N)$ or $ST \geq \tilde{\Omega}(\epsilon^2 KN)$, where $K$ is the number of keys.

## 5    Model of Computation

**Positive Results.** In the time/space trade-offs of Hellman, of Fiat and Naor, and of this paper, an algorithm uses "time $T$" and "space $S$" if it runs in time at most $T$ (in a RAM model), uses at most $S$ bits of space, and works correctly upon receiving $S$ bits of advice, in the form of an $S$-bit data structure that dominates the space requirement of the algorithm.

The "advice," in turn, can be computed in uniform time $\tilde{O}(N)$. In the work of Hellman and of Fiat and Naor, one cannot hope, in general, to have processing time significantly smaller than $N$ in order to generate the data structure used by the algorithm. Otherwise, one would have a uniform algorithm that inverts an arbitrary one-way permutation (or function) in time noticeable smaller than $N$, which is impossible relative to a random permutation (or function) oracle.

In our paper, the data structure we use is also easily pre-computable in time $\tilde{O}(N)$. Pre-processing time significantly smaller than $\epsilon N$ should not be expected, because then we would have a uniform algorithm to invert a random function on an $\epsilon$ fraction of inputs in time significantly smaller than $\epsilon N$. With some care, our data structure can indeed be pre-computed using optimal uniform time $\tilde{O}(\epsilon N)$. We, however, do not describe it here for the sake of simplicity.

**Negative Results.** When we show that a particular combination of space $S$ and time $T$ is not achievable, our result rules out non-uniform algorithms that make at most $T$ oracle queries to the function (or generator) oracle, and which receive at most $S$ bits of advice. The actual space used by the algorithm, as well as the complexity of the computations performed between oracle queries, can be unbounded. Likewise, the non-uniform advice can have arbitrary complexity.

## Acknowledgements

We would like to thank Daniel Wichs for suggesting the study of the common random string model, and Scott Aaronson, Cynthia Dwork, Omer Reingold, Udi Wieder and Hoeteck Wee for pointers to the literature. We also thank the anonymous reviewers for helpful comments and pointing us to [16,13].

## References

1. Aaronson, S.: Lower bounds for local search by quantum arguments. SIAM Journal of Computing 35(4), 804–824 (2006)
2. Alon, N., Goldreich, O., Håstad, J., Peralta, R.: Simple constructions of almost $k$-wise independent random variables. Random Structures and Algorithms 3(3), 289–304 (1992)
3. Andreev, A.E., Clementi, A.E.F., Rolim, J.D.P.: Optimal bounds for the approximation of boolean functions and some applications. Theoretical Computer Science 180, 243–268 (1997)
4. Barkan, E., Biham, E., Shamir, A.: Rigorous bounds on cryptanalytic time/memory tradeoffs. In: Dwork, C. (ed.) CRYPTO 2006. LNCS, vol. 4117, pp. 1–21. Springer, Heidelberg (2006)
5. Bennett, C., Bernstein, E., Brassard, G., Vazirani, U.: Strengths and weaknesses of quantum computing. SIAM Journal on Computing 26(5), 1510–1523 (1997)
6. Capalbo, M.R., Reingold, O., Vadhan, S.P., Wigderson, A.: Randomness conductors and constant-degree lossless expanders. In: Proceedings of the 34th ACM Symposium on Theory of Computing, pp. 659–668 (2002)
7. Fiat, A., Naor, M.: Rigorous time/space trade-offs for inverting functions. SIAM Journal on Computing 29(3), 790–803 (1999)

8. Gennaro, R., Trevisan, L.: Lower bounds on the efficiency of generic cryptographic constructions. In: Proceedings of the 41st IEEE Symposium on Foundations of Computer Science, pp. 305–313 (2000)

9. Goldreich, O., Levin, L.: A hard-core predicate for all one-way functions. In: Proceedings of the 21st ACM Symposium on Theory of Computing, pp. 25–32 (1989)

10. Golynski, A.: Cell probe lower bounds for succinct data structures. In: Proceedings of the 20th ACM-SIAM Symposium on Discrete Algorithms, pp. 625–634 (2009)

11. Grover, L.: A fast quantum mechanical algorithm for database search. In: Proceedings of the 28th ACM Symposium on Theory of Computing, pp. 212–219 (1996)

12. Hellman, M.: A cryptanalytic time-memory trade-off. IEEE Transactions on Information Theory 26(4), 401–406 (1980)

13. Joux, A., Lucks, S.: Improved generic algorithms for 3-collisions. In: Matsui, M. (ed.) ASIACRYPT 2009. LNCS, vol. 5912, pp. 347–363. Springer, Heidelberg (2009)

14. Ostlin, A., Pagh, R.: Uniform hashing in constant time and linear space. In: Proceedings of the 35th ACM Symposium on Theory of Computing, pp. 622–628 (2003)

15. Siegel, A.: On universal classes of extremely random constant-time hash functions. SIAM Journal of Computing 33(3), 505–543 (2004)

16. van Oorschot, P.C., Wiener, M.J.: Parallel Collision Search with Cryptanalytic Applications. Journal of Cryptology 12, 1–28 (1999)

17. Wee, H.: On obfuscating point functions. In: Proceedings of the 37th ACM Symposium on Theory of Computing, pp. 523–532 (2005)

18. Yao, A.: Coherent functions and program checkers. In: Proceedings of the 22nd ACM Symposium on Theory of Computing, pp. 84–94 (1990)

# Pseudorandom Functions and Permutations Provably Secure against Related-Key Attacks

Mihir Bellare and David Cash

Dept. of Computer Science & Engineering, University of California San Diego, USA
{mihir,cdcash}@cs.ucsd.edu
http://www.cs.ucsd.edu/users/{mihir,cdcash}

**Abstract.** This paper fills an important foundational gap with the first proofs, under standard assumptions and in the standard model, of the existence of PRFs and PRPs resisting rich and relevant forms of related-key attack (RKA). An RKA allows the adversary to query the function not only under the target key but under other keys derived from it in adversary-specified ways. Based on the Naor-Reingold PRF we obtain an RKA-PRF whose keyspace is a group and that is proven, under DDH, to resist attacks in which the key may be operated on by *arbitrary* adversary-specified group elements. Our framework yields other RKA-PRFs including a DLIN-based one derived from the Lewko-Waters PRF. We show how to turn these PRFs into PRPs (blockciphers) while retaining security against RKAs. Over the last 17 years cryptanalysts and blockcipher designers have routinely and consistently targeted RKA-security; it is important for abuse-resistant cryptography; and it helps protect against fault-injection sidechannel attacks. Yet ours are the first significant proofs of existence of secure constructs. We warn that our constructs are proofs-of-concept in the foundational style and not practical.

## 1 Introduction

Alarmed by the number of successful related-key attacks (RKAs) against real blockciphers [15,17,16,38,42,19,9,10,12,11,49,54,29,36,13,39,34], theoreticians have stepped back to ask to what extent the underlying goal of RKA-secure PRFs and PRPs is achievable at all. The question is made challenging by the unusual nature of the attack model which allows the adversary to manipulate the key. Previous works providing RKA-secure PRFs and PRPs have bypassed rather than overcome the core technical difficulties by using the ideal cipher or random oracle models, making non-standard assumptions themselves "related-key" in nature, or limiting attackers to weak classes of RKAs for which the problem disappears [5,46]. We provide a new technical approach based on which we obtain the first designs of PRFs and PRPs secure against non-trivial and application-relevant forms of RKAs under standard assumptions (DDH) and in the standard model. Our constructions are not practical, providing, instead, in-principle proofs of achievability of the goals in the classical foundational style.

T. Rabin (Ed.): CRYPTO 2010, LNCS 6223, pp. 666–684, 2010.

THE MODEL. RKAs were introduced by Biham and Knudsen [7,8,40] and formalized by Bellare and Kohno (BK) [5]. Referring to any $\phi\colon \mathcal{K} \to \mathcal{K}$ as a related-key deriving (RKD) function, the latter define what it means for a family of functions $F\colon \mathcal{K} \times \mathcal{D} \to \mathcal{R}$ to be a $\Phi$-RKA-PRF, where $\Phi$ is a class (set) of RKD functions. The game begins by picking a random challenge bit $b$, a random target key $K \in \mathcal{K}$ and, for each $L \in \mathcal{K}$, a random function $G_L\colon \mathcal{D} \to \mathcal{R}$. The adversary is allowed multiple queries to an oracle that, given a pair $(\phi, x) \in \Phi \times \mathcal{D}$, returns $F_{\phi(K)}(x)$ if $b = 1$ and $G_{\phi(K)}(x)$ if $b = 0$, and its advantage is $2 \Pr[b = b'] - 1$, where $b'$ is the bit it outputs. The definition of a family of permutations (block-cipher) $F\colon \mathcal{K} \times \mathcal{D} \to \mathcal{D}$ being a $\Phi$-RKA-PRP is analogous, the difference being that each $G_L$ is a random permutation on $\mathcal{D}$ rather than a random function. Note that when $\Phi$ consists of just the identity function, we recover the standard PRF [33] and PRP [45] notions.

GROUP-INDUCED CLASSES. We must beware of inherent limitations. It is observed in [5] that some $\Phi$ are "impossible" in the sense that *no* $F$ can be a $\Phi$-RKA-PRF or a $\Phi$-RKA-PRP. Indeed, any $\Phi$ that contains a constant function $\phi(\cdot) = C$, for some attacker-known constant $C \in \mathcal{K}$, is impossible. (For some $x$, just query the RK-oracle with $(\phi, x)$ and return 1 if the response is $F_C(x)$.) The class of all RKD functions is impossible, and so is the class of all permutations. The basic foundational question, then, is to identify *specific* classes $\Phi$, as rich, interesting and relevant as possible, for which we can prove "possibility," meaning existence of $\Phi$-RKA-PRFs and $\Phi$-RKA-PRPs. But which classes are good candidates?

BK [5] showed the (standard model) possibility of any class $\Phi$ whose member RKDs modify only the second half of the given key, and Lucks [46] gave, for the same class, an alternative construction with better concrete security. But if part of the key is unmodified, we can just use it as the "actual" key and put the rest in the input, meaning RKA-security here is for "trivial" reasons. For the proof-of-concept results in which we are interested, we seek candidate classes where the core technical difficulties cannot be bypassed in this way.

Luckily, Lucks [46] has already pinpointed a worthy target. His group-induced classes are elegant, appealing, non-trivial and application-relevant. If $(\mathcal{K}, *)$ is a group under an operation "$*$", the associated group-induced class is $\mathsf{rkd}[\mathcal{K}, *] = \{ \phi_{\Delta}^* \colon \Delta \in \mathcal{K} \}$ where $\phi_{\Delta}^*(K) = K * \Delta$ for all $K \in \mathcal{K}$. These classes are rich because all group actions are included. They also have what in [32] is called the completeness property and viewed as important to non-triviality of the class, namely that for any $K, K' \in \mathcal{K}$ there is a $\phi \in \mathsf{rkd}[\mathcal{K}, *]$ such that $\phi(K) = K'$. Security relative to these classes suffices for applications and cannot be established by tricks such as the above. The quest that emerges is to find (non-trivial) groups $(\mathcal{K}, *)$ for which we can show the possibility of $\mathsf{rkd}[\mathcal{K}, *]$, meaning exhibit $\mathsf{rkd}[\mathcal{K}, *]$-RKA-PRFs and $\mathsf{rkd}[\mathcal{K}, *]$-RKA PRPs $F\colon \mathcal{K} \times \mathcal{D} \to \mathcal{R}$ whose keyspace is $\mathcal{K}$.

PREVIOUS WORK. Results of [5] imply that ideal ciphers achieve $\mathsf{rkd}[\mathcal{K}, *]$-RKA-PRP security for *any* large enough group $(\mathcal{K}, *)$. Also, one can easily strengthen a given PRF or PRP to be a $\mathsf{rkd}[\mathcal{K}, *]$-RKA one by hashing the key with a random oracle before use [46]. However, it is unclear how to instantiate the ideal primitives here to get "real" constructions for even a *single* group [25,18]. For certain composite numbers $M$, Lucks [46] provides $\mathsf{rkd}[\mathbb{Z}_M, +]$-RKA-PRFs for the group $(\mathbb{Z}_M, +)$, where $+$ is addition modulo $M$, but the assumptions on which he bases security are not only interactive and novel but also themselves "related-key" in nature and uncomfortably close to just assuming the construct itself is secure, making the value of the proofs debatable from the point of view of security assurance. Existing PRFs such as the DDH-based one of Naor and Reingold [47] or the DLIN-based one of Lewko and Waters [43] are subject to simple attacks showing they provide no RKA-security. (Nonetheless they will be a starting point for our constructs.) Research has expanded to consider RKA-security of other primitives while leaving the goal unachieved for the more basic PRF, PRP and PRG ones [1,32].

The salient fact that emerges from this previous work is that we do not have *even a single example* of a group $(\mathcal{K}, *)$ for which we can prove the existence of a $\mathsf{rkd}[\mathcal{K}, *]$-RKA-PRF or $\mathsf{rkd}[\mathcal{K}, *]$-RKA-PRP under standard assumptions in the standard model. The reason for the lack of progress is technical obstacles. The attack models underlying standard definitions of standard primitives do not allow any key-manipulation by the adversary. This makes it unclear how one can do any reductions, which seem to require applying RKD functions to an unknown key. This difficulty is appreciated, with Goldenberg and Liskov [32, Section 4] saying "The major open problem in related-secret security is whether or not related-key secure blockciphers exist ... related-secret pseudorandom bits cannot be constructed using traditional techniques. This leaves a significant open problem ... can fundamentally new techniques be found to create related-secret pseudorandom bits?"

NEW RKA-PRFs. We fill the above gap, providing the first constructions, under the standard DDH assumption and in the standard model, of $\Phi$-RKA-PRFs where $\Phi$ is group-induced. We obtain and analyze our designs via a general framework using two new primitives which may be of independent interest, namely key-malleable PRFs and key fingerprints. However, (surprisingly) at least one of our constructions, that we call the multiplicative DDH based RKA-PRF, is compact enough to state here. Let NR: $(\mathbb{Z}_p^*)^{n+1} \times \{0,1\}^n \to \mathbb{G}$ denote the Naor-Reingold PRF [47] that given key $\mathbf{a} = (\mathbf{a}[0], \ldots, \mathbf{a}[n]) \in (\mathbb{Z}_p^*)^{n+1}$ and input $x = x[1] \ldots x[n] \in \{0,1\}^n$ returns

$$\mathrm{NR}(\mathbf{a}, x) = g^{\mathbf{a}[0] \prod_{i=1}^n \mathbf{a}[i]^{x[i]}}, \tag{1}$$

where $\mathbb{G} = \langle g \rangle$ is a group of prime order $p$. The keyspace $\mathcal{K} = (\mathbb{Z}_p^*)^{n+1}$ is a group under the operation $*$ of componentwise multiplication modulo $p$, but simple attacks [5] show that NR is not itself a $\mathsf{rkd}[\mathcal{K}, *]$-RKA-PRF. Let $h$ be a collision-resistant hash function with domain $\{0,1\}^n \times \mathbb{G}^{n+1}$ and range $\{0,1\}^{n-2}$. Given key $\mathbf{a}$ and input $x$, our construct $F$: $(\mathbb{Z}_p^*)^{n+1} \times \{0,1\}^n \to \mathbb{G}$ returns

$$F(\mathbf{a}, x) = \mathrm{NR}(\mathbf{a}, 11 \| h(x, (g^{\mathbf{a}[0]}, g^{\mathbf{a}[0]\mathbf{a}[1]}, \ldots, g^{\mathbf{a}[0]\mathbf{a}[n]}))),$$

where "$\|$" denotes concatenation. Theorem 3 says that $F$ is a $\mathsf{rkd}[(\mathbb{Z}_p^*)^{n+1}, *]$-RKA-PRF under the DDH assumption. The difficulty such a proof had to overcome was how the "simulator," given $\mathbf{d}$, can answer queries for $F$ on keys of the form $\mathbf{a} * \mathbf{d}$ without itself knowing $\mathbf{a}$ and without contradicting RKA security by enabling an attack.

This and other results are obtained via a general framework hinging on two new primitives. We call a PRF $M: \mathcal{K} \times \mathcal{D} \rightarrow \mathcal{R}$ key-malleable relative to a class $\Phi$ of RKD functions on $\mathcal{K}$ if there is an efficient algorithm that given $(\phi, x) \in \Phi \times \mathcal{D}$ and oracle access to $M_K$ returns $M_{\phi(K)}(x)$. That this could be useful for building a $\Phi$-RKA-PRF is, on the one, hand, intuitive, because it allows us to simulate an oracle for $M(\phi(K), \cdot)$ via an oracle for $M(K, \cdot)$. But it is, on the other hand, counter-intuitive, because the same property immediately gives rise to an attack showing that $M$ is not a $\Phi$-RKA-PRF! Something else is necessary. This turns out to be the new concept of a key fingerprint, a vector $\mathbf{w}$ over $\mathcal{D}$ that uniquely identifies a key in the sense that for all $(\phi, \phi', K) \in \Phi \times \Phi \times \mathcal{K}$ we have $M_{\phi(K)}(\mathbf{w}) \neq M_{\phi'(K)}(\mathbf{w})$ whenever $\phi \neq \phi'$, where we have extended $M$ to vector second arguments on which it operates componentwise. Given $M, \mathbf{w}$ and a collision-resistant hash function, our general construction shows how to build $F: \mathcal{K} \times \mathcal{D} \rightarrow \mathcal{R}$ that we can show is a $\Phi$-RKA-PRF (cf. Theorem 1). The multiplicative DDH based RKA-PRF noted above is obtained by showing that NR is a key-malleable PRF relative to $\mathsf{rkd}[(\mathbb{Z}_p^*)^{n+1}, *]$ and then finding a key fingerprint for it. It is interesting that we turn malleability [28], typically viewed as a "bad" property, into a "good" property that we can exploit.

Two more constructs emanate from this framework. There are groups where DDH is easy but the Decision Linear (DLIN) problem of [22] still seems hard. Lewko and Waters [43] provide a DLIN-based analogue of the Naor-Reingold PRF, commenting that they know of no "closed-form" rendition of it akin to the above Equation (1) for NR. Using matrices, we provide in Equation (21) such a closed-form, and then, restricting attention to invertible matrices and slightly modifying the function, we obtain in Equation (22) a PRF that we can show is key-malleable and admits a key fingerprint. Our framework then yields a DLIN-based RKA-PRF [4].

The group $(\mathbb{Z}_p^*)^{n+1}$ underlying our multiplicative DDH-based RKA-PRF is, as the name indicates, multiplicative. Providing a DDH-based $\mathsf{rkd}[\mathbb{Z}^{n+1}, *]$-RKA-PRF where $*$ is componentwise addition modulo $p$ is more difficult. We provide in [4] a solution that involves first modifying the Naor-Reingold PRF and then applying our framework. However, the running time of our reduction is exponential in the input size. Theoretically, this means we must assume hardness of DDH against exponential-time algorithms. In practice, one can get security by using larger groups. This situation parallels that for the BB IBE scheme [20].

FROM RKA-PRFS TO RKA-PRPS. Practical interest centers on RKA-secure blockciphers, meaning PRPs, and the constructions above are RKA-PRFs. It is not clear how one might modify the constructions to get RKA-PRPs. We use a different approach. Using deterministic extractors [24,30,26], we convert our $\Phi$-RKA-PRFs into $\Phi$-RKA-PRGs with bitstring outputs. When these are used as key-derivation functions to key an ordinary (not RKA) PRP, we obtain a $\Phi$-RKA-PRP. (This second, composition step extends similar ones from [46,32]).

For each class $\Phi$ for which we have a $\Phi$-RKA-PRF, this not only yields a CPA-secure $\Phi$-RKA-PRP but even a CCA-secure one.

RELATED WORK AND TECHNIQUES. Based on the Boneh-Boyen short signature scheme [21], Dodis and Yampolskiy [27] define a PRF BBDY: $\mathbb{Z}_p \times S \to \mathbb{G}$ via $\text{BBDY}(k, x) = \mathbf{e}(g, g)^{1/(k+x)}$, where $\mathbf{e}: \langle g \rangle \times \langle g \rangle \to \mathbb{G}$ is a bilinear map and $S \subseteq \mathbb{Z}_p$. This had seemed to us promising towards building a $\text{rkd}[\mathbb{Z}_p, +]$-RKA-PRF, but (disappointingly) did not lead there. To begin with, BBDY is easily shown by attack to not itself be a $\text{rkd}[\mathbb{Z}_p, +]$-RKA-PRF. (Adding 1 to $k$ or to $x$ yields the same outcome.) By exploiting the symmetry between $k$ and $x$ and using the composition paradigm, it turns out one can show how to construct a $\text{rkd}[\mathbb{Z}_p, +]$-RKA-PRF if BBDY was a (plain) PRF, but *only if the input domain $S$ was equal to $\mathbb{Z}_p$*. The problem is that the q-DBDHI-based proof of [27,21] requires $S$ to be "small" and in particular delivers nothing at all when $S = \mathbb{Z}_p$. We comment that there is no attack showing BBDY is not a PRF when $S = \mathbb{Z}_p$ and one might prove this in the generic model, but there seems little reason to pursue a generic group model solution when we already have a standard model, DDH-based solution. (In fact, since DDH is hard in the generic group model [53], our results already imply a generic model solution anyway.)

RKA-security is much easier for randomized primitives than deterministic ones. From the ElGamal scheme over a group of prime order $p$, one can easily get a (randomized) $\text{rkd}[\mathbb{Z}_p, +]$-RKA-CPA-secure DDH-based symmetric encryption scheme. Applebaum [1] presents a more efficient $\text{rkd}[\{0, 1\}^n, \oplus]$-RKA-CPA-secure (still randomized) symmetric encryption scheme assuming hardness of the LPN problem. There seems to be no simple way, from these techniques, to get the full-fledged group-induced RKA-PRFs that we target, where the computation is deterministic. That the deterministic case is more difficult than the randomized one is not surprising or unusual. In analogy, DDH based injective trapdoor functions [48] were discovered much later than DDH-based public-key encryption schemes.

Goldenberg and Liskov [32] broaden the scope to consider related-secret security. As with Lucks [46] they can, via composition, reduce the design of $\Phi$-RKA-PRFs to the design of $\Phi$-RKA-PRGs, but provide no new constructions of the latter and hence of the former. They have negative results indicating the difficulty of getting these for non-trivial classes $\Phi$, and comment [32, Section 1] that "This leads us to the conclusion that if related-secret pseudorandomness (including related-key blockciphers) are possible, they must be proven either based on other related-secret pseudorandomness assumptions, or a dramatically new way of creating pseudorandomness from hardness must be developed." Our results are answers to these questions, showing that one can in fact obtain related-key pseudorandomness under standard assumptions. (Our RKA-PRFs of course directly yield RKA-PRGs.) Their negative results are in a limited model of computation and do not apply in our context.

CONTEXT. Conceived with the goal of studying the strength of blockcipher key-schedules [7,8,40], RKAs quickly became mainstream. RKA-security is viewed as necessary for the collision-resistance of blockcipher-based compression functions [50]. (But one should note that this view has no formal justification.) RKA-resistance was a stated design goal of AES and remains so for other modern ciphers. A successful RKA is universally viewed by cryptanalysts as a break of the cipher. The recent attention-grabbing attacks on AES-192 and AES-256 [17,16,15] were RKAs, and far from unique in this regard: a look at the literature shows that RKAs abound [38,42,19,9,10,12,11,49,54,29,36,13,39,34]. Several higher-level cryptographic constructs, including HMAC [3,2], the 3GPP confidentiality and integrity algorithms f8,f9 [35], and RMAC [37,41], use related keys and thus rely for their (standard, not RKA) security on RKA-security of the underlying compression function or blockcipher.

The most direct use of RKA-security is for very cheap, simple and natural ways to rekey or tweak block ciphers. Subkeys of $K$ for use with modes of operation of a blockcipher $E$ might be derived in standard usage via $E_K(\Delta_1), E_K(\Delta_2), \ldots$ where $\Delta_1, \Delta_2, \ldots$ are constants. If $E$ is a RKA-PRP one can just use instead $K * \Delta_1, K * \Delta_2, \ldots$, where $*$ is a group operation, saving many blockcipher operations. On the other hand if $E$ is a $\mathsf{rkd}[\mathcal{K}, *]$-RKA-PRP, then $F_K^T(x) = E_{K*T}(x)$ is shown in [5] to be a tweakable blockcipher, a primitive that has proven to be of great importance both conceptually and in applications [44,51].

More designs would probably use related keys if it were possible to do so safely. Non-expert (in practice, most!) designers do it anyway, making RKA-security, in the words of Biryukov, Dunkelman, Keller, Khovratovich and Shamir [15], central to abuse-resistant cryptography.

Beyond this, RKA-security provides resistance to fault injection attacks [23,14] where the attacker can inject faults that change bits of a hardware-stored key and observe the outputs of the cryptographic primitive under the modified key, putting RKAs under the umbrella of sidechannel attacks. This sidechannel connection is captured by the tamper-proof security model of Gennaro, Lysyanskaya, Malkin, Micali and Rabin [31]. (They were apparently not aware of the prior model of [5] and the cryptanalytic literature on RKAs. We hope our current paper helps connect these two lines of work.)

Overall, the motivation for the theoretical study of RKA-security is not just powerful but unusual in coming from so many different parts of cryptography, namely foundations, cryptanalysis, protocol design and resistance to sidechannel attacks.

## 2    Basic Definitions

A family of functions $F \colon \mathcal{K} \times \mathcal{D} \to \mathcal{R}$ takes a key $K \in \mathcal{K}$ and input $x \in \mathcal{D}$ and returns an output $F_K(x) = F(K, x) \in \mathcal{R}$. Let $\mathsf{FF}(\mathcal{K}, \mathcal{D}, \mathcal{R})$ be the set of all families of functions $F \colon \mathcal{K} \times \mathcal{D} \to \mathcal{R}$. For sets $X, Y$ let $\mathsf{Fun}(X, Y)$ be the set of all functions mapping $X$ to $Y$. If $S$ is a (finite) set then $s \xleftarrow{\$} S$ denotes the operation of picking $s$ from $S$ at random and $|S|$ is the size of $S$. We denote by $y \xleftarrow{\$} A(x_1, x_2, \ldots)$ the operation of running randomized algorithm $A$ on inputs $x_1, x_2, \ldots$ and fresh coins and letting $y$ denote the output. If $\mathbf{v}$ is a vector then $|\mathbf{v}|$ denotes the number of its coordinates and $\mathbf{v}[i]$ denotes its $i$-th coordinate, meaning $\mathbf{v} = (\mathbf{v}[1], \ldots, \mathbf{v}[|\mathbf{v}|])$. A (binary) string $x$ is identified with a vector over $\{0, 1\}$ so that $|x|$ is its length and $x[i]$ is its $i$-th bit. If $F\colon \mathcal{K} \times \mathcal{D} \to \mathcal{R}$ is a family of functions and $\mathbf{x}$ is a vector over $\mathcal{D}$ then $F(K, \mathbf{x})$ denotes the vector $(F(K, \mathbf{x}[1]), \ldots, F(K, \mathbf{x}[|\mathbf{x}|]))$. Read the term "efficient" as meaning "polynomial-time" in the natural asymptotic extension of our concrete framework.

GAMES. Some of our definitions and proofs are expressed via code-based games [6]. Recall that such a game —see Fig. 1 for an example— consists of an (optional) INITIALIZE procedure and procedures to respond to adversary oracle queries. A game $G$ is executed with an adversary $A$ as follows. First, INITIALIZE (if present) executes. Then $A$ executes, its oracle queries being answered by the corresponding procedures of $G$. When $A$ terminates, its output, denoted $G^A$, is called the output of the game, and we let "$G^A \Rightarrow 1$" denote the event that this game output takes value 1. Boolean flags are assumed initialized to false. The running time of an adversary by convention is the worst case time for the execution of the adversary with any of the games defining its security, so that the time of the called game procedures is included. When (as often) we describe a game in text and say the game "begins" by doing something, we are describing how INITIALIZE works.

PRFs. The advantage of an adversary $A$ in attacking the (standard) prf security of a family of functions $F\colon \mathcal{K} \times \mathcal{D} \to \mathcal{R}$ is defined via

$$\mathbf{Adv}_F^{\mathrm{prf}}(A) = \Pr\left[\mathrm{PRFReal}_F^A \Rightarrow 1\right] - \Pr\left[\mathrm{PRFRand}_F^A \Rightarrow 1\right]. \qquad (2)$$

Game $\mathrm{PRFReal}_F$ begins by picking $K \xleftarrow{\$} \mathcal{K}$ and responds to oracle query $\mathrm{FN}(x)$ via $F(K, x)$. Game $\mathrm{PRFRand}_F$ begins by picking $f \xleftarrow{\$} \mathsf{Fun}(\mathcal{D}, \mathcal{R})$ and responds to oracle query $\mathrm{FN}(x)$ via $f(x)$.

RKA-PRFs. We recall definitions from [5]. Let $F\colon \mathcal{K} \times \mathcal{D} \to \mathcal{R}$ be a family of functions and $\Phi \subseteq \mathsf{Fun}(\mathcal{K}, \mathcal{K})$. The members of $\Phi$ are called RKD (related-key deriving) functions. An adversary is said to be $\Phi$-restricted if its oracle queries $(\phi, x)$ satisfy $\phi \in \Phi$. The advantage of a $\Phi$-restricted adversary $A$ in attacking the prf-rka security of $F$ is defined via

$$\mathbf{Adv}_{\Phi,F}^{\mathrm{prf\text{-}rka}}(A) = \Pr\left[\mathrm{RKPRFReal}_F^A \Rightarrow 1\right] - \Pr\left[\mathrm{RKPRFRand}_F^A \Rightarrow 1\right]. \quad (3)$$

Game $\mathrm{RKPRFReal}_F$ begins by picking $K \xleftarrow{\$} \mathcal{K}$ and responds to oracle query $\mathrm{RKFN}(\phi, x)$ via $F(\phi(K), x)$. Game $\mathrm{RKPRFRand}_F$ begins by picking $K \xleftarrow{\$} \mathcal{K}$ and $G \xleftarrow{\$} \mathsf{FF}(\mathcal{K}, \mathcal{D}, \mathcal{R})$, and responds to oracle query $\mathrm{RKFN}(\phi, x)$ via $G(\phi(K), x)$.

CR HASH FUNCTIONS. The advantage of $C$ in attacking the cr (collision-resistance) security of $H\colon \mathcal{D} \to \mathcal{R}$ is

$$\mathbf{Adv}_H^{\mathrm{cr}}(C) = \Pr\left[x \neq x' \text{ and } H(x) = H(x')\right]$$

where the probability is over $(x, x') \xleftarrow{\$} C$. For simplicity and to better reflect practice, we view hash functions as unkeyed. This means there always *exists* an efficient $C$ whose cr-advantage is 1, but that does not mean we can find it, and our results remain meaningful because the proofs give *explicit* constructions of cr-adversaries from other adversaries [52]. We could extend our treatment to let hash functions be families, which would be more rigorous. We can't make the hash key part of the PRF key because then it would be subject to the RKA, but since its secrecy is not needed for security, we can make it a public parameter. Thus, keyed hash functions require an extended syntax for function families in which functions in the family depended on a public parameter, and we have chosen to avoid this.

# 3  Constructions of RKA-PRFs and RKA-PRPs

In this section we describe and analyze our RKA-PRF constructions. We begin by defining the key-malleability and key fingerprint notions on which the general construction is based. Theorem 1 states the general construction and proves its security. Then we show how to instantiate the general construction to obtain DDH based RKA-PRFs for group-induced classes as well as other RKA-PRFs.

KEY-MALLEABILITY. Suppose $M\colon \mathcal{K} \times \mathcal{D} \to \mathcal{R}$ is a family of functions and $\Phi \subseteq \mathsf{Fun}(\mathcal{K}, \mathcal{K})$ is a set of RKD functions. Suppose T is a deterministic algorithm that given an oracle $f\colon \mathcal{D} \to \mathcal{R}$ and inputs $(\phi, x) \in \Phi \times \mathcal{D}$ returns a point $\mathsf{T}^f(\phi, x) \in \mathcal{R}$. We say that T is a *key-transformer* for $(M, \Phi)$ if it satisfies two conditions. The first, called *correctness*, asks that $M(\phi(K), x) = \mathsf{T}^{M(K, \cdot)}(\phi, x)$ for every $(\phi, K, x) \in \Phi \times \mathcal{K} \times \mathcal{D}$. This is a relatively straightforward condition saying that one can compute $M(\phi(K), x)$ from $\phi, x$ if one has an oracle for $M(K, \cdot)$. The second condition, called *uniformity*, is more subtle. Roughly, it says that if the oracle provided to T is random then the outputs of T on any input sequence $(\phi_1, x_1), \ldots, (\phi_q, x_q)$ are uniformly and independently distributed *as long as $x_1, \ldots, x_q$ are distinct*. Formally, game KTReal$_\mathsf{T}$ begins by picking $f \xleftarrow{\$} \mathsf{Fun}(\mathcal{D}, \mathcal{R})$ and responds to oracle query KTFN$(\phi, x)$ via $\mathsf{T}^f(\phi, x)$ while game KTRand$_\mathsf{T}$ makes no initializations and responds to oracle query KTFN$(\phi, x)$ by picking and returning a random point in $\mathcal{R}$. Let us say a $\Phi$-restricted adversary is *unique input* if, in its oracle queries $(\phi_1, x_1), \ldots, (\phi_q, x_q)$, the points $x_1, \ldots, x_q$ are always distinct, where by "always" we mean with probability one regardless of how oracle queries are answered and what are the coins of the adversary. The uniformity requirement is that

$$\Pr\left[\mathrm{KTReal}_\mathsf{T}^U \Rightarrow 1\right] = \Pr\left[\mathrm{KTRand}_\mathsf{T}^U \Rightarrow 1\right] \tag{4}$$

for every unique-input $\Phi$-restricted adversary $U$ against the uniformity of T. We say $M$ is $\Phi$-*key-malleable* if there exists an efficient key transformer for $(M, \Phi)$.

That key-malleability might be useful to obtain RKA-PRFs is, on the one hand, intuitive, because the correctness property clearly allows us to simulate queries to $M(\phi(K), \cdot)$ via queries to $M(K, \cdot)$. It is, on the other hand, counter-intuitive, because the same correctness property immediately yields an attack showing that $M$ is *not* a $\Phi$-RKA-PRF as long as $\Phi$ contains the identity function id and a function $\phi$ satisfying $\phi(K) \neq K$ for all $K \in \mathcal{K}$, conditions met by any group-induced $\Phi$. Indeed, consider $\Phi$-restricted adversary $A$ that, for some $x \in \mathcal{D}$, makes query $y \leftarrow \text{RKFN}(\phi, x)$. Then it runs $\mathsf{T}$ on inputs $\phi, x$ to get an output $z$, answering any oracle query $w$ made in this computation by $\text{RKFN}(\text{id}, w)$. It returns 1 if $y = z$ and 0 otherwise. Correctness says that $A$ always returns 1 in game $\text{RKPRFReal}_M$. But the assumption on $\phi$ implies that $A$ returns 1 with probability at most $1/|\mathcal{R}|$ in game $\text{RKPRFRand}_M$. So $\mathbf{Adv}^{\text{prf-rka}}_{M,\Phi}(A)$ is almost 1.

Although a key-malleable $M$ is not a $\Phi$-RKA-PRF, one can show that it is an RKA-PRF versus unique-input adversaries. (The adversary of the above attack need not be unique-input.) This leaves two questions. The first is how to bridge the gap to arbitrary adversaries, which we do via the concept of key fingerprints discussed below. The second is how to obtain key-malleable PRFs, which we will do later via the Naor-Reingold [47] and Lewko-Waters [43] constructs.

KEY FINGERPRINTS. Suppose $M \colon \mathcal{K} \times \mathcal{D} \to \mathcal{R}$ is a family of functions and $\Phi \subseteq \mathsf{Fun}(\mathcal{K}, \mathcal{K})$ is a set of RKD functions. Let $\mathbf{w}$ be vector over $\mathcal{D}$ and let $m = |\mathbf{w}|$. We say that $\mathbf{w}$ is a *key fingerprint* for $(M, \Phi)$ if

$$(M(\phi(K), \mathbf{w}[1]), \ldots, M(\phi(K), \mathbf{w}[m]))$$
$$\neq (M(\phi'(K), \mathbf{w}[1]), \ldots, M(\phi'(K), \mathbf{w}[m])) \tag{5}$$

for all $K \in \mathcal{K}$ and all distinct $\phi, \phi' \in \Phi$.

Let's call a class $\Phi \subseteq \mathsf{Fun}(\mathcal{K}, \mathcal{K})$ of RKD functions *claw-free* if $\phi(K) \neq \phi'(K)$ for every key $K \in \mathcal{K}$ and every distinct $\phi, \phi' \in \Phi$ [46,5]. We note that if $(M, \Phi)$ has a key fingerprint then it follows automatically that $\Phi$ is claw-free. Indeed, if there is a $K$ and $\phi, \phi'$ such that $\phi(K) = \phi'(K)$ then there can be no $\mathbf{w}$ for which Equation (5) is true. We will use this frequently below.

We say that $\mathbf{w}$ is a *strong key fingerprint* for $(M, \Phi)$ if

$$(M(K, \mathbf{w}[1]), \ldots, M(K, \mathbf{w}[m])) \neq (M(K', \mathbf{w}[1]), \ldots, M(K', \mathbf{w}[m])) \tag{6}$$

for all distinct $K, K' \in \mathcal{K}$. If $\Phi$ is claw-free then a strong key fingerprint for $(M, \Phi)$ is also a key fingerprint for $(M, \Phi)$, which we will use in analyzing our constructs. If $\Phi$ is complete —recall this means that for every $K, K' \in \mathcal{K}$ there is a $\phi \in \Phi$ such that $\phi(K) = K'$— then any key fingerprint for $(M, \Phi)$ is also a strong key fingerprint for $(M, \Phi)$. In general, however, the existence of a key fingerprint may not imply the existence of a strong key fingerprint.

CONSTRUCTION. Let $M \colon \mathcal{K} \times \mathcal{D} \to \mathcal{R}$ be a key-malleable family of functions and $\mathsf{T}$ a key transformer for $(M, \Phi)$. Let $\mathbf{w} \in \mathcal{D}^m$ be a key-fingerprint for $(M, \Phi)$. We say that a point $w \in \mathcal{D}$ is a *possible oracle query* for $\mathsf{T}$ relative to $(M, \Phi, \mathbf{w})$ if there exists $(f, \phi, i) \in \mathsf{Fun}(\mathcal{D}, \mathcal{R}) \times \Phi \times \{1, \ldots, m\}$ such that the computation $\mathsf{T}^f(\phi, \mathbf{w}[i])$ makes oracle query $w$. We let $\mathsf{Qrs}(\mathsf{T}, M, \Phi, \mathbf{w})$ be the set of all possible oracle queries $w$ for $\mathsf{T}$ relative to $(M, \Phi, \mathbf{w})$. Let $\overline{\mathcal{D}} = \mathcal{D} \times \mathcal{R}^m$. A hash function $H$ with domain $\overline{\mathcal{D}}$ is said to be *compatible* with $(\mathsf{T}, M, \Phi, \mathbf{w})$ if its range is $\mathcal{D} \setminus \mathsf{Qrs}(\mathsf{T}, M, \Phi, \mathbf{w})$. That is, possible oracle queries of $\mathsf{T}$ relative to $(M, \Phi, \mathbf{w})$ are not allowed to be outputs of $H$. With this, we can say what are the ingredients of our construction of a $\Phi$-RKA-PRF: (1) a $\Phi$-key-malleable PRF, meaning a family of functions $M\colon \mathcal{K} \times \mathcal{D} \to \mathcal{R}$ such that, on the one hand, $M$ is a PRF and, on the other hand, there exists a key transformer $\mathsf{T}$ for $(M, \Phi)$; (2) a key fingerprint $\mathbf{w}$ for $(M, \Phi)$; and (3) a collision-resistant hash function $H\colon \overline{\mathcal{D}} \to \mathcal{D} \setminus \mathsf{Qrs}(\mathsf{T}, M, \Phi, \mathbf{w})$ that is compatible with $(\mathsf{T}, M, \Phi, \mathbf{w})$. We combine them to build $F\colon \mathcal{K} \times \mathcal{D} \to \mathcal{R}$ that on input $K, x$ computes $\overline{\mathbf{w}} \leftarrow M(K, \mathbf{w})$ —recall that, as per our notational conventions, $M(K, \mathbf{w})$ is the vector whose $i$-th component is $M(K, \mathbf{w}[i])$ for $1 \le i \le m$— and then returns $M(K, H(x, \overline{\mathbf{w}}))$. The following theorem says that $F$ is a $\Phi$-RKA-PRF assuming $M$ is a PRF and $H$ is collision-resistant. No assumptions are made on $\Phi$ beyond those implied by the conditions stated here.

**Theorem 1.** *Let* $M\colon \mathcal{K} \times \mathcal{D} \to \mathcal{R}$ *be a family of functions and* $\Phi \subseteq \mathsf{Fun}(\mathcal{K}, \mathcal{K})$ *a class of RKD functions. Let* $\mathsf{T}$ *be a key-transformer for* $(M, \Phi)$ *making* $Q_\mathsf{T}$ *oracle queries, and let* $\mathbf{w} \in \mathcal{D}^m$ *be a key fingerprint for* $(M, \Phi)$. *Let* $\overline{\mathcal{D}} = \mathcal{D} \times \mathcal{R}^m$ *and let* $H\colon \overline{\mathcal{D}} \to S$ *be a hash function that is compatible with* $(\mathsf{T}, M, \Phi, \mathbf{w})$, *so that* $S = \mathcal{D} \setminus \mathsf{Qrs}(\mathsf{T}, M, \Phi, \mathbf{w})$. *Define* $F\colon \mathcal{K} \times \mathcal{D} \to \mathcal{R}$ *by*

$$F(K, x) = M(K, H(x, M(K, \mathbf{w}))) \tag{7}$$

*for all* $K \in \mathcal{K}$ *and* $x \in \mathcal{D}$. *Let* $A$ *be a* $\Phi$-*restricted adversary against the prf-rka security of* $F$ *that makes* $Q_A \le |S|$ *oracle queries. Then we can construct an adversary* $B$ *against the prf-security of* $M$ *and an adversary* $C$ *against the cr-security of* $H$ *such that*

$$\mathbf{Adv}_{\Phi, F}^{\mathrm{prf\text{-}rka}}(A) \le \mathbf{Adv}_M^{\mathrm{prf}}(B) + \mathbf{Adv}_H^{\mathrm{cr}}(C) . \tag{8}$$

*Adversary* $B$ *makes* $(m+1) \cdot Q_\mathsf{T} Q_A$ *oracle queries, and* $B$ *and* $C$ *have the same running time as* $A$. ∎

*Proof (Theorem 1).* We use the game sequence of Fig. 1, in the analysis below abbreviating by $W_i$ the event "$G_i^A \Rightarrow 1$". We assume (wlog) that $A$ never repeats an oracle query. Game $G_0$ simply instantiates game $\mathrm{RKPRFReal}_F$ of the definition of Section 2 with our construction $F$, so

$$\Pr\left[\mathrm{RKPRFReal}_F^A \Rightarrow 1\right] = \Pr[W_0] . \tag{9}$$

Game $G_1$, which does not include the boxed code, introduces some book-keeping, keeping track of hash values in a set $D$ and setting a flag bad to true if it ever sees a repeat. The book-keeping does not affect the values returned by RKFN so

$$\Pr[W_1] = \Pr[W_0] . \tag{10}$$

Game $G_2$ adds the boxed code which "corrects" a hash value repetition by picking instead a value that, being drawn from $S \setminus D$, will not repeat any previous one. The addition of this "artificial" step, leading to a game different from the "real" one, is to ensure that the values of $h$ on which $\mathsf{T}^f(\phi, h)$ is later called (lines 37,47,57) are distinct, putting us in a position to exploit the uniformity of $\mathsf{T}$ and replace the outputs by random values. This, however, is some distance away. For the moment we observe that games $G_1, G_2$ are identical until bad —differ only

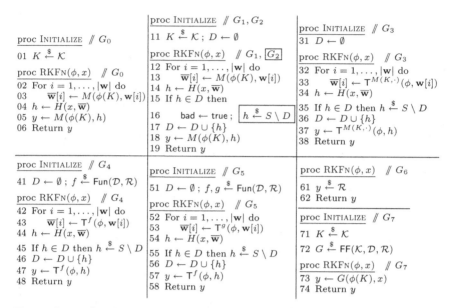

**Fig. 1.** Games for the proof of Theorem 1. Game $G_2$ includes the boxed code and game $G_1$ does not.

in code following the setting of bad to true— and hence the fundamental lemma of game playing [6] implies that

$$\Pr[W_1] \leq \Pr[W_2] + \Pr[B_1] \tag{11}$$

where $B_1$ denotes the event that the execution of $A$ with game $G_1$ sets the flag bad to true. Making crucial use of the assumption that $\mathbf{w}$ is a key fingerprint for $(M, \Phi)$, we design adversary $C$ attacking the cr-security of $H$ such that

$$\Pr[B_1] \leq \mathbf{Adv}_H^{\mathrm{cr}}(C) . \tag{12}$$

Adversary $C$ begins by picking $K \xleftarrow{\$} \mathcal{K}$ and initializing a counter $j \leftarrow 0$. It then runs $A$. When the latter makes a RKFN-query $(\phi, x)$, adversary $C$ responds via

For $i = 1, \ldots, |\mathbf{w}|$ do $\overline{\mathbf{w}}[i] \leftarrow M(\phi(K), \mathbf{w}[i])$
$j \leftarrow j + 1$ ; $\phi_j \leftarrow \phi$ ; $x_j \leftarrow x$ ; $\overline{\mathbf{w}}_j \leftarrow \overline{\mathbf{w}}$ ; $h_j \leftarrow H(x, \overline{\mathbf{w}})$ ; $y \leftarrow M(\phi(K), h)$
Return $y$

When $A$ halts, $C$ searches for $a, b$ satisfying $1 \leq a < b \leq j$ such that $h_a = h_b$ and, if it finds them, outputs $(x_a, \overline{\mathbf{w}}_a), (x_b, \overline{\mathbf{w}}_b)$ and halts. Towards justifying Equation (12) the main question is, why are $(x_a, \overline{\mathbf{w}}_a), (x_b, \overline{\mathbf{w}}_b)$ distinct? The assumption that $A$ never repeats an oracle query means that $(\phi_a, x_a) \neq (\phi_b, x_b)$. Now consider two cases. First, if $\phi_a = \phi_b$ then we must have $x_a \neq x_b$ whence of course $(x_a, \overline{\mathbf{w}}_a) \neq (x_b, \overline{\mathbf{w}}_b)$. Second, if $\phi_a \neq \phi_b$ then the assumption that $\mathbf{w}$ is a key fingerprint for $(M, \Phi)$ means, by Equation (5), that $\overline{\mathbf{w}}_a \neq \overline{\mathbf{w}}_b$ and again $(x_a, \overline{\mathbf{w}}_a) \neq (x_b, \overline{\mathbf{w}}_b)$.

In game $G_3$, we use the key transformer T, given by the assumed $\Phi$-key-malleability of $M$, to compute $M(\phi(K), \cdot)$ via oracle calls to $M(K, \cdot)$, both at line 33 and at line 37. The correctness property of the key transformer implies

$$\Pr[W_2] = \Pr[W_3] . \tag{13}$$

Game $G_4$ replaces the oracle given to T by a random function. We design adversary $B$ attacking the prf-security of $M$ such that

$$\Pr[W_3] - \Pr[W_4] \leq \mathbf{Adv}_M^{\mathrm{prf}}(B) . \tag{14}$$

This is possible because the games make only oracle access to $M(K, \cdot)$ and $f$, respectively. In detail, adversary $B$ runs $A$. When the latter makes a RKFN-query $(\phi, x)$, adversary $B$ responds via

For $i = 1, \ldots, |\mathbf{w}|$ do $\overline{\mathbf{w}}[i] \leftarrow \mathsf{T}^{\mathrm{FN}}(\phi, \mathbf{w}[i])$ ; $h \leftarrow H(x, \overline{\mathbf{w}})$ ; $y \leftarrow \mathsf{T}^{\mathrm{FN}}(\phi, h)$
Return $y$

where FN is $B$'s own oracle. When $A$ halts, $B$ halts with the same output. Then

$$\Pr\left[ \mathrm{PRFReal}_M^B \Rightarrow 1 \right] = \Pr[W_3] \quad \text{and} \quad \Pr\left[ \mathrm{PRFRand}_M^B \Rightarrow 1 \right] = \Pr[W_4]$$

so Equation (14) follows from Equation (2).

Rather than return $y = \mathsf{T}^f(\phi, h)$ as at lines 47,48, we would like to pick and return a random $y$, as at lines 61,62 of game $G_6$, saying this makes no difference by the uniformity of T. But we have to be careful, because line 47 is not the only place $f$ is used in $G_4$. Oracle $f$ is also being queried in the computation $\mathsf{T}^f(\phi, \mathbf{w}[i])$ at line 43, and if a $f$-query made here equals an input $h$ at line 47, then it is unclear we can argue randomness of the line 47 output $y$ based on the uniformity of T. The assumed compatiblity of $H$ with $(\mathsf{T}, M, \Phi, \mathbf{w})$ comes to the rescue. It says the queries to $f$ in the computation $\mathsf{T}^f(\phi, \mathbf{w}[i])$ at line 43, which fall within the set $\mathsf{Qrs}(\mathsf{T}, M, \Phi, \mathbf{w})$, are not in the set $S$ that is the range of $H$. Thus, the calls to $f$ at lines 43 and 47 can be answered with different, independent random functions without affecting the distribution of the procedure output. In other words, considering game $G_5$, which switches $f$ to $g$ at line 53 but not at line 57, the compatibility of $H$ with $(\mathsf{T}, M, \Phi, \mathbf{w})$ implies that

$$\Pr[W_4] = \Pr[W_5] . \tag{15}$$

We will now exploit the uniformity of T to show that

$$\Pr[W_5] = \Pr[W_6] . \tag{16}$$

To do this we design *unique-input* $\Phi$-restricted adversary $U$ against the uniformity of T such that

$$\Pr\left[ \mathrm{KTReal}_M^U \Rightarrow 1 \right] = \Pr[W_5] \quad \text{and} \quad \Pr\left[ \mathrm{KTRand}_M^U \Rightarrow 1 \right] = \Pr[W_6] . \tag{17}$$

Equation (16) follows from Equation (4). Adversary $U$ begins by initializing set $D \leftarrow \emptyset$ and picking $g \xleftarrow{\$} \mathsf{Fun}(\mathcal{D}, \mathcal{R})$. (Adversary $U$ of the uniformity condition is not required to be efficient so picking $g$ like this is okay but in any case we could make $U$ efficient if we liked by simulating $g$ via lazy sampling rather than picking it upfront.) It then runs $A$. When the latter makes a RKFN-query $(\phi, x)$, adversary $U$ responds via

> For $i = 1, \ldots, |\mathbf{w}|$ do $\overline{\mathbf{w}}[i] \leftarrow \mathsf{T}^g(\phi, \mathbf{w}[i])$
> $j \leftarrow j + 1$ ; $\phi_j \leftarrow \phi$ ; $h_j \leftarrow H(x, \overline{\mathbf{w}})$ ; If $h_j \in D$ then $h_j \xleftarrow{\$} S \setminus D$
> $y \leftarrow \mathrm{KTFN}(\phi_j, h_j)$ ; Return $y$

where KTFN is $U$'s own oracle. The delicate question is, why is $U$ unique-input? The boxed code introduced at line 16, carried through to line 55, and reflected by the "If" statement in the code for $U$ above, ensures that $h_1, \ldots, h_j$ are all distinct at the end of $U$'s computation as long as $Q_A \leq |S|$, which the theorem assumed. Equation (17) follows.

The claw-freeness of $\Phi$ —recall this follows from the assumption that $(M, \Phi)$ has a key fingerprint— implies that if $(\phi, x) \neq (\phi', x')$ then $(\phi(K), x) \neq (\phi'(K), x')$. This together with the assumption that $A$ does not repeat an oracle query imply

$$\Pr[W_6] = \Pr[W_7] = \Pr\left[ \mathrm{RKPRFRand}_F^A \Rightarrow 1 \right]. \qquad (18)$$

Equation (8) follows from Equations (9), (10), (11), (12), (13), (14), (15), (16), (18), (3).

OUR MULTIPLICATIVE DDH-BASED RKA-PRF. We instantiate our general construction to get a DDH-based $\Phi$-RKA-PRF where $\Phi$ is group induced. Let $\mathbb{G}$ be a (multiplicatively written) group of prime order $p$, and let $g \in \mathbb{G}$ be an arbitrary generator of $\mathbb{G}$. The classic Naor-Reingold [47] PRF NR: $\mathbb{Z}_p^{n+1} \times \{0, 1\}^n \to \mathbb{G}$ is defined via

$$\mathrm{NR}(\mathbf{a}, x) = g^{\mathbf{a}[0] \prod_{i=1}^n \mathbf{a}[i]^{x[i]}} \qquad (19)$$

for all $\mathbf{a} \in \mathbb{Z}_p^{n+1}$ and $x \in \{0, 1\}^n$. Recall the advantage of an adversary $B$ against the DDH problem in $\mathbb{G}$ is

$$\mathbf{Adv}_{\mathbb{G}}^{\mathrm{ddh}}(B) = \Pr\left[ B(g^a, g^b, g^{ab}) \Rightarrow 1 \right] - \Pr\left[ B(g^a, g^b, g^c) \Rightarrow 1 \right],$$

where the probabilities are over $a, b, c \xleftarrow{\$} \mathbb{Z}_p^*$. The following result of [47] says that NR is a PRF if DDH is hard in $\mathbb{G}$.

**Lemma 2.** [47] *Let* $\mathbb{G} = \langle g \rangle$ *be a group of prime order $p$ and* NR: $\mathbb{Z}_p^{n+1} \times \{0, 1\}^n \to \mathbb{G}$ *the family of functions defined via Equation (19). Let $A$ an adversary against the prf-security of* NR *that makes $Q$ oracle queries. Then we can construct an adversary $B$ against the DDH problem in $\mathbb{G}$ such that*

$$\mathbf{Adv}_{\mathrm{NR}}^{\mathrm{prf}}(A) \leq n \cdot \mathbf{Adv}_{\mathbb{G}}^{\mathrm{ddh}}(B). \qquad (20)$$

*The running time of $B$ is that of $A$ plus the time required for $\mathcal{O}(Q)$ exponentiations in $\mathbb{G}$.* ∎

Group-induced class: Define operation $*$ by $\mathbf{a} * \mathbf{d} = (\mathbf{a}[0]\mathbf{d}[0], \ldots, \mathbf{a}[n]\mathbf{d}[n])$ where operations on components are multiplications modulo $p$. Then the set $\mathcal{K} = (\mathbb{Z}_p^*)^{n+1}$ is a group under $*$. Let $\phi_{\mathbf{d}}^*: \mathcal{K} \to \mathcal{K}$ be defined by $\phi_{\mathbf{d}}^*(\mathbf{a}) = \mathbf{a} * \mathbf{d}$

for all $\mathbf{a}, \mathbf{d} \in \mathcal{K}$. Let $\varPhi = \mathsf{rkd}[(\mathbb{Z}_p^*)^{n+1}, *]$ be the class of all $\phi_{\mathbf{d}}^*$ as $\mathbf{d}$ ranges over $\mathcal{K}$. This class is group-induced, the group being $(\mathcal{K}, *)$.

Key malleability: We claim that NR is $\varPhi$-key-malleable. The key-transformer $\mathsf{T}$, given oracle $f \colon \{0,1\}^n \to \mathbb{G}$ and inputs $\phi_{\mathbf{d}}^*, x$, returns $f(x)^{\mathbf{d}[0] \prod_{i=1}^n \mathbf{d}[i]^{x[i]}}$. Correctness holds because

$$\mathsf{T}^{\mathrm{NR}(\mathbf{a}, \cdot)}(\phi_{\mathbf{d}}^*, x) = \mathrm{NR}(\mathbf{a}, x)^{\mathbf{d}[0] \prod_{i=1}^n \mathbf{d}[i]^{x[i]}} = \mathrm{NR}(\mathbf{a} * \mathbf{d}, x) \,.$$

In game $\mathrm{KTReal}_{\mathrm{NR}}$, the responses received by unique-input, $\varPhi$-restricted adversary $U$ to KTFN-queries $(\phi_{\mathbf{d}_1}^*, x_1), \ldots, (\phi_{\mathbf{d}_q}^*, x_q)$ are $f(x_1)^{\mathbf{d}_1[0] \prod_{i=1}^n \mathbf{d}_1[i]^{x[i]}}, \ldots,$ $f(x_q)^{\mathbf{d}_q[0] \prod_{i=1}^n \mathbf{d}_q[i]^{x[i]}}$ where $f \xleftarrow{\$} \mathsf{Fun}(\{0,1\}^n, \mathbb{G})$ was chosen by the game. Since $x_1, \ldots, x_q$ are distinct and the exponents are non-zero, these responses are randomly and independently distributed over $\mathbb{G}$. We have verified the uniformity condition.

Key fingerprint: For $i = 1, \ldots, n$ let $\mathbf{w}[i] = 0^{i-1} \| 1 \| 0^{n-i}$ be the string that is all zeros except at position $i$, where it has a one. Let $\mathbf{w}[0] = 0^n$. We claim that $\mathbf{w}$ is a strong key fingerprint for $(\mathrm{NR}, \varPhi)$. To see this, first note that $(\mathrm{NR}(\mathbf{a}, \mathbf{w}[0]), \mathrm{NR}(\mathbf{a}, \mathbf{w}[1]) \ldots, \mathrm{NR}(\mathbf{a}, \mathbf{w}[n])) = (g^{\mathbf{a}[0]}, g^{\mathbf{a}[0]\mathbf{a}[1]}, \ldots, g^{\mathbf{a}[0]\mathbf{a}[n]})$. Now if $\mathbf{a}, \mathbf{a}' \in \mathcal{K}$ are distinct keys and $\mathbf{a}[0] \neq \mathbf{a}'[0]$ then $g^{\mathbf{a}[0]} \neq g^{\mathbf{a}'[0]}$. On the other hand if $\mathbf{a}[0] = \mathbf{a}'[0]$ and $\mathbf{a}[i] \neq \mathbf{a}'[i]$ for some $i > 0$, then $g^{\mathbf{a}[0]\mathbf{a}[i]} \neq g^{\mathbf{a}'[0]\mathbf{a}'[i]}$. The claim follows from the definition of Equation (6) with $M = \mathrm{NR}$. Since $\varPhi$ is claw-free, $\mathbf{w}$ is also a key fingerprint for $(\mathrm{NR}, \varPhi)$, satisfying Equation (5) with $M = \mathrm{NR}$.

Compatible hash function: The set of possible oracle queries of $\mathsf{T}$ relative to $(\mathrm{NR}, \varPhi, \mathbf{w})$ is $\mathsf{Qrs}(\mathsf{T}, \mathrm{NR}, \varPhi, \mathbf{w}) = \{\, \mathbf{w}[i] \ : \ 0 \leq i \leq n \,\}$ because on inputs $\phi, x$ the only oracle query made by $\mathsf{T}$ is $x$ itself. Let $\overline{\mathcal{D}} = \{0,1\}^n \times \mathbb{G}^{n+1}$. If $h \colon \overline{\mathcal{D}} \to \{0,1\}^{n-2}$ is collision resistant, then $H \colon \overline{\mathcal{D}} \to \{0,1\}^n \setminus \mathsf{Qrs}(\mathsf{T}, \mathrm{NR}, \varPhi, \mathbf{w})$ defined by $H(x, \mathbf{z}) = 11 \| h(x, \mathbf{z})$ is collision resistant and compatible with $(\mathsf{T}, \mathrm{NR}, \varPhi, \mathbf{w})$ because all members of $\mathsf{Qrs}(\mathsf{T}, \mathrm{NR}, \varPhi, \mathbf{w})$ have Hamming weight at most 1 while outputs of $H$ have Hamming weight at least 2.

We have all the ingredients. The following theorem combines the above with Theorem 1 and Lemma 2 to present our DDH-based $\varPhi$-RKA-PRF for group-induced $\varPhi$ and specify its security.

**Theorem 3.** *Let $\mathbb{G} = \langle g \rangle$ be a group of prime order $p$ and* $\mathrm{NR} \colon \mathbb{Z}_p^{n+1} \times \{0,1\}^n \to \mathbb{G}$ *the family of functions defined via Equation (19). Let $\overline{\mathcal{D}} = \{0,1\}^n \times \mathbb{G}^{n+1}$ and let $h \colon \overline{\mathcal{D}} \to \{0,1\}^{n-2}$ be a hash function. Define $F \colon (\mathbb{Z}_p^*)^{n+1} \times \{0,1\}^n \to \mathbb{G}$ by*

$$F(\mathbf{a}, x) = \mathrm{NR}(\mathbf{a}, 11 \| h(x, (g^{\mathbf{a}[0]}, g^{\mathbf{a}[0]\mathbf{a}[1]}, \ldots, g^{\mathbf{a}[0]\mathbf{a}[n]}))) $$

*for all $\mathbf{a} \in (\mathbb{Z}_p^*)^{n+1}$ and $x \in \{0,1\}^n$. Let $\varPhi = \mathsf{rkd}[(\mathbb{Z}_p^*)^{n+1}, *]$ where $*$ is the operation of component-wise multiplication modulo $p$. Let $A$ be a $\varPhi$-restricted adversary against the prf-rka security of $F$ that makes $Q_A \leq 2^{n-2}$ oracle queries. Then we can construct an adversary $B$ against the DDH problem in $\mathbb{G}$ and an adversary $C$ against the cr-security of $h$ such that*

$$\mathbf{Adv}_{\varPhi, F}^{\mathrm{prf\text{-}rka}}(A) \leq n \cdot \mathbf{Adv}_{\mathbb{G}}^{\mathrm{ddh}}(B) + \mathbf{Adv}_h^{\mathrm{cr}}(C) \,.$$

*The running time of $B$ is that of $A$ plus the time required for $\mathcal{O}(nQ)$ exponentiations in $\mathbb{G}$. $C$ has the same running time as $A$.* ∎

A DLIN-BASED RKA-PRF. There are groups where DDH is easy but the DLIN problem [22] still seems hard, which motivated Lewko and Waters [43] to find a DLIN-based PRF. In the same vein, we seek a DLIN-based RKA-PRF.

Let $\mathbb{G} = \langle g \rangle$ be a group of prime order $p$. Lewko and Waters [43] describe their DLIN-based PRF as having key a randomly chosen tuple $(y_0, z_0, y_1, z_1, w_1, v_1,$ $\ldots, y_n, z_n, w_n, v_n) \in \mathbb{Z}_p^{4n+2}$ and then on input $x \in \{0,1\}^n$ computing its output as follows. Set $a \leftarrow y_0$ ; $b \leftarrow z_0$ and then for $i = 1, \ldots, n$ execute "If $x[i] = 1$ then $a \leftarrow ay_i + bz_i$ ; $b \leftarrow aw_i + bv_i$." Finally, return $g^a$. Lewko and Waters [43, Section 1] comment that "the additional complexity required to accomodate the weaker assumptions means that our functions can no longer be described by closed-form formulas like ...," referring, in the "...," to the formula for NR that we have given as Equation (19). We provide such a closed-form formula based on matrices. (This will put us in a position, via a slight modification of the construction, to apply Theorem 1 and obtain a RKA-PRF.) Let $AL_2(p)$ denote the set of all 2 by 2 matrices over $\mathbb{Z}_p$. If $\mathbf{M} \in AL_2(p)$ and $b \in \{0,1\}$ then $\mathbf{M}^b$ is the identity matrix if $b = 0$ and is of course just $\mathbf{M}$ if $b = 1$. If $\mathbf{u} = (\mathbf{u}[1], \mathbf{u}[2])$ is a 2-vector over $\mathbb{Z}_p$ then $\mathbf{u} \cdot \mathbf{M}$ denotes the 2-vector obtained by the vector-matrix product in which $\mathbf{u}$ is viewed as a 1 by 2 matrix. We define LW: $AL_2(p)^{n+1} \times \{0,1\}^n \to \mathbb{G}$ via

$$\mathrm{LW}(\mathbf{A}, x) = g^{\mathbf{y}[1]} \quad \text{where} \quad \mathbf{y} = (1,0) \cdot \mathbf{A}[0] \prod_{i=1}^n \mathbf{A}[i]^{x[i]} \quad (21)$$

for all $\mathbf{A} \in AL_2(p)^{n+1}$ and $x \in \{0,1\}^n$. Here the key is an $(n+1)$-vector $\mathbf{A} = (\mathbf{A}[0], \ldots, \mathbf{A}[n])$ of 2 by 2 matrices over $\mathbb{Z}_p$. The formula left-multiplies the matrix product by the 2-vector $(1,0)$ to get a 2-vector $\mathbf{y}$ whose first component $\mathbf{y}[1]$ becomes the exponent to which $g$ is raised to get the function output. We claim LW is exactly the function described by the code above. (To verify this it helps to recall that matrix multiplication is associative. Strictly speaking the LW key is longer, being $4n+4$ elements of $\mathbb{Z}_p$, but the second row of $\mathbf{A}[0]$ is effectively unused due to the product with $(1,0)$ so the effective key is $4n + 2$ points in $\mathbb{Z}_p$, as in the original construct.) Comparing with Equation (19), the closed-form formulation of Equation (21) makes clearer how LW is an analogue of NR.

To obtain a key-malleable PRF admitting a key fingerprint, we need two modifications. (The modifications are in fact to get the key fingerprint, not the key malleability.) First, we restrict the keyspace, drawing the matrices from $GL_2(p) \subset AL_2(p)$ rather than $AL_2(p)$, where $GL_2(p)$ is the set of invertible matrices in $AL_2(p)$, usually referred to as the general linear group. Second, if $\mathbf{y}[1] = 0$, we use $\mathbf{y}[2]$, which we will be able to guarantee is not 0 in this case, in its place. In detail, define LW*: $GL_2(p)^{n+1} \times \{0,1\}^n \to \mathbb{G}$ via

$$\mathrm{LW}^*(\mathbf{A}, x) = \begin{cases} g^{\mathbf{y}[1]} & \text{if } \mathbf{y}[1] \neq 0 \\ g^{\mathbf{y}[2]} & \text{otherwise} \end{cases} \quad \text{where} \quad \mathbf{y} = (1,0) \cdot \mathbf{A}[0] \prod_{i=1}^n \mathbf{A}[i]^{x[i]} \quad (22)$$

for all $\mathbf{A} \in GL_2(p)^{n+1}$ and $x \in \{0,1\}^n$. In [4] we show how to apply our framework of Theorem 1 with LW* in the role of $M$, obtaining our DLIN-based RKA-PRF.

AN ADDITIVE DDH-BASED RKA-PRF. The group $(\mathbb{Z}_p^*)^{n+1}$ of our multiplicative DDH-based RKA-PRF is multiplicative. Our final construct is a $\mathsf{rkd}[\mathbb{Z}_p^n, +]$-RKA-PRF, still DDH-based, where $+$ is componentwise addition, so that the group is additive. This takes more work than the multiplicative construct. We were not able to work with NR itself but had to slightly modify it. The key transformer and proof of uniformity are more complex. The materiel is in [4].

FROM RKA-PRFs TO RKA-PRPs. Cryptanalytic interest has mostly been in RKA-secure blockciphers, meaning, families of permutations. It is not clear how one might directly modify the constructions of Section 3, which are families of functions, to make them families of permutations. We use, instead, a simple but powerful composition approach that produces a $\Phi$-RKA-PRP from a given $\Phi$-RKA-PRG and an ordinary PRP. We obtain appropriate $\Phi$-RKA-PRGs by combining our $\Phi$-RKA-PRFs with deterministic extractors [26,30,24]. This approach not only yields RKA-secure PRPs under chosen-plaintext attack (CPA) but even under chosen-ciphertext attack (CCA). In [4] we provide formal definitions for PRPs, RKA-PRPs and RKA-PRGs and state and prove the composition results.

## Acknowledgments

We thank Xavier Boyen and Damien Vergnaud for explaining that the q-BDHI-based proof of PRF-security of BBDY from [27,21] only works when the domain of the function is small, rendering abortive our attempts to obtain RKA-security via this function. We thank Mira Belenkiy for suggesting that BBDY might help for RKA-security in the first place. We thank Mira and Tolga Acar for discussions on the practical relevance of RKAs that rekindled the first author's interest in this area.

## References

1. Applebaum, B.: Fast cryptographic primitives based on the hardness of decoding random linear code. Technical Report TR-845-08, Princeton University (2008)
2. Bellare, M.: New proofs for NMAC and HMAC: Security without collision-resistance. In: Dwork, C. (ed.) CRYPTO 2006. LNCS, vol. 4117, pp. 602–619. Springer, Heidelberg (2006)
3. Bellare, M., Canetti, R., Krawczyk, H.: Keying hash functions for message authentication. In: Koblitz, N. (ed.) CRYPTO 1996. LNCS, vol. 1109, pp. 1–15. Springer, Heidelberg (1996)
4. Bellare, M., Cash, D.: Pseudorandom functions and permutations provablysecure against related key attacks. Cryptology ePrint Archive (2010) (full version of this abstract)
5. Bellare, M., Kohno, T.: A theoretical treatment of related-key attacks: RKA-PRPs, RKA-PRFs, and applications. In: Biham, E. (ed.) EUROCRYPT 2003. LNCS, vol. 2656, pp. 491–506. Springer, Heidelberg (2003)
6. Bellare, M., Rogaway, P.: The security of triple encryption and a framework for code-based game-playing proofs. In: Vaudenay, S. (ed.) EUROCRYPT 2006. LNCS, vol. 4004, pp. 409–426. Springer, Heidelberg (2006)

7. Biham, E.: New types of cryptoanalytic attacks using related keys (extended abstract). In: Helleseth, T. (ed.) EUROCRYPT 1993. LNCS, vol. 765, pp. 398–409. Springer, Heidelberg (1994)

8. Biham, E.: New types of cryptanalytic attacks using related keys. Journal of Cryptology 7(4), 229–246 (1994)

9. Biham, E., Dunkelman, O., Keller, N.: Related-key boomerang and rectangle attacks. In: Cramer, R. (ed.) EUROCRYPT 2005. LNCS, vol. 3494, pp. 507–525. Springer, Heidelberg (2005)

10. Biham, E., Dunkelman, O., Keller, N.: A related-key rectangle attack on the full KASUMI. In: Roy, B. (ed.) ASIACRYPT 2005. LNCS, vol. 3788, pp. 443–461. Springer, Heidelberg (2005)

11. Biham, E., Dunkelman, O., Keller, N.: Related-key impossible differential attacks on 8-round AES-192. In: Pointcheval, D. (ed.) CT-RSA 2006. LNCS, vol. 3860, pp. 21–33. Springer, Heidelberg (2006)

12. Biham, E., Dunkelman, O., Keller, N.: A simple related-key attack on the full SHACAL-1. In: Abe, M. (ed.) CT-RSA 2007. LNCS, vol. 4377, pp. 20–30. Springer, Heidelberg (2006)

13. Biham, E., Dunkelman, O., Keller, N.: A unified approach to related-key attacks. In: Nyberg, K. (ed.) FSE 2008. LNCS, vol. 5086, pp. 73–96. Springer, Heidelberg (2008)

14. Biham, E., Shamir, A.: Differential fault analysis of secret key cryptosystems. In: Kaliski Jr., B.S. (ed.) CRYPTO 1997. LNCS, vol. 1294, pp. 513–525. Springer, Heidelberg (1997)

15. Biryukov, A., Dunkelman, O., Keller, N., Khovratovich, D., Shamir, A.: Key recovery attacks of practical complexity on AES variants with up to 10 rounds. In: Gilbert, H. (ed.) EUROCRYPT 2010. LNCS, vol. 6110, pp. 299–319. Springer, Heidelberg (2010)

16. Biryukov, A., Khovratovich, D.: Related-key cryptanalysis of the full AES-192 and AES-256. In: Matsui, M. (ed.) ASIACRYPT 2009. LNCS, vol. 5912, pp. 1–18. Springer, Heidelberg (2009)

17. Biryukov, A., Khovratovich, D., Nikolic, I.: Distinguisher and related-key attack on the full AES-256. In: Halevi, S. (ed.) CRYPTO 2009. LNCS, vol. 5677, pp. 231–249. Springer, Heidelberg (2009)

18. Black, J.: The ideal-cipher model, revisited: An uninstantiable blockcipher-based hash function. In: Robshaw, M.J.B. (ed.) FSE 2006. LNCS, vol. 4047, pp. 328–340. Springer, Heidelberg (2006)

19. Blunden, M., Escott, A.: Related key attacks on reduced round KASUMI. In: Matsui, M. (ed.) FSE 2001. LNCS, vol. 2355, pp. 277–285. Springer, Heidelberg (2002)

20. Boneh, D., Boyen, X.: Efficient selective-ID secure identity based encryption without random oracles. In: Cachin, C., Camenisch, J.L. (eds.) EUROCRYPT 2004. LNCS, vol. 3027, pp. 223–238. Springer, Heidelberg (2004)

21. Boneh, D., Boyen, X.: Short signatures without random oracles. In: Cachin, C., Camenisch, J.L. (eds.) EUROCRYPT 2004. LNCS, vol. 3027, pp. 56–73. Springer, Heidelberg (2004)

22. Boneh, D., Boyen, X., Shacham, H.: Short group signatures. In: Franklin, M. (ed.) CRYPTO 2004. LNCS, vol. 3152, pp. 41–55. Springer, Heidelberg (2004)

23. Boneh, D., DeMillo, R.A., Lipton, R.J.: On the importance of checking cryptographic protocols for faults (extended abstract). In: Fumy, W. (ed.) EUROCRYPT 1997. LNCS, vol. 1233, pp. 37–51. Springer, Heidelberg (1997)

24. Canetti, R., Friedlander, J.B., Konyagin, S.V., Larsen, M., Lieman, D., Shparlinski, I.: On the statistical properties of Diffie-Hellman distributions. Israel J. Math. 120, 23–46 (2000)
25. Canetti, R., Goldreich, O., Halevi, S.: The random oracle methodology, revisited (preliminary version). In: 30th ACM STOC, pp. 209–218. ACM Press, New York (May 1998)
26. Chevalier, C., Fouque, P.-A., Pointcheval, D., Zimmer, S.: Optimal randomness extraction from a Diffie-Hellman element. In: Joux, A. (ed.) EUROCRYPT 2009. LNCS, vol. 5479, pp. 572–589. Springer, Heidelberg (2010)
27. Dodis, Y., Yampolskiy, A.: A verifiable random function with short proofs and keys. In: Vaudenay, S. (ed.) PKC 2005. LNCS, vol. 3386, pp. 416–431. Springer, Heidelberg (2005)
28. Dolev, D., Dwork, C., Naor, M.: Nonmalleable cryptography. SIAM Journal on Computing 30(2), 391–437 (2000)
29. Dunkelman, O., Keller, N., Kim, J.: Related-key rectangle attack on the full SHACAL-1. In: Biham, E., Youssef, A.M. (eds.) SAC 2006. LNCS, vol. 4356, pp. 28–44. Springer, Heidelberg (2007)
30. Fouque, P.-A., Pointcheval, D., Stern, J., Zimmer, S.: Hardness of distinguishing the MSB or LSB of secret keys in Diffie-Hellman schemes. In: Bugliesi, M., Preneel, B., Sassone, V., Wegener, I. (eds.) ICALP 2006. LNCS, vol. 4052, pp. 240–251. Springer, Heidelberg (2006)
31. Gennaro, R., Lysyanskaya, A., Malkin, T., Micali, S., Rabin, T.: Algorithmic tamper-proof (ATP) security: Theoretical foundations for security against hardware tampering. In: Naor, M. (ed.) TCC 2004. LNCS, vol. 2951, pp. 258–277. Springer, Heidelberg (2004)
32. Goldenberg, D., Liskov, M.: On related-secret pseudorandomness. In: Micciancio, D. (ed.) TCC 2010. LNCS, vol. 5978, pp. 255–272. Springer, Heidelberg (2010)
33. Goldreich, O., Goldwasser, S., Micali, S.: How to construct random functions. Journal of the ACM 33, 792–807 (1986)
34. Hong, S., Kim, J., Lee, S., Preneel, B.: Related-key rectangle attacks on reduced versions of SHACAL-1 and AES-192. In: Gilbert, H., Handschuh, H. (eds.) FSE 2005. LNCS, vol. 3557, pp. 368–383. Springer, Heidelberg (2005)
35. Iwata, T., Kohno, T.: New security proofs for the 3GPP confidentiality and integrity algorithms. In: Roy, B., Meier, W. (eds.) FSE 2004. LNCS, vol. 3017, pp. 427–445. Springer, Heidelberg (2004)
36. Jakimoski, G., Desmedt, Y.: Related-key differential cryptanalysis of 192-bit key AES variants. In: Matsui, M., Zuccherato, R.J. (eds.) SAC 2003. LNCS, vol. 3006, pp. 208–221. Springer, Heidelberg (2004)
37. Jaulmes, É., Joux, A., Valette, F.: On the security of randomized CBC-MAC beyond the birthday paradox limit: A new construction. In: Daemen, J., Rijmen, V. (eds.) FSE 2002. LNCS, vol. 2365, pp. 237–251. Springer, Heidelberg (2002)
38. Kelsey, J., Schneier, B., Wagner, D.: Related-key cryptanalysis of 3-WAY, Biham-DES, CAST, DES-X, NewDES, RC2, and TEA. In: Han, Y., Okamoto, T., Qing, S. (eds.) ICICS 1997. LNCS, vol. 1334, pp. 233–246. Springer, Heidelberg (1997)
39. Kim, J., Hong, S., Preneel, B.: Related-key rectangle attacks on reduced AES-192 and AES-256. In: Biryukov, A. (ed.) FSE 2007. LNCS, vol. 4593, pp. 225–241. Springer, Heidelberg (2007)
40. Knudsen, L.R.: Cryptanalysis of LOKI91. In: Zheng, Y., Seberry, J. (eds.) AUSCRYPT 1992. LNCS, vol. 718, pp. 196–208. Springer, Heidelberg (1993)
41. Knudsen, L.R., Kohno, T.: Analysis of RMAC. In: Johansson, T. (ed.) FSE 2003. LNCS, vol. 2887, pp. 182–191. Springer, Heidelberg (2003)

42. Ko, Y., Hong, S., Lee, W., Lee, S., Kang, J.-S.: Related key differential attacks on 27 rounds of XTEA and full-round GOST. In: Roy, B., Meier, W. (eds.) FSE 2004. LNCS, vol. 3017, pp. 299–316. Springer, Heidelberg (2004)

43. Lewko, A.B., Waters, B.: Efficient pseudorandom functions from the decisional linear assumption and weaker variants. In: Al-Shaer, E., Jha, S., Keromytis, A.D. (eds.) ACM CCS 2009, pp. 112–120. ACM Press, New York (November 2009)

44. Liskov, M., Rivest, R.L., Wagner, D.: Tweakable block ciphers. In: Yung, M. (ed.) CRYPTO 2002. LNCS, vol. 2442, pp. 31–46. Springer, Heidelberg (2002)

45. Luby, M., Rackoff, C.: How to construct pseudorandom permutations from pseudorandom functions. SIAM Journal on Computing 17(2) (1988)

46. Lucks, S.: Ciphers secure against related-key attacks. In: Roy, B., Meier, W. (eds.) FSE 2004. LNCS, vol. 3017, pp. 359–370. Springer, Heidelberg (2004)

47. Naor, M., Reingold, O.: Number-theoretic constructions of efficient pseudo-random functions. Journal of the ACM 51(2), 231–262 (2004)

48. Peikert, C., Waters, B.: Lossy trapdoor functions and their applications. In: Ladner, R.E., Dwork, C. (eds.) 40th ACM STOC, pp. 187–196. ACM Press, New York (2008)

49. Phan, R.C.-W.: Related-key attacks on triple-DES and DESX variants. In: Okamoto, T. (ed.) CT-RSA 2004. LNCS, vol. 2964, pp. 15–24. Springer, Heidelberg (2004)

50. Preneel, B., Govaerts, R., Vandewalle, J.: Hash functions based on block ciphers: A synthetic approach. In: Stinson, D.R. (ed.) CRYPTO 1993. LNCS, vol. 773, pp. 368–378. Springer, Heidelberg (1994)

51. Rogaway, P.: Efficient instantiations of tweakable blockciphers and refinements to modes OCB and PMAC. In: Lee, P.J. (ed.) ASIACRYPT 2004. LNCS, vol. 3329, pp. 16–31. Springer, Heidelberg (2004)

52. Rogaway, P.: Formalizing human ignorance. In: Nguyên, P.Q. (ed.) VIETCRYPT 2006. LNCS, vol. 4341, pp. 211–228. Springer, Heidelberg (2006)

53. Shoup, V.: Lower bounds for discrete logarithms and related problems. In: Fumy, W. (ed.) EUROCRYPT 1997. LNCS, vol. 1233, pp. 256–266. Springer, Heidelberg (1997)

54. Zhang, W., Wu, W., Zhang, L., Feng, D.: Improved related-key impossible differential attacks on reduced-round AES-192. In: Biham, E., Youssef, A.M. (eds.) SAC 2006. LNCS, vol. 4356, pp. 15–27. Springer, Heidelberg (2007)

# Secure Two-Party Quantum Evaluation of Unitaries against Specious Adversaries

Frédéric Dupuis[1,*], Jesper Buus Nielsen[2], and Louis Salvail[3,**]

[1] Institute for Theoretical Physics, ETH Zurich, Switzerland
dupuis@phys.ethz.ch
[2] DAIMI, Aarhus University, Denmark
jbn@cs.au.dk
[3] Université de Montréal (DIRO), QC, Canada
salvail@iro.umontreal.ca

**Abstract.** We describe how any two-party quantum computation, specified by a unitary which simultaneously acts on the registers of both parties, can be privately implemented against a quantum version of classical semi-honest adversaries that we call specious. Our construction requires two ideal functionalities to garantee privacy: a private SWAP between registers held by the two parties and a classical private AND-box equivalent to oblivious transfer. If the unitary to be evaluated is in the Clifford group then only one call to SWAP is required for privacy. On the other hand, any unitary not in the Clifford requires one call to an AND-box per R-gate in the circuit. Since SWAP is itself in the Clifford group, this functionality is universal for the private evaluation of any unitary in that group. SWAP can be built from a classical bit commitment scheme or an AND-box but an AND-box cannot be constructed from SWAP. It follows that unitaries in the Clifford group are to some extent the easy ones. We also show that SWAP cannot be implemented privately in the bare model.

## 1 Introduction

In this paper, we address the problem of privately evaluating some unitary transform $U$ upon a joint quantum input state held by two parties. Since unitaries model what quantum algorithms are implementing, we can see this problem as a natural extension of secure two-party evaluation of functions to the quantum realm. Suppose that a state $|\phi_{\text{in}}\rangle \in \mathcal{A} \otimes \mathcal{B}$ is the initial shared state where Alice holds register $\mathcal{A}$ and Bob holds register $\mathcal{B}$. Let $U \in \text{U}(\mathcal{A} \otimes \mathcal{B})$ be some unitary transform acting upon $\mathcal{A}$ and $\mathcal{B}$. What cryptographic assumptions are needed for a private evaluation of $|\phi_{\text{out}}\rangle = U|\phi_{\text{in}}\rangle$ where *private* means that each player learns no more than in the ideal situation depicted in Fig. 1? Of course, answers to this question depend upon the adversary we are willing to tolerate.

* Supported by Canada's NSERC Postdoctoral Fellowship Program.
** Supported by Canada's NSERC discovery grant, MITACS, and the QuantumWorks networks(NSERC).

T. Rabin (Ed.): CRYPTO 2010, LNCS 6223, pp. 685–706, 2010.

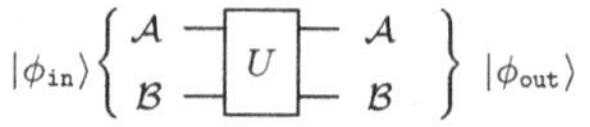

In [17], it was shown that unitaries cannot be used to implement classical cryptographic primitives. Any non-trivial primitive implemented by unitaries will necessarily leak information toward one party. Moreover, this leakage is available to a weak class of adversaries that can be interpreted as

**Fig. 1.** Ideal Functionality for unitary $U$

the quantum version of classical semi-honest adversaries. It follows that quantum two-party computation of unitaries cannot be used to implement classical cryptographic primitives. This opens the possibility that the cryptographic assumptions needed for private evaluations of unitaries are weaker than for their classical counterpart. So, what classical cryptographic assumptions, if any, are required to achieve privacy in our setting? Are there unitaries more difficult to evaluate privately than others?

In this work, we answer these questions against a class of weak quantum adversaries, called *specious*, related to classical semi-honest adversaries. We say that a quantum adversary is specious if at any step during the execution of a protocol, it can provide a judge with some state that, when joined with the state held by the honest player, will be indistinguishable from a honest interaction. In other words, an adversary is specious if it can pass an audit with success at any step. Most known impossibility proofs in quantum cryptography apply when the adversary is restricted to be specious. Definitions similar to ours have been proposed for the quantum setting and usually named semi-honest. However, translating our definition to the classical setting produces a strictly stronger class of adversaries than semi-honest[1] which justifies not adopting the term *semi-honest*. We propose the name *specious* as the core of the definition is that the adversary must appear to act honestly.

*Contributions.* First, we define two-party protocols for the evaluation of unitaries having access to oracle calls. This allows us to consider protocols with security relying on some ideal functionalities in order to be private. We then say that a protocol is in the *bare model* if it does not involve any call to an ideal functionality. We then formally define what we mean by specious adversaries. Privacy is then defined via simulation. We say that a protocol for the two-party evaluation of unitary $U$ is private against specious adversaries if, for any joint input state and at any step of the protocol, there exists a simulator that can reproduce the adversary's view having only access to its own part of the joint input state. Quantum simulation must rely on a family of simulators for the view of the adversary rather than one because quantum information does not accumulate but can vanish as the protocol evolves. For instance, consider the

---

[1] As an example, assume there exist public key cryptosystems where you can sample a public key without learning the secret key. Then this is a semi-honest oblivious transform: The receiver, with choice bit $c$, samples $pk_c$ in the normal way and learns its corresponding secret key and samples $pk_{1-c}$ without learning its secret key. He sends $(pk_0, pk_1)$. Then the sender sends $(E_{pk_0}(m_0), E_{pk_1}(m_1))$ and the receiver decrypts $E_{pk_c}(m_c)$. This is not secure against a specious adversary who can sample $pk_{1-c}$ along with its secret key $sk_{1-c}$ and then delete $sk_{1-c}$ before the audit.

trivial protocol that let Alice send her input register to Bob so that he can apply locally $|\phi_{\text{out}}\rangle = U|\phi_{\text{in}}\rangle$ before returning her register. The final state of such a protocol is certainly private, as Bob cannot clone Alice's input and keep a copy, yet at some point Bob had access to Alice's input thus violating privacy. No simulator can possibly reproduce Bob's state after he received Alice's register without having access to her input state.

Second, we show that no protocol can be shown statistically private against specious adversaries in the bare model for a very simple unitary: the swap gate. As the name suggests, the swap gate simply permutes Alice's and Bob's input states. Intuitively, the reason why this gate is impossible is that at some point during the execution of such protocol, one party that still has almost all its own input state receives a non-negligible amount of information (in the quantum sense) about the other party's input state. At this point, no simulator can possibly re-produce the complete state held by the receiving party since a call to the ideal functionality only provides access to the other party's state while no call to the ideal functionality only provides information about that party's own input. Therefore, any simulator cannot re-produce a state that contains information about the input states of both parties. It follows that cryptographic assumptions are needed for the private evaluation of unitaries against specious adversaries. On the other hand, a classical bit commitment is sufficient to implement the swap privately in our model.

Finally, we give a very simple protocol for the private evaluation of any unitary based on ideas introduced by [7,6] in the context of fault tolerant quantum computation. Our construction is similar to Yao's original construction in the classical world[22,9]. We represent any unitary $U$ by a quantum circuit made out of gates taken from the universal set $\mathcal{UG} = \{X, Y, Z, \text{CNOT}, \text{H}, \text{P}, \text{R}\}$ [13]. The protocol evaluates each gate of the circuit upon shared encrypted input where the encryption uses the Pauli operators $\{X, Y, Z\}$ together with the identity. In addition to the Pauli gates $X, Y$, and $Z$, gates CNOT, H, and P can easily be performed over encrypted states without losing the ability to decrypt. Gates of that kind belong to what is called the *Clifford group*. The CNOT gate is the only gate in $\mathcal{UG}$ acting upon more than one qubit while the R-gate is the only one that does not belong to the Clifford group. In order to evaluate it over an encrypted state while preserving the ability to decrypt, we need to rely upon a classical ideal functionality computing securely an additive sharing for the AND of Alice's and Bob's input bits. We call this ideal functionality an AND-box. Upon input $x \in \{0, 1\}$ for Alice and $y \in \{0, 1\}$ for Bob, it produces $a \in_R \{0, 1\}$ and $b \in \{0, 1\}$ to Alice and Bob respectively such that $a \oplus b = x \wedge y$. An AND-box can be obtained from any flavor of oblivious transfer and is defined the same way than an NL-box[14,15] without the property that its output can be obtained before the input of the other player has been provided to the box (i.e., NL-boxes are non-signaling). The *equivalence* between AND-boxes, NL-boxes, and oblivious transfer is discussed in [21]. At the end of the protocol, each part of the shared key allowing to decrypt the output must be exchanged in a fair way. For this task, Alice and Bob rely upon an ideal swap functionality called

SWAP. The result is that any $U$ can be evaluated privately upon any input provided Alice and Bob have access to one AND-box per R-gate and one call to the an ideal swap. If the circuit happens to have only gates in the Clifford group then only one call to an ideal swap is required for privacy. In other words, SWAP is universal for the private evaluation of circuits in the Clifford group (i.e., those circuits having no R-gate) and itself belongs to that group (SWAP is not a classical primitive). To some extent, circuits in the Clifford group are the *easy* ones. Privacy for circuits containing R-gates however needs a classical cryptographic primitive to be evaluated privately by our protocol. It means that AND-boxes are universal for the private evaluation of any circuit against specious adversaries. We don't know whether there exist some unitary transforms that are universal for the private evaluation of any unitary against specious adversaries.

*Previous works.* All impossibility results in quantum cryptography we are aware of apply to classical primitives. In fact, the impossibility proofs usually rely upon the fact that an adversary with a seemingly honest behavior can force the implementation of classical primitives to behave quantumly. The result being that implemented that way, the primitive must leak information to the adversary. This is the spirit behind the impossibility of implementing oblivious transfer securely using quantum communication[10]. In that same paper the impossibility of any one-sided private evaluation of non-trivial primitives was shown. All these results can be seen as generalizations of the impossibility of bit commitment schemes based on quantum communication[11,12]. The most general impossibility result we are aware of applies to any *non-trivial* two-party classical function[17]. It states that it suffices for the adversary to *purify* its actions in order for the quantum primitive to leak information. An adversary purifying its actions is specious as defined above. None of these impossibility proofs apply to quantum primitives characterized by some unitary transform applied to joint quantum inputs. Blind quantum computation is a primitive that shows similarities to ours. In [4], a protocol allowing a client to get its input to a quantum circuit evaluated blindly has been proposed. The security of their scheme is unconditional while in our setting almost no unitary allows for unconditional privacy.

An unpublished work of Smith[19] shows how one can devise a private protocol for the evaluation of any unitary that seems to remain private against all quantum adversaries. However, the techniques used require strong cryptographic assumptions like homomorphic encryption schemes, zero-knowledge and witness indistinguishable proof systems. The construction is in the spirit of protocols for multiparty quantum computation[3,5] and fault tolerant quantum circuits[18,1]. Although our protocol only guarantees privacy against specious adversaries, it is obtained using much weaker cryptographic assumptions.

## 2    Preliminaries

The $N$-dimensional complex Euclidean space (i.e., Hilbert space) will be denoted by $\mathcal{H}_N$. We denote quantum registers using calligraphic typeset $\mathcal{A}$. As usual, $\mathcal{A} \otimes \mathcal{B}$

denotes the space of two such quantum registers. We write $\mathcal{A} \approx \mathcal{B}$ when $\mathcal{A}$ and $\mathcal{B}$ are such that $\dim(\mathcal{A}) = \dim(\mathcal{B})$. A register $\mathcal{A}$ can undergo transformations as a function of time; we denote by $\mathcal{A}_i$ the state of space $\mathcal{A}$ at time $i$. When a quantum computation is viewed as a circuit accepting input in $\mathcal{A}$, we denote all wires in the circuit by $w \in \mathcal{A}$. If the circuit accepts input in $\mathcal{A} \otimes \mathcal{B}$ then the set of all wires is denoted $w \in \mathcal{A} \cup \mathcal{B}$.

The set of all linear mappings from $\mathcal{A}$ to $\mathcal{B}$ is denoted by $\mathrm{L}(\mathcal{A}, \mathcal{B})$ while $\mathrm{L}(\mathcal{A})$ stands for $\mathrm{L}(\mathcal{A}, \mathcal{A})$. To simplify notation, for $\rho \in \mathrm{L}(\mathcal{A})$ and $M \in \mathrm{L}(\mathcal{A}, \mathcal{B})$ we write $M \cdot \rho$ for $M\rho M^\dagger$. We denote by $\mathrm{Pos}(\mathcal{A})$ the set of positive semi-definite operators in $\mathcal{A}$. The set of positive semi-definite operators with trace 1 acting on $\mathcal{A}$ is denoted $\mathrm{D}(\mathcal{A})$; $\mathrm{D}(\mathcal{A})$ is the set of all possible quantum states for register A. An operator $A \in \mathrm{L}(\mathcal{A}, \mathcal{B})$ is called a *linear isometry* if $A^\dagger A = \mathbb{1}_\mathcal{A}$. The set of unitary operators (i.e., linear isometries with $\mathcal{B} = \mathcal{A}$) acting in $\mathcal{A}$ is denoted by $\mathrm{U}(\mathcal{A})$. The identity operator in $\mathcal{A}$ is denoted $\mathbb{1}_\mathcal{A}$ and the completely mixed state in $\mathrm{D}(\mathcal{A})$ is denoted $\mathbb{I}_\mathcal{A}$. For any positive integer $N > 0$, $\mathbb{1}_N$ and $\mathbb{I}_N$ denote the identity operator respectively the completely mixed state in $\mathcal{H}_N$. When the context requires, a pure state $|\psi\rangle \in \mathcal{AB}$ will be written $|\psi\rangle^{\mathcal{AB}}$ to make explicit the registers in which it is stored.

A linear mapping $\Phi : \mathrm{L}(\mathcal{A}) \mapsto \mathrm{L}(\mathcal{B})$ is called a *super-operator* since it belongs to $\mathrm{L}(\mathrm{L}(\mathcal{A}), \mathrm{L}(\mathcal{B}))$. $\Phi$ is said to be *positive* if $\Phi(A) \in \mathrm{Pos}(\mathcal{B})$ for all $A \in \mathrm{Pos}(\mathcal{A})$. The super-operator $\Phi$ is said to be *completely positive* if $\Phi \otimes \mathbb{1}_{\mathrm{L}(\mathcal{Z})}$ is positive for every choice of the Hilbert space $\mathcal{Z}$. A super-operator $\Phi$ can be physically realized or is *admissible* if it is completely positive and preserves the trace: $\mathrm{tr}(\Phi(A)) = \mathrm{tr}(A)$ for all $A \in \mathrm{L}(\mathcal{A})$. We call such a super-operator a *quantum operation*. Another way to represent any quantum operation is through a linear isometry $W \in \mathrm{L}(\mathcal{A}, \mathcal{B} \otimes \mathcal{Z})$ such that $\Phi(\rho) = \mathrm{tr}_\mathcal{Z}(W \cdot \rho)$, for some extra space $\mathcal{Z}$. Any such isometry $W$ can be implemented by a physical process as long as the resource to implement $\mathcal{Z}$ is available. This is just a unitary transform in $\mathrm{U}(\mathcal{A} \otimes \mathcal{Z})$ where the system in $\mathcal{Z}$ is initially in known state $|0_\mathcal{Z}\rangle$.

For two states $\rho_0, \rho_1 \in \mathrm{D}(\mathcal{A})$, we denote by $\Delta(\rho_0, \rho_1)$ the trace norm distance between $\rho_0$ and $\rho_1$: $\Delta(\rho_0, \rho_1) := \frac{1}{2}\|\rho_0 - \rho_1\|$. If $\Delta(\rho_0, \rho_1) \le \varepsilon$ then any quantum process applied to $\rho_0$ behaves exactly as for $\rho_1$ except with probability at most $\varepsilon$ [16].

Let $X, Y$, and $Z$ be the three non-trivial one-qubit Pauli operators. The Bell measurement is a complete orthogonal measurement on two qubits made out of the measurement operators $\{|\Psi_{x,y}\rangle\langle\Psi_{x,y}|\}_{x,y \in \{0,1\}}$ where $|\Psi_{x,y}\rangle := \frac{1}{\sqrt{2}}(|0, x\rangle + (-1)^y|1, \overline{x}\rangle)$. We say that the outcome of a Bell measurement is $(x, y) \in \{0, 1\}^2$ if $|\Psi_{x,y}\rangle\langle\Psi_{x,y}|$ has been observed. The quantum one-time-pad is a perfectly secure encryption of quantum states[2]. It encrypts a qubit $|\psi\rangle$ as $X^x Z^z|\psi\rangle$, where the key is two classical bits, $(x, z) \in \{0, 1\}^2$ and $X^0 Z^0 = \mathbb{1}$, $X^0 Z^1 = Z$, $X^1 Z^0 = X$ and $X^1 Z^1 = Y$ are the Pauli operators.

## 2.1   Modeling Two-Party Strategies

Consider an interactive two-party strategy $\Pi^{\mathcal{O}}$ between parties $\mathscr{A}$ and $\mathscr{B}$ and oracle calls $\mathcal{O}$. $\Pi^{\mathcal{O}}$ can be modeled by a sequence of quantum operations for each player together with some oracle calls also modeled by quantum operations. Each quantum operation in the sequence corresponds to the action of one party at a certain step of the strategy. The following definition is a straightforward adaptation of $n$-turn interactive quantum strategies as described in [8]. The main difference is that here, we provide a joint input state to both parties and that quantum transmissions taking place during the execution is modeled by a quantum operation; one that is moving a state on one party's side to the other party.

**Definition 2.1.** *A $n$-step two party strategy with oracle calls denoted $\Pi^{\mathcal{O}} = (\mathscr{A}, \mathscr{B}, \mathcal{O}, n)$ consists of:*

1. *input spaces $\mathcal{A}_0$ and $\mathcal{B}_0$ for parties $\mathscr{A}$ and $\mathscr{B}$ respectively,*

2. *memory spaces $\mathcal{A}_1, \ldots, \mathcal{A}_n$ and $\mathcal{B}_1, \ldots, \mathcal{B}_n$ for $\mathscr{A}$ and $\mathscr{B}$ respectively,*

3. *an n-tuple of quantum operations $(\mathscr{A}_1, \ldots, \mathscr{A}_n)$ for $\mathscr{A}$, $\mathscr{A}_i : \mathrm{L}(\mathcal{A}_{i-1}) \mapsto \mathrm{L}(\mathcal{A}_i)$, $(1 \leq i \leq n)$,*

4. *an n-tuple of quantum operations $(\mathscr{B}_1, \ldots, \mathscr{B}_n)$ for $\mathscr{B}$, $\mathscr{B}_i : \mathrm{L}(\mathcal{B}_{i-1}) \mapsto \mathrm{L}(\mathcal{B}_i)$, $(1 \leq i \leq n)$,*

5. *memory spaces $\mathcal{A}_1, \ldots, \mathcal{A}_n$ and $\mathcal{B}_1, \ldots, \mathcal{B}_n$ can be written as $\mathcal{A}_i = \mathcal{A}_i^{\mathcal{O}} \otimes \mathcal{A}_i'$ and $\mathcal{B}_i = \mathcal{B}_i^{\mathcal{O}} \otimes \mathcal{B}_i'$, $(1 \leq i \leq n)$, and $\mathcal{O} = (\mathcal{O}_1, \mathcal{O}_2, \ldots, \mathcal{O}_n)$ is an n-tuple of quantum operations: $\mathcal{O}_i : \mathrm{L}(\mathcal{A}_i^{\mathcal{O}} \otimes \mathcal{B}_i^{\mathcal{O}}) \mapsto \mathrm{L}(\mathcal{A}_i^{\mathcal{O}} \otimes \mathcal{B}_i^{\mathcal{O}})$, $(1 \leq i \leq n)$.*

*If $\Pi = (\mathscr{A}, \mathscr{B}, n)$ is a n-turn two-party protocol then the final state of the interaction upon input state $\rho_{\mathrm{in}} \in \mathrm{D}(\mathcal{A}_0 \otimes \mathcal{B}_0 \otimes \mathcal{R})$, where $\mathcal{R}$ is a system of dimension $\dim \mathcal{R} = \dim \mathcal{A}_0 \dim \mathcal{B}_0$, is:*

$$[\mathscr{A} \circledast \mathscr{B}](\rho_{\mathrm{in}}) := (\mathbb{1}_{\mathrm{L}(\mathcal{A}_n' \otimes \mathcal{B}_n' \otimes \mathcal{R})} \otimes \mathcal{O}_n)(\mathscr{A}_n \otimes \mathscr{B}_n \otimes \mathbb{1}_{\mathcal{R}})$$
$$\ldots (\mathbb{1}_{\mathrm{L}(\mathcal{A}_1' \otimes \mathcal{B}_1' \otimes \mathcal{R})} \otimes \mathcal{O}_1)(\mathscr{A}_1 \otimes \mathscr{B}_1 \otimes \mathbb{1}_{\mathcal{R}})(\rho_{\mathrm{in}}) .$$

Step $i$ of the strategy corresponds to the actions of $\mathscr{A}_i$ and $\mathscr{B}_i$ followed by the oracle call $\mathcal{O}_i$.

Note that we consider input states defined on the input systems together with a reference system $\mathcal{R}$; this allows us to show the correctness and privacy of the protocol not only for pure inputs, but also for inputs that are entangled with a third party. This is the most general case allowed by quantum mechanics.

A two-party strategy is therefore defined by quantum operation tuples $(\mathscr{A}_1, \ldots, \mathscr{A}_n)$, $(\mathscr{B}_1, \ldots, \mathscr{B}_n)$, and $(\mathcal{O}_1, \ldots, \mathcal{O}_n)$. These operations also define working spaces $\mathcal{A}_0, \ldots, \mathcal{A}_n, \mathcal{B}_0, \ldots, \mathcal{B}_n$ together with the input-output spaces to the oracle calls $\mathcal{A}_i^{\mathcal{O}}$ and $\mathcal{B}_i^{\mathcal{O}}$ for $1 \leq i \leq n$.

A *communication oracle* from Alice to Bob is modeled by having $\mathcal{A}_i^{\mathcal{O}} \approx \mathcal{B}_i^{\mathcal{O}}$ and letting $\mathcal{O}_i$ move the state in $\mathcal{A}_i^{\mathcal{O}}$ to $\mathcal{B}_i^{\mathcal{O}}$ and erase $\mathcal{A}_i^{\mathcal{O}}$. Similarly for communication in the other direction. We define a *bare model* protocol to be one which only uses communication oracles.

# 3   Specious Quantum Adversaries

## 3.1   Protocols for Two-Party Evaluation

Let us consider two-party protocols for the quantum evaluation of unitary transform $U \in U(\mathcal{A}_0 \otimes \mathcal{B}_0)$ between parties $\mathcal{A}$ and $\mathcal{B}$ upon joint input state $\rho_{\mathrm{in}} \in D(\mathcal{A}_0 \otimes \mathcal{B}_0 \otimes \mathcal{R})$:

**Definition 3.1.** *A two-party protocol $\Pi_U^{\mathcal{O}} = (\mathcal{A}, \mathcal{B}, \mathcal{O}, n)$ for $U \in U(\mathcal{A}_0 \otimes \mathcal{B}_0)$ is an $n$–step two-party strategy with oracle calls, where $\mathcal{A}_n \approx \mathcal{A}_0$ and $\mathcal{B}_n \approx \mathcal{B}_0$. It is said to be $\varepsilon$–correct if*

$$\Delta\left([\mathcal{A} \circledast \mathcal{B}](\rho_{\mathrm{in}}), (U \otimes \mathbb{1}_{\mathcal{R}}) \cdot \rho_{\mathrm{in}}\right) \leq \varepsilon \quad \text{for all } \rho_{\mathrm{in}} \in D(\mathcal{A}_0 \otimes \mathcal{B}_0 \otimes \mathcal{R}) \ .$$

*We denote by $\Pi_U$ a two-party protocol in the* bare model *where, without loss of generality, we assume that $\mathcal{O}_{2i+1}$ ($0 \leq i \leq \lfloor \frac{n}{2} \rfloor$) implements a communication channel from $\mathcal{A}$ to $\mathcal{B}$ and $\mathcal{O}_{2i}$ ($1 \leq i \leq \lfloor \frac{n}{2} \rfloor$) implements a communication channel from $\mathcal{B}$ to $\mathcal{A}$. Communication oracles are said to be* trivial.

In other words, a two-party protocol $\Pi_U^{\mathcal{O}}$ for unitary $U$ is a two-party interactive strategy where, at the end, the output of the computation is stored in the memory of the players. $\Pi_U^{\mathcal{O}}$ is correct if, when restricted to the output registers (and $\mathcal{R}$), the final quantum state shared by $\mathcal{A}$ and $\mathcal{B}$ is $(U \otimes \mathbb{1}_{\mathcal{R}}) \cdot \rho_{\mathrm{in}}$.

As it will become clear when we discuss privacy in Sect. 3.3, we need to consider the joint state at any step during the evolution of the protocol:

$$\rho_1(\rho_{\mathrm{in}}) := (\mathbb{1}_{L(\mathcal{A}_1' \otimes \mathcal{B}_1' \otimes \mathcal{R})} \otimes \mathcal{O}_1)(\mathcal{A}_1 \otimes \mathcal{B}_1 \otimes \mathbb{1}_{L(\mathcal{R})})(\rho_{\mathrm{in}}),$$
$$\rho_{i+1}(\rho_{\mathrm{in}}) := (\mathbb{1}_{L(\mathcal{B}_{i+1}' \otimes \mathcal{A}_{i+1}' \otimes \mathcal{R})} \otimes \mathcal{O}_{i+1})(\mathcal{A}_{i+1} \otimes \mathcal{B}_{i+1} \otimes \mathbb{1}_{L(\mathcal{R})})(\rho_i(\rho_{\mathrm{in}})) \ , \quad (1)$$

for $1 \leq i < n$. We also write the final state of $\Pi_U^{\mathcal{O}}$ upon input state $\rho_{\mathrm{in}}$ as $\rho_n(\rho_{\mathrm{in}}) = [\mathcal{A} \circledast \mathcal{B}](\rho_{\mathrm{in}})$.

## 3.2   Modeling Specious Adversaries

Intuitively, a specious adversary acts in any way apparently indistinguishable from the honest behavior, in the sense that no audit can distinguish the behavior of the adversary from the honest one.

More formally, a specious adversary in $\Pi_U^{\mathcal{O}} = (\mathcal{A}, \mathcal{B}, \mathcal{O}, n)$ may use an arbitrary large quantum memory space. However, at any step $1 \leq i \leq n$, the adversary can transform its own current state to one that is indistinguishable from the honest joint state. These transforms are modeled by quantum operations, one for each step of the adversary in $\Pi_U^{\mathcal{O}}$, and are part of the adversary's specification. We denote by $(\mathcal{T}_1, \ldots, \mathcal{T}_n)$ these quantum operations where $\mathcal{T}_i$ produces a valid transcript at the end of the $i$–th step.

Let $\tilde{\mathcal{A}}$ and $\tilde{\mathcal{B}}$ be adversaries in $\Pi_U^{\mathcal{O}}$. We denote by $\Pi_U^{\mathcal{O}}(\tilde{\mathcal{A}}) = (\tilde{\mathcal{A}}, \mathcal{B}, \mathcal{O}, n)$ and $\Pi_U^{\mathcal{O}}(\tilde{\mathcal{B}}) = (\mathcal{A}, \tilde{\mathcal{B}}, \mathcal{O}, n)$ the resulting $n$–step two-party strategies. We denote by $\tilde{\rho}_i(\tilde{\mathcal{A}}, \rho_{\mathrm{in}})$ the state defined in (1) for protocol $\Pi_U^{\mathcal{O}}(\tilde{\mathcal{A}})$ and similarly by $\tilde{\rho}_i(\tilde{\mathcal{B}}, \rho_{\mathrm{in}})$ that state for protocol $\Pi_U^{\mathcal{O}}(\tilde{\mathcal{B}})$.

Adding the possibility for the adversary to be $\varepsilon$-*close* to honest, we get the following definition:

**Definition 3.2.** *Let* $\Pi_U^{\mathscr{O}} = (\mathscr{A}, \mathscr{B}, \mathscr{O}, n)$ *be an* $n$-*step two-party protocol with oracle calls for* $U \in \mathrm{U}(\mathcal{A}_0 \otimes \mathcal{B}_0)$. *We say that:*

- $\mathscr{A}$ *is* $\varepsilon$-specious if $\Pi_U^{\mathscr{O}}(\tilde{\mathscr{A}}) = (\tilde{\mathscr{A}}, \mathscr{B}, \mathscr{O}, n)$ *is an* $n$-*step two-party strategy with* $\tilde{\mathcal{A}}_0 = \mathcal{A}_0$ *and there exists a sequence of quantum operations* $(\mathscr{T}_1, \ldots, \mathscr{T}_n)$ *such that:*
  1. *for every* $1 \leq i \leq n$, $\mathscr{T}_i : \mathrm{L}(\tilde{\mathcal{A}}_i) \mapsto \mathrm{L}(\mathcal{A}_i)$,
  2. *for every input state* $\rho_{\mathrm{in}} \in \mathrm{D}(\mathcal{A}_0 \otimes \mathcal{B}_0 \otimes \mathcal{R})$, *and for all* $1 \leq i \leq n$,

$$\Delta\left( (\mathscr{T}_i \otimes \mathbb{1}_{\mathrm{L}(\mathcal{B}_i \otimes \mathcal{R})}) \left( \tilde{\rho}_i(\tilde{\mathscr{A}}, \rho_{\mathrm{in}}) \right), \rho_i(\rho_{\mathrm{in}}) \right) \leq \varepsilon .$$

- $\mathscr{B}$ *is* $\varepsilon$-specious if $\Pi_U^{\mathscr{O}}(\tilde{\mathscr{B}}) = (\mathscr{A}, \tilde{\mathscr{B}}, \mathscr{O}, n)$ *is a* $n$-*step two-party strategy with* $\tilde{\mathcal{B}}_0 = \mathcal{B}_0$ *and there exists a sequence of quantum operations* $(\mathscr{T}_1, \ldots, \mathscr{T}_n)$ *defined as before with* $\mathcal{B}_i, \tilde{\mathcal{B}}_i$, *and* $\tilde{\rho}_i(\tilde{\mathscr{B}}, \rho_{\mathrm{in}})$ *replacing* $\mathcal{A}_i, \tilde{\mathcal{A}}_i$, *and* $\tilde{\rho}_i(\tilde{\mathscr{A}}, \rho_{\mathrm{in}})$ *respectively.*

*If a party is* $\varepsilon(m)$-*specious with* $\varepsilon(m)$ *negligible for* $m$ *a security parameter then we say that this party is* statistically specious.

### 3.3   Privacy

Privacy for $\Pi_U^{\mathscr{O}}$ is defined as the ability for a simulator, having only access to the adversary's input and the ideal functionality $U$, to reproduce the state of the adversary at any step in the execution of $\Pi_U^{\mathscr{O}}$. Our definition is similar to the one introduced in [20] for statistical zero-knowledge proof systems.

A simulator for an adversary in $\Pi_U^{\mathscr{O}}$ is represented by a sequence of quantum operations $(\mathscr{S}_i)_{i=1}^n$, where $\mathscr{S}_i$ re-produces the view of the adversary after step $i$. $\mathscr{S}_i$ initially receives the adversary's input and has access to the ideal functionality for $U$ evaluated upon the joint input of the adversary and the honest player. Because of no-cloning, a simulator calling $U$ loses its input, and the input might be required to simulate e.g. early steps in the protocol, so we have to allow that $\mathscr{S}_i$ does not call $U$. For this purpose we introduce a bit $q_i \in \{0, 1\}$. When $q_i = 0$, $\mathscr{S}_i$ does not call $U$ and when $q_i = 1$, $\mathscr{S}_i$ must first call the ideal functionality $U$ before performing some post-processing. More precisely,

**Definition 3.3.** *Let* $\Pi_U^{\mathscr{O}} = (\mathscr{A}, \mathscr{B}, \mathscr{O}, n)$ *be an* $n$-*step two-party protocol for* $U \in \mathrm{D}(\mathcal{A}_0 \otimes \mathcal{B}_0)$. *Then,*

- $\mathscr{S}(\tilde{\mathscr{A}}) = \langle (\mathscr{S}_1, \ldots, \mathscr{S}_n), q \rangle$ *is a simulator for adversary* $\tilde{\mathscr{A}}$ *in* $\Pi_U^{\mathscr{O}}$ *if it consists of:*
  1. *a sequence of quantum operations* $(\mathscr{S}_1, \ldots, \mathscr{S}_n)$ *where for* $1 \leq i \leq n$, $\mathscr{S}_i : \mathrm{L}(\mathcal{A}_0) \mapsto \mathrm{L}(\tilde{\mathcal{A}}_i)$,
  2. *a sequence of bits* $q \in \{0, 1\}^n$ *determining if the simulator calls the ideal functionality at step* $i$: $q_i = 1$ *iff the simulator calls the ideal functionality.*

– *Similarly,* $\mathscr{S}(\tilde{\mathscr{B}}) = \langle (\mathscr{S}_1, \ldots, \mathscr{S}_n), q' \rangle$ *is a simulator for adversary* $\tilde{\mathscr{B}}$ *in* $\Pi_U^{\mathcal{O}}$ *if it satisfies conditions 1 and 2 above with* $q', \mathcal{B}_0, \mathcal{B}_i,$ *and* $\tilde{\mathcal{B}}_i$ *replacing* $q, \mathcal{A}_0, \mathcal{A}_i,$ *and* $\tilde{\mathcal{A}}_i$ *respectively.*

Given an input state $\rho_{\text{in}} \in \text{D}(\mathcal{A}_0 \otimes \mathcal{B}_0 \otimes \mathcal{R})$, we define the $\tilde{\mathscr{A}}$'s respectively $\tilde{\mathscr{B}}$'s simulated views as:

$$\nu_i(\tilde{\mathscr{A}}, \rho_{\text{in}}) := \text{tr}_{\mathcal{B}_0}\left( (\mathscr{S}_i \otimes \mathbb{1}_{\text{L}(\mathcal{B}_0 \otimes \mathcal{R})}) ((U^{q_i} \otimes \mathbb{1}_{\mathcal{R}}) \cdot \rho_{\text{in}}) \right) ,$$

$$\nu_i(\tilde{\mathscr{B}}, \rho_{\text{in}}) := \text{tr}_{\mathcal{A}_0}\left( (\mathbb{1}_{\text{L}(\mathcal{A}_0 \otimes \mathcal{R})} \otimes \mathscr{S}_i) \left( (U^{q'_i} \otimes \mathbb{1}_{\mathcal{R}}) \cdot \rho_{\text{in}} \right) \right) .$$

We say that protocol $\Pi_U^{\mathcal{O}}$ is private against specious adversaries if there exits a simulator for the view at any step of any such adversary. In more details,

**Definition 3.4.** *Let* $\Pi_U^{\mathcal{O}} = (\mathscr{A}, \mathscr{B}, \mathcal{O}, n)$ *be a protocol for* $U \in \text{U}(\mathcal{A}_0 \otimes \mathcal{B}_0)$ *and let* $0 \leq \delta \leq 1$. *We say that* $\Pi_U^{\mathcal{O}}$ *is* $\delta$–*private against* $\varepsilon$–*specious* $\tilde{\mathscr{A}}$ *if there exists a simulator* $\mathscr{S}(\tilde{\mathscr{A}})$ *such that for all input states* $\rho_{\text{in}} \in \text{D}(\mathcal{A}_0 \otimes \mathcal{B}_0 \otimes \mathcal{R})$ *and for all* $1 \leq i \leq n$, $\Delta\left( \nu_i(\tilde{\mathscr{A}}, \rho_{\text{in}}), \text{tr}_{\mathcal{B}_i}(\tilde{\rho}_i(\tilde{\mathscr{A}}, \rho_{\text{in}})) \right) \leq \delta$. *Similarly, we say that* $\Pi_U$ *is* $\delta$–*private against* $\varepsilon$–*specious* $\tilde{\mathscr{B}}$ *if there exists a simulator* $\mathscr{S}(\tilde{\mathscr{B}})$ *such that for all input states* $\rho_{\text{in}} \in \text{D}(\mathcal{A}_0 \otimes \mathcal{B}_0 \otimes \mathcal{R})$ *and for all* $1 \leq i \leq n$, $\Delta\left( \nu_i(\tilde{\mathscr{B}}, \rho_{\text{in}}), \text{tr}_{\mathcal{A}_i}(\tilde{\rho}_i(\tilde{\mathscr{B}}, \rho_{\text{in}})) \right) \leq \delta$. *Protocol* $\Pi_U^{\mathcal{O}}$ *is* $\delta$–*private against* $\varepsilon$–*specious adversaries if it is* $\delta$–*private against both* $\tilde{\mathscr{A}}$ *and* $\tilde{\mathscr{B}}$. *For* $\gamma > 0$, *if* $\Pi_U^{\mathcal{O}}$ *is* $2^{-\gamma m}$–*private for* $m \in \mathbb{N}^+$ *a security parameter then we say that* $\Pi_U^{\mathcal{O}}$ *is statistically private.*

We show next that for some unitary, statistical privacy cannot be satisfied by any protocol in the bare model.

# 4    Unitaries with No Private Protocols

In this section, we show that no statistically private protocol for the swap gate exists in the bare model. The swap gate, denoted SWAP, is the following unitary transform:

$$\text{SWAP} : |\phi_A\rangle^{\mathcal{A}_0} |\phi_B\rangle^{\mathcal{B}_0} \mapsto |\phi_B\rangle^{\mathcal{A}_0} |\phi_A\rangle^{\mathcal{B}_0} ,$$

for any one qubit states $|\phi_A\rangle \in \mathcal{A}_0$ and $|\phi_B\rangle \in \mathcal{B}_0$ (i.e., $\dim(\mathcal{A}_0) = \dim(\mathcal{B}_0) = 2$). Notice that SWAP is in the Clifford group since it can be implemented with three CNOT gates. It means that universality is not required (gates in the Clifford groups are not universal for quantum computation) for a unitary to be impossible to evaluate privately. The impossibility of SWAP essentially follows from no cloning.

**Theorem 4.1 (Impossibility of swapping).** *There is no correct and statistically private two-party protocol* $\Pi_{\text{SWAP}} = (\mathscr{A}, \mathscr{B}, \mathcal{O}, n(m))$ *in the bare model.*

*Proof.* Suppose that there exists an $\varepsilon$-correct, $\varepsilon$-private protocol in the bare model for SWAP for sufficiently small $\varepsilon$; we will show that this implies that one of the two players must *lose* information upon receiving a message, which is clearly impossible.

We will consider the following particular pure input state: $|\varphi\rangle := |\Psi_{0,0}\rangle^{\mathcal{A}_0 \mathcal{R}_\mathcal{A}} \otimes |\Psi_{0,0}\rangle^{\mathcal{B}_0 \mathcal{R}_\mathcal{B}}$, a maximally entangled state between $\mathcal{A}_0 \otimes \mathcal{B}_0$ and the reference system $\mathcal{R}_\mathcal{A} \otimes \mathcal{R}_\mathcal{B}$ that is broken down into two subsystems for convenience. Furthermore, we will consider the "purified" versions of the honest players for this protocol; in other words, we will assume that the super-operators $\mathscr{A}_1, \ldots, \mathscr{A}_n$ and $\mathscr{B}_1, \ldots, \mathscr{B}_n$ are in fact linear isometries and that therefore the players never discard any information unless they have to send it to the other party. The global state $\rho_i(\varphi)$ after step $i$ is therefore a pure state on $\mathcal{A}_i \otimes \mathcal{B}_i \otimes \mathcal{R}_\mathcal{A} \otimes \mathcal{R}_\mathcal{B}$.

After step $i$ of the protocol (i.e., after the $i$th message has been sent), Alice's state must either depend only on her own original input (if $q_i = 0$ for her simulator), or on Bob's original input (if $q_i = 1$). More precisely, by the definition of privacy (Definition 3.4), we have that

$$\Delta\left(\nu_i(\mathscr{A}, \varphi), \operatorname{tr}_{\mathcal{B}_i}[\rho_i(\varphi)]\right) \leq \varepsilon \ ,$$

where $\nu_i(\mathscr{A}, \varphi)$ is $\mathscr{A}$'s simulated view after step $i$ and $\rho_i(\varphi)$ is the global state in the real protocol after step $i$. Now, suppose that $q_i = 0$, and let $|\xi\rangle \in \mathcal{A}_i \otimes \mathcal{R}_\mathcal{A} \otimes \mathcal{R}'_\mathcal{B} \otimes \mathcal{Z}$ be a purification of $\nu_i(\mathscr{A}, \varphi)$ with $\mathcal{Z}$ being the purifying system, and $\mathcal{R}_\mathcal{B}$ renamed for upcoming technical reasons. The pure state $|\xi\rangle \otimes |\Psi_{0,0}\rangle^{\mathcal{R}_\mathcal{B} \mathcal{B}_0}$ has the same reduced density matrix as $\nu_i(\mathscr{A}, \varphi)$ on $\mathcal{A}_i \otimes \mathcal{R}_\mathcal{A} \otimes \mathcal{R}_\mathcal{B}$. Hence, by Uhlmann's theorem, there exists a linear isometry $V : \mathcal{B}_i \to \mathcal{B}_0 \otimes \mathcal{Z} \otimes \mathcal{R}'_\mathcal{B}$ such that

$$V \nu_i(\mathscr{A}, \varphi) V^\dagger = |\xi\rangle\langle\xi| \otimes |\Psi_{0,0}\rangle\langle\Psi_{0,0}|^{\mathcal{B}_0 \mathcal{R}_\mathcal{B}}$$

and hence

$$\Delta\left(V \rho_i(\varphi) V^\dagger, |\xi\rangle\langle\xi| \otimes |\Psi_{0,0}\rangle\langle\Psi_{0,0}|^{\mathcal{B}_0 \mathcal{R}_\mathcal{B}}\right) \leq \sqrt{2\varepsilon} \ .$$

This means that if $q_i = 0$, then Bob is still capable of reconstructing his own input state after step $i$ by applying $V$ to his working register. Clearly, this means that $q'_i = 0$ (i.e., Bob's simulator must also not call SWAP), and therefore, by the same argument, Alice must also be able to reconstruct her own input with an isometry $V_A : \mathcal{A}_i \to \mathcal{B}_0 \otimes \mathcal{Z} \otimes \mathcal{R}'_\mathcal{A}$. The same argument also holds if $q_i = 1$: we then conclude that $q'_i = 1$ and that Alice and Bob must have each other's inputs; no intermediate situation is possible. We conclude that, at every step $i$ of the protocol, $q_i = q'_i$.

Now, before the protocol starts, Alice must have her input, and Bob must have his, hence, $q_0 = q'_0 = 0$. At the end, the two inputs must have been swapped, which means that $q_n = q'_n = 1$; there must therefore be a step $k$ in the protocol after which the two inputs are swapped but not before, meaning that $q_k = 1$ and $q_{k-1} = 0$. But at each step, only one player receives information, which means that at this step $k$, the player who received the message must lose the ability to reconstruct his own input, which is clearly impossible. $\qquad\square$

Using this line of reasoning, Theorem 4.1 can be extended to apply to any protocol for almost any unitary preventing both parties to recover their input states from its output.

**Sufficient Assumptions for Private SWAP.** A private protocol for SWAP in the bare model would exist if the players could rely on special relativity and a lower bound on their separation in space: they simply send their messages simultaneously. The fact that messages cannot travel faster than the speed of light ensures that the messages are independent of each other. It is also straightforward to devise a private protocol for SWAP based on commitment schemes. $\mathscr{A}$ sends one half EPR-pair to $\mathscr{B}$ while keeping the other half. $\mathscr{A}$ then teleports (without announcing the outcome of the measurement) her register and commits on the outcome of the Bell measurement. $\mathscr{B}$ sends his register to $\mathscr{A}$ before she opens her commitment.

# 5   The Protocol

We now describe a private protocol for the two-party evaluation of any unitary $U \in \mathrm{U}(\mathcal{A}_0 \otimes \mathcal{B}_0)$ denoted by $P_U^{\mathcal{O}} = (\mathscr{A}^*, \mathscr{B}^*, \mathcal{O}, n_U + 1)$ where $U$ is represented by a circuit $C_U$ with $u$ gates in $\mathcal{UG}$. We slightly abuse the notation with respect to the parameter $n_U + 1$. Given circuit $C_U$, we let $n_U$ be the number of oracle calls (including calls to communication oracles). Setting the last parameter to $n_U + 1$ instead of $n_U$ comes from the fact that in our protocol, $\mathscr{A}^*$ and $\mathscr{B}^*$ have to perform a last operation each in order to get their outcome. These last operations do not involve a call to any oracle. Let $G_j$ be the $j$-th gate in $C_U = G_u G_{u-1} \ldots G_1$. The protocol is obtained by composing sub-protocols for each gate similarly to well-known classical constructions[22,9]. Notice that $P_U^{\mathcal{O}}$ will not be presented in the form of Definition 3.1. $\mathscr{A}^*$ is not necessarily sending the first and the last messages. This can be done without consequences since we provide a simulation for each step where a message from the honest party is received or the output of a call to an ideal functionality is available. Putting $P_U^{\mathcal{O}}$ in the standard form of Definition 3.1 is straightforward and changes nothing to the proof of privacy.

The evaluation of each gate is performed over shared encrypted states. Each wire in $C_U$ will be updated from initially holding the input $\rho_{\mathrm{in}} \in \mathrm{D}(\mathcal{A}_0 \otimes \mathcal{B}_0 \otimes \mathcal{R})$ to finally holding the output $(U \otimes \mathbb{1}_{\mathcal{R}}) \cdot \rho_{\mathrm{in}} \in \mathrm{D}(\mathcal{A}_0 \otimes \mathcal{B}_0 \otimes \mathcal{R})$. The state of wires $\mathtt{w} \in \mathcal{A}_0 \cup \mathcal{B}_0$ after the evaluation of $G_j$ are stored at $\mathscr{A}^*$'s or $\mathscr{B}^*$'s according if $\mathtt{w} \in \mathcal{A}_0$ or $\mathtt{w} \in \mathcal{B}_0$. The shared encryption keys for wire $\mathtt{w} \in \mathcal{A}_0 \cup \mathcal{B}_0$ updated after the evaluation of $G_j$ are denoted by $K_{\mathscr{A}^*}^j(\mathtt{w}) = (X_{\mathscr{A}^*}^j(\mathtt{w}), Z_{\mathscr{A}^*}^j(\mathtt{w})) \in \{0,1\}^2$ and $K_{\mathscr{B}^*}^j(\mathtt{w}) = (X_{\mathscr{B}^*}^j(\mathtt{w}), Z_{\mathscr{B}^*}^j(\mathtt{w})) \in \{0,1\}^2$ for $\mathscr{A}^*$ and $\mathscr{B}^*$ respectively and are held privately in internal registers of each party.

The final phase of the protocol is where a call to an ideal functionality is required. $\mathscr{A}^*$ and $\mathscr{B}^*$ exchange their own part of each encryption key for the other party's wires. In order to do this, the *key-releasing phase* invokes an ideal SWAP-gate as functionality: $\mathcal{O}_{n_U} : \mathrm{L}(\mathcal{A}_{n_U}^{\mathcal{O}} \otimes \mathcal{B}_{n_U}^{\mathcal{O}}) \mapsto \mathrm{L}(\mathcal{A}_{n_U}^{\mathcal{O}} \otimes \mathcal{B}_{n_U}^{\mathcal{O}})$, where

$\mathcal{O}_{n_U}(\rho) := \mathsf{SWAP} \cdot \rho$. Upon joint input state $\rho_{\mathrm{in}} \in \mathrm{D}(\mathcal{A}_0 \otimes \mathcal{B}_0 \otimes \mathcal{R})$, protocol $P_U^{\mathcal{O}(U)}$ runs the following phases:

**Initialization:** We assume that $\mathscr{A}^*$ and $\mathscr{B}^*$ have agreed upon a description of $U$ by a circuit $C_U$ made out of $u$ gates $(G_1, \ldots, G_u)$ in $\mathcal{UG}$. For all wires $\mathsf{w} \in \mathcal{A}_0 \cup \mathcal{B}_0$, $\mathscr{A}^*$ and $\mathscr{B}^*$ set their initial encryption keys as $K^0_{\mathscr{A}^*}(\mathsf{w}) = (X^0_{\mathscr{A}^*}(\mathsf{w}), Z^0_{\mathscr{A}^*}(\mathsf{w})) := (0,0)$ and $K^0_{\mathscr{B}^*}(\mathsf{w}) = (X^0_{\mathscr{B}^*}(\mathsf{w}), Z^0_{\mathscr{B}^*}(\mathsf{w})) := (0,0)$ respectively.

**Evaluation:** For each gate number $1 \le j \le u$, $\mathscr{A}^*$ and $\mathscr{B}^*$ evaluate $G_j$ as described in details below. This evaluation results in shared encryption under keys $K^j_{\mathscr{A}^*}(\mathsf{w}) = (X^j_{\mathscr{A}^*}(\mathsf{w}), Z^j_{\mathscr{A}^*}(\mathsf{w}))$ and $K^j_{\mathscr{B}^*}(\mathsf{w}) = (X^j_{\mathscr{B}^*}(\mathsf{w}), Z^j_{\mathscr{B}^*}(\mathsf{w}))$ for all wires $\mathsf{w} \in \mathcal{A}_0 \cup \mathcal{B}_0$, which at that point hold a shared encryption of $((G_j G_{j-1} \ldots G_1) \otimes \mathbb{1}_{\mathcal{R}}) \cdot \rho_{\mathrm{in}}$. Only the evaluation of the R-gate requires a call to an ideal functionality (i.e., an AND-BOX).

**Key-Releasing:** Let $\mathcal{A}^{\mathcal{O}}_{n_U}$ and $\mathcal{B}^{\mathcal{O}}_{n_U}$ be the set of registers holding respectively $K^u_{\mathscr{A}^*}(\mathsf{w}) = (X^u_{\mathscr{A}^*}(\mathsf{w}), Z^u_{\mathscr{A}^*}(\mathsf{w}))$ for $\mathsf{w} \in \mathcal{B}_0$ and $K^u_{\mathscr{B}^*}(\mathsf{w}) = (X^u_{\mathscr{B}^*}(\mathsf{w}), Z^u_{\mathscr{B}^*}(\mathsf{w}))$ for $\mathsf{w} \in \mathcal{A}_0$. We assume w.l.g that dimensions of both sets of registers are identical[2]:

1. $\mathscr{A}^*$ and $\mathscr{B}^*$ run the ideal functionality for the SWAP-gate upon registers $\mathcal{A}^{\mathcal{O}}_{n_U}$ and $\mathcal{B}^{\mathcal{O}}_{n_U}$.
2. $\mathscr{A}^*$ applies the decryption operator $K_{\mathscr{A}^*}(\mathsf{w}) = (X^u_{\mathscr{A}^*}(\mathsf{w}) \oplus X^u_{\mathscr{B}^*}(\mathsf{w}), Z^u_{\mathscr{A}^*}(\mathsf{w}) \oplus Z^u_{\mathscr{B}^*}(\mathsf{w}))$ to each of her wires $\mathsf{w} \in \mathcal{A}_0$.
3. $\mathscr{B}^*$ applies the decryption operator for key $K_{\mathscr{B}^*}(\mathsf{w}) = (X^u_{\mathscr{A}^*}(\mathsf{w}) \oplus X^u_{\mathscr{B}^*}(\mathsf{w}), Z^u_{\mathscr{A}^*}(\mathsf{w}) \oplus Z^u_{\mathscr{B}^*}(\mathsf{w}))$ to each of his wires $\mathsf{w} \in \mathcal{B}_0$.

**Swapping for Key-Releasing.** Notice that the key-releasing phase only uses the SWAP-gate with classical input states. The reader might therefore wonder why this functionality is defined quantumly when a classical swap would work equally well. The reason is that, perhaps somewhat surprisingly, a classical swap is a potentially stronger primitive than a quantum swap. From a classical swap one can build a quantum swap by encrypting the quantum states with classical keys, exchange the encrypted states using quantum communication, and then using the classical swap to exchange the keys. Obtaining a classical swap from a quantum one, however, is not obvious. Suppose that registers $\mathcal{A}$ and $\mathcal{B}$ should be swapped classically while holding quantum states beforehand. These registers could be entangled with some purification registers before being swapped. Using a quantum swap between $\mathcal{A}$ and $\mathcal{B}$ will always leave these registers entangled with the purification registers until they become measured while a classical swap will ensure that $\mathcal{A}$ and $\mathcal{B}$ become unentangled with the purification registers after its invocation. In other words, a classical swap could prevent an adversary from exploiting entanglement in his attack.

---

[2] Otherwise, add enough registers initially in state $|0\rangle$ to the smaller set.

**The Ideal AND-Box Functionality.** As we are going to see next, a call to an ideal AND-box is required during the evaluation of the R-gate. Unlike the ideal SWAP used for key-releasing, the AND-box will be modeled by a purely classical primitive denoted AND-BOX. This is required for privacy of our protocol since any implementation of it by some unitary will necessarily leak[17]. The quantum operation implementing it will first measure the two one-qubit input registers in the computational basis in order to get classical inputs $x, y \in \{0,1\}$ for $\mathscr{A}^*$ and $\mathscr{B}^*$ respectively. The classical output bits are then set to $a \in_R \{0,1\}$ for $\mathscr{A}^*$ and $b = a \oplus xy$ for $\mathscr{B}^*$.

## 5.1 Computing over Encrypted States

Before the execution of $G_{j+1}$ in $C_U$, $\mathscr{A}^*$ and $\mathscr{B}^*$ share an encryption of $\rho_j = ((G_j \cdot G_{j-1} \cdot \ldots \cdot G_1) \otimes \mathbb{1}_{\mathcal{R}}) \cdot \rho_{\text{in}}$ in registers[3] holding wires $\mathtt{w} \in \mathcal{A}_0 \cup \mathcal{B}_0$. Each wire $\mathtt{w} \in \mathcal{A}_0 \cup \mathcal{B}_0$ is encrypted by a shared quantum one-time pad as

$$\left( \left( \bigotimes_{\mathtt{w} \in \mathcal{A}_0 \cup \mathcal{B}_0} X^{X^j_{\mathscr{A}^*}(\mathtt{w}) \oplus X^j_{\mathscr{B}^*}(\mathtt{w})} Z^{Z^j_{\mathscr{A}^*}(\mathtt{w}) \oplus Z^j_{\mathscr{B}^*}(\mathtt{w})} \right) \otimes \mathbb{1}_{\mathcal{R}} \right) \cdot \rho_j , \qquad (2)$$

where $K^j_{\mathscr{A}^*}(\mathtt{w}) := (X^j_{\mathscr{A}^*}(\mathtt{w}), Z^j_{\mathscr{A}^*}(\mathtt{w})) \in \{0,1\}^2$ and $K^j_{\mathscr{B}^*}(\mathtt{w}) := (X^j_{\mathscr{B}^*}(\mathtt{w}), Z^j_{\mathscr{B}^*}(\mathtt{w})) \in \{0,1\}^2$ are two bits of secret keys for $\mathscr{A}^*$ and $\mathscr{B}^*$ respectively. In other words, wires $\mathtt{w} \in \mathcal{A}_0 \cup \mathcal{B}_0$ are encrypted by $X^x Z^z$ where $x = X^j_{\mathscr{A}^*}(\mathtt{w}) \oplus X^j_{\mathscr{B}^*}(\mathtt{w})$ and $z = Z^j_{\mathscr{A}^*}(\mathtt{w}) \oplus Z^j_{\mathscr{B}^*}(\mathtt{w})$ are additive sharings for the encryption of $\mathtt{w}$. Then, evaluating $G_{j+1}$ upon state (2) will produce a new sharing $K^{j+1}_A(\mathtt{w}) := (X^{j+1}_A(\mathtt{w}), Z^{j+1}_A(\mathtt{w}))$ and $K^{j+1}_B(\mathtt{w}) := (X^{j+1}_B(\mathtt{w}), Z^{j+1}_B(\mathtt{w}))$ for the encryption of state $\rho_{j+1} = (G_{j+1} \otimes \mathbb{1}_{\mathcal{R}}) \cdot \rho_j$. In the following, we describe how to update the keys for the wires involved in the current gate to be evaluated—all other wires retain their previous values.

## 5.2 Evaluation of Gates in the Pauli and Clifford Groups

Non-trivial Pauli gates (i.e., $X, Y$, and $Z$) can easily be computed on encrypted quantum states since they commute or anti-commute pairwise. Let $G_{j+1} \in \{X, Y, Z\}$ be the Pauli gate to be executed on wire $\mathtt{w}$. It means that up to an irrelevant phase factor, it suffices for the owner of $\mathtt{w}$ to apply $G_{j+1}$ without the need for neither party to update their shared keys, i.e., $K^{j+1}_{\mathscr{A}^*}(\mathtt{w}) := K^j_{\mathscr{A}^*}(\mathtt{w})$ and $K^{j+1}_{\mathscr{B}^*}(\mathtt{w}) := K^j_{\mathscr{B}^*}(\mathtt{w})$.

Now, suppose that $G_{j+1} \in \{\mathsf{H}, \mathsf{P}\}$. Each of these one-qubit gates applied upon wire $\mathtt{w}$ will be computed by simply letting the party owning $\mathtt{w}$ apply $G_{j+1}$. Encryption keys are updated locally as:

$$\mathsf{H} : K^{j+1}_{\mathscr{A}^*} = (X^{j+1}_{\mathscr{A}^*}(\mathtt{w}), Z^{j+1}_{\mathscr{A}^*}(\mathtt{w})) := (Z^i_{\mathscr{A}^*}(\mathtt{w}), X^j_{\mathscr{A}^*}(\mathtt{w})) ,$$
$$K^{j+1}_{\mathscr{B}^*} = (X^{j+1}_{\mathscr{B}^*}(\mathtt{w}), Z^{j+1}_{\mathscr{B}^*}(\mathtt{w})) := (Z^j_{\mathscr{B}^*}(\mathtt{w}), X^j_{\mathscr{B}^*}(\mathtt{w})) ,$$

---

[3] To ease the notation in the following, we assume $\rho_j \in \mathrm{D}(\mathcal{A}_0 \otimes \mathcal{B}_0)$ rather than in $\mathrm{D}(\mathcal{A}_0 \otimes \mathcal{B}_0 \otimes \mathcal{R})$. It is easy to see that this can be done without loss of generality.

$$\mathsf{P}: K_{\mathscr{A}*}^{j+1} = (X_{\mathscr{A}*}^{j+1}(\mathtt{w}), Z_{\mathscr{A}*}^{j+1}(\mathtt{w})) := (X_{\mathscr{A}*}^{j}(\mathtt{w}), X_{\mathscr{A}*}^{j}(\mathtt{w}) \oplus Z_{\mathscr{A}*}^{j}(\mathtt{w})) \ ,$$

$$K_{\mathscr{B}*}^{j+1} = (X_{\mathscr{B}*}^{j+1}(\mathtt{w}), Z_{\mathscr{B}*}^{j+1}(\mathtt{w})) := (X_{\mathscr{B}*}^{j}(\mathtt{w}), X_{\mathscr{B}*}^{j}(\mathtt{w}) \oplus Z_{\mathscr{B}*}^{j}(\mathtt{w})) \ .$$

Any one-qubit gate in the Clifford group can be implemented the same way using their own commutation relations with the Pauli operators used for encryption. A CNOT-gate on local wires can be evaluated in a similar way. That is, whenever both wires $\mathtt{w}$ and $\mathtt{w}'$ feeding the CNOT belong to the same party. Assume that $\mathtt{w}$ is the control wire while $\mathtt{w}'$ is the target and that $\mathscr{A}^*$ holds them both. Then, $\mathscr{A}^*$ simply applies CNOT on wires $\mathtt{w}$ and $\mathtt{w}'$. Encryption keys are updated as:

$$\mathsf{CNOT}: K_{\mathscr{A}*}^{j+1}(\mathtt{w}) = (X_{\mathscr{A}*}^{j+1}(\mathtt{w}), Z_{\mathscr{A}*}^{j+1}(\mathtt{w})) := (X_{\mathscr{A}*}^{j}(\mathtt{w}), Z_{\mathscr{A}*}^{j}(\mathtt{w}) \oplus Z_{\mathscr{A}*}^{j}(\mathtt{w}')) \ ,$$

$$K_{\mathscr{A}*}^{j+1}(\mathtt{w}') = (X_{\mathscr{A}*}^{j+1}(\mathtt{w}'), Z_{\mathscr{A}*}^{j+1}(\mathtt{w}')) := (X_{\mathscr{A}*}^{j}(\mathtt{w}') \oplus X_{\mathscr{A}*}^{j}(\mathtt{w}), Z_{\mathscr{A}*}^{j}(\mathtt{w}')) \ ,$$

$$K_{\mathscr{B}*}^{j+1}(\mathtt{w}) := K_{\mathscr{B}*}^{j}(\mathtt{w}) \text{ and } K_{\mathscr{B}*}^{j+1}(\mathtt{w}') := K_{\mathscr{B}*}^{j}(\mathtt{w}') \ .$$

When $\mathscr{B}^*$ holds both wires, the procedure is simply performed with the roles of $\mathscr{A}^*$ and $\mathscr{B}^*$ reversed.

**Nonlocal CNOT.** We now look at the case where $G_{j+1} = \mathsf{CNOT}$ upon wires $\mathtt{w}$ and $\mathtt{w}'$, one of which is owned by $\mathscr{A}^*$ while the other is owned by $\mathscr{B}^*$. In this case, interaction is unavoidable for the evaluation of the gate. Let us assume w.l.g that $\mathscr{A}^*$ holds the control wire $\mathtt{w}$ while $\mathscr{B}^*$ holds the target wire $\mathtt{w}'$ (i.e., $\mathtt{w} \in \mathcal{A}_0$ and $\mathtt{w}' \in \mathcal{B}_0$). We start from a construction introduced in [7] in the context of fault tolerant quantum computation.

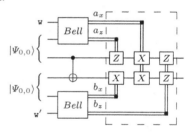

The idea behind the sub-protocol is depicted in Fig. 2. The effect of the Bell measurement is to *teleport* the input state of wires $\mathtt{w}$ and $\mathtt{w}'$ *through* the CNOT-gate[7]. The input to the CNOT appearing in the circuit of Fig. 2 is independent of both input wires $\mathtt{w}$ and $\mathtt{w}'$ (they are just two half EPR-pairs).

The sub-protocol for the evaluation of CNOT simply consists in executing the circuit of Fig. 2 without the decryption part (i.e., the part inside the dotted rectangle). The state

**Fig. 2.** Evaluation of CNOT

$|\xi\rangle := (\mathbb{1}_A \otimes CNOT \otimes \mathbb{1}_B)|\Psi_{0,0}\rangle|\Psi_{0,0}\rangle$ can be prepared by one party. We let the holder of the *control wire* (i.e., $\mathscr{A}^*$ in Fig. 2) prepare $|\xi\rangle$ before sending its two rightmost registers to the other party. The decryption in the dotted-rectangle is used to update the encryption keys according to the measurement outcomes $(a_x, a_z, b_x, b_z)$:

$$\mathsf{CNOT}: K_{\mathscr{A}*}^{j+1}(\mathtt{w}) := (X_{\mathscr{A}*}^{j}(\mathtt{w}) \oplus a_x, Z_{\mathscr{A}*}^{j}(\mathtt{w}) \oplus a_z) \ ,$$

$$K_{\mathscr{B}*}^{j+1}(\mathtt{w}) := (X_{\mathscr{B}*}^{i}(\mathtt{w}), Z_{\mathscr{B}*}^{j}(\mathtt{w}) \oplus b_z) \ ,$$

$$K_{\mathscr{A}*}^{j+1}(\mathtt{w}') := (X_{\mathscr{A}*}^{j}(\mathtt{w}') \oplus a_x, Z_{\mathscr{A}*}^{j}(\mathtt{w}')) \ ,$$

$$K_{\mathscr{B}*}^{j+1}(\mathtt{w}') := (X_{\mathscr{B}*}^{j}(\mathtt{w}') \oplus b_x, Z_{\mathscr{B}*}^{j}(\mathtt{w}') \oplus b_z) \ .$$

As for all previous gates, the key updating phase is performed locally without the need for communication.

## 5.3   Evaluation of the R-Gate

The only gate left in $\mathcal{UG}$ is $G_{j+1} := \mathsf{R}$. We assume without loss of generality that $\mathscr{A}^*$ owns wire $\mathbf{w}$ upon which $\mathsf{R}$ is applied (i.e., $\mathbf{w} \in \mathcal{A}_0$). The subprotocol needs a call to an ideal AND-BOX in order to guarantee privacy during the key updating process. Observe first that the R-gate commutes with Pauli encryption operator $Z$. It means that applying the R-gate upon a state encrypted with $Z$ produces the correct output state still encrypted with $Z$. However, the equality $\mathsf{R} \cdot X = e^{-i\pi/4} Y \mathsf{P} \cdot \mathsf{R}$ tells us that a P-gate should be applied for the decryption of the output when the input has been encrypted using $X$. This breaks the invariant that wires after each gate are all encrypted by Pauli operators. We remove the P-gate by converting it into a sequence of Pauli operators.

Ignoring an irrelevant global phase, the result of applying $\mathsf{R}$ on wire $\mathbf{w}$ is

$$\mathsf{R} Z^{Z^j_{\mathscr{A}*}(\mathbf{w}) \oplus Z^i_{\mathscr{B}*}(\mathbf{w})} X^{X^j_{\mathscr{A}*}(\mathbf{w}) \oplus X^j_{\mathscr{B}*}(\mathbf{w})} =$$

$$Z^{Z^j_{\mathscr{A}*}(\mathbf{w}) \oplus Z^j_{\mathscr{B}*}(\mathbf{w}) \oplus X^j_{\mathscr{A}*}(\mathbf{w}) \oplus X^j_{\mathscr{B}*}(\mathbf{w})} X^{X^j_{\mathscr{A}*}(\mathbf{w}) \oplus X^j_{\mathscr{B}*}(\mathbf{w})} \mathsf{P}^{X^j_{\mathscr{A}*}(\mathbf{w}) \oplus X^j_{\mathscr{B}*}(\mathbf{w})} \mathsf{R} \ , \tag{3}$$

To remove the P-gate, we let each party remove his part of $\mathsf{P}^{X^j_{\mathscr{A}*}(\mathbf{w}) \oplus X^j_{\mathscr{B}*}(\mathbf{w})}$ in a private interactive process. To do this, $\mathscr{A}^*$ picks random bits $r$ and $r'$, and $\mathscr{B}^*$ picks random bits $s$ and $s'$. $\mathscr{A}^*$ applies the operator $X^r Z^{r'} \mathsf{P}^{X^j_{\mathscr{A}*}(\mathbf{w})}$ and sends the resulting quantum state to $\mathscr{B}^*$. $\mathscr{B}^*$

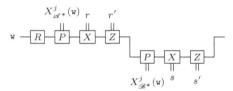

**Fig. 3.** Implementation of the R-gate

applies the operator $X^s Z^{s'} \mathsf{P}^{X^j_{\mathscr{B}*}(\mathbf{w})}$ and sends the result back to $\mathscr{A}^*$. The resulting protocol is shown in Fig. 3. It starts with $\mathscr{A}^*$ applying $\mathsf{R}$ upon the encrypted state before the one-round interactive process described above starts.

After $\mathscr{A}^*$'s application of $\mathsf{R}$, the resulting state is as described on the right-hand side of (3). At the end of the process (i.e., circuit of Fig. 3), the encryption becomes:

$$Z^{s'} X^s \mathsf{P}^{X^j_{\mathscr{B}*}(\mathbf{w})} Z^{r'} X^r \mathsf{P}^{X^j_{\mathscr{A}*}(\mathbf{w})}$$

$$Z^{Z^j_{\mathscr{A}*}(\mathbf{w}) \oplus Z^j_{\mathscr{B}*}(\mathbf{w}) \oplus X^j_{\mathscr{A}*}(\mathbf{w}) \oplus X^j_{\mathscr{B}*}(\mathbf{w})} X^{X^j_{\mathscr{A}*}(\mathbf{w}) \oplus X^j_{\mathscr{B}*}(\mathbf{w})} \mathsf{P}^{X^j_{\mathscr{A}*}(\mathbf{w}) \oplus X^j_{\mathscr{B}*}(\mathbf{w})} \ . \tag{4}$$

Now, we use the fact that $Z$ commutes with $\mathsf{P}$ and $\mathsf{P} \cdot X = XZ \cdot \mathsf{P}$. In addition, since for $a, b \in \{0, 1\}$, $\mathsf{P}^{a+b} = Z^{ab}\mathsf{P}^{a \oplus b}$ we re-write (4) as

$$Z^{s' \oplus r' \oplus X^j_{\mathscr{A}*}(\mathbf{w}) \oplus X^j_{\mathscr{B}*}(\mathbf{w}) \oplus Z^j_{\mathscr{A}*}(\mathbf{w}) \oplus Z^j_{\mathscr{B}*}(\mathbf{w}) \oplus (r \oplus X^j_{\mathscr{A}*}(\mathbf{w})) \cdot X^j_{\mathscr{B}*}(\mathbf{w})}$$

$$X^{s \oplus r \oplus X^j_{\mathscr{A}*}(\mathbf{w}) \oplus X^j_{\mathscr{B}*}(\mathbf{w})} \ . \tag{5}$$

Encryption (5) is not a proper additive sharing since the $Z$-operator depends on $(r \oplus X^j_{\mathscr{A}*}(\mathbf{w})) \cdot X^j_{\mathscr{B}*}(\mathbf{w})$; the logical AND between a value known only by $\mathscr{A}^*$ (i.e., $r \oplus X^j_{\mathscr{A}*}(\mathbf{w})$) and a value known only by $\mathscr{B}^*$ (i.e., $X^j_{\mathscr{B}*}(\mathbf{w})$).

To get back to an additive sharing, $\mathscr{A}^*$ and $\mathscr{B}^*$ can simply call the AND-BOX once with inputs $r \oplus X^j_{\mathscr{A}^*}(\mathtt{w})$ and $X^j_{\mathscr{B}^*}(\mathtt{w})$ respectively as depicted in Fig. 4. After this, $\mathscr{A}^*$ and $\mathscr{B}^*$ share a proper encryption of the resulting state. The new encryption key for $\mathscr{A}^*$'s wire $\mathtt{w}$ becomes:

**Fig. 4.** $\alpha \oplus \beta = (r \oplus X^j_{\mathscr{A}^*}(\mathtt{w})) \cdot X^j_{\mathscr{B}^*}(\mathtt{w})$ from an AND-BOX.

$$\mathsf{R}: K^{j+1}_{\mathscr{A}^*}(\mathtt{w}) := (r \oplus X^j_{\mathscr{A}^*}(\mathtt{w}), r' \oplus \alpha \oplus Z^j_{\mathscr{A}^*}(\mathtt{w}) \oplus X^j_{\mathscr{A}^*}(\mathtt{w})) ,$$
$$K^{j+1}_{\mathscr{B}^*}(\mathtt{w}) := (s \oplus X^j_{\mathscr{B}^*}(\mathtt{w}), s' \oplus \beta \oplus Z^j_{\mathscr{B}^*}(\mathtt{w}) \oplus X^j_{\mathscr{B}^*}(\mathtt{w})) .$$

### 5.4    On the Necessity of Swapping Privately

One may ask whether relying upon SWAP is necessary for the protocol to be private against specious adversaries. For instance, what would happen if one party announces the encryption keys before the other party? We now show that as soon as one party gets the other party's decryption key before having announced its own, a specious adversary can break privacy.

Consider the protocol for a quantum circuit made out of one single CNOT-gate. Suppose that $\mathscr{A}^*$ holds the control wire $\mathtt{w}$ while $\mathscr{B}^*$ holds the target wire $\mathtt{w}'$. Suppose also the key-releasing phase first asks $\mathscr{B}^*$ to announce the encryption keys $K_{\mathscr{B}^*}(\mathtt{w})$ before $\mathscr{A}^*$ announces $K_{\mathscr{A}^*}(\mathtt{w}')$. Suppose $\tilde{\mathscr{A}}$'s input state is $|0\rangle$.

The adversary $\tilde{\mathscr{A}}$ can now act as follows. $\tilde{\mathscr{A}}$ runs the protocol for CNOT without performing the Bell measurement until she receives the encryption key $b_z$ from $\mathscr{B}^*$. Clearly, $\tilde{\mathscr{A}}$'s behavior is specious up to that point since she could re-produce the honest state by just applying the Bell measurement on her input state . However, given $b_z$ she could also in principle compute the CNOT upon any input state of her choice. This means that the state she holds after $b_z$ has been announced and before applying her Bell measurement contains information about $\mathscr{B}^*$'s input. On the one hand, when $\tilde{\mathscr{A}}$'s input state is $|0\rangle$ no information whatsoever on $\mathscr{B}^*$'s input state should be available to her (i.e., in this case CNOT behaves like the identity). On the other hand, had her input state been $|-\rangle$, information about $\mathscr{B}^*$'s state would have become available since the control and target wires exchange their roles when the input states are in the Hadamard basis. However, when $\tilde{\mathscr{A}}$'s input state is $|0\rangle$, any simulation of her view can only call the ideal functionality with input state $|0\rangle$. It follows that no simulator can reproduce $\tilde{\mathscr{A}}$'s state right after the announcement of $b_z$.

## 6    Proof of Privacy

**Privacy of the Evaluation Phase.** We start by showing privacy of protocol $P^{\mathcal{O}}_U = (\mathscr{A}^*, \mathscr{B}^*, n_U + 1)$ at all steps $1 \leq i \leq n_U - 1$ occurring during the *evaluation phase* of quantum circuit $C_U$ implementing $U$ with $u$ gates in $\mathcal{UG}$. The last step of the evaluation phase is $n_U - 1$ since only one oracle call is left to complete the execution. This phase is the easy part of the simulation since all transmissions are independent of the joint input state $\rho_{\mathrm{in}} \in \mathrm{D}(\mathcal{A}_0 \otimes \mathcal{B}_0 \otimes \mathcal{R})$. The lemma below can easily be proven and provides a perfect simulation of any adversary's view generated during the evaluation of any gate in $C_U$. No call to the ideal functionality for $U$ is required.

**Lemma 6.1.** $P_U^{\mathcal{O}} = (\mathscr{A}^*, \mathscr{B}^*, n_U + 1)$ *admits a simulator* $\mathscr{S}(\tilde{\mathscr{A}})$ *for any adversary* $\tilde{\mathscr{A}}$ *(not necessarily specious) that does not call the ideal functionality for* $U \in \mathrm{U}(\mathcal{A}_0 \otimes \mathcal{B}_0)$ *such that for any joint input state* $\rho_{\mathrm{in}} \in \mathrm{D}(\mathcal{A}_0 \otimes \mathcal{B}_0 \otimes \mathcal{R})$, *every* $1 \leq i \leq n_U - 1$:
$$\Delta\left(\nu_i(\tilde{\mathscr{A}}, \rho_{\mathrm{in}}), \mathrm{tr}_{\mathcal{B}_i}\left(\tilde{\rho}_i(\tilde{\mathscr{A}}, \rho_{\mathrm{in}})\right)\right) = 0 .$$

*The same holds against any adversary* $\tilde{\mathscr{B}}$.

**Privacy of the Key-Releasing Phase.** Before proving privacy of the key-releasing phase, we need the following lemma establishing that at the end of the protocol, specious adversaries must leave their extra working registers (used to implement the attack) independent of the joint input state. In other words, no extra information is available to the adversary at the very end of any correct protocol. Hence, if the adversary can break the privacy of a protocol, then he must "rush" to do so before the last step.

**Lemma 6.2 (Rushing Lemma).** *Let* $\Pi_U^{\mathcal{O}} = (\mathscr{A}, \mathscr{B}, n)$ *be a correct protocol for the two party evaluation of* $U$. *Let* $\tilde{\mathscr{A}}$ *be any* $\varepsilon$-*specious adversary in* $\Pi_U^{\mathcal{O}}$. *Then, there exists an isometry* $T : \tilde{\mathcal{A}}_n \to \mathcal{A}_n \otimes \widehat{\mathcal{A}}$ *and a mixed state* $\tilde{\varrho} \in \mathrm{D}(\widehat{\mathcal{A}})$ *such that for all joint input states* $\rho_{\mathrm{in}} \in \mathrm{D}(\mathcal{A}_0 \otimes \mathcal{B}_0 \otimes \mathcal{R})$,
$$\Delta\left((T \otimes \mathbb{1}_{\mathcal{B}_n \otimes \mathcal{R}}) \cdot \left([\tilde{\mathscr{A}} \circledast \mathscr{B}](\rho_{\mathrm{in}})\right), \tilde{\varrho} \otimes (U \otimes \mathbb{1}_{\mathcal{R}}) \cdot \rho_{\mathrm{in}}\right) \leq 12\sqrt{2\varepsilon} .$$

*The same also applies to any* $\varepsilon$-*specious adversary* $\tilde{\mathscr{B}}$.

*Proof.* We shall only prove the statement for an $\varepsilon$-specious $\tilde{\mathscr{A}}$; the statement for an $\varepsilon$-specious $\tilde{\mathscr{B}}$ is identical. Furthermore, by convexity, it is sufficient to prove the theorem for pure $\rho_{\mathrm{in}}$.

Consider any pair of pure input states $|\psi_1\rangle$ and $|\psi_2\rangle$ in $\mathcal{A}_0 \otimes \mathcal{B}_0 \otimes \mathcal{R}$. Now, let $\mathcal{R}' := \mathcal{R} \otimes \mathcal{R}_2$, where $\mathcal{R}_2 = \mathrm{span}\{|1\rangle, |2\rangle\}$ represents a single qubit, and define the state $|\psi\rangle := \frac{1}{\sqrt{2}}(|\psi_1\rangle|1\rangle + |\psi_2\rangle|2\rangle) \in \mathcal{A}_0 \otimes \mathcal{B}_0 \otimes \mathcal{R}'$. Note that $\mathrm{tr}_{\mathcal{R}_2}(|\psi\rangle\langle\psi|) = \frac{1}{2}|\psi_1\rangle\langle\psi_1| + \frac{1}{2}|\psi_2\rangle\langle\psi_2|$. Due to the correctness of the protocol and to the speciousness of $\tilde{\mathscr{A}}$, there exists a quantum operation $\mathscr{T}_n : \mathrm{L}(\tilde{\mathcal{A}}_n) \to \mathrm{L}(\mathcal{A}_n)$ such that
$$\Delta\left((\mathscr{T}_n \otimes \mathbb{1}_{\mathrm{L}(\mathcal{B}_n \otimes \mathcal{R}')})([\tilde{\mathscr{A}} \circledast \mathscr{B}](|\psi\rangle\langle\psi|)), (U \otimes \mathbb{1}_{\mathcal{R}'}) \cdot |\psi\rangle\langle\psi|\right) \leq 2\varepsilon .$$

Now, consider any isometry $T : \tilde{\mathcal{A}}_n \to \mathcal{A}_n \otimes \widehat{\mathcal{A}}$ such that $\mathscr{T}_n(\sigma) = \mathrm{tr}_{\widehat{\mathcal{A}}}(T\sigma T^\dagger)$ for every $\sigma \in \mathrm{L}(\tilde{\mathcal{A}}_n)$ — in other words, any operation that implements $\mathscr{T}_n$ while keeping any information that would otherwise be destroyed in $\widehat{\mathcal{A}}$. By Uhlmann's theorem, there must exist a state $\tilde{\varrho} \in \mathrm{D}(\widehat{\mathcal{A}})$ such that
$$\Delta\left((T \otimes \mathbb{1}_{\mathcal{B}_n \otimes \mathcal{R}'}) \cdot \left([\tilde{\mathscr{A}} \circledast \mathscr{B}](|\psi\rangle\langle\psi|)\right), \tilde{\varrho} \otimes ((U \otimes \mathbb{1}_{\mathcal{R}'}) \cdot |\psi\rangle\langle\psi|)\right) \leq 2\sqrt{2\varepsilon} .$$

Now, the trace distance is monotonous under completely positive, trace non-increasing maps. In particular, we can apply the projector $P_1 = \mathbb{1}_{L(\mathcal{A}_n \otimes \mathcal{B}_n \otimes \mathcal{R})} \otimes |1\rangle\langle 1|$ to both states in the above trace distance and the inequality will still hold. In other words, we project both states onto $|1\rangle$ on $\mathcal{R}_2$, thereby turning $|\psi\rangle\langle\psi|$ into $\frac{1}{2}|\psi_1\rangle\langle\psi_1|$. Factoring out the $\frac{1}{2}$, we get that

$$\Delta\left((T \otimes \mathbb{1}_{\mathcal{B}_n \otimes \mathcal{R}}) \cdot \left([\tilde{\mathcal{A}} \circledast \mathcal{B}](|\psi_1\rangle\langle\psi_1|)\right), \tilde{\varrho} \otimes ((U \otimes \mathbb{1}_{\mathcal{R}}) \cdot |\psi_1\rangle\langle\psi_1|)\right) \leq 4\sqrt{2\varepsilon} \ .$$

Likewise, projecting onto $|2\rangle$ yields

$$\Delta\left((T \otimes \mathbb{1}_{\mathcal{B}_n \otimes \mathcal{R}}) \cdot \left([\tilde{\mathcal{A}} \circledast \mathcal{B}](|\psi_2\rangle\langle\psi_2|)\right), \tilde{\varrho} \otimes ((U \otimes \mathbb{1}_{\mathcal{R}}) \cdot |\psi_2\rangle\langle\psi_2|)\right) \leq 4\sqrt{2\varepsilon} \ .$$

Our only problem at this point is that $\tilde{\varrho}$ in principle depends on $|\psi_1\rangle$ and $|\psi_2\rangle$. However, repeating the above argument with $|\psi_1\rangle$ and $|\psi_3\rangle$ for any $|\psi_3\rangle$ will yield a $\tilde{\varrho}'$ with

$$\Delta\left((T \otimes \mathbb{1}_{\mathcal{B}_n \otimes \mathcal{R}}) \cdot \left([\tilde{\mathcal{A}} \circledast \mathcal{B}](|\psi_1\rangle\langle\psi_1|)\right), \tilde{\varrho}' \otimes ((U \otimes \mathbb{1}_{\mathcal{R}}) \cdot |\psi_1\rangle\langle\psi_1|)\right) \leq 4\sqrt{2\varepsilon}$$

and hence, by the triangle inequality, $\Delta(\tilde{\varrho}, \tilde{\varrho}') \leq 8\sqrt{2\varepsilon}$. Therefore, for any state $|\varphi\rangle \in \mathcal{A}_0 \otimes \mathcal{B}_0 \otimes \mathcal{R}$, there exists a state $\tilde{\rho} \in \hat{\mathcal{A}}$ with $\Delta(\tilde{\rho}, \tilde{\varrho}) \leq 8\sqrt{2\varepsilon}$ such that

$$\Delta\left((T \otimes \mathbb{1}_{\mathcal{B}_n \otimes \mathcal{R}}) \cdot \left([\tilde{\mathcal{A}} \circledast \mathcal{B}](|\varphi\rangle\langle\varphi|)\right), \tilde{\rho} \otimes ((U \otimes \mathbb{1}_{\mathcal{R}}) \cdot |\varphi\rangle\langle\varphi|)\right) \leq 4\sqrt{2\varepsilon} \ .$$

The lemma then follows by the triangle inequality:

$$\Delta\left((T \otimes \mathbb{1}_{\mathcal{B}_n \otimes \mathcal{R}}) \cdot \left([\tilde{\mathcal{A}} \circledast \mathcal{B}](|\varphi\rangle\langle\varphi|)\right), \tilde{\varrho} \otimes ((U \otimes \mathbb{1}_{\mathcal{R}}) \cdot |\varphi\rangle\langle\varphi|)\right)$$
$$\leq \Delta\left((T \otimes \mathbb{1}_{\mathcal{B}_n \otimes \mathcal{R}}) \cdot \left([\tilde{\mathcal{A}} \circledast \mathcal{B}](|\varphi\rangle\langle\varphi|)\right), \tilde{\rho} \otimes ((U \otimes \mathbb{1}_{\mathcal{R}}) \cdot |\varphi\rangle\langle\varphi|)\right) + \Delta(\tilde{\rho}, \tilde{\varrho})$$
$$\leq 4\sqrt{2\varepsilon} + 8\sqrt{2\varepsilon} = 12\sqrt{2\varepsilon} \ . \qquad \square$$

In order to conclude the privacy of $P_U^{\mathcal{O}}$, families $\mathscr{S}(\tilde{\mathcal{A}})$ and $\mathscr{S}(\tilde{\mathcal{B}})$ need one more simulator each: $\mathscr{S}_{n_U} \in \mathscr{S}(\tilde{\mathcal{A}})$ and $\mathscr{S}'_{n_U} \in \mathscr{S}(\tilde{\mathcal{B}})$ corresponding to the simulation of the key-releasing phase. This time, these simulators need to query the ideal functionality for $U$ and also need the adversary to be specious. We show that privacy of the key-releasing phase follows from the "Rushing Lemma" (Lemma 6.2). This is the role of the ideal SWAP to make sure that before the adversary gets the output of the computation, the information needed by the honest player to recover its own output has been given away by the adversary.

It should be mentioned that we're not explicitly simulating the final state of the adversary since simulating the SWAP allows also to get $\tilde{\mathcal{A}}$'s final state by simply adding $\tilde{\mathcal{A}}$'s last quantum operation to the simulated view. We therefore set step $n_U$ in $P_U^{\mathcal{O}}$ to be the step reached after the call to SWAP. This abuses the notation a bit since after SWAP, $\tilde{\mathcal{A}}$ and $\mathcal{B}^*$ must each apply a final quantum operation with no more oracle call. We'll denote by $\tilde{\mathcal{A}}_{n_U+1}$ and $\mathcal{B}^*_{n_U+1}$ these last operations allowing to reconstruct the output of the computation (no comunication).

**Lemma 6.3.** *For any $\varepsilon$-specious quantum adversary $\tilde{\mathscr{A}}$ against $P_U^{\mathcal{O}} = (\mathscr{A}^*, \mathscr{B}^*, n_U + 1)$, there exist simulators $\mathscr{S}_{n_U} \in \mathscr{S}(\tilde{\mathscr{A}})$ such that for all $\rho_{\mathrm{in}} \in \mathrm{D}(\mathcal{A}_0 \otimes \mathcal{B}_0 \otimes \mathcal{R})$,*

$$\Delta\left(\nu_{n_U}(\tilde{\mathscr{A}}, \rho_{\mathrm{in}}), \mathrm{tr}_{\mathcal{B}_{n_U}}\left(\tilde{\rho}_{n_U}(\tilde{\mathscr{A}}, \rho_{\mathrm{in}})\right)\right) \le 24\sqrt{2\varepsilon} \ .$$

*Simulator $\mathscr{S}_{n_U}$ calls the ideal functionality for $U$ and can be used directly to simulate step $n_U + 1$ as well. The same holds for adversary $\tilde{\mathscr{B}}$.*

*Proof (sketch).* We only prove privacy against adversary $\tilde{\mathscr{A}}$, privacy against $\tilde{\mathscr{B}}$ follows directly since the key-releasing phase is symmetric. The idea behind the proof is to run $\tilde{\mathscr{A}}$ and $\mathscr{B}^*$ upon a dummy joint input state until the end of the protocol. Since the adversary is specious, it can re-produce the honest state at the end. The Rushing Lemma tells us that at this point, the output of the computation is essentially in tensor product with all the other registers. Moreover, the state of all other registers is independent of the input state upon which the protocol is executed. The *dummy output* can then be replaced by the output of the ideal functionality for $U$ before $\tilde{\mathscr{A}}$ goes back to the stage reached just after SWAP.

More formally, we define a simulator $\mathscr{S}_{n_U} \in \mathscr{S}(\tilde{\mathscr{A}})$ producing $\tilde{\mathscr{A}}$'s view just after the call to SWAP. Let $\tilde{\mathscr{A}}_{\mathsf{SWAP}} \in \mathrm{L}(\mathcal{A}_0, \tilde{\mathcal{A}}_{n_U})$ and $\mathscr{B}^*_{\mathsf{SWAP}} \in \mathrm{L}(\mathcal{B}_0, \tilde{\mathcal{B}}_{n_U})$ be the quantum operations run by $\tilde{\mathscr{A}}$ and $\mathscr{B}^*$ respectively until after SWAP is executed. Notice that at this point, $\tilde{\mathscr{A}}$'s and $\mathscr{B}^*$'s registers do not have any further oracle registers since no more communication or oracle call will take place. Let $\tilde{A}_{n_U} \in \mathrm{L}(\tilde{\mathcal{A}}_{n_U}, \tilde{\mathcal{A}}_{n_U+1} \otimes \mathcal{Z})$ be the isometry implementing $\tilde{\mathscr{A}}$'s last quantum operation taking place after the call to SWAP (and producing her final state) and let $B_{n_U} \in \mathrm{L}(\mathcal{B}_{n_U}, \mathcal{B}_{n_U+1} \otimes \mathcal{W})$ be the isometry implementing $\mathscr{B}^*$'s last quantum operation. Finally, let $T \in \mathrm{L}(\tilde{\mathcal{A}}_{n_U+1}, \mathcal{A}_{n_U+1} \otimes \hat{\mathcal{A}})$ be the isometry implementing $\mathscr{T}_{n_U+1}$ as defined in Lemma 6.2 (i.e., the transcript produced at the very end of the protocol). As usual, let $\rho_{\mathrm{in}} \in \mathrm{D}(\mathcal{A}_0 \otimes \mathcal{B}_0 \otimes \mathcal{R})$ be the joint input state. The simulator $\mathscr{S}_{n_U}$ performs the following operations:

1. $\mathscr{S}_{n_U}$ generates the quantum state $\sigma(\phi^*) = [\tilde{\mathscr{A}}_{\mathsf{SWAP}} \circledast \mathscr{B}^*_{\mathsf{SWAP}}](|\phi^*\rangle\langle\phi^*|) \in \mathrm{D}(\tilde{\mathcal{A}}_{n_U} \otimes \mathcal{B}_{n_U})$ implementing $\tilde{\mathscr{A}}$ interacting with $\mathscr{B}^*$ until SWAP is applied. The execution is performed upon a predetermined (dummy) arbitrary input state $|\phi^*\rangle \in \mathcal{A}_0 \otimes \mathcal{B}_0$.
2. $\mathscr{S}_{n_U}$ sets $\sigma'(\phi^*) = (T\tilde{A}_{n_U} \otimes B_{n_U}) \cdot \sigma(\phi^*) \in \mathrm{D}(\mathcal{A}_{n_U+1} \otimes \mathcal{B}_{n_U+1} \otimes \mathcal{Z} \otimes \hat{\mathcal{A}} \otimes \mathcal{W})$.
3. $\mathscr{S}_{n_U}$ replaces register $\mathcal{A}_{n_U+1} \approx \mathcal{A}_0$ by $\mathscr{A}^*$'s output of the ideal functionality for $U$ evaluated upon $\rho_{\mathrm{in}}$. That is, $\mathscr{S}_{n_U}$ generates the state $\sigma'(\rho_{\mathrm{in}}) = (U \otimes \mathbb{1}_{\mathcal{R}}) \cdot \rho_{\mathrm{in}} \otimes \mathrm{tr}_{\mathcal{A}_{n_U+1}\mathcal{B}_{n_U+1}}(\sigma'(\phi^*)) \in \mathrm{D}(\mathcal{A}_{n_U+1} \otimes \mathcal{B}_{n_U+1} \otimes \mathcal{R} \otimes \mathcal{Z} \otimes \hat{\mathcal{A}} \otimes \mathcal{W})$.
4. $\mathscr{S}_{n_U}$ finally sets $\nu_{n_U}(\tilde{\mathscr{A}}, \rho_{\mathrm{in}}) = \mathrm{tr}_{\mathcal{B}_{n_U+1}\mathcal{W}}((T\tilde{A}_{n_U} \otimes \mathbb{1}_{\mathcal{B}_{n_U+1}\mathcal{R}})^\dagger \cdot \sigma'(\rho_{\mathrm{in}})) \in \mathrm{D}(\tilde{\mathcal{A}}_{n_U} \otimes \mathcal{R})$.

Notice that execution of the ideal SWAP ensures that the keys swapped are independent of each other and of the joint input state $\rho_{\mathrm{in}}$. This is because for any input state, all these keys are uniformly distributed bits if they are outcomes of Bell measurements and otherwise are set to 0. By Lemma 6.2 and the fact that $\tilde{\mathscr{A}}$ is $\varepsilon$–specious, we have:

$$\Delta\left(\operatorname{tr}_{\mathcal{Z}\hat{A}\mathcal{W}}\left(\sigma'(\phi^*)\right),\tilde{\varrho}\otimes U\cdot|\phi^*\rangle\langle\phi^*|\right)\le 12\sqrt{2\varepsilon}\text{ and}$$

$$\Delta\left(\left(\mathscr{T}_{n_U+1}\otimes\mathbb{1}_{\mathrm{L}(\mathcal{B}_{n_U+1})}\right)\left([\tilde{\mathscr{A}}\circledast\mathscr{B}^*](\rho_{\mathrm{in}})\right),\tilde{\varrho}\otimes U\cdot\rho_{\mathrm{in}}\right)\le 12\sqrt{2\varepsilon}\ .$$

It follows using the triangle inequality that,

$$\Delta\left(\left(\mathscr{T}_{n_U+1}\otimes\mathbb{1}_{\mathrm{L}(\mathcal{B}_{n_U+1})}\right)\left([\tilde{\mathscr{A}}\circledast\mathscr{B}^*](\rho_{\mathrm{in}})\right),\operatorname{tr}_{\mathcal{Z}\hat{A}\mathcal{W}}\left(\sigma'(\rho_{\mathrm{in}})\right)\right)\le 24\sqrt{2\varepsilon}\ . \quad (6)$$

Using the fact that isometries cannot increase the trace-norm distance and that $(T\tilde{A}_{n_U})^\dagger$ allows $\tilde{\mathscr{A}}$ to go back from the end of the protocol to the step reached after SWAP, we get from (6) that

$$\Delta\left(\nu_{n_U}(\tilde{\mathscr{A}},\rho_{\mathrm{in}}),\operatorname{tr}_{\mathcal{B}_{n_U}}\left(\tilde{\rho}_{n_U}(\tilde{\mathscr{A}},\rho_{\mathrm{in}})\right)\right)=$$
$$\Delta\left(\left(\mathscr{T}_{n_U+1}\otimes\mathbb{1}_{\mathrm{L}(\mathcal{B}_{n_U+1})}\right)\left([\tilde{\mathscr{A}}\circledast\mathscr{B}^*](\rho_{\mathrm{in}})\right),\operatorname{tr}_{\mathcal{Z}\hat{A}\mathcal{W}}\left(\sigma'(\rho_{\mathrm{in}})\right)\right)\le 24\sqrt{2\varepsilon}\ .$$

The proof of the statement follows.    □

# 7    Main Result and Open Questions

Putting Lemma 6.1 and Lemma 6.3 together gives the desired result:

**Theorem 7.1 (Main Result).** *Protocol $P_U^{\mathcal{O}}$ is statistically private against any statistically specious quantum adversary and for any $U\in\mathrm{U}(\mathcal{A}_0\otimes\mathcal{B}_0)$. If $U$ is in the Clifford group then the only non-trivial oracle call in $\mathcal{O}$ is one call to an ideal SWAP. If $U$ is not in the Clifford group then $\mathcal{O}$ contains an additional oracle call to* AND-BOX *for each* R*-gate in the circuit for $U$.*

It should be mentioned that it is not too difficult to modify our protocol in order to privately evaluate quantum operations rather than only unitary transforms. Classical two party computation together with the fact that quantum operations can be viewed as unitaries acting in larger spaces can be used to achieve this extra functionality. Privacy can be preserved by keeping these extra registers encrypted after the execution of the protocol. We leave this discussion to the full version of the paper.

A few interesting questions remain open:

– It would be interesting to know whether there exists a unitary transform that can act as a universal primitive for private two-party evaluation of unitaries. This would allow to determine whether classical cryptographic assumptions are required for this task.
– Finally, is there a way to compile quantum protocols secure against specious adversaries into protocols secure against arbitrary quantum adversaries? An affirmative answer would allow to simplify greatly the design of quantum protocols. Are extra assumptions needed to preserve privacy against any adversary?

# Acknowledgements

The authors would like to thank the referees for their comments and suggestions. We would also like to thank Thomas Pedersen for numerous helpful discussions in the early stage of this work.

# References

1. Aharonov, D., Ben-Or, M.: Fault-tolerant quantum computation with constant error. In: 29th Annual ACM Symposium on Theory of Computing (STOC), pp. 176–188 (1997)
2. Ambainis, A., Mosca, M., Tapp, A., de Wolf, R.: Private quantum channels. In: 41st Annual IEEE Symposium on Foundations of Computer Science (FOCS), pp. 547–553 (2000)
3. Ben-Or, M., Crépeau, C., Gottesman, D., Hassidim, A., Smith, A.: Secure multiparty quantum computation with (only) a strict honest majority. In: 47th Annual IEEE Symposium on Foundations of Computer Science (FOCS), pp. 249–260 (2006)
4. Broadbent, A., Fitzsimons, J., Kashefi, E.: Universal blind quantum computation (December 2009), http://arxiv.org/abs/0807.4154
5. Crépeau, C., Gottesman, D., Smith, A.: Secure multi-party quantum computation. In: 34th Annual ACM Symposium on Theory of Computing (STOC), pp. 643–652 (2002)
6. Gottesman, D., Chuang, I.L.: Demonstrating the viability of universal quantum computation using teleportation and single-qubit operations. Nature 402, 390–393 (1999)
7. Gottesman, D., Chuang, I.L.: Quantum teleportation is a universal computational primitive (August 1999), http://arxiv.org/abs/quant-ph/9908010
8. Gutoski, G., Watrous, J.: Quantum interactive proofs with competing provers. In: Diekert, V., Durand, B. (eds.) STACS 2005. LNCS, vol. 3404, pp. 605–616. Springer, Heidelberg (2005)
9. Kilian, J.: Founding cryptography on oblivious transfer. In: 20th Annual ACM Symposium on Theory of Computing (STOC), pp. 20–31 (1988)
10. Lo, H.-K.: Insecurity of quantum secure computations. Physical Review A 56(2), 1154–1162 (1997)
11. Lo, H.-K., Chau, H.F.: Is quantum bit commitment really possible? Physical Review Letters 78, 3410–3413 (1997)
12. Mayers, D.: Unconditionally secure quantum bit commitment is impossible. Physical Review Letters 78, 3414–3417 (1997)
13. Nielsen, M.A., Chuang, I.L.: Quantum Computation and Quantum Information. Cambridge University Press, Cambridge (2000)
14. Popescu, S., Rohrlich, D.: Quantum nonlocality as an axiom. Foundations of Physics 24(3), 379–385 (1994)
15. Popescu, S., Rohrlich, D.: Causality and nonlocality as axioms for quantum mechanics. In: Symposium on Causality and Locality in Modern Physics and Astronomy: Open Questions and Possible Solutions (1997), http://arxiv.org/abs/quant-ph/9709026
16. Renner, R., König, R.: Universally composable privacy amplification against quantum adversaries. In: Kilian, J. (ed.) TCC 2005. LNCS, vol. 3378, pp. 407–425. Springer, Heidelberg (2005)

17. Salvail, L., Sotáková, M., Schaffner, C.: On the power of two-party quantum cryptography. In: Matsui, M. (ed.) ASIACRYPT 2009. LNCS, vol. 5912, pp. 70–87. Springer, Heidelberg (2009)
18. Shor, P.W.: Fault-tolerant quantum computation. In: 37th Annual IEEE Symposium on Foundations of Computer Science (FOCS), pp. 56–65 (1996)
19. Smith, A.: Techniques for secure distributed computing with quantum data. Presented at the Field's institute Quantum Cryptography and Computing Workshop (October 2006)
20. Watrous, J.: Limits on the power of quantum statistical zero-knowledge. In: 43rd Annual IEEE Symposium on Foundations of Computer Science (FOCS), pp. 459–468 (2002)
21. Wolf, S., Wullschleger, J.: Oblivious transfer and quantum non-locality. In: International Symposium on Information Theory (ISIT 2005), pp. 1745–1748 (2005)
22. Yao, A.: How to generate and exchange secrets. In: 27th Annual IEEE Symposium on Foundations of Computer Science (FOCS) (1986)

# On the Efficiency of Classical and Quantum Oblivious Transfer Reductions

Severin Winkler[1] and Jürg Wullschleger[2]

[1] ETH Zurich, Switzerland
swinkler@ethz.ch
[2] University of Bristol, United Kingdom
j.wullschleger@bristol.ac.uk

**Abstract.** Due to its universality oblivious transfer (OT) is a primitive of great importance in secure multi-party computation. OT is impossible to implement from scratch in an unconditionally secure way, but there are many reductions of OT to other variants of OT, as well as other primitives such as noisy channels. It is important to know how efficient such unconditionally secure reductions can be in principle, i.e., how many instances of a given primitive are at least needed to implement OT. For perfect (error-free) implementations good lower bounds are known, e.g. the bounds by Beaver (STOC '96) or by Dodis and Micali (EUROCRYPT '99). However, in practice one is usually willing to tolerate a small probability of error and it is known that these *statistical* reductions can in general be much more efficient. Thus, the known bounds have only limited application. In the first part of this work we provide bounds on the efficiency of secure (one-sided) two-party computation of arbitrary finite functions from distributed randomness in the statistical case. From these results we derive bounds on the efficiency of protocols that use (different variants of) OT as a black-box. When applied to implementations of OT, our bounds generalize known results to the statistical case. Our results hold in particular for transformations between a finite number of primitives and for *any* error. Furthermore, we provide bounds on the efficiency of protocols implementing Rabin OT.

In the second part we study the efficiency of quantum protocols implementing OT. Recently, Salvail, Schaffner and Sotakova (ASIACRYPT '09) showed that most classical lower bounds for *perfectly* secure reductions of OT to distributed randomness still hold in a quantum setting. We present a statistically secure protocol that violates these bounds by an arbitrarily large factor. We then present a weaker lower bound that *does* hold in the statistical quantum setting. We use this bound to show that even quantum protocols cannot extend OT. Finally, we present two lower bounds for reductions of OT to commitments and a protocol based on string commitments that is optimal with respect to both of these bounds.

**Keywords:** Unconditional Security, Oblivious Transfer, Lower Bounds, Quantum Cryptography, Two-Party Computation.

T. Rabin (Ed.): CRYPTO 2010, LNCS 6223, pp. 707–723, 2010.
© International Association for Cryptologic Research 2010

# 1    Introduction

Secure multi-party computation allows two or more distrustful players to jointly compute a function of their inputs in a secure way [48]. Security here means that the players compute the value of the function correctly without learning more than what they can derive from their own input and output.

A primitive of central importance in secure multi-party computation is *oblivious transfer* (OT), as it is sufficient to execute any multi-party computation securely [25,27]. The original form of OT $((\frac{1}{2})\text{-RabinOT}^1)$ has been introduced by Rabin in [35]. It allows a sender to send a bit $x$, which the receiver will get with probability $\frac{1}{2}$. Another variant of OT, called one-out-of-two bit-OT $(\binom{2}{1}\text{-OT}^1)$ was defined in [23] (see also [39]). Here, the sender has two input bits $x_0$ and $x_1$. The receiver gives as input a choice bit $c$ and receives $x_c$ without learning $x_{1-c}$. The sender gets no information about the choice bit $c$. Other important variants of OT are $\binom{n}{t}\text{-OT}^k$ where the inputs are strings of $k$ bits and the receiver can choose $t < n$ out of $n$ secrets and $(p)\text{-RabinOT}^k$ where the inputs are strings of $k$ bits and the erasure probability is $p \in [0,1]$.

If the players have access to noiseless (classical or quantum) communication only, it is impossible to implement unconditionally secure OT, i.e. secure against an adversary with unlimited computing power. It has been shown in [13] that $(p)\text{-RabinOT}^k$ and $\binom{2}{1}\text{-OT}^1$ are equally powerful, i.e., one can be implemented from the other. Numerous reductions between different variants of $\binom{n}{1}\text{-OT}^k$ are known as well: $\binom{2}{1}\text{-OT}^k$ can be implemented from $\binom{2}{1}\text{-OT}^1$ [5,15,9,8], and $\binom{n}{1}\text{-OT}^k$ can be implemented from $\binom{2}{1}\text{-OT}^{k'}$ [7,9,21,44]. There has also been a lot of interest in reductions of OT to weaker primitives. It is known that OT can be realized from noisy channels [12,14,18,47], noisy correlations [42,33], or weak variants of OT [12,10,20,8,19,46].

In the quantum world, it has been shown in [6,49,17,38] that OT can be implemented from black-box commitments, something that is impossible in the classical setting.

Given these positive results it is natural to ask how efficient such reductions can be in principle, i.e., how many instances of a given primitive are needed to implement OT.

## 1.1    Previous Results

In the classical setting, several lower bounds for OT reductions are known. The first impossibility result for unconditionally secure reductions of OT has been presented in [2]. There it has been shown that the number of $\binom{2}{1}\text{-OT}^1$ cannot be *extended*[1], i.e., there does not exist a protocol using $n$ instances of $\binom{2}{1}\text{-OT}^1$ that perfectly implements $m > n$ instances. Lower bounds for the number of instances of OT needed to perfectly implement other variants of OT have been presented in [21] (see also [31]) and generalized in [44,43]. These bounds apply to both the semi-honest (where dishonest players follow the protocol) and the malicious

---

[1] Note that in the computational setting, OT *can* be extended, see [2,26].

(where dishonest players behave arbitrarily) model. If we restrict ourselves to the malicious model these bounds can be improved, as shown in [28]. Lower bounds on the number of ANDs needed to implement general functions have been presented in [4].

All these results only consider *perfect* protocols and do not give much insight into the case of statistical implementations. As pointed out in [28], their result *only* applies to the perfect case, because there is a statistical protocol that is more efficient [16]. The bounds for perfect and statistical protocols can in fact be *very* far apart, as shown in [4]: The amount of OTs needed to compute the equality function is exponentially bigger in the perfect case than in the statistical case. Therefore, it is not true in general that a bound in the perfect case implies a similar bound in the statistical case.

So far very little is known in the statistical case. In [1] a proof sketch of a lower bound for statistical implementations of $\binom{2}{1}$-$\mathsf{OT}^k$ has been presented. However, this result only holds in the asymptotic case, where the number $n$ of resource primitives goes to infinity and the error goes to zero as $n$ goes to infinity. In [4] a non-asymptotic lower bound on the number of ANDs needed for one-sided secure computation of arbitrary functions with *boolean* output has been shown. This result directly implies lower bounds for protocols that use $\binom{n}{t}$-$\mathsf{OT}^k$ as a black-box. However, besides being restricted to boolean-valued functions this result is not strong enough to show optimality of several known reductions and it does not provide bounds for reductions to randomized primitives such as $(\frac{1}{2})$-$\mathsf{RabinOT}^1$.

In the quantum setting almost all negative results known show that a certain primitive is impossible to implement from scratch. Commitment has been shown to be impossible in the quantum setting in [32,30]. Using a similar proof, it has been shown in [29] that general one-sided two-party computation and in particular oblivious transfer are also impossible to implement securely in the quantum setting.

To our knowledge, the only lower bounds for quantum protocols where the players have access to resource primitives (such as different variants of OT) have been presented in [36] where Theorem 4.7 shows that important lower bounds for classical protocols also apply to *perfectly* secure quantum reductions.

### 1.2   Contribution

*Classical Reductions.* In Section 2 we consider statistically secure protocols in the semi-honest model that compute a function between two parties from trusted randomness distributed to the players. We provide two bounds on the efficiency of such reductions that allow in particular to derive bounds on the minimal number of $\binom{n}{t}$-$\mathsf{OT}^k$ or $(p)$-$\mathsf{RabinOT}^k$ needed to compute any given function securely. Our bounds do not involve any asymptotics, i.e., we consider a finite number of resource primitives and our results hold for *any* error.

In Section 2.3 we provide an additional bound for the special case of statistical implementations of $\binom{n}{1}$-$\mathsf{OT}^k$. Note that for implementations of OT bounds in the semi-honest model imply similar bounds in the malicious model[2]. The bounds for implementations of $\binom{n}{1}$-$\mathsf{OT}^k$ (Theorem 3) imply the following corollary that gives a general bound on the conversion rate between different variants of OT.

**Corollary 1.** *For any reduction that implements $M$ instances of $\binom{N}{1}$-$\mathsf{OT}^K$ from $m$ instances of $\binom{n}{1}$-$\mathsf{OT}^k$ in the semi-honest model with an error of at most $\varepsilon$, we have*

$$\frac{m}{M} \geq \max\left(\frac{(N-1)K}{(n-1)k}, \frac{K}{k}, \frac{\log N}{\log n}\right) - 7NK \cdot (\varepsilon + h(\varepsilon)).$$

Corollary 1 generalizes the lower bounds from [21,44,43] to the statistical case and is strictly stronger than the impossibility bounds from [1]. If we let $M = m + 1$, $N = n = 2$ and $K = k = 1$, we obtain a stronger version of Theorem 3 from [2] which states that OT cannot be extended.

In the full version of this paper [40], we also derive new bounds in the statistical case for protocols implementing $(p)$-$\mathsf{RabinOT}^k$, and show that our bounds imply bounds for implementations of oblivious linear function evaluation (OLFE).

Our lower bounds show that the following protocols are (close to) optimal in the sense that they use the minimal number of instances of the given primitive.

– The protocol in [9,21] which uses $\frac{N-1}{n-1}$ instances of $\binom{n}{1}$-$\mathsf{OT}^k$ to implement $\binom{N}{1}$-$\mathsf{OT}^k$ is optimal.
– The protocol in [44] which uses $t$ instances of $\binom{n}{1}$-$\mathsf{OT}^{kn^{t-1}}$ to implement $\binom{n^t}{1}$-$\mathsf{OT}^k$ is optimal.
– In the semi-honest model, the trivial protocol that implements $\binom{2}{1}$-$\mathsf{OT}^k$ from $k$ instances of $\binom{2}{1}$-$\mathsf{OT}^1$ is optimal. In the malicious case, the protocol in [16] uses asymptotically (as $k$ goes to infinity) the same amount of instances and is therefore asymptotically optimal.
– The protocol in [37] that implements $\binom{2}{1}$-$\mathsf{OT}^k$ from $(\frac{1}{2})$-$\mathsf{RabinOT}^1$ in the malicious model is asymptotically optimal.

*Quantum Reductions.* While previous result show that quantum protocols show similar limits as classical protocols for reductions between different variants of oblivious transfer, we present in Section 3.1 a statistically secure protocol that violates the classical bounds and the bound for perfectly secure quantum protocols by an arbitrarily large factor. More precisely, we prove that, in the quantum setting, string oblivious transfer can be reversed much more efficiently than by any classical protocol.

**Theorem 4.** *There exists a protocol that implements $\binom{2}{1}$-$\mathsf{OT}^{k'}$ with an error $\varepsilon$ from $\kappa = O(\log 1/\varepsilon)$ instances of $\binom{2}{1}$-$\mathsf{OT}^k$ in the opposite direction where $k' = \Omega(k)$ if $k = \Omega(\kappa)$.*

---

[2] For implementations of OT (and any other so-called deviation revealing functionality) security in the malicious model implies security in the semi-honest model [34]. In [40] we show this implication for $\binom{n}{1}$-$\mathsf{OT}^k$ and $(p)$-$\mathsf{RabinOT}^k$ with explicit bounds on the simulation errors.

For classical and perfect quantum protocols $k'$ is essentially upper bounded by $\kappa$. In Theorem 5 we show that a weaker lower bound for quantum reductions holds also for quantum protocols in the statistical setting. Theorem 5 implies that quantum protocols cannot extend oblivious transfer, i.e., we show that there exists a constant $c > 0$ such that any quantum reduction of $m + 1$ instances of $\binom{2}{1}$-OT$^1$ to $m$ instances of $\binom{2}{1}$-OT$^1$ must have an error of at least $\frac{c}{m}$.

Furthermore, Theorem 5 implies a lower bound for reductions between different variants of OT.

**Corollary 2.** *For any quantum reduction that implements* $\binom{2}{1}$-OT$^K$ *from* $m$ *instances of* $\binom{n}{1}$-OT$^k$ *with an error smaller than* $\varepsilon$, *we have*

$$m \geq \frac{K}{2nk + 2\log n} - 3K\sqrt{\varepsilon} - 13h(\sqrt{\varepsilon}) .$$

Finally, we also derive a lower bound on the number of commitments (Theorem 7) and on the total number of bits the players need to commit to (Theorem 6) in any $\varepsilon$-secure implementation of $\binom{2}{1}$-OT$^k$ from commitments.

**Corollary 3.** *A protocol that implements* $\binom{2}{1}$-OT$^k$, *using commitments only, with an error of at most* $\varepsilon$ *must use at least* $\log(1/\varepsilon) - 6$ *commitments and needs to commit to at least* $k/2 - 12k\sqrt{\varepsilon} - 7h(\sqrt{\varepsilon})$ *bits in total.*

Corollary 3 implies that bit commitments cannot be extended. More precisely, there exists a constant $c > 0$ such that any protocol that implements $m + 1$ bit commitments out of $m$ bit commitments must have an error of at least $\frac{c}{m}$. Finally, in Section 8 we show that there exists a protocol that is essentially optimal with respect to Corollary 3. We use the protocol from [6,17], but let the receiver commit to blocks of measurements at once, to prove the following theorem.

**Theorem 8.** *There exists a quantum protocol that implements* $\binom{2}{1}$-OT$^k$ *with an error of at most* $\varepsilon$, *using* $\kappa = O(\log 1/\varepsilon)$ *commitments to strings of size* $b$, *where* $\kappa b = O(k + \log 1/\varepsilon)$.

All proofs are in the full version of this work [40].

### 1.3   Notation

We use calligraphic letters to denote sets. We denote the distribution of a random variable $X$ over $\mathcal{X}$ by $P_X$. A conditional distribution $P_{X|Y}(x,y)$ over $\mathcal{X} \times \mathcal{Y}$ defines for every $y \in \mathcal{Y}$ a distribution $P_{X|Y=y}$. $P_{X|Y}$ can be seen as a randomized function that has input $y$ and output $x$. The *conditional Shannon entropy* of $X$ given $Y$ is defined as[3]

$$\mathrm{H}(X \mid Y) := -\sum_{x,y} P_{XY}(x,y) \log P_{X|Y}(x,y) ,$$

---

[3] All logarithms are binary, and we use the convention that $0 \cdot \log 0 = 0$.

and the *mutual information* of $X$ and $Y$ as $I(X;Y) = H(X) - H(X \mid Y)$. We use the notation $h(p) = -p \log p - (1-p) \log(1-p)$ for the binary entropy function. Furthermore, we write $[k]$ to denote the set $\{1, \ldots, k\}$. If $x = (x_1, \ldots, x_n)$ and $T := \{i_1, \ldots, i_k\} \subseteq [n]$, then $x|_T$ denotes the substring $(x_{i_1}, x_{i_2}, \ldots, x_{i_k})$ of $x$. If $x, y \in \{0,1\}^n$, then $x \oplus y$ denotes the bitwise XOR of $x$ and $y$.

### 1.4 Primitives and Randomized Primitives

In the following we consider two-party primitives that take inputs $x$ from Alice and $y$ from Bob and outputs $\bar{x}$ to Alice and $\bar{y}$ to Bob, where $(\bar{x}, \bar{y})$ are distributed according to $P_{\bar{X}\bar{Y}|XY}$. For simplicity, we identify such a primitive with $P_{\bar{X}\bar{Y}|XY}$. If the primitive has no input and outputs values $(u, v)$ distributed according to $P_{UV}$, we may simply write $P_{UV}$. If the primitive is deterministic and only Bob gets an output, i.e., if there exists a function $f : \mathcal{X} \times \mathcal{Y} \to \mathcal{Z}$ such that $P_{\bar{X}\bar{Y}|X=x,Y=y}(\perp, f(x,y)) = 1$ for all $x, y$, then we identify the primitive with the function $f$.

Examples of such primitives are $\binom{n}{t}$-$\mathsf{OT}^k$, $(p)$-$\mathsf{RabinOT}^k$, $\mathsf{EQ}_n$ and $\mathsf{IP}_n$.

- $\binom{n}{t}$-$\mathsf{OT}^k$ is the primitive where Alice has an input $x = (x_0, \ldots, x_{n-1}) \in \{0,1\}^{k \cdot n}$, and Bob has an input $c \subseteq \{0, \ldots, n-1\}$ with $|c| = t$. Bob receives $y = x|_c \in \{0,1\}^{tk}$.
- $(p)$-$\mathsf{RabinOT}^k$ is the primitive where Alice has an input $x \in \{0,1\}^k$. Bob receives $y$ which is equal to $x$ with probability $p$ and $\Delta$ otherwise.
- The *equality* function $\mathsf{EQ}_n : \{0,1\}^n \times \{0,1\}^n \to \{0,1\}$ is defined as $\mathsf{EQ}_n(x,y) = 1$ if $x = y$ and $\mathsf{EQ}_n(x,y) = 0$ otherwise.
- The *inner product modulo two* function $\mathsf{IP}_n : \{0,1\}^n \times \{0,1\}^n \to \{0,1\}^n$ is defined as $\mathsf{IP}_n(x,y) = \oplus_{i=1}^n x_i y_i$.

We often allow a protocol to use a primitive $P_{UV}$ that does not have any input. This is enough to model reductions to $\binom{n}{t}$-$\mathsf{OT}^k$ and $(p)$-$\mathsf{RabinOT}^k$, since these primitives are equivalent to distributed randomness $P_{UV}$, i.e., there exist two protocols that are secure in the semi-honest model: one that generates the distributed randomness using *one* instance of the primitive, and one that implements *one* instance of the primitive using the distributed randomness as input to the two parties. The fact that $\binom{2}{1}$-$\mathsf{OT}^1$ is equivalent to distributed randomness has been presented in [6,3]. The generalization to $\binom{n}{t}$-$\mathsf{OT}^k$ is straightforward. The randomized primitives are obtained by simply choosing all inputs uniformly at random. For $(p)$-$\mathsf{RabinOT}^k$ the implementation is straightforward. Hence, any protocol that uses some instances of $\binom{n}{t}$-$\mathsf{OT}^k$ or $(p)$-$\mathsf{RabinOT}^k$ can be converted into a protocol that only uses a primitive $P_{UV}$ without any input.

## 2 Lower Bounds for Classical Two-Party Computation

### 2.1 Protocols and Security in the Semi-honest Model

We will consider the *semi-honest model*, where both players behave honestly, but may save all the information they get during the protocol to obtain extra information about the other player's input or output. A protocol securely implements $P_{\bar{X}\bar{Y}|XY}$ with an error of $\varepsilon$, if the entire view of each player can be simulated[4] with an error of at most $\varepsilon$ in an ideal setting, where the players only have black-box access to the primitive $P_{\bar{X}\bar{Y}|XY}$. Note that this simulation is not allowed to change neither the input nor the output. (See the full version [40] for a formal definition.) This definition of security follows Definition 7.2.1 from [24], but is adapted to the case of computationally unbounded adversaries and statistical indistinguishability.

## 2.2 Lower Bounds for Secure Function Evaluation

We will now give lower bounds for $\varepsilon$-secure implementations of functions $f : \mathcal{X} \times \mathcal{Y} \to \mathcal{Z}$ from a primitive $P_{UV}$ in the semi-honest model. A function $f$ has no redundant inputs for Alice if

$$\forall x \neq x' \in \mathcal{X} \; \exists y \in \mathcal{Y} : \; f(x,y) \neq f(x',y) . \tag{2.1}$$

Clearly, a function $f$ can be computed from a primitive $P_{UV}$ with an error $\varepsilon$ in the semi-honest model if and only if the function $f'$ obtained by combining all redundant inputs for Alice can be computed with the same error.

Let Alice's and Bob's inputs $X$ and $Y$ be independent and uniformly distributed and let $M$ be the whole communication in the protocol. Loosely speaking, Alice must enter (almost) all the information about $X$ into the protocol as follows: If Bob's input is $y$, then he must be able to compute $f(X,y)$. But, as Alice must not learn $y$, she has to enter all information about $f(X,y)$ into the protocol independent of Bob's input. Thus, Alice must input all information about $f(X,y)$ into the protocol for all $y$. If $f$ satisfies (2.1), then $\{f(x,y) : y \in \mathcal{Y}\}$ allows to compute $x$. Thus, Alice must enter all information about $X$ into the protocol. More precisely, it can be shown that

$$H(X \mid UM, Y = y) \leq (3|\mathcal{Y}| - 2)(\varepsilon \log |\mathcal{Z}| + h(\varepsilon)) .$$

Since the protocol is secure against Bob, one can prove that for all $y$

$$\mathrm{H}(X \mid VM, Y = y) \geq \mathrm{H}(X \mid f(X,y)) - \varepsilon \log |\mathcal{X}| - h(\varepsilon) .$$

The following theorem that gives a lower bound on the conditional entropy of $P_{UV}$ can then be obtained from these two inequalities.

**Theorem 1.** *Let $f : \mathcal{X} \times \mathcal{Y} \to \mathcal{Z}$ be a function that satisfies (2.1). Let a protocol having access to $P_{UV}$ be an $\varepsilon$-secure implementation of $f$ in the semi-honest model. Then*

$$\mathrm{H}(U \mid V) \geq \max_{y} \mathrm{H}(X \mid f(X,y)) - 3|\mathcal{Y}|(\varepsilon \log |\mathcal{Z}| + h(\varepsilon)) - \varepsilon \log |\mathcal{X}| .$$

---

[4] The simulation is not required to be efficient.

Note that for some functions the bound of Theorem 1 can be improved by maximizing over all restrictions of the function $f$, i.e., over all functions $f'(x, y)$ : $\mathcal{X}' \times \mathcal{Y}' \to \mathcal{Z}'$ where $\mathcal{X}' \subset \mathcal{X}$, $\mathcal{Y}' \subset \mathcal{Y}$ and $\mathcal{Z}' \subset \mathcal{Z}$ with $f'(x, y) = f(x, y)$ that still satisfy condition (2.1).

Any lower bound for $f'$ implies a lower bound for $f$. The following corollaries follow immediately from Theorem 1.

**Corollary 4.** *Let a protocol having access to $P_{UV}$ be an $\varepsilon$-secure implementation of $\binom{n}{t}$-OT$^k$ in the semi-honest model. Then*

$$H(U \mid V) \geq (n - t)k - 3\lceil n/t \rceil (\varepsilon t k + h(\varepsilon)) - \varepsilon n k .$$

**Corollary 5.** *Let a protocol having access to $P_{UV}$ be an $\varepsilon$-secure implementation of $EQ_n$ in the semi-honest model. Then*

$$H(U|V) \geq \max_{0 < k \leq n} \left( (1 - \varepsilon)k - 3 \cdot 2^k(\varepsilon + h(\varepsilon)) - 1 \right) .$$

There exists a secure reduction of $EQ_n$ to $EQ_k$ [4]: Alice and Bob compare $k$ inner products of their inputs with random strings using $EQ_k$. This protocol is secure in the semi-honest model with an error[5] of at most $2^{-\kappa}$. Since there exists a circuit to implement $EQ_k$ with $k$ XOR and $k$ AND gates, it follows from [25] that $EQ_k$ can be securely implemented using $k$ instances to $\binom{4}{1}$-OT$^1$ or $3k$ instances of $\binom{2}{1}$-OT$^1$ in the semi-honest model. Since $m$ instances of $\binom{2}{1}$-OT$^1$ are equivalent to a primitive $P_{UV}$ with $H(U|V) = m$, the bound of Corollary 5 is optimal up to a factor of 3. This implies that the term $|\mathcal{Y}|$ in the statement of the bound given in Theorem 1 cannot be reduced significantly, i.e., it is not possible to replace $|\mathcal{Y}|$ with $\log |\mathcal{Y}|$ for example.

**Corollary 6.** *Let a protocol having access to a primitive $P_{UV}$ be an $\varepsilon$-secure implementation of the inner product function $IP_n$ in the semi-honest model. Then* $H(U|V) \geq n - 1 - 4n(\varepsilon + h(\varepsilon))$.

If $\varepsilon + h(\varepsilon) \leq 1/8$, then it immediately follows from Corollary 6 that we need at least $n/2 - 1$ calls to $\binom{2}{1}$-OT$^1$ to compute $IP_n$ with an error of at most $\varepsilon$. From the protocol presented in [4] we know that there exists a perfectly secure protocol that computes $IP_n$ from $n$ instances of $\binom{2}{1}$-OT$^1$. Therefore, the bound is optimal up to a factor of 2.

For our next lower-bound, the function $f : \mathcal{X} \times \mathcal{Y} \to \mathcal{Z}$ must satisfy the following property. There exist $y_1 \in \mathcal{Y}$ such that

$$\forall x \neq x' \in \mathcal{X} : f(x, y_1) \neq f(x', y_1) , \tag{2.2}$$

and $y_2 \in \mathcal{Y}$ such that

$$\forall x, x' \in \mathcal{X} : f(x, y_2) = f(x', y_2) . \tag{2.3}$$

---

[5] Note that our security definition is different from the one used in [4].

Let Alice's input $X$ be uniformly distributed. Loosely speaking, the security of the protocol implies that the communication gives (almost) no information about Alice's input $X$ if Bob's input is $y_2$. But the communication must be (almost) independent of Bob's input, otherwise Alice could learn Bob's input. Thus, Alice's input $X$ is uniform with respect to the whole communication even when Bob's input is $y_1$. Let now Bob's input be fixed to $y_1$ and let $M$ be the whole communication. Then the following lower bound can be proved using the given intuition.

$$H(f(X, y_1) \mid M) \geq \log |\mathcal{X}| - 6\varepsilon \log |\mathcal{X}| - 6h(\varepsilon) .$$

As Bob must be able to compute the correct output, one can show that

$$H(f(X, y_1) \mid VM) \leq \varepsilon \log |\mathcal{X}| + h(\varepsilon) .$$

The following lower bound on the mutual information of $P_{UV}$ can be obtained from these two inequalities.

**Theorem 2.** *Let $f : \mathcal{X} \times \mathcal{Y} \to \mathcal{Z}$ be a function that satisfies (2.2) and (2.3). Then for any protocol that implements $f$ from a primitive $P_{UV}$ with an error of at most $\varepsilon$ in the semi-honest model*

$$I(U; V) \geq \log |\mathcal{X}| - 7\varepsilon \log |\mathcal{X}| - 7h(\varepsilon) .$$

Since properties (2.2) and (2.3) can be satisfied by restricting Alice's input in $\binom{n}{t}$-$\mathsf{OT}^k$, we obtain the following corollary.

**Corollary 7.** *Let a protocol having access to $P_{UV}$ be an $\varepsilon$-secure implementation of $\binom{n}{t}$-$\mathsf{OT}^k$ in the semi-honest model where $t \leq \lfloor n/2 \rfloor$. Then*

$$I(U; V) \geq tk - 7\varepsilon tk - 7h(\varepsilon) .$$

We further generalize Theorem 2 to arbitrary functions $f : \mathcal{X} \times \mathcal{Y} \to \mathcal{Z}$ in [40]. In the case of perfect implementations the bound $H(U) = H(U|V) + I(U; V) \geq \log |\mathcal{X}|$ follows from Theorem 1 and the generalization of Theorem 2. From this bound we get that any perfectly secure protocol needs at least $\log |\mathcal{X}|$ instances of $\binom{2}{1}$-$\mathsf{OT}^1$ to implement a function $f : \mathcal{X} \times \mathcal{Y} \to \mathcal{Z}$, which implies Theorem 4.11 from [4].

## 2.3   Lower Bounds for Protocols Implementing OT

$\binom{2}{1}$-$\mathsf{OT}^1$ can be implemented from one instance of $\binom{2}{1}$-$\mathsf{OT}^1$ in the opposite direction [45]. Therefore, it follows immediately from Corollary 4 that for any $\varepsilon$-secure reduction of $\binom{2}{1}$-$\mathsf{OT}^1$ to $P_{UV}$, we must also have

$$H(V \mid U) \geq 1 - 5(\varepsilon + h(\varepsilon)) ,$$

since any violation of this bound could be used to construct a violation of the bound from Corollary 4. This bound can be generalized to $n > 0$. Together with the bounds from Theorem 1 and 2 we get the following theorem.

**Theorem 3.** *Let a protocol having access to $P_{UV}$ be an $\varepsilon$-secure implementation of $m$ instances of $\binom{n}{1}$-$OT^k$ in the semi-honest model. Then*

$$H(U \mid V) \geq m(n-1)k - 4n(\varepsilon mk + h(\varepsilon)),$$
$$H(V \mid U) \geq m\log n - m(4\log n + 7)(\varepsilon + h(\varepsilon)),$$
$$I(U;V) \geq mk - 7\varepsilon mk - 7h(\varepsilon) .$$

The statement of Corollary 1 follows from the fact that $m$ instances of $\binom{n}{1}$-$OT^k$ are equivalent to a primitive $P_{UV}$ with $H(U \mid V) = m(n-1)k$, $I(U;V) = mk$ and $H(V \mid U) = m\log n$.

In the full version of this paper [40], we show that the bounds of Theorem 1-3 can be generalized to the monotones from [43]. Furthermore, we derive new bounds for protocols implementing $(p)$-RabinOT$^k$, and show that our bounds imply bounds for implementations of oblivious linear function evaluation (OLFE).

# 3    Quantum Reductions

## 3.1    Reversing String OT Efficiently

As the bounds of the last section generalize the known bounds for perfect implementations of OT from [2,21,44,43] to the statistical case, it is natural to ask whether similar bounds also hold for quantum protocols, i.e., if the bounds presented in [36] can be generalized to the statistical case. We give a negative answer to this question by presenting a statistically secure quantum protocol that violates these bounds.

$\binom{2}{1}$-OT$^k$ can be implemented from $m = O(k + \kappa)$ bit commitments with an error of $2^{-\Omega(\kappa)}$ [6,49,17]. In the protocol, Alice sends $m$ BB84-states to Bob who measures them either in the computational or in the diagonal basis. To ensure that he really measures Bob has to commit to the basis he has measured in and the measurement outcome for every qubit received. Alice then asks Bob to open a small subset $\mathcal{T}$ of size $\alpha m$ of these pairs of commitments. OT can then be implemented using further classical processing. (See [17] for a complete description of the protocol.) This protocol implements oblivious transfer that is statistically secure in the quantum *universal composability* model [38]. Obviously the $m$ instances of bit commitments can be replaced by a single functionality, denoted by $\mathcal{F}_{MCOM}^{A \to B, m}$, which allows one player to commit to a bit string of length $m$ and later open an arbitrary substring. The following protocol implements $\mathcal{F}_{MCOM}^{A \to B, k}$ from the oblivious transfer functionality $\mathcal{F}_{OT}^{A \to B, k}$ (see [38] for a definition of $\mathcal{F}_{OT}^{A \to B, k}$).

---

**Inputs:** Alice has an input $b = (b_1, \ldots, b_k) \in \{0, 1\}^k$ in Commit. Bob has an input $T \subseteq [k]$ in Open.

Commit($b$):

For all $1 \leq i \leq \kappa$:

1. Alice and Bob invoke $\mathcal{F}_{OT}^{A \to B, k}$ with random inputs $x_0^i, x_1^i \in \{0, 1\}^k$ and $c^i \in_R \{0, 1\}^k$.
2. Bob receives $y^i = x_{c^i}^i$ from $\mathcal{F}_{OT}^{A \to B, k}$.
3. Alice sends $m^k := x_0^i \oplus x_1^i \oplus b$ to Bob.

Open(T):

1. Alice sends $b|_T$, $T$ and $x_0^i|_T, x_1^i|_T$ for all $1 \leq i \leq \kappa$ to Bob.
2. If $m^i|_T = x_0^i|_T \oplus x_1^i|_T \oplus b^i|_T$ and $y^i|_T = x_c^i|_T$ for all $1 \leq i \leq \kappa$, Bob accepts and outputs $b_T$, otherwise he rejects.

---

**Lemma 1.** *There exists a protocol that is statistically secure and universally composable that realizes $\mathcal{F}_{MCOM}^{A \to B, k}$ with an error of $2^{-\kappa/2}$ using $\kappa$ instances of $\mathcal{F}_{OT}^{A \to B, k}$.*

Since any protocol that is also statistically secure in the classical universal composability model [11] is also secure in the quantum universal composability model [38], we get, together with the proofs from [17,38], the following theorem.

**Theorem 4.** *There exists a protocol that implements $\binom{2}{1}$-$OT^{k'}$ with an error $\varepsilon$ from $\kappa = O(\log 1/\varepsilon)$ instances of $\binom{2}{1}$-$OT^k$ in the opposite direction where $k' = \Omega(k)$ if $k = \Omega(\kappa)$.*

Since we can choose $k \gg \kappa$, this immediately implies that the bound of Corollary 4 does not hold for quantum protocols. Similar violations can be shown for the other two lower bounds given in Theorem 7. For example, statistically secure and universally composable[6] commitments can be implemented from shared randomness $P_{UV}$ that is distributed according to $(p)$-RabinOT at a rate of $H(U \mid V) = 1 - p$ [41]. Using Theorem 8, one can implement $\mathcal{F}_{OT}^{B \to A, k}$ with $k \in \Omega(n(1 - p))$ from $n$ copies of $P_{UV}$. Since $I(U; V) = p$, quantum protocols can also violate the bound of Corollary 7.

It has been an open question whether noiseless quantum communication can increase the commitment capacity [41]. Our example implies a positive answer to this question.

## 3.2    Lower Bounds

The protocols presented in the previous section prove that the known impossibility results for perfectly secure oblivious transfer reductions from [36] do not

---

[6] Stand-alone statistically secure commitments based on stateless two-party primitives are universally composable [22].

hold for statistically secure quantum protocols. Thus, it is natural to ask whether quantum protocols can even extend oblivious transfer or, more generally, how efficient statistically secure quantum protocols can be. In this section we prove an impossibility result that holds for statistically secure quantum protocols and that implies in particular that also quantum protocols *cannot* extend OT. Since, in contrast to the classical case, security against semi-honest adversaries can be trivially achieved in the quantum setting, we consider in the following protocols that are secure against malicious adversaries in the stand-alone model. A protocol is an $\varepsilon$-secure implementation of OT if for any adversary attacking the protocol (real setting), there exists a simulator using the ideal OT (ideal setting) such that for all inputs of the honest players the real and the ideal setting can be distinguished with an advantage of at most $\varepsilon$.

In the following we will give two lower bounds for quantum protocols that implement $\binom{2}{1}$-$\mathsf{OT}^k$ using a trusted resource such as trusted randomness distributed to the players or a bit commitment functionality. Our proofs use similar techniques as the impossibility results in [32,30,29]. First, the protocol is replaced by a purified version of the protocol that is equivalent in a certain sense. In particular the purified version has the same security properties as the original protocol and the impossibility of the former implies the impossibility of the latter. In this protocol the players defer all of their measurements to the very end of the protocol. See [32,30,29] for details.

We use the notation $\rho^{AB}$ for a state in the Hilbert space $\mathcal{H}_A \otimes \mathcal{H}_B$, and $\rho^A := \mathrm{tr}_B(\rho^{AB})$. The *conditional von Neumann entropy* is defined as $H(A \mid B)_\rho := H(\rho^{AB}) - H(\rho^B)$, where $H(\rho) := \mathrm{tr}(-\rho \log(\rho))$.

We first consider protocols where the players have access to a primitive that generates a pure state $|\psi\rangle^{ABE}$, distributes registers $A$ and $B$ to Alice and Bob respectively and keeps the purification in its register $E$.

Let Alice choose her inputs $X_0$ and $X_1$ uniformly at random and let Bob's input be $c$. When Alice and Bob execute the purified protocol honestly the final state just before the honest players perform their measurements is a pure state $|\rho\rangle_c^{ABE}$, where A and B are the registers of Alice and Bob and E is the register of the trusted resource.

Loosely speaking, security for Alice guarantees that Bob has (almost) no information about $X_0$ if $c = 1$, i.e., the entropy $H(X_0 \mid B)_{\rho_1}$ is almost maximal. On the other hand, Alice must not be able to learn Bob's choice bit. Therefore, we have $\rho_0^A \approx \rho_1^A$. As shown in [32,30,29], this implies that there exists a unitary on system $BE$ that transforms $|\rho\rangle_1^{ABE}$ into a state close to $|\rho\rangle_0^{ABE}$. Since Bob can learn $X_0$ if $c = 0$, this implies that $H(X_0 \mid BE)_{\rho_1}$ is small. Using these two facts, one can then prove the following lower bound on the entropy of $E$.

**Theorem 5.** *To implement one instance of $\binom{2}{1}$-$\mathsf{OT}^k$ over strings of size $k$ with an error of at most $\varepsilon$ from a primitive $|\psi\rangle^{ABE}$ with a quantum protocol we need*

$$2H(E)_\psi \geq (1 - 21\varepsilon - 2\sqrt{\varepsilon}) \cdot k - 11h(\varepsilon) - 2h(\sqrt{\varepsilon}) .$$

A classical primitive $P_{UV}$ can be modeled by the quantum primitive

$$|\psi\rangle^{ABE} = \sum_{u,v} \sqrt{P_{UV}(u,v)} \cdot |u,v\rangle^{AB} \otimes |u,v\rangle^E$$

that distributes the values $u$ and $v$ and keeps the purification in its register $E$. Therefore, we get the following corollary from Theorem 5.

**Corollary 8.** *To implement one instance of $\binom{2}{1}$-$OT^k$ with an error of at most $\varepsilon$ from $P_{UV}$ with a quantum protocol, we need*

$$2H(UV) \geq (1 - 21\varepsilon - 2\sqrt{\varepsilon}) \cdot k - 11h(\varepsilon) - 2h(\sqrt{\varepsilon}) .$$

Since $m$ instances of $\binom{2}{1}$-$OT^k$ can be implemented from shared randomness with $H(UV) = 2k + 1$ we get the following corollary.

**Corollary 9.** *To implement one instance of $\binom{2}{1}$-$OT^k$ with an error of at most $\varepsilon$ from $n$ instances of $\binom{2}{1}$-$OT^{k'}$ in either direction with a quantum protocol, we need*

$$2n(2k' + 1) \geq (1 - 21\varepsilon - 2\sqrt{\varepsilon}) \cdot k - 11h(\varepsilon) - 2h(\sqrt{\varepsilon}) .$$

Next, we present a bound for implementations of $\binom{2}{1}$-$OT^k$ from commitments. We can model black-box commitments by a trusted functionality that receives bits over a classical channel and stores them in a register $E$. When the committer sends the open command, the functionality sends the bits to the receiver. We can replace the two classical channels with a quantum channel where the players measure the qubits when sending and after receiving them. These measurements can then be purified by the players. The following bound can be obtained by adapting the proof of Theorem 5 to this scenario.

**Theorem 6.** *To implement a $\binom{2}{1}$-$OT^k$ with an error of at most $\varepsilon$ we need to commit to at least $(1 - 21\varepsilon - 2\sqrt{\varepsilon})k/2 - 6h(\varepsilon) - h(\sqrt{\varepsilon})$ bits in total.*

From Corollary 9 and Theorem 6 follows that OTs and commitments cannot be extended by quantum protocols.

**Corollary 10.** *Any quantum protocol that implement $m+1$ instances of $\binom{2}{1}$-$OT^1$ from $m$ instances of $\binom{2}{1}$-$OT^1$ must have an error of at least $\frac{5 \cdot 10^{-6}}{m}$ for any $m > 0$.*

**Corollary 11.** *Any quantum protocol that implements $m + 1$ bit commitments out of $m$ commitments must have an error of at least $\frac{10^{-9}}{m}$ for any $m > 0$.*

Next, we give an additional lower bound for reductions of OT to commitments that shows that the *number* of commitments (of arbitrary size) used in any $\varepsilon$-secure protocol must be at least $\Omega(\log(1/\varepsilon))$. We model the commitments as before, but store the commitments of Alice and Bob separately in $E_A$ and $E_B$. The proof idea is the following: We let the adversary guess a subset $\mathcal{T}$ of commitments that he will be required to open during the protocol. He honestly executes all commitments in $\mathcal{T}$, but cheats in all others. If the adversary guesses $\mathcal{T}$ right, he is able to cheat in the same way as in any protocol that does not use any commitments.

**Theorem 7.** *Any quantum protocol that implements $\binom{2}{1}$-$OT^k$ using $\kappa$ commitments (of arbitrary length) must have an error of at least $2^{-\kappa}/36$.*

### 3.3  Reduction of OT to String-Commitments

The protocol we described in Section 3.1 uses $m = O(k + \kappa)$ commitments to 2 bits to implement $\binom{2}{1}$-$OT^k$ with an error of $2^{-\Omega(\kappa)}$. If $k = \omega(\kappa)$ this it is not optimal with respect to Theorem 7. We will now show how to construct a protocol that is optimal with respect to the lower bounds of both Theorem 6 and Theorem 7. We modify the protocol by grouping the $m$ pairs into $\kappa$ blocks of size $b := m/\kappa$. We let Bob commit to the blocks of $b$ pairs of values at once. The subset $\mathcal{T}$ is now of size $\alpha\kappa$, and defines the blocks to be opened by Bob. If Bob is able to open all commitments in $\mathcal{T}$ correctly, then with high probability, he must have correctly measured almost all qubits. We only need to estimate the error probability of the sampling strategy that corresponds to the new checking procedure which Alice applies and apply the proof of [17] to get the following theorem.

**Theorem 8.** *There exists a quantum protocol that implements $\binom{2}{1}$-$OT^k$ with an error of at most $\varepsilon$ out of $\kappa = O(\log 1/\varepsilon)$ commitments of size $b$, where $\kappa b = O(k + \log 1/\varepsilon)$.*

Using Theorem 8, it can be shown that string-commitments cannot be extended.

**Corollary 12.** *Let $m > 0$. If there exists a (quantum) protocol that implements string commitments of length $m' + 1$ out of string commitments of length $m'$ for all $m' > m$ with an error of at most $\varepsilon$, then there exists a constant $c > 0$ such that $\varepsilon \geq \frac{c}{m}$.*

## 4  Conclusions

The main contribution of this work are impossibility proofs for statistical oblivious transfer reductions. In the classical case we have generalized several known lower bounds for perfect reductions to statistical security. In the quantum case we have shown that the known bound for perfect reductions does not apply to statistical reductions, and have presented a new bound that *does* hold in the statistical quantum setting. Our bounds imply several important impossibility results, for example, that OT cannot be extended, neither in the classical nor in the quantum setting.

There are many interesting open questions. For example, it is not known whether more than two instances of $\binom{2}{1}$-$OT^1$ can be implemented (in the classical or the quantum setting) from two instances of $\binom{2}{1}$-$OT^\ell$, one in each direction.

## Acknowledgments

We thank Esther Hänggi, Thomas Holenstein and Stephanie Wehner for helpful discussions, and the referees for their useful comments. This work was funded by the Swiss National Science Foundation (SNSF) and the U.K. EPSRC, grant EP/E04297X/1. Part of this work was done while JW was visiting McGill University.

## References

1. Ahlswede, R., Csiszar, I.: On oblivious transfer capacity. In: IEEE Information Theory Workshop on Networking and Information Theory, ITW 2009, December 10, pp. 1–3 (2009)
2. Beaver, D.: Correlated pseudorandomness and the complexity of private computations. In: STOC 1996: Proceedings of the 28th Annual ACM Symposium on Theory of Computing, pp. 479–488. ACM Press, New York (1996)
3. Beaver, D.: Precomputing oblivious transfer. In: Coppersmith, D. (ed.) CRYPTO 1995. LNCS, vol. 963, pp. 97–109. Springer, Heidelberg (1995)
4. Beimel, A., Malkin, T.: A quantitative approach to reductions in secure computation. In: Naor, M. (ed.) TCC 2004. LNCS, vol. 2951, pp. 238–257. Springer, Heidelberg (2004)
5. Bennett, C.H., Brassard, G., Robert, J.M.: Privacy amplification by public discussion. SIAM Journal on Computing 17(2), 210–229 (1988)
6. Bennett, C.H., Brassard, G., Crépeau, C., Skubiszewska, M.H.: Practical quantum oblivious transfer. In: Feigenbaum, J. (ed.) CRYPTO 1991. LNCS, vol. 576, pp. 351–366. Springer, Heidelberg (1992)
7. Brassard, G., Crépeau, C., Robert, J.M.: Information theoretic reductions among disclosure problems. In: Proceedings of the 27th Annual IEEE Symposium on Foundations of Computer Science (FOCS 1986), pp. 168–173 (1986)
8. Brassard, G., Crépeau, C., Wolf, S.: Oblivious transfers and privacy amplification. Journal of Cryptology 16(4), 219–237 (2003)
9. Brassard, G., Crépeau, C., Santha, M.: Oblivious transfers and intersecting codes. IEEE Transactions on Information Theory 42(6), 1769–1780 (1996)
10. Cachin, C.: On the foundations of oblivious transfer. In: Nyberg, K. (ed.) EUROCRYPT 1998. LNCS, vol. 1403, pp. 361–374. Springer, Heidelberg (1998)
11. Canetti, R.: Universally composable security: A new paradigm for cryptographic protocols. In: FOCS 2001, pp. 136–145 (2001)
12. Crépeau, C., Kilian, J.: Achieving oblivious transfer using weakened security assumptions (extended abstract). In: Proceedings of the 29th Annual IEEE Symposium on Foundations of Computer Science (FOCS 1988), pp. 42–52 (1988)
13. Crépeau, C.: Equivalence between two flavours of oblivious transfers. In: Pomerance, C. (ed.) CRYPTO 1987. LNCS, vol. 293, pp. 350–354. Springer, Heidelberg (1988)
14. Crépeau, C., Morozov, K., Wolf, S.: Efficient unconditional oblivious transfer from almost any noisy channel. In: Blundo, C., Cimato, S. (eds.) SCN 2004. LNCS, vol. 3352, pp. 47–59. Springer, Heidelberg (2005)
15. Crépeau, C., Santha, M.: On the reversibility of oblivious transfer. In: Davies, D.W. (ed.) EUROCRYPT 1991. LNCS, vol. 547, pp. 106–113. Springer, Heidelberg (1991)

16. Crépeau, C., Savvides, G.: Optimal reductions between oblivious transfers using interactive hashing. In: Vaudenay, S. (ed.) EUROCRYPT 2006. LNCS, vol. 4004, pp. 201–221. Springer, Heidelberg (2006)

17. Damgård, I., Fehr, S., Lunemann, C., Salvail, L., Schaffner, C.: Improving the security of quantum protocols via commit-and-open. In: Halevi, S. (ed.) CRYPTO 2009. LNCS, vol. 5677, pp. 408–427. Springer, Heidelberg (2009)

18. Damgård, I., Fehr, S., Morozov, K., Salvail, L.: Unfair noisy channels and oblivious transfer. In: Naor, M. (ed.) TCC 2004. LNCS, vol. 2951, pp. 355–373. Springer, Heidelberg (2004)

19. Damgård, I., Fehr, S., Salvail, L., Schaffner, C.: Oblivious transfer and linear functions. In: Dwork, C. (ed.) CRYPTO 2006. LNCS, vol. 4117, pp. 427–444. Springer, Heidelberg (2006)

20. Damgård, I., Kilian, J., Salvail, L.: On the (im)possibility of basing oblivious transfer and bit commitment on weakened security assumptions. In: Stern, J. (ed.) EUROCRYPT 1999. LNCS, vol. 1592, pp. 56–73. Springer, Heidelberg (1999)

21. Dodis, Y., Micali, S.: Lower bounds for oblivious transfer reductions. In: Stern, J. (ed.) EUROCRYPT 1999. LNCS, vol. 1592, pp. 42–55. Springer, Heidelberg (1999)

22. Dowsley, R., van de Graaf, J., Müller-Quade, J., Nascimento, A.C.A.: On the composability of statistically secure bit commitments. Cryptology ePrint Archive, Report 2008/457 (2008)

23. Even, S., Goldreich, O., Lempel, A.: A randomized protocol for signing contracts. Commun. ACM 28(6), 637–647 (1985)

24. Goldreich, O.: Foundations of Cryptography. Basic Applications, vol. II. Cambridge University Press, Cambridge (2004)

25. Goldreich, O., Vainish, R.: How to solve any protocol problem - an efficiency improvement. In: Pomerance, C. (ed.) CRYPTO 1987. LNCS, vol. 293, pp. 73–86. Springer, Heidelberg (1988)

26. Ishai, Y., Kilian, J., Nissim, K., Petrank, E.: Extending oblivious transfers efficiently. In: Boneh, D. (ed.) CRYPTO 2003. LNCS, vol. 2729, pp. 145–161. Springer, Heidelberg (2003)

27. Kilian, J.: Founding cryptography on oblivious transfer. In: Proceedings of the 20th Annual ACM Symposium on Theory of Computing (STOC 1988), pp. 20–31. ACM Press, New York (1988)

28. Kurosawa, K., Kishimoto, W., Koshiba, T.: A combinatorial approach to deriving lower bounds for perfectly secure oblivious transfer reductions. IEEE Transactions on Information Theory 54(6), 2566–2571 (2008)

29. Lo, H.K.: Insecurity of quantum secure computations. Physical Review A 56, 1154 (1997)

30. Lo, H.K., Chau, H.F.: Is quantum bit commitment really possible? Physical Review Letters 78, 3410–3413 (1997)

31. Maurer, U.: Information-theoretic cryptography. In: Wiener, M. (ed.) CRYPTO 1999. LNCS, vol. 1666, pp. 47–64. Springer, Heidelberg (1999)

32. Mayers, D.: Unconditionally secure quantum bit commitment is impossible. Physical Review Letters 78, 3414–3417 (1997)

33. Nascimento, A., Winter, A.: On the oblivious transfer capacity of noisy correlations. In: Proceedings of the IEEE International Symposium on Information Theory, ISIT 2006 (2006)

34. Prabhakaran, M., Rosulek, M.: Cryptographic complexity of multi-party computation problems: Classifications and separations. In: Wagner, D. (ed.) CRYPTO 2008. LNCS, vol. 5157, pp. 262–279. Springer, Heidelberg (2008)

35. Rabin, M.O.: How to exchange secrets by oblivious transfer. Tech. Rep. TR-81, Harvard Aiken Computation Laboratory (1981)
36. Salvail, L., Schaffner, C., Sotáková, M.: On the power of Two-party quantum cryptography. In: Matsui, M. (ed.) ASIACRYPT 2009. LNCS, vol. 5912, pp. 70–87. Springer, Heidelberg (2009)
37. Savvides, G.: Interactive Hashing and reductions between Oblivious Transfer variants. Ph.D. thesis, McGill University, Montréal (2007)
38. Unruh, D.: Universally composable quantum multi-party computation. In: Gilbert, H. (ed.) EUROCRYPT 2010. LNCS, vol. 6110, pp. 486–505. Springer, Heidelberg (2010)
39. Wiesner, S.: Conjugate coding. SIGACT News 15(1), 78–88 (1983)
40. Winkler, S., Wullschleger, J.: On the efficiency of classical and quantum oblivious transfer reductions. Cryptology ePrint Archive, Report 2009/508 (2009)
41. Winter, A., Nascimento, A.C.A., Imai, H.: Commitment capacity of discrete memoryless channels. In: IMA Int. Conf., pp. 35–51 (2003)
42. Wolf, S., Wullschleger, J.: Zero-error information and applications in cryptography. In: Proceedings of 2004 IEEE Information Theory Workshop, ITW 2004 (2004)
43. Wolf, S., Wullschleger, J.: New monotones and lower bounds in unconditional two-party computation. IEEE Transactions on Information Theory 54(6), 2792–2797 (2008)
44. Wolf, S., Wullschleger, J.: New monotones and lower bounds in unconditional two-party computation. In: Shoup, V. (ed.) CRYPTO 2005. LNCS, vol. 3621, pp. 467–477. Springer, Heidelberg (2005)
45. Wolf, S., Wullschleger, J.: Oblivious transfer is symmetric. In: Vaudenay, S. (ed.) EUROCRYPT 2006. LNCS, vol. 4004, pp. 222–232. Springer, Heidelberg (2006)
46. Wullschleger, J.: Oblivious-transfer amplification. In: Naor, M. (ed.) EUROCRYPT 2007. LNCS, vol. 4515, pp. 555–572. Springer, Heidelberg (2007)
47. Wullschleger, J.: Oblivious transfer from weak noisy channels. In: Reingold, O. (ed.) TCC 2009. LNCS, vol. 5444, pp. 332–349. Springer, Heidelberg (2009)
48. Yao, A.C.: Protocols for secure computations. In: Proceedings of the 23rd Annual IEEE Symposium on Foundations of Computer Science (FOCS 1982), pp. 160–164 (1982)
49. Yao, A.C.C.: Security of quantum protocols against coherent measurements. In: STOC 1995: Proceedings of the 27th Annual ACM Symposium on Theory of Computing, pp. 67–75. ACM Press, New York (1995)

# Sampling in a Quantum Population, and Applications

Niek J. Bouman and Serge Fehr

Centrum Wiskunde & Informatica (CWI), Amsterdam, The Netherlands
{n.j.bouman,s.fehr}@cwi.nl

**Abstract.** We propose a framework for analyzing classical sampling strategies for estimating the Hamming weight of a large string from a few sample positions, when applied to a multi-qubit quantum system instead. The framework shows how to interpret the result of such a strategy and how to define its accuracy when applied to a quantum system. Furthermore, we show how the accuracy of any strategy relates to its accuracy in its classical usage, which is well understood for the important examples. We show the usefulness of our framework by using it to obtain new and simple security proofs for the following quantum-cryptographic schemes: BB84 quantum-key-distribution, and quantum oblivious-transfer from bit-commitment.

## 1 Introduction

Sampling allows to learn some information on a large population by merely looking at a comparably small number of individuals. For instance it is possible to predict the outcome of an election with very good accuracy by analyzing a relatively small subset of all the votes. In this work, we study sampling in a *quantum* population: we want to learn information about a large quantum state by measuring only a small part. Specifically, we investigate the quantum version of the following classical sampling problem (and of variants thereof). Given a bit-string $q = (q_1, \dots, q_n) \in \{0,1\}^n$ of length $n$, the task is to estimate the Hamming weight of $q$ by sampling and looking at only a few positions within $q$. This classical sampling problem is well understood. For instance, the following particular *sampling strategy* works well: sample (with or without replacement) a linear number of positions uniformly at random, and compute an estimate for the Hamming weight of $q$ by scaling the Hamming weight of the sample accordingly; Hoeffding's bounds guarantee that the estimate is close to the real Hamming weight except with small probability. In particular, one can use a sampling strategy to *test* whether $q$ is close to the all-zero string $(0, \dots, 0)$ by looking only at a relatively small number of positions, where the test is accepted if and only if all the sample positions are zero, i.e., the estimated Hamming weight vanishes.

In the quantum version of the sampling problem from above, the string $q$ is replaced by a $n$-qubit quantum system $A$. Obviously, a sampling strategy from the classical can be *applied* to the quantum setting as well: pick a sample of

T. Rabin (Ed.): CRYPTO 2010, LNCS 6223, pp. 724–741, 2010.

qubit positions within $A$, measure (in the computational basis) these sample positions, and compute the estimate as dictated by the sampling strategy from the observed values (i.e., typically, scale the Hamming weight of the measured sample appropriately). However, due to the special nature of quantum states, it is not clear and to the best of our knowledge so far not well understood, how to formally *interpret* the computed estimate. Simply extending the classical results in a straightforward way to the quantum setting does not work due to several reasons (e.g., one reason being that it is not clear what the Hamming weight of a quantum state should be).

In this work, we present a framework that addresses the above and fully characterizes the behavior of a classical sampling strategy when applied to a quantum population, i.e., to a $n$-qubit system or, more general, to $n$ copies of an arbitrary "atomic" system. Our framework incorporates the following. First, we specify an abstract property on the state of $A$ (after the measurements done by the sampling strategy), with the intended meaning that this is the property one should conclude from the outcome of the sampling strategy when applied to $A$. We also demonstrate that this property has useful consequences: specifically, that a suitable measurement will lead to a high-entropy outcome; this is handy in particular for quantum-cryptographic purposes. Then, we define a meaningful measure, sort of a "quantum error probability" (although technically speaking it is not a probability), that tells how reliable it is to conclude the specified property from the outcome of the sampling strategy. Finally, we show that for *any* sampling strategy, the quantum error probability of the strategy, as we define it, is bounded by the square-root of its classical error probability. This means that in order to understand how well a sampling strategy performs in the quantum setting, it suffices to analyze it in the classical setting, which is typically much simpler. Furthermore, for typical sampling strategies, like when picking the sample uniformly at random, there are well-known good bounds on the classical error probability.

We demonstrate the usefulness of our framework by means of two applications. Our applications do not constitute actual new results, but they provide new and simple(r) proofs for known results, both in the area of quantum cryptography. We take this as strong indication for the usefulness of the framework, and that the framework is likely to prove valuable in other applications as well.

The first application is to quantum key-distribution (QKD). We show how our framework for analyzing sampling strategies in the quantum setting leads to a conceptually very simple and easy-to-understand security proof for the BB84 QKD scheme.[1] The main idea behind the proof is that the checking phase of the BB84 scheme can be viewed as executing a specific sampling strategy. From the framework, it then follows that the raw key has high min-entropy from the adversary's point of view, and the proof is concluded by applying the privacy amplification theorem.

---

[1] Actually, we prove security for an entanglement-based version of BB84 that implies security for the original BB84 scheme.

QKD schemes initially came without security proofs, and proving QKD schemes rigorously secure turned out to be an extremely challenging and subtle task. Nowadays, though, the security of QKD schemes is better understood, and we know of different ways of proving, say, BB84 secure, ranging from Shor and Preskill's proof based on quantum error-correcting codes [9] to Renner's approach using a quantum De Finetti theorem which allows to reduce security against general attacks to security against the much weaker class of so-called collective attacks [7]. Nonetheless, we think that our proof is interesting because of the following reasons. It provides an *explicit* expression for the security of the scheme, given in terms of an easy-to-compute function of the observed error-rate, the parameters of the code used to do error correction, and the number of extracted key-bits (and the parameters of the scheme). This is in contrast to most proofs in the literature which merely provide an asymptotic analysis. Furthermore, the proof is technically very accessible (e.g. compared to quantum-De-Finetti-based proofs) and as such for instance particularly well-suited for teaching. Finally, it does not require any "symmetrization of the qubits" (e.g. by applying a random permutation) from the protocol, and it gives a *direct* security proof, rather than a reduction to the security against collective attacks.

The second application is to quantum oblivious transfer (QOT). It is well known that QOT is not possible from scratch, but one can build a secure QOT scheme when given a bit-commitment (BC) primitive "for free". Also for this cryptographic primitive, our framework allows for a simple and easy-to-understand security proof. Due to space restriction, this second application is only given in the full version [2] of this paper. The security of QOT (when given bit commitments) has also recently been rigorously proven in [4]. Although at the technical level similar ideas are used, our work distinguishes from [4] in that we introduce and rigorously study the concept of a general sampling strategy. This not only gives a nice framework and makes the security of QOT easier to understand, but it also opens the door for other applications (as we demonstrate).

We find it particularly interesting that with our framework, the protocols for QKD and QOT can be prover secure by means of very similar techniques, even though they implement fundamentally different cryptographic primitives, and are intuitively secure due to different reasons.

## 2    Notation, Terminology, and Some Tools

*Strings and Hamming Weight.* Throughout the paper, $\mathcal{A}$ denotes some fixed finite alphabet with $0 \in \mathcal{A}$. It is safe to think of $\mathcal{A}$ as $\{0, 1\}$, but our claims also hold for larger alphabets. For a string $\boldsymbol{q} = (q_1, \ldots, q_n) \in \mathcal{A}^n$ of arbitrary length $n \geq 0$, the *Hamming weight* of $\boldsymbol{q}$ is defined as: $\mathrm{wt}(\boldsymbol{q}) := |\{i \in [n] : q_i \neq 0\}|$, where $[n]$ is a short hand for $\{1, \ldots, n\}$. The *relative* Hamming weight of $\boldsymbol{q}$ is defined as $\omega(\boldsymbol{q}) := \mathrm{wt}(\boldsymbol{q})/n$. By convention, the relative Hamming weight of the empty string $\perp$ is set to $\omega(\perp) := 0$. For a subset $J \subset [n]$, we write $\boldsymbol{q}_J := (q_i)_{i \in J}$ for the restriction of $\boldsymbol{q}$ to the positions $i \in J$.

*Quantum Systems and States.* We assume the reader to be familiar with the basic concepts of quantum information theory; we merely fix some specific terminology and notation here.

By default, we write $\mathcal{H}_A$ for the state space of system $A$, and $\rho_A$ for the density matrix and $|\varphi_A\rangle$ for the state vector (in case of a pure state) describing the state of $A$. To simplify language we are sometimes a bit sloppy in distinguishing between a quantum system, its state, and the state vector or density matrix describing the state. A *qubit* is a quantum system $A$ with state space $\mathcal{H}_A = \mathbb{C}^2$. The *computational basis* $\{|0\rangle, |1\rangle\}$ (for a qubit) is given by $|0\rangle = \binom{1}{0}$ and $|1\rangle = \binom{0}{1}$, and the *Hadamard basis* by $H\{|0\rangle, |1\rangle\} = \{H|0\rangle, H|1\rangle\}$, where $H$ denotes the 2-dimensional *Hadamard matrix* $H = 2^{-1/2}\left(\begin{smallmatrix} 1 & 1 \\ 1 & -1 \end{smallmatrix}\right)$. The state space of an $n$-qubit system $A = A_1 \cdots A_n$ is given by $\mathcal{H}_A = (\mathbb{C}^2)^{\otimes n} = \mathbb{C}^2 \otimes \cdots \otimes \mathbb{C}^2$. For $\boldsymbol{x} = (x_1, \ldots, x_n)$ and $\boldsymbol{\theta} = (\theta_1, \ldots, \theta_n)$ in $\{0,1\}^n$, we write $|\boldsymbol{x}\rangle$ for $|\boldsymbol{x}\rangle = |x_1\rangle \cdots |x_n\rangle$ and $H^{\boldsymbol{\theta}}$ for $H^{\boldsymbol{\theta}} = H^{\theta_1} \otimes \cdots \otimes H^{\theta_n}$, and thus $H^{\boldsymbol{\theta}}|\boldsymbol{x}\rangle$ for $H^{\boldsymbol{\theta}}|\boldsymbol{x}\rangle = H^{\theta_1}|x_1\rangle \cdots H^{\theta_n}|x_n\rangle$. Finally, we write $\{|0\rangle, |1\rangle\}^{\otimes n} = \{|\boldsymbol{x}\rangle : \boldsymbol{x} \in \{0,1\}^n\}$ for the computational basis on an $n$-qubit system, and $H^{\boldsymbol{\theta}}\{|0\rangle, |1\rangle\}^{\otimes n} = \{H^{\boldsymbol{\theta}}|\boldsymbol{x}\rangle : \boldsymbol{x} \in \{0,1\}^n\} = H^{\theta_1}\{|0\rangle, |1\rangle\} \otimes \cdots \otimes H^{\theta_n}\{|0\rangle, |1\rangle\}$ for the basis that is made up of the computational basis on the subsystems $A_i$ with $\theta_i = 0$ and of the Hadamard basis on the subsystems $A_i$ with $\theta_i = 1$. To simplify notation, we will sometimes abuse terminology and speak of the basis $\boldsymbol{\theta}$ when we actually mean $H^{\boldsymbol{\theta}}\{|0\rangle, |1\rangle\}^{\otimes n}$.

*Measuring* a system $A$ in basis $\{|i\rangle\}_{i \in I}$, where $\{|i\rangle\}_{i \in I}$ is an orthonormal basis of $\mathcal{H}_A$, means applying the measurement described by the projectors $\{|i\rangle\langle i|\}_{i \in I}$, such that outcome $i \in I$ is observed with probability $p_i = \mathrm{tr}(|i\rangle\langle i|\rho_A)$ (respectively $p_i = |\langle i|\varphi_A\rangle|^2$ in case of a pure state). If $A$ is a subsystem of a bipartite system $AB$, then it means applying the measurement described by the projectors $\{|i\rangle\langle i| \otimes \mathbb{I}_B\}_{i \in I}$, where $\mathbb{I}_B$ is the identity operator on $\mathcal{H}_B$.

We measure closeness of two states $\rho$ and $\sigma$ by their *trace distance*: $\Delta(\rho, \sigma) := \frac{1}{2}\mathrm{tr}|\rho - \sigma|$, where for any square matrix $M$, $|M|$ denotes the positive-semi-definite square-root of $M^\dagger M$. For *pure* states $|\varphi\rangle$ and $|\psi\rangle$, the trace distance of the corresponding density matrices coincides with $\Delta(|\varphi\rangle\langle\varphi|, |\psi\rangle\langle\psi|) = \sqrt{1 - |\langle\varphi|\psi\rangle|^2}$. If the states of two systems $A$ and $B$ are $\epsilon$-close, i.e. $\Delta(\rho_A, \rho_B) \leq \epsilon$, then $A$ and $B$ cannot be distinguished with advantage greater than $\epsilon$; in other words, $A$ behaves exactly like $B$, except with probability $\epsilon$.

*Classical and Hybrid Systems (and States).* Subsystem $X$ of a bipartite quantum system $XE$ is called *classical*, if the state of $XE$ is given by a density matrix of the form $\rho_{XE} = \sum_{x \in \mathcal{X}} P_X(x)|x\rangle\langle x| \otimes \rho_E^x$, where $\mathcal{X}$ is a finite set of cardinality $|\mathcal{X}| = \dim(\mathcal{H}_X)$, $P_X : \mathcal{X} \to [0,1]$ is a probability distribution, $\{|x\rangle\}_{x \in \mathcal{X}}$ is some fixed orthonormal basis of $\mathcal{H}_X$, and $\rho_E^x$ is a density matrix on $\mathcal{H}_E$ for every $x \in \mathcal{X}$. Such a state, called *hybrid* or *cq* (for classical-quantum) state, can equivalently be understood as consisting of a *random variable* $X$ with distribution $P_X$, taking on values in $\mathcal{X}$, and a system $E$ that is in state $\rho_E^x$ exactly when $X$ takes on the value $x$. This formalism naturally extends to two (or more) classical systems $X$, $Y$ etc.

If the state of $XE$ satisfies $\rho_{XE} = \rho_X \otimes \rho_E$, where $\rho_X = \mathrm{tr}_E(\rho_{XE}) = \sum_x P_X(x)|x\rangle\langle x|$ and $\rho_E = \mathrm{tr}_X(\rho_{XE}) = \sum_x P_X(x)\rho_E^x$, then $X$ is *independent* of $E$, and thus no information on $X$ can be obtained from system $E$. Moreover, if $\rho_{XE} = \frac{1}{|\mathcal{X}|}\mathbb{I}_X \otimes \rho_E$, where $\mathbb{I}_X$ denotes the identity on $\mathcal{H}_X$, then $X$ is *random-and-independent* of $E$. This is what is aimed for in quantum cryptography, when $X$ represents a classical cryptographic key and $E$ the adversary's potential quantum information on $X$.

It is not too hard to see that for two hybrid states $\rho_{XE}$ and $\rho_{XE'}$ with the same (distribution of) $X$, the trace distance between $\rho_{XE}$ and $\rho_{XE'}$ can be computed as $\Delta(\rho_{XE}, \rho_{XE'}) = \sum_x P_X(x)\Delta(\rho_E^x, \rho_{E'}^x)$.

*Min-Entropy and Privacy Amplification.* We make use of Renner's notion of the *conditional min-entropy* $\mathrm{H}_{\min}(\rho_{XE}|E)$ of a system $X$ conditioned on another system $E$ [7]. Although the notion makes sense for arbitrary states, we restrict to hybrid states $\rho_{XE}$ with classical $X$. If the hybrid state $\rho_{XE}$ is clear from the context, we may write $\mathrm{H}_{\min}(X|E)$ instead of $\mathrm{H}_{\min}(\rho_{XE}|E)$. The formal definition is not very relevant to us, we merely rely on some elementary properties. For instance, the *chain rule* guarantees that $\mathrm{H}_{\min}(X|YE) \geq \mathrm{H}_{\min}(XY|E) - \log(|\mathcal{Y}|) \geq \mathrm{H}_{\min}(X|E) - \log(|\mathcal{Y}|)$ for classical $X$ and $Y$ with respective ranges $\mathcal{X}$ and $\mathcal{Y}$. Note that throughout this paper, log denotes the *binary* logarithm (we write ln for the *natural* logarithm). Furthermore, it holds that if $E'$ is obtained from $E$ by measuring (part of) $E$, then $\mathrm{H}_{\min}(X|E') \geq \mathrm{H}_{\min}(X|E)$.

Finally, we make use of Renner's privacy amplification theorem [8,7], as given below. Recall that a function $g : \mathcal{R} \times \mathcal{X} \to \{0,1\}^\ell$ is called a *universal* (hash) function, if for the random variable $R$, uniformly distributed over $\mathcal{R}$, and for any distinct $x, y \in \mathcal{X}$: $\Pr[g(R,x) = g(R,y)] \leq 2^{-\ell}$.

**Theorem 1 (Privacy amplification).** *Let $\rho_{XE}$ be a hybrid state with classical $X$. Let $g : \mathcal{R} \times \mathcal{X} \to \{0,1\}^\ell$ be a universal hash function, and let $R$ be uniformly distributed over $\mathcal{R}$, independent of $X$ and $E$. Then $K = g(R, X)$ satisfies*

$$\Delta\left(\rho_{KRE}, \frac{1}{|K|}\mathbb{I}_K \otimes \rho_{RE}\right) \leq \frac{1}{2} \cdot 2^{-\frac{1}{2}(\mathrm{H}_{\min}(X|E)-\ell)} .$$

Informally, Theorem 1 states that if $X$ contains sufficiently more than $\ell$ bits of entropy when given $E$, then $\ell$ nearly random-and-independent bits can be extracted from $X$.

## 3   Sampling in a Classical Population

As a warm-up, and in order to study some useful examples and introduce some convenient notation, we start with the classical sampling problem, which is rather well-understood.

### 3.1   Sampling Strategies

Let $\boldsymbol{q} = (q_1, \ldots, q_n) \in \mathcal{A}^n$ be a string of given length $n$. We consider the problem of estimating the relative Hamming weight $\omega(\boldsymbol{q})$ by only looking at a substring

$\boldsymbol{q}_t$ of $\boldsymbol{q}$, for a small subset $t \subset [n].^2$ Actually, we are interested in the equivalent problem of estimating the relative Hamming weight $\omega(\boldsymbol{q}_{\bar{t}})$ of the *remaining* string $\boldsymbol{q}_{\bar{t}}$, where $\bar{t}$ is the complement $\bar{t} = [n] \backslash t$ of $t$.$^3$ A canonical way to do so would be to sample a uniformly random subset (say, of a certain small size) of positions, and compute the relative Hamming weight of the sample as estimate. Very generally, we allow any strategy that picks a subset $t \subset [n]$ according to some probability distribution and computes the estimate for $\omega(\boldsymbol{q}_{\bar{t}})$ as some (possibly randomized) function of $t$ and $\boldsymbol{q}_t$, i.e., as $f(t, \boldsymbol{q}_t, s)$ for a *seed* $s$ that is sampled according to some probability distribution from a finite set $\mathcal{S}$. This motivates the following formal definition.

**Definition 1 (Sampling strategy).** *A sampling strategy $\Psi$ consists of a triple $(P_T, P_S, f)$, where $P_T$ is a distribution over the subsets of $[n]$, $P_S$ is a (independent) distribution over a finite set $\mathcal{S}$, and $f$ is a function*

$$f : \left\{ (t, v) : t \subset [n], v \in \mathcal{A}^{|t|} \right\} \times \mathcal{S} \to \mathbb{R}.$$

We stress that a sampling strategy $\Psi$, as defined here, specifies how to choose the sample subset as well as how to compute the estimate from the sample (thus a more appropriate but lengthy name would be a "sample-and-estimate strategy").

*Remark 1.* By definition, the choice of the seed $s$ is specified to be independent of $t$, i.e., $P_{TS} = P_T P_S$. Sometimes, however, it is convenient to allow $s$ to depend on $t$. We can actually do so without contradicting Definition 1. Namely, to comply with the independence requirement, we would simply choose a (typically huge) "container" seed that contains a seed for every possible choice of $t$, each one chosen with the corresponding distribution, and it is then part of $f$'s task, when given $t$, to select the seed that is actually needed out of the container seed.$^4$

A sampling strategy $\Psi$ can obviously also be used to *test* if $\boldsymbol{q}$ (or actually $\boldsymbol{q}_{\bar{t}}$) is close to the all-zero string $0 \cdots 0$: compute the estimate for $\omega(\boldsymbol{q}_{\bar{t}})$ as dictated by $\Psi$, and *accept* if the estimate vanishes and else *reject*.

We briefly discuss a few example sampling strategies (two more examples, including random sampling *with* replacement, can be found in the full version [2].

*Example 1 (Random sampling* without *replacement).* In random sampling without replacement, $k$ *distinct* indices $i_1, \ldots, i_k$ within $[n]$ are chosen uniformly at random, where $k$ is some parameter, and the relative Hamming weight of $\boldsymbol{q}_{\{i_1, \ldots, i_k\}}$ is used as estimate for $\omega(\boldsymbol{q}_{\bar{t}})$. Formally, this sampling strategy is given

---

$^2$ More generally, we may consider the problem of estimating the Hamming *distance* of $\boldsymbol{q}$ to some arbitrary *reference string* $\boldsymbol{q}_\circ$; but this can obviously be done simply by estimating the Hamming weight of $\boldsymbol{q}' = \boldsymbol{q} - \boldsymbol{q}_\circ$.

$^3$ In our applications, the sampled positions within $\boldsymbol{q}$ will be *discarded*, and thus we will be interested merely in the remaining positions.

$^4$ Alternatively, we could simply drop the independence requirement in Definition 1; however, we feel it is conceptually easier to think of the seed as being independently chosen.

by $\Psi = (P_T, P_S, f)$ where $P_T(t) = 1/\binom{n}{k}$ if $|t| = k$ and else $P_T(t) = 0$, $\mathcal{S} = \{\perp\}$ and thus $P_S(\perp) = 1$, and $f(t, \boldsymbol{q}_t, \perp) = \tilde{f}(t, \boldsymbol{q}_t) = \omega(\boldsymbol{q}_t)$.    ◇

*Example 2 (Uniformly random subset sampling).* The sample set $t$ is chosen as a uniformly random subset of $[n]$, and the estimate is computed as the relative Hamming weight of the sample $\boldsymbol{q}_t$: $P_T(t) = 1/2^n$ for any $t \subseteq [n]$, and $\mathcal{S} = \{\perp\}$ and $f(t, \boldsymbol{q}_t, \perp) = \tilde{f}(t, \boldsymbol{q}_t) = \omega(\boldsymbol{q}_t)$.    ◇

The next example is a somewhat unnatural and in some sense non-optimal sampling strategy, but it will be of use for the QKD proof in Section 5.

*Example 3 (Pairwise one-out-of-two sampling, using only part of the sample).* For this example, it is convenient to consider the index set from which the subset $t$ is chosen, to be of the form $[n] \times \{0, 1\}$. Namely, we consider the string $\boldsymbol{q} \in \mathcal{A}^{2n}$ to be indexed by *pairs* of indices, $\boldsymbol{q} = (q_{ij})$, where $i \in [n]$ and $j \in \{0, 1\}$; in other words, we consider $\boldsymbol{q}$ to consist of $n$ pairs $(q_{i0}, q_{i1})$. The subset $t \subset [n] \times \{0, 1\}$ is chosen as $t = \{(1, j_1), \dots, (n, j_n)\}$ where every $j_k$ is picked independently at random in $\{0, 1\}$. In other words, $t$ selects one element from each pair $(q_{i0}, q_{i1})$. Furthermore, the estimate for $\omega(\boldsymbol{q}_{\bar{t}})$ is computed from $\boldsymbol{q}_t$ as $f(t, \boldsymbol{q}_t, s) = \omega(\boldsymbol{q}_s)$ where the seed $s$ is a random subset $s \subset t$ of size $k$.    ◇

### 3.2    The Error Probability

We formally define a measure that captures for a given sampling strategy how well it performs, i.e., with what probability the estimate, $f(t, \boldsymbol{q}_t, s)$, is how close to the real value, $\omega(\boldsymbol{q}_{\bar{t}})$. For the definition and for later purposes, it will be convenient to introduce the following notation. For a given sampling strategy $\Psi = (P_T, P_S, f)$, consider arbitrary but fixed choices for the subset $t \subset [n]$ and the seed $s \in \mathcal{S}$ with $P_T(t) > 0$ and $P_S(s) > 0$. Furthermore, fix an arbitrary $\delta > 0$. Define $B_{t,s}^{\delta}(\Psi) \subseteq \mathcal{A}^n$ as

$$B_{t,s}^{\delta}(\Psi) := \{\boldsymbol{b} \in \mathcal{A}^n : |\omega(\boldsymbol{b}_{\bar{t}}) - f(t, \boldsymbol{b}_t, s)| < \delta\},$$

i.e., as the set of all strings $\boldsymbol{q}$ for which the estimate is $\delta$-close to the real value, assuming that subset $t$ and seed $s$ have been used. To simplify notation, if $\Psi$ is clear from the context, we simply write $B_{t,s}^{\delta}$ instead of $B_{t,s}^{\delta}(\Psi)$. By replacing the specific values $t$ and $s$ by the corresponding (independent) random variables $T$ and $S$, with distributions $P_T$ and $P_S$, respectively, we obtain the *random variable* $B_{T,S}^{\delta}$, whose range consists of subsets of $\mathcal{A}^n$. By means of this random variable, we now define the *error probability* of a sampling strategy as follows.

**Definition 2 (Error probability).** *The* (classical) error probability *of a sampling strategy $\Psi = (P_T, P_S, f)$ is defined as the following value, parametrized by $0 < \delta < 1$:*

$$\varepsilon_{\text{class}}^{\delta}(\Psi) := \max_{\boldsymbol{q} \in \mathcal{A}^n} \Pr\left[\boldsymbol{q} \notin B_{T,S}^{\delta}(\Psi)\right].$$

By definition of the error probability, it is guaranteed that for any string $q \in \mathcal{A}^n$, the estimated value is $\delta$-close to the real value except with probability at most $\varepsilon_{\text{class}}^\delta(\Psi)$. When used as a sampling strategy to test closeness to the all-zero string, $\varepsilon_{\text{class}}^\delta(\Psi)$ determines the probability of accepting even though $q_{\bar{t}}$ is "not close" to the all-zero string, in the sense that its relative Hamming weight exceeds $\delta$. Whenever $\Psi$ is clear from the context, we will write $\varepsilon_{\text{class}}^\delta$ instead of $\varepsilon_{\text{class}}^\delta(\Psi)$.

Below, we analyze the error probability of the sampling strategy discussed in Example 3, because this sampling strategy is used in our QKD security proof (Section 5). The error probabilities of the other examples can be found in the full version of this paper [2]. To bound the error probability, we use Hoeffding's inequality [5]. The following theorem summarizes this inequality, tailored to our needs.

**Theorem 2 (Hoeffding).** *Let $b \in \{0,1\}^n$ be a bit string with relative Hamming weight $\mu = \omega(b)$. Let the random variables $X_1, X_2, \ldots, X_k$ be obtained by sampling $k$ random entries from $b$ with replacement, i.e., the $X_i$'s are independent and $P_{X_i}(1) = \mu$. Furthermore, let the random variables $Y_1, Y_2, \ldots, Y_k$ be obtained by sampling $k$ random entries from $b$ without replacement. Then, for any $\delta > 0$, the random variables $\bar{X} := \frac{1}{k}\sum_i X_i$ and $\bar{Y} := \frac{1}{k}\sum_i Y_i$ satisfy*

$$\Pr\big[|\bar{Y} - \mu| \geq \delta\big] \leq \Pr\big[|\bar{X} - \mu| \geq \delta\big] \leq 2\exp(-2\delta^2 k)\,.$$

*Error Probability of Example 3.* For $\mathcal{A} = \{0,1\}$, a bound on the error probability $\varepsilon_{\text{class}}^\delta$ is obtained as follows. Let $q$ be arbitrary, indexed as discussed earlier. First, we show that $\omega(q_{\bar{T}})$ is likely to be close to $\omega(q_T)$. For this, consider the pairs $(q_{i0}, q_{i1})$ for which $q_{i0} \neq q_{i1}$. Let there be $\ell$ such pairs (where obviously $\ell \leq n$.) We denote the restrictions of $q_T$ and $q_{\bar{T}}$ to these indices $i$ with $q_{i0} \neq q_{i1}$ by $\tilde{q}_T$ and $\tilde{q}_{\bar{T}}$, respectively. It is easy to see that $\text{wt}(\tilde{q}_T) + \text{wt}(\tilde{q}_{\bar{T}}) = \ell$. It follows that for any $\epsilon > 0$ we have

$$\Pr\big[|\omega(q_{\bar{T}}) - \omega(q_T)| \geq \epsilon\big] = \Pr\big[|\text{wt}(q_T) - \text{wt}(q_{\bar{T}})| \geq n\epsilon\big]$$

$$= \Pr\big[|\text{wt}(\tilde{q}_T) - \text{wt}(\tilde{q}_{\bar{T}})| \geq n\epsilon\big] = \Pr\big[|2\text{wt}(\tilde{q}_T) - \ell| \geq n\epsilon\big]$$

$$\leq 2\exp\left(-2\left(\tfrac{n\epsilon}{2\ell}\right)^2 \ell\right) = 2\exp\left(-\tfrac{n\epsilon^2}{2} \cdot \tfrac{n}{\ell}\right) \leq 2\exp\left(-\tfrac{1}{2}\epsilon^2 n\right)\,,$$

where the third equality follows from replacing $\text{wt}(\tilde{q}_{\bar{T}})$ by $\ell - \text{wt}(\tilde{q}_T)$, and the first inequality follows from Hoeffding's inequality (as each entry of $\text{wt}(\tilde{q}_T)$ is 0 with independent probability $\frac{1}{2}$).

Furthermore, for any $\gamma > 0$ we have the following relation involving $q_S$:

$$\Pr\big[|\omega(q_T) - \omega(q_S)| \geq \gamma\big] \leq 2\exp\left(-2k\gamma^2\right)\,,$$

which follows from directly applying Hoeffding's inequality. Applying the union bound and letting $\delta = \epsilon + \gamma$, we obtain

$$\varepsilon_{\text{class}}^\delta = \Pr\big[|\omega(q_{\bar{T}}) - \omega(q_S)| \geq \delta\big] < 2\min_{\epsilon \in (0,\delta)}\left[\exp\left(-\tfrac{1}{2}\epsilon^2 n\right) + \exp\left(-2k(\delta - \epsilon)^2\right)\right]$$

$$\leq 4\exp\left(-\frac{2kn\delta^2}{(2\sqrt{k} + \sqrt{n})^2}\right) \leq 4\exp\left(-\tfrac{1}{3}\delta^2 k\right)\,,$$

where the last line follows from choosing $\epsilon$ such that the two exponents coincide, and from doing some simplifications while assuming $k \leq n/2$.

## 4 Sampling in a *Quantum* Population

We now want to study the behavior of a sampling strategy when applied to a quantum population. More specifically, let $A = A_1 \cdots A_n$ be an $n$-partite quantum system, where the state space of each system $A_i$ equals $\mathcal{H}_{A_i} = \mathbb{C}^d$ with $d = |\mathcal{A}|$, and let $\{|a\rangle\}_{a \in \mathcal{A}}$ be a fixed orthonormal basis of $\mathbb{C}^d$. We allow $A$ to be entangled with some additional system $E$ with arbitrary finite-dimensional state-space $\mathcal{H}_E$. We may assume the joint state of $AE$ to be pure, and as such be given by a state vector $|\varphi_{AE}\rangle \in \mathcal{H}_A \otimes \mathcal{H}_E$; if not, then it can be purified by increasing the dimension of $\mathcal{H}_E$.

Similar to the classical sampling problem of testing closeness to the all-zero string, we can consider here the problem of testing if the state of $A$ is close to the all-zero *reference state* $|\varphi_A^\circ\rangle = |0\rangle \cdots |0\rangle$ by looking at, which here means *measuring*, only a few of the subsystems of $A$. More generally, we will be interested in the sampling problem of estimating the "Hamming weight of the state of $A$", although it is not clear at the moment what this should mean. Actually, like in the classical case, we are interested in testing closeness to the all-zero state, respectively estimating the Hamming weight, of the *remaining subsystems* of $A$.

It is obvious that a sampling strategy $\Psi = (P_T, P_S, f)$ can be applied in a straightforward way to the setting at hand: sample $t$ according to $P_T$, measure the subsystems $A_i$ with $i \in t$ in basis $\{|a\rangle\}_{a \in \mathcal{A}}$ to observe $q_t \in \mathcal{A}^{|t|}$, and compute the estimate as $f(t, q_t, s)$ for $s$ chosen according to $P_S$ (respectively, for testing closeness to the all-zero state, accept or reject depending on the value of the estimate). However, it is a-priori *not* clear, how to interpret the outcome. Measuring a random subset of the subsystems of $A$ and observing 0 all the time indeed seems to suggest that the original state of $A$, and thus the remaining subsystems, must be in some sense close to the all-zero state; but in what formal sense is this true? And what can we conclude about the remaining state in case of a general sampling strategy for estimating the (relative) Hamming weight?

We give in this section a rigorous analysis of sampling strategies when applied to a $n$-partite quantum system $A$. Our analysis completely answers above questions. Later in the paper, we demonstrate the usefulness of our analysis of sampling strategies for studying and analyzing quantum-cryptographic schemes.

### 4.1 Analyzing Sampling Strategies in the Quantum Setting

We start by suggesting the property on the remaining subsystems of $A$ that one should expect to be able to conclude from the outcome of a sampling strategy. A somewhat natural approach is as follows.

**Definition 3.** *For system $AE$, and similarly for any subsystem of $A$, we say that the state $|\varphi_{AE}\rangle$ of $AE$ has* relative Hamming weight $\beta$ within $A$ *if it is of the form $|\varphi_{AE}\rangle = |b\rangle|\varphi_E\rangle$ with $b \in \mathcal{A}^n$ and $\omega(b) = \beta$.*

Now, given the outcome $f(t, \boldsymbol{q}_t, s)$ of a sampling strategy when applied to $A$, we want to be able to conclude that, up to a small error, the state of the remaining subsystem $A_{\bar{t}}E$ is a *superposition* of states with relative Hamming weight close to $f(t, \boldsymbol{q}_t, s)$ within $A_{\bar{t}}$. To analyze this, we extend some of the notions introduced in the classical setting. Recall the definition of $B^\delta_{t,s}$, consisting of all strings $\boldsymbol{b} \in \mathcal{A}^n$ with $|\omega(\boldsymbol{b}_{\bar{t}}) - f(t, \boldsymbol{b}_t, s)| < \delta$. By slightly abusing notation, we extend this notion to the quantum setting and write

$$\mathrm{span}(B^\delta_{t,s}) := \mathrm{span}(\{|\boldsymbol{b}\rangle : \boldsymbol{b} \in B^\delta_{t,s}\}) = \mathrm{span}(\{|\boldsymbol{b}\rangle : |\omega(\boldsymbol{b}_{\bar{t}}) - f(t, \boldsymbol{b}_t, s)| < \delta\}).$$

Note that if the state $|\varphi_{AE}\rangle$ of $AE$ happens to be in $\mathrm{span}(B^\delta_{t,s}) \otimes \mathcal{H}_E$ for some $t$ and $s$, and if exactly these $t$ and $s$ are chosen when applying the sampling strategy to $A$, then *with certainty* the state of $A_{\bar{t}}E$ (after the measurement) is in a superposition of states with relative Hamming weight $\delta$-close to $f(t, \boldsymbol{q}_t, s)$ within $A_{\bar{t}}$, regardless of the measurement outcome $\boldsymbol{q}_t$.

Next, we want to extend the notion of error probability (Definition 2) to the quantum setting. For this, we consider the *hybrid* system $TSAE$, consisting of the classical random variables $T$ and $S$ with distribution $P_{TS} = P_T P_S$, describing the choices of $t$ and $s$, respectively, and of the actual quantum systems $A$ and $E$. The state of $TSAE$ is given by

$$\rho_{TSAE} = \sum_{t,s} P_{TS}(t, s)|t, s\rangle\langle t, s| \otimes |\varphi_{AE}\rangle\langle\varphi_{AE}|.$$

Note that $TS$ is independent of $AE$: $\rho_{TSAE} = \rho_{TS} \otimes \rho_{AE}$; indeed, in a sampling strategy $t$ and $s$ are chosen independently of the state of $AE$. We compare this *real* state of $TSAE$ with an *ideal* state which is of the form

$$\tilde{\rho}_{TSAE} = \sum_{t,s} P_{TS}(t, s)|t, s\rangle\langle t, s| \otimes |\tilde{\varphi}^{ts}_{AE}\rangle\langle\tilde{\varphi}^{ts}_{AE}| \quad \text{with}$$

$$|\tilde{\varphi}^{ts}_{AE}\rangle \in \mathrm{span}(B^\delta_{t,s}) \otimes \mathcal{H}_E \;\; \forall\, t, s \tag{1}$$

for some given $\delta > 0$. Thus, $T$ and $S$ have the same distribution as in the real state, but here we allow $AE$ to depend on $T$ and $S$, and for each particular choice $t$ and $s$ for $T$ and $S$, respectively, we require the state of $AE$ to be in $\mathrm{span}(B^\delta_{t,s}) \otimes \mathcal{H}_E$. Thus, in an "ideal world" where the state of the hybrid system $TSAE$ is given by $\tilde{\rho}_{TSAE}$, it holds *with certainty* that the state $|\psi_{A_{\bar{t}}E}\rangle$ of $A_{\bar{t}}E$, after having measured $A_t$ and having observed $\boldsymbol{q}_t$, is in a superposition of states with relative Hamming weight $\delta$-close to $\beta := f(t, \boldsymbol{q}_t, s)$ within $A_{\bar{t}}$. We now define the quantum error probability of a sampling strategy by looking at how far away the closest ideal state $\tilde{\rho}_{TSAE}$ is from the real state $\rho_{TSAE}$.

**Definition 4 (Quantum error probability).** *The* quantum error probability *of a sampling strategy* $\Psi = (P_T, P_S, f)$ *is defined as the following value, parametrized by $0 < \delta < 1$:*

$$\varepsilon^\delta_{\mathrm{quant}}(\Psi) = \max_{\mathcal{H}_E} \max_{|\varphi_{AE}\rangle} \min_{\tilde{\rho}_{TSAE}} \Delta(\rho_{TSAE}, \tilde{\rho}_{TSAE}),$$

*where the first* max *is over all finite-dimensional state spaces* $\mathcal{H}_E$, *the second* max *is over all state vectors* $|\varphi_{AE}\rangle \in \mathcal{H}_A \otimes \mathcal{H}_E$, *and the* min *is over all ideal states* $\tilde{\rho}_{TSAE}$ *as in* (1).[5]

As with $B_{t,s}^\delta$ and $\varepsilon_{\text{class}}^\delta$, we simply write $\varepsilon_{\text{quant}}^\delta$ when $\Psi$ is clear from the context. We stress the meaningfulness of the definition: it guarantees that on average over the choice of $t$ and $s$, the state of $A_{\bar{t}}E$ is $\varepsilon_{\text{quant}}^\delta$-close to a superposition of states with Hamming weight $\delta$-close to $f(t, \mathbf{q}_t, s)$ within $A_{\bar{t}}$, and as such it *behaves* like a superposition of such states, except with probability $\varepsilon_{\text{quant}}^\delta$. We will argue below and demonstrate in the subsequent sections that being (close to) a superposition of states with given approximate (relative) Hamming weight has some useful consequences.

*Remark 2.* Similarly to footnote 2, also here the results of the section immediately generalize from the all-zero reference state $|0\rangle \cdots |0\rangle$ to an arbitrary reference state $|\varphi_A^\circ\rangle$ of the form $|\varphi_A^\circ\rangle = U_1|0\rangle \otimes \cdots \otimes U_n|0\rangle$ for unitary operators $U_i$ acting on $\mathbb{C}^d$. Indeed, the generalization follows simply by a suitable change of basis, defined by the $U_i$'s. Or, in the special case where $\mathcal{A} = \{0, 1\}$ and

$$|\varphi_A^\circ\rangle = H^{\hat{\boldsymbol{\theta}}}|\hat{\boldsymbol{x}}\rangle = H^{\hat{\theta}_1}|\hat{x}_1\rangle \otimes \cdots \otimes H^{\hat{\theta}_n}|\hat{x}_n\rangle$$

for a fixed reference basis $\hat{\boldsymbol{\theta}} \in \{0, 1\}^n$ and a fixed reference string $\hat{\boldsymbol{x}} \in \{0, 1\}^n$, we can, alternatively, replace in the definitions and results the computational by the Hadamard basis whenever $\hat{\theta}_i = 1$, and speak of the (relative) Hamming distance to $\hat{\boldsymbol{x}}$ rather than of the (relative) Hamming weight.

## 4.2   The Quantum vs. the Classical Error Probability

It remains to discuss how difficult it is to actually *compute* the quantum error probability for given sampling strategies, and how the *quantum* error probability $\varepsilon_{\text{quant}}^\delta$ relates to the corresponding *classical* error probability $\varepsilon_{\text{class}}^\delta$. To this end, we show the following simple relationship between $\varepsilon_{\text{quant}}^\delta$ and $\varepsilon_{\text{class}}^\delta$.

**Theorem 3.** *For any sampling strategy $\Psi$ and for any $\delta > 0$:*

$$\varepsilon_{\text{quant}}^\delta(\Psi) \leq \sqrt{\varepsilon_{\text{class}}^\delta(\Psi)}.$$

As a consequence of this theorem, it suffices to analyze a sampling strategy in the classical setting, which is much easier, in order to understand how it behaves in the quantum setting. In particular, sampling strategies that are known to behave well in the classical setting, like Example 1 to 3, are also automatically guaranteed to behave well in the quantum setting. We will use this for our applications.

Our bound on $\varepsilon_{\text{quant}}^\delta$ is in general tight. Indeed, in [2] we show tightness for an explicit class of sampling strategies, which e.g. includes Example 1 and Example 3. Here, we just mention the tightness result.

---

[5] It is not too hard to see, in particular after having gained some more insight via the proof of Theorem 3 below, that these min and max exist.

**Proposition 1.** *There exist natural sampling strategies for which the inequality in Theorem 3 is an equality.*

*Proof (of Theorem 3).* We need to show that for any $|\varphi_{AE}\rangle \in \mathcal{H}_A \otimes \mathcal{H}_E$, with arbitrary $\mathcal{H}_E$, there exists a suitable ideal state $\tilde{\rho}_{TSAE}$ with $\Delta(\rho_{TSAE}, \tilde{\rho}_{TSAE}) \leq (\varepsilon_{\text{class}}^\delta)^{1/2}$. We construct $\tilde{\rho}_{TSAE}$ as in (1), where the $|\tilde{\varphi}_{AE}^{ts}\rangle$'s are defined by the following decomposition.

$$|\varphi_{AE}\rangle = \langle\tilde{\varphi}_{AE}^{ts}|\varphi_{AE}\rangle|\tilde{\varphi}_{AE}^{ts}\rangle + \langle\tilde{\varphi}_{AE}^{ts\perp}|\varphi_{AE}\rangle|\tilde{\varphi}_{AE}^{ts\perp}\rangle,$$

with $|\tilde{\varphi}_{AE}^{ts}\rangle \in \text{span}(B_{t,s}^\delta) \otimes \mathcal{H}_E$, $|\tilde{\varphi}_{AE}^{ts\perp}\rangle \in \text{span}(B_{t,s}^\delta)^\perp \otimes \mathcal{H}_E$ and $|\langle\tilde{\varphi}_{AE}^{ts}|\varphi_{AE}\rangle|^2 + |\langle\tilde{\varphi}_{AE}^{ts\perp}|\varphi_{AE}\rangle|^2 = 1$. In other words, $|\tilde{\varphi}_{AE}^{ts}\rangle$ is obtained as the re-normalized projection of $|\varphi_{AE}\rangle$ into $\text{span}(B_{t,s}^\delta) \otimes \mathcal{H}_E$. Note that $|\langle\tilde{\varphi}_{AE}^{ts\perp}|\varphi_{AE}\rangle|^2$ equals the probability $\Pr[Q \notin B_{t,s}^\delta]$, where the random variable $Q$ is obtained by measuring subsystem $A$ of $|\varphi_{AE}\rangle$ in basis $\{|a\rangle\}_{a\in\mathcal{A}}^{\otimes n}$. Furthermore,

$$\sum_{t,s} P_{TS}(t,s)\,|\langle\tilde{\varphi}_{AE}^{ts\perp}|\varphi_{AE}\rangle|^2 = \sum_{t,s} P_{TS}(t,s)\Pr[Q \notin B_{t,s}^\delta] = \Pr[Q \notin B_{T,S}^\delta]$$

$$= \sum_q P_Q(q)\Pr[q \notin B_{T,S}^\delta],$$

where by definition of $\varepsilon_{\text{class}}^\delta$, the latter is upper bounded by $\varepsilon_{\text{class}}^\delta$. From elementary properties of the trace distance, and using Jensen's inequality, we can now conclude that

$$\Delta(\rho_{TSAE}, \tilde{\rho}_{TSAE}) = \sum_{t,s} P_{TS}(t,s)\Delta(|\varphi_{AE}\rangle\langle\varphi_{AE}|, |\tilde{\varphi}_{AE}^{ts}\rangle\langle\tilde{\varphi}_{AE}^{ts}|)$$

$$= \sum_{t,s} P_{TS}(t,s)\sqrt{1 - |\langle\tilde{\varphi}_{AE}^{ts}|\varphi_{AE}\rangle|^2} = \sum_{t,s} P_{TS}(t,s)|\langle\tilde{\varphi}_{AE}^{ts\perp}|\varphi_{AE}\rangle|$$

$$\leq \sqrt{\sum_{t,s} P_{TS}(t,s)|\langle\tilde{\varphi}_{AE}^{ts\perp}|\varphi_{AE}\rangle|^2} \leq \sqrt{\varepsilon_{\text{class}}^\delta},$$

which was to be shown. □

As a side remark, we point out that the particular ideal state $\tilde{\rho}_{TSAE}$ constructed in the proof minimizes the distance to $\rho_{TSAE}$; this follows from the so-called Hilbert projection theorem.

## 4.3   Superpositions with a Small Number of Terms

We give here some argument why being (close to) a superposition of states with a given approximate Hamming weight may be a useful property in the analyses of quantum-cryptographic schemes. For simplicity, and since this will be the case in our applications, we now restrict to the binary case where $\mathcal{A} = \{0, 1\}$. Our argument is based on the following lemma, which follows immediately from Lemma 3.1.13 in [7]; we give a direct proof of Lemma 1 in the full version [2].

Informally, it states that measuring (part of) a *superposition* of a small number of orthogonal states produces a similar amount of uncertainty as when measuring the *mixture* of these orthogonal states.

**Lemma 1.** *Let $A$ and $E$ be arbitrary quantum systems, let $\{|i\rangle\}_{i \in I}$ and $\{|w\rangle\}_{w \in \mathcal{W}}$ be orthonormal bases of $\mathcal{H}_A$, and let $|\varphi_{AE}\rangle$ and $\rho_{AE}^{\text{mix}}$ be of the form*

$$|\varphi_{AE}\rangle = \sum_{i \in J} \alpha_i |i\rangle |\varphi_E^i\rangle \in \mathcal{H}_A \otimes \mathcal{H}_E \quad and \quad \rho_{AE}^{\text{mix}} = \sum_{i \in J} |\alpha_i|^2 |i\rangle\langle i| \otimes |\varphi_E^i\rangle\langle\varphi_E^i|$$

*for some subset $J \subseteq I$. Let $\rho_{WE}$ and $\rho_{WE}^{\text{mix}}$ describe the hybrid systems obtained by measuring subsystem $A$ of $|\varphi_{AE}\rangle$ and $\rho_{AE}^{\text{mix}}$, respectively, in basis $\{|w\rangle\}_{w \in \mathcal{W}}$ to observe outcome $W$. Then, $\mathrm{H}_{\min}(\rho_{WE}|E) \geq \mathrm{H}_{\min}(\rho_{WE}^{\text{mix}}|E) - \log |J|$.*

We apply Lemma 1 to an $n$-qubit system $A$ where $|\varphi_{AE}\rangle$ is a superposition of states with relative Hamming weight $\delta$-close to $\beta$ within $A$:[6]

$$|\varphi_{AE}\rangle = \sum_{\substack{b \in \{0,1\}^n \\ |\omega(b) - \beta| \leq \delta}} |b\rangle |\varphi_E^b\rangle .$$

It is well known that $\left|\{b \in \{0,1\}^n : |\omega(b) - \beta| \leq \delta\}\right| \leq 2^{\mathrm{h}(\beta+\delta)n}$ for $\beta + \delta \leq \frac{1}{2}$, where $\mathrm{h}(p) := -\big(p\log(p) + (1-p)\log(1-p)\big)$ denotes the binary entropy function.

Since measuring qubits within a state $|b\rangle$ in the *Hadamard* basis produces uniformly random bits, we can conclude the following.

**Corollary 1.** *Let $A$ be an $n$-qubit system, let the state $|\varphi_{AE}\rangle$ of $AE$ be a superposition of states with relative Hamming weight $\delta$-close to $\beta$ within $A$, where $\delta + \beta \leq \frac{1}{2}$, and let the random variable $X$ be obtained by measuring $A$ in basis $H^{\theta}\{|0\rangle, |1\rangle\}^{\otimes n}$ for $\theta \in \{0,1\}^n$. Then*

$$\mathrm{H}_{\min}(X|E) \geq \mathrm{wt}(\theta) - \mathrm{h}(\beta + \delta)n .$$

Consider now the following quantum-cryptographic setting. Bob prepares and hands over to Alice an $n$-qubit quantum system $A$, which ought to be in state $|\varphi_A^\circ\rangle = |0\rangle \cdots |0\rangle$. However, since Bob might be dishonest, the state of $A$ could be anything, even entangled with some system $E$ controlled by Bob. Our results now imply the following: Alice can apply a suitable sampling strategy to convince herself that the joint state of the remaining subsystem of $A$ and of $E$ is (close to) a superposition of states with bounded relative Hamming weight. From Corollary 1, we can then conclude that with respect to the min-entropy of the measurement outcome, the state of $A$ behaves similarly to the case where Bob honestly prepares $A$ to be in state $|\varphi_A^\circ\rangle$. By Remark 2, i.e., by doing a suitable change of basis, the same holds if $|\varphi_A^\circ\rangle = H^{\hat{\theta}}|\hat{x}\rangle$ for arbitrary fixed $\hat{\theta}, \hat{x} \in \{0,1\}^n$, where $\mathrm{wt}(\theta)$ is replaced by the Hamming distance between $\theta$ and $\hat{\theta}$. We will make use of this in the application in the upcoming section.

---

[6] System $A$ considered here corresponds to the subsystem $A_{\bar{t}}$ in the previous section, after having measured $A_t$ of the ideal state.

# 5    Application: Quantum Key Distribution (QKD)

In quantum key distribution (QKD), Alice and Bob want to agree on a secret key in the presence of an adversary Eve. Alice and Bob are assumed to be able to communicate over a quantum channel and over an authenticated classical channel. Eve may eavesdrop the classical channel (but not insert or modify messages), and she has full control over the quantum channel. The first and still most prominent QKD scheme is the famous BB84 QKD scheme due to Bennett and Brassard [1].

In this section, we show how our sampling-strategy framework leads to a simple security proof for the BB84 QKD scheme. Proving QKD schemes rigorously secure is a highly non-trivial task, and as such our new proof nicely demonstrates the power of the sampling-strategy framework. Furthermore, our new proof has some nice features. For instance, it allows us to explicitly state (a bound on) the error probability of the QKD scheme for any given choices of the parameters. Additionally, our proof does not seem to take unnecessary detours or to make use of "loose bounds", and therefore we feel that the bound on the error probability we obtain is rather tight (although we have no formal argument to support this). Our proof strategy can also be applied to other QKD schemes that are based on the BB84 encoding. For example, Lo *et al.*'s QKD scheme[7] [6] can be proven secure by following exactly our proof, except that one needs to analyze a slightly different sampling strategy. On the other hand, it is yet unknown whether our framework can be used to prove e.g. the six-state QKD protocol [3] secure.

As a matter of fact, the QKD scheme we analyze is an entanglement-based version of the BB84 scheme. However, it is very well known and not too hard to show that security of the entanglement-based version implies security of the original BB84 QKD scheme.

The entanglement-based QKD scheme, QKD, is parametrized by the total number $n$ of qubits sent in the protocol and the number $k$ of qubits used to estimate the error rate of the quantum channel (where we require $k \leq n/2$). Additional parameters, which are determined during the course of the protocol, are the observed error rate $\beta$ and the number $\ell \in \mathbb{N} \cup \{0\}$ of extracted key bits. QKD makes use of a universal hash function $g : \mathcal{R} \times \{0,1\}^{n-k} \to \{0,1\}^\ell$ and a linear binary error correcting code of length $n - k$ that allows to correct up to a $\beta'$-fraction of errors (except maybe with negligible probability) for some $\beta' > \beta$. The choice of how much $\beta'$ exceeds $\beta$ is a trade-off between keeping the probability that Alice and Bob end up with different keys small and increasing the size of the extractable key. We will write $m$ for the bit size of the syndrome of this error-correcting code. Protocol QKD can be found below.

It is not hard to see that $k = \hat{k}$ except with negligible probability (in $n$). Furthermore, if no Eve interacts with the quantum communication in the qubit distribution phase then $\boldsymbol{x} = \boldsymbol{y}$ in case of a noise-free quantum channel, or more generally, $\omega(\boldsymbol{x} - \boldsymbol{y}) \approx \phi$ in case the quantum channel is noisy and introduces

---

[7] In this scheme, Alice and Bob bias the choice of the bases so that they measure a bigger fraction of the qubits in the same basis.

---

**Protocol** QKD

1. *(Qubit distribution)* Alice prepares $n$ EPR pairs of the form $(|0\rangle|0\rangle + |1\rangle|1\rangle)/\sqrt{2}$, and sends one qubit of each pair to Bob, who confirms the receipt of the qubits. Then, Alice picks random $\boldsymbol{\theta} \in \{0,1\}^n$ and sends it to Bob, and Alice and Bob measure their respective qubits in basis $\boldsymbol{\theta}$ to obtain $\boldsymbol{x}$ on Alice's side and $\boldsymbol{y}$ on Bob's side.
2. *(Error estimation)* Alice chooses a random subset $s \subset [n]$ of size $k$ and sends it to Bob. Then, Alice and Bob exchange $\boldsymbol{x}_s$ and $\boldsymbol{y}_s$ and compute $\beta := \omega(\boldsymbol{x}_s \oplus \boldsymbol{y}_s)$.
3. *(Error correction)* Alice sends the syndrome $syn$ of $\boldsymbol{x}_{\bar{s}}$ to Bob with respect to a suitable linear error correcting code (as described above). Bob uses $syn$ to correct the errors in $\boldsymbol{y}_{\bar{s}}$ and obtains $\hat{\boldsymbol{x}}_{\bar{s}}$. Let $m$ be the bit-size of $syn$.
4. *(Key distillation)* Alice chooses a random seed $r$ for a universal hash function $g$ with range $\{0,1\}^\ell$, where $\ell$ satisfies $\ell < (1-\mathrm{h}(\beta))n - k - m$ (or $\ell = 0$ if the right-hand side is not positive), and sends it to Bob. Then, Alice and Bob compute $\boldsymbol{k} := g(r, \boldsymbol{x}_{\bar{s}})$ and $\hat{\boldsymbol{k}} := g(r, \hat{\boldsymbol{x}}_{\bar{s}})$, respectively.

---

an error probability $0 \le \phi < \frac{1}{2}$. It follows that $\beta \approx \phi$, so that using an error correcting code that approaches the Shannon bound, Alice and Bob can extract close to $(1 - 2\mathrm{h}(\phi))(n - k)$ bits of secret key, which is positive for $\phi$ smaller than approximately 11%. The difficult part is to prove security against an active adversary Eve. We first state the formal security claim.

Note that we cannot expect that Eve has (nearly) no information on $\boldsymbol{K}$, i.e. that $\Delta\big(\rho_{KE}, \frac{1}{|K|}\mathbb{I}_K \otimes \rho_E\big)$ is small, since the bit-length $\ell$ of $\boldsymbol{K}$ is not fixed but depends on the course of the protocol, and Eve can influence and thus obtain information on $\ell$ (and thus on $\boldsymbol{K}$). Theorem 4 though guarantees that the bit-length $\ell$ is the *only* information Eve learns on $\boldsymbol{K}$, in other words, $\boldsymbol{K}$ is essentially random-and-independent of $E$ when given $\ell$.

**Theorem 4 (Security of** QKD**).** *Consider an execution of* QKD *in the presence of an adversary Eve. Let $\boldsymbol{K}$ be the key obtained by Alice, and let $E$ be Eve's quantum system at the end of the protocol. Let $\tilde{\boldsymbol{K}}$ be chosen uniformly at random of the same bit-length as $\boldsymbol{K}$. Then, for any $\delta$ with $\beta + \delta \le \frac{1}{2}$:*

$$\Delta\big(\rho_{KE}, \rho_{\tilde{K}E}\big) \le \frac{1}{2} \cdot 2^{-\frac{1}{2}\big((1-\mathrm{h}(\beta+\delta))n-k-m-\ell\big)} + 2\exp\big(-\tfrac{1}{6}\delta^2 k\big).$$

From an application point of view, the following question is of interest. Given the parameters $n$ and $k$, and given a course of the protocol with observed error rate $\beta$ and where an error-correcting code with syndrome length $m$ was used, what is the maximal size $\ell$ of the extractable key $\boldsymbol{K}$ if we want $\Delta(\rho_{KE}, \rho_{\tilde{K}E}) \le \epsilon$ for a given $\epsilon$? From the bound in Theorem 4, it follows that for every choice of $\delta$ (with $\beta + \delta \le \frac{1}{2}$), one can easily compute a possible value for $\ell$ simply by solving for $\ell$. In order to compute the optimal value, one needs to maximize $\ell$ over the choice of $\delta$.

The formal proof of Theorem 4 is given below. Informally, the argument goes as follows. The error estimation phase can be understood as applying a sampling strategy. From this, we can conclude that the state from which the raw key,

$\boldsymbol{x}_{\bar{s}}$, is obtained, is a superposition of states with bounded Hamming weight, so that Corollary 1 guarantees a certain amount of min-entropy within $\boldsymbol{x}_{\bar{s}}$. Privacy amplification then finishes the proof.

To indeed be able to model the error estimation procedure as a sampling strategy, we will need to consider a modified but *equivalent* way for Alice and Bob to jointly obtain $\boldsymbol{x}_s$ and $\boldsymbol{y}_s$ from the initial joint state, which will allow them to obtain the XOR-sum $\boldsymbol{x}_s \oplus \boldsymbol{y}_s$, and thus to compute $\beta$, *before* they measure the remaining part of the state, whose outcome then determines $\boldsymbol{x}_{\bar{s}}$. This modification is based on the so-called CNOT operation, $U_{\text{CNOT}}$, acting on $\mathbb{C}^2 \otimes \mathbb{C}^2$, and its properties that

$$U_{\text{CNOT}}(|b\rangle|c\rangle) = |b\rangle|b \oplus c\rangle \quad \text{and} \quad U_{\text{CNOT}}(H|b\rangle H|c\rangle) = H|b \oplus c\rangle H|c\rangle, \quad (2)$$

where the first holds by definition of $U_{\text{CNOT}}$, and the second is trivial to verify.

*Proof.* Throughout the proof, we use capital letters, $\boldsymbol{\Theta}$, $\boldsymbol{X}$ etc. for the *random variables* representing the corresponding choices of $\boldsymbol{\theta}$, $\boldsymbol{x}$ etc. in protocol QKD. Let the state, shared by Alice, Bob and Eve right after the quantum communication in the qubit distribution phase, be denoted by $|\psi_{ABE_\circ}\rangle$;[8] without loss of generality, we may indeed assume the shared state to be pure. For every $i \in [n]$, Alice and Bob then measure the respective qubits $A_i$ and $B_i$ from $|\psi_{ABE_\circ}\rangle$ in basis $\Theta_i$, obtaining $X_i$ and $Y_i$. This results in the hybrid state $\rho_{\boldsymbol{\Theta}\boldsymbol{X}\boldsymbol{Y}E_\circ}$. For the proof, it will be convenient to introduce the additional random variables $\boldsymbol{W} = (W_1, \ldots, W_n)$ and $\boldsymbol{Z} = (Z_1, \ldots, Z_n)$, defined by

$$W_i := \begin{cases} X_i \text{ if } \Theta_i = 0 \\ Y_i \text{ if } \Theta_i = 1 \end{cases} \quad \text{and} \quad Z_i := X_i \oplus Y_i. \quad (3)$$

Note that, when given $\boldsymbol{\Theta}$, the random variables $\boldsymbol{W}$ and $\boldsymbol{Z}$ are uniquely determined by $\boldsymbol{X}$ and $\boldsymbol{Y}$ *and vice versa*, and thus we may equivalently analyze the hybrid state $\rho_{\boldsymbol{\Theta}\boldsymbol{W}\boldsymbol{Z}E_\circ}$.

For the analysis, we will consider a slightly *different* experiment for Alice and Bob to obtain the very *same* state $\rho_{\boldsymbol{\Theta}\boldsymbol{W}\boldsymbol{Z}E_\circ}$; the advantage of the modified experiment is that it can be understood as a sampling strategy. The modified experiment is as follows. First, the CNOT transformation is applied to every qubit pair $A_iB_i$ within $|\psi_{ABE_\circ}\rangle$ for $i \in [n]$, such that the state $|\varphi_{ABE_\circ}\rangle = (U_{\text{CNOT}}^{\otimes n} \otimes \mathbb{I}_{E_\circ})|\psi_{ABE_\circ}\rangle$ is obtained. Next, $\boldsymbol{\Theta}$ is chosen at random as in the original scheme, and for every $i \in [n]$ the qubit pair $A_iB_i$ of the transformed state is measured as in the original scheme depending on $\Theta_i$; however, if $\Theta_i = 0$ then the resulting bits are denoted by $W_i$ and $Z_i$, respectively, and if $\Theta_i = 1$ then they are denoted by $Z_i$ and $W_i$, respectively, such that which bit is assigned to which variable depends on $\Theta_i$. This is illustrated in Figure 1 (left and middle), where light and dark colored ovals represent measurements in the computational and Hadamard basis, respectively. It now follows immediately from the properties (2)

---

[8] $E_\circ$ represents Eve's quantum state just after the quantum communication stage, whereas $E$ represents Eve's entire state at the end of the protocol (i.e., her quantum information and all classical information gathered during execution of QKD).

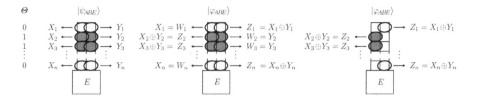

**Fig. 1.** Original and modified experiments for obtaining the same state $\rho_{\Theta WZE_\circ}$

of the CNOT transformation and from the relation (3) between $\boldsymbol{X}, \boldsymbol{Y}$ and $\boldsymbol{W}, \boldsymbol{Z}$ that the state $\rho_{\Theta WZE_\circ}$ (or, equivalently, $\rho_{\Theta XYE_\circ}$) obtained in this modified experiment is exactly the same as in the original.

An additional modification we may do without influencing the final state is to *delay* some of the measurements: we assume that first the qubits are measured that lead to the $Z_i$'s, and only at some later point, namely after the *error estimation* phase, the qubits leading to the $W_i$'s are measured (as illustrated in Figure 1, right). This can be done since the relative Hamming weight of $X_S \oplus Y_S$ for a random subset $S \subset [n]$ (of size $k$) can be computed given $\boldsymbol{Z}$ alone.

The crucial observation is now that this modified experiment can be viewed as a particular sampling strategy $\Psi$, as a matter of fact as the sampling strategy discussed in Example 3, being applied to systems $A$ and $B$ of the state $|\varphi_{ABE_\circ}\rangle$. Indeed: first, a subset of the $2n$ qubit positions is selected according to some probability distribution, namely of each pair $A_iB_i$ one qubit is selected at random (determined by $\Theta_i$). Then, the selected qubits are measured to obtain the bit string $\boldsymbol{Z} = (Z_1, \ldots, Z_n)$. And, finally, a value $\beta$ is computed as a (randomized) function of $\boldsymbol{Z}$: $\beta = \omega(\boldsymbol{Z}_S)$ for a random $S \subset [n]$ of size $k$. We point out that here the reference basis (as explained in Remark 2) is not the computational basis for all qubits, but is the Hadamard basis on the qubits in system $A$ and the computational basis in system $B$; however, as discussed in Remark 2, we may still apply the results from Section 4 (appropriately adapted).

It thus follows that for any fixed $\delta > 0$, the remaining state, from which $\boldsymbol{W}$ is then obtained, is (on average over $\Theta$ and $S$) $\varepsilon^\delta_{\text{quant}}$-close to a state which is (for any possible values for $\Theta, \boldsymbol{Z}$ and $S$) a superposition of states with relative Hamming weight in a $\delta$-neighborhood of $\beta$. Note that the latter has to be understood with respect to the fixed reference basis (i.e., the Hadamard basis on $A$ and the computational basis on $B$). In the following, we assume that the remaining state *equals* such a superposition, but we remember the error

$$\varepsilon^\delta_{\text{quant}} \leq \sqrt{\varepsilon^\delta_{\text{class}}} \leq 2\exp\left(-\tfrac{1}{6}\delta^2 k\right).$$

where the bound on $\varepsilon^\delta_{\text{class}}$ was derived in Section 3.2.

Recall that $\boldsymbol{W}$ is now obtained by measuring the remaining qubits; however, the basis used is opposite to the reference basis, namely the computational basis on the qubits $A_i$ and the Hadamard basis on the qubits $B_i$. Hence, by Corollary 1 (and the subsequent discussion) we get a lower bound on the min-entropy of $\boldsymbol{W}$:

$$\mathrm{H}_{\min}(\boldsymbol{W}|\boldsymbol{\Theta}\boldsymbol{Z}SE_\circ) \geq (1 - \mathrm{h}(\beta + \delta))n \,.$$

Since $\boldsymbol{W}$ is uniquely determined by $\boldsymbol{X}$ (and vice versa) when given $\boldsymbol{\Theta}$ and $\boldsymbol{Z}$, the same lower bound also holds for $\mathrm{H}_{\min}(\boldsymbol{X}|\boldsymbol{\Theta}\boldsymbol{Z}SE_\circ)$. Note that in QKD, the $k$ qubit-pairs that are used for estimating $\beta$ are not used anymore in the key distillation phase, so we are actually interested in the min-entropy of $\boldsymbol{X}_{\bar{S}}$. Additionally, we should take into account that Alice sends an $m$-bit syndrome $SYN$ during the error correction phase. Hence, by using the chain rule, we obtain

$$\mathrm{H}_{\min}(\boldsymbol{X}_{\bar{S}}|\boldsymbol{\Theta}\boldsymbol{Z}\boldsymbol{X}_S SYN E_\circ) \geq (1 - \mathrm{h}(\beta + \delta))n - k - m.[9]$$

Finally, we apply privacy amplification (Theorem 1) to conclude the proof. □

# References

1. Bennett, C.H., Brassard, G.: Quantum cryptography: Public key distribution and coin tossing. In: Proceedings of IEEE International Conference on Computers, Systems, and Signal Processing, pp. 175–179 (1984)
2. Bouman, N.J., Fehr, S.: Sampling in a quantum population, and applications (2009), http://arxiv.org/abs/0907.4246
3. Bruss, D.: Optimal eavesdropping in quantum cryptography with six states. Physical Review Letters 81, 3018 (1998), http://arxiv.org/abs/quant-ph/9805019
4. Damgård, I., Fehr, S., Lunemann, C., Salvail, L., Schaffner, C.: Improving the security of quantum protocols via commit-and-open. In: Halevi, S. (ed.) CRYPTO 2009. LNCS, vol. 5677, pp. 408–427. Springer, Heidelberg (2009)
5. Hoeffding, W.: Probability inequalities for sums of bounded random variables. Journal of the American Statistical Association 58(301), 13–30 (1963)
6. Lo, H.K., Chau, H.F., Ardehali, M.: Efficient quantum key distribution scheme and a proof of its unconditional security. J. Cryptol. 18(2), 133–165 (2005)
7. Renner, R.: Security of Quantum Key Distribution. Ph.D. thesis, ETH Zürich (Switzerland) (September 2005), http://arxiv.org/abs/quant-ph/0512258
8. Renner, R.S., König, R.: Universally composable privacy amplification against quantum adversaries. In: Kilian, J. (ed.) TCC 2005. LNCS, vol. 3378, pp. 407–425. Springer, Heidelberg (2005)
9. Shor, P.W., Preskill, J.: Simple Proof of Security of the BB84 Quantum Key Distribution Protocol. Phys. Rev. Lett. 85, 441–444 (2000)

---

[9] Probably, it is possible to prove the lower bound: $(1 - \mathrm{h}(\beta + \delta))(n - k) - m$ using a different sampling strategy. However, for that case the error probability of the related classical sampling strategy becomes harder to analyze. We have chosen for the current proof strategy and bound for the sake of simplicity.

# Author Index